Austin
June 2012
Robles Jct., AZ

Cultivated vegetables of the world: a multilingual onomasticon

Dedicated to Sandra E. Kays

Cultivated vegetables of the world: a multilingual onomasticon

Stanley J. Kays

University of Georgia

Wageningen Academic
Publishers

ISBN: 978-90-8686-164-4
e-ISBN: 978-90-8686-720-2
DOI: 10.3921/978-90-8686-720-2

First published, 2011

Table of contents

Introduction

With the human population in many areas of the world already above the carrying capacity of the land, food availability and advances in productivity are becoming increasingly critical. It is predicted that the world population will reach 9 billion people by 2050, which represents a dramatic increase in the number of people to feed, with increasingly scarce crop land, water, fertilizer, and fuel. Vegetables make up a major portion of the diet of humans, providing not only calories but essential vitamins, fiber, and minerals and will play an increasingly important role in food availability. Of the over 400 vegetable crops that are grown commercially, the genetic diversity in critical attributes is exceptional. The edible portion of vegetables is comprised of an array of morphological parts [e.g. leaves, stems, lateral buds, young plants, shoots, petioles, flowers, flower buds, fruit (immature to mature), seeds, seedlings, roots, tubers, corms, bulbs, rhizomes]. Vegetables vary widely in the length of time required from planting to harvest, postharvest longevity, and day length, temperature, water, fertility, and pesticide requirements. They provide a diverse range of tastes, aromas, textures, colors, and nutritional attributes, greatly increasing the variety in the foods we eat and satisfying a myriad of personal preferences. The length of time required to produce a crop (i.e. planting to harvest) varies from days to several years and the length of optimum harvest maturity from hours (gerkins – *Cucumis sativus* L.) to years (giant taro – *Alocasia macrorrhiza* (L.) Schott). The fact that short production cycle vegetables allow multiple cropping and a significant volume of the vegetables grown worldwide are produced on small plots, militates against accurate production statistics, preventing a clear understanding and appreciation of the value of these crops to the world food supply.

Vegetables are a critical component in the health of humans. While the nutritional importance of vegetables has long been recognized within the nutrition and medical communities, there is an increasing awareness among the general public of the health advantages of increased vegetable consumption. In many areas of the world, increased consumption of vegetables has been stimulated not only by a growing understanding of the health implications but an increasing variety of vegetables available.

The introduction of new crops via Columbian and Eurasian exchanges has had an immeasurable impact on world food availability and in turn on population growth and the history of mankind (Diamond, 2005; Gunn, 2003). The introduction of plants between the New and Old World since the latter part of the 13th century has been described as "one of the most important aspects of the history of life on this planet since the retreat of the continental glaciers" (Crosby, 1972). In the 16th and 17th centuries, ship captains and diplomats routinely collected and sent unique plants to their home countries. This was especially true for several European countries. Plant exploration subsequently became officially supported by various governments and combined with the efforts of private and commercial plant collectors and botanical gardens, many exotic new species were introduced around the world. Expeditions by Joseph Banks, Joseph Hooker, Nathaniel Ward, David Fairchild, Frank N. Meyer, N.I. Vavilov and others provided a wealth of new species. Initially the focus was on finding superior cultivars; however, the emphasis shifted in the mid-20th century toward collecting germplasm with the focus now on unique genes (Harlan, 1975).

The introduction of a new species, however, does not equate to adoption. In many instances, even with ornamentals, new introductions represented little more than curiosities, seldom gaining wide spread popularity. What then drives the adoption of new crops? Demand is the key factor and the ease and avenue for creating demand differ widely among types of crops. With many agronomic (sunflower – *Helianthus annuus* L.) and industrial (longleaf pine – *Pinus palustris* Miller) crops, government and business interests create the demand and contract production with growers, guaranteeing purchase. Vegetable crops, on-the-other-hand, generally involve much lower production volumes and with successful introduction, demand increases slowly until the positive attributes of the crop become common knowledge. Gourmet restaurants, looking for unique entrees, represent one recent means by which the demand for new vegetables has increased. More importantly, increased exposure to new crops and the substantially greater variety of vegetable crops available in many local markets is due to immigration and, to a lesser extent, international travel. Cuban, Thai, Indian, Filipino, Chinese, Vietnamese and Cambodian immigrants arriving in South Florida in the later half of the 20th century brought with them an existing demand for specific vegetables that were an essential component of everyday live in their countries of origin. Crops, such as bitter melon – *Momordica charantia* L., chayote – *Sechium edule* (Jacq.) Swartz., guar – *Cyamopsis tetragonoloba* (L.) Taubert, luffa – *Luffa aegyptiaca* P. Miller, pea eggplant – *Solanum torvum* Swartz, pointed gourd – *Trichosanthes dioica* Roxb., tindora – *Coccinia grandis* (L.) Voigt, and winged bean – *Psophocarpus tetragonolobus* (L.) DC., began being grown locally and offered in ethnic markets (Lamberts, 1992). Some subsequently spread to more traditional markets and wider use. Similar stories are found in hundreds of locations around the world where immigrants congregate.

With the rapidly growing world population, vegetables have a critical role to play. Vegetable research is increasingly focusing on elevating productivity (kg/ha/day) and broadening diversity, while reducing inputs of water, fertilizer, fuel and pesticides.

Purpose of the book

The purpose of this book is to facilitate accuracy in communication among individuals working in agriculture, commerce, regulation, and other fields dealing with vegetable crops; encourage consistency in terminology; and better understand the extent and diversity of vegetable production and utilization worldwide. The book represents an essential reference source for research scientists to gardeners, with the information presented in a concise and readily accessible format. It provides up-to-date scientific names, synonyms, and vernacular names for the commercially cultivated vegetable crops of economic importance worldwide (402 crops), in addition to information on the plant parts utilized and their method of preparation. Vernacular names from 370 languages are presented along with information on each of the languages.

Increasing global economic interdependence and trade in agricultural products makes precise communication among individuals utilizing different languages essential. There currently is a wide range of vegetables shipped around the world as seasonal, economic and other forces are shifting markets from exclusively local toward global. This is especially so for crops that cannot be grown in the importing country or have a limited season of availability due to climatic conditions. Vegetables that are more probable candidates for import/export markets typically have sufficient postharvest longevity, allowing them to reach retail markets within the importing country in excellent condition.

Increased international exchange of agricultural products, as well as travel, too frequently results in confusion concerning the identity of many crops. Likewise, when new crops are introduced to an area, the common name used by the group introducing the crop is often maintained. In some cases, different names are utilized in neighboring markets run by the same or other ethnic groups. When the large number of vegetable crops is combined with the extremely large number of languages worldwide, the number of common names is in the tens of thousands. As a consequence, starting with a common name in an unfamiliar language, it can be difficult, if not impossible, to determine the crop in question. The information in this text allows indentifying such crops and is therefore of use to university and governmental researchers, libraries worldwide, agricultural organizations, agricultural scientists, embassies, international travelers, vegetable growers, shippers, packers, produce buyers, grocery store managers, gourmet restaurants, chefs, and gardeners.

Linguistic value of common names

The study of the origin of cultivated plants and the forces driving their introduction into other areas of the world utilizes an interdisciplinary approach that may involve genetics, linguistics, history, anthropology, archaeology, ethnology, economics, geography, food science, horticulture, plant pathology, climatology, and disciplines studying human behavior. Linguistic evidence, in some cases, can be exceptionally helpful in identifying the timing and therefore sequence of crop domestication and/or introduction into a particular geographic location (Alpern, 2008). The origins of taro (Williamson, 1993) and maize (Blench *et al.*, 1994) in West Africa were inferred from the distribution of the vernacular names among local languages and historical data. The names for maize in southern Nigeria indicated a northern, trans-Saharan origin instead of introduction by the Portuguese. According to Diamond (2005), the names for crops in the Niger-Congo language family in southern Nigeria can be separated into three groups; (1) those in which the words for a particular crop are very similar in all of the languages due to the crop probably being first domesticated there; (2) those in which the names are consistent within only a small subgroup of southern Nigerian languages and probably represented crops of Indonesian origin; and (3) names that are inconsistent among the languages and are thought to represent New World introductions. For a number of crops that are more recent introductions around the world, the vernacular name across many languages is fairly consistent. When tomato (*Lycopersicon esculentum* var. *esculentum* P. Miller) is contrasted with onion (*Allium cepa* L. Cepa group), a crop in which wild species are relatively pandemic, there is a tremendous range in the names of the latter across languages. Comparing the names for tomato/onion across several languages illustrates the point: (AFAR) tamaàtim/bàsal; (AFRIKAANS) tamatie/ui; (ALBANIAN, TOSK) domate/qepë; (AMHARIC) tomatim/gäyy šənkurt; (ARABIC) tamatum/bassal; (ARMENIAN) tomat/soch; (BAFUT) tòmátò/ángyɛ̌sə̀; (BAOULÉ) tomati/djaba; (BASQUE) tomate/kipula; (BEMBA) matimati/kanyense; (BEROM) tùmatùr/àlbásà; (BÉTÉ) tomaté/djaba; (BRETON) tomatez/ognon; (BULGARIAN) domat/luk; (CAROLINIAN) tumóótis/subwuyas; (CATALAN-VALENCIAN-BALEAR) tomatera/ceba; (CEBUANO) kamatis/sibuyas; (CHEROKEE) tamahli/svgi; (DAGAARE, SOUTHERN) tomati/zanmē; (DANISH) tomat/kepa-løg; (DENDI) tìmátì/albasa; (DUTCH) tomaat/ui; (ENGLISH) tomato/onion; (ESTONIAN) tomat/sibul; (ÉWÉ) tomatos/sabala; (FINNISH) tomaatit/sipuli; (FRENCH) tomate/oignon; and (GIANGAN) kamatis/sibuyas.

Derivation of common names

The derivation of common names occasionally represents an intriguing story. For example, the name for *Helianthus tuberosus* L. in a cross-section of European countries is "topinambur" [or a close variation thereof (Kays and Nottingham, 2007)] the use of which began in France in the early 1600s. A French expedition to South America returned to Paris in 1613 with six members of a warlike tribe of the Guaranis race, known as the Topinambous, who were taken from the Isle de S. Luiz de Maranhão in Brazil (De Malherbe, F., 1862). These exotic looking natives created a great attraction that more or less coincided with the introduction and sale in Parisian markets of the equally exotic and somewhat uncouth looking tubers of *H. tuberosus*. Street hawkers adopted the term topinambou to draw attention to their new offering (Salaman, 1940).

Buck (1949) lists the synonyms for the modern English word "cabbage", describing the apparent derivation of names in a cross-section of Indo-European languages. The word appears to have evolved from the Greek kavλós (stalk) and subsequently caulis (stalk) in Latin. The subsequent similarity in names is quite striking: cavolo (Italian), col (French), col (Spanish), cāl (Old Irish), cabāiste (New Irish), cawl (Welsh), kaolenn (Breton), kāl (Old Norse), kaal (Danish), kål (Swedish), cawel (Old English), cole, caboche (Middle English), cabbage (New English), kool (Dutch), cōl (Old High German), kōl (Middle High German), and kohl, kraut (New High German). English words sauerkraut and coleslaw evolved from several of these words. In Eastern Europe, the names for cabbage appear to have been derived from 'caput' (head) to give the following: kopustas (Lithuanian), kāposts (Lettish), kupus (Serbo-Croatian), kapusta (Czech), kapusta (Polish), and kapusta (Russian). For many vegetables, however, there is relatively little information on the derivation of their names.

The first usage of a word for cabbage in the English language was in 1440 (Royal Society, 1790) "Take cabaches and cut hom on foure..and let hit boyle." In 1495 Caxton (Jerome, *et al.*, 1495) wrote "He laboured the gardins, sewe the seedes for cabochis, and colewortes" and subsequently in 1580 Baret wrote "Cabage, or colewoort, *brassica*. Cabage, or cole cabege, *brassica capitata*." The word cabbage appears to have evolved from colewort (Simpson and Weiner, 1989).

Using the text

The publication contains of the following information:

- Common Names of Individual Crops
  a. Listed by division, family, genus and species (Chapter 1).
  b. Listed alphabetically (Chapter 2) with each of the common names given a numerical designation for the family and species, followed by the language.
- Scientific Names and Synonyms
  Chapter 3 lists the correct scientific name (in bold with the family and species numerically keyed back to the taxonomic list) along with synonyms that are followed by the correct genus and species.
- Crops listed by Plant Part Consumed (Chapter 4)
- Crops Listed by Method of Preparation (Chapter 5)
- Languages
  a. Language critiques – A short synopsis is given for each language listed (Chapter 6) that includes language synonyms, number of speakers, geographical location where spoken, language family classification, and dialects.
  b. Language synonyms – Since many languages have more than one name, alternative names are also listed and keyed back to the appropriate name (Chapter 7).
  c. Dialects – A number of languages are known by their dialect name. Chapter 8 keys the dialects to the language name utilized in the text.
- References
- Appendices
  1. Diacritics.
  2. Unique letters/symbols and alphabetization.

Starting with a common name and not knowing the language or scientific name of the plant, it is difficult to determine the species, and hence alternative names in the same or other languages using a taxonomic list alone. The information presented in this book evolved from a journal publication (Kays and Silva Dias, 1995) covering 15 languages. The current list greatly expands the amount of information presented and has transformed it into a much more readily accessible format.

To find a crop starting with the scientific name (e.g. *Aalius androgynus* Kuntze), look the binomial up initially in Chapter 3 to ascertain if it is correct or a synonym. In this case, *Aalius androgynus* Kuntze is a synonym for *Sauropus androgynus* (L.) Merrill that is followed by a numerical code (39.5) that keys back to Chapter 1. The number 39.5 indicates family 39 (Euphorbiaceae) and crop number 5 under that family. Starting with Division Sphendophyta, the families are listed in numerical order.

To find the scientific name of a crop starting with a common name, look up the common name which is arranged alphabetically in Chapter 2. For example, "kangkong" is followed by a numerical code (36.2). In Chapter 1, 36.2 corresponds to the family 36 (Convolvulaceae) and crop 2 (*Ipomoea aquatica* Forssk.). Looking over the names for the crop in different languages, "kangkong" is found to be a vernacular name for the crop in Tagalog, Malay, Ilocano, Pangasinan, Tok Pisin, Dutch and others.

To find all of the crops that are grown for a specific edible part (e.g. tubers), use Chapter 4. Under the heading of "Tuber" the individual crops are listed by their English (Chapter 4a) and scientific names (Chapter 4b). The edible part (EP) is also indicated in Chapter 1, to the right of the scientific name. When an additional part (or parts) of the plant has a secondary use, it is designated with a superscript "s" (e.g. root^{s}).

The methods of preparation (P) have been separated in Chapter 5 into the following classes (R - raw, C - cooked, P - preserved). Preservation includes pickled, salted, jellied, or preserved by other methods, excluding canning and freezing. When more than one method of preparation is employed, they are presented in the general sequence of frequency of use. As with the edible plant parts (Chapter 4), the plants categorized under each method(s) of preparation are listed by their English (Chapter 5a) and scientific names (Chapter 5b).

Crops

"Vegetable crop" is a somewhat nebulous term that usually denotes a plant cultivated for an edible part or parts and generally excludes sweet fruits and dried seeds. Vegetable is not a scientific term but rather one that is based largely on culinary and cultural traditions. Vegetables can be consumed raw, cooked, or preserved in some manner and prepared in innumerable of ways, though generally not as desserts. However, exceptions are common; for example, rhubarb (*Rheum rhabarbarum* L.), considered a vegetable, is almost invariably served as a dessert. Likewise, certain members of the Cucurbitaceae, such as watermelon (*Citrullus lanatus* (Thunb.) Natsum & Nakai) and cantaloupe (*Cucumis melo* L. Cantaloupensis group), are technically sweet fruits; however, their culture is essentially the same as a number of other members of the Cucurbitaceae that are vegetables and as a consequence, are commonly grouped together as vegetables. Therefore, what is considered a vegetable varies somewhat among the diverse range of individuals (e.g. scientists, government officials, gardeners, chefs) dealing with this food group.

In the text, vegetables are crops that are: (1) herbaceous or largely herbaceous plants or plant parts; (2) consumed by humans; (3) eaten raw, cooked, or preserved; and 4) typically eaten as part of the main course or appetizer, rather than as a dessert (Chittenden, 1951; Fritz *et al.*, 1989; Gove, 1971; Huxley, 1992; Parker, 1989; Seymour, 1990; Soule, 1985). The list of vegetables was further restricted by excluding: (1) non-cultivated species; (2) lower organisms (e.g. fungi); (3) most trees and woody shrubs; and (4) plants grown in or gathered from salt water (e.g. sea weeds). The list has several exceptions that were included: (1) banana inflorescence (*Musa paradisiaca* L. var. *paradisiaca*) and plantain fruit (*M. paradisiaca* L. var. *paradisiaca*) that are from herbaceous trees; (2) several species that are traditionally utilized as condiments but are consumed in sufficient quantities in salads in some areas of the world (e.g. basil or mint); (3) several species that could be considered woody shrubs but are cultivated as leafy vegetables, such as *Cnidoscolus aconitifolius* (Miller) Johnston and *C. chayamansa* McVaughn or as a root crop, *Manihot esculenta* Crantz.; and (4) a few that are sufficiently sweet as to be considered a dessert fruit.

The list of crops was compiled with the help of horticultural and botanical authorities from over 40 countries, representing diverse climatic and ethnic regions.

Scientific names

Modern plant taxonomy began with publication of Species Plantarum by Linnaeus (Von Linné, 1753). While there was some initial resistance to acceptance of the binomial system established by Linnaeus (e.g. "... attempts to reduce the names of plants to a ſyſtem, has rendered the ſtudy more difficult and more ſubject to error, than it would have been if the Student had only uſed his ſight for the diſtinguiſhing of plants, and his memory for regiſtring them." Brookes, 1763), it fairly rapidly became the universal standard.

Plant taxonomy is based on a hierarchy of categories to which taxa (plants) are assigned. The hierarchy, established by the International Code of Botanical Nomenclature (McNeil *et al.*, 2005), includes the order, family, genus, and species with each progressive category being more specific. Families are groups of plants that have distinctive reproductive and vegetative features and are subdivided into genera and species. Genera represent a category in which the included species have more characteristics in common than species of other genera and therefore represent an aggregation of closely related species. Individual species are characterized by distinct morphological differences from other closely related species and can be further subdivided into subspecies, variety and form. These latter categories are used to characterize variation within a population of plants of the same species.

A scientific name is not complete without indicating the author or authors who first described the taxon. These names are critical for accuracy and assigning priority. Authors who have named a number of taxa often have their names abbreviated. For example, Linnaeus is shortened to L. and A.P. de Candolle to DC. (see Brummitt and Powell, 1992 for a list of these abbreviations).

There are several variations in how the author is cited after the binomial that concur with established rules. For example in the binomial *Centella asiatica* (L.) Urban, the original author (Linnaeus) is listed in parenthesis when a later author (Urban) transferred the original species to another genus or rank. When the author designation has an "ex" connecting two authors (e.g. *Allium tuberosum* Rottler, ex Sprengel), the first author (Rottler) initially proposed the name but did not validly publish it; the second author (Sprengel) validated the name while citing the first author.

Synonyms

In the early days of plant classification, due in part to difficulties in communication, it was not uncommon for a specific taxon to be assigned a Latin binomial by more than one individual. A synonym is created when a taxon is given different scientific names by separate authors. Synonyms can greatly complicate the identification and communication of a species of interest. According to the International Code of Botanical Nomenclature (McNeill *et al.*, 2005), the first author naming the taxon has priority. Subsequent names are deemed synonyms and their use is incorrect. Synonyms are always referred to as a synonym of a particular taxon (e.g. *Acmella caulorrhiza* Delile is a synonym of *Acmella oleracea* (L.) R.K. Jansen). In the text, finding the correct vegetable crop in the botanical list (Chapter 1) is difficult, if not impossible, when starting with the incorrect scientific name. Therefore, synonyms for the crops are listed in Chapter 3, with the correct binomial indicated in bold, followed by a number that indicates the family and the specific crop within the family in Chapter 1 (e.g. 24.1. indicates the 24th family and the 1st crop).

Changes in scientific names

Changes in nomenclature occur due to the fact that the classification of a taxon is not absolutely fixed. While there are specific rules with regard to assigning binomials, there remains considerable debate on the phylogenetic level. Some taxonomists are more inclined to separate groups of similar plants into discrete species while others are more inclined to lump them together into one or a small number of species. The advent of molecular genetics and the emerging use of molecular evidence and cladistic analysis have further stimulated proposals for changes in existing classifications. For example, molecular evidence supports the tomato (*Lycopersicon esculentum* var. *esculentum* P. Miller) belonging in the Solanum genus (e.g. *Solanum lycopersicon* Blanco). As a consequence, there is a continuing debate between developing a more phytogenetically meaningful classification and that of preserving nomenclature that has been used, in some cases, for a century or more.

Another taxonomic problem is that a number of cultivated species have undergone considerable genetic manipulation by plant breeders. In certain instances, there are new forms of plants produced that represent hybrids between two species. These are generally denoted, as in the edible chrysanthemum, with the binomial for each species separated by an × sign (e.g. *Chrysanthemum coronarium* L. × *Chrysanthemum morifolium* Ramat.). Plant breeders continue to produce a diverse selection of cultivated plants that are termed "cultivars", generally abbreviated as "cv". A cultivar is a cultivated variety of a species that had been selected for specific desirable characteristics (e.g. fruit size, shape, color, flavor, postharvest longevity) that are genetically fixed with the offspring retaining the plant's unique characteristics from one generation to the next.[1] Cultivars are reproduced under carefully controlled conditions, especially when they are a cross-pollinated species, and many are now routinely patented to prevent unwarranted sale by competitors.

Although the practice is declining, the term "variety" continues to be widely and incorrectly applied to cultivars. Variety is a taxonomic category and not an agricultural term. Cultivar names are not in Latin, begin with a capital letter and can be listed after the scientific name (e.g. *Zea mays* L. ssp. *mays* cv. Silver Queen).

In the taxonomic treatment herein, several closely related species in which the edible parts are very similar are grouped under the genus name (e.g. Dendrocalamus, Amaranthus) and the common names are listed collectively. When distinct names for particular species are available, they are indicated using a numerical superscript at the end of the common name indicating the species [e.g. *Dendrocalamus* spp. (e.g. *D. asper*[1], *D. brandisii*[2], *D. giganteus*[3], *D. hamiltonii*[4], *D. latiflorus*[5], *D. membranaceus*[6], *D. strictus*[7]) or *Amaranthus* spp. (*A. albus*[1], *A. caudatus*[2], *A. cruentus*[3], *A. dubius*[4], *A. hybridus*[5], *A. lividus*[6], *A. mangostanus*[7], *A. spinosus*[8], *A. sylvestris*[9], *A. tricolor*[10], *A. viridis*[11])]. Therefore "sraha nsoe[8]" is an Akan name for *Amaranthus spinosus* L.

With some species, breeders have created such diversity in morphology that they are segregated using group categories (e.g. *Allium cepa* L. Cepa group, *Allium cepa* L. Aggregatum group). While several authors have separated crops within the Brassicaceae into groups (e.g. *Brassica oleracea* Botrytis group vs. *B. oleracea* L. var. *botrytis* L.), the older, more widely adopted botanical variety designation has been utilized in the text. In *B. oleracea*, this treatment was followed, taking into consideration the morphological diversity and cultural traditions associated with the kales and other compacted rosette coles like tronchuda cabbages (*Brassica oleracea* L. var. *costata* DC.).

---

[1] With the exception of hybrids.

The taxonomy within some of the Cucurbitaceae is also a problem, with distinctions among some crop types occasionally being nebulous at best. For example, pumpkins can be found in both *Cucurbita maxima* Duch. and *C. moschata* Duch. ex Poir. Therefore common names, especially more generic ones, may be assigned to multiple species. In the Cucurbitaceae, classification is further complicated by hybridization among some species.

Taxonomic sources

Taxonomic divisions follow Raven, Evert, and Eichhorn (1992) in the text, with classification terminology within parentheses following Cronquist (1988). Alternate family names, that have the correct ending (*-aceae*), have been sanctioned by the Code [e.g. Apiaceae (Umbelliferae), Asteraceae (Compositae), Brassicaceae (Cruciferae), Fabaceae (Leguminosae), Lamiaceae (Labiatae)] and adopted by many taxonomists. The newer version has been used in Chapter 1. Genus and species names, where applicable, largely follow the USDA-ARS Germplasm Resources Information Network (GRIN) of the National Plant Germplasm System (NPGS) treatment <http://www.ars-grin.gov/cgi-bin/npgs/html/taxecon.pl>. In several instances where large numbers of species have been used similarly (e.g. amaranths, lily), the group is condensed under the genus, with a representative list of the species grown, though not as individual crops. Where distinct differences are evident among subspecies, the nomenclature used in the primary country of utilization is adopted. For example with *Brassica juncea* (L.) Czern., a vegetable that originated in China and has undergone extensive morphological manipulation in breeding programs there, the classification system presented by Yang *et al.*, 1988 is adopted.

Common names

Sources

A publication of this nature would be impossible without the help of a number of horticultural and botanical authorities from around the world (see Acknowledgements). Vernacular names have also been gleaned from an extremely wide range of published sources.

Alphabetization

In a number of instances (e.g. some of the African languages), the 26 Latin characters used as the basis for the English language are not adequate for creating all of the sounds required. As a consequence, it has been essential to include several AfroRoman characters as well as certain other unique characters from other languages. Where multiple languages are used, even when using the same alphabet, the appropriate alphabetical order is not always clear. For example, in French é and ô are simply accented letters while á, í, and ö in Icelandic are distinct letters. The common names listed herein (e.g. Chapter 2) are alphabetized in the following order: a/ɑ, æ/ae, b/ɓ, c/ɔ, d/ɗ/ɖ, e/ɛ/ʃ/ə, f, ƒ, g, h/ɣ, i/ı, j, k/ƙ, l, m, n/ɲ/ŋ, o, p, q, r, s, ß, t, u, v, w, x, y/ƴ, z, δ, η. A / between letters indicates that they are handled as equivalents with regard to alphabetization. Diacritics (see Appendix I) are ignored during alphabetization. For a more detailed explanation of non-Latin characters see the Appendix II.

The crop names are derived from a number of alphabets (e.g. Hebrew, Greek, Cyrillic, Armenian, Arabic) and logograms (e.g. Chinese, Korean, Japanese). Presenting the names in a uniform text necessitates transcription or transliteration of unique characters into the Latin alphabet (Romanization) or in some instances, AfroRoman characters. Conversion can vary among individuals, resulting in slight differences in spelling for a specific word. Alphabet and logogram alterations come from the source of the names and are accepted as received. When variation in spelling is present between two sources, both versions are included.

Most appropriate name

When possible, the most widely used common name has been listed first for each language, followed by less frequently used names. However, the source for some languages did not indicate the preferred name; as a consequence, in a number of instances, accurately prioritizing the names is not possible. In others several names are comparable in the frequency of use. The priority for the names may also differ among countries using the same language. For example, Crioulo is a Portuguese Creole language that is spoken in Cape Verde as well as certain countries in West Africa (e.g. Guinea-Bissau, Senegal). Many Crioulo names differ markedly with location. The situation is made even more complex when the most appropriate common name varies with location within a country as in the case of a number of Spanish names for vegetables in Mexico. In some instances, due to the complexity of distinguishing the most appropriate name for individual locations, the common names are simply listed. As a consequence, the first name listed may not necessarily be the most frequently used.

When foreign names are used locally ("loan words"), they have been included [e.g. "aubergine" for *Solanum melongena* L. in England versus "eggplant" in the United States]. The occurrence of loan words is fairly common, especially with more recent introductions. For example oca, a Quechua word from the Cusco region of the Andes for *Oxalis tuberosa* Molina, has been adopted by several languages [oka (Czech), oka (Danish), oca (English), oca (Portuguese), oca (Spanish), oka (Vlaams)] even though the crop is not grown or consumed in most locations where the languages are spoken. The absence of a name for a crop in the language, the willingness to adopt loan words, and the relative ease of pronounciation by speakers of the language adopting the name are factors affecting adoption.

Many of the crops listed lack names in one or more languages. Likewise, the presence of a name in a language does not necessarily indicate that the crop is grown in the country where the language is spoken.

Number of names

The list does not include all possible common names for a crop within a language and the level of thoroughness varies somewhat among languages. Hung *et al.* (1992), for example, lists 34 common names for tomato in Chinese. While the Chinese characters for a vegetable name are uniform across dialects, the pronunciation of the name can vary widely depending upon the dialect. Therefore, the preferred common name can vary with dialect, as well as with location. Including every name for each language was simply not feasible due to the number of vegetable crops. What has been included, in the case of 8 Chinese dialects covered, is the name designated by Hung *et al.* (1992) as the most widely used. While this name may not be the most appropriate across all of the dialects and/or locations in which Chinese is spoken, a more detailed treatment was not within the scope of the current work.

Lack of a common name

There is an extremely wide range in the number of common names listed across languages. Inuktitut and Aleut, for example, are used in the very northern latitudes where the growing season and/or temperature tremendously limit the vegetable species that can be successfully cultivated. As a consequence, there are very few Eskimo names for vegetables. In contrast, speakers of certain other languages very readily coin names for new crops, even those that are not grown or, in some cases, even known where the language is spoken. In some instances, the new name simply represents the name used in the country of the plant's origin. This is often the case when the language in question is on the wane and the names for new crops are generally derived from the replacement language (e.g. English names for vegetables being used by many speakers of Native American Indian languages).

The existence of a name in a particular language does not necessarily mean a vegetable is cultivated in the area where the language is spoken. Conversely, the absence of a name does not necessarily mean that the crop is not cultivated. As indicated earlier, the existing collection of vernacular names is a "work in progress" and there are numerous additional names and languages that would be advantageous to eventually include.

Generic terms

Generic names, names that refer to a general kind or group of plants or plant products, are widely used in cultivated crops. Unfortunately in some instances, generic names can create a problem due to their lack of specificity. Bean or beans, for example, can be correctly applied to the edible fruit or seeds of a number of legumes that are used as vegetables. So when one refers to a bean, is it a green, lima, sword, jack or other type of bean? Likewise, corn (maize) can be equally applied to sweet corn, field corn, popcorn and other forms. The term corn is also used for wheat (*Triticum aestivum* L.) and oats (*Avena sativa* L.) in Europe and occasionally for certain other cereals. Generic names/terms are commonly used in a number of languages. In some cases, they indicate all forms that are not distinguished using more specific names. A similar situation occurs for the name used for both banana and plantain in some languages, where the two plant types are not distinguished. Likewise, the generic term "mint" includes *Mentha spicata* L., *Mentha pulegium* L. and a number of wild mint species. Spinach is the English name for *Spinacia oleracea* L.; however, spinach is also used for other leafy greens such as Amaranthus spp. in Africa and *Basella alba* L. in parts of Asia. In such cases, it is important to be cognizant of the setting in which the name is used. The potential for confusion when using generic names is a primary reason to know and, when appropriate, use the scientific name for a crop in question.

Dried crops

Vegetables are routinely thought of as relatively succulent, fresh products; however, in some instances, they are dried (e.g. beans, corn, peas, radish leaves) or altered in some manner before using, often to increase the postharvest longevity of the product. A number of the legumes are consumed as both fresh and dried products (e.g. cowpeas, *Vigna unguiculata* (L.) Walp. ssp. *unguiculata*). Dried seed is the prevalent form in which cowpeas are utilized worldwide, though in the United States the seeds are generally harvested and used prior to significant drying in the field or after harvest.

Languages

While there are over 6800 languages worldwide (Gordon, 2005), 96% percent of the people of the world speak only 4% of the languages. In contrast, 25% of the languages are spoken by fewer than 1000 people. It is currently estimated that a language becomes extinct every two weeks. Sabah and Sarawak in Malaysia are now extinct and there are similar losses elsewhere. In the United States, the number of monolingual speakers for many Native American languages has declined markedly [e.g. Kiowa Apache (18), Caddo (25), Kickapoo (6)] with the remaining speakers generally limited to older adults (Gordon, 2005). The death of languages results in a loss in language diversity, ethnic identity, a repository of history, a part of human knowledge, and simply interesting subjects in their own right (Crystal, 2000). While a few previously undiscovered languages are identified each year, they do not numerically compensate for those lost and are often also in a precarious position with regard to their own future.

Language families

A language family is a group of related languages that have evolved from a common ancestor (i.e. a proto-language). The objective herein was to include as many language families in the text as possible, and while the number of families less than desired, the text does provide the most detailed cross-section of families to date. Languages families and the number of languages within each are: Afro-Asiatic (14), Algic (4), Altaic (9), Austro-Asiatic (5), Austronesian (61), Dravidian (6), Eskimo-Aleut (3), Hmong-Mien (2), Indo-European (69), Indo-Iranian (1), Iroquoian (2), Kartvelian (1), Khoisan (2), Kiowa Tanoan (1), Misumalpan (1), Muskogean (2), Na-Dene (2), Northeast Caucasian (1), Niger-Congo (128), Nilo-Saharan (6), Quechuan (1), Sino-Tibetan (18), Siouan (2), Tai-Kadai (4), Uralic (3), Uto-Aztecan (3), language isolates (4), Creole (8), mixed (1), unclassified (6).

Individual languages

In Chapter 6, a very brief critique of each of the languages is presented that includes the family and further sub-categorization, synonyms, number of speakers, and locations where spoken. Individual languages in the text are: Abua, Adamawa Fulfulde, Afar, Afrikaans, Agta, Aja, Akan, Akan-Akuapem, Akan-Asante, Akan-Fante, Akha, Aklanon, Akpa, Akwa, Aleut, Amharic, Ando, Anuak, Anyin, Arabic, Armenian, Assamese, Baatonum, Badaga, Bafia, Bafut, Baga, Balanta-Kentohe, Balti, Balkan Romani, Bamanankan, Bamileke, Baoulé, Basque, Bassari, Belarusan, Bemba, Bembé, Bengali, Berom, Bété, Bete-Bendi, Binukid, Blackfoot, Bodo, Bokyi, Bontoc Central, Bosnian, Botolan Sambal, Breton, Budik, Bughotu, Buhid, Bulgarian, Bullom So, Bulu, Burmese, Butbut Kalinga, Butuanon, Cajun French, Cape Verde Crioulo, Car Nicobarese, Carolinian, Casiguran Dumagat Agta, Catalan-Valencian-Balear, Cebaara Senoufo, Cebuano, Central Bicolano, Central Khmer, Central Palawano, Central Sama, Chechen, Cherokee, Chickasaw, Chuwabu, Comanche, Croatian, Cusco Quechua, Czech, Dagbani, Dakota, Dangme, Dangme-Krobo, Danish, Dendi, Digo, Ditammari, Doondo, Dutch, Eastern Canadian Inuktitut, Eastern Farsi, Eastern Maninkakan, Eastern Panjabi, Ede Nago, Edo, Efik, English, Estonian, Éwé, Ewondo, Fijian, Filipino, Finnish, Fon, French, Fulfulde, Ga, Gaddang, Galician, Gàn, Ganda, Gbaya, Gbe-Vhe, Gen, Georgian, German, Giangan, Gikuyu, Gola, Gonja, Gourmanchéma, Gourmantché, Greek, Gujarati, Gun, Guro, Haitian French Creole, Hakka, Hanunoo, Harari, Harsūsi, Hausa, Hawaiian, Hebrew, Herero, Hiligaynon, Hindi, Hindustani, Hmong, Daw, Hmong-Njua, Hopi, Hungarian, Ibaloi, Ibanag, Icelandic, Iceve-Maci, Ifugao, Igbo, Ilocano, Indonesian, Irish Gaelic, Isnag, Italian, Itawit, Ivatan, Izon, Japanese, Javanese, Jicarilla Apache, Jola-Fonyi, Jula, Junkun Takum, Kabiyé, Kalagan, Kankanaey, Kannada, Kanuri, Karen, Kashmiri, Kati, Kazakh, Kiowa, Kirghiz, Kiribati, Konkani, Konkomba, Kono, Koongo, Korean, Koro, Krio, Kung-Ekoka, Kuranko, Kwangali, Kwanyama, Kwéyòl, Ladino, Lahu, Lakota, Lao, Latvian, Limba, Lisu, Lithuanian, Loko, Lower Sorbian, Luo, Luxenbourgeois, Luyia, Macedonian, Madura, Maguindanao, Maithili, Makhuwa, Malagasy, Malawi Lomwe, Malay, Malayalam, Malimba, Maltese, Mandarin, Manding-Bambara, Manding-Mandinka, Manding-Maninka, Mandinka, Mandjak, Maninka, Mankanya, Manobo, Manu'a Samoan, Manx Gaelic, Maori, Maranao, Marathi, Marinduque, Marquesan, Marwari, Mehri, Mende, Micmac, Min Bei, Min Nan, Minangkabau, Mizo, Mohawk, Mongolian, Mòoré, Morisyen, Mundari, Muskogee, Nama-Damara, Navajo, Ndali, Ndonga, Nepali, Newar, Ngizim, Ngundi, Niue, Northern Azerbaijani, Northern Kissi, Northern Kurdish, Northern Pashto, Northern Sotho, Northern Thai, Norwegian, Nupe-Nupe-Tako, Nyakyusa-Ngonde, Nyanja, Ogbia, Ogbia-Agholo, Oriya, Pali, Pampangan, Pamplona Atta, Pangasinan, Plains Cree, Polish, Portuguese, Potawatomi, Pulaar, Pular, Pumé, Rarotongan, Romanian, Ronga, Russian, Samoan, Sanga, Sanskrit, Santali, Sarangani Blaan, Scots, Scottish

Gaelic, Sena, Serbo-Croatian, Serer-Sine, Shan, Shona, Sidamo, Sinhala, Slovak, Slovenian, Somali, Songhay, Southern Bobo Madaré, Southern Dagaare, Southern Kisi, Southern Paiute, Southern Sotho, Spanish, Sranan, Stellingwerfs, Subanon, Sumo-Mayangna, Sunda, Susu, Swahili, Swedish, Tagalog, Tagbanwa, Tajiki, Tamil, Tatar, Tausug, Teke, Telugu, Tem, Thai, Themne, Tibetan, Tigrigna, Tinà Sambal, Tiruray, Tiv, Tok Pisin, Tonga, Tongan, Tosk Albanian, Tsonga, Tswana, Tulu, Tumbuka, Tupuri, Turkish, Turkmen, Twi, Ukrainian, Upper Guinea Crioulo, Upper Sorbian, Urdu, Uyghur, Uzbek, Vai, Venda, Vietnamese, Vlaams, Waama, Waci Gbe, Wamey, Wáray-Wáray, Welsh, West Central Oromo, West Greenlandic Inuktitut, Western Farsi, Western Maninkakan, Wolof, Wú, Xhosa, Xiang, Yala, Yalunka, Yao, Yiddish, Yom, Yoruba, Yue, Zarma, Zulu, Zuni.

The language names used correspond to the preferred names identified in *Ethnolofue Languages of the World* (Gordon, 2005). If a language of interest is not found listed above, consult Chapter 8 that lists synonyms.

Language critiques

A short critique of each of the languages is provided in Chapter 6. It includes synonyms for the language, number of speakers, geographical locations where the language is spoken, language family classification and dialects. Each language has a three letter ISO/DIS 639-3 identity code, established by the International Organization for Standardization. The three letter identifier provides a condensed means of enumerating living, extinct, ancient and other languages and has many uses, in particular in computer systems. In instances within the text where the identity of a language is not definitive, a temporary code in italics has been assigned.

Language synonyms

Many languages have a number of synonyms for individual vegetables. For example, Gourma, Gourmantche, Gurma, Migulimancema, Goulmacema, Gulmancema, Gulimancema, Gourmantché are synonyms for Gourmanchéma. In some instances, a synonym or synonyms are currently or were previously widely used. When reviewing earlier literature synonyms are routinely encountered. See Chapter 7 for an alphabetical list of language names and synonyms with the latter keyed to the name utilized in Chapter 1 and 2 of this text.

Dialects

In some instances, certain languages are identified by their dialect name. Chapter 8 lists the dialects of the languages covered in the text and keys back to the correct language name.

Acknowledgements

A publication of this nature would be impossible without the help of a number of plant scientists from around the world. In many instances, the compilation of the names for the languages entailed considerable time and effort and I am deeply indebted to these individuals. In particular, I would like to acknowledge the significant contributions made by I.C. Silva Dias, Chris Starr and Betty Schroeder. Other individuals that have contributed to either compiling the list of cultivated vegetables or providing common names for the manuscript are: Français Acadien, F. Adamicki, Saida Agaueva, S. Alexander, Y.M. Allain, A.M. Allam, Noor ul Amin, A. Andásfalvy, J. Apelaud, Gardar R.Arnason, D. Astley, Khin Swe Swe Aye, B.Bar-Tel, H. Baugerød, Jannick Bayon, A.S. Bindra, Sistske A.Bloemhoff, Conrad Bonsi, D.R. Caccioni, A. Casas, Mihaela Cerne, Alfonso Cerrate, Satish Chandra, N.J. Chen, Lee Kiang Chua, A.F. Coertze, Constantinos Dagras, M.J. Day, Elias Dekazos, P.H.M. Dekker, A.P.M. deu Nijs, Ivo Djinovic, C.C. Doyle Hugues Dupriez, Z-ca Dyrektora, Charles Ellis, Elazar Fallik, M.E.S.Ferreira, K.J. Fisher, D. Fritz, Richard Gáborjányi, Alex Garcia-Lagunzad, Jr., Tatjana A. Garnileuko, J.B. George, David E. Giannasi, Valdimir I. Gontcharenok, Noel Govinden, M.Y.D. Gumedzoe, O.V. Hansson, S.S. Harjadi, Styati Harjadi, R.M.R. Harun, H. Haung, Maret Herr, Hikmat Hisoriev, Ole Høst, Marcel Houinato, L. Hung, F. Serracino Inglott, Sabitrie Jairam-Doerga, Hans Arne Jensen, S.B. Jones, J.B.D.S. Kahandawels, S. Kamimura, George Kessler, S. Ketsa, Shahida Khalid, William Kisaalita, M.H. Kistner, Jared S. Klein, S. Klug-Andersen, Michael Kolomiets, Dimitri Korulkin, Louise B. Koumouno, G. Kuo, Lydia A. Ladape, E.W. Laubscher, Stephen Lewthwaite, Marcela Libomobo, M.E. Ling, Xingui Liu, S. Lopex, L. López-Bellido, Risto Lozanovski, G.H. Massiha, Alain Mbaye, René Bikomo Mbonomo, Z. Milakovic, Keigo Minami, Richard Molinar, E. Moschini, E. Müller, Esther Murugi Kahangi, Robert Mwanga, Rupert Nelson, Torsten Nilsson, Stacia Nordin, Hasim Ogut, C.M. Olympios, I.C. Onwueme, Arija Ozola, K.H.S. Peiris, J.Y. Peron, Raili Pessala, Uzi Plitmann, A. Posthoff, C.S. Prakash, S. Qaiser Abbas, Lila Raeva, E.T. Rasco, Jr., Valerijus Rašomavičius, Nerius I. Roperos, Pierre Michel Saint-Clair, H.O. Saxena, A. Shadeque, João C. Silva Dias, Brice Sinsin, Otto Soemarwoto, C. Zana Somda, Ki-Cheol Son, Etien Dibié Théodore, R. Tillaev, H.D. Tindall, R.T. Toledo, T.S. Tsendeekhuu, Özlem Tuncay, Yüksel Tüzel, J. Van Mullen, A. Vanachter, G. Varga, J.J. Vasquez, Japaridze Vazha, Qiao-mei Wang, Yan Wang, D. Wei, J. Williamson, K.F. Lan Chow Wing, Ding-hua Wu, Joan Vallès Xirau, S. Yui, Václav Zelený.

Chapter 1. Common names of the cultivated vegetable crops of the world, listed by division, family, genus and species

Division Sphendophyta[1]

1. Equisetaceae

1.1. *Equisetum arvense* L. (EP)[2] young strobili; (P)[3] C,P.

(AFRIKAANS) perdestertvaring; (ARABIC) el-konbath, zail el-hosam; (BELARUSAN) hvost poljave; (CHINESE, MANDARIN) wen jing; (CHINESE, WÚ) wen jing; (CZECH) přeslička rolní; (DANISH) ager-padderokk; (DUTCH) heermoes; (ENGLISH) horsetail, field horsetail; (ESTONIAN) osi; (FINNISH) peltokorte; (FRENCH) prêle des champs; (GERMAN) Zinnkraut, Acker-Schachtelhalm; (GREEK) lippoyris, dolycombi; (HEBREW) shvatvat, shvatvat ha-sadè; (HUNGARIAN) mezei zsurló; (ICELANDIC) kloelfting; (ITALIAN) coda cavallina; (JAPANESE) sugina, tsukushi; (KALAGAN) mangobyon; (KAZAKH) hvosch; (LATVIAN) tīruma kosa; (LITHUANIAN) dirvinis asiūklis; (MACEDONIAN) polska pršlika; (MALTESE) denb iz-ziemel; (MONGOLIAN) khuduugiin shivlee; (NEPALI) aankhle jhar, kurkure jhar; (NORWEGIAN) åkersnelle; (POLISH) skrzyp polny; (PORTUGUESE) cavalinha, cavalinha dos campos, pinheira, cauda de cavalo; (SERBO-CROATIAN) preslica, poljska; (SPANISH) cola de caballo, equiseto menor; (STELLINGWERFS) lidruske, peerdestat, peerds, peers, robol, lidrus; (SWEDISH) fråken; (TAJIKI) chilbughum; (TURKISH) tarla atkuyruğu, at kuyruğu, çam otu, çiğçiğ, ekleme; (TURKMEN) mejdan nwoşy; (URDU) mathi arwin; (VLAAMS) heermoes, akkerpaardestaart; (WELSH) marchrawn-yr-ardir.

Division Pterophyta

2. Dennstaedtiaceae

2.1. *Pteridium aquilinum* (L.) Kuhn (EP) immature fronds; (P) C.

(AFRIKAANS) adelaarsvaring; (ARABIC) sarkhas; (ASSAMESE) dhekia; (BELARUSAN) arljak zvanchine, paparats; (BLAAN, SARANGANI) saphong; (BONTOC, CENTRAL) anamam, anam-am; (CEBUANO) agsam; (CHINESE, GÀN) jue; (CHINESE, MANDARIN) huo jue cai; (CHINESE, MIN BEI) jue cai; (CHINESE, WÚ) jue; (CHINESE, YUE) jue cai; (CRIOULO, CAPE VERDE) feto; (CZECH) hasivka orličí; (DANISH) ornebregne; (DIGO) mvwiko; (DUTCH) adelaarsvaren; (ENGLISH) bracken, bracken fern, pasture brake, eagle fern; (ESTONIAN) kilpjakg; (FINNISH) sananjalka; (FRENCH) ptèride â l'aigle, ptéris aigle, fougere aigle; (GAELIC, IRISH) raithneach; (GBAYA) ndélé; (GERMAN) Adlerfarn; (GREEK) fteri; (HANUNOO) sunkuk buladlad, tuyapak; (HEBREW) evronit ha-nesher; (HUNGARIAN) sasharaszt; (IBALOI) alam-am; (ICELANDIC) arnarburkni; (IFUGAO) apatapat an tawayan, tayawan; (ITALIAN) felce aquilina; (JAPANESE) warabi; (JAVANESE) pakis gemblung; (KALAGAN) tollobang, tollobong; (KANKANAEY) annapat; (KISSI, NORTHERN) kumtõ; (KONO) kumbi, kumbili; (KOONGO) koungou; (LAHU) da_{v} ka; (LATVIAN) parastā ērgļpaparde, jāņa paparde; · (LITHUANIAN) stambialapis šakya; (MACEDONIAN) obična orlova paprat; (MALAY) pakis gila; (MALTESE) felci; (MAORI)

---

[1] Taxonomic divisions follow Raven *et al.*, 1992.

[2] EP = Edible part of the plant.

[3] P = Method of preparation: C-cooked, R-raw, P-processed (e.g. pickled, salted).

raarahu, rahurahu, maakaka, manehu, rarauhe, taakaka; (MENDE) gbũhɔ̃; (MONGOLIAN) egel bavran; (NEPALI) unyu; (NORWEGIAN) einstape; (PANJABI, EASTERN) kakhash, deo, kakei, lungar; (POLISH) orlica pospolita, paproć; (PORTUGUESE) feto ordinário, feto fêmea, samambaia do campo; (QUECHUA, CUSCO) chakcchaca; (ROMANIAN) ferigă mare, paproi; (RUSSIAN) arljak ohyknovennyi, paparatnik; (SERBO-CROATIAN) bujad, obična; (SPANISH) helecho comun; (SUNDA) andam dangdeur, paku geulis; (TAJIKI) pharn; (TAMIL) parnai; (THAI) kut kia, kut kin, chon yai; (TURKISH) kartal eğreltisi, eğrelti, kizilot, papra, ifteri otu; (VIETNAMESE) guy[ees]t; (VLAAMS) adelaarsvaren; (WELSH) cyffrhedyn.

3. Osmundaceae

3.1. *Osmunda cinnamomea* L. (EP) immature fronds; (P) C.

(ARABIC) el-khenshar alkarfy; (CHINESE, MANDARIN) fen zhu zi qi; (CHINESE, WÚ) fen zhu zi qi; (CZECH) podezřeň; (ENGLISH) cinnamon fern; (FARSI, WESTERN) darchin; (FRENCH) osmonde cannelle; (GERMAN) Zimtrispenfarn; (GREEK) osmounda to kinnamomios; (HEBREW) sarach hakinamon; (HMONG DAW) kaus suab; (JAPANESE) yamadori zenmai; (LATVIAN) papardes; (MALTESE) bastun talkanella; (NEPALI) rani unyu; (POLISH) paproć cynamonowa; (PORTUGUESE) feto cinanono, osmunda; (SPANISH) osmunda; (TAJIKI) dorchin, dolchin; (TURKISH) tarçin eğreltisi, tarçini kral eğreltisi; (VLAAMS) kaneelvaren.

3.2. *Osmunda japonica* Thunb. (EP) immature frond; (P) C.

(CHINESE, MANDARIN) zi qi; (CHINESE, WÚ) zi qi; (ENGLISH) Japanese flowering fern; (FRENCH) osmonde du Japon; (HEBREW) sarach yapani; (JAPANESE) zenmai; (NEPALI) rani unyu; (POLISH) paproć japońska kwitnaca; (PORTUGUESE) osmundo do Japão; (SPANISH) osmunda japonesa; (TURKISH) Japon çiçekli, eğrelti, cayir sedef olumsu; (URDU) japani phol wala fern; (VLAAMS) Japanse bloeiende varen.

4. Parkeriaceae

4.1. *Ceratopteris thalictroides* (L.) Brongn. (EP) young leaves; (P) C.

(AFRIKAANS) watervaring; (ARABIC) el-khenshar almaee; (CHINESE, MANDARIN) shui jue; (CHINESE, MIN BEI) wu ma jue; (CHINESE, WÚ) shui jue; (CHINESE, YUE) wu mao jue, long chuan jue; (CRIOULO, CAPE VERDE) feto d'água; (CZECH) rohatec žluťuchový; (ENGLISH) water fern, watersprite, floating stag's horn, pod fern, oriental water fern; (FILIPINO) pakong-roman; (FINNISH) riisisaniainen; (GREEK) fteri to ithrohari; (HANUNOO) banawug sapat; (HEBREW) sarach hamaim; (HMONG DAW) kaus suab; (IFUGAO) bakkaku; (INDONESIAN) pakurioh, pakis rawa, paku tespong, paku cai; (ITALIAN) paku roman; (JAPANESE) mizu-warabi; (LAO) kok karn pu; (LATVIAN) papardes; (MACEDONIAN) vodna paprat; (MALAY) paku ruan, paku roman, sayur kodok; (MARANAO) pako a paletaw; (NEPALI) unyu; (POLISH) paproć wodna; (PORTUGUESE) feto aquático, mururé; (SAMBAL, BOTOLAN) pako buaya; (TAGALOG) makahirak-hirak, pakong sungai, pakong tubig, pakong-roman, saguray; (THAI) phak khaakhiat, phak kuutnam, phak kuut kao kwuang; (TURKISH) boynuz eğrelti, boynuz eğrekti, su eğreltisi; (URDU) pani ka fern; (VIETNAMESE) r[as]ng g[aj]c nai, quy[ees]t g[aj]c nai, rau c[aaf]n tr[oo]i; (VLAAMS) watervaren; (WELSH) marchrhedyn-y-dwr.

5. Polypodiaceae

5.1. *Diplazium esculentum* (Retz.) Swartz (EP) young leaves; (P) C.

(ARABIC) el-klenshar alkhamel; (BUHID) tigafun; (CHINESE, MANDARIN) cai jue; (CHINESE, WÚ) shuang gai jue; (DANISH) spiselig fjerbregne; (ENGLISH) vegetable fern; (HMONG DAW) kaus suab; (IFUGAO) appaku, pako a saladeng; (INDONESIAN) puka beunteur, paku sayur, paku tanjung; (JAPANESE) kuware-shida; (JAVANESE) pakis ijo, pakis wilis; (KABIYÉ) amalore; (KANKANAEY) appaku, pakpakong; (MACEDONIAN) zelenčukova paprat, paku tanjong, paku benar; (MALAY) paku, pakis; (NEPALI) nigurio, neuro; (POLISH) paproć warzywna; (PORTUGUESE) feto; (SUNDA) paku beunyeur, paku leuheur; (TAGALOG) pako, tagabas, pakong-kinakain; (THAI) phak kuut, hasdam, kuut khue; (TURKISH) sebze eğreltisi, yenebilir diplazyum; (URDU) khana ka fern; (VLAAMS) pako; (WÁRAY-WÁRAY) pako.

Division Anthophyta (Magnoliophyta)[4]

Class Monocotyledons (Lilopsida)

6. Alismataceae

6.1. *Sagittaria sagittifolia* L. subsp. *leucopetala* (Miq.) Hartog (EP) corms; (P) C.

(ARABIC) kewi; (BICOLANO, CENTRAL) tikog; (CEBUANO) gauai-gauai; (CHINESE, MANDARIN) ye ci qu; (CHINESE, WÚ) ye ci gu; (ENGLISH) Chinese arrowhead, arrowhead, arrow-weed, swamp potato; (FRENCH) fléchière commune, sagittaire de Chine; (GERMAN) gemeines Pfeilkraut; (HEBREW) hatẕetẕon tlat-ʿalim; (INDONESIAN) bea-bea; (ITALIAN) erba saetta chinese; (JAPANESE) kuwai, shiro-guwai; (KHMER, CENTRAL) slök lumpaèng; (LAO) phak sôb; (MACEDONIAN) trolisten strelolist; (MALAY) ubi keladi, keladi chabang; (POLISH) uszyca trójlistna; (PORTUGUESE) erva frecha chinesa; (SPANISH) flecha chinesa, saeta chinesa; (SUNDA) eceng genjer; (THAI) khakhiat, taokiat, phakkhangkai; (TURKISH) çin su oku; (URDU) chini sir naza; (VIETNAMESE) rau m[as]c, t[uwf] c[oo], c[ur] ch[os]c; (VLAAMS) kuwai.

6.2. *Sagittaria sagittifolia* L. subsp. *sagittifolia* (EP) corms; (P) C.

(AFRIKAANS) lansblaar; (ARABIC) raas el-saham; (BELARUSAN) shalnikavaja strelkslist; (BENGALI) chhoto-kut, muya muya; (BICOLANO, CENTRAL) lagbang, tikog; (CEBUANO) longiloba, tikog; (CHINESE, GÀN) ci gu; (CHINESE, MANDARIN) ci qu; (CHINESE, MIN BEI) shu zai yu; (CHINESE, WÚ) ci gu; (CHINESE, YUE) ci gu, wu yu; (CREOLE, HAITIAN FRENCH) malanga; (CZECH) šípatka střelolistá, šípatka vodní; (DUTCH) pijlkruid; (ENGLISH) arrowhead; (ESTONIAN) kõõlusleht; (FILIPINO) tikog; (FINNISH) pystykeiholehti; (FRENCH) fléchière, flèche d'eau, sagittaire; (GAELIC, IRISH) rinn; (GERMAN) echtes Pfeilkraut; (GREEK) saggitaria to velofyllos; (HEBREW) hatseson, hatzetzon hitzi, haẕeẕon; (HUNGARIAN) nyílfű; (ICELANDIC) örvarblad; (INDONESIAN) bia-bia; (ITALIAN) sagittaria commune, erba saetta, barba silvuna; (JAPANESE) kuwai; (KAZAKH) strelolist strelolistyj; (LATVIAN) parastā bultene; (LITHUANIAN) strėlialapė papliauška; (MACEDONIAN) običen strelolist; (MALAY) ubi keladi; (NORWEGIAN) pilblad; (POLISH) strzałka wodna, uszyca; (PORTUGUESE) seta, erva frecha, chapéu de couro; (POTAWATOMI) wabasi'binik; (SPANISH) saeta de agua, flecha de agua;

[4] Classification terminology within parentheses follows Cronquist, 1988.

(TAGALOG) gáuai-gáuai; (TURKISH) adi suoku, suotu, adi su oku; (URDU) sir naza; (VLAAMS) pijlkruid; (WÁRAY-WÁRAY) gaway, gaway-gaway; (WELSH) saethlys.

7. Alliaceae

7.1. *Allium ampeloprasum* L. var. *ampeloprasum* (EP)[5] bulb and leaves[s]; (P) C,R.

(ARABIC) tum-zu el-raas; (BELARUSAN) tsabulevaja; (BULGARIAN) biseren luk; (CATALAN-VALENCIAN-BALEAR) all porret; (CHINESE, MANDARIN) da tou suan, nan ou suan,; (CHINESE, WÚ) da suan, da tou suan; (CHINESE, YUE) da suan, suan; (CZECH) pór zahradní, pór; (DAGAARE, SOUTHERN) zanmē-var; (DANISH) perleleg, orientalsk løg; (DUTCH) wilde prei, wild look; (ENGLISH) great-headed garlic, levant garlic, wild leek; (ESTONIAN) pärlsibul; (FRENCH) ail d'orient, ail gros, poireau d'été; (GERMAN) Perllauch, Ackerknoblauch; (GIKUYU) gĩtũngũrũ; (GREEK) ampelopraso; (HEBREW) shum gavoha; (HUNGARIAN) francia hagyma, gyöngyhagyma; (INDONESIAN) bawang prei; (ITALIAN) porrandello, porraccio, porro selvatico; (LATVIAN) sīpoli, puravi; (MACEDONIAN) golemoglavest luk; (MALTESE) kurrat slavagg; (MONGOLIAN) songino; (NORWEGIAN) perleløk; (POLISH) czosnek wielocebulowy; (PORTUGUESE) alho porro bravo, alho gigante; (PULAR) ladji; (ROMANIAN) ceapă margaritar; (RUSSIAN) luk vinogradnyj; (SERBO-CROATIAN) biser luk; (SERER-SINE) ladj; (SLOVAK) pór; (SLOVENIAN) srebrna čebulica; (SPANISH) puerro agreste, puerro salvage, puerro silvestre, cebollino; (SWEDISH) pärllök; (TAGALOG) kutsai; (TAJIKI) siri tez; (TURKISH) yaz sarmisaği, kaya sarimsaği, yalanci pirasa, bağ pirasasi, yaz sarmaşiği; (TURKMEN) ýapyşak şogan; (URDU) bara sir wala lasan; (VLAAMS) parelui; (WELSH) nionyn dodwy, craf-y-geifr.

7.2. *Allium cepa* L. var. *aggregatum* G. Don (EP) pseudostem and leaves; (P) R,C.

(AFRIKAANS) salot, aanteelui; (ALBANIAN, TOSK) qepë shalotë; (ARABIC) bassal el shallut; (ASSAMESE) asomiya piyaj; (BASSARI) diàɓá; (BELARUSAN) tsabulja shalot; (BEMBÉ) bala; (BENGALI) ghundhun; (BICOLANO, CENTRAL) cebollas, sibolyino, sibulyas; (BINUKID) salaysay; (BONTOC, CENTRAL) idaya; (BRETON) chalotez; (BULGARIAN) šalot; (BULLOM SO) sibɔla; (BURMESE) kyet-thun-u-galay; (CATALAN-VALENCIAN-BALEAR) ascalunia; (CEBUANO) lasona, lansona; (CHICKASAW) ato̱falaa' himona', bala' albilha, itti' tiyya', bala' sinti, bala' okchemali', balokchamali'; (CHINESE, GÀN) xi xiang cong; (CHINESE, MANDARIN) fen nie yang cong, hu ts'ung; (CHINESE, MIN BEI) xiang cong; (CHINESE, MIN NAN) chhang, chhang-á, lō-giō; (CHINESE, WÚ) fen nie cong, fen cong; (CHINESE, YUE) fen cong, hu cong, xiang cong; (CREOLE, HAITIAN FRENCH) echalott, echalot; (CZECH) cibule šalotka množilka, šalotka; (DAGAARE, SOUTHERN) zanmē; (DANGME) sabola; (DANGME-KROBO) samana, samanatʃupa; (DANISH) skalotteløg, kartoffelløg; (DUTCH) sjalot; (EDE NAGO) alubosa; (ENGLISH) shallot, potato onion, multiplier onion; (ESTONIAN) šalott; (FILIPINO) sibuyas-tagalog; (FINNISH) ryvässipuli, salotten sipuli, salottisipuli; (FRENCH) échalote, oignon-patate, ail stérile; (FULFULDE, ADAMAWA) lawasi, lawasīji; (GA) sabola; (GAELIC, IRISH) seallóid; (GBE-VHE) sabala; (GERMAN) Eschlauch; (GIKUYU) gĩtũngũrũ; (GOURMANCHÉMA) djaba nachine; (GOURMANTCHÉ) salbissa; (GREEK) kremmydi to askoclonio; (HANUNOO) lasuna; (HAUSA) àlbásàà, àlbásàà mai-gora, àlbásàà mai-kara, àlbásàà mai-wuya, ƙwàn kádàà, ramuza, safa, shafa, safa mai-gora, shalamta, shalanta, tsattsafi, yár bààkín kòògíí, yár bààkín ràáfíí, yár làmbúú, zalbo; (HEBREW) shum ashkelon; (HINDI) gandana; (HUNGARIAN) mogyoróhogyma, aslottahagyma; (ICELANDIC) skalotlaukur; (IFUGAO) amput an dumalom, danggu, pumalapaggar danggu; (INDONESIAN) bawang merah, bawang abang mirah; bawang

[5] A superscript *s* indicates a plant part that has a secondary use.

acar, dasun merah, berambang, brambang; (ITALIAN) scalogno; (ITAWIT) lasuna; (JAPANESE) sharotto; (JAVANESE) brambang, brambang abang, bawang merah; (JOLA-FONYI) eladje; (JULA) djaba; (KHMER, CENTRAL) khtüm krâhâàm; (KIRGHIZ) kyrk muun pijaz; (KIRIBATI) te anian; (KONKOMBA) tčèbènur; (KRIO) shilɔt, lif yabas; (LAO) hoom bwàx; (LATVIAN) dārza sīpols, sīpols loki, šalotes sīpoli; (LUO) atung gula; (MACEDONIAN) skalunka; (MALAY) daun bawang, bawang merah, bawang kecil; (MANDING-MANDINKA) catinhá-ô, jabô; (MALTESE) ix-xalloti; (MENDE) yabasi-laa; (MOHAWK) anonkshera; (MÒORÉ) gãbdo, albaharé; (NEPALI) pyaz; (NGUNDI) bala; (NORWEGIAN) sjalottløk, potetlåk; (PANGASINAN) lasona; (PANJABI, EASTERN) gandana, gandhan, piäz, gǣda; (POLISH) cebula kartoflanka; (PORTUGUESE) chalota, cebolinha-de-bubinhos, cebolinha; (ROMANIAN) ceapă esalota; (RUSSIAN) luk-šalot; (SERBO-CROATIAN) kozjak; (SINHALA) rathu loonu; (SLOVAK) šarlotka; (SLOVENIAN) šalotka; (SONGHAY) àlbésel; (SORBIAN, UPPER) šalota, worješenc; (SPANISH) chalote, escaluña, ascalonia; (STELLINGWERFS) sjelot, slotten; (SUBANON) genda; (SUMO-MAYANGNA) onyan; (SUNDA) bawang acar, bawang beureum; (SWAHILI) vitunguu shaloti; (SWEDISH) scharlottenlök; (TAGALOG) sibuyas talalog, bawang-pula, lukyu, sibuyas, sibuyas-Tagalog; (TAJIKI) pioz; (TEM) alebassa; (THAI) hom, hom-daeng, hom-lek; (TOK PISIN) lip anian; (TURKISH) yabani sarmisak, askalon; (TURKMEN) adaty şogan; (TWI) gyeene; (UKRAINIAN) šalot; (URDU) piaz; (VIETNAMESE) h[af]nh ta, h[af]nh t[aw]m, h[af]nh n[es]n; (VLAAMS) sjalot; (WÁRAY-WÁRAY) sibuyas-bombay; (WELSH) sibwn; (WOLOF) soblé, soblé futô; (XHOSA) uhlobo oluthile lwetswele; (YORUBA) alubosa, àlubọ̀sà eléwé, àlubọ̀sà onịṣu.

7.3. *Allium cepa* L. var. *cepa* (EP) bulb; (P) R,C.

Dialects of Luyia (Bukusu[1], Itsukha[2], Kabarasi[3], Marama[4])

(ABUA) áyo; (AFAR) bàsal; (AFRIKAANS) ui; (AKPA) árùbásā; (ALBANIAN, TOSK) qepë; (ALEUT) kianartok; (AMHARIC) gäyy šənkurt; (APACHE, JICARILLA) ttōshchi̱nii, ttōhchinií; (ARABIC) bassal; (ARMENIAN) soch; (ASSAMESE) piyaj, piyaz, ponoru; (AZERBIAJANI) sogan; (BAATONUM) wuasaku; (BADAGA) uḷḷi, bengue, benguve; (BAFUT) ángyɛ̌sə̀; (BALTI) tsong; (BAOULÉ) djaba, djé n'dé; (BASQUE) kipula, tipula; (BASSARI) diàɓá; (BELARUSAN) tsabulji zepchatyi, tsabulji repchatui; (BEMBA) kanyense; (BEMBÉ) bala; (BENGALI) palandu, piyaj, piaj, pyanj; (BEROM) àlbásà; (BÉTÉ) djaba; (BETE-BENDI) à-làbásyà; (BICOLANO, CENTRAL) sibuyas; (BINUKID) ganda; (BLACKFOOT) pisatsiinikimm; (BODO) sambram gøjah; (BOKYI) eyeb; (BONTOC CENTRAL) sibuyas; (BOSNIAN) luk, zvibel, sogan; (BRETON) ognon; (BUDIK) dyaɓá; (BULGARIAN) luk; (BULLOM SO) sibɔla, yábas; (BULU) anyon; (BURMESE) kyet-thun, kyet-th-wonini; (CAJUN FRENCH) oignon; (CAROLINIAN) subwuyas; (CATALAN-VALENCIAN-BALEAR) ceba; (CEBUANO) sibuyas; (CHECHEN) xox; (CHEROKEE) svgi; (CHICKASAW) ato̱falla'a', ato̱falla'ha'a, ato̱falaa', ato̱falla'a' homma', ato̱falla'a' lakna', ato̱falla'a' tohbi'; (CHINESE, GÀN) cong tou; (CHINESE, HAKKA) ts'oūng-t'êou, ts'oūng-tzè; (CHINESE, MANDARIN) yang cong, congtou; (CHINESE, MIN BEI) fan zai cong; (CHINESE, MIN NAN) chhang-thâu, iû-chhang; (CHINESE, XIANG) yuan cong; (CHINESE, WÚ) yan cong; (CHINESE, YUE) yang cong, cong tou; (COMANCHE) kuukanaruhkaʔ, kuuka̱; (CREE, PLAINS) wihcekaskosiy, wâpiskaskosiy; (CREOLE, HAITIAN FRENCH) zonyon; (CRIOULO, CAPE VERDE) cabola; (CROATIAN) luk; (CZECH) cibule kuchyňská, cibale; (DAGAARE, SOUTHERN) zanmɛ̄; (DAGBANI) gab'nyule; (DAKOTA) pśiŋ; (DANGME-KROBO) samana; (DANISH) kepa-løg; (DENDI) albasa; (DIGO) chitunguu madzi; (DUTCH) ui; (EDE NAGO) alubosa gambari; (EDO) alubarha; (EFIK) òyím, òyím ḿbàkárá; (ENGLISH) onion, common onion, bulb onion; (ESTONIAN) sibul; (ÉWÉ) sabala; (FARSI, EASTERN) piyâz; peaj; (FARSI, WESTERN) piyâz, piaz; (FILIPINO) sibuyas; (FINNISH) sipuli, ruokasipuli; (FON) ayomasa; (FRENCH) oignon; (FULFULDE, ADAMAWA) albacce, albasa, albasal, tingyeje gorɗe, tingyēre, tingyere, tinyēre, tinyēre, tinggēre, toiyeere, kárán àlbásàà, katakata, láwàshìì, wúyàn mákwárwáá; (GA) blɔfo sabola; (GADDANG) danggu; (GAELIC, IRISH) oinniún; (GAELIC, MANX) unnish; (GAELIC, SCOTTISH) uinnean; (GALICIAN) cebola; (GANDA) butungulu; (GBE, WACI) sabule; (GEN) sabulevi; (GEORGIAN) khakhvi, chachvi, háhvi; (GERMAN) Speisezwiebel, Küchenzwiebel; (GIANGAN) sibuyas; (GIKUYU) gĩtũngũrũ-kĩa-ngwacĩ; (GONJA) gébú, kàlèbásà; (GOURMANCHÉMA) djaba napaan; (GOURMANTCHÉ) djaba; (GREEK) kremmy di to koino, kremmy di, kremmydi; (GUJARATI) dungari, dungli, kando; (GUN) ganganlubassa; (HARARI) šənkurta; (HAUSA) albasa, albása, àlbásàà, àlbásàà gudàajíí, àlibácá, gingen tasa, gùdán àlbásàà, ƙuduru; (HEBREW) batzal ha-ginna; (HERERO) !kharo-i, onjanga; (HILIGAYNON) sibúyas; (HINDI) piaz, piyaz, piaj, pyāj; (HINDUSTANI) piyáz, basal, gaṭṭhá; (HMONG DAW) dos; (HOPI) siiwi; (HUNGARIAN) vöröshagyma, vörösjagyma; (IBALOI) shanghu; (IBANAG) sibuyas, sibolyas; (ICELANDIC) höfuđlaukur, rauolaukur; (IFUGAO) sibuyas; (IGBO) yabasi, àyàbaasị, ayo, yàbaàsi; (ILOCANO) lasona, sibulyas, cebollas, siboyas; (INDONESIAN) bawang manis, bawang lampoeng, bawang Bombay, bawang merah, bawang besar; (INUKTITUT, WEST GREENLANDIC) inanitsoq, uvanitsoq; (ITALIAN) cipolla; (ITAWIT) seboliat; (IVATAN) buliyas; (IZON) ụtụta, yabásị; (JAPANESE) tamanegi, tamane; (JUNKUN TAKUM) pádòn, ribásá; (KABIYÉ) cahoun; (KANNADA) eerulli, irulli, nirulli, neeruli, neerulli, ullaagedde; (KANURI) lebasar, lùwásàr; (KASHMIRI) prān, prāna-bagāra; (KATI) trashtu; (KAZAKH) luk repcatyj, piyaz; (KHMER, CENTRAL) khtüm barang; (KIRGHIZ) pijaz tup, piyaz; (KIRIBATI) te anian; (KONKANI) kando, piyaavu; (KOREAN) dungulpha, yangp'a; (KORO) àl̀pása; (KRIO) yabas, yabasi, alibosa; (KURDISH, NORTHERN) pyāz; (KWANGALI) uriri; (KWÉYÒL) lonyon, gwo-lòyon; (LADINO) sevóya; (LAHU) shu_; (LAKOTA) pśiŋ (LAO) bwàx fàlangx; (LATVIAN) sīpols; (LUXENBOURGEOIS) Ënn; (LIMBA) oyaba; (LITHUANIAN) valgomasis svoqūnas; (LOMWE, MALAWI) anyenzi; (LUO) atung gula; (LUYIA) bitungulu[1], shitunguo[3], eshitukwo[2,4]; (MACEDONIAN) običen kromid; (MAITHILI) peāju; (MALAGASY) tongolo; (MALAY) bawang, besar, bawang besar; (MALAYALAM) iravengaayam, ulli, eerulli, bavang, chuvannaulli; (MANDING-BAMBARA) diaba; (MANDING-MANDINKA) jabbo; (MANINKAKAN, WESTERN) ladio; (MALTESE) il-basal; (MAORI) aniana; (MARANAO) lasona, siboyas, tukupun; (MARATHI) kanda, kandaa; (MEHRI) bəṣəlēt, bəṣāl; (MENDE) yabas, yabasi, yawɛ, siba; (MICMAC) ăpsāāsisk; (MIZO) purun, purun sen; (MOHAWK) anonk, anonkshera; (MÒORÉ) zãyõ, gãbdo, alba; (MORISYEN) zonion; (MUNDARI) peāju; (MUSKOGEE) tafvmpe; (NAMA-DAMARA) onyanga; (NAVAJO) tł'ohchin; (NDONGA) nyanga; (NEPALI) pyaz; (NEWAR) chāp; (NGIZIM) àlbásâr̃; (NIUE) aniani; (NORWEGIAN) kepaløk, vanlig kepaløk, løk, matløk, rødløk; (NUPE-NUPE-TAKO) àlùbăsa, lùbăsa; (OGBIA) ə̀yò; (OGBIA-AGHOLO) ə̀yò; (ORIYA) peas, ulli; (OROMO, WEST CENTRAL) kulobi dima, shunkurti; (PAIUTE, SOUTHERN) sevoy′u; (PALI) palaṇḍu, sukandako; (PAMPANGAN) sibuyas; (PANGASINAN) sibúyas; (PANJABI, EASTERN) ganda, piaz, payaz; (PASHTO, NORTHERN) pyâz, piyāz; (POLISH) cebula, cebula jadalna; (PORTUGUESE) cebola; (POTAWATOMI) shakwesh, skakwe'shak; (PULAAR) wasalde, soblere; (QUECHUA, CUSCO) chabura; (ROMANIAN) ceapă ceaclama, ceapa cultivată, pentrubulb; (RUSSIAN) luk repčatyj, louk repčhatyi, louk zepchatnyi, louk repka; (SAMOAN) aniani; (SANGA) aburuma; (SANSKRIT) palaandu, nripakanda, palandu, raktakanda, deerghapathra, durghandh, mahaakanda, nripapriya, nripavhaya, nripeshtha, raajapalaandu, raajapriya, raajeshtha, rakthakanda, rochaka, yavaneshttha; (SANTALI) peaj, piạj; (SCOTS) ingan; (SENOUFO, CEBAARA) djabahou; (SERBO-CROATIAN) luk, crniluk; (SERER-SINE) soblé; (SHONA) hanyanisi, nyanyisi; (SIDAMO) tuma, tuma shunkurta; (SINHALA) luku loonu; (SLOVAK) cibul'a kychyňská; (SLOVENIAN) čebula, navadna čebula; (SOMALI) basal; (SORBIAN, LOWER) cybula; (SORBIAN, UPPER) cybla; (SOTHO, NORTHERN) eie; (SOTHO, SOUTHERN) eie; (SPANISH) cebolla; (SRANAN) ayun; (STELLINGWERFS) siepel, oje, oije; (SUNDA) bawang Bombay; (SWAHILI) kitunguu maji, kitunguu; (SWEDISH) lök, matlök; (TAGALOG) sibuyas bombay, sibuyas buyas, sibuyas; (TAJIKI) pioz; (TAMIL) vengaayam eerulli, periya venkayam, ira-vengayam, irulli, vella-vengayam, vengayam, erangayam, iravengaayam; (TATAR) sugan; (TAUSUG) sibuyas; (TELUGU) neerulli, nirulli, vulli,

ulli, gaddalu, urigaddalu, erragaddalu, erra ulligadda, vulligaddalu, vullipayalu; (THAI) hom, hom-yai, hom-huayai; (THEMNE) ka-yaba, *koe*-yaba, *ta*-yaba; (TIBETAN) tsōn; (TIV) a̲la̲busa̲; (TOK PISIN) anian; (TSWANA) eie; (TULU) neerullikande; (TURKISH) soğan; (TURKMEN) sogan; (TWI) anyau; (UKRAINIAN) cibulja; (URDU) piaz; (UYGHUR) piyaz; (UZBEK) piyoz, piëz; (VAI) siba; (VENDA) nyanìsa; (VIETNAMESE) cu hành, h[af]nh c[ur], h[af]nh t[aa]y; (VLAAMS) ui, ajuin, ai; (WAMEY) dyaba; (WÁRAY-WÁRAY) sibuyas; (WELSH) nionyn; (WOLOF) soblé; (XHOSA) itswede; (YALA) èjím, yalibáca; (YIDDISH) tsibehleh; (YOM) amisambɛɛ; (YORUBA) àlubọsà, àlubọsà gambari, àlubọsà keta, ẹlubásà; (ZULU) u-anyanisi; (ZUNI) moqqi.

7.4. *Allium chinense* G. Don. (EP) bulb; (P) C.

(ARABIC) tum el-khabazeen; (CHINESE, GÀN) qiao tou; (CHINESE, MANDARIN) xie; (CHINESE, MIN BEI) qiao tou; (CHINESE, XIANG) qiao tou; (CHINESE, WÚ) cong, xie; (CHINESE, YUE) qiao, qiao tou; (DANISH) rakkyo; (ENGLISH) rakkyo, Baker's garlic; (FRENCH) echalotte chinoise, rakkyo; (HEBREW) shum sini; (HMONG-DAW) dos; (HMONG-NJUA) dlog; (HUNGARIAN) iyabasi; (INDONESIAN) bawang ganda; (JAPANESE) rakkyou; (MALAY) lokyo; (POLISH) szczypiorek chiński; (PORTUGUESE) chalota chinesa, cebolinha para picles; (SPANISH) chalote chinesa; (THAI) krathiam-chin, hom-prang, hom-paenyuak; (VIETNAMESE) ki[eej]u.

7.5. *Allium fistulosum* L. (EP) pseudostem and leaves; (P) R,C.

(AFRIKAANS) bindui, soepui; (ALBANIAN, TOSK) qepë dimrore; (ARABIC) bassal el-mustatere, bassal el-ankudy; (ASSAMESE) Welsh onion; (BASQUE) tipulatxi; (BELARUSAN) tasbulja repchataja, tsabulja tabchataja, tsabulja batun; (BRETON) sivolez, krak-sivolez; (BULGARIAN) zimen luk; (CHINESE, MANDARIN) cong, ts'ung; (CHINESE, MIN NAN) tōa-chhang; (CHINESE, WÚ) cong, da guan cong, da cong; (CHINESE, YUE) da cong, mu cong, shui cong; (CRIOULO, CAPE VERDE) cabolinha; (CZECH) cibule zimní, cibulc sibiřská, ošlejch; (DANISH) pibe-løg; (DUTCH) grof bieslook; (ENGLISH) Welsh onion, Japanese bunching onion; (ESTONIAN) talisibul; (FINNISH) pillisipuli, talvispuli; (FRENCH) ciboulette, cive, ail-ciboule, ciboule; (GEORGIAN) tshlacvi; (GERMAN) Winterzwiebel; (GONJA) gebu, kalebasa; (GREEK) kremmydi to syriggodes, pikpo kremmydi; (HEBREW) batzal yapani; (HMONG DAW) dos; (HUNGARIAN) metélőhagyma, téli hagyma; (IFUGAO) buyah, buyas; (INDONESIAN) bawang daun, bawang daven, bawang bakung, bawang semprong, bawang sop, bawang benggala; (INUKTITUT, EASTERN CANADIAN) uaniujaq; (ITALIAN) cipolla d'Inverno, cipolletta; (JAPANESE) negi; (JAVANESE) bawang oncang; (KAZAKH) luk batun; (KHMER, CENTRAL) khtüm sânlök; (KIRGHIZ) batun pijazi; (KIRIBATI) te anian; (KOREAN) pha; (LATVIAN) stobrainais sīpols, batūnsīpoli; (LITHUANIAN) tuščialaiškis česnakas; (LUO) atung gula; (MACEDONIAN) zimski kromid; (MALAY) daun bawang; (MORISYEN) lake zonion; (NEPALI) pyaz; (NORWEGIAN) pipeløk; (PALAWANO, CENTRAL) tawar; (POLISH) cebula siedmiolatka; (PORTUGUESE) cebolinha, cebolinha comum, cebolinha de todo o ano, cebolinha verde; (ROMANIAN) ceapă de tuns; (RUSSIAN) luk batun; (SERBO-CROATIAN) aršlama; (SLOVAK) cibul'a zimná; (SLOVENIAN) zimski luk, qulë dimrore; (SPANISH) cebolleta, cebolleta común, cebollino de verdeo común; (SUNDA) bawang bakung, bawang sop, bawang cina, bawang daun; (SWEDISH) piplök, sallada lök; (TAGALOG) nebuka; (THAI) hom-ton, hom-chin; (TURKISH) kiş soğani, gal soğani; (UKRAINIAN) cibulja-batun; (VIETNAMESE) h[af]nh hoa, h[af]nh h[uw][ow]ng; (VLAAMS) Japansc bosui, bieslook; (WELSH) nionyn cymru; (YIDDISH) tsibekhes; (ZARMA) albasan.

7.6. *Allium kurrat* Schweinf. ex K. Krause (EP) pseudostem and bulb; (P) C,R.

(ARABIC) kurrat; (CHINESE, WÚ) nan ou suan; (ENGLISH) kurrat; (FRENCH) kurrat; (GIKUYU) gĩtũngũrũ; (HEBREW) karrash, kresha, shum kurrat; (HUNGARIAN) kurrat; (INDONESIAN) prei; (JAPANESE) kratto; (POLISH) por dziki, kurat; (PORTUGUESE) kurrat; (SPANISH) kurrat; (VLAAMS) kurrat.

7.7. *Allium macrostemon* Bunge (EP) leaves; (P) C,R.

(CHINESE, WÚ) xie bai; (ENGLISH) Japanese garlic; (FRENCH) ail du Japon, nobiru; (GREEK) kineziko skordo; (ITALIAN) aglio giapponese; (JAPANESE) nobiru; (PORTUGUESE) alho do Japão; (SPANISH) ajo del Japon; (TURKISH) Japon sarmisaği, sarmisak; (URDU) japani lasan; (VLAAMS) nobiru.

7.8. *Allium porrum* L. (EP) pseudostem and leaves; (P) C,R.

(AFRIKAANS) prei; (ALBANIAN, TOSK) presh, pras; (ARABIC) kurrat; (ASSAMESE) leek; (BASQUE) porru; (BELARUSAN) poréi; (BENGALI) paru; (BOSNIAN) praziluk, prasa, poriluk; (BRETON) pour; (BULGARIAN) praz; (CATALAN-VALENCIAN-BALEAR) porro; (CEBUANO) sibuyas-bisaya, sibuyas-Bisaya, sibuyas-litmo; (CHINESE, HAKKA) kiòu-ts'ói; (CHINESE, MANDARIN) jiu cong, nan ou ts'ung; (CHINESE, MIN NAN) kú-chhài, kú, soàn-á; (CHINESE, WÚ) jiu cong; (CREE, PLAINS) wehcekaskosiy; (CREOLE, HAITIAN FRENCH) poro, powo; (CROATIAN) poriluk; (CZECH) pór zahradní setý, pór; (DAGAARE, SOUTHERN) zanmē-var; (DAKOTA) pšiŋ; (DANISH) porre; (DUTCH) prei; (ENGLISH) leek; (ESTONIAN) porrulauk; (FARSI, EASTERN) gandana; (FARSI, WESTERN) tarché farangy; (FILIPINO) kuchai, leek; (FINNISH) purjo; (FRENCH) poireau, porreau; (GAELIC, IRISH) cainneann; (GAELIC, SCOTTISH) cainneann, creamh-gàrraidh; (GALICIAN) porreta; (GEORGIAN) prassa, prása; (GERMAN) Porree; (GIKUYU) gĩtũngũrũ; (GOURMANCHÉMA) djaba vortz; (GREEK) praso; (HAUSA) àlbásàà-mai-kara; (HEBREW) karrash, kresha, shum hakarash; (HINDI) kirath; (HINDUSTANI) gandáná, piyáz ki taran kadarak̦ht; (HMONG DAW) dos ntaj; (HUNGARIAN) póréhagyma; (ICELANDIC) blađylaukur; (ILOCANO) leek; (INDONESIAN) bawang prei, bawang péré, bawang sayuran; (ITALIAN) porro, porretta; (JAPANESE) liiki, niranegi; (JAVANESE) bawang sayuran, bawang pre, pre; (KASHMIRI) godu; (KAZAKH) luk porej; (KHMER, CENTRAL) khtüm khchâl; (KIRGHIZ) porej pijazi; (KIRIBATI) rike; (KWÉYÒL) powo; (LADINO) prása, puerro; (LAKOTA) pśiŋ; (LAO) pèènz fàlangx; (LATVIAN) puravi; (LITHUANIAN) poras; (LUXENBOURGEOIS) porett; (MACEDONIAN) običen praz; (MALAGASY) tongolobe; (MALAY) lik; (MALTESE) kurrat; (MAORI) riiki; (MARANAO) genda; (MORISYEN) pwaro; (NGUNDI) poilo, poireau; (NORWEGIAN) purre; (PAMPANGAN) kuse; (PANGASINAN) kusay; (PASHTO, NORTHERN) gandana; (POLISH) por, warzywo; (PORTUGUESE) alho porro, alho francês; (ROMANIAN) praz; (RUSSIAN) louk parei, luk porej; (SCOTS) foos; (SERBO-CROATIAN) praziluk; (SIDAMO) shunkurta; (SINHALA) leek; (SLOVAK) pór; (SLOVENIAN) pór, pór pravý; (SORBIAN, LOWER) porej; (SORBIAN, UPPER) porej; (SPANISH) ajo porro, ajo kiras; (SUNDA) bawang pere, bawang parai, bawang bali, bawang pire; (SWAHILI) vitunguu liki, namna ya kitunguu kidogo; (SWEDISH) purjo lök, purjo; (TAGALOG) kutsai, leek; (TEKE) poilo, poireau; (THAI) krathiam-ton, krathiam-bai; (TIBETAN) ñũqtsoon; (TURKISH) pirasa; (UKRAINIAN) cibulja-porej; (VIETNAMESE) lò chân lông, t[or]i t[aa]y; (VLAAMS) prei; (WELSH) cenhinen; (WOLOF) poro; (XHOSA) iliki, uhlobo oluthile lwetswele; (YIDDISH) pureh-tsibehleh.

7.9. *Allium × proliferum* (Moench) Schrad. ex Willd (EP) aerial bulbs; (P) R,C.

a. tree onions

(AFRIKAANS) Egiptiese aanteelui; (ARABIC) bassal el-masry; (BULU) anyon; (BURMESE) kyet-thun-maik; (CHINESE, MANDARIN) ding sheng yang cong; (CHINESE, WÚ) ding sheng yang cong; (CREOLE, HAITIAN FRENCH) zonyon vèt; (DANISH) etage-løg; (DUTCH) egyptische ui; (ENGLISH) tree onion, Egyptian onion; (ESTONIAN) rindeline sibul; (ÉWÉ) sabala; (FINNISH) ilmasipuli; (FRENCH) oignon d'Egypte, oignon catawissa; (GA) blɔfo sabola; (GERMAN) Etagenzwiebel; (GOURMANTCHÉ) djaba; (GREEK) aigyptiako kremmydi; (HEBREW) batzal mitzree, baẕal; (HUNGARIAN) fiókhagyma, rokambol hagyma; (INDONESIAN) bawang putih; (ITALIAN) cipolla di Egitto; (JOLA-FONYI) ediaba ou élignon; (KABIYÉ) cahou; (KAZAKH) luk mnogojarusnyj; (KIRGHIZ) kep tepkičtuupijaz; (LATVIAN) sīpols, daudzposmu sīpoli; (LITHUANIAN) daugiaaukštis svoqūnas; (MACEDONIAN) egipatski kromid; (MANINKAKAN, WESTERN) diabo; (NORWEGIAN) luftløk; (POLISH) cebula wielopiętrowa; (PORTUGUESE) cebola do Egipto, cebola egípcia; (PULAR) bassaleé; (ROMANIAN) ceapă de egipt; (RUSSIAN) luk mnogojarusnyj; (SERBO-CROATIAN) egipatski luk; (SLOVAK) cibul'a poschodova; (SLOVENIAN) egiptovski luk; (SPANISH) cebolla del Egipto; (SWEDISH) matlök, kepalök; (TAJIKI) pioz; (TURKISH) misir soğani; (TWI) anyau; (UKRAINIAN) bagatojarusna cibulja; (URDU) misri piaz; (VLAAMS) boomui, Egyptische ui; (WÁRAY-WÁRAY) ahos; (WELSH) nionyn prew.

b. turfed stone leek (EP) leaves; (P) C.

(Previously classified as *Allium × wakegi* Araki and is now considered a subgroup of *Allium × proliferum.*)

(CHINESE, MANDARIN) da gaun cong; (CHINESE, WÚ) cong; (ENGLISH) turfed stone leek; (FRENCH) ciboule, oignon d'Hiver; (GREEK) allion to kollon; (HUNGARIAN) téli hagyma; (ICELANDIC) sigurlaukur; (ITALIAN) cipolla d'Inverno; (JAPANESE) wakegi; (PORTUGUESE) cebolinha verde; (SPANISH) cebolleta; (VLAAMS) grove bieslook, stengelui.

7.10. *Allium sativum* L. (EP) bulb and leaves[S]; (P) R,C.

(AFAR) tooma; (AFRIKAANS) knoffel; (ALBANIAN, TOSK) hughër, hudhër, hudrë; (AMHARIC) näč̣č̣ šənnurt; (ARABIC) tum, saum; (ARMENIAN) schtor; (ASSAMESE) nohoru, nohoyo; (AZERBAIJANI, NORTHERN) sarımsaq; (BADAGA) beḷḷe benguve; (BASQUE) barakatz; (BELARUSAN) chesnok agorodnaja; (BEMBÉ) ailli; (BENGALI) rashun, lashan, lashun, rasun, lusun; (BEROM) ǹgyùsmà; (BICOLANO, CENTRAL) bawang; (BINUKID) ahus, ganda; (BODO) sambram gufut; (BONTOC CENTRAL) bawang; (BOSNIAN) bijeli luk, češnjak; (BRETON) kignen; (BULGARIAN) češan; (BURMESE) kyet-thun-byu; (CAJUN FRENCH) zail; (CATALAN-VALENCIAN-BALEAR) all; (CEBUANO) ahos; (CHECHEN) saaramsiaq, seeramsiaq; (CHICKASAW) kosoma'si'; (CHINESE, GÀN) da siam; (CHINESE, HAKKA) són-t́zè, són-t'êou; (CHINESE, MANDARIN) da suan, siao suan; (CHINESE, MIN BEI) suan; (CHINESE, MIN NAN) soàn-thâu, soàn-á; (CHINESE, WÚ) da suan, suan; (CHINESE, YUE) suan, ola suan; (CREOLE, HAITIAN FRENCH) lay, laj; (CRIOULO, CAPE VERDE) adjo; (CROATIAN) cešnjak; (CZECH) česnek kuchyňský, česnek kuchyňský obecný; (DAGAARE, SOUTHERN) laï; (DAKOTA) pšiŋ ikceka; (DANGME-KROBO) samana tʃupã; (DANISH) hvid-løg, slangehvidløg; (DENDI) tafanua; (DIGO) chitunguu saumu; (DUTCH) knoflook; (EDE NAGO) ayu; (EFIK) òyím; (ENGLISH) garlic; (ESTONIAN) küüslauk; (ÉWÉ) ayo; (FARSI, EASTERN) sêr; (FARSI, WESTERN) sir, seer; (FIJIAN) lesun; (FILIPINO) bawang; (FINNISH) valksipuli; (FON) ayo; (FRENCH) ail, ail ordinaire, ail blanc; (FULFULDE, ADAMAWA) albatje dadi, albatje darri, arngalore, arngalāre, arngalāje; (GA) aya, ayo; (GADDANG) amfut; (GAELIC, IRISH) gairleog; (GAELIC, MANX) garleid; (GAELIC, SCOTTISH) creamh; (GBE, WACI) ayo; (GIANGAN) bawang; (GALICIAN) allo; (GEN) ayo; (GEORGIAN) niori; (GERMAN) Knoblauch; (GIKUYU) saumu, gĩtũngũrũ-kĩa-ngwacĩ; (GOURMANCHÉMA) lĩng; (GOURMANTCHÉ) salbissa-djãg; (GREEK) skordo; (GUJARATI) lasan, shunam; (GUN) ayo; (HANUNOO) bawan malagti; (HARARI) tumma; (HAUSA) tafannuwa, tafárnúúwáá; (HEBREW) shum ha-ginna, shum hagina; (HILIGAYNON) áhos; (HINDI) lehsoon, lahsan, lehsun, lah.sun; (HINDUSTANI) lahsan, lahsan ká jawá; (HMONG DAW) qej; (HUNGARIAN) fokhagyma; (IBANAG) bawang, ahus; (ICELANDIC) hvitlaukur; (IFUGAO) bawang; (IGBO) àyo, ayón, ayún; (ILOCANO) bawang, ahos; (INDONESIAN) bawang putih, dasun, lasoen; (ITALIAN) aglio, aglio domestico; (ITAWIT) acut; (JAPANESE) ninniku; (JAVANESE) bawang; (JUNKUN TAKUM) adong shinga; (KANNADA) bellulli belluli; (KANURI) kàálmú, kagalmu, kə̀ngálmú; (KASHMIRI) rahun; (KAZAKH) chesnok, sarımsaq; (KHMER, CENTRAL) khtüm sâå; (KIRGHIZ) sarımsak, sarymsak; (KIRIBATI) karike; (KOREAN) manul; (KWÉYÒL) lai, lay; (LADINO) ájo; (LAHU) hpui$^{\vee}$ shu_; shu_hpu; (LAO) kath ’ièm; (LATVIAN) ķīploki; (LITHUANIAN) valgomasis česnakas; (LUO) atung gula; (LUXENBOURGEOIS) Knueweléck; (MACEDONIAN) običen luk; (MAITHILI) lahsun; (MALAGASY) tongologasy; (MALAY) bawang putih; (MALAYALAM) veluthulli, vellulli, tellagadda; (MALTESE) tewm tal-icheli; (MANDING-BAMBARA) tumé, layi; (MANINKAKAN, WESTERN) ladio; (MARANAO) lasona a tukapun, sakorab; (MARATHI) lasunas, lasun, lasuna; (MEHRI) t̠əmēt, t̠əm; (MIZO) purun var; (MONGOLIAN) tarimal songino; (MÒORÉ) aye, laï; (MORISYEN) lay; (MUNDARI) rasuni; (MUSKOGEE) tafvmpe-vhake; (NEPALI) lasun; (NEWAR) phakan, hmukan, tawo; (NGUNDI) ailli; (NIUE) kāliki; (NORWEGIAN) kvitløk; (ORIYA) rasun; (OROMO, WEST CENTRAL) kulobi adi; (PALI) mahākando; (PAMPANGAN) bawang; (PANGASINAN) báwañg; (PANJABI, EASTERN) thom, lassan, lussun, garlic; (PASHTO, NORTHERN) huz̲ha, ūga; (POLISH) czosnek, czosnek pospolity; (PORTUGUESE) alho, alho comum, alho porro; (ROMANI, BALKAN) sir; (ROMANIAN) usturoi cultivat, ai; (RUSSIAN) česnok, louk chesnok, čhesnok, palevoi louk; (SANGA) aburuma nẹẹra; (SANSKRIT) lashuna, arishtha, laseena, ugragandha, bhutagna, mahasuda; (SANTALI) rạsun; (SERBO-CROATIAN) češnjak; (SERER-SINE) ladj; (SINHALA) sudu loonu; (SLOVAK) cesnak; (SLOVENIAN) česen, cesnak kuchynský; (SOMALI) ton; (SORBIAN, UPPER) kobołk; (SOTHO, NORTHERN) konofele; (SOTHO, SOUTHERN) konofole; (SPANISH) ajo, ajo común; (SUMO-MAYANGNA) kyalic; (SUNDA) bawang putih, bawang bodas; (SWAHILI) kitunguu saumu; (SWEDISH) vitlök; (TAGALOG) bawany, bawang, bawang-puti; (TAJIKI) sir; (TAMIL) vallai-pundu, vella poodu, poodu; (TATAR) sarımsak; (TAUSUG) bawang; (TELUGU) vellulli, velluli, tellagadda; (THAI) krathiam, hom-tiam; (TIBETAN) qcqpa; (TIV) nu̲man; (TOK PISIN) galik; (TSWANA) konofole; (TURKISH) sarmisak; (TURKMEN) sarymsak şogan, sarymsak; (UKRAINIAN) časnik; (URDU) lasan; (UYGHUR) samsaq; (UZBEK) sarimsock piyoz, sarimsok piez, sarimsoq; (VIETNAMESE) t[or]i; (VLAAMS) knoflook; (WELSH) garlleg; (WOLOF) ladji, laji; (XHOSA) ikonfile, ikonofile; (YIDDISH) knubel; (YOM) kafanufa; (YORUBA) ayu, àlubọ́sà, ayón; (ZARMA) tafarnuwa (ZULU) ugaliki, ishaladi lenyoka; (ZUNI) moqqi: techi.

7.11. *Allium schoenoprasum* L. (EP) leaves; (P) R.

(AFRIKAANS) grasiu; (ALBANIAN, TOSK) presh kulmak; (ARABIC) bassal el-shifee; (ARMENIAN) soch; (ASSAMESE) chive; (AZERBIAJANI) sogan; (BELARUSAN) tsabulja skarada, threebulka; (BEMBÉ) ciboula; (BULGARIAN) rjazanec; (CATALAN-VALENCIAN-BALEAR) ceballot; (CHINESE, MANDARIN) xi xiang cong, hsi hsiang ts’ung; (CHINESE, WÚ) xi xiang cong; (CREOLE, HAITIAN FRENCH) siv; (CROATIAN) luk vlasac; (CZECH) pažitka pobřežní, pažitka zahradni; (DANISH) pur-løg; (DUTCH) bieslook; (ENGLISH) chive; (ESTONIAN) murulauk; (FARSI, EASTERN) gandana; (FARSI, WESTERN) peyazcheh;

(FINNISH) ruohosipuli; (FRENCH) ciboulette, civette, fines herbes; (GAELIC, MANX) shoushan; (GALICIAN) cebolino, cebolo; (GEORGIAN) chachvi; (GERMAN) Schnittlauch; (GREEK) schinoprasson, allion to sholnoproeso; (HEBREW) shum dak; (HINDI) cives; (HMONG DAW) dos ntaj; (HUNGARIAN) metélőhagyma; (ICELANDIC) graslaukur; (INDONESIAN) kucai, langkio, ganda isi, lokio, bawang langkio; (ITALIAN) cipollina; (JAPANESE) asatsuki, ezonegi; (KAZAKH) shnittluk; (KHMER, CENTRAL) kukiiw; (KIRGHIZ) šnitt pijazi; (KWÉYÒL) lonyon, ti-lòyon; (LAO) hoom pèèn; (LATVIAN) maurloki, mourloki; (LITHUANIAN) laiškinis česnakas; (LUO) atung gula; (LUXENBOURGEOIS) Bratzelen; (MACEDONIAN) kromid rezanec; (MALAY) kucai; (MALTESE) basal tal-qsari; (MONGOLIAN) sibir songino; (MORISYEN) lake zonion; (NEPALI) chhyapi; (NGUNDI) ciboula; (NORWEGIAN) gresløk, grasløk, sibirgrasløk; (POLISH) szczypiorek; (PORTUGUESE) cebolinho, cebolinha miúda, cebolinha galega, ceboleta de França; (ROMANIAN) pur, ceară-puri, arpagic; (RUSSIAN) šnittluk, louk-scarada, louk-rezanets, šnitt-louk; (SCOTS) sithes; (SERBO-CROATIAN) vlašac; (SLOVAK) pažitka; (SLOVENIAN) drobnjak, pažíka; (SORBIAN, UPPER) cyblink; (SPANISH) cebollino, cebollino francés; (STELLINGWERFS) loek; (SUNDA) langkio; (SWAHILI) vitunguu saumu; (SWEDISH) gräslök; (TAGALOG) chives; (TAJIKI) sirac; (THAI) kuichai-farang, kuichai-baiklom, hom-paen; (TURKISH) başsarimsak, küçük taze soğan, sivet, sibulet, soğancik; (VIETNAMESE) he; (VLAAMS) fijn bieslook, kleine bieslook; (WELSH) cennin syfi.

7.12. *Allium scorodoprasum* L. (EP) leaves and bulb; (P) C,R.

(AFRIKAANS) reuseknoffel; (ARABIC) tum el-emlak; (BULGARIAN) rokambol; (CHINESE, MANDARIN) xiao suan; (CHINESE, WÚ) xiao suan; (CZECH) česnek ořešec, paličak; (DANISH) skov-løg; (DUTCH) slangenlook; (ENGLISH) giant garlic, sand leek; (FINNISH) käärmeenlaukka, helmivalkosipuli; (FRENCH) ail rocambole; (GERMAN) falscher Schlangen(-Knob)lauch, Graslauch, Rockenbolle, Sandlauch; (GREEK) allion to skorodoprolso; (HMONG DAW) qej ntyw; (HUNGARIAN) kígyóhagyma, rokambol; (ITALIAN) aglio romano; (LATVIAN) maurloki, kiploksīpoli; (LITHUANIAN) porinis česnakas; (MACEDONIAN) misirski kromid; (MALTESE) tewm ta'ras hoxna; (MONGOLIAN) mangir, khumkhell; (NORWEGIAN) bendelløk; (POLISH) czosnek wężowy, wężyk; (PORTUGUESE) alho grosso de Espanha, alho gigante, alho rocambole, alho vermelho; (ROMANIAN) robambole; (RUSSIAN) ispanskij česnok; (SLOVAK) cesnak zubatý; (SLOVENIAN) lucka; (SPANISH) ajo ramano; (SWEDISH) spansk hvitlök; (TAJIKI) sirpioz; (TURKISH) iri sarimsak, tas sarmisaği, rokambak; (URDU) bara lasan; (VLAAMS) graslook.

7.13. *Allium tuberosum* Rottler, ex Sprengel (EP) leaves, immature flowers; (P) C,R.

(AFRIKAANS) sjinese prei; (ARABIC) kurrat seeny; (ASSAMESE) China leek; (BINUKID) ganda; (BURMESE) gyu-myit; (CEBUANO) ganda; (CHINESE, GÀN) jin cai; (CHINESE, MANDARIN) jiu cai; (CHINESE, MIN BEI) jiu cai; (CHINESE, WÚ) jiu cai; (CHINESE, YUE) jiu, jiu cai, yu tou; (CZECH) česnek vonný; (DANISH) kina-løg; (ENGLISH) Chinese chive, Chinese leek, oriental garlic; (FILIPINO) kuchai; (FINNISH) kiinansipuli; (FRENCH) ail civette de Chine, ail chinoise, ciboulette chinoise; (GERMAN) Chinalauch; (HEBREW) shum ha-peka'ot; (HMONG DAW) dos ntxw; (IFUGAO) amput di imayyaw, gamput di gimayyaw; (INDONESIAN) kucai, bawang prey, sigando, ganda kawat, bawang kucai; (ITALIAN) cipollina chinese; (ITAWIT) cuchay; (JAPANESE) nira; (JAVANESE) kucai, pucai; (KHMER, CENTRAL) kachaay; (MACEDONIAN) kineski luk; (MALAY) kuchai; (MALTESE) kurrat ciniz; (MONGOLIAN) ankhil songino; (MORISYEN) lake lay; (NEPALI) dundu; (NORWEGIAN) kinagrasløk; (PAMPANGAN) kutse; (POLISH) czosnek bulwiasty; (PORTUGUESE) alho chinês, nirá; (SPANISH) cive chino, cebollino de la China; (SUNDA) bawang kucai, bawang ganda; (SWEDISH) kinesisk lök; (TAGALOG) kutsay;

(TAGBANWA) ganda; (THAI) kuichai, hom-paen; (VIETNAMESE) h[ej]; (VLAAMS) Chinese bieslook; (WELSH) cenhinen china.

7.14. *Allium victorialis* L. (EP) bulb, leaves[S]; (P) C.

(ARABIC) bassal zu-el-gazoor; (CHINESE, GÀN) da cong; (CHINESE, MIN BEI) mai cong, cai cong, shui cong; (CHINESE, WÚ) ge cong; (CZECH) česnek hadí; (DANISH) allemandsharnisk; (ENGLISH) longroot onion, victory onion; (FINNISH) voitonlaukka; (FRENCH) ail de cerf; (GREEK) allion to nikiforo; (HUNGARIAN) győzedelmes hagyma; (JAPANESE) gyouja ninniku; (MONGOLIAN) jalatiin songino; (POLISH) czosnek siatkowany; (TURKISH) geyik sarmisaği, sarimsak; (VLAAMS) gyo-janinniku; (WELSH) nionyn gwraidd hir.

8. Araceae

8.1. *Alocasia macrorrhizos* (L.) G. Don (EP) corm, immature leaves[S], petioles[S]; (P) C.

(ARABIC) el-qulqass, el-emlak; (ASSAMESE) man kachu, boro mankachv; (BENGALI) mankachu; (BICOLANO, CENTRAL) talipan; (BINUKID) bagyang; (BONTOC, CENTRAL) aba-aba, bilbila; (BURMESE) pein-u, pein-mohawaya; (CAROLINIAN) filé; (CEBUANO) badiang, bagiang, bagyang, biga, galiang, gallan, galyang, ragiyang, talyang; (CHINESE, MANDARIN) hai yu; (CHINESE, WÚ) hai yu; (CREOLE, HAITIAN FRENCH) mazumbel, mazoumbèl; (CZECH) alokázie; (DAGAARE, SOUTHERN) kũdbari; (DANISH) kæmpetaro; (ENGLISH) giant taro, alocasia, giant alocasia, elephant ear; (ÉWÉ) koko; (FIJIAN) via, viadranu; (FILIPINO) biga; (FINNISH) jättialokasia; (FRENCH) taro géant, grande tayove; (GOURMANTCHÉ) ulũebi-tũba; (HAWAIIAN) ʻape; (HEBREW) alocasia; (HINDI) mankind, mankanda, hai yu; (HMONG DAW) qes tsw ha; (HUNGARIAN) táró; (IBANAG) aba; (IFUGAO) bilan mumpaha, helan mumpahag; (ILOCANO) bira, biga, sininaba; (INDONESIAN) bira, sente, lompong, birah; (ITALIAN) apé; (ITAWIT) bia; (JAPANESE) dokuimo, kuwazu-imo; (JAVANESE) sente, mael; (KANKANAEY) bila-bila; (KANNADA) marasa kaage, maanka, marasanige, mundigida, mundi gedde; (KHMER, CENTRAL) k'da:t haôra; (KIRIBATI) kabe; (LAO) kaph'uk; (LATVIAN) lielsakņu alokāzija; (LUO) opele; (MALAY) birah nĕgĕri, keladi sebaring; (MARANAO) lepeng, tangila a elepanti; (MARATHI) kaamsaalu, alu; (MARQUESAN) kape; (MÒORÉ) wobg-tuba; (NIUE) kape; (PAMPANGAN) biga, gandus; (PANJABI, EASTERN) pahärrhä arbi; (POLISH) alokazja, zakleśń olbrizymia; (PORTUGUESE) inhame gigante, inhame monstro, inhame taro; (RAROTONGAN) kape; (SAMOAN) ta'amu; (SANSKRIT) hasthi karni, kaasaalu; (SPANISH) cará tayá, malangá; (SUNDA) sente, taleus sente, bira; (SWAHILI) magimbi; (TAGALOG) gabing San Fernando, biga, badiang, malabiga; (TAMIL) merukankilangu; (TELUGU) chara kanda, sara chema; (THAI) kradaat, kradaat dam, kradatdam, horaa, hora; (TOK PISIN) abir, pia, via; (TONGAN) kape; (TURKISH) alokasye; (URDU) mankanda; (VIETNAMESE) khoai s[as] p, r[as]y, r[as]y [aw]n; (VLAAMS) reuzetaro; (WÁRAY-WÁRAY) badyáng.

8.2. *Amorphophallus konjac* Koch. (EP) corm; (P) C.

(*A. konjac* is now taxonomically considered by many as *A. paeoniifolius* var. *campanulatus*)

(BUHID) lungun; (ENGLISH) konjac, elephant-foot yam, snake palm, voodoo lily; (HANUNOO) bangak daraka, bangak iruy, buruy; (JAPANESE) konnyaku; (TAGALOG) pungapung; (VIETNAMESE) khoai n[uw]a.

8.3. *Amorphophallus paeoniifolius* (Dennst.) Nicolson var. *campanulatus* (Decne.) Sivad. (EP) corm; (P) C.

(ARABIC) batata el-feel; (ASSAMESE) ol kochu, ol; (BENGALI) ol, pharashbin; (BICOLANO, CENTRAL) alupahi, bagong; (BUHID) bangak, lungun; (BURMESE) wa; (CEBUANO) anto, oroy, pamangkilon; (CHINESE, MANDARIN) bai ban mo; (CHINESE, WÚ) bai ban mo yu; (CZECH) zmijovec zvonovitý; (DAGAARE, SOUTHERN) wēnaangnié; (DANISH) elefantyams; (ENGLISH) elephant yam, telinga potato, telugo potato; (FIJIAN) daiga; (FILIPINO) pungapung; (FRENCH) kouniak d'annam, pomme de terre de télinga; (GOURMANCHÉMA) makainēs; (GOURMANTCHÉ) apara; (GREEK) elephantini dioscorea; (GUJARATI) suran; (HEBREW) amorpha; (HINDI) suran, jangli suran, jimikand, jimmy kand, zaminkand, aula, kanda, karodkand, sooran, zameenkand; (IBANAG) bagabag, bagang; (ILOCANO) tigi-nga magmanto; (INDONESIAN) suweg, walur; (ITAWIT) bagabag; (JAPANESE) konjac, konniaku; (JAVANESE) suweg, walur; (KALAGAN) tokgang; (KANNADA) suvarna gedda, suvarna gedde, suvarna-guddab, soorana gedde, suvarna gedde, choorana; choorna gedde, kanda gedde, panjaragedde; (KHMER, CENTRAL) toal; (KONKANI) soorna, suma; (LAHU) bvuh¯ meu$^{\vee}$ ceh$_{\vee}$; (LAO) duk düa, kabuki; (LOMWE, MALAWI) chilazi; (MALAY) loki, chena, karrk-kavanai, ubi kekek; (MALAYALAM) kachil, kizhanna, cena, karunakarang; (MANU'A SAMOAN) teve; (MARANAO) delas, saging a riya sa kalasan; (MARATHI) suran; (MARQUESAN) teve; (MÒORÉ) piēcē; (MORISYEN) sourann; (NEPALI) wol; (NIUE) teve; (ORIYA) olua, farasi, owa, samba, simba; (PAMPANGAN) tokod-banuwa; (PANJABI, EASTERN) jimikand, zimikand; (POLISH) dziwidło dzwonkowate; (PORTUGUESE) jararaca mirim, batata de telinga; (RAROTONGAN) teve; (SAMOAN) teve; (SANSKRIT) suran, suranah, arsaghna, kunda, soorana, ashoghna, bahukanda, durnamari, kanda, kandala, kandanaayaka, kandarha, kandashoorana, kandi, kandoola, kandvardhana, kanthala, olla, rutchyakanda, soornaka, sthoolakandaka, sukandi, suvithra, theevrakantha, vathaari; (SANTALI) ol, pinḍe; (SPANISH) patata de Telinga; (SUNDA) lleus, suweg; (TAGALOG) pañgapoñg, ápon, pungapung, apon, apung-apong, pungapong; (TAJIKI) batat, kartoshkai shirin; (TAMIL) chenai kizhangu, kanaikizhangu, karanaippala, karakarunai, karnaikkizhangu,. karunaikkalang, chena, kachul, karanai-kilangu, karuna-kalang; (TELUGU) senai kizhangu, mashikanda, potikunda, manchi-kunda-guddae, kanda, alukamu, daradakandagadda, ghemikanda, kandagodda, manchikanda, naati kanda, oorakanda, paati kanda, potikanda; (THAI) buk, ukkhungkhok, man-suran; (TONGAN) teve; (TULU) parinki gadde, pariṅkigaḍḍè, sūraṇa; (TURKISH) amorfotallus; (URDU) hati yam, jangli suran; (VIETNAMESE) khoai n[uw]a, n[uw]a chu[oo]ng; (VLAAMS) olifantspoot.

8.4. *Colocasia esculenta* (L.) Schott (EP) corm, immature leaves[S]; (P) C.

[1]Itsukha dialect of Luyia.

(AFRIKAANS) amadumbi; (AGTA) ganet; (ALBANIAN, TOSK) kolokasie; (AMHARIC) gudäre, gʷədäre; (ANDO) mangani; (ARABIC) qolqas; (ASSAMESE) kochu; (BAATONUM) man kani; (BALANTA-KENTOHE) táem; (BAMANANKAN) diabéré, dàbéré, jabéré; (BAOULÉ) manganni, coco; (BASSARI) a-néwùrè, ɛ-ɗéwùrè, ɔ-léwùrè, guge; (BEMBÉ) talo; (BENGALI) kachu; (BEROM) jwɛ, jwɛ-pwɛŋ, jwɛ-sunàŋ; (BÉTÉ) koko; (BICOLANO, CENTRAL) natong, gabi, linsa, apay-ingkato, gabe, na-tong, tangoy; (BINUKID) suli, tuved, dagmai, kimpoi, lagbai; (BODO) tahdun, basor; (BONTOC, CENTRAL) amowang, baklag, lakat, pising; (BUDIK) ɛ-ɓɛn, ɛ-mɛn; (BUHID) alufa, amle malagsi, amle malayong, ayuskus, bage, fakli, inamlong, sapnuan, siggalfut, simbung, surnawi, turenduy; (BULGARIAN) kolokazija; (BULLOM SO) ɛ-lɛba; (BULU) atu; (BUTUANON) karlan; (CAROLINIAN) oot, otogúúh, otorhall, woot; (CEBUANO) abalong, butig, dagmay, gaway, kimpoy, lagbay; (CHINESE, GÀN) yu; (CHINESE, MANDARIN) yu, yü tou; (CHINESE, MIN BEI) yu tou; (CHINESE, MIN NAN) ō˙, ō˙-á; (CHINESE, WÚ) yu, yu nai, mao yu, yu tou; (CHINESE, XIANG) yu zi; (CHINESE, YUE) yu, yu tou; (CREOLE, HAITIAN FRENCH) mazumbel, mazoumbèl, tawo; (CRIOULO, CAPE VERDE) ihname, nhami; (CZECH) taro, kolokasie; (DAGAARE, SOUTHERN) pièrè; (DAGBANI) koúku; (DANISH) taro, kokosyams; (DITAMMARI) yakotanko; (DUTCH) taro; (EDE NAGO) ikoko; (EDO) íyòkhó; (EFIK) àta ḿkpọ̀ñ; (ENGLISH) taro, dasheen, cocoyam, eddoe; (ESTONIAN) taro; (ÉWÉ) mankani, asɔvle, koko, ulẽ; (FIJIAN) dalo; (FILIPINO) gabi; (FINNISH) taaro; (FON) glen; (FRENCH) taro, aronille; (FULFULDE) bonntoore, goojare, meeroore, tanndawre; (FULFULDE, ADAMAWA) tandaure, tandauje; (GA) kratse, kooloo amaɔkani; (GADDANG) tafal; (GBE, WACI) mankanlin; (GEN) mankanlin; (GERMAN) Taro; (GIKUYU) ndũũma; (GOLA) duu; (GOURMANTCHÉ) apara; (GREEK) kolokasi, kolokasia edodimos; (GUJARATI) alvi, alavu; (GUN) glen; (GURO) kólúo; (HANUNOO) badyan; (HAUSA) gwāza, guaza, gwaázaá, gwáázáá, kunnen jaakii, tagwamba; (HAWAIIAN) kalo; (HEBREW) kulkas, colocasia; (HILIGAYNON) gábi; (HINDI) akhi dhueva, arvi, kachalu, ghuiya, ghuiyan, ashukachu, auri, avois, ghoya, ghuya, ghwiya, gorikachu, jangli ghuya, kachu; (HINDUSTANI) arui; (HMONG DAW) qos tsw ha, qos tsov; (HMONG-NJUA) kaav ywj; (HUNGARIAN) táró, kokoyam, Egyiptomi taro, kalo; (IBALOI) aba, pising; (IBANAG) gavi; (IFUGAO) lubingan, bilaan aba, hukhukin di gayang, laat an aba, laat an banig, palaganggan latod; (IGBO) akasi, ede, édè, édè nkiti, édè ọ́kụ̄, édè ōsị́lị́, édè óyìbó, ogu édè, ńkàsì, àkàsì, édè éfù, édè ọ́bà; (ILOCANO) aba, aua, awa; (INDONESIAN) talas, talas bogor, keladi, bentul; (ITALIAN) colocasia, aro di Egitto; (ITAWIT) atang; (IVATAN) bola, sudi, yasi; (IZON) òdú, lòdú, odú, lòdú; (JAPANESE) satoimo, imo; (JAVANESE) tales, linjal, gelo; (JOLA-FONYI) usub; (JULA) mangani; (KABIYÉ) gbagbeng, kpero; (KANKANAEY) pising; (KANNADA) kesavina gedde, kesave dantu, kachchi, shamagadde, alavi, avigedde, keeshina gedde, kesavu, saave gedde, shaame gedde, shaave gedde; (KHMER, CENTRAL) tra:w; (KIRIBATI) te taororo; (KISI, SOUTHERN) ouai; (KISSI, NORTHERN) uai, wai, wayelẽ, waye-pɔmbɔe; (KONKANI) vaenti; (KONO) g-busi; (KOONGO) talo; (KOREAN) toran; (KRIO) krach-koko, koko; (KWÉYÒL) dachinn; (LAHU) peh$^{\vee}$ shi_; (LAO) bo:n, phüak; (LATVIAN) taro; (LOKO) k-pohi; (LUO) opele; (LUYIA) likhwana[1]; (MAITHILI) arokõch, arubĩ, pēchī; (MALAGASY) anantsaonjo; (MALAY) keladi, birah keladi, keladi China; (MALAYALAM) chembu, chempakizhanna, kaladi; (MALIMBA) yabɛrɛ; (MANDING-MANDINKA) jabero; (MANDINKA) oosoo, yabéré, fira-koko; (MANDJAK) umpita; (MANINKAKAN, EASTERN) fira-koko, yabɛre; (MAORI) taro; (MARANAO) dalog; (MARATHI) alu, aaloo, chempu, ran aalu; (MARQUESAN) ta'o; (MENDE) koko, kpoji; (MIZO) telhawng chi khat; (MÒORÉ) kũdbade; (MORISYEN) arouil violet, bred sonz; (NEPALI) pindalu; (NIUE) talo; (NORWEGIAN) taro; (NYANJA) ntembe; (ORIYA) saru; (PALAWANO, CENTRAL) lapung, talus-talus; (PAMPANGAN) gandos, gandus; (PANJABI, EASTERN) arvi, arbi; (POLISH) kolokazja, kleśnica, kolokazja jadalna; (PORTUGUESE) inhame, inhame branco, taro, taioba; (PULAAR) djáberè, iabéré-koko, jabéré-koko, jaibereh yabɛrɛ-ŋanya; (PUMÉ) t^{h}ada; (QUECHUA, CUSCO) mota, ñua; (ROMANIAN) colocazia; (RUSSIAN) taro; (SAMBAL, BOTOLAN) batlingan, tau; (SAMBAL, TINÀ) balingan, lapa, luko; (SAMOAN) talo; (SANSKRIT) alooka, aalooki, aashukachu, kachhi, kachhu, kachwi, katchu; (SANTALI) picki; (SENOUFO, CEBAARA) manganouwi; (SERBO-CROATIAN) taro; (SHONA) dumbe; (SINHALA) dehi ala; (SLOVAK) taro; (SPANISH) colocasia, malanga, yame de Canarias, alcocaz; (SUMO-MAYANGNA) pailumakka; (SUNDA) bolang, taleus balang; (SUSU) bari, bare; (SWAHILI) magimbi; (SWEDISH) Egyptisk taro, taro; (TAGALOG) gabi, lagbay, hupi, taro; (TAGBANWA) tales, gabi; (TAMIL) sambu, shana dumpa, sapan kizhangu, seppan-kizhangu, shamakkizhangu, shemakkalenga; (TELUGU) chemadumpalu, chaamakoora, chamadumda, chemagadda, chema, chamadumpa, chamakoora, shamakoora, shamathumpa; (TEM) kóu; (THAI) phuak, bon-nam, tun; (THEMNE) ɛ-koko-ɛ-nɛthi, ɛ-koko, *an*-koko, *a*-koko-*a*-nɛth, a-rɛŋ; (TIV) mọndọ; (TOK PISIN) anega, ba biloun; (TONGA) masimbi; (TONGAN) talo; (TULU) thevu; (TURKISH) fil kulaği, göleğez, gövelez, kolokaz, karayip lahanasi, kolokasya, güleis; (TWI) borɔbɛ; (VIETNAMESE) khoai-mon, khoai n[uw][ows]c, m[oo]n n[uw][ows]c, khoai s[oj]; (VLAAMS) taro; (WAAMA) kokoobu; (WAMEY) va-ngond; (WÁRAY-

WÁRAY) gawáy; (XHOSA) isityalo esinegaqa elityiwayo elifana nebhatata; (YAO) koko, lombo; (YOM) ikoko, kpelekpelei; (YORUBA) ikoko, kok, iṣu kókò, kókò, kókò funfun, kókò pupa.

8.5. *Colocasia gigantea* (Blume) Hook. f. (EP) petioles; (P) C.

(CHINESE, MIN NAN) ō͘-hôai; (ENGLISH) giant elephant's ear; (VIETNAMESE) bạc hà, dọc mùng.

8.6. *Cyrtosperma chamissonis* (Schott) Merr. (EP) corm; (P) C.

(AFRIKAANS) olifantsoor, taro; (ARABIC) qolqas el-mustaukat; (BINUKID) binagyang; (CAROLINIAN) bwula; (CZECH) maota; (DANISH) sumptaro; (ENGLISH) giant swamp taro; (FIJIAN) via kana; (FRENCH) faux-taro; (JAPANESE) te babai; (KIRIBATI) te babai, te papai; (MALAY) keladi air; (MALTESE) ghorghas; (POLISH) kolokazja błotna; (PORTUGUESE) falso taro; (RAROTONGAN) puraka; (SAMOAN) pula'a; (SANSKRIT) kachu; (SINHALA) gahala; (SPANISH) falso taro; (TONGAN) via; (VLAAMS) moerastaro; (WÁRAY-WÁRAY) palawan.

8.7. *Cyrtosperma merkusii* (Hassk.) Schott (EP) corm; (P) C.

(*C. merkusii* is now taxonomically considered by many as *Cyrtosperma chamissonis*)

(AKLANON) palawan; (BICOLANO, CENTRAL) galiang, galyang; (CEBUANO) palauan, palau, galyan, palaw; (ENGLISH) gallan; (FILIPINO) palauan; (FRENCH) siguine; (MALAY) geli-geli, keladi pari, birah hutan; (MANOBO) binagyang; (POLISH) kolokazja filipińska; (SAMOAN) pala'a; (SPANISH) gallan; (THAI) kli; (VIETNAMESE) h[aj]t cong; (VLAAMS) gallan; (WÁRAY-WÁRAY) palawan.

8.8. *Xanthosoma brasiliense* (Desf.) Engler (EP) immature leaves; (P) C.

(ARABIC) sabanekh; (ASSAMESE) dohi kochu; (DANISH) belembe; (ENGLISH) tannier spinach, calalou; (FRENCH) calalou, belembe, quélimbé; (HEBREW) xanthosoma; (LUO) opele; (POLISH) ksantosoma brazylijska; (PORTUGUESE) belembe, calalu; (SPANISH) belembe, calalú, kelembé, quelembe; (TAGALOG) San Fernando gabi; (VLAAMS) belembe, calalou.

8.9. *Xanthosoma sagittifolium* (L.) Schott (EP) corm and young leaves[S]; (P) C.

(AKAN-AKUAPEM) amankani, mankani, amankani antwibo, amankani fita, amankani fufu, amankani kokoo, amankani pa; (AKAN-ASANTE) amankani, antwibo; (ASSAMESE) dohi kochu; (BEMBÉ) makoutou ma nzaou; (BÉTÉ) ptê; (BINUKID) lutya; (BULLOM SO) ɛ-lɛba, koku-lɛ; (BULU) ekabé, lombo; (CHINESE, MANDARIN) ya yu; (CHINESE, WÚ) ya yu; (CREOLE, HAITIAN FRENCH) karaib, malanga karayib; (CRIOULO, CAPE VERDE) inhame, malanca; (DANGME) amankani; (DANGME-KROBO) amãkãni, amankani hiɔ, amankani tchu, amankanidzẽ; (DANISH) tannia; (DUTCH) tajer; (EDO) íyòkhó àkàrá; (EFIK) àta ḿkpọ̀ñ mbàkárá, idídùòt ḿkpọ̀ñ mbàkárá; (ENGLISH) tannia, yellow yautia, yellow ocumo; (ÉWÉ) kontomne, amankani, mankani; (EWONDO) lombo; (FIJIAN) tannia; (FILIPINO) yautia; (FINNISH) kaakaotaaro; (FON) tevi gami; (FRENCH) yautia des anglo saxons, tannie, tanier, chou caraibe, mangaride, taye, tayotte; (GA) amaŋkani, amankani; (GBAYA) wa macaba; (GEN) makalɛ; (GERMAN) Goldnarbe, Okumo, Tannia, Yautia; (GOLA) duu; (HAUSA) gwaàzaàmaì goòraá, gwaàzaá; (HEBREW) xanthosoma; (HINDUSTANI) lomka taya; (HUNGARIAN) tánia; (IGBO) akasi-oyibo, ede-beké, édè eko, édè ékò, àkàsị óyìbó, édè árọ̀, édẹ bèkẹ̀è, édè ọ̄kpọ́rọ́, ǹkàsí bèkéè; (INDONESIAN) kimpul, bentul, linjik, talas belitung; (IZON) bèké òdù; (JAPANESE) yautea; (JAVANESE) bentul, linjik, jokotowo, kimpul; (KABIYÉ) kpekpèou; (KISSI, NORTHERN) waye-bɛnde, wayele, wayilẽ; (KONO) koko; (KOONGO) makoutou ma nzaou; (KRIO) koko; (KWÉYÒL) chou; (LOKO) koko; (LUO) opele; (MALAY) keladi betawi; (MALIMBA) kogo, (MANDING-MANDINKA) diabereen puté, gbéne-uku, yabɛre; (MENDE) koko; (MORISYEN) arouil kari; (NIUE) pulaka; (NORWEGIAN) okumo; (POLISH) ksantosoma fiołkowa; (PORTUGUESE) mangará mirim, magarás, mangarito, taioba, qualuza, mafafa; (PULAAR) jabɛrɛ, jabɛrɛ-koko, yabɛrɛ; (QUECHUA, CUSCO) mairino; (SAMOAN) ta'amu papalagi; (SINHALA) kiri ala; (SPANISH) yautía bravia, mangaras, mangarito, malanga amarilla, yautía, cogollos; (SUMO-MAYANGNA) wilis; (SUNDA) kimpul, bentul, taleus hideung, taleus balitung, taleus lahun indung; (SUSU) koko; (TAGALOG) yautia; (THEMNE) ɛ-koko, ɛ-kol, ɛ-gbaŋkaŋ, *an*-ghaŋkaŋ, *an*-koko, *an*-kol, *am*-bɛroŋ, *an*-yabɛrɛ, amankani; (TURKISH) ok biçimli altin göbek; (TWI) amankani; (VAI) k-posi; (VLAAMS) tajerblad.

8.10. *Xanthosoma violaceum* Schott (EP) corm; (P) C.

(CEBUANO) cebu-gabi, tumon; (ENGLISH) Indian kale, blue taro, primrose malanga; (FILIPINO) cebu-yautia; (HANUNOO) singapur; (IFUGAO) bilan ngalngalitan, hilan nagalgaletan; (INDONESIAN) talas belitung, kimpul; (JAVANESE) kimpul; (LAO) th'u:n; (MALAY) keladi hitam, birah hitam, keladi kelamino; (MARANAO) palaw; (SUNDA) kimpul; (TAGALOG) gabing-cebu, gabing-Cebu, tumon; (THAI) kradat-dam; (VIETNAMESE) khoai s[as]p, ho[af]ng thu;

9. Cannaceae

9.1. *Canna indica* L. (EP) rhizomes; (P) C.

(AFRIKAANS) styselkanna; (AGTA) sagu; (AKAN-AKUAPEM) aburobia; (AKAN-ASANTE) akondogu, aburobia, ahabia; (AKLANON) tikas-tikas; (AKWA) mbobo-ikokomo; (ALBANIAN, TOSK) kanë; (ARABIC) kanna; (ASSAMESE) parijat; (BAATONUM) irobinu; (BAOULÉ) woko; (BASSARI) ɔ-ngànɛ́r; (BICOLANO, CENTRAL) bangali, batag-batag; (BUDIK) ma-ngalàl; (BUHID) talapias; (BULLOM SO) k-pɔwe, yumɛn; k-pɔklɔ-lɛ; (BURMESE) adalut, butsarana; (CEBUANO) balunsaying, kolintasan, tampuranga, tapuranga; (CHINESE, MANDARIN) shi yong mei ren jiao; (CHINESE, WÚ) jiao yu, shi yong mei ren jiao; (CHINESE, YUE) jia yu; (CRIOULO, CAPE VERDE) coqueirinho-de-jardim, cana-de-india; (CZECH) dosna jedlá, kana jedlá; (DANGME) blɛfo-tobi; (DANGME-KROBO) blɛfo-tobi; (DANISH) spiselig kanna; (DENDI) iro kize; (DUTCH) eetbaar bloemriet, Indisch bloemriet, ganjong; (EDE NAGO) okun; (EDO) èbèsälébō, (EFIK) n̈kwà édọ́n̈; (ENGLISH) edible canna, Indian canna, Queensland arrowroot, Indian beed, Indian reed, Indian shot; (ESTONIAN) kanna; (ÉWÉ) tovi aku, aku; (FILIPINO) canna, kakuintasan; (FINNISH) kanna; (FRENCH) balisier comestible, achira, canna d'Inde, canna; (GA) ahabia, ahabía; (GEN) atuklwi; (GERMAN) eßbares Blumenrohr, Indisches Blumenrohr, Achira, Kapocho; (GREEK) marantia kalamoidis; (GUN) alodan; (GURO) torowuin; (HANUNOO) kulintas mabiru, kulintas malagti; (HAUSA) baƙaleƙale, baƙalele, báƙáleƙáre, gwangwama; (HEBREW) kanna hodit, canna hodi; (HINDI) sabbajaya, sarvajjya; (HMONG-DAW) qhau liab; (HMONG-NJUA) nplooj nre lab; (HUNGARIAN) kanna, virágnád; (IFUGAO) gullu-gullu, kiwingan, ligyu an palawel, ullu-uIlu; (IGBO) abereka mwọ; (INDONESIAN) ganyong, tasbih, laos mekah, laos jambe, kanas; (ITALIAN) canna commestible, canna dolce, canna edule, canna d'India; (IVATAN) lasa; (JAPANESE) shokuyou-kanna; (JAVANESE) ganyong, lembong nyidra, senitra, buah tasbeh, ubi pikul; (KANNADA) kaabaale, chare gundina gida, hoodingana, hoodingala, kaalahoo, kaela hoo, kare gulaganji, kyana gida, nirghandha kusuma, swarajaya; (KHMER, CENTRAL) ché:k té:hs; (KISSI, NORTHERN) g-bclɔ̃, g-bɔlɔ̃kpɔ̃-siaŋa; (KONKANI) kaelaaphoola; (KONO) g-bɔɛ, g-bɔ-mɛsɛ, g-bɔyɛ; (KRIO) ɔkpɔlɔ-bid; (LAO) kwàyz ké: so:n, kwàyz ph'uuttha so:n; (LITHUANIAN)

Indinė kana; (LOKO) fɔrɔ, tɔmbega; (MALAY) ubi gereda, kenyong, zemba, daun tasbeh, ganjong, pisang sebiak; (MALAYALAM) kaattuvaala, kaattubaala; (MANDING-MANDINKA) fɔrɔndɔ-jɔ̃, bambru, nigoko; (MARANAO) korintas sa kalasan, maorintang; (MARATHI) daevakeli, kardali; (MENDE) fɔlo, fɔlɔ-gbama, giema-vɛlɔ, tomboya-vɛlɔ, ngiye ma vɔlɔ, tombo ya vɔlɔ; (MORISYEN) korn flor; (NEPALI) sarbada; (NORWEGIAN) kanna, canna; (PALAWANO, CENTRAL) cupialo; (POLISH) paciorecznik indyjski; (PORTUGUESE) cana comestível, cana da Índia, birú manso, merú; (PULAAR) fɔrɔndɔ; (RAROTONGAN) nuaenga; (ROMANIAN) cana; (SAMOAN) fagamanu, fanamanu, fa'i masoa, māsoā; (SANSKRIT) sarvajaya, daevakili, kamaakshi, krishnathaamara, shilarambha, vanakadali; (SERBO-CROATIAN) kana; (SINHALA) buth sarana; (SLOVAK) kanna; (SPANISH) achera, achira, caña comestible, caña de las Indias; (SUBANON) kulintas; (SUNDA) ganyong, ganyol, sebe; (SUSU) bɔhɔri, fɔrɔn-dai; (SWAHILI) kana; (SWEDISH) kanna; (TAGALOG) tikas-tikas, kukuwintasan, kakuwintasan, kulintas, kuwintas-kuwintasan, saging-saging, tikas, tikis-tikis, tukas-tukas; (TAMIL) kalvaalai, kalvalaimani, kundimani, puvalai, siramalai, shivamaale chedy; (TELUGU) guruginja, krishnathaamara, metta thaamara; (THAI) phuttharaksa, phutthason; (THEMNE) *an*-pɔlɔ; (TONGAN) misimisi; (TULU) kaelapola; (TURKISH) kana, ararot, hint kamişi; (TWI) aburobia, ahabia; (URDU) khana wala canna; (VAI) fɔle; (VIETNAMESE) chu[oos]i hoa, dong ri[eef]ng, khoai dao; (VLAAMS) indisch bloemriet; (WAMEY) u-ngwar; (WÁRAY-WÁRAY) kuwintáson; (YORUBA) ìdò, okun.

10. Cyperaceae

10.1. *Cyperus esculentus* L. (EP) tubers; (P) C,R.

(AFRIKAANS) geeluintjie, uintjie, hoenderuintjie, patrysuintjie; (AJA) efio; (AKAN-AKUAPEM) atadwe, atadwe awehea, atadwe fufuu, atadwe hene, hene, atadwe tsintsin, atadwe tuntum; (ALBANIAN, TOSK) mandër, manëz; (ARABIC) hab-el-samar, saad; (BAATONUM) tantan kpwen; (BAMANANKAN) ntogon, ntô, ntogô, togô, njoro, nton, n-togon, togon-kayogo, togon; (BAOULÉ) atadjô, n'gbomi; (BASSARI) dimori; (BEROM) pyò; (BÉTÉ) tjongô; (CHINESE, MANDARIN) yang di li; (CHINESE, WÚ) yang di li, suo cao; (CHUWABU) ndawe; (CRIOULO, CAPE VERDE) vista; (CZECH) zelenošáchor jedlý, zemní mandle, šachor jedlý; (DAGAARE, SOUTHERN) sicimè; (DAGBANI) nansaga, nansacha; (DANGME) fie; (DANGME-KROBO) fɛ; (DANISH) jordmandel; (DENDI) gaasu; (DITAMMARI) yanaacɛmmra; (DUTCH) aardmandel; (EDE NAGO) ofio; (EFIK) ísíp akṙâ; (ENGLISH) chufa, rushnut, earth almond, nut grass, ground-almond, tiger-nuts; (ÉWÉ) efio, fie, fie ge, fie yibɔ, fio; (FINNISH) maakastanja; (FON) fio; (FRENCH) amande de terre, souchet comestible, chufa; (FULFULDE, ADAMAWA) ayaare, waccuure; (GA) ataŋme; (GERMAN) Erdmandel; (GOURMANCHÉMA) venanachine; (GOURMANTCHÉ) hãnti; (GUN) fio; (HAUSA) árìgízà, ayáá, áyaá, àyaà-áyaà, áyaá rigiza, ba-dusaya, bákár-áyaá, chiso, cizo, guzu-guzu, jar áyaá, jiji, rìgízà, haya, bakar; (HEBREW) gome neëkhal, gomeh; (HERERO) okatjako; (HINDI) chichoda, chichada; (HUNGARIAN) palka, éti palka, mandulakávé; (ITALIAN) zigolo dolce; (JOLA-FONYI) épaympay; (JULA) tjongô; (KABIYÉ) amo; (KANNADA) nela baadaami hullu, musthe, thunga masthe; (KANURI) nebu, nufu; (KONKOMBA) idabongmale, idabunyal amo, indabongmale; (KRIO) kramanti, gramantin, gramanti; (LATVIAN) čufa, zemesmandele; (LITHUANIAN) valgomoji viksvuolė; (MALTESE) habb ghaziz; (MANDING-MANDINKA) n'chongo, n'chongo ba; (MANINKAKAN, EASTERN) toki; (MANINKAKAN, WESTERN) toki; (MONGOLIAN) sakhal öbs, ölön; (MÒORÉ) kiogo; (NAMA-DAMARA) !hân, !han-i; (NGUNDI) ntsantsaku; (PANJABI, EASTERN) kaseru, dila; (POLISH) migdał ziemny, cibora jadalna, gatunek cibory; (PORTUGUESE) chufa, junça doce, cipero comestível, juncinha, iririca-amarela; (QUECHUA, CUSCO) chufa; (ROMANIAN) căprişor, căprişor de mîncare; (RONGA) tirrica amarela; (RUSSIAN) mindal zemljanio, kofe mindalnyi, oreh zemljanoi, syt sedobnaja; (SANSKRIT) bhadramustha, chichotaka, kaseruka, kshudramustha, sugandhi, sukanda, sukareshta; (SENOUFO, CEBAARA) nouguele;

(SERBO-CROATIAN) zanovet; (SERER-SINE) ról; (SHONA) ndeka, ndekeret, tsuwa; (SLOVAK) zanovät, čilimník; (SONGHAY) hántí; (SPANISH) chufa; (SWAHILI) ndago; (SWEDISH) ginst; (TAMIL) palkorai; (TEM) amo, ano, mŏĕ; (THEMNE) *an*-soi, *an*-kolma-*a*-ro-gban; (TIGRIGNA) kwenti; (TIV) ìhọ̀họ̀; (TURKISH) sari topalak, abdülleziz, topoalağ, yer baddemi, kösnü otu, Venüs otu, şehvet otu, mablüloziz; (TWI) atajwe; (URDU) chi choda; (VLAAMS) aardamandel; (WOLOF) nder, nâgaro; (YOM) perempe; (YORUBA) ofio, ofio omu, imùmú òfio õmu, ẹrũnshà.

10.2. *Eleocharis dulcis* (Burm. f.) Trin. ex Henschel (EP) corms; (P) C,R.

(ARABIC) el-kustnaa, el-maee; (ASSAMESE) pani shingori; (BICOLANO, CENTRAL) apulid, buslig; (BONTOC, CENTRAL) tebtebbek; (CEBUANO) buslig, kalangub; (CHINESE, GÀN) bi qi; (CHINESE, MANDARIN) bi qi; (CHINESE, MIN BEI) wei li; (CHINESE, XIANG) qi mi, di li; (CHINESE, WÚ) bi ji, di li, ma ti; (CHINESE, YUE) ma ti; (CZECH) bahnička hlíznatá, vodní kaštan; (DANISH) kinavandkastanie; (ENGLISH) Chinese water chestnut, water chestnut, matai; (ESTONIAN) vesipähkel; (FILIPINO) apulid; (FINNISH) vesikastanja; (FRENCH) châtaigne d'eau; (GREEK) kastano idrohars; (HEBREW) bitẕöni matok, armon maim; (HILIGAYNON) tayok-tayok; (HUNGARIAN) csetkáka; (IBANAG) pagappo, pagippi; (ILOCANO) buslig; (INDONESIAN) teki; (JAPANESE) ohkuro guai; (JAVANESE) peperetan; (KHMER, CENTRAL) mëm phlông; (MALAY) tike; (MANU'A SAMOAN) utuutu; (MONGOLIAN) gurvalj; (NORWEGIAN) knollsivaks; (PANGASINAN) apoler, apular; (POLISH) ponikło bulwowe, słodkie; (PORTUGUESE) castanheiro de água; (SPANISH) cabezas de negrito; (SUNDA) babawangan; (TAGALOG) apúlid, buslig, apulid, apulid-tsina, potok, sibo-siboyasan; (TAJIKI) shohbulut; (THAI) haeo, haeo-chin; (THEMNE) a-kɔlma-ka-tar; (TONGAN) kuta; (TURKISH) su kestanesi, sandalye sazi; (VIETNAMESE) m[ax] th [af]y, n[aw]n ng[oj]t, n[aw]ng; (VLAAMS) waterkastanje; (WELSH) castawwydden dwr.

10.3. *Eleocharis kuroguwai* Ohwi (EP) corms; (P) C,R.

(CHINESE, WÚ) yie bi ji; (ENGLISH) wild water chestnut; (FINNISH) vesikastanja; (FRENCH) pi tsi; (GREEK) agrio-nerokàstano; (HUNGARIAN) csetkáka; (JAPANESE) kuro-guwai; (MONGOLIAN) gurvalj; (POLISH) ponikło kuro; (TAJIKI) shohbulut dashti; (TURKISH) yabani su kestanesi; (VLAAMS) kuro-guwai.

11. Dioscoreaceae

11.1. *Dioscorea alata* L. (EP) tuber; (P) C.

(AFRIKAANS) jam; (AKAN-AKUAPEM) ad-ammã-wo-ba, adjugo somadrewe, afasew, afasew anamasu, afasew apuka, afasew kani, afasew pa, afasew ɔdepa; (AKAN-FANTE) ɛdwew; (ALBANIAN, TOSK) dioskore; (AMHARIC) yäsəkk^{w} ar dəbbəč; (ARABIC) batata maee; (ASSAMESE) kathalu, kath alu, katalu, ractaguranialu; (BAATONUM) sàkuru; (BALANTA-KENTOHE) bosétigyé; (BAMANANKAN) danda ba, dabda ba, gua, guagara, guara, massa ku, uagara; (BAOULÉ) bêtai bêtai, n'ziwa, n'za, aken zaya, alua ualé, bété bété, wété wété, bété bété alengbe, bété bété kpa; (BEMBA) mûmbu; (BEMBÉ) mbala ngun; (BENGALI) chupri alu, khamalu; (BEROM) dém pwɛŋ, dém sunàŋ, kit jey, kit kácìk; (BÉTÉ) onhou gni; (BINUKID) uri; (BONTOC, CENTRAL) ang-angel, angel, ipoy, obi, tugi; (BUGHOTU) uvi; (BUHID) alfuyuan, alukha, barangse, bugraya, bugsi, fuluy, kasudu, lugiman, sinarafung, ubi fufu, ubi kasure, ubi malayung, ubi maraska, urabi; (BULU) engôm; (CAROLINIAN) nikka, likka, lóóghu; (CEBUANO) kinampai, kinampay; (CHINESE, HAKKA) voú-tzè; (CHINESE, MANDARIN) da shu; (CHINESE, MIN NAN) chî, chû; (CHINESE, XIANG) jiao ban shu; (CHINESE, WÚ) da shu, shan yao, liang gua; (CHINESE, YUE) da shu, can shu, tian shu, shu yu, sha ge, di qua, tan ge; (CREOLE, HAITIAN FRENCH) yanm;

(CZECH) bílý jam, batat, sladkyy brambor; (DAGAARE, SOUTHERN) gnié; (DANGME-KROBO) alamoa, alamoa gaga, alamoa kani, alamoa gu, alamoa ku, alamoa puka, alamoa sale, alamoa tun, alamoa tun gaga, asete alamoa; (DENDI) dundu hariya; (DIGO) mriga yere; (DITAMMARI) yanwà; (EDE NAGO) ewura duduku; (EDO) ígìorùa, udin; (EFIK) èbìgè, èbìgè ǹtán, èbìgè òsúkpà, n̈poyo èìgè, ọsoñ-ikpọ̀k èbìgè; (ENGLISH) water yam, greater yam, winged yam, asiatic yam, common yam, humped yam, pondichery potato; (ESTONIAN) jamss; (ÉWÉ) téklo, afasi, adzigo, adzugo, adzugo bolobolo, adzugo dzedze, adzugo kpuka, adzugo nkani; (FARSI, WESTERN) sibezamihic hendie; (FIJIAN) uvi; (FILIPINO) ubi; (FINNISH) isojamssi; (FON) tevi; (FRENCH) igname de Chine, igname ailée, pacala; (FULFULDE, ADAMAWA) ɓulumwol, ɓulumji; (GA) afaseo, afasέo, afasó, dãna; (GEN) ete; (GERMAN) Yamswurzel, Wasseryam; (GONJA) kuboru; (GREEK) aioskorea ptery-gioforos; (GUJARATI) goradu, ratalu; (HANUNOO) pappak upa; (HAUSA) doya, jikin mutun, sakataa, sakatáá; (HAWAIIAN) uhi; (HINDI) rataaloo, ratalu, chupri alu, chupri aalu, khamalu; (HINDUSTANI) ratálú, zamínqand; (HUNGARIAN) jamszgyökér; (IBANAG) uvi; (IFUGAO) luktu, ubin oongal; (IGBO) ji-mbo, abina, ji, jí m̀bàlà m̀bàlà, m̀bàlà, jí àbànà, mbwẹde, ngbẹde, onoko, jí àbàlà, jí m̀bàlà àdàkà, jí m̀bàlà m̀me; (INDONESIAN) uwi singapur, uwi, huwi, ubi manis; (ITALIAN) igname; (IZON) ìyọ́rọ̀bùrú; (JAPANESE) daijo; (JAVANESE) uwi kata, jebubug basu, jebubug endog, gembolo, uwi gandul; (JOLA-FONYI) é kama, e kama; (JULA) kou; (KABIYÉ) héyé, afasse; (KANNADA) kunti genasu, thunagenasu, dappa genasu, hebbu genasu, mudigenasu, nurele genasu; (KHMER, CENTRAL) dâmlô:ng chiem moan, dâmlô:ng phluk; (KIRIBATI) te iam; (KONKANI) kona; (KRIO) wata-yam, watayams, ishu, agbana; (KWÉYÒL) kanm, yanm; (LAHU) meu_ ce$^{\vee}$ ba$_{\vee}$ te; (LAO) man hlièmx; (LITHUANIAN) sparnuotoji dioskorėja; (LOMWE, MALAWI) chilazi; (LUO) obato; (MAITHILI) khamhāru, ratāru; (MALAY) ubi kemali, ubi kipas, pokok ubi, ubi tiyang; (MALAYALAM) kachil, aachilkkilangu, kaavattu, peruvallikkilannu; (MANDING-MANDINKA) balantaɲambi, massaku, ɲambí-ba; (MANINKAKAN, WESTERN) gbara-gué; (MAORI) uwhi, uhi, uwhikaaho, ngangarangi; (MARANAO) awi, awi a kakowak; (MARATHI) goradu, chipari-aalu, goradu, khanphal, pin daalu; (MARQUESAN) puauhi; (MENDE) nja-mbole; (MUSKOGEE) vhv-cvmpv; (NDALI) fitugu; (NEPALI) tarul; (NGUNDI) ngumvu; (NICOBARESE, CAR) si-në-meh, tö-ki-nī-ön, ta-ū-lö, ta-hā-ngen; (NIUE) ufi; (NORWEGIAN) kjempeyams, yams; (NUPE-NUPE-TAKO) eci; (NYAKYUSA-NGONDE) fitugu; (NYANJA) cilizi, mpama; (ORIYA) desi ba khamba alu; (OROMO, WEST CENTRAL) hanchote, dinicha, hochino; (PANJABI, EASTERN) ratalu; (POLISH) pochrzyn skrzydlaty; (PORTUGUESE) cará de Angola, cará branco, inhame bravo, cará da terra, cará; (PULAAR) pulóók; (PUMÉ) para; (QUECHUA, CUSCO) ticuna; (RAROTONGAN) u'i, ui parai; (ROMANIAN) batată, cartof brambor; (RUSSIAN) batat, patat, sladkyi kartofel; (SAMA, CENTRAL) mapiya; (SANSKRIT) dandaalu, pindaalu; (SENOUFO, CEBAARA) natogue; (SERBO-CROATIAN) jams; (SERER-SINE) an; (SIDAMO) ma t̲a t̲êssha; (SINHALA) hingurala; (SLOVAK) jams; (SPANISH) ñame de agua, ñame grande, ñame branco, cará branco; (SRANAN) jamsi; (SUNDA) huwi buwah, huwi nagri, huwi gandul, huwi upas, huwi blicik, huwi; (SUSU) khabi-gboueli, khabi-gbueli; (SWAHILI) kiazi kikuu, viazi vikuu; (SWEDISH) jams; (TAGALOG) ubi; (TAGBANWA) ubi; (TAMIL) kizhangu, perumvalli kizhangu, mullu-valli, siruvalli, ilangilangu, kaachikkizhngu, kayavalli, peruvalli, yaemskelung; (TELUGU) pendalam, pandalamu, pendalamu, daeshavaali pendalam, gadinidonda pendalamu, gunapendalamu, naarathega, niluvu pendalam, yadduthoka dumpa; (THAI) man-liam, man-sao, man-thu; (THEMNE) ɛ-wata-yams, *a*-wata-yams; (TIV) iyou̯gh; uhlobo oluthile olukhulu lwebhatata, àgbò, àgbò tsábàgú; (TOK PISIN) yam tru; (TONGA) ciyao; (TONGAN) 'ufi; (TULU) tunakereng, tuppekerengu; (TUMBUKA) ciyao; (TURKISH) kanatli yam, büyük yam; (TWI) afaseɛ, afasew; (URDU) pani yam, chini yam; (VIETNAMESE) c[ur] c[as]i, c[ur] m[owx], khoai v[aj]c; (VLAAMS) yam, igname; (WAAMA) wɔɔre; (WÁRAY-WÁRAY) ubi; (WOLOF) añ, ɲam-ba, ku-ɲambo, ɲamba ba; (YAO) lipeta; (YOM) nur; (YORUBA) ewura duduku, kiamfa, arunfanfan, atti, duduku, ègbódo, ewùrà, ewùrà funfun; (ZULU) uhlobo lukamanga.

11.2. *Dioscorea bulbifera* L. (EP) tubers; (P) C.

(AFRIKAANS) jam, aartappeljam; (AKAN-AKUAPEM) akam, akam moto; (AKAN-ASANTE) akaï, akam; (AKAN-FANTE) man; (ALBANIAN, TOSK) dioskore; (ASSAMESE) goch alu, kathalu; (BAATONUM) mɔkoro; (BADAGA) eresingi, eresinge; (BALANTA-KENTOHE) busu bulé, n-dome; (BAMANANKAN) danda, dâda, dana, kamu; (BAOULÉ) bronu kaha, ka; (BASSARI) ɔ-pɔngwón, a-fɔngwón, nubulentschor; (BEMBA) mûmbu; (BENGALI) banalu, suaralu, pitaalu, chupri alu, chamalu, gaichaalu, kukuralu; (BEROM) edɔ̀ŋ, etɔt, kwáŋ, yɔ̀ŋ; (BICOLANO, CENTRAL) oro-ablao, pulugan; (BINUKID) uri; (BONTOC, CENTRAL) lugtun di pinading; (BUDIK) ga-pángòl; (BUGHOTU) uvi; (BUHID) bed, bel, ligayan; (CAROLINIAN) nikka, likka, lóóghu; (CHINESE, HAKKA) voú-tzè; (CHINESE, MANDARIN) huang du; (CHINESE, MIN NAN) chî, chû; (CHINESE, WÚ) huang du; (CRIOULO, UPPER GUINEA) genebra; (CZECH) jam čibulkatý, vzduš; (DAGAARE, SOUTHERN) gnié; (DAGBANI) friguma; (DANGME-KROBO) akam, akom; (DANISH) yams; (DENDI) dundu mbissa; (DIGO) mriga yere; (EDE NAGO) emina; (ENGLISH) potato yam, aerial yam, bulb-bearing yam, bulbil-bearing yam; (ESTONIAN) jamss; (ÉWÉ) etetsetse, gbe-tɔ; (FIJIAN) kailegaga, kaile; (FILIPINO) ubi-ubihan; (FON) jite sɛnde; (FRENCH) igname pousse en l'air, igname bulbifere, igname patate, pomme de l'air; (FULFULDE, ADAMAWA) ɓulumwol, ɓulumji; (GEN) jite; (GERMAN) Knollen-Yam; (GIANGAN) dadakan; (GREEK) glykopatata; (GUJARATI) goradu, ratalu, salvinavelya, suariya, varahikanda; (HANUNOO) batyay; (HAUSA) doya, dóóya, dóóyan ítaacéé, dóóyar-bísà, dóóyar-bisa; (HAWAIIAN) hoi; (HINDI) rataaloo, ratalu, suaralu, pitaalu, agitha, karukanda, bhiruvolikanda, genth, genthi, gethi, genttee, karawakand, suvariyaa kand, zaminkand; (HINDUSTANI) ratálú, zamínqand; (HUNGARIAN) jamszgyökér; (IGBO) adụ, ji, àbànà ófị̄á, àbànà ófị̄á baba, àbànà ófị̄á gbara, àbànà ófị̄á mbaba, ádụ̀; (ILOCANO) aribukbuk; (INDONESIAN) ubi gantung, ubi buah; (ITALIAN) igname; (JAPANESE) kashû-imo, niga-gashû, maruba-dokoro; (JAVANESE) jebubug basu; (JOLA-FONYI) diéban, éborimbor, kanum, karamba; (KABIYÉ) dihere, nbanioke; (KANNADA) kunti genasu, heggenasu, bellarai, inasara, nirbatte, tabinere, handi gedde, kaikaraande, kaimoode gedde, kunta genasu; (KHMER, CENTRAL) dâmlô:ng dü:hs, dâmlô:ng sdâm préi; (KIRIBATI) te iam; (KISSI, NORTHERN) kule-yondolẽ; (KONKANI) karando; (KONO) kuyawa; (KRIO) ɛmina, ishu; (KWÉYÒL) yanm; (LAHU) la^{V} hk' a^{V}; (LAO) man pauz, hwà 'i: mu:z; (LITHUANIAN) gumbinė dioskorėja; (LOKO) m-bo boni, n-gengange; (LOMWE, MALAWI) chilazi; (LUO) obato; (MALAY) ubi atatus, ubi atas, ubi china, ubi kestali; (MALAYALAM) kachil, kattukachil, kanji, kaattu kaachil, mukkilangu; (MALIMBA) kəyoro; (MANDING-MANDINKA) bayulo, kama, kamao, tubaabuɲambi, camó dandandim, dandaõ, danda; (MANINKAKAN, EASTERN) dâda, dana, dan-dan, guinfiné, digegbe; (MANINKAKAN, WESTERN) danda; (MANKANYA) catoco; (MANU'A SAMOAN) soi; (MAORI) uwhi, uhi, uwhikaaho, ngangarangi; (MARANAO) leteng; (MARATHI) dukarakanda, manakund, karuka-rinda, gathalu, goradu, mataru, kadoo karaandaa, konaphala, kukarkand, mibaelikand, varaahi; (MARQUESAN) hoi; (MENDE) fɔli, mbole, ngawu, puli; (MUSKOGEE) vhv-cvmpv; (NDALI) fitugu; (NEPALI) gittha; (NGUNDI) ngamba; (NICOBARESE, CAR) si-në-meh, tö-ki-nī-ön, ta-ū-lö, ta-hā-ngen; (NIUE) ufi, hoi; (NORWEGIAN) luftyams, yams; (NUPE-NUPE-TAKO) eci; (NYAKYUSA-NGONDE) fitugu; (NYANJA) cilizi, mpama; (ORIYA) desi ba khamba alu, pita alu, pittale, gourobi; (OROMO, WEST CENTRAL) hanchote, dinicha, hochino; (PALAWANO, CENTRAL) ablao; (PANJABI, EASTERN) ratalu, zaminkhand; (POLISH) pochrzyn ziemniaczany; (PORTUGUESE) inhame de bolbos aéreos, inhame de rama, cará de sapateiro, cará aereo; (PULAAR) pulóók, púri, puri-balé; (PUMÉ) para; (RAROTONGAN) oi, ui; (SAMBAL, BOTOLAN) bukyot; (SAMOAN) soi, ufi sina, ufi soi; (SANSKRIT) varahi, varahikanda, varaahikanda, amlardraka, sarpakkya, amritha, badarakachha, balya, bilvamoola, brahmaputhri, brhamikanda, charmakaraluka, ghrishti, kanya, kaumaari, krodakanya, krodi, kushthanaashaka, madhaveshtagrishtika, magadhi, mahaaveerya, mahaushadha, shambarakanda, shukari, sukandaka, thrinethra, vanamaalini, vanavaasi, vaaraahi, vishvaksenapriya, virdhida, vyadhihanta; (SANTALI) beṅgo̱ nā̱ṛī̱; (SERBO-CROATIAN) jams; (SERER-SINE) ań, sipa; (SIDAMO) ma t̲a t̲êssha; (SINHALA) undala; (SLOVAK) jams; (SPANISH) ñame del aire, ñame congo; (SUBANON) glingu gubi menaik, panubulan; (SUNDA) huwi buah; (SUSU) dané; (SWEDISH) jams; (TAGALOG) bayag-toro, ubi-ubihan, ubi-ubingan, bayag-kabayo, utong-utongan; (TAGBANWA) butuan; (TAMIL) kodikilangu, pannukilangu, kizhangu, pannukkizhangu, kaattukkizhangu, kaivalli kodi, kavala-kodi, kattukkilangu, thavaikachchu, venilai valli, vettilaivalli; (TEKE) akô; (TELUGU) chedupaddu-dumpa, chedupaddudumpa, malakakayapendalamu, malaakaakaayapendalamu, pendalam, adairdumpa, adavidumpatige, aretegalu, avatengatige, chenchudampa, yellagadda; (TEM) agbanio; (THAI) wan-prachim, hampao, man-soen; (THEMNE) *am*-boŋk, *ra*-boŋk, *an*-wun; (TIV) ájíē, íjémbēàóndó, íjíē, íjíē shìbèrá, iyo̭ugh; uhlobo oluthile olukhulu lwebhatata; (TONGA) ciyao; (TONGAN) hoi; (TULU) ballisoorala; (TUMBUKA) ciyao; (TURKISH) yumrulu yam, hava patatesi; (TWI) akan, akanmmto, akqmbçowoma, ananseakán; (URDU) ratalu, zaminekand; (VIETNAMESE) c[ur] d[aj] i, khoai ch[oof]i, khoai d[as]i; (VLAAMS) aardappelyam; (WAMEY) ɓẹ-ngwaf u-nad̈, tẹmp-tẹmp; (WÁRAY-WÁRAY) abubu sa tawo; (WOLOF) ań, ngolgol; (YAO) lipeta; (YORUBA) emina, ẹmina, djiteodé, guite sindé, guite sodé, aparia iṣu, ẹmina-esi, èsúrú, ẹwura-esi.

11.3. *Dioscorea cayenensis* Lam. (EP) tuber; (P) C.

(AKAN-AKUAPEM) akwakwɔ-bedina, aniwa-aniwa, *n*-kani, n-kani-fufu, n-kani-hene, n-kani-pa, nkuku; (AKAN-ASANTE) afũũ, afũũ-ba, nkontia; (ALBANIAN, TOSK) dioskore; (ARABIC) batata safraa, batata swdaa; (BAMANANKAN) dãnda, fasaka; (BAOULÉ) klinglin; (BASSARI) kh̃áɓ, ɔ-ŋáɓ; (BEMBA) mûmbu; (BÉTÉ) iberou gni; (BUDIK) ɛ-gɛ̀b; (BULU) andia; (CHINESE, HAKKA) voú-tzè; (CHINESE, MANDARIN) fei zhou shan yao; (CHINESE, MIN NAN) chî, chû; (CHINESE, WÚ) fei zhou shan yao; (CREOLE, HAITIAN FRENCH) yanm jonn; (CZECH) ný jam; (DAGAARE, SOUTHERN) gnié; (DANGME-KROBO) kani, kani-gu, kani-hini, kani-kuku, kani-tʃu, otri-kani; (DANISH) yams; (DIGO) mriga yere; (EDE NAGO) pansaga; (EDO) íkpèn, òkpòdọ̀ghòn; (EFIK) àkpàná, àkpàná-ntàntà, èfián; (ENGLISH) yellow yam, negro yam, yellow Guinea yam; (ESTONIAN) jamss; (ÉWÉ) ete, kanyi, dzafutu, dzoboli, kani, kani gu, kani hene, kani kuku, kani tchu, kani yie, kpasa, n-kani, otre-kani, avete, boto, dza, loboko; (FRENCH) igname de la Guinée, igname jaune, igname pays-nègre; (GA) ɔkani, nkani; (GERMAN) Guinea Yamswurzel, Cayenne Yamswurzel; (GOURMANTCHÉ) adja-bouinga; (GREEK) dioscorea, enaerios, kitrini, maúpy, pikri; (GUN) aglandan; (HAUSA) doóyàr kúdù; (HINDUSTANI) ratálú, zamín̲qand; (HMONG DAW) qos daj, qos liab; (HUNGARIAN) jamszgyökér; (IGBO) ji-okwu; (ITALIAN) igname; (IZON) òwéí bùrú; (JULA) kou; (KABIYÉ) koukou; (KIRIBATI) te iam; (KRIO) agbani-yam, agbaniyams, akpani, agbani; (KWÉYÒL) yanm; (LOMWE, MALAWI) chilazi; (LUO) obato; (MANDING-MANDINKA) ɲambi; (MAORI) uwhi, uhi, uwhikaaho, ngangarangi; (MENDE) mbole-gbɔlu, mbole-gbɔlɔ; (NDALI) fitugu; (NORWEGIAN) yams; (NYAKYUSA-NGONDE) fitugu; (NYANJA) cilizi, mpama; (POLISH) pochrzyn żółty; (PORTUGUESE) cará do Pará; (PUMÉ) para; (SENOUFO, CEBAARA) wakourouhou; (SERBO-CROATIAN) jams; (SLOVAK) jams; (SPANISH) affoo, ñame amarillo, ñame negro, ñame de Guinea; (SWEDISH) jams; (THEMNE) ma-nəi, *an*-nəi, ma-nai, *an*-nai, *a*-mas, *an*-ku, *a*-nyame, *a*-nyams-*a*-yin; (TIV) íyogh, nwángē; (TONGA) ciyao; (TUMBUKA) ciyao; (TWI) nkani, afu; (URDU) pela yam; (VLAAMS) gele yam; (WAMEY) u-nkwav; (YAO) lipeta, (YORUBA) pansaga, àgándán, aginipa, apèpe, ìgángán, ìgángán alo, àlọ̄.

11.4. *Dioscorea dumetorum* (Kunth) Pax (EP) tuber; (P) C.

(AFRIKAANS) bitter jam; (AKAN-AKUAPEM) akori ekyi, nkanfo, oworo-woro, yaw serewin; (AKAN-ASANTE) nkanfoɔ; (ALBANIAN, TOSK) dioskore; (ARABIC) batata mura; (ASSAMESE) tita alu; (BEMBA) mûmbu; (BUDIK) ɛ-dyíɗy, gi-nyíɗy; (BULU) andia; (CAROLINIAN) nikka, likka, lóóghu; (CHINESE, HAKKA) voú-tzè; (CHINESE, MIN NAN) chî, chû; (CHINESE, WÚ) ku shu; (DAGAARE, SOUTHERN) gnié; (DANGME-KROBO) kamfo, nkampo; (DANISH) yams; (DENDI) dundu kiré; (DIGO) mriga, mriga yere; (EDO) ùfúa, èmówè; (EFIK) afia edidia, àfia èdidià iwá, àfia iwá, àkpàná, eba edi, èdèm ìnày, èdiá abañ, èdidià ìwá; (ENGLISH) bitter yam, cluster yam, trifoliate yam; (ESTONIAN) jamss; (ÉWÉ) dekute, adekute; (FON) gudu gudu; (FRENCH) igname sauvage; (GA) ɔkamfo, ɔkafo, akori ekyi, nkampo, oworo-woro; (GERMAN) Wild Yamswurzel; (GOURMANCHÉMA) kũ; (GOURMANTCHÉ) adja-noũga; (GUN) suruma; (HAUSA) koósán-roógoò, gursami; (HINDUSTANI) ratálú, zamíṉqand; (HUNGARIAN) jamszgyökér; (IGBO) ọna, ọ̀nà, ọ̀nụ̀, ádụ̀, akpana, atọka, eba edi, obubit iwa; (ITALIAN) igname salvatico; (KABIYÉ) tchevèou, lala; (KIRIBATI) te iam; (KRIO) ɛsuru-yams, esuru; (KWÉYÒL) yanm; (LOKO) n-gengange; (LUO) obato; (MAORI) uwhi, uhi, uwhikaaho, ngangarangi; (NDALI) fitugu; (NGUNDI) tichunia; (NORWEGIAN) yams; (NYAKYUSA-NGONDE) fitugu; (NYANJA) cilizi, mpama; (POLISH) poschzyn gorzki; (PORTUGUESE) inhame amargo, inhame bravo; (PULAAR) mabaia; (PUMÉ) para; (SERBO-CROATIAN) jams; (SLOVAK) jams; (SPANISH) ñame amargo, ñame de tres hojas; (SWEDISH) jams; (TIV) ínímbē, anembe; (TONGA) ciyao; (TUMBUKA) ciyao; (TWI) nkamfo; (URDU) karwa yam; (VLAAMS) bittere yam; (YAO) lipeta; (YORUBA) itakun iṣu, èsúru.

11.5. *Dioscorea esculenta* (Lour.) Burkill (EP) tuber; (P) C.

(ALBANIAN, TOSK) dioskore; (ASSAMESE) moa alu; (BAOULÉ) lopka, blofouè douo, brofue duo; (BEMBA) mûmbu; (BENGALI) suthni, susnialu; (BICOLANO, CENTRAL) kamiging, baribaran; (BINUKID) buley, uri; (BUGHOTU) uvi; (BUHID) bel, bong; (BULU) andia; (CAROLINIAN) nikka, likka, lóóghu; (CEBUANO) tam-is; (CHINESE, HAKKA) voú-tzè; (CHINESE, MANDARIN) tian shu; (CHINESE, MIN NAN) chî, chû; (CHINESE, WÚ) tian shu; (CZECH) malý jam; (DAGAARE, SOUTHERN) gnié; (DANGME-KROBO) biɛfo hiɛ; (DANISH) yams; (DIGO) mriga yere; (ENGLISH) lesser yam, Chinese yam, Asiatic yam; (ESTONIAN) jamss; (ÉWÉ) anago-tɛ, yevu-tɛ, anago té, ete (yovote); (FIJIAN) kawai; (FILIPINO) tugi; (FINNISH) pikkujamssi; (FRENCH) igname des blancs, igname patate, igname pas possible, igname de Chine; (GA) ant-wakɔle; (GADDANG) luttu; (GERMAN) Yamswurzel; (GOURMANCHÉMA) kũ; (GOURMANTCHÉ) anoũga; (GREEK) eloasson dioscorea; (HANUNOO) basun, borot, burut, guhayan; (HILIGAYNON) lupisan, tugi; (HINDUSTANI) ratálú, zamíṉqand; (HUNGARIAN) jamszgyökér; (IBANAG) a-neg, aneg, luttu; (IFUGAO) lugtus puggupug, luttu; (IGBO) ji; (ILOCANO) boga, tugi; (INDONESIAN) gembili, ubi gembili, ubi aung, ubi landak; (ITALIAN) igname; (ITAWIT) aneg; (IVATAN) dukay; (JAPANESE) hari-imo; (JAVANESE) gembili, sudo, gembolo, ubi aung, ubi gembili; (JULA) kou fetini; (KABIYÉ) hèyè; (KANNADA) mullu genasu, see genasu; (KHMER, CENTRAL) dâmlô:ng sya; (KIRIBATI) te iam; (KWÉYÒL) yanm; (LAO) hwà katha:d, man ‘o:nz; (LATVIAN) ēdamā diskoreja; (LUO) obato; (MALAY) ubi torak, kembili, kemarung; (MAORI) uwhi, uhi, uwhikaaho, ngangarangi; (MARATHI) kangar; (MÒORÉ) kũ; (MORISYEN) kambar beti; (MUSKOGEE) vhv-cvmpv; (NDALI) fitugu; (NEPALI) tarul; (NICOBARESE, CAR) si-në-meh, tö-ki-nī-ön, ta-ū-lö, ta-hā-ngen; (NIUE) ufi, ufilei; (NORWEGIAN) yams; (NYAKYUSA-NGONDE) fitugu; (NYANJA) cilizi, mpama; (POLISH) pochrzyn jadalny; (PORTUGUESE) inhame de São Tomé; (PULAAR) pulóók; (PUMÉ) para; (SAMBAL, BOTOLAN) baugan, bawugan, kopastak; (SAMOAN) ufi lei; (SANSKRIT) madhvaluka, munjaatha; (SENOUFO, CEBAARA) fogbogui dangarou; (SERBO-CROATIAN) jams; (SINHALA) kukulala; (SLOVAK) jams; (SPANISH) ñame papa, ñame pequeño; (SUNDA) huwi kamayung, huwi cucuk, huwi butul, huwi landak, huwi taropong, huwi ceker, huwi jahe, huwi kawoyang; (SWEDISH) jams; (TAGALOG) apali, tongo, tugi, tungo; (TAMIL) musilam valli kilangu, siruvalli kilangu, siruvalli kizhangu; (TELUGU) silakadom, tivvitiga, tippa tiga; (THAI) man-musua, man-chuak; (TOK PISIN) mami; (TONGA) ciyao; (TONGAN) ʻufilei; (TUMBUKA) ciyao; (TURKISH) çin patatesi, yenilen yam, hint yerelmasi; (TWI) ɔdebɔduanan; (VIETNAMESE) c[ur] t[uwf], khoai t[uwf], t[uwf] gai; (VLAAMS) kleine yam; (YAO) lipeta; (YORUBA) iṣu, iṣu àlùbọ́sà, iṣu ànàmọ́, odunkun.

11.6. *Dioscorea oppositifolia* L. (EP) tuber; (P) C.

Dialects of Luyia (Bukusu[1], Maragoli[2])

(AFRIKAANS) sjinese jam; (ALBANIAN, TOSK) dioskore; (ARABIC) batata seeny; (BEMBA) mûmbu; (CHINESE, HAKKA) voú-tzè; (CHINESE, MANDARIN) pu tong shan yao; (CHINESE, MIN BEI) xiao shu, huai shan, jia shan yao; (CHINESE, MIN NAN) chî, chû; (CHINESE, WÚ) pu tong shan yao; bai shao, shan shu, shu yu; (CHINESE, YUE) jia shan yao, huai shan; (CZECH) jam čínský; (DAGAARE, SOUTHERN) gnié; (DANISH) yams; (DIGO) mriga yere; (DUTCH) eetbare yamswortel; (ENGLISH) Chinese yam, cinnamon yam, betel yam; (ESTONIAN) jamss; (ÉWÉ) ete; (FINNISH) jamssi; (FRENCH) igname de Chine; (GERMAN) Chinesische Yamswurzel, eßbare Yamswurzel; (GIKUYU) gĩkwa; (HINDUSTANI) ratálú, zamínqand; (HMONG DAW) qos; (HUNGARIAN) jámszgyökér; (ITALIAN) igname Chinese; (JAPANESE) yamaimo; (JOLA-FONYI) é borimbor; (KANNADA) aenna soora, bellaraai, hakki genasu, imsara, inaasara, kaadu genasu, neerbatte, thabinera; (KIRIBATI) te iam; (KWÉYÒL) yanm; (LATVIAN) batātu dioskoreja; (LITHUANIAN) batatinė dioskorėja; (LUO) obato; (LUYIA) liliakhunyu[1], litugu[2], lirungu[2]; (MAORI) uwhi, uhi, uwhikaaho, ngangarangi; (NDALI) fitugu; (NORWEGIAN) kinayams, yams; (NYAKYUSA-NGONDE) fitugu; (NYANJA) cilizi, mpama; (POLISH) pochrzyn chińsik, ziemniak chińsik; (PORTUGUESE) inhame da China, cará amarelo; (PUMÉ) para; (SANSKRIT) aamlaardraka, sarpakhya, varaahi; (SERBO-CROATIAN) jams; (SLOVAK) jams; (SPANISH) ñame de la China; (SWEDISH) jams; (TAMIL) kavalla-kodi, kaivalli kodi, mahaparuvallikkodi, malayan kizhangukkodi, verrolaivalli, vettiiaivalli kizhangu; (TELUGU) adavidumpathige, adaviyatagathige, arethegalu, arithige, athyaga, avathengathige, chenchudumpa, yella gadda; (TONGA) ciyao; (TUMBUKA) ciyao; (TURKISH) çin patatesi, hintyerelmasi; (URDU) aloo yam, barahi kund; (VLAAMS) aardappelyam; (YAO) lipeta.

11.7. *Dioscorea rotundata* Poiret (EP) tuber; (P) C.

(AKAN-AKUAPEM) ade, akimbayeray, akuku, akwakor, aso bayerɛ, ayebir, ɔde pa, osibã, pona, tonto; (AKAN-FANTE) ayebir, pona; (ALBANIAN, TOSK) dioskore; (ARABIC) batata beda; (ASSAMESE) kotahi alu; (BAOULÉ) kiri kiri, kuana, sopéré; (BASSARI) lili; (BEMBA) mûmbu; (BEROM) kit ryáŋ; (BULU) andia; (CHINESE, HAKKA) voú-tzè; (CHINESE, MANDARIN) yuan xing shu yu; (CHINESE, MIN BEI) da shu, tian shu; (CHINESE, MIN NAN) chî, chû; (CHINESE, WÚ) yuan xing shu yu; (CHINESE, YUE) shen shu, tian shu; (DAGAARE, SOUTHERN) gnié; (DAGBANI) bamegu, dakpan, gungunsale, gunuukple, kpasajɔ, larbakɔ, kpeney, lili, perenga, kprenga, sagalanga, sanaraje, ziglanbo, zon, zugulangbɔ, zulanbo; (DANGME) yɛ; (DANGME-KROBO) ade, akim bali, akwakor, akwakor pechiwa, akwakwaleekwa, deeka, hiɛ, jeow, kate, kentem, kumiyo, mãade, mali, obobi, odornor, osuban; (DANISH) yams; (DIGO) mriga yere; (EDO) èmọ́wè, ígìerù, ómí, ómí funfun, órí; (EFIK) afia oko, àfia òkò, òkò, èkà biâ, ekefià, èkò, oko, okpo, ọ̀kpọ́, okpo uman, ọ̀kpọ́ úmàn íwā, okpuru, otuk okpo; (ENGLISH) white Guinea yam, Guinea yam; (ESTONIAN) jamss; (ÉWÉ) ete (fufute), ade, akuku, dreeka, dze-tɛ, ge-tɛ, klewu, kokolimakoe, nkam, sonka, sunka, tɛ, tɛ-dze, te-gba; (FILIPINO) white Guinea yam; (FINNISH) valkoinen guineanjamssi; (FRENCH) igname de la Guinée, igname blanche; (GA) dãna, male, tonto, yɛlɛ; (GERMAN) Guinea Yamswurzel; (GONJA) kujɔ, ajɔ; (GOURMANTCHÉ) afuua-noũga; (GREEK) lefki dioacorea; (HAUSA) dóóya; (HINDUSTANI) ratálú, zamínqand; (HUNGARIAN) jamszgyökér; (IGBO) ji-ocha, àbí̩, jí àbí, àgà, jí àgà, ala, alafolo, jí alafolo alẹfulu, asokolo, awudo, jí awudo, asuku, ẹkpẹ, jí àyọ̀bè, jí ọ́chá, jí íkē, jí òkò jí-akero, ólū, omi; (ITALIAN) igname; (KIRIBATI) te iam; (KONKOMBA) gbelenga; (KRIO) wet-yam, wait-yam; (KWÉYÒL) yanm; (LOMWE, MALAWI) chilazi; (LUO) obato; (MAORI) uwhi, uhi, uwhikaaho, ngangarangi; (MENDE) mbole-gowe, mbole-gowɔ; (MORISYEN) kambar douri; (NDALI) fitugu; (NORWEGIAN) yams; (NYAKYUSA-NGONDE) fitugu; (NYANJA) cilizi, mpama; (POLISH) pochrzyn biały; (PORTUGUESE) inhame da Guiné branco; (PULAAR) kapé; (PUMÉ) para; (SERBO-CROATIAN) jams; (SINHALA) suduwelala; (SLOVAK) jams; (SPANISH) ñame Guieno blanco; (SUSU) khabi; (SWEDISH) jams; (TAGALOG) nami; (TEKE) akô; (THEMNE) ɛ-yams-ɛ-fɛra, *an*-won, *a*-yams-*a*-fɛra; (TIV) nwángē; (TONGA) ciyao; (TUMBUKA) ciyao; (URDU) safad yam; (VLAAMS) witte yam; (YAO) lipeta; (YORUBA) ààlàoko, àbàjẹ, aga, agake, agẹmọkun, aginni, agogo, aladẹ, alo, àlọ̀lọ, ayin, bọki, apepe, awure.

11.8. *Dioscorea trifida* L. f. (EP) tuber; (P) C.

(ALBANIAN, TOSK) dioskore; (CHINESE, HAKKA) voú-tzè; (CHINESE, MIN NAN) chî, chû; (CHINESE, WÚ) shan ye shu yu; (CZECH) kaš-kaš jam; (DAGAARE, SOUTHERN) gnié; (DANISH) yams; (DIGO) mriga yere; (ENGLISH) cush-cush yam, Indian yam; (ESTONIAN) jamss; (FINNISH) kuskusjamssi; (FRENCH) igname couche-couche, igname douce, igname de la Martinique; (GREEK) tropiki amerikaniki dioscorea; (HINDUSTANI) ratálú, zamíṉqand; (HUNGARIAN) jamszgyökér; (ITALIAN) igname couche-couche; (KIRIBATI) te iam; (KWÉYÒL) yanm; (LUO) obato; (MAORI) uwhi, uhi, uwhikaaho, ngangarangi; (NDALI) fitugu; (NEPALI) vyakur; (NORWEGIAN) yams; (NYAKYUSA-NGONDE) fitugu; (NYANJA) cilizi, mpama; (POLISH) pochrzyn trójdzielny; (PORTUGUESE) cará doce, cará mimoso; (PUMÉ) čaramekə; (SERBO-CROATIAN) jams; (SLOVAK) jams; (SPANISH) ñame morado, ñame yampí, ñame mapuey, ñame couche-couche; (SRANAN) napi; (SUMO-MAYANGNA) usi; (SWEDISH) jams; (TONGA) ciyao; (TUMBUKA) ciyao; (TURKISH) üç yarikli yam; (VLAAMS) cush-cush yam; (YAO) lipeta.

12. Iridaceae

12.1. *Tigridia pavonia* (L. f.) Ker.-Gawl. (EP) bulb; (P) C.

(CHINESE, WÚ) hu pi hua; (CZECH) tygřice; (DANISH) tigerblomst; (DUTCH) tijgerbloem; (ENGLISH) maravilla bulb, common tiger flower; (FRENCH) iris, lis de tigre, oeil de paon, cacomite; (FINNISH) tiikerikukka; (GERMAN) echte Tigerblume, Tigerlilie; (HUNGARIAN) tigrisvirág; (ICELANDIC) tigrisblom; (NORWEGIAN) tigerblomst; (PORTUGUESE) flor de lis, tigrídia, açucena de Trinidade, flor tigre; (SPANISH) flor de la maravilla; hierba de Trinidad; (TURKISH) kaplan çiçeği, kaplan zambaği, adi kaplan çiçeği; (VLAAMS) maravilla knol.

13. Liliaceae

13.1. *Asparagus acutifolius* L. (EP) shoot (spear); (P) C.

(AFRIKAANS) wilde aspersie; (ARABIC) aspargus barry; (ASSAMESE) bon satamul; (BELARUSAN) lilanaja; (CHINESE, MANDARIN) xi zhen song ye; (CHINESE, WÚ) ye lu sun, xi zhen song ye; (DUTCH) spitsbladige asperge; (ENGLISH) wild asparagus; (FRENCH) asperge sauvage; (GERMAN) spitzblättriger Spargel; (GREEK) agrio sparaggi; (HEBREW) asparag mehudad; (HINDI) shatawar; (HUNGARIAN) vad spárga; (ITALIAN) asparago pungente; (KABIYÉ) kozengobouya; (MACEDONIAN) diva špargla; (MALTESE) sprag salvagg; (NEPALI) ban kurilo; (NORWEGIAN) asparges; (POLISH) szparag dziki; (PORTUGUESE) espargo silvestre, espargo bravo; (SERER-SINE) ngol a sav; (SPANISH) espárrago amarguero; (SWEDISH) sparris; (TAJIKI) sarsabili dashti; (TURKISH) yabani kuşkonmaz, aci ot, kirgin otu, tilkişen, sivri yaprakli-çalimsi kuşkonmaz; (URDU) spatawar jangli asparagus; (VLAAMS) wilde asperge; (WELSH) merllys gwylt; (WOLOF) fir i buki, ñarara.

13.2. *Asparagus officinalis* L. (EP) shoot (spear); (P) C.

(AFRIKAANS) aspersie, makaspersie; (ALBANIAN, TOSK) sparengj, shpargull; (ARABIC) holgun; (ARMENIAN) spevek; (ASSAMESE) satamul; (AZERBIAJANI) merečejud; (BASQUE) frantsesporru; (BELARUSAN) sparzha leuavaja; (BENGALI) hillua; (BRETON) aspergez; (BULGARIAN) asperža; (BURMESE) ka-nyut; (CATALAN-VALENCIAN-BALEAR) espàrrec, esparrech; (CHINESE, HAKKA) liôung-sī-ts'ói; (CHINESE, MANDARIN) lu sun, lo sun; (CHINESE, MIN NAN) lô-sún; (CHINESE, WÚ) lu sun, shi diao bai; (CHINESE, YUE) lu sun, shi diao bai; (CREOLE, HAITIAN FRENCH) aspèl, aspèj; (CRIOULO, CAPE VERDE) espargo; (CROATIAN) šparoga; (CZECH) chřest obecný, chřest, chřest lekařsky, špargl; (DANISH) asparges; (DUTCH) asperge; (ENGLISH) asparagus; (ESTONIAN) aspar, spargel; (FARSI, WESTERN) marchobch, halgun; (FILIPINO) asparagus; (FINNISH) parsa, ruoka parsa; (FRENCH) asperge; (GAELIC, MANX) asparag; (GAELIC, SCOTTISH) creamh, na muice fiadhaich; (GALICIAN) espargo, espargueira, esparragèra; (GEORGIAN) satatsuri; (GERMAN) Spargel; (GIKUYU) rũreru; (GREEK) sparangi; (HEBREW) asparag refuyi, asparag refui, asparagus refuee; (HINDI) nagdour, nag-down, halyum; (HINDUSTANI) nágdauná; (HUNGARIAN) spárga; (ICELANDIC) spergill; (INDONESIAN) asparagus, spèrsis, sepersi, sapersis, saperis, akar parsi; (ITALIAN) asparago; (JAPANESE) asuparagasu, matsubaudo; (KAZAKH) lopukh; (KHMER, CENTRAL) tum'-peang barang; (KOREAN) jakbicharu; (LADINO) sparanga; (LATVIAN) sparģelis; (LITHUANIAN) vaistinis smidras; (MACEDONIAN) obična špargla; (MALAY) asparagus; (MALTESE) l-sprag; (MARANAO) asparagos; (MICMAC) egadagwŏdŭmŭgāwā', egadâkŭnckeech tŭlā'; (MONGOLIAN) emiin khereen nuden; (MORISYEN) asperz; (NEPALI) kurilo; (NORWEGIAN) matasparges, asparges; (POLISH) szparag, szparag lekarski; (PORTUGUESE) espargo, aspargo; (ROMANIAN) sparanghel; (RUSSIAN) sparzha obyknovennaja, sparzha sedovnaja, sparzha ovoshtnaja, sparžha aptečnhaja; (SERBO-CROATIAN) špargla; (SERER-SINE) dug o môn; (SLOVAK) špargel'a; (SLOVENIAN) špargelj; (SORBIAN, LOWER) gromak; (SORBIAN, UPPER) hromak, spargl; (SPANISH) espárrago; (SUNDA) asparageus, spersis; (SWAHILI) asparagasi; (SWEDISH) sparris, spärgel; (TAGALOG) asparagus, asparago; (TAJIKI) sarsabili davoi; (THAI) nomai-farang; (TIBETAN) kūūtsɛɛ; (TURKISH) kuşkonmaz; (TURKMEN) dermanlyk sersepil; (UKRAINIAN) sparža; (URDU) asparagus; (UZBEK) sparja, kuš konmas; (VIETNAMESE) m[aw]ng t[aa] y; (VLAAMS) asperge; (WELSH) merllys; (WOLOF) yar u golo; (XHOSA) imvane elinywayo; (YIDDISH) spamridgeh.

13.3. *Hemerocallis* spp. (EP) flower; (P) C

(e.g. *H. flava, H. fulva, H. minor* – many of the edible species have specific names in Chinese and Japanese)

(AFRIKAANS) daglelie; (ARABIC) el-sawsan, el-naharry; (CAROLINIAN) ghiyobw, liiriyo; (CATALAN-VALENCIAN-BALEAR) lliri; (CHICKASAW) lili'; (CHINESE, GÀN) huang hua cai; (CHINESE, HAKKA) pâc-hăp-fā; (CHINESE, MANDARIN) jin zhen cai, chin-chên-ts'ao, ch'ien-yieh-hsüan-ts'ao, chin-hgüan-ts'ao; (CHINESE, MIN BEI) jin zhen cai; (CHINESE, XIANG) huang hua cai; (CHINESE, WÚ) ni meng suan cao, xuan cao, jin zhen cai; (CHINESE, YUE) huang hua cai, jin zhen cai; (CZECH) denivka; (DAKOTA) mnaȟcaȟca; (DANISH) daglilje; (DUTCH) daglelie; (ENGLISH) day lily; (ESTONIAN) päevaliiliad; (FARSI, WESTERN) soosan; (FINNISH) päivänliljat; (FRENCH) hémérocalle; (GERMAN) Taglilie; (GREEK) krimos imeras; (HEBREW) shoshan yom, hemerocallis; (HINDUSTANI) gul i sausan; (HUNGARIAN) sárgaliliom, tűzliliom; (ICELANDIC) dagilja; (INDONESIAN) sedep malem; (ITALIAN) giglio; (JAPANESE) kanzo, kinshinsai, wasure-gusa, shina-kanzô, yabu-kanzô, ezo-zentei-ka, hosoba-kisuge; (KAZAKH) it shu; (KRIO) lili; (LATVIAN) dienziedis, dzeltenā dienziede, rūsganā dienziede; (LITHUANIAN) vienadienė; (MACEDONIAN) bogorodično cveke; (MICMAC) pagose, moojelăbĕch'; (NORWEGIAN) daglilje;

(POLISH) liliowiec żółty; (PORTUGUESE) hemerocalis; (SORBIAN, UPPER) lilija; (SPANISH) azucena; (SWAHILI) au zuri, jamaii ya yungiyungi; (SWEDISH) daglilja; (TAJIKI) savsan; (TIV) i̱ga̱so alôm lôhô; (TURKISH) kaya çiçeği, güngüzeli; (TURKMEN) gemerokallis; (URDU) din ki lili; (VLAAMS) daglelies; (WELSH) liliawir dydd.

13.4. *Lilium* spp. (EP) bulb; (P) C.

(e.g. *L. auratum, L. brownii, L. concolor, L. cordifolium, L. japonicum, L. lancifolium, L. leichtlinii, L. longiflora, L. maximowiczii, L. tigrinum,* etc. – many of the edible species have specific names in Chinese and Japanese)

(AFRIKAANS) lelie; (ALBANIAN, TOSK) zambak; (ARABIC) el-sawsan; (ASSAMESE) lily; (CHINESE, MANDARIN) bai he; (CHINESE, MIN NAN) pek-ha̍p, pek-ha̍p-hoe; (CHINESE, WÚ) bai he shu, bai he; (CRIOULO, CAPE VERDE) lírio; (CZECH) lilie; (DANISH) lilje; (DUTCH) lelie; (ENGLISH) lily; (ESTONIAN) liilia; (FRENCH) lis; (GAELIC, MANX) lilee; (GERMAN) Lilie; (GREEK) krimos, liliou; (HEBREW) shoshan; (HUNGARIAN) liliom; (ICELANDIC) lilja; (ITALIAN) giglio; (JAPANESE) yurine, teppô-yurine, oni-yurine, sasa-yurine, hime-yurine, kibana-no-oni-yurine, yama-yurine, hakata-yurine, ko-oni-yurine; (LATVIAN) lilija; (LITHUANIAN) lelija; (LUXENBOURGEOIS) Liljen; (MACEDONIAN) lilija; (MALTESE) gilju; (MONGOLIAN) saraana; (NORWEGIAN) lilje; (POLISH) lilia; (PORTUGUESE) lírios, açucenas; (ROMANIAN) crin; (RUSSIAN) lilija; (SAMOAN) lili; (SERBO-CROATIAN) liljan, krin; (SLOVAK) l'alia; (SPANISH) azucena, lirio; (SWEDISH) liljor, lilja; (TAJIKI) savsan; (TIBETAN) qūmüü; (TURKISH) zambak; (URDU) lili; (VLAAMS) lelie; (WELSH) alaw, lili; (YIDDISH) lileeah.

13.5. *Muscari comosum* (L.) Mill. (EP) bulb; (P) C.

(ARABIC) el-sawsan, el-masamy; (CZECH) modravec chocholatý, modřenec chocholatý; (DANISH) perlehyacinth, druehyacinth; (DUTCH) pluimhyacint, druifhyacint, blauwe druifjes; (ENGLISH) tuffed hyacinth, tassel grapehyacinth; (ESTONIAN) hüatsint; (FINNISH) töyhtöhelmililja; (FRENCH) muscari à toupet, jacinthe musquée; (GERMAN) schopfige Traubenhyazinthe; (GREEK) fountotos yakinthos; (HEBREW) mez̤üyaz̤ot, mez̤ilot, kadan; (HUNGARIAN) fürtös gyöngyike; (ITALIAN) cipollaccia, giacinto del pennacchio, muscari comune; (MACEDONIAN) obična kukavičino; (MALTESE) basal il-hniezer; (NORWEGIAN) perleblomst; (POLISH) szafirek miękkolistny; (PORTUGUESE) jacinto das searas, jacinto bravo, muscari; (SPANISH) cebolla de lagarto, nazarena de penacho, muscari, jacinto racimoso; (SWEDISH) plym hyacint; (TAJIKI) sumbul; (TURKISH) perçemli misksoğani, misksoğani, araptaşaği; (VLAAMS) kuifhyacint.

14. Limnocharitaceae

14.1. *Limnocharis flava* (L.) Buchenau (EP) young leaves, petioles and floral shoots; (P) C.

(ARABIC) el-makmalia, el-safrae; (CHINESE, MANDARIN) shi yong lan hua jiao; (ENGLISH) yellow velvetleaf, sawah lettuce; (INDONESIAN) genjer; (MALAY) jinjir paku rawan; (POLISH) aksamitka żółta; (PORTUGUESE) limnocares, golfo, barata; (SPANISH) hoja de buitre; (TURKISH) batak seven, sari batakseven; (VLAAMS) geel fluweelblad.

15. Marantaceae

15.1. *Calathea allouia* (Aubl.) Lindley (EP) tubers; (P) C.

(ARABIC) topia; (CHINESE, WÚ) shi yong lan hua jiao; (DANISH) topi; (ENGLISH) topee, leren, sweet corn root, topee tambu, Guinea arrowroot, sweetcorn tuber; (FRENCH) topi, alléluia, cucuma d'Amerique; (HEBREW) calathea; (POLISH) ostrzeszyn jadalny; (PORTUGUESE) tupinambor; (PUMÉ) hoičo; (QUECHUA, CUSCO) cui-cui, dale-dale, toyita; (SPANISH) tupinambur, topitambou, lairén, alleluia, laren, leren; (TURKISH) kalatea; (VIETNAMESE) c[ur] l[uf]n, dong t[is]a; (VLAAMS) topee.

15.2. *Maranta arundinacea* L. (EP) rhizomes; (P) C.

(AFRIKAANS) araroet, maranta; (ASSAMESE) tora alu; (BENGALI) ararut; (BICOLANO, CENTRAL) sago, sagu, uraru; (BONTOC, CENTRAL) galamaka, pikaw; (CAROLINIAN) aruuru, mógomóg, móghomógh; (CHINESE, MANDARIN) zhu yu; (CHINESE, MIN NAN) ngāu; (CHINESE, WÚ) zhu yu; (CZECH) maranta třtinovitá; (DANISH) maranta, vestindisk pilrod; (DUTCH) bamboeachtige pijlwortel; (ENGLISH) West Indian arrowroot, Bermuda arrowroot, St. Vincent arrowroot, arrowroot; (ESTONIAN) aroruut; (FILIPINO) araru'; (FINNISH) nuolijuuri; (FRENCH) marante en forme de bambou, arrowroot des Antilles, marante, herbe aux flèches, arruruz; (GERMAN) Pfeilwurz, schilfartige Pfeilwurz; (GREEK) marocntia kalamoidis, marantamylon; (GUJARATI) ararut; (HANUNOO) uraru; (HEBREW) maranta; (HINDI) tavkil, arārot, tikhor; (IBANAG) araru; (IFUGAO) unig; (ILOCANO) sagu; (INDONESIAN) ararut, arerut, araut, sagu belanda, segu betawi, garut, leret, sagu; (ITALIAN) fecola a froma di bambù, maranta; (ITAWIT) araru; (IVATAN) bay, vay; (JAPANESE) kuzu-ukon; (JAVANESE) garut, nggarut, angkrik, arus, jelarut, waerut; (KANKANAEY) galamaka; (KANNADA) tavaksha, kuvehittu, kove hittu; (KHMER, CENTRAL) saku; (KRIO) ararut; (LAHU) a⁻ hpa$_\wedge$ meu$_{_}$; (LAO) sa:kh'u:; (LITHUANIAN) Bermudinė maranta; (MALAY) ubi garut, sagu belanda, ararut, berolu; (MARANAO) obi bamban; (MARATHI) tavkil, aara root; (MIZO) thlaizung ei mi; (MORISYEN) arrou-rout; (MUSKOGEE) helv; (NORWEGIAN) pilrotmaranta, maranta; (POLISH) maranta trzcinowata; (PORTUGUESE) araruta, maranta; (PUMÉ) kaip^{h}unda; (QUECHUA, CUSCO) araruta, yuquilla; (RAROTONGAN) tiare aru ra; (SAMBAL, BOTOLAN) sago; (SAMOAN) māosā; (SANSKRIT) thavaksheera, thungaksheeri; (SINHALA) hulankeeriya; (SPANISH) maranta, cará maco, arrurruz arundinácco; (SUNDA) larut, patat sagu, harut, daun lipung; (SWEDISH) arrow rot; (TAGALOG) urard, aroru, aru-aru, araro, aruru; (TAGBANWA) aru-aru; (TAMIL) araruttukkilangu, kuvamavv, aruruttukizhangu; (TELUGU) palagunda; (TONGAN) misimisi koka'anga; (TURKISH) kamişsi ararot, ararot; (VIETNAMESE) khoai-mi-tinh, ho[af]ng tinh, c[ur] dong, hu[yf]nh tinh; (VLAAMS) West-indische pijlwortel; (WÁRAY-WÁRAY) aloró.

16. Musaceae

16.1. *Musa × paradisiaca* L. var. *paradisiaca* (EP) a. fruit, b. flower bud; (P) a. C,R, b. C.

Plantain is considered a vegetable as are banana flower buds. A number of languages do not distinguish between banana and plantain, using only a generic term. The following names can be either or both. Dialects of Luyia (Bukusu[1], Itsukha[2], Kabarasi[3]).

a. fruit

(AFRIKAANS) piesang; (AGTA) biget; (AGTA, CASIGURAN DUMAGAT) abat, dupat; (AJA) mandan; (ALBANIAN, TOSK) banane; (AMHARIC) muz; (ANDO) banda; (ANUAK) b̭̭̄ḽ̄, b̭̭́lé; (ARABIC) mooz el-gana; (ASSAMESE) kol; (ATTA, PAMPLONA) dupa, dupag; (AZERBAIJANI, NORTHERN) banan; (BADAGA) ba:caṉṉu, ba:ḷe, ba:ḷe haṉṉu; (BAFUT) àmbɔ́, àmbótsu, bànánà, mùnggọ, nènggọ; (BAGA) a-babani; (BAMANANKAN) namassa; (BAOULÉ) manda aliê, aliè manda; (BASSARI) banana; (BEMBA) nkonde; (BEMBÉ) dikondo; (BÉTÉ) batei; (BETE-BENDI) kí-gwā, kí-mbɔ̄, kí-gwàā; (BICOLANO, CENTRAL) latunda, sab-a, saging-sab-a; (BINUKID) amas, bulig, bunguran, saging, televerng, tundang; (BODO) laihfang, sinisamfa; (BOKYI) kenkwa; (BONTOC, CENTRAL) poso; (BUGHOTU) vudi; (BUHID) punti, yakyaw; (BULLOM SO) puloth, santumir, bana, bàná, bàná-bombom-dɛ, bàná-bumbum, bàná-katul-lɛ, bàná-lɛ, kpende-mane-lɛ, sintem; (BULU) ekon; (BURMESE) nget-pyaw; (CAJUN FRENCH) banane; (CAROLINIAN) uurh, wiisch; (CATALAN-VALENCIAN-BALEAR) plàtan, banana; (CEBUANO) turdan, kardaba; (CHECHEN) banaan; (CHINESE, HAKKA) kioūmg-tsiāo; (CHINESE, MIN NAN) chio, keng-chio, kin-chio; (CHINESE, WÚ) ba jiao ke; (COMANCHE) aakaaʔ; (CREE, PLAINS) wâkâs; (CREOLE, HAITIAN FRENCH) bannann fran, gwo bannann; (CRIOULO, UPPER GUINEA) bananeira, banana anã, banana de cubissique, banana maçã, banana ouro, banana pom, banana prata; (CZECH) banánovník zeleninový; (DAGAARE, SOUTHERN) boordiē; (DAKOTA) iyaġezizi; (DANGME-KROBO) blɔfo kodu, kodu, mãdãa, mãnãa; (DANISH) banan, pisang; (DIGO) mgomba, mkoo; (DUTCH) banaan, bakbanaan; (EDE NAGO) ɔgede agbagba; (EDO) ọghẹdẹ, ọghẹdebo; (EFIK) adabri, akpanan-ekọp, anwang-ébát, ébát-iba, m̀bòró; (ENGLISH) plantain; (ESTONIAN) jahubanaan; (ÉWÉ) abladzo; (FARSI, EASTERN) kêle; (FARSI, WESTERN) barhang, bartang; (FIJIAN) vudi; (FINNISH) jauhobanaani; (FON) kokwe sozu; (FRENCH) plantain, banane à cuire, bananier plantain; (FULFULDE, ADAMAWA) kondong, kondongji, ayabaje; (GA) amadaa; (GADDANG) abat, bat; (GAELIC, MANX) duilag, pherick; (GBE-VHE) abladzo, akɔdu, akɔdua-zẽ, akɔdukɔkɔ, akɔdukpe, ewe-kɔdu, kɔdu, yevu kɔdu; (GEN) amandan; (GERMAN) Kochbanane, Mehlbanane, Plantainbanane, Wegerich; (GIKUYU) marĩgũ; (GOLA) banda, banda-*g*bɛgbɛ, konde; (GONJA) bordie, kayeba; (GOURMANTCHÉ) ibourdi; (GREEK) idos mpanana; (HANUNOO) sab-a, sagin balayan, sagin binalatan, sagin buyun, sagin pinakul mukay, saging-sab-a; (HARARI) mūz, gē mūz; (HAUSA) agade, ayaba, àyàbà, àyàbàr tuùraàwaá, àyàbàr dáaji; (HAWAIIAN) mai'a; (HEBREW) banana, moz; (HERERO) banana-i; (HILIGAYNON) ságing; (HINDI) kela, kaela, amrit, kachkula, maozkula; (HINDUSTANI) bakwa, kelá, kadalí, gahr; (HUNGARIAN) banán, pizáng; (IBALOI) balat, doman, soging; (IBANAG) balat, damilig, damilik, dufo, duho, dupo; (ICELANDIC) mjölbanani; (ICEVE-MACI) bànánà; (IFUGAO) abat, balat, balatan dipping; (IGBO) abị rị ka, elili ji okō, ẹze ọgẹde ji oko, ji okō, ọ̀gèdè, ọ̀gèdè ñkịtị, unè, unèlè, unèlò ọcha, òmèjl, ukam, unè, ípà, ulè, àbìrìka unèr̃è, unèr̃è, abereka, ebegehi, unere, ọgede, ọgede ojoko, abrika; (ILOCANO) banangar, latundal, tukol, apugin; (INDONESIAN) pisang, pisang tanduk; (ITALIAN) banana da cuocere, banana verde, piantaggine, banano plantain; (ITAWIT) bahat, bat; (IVATAN) gayod, soteen, soteen gayod, vineveh, vinivih, vinivik; (IZON) beiba, oyinbó, beribă, apáḷị́ béribá, abángha, beké abangha, berebe-ghịsa; (JAVANESE) gedang sungu; (JOLA-FONYI) banana; (JULA) brana soussouta; (KABIYÉ) iyabiye; (KALAGAN) balat; (KANKANAEY) baat, bagat; (KANNADA) bale, baale, budibaale, chandrabaale, elebaale, elakkibaale, haavubaale, hombaale, jenubaale, gulurbaale, kadali, kandu, kattubaale, madarangabaale, mavuju, pachhabaale, puttabaale, raajabaale, rambha, rasabaale; (KANURI) àyàwà; (KHMER, CENTRAL) ché:k; (KIRIBATI) te banana, te touru, te rarai; (KISSI, NORTHERN) *g*bɔlɔ, *g*bɔlɔ-bana, *g*bɔlɔ-bendo, *g*bɔlɔ-mɔyo, *g*bɔlɔ-pɔmbɔ, *k*pɔlɔ; (KONKANI) kaeli; (KONKOMBA) ayaba; (KONO) kasawane, kondeke, maa, seminga; (KRIO) plantin, kɔ́tɔ̀kwakyi, ɔ̀ŋgùgù dɔ̀ńdè, kraykray banana; (KURDISH, NORTHERN) moz; (KWANGALI) otjipanana; (KWÉYÒL) bannann, banann; (LADINO) banana; (LAHU) a⁻ pawᵛ hpui ma; (LAO) kwàyz khauz; (LIMBA) hobana, bana, bana-batogoe; (LITHUANIAN) tikrasis bananas; (LOKO) maa, maa-kotɔpɔ-gutugutu, maa-wawa; (LOMWE, MALAWI) matoki; (LUO) abolo; (LUXENBOURGEOIS) Banan; (LUYIA) litore[1,3], liremwa[2]; (MALAY) pisang; (MALAYALAM) bala, ettakkaaya, ettavaala, kaadalam, kaashthhila, karinkadali, kunnan, pichha, vannan, vala; (MANDING-MANDINKA) banana; kyɛmɛ, namassa; (MANDJAK) messente, messente mlame; (MANOBO) amas, bulig, sahing; (MANU'A SAMOAN) fa'i; (MARANAO) amas, borongan, kantong, manggai, pakal, tadyaw, tendok; (MARATHI) kaela, kadali, sonakaela; (MARQUESAN) meika, mei'a; (MENDE) maa, mana, *m*bana, jelegboi, jɛmiŋga, konde, mana-wa, mɛnde-hedi, momɔŋa, sɛlɛ, sɛlɛ-guti-guti, tɛndɛ; (MOHAWK) tyotahyàkton; (MÒORÉ) barãde; (MUNDARI) kadal; (MUSKOGEE) welvntvlo; (NAMA-DAMARA) ebanana; (NDONGA) sibanana; (NEPALI) kera; (NEWAR) mvāc; (NGUNDI) dikondo; (NICOBARESE, CAR) tai-nyuk-ngö; (NIUE) laumalika, futi; (NORWEGIAN) kokebanan, banan; (NUPE-NUPE-TAKO) yàbákó; (OGBIA) ɔ́ɓààmí̩, oɓá̩mé̩, o̩fo̩ĩ, ɔ́fɔ́ín, ɔ́ká̩í̩; (OGBIA-AGHOLO) ɔ́ka̩í̩, òsókpó; (OROMO, WEST CENTRAL) musa, kocho, werke; (PALI) kadalī, kadalīkā, moco, rumbhā; (PAMPANGAN) saguin a latondan; (PANGASINAN) punti, seba; (PASHTO, NORTHERN) kela; (POLISH) banan zwyczajny, banan; (PORTUGUESE) banana pão, pacova, pacoba, banana prata; (PULAAR) banaana, banana, banana dánès, banana gábu, banana mánea; kɔntɔmbɔlɔ̃; (RAROTONGAN) meiki; (ROMANIAN) banan, bananier; (RUSSIAN) banan; (SAMBAL, BOTOLAN) hapa, haa; (SAMBAL, TINÀ) haa, haba; (SAMOAN) tōfa'i, fa'i, fa'i samoa, fa'i papalagi; (SANGA) ayaba; (SANSKRIT) kadali, rambha, alabu, ambusaara, amshumathphala, baalakapriya, bhaanuphala, charmannvati, deerghapathra, guchhandathika, guchhaphala, hastivishaani, kashthhila, mocha, nagaranshadhi, nisara, raajeshta, radraka, sakrithphalaa, sukumaara, suphala, thanthuvigraha, thathpathri, urustambhhaa, vaarana, vanabhhushaa, vanalakshmi, varanabusa, varanavallabha, varariksha; (SCOTS) bananie; (SENOUFO, CEBAARA) barida souwi; (SERBO-CROATIAN) banana; (SERER-SINE) banana; (SHONA) bhanana, dzoro, figu, hobo, hova; (SIDAMO) mûse, mûze; (SINHALA) malu kesel; (SLOVAK) banán; (SORBIAN, LOWER) banana; (SORBIAN, UPPER) banana; (SOTHO, NORTHERN) panama; (SOTHO, SOUTHERN) panama; (SPANISH) llantén, plátano de cocinar, plãtano macho, platano; (SUMO-MAYANGNA) waka, waki; (SUNDA) cau tanduk, cau galek; (SUSU) dugu-fui, kontubulogue, ndunké, santuméri, banane, kɔntɔmbolonyi; (SWAHILI) ndizi, mgomba; (SWEDISH) banan; (TAGALOG) saping na saba, saging, latundan, tordan, letondal; (TAGBANWA) saging, punto, punti, puntik; (TAJIKI) zuf; (TAMIL) vazhai, ambanam, angusam, arambai, arayakkommai, arbaruttam, aresigam, asogam, irasandalai, kadali, kavar, kavargali, mandan, mamarai valai, namavalai, pachilandai, paivalai, pidagadali, puvalai, puvanvali, sami, segili, sevali, sugandham, thuraivalai, udiranvalai, valai; (TAUSUG) busag, sahing, saing; (TELUGU) kadali, amrithapany, ananthi, anathi, anthi, aratthi, ariti chettu, batila, bontharathi, chakrakaeli, desavala chakrakaeli, errachakrakaeli, errarati, kadalamu, karpoora chakravkaeli, kommanatti, kommu ariti, kommaratti, naatabam, nallarati, pachharati, rambha, sugandhhalu, thatachhadamu, theneyarathi; (THAI) kluai; (THEMNE) *a*-bana, *ma*-bana, *an*-kuntibi, *a*-manawa, *ɛ*-plantan, *ɛ*-planti, *ɛ*-santi; (TIBETAN) qera; (TIV) a̱yaba, kô̩rko̩mbo̩, a̱ya̱ba̱; (TONGAN) fusi; (TSWANA) panama; (TULU) bārè, kadoli, baare; (TURKISH) inciriadem, cennetinciri; (TWI) ɔborɔde; (URDU) kala; (VAI) bana, bana-baba; (VENDA) muomva; (VIETNAMESE) chu[oos]i; (VLAAMS) djantoeng, entoet, ontong, ongkol; (WÁRAY-WÁRAY) saging, baloy, costa, inaloaba, kadulsi, kahapog, pakol, paguhâ, pinipita, sabá, tinagakan, tindok, tinumbaga; (XHOSA) ibhanana, isiqhamo esithile esifana kanye nebhanana; (YALA) wɔ̀gbɔ̀, wɔ̀gbɔ̀-àlèlè, ikpāmásányā, wògbùɔ̀; (YORUBA) ɔgede agbagba, ọ̀gẹ̀dẹ̀; (ZULU) ubhanana, u(lu)khova; (ZUNI) z/u-piya.

b. flower bud

(AFRIKAANS) piesangblom; (AMHARIC) muz abäba; (ARABIC) zaher el-mooz; (ASSAMESE) kol dil; (BINUKID) pusu'; (BULU) ajôé; (CAJUN FRENCH) bouton de banane; (CATALAN-VALENCIAN-BALEAR) banana flor; (CHINESE, WÚ) xiang jiao hua; (CZECH) banánovník ovocný;

(DAGAARE, SOUTHERN) baarna; (DANISH) bananblomsterstand, blomst banan; (DUTCH) bloesem banaan; (ENGLISH) banana inflorescence; (ÉWÉ) akodu; (FIJIAN) kera phul; (FILIPINO) puso; (FINNISH) banaani; (FRENCH) banane à curie, fleur banane; (GA) akwadu; (GERMAN) Bananenblüte; (GIKUYU) kĩongoro; (GOURMANCHÉMA) badnã; (GOURMANTCHÉ) bananni; (GREEK) mausa anthousa; (HMONG DAW) pob tsawb; (HUNGARIAN) banán virág; (INDONESIAN) jantung pisang; (ITALIAN) flore banana; (LATVIAN) banāni; (LUO) abolo; (MAKHUWA) magonho; (MALAY) jantung pisang; (MORISYEN) baba banann, baba fig; (NEPALI) kerako phool; (NORWEGIAN) bananblomst; (POLISH) banan rajski; (PORTUGUESE) flôr de bananeira; (PULAR) bananayé; (SAMOAN) fa'i fuga; (SINHALA) kesel muwa; (SPANISH) flor de plátano; (SWEDISH) banan; (TAGALOG) puso ug saging; (TWI) kwadu; (URDU) kala; (VLAAMS) pisangbloem; (YIDDISH) bahnahna blum; (ZUNI) z/u-piya /uteya.

17. Poaceae (Gramineae)*[6]

17.1. *Bambusa* spp. (EP) young shoots; (P) C.

(e.g. [1]*B. arundinacea,* [2]*B. beecheyana,* [3]*B. blumeana,* [4]*B. oldhamii,* [5]*B. vulgaris,* [6]*B. bambos)*

(ABUA) àgàràbà[3]; (AFAR) kàsab; (AFRIKAANS) bamboesspruite, bamboeslote; (AKHA) a cu[1], za ju[1], za shaw[5], ya shu[5]; (ALBANIAN, TOSK) bambus; (AMHARIC) qärqäḥa[1], šämäqqo, šəmäl; (ARABIC) crasen el-kharazan; (ASSAMESE) bah gaj; (BADAGA) bidil[1], bidilu[1], tabare, me:de; (BAGA) ko-tatami; (BAOULÉ) m'blangblô[5], féfé[5]; (BEMBA) lusengu; (BEMBÉ) mumbanbou; (BENGALI) bans[1], behar[1], behurbans[1], katausi[1], basini bans[5]; (BÉTÉ) paihoun[5]; (BICOLANO, CENTRAL) kabaloan[5], dugian, kabugawan, kawayan-totoo, marurugi, rugian; (BINUKID) davung, kering[5]; (BODO) meoa, oa, oua; (BONTOC, CENTRAL) bengwil; (BUGHOTU) duee, gau, gauratu, kadolulu, ruavatu; (BUHID) ararang; (BULLOM SO) kana-lɛ[3], wus-lɛ[3]; (BURMESE) hmyit, kay-kat-wa[1,6], wanet[5]; (BELARUSAN) zlanavaja; (CAROLINIAN) bwaay, ligippwaay, lighippwáy; (CATALAN-VALENCIAN-BALEAR) bambú; (CEBUANO) batakan[3], kaawno, pawa; (CHICKASAW) oskapi'; (CHINESE, HAKKA) tchôuc-tzè; (CHINESE, MANDARIN) wu jiao lu zhu; (CHINESE, MIN NAN) tek-á, tek-sún, tang-sún, lėk-tek-sún, sún; (CHINESE, WÚ) diao si qiu zhu, pa dou dian zhu, chang zhi zhu; (CREOLE, HAITIAN FRENCH) bonbou, banbou; (CRIOULO, CAPE VERDE) carisso-da-guiné, bamboo-grande, bambu; (CZECH) bambusové výhonky; (DAGAARE, SOUTHERN) tára-dar; (DANISH) bambusskud; (DUTCH) bamboespruiten; (ENGLISH) bamboo shoots, spiny bamboo[1,3,6], throny bamboo[1,3,6], common bamboo[5], golden bamboo[5], tiger bamboo[5]; (ESTONIAN) bambus; (ÉWÉ) pamplo, pambro; (FARSI, WESTERN) kkeyzaran; (FIJIAN) bas; (FILIPINO) lahong; (FINNISH) bambu; (FON) dawe; (FRENCH) bambou, pousses de bambou, grand bambou[5]; (FULFULDE, ADAMAWA) kewal, kewe; (GAELIC, SCOTTISH) cuile; (GALICIAN) bambueira; (GEORGIAN) bambúk; (GERMAN) Bambus, Bambussproßen, Bambusspitzen; (GIKUYU) mũrangi; (GOLA) sen, seni; (GOURMANCHÉMA) sucré-dow, ŋmàlù[5]; (GOURMANTCHÉ) moũkardarma; (GUJARATI) toncor[1], wans[1]; (GURO) balé[3]; (HARARI) qärqäḥa; (HAUSA) gora; (HEBREW) ḫizran, chizran, bamboo; (HILIGAYNON) támbu'; (HINDI) bans[1], kanta bans[1], katang bans[1]; (HINDUSTANI) báṉs, bāns; (HMONG DAW) ntsuag xyoob, tsong[1]; (HOPI) paqavi; (HUNGARIAN) bambuszrügy; (IBANAG) pasingan; (IFUGAO) awayan an pallotan, kawayan an paluttan; (IGBO) ọtọsi; (ILOCANO) ka-wayan-siitan[3], kawayan-siitan[3], kawayan-nga-bayog, siitan; (INDONESIAN) rebung, bambu duri[1,3], bamboo duri[6], bambu kuning[5], bambu ampel[5], domar[5]; (ITALIAN) bambú, getti di bambú; (ITAWIT) pasingan; (IZON) igbon ịkịrai[5]; (JAPANESE) takenoko, diasan-chiku, ryoku-chiku; (JAVANESE) pring ori[1,6], pring

[6] Names marked with an asterisk have been changed under article 18 of the International Code of Botanical Nomenclature. Either name is correct.

gesing[3]; (JUNKUN TAKUM) vyo; (KABIYÉ) soumou; (KANKANAEY) kawwayan; (KANNADA) andebidiru[1], bidungalu[1], dongi[1], habbidiru[1], hebbidiru[1], hebbidru[1]; (KAREN) waa-chuu[1], waa baw[5]; (KASHMIRI) baĩs; (KHMER, CENTRAL) rüssèi khléi[1,6], rüssèi préi[1,6], rüssèi roliëk[3], rüssèi kaèw[5]; (KIRIBATI) te kaibaba, te kaipapa; (KISSI, NORTHERN) pilanda[3]; (KONKANI) conopui[1], kanaki[1], vellu[1]; (KONO) siminɛ[3]; (KRIO) kɛn; (LAHU) va$^{\vee}$ shi, yaw ceh[1], za khao[5]; (LAO) phaix pa:x[1], phaix ba:nz[3], s'a:ng kh'am'[5], phaix pa:x[6]; (LATVIAN) bambusi, bambuss, bambusniedre; (LIMBA) barãʔ[3]; (LISU) mai hio[1]; (LITHUANIAN) bambukas; (LOKO) sii[3]; (LOMWE, MALAWI) nsengere; (LUO) keo; (LUXENBOURGEOIS) Bambus; (MALAY) rebung, buloh duri[3], buloh sikai[3], buloh minyak[5], buloh kuning[5], tamelang[5]; (MALAYALAM) illi[1], karmmaram[1], mulla[1], mungil[1]; (MANDING-MANDINKA) bɔɔ[3]; (MANOBO) kewayan; (MARANAO) tanilang; (MARATHI) kallak[1], mandga[1], bambu[1], kalaka[5]; (MARWARI) abāns, bānsdo; (MENDE) semi[3]; (MIZO) mau, chal, rua; (MÒORÉ) tãwiga; (MORISYEN) leker bambou; (MUNDARI) kāril, hanḍuā-utu, māt', mād'; (NEPALI) bans tama; (NEWAR) pat; (NGUNDI) mumbanbou; (NIUE) kaho, papalagi; (NORWEGIAN) bambusskudd, bambus; (NUPE-NUPE-TAKO) gusa, èba; (NYAKYUSA-NGONDE) lulasi[5]; (NYANJA) nsungwi[5]; (ORIYA) bendo[1], beudobaunso[1], kontabanso[1], kontobanso[1], sundrogai[5]; (OROMO, WEST CENTRAL) leman, simalla; (PALI) kisalayaṁ; (PAMPANGAN) baugin; (PANGASINAN) kawayán, kilíng, kawayan-tuwa; (PANJABI, EASTERN) magae[1], magar bans[1], nal[1]; (POLISH) bambus; (PORTUGUESE) bambú, rebentos de bambú, brotos de bambú; (PULAAR) kɛvɛ[3], kewal; (RAROTONGAN) koe; (ROMANIAN) bambus; (SAMBAL, BOTOLAN) kawayan mantug; (SAMBAL, TINÀ) kawayan-potog; (SAMOAN) ofe fiti, ofe papalagi; (SANSKRIT) bahupallava vansa[1], vansh[1]; (SANTALI) helta, jạil, kạril; (SENOUFO, CEBAARA) tacaltique[5]; (SERER-SINE) i n'gol[3]; (SHAN) mai-sang-ram[1]; (SHONA) musengere, mushenjere; (SLOVENIAN) bambusovité; (SOMALI) jara'; (SORBIAN, UPPER) bambus; (SOTHO, NORTHERN) pampu; (SOTHO, SOUTHERN) leqala; (SPANISH) bambú, brotes de bambú; (SUMO-MAYANGNA) itikna; (SUNDA) haur cucuk[3]; (SUSU) tatami[5]; (SWAHILI) mianzi, muanzi; (SWEDISH) bamboo skott, bambu; (TAGALOG) labong, tambo, kauayan-tinik[3], kauayan-kiling[5], Indian bamboo[6], kawayan-tinik, kawayan-totoo, patung; (TAGBANWA) kawayan; (TAJIKI) nay; (TAMIL) mullumungil[1], periya mungili[1], peru varaimungil[1], mungil[1,5], ponmungil[5]; (TELUGU) bongudu[1], bongu veduru[1], mullabongu[1], mullaveduru[1], mullavendru[1], pentiveduru[1], veduru[1], vedru[5]; (THAI) phai[1], phai-pa[1], phai-pah[1], phai-nam[1], mai-phai[1], phai paa[1], phai-sisuk[3], phai-luang[5], phai-luong[5], phai lueang[5], phai-ngachang[5], phai-nam[6]; (THAI, NORTHERN) chan-kham[5]; (THEMNE) *ka*-sul[3]; (TIBETAN) ñūqtsə; (TIV) sôngough; (TSWANA) bambusa; (TUMBUKA) musyombe[5]; (TURKISH) bambu kamişi, bambu; (URDU) bans ki shakhan, bansa[1]; (VAI) kenye, seni[3], senye[3]; (VENDA) mushaka tembwe; (VIETNAMESE) tre gai r [uw]ng[1,6], tre gai[3], tre gai[3], tre l[af] ng[af][6]; (VLAAMS) bamboe, bamboescheuten; (WÁRAY-WÁRAY) tangnan; (XHOSA) uqalo, umthi webhambu; (YALUNKA) tatami-na[3]; (YAO) mlasi[5]; (YIDDISH) yamsh shpratsen; (ZULU) uqalo.

17.2. *Dendrocalamus* spp. (EP) young shoots; (P) C.

(e.g. [1]*D. asper*, [2]*D. brandisii*, [3]*D. giganteus*, [4]*D. hamiltonii*, [5]*D. latiflorus*, [6]*D.membranaceus*, [7]*D. strictus*)

(AFAR) kàsab; (AKHA) ma bu[2], za pi-euh[3], za-jeh[3], ya baw[4], za baw[4], za cui[6]; (AMHARIC) šämäqqo, šəmäl; (ASSAMESE) bah gaj[5]; (BEMBA) lusengu; (BICOLANO, CENTRAL) botong[1], (BINUKID) davung; (BODO) meoa, oa, oua; (BUGHOTU) duee, gau, gauratu, kadolulu, ruavatu; (BURMESE) kya-lo-wa[2], wabo[2], hmyit[5], wani[5], waya[6], wamu[6], wapyu[6], myinwa[7]; (CAROLINIAN) bwaay, ligippwaay, lighippwáy; (CATALAN-VALENCIAN-BALEAR) bambú; (CHICKASAW) oskapi'; (CHINESE, HAKKA) tchôuc-tzè; (CHINESE, MANDARIN) ma zhu[5]; (CHINESE, MIN NAN) tek-á, tek-sún, tang-sún, lėk-tek-sún, sún; (CHINESE, WÚ) ma zhu, liang shan ci zhu; (CHINESE, YUE) ma zhu sun[5]; (CREOLE, HAITIAN FRENCH) bonbou, banbou; (DANISH) bambusskud; (DUTCH) bamboespruiten; (ENGLISH) bamboo shoots, giant bamboo[1], Taiwan giant bamboo[5], ma bamboo[5], waya[6], male bamboo[7], solid bamboo[7]; (ESTONIAN) bambus; (ÉWÉ) pamplo, pambro; (FARSI, WESTERN) kkeyzaran; (FILIPINO) labong; (FRENCH) pousses de bambou; (FULFULDE, ADAMAWA) kewal, kewe; (GALICIAN) bambueira; (GEORGIAN) bambúk; (GERMAN) Bambus; (GIKUYU) mũrangi; (GOURMANCHÉMA) sucré-dow; (GREEK) dendrokàloomos the platyanthos[5]; (HARARI) qärqäḥa; (HAUSA) gora; (HEBREW) chizran; (HINDI) bāns; (HINDUSTANI) báṉs; (HMONG DAW) dong ngeh[3], ntsuag xyoob[5], xyong[6]; (HOPI) paqavi; (HUNGARIAN) bambuszrügy; (INDONESIAN) bambu rebung[1], bambu betung[1], bambu taiwan[5]; (ITALIAN) bambú; (JAPANESE) takenoko, machiku[5]; (JAVANESE) pring petung[1], jajang betung[1], deling petung[1]; (KAREN) waa-khlue[2], waa-kahlue-pho[2], waa-khru[2], waa-kwaa[3], waa-klu[4], waa-gler[4], waa-mee[6], wa-mu[6], wa-mi[6]; (KASHMIRI) baĩs; (KIRIBATI) te kaibaba[5], te kaipapa[5]; (LAHU) meh va bvuh-eh[2], va bvuh-eh[2], mai pao[3], mai pok[3], va sha[4], va k'aw[4], va meu[5], va meu hpu[5], naw van[6]; (LAO) hok[1], hôk[2], hôk hnhaix[6], s'a:ng[6,7]; (LATVIAN) bambusi, bambuss, bambusniedre; (LISU) ma baw da ma[2], ma khaw neh[3], ma khaw neh[4], ma khaw neh[6]; (LOMWE, MALAWI) nsengere; (LUO) keo; (LUXENBOURGEOIS) Bambus; (MALAY) rebung, buloh beting[1], buloh betong[1], buloh panching[1], buloh batu[7]; (MARWARI) abāns, bānsdo; (MIZO) mau, chal, rua; (MORISYEN) leker bamboo; (MUNDARI) kāril, hanḍuā-utu, māt', mād'; (NEPALI) bans tama; (NEWAR) pat; (NIUE) kaho; (NORWEGIAN) bambusskudd; (NUPE-NUPE-TAKO) gusa, èba; (OROMO, WEST CENTRAL) leman, simalla; (PALI) kisalayaṁ; (POLISH) bambus; (PORTUGUESE) bamboo, broto de bambú, bambú gigante[3]; (PULAAR) kewal; (SANTALI) helta, jạil, kạril; (SHAN) mai-puk[2], mai-hok[4], mai-hok-khu[4], mai pok[5], mai-sang[6]; (SHONA) musengere, mushenjere; (SOMALI) jara'; (SORBIAN, UPPER) bambus; (SOTHO, NORTHERN) pampu; (SOTHO, SOUTHERN) leqala; (SPANISH) bambú; (SUNDA) bitung[1], awi bitung[1], awi iwung[1]; (SWAHILI) mianzi, muanzi; (SWEDISH) bamboo skott; (TAGALOG) labong, bukawe[1], botong[5]; (TAJIKI) nay; (THAI) phai-tong[1], mai-bongyai[2], mai-sang-mon[2], phai bong yai[2], mai-po[2,3], phai-pao[3], pai-poh[3], pai[3], phai-nual-yai[4], pai-nuan-yai[4], pai-nual[6], mai-nuan[6], pai-naan[6], mai-lai-lo[6], pai-sang[6,7]; (THAI, NORTHERN) phai- sangyen[2], phai-zangkum[5], pai-hok[4], mai-hok[4], mai-po[4], mai-phieo[4], saang doi[6], phai-sangnuan[6], phai-nuan[6], phai-sangdoi[6]; (TIBETAN) ñūqtsə̂; (TIV) sôngough; (TSWANA) bambusa; (TURKISH) dev bambu; (URDU) bans ki shakhan; (VENDA) mushaka tembwe; (VIETNAMESE) manh tong[1], m[aj]nh t[oo]ng hoa to[5], tre ta[uf][5], t[aaf] m v[oo]ng[7]; (VLAAMS) bamboescheuten; (WÁRAY-WÁRAY) patóng; (XHOSA) umthi webhambu, uqalo; (YIDDISH) yamsh shpratsen; (ZULU) uqalo.

17.3. *Pennisetum purpureum* Schumacher (EP) young spears; (P) C.

(AFRIKAANS) olifantsgras; (AKAN-ASANTE) akokɔ ani, anan hwerew, hwediɛ; (ANYIN) chelié; (ARABIC) osheb el-feel; (BAOULÉ) n'dê, nè; (BEROM) eromo; (BÉTÉ) gako; (BUHID) lintukay; (BULLOM SO) ŋà; (CEBUANO) darawi, daray, handalawi; (CHINESE, WÚ) lang wei cao; (CREOLE, HAITIAN FRENCH) zèb éleéfan, zèb elefan; (CZECH) sloní tráva; (DAGAARE, SOUTHERN) kĩr; (DANISH) elefantgræs, afrikansk elefantgræs; (DUTCH) olifantsgras; (EDE NAGO) ikɛn; (EDO) ọghọdọ́gbọ; (EFIK) m̀bíit, ḿbọ́kọ́k ékpò; (ENGLISH) elephant grass, napier grass; (ÉWÉ) eko; (EWONDO) sissongho, essong; (FINNISH) elefanttiheinä; (FON) likun; (FRENCH) herbe a éléphant, napier, fausse canne à sucre, sissongo; (FULFULDE) gawri ngabbu, toloore; (FULFULDE, ADAMAWA) toll ore; (GA) g'la, adaí, glầ; (GBE-VHE) adá; (GERMAN) Elefantengras; (GIANGAN) lagoli; (GOURMANCHÉMA) yapron; (GOURMANTCHÉ) okan-anlo; (GREEK) penniseton the porphyroun; (GUN) kusu; (GURO) dia, dian voli; (HANUNOO) ikug-idu; (HAUSA) daáwàr kádàà, kyambama, yambámáá; (HEBREW) zifnoẕa argemanit, zifnotza; (IGBO) àchàlà, àchàrà, àchàrà mili, ùkpò ukwu; (INDONESIAN) rumput gajah; (ITALIAN) erba di napier, erba elefantina, penniseto rosso; (IZON) ịzọ́n úsí, osĭ, usĭ; (JULA) tchekala, ouélé, gbôgôh; (KANNADA) aane hullu, dappa naepear hullu; (KISSI, NORTHERN) chɛŋgio; (KONO) fa, fa-wa; (KRIO) bush-shuga-kɛn; (KURANKO) molike; (LITHUANIAN) dramblinė soruolė; (LOKO) *n*gara; (LOMWE, MALAWI) udzu; (LUO) ogada; (MAKHUWA) churchuru marrakele; (MALAY) rumput gajah; (MANDING-MANDINKA) moloko, mɔlɔkɔ-yɔ̃; (MENDE) *n*gala, *m*bowi-hei, ngɔngɔi; (MONGOLIAN) ödölj; (MÒORÉ) kibgo; (NIUE) mōtie alefane; (NORWEGIAN) elefantgras; (PANJABI, EASTERN) kaē, sarkǣdä; (POLISH) proso perłowe afrykańskie; (PORTUGUESE) capim elefante, capim da Rodésia, capim napier; (PULAAR) tambɛn; (SENOUFO, CEBAARA) lôhô sougueul, lôhôtrihi, lôhôchaune; (SPANISH) pasto elefante, pasto napier, hierba elefante; (SUBANON) gulalay; (SUNDA) jukut gajah; (SWAHILI) majani ya tembo; (TAGALOG) buntot-pusa, darawi; (TAJIKI) filalaf; (THEMNE) *an*-lal, *ka*-staf; (TIV) awo, uwua nor; (TURKISH) filotu, erguvani zencidarisi, hint darisi; (TWI) akokɔ-ani, ananghwerew hwedie; (URDU) hati ghass; (VLAAMS) olifantsgras; (YALUNKA) kulon-na, kulu-na; (YORUBA) eèsún, ikẹn.

17.4. *Phyllostachys* spp. (EP) young shoots; (P) C.

(e.g. *P. edulis, P. dulcis, P. makinoi, P. nuda, P. pubescens, P. viridis*)

(AFAR) kàsab; (AMHARIC) šämäqqo, šəmäl; (ASSAMESE) bah gaj; (BEMBA) lusengu; (BEMBÉ) moutsoungou; (BINUKID) davung; (BODO) meoa, oa, oua; (BUGHOTU) duee, gau, gauratu, kadolulu, ruavatu; (BURMESE) hmyit; (CAROLINIAN) bwaay, ligippwaay, lighippwáy; (CATALAN-VALENCIAN-BALEAR) bambú; (CHICKASAW) oskapi'; (CHINESE, MANDARIN) gang zhu, tan-chu, k'u-chu; (CHINESE, MIN NAN) tek-á, tek-sún, tang-sún, le̍k-tek-sún, sún; (CHINESE, WÚ) gang zhu shu; (CHINESE, YUE) mao zhu sun; (CREOLE, HAITIAN FRENCH) bonbou; (CRIOULO, CAPE VERDE) bambu; (CZECH) fylostachys; (DANISH) bambusskud; (DUTCH) bamboesspruiten; (ENGLISH) bamboo shoots; (ESTONIAN) bambus; (ÉWÉ) pamblo, pambro; (FARSI, WESTERN) kkeyzaran; (FRENCH) pousses de bambou; (FULFULDE, ADAMAWA) kewal, kewe; (GALICIAN) bambueira; (GEORGIAN) bambúk; (GERMAN) Bambus; (GIKUYU) mũrangi; (GOURMANCHÉMA) sucré-dow; (GREEK) phyllostochis; (HARARI) qärqäḥa; (HAUSA) gora; (HEBREW) chizran, bamboo; (HINDI) bāns; (HINDUSTANI) báns̱; (HMONG DAW) ntsuag xyoob; (HOPI) paqavi; (HUNGARIAN) bambuszrügy; (INDONESIAN) rebung; (ITALIAN) bambú; (JAPANESE) takenoko, hotei-chiku, kei-chiku, kuro-chiku, ma-dake; (KAREN) mep-we; (KASHMIRI) baĩs; (KIRIBATI) te kaibaba, te kaipapa; (LATVIAN) bambusi, bambuss, bambusniedre; (LOMWE, MALAWI) nsengere; (LUO) keo; (LUXENBOURGEOIS) Bambus; (MALAY) rebung; (MARWARI) abāns, bānsdo; (MIZO) mau, chal, rua; (MUNDARI) kāril, hanḍuā-utu, māt', mād'; (NEWAR) pat; (NGUNDI) muntsié; (NIUE) kaho; (NORWEGIAN) bambusskudd; (NUPE-NUPE-TAKO) gusa, èba; (OROMO, WEST CENTRAL) leman, simalla; (PALI) kisalayaṁ; (POLISH) bambus; (PORTUGUESE) bambú; (PULAAR) kewal; (SANTALI) helta, jạil, kạril; (SHAN) mai-pang-puk; (SHONA) chiwere tsanga, mapfup, musengere, mushenjere; (SOMALI) jara'; (SPANISH) bambú; (SORBIAN, UPPER) bambus; (SOTHO, NORTHERN) pampu; (SOTHO, SOUTHERN) leqala; (SWAHILI) mianzi, muanzi; (SWEDISH) bamboo skott; (TAGALOG) labong; (TAJIKI) nay; (THAI) paang-puk; (TIBETAN) ñũqtsә̂; (TIV) sôngough; (TSWANA) bambusa; (TURKISH) hint kamişi; (URDU) bans ki shakhan; (VENDA) mushaka tembwe; (VLAAMS) bamboescheuten; (XHOSA) umthi webhambu, uqalo; (YIDDISH) yamsh shpratsen; (ZULU) uqalo.

17.5. *Saccharum edule* Hassk. (EP) immature inflorescence; (P) C.

This species is rare and is thought to be a variant of *S. officinarum*. [1]Denotes names for *S. edule*; the remainding names are simply generic for sugarcane. In some instances (e.g. English), the name refers to the inflorescence of sugarcane. Dialects of Luyia (Bukusu[2], Kabarasi[3], Marama[4]).

(AFAR) sokkarqaàla; (AFRIKAANS) aartappel, suikerriet; (AGTA) talad; (AGTA, CASIGURAN DUMAGAT) unat; (AKAN-ASANTE) ahwereɛ; (AKAN-FANTE) ahwere; (AKPA) ànggbó ibó;

an mumbola, gokbilang; (IGBO) ọka; (ILOCANO) mais; (INDONESIAN) jagung manis, jagung; (ITALIAN) mais zuccherino, mais dolce, granoturco da zucchero; (ITAWIT) mait; (IZON) aká; (JAPANESE) tomorokoshi; (JAVANESE) jagung, janggel; (JOLA-FONYI) koumboha; (JULA) kaba; (KABIYÉ) wamla; (KALINGA, BUTBUT) kaaw; (KANNADA) mekkejola, goinjol, musukojola, makkai, jola, mekke jola, musukina jola, dodda jola, govinajola, hallina jola, kundige jola, meksikan jola; (KASHMIRI) dŏda-hĕdur; maköyü, (KATI) jigor, jawār; (KAZAKH) sacharnaja kukuruza, zhügeri; (KIOWA) ei-t'Hî-dα; (KIRGHIZ) kanttuu žugeru, jügörü; (KIRIBATI) kaon; (KONKANI) maeo, zonallo, zondlo, jolu; (KRIO) kɔn; (KURDISH, NORTHERN) gēnmēsāmi; (KWANGALI) omiriva; (KWÉYÒL) mi, gwenn; (LADINO) maiz, misir; (LAHU) sha ma; (LATVIAN) kukurūza, cucurkukurūza; (LITHUANIAN) paprastasis kukurūzas, daržo kukurūzai; (LOMWE, MALAWI) mondokwa - dowe, nahebwe; (LUO) nywagi; (LUXENBOURGEOIS) Mais; (MACEDONIAN) slatka pčenka; (MALAY) jagong; (MALAYALAM) cholam, makka cholam, ponthicholam; (MALTESE) qamh ir-rum; (MANINKAKAN, WESTERN) toubanio; (MANOBO) hapey, sanley, serat, sileban; (MAORI) kaanga, koopkipaki, paratee; (MARANAO) bito-an, buto sa a-i, kamais; (MARATHI) maka, makai, makka, guta, buta; (MEHRI) barr, bər; (MENDE) nyɔ; (MICMAC) peās'kŭmŭn; (MOHAWK) onenhste; (MONGOLIAN) erdene shish; (MÒORÉ) kãmãna; (MORISYEN) may dou; (MUNDARI) makāi, jonṛa; (MUSKOGEE) vce; (NAMA-DAMARA) -kula; (NAVAJO) naadą́ą́'; (NDONGA) epungu; (NEPALI) makai; (NEWAR) wā, pu-wā; (NGIZIM) másáȓmì; (NGUNDI) sangou; (NICOBARESE, CAR) peòk; (NIUE) ahi, ahi taina, hana; (NORWEGIAN) sukkermais, mais; (NUPE-NUPE-TAKO) kàba; (NYAKYUSA-NGONDE) ifirombe; (NYANJA) cimanga; (ORIYA) buta, maka, mokka; (OROMO, WEST CENTRAL) bekkolo; (PAIUTE, SOUTHERN) ung-weev', komee'; (PAMPANGAN) mais; (PANJABI, EASTERN) barajaur, chhale, juar, kukri, kuthi, makai, makki; (PASHTO, NORTHERN) jnār, ghat jnār; (POLISH) kukurydza cukrowa[1]; (PORTUGUESE) milho doce[1], milho; (POTAWATOMI) wishpumnuk, mandamin, ndamnuk, ndamen; (PULAAR) makka; (PULAR) makka; (ROMANI, BALKAN) misiri; (ROMANIAN) porumb zaharat; (RUSSIAN) sacharnaja kukuruza; (SAMBAL, BOTOLAN) maih; (SAMOAN) sana; (SANSKRIT) yavanala, makaaya, kandaja, mahaakaaya, samputantastha, shikhalu, vavanala; (SANTALI) joṇḍra; (SENA) mapiramanga; (SENOUFO, CEBAARA) padegue; (SERBO-CROATIAN) kukuruz sečerac; (SHONA) bonore, chibage, chibahwe, chibarwe, chibere; (SIDAMO) badalco; (SINHALA) bada iringu; (SLOVAK) kukurica cukrová; (SLOVENIAN) sladka koruza; (SORBIAN, LOWER) kukurica, majs; (SORBIAN, UPPER) kukurica, kurjace woko, majs; (SOTHO, NORTHERN) mahea; (SOTHO, SOUTHERN) poone; (SPANISH) maíz dulce[1], maíz azucarado[1], maíz tierno[1]; (STELLINGWERFS) mais; (SUMO-MAYANGNA) am, ama; (SUNDA) jagong; (SWAHILI) mhindi, punje, mafaka, ngano, gunzi, kigunzi; (SWEDISH) sockermajs[1]; (TAGALOG) mais; (TAGBANWA) mais; (TAJIKI) juvorimakka; (TAMIL) makka-cholam, makka sholam, turakacholam, mutthucholam, turaka cholam; (TATAR) kukuruz; (TAUSUG) mais, gandum; (TELUGU) mokka-janna, makka jonnalu, makkajounalu, mokkajanna, makkazonnalu, mokka, mosanam, zonaloo; (THEMNE) ta-mank; (TIBETAN) āshom; (TIRURAY) kamais; (TIV) i̱ku̱lekẹ, anagajia; (TONGA) cingoma; (TSWANA) mmidi; (TULU) jola; (TURKISH) tatli misir, misir; (TURKMEN) mekgejöwen, adaty mekgejöwen; (UKRAINIAN) cukrova kukurudza[1]; (URDU) makai; (UYGHUR) qonaq; (UZBEK) makkajo'xori, maccadjukhory; (VENDA) ḽivhele, gurùnngu, vandisi, mbala-makole, mutonga, vele; (VLAAMS) suikermais[1]; (WAAMA) manaso tɛɛriya; (WÁRAY-WÁRAY) maís; (WOLOF) mbox; (XHOSA) umbona; (YAO) imanga; (YIDDISH) kukuruzeh; (YOM) paazaa; (YORUBA) agbado; (ZARMA) kolkoti; (ZULU) ummbila, umumbu; (ZUNI) atowa.

17.8. *Zizania latifolia* (Griseb.) Turcz. ex Stapf. (EP) swollen shoot/stem; (P) C.

(ARABIC) kharazan el-maa; (BELARUSAN) tsatsanija shirakolistnaja; (CHINESE, MANDARIN) jiao bai; (CHINESE, WÚ) jiao bai; (DANISH) vildris; (ENGLISH) water bamboo, cobo, Manchurian wild rice, Manchurian water rice; (FRENCH) riz sauvage, riz sauvage de Mandchourie; (GREEK) zizauia the elovios; (HUNGARIAN) tuszkarora; (ITALIAN) riso d'acqua, riso d'America, riso indiano, tuscarora; (JAPANESE) makomo; (LITHUANIAN) vandeninė zizanija; (LUYIA) amatayia kokhumwalo; (MONGOLIAN) örgön navchit tutarga; (POLISH) zizania wodna, dziki ryż; (PORTUGUESE) arroz bravo; (RONGA) batata rcno, chizambana; (SPANISH) arroz silvestre; (TAJIKI) nai obi; (THAI) nomai-nam, kapek; (TURKISH) kanada çeltiği, yabani çeltik; (URDU) pani ka bans; (VIETNAMESE) c[ur] ni[eex]ng, ni[ee]ng ni[eex]ng, l[us]a mi[ee]u; (VLAAMS) waterbamboe.

18. Pontederiaceae

18.1. *Monochoria hastata* (L.) Solms (EP) young leaves; (P) C.

(CEBUANO) kasal-kasal, payaw-payaw, gabi-gabi, kosol-kosol; (CHINESE, WÚ) jian ye yu jiu hua; (ENGLISH) hastate-leaved pondweed, arrow-leaved monochoria; (FILIPINO) gabi-gabihan; (HUNGARIAN) vizijátszint; (INDONESIAN) eceng gede, bia-bia, wewehan; (JAPANESE) nan'yo-mizu-aoi; (KHMER, CENTRAL) chrach; (LAO) 'ii hin; (MALAY) chacha layar, kangkong air; (PORTUGUESE) monocória; (TAGALOG) gabi-gabihan; (TELUGU) neeru thaamara; (THAI) phaktop-thai, phaktop, phakpong; (TOK PISIN) maoa; (URDU) sir naza, pati talabi khoodro poda; (VIETNAMESE) rau m[as]c.

18.2. *Monochoria vaginalis* (Brum. f.) C. Presl. (EP) young leaves; (P) C.

(BENGALI) nukha, nanka; (CHINESE, WÚ) ya she cao; (ENGLISH) oval-leaved pondweed, oval-leaved monochoria, pickerel weed; (FILIPINO) gabing-uak; (ILOCANO) bilagut; (INDONESIAN) jagung manis,eceng padi, eceng leutik, wewehan; (KHMER, CENTRAL) chrach; (LAHU) hpa^ li⁻; (MALAY) kelayar, keladi agak; (MALAYALAM) kakapola; (PORTUGUESE) monocória, udoendo; (SANSKRIT) indivarah; (TAGALOG) biga-bigaan, gabing uwak; (TELUGU) nirkancha, nirokancha; (THAI) khakhiat, ninlabon, phakhin; (URDU) bazvi pati khoodro poda; (VIETNAMESE) rau ch[os] c, rau m[as]c l[as] thon.

19. Taccaceae

19.1. *Tacca leontopetaloides* (L.) O. Kuntze (EP) rhizome, (P) C.

(ARABIC) gezer sahmy hindy; (BASSARI) *a*-tiíti *ɓi* yil; (BUDIK) gi-ngùrkùr; (CEBUANO) tayobong, kanobong; (CHINESE, GÀN) qi zi yu; (CHINESE, MANDARIN) ju ruo shu; (CHINESE, MIN BEI) mo yu, mi guo; (CHINESE, XIANG) mo yu; (CHINESE, WÚ) shan da huang, ju ruo shu; (CHINESE, YUE) mo yu, gou gua yu; (CZECH) taka; (DANISH) ostindisk pilrod; (DIGO) ulanga; (ENGLISH) East Indian arrowroot, Fiji arrowroot, Tahiti arrowroot, Polynesian arrowroot; (FIJIAN) yabia; (FILIPINO) gau-gau; (FINNISH) nuolijuuri; (FRENCH) hydropire, arrowroot de Tahiti, arrowroot polynésien; (FULFULDE, ADAMAWA) duɓɓi waandu, labbo fowru; (GA) ɔmɔkç, anu akatiɔwia; (GBE-VHE) dzogbenyabo; (GREEK) arrarouti anatolir kon indion; (HANUNOO) kanubun; (HAUSA) gaatarin zoomoo, giginyar birii, masara maciji, sandan birii, sandan dutse, sandan yan bori, taara-yaara, yara; (HAWAIIAN) pia; (ILOCANO) panarien, panariyen, pannariean; (INDONESIAN) gadung tikus, gadung laut, iles-iles, likir, leker, kacunda, taka laut, kecondang; (JAPANESE) makomo; (JAVANESE) iles-iles, kecondang, condang, kemendulan, ceker ayam; (KANNADA) handee gedde, daeva kaanda; (KIRIBATI) makemake; (LOKO) bakeyowi; (MALAY) lukeh, poko lukeh; (MALAYALAM) chanekizhana; (MANDING-BAMBARA) bua-bayayo, suruku katama; (MANU'A SAMOAN) masoa, pia; (MARQUESAN) pia; (MENDE) buai ŋgīli; (MÒORÉ) bolbogho, luré; (NIUE) pia; (POLISH) krąpiel pierzastodzielna; (PORTUGUESE) araruta do Tahiti;

(RAROTONGAN) pia; (SAMBAL, TINÀ) tambobon; (SAMOAN) māsoā, pia; (SANTALI) boṅga ṭaini; (SERER-SINE) *n*dof *n*diãng, *n*dof rôg, ka kéna, lar; (SPANISH) yabia; (SUNDA) kumis ucing, kacondang, kotok bongkok; (TAGALOG) yabyaban, gaw-gaw, gawgaw; (TAMIL) kaattu karunai, karachunai, periyakarumeikkilhaugu; (TELUGU) chanda, kanda, peddakandagadda, ritta kanda; (TEM) ludju; (THAI) thaoyaimom, buk-ro; (TIV) amara, gbache; (TONGAN) mahoa'a; (TURKISH) takka, tahiti ararotu; (TWI) akoropaa, aborobaạ; (URDU) arrarot; (VIETNAMESE) b[aj]ch tinh, c[ur] n[uw]a; (VLAAMS) oost-indische pijlwortel; (WOLOF) bara, èt i buki, lar; (YORUBA) akana maigbo, ọlọ̀ pàpà nìràgà.

20. Zingiberaceae

20.1. *Alpinia galanga* (L.) Sw. (EP) floral sprouts and flowers, tender shoots[S], rhizomes[S]; (P) C,R,P.

(ARABIC) el-galangal, el-adkham; (BENGALI) barakulanjar, kulanjan; (BICOLANO, CENTRAL) langkawas, palla, langkuas; (BURMESE) padagoji; (CHINESE, MANDARIN) hong dou kou; (CHINESE, WÚ) hong dou kou, dou kou; (CZECH) galgán velký; (DANISH) stor galanga; (ENGLISH) greater galangal, galangal, Java galangal, Siamese galangal, mongoose plant, monkey plant; (FILIPINO) lankau'as; (FINNISH) galangajuuri; (FRENCH) galanga, galanga de l'Inde, galanga majeur, herbe indienne, grand galanga; (GERMAN) Galanga, Galgant, großer Galgant; (GUJARATI) kulinjan; (HEBREW) alpinia; (HINDI) kulinjan, kulanjan, barakulanjar, mahaabharee, bharakalijan, barakulanjan, kuljan; (HINDUSTANI) kulanjan, laos; (INDONESIAN) langkuas, lengkuas; (ITALIAN) galanga; (JAPANESE) nankyo; (JAVANESE) laos; (KANNADA) dumparaasme, dumparasmi, dumbarasme, doddadumprashme, dooparaasme, dumbaraasme, dumbraashta, gandhamoola, kulinjana, raasmi, raeshime gadde, sugandhavaachi; (KHMER, CENTRAL) remdéng; (LAO) kha:x ta: dé:ng; (MALAY) lengkuas; (MALAYALAM) aratta, perasatta, kol-inji, peraratta; (MARATHI) kolirlja, koshtkulinjan, koshtkulayan; (NEPALI) seto bojho; (POLISH) alpinia galgant wielki, kardamon dziki; (PORTUGUESE) galanga maior; (SANSKRIT) malayavacha, kulajan, kulanja, raasanaa, aelaaparni, aruna, dhumala, gandhamoola, gandhavaaruni, kapidruma, korajaa, kulanjana, mahaabhareevacha, naakuli, paatala, purusha, raktharenu, rakthapushpa, sugandha, sugandhavacha, sugandhayoga, theekshna moola, ugragandha; (SPANISH) galanga, galanga mayor; (SUNDA) laja; (TAMIL) peraratthai, pera-rattai, aanandam, aratthai, ardubam, atthumam, gorararatthaayi, kandanaaguliyan, ormarundu, satthiradji, sugandam, titthiram, tumbaraattagam; (TELUGU) peddadumparashtram, peddadumparaashtrakmu, dumparaashtrakamu, kachoramu; (THAI) kha; (TURKISH) büyük galanga, havlican; (URDU) khoosjan, khalanj, kalanjan; (VIETNAMESE) ri[eef]ng n[ees]p, s[wo]n n[aj]i; (VLAAMS) galangawortel.

20.2. *Curcuma longa* L. (EP) rhizome; (P) C.

(AFAR) hùrud, ùrud; (AFRIKAANS) kurkuma, borrie; (AKAN-ASANTE) gangamu; (AKLANON) lampuyang; (ALBANIAN, TOSK) kurkumë; (AMHARIC) hərd, ərd, zənğəbəl; (ARABIC) kur kum, zangabeel asfer; (ASSAMESE) halodhi; (BAATONUM) yorutambu; (BALTI) yung; (BENGALI) haldi, halada, halud; (BINUKID) kelawag; (BODO) haldøi; (BUHID) dilaw; (BURMESE) nanwin; (CEBUANO) dulaw, duwaw, kalabaga, kalauag, kinamboi, lumpuyang, kalawag, kinamboy, tamangyan; (CHINESE, MANDARIN) jiang huang; (CHINESE, WÚ) jiang huang; (CHINESE, YUE) yu jin; (CRIOULO, UPPER GUINEA) açafron; (CZECH) kurkuma dlouhá, žlutý zázvor; (DANISH) gurkemeje, kurkuma; (DUTCH) geelwortel; (ENGLISH) turmeric, yellow ginger; (ESTONIAN) kurkum; (FIJIAN) haldi, cago; (FILIPINO) luyang dilaw; (FINNISH) maustekurkuma; (FON) can koko; (FRENCH) curcuma, curcuma longue, curry, safran des Indes; (FULFULDE, ADAMAWA) mangina, magina, gangamuki; (GA) kakaotsofa tsuru; (GAELIC, SCOTTISH) dinnsear; (GEN) dotɛ jɛn; (GERMAN) lange Kurkuma, Gelbwurzel; (GIANGAN) salampauyan; (GREEK) kourkoumi hie long, chysorriza; (GUJARATI) haldi, halada; (GUN) dotɛ jɛn; (HANUNOO) alumpan; (HAUSA) gàngàmóó, kurgum, magina, mangina, tírìì, túrìì, zabibi; (HAWAIIAN) ʻolena; (HEBREW) kurkum; (HILIGAYNON) dulaw; (HINDI) haldi, halada, adarak; (HINDUSTANI) hardi, pitras hardi; (HMONG DAW) qhiav qus; (HUNGARIAN) kurkuma, gyökér; (IBALOI) kuhig; (IBANAG) kunik; (ICELANDIC) ukon; (IGBO) uchiri-ilu; (ILOCANO) kunig, kuliaw, kalawag, kuliyaw, kulyaw; (INDONESIAN) kunyit, temu kunyit, temu kuning, kunir, kumkuma, kuma-kuma, kungkuma, kurkuma, komkoma; (ITALIAN) curcuma lunga, zafferano delle Indie, radice gialla; (ITAWIT) kunig, kunik; (IVATAN) hama; (JAVANESE) kunir, kunir bentis, temu kunir; (KALINGA, BUTBUT) unig; (KANNADA) arisina, arishina; (KHMER, CENTRAL) rômiêt, lômiêt; (LAHU) mi$_v$ shi; (LAO) khminz khünz; (LATVIAN) kulitvētā kurkuma; (MAITHILI) haradi, haraid; (MALAY) kunyit, temu kunyit; (MALAYALAM) manjal; (MANOBO) kalawag; (MARANAO) kalawag; (MARATHI) haldi, halad, halada; (MARQUESAN) ʻeka, ʻena; (MENDE) *n*dɔo-bundɔ; (MIZO) sawhthing; (MORISYEN) safran; (MUNDARI) sāsāng; (NEPALI) besar; (NEWAR) mā-haraḍī, mi-haraḍĭ; (NICOBARESE, CAR) ma-nū-lö; (NIUE) ago; (NORWEGIAN) gurkemeie; (ORIYA) haldi; (PALI) haliddā; (PAMPANGAN) angay, ange, kulalo, pangas, añgai, pañgas; (PANGASINAN) kulalaw; (PANJABI, EASTERN) haldi, hald, bassar; (PASHTO, NORTHERN) adrak; (POLISH) ostryż długi, kurkuma długa; (PORTUGUESE) açafrão da Índia, curcuma, gengibre dourada; (RAROTONGAN) rea; (ROMANIAN) ghimber galben, curcuma; (SAMBAL, BOTOLAN) kalawag; (SAMOAN) ago, ano, sama; (SANSKRIT) haridra; (SANTALI) gāṭ, sasaṅ; (SERBO-CROATIAN) žutnjak, zerdečaf; (SINHALA) kaha; (SLOVAK) kurkuma; (SOMALI) hurūd; (SPANISH) cúrcuma larga, cúrcuma longa, azafrán de la India; (SUBANON) lawag; (SUMO-MAYANGNA) azáfran; (SUNDA) koneng, koneng temen, kunyir; (SWAHILI) binzari, manjano, tangawizi, nguvu, bidii, juhudi; (SWEDISH) gurkmeja; (TAGALOG) luyang dilaw, dilaw, dilaw-pula, kilaw, luyang-dilaw; (TAGBANWA) kunit; (TAJIKI) kijkita; (TAMIL) peraratthei-manjal, manjal; (TELUGU) pasupu; (THAI) khamin, khamin-chan; (THAI, NORTHERN) khamin-kaeng; (TIRURAY) kalawag sangir; (TONGAN) ango; (TULU) ǰaḷu, ānemaṅǰaḷu, kaččūra, kūvè; (TURKISH) zerdeçal, zerdeçal hint zafarni, sariboya, zerdeçav, zerdeçöp, safran kükü; (URDU) haldi; (VIETNAMESE) ngh[eej], u[aas] t kim; (VLAAMS) geelwortel, kocnjit; (WÁRAY-WÁRAY) duláw, dulaw, tanmanan.

20.3. *Curcuma mangga* Valeton & Van Zijp (EP) rhizomes, young shoots; (P) R,C.

(CHINESE, WÚ) jiang huang; (ENGLISH) mango ginger; (HUNGARIAN) kurkuma; (INDONESIAN) temu lalab, temu mangga, kuning pari; (JAPANESE) temu mangga; (JAVANESE) temu mangga, temu glenyeh, temu putih, temu bajangan, temu lalab, temu poh; (MALAY) temu pauh; (SUNDA) koneng manggah, koneng joho, koneng lalab, koneng pari; (THAI) khamin-khao.

20.4. *Curcuma zedoaria* (Christm.) Roscoe (EP) rhizomes; (P) C.

(AFRIKAANS) Japanese wilde gemmer; (AKLANON) lampoyang; (ARABIC) gezer sahmy hindy; (ASSAMESE) keturi; (BENGALI) kachura; (BICOLANO, CENTRAL) tamahilan, tamaylan; (BONTOC, CENTRAL) koneg; (BURMESE) thanuwen; (CEBUANO) alimpuyas, alimpuyang; (CHINESE, WÚ) eshu; (CHINESE, YUE) peng e shu; (CZECH) kurkuma; (DANISH) zedoar; (ENGLISH) long zedoary, round zedoary, zedoary; (FILIPINO) luya-luyahan; (FINNISH) nuolijuuri; (FRENCH) zédoaire, curcuma zédoaire; (GERMAN) Zitwer; (GREEK) to indiko arrapoúte; (GUJARATI) kachura; (HANUNOO) alumpan budak; (HEBREW) kurkum zedoar; (HINDI) kachura, kachoora, aamaada, kaalihaladi; (HUNGARIAN) citvor; (IFUGAO) unig, unig an nahamad; (ILOCANO) konik, langguwas, langkuwas; (INDONESIAN) temu putih, temulawak; (ITALIAN) zedoaria; (JAPANESE) gajutsu; (JAVANESE) temu putih; (KANKANAEY) kunig, laylaya; (KANNADA) kachura, ambe haldi, kaadu arisina, kachora; (KHMER, CENTRAL) prâtiël prèah 'ângkaôl; (KONKANI) aamba haldi; (LAO) khminz khünz; (LITHUANIAN) geltonoji ciberžolė; (MALAY) temu kuning, temu lawak, kunchur; (MALAYALAM) kaattumannaal, kachholam, kachhurikizhanna, pulakizhanna; (MARANAO) lempoyan; (MARATHI) kachura, narakachora, kachari, kachora; (PAMPANGAN) tamo; (POLISH) ostryż plamisty; (PORTUGUESE) zedoária, zedoeira; (QUECHUA, CUSCO) kisalor; (SAMBAL, TINÀ) ganda (SANSKRIT) kachoora, karchoora, draavida, durlabha, gandhamoolaka, gandhasara, jatala, kalpaka, karshya, mukhya, narkachoor, sathi, shathi, vedhya; (SPANISH) cedoaria, cedoaria larga, cedoaria redonda; (SUBANON) tampoyang; (SUNDA) koneng tegal; (TAGALOG) barak, bolon, bulon, luya-Iuyahan, tamahiba, tamo, tamo-kansi; (TAMIL) kichili-kizhanghv, kachora, kasthoori-manjil, kichhilik-kishangu, pulankilhangu; (TELUGU) kachoram, kichhiligaddalu, adavipasuvu, kasthooripasupu; (THAI) khamin-khun, khamin-oi; (TULU) aaane manjal; (TURKISH) ceduvar, zidvar, zalumba kökü, zurumbaz, cedvar, kara çavdar; (VIETNAMESE) ngh[eej] den, nga tru[aaj]t, ng[ar]i t[is]m; (VLAAMS) zeddarwortel; (WÁRAY-WÁRAY) nuláw-duláw.

20.5. *Zingiber mioga* (Thunb.) Roscoe (EP) rhizomes, tender shoots[S], leaves[S] and flowers[S]; (P) C.

(ARABIC) zangabeel; (ASSAMESE) Japani ada; (CHINESE, MANDARIN) rang he; (CHINESE, WÚ) rang he, ming he; (CHINESE, YUE) ye jiang; (CZECH) zázvorovník pravý; (ENGLISH) Japanese wild ginger, mioga ginger; (FRENCH) zédoaire; (GERMAN) Zedoarwurzel, zittwer Kukumc, Japaningwer; (GREEK) ziggiveri mióga; (HEBREW) zangvil, ginger; (HMONG DAW) qhiav qus; (HUNGARIAN) gyömbér; (ITALIAN) zenzero; (JAPANESE) myouga, mioga; (LITHUANIAN) Japoninis imbieras; (MALAY) temu kuning; (POLISH) imbir japoński, imbir mioga; (PORTUGUESE) gengibre, miogá; (SPANISH) jengibre; (URDU) adrak, soant; (VLAAMS) Japanse wilde gember.

20.6. *Zingiber officinale* Roscoe (EP) rhizomes and tender shoots[S]; (P) R,C,P.

(AFRIKAANS) gemmer; (AGTA) loyi; (AKAN-ASANTE) akakadur, tsintsimir, tsintsinmin; (ALBANIAN, TOSK) xhenxhefil; (ARABIC) alzangabeel; (ASSAMESE) ada; (AZERBAIJANI, NORTHERN) zəncəfil; (BADAGA) ingu, inji, sukku; (BALTI) cazga; (BAOULÉ) asiê ounsa; (BASSARI) ɔ-ngı̧́n ɔnd ɓi- ɗyàr, afu; (BEMBÉ) tangawissi, moungountsimbala; (BENGALI) ada; (BEROM) syàtáà dángás; (BÉTÉ) gnamankou; (BICOLANO, CENTRAL) laya, lay-a, luy-a; (BINUKID) luya, mererasen; (BOBO MADARÉ, SOUTHERN) dùgùmà nyamugu; (BODO) haijeng; (BONTOC CENTRAL) laya; (BOSNIAN) džindžer, džumbir; (BUDIK) nyámákù; (BUGHOTU) konggu; (BUHID) ria, rid; (BULLOM SO) wischa, lone, loni, saŋk-ɛ; (BURMESE) gyin; (BUTUANON) luy-a; (CAROLINIAN) gúrhúl, ghuschél; (CATALAN-VALENCIAN-BALEAR) gingebre; (CEBUANO) luy-a; (CHECHEN) zamzhaabil, waambar; (CHINESE, GÀN) sheng jiang; (CHINESE, HAKKA) kiōng-mâ; (CHINESE, MANDARIN) jiang; (CHINESE, MIN BEI) sheng jiang, mu jiang; (CHINESE, MIN NAN) chhe-kiu, kiu; (CHINESE, WÚ) jiang, shen jiang; (CHINESE, YUE) sheng jiang; (CREE, PLAINS) manito oskatask; (CREOLE, HAITIAN FRENCH) jinjanm, jenjanm; (CRIOULO, CAPE VERDE) gengibre; (CRIOULO, UPPER GUINEA) gengipe; (DAGAARE, SOUTHERN) yamankon; (DAGBANI) sakarra, tschibilli; (DAKOTA) aġuyapiicahi ġi; (DANGME-KROBO) odzahwi; (DANISH) ingefær; (DIGO) tangawizi; (DUTCH) gember, amoom; (ENGLISH) ginger, common ginger; (ESTONIAN) ingver; (ÉWÉ) dote, gumetakui, agometaku; (FARSI, WESTERN) zenjebil, zanjebeel; (FIJIAN) aadi; (FILIPINO) luya; (FINNISH) inkivääri; (FRENCH) amome, gingembre; (GA) eŋelfɛl, kàkao tsofa, kakatcofa; (GAELIC, MANX) jinshar; (GAELIC, SCOTTISH) dinnsear; (GALICIAN) sinxebra, xinxebra; (GBE-VHE) abrɔfo-yisa, ŋkrabodo, ŋkrabrɔ, ŋkrama; (GERMAN) Ingwer-Gewürz, Kardamom; (GIANGAN) riya; (GOLA) *m*bada-nyum, genja, nyama, nyama-ko-kɛmpu; (GOURMANCHÉMA) sakadrou; (GOURMANTCHÉ) igoũo kambi; (GREEK) ziggivery to pharmakeytiko; (GUJARATI) adu; (HANUNOO) Ii-a tamlan; (HARARI) zänğäbīl; (HAUSA) citta, cìttar àhò, cìttár àhóó, cìttár-fo, cìttár-jábà, cìttár-mai-ƙwaya, cìttár-mai-yáátsàà, sakanjabir, sakan-zabur, zànzàbîr; (HEBREW) zangvil, ginger; (HILIGAYNON) lúy'a; (HINDI) adrakh, adrak, ada, sandhi, sonta; (HINDUSTANI) adrak, son̠th; (HMONG DAW) qhiav; (HUNGARIAN) gyömbér; (IBALOI) agat; (IBANAG) laya; (ICELANDIC) engifer; (IFUGAO) kallawagan puupu, laiya, layya; (IGBO) jinjà, ukpara-nsize; (ILOCANO) basing, laya, baseng, layya; (INDONESIAN) jahe, jae, lia, alia, halia; (ITALIAN) zenzero officinale; (IVATAN) anaha, naha; (JAPANESE) shouga; (JAVANESE) jae, jahe; (JULA) gnamankou; (KABIYÉ) wissikoye; (KALINGA, BUTBUT) ladja; (KANNADA) shunti, alla, hasisunti, aardraka, hasi shunti, khaara genasu, vanashunti; (KANURI) njittá gòló, tààjiwâr; (KASHMIRI) adrakh, ödürü; (KHMER, CENTRAL) khnhei, khnhei phlung; (KISSI, NORTHERN) kisà-dilɛndɛ, kisà-*k*pɛŋgilo; (KONKANI) alem, alle, sonti; (KONKOMBA) idčin benul, kelem bwapà; (KONO) *n*du-nyamoɛ, nyamoɛ; (KRIO) jinja; (KURANKO) nyamoku, nyamɔŋhɛ; (KWÉYÒL) jenjanm, jenjamb; (LADINO) jinjibrè, jenjibre; (LAHU) chu$^{\vee}$ pi^{-}; (LAO) khi:ng; (LATVIAN) ārstniecības ingvēras; (LITHUANIAN) tikrasis imbieras; (LOKO) kihyɛ, lɔogihye; (MAITHILI) ād; (MALAY) halia, haliya, jahi, atujapi; (MALAYALAM) inji, andrakam, chinchatakam, chinchiver, chukku, sringivaeram, tinkshottham; (MANDING-BAMBARA) dugukoro niamaku, niamaku, nyamaku; (MANDING-MANDINKA) duguru-nyamuŋhu, linkkwum, nyamuŋhu; (MANDING-MANINKA) niamaku; (MANOBO) marerasan, loya pagirisun; (MARATHI) ale, alen, sunte; (MENDE) *n*dɔo-kijɛ, *n*dɔo-gijɛ, kijɛ; (MÒORÉ) yamaku, yãmãkou; (MORISYEN) zenzam; (MUNDARI) ādi, ādrak; (NEPALI) aduwa; (NEWAR) suthi; (NGIZIM) cìitá áfó; (NGUNDI) tangawissi, moungountsimbala; (NORWEGIAN) ingefær; (ORIYA) ada; (OROMO, WEST CENTRAL) gijimbilla, jijimbilla; (PALI) addakaṁ, siṅgiveraṁ; (PAMPANGAN) laya; (PANGASINAN) agát, agat; (PANJABI, EASTERN) adrak, udruk; (PASHTO, NORTHERN) sund̲; (POLISH) imbir lekarsik; (PORTUGUESE) gengibre, gengibre amarelo, gengibre das boticas; (PULAAR) niamaku-limban; (RAROTONGAN) kopekai; (ROMANIAN) ghimber; (SAMBAL, BOTOLAN) alomangi, laiya, layya, naupak; (SAMBAL, TINÀ) laiyal, lay-a, layal; (SAMOAN) 'ava pui; (SANSKRIT) adraka, aardrakah, shunti, anupama, apakrishnaka, aardrashaka, chandrakhya, gulma moola, kandara, katubhadra, katukkata, machhaka, mahinja, moolaja, rahuchhana, saikatheshtha, sharanga, shrigavera, sushkaka, vara, nagaraja, mahaushad, vishvabheshaj; (SANTALI) ada, adhe̱; (SCOTS) ginge; (SENOUFO, CEBAARA) siguane; (SERBO-CROATIAN) gingibar; (SERER-SINE) ginger; (SIDAMO) janjiwêlo; (SINHALA) inguru; (SLOVAK) zázvor; (SOMALI) sinjibil; (SORBIAN, UPPER) himbjer, ingwer; (SPANISH) jengibre, jengibre oficinal; (SUMO-MAYANGNA) sinsa, marid puluni; (SUNDA) jahe; (SUSU) niohomi, kamanghi, kisà-khɔŋgbe, kisànyi, nyɔkhɔmi; (SWAHILI) tangawizi; (SWEDISH) inge fära; (TAGALOG) luya; (TAGBANWA) luya; (TAJIKI) zanjabil; (TAMIL) allam, inji, sukku, artthiragam, atthiradam, kodatharam, kulu mamulam, maruppu, sangai, sigaram, singavaeram, singivaeram, sundi, ubugallam, verrkkombu; (TAUSUG) lu-uya; (TELUGU) allam, allamu, sonthi, sonti, aardrakamu, mahaushadamu, shunthi, sringabaeramu, (TEM) ata rrer, atarraga; (THEMNE) ta-sɛŋa, *ta*-seŋa, *ta*-seŋa-*ta*-thənk; (THAI) khing, khing-daeng; (TIBETAN) q̱əmuù; (TIRURAY) fute giya, giya, kasumba giya; (TIV) shi̱ta, se̱ta; (TOK PISIN) kawawar, kawawari; (TULU) šuṇṭhi, sunthi, sunti, pajasunti; (TURKISH) zencefil, hakiki zencefil; (TWI) akekaduru, kakaduru, akeka, kekeduru; (URDU) adrak, soant; (VAI) falo-nyan, nyan-mɛsɛ; (VIETNAMESE) g[uwf]ng, sinh kh [uw][ow]ng; (VLAAMS) gember; (WAMEY) *a*-mbir; (WÁRAY-WÁRAY) luyá, luy-a (WELSH) sinsir; (WOLOF) dinjar, dinjer, jinjer, ginger; (XHOSA) ijinja; (YALUNKA) nyakhamu-na; (YIDDISH) ingber; (YORUBA) atalẹ̀, ewere; (ZARMA) citta; (ZULU) ujinja.

Division Anthophyta (Magnoliophyta)

Class Diocotyledons (Magniliopsida)

21. Acanthaceae

21.1. *Justicia insularis* T. Anderson (EP) young shoots, leaves, roots[S]; (P) R,C.

(ASSAMESE) bahok tita; (CHINESE, MANDARIN) jou yeh jue chuang; (CHINESE, WÚ) jue chuang; (ENGLISH) tettu; (ÉWÉ) dzivi-dzivi; (HUNGARIAN) akantusz; (ITALIAN) carmantina; (MENDE) tɛvɛ, tɛyɛ; (PORTUGUESE) justícia; (URDU) teto; (WAAMA) muntun; (YORUBA) tettu, esiri.

21.2. *Rungia klossii* S. Moore (EP) leaves; (P) C.

(CHINESE, WÚ) hai er cao; (DANISH) rungia; (DUTCH) rungia; (ENGLISH) rungia; (FRENCH) rungia; (GERMAN) Rungia; (ITALIAN) rungia; (PORTUGUESE) rungia; (SPANISH) rungia; (TOK PISIN) kenkaba, tani, moku, ainbe.

22. Aizoaceae

22.1. *Mesembryanthemum crystallinum* L. (EP) leaves; (P) R.

(AFRIKAANS) ysplant, braksiaai, noeta, nuta, soutslaai, volstruisslaai; (ARABIC) napat althalg; (ASSAMESE) mesembryanthemum; (CHINESE, MANDARIN) bing hua; (CHINESE, WÚ) bing hua; (CHINESE, YUE) bing ye ri shong hua; (CZECH) kosmatec krystalový, kosmatec křištálový; (DANISH) is-middagsblomst; (DUTCH) ijskruid, ijsplantje; (ENGLISH) ice plant; (ESTONIAN) kristaljas keskpäevalill; (FINNISH) jääpäivikki; (FRENCH) ficoïde, ficoïde cristalline, ficoïde glaciale, herbe à la glace; (GERMAN) gemeines Eiskraut, Eisblume; (GREEK) mesimvriathemo to krystallino; (HEBREW) ahal ha-gevishim, ahal hagvishim; (HUNGARIAN) kristályvirág; (ITALIAN) erba cristallina; (LATVIAN) pusdienziedes; (LITHUANIAN) krištolinė pluoštagėlė; (MALTESE) cristallina; (NORWEGIAN) is-middagsblomst; (POLISH) krysztłka lśniąca; (PORTUGUESE) folha de gelo, flor de gelo, erva de orvalho; (SPANISH) escarchada, algazul; (TURKISH) buz çiçeği, buz otu; (VLAAMS) ijskruid; (WELSH) plawhigyn-ia.

22.2. *Tetragonia tetragoniodes* (Pallas) O. Kuntze (EP) tender shoots and leaves; (P) C,R.

(AFRIKAANS) Nieu-Seelandse spinasie; (ARABIC) is banekh; (ASSAMESE) New Zealand spinach; (BULGARIAN) Novozelandski spanak; (CHINESE, MANDARIN) fan xing, yeung po ts'ai; (CHINESE, WÚ) fan xing; (CHINESE, YUE) fan xing; (CZECH) novozélandský špenát, čtyřboč rozložita, čtyřboč; (DANISH) Newzealandsk spinat; (DUTCH) Nieuw Zeelandse spinazie, Neuseeland-spinat; (ENGLISH) New Zealand spinach, warrigal cabbage; (ESTONIAN) uus-meremaa spinat; (FINNISH) lamopinaatti; (FRENCH) êpinard de la Nouvellle Zêlande, tétragonne, tétragonne cornue, êpinard d'été; (GERMAN) Neuzeeländer Spinat; (GOURMANCHÉMA) epana; (GREEK) spanaki Neas Zilandias; (HEBREW) revuah, revuʻa seru'a, tered new-zelandy; (HINDI) katela pulak; (HUNGARIAN) újzélandi spenót, iy-zélandi spenót; (ICELANDIC) nysjalenskt spinat; (INDONESIAN) bayam new zelan, kabak; (ITALIAN) tetragonia della Nuova Zelanda, spinacio di Nuova Zelanda; (JAPANESE) tsuruna; (KANNADA) chikesoppu; (LATVIAN) jaunzēlandes spināts, jaunzelandes spināti; (LITHUANIAN) Naujosios Zelandijos špinokas; (MACEDONIAN) novozelandski spanak; (MALAY) kabak; (MALTESE) spinaci ta' Nuova Zelanda; (MAORI) kookihi, rengamutu, rengarenga, tuutae-ika-moana; (MORISYEN) zepinar; (NORWEGIAN) bladtetragonia, Nyseelandsk spinat, nyzealandspinat; (POLISH) szpinak nowozelandzki; (PORTUGUESE) espinafre da Nova Zelândia, tetragónia; (ROMANIAN) spanak de Nova Zeelandă, spanacul de Nova Zeelandă; (RUSSIAN) novozelandskij špinat; (SERBO-CROATIAN) Novozelandski špinat; (SLOVAK) špenát Novozélandksý; (SLOVENIAN) Novozelandska špinača; (SPANISH) espinaca de Nueva Zelandia; (SUNDA) kabak; (SWEDISH) Nyzeeländsk spendt; (TAGALOG) ispinaka; (TURKISH) solmazçiçek, Yeni Zelanda ispanaği; (URDU) katela pulak; (VIETNAMESE) d[eef]n t[aa]y, gi[eef]n t[aa]y; (VLAAMS) vierhoruk, Nieuwzeelandse spinazie; (WELSH) pigoglys.

23. Amaranthaceae

23.1. *Alternanthera philoxeroides* (C. Martius) Griseb. (EP) young tops; (P) R,C.

(ASSAMESE) matikanduri sak; (CHINESE, WÚ) xi han lian zi cao; (CHINESE, YUE) pang qi ju, kong xin xian; (DANISH) papegocjoblad; (ENGLISH) alligator weed; (FRENCH) alternanthère; (GERMAN) Pagageienblatt; (HEBREW) beetzan; (INDONESIAN) tolod, jukut kriminil; (ITALIAN) alternantera; (JAVANESE) krokot; (KANNADA) dodda honagone, mosale honagone; (LAHU) a^{-} hpe$_{\wedge}$ g'aw$^{\vee}$; (NEPALI) jala jambu; (PORTUGUESE) alternantera; (SPANISH) alternantera, lagunilla; (SWAHILI) mchicha; (THAI) phakpet-nam; (WOLOF) mbum bu gôr.

23.2. *Alternanthera sessilis* (L.) R. Br. ex DC. (EP) young tops; (P) R,C.

(AJA) gomi; (AKAN) abanase-abanase; (ASSAMESE) matikanduri sak; (BENGALI) sanchi; (CHINESE, MANDARIN) mau-t'ien-hsing; (CHINESE, WÚ) lian zi cao; (DUTCH) garnalenmoes; (EDE NAGO) shawewe; (ENGLISH) sessile alternanthera; (FILIPINO) bunga-bunga; (FON) agwe ayigbaton; (FRENCH) alternanthère sessile, magloire; (GBE-VHE) tosre; (GBE, WACI) tɔslwɛn; (GEN) aditi; (GUJARATI) jalajambo, pani ni bhaji; (GUN) agwe ayigbaton; (HEBREW) beetzan machseef; (HINDI) jal jimbua, guru bhaji, kanchari; (INDONESIAN) krokot, krema, keremek, daun tol okeremak, daun kermak, bayam putih, daun rusa, sayur udang; (ITALIAN) alternantera sessile; (JAPANESE) tsuro-no-geitô; (JAVANESE) bayem kremah, matean, keremi, keremah; (KABIYÉ) sowatou; (KANNADA) honagone soppu, honagonne, nasarchang, honagonne soppu; (KISI, SOUTHERN) mého; (LAHU) a^{-} hpe$_{\wedge}$ g'aw$^{\vee}$; (MALAY) bayam pasir, bayam tanah; (MALAYALAM) koluppa, ponnaganni keeray; (MANINKAKAN) missinikoumbré; (MANU'A SAMOAN) vao sosolo; (MAORI) maahuri, nahui; (MARATHI) kanchari, laanchari; (MENDE) *n*datawulo; (MORISYEN) bred ambalaz; (NEPALI) bhiringi saag; (PORTUGUESE) alternantera séssil; (SAMOAN) vao sosolo; (SANSKRIT) mathsyaakshi, gidjihra, giojihra, matsyaksi, lonika, meenaakshi, putthur; (SPANISH) alternantera sesil, sanguinaria, santoma cimarrona; (SUNDA) keremek, daun tolod, tolod, letah hayam; (SWAHILI) mchicha; (TAMIL) kodupai, ponnanganni-keeray, ponnagannikkirai, ponnanganni, ponnaankannikkeerai; (TELUGU) madanaganti, ponnaganti koora, ponnaganta kura, ponnagantikura; (TONGAN) fisi'i'ano; (TULU) ponnāṅkaṇṇi, vaṇaṅṅana; (URDU) kanchari shalanchi; (YOM) dafaar gbanda; (YORUBA) shawewe.

23.3. *Amaranthus* spp. (EP) tender shoots, leaves and sprouted seeds; (P) C,R.

A number of languages have specific names for each species. When available, the species is indicated with a numerical superscript. Names without a numerical designation represent either generic terms or when a species designation was not available. Cultivated species include: *A. blitum* subsp. *oleraceus*[1], *A. caudatus*[2], *A. cruentus*[3], *A. dubius*[4], *A. hybridus*[5], *A. mangostanus*[6], *A. retroflexus*[7], *A. spinosus*[8], *A. sylvestris*[9], *A. tricolor*[10], and *A. viridis*[11]. *A. lividis* and *A. oleracea* are synonyms of *A. blitum* subsp. *oleraceus*[1]. Dialects of Luyia (Bukusu[z], Marama[y], Tachoni[x]).

(AFRIKAANS) marog, amaranthus, hanekam, gewone misbredie, wilde-ertappel, hanekam, transvaalbossie, kalkoenslurp, wildegrenadella; (AJA) bura bura ba; (AKAN) nantwi bin[8], sraha nkasee[8], sraha nsoe[8]; (AKAN-AKUAPEM) asibe[8], esibe[8], nantwi bin[8], sraha nkaseɛ[8], sarha nsoe[8], ɛfan[5], fan[5]; (AKAN-ASANTE) famufuéué[11], fankokoré[11], ɛfan[11], efai[1], ɔtofammẽn; (AKLANON) kulitis, kulyapa, hagok-hok, tagughug; (AKWA) lendebê[1]; (ALBANIAN, TOSK) nenë; (AMHARIC) lambut, aluma; (ARABIC) napat orguani; (ASSAMESE) khutura sak, lal sak, ranga sak; (BAATONUM) tɛtɛ; (BADAGA) ki:re, nela ki:re, muḷ ki:re; (BAMANANKAN) boron[5], boron gbé[5], boron ulé[5], moron[5], moron dyé[5], ngoroba blé[8]; (BAMILEKE) ndjap; (BAOULÉ) klô afani; (BELARUSAN) aksamitic bely, aksamitic hvastaty; (BEMBÉ) bari[1]; (BENGALI) lal shag, natey shag, shada natey, anantamul[3], cholai[3], chuamarasa[3], ban note[1,11], ban natey, bathu[5], chuko[5], chuka, gobra note[1], note sriak[1], chanpanoteya[10], lal sag, shadanatey, dengo danta, kantamaris[8], kantanote[8], kanta notya[8], dengua[10], lal-shak[10], raknashak[10]; (BEROM) enɛ̀p[5]; (BÉTÉ) lala gnini; (BICOLANO, CENTRAL) kilitis, kadyapa, kudyapa, uray; (BINUKID) akum, belesagan; (BONTOC CENTRAL) baoan, tadtad, bawan, bawan, pakhat, segang, sibsibit, sibsifit, sisiitan, gogowat, popowegan, sakit, tagtagu; (BUHID) borarad, kudyapa, makma madugi; (BULLOM SO) bɛntisɔk-lɛ[11]; (BULU) folong, foloñ; (BURMESE) hyin-nu-nwe; (CEBUANO) harum, colitis, kulitis, tilitis, halom, harum, haum, iting-iting, kalitis, koyapa, kudyapa; (CHINESE, GÀN) xin cai; (CHINESE, MANDARIN) xian cai; (CHINESE, MIN BEI) xing cai; (CHINESE, MIN NAN) hēng-chhài; (CHINESE, WÚ) mi xian, fan sui xian, lii sui xian, xian cai; (CHINESE, YUE) xian cai; (CHUWABU) nyewe; (CRIOULO, CAPE VERDE) bredo, rabo-degalo, bredo-matcho; (CRIOULO, UPPER GUINEA) bredo[11], bredo femea[1]; (CZECH) laskavec; (DAGAARE, SOUTHERN) fokiolo; (DAGBANI) boboroa[11]; (DANGME) anago mio[8]; (DANGME-KROBO) sɔbuɛ[5]; (DANISH) amarant; (DENDI) aleefɔ; (DIGO) mchichá; (DUTCH) amarant; (EDE NAGO) tɛtɛ n'la, tɛtɛ kereke; (EDO) èbáfọ̃[5]; (EFIK) ìnyáñ áfiá[5], okpu-utere[4], ìnyàñ[1]; (ENGLISH) amaranthus, pigweed, tampala, love-lies-bleeding[3,10], tassel flower[3], prince's feather[5], spleen amaranth[5], red amaranth[5], tumble pigweed, livid pigweed, smooth pigweed, African spinach[3,11], Indian spinach[3], prickly amaranth[8], spiny pigweed[8], thorny amaranth[8], spiny amaranth[8], Chinese spinach[10], fountain plant[10], garden amaranth[10], weeping-willow leafed amaranth[10], Joseph's coat[10], green amaranth[11], local tete[11], trailing amaranth[1], purple amaranth[1], wild blite[1,7], careless weed[7], red root amaranth[7], rough green pigweed[7]; (ÉWÉ) fotété, madzẽ[5], matoŋui[8], sɛnutsõe[8]; (EWONDO) zom bikabeli, ebienlen, folon; (FIJIAN) churaiya; (FILIPINO) kulitis; (FINNISH) revonhäntä; (FON) fɔtɛtɛ, tɛtɛ moto; (FRENCH) amaranthe, amarante, amarante blanche, amarante sauvag[1], amarante verte[11], épinard vert[11], épinard du Cong[11]; (FULFULDE) adjaka, hako ndiam; (FULFULDE-ADAMAWA) haako ndiyam[8], haako-ndiyam[11], rukuɓuho[11]; (GA) glin, awusaŋaei, soaga-nmei[8], slaha- ŋmei[8]; (GADDANG) allay; (GAELIC, MANX) lus ny gfaih; (GALICIAN) bledo, bredo; (GBE-VHE) ama, matonui[8]; (GBE, WACI) tɛtɛma, nangbanjɛn; (GEN) fɔtɛtɛ, tɛtɛ wuli wuli; (GERMAN) Amarant, Fuchsschwanz, weißer Amarant, bleifarbiger Amarant, bastard Amarant; (GIKUYU) terere; (GOLA) mando[5], n-dolo[5]; (GOURMANCHÉMA) egéanchio; (GREEK) amaranthos; (GUJARATI) tandaljo, tandalja, parwar, dambho, kantalo dambho, kantalo dhimdo[8], kantanu dant[8], chuko[3], rajagaro[3,5], rajgaroo, lambho, adabav[1], tandalajo[1], tandulja[1], tandaljo[8], cholai, dhinmado[11], adabau lambho[10]; (GUN) fɔtɛtɛ, tɛtɛ moto; (HANUNOO) gindayuhan barusak, karubadha, kudyapa, kulitis tagabas; (HAUSA) alayafou, rukuɓu[8,11], cilli[8], námíjìn gaásàyaá[8], námíjìn-zaákiíbánzaá[8], malan-kotshi[11], námíjìn-gaásàyaá[11], námíjìn-zaákií-bánzaá[11], rùb-dà-túkúnyá[11], taatsunyaa[11], zaákiíbánzaá zarangade[11], lábàshií[11], malan kochi[11]; (HEBREW) yarbuz, yarbooz; (HINDI) chaulai[3], chuamarsa[5], ganhar[5], kalgaghasa[5], cholai[8], kaantenatia[8], adharanga[10], chowlaayi[10], laalnatiya[10], laalsag[10], raajkiri[10], marasa[1], vavad[1], chaulai sag, ram dana, lal sag[10], rajgira[3], ramdana[3], chua[5], chaulai, choli, chumli sag, lalnatiya[10], rajgira, chaulaie, ramdana, marsa, chotti chaulai[1], janumarak, kantabhaji[8], kantanatia[8], kanta nutiya, sadabatuta[1], kataili-chaulai[8]; (HINDUSTANI) cauraiya; (HMONG DAW) txhuv ntuj; (HOPI) poosiw, komo, wiiwa; (HUNGARIAN) disznóparéj, paréj; (IBALOI) baudan; (IBANAG) sitan; (ICELANDIC) skrauthali; (IFUGAO) alayon, alayyon, alayyon an mumbola, alayyon di tindalo, galayyong an napagatan, pikakku; (IGBO) ìnìne-ógwū[8], nnuno uku[8], ìnìnè ójíō[11], memomene[11], ìnìnè ófīá óló[11], ìnìnèḿbèkwú[11], ìnìnèḿmé[11], ìnìnè ḿmē[1]; (ILOCANO) jyabtibg, kalunai, kuanton, kalunay, kuwanton, kuwantung, siitan; (INDONESIAN) bayam cabut, bayam tur, bayam liar, bayem, bayam, bayam selasih[5], bayam itik[1], bayam monyet[1], bayam putih[1], bayam duri[8], bayam hutan[8], bayaduri[8], bayam merah[10], bayam gelatik[10], bayam kakap[10]; (ITALIAN) amaranto; (ITAWIT) nasi, nasit, ollay; (IVATAN) kilitis; (IZON) ìnìnáịn[8], ìnìnàín[11], níníyà[11]; (JAPANESE) hiyu, inu-hime-shiro-biyu, senninkoku, ha-geitô, honaga-ao-geitô, inu-biyu, seiyô-ha-geito, ao-biyu; (JAVANESE) bayem keyong[5], bayem abang[5,10], bayem lemah[1,10], bayem menir[1], bayem itik[1], bayem dempo[1], bayem sapi[1], bayem raja[8], bayem eri[8], cikron[8], bayem sekul[10], bayem ringgit[10], bayem siti[10], bayem abrit[10], bayem kakap[10]; (JOLA-FONYI) é badbaday, ébayibadjay[11]; (JULA) brombou; (KANKANAEY) adtad, bawan; (KANNADA) chilikesoppu[3], keelu harive[3], keere soppu[5,11], raajgeer[3], raamadana sirukeere[3], mullu keere soppu[8], mullu dantu[8], mullu gorakoochi[8], mullu harive soppu[8], chelakeerae soppu[10,11], dagglarive soppu[11], dagglisoppu[11], kempu dantu soppu[1,10], kadu padayala, chilikaesoppu, chilikeerae soppu, chikkere soppu, chilikiraesoppu[11], kirae soppu, daglisoppu[11], mulladantu[8], mulluharivesoppu[8], mullukiresoppu, raajanea[5], rajakere[5], dantina soppu[10], chikkeere soppu[10], dantu[10], dantu soppu[10], geerakasaale[10], keeraka saale[10], kempudantu soppu[10], kempuharive[10], naeneyakki gida[10]; (KANURI) lekir[11]; (KASHMIRI) bustān afrōz[8], ganahār[10], lisa, mawal[8]; (KHMER, CENTRAL) phtii; (KISSI, NORTHERN) fondulo[1,5], fondoulo[1,5], fondulo sama[5], fondoulo saman[5], fondulo sankura[5], fondoulo sankoura[5], fondolo-ungu[8], fon'dolo-oungou[8], fondolo-bɛndoŋ; (KONKANI) dentto[8]; (KONO) bɔbɔɛ[5], bɔbɔ-wa[5], sɔndi[8,11]; (KRIO) grins[5], grin[5], red green[4]; (LAHU) zuh^{-} mi$^{\vee}$[2], hkui$^{\vee}$ chu$^{\vee}$g'aw$^{\vee}$[8]; (LAO) hôm; (LATVIAN) amaranti; (LITHUANIAN) burnotis; (LOKO) hɔndi[5]; (LOMWE, MALAWI) naminya; (LUO) buga; (LUYIA) lidodo[x], emboka[z], litoto[z], tsimboka[y]; (MACEDONIAN) štir; (MAGUINDANAO) akum, bilanamanut; (MAITHILI) gen-hārī, rattā sāg; (MAKHUWA) injewe, enywe, ntchitcha; (MALAY) bayam, bayam hijau, bayam merah; (MALAYALAM) mullanchira[8,11], ceruchira[8], cerchiraa[11], lhanamat[11], cheera[10], mulla-cheera[8]; (MALIMBA) gbuhaŋga[5], bɔren ɔre-ba[5]; (MALTESE) denb id-dib; (MANDING-MANDINKA) bóró[5], bɔrəbɔrɛ[5], bulunonima[8], boromonema[11], bɔrɔ̃[5], bɔrɔ̃-m-bɛbɛ[5]; (MANINKAKAN, EASTERN) boron[5], moron[5], bɔrəbɔrɛ[5], bɔrɔ̃[5], bɔrɔ̃-m-bɛdɛ[5], boromonema[1]; (MANINKAKAN, WESTERN) boron gbé[5], boron oulé[5], boulounonima[8]; (MANOBO) ogom; (MANU'A SAMOAN) autasi; (MARANAO) birdo, kadyapa, katebek, morado, pagang a manok, dila a manok; (MARATHI) raajagira[3], chanalai, kantebhaji[8], thanduliya[8], chavalaayi[10], ranmaath[10], thambada maath[10], thandulajaa[10], bhaji[1], dant[1], maatha[1], pokala[1], tambadamaath[1], parwar, pokal, tambda math[10], kaholabhaji[3], kahola-bhaji[3], rajgira[5], tambdamath[1], lhanamat[11], math[10], ranmat[10], pokla, chavlai, kantemath[8], tandulja[8]; (MARWARI) rajgira; (ME'EN) katila; (MENDE) hɔndi[4,5,8,11], hɔndi-gbogboil[5], hɔndi-gowuli[5], hɔndi-guli[5], ta-hɔndi[8,11]; (MÒORÉ) lisambo, lisama, bolinbola; (MORISYEN) bred malbar; (NEPALI) latte; (NEWAR) cabakan; (NGUNDI) bari[1]; (NIUE) katule; (NORWEGIAN) amarant, buskamarant; (ORIYA) lal khada[10], balal sag, ba lal saga, prakar khada, kosala saga[1], leutia[1], leotia, champa leutia, kantamarisha, kantaneutia[8]; (PALI) appamāriso, taṇḍulīyo, taṇḍuleyyo; (PAMPANGAN) ayantoto, aynatoto; (PANJABI, EASTERN) parmal, chulai, lal sag[10]; (POLISH) szarłat zwisły; (PORTUGUESE) amarantos, bredos, carurú; (PULAAR) boroboro[5], bórborò-dóròò[11], bóròborò-déo[11], nafa-nafa[11]; (QUECHUA, CUSCO) achis, achita, ratago, ataco-casha; (RONGA) m'boa, tseque, m'bowa; (SAMBAL, BOTOLAN) aya, aya mantug, aya-kabayo, tiliti; (SAMOAN) 'au tasi, totoe; (SANSKRIT) raajagiri[3], raajaadri[3], raajashakini[3], thanduliya[8,10,11], alpamaareesha[8,10], bahuveerya[8], bhandira[8], ghanasvana[8], granthila[8], kandaka maareesha[8], kandera[8], meghanada[8], pathyashaka[8], sphurjathu[8], sushaka[8], savnitavhaya[8], thandula[8], thandulanama[8], thanduleraka[8], thandulibeeja[8], vira[8], vishaghna[8,10], raama seethalika[10], kumaarajeeva[10], maareesha[1,10], meghanaad[10], bashpaka[1], maarsha[1], rajagiri[3,5], rahadri[5], alpamarisha[1,8], marisha[10], marish[10], mekanada[10], tandula[8], tanduliya[10,11], vishaghana[11], rajashakini[5]; (SANTALI) gạndhạri aṛaḱ, jạmi, aṛaḱ; (SENA) bonongue; (SENOUFO, CEBAARA) dasouvigui, dasouyigui; (SERBO-CROATIAN) šćir, bijeli, zeleni, križani; (SERER-SINE) dahdir, dahdir gor[8]; (SHONA) mboamanga, mboa, mowa; (SINHALA) thampala, mchicha; (SLOVENIAN) láskavec; (SONGHAY) walia bocio[11]; (SPANISH) amarantos, bledos blancos; (SUBANON) pugyapa; (SUNDA) senggang bener[5], senggang beureum[5], senggang bodas[5], bayem salasi[5], bayem lemah[1], bayem hayam[1], bayem bodas[1], bayem tanah[1], bayem tai sapi[1], senggang itik[1], senggang cucuk[8], bayem cucuk[8], senggang beureum[10]; (SWAHILI) mchicha; (SWEDISH) amarant; (TAGALOG) kulitis, uray, bayambang, colitis, oort, oray, uri, halom, halon, halon, kalumpay, kilitis, kolitis, kolitis, kolitis, kuletos, kulitis, kulitis-na-babae, oray, uray, uray, bonga-bonga, bunga-bunga, talilong; (TAGBANWA) kaliyaba, kanriyuwan, kanroyuwan, samsam; (TAJIKI) amarant; (TAMIL) pungikeerai[3,5], mullukeerai[8], keerai[10], serikeerai[10], thandukkeerai[10], kuppaikkeerai[11], keeraithand[1], thankeerai[1], pungikeerai, sirukeerai[3], kaninee[5], mulaikeerai[5], malaikkeerai[8], mullukkirai, mud-keerai[8], aarumatha-thandu[1], aarumaatha thandu[1], aoraumaathathandu, kiraitandu, arai keerai[10], mulakkeerai[10], siru keerai, kaattu siru keerai, thandu keerai[10], kappai keerai[11], sinna keerai[11], sira keeraipullu[11]; (TAUSUG) gitin-giting, taikuda, giting-giting, haIunapa; (TEKE) ambouo-babié[1]; (TELUGU) netakoora[3], mulluthota koora[8], erramullugoranta[8], mundlatotakoora[8], nalladoggali[8], dantu koora[5,10], erra doggali koora[10], karnulu[10], perugu thotakoora[10], chilaka thota koora[11], erratotakura[1], peddathotakura[1], thellatotakura[1], thotakura[1], erratoakura[10], keikera, tota kura, totakura[10], peddatotakuru[1,3], pedda totakura, doggalikura[10], ettamulugoranta, mullatotakura[8], mundla kura[8], chailaka thota-kura[11]; (THAI) phakkhom-suan; (THEMNE) ka-bonthila, *a*-bonthila[4,8], *a*-niŋkuna[8,11], *an*-senduku[11], *a*-senduku[8]; (TIV) kashia[5], aleifu buter[10]; (TOK PISIN) aopa; (TONGAN) longolongo'uha, tupu'a; (TULU) padpe[10], nekkilụpadpè, doggoli padpè, padpè, mulḷapadpè[8], mulḷupadpè[8]; (TURKISH) horozibiği, çiçek, amarant; (TURKMEN) hulpa; (TWI) efan, srahasoe, nanćwibin, srahan kasɛe; (URDU) choulai, lal sag; (UZBEK) gultodjikhuroz; (VAI) lolo[5]; (VENDA) vowa; (VIETNAMESE) rao den do, rau d[eef]n; (VLAAMS) amarant; (WÁRAY-WÁRAY) harum, talitis, kulitis, sumpa, iginsumpa, tilites; (WELSH) math o blawhigyn; (WOLOF) ubum i ker, mboum bou gor[8]; mboum i keur[5], mbum bu gor[8], mbum i gor[8], mbum bu digèn[1,11], mbum bu levet[11], mbum i kör[11]; (YOM) bɛɛ kumpeeyu; (YORUBA) tɛtɛ n'la, tɛtɛ kereke, tẹ̀tẹ̀[4,11], olorungbin[4], ẹfọ tẹ̀tẹ̀[8], tẹ̀tẹ̀ ẹ̀gún[8], tẹ̀tẹ̀ eléègún[8], tẹ̀tẹ̀ atẹle danji[11], tẹ̀tẹ̀ àtètèdáyé[11], tẹ̀tẹ̀ kékeré[11]; (ZARMA) cappata[8], tchapata.

23.4. *Celosia* spp. (EP) leaves and tender shoots; (P) C.

Several languages have individual names for each species. When available, the species is indicated with a numerical superscript. Names without a numerical designation represent either generic terms or when a species designation was not available. Principal cultivated vegetable species are *C. argentea* var. *cristata*[1] and *C. trigyba*[2].

(AFRIKAANS) hanekam; (AJA) sɔma; (AKLANON) hagok-hok, tagughug; (ARABIC) katifa orf el-deek; (ASSAMESE) kukura joba; (BAMANANKAN) ngo ban ku[1]; (BENGALI) swetmurga, swetmurgha, safed-morugphul, lalmurga, huldimurga; (BICOLANO, CENTRAL) palung-palungan, papaknongan manok; (CEBUANO) kudyapa, padong-padongan; (CHINESE, MANDARIN) ji guan hua, quing hsiang; (CHINESE, WÚ) qing xing, ji quan hua; (CHINESE, YUE) ji guang hua; (CZECH) celosie, kohoutí hřeben; (DAGAARE, SOUTHERN) kankir; (DANGME-KROBO) tɔbɔ[1]; (DANISH) hanekam; (DUTCH) hanekam; (EDE NAGO) shɔkɔt; (EDO) àbórrà[1]; (ENGLISH) cockscomb[1,2], silver-spiked cockscomb[1], quail grass[1], celosia[1], Lagos spinach[1], soko[1], cock's comb[1,2], silver spinach[2], woolflower[2]; (ESTONIAN) robirohi; (EWONDO) nkan; (FARSI, WESTERN) goal halva; (FILIPINO) palong-manok; (FINNISH) kukonjarja; (FON) sɔma; (FRENCH) célosie[1,2], célosie argentée[1], crête de coq[1,2]; (GA) too lilɛi, shɔkɔ; (GAELIC, MANX) lvs ny meeyl; (GBE, WACI) yakiporiduure; (GEN) avunvɔ; (GERMAN) Hahnenkamm; (GOURMANCHÉMA) etiantio; (GREEK) kilosia; (GUJARATI) mora shikha[1], lambadi, lapadi; (HANUNOO) kalumbad, katumbad, luy manuk, palung-manok; (HAUSA) farar áláyyafóo[1], riímií[1]; (HEBREW) kelosia, celosia; (HINDI) sufaid murga, sarwari, sarwani, lal murghka[1], kokan[1], pile murghka[1], silara, debkoti, survari; (HMONG-DAW) zaub luag qab; (HMONG-NJUA) paaj qab ib; (HUNGARIAN) magyarparéj; (ICELANDIC) skrautkambur; (IGBO) shọkọtọ-yọkọtọ; (ILOCANO) lofoy, sansandok, taptapingar; (INDONESIAN) boroco, cuca, jengger ayam, bayam ekor belanda, bayam daun, bayam petik; (ISNAG) laylaya; (ITALIAN) cresta di gallo; (ITAWIT) allalaya; (IVATAN) taruyin-pabo; (JAPANESE) no-geito, keitó; (JAVANESE) bayem cenggeng, jengger ayam, jengger rumpuk; (KANNADA) amaesoppu, anne soppu; (KONKANI) kurdu, kurduqui bagi; (KRIO) shɔkɔtɔ-yɔkɔtɔ[1]; (LAHU) g'a^ na_ tzuh ve^, g'a^ hpfuh tzuh ve^; (LATVIAN) sekstainā celozija; (LITHUANIAN) celiozija; (MAKHUWA) enyewe, n'hampara; (MALAY) balung ayam; (MALAYALAM) ceruchira; (MALTESE) ghallet is-serduq; (MARANAO) pagang-a-manok; (MARATHI) kurdu, kurada, mayurshikra[1]; (MENDE) kikpɔi[1], kipɔi[1], kpange[1], yɔgɔtɔ[1]; (MÒORÉ) bougma, zikoom, nikièm-zu-belog; (NEPALI) sarwari saag; (NORWEGIAN) hanekam; (NYANJA) ndangale; (ORIYA) gangachulia; (PANGASINAN) sumpi-sumping; (PANJABI, EASTERN) sarwali, sil, sarpankha, chilchil, kûkurh kulgi, kûku kulga; (POLISH) grzebionatka właściwa, celozja; (PORTUGUESE) crista de galo[1], celosia[1], amaranto branco[1], bóróbóró déo[2]; (SAMOAN) māmoe; (SANSKRIT) vitunna[1], mayura-shikha[1]; (SANTALI) sirgiṭ aṛak̇; (SHONA) nyamugjakara; (SPANISH) cresta de gallo; (SUBANON) mandyana tubig; (SUNDA) baroko, sangseri, jawer hayam, jawer kotok, bayem biludu, rara ayam; (SWAHILI) mchicha[1,2]; (SWEDISH) tuppkam; (TAGALOG) buntot-pusa, kadayuhan, kindayohan, palong-manok, palung-palungan; (TAGBANWA) tagogtog, tagughug; (TAJIKI) toji khurus; (TAMIL) panna, pannakeerai; (TAUSUG) daling-manuk, parang; (TELUGU) gurugu, gulugkura, panchechettu, kodijuttutotakura[1]; (THEMNE) *an*-lami, *an*-kɔkɔrɔ, ɛ-kẹɔkɔrɔ; (TIV) igyór[1]; (TURKISH) selezya, horozibiği; (TURKMEN) täjihoraz; (TWI) nkawudua; (URDU) sarwari, sarwali, lal murghaka, pile murgha; (VLAAMS) hanekam; (WÁRAY-WÁRAY) surangga, surang-surang, lagiknay; (YOM) nyaʊu nyanµ; (YORUBA) shɔkɔt[1], ṣọkọ[1], sọkọ yọkọtọ[1], sọkọ yọkọtọ sọkọtọ[1], pupa, ajefowo, alari funfun, alari, pupa, osun[1].

24. Apiaceae (Umbelliferae)*

24.1. *Angelica archangelica* L. (EP) tender shoots and leaves; (P) C,R.

(AFRIKAANS) engelwortel, angelica; (ARABIC) hashesha almalak; (ASSAMESE) angelica; (BELARUSAN) parasonistaja, dudnik lekavy; (CHINESE, MANDARIN) bai zhi; (CHINESE, WÚ) bai zhi, dang qui; (CHINESE, YUE) yuan dem qui; (CZECH) andělika lékařská, děhel; (DANISH) angelik, kvan; (DUTCH) engelwortel; (ENGLISH) angelica; (ESTONIAN) kikkaputk; (FARSI, WESTERN) sonbol; (FINNISH) väinönputki; (FRENCH) angélique; (GAELIC, MANX) lus yn ainle; (GALICIAN) anxélica; (GERMAN) Engelwurz, Brustwurz, Angelika; (GREEK) aggeliki, archaggeliki; (HUNGARIAN) angyalgyökér; (ICELANDIC) hvömm, ætihvönn; (ITALIAN) angelica, arcangelica; (LITHUANIAN) šventagaršvė; (MACEDONIAN) obična angelika; (MONGOLIAN) tsoorgono, balchirgana; (NORWEGIAN) kvann, fjellkvann, strandkvann; (POLISH) arcydzięgiel, litwor lekarski; (PORTUGUESE) angélica, erva do Espírito Santo; (ROMANIAN) anghelică, angelică; (RUSSIAN) dudnik, angelika; (SERBO-CROATIAN) andjelika; (SPANISH) angélica, hierba del Espíritu Santo; (SWEDISH) angelika; (TURKISH) melekotu; (VLAAMS) grote engelwortel; (WELSH) llys yr angel.

24.2. *Angelica keiskei* (Miq.) Koidz. (EP) tender shoots and leaves; (P) C.

(AFRIKAANS) Japanese engelwortel; (ASSAMESE) Japani angelica; (CHINESE, MANDARIN) ming ri cao; (CHINESE, WÚ) ming ri cao; (DANISH) ashitaba; (ENGLISH) Japanese angelica; (GREEK) aggeliki iaponiki; (JAPANESE) ashitaba; (MACEDONIAN) japonska angelika; (MONGOLIAN) tsoorgono, balchirgana; (POLISH) arcydzięgiel japoński; (VLAAMS) Japanse engelwortel.

24.3. *Anthriscus cerefolium* (L.) Hoffm. (EP) leaves; (P) R,C.

(AFRIKAANS) kerwil; (ARABIC) makdoonis afrangyy; (BELARUSAN) markonik tsymanechnalisny; (BULGARIAN) kervel; (CATALAN-VALENCIAN-BALEAR) cerefoll; (CHINESE, MANDARIN) xi ye xinag cai, hsüch wei ts 'ai; (CHINESE, YUE) xie wie cai; (CZECH) kerblík třebule; (DANISH) kørvel, have-kørvel; (DUTCH) kervel; (ENGLISH) chervil; (ESTONIAN) harakputk, varesputk; (FARSI, WESTERN) jafari farangi; (FINNISH) maustekirveli; (FRENCH) cerfeuil; (GAELIC, MANX) constag; (GALICIAN) pirifol; (GERMAN) Garten-Kerbel, Garten-Kerbelkraut; (GREEK) skatziki selino; (HEBREW) seyagit; (HUNGARIAN) zamatos turbolya, illatos turbolya; (ICELANDIC) garđakerfill; (ITALIAN) cerfoglio, cerefoglio; (LATVIAN) kārvele; (LITHUANIAN) karvelinis builis; (MACEDONIAN) obična krasnica; (MONGOLIAN) khashilj; (NORWEGIAN) kjørvel, hagekjørvel; (POLISH) trybula ogrodowa; (PORTUGUESE) cerefólio, cerefolho; (ROMANIAN) asmacius; (RUSSIAN) kervel'; (SERBO-CROATIAN) krasulica, krasuljica, obična; (SLOVAK) trebul'ka pravá siata; (SLOVENIAN) krebuljica; (SORBIAN, UPPER) korbjel; (SPANISH) cerefollo, cerefolio, salsa de pastor, perifollo, perfollo; (SWEDISH) körvel; (TURKISH) bahçe frenk maydanozu; (TURKMEN) galyň ýaprakly kupyr; (VLAAMS) kervel; (WELSH) y-berllys.

24.4. *Apium graveolens* var. *dulce* (Mill.) Pers. (EP) petioles, leaves[s]; (P) R,C.

(AFRIKAANS) seldery; (ALBANIAN, TOSK) selinë; (ARABIC) karafs; (ASSAMESE) celery; (AZERBAIJANI, NORTHERN) kərəviz; (BADAGA) nu:l ko:l; (BENGALI) chandani, chanv, randhwni; (BRETON) ach; (BULGARIAN) drăžkova celina; (BULU) seleri; (BURMESE) ta-yok-nannan; (BELARUSAN) selderay karanjaplodny; (CAJUN FRENCH) celery; (CATALAN-VALENCIAN-BALEAR) api; (CEBUANO) kinsay; (CHINESE, GÀN) xiang qin, han qin; (CHINESE, HAKKA) k'iôun- ts'ói; (CHINESE, MANDARIN) qin cai, ch'in ts'ai; (CHINESE, MIN NAN) khîn-chhài, khûn-chhài; (CHINESE, WÚ) qin cai; (CHINESE, YUE) xiang qin, qin cai, han qin; (COMANCHE) paʔpuhi tuhkaʔ; (CREOLE, HAITIAN FRENCH) séléri, seleri; (CRIOULO, CAPE VERDE) aipo; (CROATIAN) celer bjelaš; (CZECH) miřík celer kořenový, bulvový, petržel natova, celer řapíkový; (DAGAARE, SOUTHERN) cilri; (DANISH) bledselleri, bladselleri; (DUTCH) bleekselderij; (ENGLISH) celery, blanched celery, stalk celery, leaf celery; (ESTONIAN) varsseller, seller; (FARSI, WESTERN) karaf; (FILIPINO) kinchai; (FINNISH) lehtiselleri; (FRENCH) céleri à côtés, céleri à branchir, céleri rave; (GAELIC, IRISH) soilire; (GAELIC, MANX) smullish; (GAELIC, SCOTTISH) soilire; (GEORGIAN) niakhuri; (GERMAN) Bleichsellerie, Stangensellerie; (HEBREW) selery, karpas reḥani, karpas reihani; (HILIGAYNON) kíntsay; (HINDI) ajmod, ajmud, bari ajmod; (HINDUSTANI) supuwiri, ajwáyan khurásání; (HMONG DAW) zaub txhwb; (HUNGARIAN) zeller, halványitó zeller; (ICELANDIC) stilkselleri; (INDONESIAN) seledri, selederi, seladeri, seldri, daun sup; (ITALIAN) sedano, sedano da coste; (JAPANESE) serurii, serori; (JAVANESE) sederi, seledri, (KANNADA) selerina; (KAZAKH) selderej, selderey; (KHMER, CENTRAL) chii tang' 'aô, kiën chhaay; (KIRGHIZ) selderey; (KWÉYÒL) sélwi; (LADINO) apio, apyo; (LAO) s'ii s'aangz; (LATVIAN) selerija, salàtu selerija; (LITHUANIAN) valgomasis salieras; (LUXENBOURGEOIS) Zelleri; (MACEDONIAN) običen kerevis; (MALAY) sadri, daun saderi, selderi, seladeri; (MALTESE) krafes; (MARATHI) aljmoda; (MÒORÉ) celeri; (MORISYEN) celeri; (NEPALI) celery; (NORWEGIAN) stilkselleri, stangselleri, selleri; (PAMPANGAN) kinsay, kinse, kintse; (PANGASINAN) kínchay; (PANJABI, EASTERN) khuräsäni ajwain, kornauli; (POLISH) seler liściowy, pietruszka lisciowa; (PORTUGUESE) aipo, apio hortense, aipo de talos, aipo da agua, salsão; (ROMANIAN) pătrunjel cutivat, pentru frunze, telină pentru petiol; (RUSSIAN) petrushka listovaja, selderej čereškovyj; (SAMBAL, TINÀ) kintsay; (SANSKRIT) ajmoda, andhpatrika; (SERBO-CROATIAN) celer bjelaš; (SLOVENIAN) belušna zelena, zeler; (SORBIAN, LOWER) měrik; (SORBIAN, UPPER) selerij; (SOTHO, NORTHERN) seleri; (SOTHO, SOUTHERN) seleri; (SPANISH) apio blanco, apio autoblanqueable, apio de tallo; (STELLINGWERFS) greunte, gruunte, selderie, selderij; (SUNDA) saledri, saladri, sadri; (SWAHILI) namna ya mboga, seleri; (SWEDISH) stjälkselleri, blekselleri, selleri; (TAGALOG) celery, kinchai, kinintsai, kinsay, gitsay, kinintsay, kintsay; (TAJIKI) karafs; (TAMIL) celery-keerai; (TATAR) sel'derey; (THAI) khunchai, phakpum, phakkhaopun; (TIBETAN) chīntshɛɛ; (TSWANA) morogo wa seleri; (TURKISH) sap kereviz, kereviz, yaprak; (TURKMEN) ysly garawuk, seldereÿ; (URDU) ajmod, krafs romi; (UYGHUR) chingsäy; (UZBEK) sel'derey, selderey; (VIETNAMESE) rau c[aaf]n t[aa]y; (VLAAMS) bleekselderij; (WELSH) helogan; (XHOSA) iseleri, uhlobo oluthile lwemifuno ntlobo; (YIDDISH) selereeah; (ZULU) useleri.

24.5. *Apium graveolens* var. *rapaceum* (Mill.) Gaudin (EP) root, leaves[S]; (P) C,R.

(AFRIKAANS) knolseldery; (ALBANIAN, TOSK) selinë; (ARABIC) cherafes; (ARMENIAN) nechur; (BELARUSAN) ochrochuschi; (BULGARIAN) korenoplodna celina; (CATALAN-VALENCIAN-BALEAR) api; (CHINESE, MANDARIN) gen qin cai, ch'in; (CHINESE, WÚ) gen qin cai; (CHINESE, YUE) kui gen qin; (CROATIAN) celer korijenaš; (CZECH) petrzel kadeřava, celer bulvový; (DANISH) knoldselleri; (DUTCH) knolselderij; (ENGLISH) celeriac, turnip-rooted celery; (ESTONIAN) juurseller; (FINNISH) mukulaselleri; (FRENCH) céleri tubéreux, céleri rave; (GALICIAN) apeo; (GEORGIAN) ochrochuschi; (GERMAN) Knollen-Sellerie; (GREEK) selino, fymatorrizon selino; (HEBREW) selery, karpas reḥani, carpas; (HINDI) ajmud; (HUNGARIAN) gumós zeller; (ICELANDIC) hnudselleri; (INDONESIAN) umbi seledri, seledri; (ITALIAN) sedano-rapa; (JAPANESE) seruriatku, orandamitsuba; (KIRGHIZ) sel'derej; (KOREAN) batminari; (LATVIAN) selerijas, saknu selerija; (LITHUANIAN) valgomasis salieras; (MACEDONIAN) korenov kereviz; (MALAY) selderi; (NEPALI) jwano; (NORWEGIAN) knollselleri, hageselleri, sellerirot, selleri; (POLISH) pietruszka lisciowa fryzowana, seler korzeniowy; (PORTUGUESE) aipo de raíz, aipo de cabeça, aipo nabo, aipo; (ROMANIAN) pătrunjel cu frunză creată, telina cultivată pentru rădăcini; (RUSSIAN) petrushka kudrjavaja, kornevoj sel'derej; (SERBO-CROATIAN) celer korjenaš; (SLOVAK) zeler bul'vový; (SLOVENIAN) gomoljna zelena; (SPANISH) apio nabo, apirrábano; (SWEDISH) rotselleri; (TAJIKI) karafs; (TURKISH) kök kereviz, kereviz; (UKRAINIAN) selera; (VIETNAMESE) rau cän tây; (VLAAMS) knolselderij.

24.6. *Arracacia xanthorrhiza* Bancroft (EP) root; (P) C.

(CHINESE, MANDARIN) ya lai jia li ya; (CHINESE, WÚ) ya lai jia li ya; (ENGLISH) arracacha, Peruvian carrot, white carrot, Peruvian parsnip; (FINNISH) perunpalsternakka; (FRENCH) arracacha, paneme, pomme de terre céléri; (GERMAN) Arrakatascha; (NORWEGIAN) arracacharot; (POLISH) ziemniara jadalna; (PORTUGUESE) arracacha mandioquinha, mandioquinha salsa, cenoura do Perú; (QUECHUA, CUSCO) rakcacha, huisampilla; (SPANISH) arracacha; (VLAAMS) arracacha.

24.7. *Centella asiatica* (L.) Urban (EP) leaves and stolons[S]; (P) R,C.

(AFRIKAANS) varkoortjies; (ASSAMESE) manimuni; (BADAGA) kudure ko:ku; (BENGALI) thol-kuri; (BICOLANO, CENTRAL) sora-sora; (BONTOC, CENTRAL) pispising, sipsipping; (BURMESE) myin-khwa; (CEBUANO) hahang-halo, yahong-yahong, hahanghalo; (CHINESE, MANDARIN) ji xue cao; (CHINESE, WÚ) ji xue cao; (DANISH) asiatisk centella; (EDE NAGO) ewe karo; (ENGLISH) asiatic pennywort, wild violet, Indian pennywort; (FILIPINO) takio-kohol; (FINNISH) sammakonputki; (FON) adodowe; (FRENCH) ecuelle d'eau, hydrocotyle asiatique; (GREEK) ydrokotyli asiatiki; (GURO) palétrui; (HANUNOO) dulud; (HEBREW) saflila asiatit; (HINDI) brahma-manduki, khvlakhudi, mandookparni, brahmamandooki, dhulakudi, khulakhudi; (IFUGAO) unappa; (ILOCANO) anggalipungan; (INDONESIAN) daun kaki kuda, pegagan, penggaga, antanan; (IVATAN) kalabasan-karam, tagaditak; (JAPANESE) tsubo-kusa, gotukola; (JAVANESE) ganggagan, kerok batok, calingan rambat, rendeng, cowek gompel, gagan-gagan; (JOLA-FONYI) lèv; (KANNADA) braahmi soppu, elikivi balli, elikivi soppu, gadde oraga, ondelaga, saraswathi soppu, vandelaga, uraga; (KONKANI) ekapaani, braahmi; (LAHU) hpa^ naweh_; (MALAY) pegaga; (MALAYALAM) kodangal, kodagan, kutakam, muthal; (MANU'A SAMOAN) tago; (MARANAO) tangila a lopa; (MARATHI) ekapaani, karinga, karivana, braahmi, karivanaa; (NEPALI) ghod tapre; (NIUE) tono, tonu; (PAMPANGAN) balubagdais; (POLISH) wąkrotka zwyczajna, azjatycka pępownica; (PORTUGUESE) cairussú, donheiro em penca, pata de cavalo, pé de cavalo; (RAROTONGAN) kapukapu; (SAMBAL, BOTOLAN) tulin bakolaw; (SAMOAN) togo; (SANSKRIT) mandukaparni, braahmi, mandookaparni, bhekaparni, bheki, brahma chaarini, brahmamandooki, darduchhada, divya, kapotavankaa, mahaushadhi, mandookaparnika, mandooki, supriya, thvashti; (SINHALA) gotukola; (SPANISH) hierba de clavo; (SUBANON) panggaga, panggawa; (SUNDA) antanan gede, antanan biasa, antanan rambat, kurawet galeng; (TAGALOG) takip-kohol, tapingan-daga, takip-kuhol, takip-kuto, takip-suso, tengang-daga; (TAMIL) vallarei, vallaarai, babassa, vallari, vallareekeeri; (TAUSUG) panggaga; (TELUGU) brahmi, saraswataku, saraswathi aaku, babassa, bekaparnamu, bokkudu, mandooka braahmi, vellaaraku; (THAI) bua bok; (TONGAN) tono; (TULU) thimare; (URDU) brahma-manduki; (VLAAMS) gutukola; (WÁRAY-WÁRAY) takup-suso, yahong-yahong; (WELSH) llysiaur geiniog asia; (YALUNKA) nyeretul-la; (YORUBA) ewe karo.

24.8. *Chaerophyllum bulbosum* L. (EP) roots; (P) C.

(ARABIC) almakdoonis zu el-gzoor, el-liftia; (ARMENIAN) šušanbandžar; (BELARUSAN) tsymjanets clubnjanosny; (BULGARIAN) klubenoviden kervel; (CHINESE, MANDARIN) pu qin, ta kên hsuch wei ts'ai; (CHINESE, WÚ) pu qin, xi ye qin; (CHINESE, YUE) xi ye qin; (CZECH) krabilice hlíznatá; (DANISH) knoldet hulsvøb, kjorvel roe; (DUTCH) knoll-kervel, knolribzand; (ENGLISH) tuberous chervil, turnip-rooted chervil, parsnip-chervil; (FINNISH) mukulakìrveli; (FRENCH) cerfeuil tubéreux, chérophylle bulbeux; (GALICIAN) coendro, xendro; (GERMAN) knolliger Kälberkropf, Kerbeirübe, Knollenkerbel, Rübe-Kälberkropf; (GREEK) kondylodis, anthriskos to chaerephyllon; (HUNGARIAN) gumós baraboly, czemegebaraboly; (ITALIAN) cerfoglio bulboso; (JAPANESE) chuubaa-charilu, chabiru ninjin; (LATVIAN) smaržīgā kārvele; (LITHUANIAN) gumbuotasis gurgždis; (MACEDONIAN) krtulesta krasulka; (NORWEGIAN) knollkjeks, kjeks; (POLISH) swierząbek bukwiasty; (PORTUGUESE) cerefolho tuberoso, cerefolho bolboso, batata baronesa; (ROMANIAN) baraboi; (RUSSIAN) kervel' korneplodnyj; (SERBO-CROATIAN) pitoma zvolina, krabljica, repasta; (SLOVAK) krkoška hl'uznatá; (SLOVENIAN) kosarnik; (SPANISH) perifollo tuberoso, perifollo bulboso; (SWEDISH) knölflocka; (TURKISH) yumrulu frenk maydanozu, hirhindilik; (TURKMEN) düýpli buten; (VLAAMS) knolkervel; (WELSH) perllys y perthi.

24.9. *Coriandrum sativum* L. (EP) leaves and seeds; (P) R,C.

[1]Itsukha dialect of Luyia.

(AFRIKAANS) koljander; (ALBANIAN, TOSK) koriandër; (ARABIC) kuzbara; (ASSAMESE) dhonia; (BADAGA) kottamalli; (BELARUSAN) kaljandra pasjannaja; (BENGALI) dhane, lanka, marich, dhaniya; (BICOLANO, CENTRAL) kulantra; (BODO) dundia; (BURMESE) nan-nan-pin; (CHINESE, GÀN) xiang cai; (CHINESE, MANDARIN) yan sui; (CHINESE, MIN BEI) hu sui, feng chui, xiang cai; (CHINESE, WÚ) yan sui, xiang cai; (CHINESE, YUE) xiang cai, yan sui; (CRIOULO, CAPE VERDE) coentro; (CZECH) kociandr setý; (DANISH) koriander; (DUTCH) koriander; (ENGLISH) coriander, collender, cilantro; (ESTONIAN) koriander; (FARSI, WESTERN) kishniz, gishniz; (FIJIAN) dhania; (FILIPINO) kulantro; (FINNISH) korianteri; (FRENCH) coriandre; (GAELIC, MANX) lvs y chree; (GEORGIAN) kindzi; (GERMAN) Koriander; (GIKUYU) ndania; (GREEK) koriandron, kolliahdros; (GUJARATI) khonphir, kothamir, dhane; (HEBREW) kusbar, kusbara, gad ha-sade, gad hasadeh; (HINDI) dhania, dhanya, dhaniya, dhangiyā; (HMONG-DAW) zaub txhwb qaib; (HMONG-NJUA) zaub txib qab; (HOPI) kolanro, koranro, silanro; (HUNGARIAN) koriander, kerti koriander; (ILOCANO) uansay; (INDONESIAN) ketumbar, katumbar, surage; (ITALIAN) coriandolo; (JAPANESE) koendoro, korianda; (JAVANESE) katumbar, tumbar, penjelang; (KANNADA) kothambri, kotthambari, kattambri dhanya, dhanya beeja, haveeja, kothimari, kothimbri, kothumbari, saambhaara kaalu; (KASHMIRI) dāñĕwali bạrag; (KAZAKH) kinsa; (KHMER, CENTRAL) vannsuy; (KIRIBATI) korianta; (KONKANI) kotbor, kotbori, kothambar; (LADINO) kulántro, culantro; (LAO) phak ho:m pa:nx, phak ho:m po:mz; (LATVIAN) koriandrs; (LITHUANIAN) blakinė kalendra; (LUYIA) itania[1]; (MACEDONIAN) običen koriander; (MALAY) ketumbar, penjilang, wansui; (MALAYALAM) kothamalli, kotthumalli, kotthumpalari beeja; (MARANAO) magalay; (MARATHI) dhanna, dhane, kothimber, kothinbir; dhanya, khotbir, khotmir, khothmir; (MIZO) dhaniya; (MORISYEN) kotomili; (NEPALI) dhaniya; (NORWEGIAN) koriander; (ORIYA) lanka, dhania; (PANJABI, EASTERN) dhania; (POLISH) kolender siewna; (PORTUGUESE) coentro; (ROMANIAN) koriandru cultivat, coriandru; (RUSSIAN) kishnets posevnoi; (SANSKRIT) dhanya, kustumburi, kustumbri, dhanyaka, ababika, allaka, chhathra, dhaniyaka, dhanyabeeja, dhanyakam, dhana, dhanika, dhennika, hridyagandha, janapriya, kunati, kusthumburi, nisara, shakayogya, sukshmapathra, sugandhi, thumbaru, vedhaka, veshana, vithunnaka; (SANTALI) dhạniạ; (SCOTS) corrydander; (SERBO-CROATIAN) korijander; (SINHALA) koththamalli; (SLOVAK) koriander; (SORBIAN, UPPER) coriander; (SPANISH) coriandro, cilandrio, cilantro, culantro; (SUNDA) tumbar, katuncar, tuncar, tangoh; (SWAHILI) giligilani; (SWEDISH) koriander; (TAGALOG) kulantro, uan-soi, kulan tro, ongsoy, unsuy, wansuy; (TAJIKI) kashnich, dashtir; (TAMIL) kothamalli, kotthamalli, kotthamalii keerai; (TELUGU) dhaniyalu, dhaniyaalu, mirapakaya, kothimiri, danyaalu; (THAI) phakchi, phakhom-noi; (THAI, NORTHERN) phakhom; (TULU) kottambari, kothambari, kottamberi; (TURKISH) kişniş, kara kimyon; (TURKMEN) ekimlik koriandr; (URDU) dhanya; (UZBEK) cashnich; (VIETNAMESE) rau m[uf]i, ng[of] ta, rau ng[of]; (VLAAMS) koriander; (WELSH) planhigyn persawrus.

24.10. *Cryptotaenia japonica* Hassk. (EP) leaves; (P) C.

(CHINESE, MANDARIN) ya er qin, pai-chin-ts'ai; (CHINESE, WÚ) ya er qin, tian hu sui; (DANISH) mitsuba; (ENGLISH) Japanese hornwort; (FRENCH) mitsuba; (INDONESIAN) mitsuba, salderi Jepang; (ITALIAN) lereschia giapponese; (JAPANESE) mitsuba; (MALAY) salderi djepang; (POLISH) rogatek japoński sztywny; (PORTUGUESE) mitsubá; (VLAAMS) Japans hoornkruid, mitsuda.

24.11. *Daucus carota* L. subsp. *sativus* (Hoffm.) Arcang. (EP) root and leaves[s]; (P) R,C.

(AFRIKAANS) wortel, geelwortel; (ALBANIAN, TOSK) karotë; (ALEUT) aupaluktârsuk, kaimikutaujak; (AMHARIC) karot; (ARABIC) jazar; (ARMENIAN) gazar; (ASSAMESE) gajor; (AZERBIAJANI) er kekju, yerkökü; (BADAGA) ka:raṭu; (BALTI) wa la-phru, wàa-phru; (BAOULÉ) caroti; (BASQUE) pastana; (BELARUSAN) morkva, morkva dzikaja; (BENGALI) gajar; (BOSNIAN) mrkva, šargarepa; (BRETON) karotez; (BULGARIAN) morkov; (BULU) karot; (BURMESE) mone-la-oo-ni; (CAJUN FRENCH) carotte; (CAROLINIAN) kiyáárots; (CATALAN-VALENCIAN-BALEAR) pastanaga; (CHINESE, GÀN) hong luo bo; (CHINESE, HAKKA) fôung lô-p'ĕt, foûng ts'oi-t'êoi; (CHINESE, MANDARIN) hu luo bo, hu lo po; (CHINESE, MIN BEI) jin sun, hong luo bo; (CHINESE, MIN NAN) âng-chhài-thâu; (CHINESE, WÚ) ding xiang luo bu, hong luo bu, hong luo bo, hu luo bo; (CHINESE, XIANG) hong luo bo; (CHINESE, YUE) jin sun, hong luo bo; (COMANCHE) ohahpokopi, ohahtuhkapu̲; (CREE, PLAINS) oskâtâsk; (CREOLE, HAITIAN FRENCH) karott, kawōt; (CRIOULO, CAPE VERDE) cenoura; (CROATIAN) mrkva; (CZECH) mrkev, mrkev obecná, mrkev obecná setá; (DAGAARE, SOUTHERN) carroti; (DANISH) gulerod; (DIGO) karoti; (DUTCH) peen, wortel, waspeen, bospeen, winterwortel; (ENGLISH) carrot; (ESTONIAN) porgand; (FARSI, EASTERN) zardak; (FARSI, WESTERN) zardak, gazor, havich; (FIJIAN) gajar; (FILIPINO) carrot; (FINNISH) porkkana; (FRENCH) carotte; (GAELIC, IRISH) cariéad, meacan dearg; (GAELIC, MANX) carradje; (GAELIC, SCOTTISH) curran; (GALICIAN) cenoira, cenoura; (GEORGIAN) staphilo, stapilo, stapfílo; (GERMAN) Möhre, Karotte, Möhrübe, Gelbrübe, gelbe Rübe, Rüebli, Wurzel; (GIKUYU) karati; (GOURMANCHÉMA) carroti; (GOURMANTCHÉ) caroti; bompièn̍o-tibina; (GREEK) karotto, piza to aafkou to karotou; (GUJARATI) gajar; (HARSŪSI) yezár; (HAUSA) karas; (HEBREW) gezer, gezer ha-ginna, gezer hagina, gezer haginah; (HERERO) larin; (HINDI) gajar; (HINDUSTANI) gájar, zardak; (HOPI) sikyatotona; (HUNGARIAN) sárgarépa, étkezési sárgarépa, murok; (ICELANDIC) gulrot; (INDONESIAN) wortel, bortel, bortol, vortel; (INUKTITUT, EASTERN CANADIAN) amittukaaq, patiujaq, takittukaujaq, usuujaq; (ITALIAN) carota; (JAPANESE) ninjin; (JAVANESE) wortel, wertel, bortol, wertol; (JOLA-FONYI) koucarote; (JULA) caroti; (KANNADA) kempu mulangi, gajjari, manyal mulangi; (KANURI) káràs; (KASHMIRI) bāla, bulmuj, gāzürü, kāch, mōr, mormuj; (KAZAKH) morkov kulturnaja; (KHMER, CENTRAL) karôt; (KIRGHIZ) sabiz; (KOREAN) hongdangmuu; (KRIO) karɔts; (KURDISH, NORTHERN) gêzer, gwezēr; (KWANGALI) ombaruru; (KWÉYÒL) kowòt, kawòt; (LADINO) karotas, safanoyas, safanorya; (LAHU) ka$^{\vee}$ la$^{\vee}$ g'aw$^{\vee}$ ma; (LAKOTA) tiŋpsiŋlazizi; (LATVIAN) savvaļas burkāns, sējas burkāns, burkāns, burkāni; (LITHUANIAN) paprastoji morka, morkos; (LUO) karot; (LUXENBOURGEOIS) Muurt, Wuerzel; (MACEDONIAN) pitom morkov; (MAITHILI) gājar; (MALAY) lobak merah, karot, boktel; (MALAYALAM) carrot; (MALTESE) zunnarija; (MARATHI) gazara, gajar; (MICMAC) wĭsowĭbŭnĕksit; (MIZO) antam bulbal; (MOHAWK) otsinekwar ohtera; (MÒORÉ) carot; (MORISYEN) karot; (NAMA-DAMARA) okanakamudesha; (NAVAJO) chaasht'ezhiitsoh; (NDONGA) ehwa; (NEPALI) gajar; (NIUE) kālote; (NORWEGIAN) gulrot, dyrka gulrot, karotte, villgulrot; (ORIYA) gajar; (PANJABI, EASTERN) gajar, gajjar; (PASHTO, NORTHERN) gâzëra, gāzara, zardaka; (POLISH) marchew siewna, marchew zwyczajna, marchew ogrodowa; (PORTUGUESE) cenoura; (PULAAR) karot; (PULAR) karote; (ROMANI, BALKAN) morkoos; (ROMANIAN) morcov, morcov comun, morcov comestibil; (RUSSIAN) morkov agorodnaja, morkov' kul'turnaja; (SANSKRIT) shikha-mula, garijara; (SERBO-CROATIAN) mrkva; (SERER-SINE) carot; (SHONA) kerotsi; (SIDAMO) kârôte; (SLOVAK) karotka, mrkva; (SLOVENIAN) korenček; (SORBIAN, LOWER) marchwej; (SORBIAN, UPPER) morchej; (SOTHO, NORTHERN) mmetli; (SOTHO, SOUTHERN) mmetli; (SPANISH) zanahoria; (STELLINGWERFS) wottel; (SUNDA) bortol, boktel; (SWAHILI) karoti; (SWEDISH) morot; (TAGALOG) karot, remolacha; (TAJIKI) sabsi; (TAMIL) kaaret, gajarakkilangu, gajjara kilangu, karettukizhangu; (TELUGU) gajaragadda, gajjaragedda, gajjara gedda, pitakanda; (THAI) khaerot; (TIBETAN) qōŋi̲puù; (TOK PISIN) karot, yalow pela sayor; (TSWANA) mmetli; (TURKISH) havuç, pıtık otu; (TURKMEN) adaty käşir; (UKRAINIAN) stolova morkva; (URDU) gajar; (UZBEK) sabzi; (VENDA) kherò the; (VIETNAMESE) vie cu cā-röt, c[af] r[oos]t; (VLAAMS) wortelen, peen; (WELSH) moron; (WOLOF) carrot; (XHOSA) umchweli, umnqathe; (YIDDISH) mayer; (ZARMA) karoti; (ZULU) umbazi, khaloti; (ZUNI) kewats.

24.12. *Foeniculum vulgare* Mill. subsp. *vulgare* var. *azoricum* (Miller) Thell. (EP) leaf base; (P) R.

(AFRIKAANS) Florentynse vinkel; (ALBANIAN, TOSK) kopër; (ARABIC) bisbas; (ARMENIAN) chorom-samit; (ASSAMESE) sounf, guamuri; (AZERBIAJANI) razjana; (BASQUE) aneta; (BRETON) fanouilh-mor; (BULGARIAN) bolonsko; (CATALAN-VALENCIAN-BALEAR) fenoll; (CHINESE, MANDARIN) tian hui xiang, hu hsiang; (CHINESE, WÚ) tian hui xiang; (CHINESE, YUE) qin jing hui xiang; (CZECH) fenykl sladký, fenykl obecný sladký; (DANISH) knoldfennikel;

(DUTCH) knolvenkel; (ENGLISH) Florence fennel; (ESTONIAN) apteegitill; (FINNISH) salaattifenkoli; (FRENCH) fenouil à bulbe, fenouil doux, fenouil des jardins; (GALICIAN) fionllo; (GEORGIAN) cereco; (GERMAN) Knollenfenchel, Zwiebelfenchel; (GREEK) maratho, foinikoulo to glyko; (HEBREW) shumar, shumar pashut; (HUNGARIAN) édeskömény; (ICELANDIC) fennel, sigod, fennika; (INDONESIAN) das-pĕdas; (ITALIAN) finocchio domestico, finocchio doce, finocchio da seme; (JAPANESE) furohrensu-fennelu, uikyō; (KOREAN) hoehjang; (LATVIAN) saldais fenhelis; (LITHUANIAN) paprastasis pankolis; (MACEDONIAN) običen razijan; (MORISYEN) fenouy; (NEPALI) madise soonp; (NORWEGIAN) knollfennikel, fenikel, fennikel; (PANJABI, EASTERN) sonf; (POLISH) fenkuł florencki, fenkul wloski zwyczyjny; (PORTUGUESE) funcho doce, funcho de Florença, fiolho; (ROMANIAN) feniculul de florenta; (RUSSIAN) fenchel' ovoščnoj; (SERBO-CROATIAN) anason; (SLOVAK) fenukel záhradný; (SLOVENIAN) sladki komarček; (SPANISH) hinojo, hinojo de Florencia; (SWAHILI) feneli; (SWEDISH) jankal; (TAJIKI) rozionai jillonok; (TURKISH) Floransa rezenesi, tatli raziye, rezene; (UKRAINIAN) koreneva selera; (URDU) saunf; (VLAAMS) Italiaanse venkel, knolvenkel, groentevenkel; (WELSH) ffenigel florens.

24.13. *Foeniculum vulgare* Mill. subsp. *vulgare* var. *dulce* (Mill.) Batt. (EP) leaves; (P) R,C.

(AFRIKAANS) vinkel, makvinkel; (ALBANIAN, TOSK) kopër; (AMHARIC) ənsəlal, ənsolal, kämun; (ARABIC) besbes; (ASSAMESE) sounf, guamuri; (BELARUSAN) fenhel zvychainy; (BENGALI) pan mouri, mauri, panmuhori; (CATALAN-VALENCIAN-BALEAR) fonoll; (CHINESE, MANDARIN) yi da li hui xiang; (CHINESE, MIN NAN) hôe-hiu, sío hôe-hiu, sió-hôe; (CHINESE, WÚ) yi da li hua xiang, xiao hui xiang; (CRIOULO, CAPE VERDE) funtcho, erva-doce; (CZECH) fenykl obecný, fenicul obecny; (DANISH) fennikel, sødfennikel, krydderfennikel; (DUTCH) venkel; (ENGLISH) fennel, sweet fennel; (ESTONIAN) apteegitill; (FARSI, WESTERN) raziyâne; (FIJIAN) soap; (FILIPINO) anis; (FINNISH) maustefenkoli; (FRENCH) fenouil, anethdoux; (GAELIC, MANX) finnal; (GALICIAN) fionllo, fiuncho; (GEORGIAN) kama; (GERMAN) Fenchel; (GREEK) marathron, foinikoulon to koivov; (GUJARATI) variali, variari, variyali; (HEBREW) shumar, shumar pashut; (HINDI) sonf, saunf, sonp, barisaunf, sont, varyaali; (HINDUSTANI) ek pauda, saupf, soá; (HOPI) kwangwaviva; (HUNGARIAN) gumós kömény, édesköménу, magyar ánisz; (INDONESIAN) adas, adas pedas, adhas; (ITALIAN) finocchio; (JAPANESE) uikyou; (JAVANESE) adas, adas landi, adas londo; (KANNADA) dodda sompu, badisopu, dodda jeerige, sompu, badesopu; (KASHMIRI) bödiyāna; (KAZAKH) fenchel ovoshcnoj; (LADINO) finojo, finozo, chinólyio; (LAO) phak s 'i; (LITHUANIAN) paprastasis pankolis; (MACEDONIAN) običen razijan; (MALAY) adas pedas; (MALAYALAM) perum jeerakam; (MALTESE) il-buzbiez talgidra; (MARATHI) badishep; (NEPALI) madise soonp; (NIUE) taletale; (NORWEGIAN) hagefennikel, fönikkel, krydderfennikel; (ORIYA) panmahuri; (PANJABI, EASTERN) sonf; (PASHTO, NORTHERN) sona, kāgah; (POLISH) fenkuł włoski, fenkuł, koper włoski; (PORTUGUESE) funcho, erva doce; (ROMANIAN) fenicul, anacon mare, chimeu dulce, molură; (RUSSIAN) fenhel obyknovennyi, ukrop aptechnyi, ukrop voloshski; (SANSKRIT) madhurika, ahichhathra, avakapushpi, bhuripushpa, chhathra, gandhadhika, ghosha, ghoshavathi, karavi, maadhavi, madhura, madhuri, mangalya, mishreya, mishi, potika, pushpavha, sanghata pathrika, shaleya, shalina, shathapathrika, shathaprasuna, shathapushpa, sheethashiva, sugandha, supushpika, talaparni, tapasapriya, trishahara, vanapushpa; (SERBO-CROATIAN) anason, komorač; (SLOVAK) fenikel; (SONGHAY) àlkáfún; (SORBIAN, UPPER) koprica; (SPANISH) hinojo común, fenícula; (SUNDA) hades; (SWAHILI) shamari, feneli; (SWEDISH) fänkål; (TAGALOG) anis, haras; (TAJIKI) roziona; (TAMIL) shombu, perungeeragam, kaattu seeragam, sohikeerai; (TELUGU) seemajeelakarra, peddajeelakaramu, peddajeela koora, sopu; (THAI) yira; (THAI, NORTHERN) thian-klaep, phakchi-duanha; (TURKISH) rezene, dere otu; (TURKMEN) adaty arpabadýan; (URDU) saunf; (UZBEK) ok zira; (VIETNAMESE) ti[eer]uh[oof]i h[uw][ow]ng; (VLAAMS) zoete venkel, kruidige venkel; (WELSH) ffenigel; (XHOSA) isihlahla esithile esinuka kamnandi esisetyenziswa kwisaladi nakwisosi.

24.14. *Glehnia littoralis* F. Schmidt (EP) leaves, stems, roots; (P) C.

(CHINESE, MANDARIN) shan hu cai; (CHINESE, WÚ) shan hu cai, bei sha shen; (ENGLISH) coastal glehnia; (GREEK) ydrokotyli tou chortotapitos; (JAPANESE) boufuu, hama-bôfû; (VLAAMS) hama-bofu.

24.15. *Hydrocotyle sibthorpioides* Lam. (EP) young shoots and leaves; (P) R,C.

(ASSAMESE) soru manimuni; (BONTOC CENTRAL) tomtomon, kanapa; (DOONDO) moueboula; (DUTCH) waternavel; (ENGLISH) hydrocotyle, lawn pennywort; (GIANGAN) sigbut-ta-bato; (IFUGAO) unappa; (INDONESIAN) semanggi, patikan cina, salatum tikin, tikim, kaki kuda, salatun, daun pegagan, daun penggaga, antanan; (ITALIAN) soldinella sibtorpioide; (JAPANESE) chidome-gusa; (JAVANESE) semanggi, andem, katepan, rendeng, patikim, penjelangan, antanan leutik; (KALAGAN) al-allobo, enit; (KANKANAEY) pangpangdan di num-a, tomtom-en; (MALAY) pegaga embun, ulam gunung; (SUNDA) antanan beurit, antanan lembut, antanan leutik, kurawet; (TELUGU) aelika chaevi koora, babbasa; (THAI) ya-klethoi; (VIETNAMESE) rau m[as] m[ow].

24.16. *Myrrhis odorata* (L.) Scop. (EP) leaves, roots[S] and seeds[S]; (P) R.

(AFRIKAANS) mirre; (ARABIC) samag albostan; (CHINESE, MANDARIN) ou zhou mo yao; (CHINESE, WÚ) ou zhou mo yao; (CZECH) čechřice vonná; (DANISH) sødskærm, vandnavle; (DUTCH) roomse kervel; (ENGLISH) garden myrrh, sweet cicely; (ESTONIAN) mesiputk; (FILIPINO) kanapa; (FINNISH) saksankirveli; (FRENCH) cerfeuil musqué, cerfeuil anisé; (GERMAN) Süßdolde; (GREEK) myrris e eyosmos; (HUNGARIAN) gázló, mirhafű; (ICELANDIC) spanarkerfill; (INDONESIAN) antanan, patikan; (ITALIAN) mirride delle Alpi, mirride odorata; (LITHUANIAN) kvapioji garduoklė; (NORWEGIAN) spansk kjørvel; (POLISH) marchewnik anyżkowy, mirnik wonny; (PORTUGUESE) cerefolho almiscarado, cerefolho anisado; (SPANISH) perifollo oloroso, perifollo almizclado; (SWEDISH) spansk körvel; (TAJIKI) hanut, darachti hanut; (TURKISH) misk maydanozu, anason kokulu mür; (URDU) khal khurmi; (VLAAMS) roomse kervel; (WELSH) cegiden bêr.

24.17. *Oenanthe javanica* (Blume) DC. subsp. *javanica* (Batalin) H. Hara (EP) leaves and tender shoots; (P) R,C.

(ARABIC) oshb el-maa almodala; (BENGALI) pan-turasi; (BONTOC CENTRAL) alusis, alusiis, galiwi-giwan, gisgisi; (CHINESE, MANDARIN) shui qin; (CHINESE, WÚ) shui qin; (CHINESE, YUE) shui qin, chu kui; (DUTCH) torkruid; (ENGLISH) water dropwort, oriental celery, water celery, Indian pennywort; (ESTONIAN) vesiputk; (FINNISH) pahaputki; (FRENCH) oenanthe, oenanthe phellandre, persil séri; (GERMAN) Pferdesaat, Wasser-Rebendolde; (GREEK) vatrachion, ranoun koulos, sepvion oinanihi; (HEBREW) yeynit; (HUNGARIAN) mételykóró; (IFUGAO) holog, baen di lagiwi, gisgisey, holo; (INDONESIAN) piopo, bacurongi, tespong; (ITALIAN) finocchio-acquatico; (JAPANESE) seri; (JAVANESE) pampung, pampung alas, seladren; (KALAGAN) taga-opay; (KANKANAEY) alusiis; (LAO) s'ii s'aangz; (MACEDONIAN) javanski vinocvet; (MALAY) shelum, selom, piopo; (MINANGKABAU) bacarongi; (MONGOLIAN) zulzagan tsagaan; (NORWEGIAN) hestekjørvel; (POLISH) seler wodny; (PORTUGUESE) funcho aquático, funcho oriental, cicuta; (SERBO-CROATIAN) trbulja, vodena; (SPANISH) cicuta, enante, felandrio, hinojo de agua; (SUNDA) tespong, bambung, salada liangan; (TAGALOG) ongsoy; (THAI) phakchi-lom, phakan; (TOK PISIN) nase; (TURKISH) su kerevizi, önant, alt tohum; (VIETNAMESE) rau c[aaf]n, rau c[aaf]n n[uw][ows]c; (VLAAMS) waterselderij, seri; (WELSH) helogan y dwr.

24.18. *Pastinaca sativa* L. (EP) root and leaves[S]; (P) C,R.

(AFRIKAANS) witwortel; (ALBANIAN, TOSK) pastinakë; (ARABIC) gazar abiad; (ARMENIAN) varot; (ASSAMESE) parsnip; (BELARUSAN) pasternak pasjany; (BULGARIAN) paštărnak; (CATALAN-VALENCIAN-BALEAR) pastenage, xirivia; (CHINESE, MANDARIN) mei guo fang feng, mei chou fang fêng; (CHINESE, WÚ) mei qui fang feng; (CHINESE, YUE) mei zhou fang feng, bai sun; (CREE, PLAINS) wâpiski oskatask, wâpisk oskatask; (CROATIAN) pastrnjak; (CZECH) pastinák, pastinák sety, pastinák luční setý; (DAKOTA) paŋġiżiżi; (DANISH) pastinak; (DUTCH) pastinaak; (ENGLISH) parsnip; (ESTONIAN) pastinaak, moorputk; (FARSI, WESTERN) haveej-farangee; (FILIPINO) parshio; (FINNISH) palsternakka; (FRENCH) pastenaque, panais, patenais; (GAELIC, IRISH) meacan bán; (GAELIC, MANX) desmad; (GALICIAN) cherouvea; (GEORGIAN) pasternaki, dzirtetra; (GERMAN) Pastinake, Pastinak, gemeiner Pastinak; (GREEK) elaphovoskon, pastinakos e edodimos; (HINDI) gujur; (HINDUSTANI) chuqandar, gájar; (HUNGARIAN) pasztinák, paszterăk; (ICELANDIC) pastinak, nipa; (ITALIAN) pastinaca, carota rossa; (JAPANESE) paasunitpu, amerikabōfū; (KAZAKH) pasternak posevnoj; (KIRGHIZ) pasternak; (KOREAN) bangphung; (LAKOTA) paŋgihaŋska; (LATVIAN) pastinaki, pastinaks; (LITHUANIAN) paprastasis pastarnokas, pastarnokai; (MACEDONIAN) običen paškanat; (MALAY) parsnip; (MALTESE) pastinakka; (MICMAC) tĕpanā; (NEPALI) gujar; (NIUE) pāsinipi; (NORWEGIAN) pastinakk; (POLISH) pasternak zwyczajny; (PORTUGUESE) pastinaca, pastinaga, chirivia; (POTAWATOMI) babigwe'wûnûsk; (ROMANIAN) păstirnac comun, păstîrnac; (RUSSIAN) pasternak posevnoj; (SERBO-CROATIAN) pastrnjak, paškanat, obični; (SLOVAK) paštrnák; (SLOVENIAN) pastinak; (SORBIAN, UPPER) pastyrnak; (SPANISH) pastinaca, chirivía; (SWEDISH) palsternacka; (TAGALOG) parsnip; (TURKISH) yabani havuç, kültür pastinak'i; (UKRAINIAN) pasternak; (UZBEK) pasternack; (VLAAMS) pastenaak; (WELSH) panasen; (XHOSA) uhlobo oluthile lomnqathe; (YIDDISH) pasternak; (ZULU) i(li)pasinipu.

24.19. *Petroselinum crispum* (P. Miller) Nyman, ex A.W. Hill var. *crispum* (EP) leaves; (P) R,C.

(AFRIKAANS) pietersielie; (ALBANIAN, TOSK) majdanoz; (ARABIC) madanous; (ASSAMESE) parsley; (AZERBAIJANI, NORTHERN) cəfəri, ortiq, şərik; (BASQUE) perexil; (BELARUSAN) petrushka kucharavaja; (BOSNIAN) peršin; (BULGARIAN) magdanoz; (CATALAN-VALENCIAN-BALEAR) julivert; (CHECHEN) chaambuc; (CHINESE, MANDARIN) yang yan sui, ou ch'in; (CHINESE, MIN NAN) tang-o; (CHINESE, WÚ) yang yan sui, zhou ye ou qin; (CHINESE, YUE) zhou ye ou qin; (CREE, PLAINS) ohpikecikana, kistikanis; (CREOLE, HAITIAN FRENCH) pèsi; (CRIOULO, CAPE VERDE) salsa; (CROATIAN) peršin; (CZECH) petržel, petržel zahradni, petržel zahradní kadeřavá; (DANISH) kruspersille, glatbladet persille, persille; (DUTCH) peterselie; (ENGLISH) parsley; (ESTONIAN) petersell; (FARSI, WESTERN) ja'faree; (FILIPINO) parsley; (FINNISH) kähäräpersilja, persilja; (FRENCH) persil, persil commum, persil frisé; (GAELIC, IRISH) peirsil; (GAELIC, MANX) parsal, perslee; (GALICIAN) prijel, prixel; (GEORGIAN) okhrakhushi, magdonosi; (GERMAN) Petersilie; (GREEK) maintanos, petroselinon; (HEBREW) pterosilia, petroslinon ha-ginna, petroslinon hagina; (HINDI) parseley; (HINDUSTANI) ajmod, karafs i bos-táni, khurásáni ajwáin; (HMONG DAW) zaub txhwb; (HUNGARIAN) metélő petrezselyem, petrezselyem; (ICELANDIC) steinselja; (INDONESIAN) peterseli, piterseli; (ITALIAN) prezzemolo; (JAPANESE) paserii, tsumamono; (JAVANESE) patraseli, puterseli, pletik apo; (KAZAKH) petrushka; (KHMER, CENTRAL) vannsuy baraing; (KIRGHIZ) aš kok, petrushka; (KOONGO) persila; (KOREAN) gobsilminari; (KWÉYÒL) pesi; (LADINO) perishíl, prishíl, perejil, persil, prexil; (LAKOTA) śahiyela tatiŋpsiŋla; (LATVIAN) pētersīļi, pētersīlis; (LITHUANIAN) sėjamoji petražolė, petražolė; (LUXENBOURGEOIS) Péiterséileg, Zoppekraut; (MACEDONIAN) običen magdonos; (MALAY) parsli, daun pasli; (MALTESE) tursin; (MORISYEN) persi; (NGUNDI) persila; (NIUE) pasilī; (NORWEGIAN) bladpersille, persille; (POLISH) pietruszka liściowa; (PORTUGUESE) salsa,

salsa comum, salsinha, salsa frisada; (ROMANIAN) pătrunjel, pătrunjel de grădină, pătronjelul; (RUSSIAN) petrušhka, petrushka pasevnaja, petrushka sadovaja; (SERBO-CROATIAN) peršun; (SLOVAK) petržlen, petržlen záhradný; (SLOVENIAN) petršilj; (SORBIAN, UPPER) pětršilka; (SOTHO, NORTHERN) pareseli; (SOTHO, SOUTHERN) sehwete se sesweu; (SPANISH) perejil; (SUNDA) peeterseli, potrasoli, portelseli; (SWAHILI) pasilei; (SWEDISH) persilja; (TAGALOG) parsley, parsli, perehil; (TAJIKI) jafari; (TATAR) petrushka; (THAI) phakchi-farang; (TSWANA) paresele; (TURKISH) maydanoz; (TURKMEN) adaty badyýan, petruşka; (UKRAINIAN) petruška; (URDU) parseley; (UYGHUR) ashköki; (UZBEK) petrushka, petrushca, petruška; (VENDA) phasìli; (VIETNAMESE) rau m[uf]i, m[uf]i t[aa]y; (VLAAMS) peterselie, bladpeterselie, snijpeterselie, krukpeterselie; (WELSH) persli; (XHOSA) umfuno, ipasili, imifuno ethile yokuqholisa nokuhombisa ukutya; (YIDDISH) petrishkeh; (ZULU) ipasili, i(li)pasili.

24.20. *Petroselinum crispum* (P. Miller) Nyman, ex A.W. Hill var. *neapolitanum* Danert (EP) leaves; (P) R,C.

(AFRIKAANS) Italiaanse pietersielie; (ARABIC) madanous etaly; (CHINESE, MANDARIN) yi da li xiang qin; (CHINESE, WÚ) yi da li xiang qin; (CZECH) petržel kadeřava; (ENGLISH) Italian parsley; (FINNISH) silopersilja; (FRENCH) persil italienne; (GERMAN) Italienisch Petersilie; (GREEK) italiko petroselino; (HEBREW) petrosilia italkit; (HUNGARIAN) fodros petrezselyem; (ICELANDIC) rotarsteinselja; (INDONESIAN) peterséli; (ITALIAN) prezzèmolo napolitanum; (MACEDONIAN) italijanski magdonos; (MALTESE) tursin taljan; (MORISYEN) persi frize; (NORWEGIAN) persille; (POLISH) pietruszka włoska, pietruszka lisciowa fryzowana; (PORTUGUESE) salsa, salsa italiana, salsa crespa; (ROMANIAN) pătrunjei cu frunză creată; (RUSSIAN) petrushka kudrjavaja, petrushka agarodnaja; (SPANISH) perejil, perejil italiana; (TAJIKI) jafari itolioi; (TURKISH) italyan maydanozu; (VLAAMS) Italiaanse peterselie; (WELSH) persli eidal.

24.21. *Petroselinum crispum* (P. Miller) Nyman, ex A.W. Hill var. *tuberosum* (Bernh.) Mart. Crov. (EP) root and leaves[S]; (P) C,R.

(ARABIC) madanous leefty; (BELARUSAN) petrushka karanjaplodnaja; (CHINESE, MANDARIN) gen xiang qin; (CHINESE, WÚ) geng xiang qin; (CZECH) petržel kořenova, petržel zahradní kořenová; (DANISH) rodpersille; (DUTCH) knolpeterselie; (ENGLISH) turnip-rooted parsley; (FINNISH) juuripersilja; (FRENCH) persil tubéreux, persil racine; (GERMAN) Knollen-Petersillie, Wurzel-Petersillie; (GREEK) kramvi e rapys; (HEBREW) petroslinon; (HUNGARIAN) gyökérpetrezselyem; (ITALIAN) prezzèmolo tuberoso; (JAPANESE) ne-paserii; (LITHUANIAN) sėjamoji petražolė; (MACEDONIAN) korenorepkast magdonos; (NORWEGIAN) rotpersille; (POLISH) pietruszka korzeniowa; (PORTUGUESE) salsa tuberosa; (ROMANIAN) pătruniel cultivat, pentru rădăcină; (RUSSIAN) petrushka karnevaja; (SPANISH) perejil tuberoso; (SWEDISH) rotpersilja; (TURKISH) kök maydanoz; (VLAAMS) wortelpeterselie; (WELSH) persli gwrraidd naip.

24.22. *Pimpinella brachycarpa* (Komarov) Nakai (EP) young leaves; (P) C.

(CHINESE, MANDARIN) duan guo hui qin; (KOREAN) chamnamul; (TAJIKI) bedrennej; (TURKMEN) anis.

24.23. *Sium sisarum* L. (EP) roots; (P) C.

(AFRIKAANS) suikerwortel; (CHINESE, MANDARIN) shen qin; (CHINESE, WÚ) shen qin, ze qin; (CZECH) sevlák zeleninový, sevlák cukrový; (DANISH) sukkerrod; (DUTCH) zoete watereppe; (ENGLISH) skirret; (FINNISH) sokerijuuri; (FRENCH) berle à sucre, chervis, girole; (GERMAN) Zucker-Merk, Zukerwurzel, Sußwurz; (GREEK) sion to sisaron; (HUNGARIAN) keleti békakorsó; (INDONESIAN) tespong; (ITALIAN) sedanina coltivata, sisaro, shervola; (LATVIAN) platlapu cemere; (MONGOLIAN) shandalj; (NORWEGIAN) søtkjeks; (POLISH) marek kucmerka; (PORTUGUESE) sísaro, sisirão, afiou, raíz açucarada; (SPANISH) escaravia, chirivía de azúcar, escaravia tudesca, sisaro; (TAJIKI) talkha; (TURKISH) su kazayaği, yabani şeker havucu; (TURKMEN) süýji sium; skirret.

25. Aponogetonaceae

25.1. *Aponogeton distachyos* L. f. (EP) tender shoots, flower spikes; (P) C, P.

(AFRIKAANS) waterblommetjie, wateruintjie; (ARABIC) basal el-maa; (CHINESE, MANDARIN) xi wang feng, yan zi cai; (CHINESE, WÚ) xi wang feng, yan zi cai; (CZECH) kalatka; (ENGLISH) cape pondweed, cape asparagus, water onion; (FINNISH) verkkolehti; (FRENCH) asperge du Cap; (GERMAN) Kap Spargel; (GREEK) kap spargel; (GUJARATI) lapadi, lambadi; (HEBREW) shibbolet ha-mayim, shibolet hamaim; (ITALIAN) asparago delle Cabo; (POLISH) cebula wodna; (PORTUGUESE) espargo do Cabo; (SPANISH) espárrago del Cabo; (TURKISH) ikibaşlikli sukabaği, su başaği; (VLAAMS) waterui.

26. Araliaceae

26.1. *Aralia continentalis* Kitag. (EP) sprouts; (P) C.

(ENGLISH) manchurian spikenard; (KOREAN) ttangdureup; (TAJIKI) nard.

26.2. *Aralia cordata* Thunb. (EP) tender shoots; (P) C.

(CHINESE, MANDARIN) tu dang qui; (CHINESE, WÚ) tu dang qui, du huo; (CZECH) arálie srdčitá; (DANISH) udo; (ENGLISH) spikenard; (ESTONIAN) nard, araalia; (FRENCH) aralia cordiforme; (GERMAN) Bergaralie; (GREEK) nardostachis o iatamansios, nardos; (JAPANESE) udo; (MICMAC) chekawŏbe; (NORWEGIAN) aralia; (POLISH) aralia sercowata; (PORTUGUESE) aralia, aralia cordiforme; (SPANISH) aralia, zarzaparrilha; (TURKISH) Japon aralyasi, kalp yaprakli arelya; (VLAAMS) nardus.

26.3. *Aralia elata* Seeman (EP) young leaves; (P) C.

(CHINESE, MANDARIN) liao dong dang qui; (CHINESE, WÚ) liao dong hu mu; (CZECH) arálie vyšší; (DANISH) manchurisk aralie; (ENGLISH) Japanese aralia, Japanese angelica; (ESTONIAN) kikkaputk, angeelikajuur; (FINNISH) piikkiaralia; (FRENCH) angelique en arbre du Japon, aralia elevée; (GERMAN) Japanische Aralie, Japanischer Angelicabum; (GREEK) aggelika e iaponiler; (ITALIAN) aralia elevata; (JAPANESE) taranoki; (LATVIAN) augstā aralija; (NORWEGIAN) høstaralia; (POLISH) bluszczownik sercowaty; (PORTUGUESE) aralia, aralia japonesa; (SPANISH) aralia; (VLAAMS) engelenboom.

26.4. *Eleutherococcus senticosus* (Nakai) S.Y. Hu (EP) young leaves and stem (P) C

(CHINESE, MANDARIN) ci wu jia; (ENGLISH) Siberian ginseng, ussurian thorny pepperbush; (FINNISH) siperian ginseng, venäjänjuuri; (FRENCH) buisson du diable, eleutherococoque; (GERMAN) Sibirischer Ginseng, Stachelpanax, Taigawurzel, Teufelsbusch; (ITALIAN)

eleuterococco; (JAPANESE) ezo ukogi; (KOREAN) gasiogapinamu, gasiogapi; (PORTUGUESE) ginseng siberiano; (RUSSIAN) eleuterokokk koliuchii, svobodnoiagodnik koliuchii; (SPANISH) acanthopanax espinoso, arbusto del diablo, eleutherococco espinoso, ginseng espinoso, ginseng siberiano, matorral del diablo, pimiento ruso espinoso, pimiento salvaje; (SWEDISH) rysk rot.

27. Asteraceae (Compositae)[7]

27.1. *Acmella ciliata* (Kunth) Cass. (EP) young leaves; (P) C,R.

(ENGLISH) guasca; (FRENCH) herbe aux serpents; (PORTUGUESE) guasca; (SPANISH) guasca.

27.2. *Acmella oleracea* (L.) R.K. Jansen (EP) young leaves; (P) C,R.

(BENGALI) roshunia; (CHECHEN) mashahwieqorg; (CHINESE, MANDARIN) qian ri ju; (DANISH) parakarse; (DUTCH) Brazil vlakbloem; (ENGLISH) para cress, Brazil cress; (FILIPINO) biri; (FRENCH) spilanthe des lieux humides, cresson du Brésil, cresson de Para, brède mafane; (GERMAN) Sumpfparakresse, Husarenknopf; (GREEK) spilanthes o lachanodes; (HINDI) pakormul, pokarmul, ukra; (INDONESIAN) gletang, seruni, jotang; (ITALIAN) crescionne di Para; (JAPANESE) sennichi-modoki; (KANNADA) vanamugali, hemmugulu, vana; (KONKANI) acharbondi, pipulka; (MALAY) pokok getang; (MALAYALAM) kuppa manjel; (MARATHI) akkalkara, pipulka, acharbondi; (MENDE) *k*pawo; (NEPALI) lato ghans; (NORWEGIAN) parakarse; (PANJABI, EASTERN) pakarmul; (POLISH) paragorczyca; (PORTUGUESE) agrião do Brasil, agrião do mato, agrião do pará, jambu; (SANSKRIT) akalkar, maharashtri, sarahattika; (SPANISH) berro de Pará, chisaca; (TAMIL) ukra; (TELUGU) maratimogga, maratiteega, maratitige; (VLAAMS) parakers.

27.3. *Acmella paniculata* (Wall. ex DC.) R.K. Jansen (EP) young leaves and flower shoots; (P) C,R.

(CHINESE, WÚ) jin niu kou; (ENGLISH) spot flower; (INDONESIAN) jotang, jocong, daun getang; (LAO) kh'aad; (MALAY) getang, kerabu; (THAI) phakkhrat, phakphet; (VIETNAMESE) c[us]c n[us]t [as]o, c[us]c [as]o.

27.4. *Acmella uliginosa* (Sw.) Cass. (EP) young leaves and flower shoots; (P) C,R.

(BONTOC CENTRAL) biri; (ENGLISH) chimapaya; (IFUGAO) dilag-dilag; (INDONESIAN) jotang, jocong, daun getang; (LAO) kh'aad; (MALAY) getang, kerabu; (THAI) phakkhrat, phakphet; (VIETNAMESE) c[us]c n[us]t [as]o, c[us]c [as]o.

27.5. *Arctium lappa* L. (EP) root, petioles[S]; (P) C.

(BELARUSAN) lopush vjaliky; (CHECHEN) mashahwieqorg; (CHINESE, MANDARIN) niu bang; (CHINESE, WÚ) niu bang, bian fu chi; (CHINESE, YUE) niu bang; (CZECH) lopuch větší; (DANISH) japansk burre; (DUTCH) grote klis; (ENGLISH) edible burdock, great burdock; (ESTONIAN) takjas; (FINNISH) isotakiainen; (FRENCH) bardane géante, grande bardane, bardane, bardane majeure, gouteron; (GAELIC, IRISH) leadán; (GALICIAN) bardan, herba dosanores, labaza, lampaza, llabaza; (GERMAN) große Klette; (GREEK) lappa; (HUNGARIAN) közönséges bojtorján; (INDONESIAN) gobo; (INUKTITUT, WEST GREENLANDIC) nípúnertôq; (ITALIAN) bardana

[7] Names in parentheses have been changed under article 18 of the International Code of Botanical Nomenclature, 2005. Either name is correct.

maggiore, lappa bardana; (JAPANESE) gobou, (yama-)gobo; (LATVIAN) diždadži; (LITHUANIAN) didžioji varnalėša; (MACEDONIAN) golem lopušnik; (NORWEGIAN) storborre; (PAIUTE, SOUTHERN) avaw'tuh kaw'menuv; (POLISH) łopian większy; (PORTUGUESE) bardana maior, erva dos tinhosos, pergamasso, carrapicho grande, bobô, bardana; (SERBO-CROATIAN) repuh, obični; (SPANISH) bardana mayor, lampazo; (TAJIKI) mushkhor, alafi mushkhor; (TURKISH) büyük dulavratotu, dulavrat otu; (VIETNAMESE) ng[uw]u b[af]ng; (VLAAMS) kliswortel, grote klit; (WELSH) cedowrach.

27.6. *Artemisia dracunculus* L. (EP) leaves; (P) R,C.

(AFRIKAANS) dragon; (BELARUSAN) estagon; (BLACKFOOT) ka'ksimó; (CHINESE, MANDARIN) xai ye qing hao; (CHINESE, WÚ) xia ye qing hao, long hao; (CZECH) pelyněk, kozalec; (DANISH) estragon; (DUTCH) dragon; (ENGLISH) tarragon, dragon's wort; (ESTONIAN) estragon; (FINNISH) ranskalainen rakuuna; (FRENCH) estragon, tarragon; (GEORGIAN) tarkhuna; (GERMAN) Estragon; (GREEK) artemisia e drakontios; (HEBREW) laana draconit; (HUNGARIAN) tárkony; (ICELANDIC) franskt estragon; (ITALIAN) dragoncello, dragone, estragone; (KAZAKH) estragon; (LITHUANIAN) peletrūnas; (MACEDONIAN) gradinarski pelin; (MONGOLIAN) ishgen sharilj; (NORWEGIAN) estragon; (POLISH) bylica głupich; (PORTUGUESE) estragão; (ROMANIAN) pelin; (RUSSIAN) estragon; (SPANISH) estragón, dragoncillo; (SWAHILI) atimesia; (SWEDISH) dragon; (TAJIKI) tarkhun; (TURKISH) estragon, tarhun; (VLAAMS) dragon.

27.7. *Artemisia indica* Willd. var. *maximowiczii* (Nakai) H. Hara (EP) leaves; (P) C.

(CHINESE, GÀN) tong hao; (CHINESE, MANDARIN) ai; (CHINESE, MIN BEI) ai cai; (CHINESE, WÚ) ai, ku ai; (CHINESE, XIANG) ai ye; (DUTCH) alsem; (ENGLISH) mugwort; (ESTONIAN) puju, ristmadar; (FRENCH) armoise; (GERMAN) Beifuß; (HEBREW) laana indica; (ITALIAN) assenzio; (JAPANESE) yomogina, nishi-yomogi; (MONGOLIAN) enetkheg sharilj; (POLISH) estragon; (PORTUGUESE) artemísia; (SPANISH) ajenjo; (TAJIKI) javshan; (TULU) mańǰapatrè, mańǰipatrè; (URDU) afsan teen handi; (VLAAMS) yomogi.

27.8. *Bidens pilosa* L. (EP) young shoots and leaves; (P) C.

(AFRIKAANS) knapsekêrel, knapsakkêrel, knapsakkerwel, tantgesiesebloommetjie, weduvrouens; (AKAN-AKUAPEM) anansee mpaane, dwirantwi, gyinantwi; (AKAN-ASANTE) anasipagné, gilingui; (AKWA) boulangou; (AMHARIC) čəgʷägʷät, čəgogot; (BAOULÉ) adjongô, alongoa, alongoĩ, zagaĩ zagagbé; (BONTOC CENTRAL) puriket, enwad, nguwad; (CEBUANO) pisau-pisau, pisaw-pisaw; (CHINESE, MANDARIN) san yeh kuei jen tsao; (CHINESE, WÚ) shan ye qui zhen cao; (CRIOULO, CAPE VERDE) seta, agulha; (CZECH) dvouzubec chlupatý; (DANGME-KROBO) dsetʃi; (DANISH) broendsel; (DIGO) todza; (DUTCH) tarrelzaad, naaldekruid; (ENGLISH) bur marigold, beggar tick, black jack, Spanish needle; (ESTONIAN) saialill, peiulill; (ÉWÉ) adzrɔkpi; (EWONDO) ntsama; (FILIPINO) dadayem; (FON) vodunga; (FRENCH) bident pileux, bident; (GERMAN) Buschmannspfeil, Zweizahn; (GUJARATI) phutium, samara-kokadi; (GURO) gonoretti, pétéoré; (HEBREW) du-shen sair; (HERERO) outuerangama; (HUNGARIAN) farksafog; (IFUGAO) inayan di onwad, onwad, onwad di binobo; (ILOCANO) borburtak, puriket; (INDONESIAN) jaringan, caringan, hareuga; (ITALIAN) forbicina, forbicina pelosa; (IVATAN) dadayem; (JULA) moussofi; (KANKANAEY) eluad; (KAZAKH) chereda; (KISSI, NORTHERN) dada; (KONO) kandanɛ; (KRIO) agbɛri-oku, nidul-lif; (LOMWE, MALAWI) namulepo; (LUO) onot not; (MALAY) kanching baju, subang puteri batek; (MANDING-MANDINKA) sanyi; (MANINKAKAN, EASTERN) passoklo; (MAORI) koheriki; (MARANAO) kaperek, perperek a amo, ragom a amo; (ME'EN) junqqu; (MENDE) nana, tombo-maga; (MONGOLIAN) ajig; (MORISYEN) vilbag; (NDALI) kapuninga;

(NEPALI) kuro; (NGUNDI) ndabu za ngulu; (NYANJA) cinomba, canonga; (PORTUGUESE) amor de burro, pega pega; (QUECHUA, CUSCO) pacunga, pirca; (SENOUFO, CEBAARA) namyara; (SERBO-CROATIAN) dvozub; (SHONA) muvuyu, nhungunira, mhuvuyu; (SPANISH) aceitilla, cadillo negro; (SRANAN) japanesi toriman; (SUBANON) tatai; (SWAHILI) shona nguo; (TAJIKI) alafi garmi, sagakkhor; (TAMIL) mukkutthi; (TEKE) potambili; (TONGA) kabata; (TONGAN) fisi'uli; (TUMBUKA) chisokono, kabata; (TURKISH) su kildizi; (YAO) cisosoci; (YORUBA) abẹ́rẹ, abẹ́rẹ olóko, abore-ire, abore-oloka, agánran-mọyan, agan[ran]moyan, akẹsan, akẹṣin-má-sọ, kẹshin, ẹlẹshin-másọ, kete kete, eyinata.

27.9. *Cichorium endivia* L. subsp. *endiva* (EP) leaves; (P) R.

(AFRIKAANS) andvvie; (ALBANIAN, TOSK) lëqike, krele; (ARABIC) siikuri; (ASSAMESE) endive; (BASQUE) eskarola; (BENGALI) kasini, kashini; (BRETON) endivez; (BULGARIAN) salata endivija; (CATALAN-VALENCIAN-BALEAR) endivia; (CEBUANO) eskarola; (CHINESE, MANDARIN) ku qu, ch'i ts'ai; (CHINESE, WÚ) ku jii, ku cai, ku ma cai, ju mai cai; (CHINESE, YUE) tong hao, ju hua; (CREOLE, HAITIAN FRENCH) andiv; (CROATIAN) endivija; (CZECH) endivie, štěrbák; (DANISH) endivie; (DUTCH) andijvie; (ENGLISH) endive, escarole, winter endive, garden endive; (ESTONIAN) endiivia; (FILIPINO) endive, endiba; (FINNISH) endiivi; (FRENCH) chicorée endive, chicorée frisée, cichorée escarole, scarole; (GAELIC, MANX) sharrag; (GALICIAN) chicoria; (GERMAN) Endivie, Winterendivie, Endiviensalat; (GREEK) antidi, idos antidiou; (GUJARATI) kasini, kashini; (HEBREW) olesh; (HINDI) kasini, kashini; (HUNGARIAN) endívia, endívia sálata, saláta-katáng; (ICELANDIC) salatfifill; (INDONESIAN) andewi, endewi, andepi, lof brussel, lof putih; (ITALIAN) endivia, indivia, scarola; (JAPANESE) endaibu, kiku-jisha; (JAVANESE) andewi, ndewi; (KANNADA) kaasini gida; (KASHMIRI) hand; (LADINO) endivya; (LATVIAN) endāvijas; (LITHUANIAN) daržinė trūkažolė; (MACEDONIAN) cikorija endivija; (MALTESE) indivja; (MONGOLIAN) idree; (MORISYEN) andiv; (NEPALI) kesharee; (NORWEGIAN) endivie, krusendivie; (PAMPANGAN) eskarola; (POLISH) endywia; (PORTUGUESE) endívia, escarola, chicória escarole; (QUECHUA, CUSCO) achicoria; (ROMANIAN) endivia, cicoare de grădină; (RUSSIAN) endivij; (SANSKRIT) kaasani; (SERBO-CROATIAN) žucanica; (SLOVAK) endívia, čakanka štrbák; (SLOVENIAN) endivija; (SPANISH) endivia, endibia, escarola; (STELLINGWERFS) andievie, andijvie; (SUNDA) andewi; (SWAHILI) endivu; (SWEDISH) endiv sallat, endiviesallad; (TAGALOG) endiba; (TAJIKI) kosni, hundbo; (TAMIL) kaarpari sallodu, kaashini; (THAI) phakkat-foi; (TURKISH) bostan hindibasi, andiv, aci marul; (UKRAINIAN) endivìj; (URDU) kasni; (VLAAMS) andijvie; (WELSH) ysgall-y-meirch.

27.10. *Cichorium intybus* L. (EP) leaves; (P) R, C

a. chicory

(AFRIKAANS) sigorei, tjiekoriebos; (ALBANIAN, TOSK) brese, kore, radhige; (ARABIC) handab; (ARMENIAN) čharčhaduk; (ASSAMESE) chicory; (AZERBIAJANI) kasny; (BASQUE) xikore; (BELARUSAN) cichorija zvichainaja; (BENGALI) kasini; (BULGARIAN) salatna cichorija; (CATALAN-VALENCIAN-BALEAR) xicoira amarga; (CHINESE, HAKKA) foù-mâc; (CHINESE, MANDARIN) ye ku qu, k'u chu; (CHINESE, WÚ) ye ku jii, jii jii; (CHINESE, YUE) qu mai cai; (CROATIAN) radic; (CZECH) čekanka obecná setá, čekankový salat; (DANISH) cikorie, cikoriesalat; (DUTCH) cichorei, groenloof, witlof; (ENGLISH) chicory, witloof chicory, Brussels chicory, bunk, succory; (ESTONIAN) salatsigur, sigur; (FARSI, WESTERN) kasnee; (FINNISH) sikuri, salaattisikuri; (FRENCH) chicorée sauvage, chicorée sauvage ameliorée, chicorée à feuilles vertes, chicorée intybe, chicorée witloof; (GAELIC, IRISH) lus an tsiocaire; (GAELIC, MANX) lus y chugyr; (GALICIAN) chicoria; (GEORGIAN) vardakačača; (GERMAN) Chicorée, Zichorienwurzel,

Fleichchkraut, gemeine Wegwarte; (GREEK) radiki, kichoriou to entetmimenon; (HEBREW) olesh tarbuti; (HINDI) kasni, kasani, kashini, kaasani, gulhand, hinduva, suchal; (HUNGARIAN) cikória, cikóriasálata, mezei katáng; (ICELANDIC) bladsalatfifill; (INDONESIAN) witlof, cikori, andewi, andepi; (ITALIAN) cicoria comune, cicoria da taglio; (JAPANESE) kiku-nigana; (JAVANESE) andewi, ndewi; (KANNADA) chicory; (KAZAKH) cikorij; (LATVIAN) parastais cigoriņš, lapu cigorini; (LITHUANIAN) paprastoji trūkažolė; (MACEDONIAN) gorčliva endivija; (MALTESE) cikwejra; (MONGOLIAN) idree; (NORWEGIAN) sikori, salatsikori, rotsikori; (POLISH) cykoria, cykorija salatowa; (PORTUGUESE) chicória do café, chicória amarga, chicória selvagem, almeirão, chicoria; (ROMANIAN) cicoaraea de salată, cicoare; (RUSSIAN) cikornyj salat vitluf; (SANSKRIT) kaasani; (SERBO-CROATIAN) radič, ženetrga, vodopija, cigura, cikorija, obična; (SLOVAK) čakanka šalatná, čakanka obyčajná; (SLOVENIAN) radič; (SORBIAN, UPPER) cigurij, cichorija, šikorej; (SPANISH) achicoria silvestre, achicoria de raíz, achicoria de hojas verdes, achicoria de bruselas, amargón, almirón; (STELLINGWERFS) sukerei, sukeri'j; (SUNDA) andewi; (SWEDISH) cikoria, cikoria sallat, cikorisallad; (TAGALOG) chicory; (TAJIKI) chatrqoqu; (TAMIL) kaashini, kasani keerai, sekkirai, sevattu mullangi; (TELUGU) kaasni; (TURKISH) kök hindiba, sikon, çikori, radikya, adi aci marul, çilik, yabani hindiba; (TURKMEN) adaty syşratgy; (UKRAINIAN) vìtluf; (URDU) kasini; (VIETNAMESE) di[ees]p xo[aw]n; (VLAAMS) witloof, groefloof, cichorei; (WELSH) sicori; (XHOSA) itshikili; (YIDDISH) tsikoriah; (ZULU) shikoli.

b. red leafed chicory

(DANISH) rød cikorie; (DUTCH) roodloof; (ENGLISH) redleaf chicory, radicchio; (FRENCH) chicorée sauvage à feuilles rouges; (GERMAN) Radicchio, Rosettenzichorie; (GREEK) kokkinophyllom kichorion; (HUNGARIAN) vörös cikória; (ITALIAN) radicchio; (MACEDONIAN) obična cikorija; (NORWEGIAN) sikori; (POLISH) cikoria liściowa; (PORTUGUESE) chicória vermelha, chicória de folhas roxas, radichio; (SPANISH) achicoria de hoja roja, endibia acogollada roja; (VLAAMS) radicchio.

27.11. *Cirsium dipsacolepis* (Maxim.) Matsum. (EP) root; (P) P,C.

(CHINESE, MANDARIN) ju niu bang; (CHINESE, WÚ) ju niu bang, yia cong; (CHINESE, YUE) ju qu; (ENGLISH) gobouazami; (FINNISH) ohdake; (GREEK) kirsio; (HUNGARIAN) aszat; (JAPANESE) gobouazami, mori-azami, yamagobou; (LATVIAN) usnes; (MONGOLIAN) azargana; (NORWEGIAN) tistel; (POLISH) ostrożeń; (TAJIKI) morak; (TURKISH) devedikeni, kenger; (VLAAMS) yamagobou, gobouazami.

27.12. *Cosmos caudatus* Kunth (EP) leaves and young shoots; (P) R,C.

(AKLANON) turay-turay; (ASSAMESE) cosmos; (BONTOC, CENTRAL) tuktukaw; (CEBUANO) turay-turay; (CHINESE, WÚ) qiu ying; (DANISH) cosmos; (DUTCH) cosmos; (ENGLISH) cosmos; (ESTONIAN) kosmos, ilmalill; (FRENCH) cosmos; (GERMAN) Cosmos; (GREEK) kosmos; (IFUGAO) onwad, onwad di malikanu; (INDONESIAN) ulan raja, ulam raja, pelampong, kenikir, radamindang, randa midang; (ITALIAN) cosmea caudata; (JAVANESE) randa midang, kenikir, semikir, dewadaru; (MALAY) ulam raja, pelampong; (NEPALI) putali phool; (PORTUGUESE) cosmos; (SAMBAL, BOTOLAN) kadayum-dagi; (SPANISH) cosmos, estrella de mar; (SUBANON) gleli-aw; (SUNDA) randa miang, mamanggahan, kanikir; (TAGALOG) cosmos; (TAJIKI) mino; (THAI) daoruang-phama, khamhae; (TURKISH) kozmos.

27.13. *Crassocephalum biafrae* (Oliv. & Hiern) S. Moore (EP) young shoots and leaves; (P) R.

(ENGLISH) Sierra Leone bologi; (YORUBA) worowo.

27.14. *Crassocephalum crepidioides* (Benth.) S. Moore (EP) young shoots and leaves; (P) R.

(AKLANON) molo-mustasa; (AKWA) ossissibole, ossissiboni; (AMHARIC) qäsbädäğğ; (BICOLANO, CENTRAL) malid-palid; (BONTOC, CENTRAL) tabtabilokong; (BUHID) burangot, mak-mak lumak; (BULLOM SO) kikpɔ-lɛ, kumeny-ɛ; (ENGLISH) hawksbeard velvetplant; (EWONDO) abek, pisek; (GOLA) gimbo; (GREEK) katifes; (HANUNOO) apuk-apuk madapug, apuk-ili mabiru; (HAUSA) babahun, babaku, babohoh; (HMONG-DAW) chais qav; (IFUGAO) holo de babalel, holo de gangagang, holo di ameleng, kampellel, sabungay; (IGBO) alapolo, n̠tì-ènē; (ILOCANO) purpurikit; (INDONESIAN) sintrong; (JAPANESE) benibana-borogiku; (KONO) kumbɔnɛ; (KRIO) bɔlɔgi; (LOKO) gigboi, gipoi, jigboi, kigboi, kipoi; (MAKHUWA) marua; (MANDING-MANDINKA) kumbɛ; (MANINKAKAN, EASTERN) kumba; (MARANAO) gomaos, makaneb, rembayog; (MENDE) kikpɔ, kipɔ; (SUBANON) baburug; (TAGALOG) bulak-manok, kamangi, parasiyut; (TAGBANWA) rabun-rabun; (THAI) phakphet chaang, yaa kho on, phakhaan; (THEMNE) ka-lotho, *a*-bwɔ, *an*-lotho; (TOK PISIN) marago beja, yogobikabika; (VAI) gɛnɛ, kimbɔ; (VIETNAMESE) rau t[aaf]u bay; (WÁRAY-WÁRAY) mara-mara; (YORUBA) ebolo, efodu, egusi, ebòlò, ẹ̀fọ èbòlò, ẹ̀fọ ebúre.

27.15. *Cynara cardunculus* L. subsp. *cardunculus* (EP) petiole; (P) C.

(AFRIKAANS) kardoen; (ARABIC) korchef; (BASQUE) gardu; (BULGARIAN) kardon; (CATALAN-VALENCIAN-BALEAR) cart comu; (CHINESE, MANDARIN) shi yong ji, ta chii chi; (CHINESE, WÚ) shi yong ji; (CHINESE, YUE) ci cai ji; (CRIOULO, CAPE VERDE) cardo; (CZECH) kardy; (DANISH) kardon, kardone; (DUTCH) kardoen; (ENGLISH) cardoon; (FARSI, WESTERN) kangar; (FINNISH) kardoni, ruoti; (FRENCH) cardon; (GERMAN) Kardone, Kardy; (GREEK) agkinara e kaktos; (HEBREW) kinres, kinras; (HUNGARIAN) vad árticsóka, kardi; (ICELANDIC) fingrakornblom; (INDONESIAN) balakacida; (ITALIAN) cardone, carciofo; (JAPANESE) karudon; (LITHUANIAN) dvgliuotasis artišokas; (MACEDONIAN) španska artičoka; (MALTESE) qaqocc tax-xewk; (NORWEGIAN) kardonskokk, cardon, kardon; (POLISH) kard, kardy; (PORTUGUESE) cardo hortense, cardo do coalho, cardo coalhador, alcachofra brava, curdo; (ROMANIAN) cardonul; (RUSSIAN) kardon; (SERBO-CROATIAN) karda; (SLOVAK) kardy; (SLOVENIAN) kardij; (SPANISH) cardo comestible, cardo de huerta; (SWEDISH) kardon; (TURKISH) yabani enginar, kengerotu; (UKRAINIAN) kardon; (VIETNAMESE) ba cai; (VLAAMS) kardoner, kardoen.

27.16. *Cynara cardunculus* L. (EP) immature flower bud; (P) C.

(AFRIKAANS) artisjok; (ALBANIAN, TOSK) angjinare; (ARABIC) karchuf; (ARMENIAN) kangarni; (ASSAMESE) globe; (AZERBAIJANI, NORTHERN) ənginar; (BASQUE) alkatxofa; (BENGALI) hathichak, hathichoke; (BOSNIAN) artičoka; (BRETON) askol debriñ; (BULGARIAN) artišok; (CATALAN-VALENCIAN-BALEAR) carxofa; (CHINESE, MANDARIN) chao xian ji, chü chi; (CHINESE, WÚ) yang jiang, chao zian ji; (CHINESE, YUE) yang ji; (CROATIAN) articoka; (CZECH) artyčok, artyčok zeleninový; (DANISH) artiskok; (DUTCH) artisjok; (ENGLISH) globe artichoke, artichoke; (ESTONIAN) artišokk; (FINNISH) artisokka, latva-artisokka; (FRENCH) artichaut commun, artichaut; (GAELIC, IRISH) bliosán; (GAELIC, MANX) onnane, vlasstal; (GALICIAN) alcachofra; (GERMAN) Artischoke; (GREEK) agynara; (HEBREW) artishok, kinres tarbuti, cinres tarbuti; (HINDI) hathichoke, hathichak; (HINDUSTANI) ek qism ká ehhoṭá; (HUNGARIAN) árticsóka; (ICELANDIC) ætipistill; (INDONESIAN) artisyok; (ITALIAN) carciofo, articiocche; (JAPANESE) aachi chohku, chosen azami; (LADINO) enjenára, enjenera, anginara, articho; (LAKOTA) paŋgi; (LATVIAN) artišoks; (LITHUANIAN) daržovinis artišokas; (MACEDONIAN) kuglesta artičoka; (MALTESE) qaqocc; (MORISYEN) artiso; (NEPALI) bajarangee; (NORWEGIAN) artiskokk; (POLISH) karczoch zwyczajny, karczoch, artyszok; (PORTUGUESE) alcachofra; (ROMANIAN) anghinare, cardon; (RUSSIAN) artišhok; (SERBO-CROATIAN) artičoka, artišoka; (SLOVAK) artičoka; (SLOVENIAN) artičoka; (SPANISH) alcachofero, alcachofa; (SWEDISH) kronärtskocka; (TURKISH) enginar; (UKRAINIAN) artišok; (VIETNAMESE) [as]c ti s[oo], hoa dira; (VLAAMS) artisjok; (YIDDISH) artichuk.

27.17. *Doellingeria scabra* (Thunb.) Nees (EP) leaves; (P) C.

(ASSAMESE) aster; (CHINESE, MANDARIN) dong feng cai; (CHINESE, WÚ) dong feng cai, zi wan; (DANISH) aster, høstaster; (DUTCH) aster; (ENGLISH) aster; (ESTONIAN) aster; (FARSI, WESTERN) mina; (FRENCH) aster; (GAELIC, MANX) lus yn 'eail vaayl; (GERMAN) Aster; (GREEK) astir; (HEBREW) aster; (HUNGARIAN) gerebcsin; (ITALIAN) astro; (JAPANESE) shirayama-giku; (LATVIAN) asteres; (MACEDONIAN) rapavo mitrovdenče; (MONGOLIAN) gol geser; (NORWEGIAN) asters; (POLISH) aster chiński; (PORTUGUESE) aster, florinhas de San Francisco; (SPANISH) aster; (TAJIKI) momoqaimoq; (TURKISH) saraypati, yildizçiçeği, aster saraypati; (VLAAMS) zeeaster, chui, shirayama-giru; (WELSH) ser-flodau.

27.18. *Emilia sonchifolia* (L.) DC. (EP) young shoots and leaves; (P) R,C.

(AKAN-ASANTE) guakuro; (AKWA) mbono; (BEMBÉ) potosso-tari; (BENGALI) sadhimodi, sudhimudi; (BICOLANO, CENTRAL) kipot-kipot, libun; (CEBUANO) pisowa-pisowa, setim; (CHINESE, MANDARIN) yi dian hong; (CHINESE, WÚ) yi dian hong; (CRIOULO, CAPE VERDE) padja-leite, serralha; (ENGLISH) sow thistle, emilia, cupid's shaving brush; (FILIPINO) tagulinao; (FRENCH) emilie, emilie à feuilles de laiteron, herbe a lapin; (GREEK) kadros; (HANUNOO) apuk-apuk taga-bas; (HAUSA) hurahuran boka; (HILIGAYNON) mulumustasa; (HINDI) hirankhuri, hirankhuree; (IFUGAO) domolnu, dumalum, holo; (ILOCANO) lamlampaka; (INDONESIAN) kemendilan, patah kemudi, jombang, tempuh wiyang, ketumbit jantan, tetambak merah, perasi; (ISNAG) babay-elas; (ITALIAN) emilia; (IVATAN) yagod-no-kangkang; (JAPANESE) usu beninigana; (JAVANESE) kemondelan, kemandelan, kemendilan, tempuh wiyang; (KANNADA) elikivi gida, elikivisoppu, jumki hoo; (KHMER, CENTRAL) smau reang tük; (KISI, SOUTHERN) koumba; (KISSI, NORTHERN) kumba; (LAHU) g'aw$^{\vee}$ hk'a$^{\vee}$ bo$_{-}$, g'aw$^{\vee}$ hk'eh$^{\vee}$ bo$_{-}$, ha pui k'a$_{-}$ g aw$^{\vee}$; (MALAY) ketumbit jantan, setumbak merah; (MALAYALAM) mulshevi; (MANINKAKAN, EASTERN) kumbi; (MANINKAKAN, WESTERN) koumbi; (MANU'A SAMOAN) fau lele lili'i; (MARANAO) katabas sa palaw; (MARATHI) sadamandi, sadamandee; (MENDE) kipɔ; (MORISYEN) lastron; (NEPALI) mula pate; (NIUE) pupu lele; (PORTUGUESE) emilia, erva do fígado, falsa serralha; (SAMBAL, BOTOLAN) handuyong a lilinaw, tagulinaw; (SAMOAN) meleni vao, fua lele lili'i; (SANSKRIT) sasasruti, sasarutih, saasasruthi; (SANTALI) kulại lutur; (SPANISH) emilia, clavel chino, pincel de amor, hierba socialista; (SUNDA) jonge, jonghe, kemondilan; (TAGALOG) tagulinau; (TAGBANWA) panisipit balbal; (TAMIL) neichitti poondu; (THAI) hangplachon, phakdaeng, phakbang; (TONGAN) longolongo'uha; (TULU) elikevi; (TURKISH) emilya; (URDU) hirankhari; (VIETNAMESE) rau m[as] ti[as], rau chua l[ef], h[oof]ng b[oos]i th[ar] o; (VLAAMS) gewone melkdistel; (YORUBA) ọdúndún odò.

27.19. *Enydra fluctuans* Lour. (EP) young shoots and leaves; (P) R,C.

(ASSAMESE) helachi sak, manoa sak; (BULU) epinar; (CHINESE, WÚ) nui bo cai; (ENGLISH) buffalo spinach; (INDONESIAN) godobos; (JAVANESE) godobos, kokoha; (KHMER, CENTRAL) kanting ring; (LAO) bungz ping; (MALAY) chengkeru, kangkong kerbau; (MINANGKABAU) ambur; (TAGALOG) kangkong kalabaw, kangkong-kabayo, kangkong-kalabaw; (THAI) phakbung-ruam, phakbung-pling, phakpaeng; (VIETNAMESE) rau ng[oor] tr[aa]u, rau ng[oor] th[ow]m.

27.20. *Farfugium japonicum* (L.) Kitam. (EP) petiole; (P) C.

(CHINESE, MANDARIN) da wu feng cao; (CHINESE, WÚ) da wu feng cao; (DANISH) japansk foelfod; (ENGLISH) Japanese farfugium; (JAPANESE) tsuwabuki; (POLISH) tsuwabuki; (PORTUGUESE) farfúgio; (SPANISH) farfugio; (VLAAMS) tsuwabuki.

27.21. *Fedia cornucopiae* Gaertn. (EP) leaves; (P) C.

(AFRIKAANS) valeriaan; (CHINESE, MANDARIN) fei zhou ye qu; (CHINESE, WÚ) fei zhou ye ju; (CZECH) kozlik; (DANISH) algiersalat; (DUTCH) overdadige fedia; (ENGLISH) African valerian, horn of plenty; (ESTONIAN) palderjan; (FINNISH) salaattivirmakka; (FRENCH) fédia, corme d'abondance, doucette d'Alger, valériane d'Alger; (GERMAN) überreichliche Fedia, algerisches Baldrian; (GREEK) valeriani e keratoïdis, valeriana; (HUNGARIAN) afrikai macskagyökér; (ITALIAN) lattughella, cornucoppia; (MACEDONIAN) valerijana; (MALTESE) sieq il-hamiema; (POLISH) kozłek lekarski, waleriana, kozłek; (PORTUGUESE) cornucopia, alface de Alger, valeriana; (ROMANIAN) odolean, valeriana; (RUSSIAN) valeriana, valerjana, maoun; (SPANISH) cornucopia, trompetillas; (TURKISH) kediotu, dördüzotu; (VLAAMS) valeriaan; (WELSH) triaglog.

27.22. *Galinsoga parviflora* Cav. (EP) young shoots; (P) C.

(AFRIKAANS) knopkruid; (BELARUSAN) galinzoga drobnakvetkava; (CHINESE, WÚ) la zi cao; (CHUWABU) nantala; (CZECH) pěťour maloúborný; (DANISH) håret kortstråle; (DUTCH) klein knopkruid; (ENGLISH) galinsoga, guasca, yellow weed, gallant soldier, small flower galinsoga; (FINNISH) tarhasaurikki; (FRENCH) galinsoga glabre, galinsoga à petites fleurs; (GERMAN) kleinblütiges Knopfkraut, kleinblütiges Franzosenkraut; (HUNGARIAN) gombvirág; (INDONESIAN) bribil; (ITALIAN) galinsoga comune; (JAVANESE) kuningan, balakecut, galinggang, galangsi; (LITHUANIAN) smulkiažiedė galinsoga; (MACEDONIAN) sitnocvetna galinsoga; (MAKHUWA) rrerra; (NEPALI) chitlanje ghams; (NORWEGIAN) peruskjellfrø; (POLISH) żółtlica drobnokwiatowa; (PORTUGUESE) galinsoga, erva da moda, picão bravo, picao branco, fazendeiro; (QUECHUA, CUSCO) pacpa yuyo; (SERBO-CROATIAN) galinsoga, sitnocvjetna; (SHONA) nwamuna ali gone, nwamuna aligone, n'wana ali gone, mamuna ali gone, teketera; (SPANISH) galinsoga de flores pequeñas; (SWAHILI) majani ya sungura, mwingereza; (TAGALOG) galinsoga; (TURKISH) küçük yaprakli düğmeotu, düğme otu; (TURKMEN) mayday gülli galinsoga; (VLAAMS) knopkruid, guasca.

27.23. *Glebionis coronarium* (L.) C. Jeffrey (EP) leaves and tender shoots, (P) C.

(AFRIKAANS) eetbare krisant; (ALBANIAN, TOSK) lule vjeshte, krizantemë; (ASSAMESE) babori sak; (BELARUSAN) chrizantema uvenchanaja; (BENGALI) guldaudi; (CHINESE, HAKKA) k'iôuc-tā; (CHINESE, MANDARIN) ou tong hao; (CHINESE, MIN NAN) kiok, kiok-á, kiok-hoe; (CHINESE, WÚ) ou tong hao, ju hua; (CREE, PLAINS) misiwapikwan; (CZECH) kopretina; (DANISH) okseøje, krysanthemum, kæmpemargerit, vinterasters; (DUTCH) ganzebloem, chrysant; (ENGLISH) edible chrysanthemum, garland chrysanthemum, crown daisy, tangho, Shungiku, chop suey green, Japanese greens; (ESTONIAN) krüsanteem; (FILIPINO) tango; (FINNISH) vihanneskrysanteemi; (FRENCH) chrysanthème, chrysanthemum, chrysanthème des jardins (GALICIAN) ollo deboi, pampullo; (GERMAN) Wucherblume, Chrysantheme; (GREEK) chrysanthemo to stefanoto; (HEBREW) ḥarẓit, harstit; (HINDI) guldaudi; (HUNGARIAN) ehető krizantém, krizantém, téli margitvirág; (ICELANDIC) krysi; (INDONESIAN) krisan, seruni; (ITALIAN) crisantemo, margherita; (JAPANESE) shungiku, shokuyou-giku; (JAVANESE) sruni, tango, saruni walanda; (KANNADA) shevanti; (LAO) tang 'ôô; (LATVIAN) vainagu krizantēma; (LITHUANIAN) skaistažiedė; (MACEDONIAN) margaritka; (MALAY) tong ho, kek wah; (MALTESE) lellux; (MARATHI) gulesvati, tursiphal; (MONGOLIAN) nartsag, tunkhuu; (NORWEGIAN) spiselig krageblom, krysantemum; (PANJABI, EASTERN) bagaur, zaenil; (POLISH) złocień jadalny; (PORTUGUESE) crisântemo, margarida; (ROMANIAN) tufănică, crizanteme; (RONGA) thosani, uncholo (SANSKRIT) chandramallika; (SERBO-CROATIAN) hrizantema; (SLOVAK) králik indický, chryzantéma; (SPANISH) crisantemo, margarita; (SUNDA) kerisan; (TAGALOG) tango; (TAJIKI) guli dovudi; (TAMIL) shamantippu; (TELUGU) chamanti; (THAI) phaktang-o, phakkhikhwai; (TURKISH) krizantena, kasimpati; (TURKMEN) hrizantema; (URDU) gul-e-daudi; (VIETNAMESE) c[ar]i c[us]c, c[us]c t[aaf]n [oo]; (VLAAMS) chrysant; (WELSH) ffarwel haf.

27.24. *Gymnanthemum amygdalinum* (Delile) Sch. Bip. ex Walp. (EP) young shoots; (P) C.

(AFRIKAANS) bitterblaar; (AJA) aloma; (AKAN-AKUAPEM) awonẽwẽ, awonywẽnẽ, (AKAN-ASANTE) bowin, mponasere; (AKAN-FANTE) o-bɔwen(e); (AMHARIC) grawa; (ASSAMESE) pat tita; (BAMILEKE) bitalé; (BEROM) etulùp; (BULU) yôlô yôlô, metet; (CHINESE, WÚ) ku ye ban jin ju; (DANGME) agba; (DANISH) vernonie; (DENDI) suwaka; (DUTCH) bitterblad; (EDE NAGO) aluba; (EDO) óríwò; (ENGLISH) bitter leaf; (ÉWÉ) gbɔ, gbɔti, gbɔtsi, avenya, aloma; (EWONDO) metet, yoloyolo; (FON) amavivɛ; (FRENCH) vernonie, vernonia; (FULFULDE, ADAMAWA) kaɗkaɗɗe, siwaakewal; (GA) agɔaflu, akpa, tà; (GBAYA) panboka; (GBE, WACI) aloma; (GERMAN) Vernonia; (HAUSA) shiwáákáá, shìwaákaá; (HMONG DAW) hmab dib iab; (IGBO) onugbu, òlúgbù, òlúbi, òlúbù, ònúgbù, ònúbù, òlúbì, òlúgbì; (ITALIAN) vernonia; (IZON) kíríòlògbò; (KRIO) bita-lif, bittas; (MAKHUWA) navata; (MANDING-MANDINKA) kossa fina; (MENDE) nje nyani; (PORTUGUESE) pau fede, fede, vernonia, libó, sucumadeira; (PULAAR) bantara bururé, dakuna; (SHONA) tosonzoro, tson:dzoro, m'sodzo, chipanza; (SPANISH) vernonia; (SWAHILI) jani chungu; (TAJIKI) chashmi gunjishk; (TEM) tusima, tingma, tusima, aloma, aloma kulu; (THEMNE) *a*-bita-lif, *a*-dif-wir, *an*-gbɔnthɔ, ɛ-bita-lif, ɛ-dif-wir; (TIV) ityuna; (TURKISH) aci yaprak, verconia; (VLAAMS) bitterblad; (WELSH) dail surion; (YORUBA) aluba, ewúrò, ewúrò jíjẹ, ewúrò oko.

27.25. *Gynura bicolor* DC. (EP) young leaves; (P) C.

(CHINESE, MANDARIN) zi bei tian kui; (CHINESE, WÚ) zi bei tian kui, guan yin xian; (CHINESE, YUE) zi bei tian kui; (CZECH) gynura, protažitka; (DANISH) flvejlsblad; (ENGLISH) gynura; (FRENCH) gynura à deux couleurs; (HUNGARIAN) ginura; (INDONESIAN) tali jiwo; (JAPANESE) suizenjina; (LITHUANIAN) dvispalvė ginūra; (TURKISH) gynura; (VLAAMS) suizenjina.

27.26. *Helianthus tuberosus* L. (EP) tubers; (P) C,R.

(AFAR) fura; (AFRIKAANS) knolartisjok, Jerusalem artisjok; (ALBANIAN, TOSK) mollë, mollë dheu; (ARABIC) tertufa; (ASSAMESE) Jerusalem artichoke; (BELARUSAN) slanechnik clubneneosny; (BENGALI) brahmokha; (BULGARIAN) topinambur; (CATALAN-VALENCIAN-BALEAR) nyamara; (CHECHEN) balg; (CHINESE, GÀN) yang sheng iiang; (CHINESE, MANDARIN) ju yu, ngah koh yui na; (CHINESE, MIN BEI) yang iian; (CHINESE, XIANG) yang jiang; (CHINESE, WÚ) ju yu, yang jinag, gui zi jiang; (CHINESE, YUE) yang jiang, qui zi jiang; (CZECH) topinambur, topinambur hlíznatý; (DANISH) jordskok; (DUTCH) aardpeer, topinamboer; (ENGLISH) Jerusalem artichoke; (ESTONIAN) maapirn, topinambur; (FINNISH) preeria-auringonkukka, mukula-artisokka; (FRENCH) topinambour, artichaut de Jérusalem; (GERMAN) Topinambur, Erdbirne, Erdartischecke; (GREEK) kollokasi, ilianthos o koudyllorizos; (HEBREW) ḥammanit ha-pekaot, artishok yerushalmy, chamanit hapkaot; (HINDI) hatichuk, hattichok; (HINDUSTANI) háthí-pích; (HMONG DAW) qos ntoo; (HOPI) aqàwtumna; (HUNGARIAN) csicsóka; (ICELANDIC) ætifífill; (INDONESIAN) artisyok jahudi; (ITALIAN) topinambour, topinambour elianto; (JAPANESE) kikimo; (KANNADA) artichoke gedde, sooryakaanthi gedde; (KAZAKH) topinambur; (LATVIAN) topinambūrs; (LITHUANIAN) topinambas; (MACEDONIAN) krtolest sončogled; (MALTESE) articokks; (NORWEGIAN) knollsolsikke, jordskokk, knolljordskokk; (PANJABI, EASTERN) hathipich; (POLISH) topinambur, bulwa; (PORTUGUESE) topinambo, girassol tuberoso, girassol batateiro, tupinambor; (ROMANIAN) topinambur, napi porceşti; (RUSSIAN) grusha zemljanaja, kartofel loshadnnyi, padsolnechnik klubnenosnyi, topinambour; (SANSKRIT) hasthipijoo, vajrangi; (SERBO-CROATIAN) topinambur; (SLOVAK) topinambur, slnečnica topinambur; (SLOVENIAN) topinambur; (SPANISH) topinambo, pataca, aguaturma; (SWEDISH) jordärtskocka, jordskocka; (TAJIKI) vartoj, oftobparast; (THAI) thantawan-hua; (TURKISH) yer elmasi, yerelmasi; (TURKMEN) er armydy, topinambur; (UKRAINIAN) topìambur; (URDU) uroslam archichoke; (VIETNAMESE) qu[yf] d[oj]i; (VLAAMS) aardpeer, topinamboer; (WELSH) march-ysgall.

27.27. *Ixeridium dentatum* (Thunb.) Tzvelev (EP) young leaves; (P) C.

(ENGLISH) sowthistle; (KOREAN) durupnamu.

27.28. *Kalimeris yomena* (Kitam.) Kitam. (EP) leaves, young plants; (P) C.

(CHINESE, MANDARIN) chi-êrh-ch'ang; (CHINESE, WÚ) ma lan, ma lan tou; (CHINESE, YUE) ju yu; (ENGLISH) kalimeris, false aster, Japanese aster; (JAPANESE) yomena; (MACEDONIAN) zemno jabolko; (POLISH) aster jomena; (VLAAMS) yomena.

27.29. *Lactuca indica* L. (EP) leaves; (P) C.

(AFRIKAANS) Indiese slaai; (ARABIC) khass hindy; (ASSAMESE) salad pat, salad; (BENGALI) letus; (BONTOC CENTRAL) batudan; (CEBUANO) gilgiloy; (CHINESE, MANDARIN) shan wo ju; (CHINESE, WÚ) shan wo ju; (DANISH) indisk salat; (ENGLISH) Indian lettuce; (FRENCH) laitue d'Inde, laitue indienne; (GERMAN) Indischer Lattich; (GREEK) indiko marouli; (GUJARATI) salit; (HEBREW) hassa odit, chasah indianit; (HINDI) salad; (ILOCANO) gilgiloy; (INDONESIAN) daun panjang, daun lampeni, kumak, komak, umek, lampenas; (IVATAN) yayod; (JAPANESE) aki-no-nogeshi; (JAVANESE) tempuh wiyang, sawirana, sawi rana, kuban kayu rana; (KANKANAEY) gatudan; (KANNADA) lettuce; (MACEDONIAN) indiska salata; (MALAY) sawi rana, daun panjang; (MALAYALAM) lettuce; (MALTESE) il-hass; (MARATHI) salit, letus; (MONGOLIAN) ziraa; (NEPALI) jiri saag; (ORIYA) salad; (PANJABI, EASTERN) salad; (POLISH) sałata indiańska; (PORTUGUESE) alface da India; (SPANISH) lechuga de la India; (SUNDA) lampenas, lampeni, hedyo huma, langkoas, tongo; (SWAHILI) mchunga; (TAJIKI) kohu; (TAMIL) shallaathu; (TELUGU) letuse; (TURKISH) hint marulu; (VIETNAMESE) di[ees]p d[aj]i, rau b[oof] c[os]c, b[oof] c[oo]ng anh; (VLAAMS) indische sla.

27.30. *Lactuca sativa* L. var. *angustana* L.H. Bailey (EP) stem; (P) C,R.

(ARABIC) khass saki; (ASSAMESE) salad pat; (CHINESE, MANDARIN) nen jing wo ju; (CHINESE, WÚ) wo ju, nen jin wo ju; (CHINESE, XIANG) wo sun; (CHINESE, YUE) wo qu sun; (CZECH) salát chřestový; (DANISH) aspergesalat; (DUTCH) aspergesla; (ENGLISH) celtuce, stem lettuce, asparagus lettuce; (FINNISH) parsasalaatti; (FRENCH) laitue asperge, laitue asperge romain; (GERMAN) Spargelsalat; (GREEK) thridax e imeros me kotsani; (HMONG DAW) zaub; (ICELANDIC) spergilsalat; (INDONESIAN) slada, letus; (ITALIAN) lattuga asparago; (JAPANESE) kaki-jisha; (LATVIAN) kultivēti salāti; (MACEDONIAN) steblena salata; (MALTESE) hass twil; (MONGOLIAN) ziraa; (NORWEGIAN) stengelsalat; (POLISH) sałata szparagowa, głąbiki krakowskie; (PORTUGUESE) alface espargo; (SPANISH) lechuga espárrago; (SWEDISH) sparrissallat; (YIDDISH) sahlaht, shahlahtn.

27.31. *Lactuca sativa* L. var. *capitata* L. (EP) leaves; (P) R.

(AFRIKAANS) slaai, kopslaai; (ALBANIAN, TOSK) marule, marule e kultivuar; (AMHARIC) sälata; (ARABIC) salata; (ASSAMESE) salad pat, salad; (AZERBAIJANI, NORTHERN) kahı; (BASQUE) ugaza; (BELARUSAN) salata latuk, salata; (BEMBÉ) saladi; (BENGALI) kahu, salad, letus; (BRETON) letuz, lezegenn; (BULGARIAN) glavesta salata; (BULU) salad; (BURMESE) sa-lat-ywet; (CAJUN FRENCH) tête de lettue; (CATALAN-VALENCIAN-BALEAR) llatuga, enciam; (CHECHEN) koopasta; (CHICKASAW) tohi'; (CHINESE, GÀN) wo ma, sheng cai; (CHINESE, HAKKA) mâc-tzè; (CHINESE, MANDARIN) wo ju, chüan hsin wo chü; (CHINESE, XIANG) sheng cai; (CHINESE, WÚ) ma ju, jie qiu wo ju; (CHINESE, YUE) wo ju sun, wo sun; (CREE, PLAINS) nipiya ohtihipak; (CREOLE, HAITIAN FRENCH) leti; (CRIOULO, CAPE VERDE) alface; (CROATIAN) salata; (CZECH) salát hlávkovy; (DAGAARE, SOUTHERN) saladi; (DANISH) hovedsalat; (DUTCH) kropsla, botersla; (ENGLISH) head lettuce, butterhead lettuce, lettuce; (ESTONIAN) peasalat; (FARSI, WESTERN) tukm-i-kahu, kaho; (FIJIAN) lettuce; (FILIPINO) letsugas; (FINNISH) keräsalaatti; (FRENCH) laitue pommée, laitue; (GAELIC, IRISH) leitis chabáiste; (GAELIC, MANX) lvs; (GAELIC, SCOTTISH) leiteis; (GALICIAN) alface; (GEORGIAN) thaviani salata; (GERMAN) Buttersalat, Krachsalat, Kopfsalat; (GIKUYU) mboga cia gĩthũngũ; (GOURMANCHÉMA) saladi; (GOURMANTCHÉ) bompièno- pkéonfari; (GREEK) marouli, kefaloto marouli; (GUJARATI) salit; (HEBREW) ḫassa, ḫassa tarbutit, chasa tarbutit, chasah; (HINDI) kahu, salad; (HUNGARIAN) fejes saláta; (ICELANDIC) höfuđsalat; (INDONESIAN) slada krop, selada kerup, selada, slada; (ITALIAN) lattuga a cappuccio, lattuga cappuccia, lattuga a testa; (JAPANESE) letasu, tama chisha; (JAVANESE) selada, sladah; (KANNADA) lettuce; (KAZAKH) salat; (KHMER, CENTRAL) sallaat; (KIOWA) sɑhyei-'ei-goup; (KIRGHIZ) salat; (KOREAN) sangchu, selat; (KRIO) salad; (KWÉYÒL) léti; (LADINO) alechuga, lechuga, litchouga; (LAKOTA) waȟpeiŋkpajiji; (LATVIAN) lapu salāti, galvinsalāti; (LITHUANIAN) salota, salotos; (LUXENBOURGEOIS) Zalot; (MACEDONIAN) glavesta salata; (MALAY) selada bulat, salad, selada; (MALAYALAM) lettuce; (MARATHI) salit, letus, kahu; (MARINDUQUE) tawa; (MONGOLIAN) ziraa; (MÒORÉ) salad; (MORISYEN) leti; (NAVAJO) ch'il łigaaí; (NGUNDI) saladi; (NORWEGIAN) hodesalat, hagesalat, issalat; (ORIYA) salad; (PANJABI, EASTERN) sallad, salad, kahu, bæ̃d putta sullad; (POLISH) sałata głowiasta; (PORTUGUESE) alface, alface repolhuda, alface de repolho, alface francesa, repolhuda; (ROMANIAN) salată căpătină, lăptucă; (RUSSIAN) salat kočhannyi, salat kočannyj; (SERBO-CROATIAN) salata glavičasta, salata vrtna; (SLOVAK) šalát záhradný, šalát hlávkový; (SLOVENIAN) glavnata solata; (SORBIAN, UPPER) hłojčka soloteje; (SPANISH) lechuga acogollada, lechuga arrepollada; (STELLINGWERFS) slao; (SUNDA) salada, salada bogor; (SWAHILI) mchunga; (SWEDISH) huvudsallat; (TAGALOG) litsugas, letsugas; (TAMIL) shallaathu, salattu, shallathu, shatlatuvirai; (TATAR) salat; (TELUGU) letuse; (THAI) phakkat-hom, phaksalat, phakkat-kaeo; (TOK PISIN) letis; (TURKISH) kivircik, baş salata, marul; (TURKMEN) salat; (UKRAINIAN) salata golovčastij; (URDU) dudak'kahu; (UYGHUR) bir khil köktat; (UZBEK) cokhy, salat; (VIETNAMESE) rau xà lách, rau di[ees]p, x[af] l[as]ch, di[ees]p xo[aw]n; (VLAAMS) kropsla; (WELSH) letysen gron; (YIDDISH) sahlaht, shahlahtn.

27.32. *Lactuca sativa* L. var. *longifolia* Lam. (EP) leaves; (P) R.

[1]Itsukha dialect of Luyia.

(AFRIKAANS) blaarslaai; (ALBANIAN, TOSK) marule e kultivuar; (AMHARIC) sälata; (ARABIC) khass waraxi; (ASSAMESE) salad pat, salad; (AZERBAIJANI, NORTHERN) kahı; (BELARUSAN) salat latuk; (BEMBÉ) saladi; (BENGALI) kahu, salad, letus; (BICOLANO, CENTRAL) letsugas, litsuga; (BINUKID) letsugas, litsuga; (BONTOC CENTRAL) letsugas, litsuga; (BOSNIAN) salata; (BRETON) letuz; (BULGARIAN) listna salata za rjazane, marulja; (BULU) salad; (BURMESE) salat-ywet; (CAJUN FRENCH) lettue; (CATALAN-VALENCIAN-BALEAR) enciam; (CEBUANO) letsugas, litsuga; (CHECHEN) koopasta; (CHICKASAW) tohi'; (CHINESE, HAKKA) mâc-tzè; (CHINESE, MANDARIN) ye wo ju, chou yeh wo chü; (CHINESE, MIN NAN) gô (giâ)-á chhài, gô-á-chhài; (CHINESE, WÚ) chang ye wo ju, san ye wo fu; (CHINESE, YUE) juan xin wo ju, qiu ye wo ju, jie qiu wo ju, sheng cai, san ye sheng cai; (COMANCHE) puhi tuhkapu̱; (CREE, PLAINS) nipiya; (CREOLE, HAITIAN FRENCH) léti; (CRIOULO, CAPE VERDE) alface; (CROATIAN) salata; (CZECH) salát křezu, salát listový, salát římský; (DAGAARE, SOUTHERN) saladi; (DAKOTA) waȟpeyutapi; (DANISH) bindsalat, pluksalat, snitsalat, romersk salat; (DUTCH) pluksla, snijsla; (ENGLISH) leaf lettuce, romaine lettuce, lettuce; (ESTONIAN) rooma salat; (FARSI, EASTERN) kâhu; (FARSI, WESTERN) tukm-i-kahu, kâhu; (FIJIAN) lettuce; (FILIPINO) letsugas; (FINNISH) lehtisalaatti, sidesalaatti; (FRENCH) laitue romaine, laitue lombard, laitue à couper, chinchon, laitue frisée, laitue; (GAELIC, MANX) lvs; (GAELIC, SCOTTISH) leiteis; (GEORGIAN) photlovani salata, salata romeni; (GERMAN) Blattsalat, Römischer Salat, Bindersalat, Bindsalat, Kochsalat, Sommer-Endivie; (GIANGAN) letsugas, litsuga; (GIKUYU) mboga cia gĩthũngũ; (GOURMANCHÉMA) saladi; (GOURMANTCHÉ) saladi; (GREEK) marouli (salata); (GUJARATI) salit; (HAUSA) letas; (HEBREW) ḥassa, ḥassa tarbutit, chasah; (HILIGAYNON) litsúgas; (HINDI) kahu, salad; (HINDUSTANI) káhú, salad; (HMONG DAW) zaub; (HUNGARIAN) kötözoő saláta, tépösaláta, kötözosaláta, leveles saláta; (IBANAG) letsugas, litsuga; (ICELANDIC) blađsalat, romverskt salat; (IFUGAO) letsugas, litsuga; (ILOCANO) letsugas, litsuga; (INDONESIAN) selada, selada kerup, slada daun, slada; (ITALIAN) lattuga di taglio, lattuga romana; (JAPANESE) liifu letasu, koki chisha; (JAVANESE) selada, selat, sladah; (KANNADA) lettuce, salaadi soppu, salaat; (KAZAKH) listovoj salat, salat; (KHMER, CENTRAL) sallaat; (KIRGHIZ) salat; (KRIO) lɛtyus, salad; (KURDISH, NORTHERN) kāhū; (KWÉYÒL) léti; (LADINO) alechuga, lechuga, litchouga; (LAKOTA) waȟpeiŋkpajiji; (LATVIAN) lapu salāti, romiešu salāti, lapu salāti; (LITHUANIAN) salota; (LUXENBOURGEOIS) Zalot; (LUYIA) libumbe[1]; (MACEDONIAN) marula, režana, lisna salata; (MALAY) selada, salad; (MALAYALAM) lettuce; (MANINKAKAN, WESTERN) salato; (MARATHI) salit, letus, kahu; (MARINDUQUE) tawa; (MICMAC) ěskipkodŭmŭgāwā'; (MOHAWK) onerahtakate; (MONGOLIAN) ziraa; (MÒORÉ) salad; (MORISYEN) tia-mimak; (MUSKOGEE) lanē-papkv; (NAVAJO) ch'il łigaaí; (NGUNDI) saladi; (NIUE) lētisi; (NORWEGIAN) bindsalat, salat romain, bladsalat, salat, hagesalat; (NYANJA) saladi; (ORIYA) salad; (PAMPANGAN) letsugas, litsuga; (PANGASINAN) luchúgas; (PANJABI, EASTERN) sallad, salad, putta sullad, kahu; (PASHTO, NORTHERN) kâhu; (POLISH) sałata długolistna, sałata liskowa, sałata rzymska, sałata do skubania; (PORTUGUESE) alface romana, alface orelha de mula, alface de cortar, pura cortar; (PULAR) salade; (ROMANIAN) marulă, salată de foi, lăptuci de grădină; (RUSSIAN) salat latuk, salat litovol, listovoj salat, rimskij latuk; (SERBO-CROATIAN) salata romana, lisna salata, salata glavičasta; (SHONA) letisi; (SIDAMO) salâ ṯa; (SINHALA) salada kola; (SLOVAK) šalát římsky, šalát listovy, šalát záhradný; (SLOVENIAN) solata vezivka, solata berivka; (SORBIAN, UPPER) hłójčkaty solotej, solotej; (SOTHO, NORTHERN) selae; (SOTHO, SOUTHERN) selae; (SPANISH) lechuga romana, lechuga par cortar, lechuga de hojas, lechuguino; (STELLINGWERFS) slao; (SUNDA) salada, salada bogor, salada bokor; (SWAHILI) mchunga, letusi; (SWEDISH) bindsallat, plocksallat, bladsallat, romersk sallat; (TAGALOG) litsugas, letsugas; (TAMIL) shallaathu, salattu, shallathu, shatlatuvirai, salladu keerai; (TATAR) salat; (TAUSUG) letsugas, litsuga; (TELUGU) letuse; (THAI) phakkat-hom, phaksalat, phakkat-kaeo; (TIBETAN) loqɔɔ pɛ̃ɛ̃tshɛɛ; (TOK PISIN) letis; (TSWANA) selae; (TURKISH) yaprak salata, marul, kirivik salata; (TURKMEN) salat; (UKRAINIAN) salat-romen; (URDU) kahu; (UYGHUR) bir khil köktat; (UZBEK) cokhy, salat; (VENDA) lethìsi, tshilài; (VIETNAMESE) rau di[ees]p, x[af] l[as]ch, di[ees]p xo[aw]n; (VLAAMS) krulsla; (WELSH) letysen ddail; (WOLOF) salat; (XHOSA) iletusi, iletisi, umfuno obukhaphetshu onamagqabi atyiwa engaphekwanga; (YIDDISH) sahlaht, shahlahtn; (ZARMA) salaati; (ZULU) uletisi.

27.33. *Launaea taraxacifolia* (Willd.) Amin, ex C. Jeffrey (EP) leaves; (P) C.

[1]Itsukha dialect of Luyia.

(CHINESE, WÚ) ye wo ju; (CHINESE, YUE) chang ye wo ju, sheng cai, san ye sheng cai; (ENGLISH) wild lettuce; (ÉWÉ) ayoto; (GREEK) agrio marouli; (HMONG DAW) zaub ntsuab qus; (ITALIAN) launea a foglia di tarassaco; (JOLA-FONYI) esalaate; (LUYIA) likatsila[1]; (SWAHILI) mchunga; (URDU) janli salad; (YORUBA) yanrin.

27.34. *Ligularia fischeri* Turcz. (EP) leaves; (P) C

(KOREAN) gomchwi.

27.35. *Petasites japonicus* (Siebold & Zucc.) Maxim (EP) petiole; (P) C.

(CHINESE, MANDARIN) feng dou cai, feng tou tsai; (CHINESE, WÚ) feng dou cai, she tou cai'; (CZECH) devětsil japonský; (DANISH) japansk hestehov; (DUTCH) Japans hoefblad; (ENGLISH) butterbur, Japanese butterbur, sweet coltsfoot; (FINNISH) japaninruttojuuri; (FRENCH) pétasitès japonais; (GERMAN) Japanische Pestwurz; (GREEK) petasitis o iaponikos; (HUNGARIAN) acsalapu; (ICELANDIC) hestahofur, pestarrot; (INDONESIAN) daun fuki; (ITALIAN) farfaraccio giapponese; (JAPANESE) fuki; (MACEDONIAN) japonski lopuš; (NORWEGIAN) japanpestrot; (PORTUGUESE) petaside japonês; (SPANISH) fárfara japonesa; (TURKISH) Japon öksürükotu, vebaotu, öksürükotu, kişbambulu; (VLAAMS) fuki.

27.36. *Scolymus hispanicus* L. (EP) roots and leaves[s]; (P) C.

(AFRIKAANS) oesterwortel; (ARABIC) sunariysh; (BASQUE) kardaberaiska; (CATALAN-VALENCIAN-BALEAR) card bort; (CHINESE, MANDARIN) xi ban ya bo luo men shen, huang hua chi; (CHINESE, WÚ) xi ban ya bo luo men shen; (DANISH) guldtidsel; (DUTCH) spaanse schorsoneer; (ENGLISH) scolymus, golden thistle, Spanish golden thistle; (FRENCH) carduille, cardon d'Espagne, scolyme d'Espagne; (GERMAN) Spanisher Golddistel, Spanische Golddistel, Goldwurzel; (GREEK) skolymos o ispanikos; (HEBREW) ḥoaḥ sefaradi, choach spharadi; (HUNGARIAN) aranybojtorján; (ICELANDIC) gullpistill; (ITALIAN) cardo scòlimo, cardoncello salvatico; (MACEDONIAN) španski zlaten oso; (MALTESE) xewk isfar; (NORWEGIAN) gulltistel, svartrot; (POLISH) parczoch hiszpański; (PORTUGUESE) cangarinha, cardo de ouro, escolimo de Espanha; (RUSSIAN) skolimus; (SERBO-CROATIAN) dragušica, dlakava; (SPANISH) escolimo, cardillo, almirón de España; (SWEDISH) spansk gold tistle; (TURKISH) ispanyol yemlik otu, ispanya altin dikeni, sevketibostan; (VLAAMS) scolymus.

27.37. *Scolymus maculatus* L. (EP) leaves; (P) C.

(CZECH) hadi mord španělsky; (ENGLISH) spotted garden thistle; (FRENCH) scolyme tache, spine-jaune; (GERMAN) gefleckte Golddistel; (GREEK) skolymos o stiktos; (HEBREW) ḥoaḥ akod, choach akod; (ITALIAN) scardaccione macchiato; (MACEDONIAN) pegav zlaten osot; (MALTESE) xewk isfar tar-riga; (POLISH) salsefia hiszpańska wezymord czarny korzen, czarne korzonki; (PORTUGUESE) cardo branco, escólimo malhado; (ROMANIAN) spaniolă, cu rădăcini dolci; (RUSSIAN) skartsanera, sladkil koren; (SPANISH) escolimo maculado; (TURKISH) benekli altin dikeni; (VLAAMS) gevlekte tuindistel.

27.38. *Scorzonera hispanica* L. (EP) roots and leaves[S]; (P) C,R.

(AFRIKAANS) skorsenier; (ALBANIAN, TOSK) stumbill; (ARABIC) khass el-kalab; (ARMENIAN) sindz; (AZERBIAJANI) emlik; (BASQUE) sendapoki; (BELARUSAN) chernakorjan; (BULGARIAN) čeren koren; (CATALAN-VALENCIAN-BALEAR) escorcomera; (CHINESE, MANDARIN) hei pi bo luo men shen, ya ts'ung; (CHINESE, WÚ) hei pi bo luo men shen, ya cong; (CZECH) hadí mord španělský, černý kořen; (DANISH) have-skorzoner, scorzoner, skorsonerrod; (DUTCH) schorseneer; (ENGLISH) scorzonera, black salsify, common viper's-grass; (ESTONIAN) mustjuur, aed-mustjuur; (FARSI, WESTERN) sefoorjench; (FINNISH) mustajuuri; (FRENCH) scorzonère, scorzonère noire, scorzonère d'Espagne, salsifis noir; (GERMAN) Schwarzwurzel, Garten-Schwarzwurzel, Spanische Schwarzwurzel, Schorzonere; (GREEK) skorzonera e ispanlki; (HEBREW) hardufnin sefaradi, hardufnin spharadi; (HUNGARIAN) feketegyökér; (ICELANDIC) svartrot; (INDONESIAN) tanaman hitam; (ITALIAN) scorzonera, scorzonera di spagna, barba di becco nera; (JAPANESE) kikugobou, kikugobō; (LATVIAN) melnsakne; (LITHUANIAN) valgomoji gelteklė; (MACEDONIAN) crn koren, španska zmijovka; (MALTESE) bajda; (MONGOLIAN) khaviskhana; (NORWEGIAN) svartrot, skorsonerrot; (POLISH) skorzonera; (PORTUGUESE) escorcioneira, salsifi negro; (ROMANIAN) scortionera, scorzoneră hispanică, scorzoneră spaniolă, cu rădăcini dulci; (RUSSIAN) černyj koren'; (SERBO-CROATIAN) kozja brada, zmijak, zmijavka, španjolska; (SLOVAK) čierny koreň, hadí mor španielsky; (SLOVENIAN) črni koren; (SORBIAN, UPPER) zatkańca; (SPANISH) escorzonera, escorzonera de España, salsifi negro; (SWEDISH) suartrot; (TAJIKI) mumtalokh; (TURKISH) kara yemlik otu, bahçe iskorcini; (UKRAINIAN) skorzonera; (VLAAMS) schorsoneer.

27.39. *Smallanthus sonchifolius* (Poepp. & Endl.) H. Rob. (EP) roots; (P) C.

(CZECH) jakon; (ENGLISH) yacon, yacon strawberry, Bolivian sunroot; (FRENCH) yacon, poire de terre, poire de terre Cochet; (GERMAN) Erdbirne; (ITALIAN) polinia; (JAPANESE) yaakon; (PORTUGUESE) polinia; (SPANISH) arboloco, jacón, llamón, aricoma, jiquimilla; (VLAAMS) yalon.

27.40. *Sonchus asper* (L.) Hill (EP) leaves; (P) C,R.

(ENGLISH) spiny sow thistle; (FRENCH) laitron épineux; (JAVANESE) delgiyu; (THAI) phakkat-hom, phakkat-hangkai.

27.41. *Sonchus malaianus* Miquel (EP) leaves; (P) C,R.

(INDONESIAN) kumindelan, blenggi.

27.42. *Sonchus oleraceus* L. (EP) leaves; (P) C,R.

(AFRIKAANS) suidissel, sydissel, melkdissel; (ARABIC) tadkernit; (BELARUSAN) asot agorodnaja; (BONTOC, CENTRAL) gagatang, keskesteng; (BRETON) laitue sauvage; (CHINESE, GÀN) liang ma; (CHINESE, MANDARIN) ku qu cai; (CHINESE, MIN BEI) ku ju; (CHINESE, WÚ) ku cai, ku ma cai, ku ju cai; (CHINESE, YUE) ku mai, you mai; (CZECH) mléč zelinný; (DANISH) svinemælk; (DUTCH) gewone melkdistel; (ENGLISH) sowthistle, milk thistle, annual milk thistle, hare's lettuce, hare's palace, hare's thistle, milk weed, sprout thistle, turn sole; (FILIPINO) gagatano; (FINNISH) kaalivalvatti; (FRENCH) laiteron, laitron commun, laitue de lierre; (GAELIC, IRISH) bleachtán; (GERMAN) Hussarenknopf, Kohl-Gänsedistel; (GREEK) zochos; (HEBREW) meror ha-ginnot, maror aginot; (HINDI) dudhi, dodak; (HUNGARIAN) szelíd csorbóka; (INDONESIAN) jombang, tempuyung, tempuh wiyang; (ITALIAN) serralha, cardeda; (JAPANESE) nogeshi; (JAVANESE) delgiyu; (KALINGA, BUTBUT) longlong; (KANKANAEY) gagattang; (KANNADA) naayi hakkarike; (KAZAKH) osot ogorodnyj; (LATVIAN) mīkstpienes; (LITHUANIAN) daržinė pienė; (MACEDONIAN) običen mlečen osot; (MALTESE) tfief; (MAORI) rauriki, puuhaa, puuwhaa, pororua; (MONGOLIAN) sharalzagana; (NEPALI) dudhe; (NIUE) pupu lele; (NORWEGIAN) haredylle; (NYANJA) cinguwo; (PANJABI, EASTERN) sɔ̃chal; (POLISH) mlecz zwyczajny; (PORTUGUESE) serralha, serralha macia, leituga; (QUECHUA, CUSCO) achcaña; (SERBO-CROATIAN) ostak, sinji; (SPANISH) cerraja, lechuguilla silvestre; (SUNDA) camawak; (SWAHILI) mchunga; (TAJIKI) shiri kharak; (TAMIL) shakadevi, oosithagarai; (TELUGU) rathrinta; (TONGAN) longolongo'uha; (TURKISH) eşek gevreği, eşek marulu, adi eşek marulu; (TURKMEN) bakja sarygalgan; (URDU) dudi; (VIETNAMESE) rau di[ees]p d[aj]i, nh[ux] c[us]c, rau c[us]c s[uwx]a; (VLAAMS) gewone melkdistel; (WELSH) llaeth ysgallen; (WOLOF) luguǰ u valo.

27.43. *Sonchus wightianus* DC. (EP) leaves; (P) C,R.

(ENGLISH) pualele, sow thistle; (IFUGAO) lunglung an manema; (ILOCANO) lampaka; (JAVANESE) tempuyung; (NEPALI) ban sampu; (VIETNAMESE) nh[ux] c[us]c d[oof]ng.

27.44. *Struchium sparganophorum* (L.) O. Kuntze (EP) young shoots; (P) C.

(AMHARIC) märara qətal; (BEMBÉ) minkonkolo; (EDE NAGO) ewuro odo; (ENGLISH) water bitter leaf; (FRENCH) oreille mouton; (FULFULDE) soaka; (NGUNDI) mundundindundi; (PORTUGUESE) libó d'água; (SWAHILI) jani chungu; (TEKE) munkankali; (YORUBA) ewuro odo, ewurodo.

27.45. *Taraxacum officinale* G.H. Webber, ex Wiggers (EP) leaves, root[S]; (P) R,C.

(AFRIKAANS) perdeblom, molslaai, platdissel; (ARMENIAN) chatutik; (AZERBIAJANI) adžygovuk; (BASQUE) galkidea; (BELARUSAN) oduvanchik lekavy; (BENGALI) pitachumki; (BLACKFOOT) naató'sipisátssaisski; (BRETON) c'hwervizon; (BULGARIAN) radika; (CATALAN-VALENCIAN-BALEAR) pixallits, dent de lleó; (CHECHEN) baappa; (CHINESE, HAKKA) p'òu-kōung-ȳn; (CHINESE, MANDARIN) pu cong ying, pu kung ying; (CHINESE, WÚ) pu gong ying; (COMANCHE) soko sikusa; (CZECH) smetanka lékařska, smetanka; (DANISH) mælkebøtte, løvetand; (DUTCH) molsla, paardebloem; (ENGLISH) dandelion, bitterwort, blow ball flower, canker, cankerwort, crow-parsnip, dashel-flower, dentilion, dindle, doon-head-clock, fortune-teller, gowan, irish daisy, milk gowan, monks hood, one o'clocks, peasant's clock, pissabed, priest crown; (ESTONIAN) võilill; (FILIPINO) dandelion; (FINNISH) voikukka, voikukat; (FRENCH) dent de lion, pissenlit commum; (GAELIC, IRISH) caisearbhán; (GAELIC, MANX) lvs ny minnag;

(GAELIC, SCOTTISH) beàrnan-brìde; (GALICIAN) leitaruga, taráxaco, muxalan; (GEORGIAN) dzaglis-satacuri; (GERMAN) Löwenzahn, Kuhblume, gemeine Kuhblume; (GREEK) taraxakon to pharmakeftikon; (GUJARATI) pathardi; (HEBREW) shinnan refuyi, sheinan refuee; (HINDI) dulal, barau, kanphul, dudhal, baran, dudhale, dudhali, dudheli, jangli kaasani kanpool, varam; (HUNGARIAN) gyermekláncfű, pongyola pitypang; (ICELANDIC) tunfifill; (INDONESIAN) jombang, djombang; (ITALIAN) dente de leone, soffione tarassaco, radichiello; (JAPANESE) shokuyou-tanpopo, tampopo; (JAVANESE) jombang; (KANNADA) kaadu seventhi, kaadushaavanthi, daandelian hoo, mullu moolangi; (KASHMIRI) dŏdijü; (KAZAKH) oduvanchik lekarstvennyj; (KRIO) wata-bitas; (LATVIAN) arstniecības pienene, cūkpiene, pienene; (LITHUANIAN) paprastoji kiaulpienė; (LUXENBOURGEOIS) Pissblumm; (MACEDONIAN) gluvarče, obično gluvarče; (MAORI) tawao, tohetaka, tohetake, tohetea; (MARATHI) undarkani, bhathur, undara chakaan; (MEHRI) ḥəźawīt, ḥəźəwēt; (MONGOLIAN) emiin bagvaakhai; (NEPALI) tuki phool; (NORWEGIAN) løvetann; (PAIUTE, SOUTHERN) o-aw sawent; (PANJABI, EASTERN) dudli, baran, radam, dudh-batthal, kanphul, shamukei; (POLISH) mniszek pospolity, dmuchawiec, mniszek lekarski, mniszek; (PORTUGUESE) dente de leão, taraxaco, amor dos homens; (POTAWATOMI) asawa' bagwûnik, wasakûtcä'pûk; (QUECHUA, CUSCO) achicoria; (ROMANIAN) păpădie; (RUSSIAN) oduvančhik lekarstvennyj; (SANSKRIT) dugdhphenee, karnaphool, loothari, payasvinee; (SERBO-CROATIAN) maslačak, ljekoviti; (SLOVAK) regrat; (SLOVENIAN) anagrón; (SPANISH) diente de léon, taraxacónn; (SUNDA) ki baceta; (SWEDISH) maskros; (TAJIKI) qoqu, guli qahqah; (TIBETAN) qhūmaà me̱tɔɔ̀; (TURKISH) kara hindiba, adi aslanağzi, arslandisi; (TURKMEN) dermanlyk tozga; (VIETNAMESE) b[oof] c[oo]ng anh, s[uw] nha; (VLAAMS) molsla, gewone paardebloem; (WELSH) dant y lleew; (YIDDISH) loftel.

27.46. *Tragopogon porrifolius* L. (EP) root and young leaves; (P) C,R.

(AFRIKAANS) hawerwortel; (ASSAMESE) salsify; (BULGARIAN) salzifi; (CATALAN-VALENCIAN-BALEAR) barba de cabra; (CHINESE, MANDARIN) bo lou men shen, p'o lo min shin; (CHINESE, WÚ) ba luo men shen, zi yang niu bang; (CHINESE, YUE) yao yong pu gong ying; (CZECH) kozí brada fialová, salsifis; (DANISH) havrerod; (DUTCH) haverwortel, blauwe morgenster; (ENGLISH) salsify, vegetable oyster, oyster plant; (ESTONIAN) piimjuur, aed-piimjuur; (FINNISH) kaurajuuri; (FRENCH) salsifi, salsifi blanc, salsifis; (GAELIC, MANX) salsey; (GALICIAN) barba cabreira; (GERMAN) Haferwurzel, Weißwurzel, Bocksbart; (GREEK) lagolachano, tragoponon o procsophylos; (HEBREW) zekan ha-tayish, zkan-hataish; (HINDUSTANI) ek qism ká; (HUNGARIAN) saláta bakszakál, kecskedin; (ICELANDIC) hafursrot; (ITALIAN) scorzobianca, anca, barba di becco; (JAPANESE) salusifai, ba ra monjin; (LITHUANIAN) valgomasis pūtelis; (MACEDONIAN) španska kozička, španska kozja brada; (MALTESE) lehjet il-bodbod; (MONGOLIAN) jamaan sakhal; (NORWEGIAN) rotgeitskjegg, geitskjegg, havrerot; (POLISH) salsefia; (PORTUGUESE) salsifi branco, cersefi, cercifi, barba de bode; (ROMANIAN) barba caprei; (RUSSIAN) sal'sifi; (SERBO-CROATIAN) bijeli korijen, kozja brada; (SLOVAK) kozia brada pórolistná; (SLOVENIAN) beli koren; (SPANISH) salsifi blanco; (SWEDISH) haver rot; (TAJIKI) rishi buz; (TURKISH) keklikotu, tekesakali, pirasa yaprakli teke sakali; (UKRAINIAN) vìvsjanij korìn; (VLAAMS) haverwortel, echte haverwortel, paarse morgenster, witte schorseneer; (XHOSA) isityalo esithile esinengcambu efana nomnqathe etyiwa njengomfuno xa iphekiwe.

27.47. *Tragopogon pratensis* L. (EP) young roots and shoots; (P) C,R.

(BELARUSAN) kazlaborod lugavy; (CHINESE, MANDARIN) cao di bo lou men shen; (CHINESE, WÚ) cao di bo luo men shen; (CZECH) kozí brada luční; (DUTCH) gele morgenster; (ENGLISH) meadow salsify, goatsbeard; (ESTONIAN) kitse-enelas; (FINNISH) pukinparta; (FRENCH) salsifis des prés; (GERMAN) Wiesen-Bocksbart; (GREEK) tragopono o leimonios; (HUNGARIAN) réti bakszakál; (ITALIAN) salsèfrica, barba di bêcco, barba di prete; (KAZAKH) kosloborodnik lugovoj; (LATVIAN) pļavas plastbārdis; (LITHUANIAN) pievinis pūtelis; (MACEDONIAN) bel koren livadska kozja brada; (MONGOLIAN) nugiin jamaan sakhal; (NORWEGIAN) enggeitskjegg; (POLISH) salsefia łąkowa, kozibrod lakowy; (PORTUGUESE) cersefi bastardo, barba de bode; (ROMANIAN) barba caprel; (RUSSIAN) kazlabarodnik lougovol; (SERBO-CROATIAN) kozja brada, livadna; (SPANISH) barba cabruna; (SWEDISH) angshaver rot; (TAJIKI) rishi buz; (TURKISH) çayir tekesakali; (VLAAMS) gele morgenster; (WELSH) llysiau`r gwewyn.

27.48. *Youngia sonchifolia* Maxim. (EP) leaves and root; (P) C.

(KOREAN) godulbbaegi.

28. Basellaceae

28.1. *Basella alba* L. (EP) leaves and young shoots; (P) C.

(AFRIKAANS) malabarspinasie; (ASSAMESE) puroi sak; (BAATONUM) sakonu; (BEMBÉ) kimbombolo; (BENGALI) poi; (BICOLANO, CENTRAL) arogbati; (BINUKID) elugbati; (BODO) møifrai; (BUHID) alugbati; (CEBUANO) alugbati, masbati; (CHINESE, MANDARIN) hong luo kui; (CHINESE, WÚ) mu e cai, hong luo kui; (CHINESE, YUE) luo kui, teng cai, chan cai; (CRIOULO, CAPE VERDE) tinta-de-macaco; (CZECH) bazela bílá; (DANISH) malabarspinat, indisk spinat; (DUTCH) basella, postelein; (EDE NAGO) shɛ ɛjɛ shɔnrɔ; (ENGLISH) Ceylon spinach, Indian spinach, Malabar spinach, Malabar nightshade; (FILIPINO) alugbati; (FINNISH) malabarinpinaatti; (FON) jamaku; (FRENCH) baselle, baselle blanche, brette d'Angole, épinard de Malabar, brède de Malabar; (GA) shwee; (GERMAN) Indischer Spinat, Malabar Spinat; (GREEK) spanaki keïlanis; (GUN) glaswe kpajit n; (HANUNOO) talugbati; (HINDI) poi, bonpoi, laalbachlu, myal-ki-bhaji, poi, poikivel, sufedbachla; (HINDUSTANI) poi; (HMONG-DAW) hmab ntsha; (HMONG-NJUA) hmab btsha, hmaab ntshaa; (IBANAG) kuvay; (ILOCANO) ilaibakir, alogbati, ilay bakir; (INDONESIAN) gendola, gandola, genjerot, basela; (ITALIAN) basella, basella bianca, spinaccio del Malabar, spinaccio della Cina, spinaccio d'America; (ITAWIT) cubay; (JAPANESE) tsuru-murasaki; (JAVANESE) jingga, genjerot, gendolak, genjulak, gedrek, oci-oci; (JULA) pinard brou, nanborou; (KANNADA) basale, basale soppu, baayi basale, bili basale, dodda basale, hire basale, kempu baayi basale, kempu basale, pothaki; (KHMER, CENTRAL) chrâlong; (KONKANI) valchibhagi; (KRIO) brɔd-bɔlɔgi; (LAO) pang; (MAITHILI) pōrō; (MALAY) remayong, pacar gandula, gendola, tembayung; (MALAYALAM) basellakkeera; (MARATHI) poi, maayaalachivel, mayalu, wahlea; (MORISYEN) bred gandol; (NEPALI) poiko saag; (NIUE) polo fua; (NORWEGIAN) basella; (PAMPANGAN) alubati, ganana; (PANGASINAN) kabkabatiti; (POLISH) miłowój biały, bazela biała; (PORTUGUESE) bacela, basela, bertalha, bredo de Angola; (SANSKRIT) upodika, pasalaikkeerai, kalambi, madushaka, mohini, pichhila, pichhila chhada, pothaki, puthika, upodaki, upoti, valipodaki, vishala, vishvatulasi, vrishchikapriya; (SINHALA) nivithi; (SPANISH) espinaca basela, espinaca blanca, espinaca de Malabar, bretaña, libato; (SUNDA) gandola, ganola, gendolan; (SWAHILI) nderema; (TAGALOG) albugbati, alugbati, dundula, libato, grana; (TAMIL) basala, paasaangali, shivappuvaslakkeerai, vaslakkeerai; (TAUSUG) dundula; (TELUGU) batsala, alla-bachhalla, allu-batchhalla, bachhalitheega, bachhali koora, erraalubachhali, kaarubachhali, peddabachhalia, peddamattu, peddamattuniatkubatsala, polambachhali; (THAI) phakplang, phakplang-yai; phakpang; (TULU) basalè; (TURKISH) hint ispanaği, çin ispanaği; (TWI) efan; (URDU) poi'saloni palak; (VIETNAMESE) m[oof]ng t[ow]i, m[uf]ng t[ow]i; (VLAAMS) basella; (WAAMA) yɔ buta; (WÁRAY-WÁRAY) alugbati; (YORUBA) shɛ ɛjɛ shɔnrɔ, amuntutu funfun, amuntutu pupa.

28.2. *Ullucus tuberosus* Caldas (EP) tubers; (P) C.

(CHINESE, MANDARIN) zhi li shu; (CHINESE, WÚ) zhi li shu; (CHINESE, YUE) kuai gen luo kui; (DANISH) ulluco; (ENGLISH) ulluco, melluco, olluco; (FINNISH) ulluco; (FRENCH) ulluque, ulluco; (GERMAN) Knollenbaselle; (POLISH) bulwotka bulwiasta; (PORTUGUESE) uluco; (QUECHUA, CUSCO) ullucu; (SPANISH) ulluco, melloco, papa lisa, ruba, olloco, timbos; (TURKISH) ulluko; (VLAAMS) ulluco.

29. Boraginaceae

29.1. *Borago officinalis* L. (EP) petioles; (P) C.

(AFRIKAANS) bernagie, komkommerkruid; (BELARUSAN) burachnikavaja, agorochnik lekavy; (CHINESE, MANDARIN) bo li qu; (CHINESE, WÚ) liu li ju; (CZECH) brutnák obesny, brutnák lékařský; (DANISH) hjulkrone, borago; (DUTCH) bernagie, borago; (ENGLISH) borage, beebread, talewort; (ESTONIAN) kurgirohi; (FARSI, WESTERN) gaolgarzaban; (FINNISH) purasruoho, kurkkuyrtti; (FRENCH) bourrache, borrage; (GALICIAN) berralas, borraxa; (GERMAN) Boretsch, Gurkenkraut; (GREEK) vorraginon to pharmakeftico; (HEBREW) borag refuyi, borago; (HUNGARIAN) borágó; (ICELANDIC) hjolkrona; (ITALIAN) boragine, borrana; (KAZAKH) ogurechnaja trev; (LATVIAN) ārstniecības gurķene; (LITHUANIAN) vaistinė aqurklė; (MACEDONIAN) običen borago; (NORWEGIAN) agurkurt, vanlig agurkurt; (POLISH) ogórecznik lekarski; (PORTUGUESE) borragem; (QUECHUA, CUSCO) borraja; (ROMANIAN) limba mielului medicinala, boranita medicinala; (RUSSIAN) burachnik lekarstvennyi; (SPANISH) borraja, borracha; (SWEDISH) gurkört; (TAJIKI) khiorak; (TURKISH) hodan, hadun, siğirdili, burancaotu; (TURKMEN) dermanlyk buraçnik; (VLAAMS) komkommerkruid, bernagie; (WELSH) tafod yr ych.

29.2. *Symphytum officinale* L. (EP) leaves and tender shoots; (P) C.

(BRETON) skouarn-azen; louzaouenn an troc'h, louzaouenn ar c'halvez; (CZECH) kostival lékařský; (DANISH) læge-kulsukker; (DUTCH) gewone smeerwortel; (ENGLISH) common comfrey; (ESTONIAN) varemerohi; (FINNISH) rohtoraunioyrtti; (FRENCH) consoule officinale; (GERMAN) gemeiner Beinwell; (GREEK) symfyton to pharmakeftiko; (HEBREW) sinfiton refuyi; (HUNGARIAN) fekete nadálytő; (ITALIAN) consolida maggiore, orecchia d'asino; (JAPANESE) kohn furii; (LATVIAN) ārstniecības tauksakne; (LITHUANIAN) vaistinė taukė; (MACEDONIAN) običen gavez; (NORWEGIAN) legevalurt; (POLISH) żywokost lekarski; (PORTUGUESE) consolda maior, orelhas de asno, confrei; (QUECHUA, CUSCO) chuncu-chuncu; (SPANISH) consuelda mayor, sinfito mayor; (SWEDISH) akta vallört; (TURKISH) adi karakafesotu, karakafes; (VLAAMS) gewone smeerwortel; (WELSH) llysiau`r cwlwm.

29.3. *Symphytum × uplandicum* Nyman (EP) young leaves and shoots ; (P) C.

(CZECH) kostival uplandský; (DANISH) foder-kulsukker; (DUTCH) Uplandse smeerwortel; (ENGLISH) prickly comfrey, Quaker comfrey, Russian comfrey; (FINNISH) ruotsinraunioyrtti; (FRENCH) consoude voyageuse, consoude d'Upland, consoude russe; (GERMAN) Komfrey, Hochland-Beinwell, Schwedischer Beinwell; (GREEK) symfyton to rossiko; (ITALIAN) consolida azzurra; (MACEDONIAN) ruski gavez; (NORWEGIAN) mellomvalurt; (POLISH) kwakier lekarski, zywokost lekarski; (PORTUGUESE) consolda; (SPANISH) cousuelda azul; (SWEDISH) uppländsk vallört; (VLAAMS) bastaard smeerwortel.

30. Brassicaceae (Cruciferae)*

30.1. *Armoracia rusticana* P. Gaertn., B. Mey., Scherb. (EP) roots, leaves[S], sprouted seeds[S]; (P) R,C.

(AFRIKAANS) peperwortel; (ALBANIAN, TOSK) kren; (ARABIC) feegel; (ARMENIAN) bogkwairi; (ASSAMESE) bon mula; (BASQUE) bitxaleka; (BELARUSAN) kapustny, kryzakvety chrjan; (BRETON) gouez-irvin; (BULGARIAN) chrjan; (CATALAN-VALENCIAN-BALEAR) rave rustica; (CEBUANO) kamunggay; (CHINESE, MANDARIN) la gen, ma lo po; (CHINESE, WÚ) la gen; (CHINESE, YUE) da tou cai, chong cai; (CROATIAN) hren; (CZECH) křen, křen selský; (DANISH) peberrod; (DUTCH) mierikswortel; (ENGLISH) horseradish; (ESTONIAN) mädarõigas, mäeroigas; (FIJIAN) barah murai; (FILIPINO) malunggay; (FINNISH) piparjuuri; (FRENCH) cran, armoracie raifort, raifort sauvage; (GAELIC, IRISH) raidis fhiáin; (GAELIC, MANX) raghyl, feie; (GEORGIAN) pirshushkha; (GERMAN) Meerrettich; (GREEK) kochliaris e armorakia; (HEBREW) hazeret, hazereth; (HINDUSTANI) sahjaná kí jaṛ; (HUNGARIAN) torma; (ICELANDIC) piparrot; (ITALIAN) cren, barbafonte, rafano rusticano; (JAPANESE) wasabi daikon, seiyo-wasabi; (KAZAKH) chren; (KIRGHIZ) chren; (LATVIAN) mārrutki, mārutki; (LITHUANIAN) valgomasis krienas; (LUXENBOURGEOIS) Päerdsrennen; (MACEDONIAN) ren, običen ren; (MALTESE) l-gherq mustarda; (NORWEGIAN) pepperrot; (PANJABI, EASTERN) urgǣd; (POLISH) chrzan pospolity, chrzan; (PORTUGUESE) armorácia, rabanete selvagem, rábano rústico, rábano de carvalho, raiz forte; (ROMANIAN) hrean; (RUSSIAN) chren; (SERBO-CROATIAN) ren; (SLOVAK) chren; (SLOVENIAN) hren; (SORBIAN, LOWER) kśěn; (SORBIAN, UPPER) chrěn; (SPANISH) rábano rusticano, rábano picante, rábano silvestre; (SWAHILI) mronge; (SWEDISH) pepparrot, skörbjuggsört; (TAJIKI) rosani shomi; (TURKISH) bayir turpu, karaturp, at turpu, hindiba; (TURKMEN) adaty hren; (UKRAINIAN) chrìn; (UZBEK) khartoi; (VLAAMS) mierikswortel; (WELSH) rhuddygl; (YIDDISH) khrayn.

30.2. *Barbarea verna* (Miller) Ascherson (EP) leaves; (P) C,R.

(CHINESE, MANDARIN) chun shan gai; (CHINESE, WÚ) chun shan jie; (DANISH) langskulpet vinterkarse, vinterkarse; (DUTCH) vroeg barbarakruid; (ENGLISH) upland cress, winter cress; (FINNISH) krassikanankaali, maakrassi; (FRENCH) cresson de terre, barbarée printanière; (GERMAN) Garten-Kress, Amerikanische Winterkresse, Frühlings-Barbarakraut; (GREEK) varvarea e prolïmi; (HUNGARIAN) borbálafű; (ITALIAN) erba di Santa Barbara vernale, crescione inglese; (JAPANESE) fuyu-garashi; (LUXENBOURGEOIS) Cresson; (MACEDONIAN) proletna surepka; (MAORI) toi; (MONGOLIAN) sarmag; (NORWEGIAN) lang vinterkarse; (POLISH) gorczycznik wiosenny; (PORTUGUESE) agrião de horta, agriao rinchão, agriao da terra; (SPANISH) hierba de Santa Bárbara, berro de tierra, masturzo de jardin; (TAJIKI) vazirak; (TURKISH) ilkbahar kiş teresi, erkenci kis teresi, kiş teresi micanotu; (VLAAMS) vroeg barbarakruid; (WELSH) berw'r mynydd.

30.3. *Brassica carinata* A. Braun (EP) leaves; (P) C.

[1]Maragoli dialect of Luyia.

(ASSAMESE) soriah; (CZECH) brukev kýlantá, hořčice habaešská; (DANISH) abessinsk sennep; (ENGLISH) Abyssinian mustard, Ethiopian mustard; (ESTONIAN) sinep; (FARSI, WESTERN) khardal; (FILIPINO) mustasa; (FRENCH) moutarde de l'Ethiopie, moutarde éthiopienne; (GERMAN) Aethiopischer Senf; (GREEK) avyssiniakos sinapis; (HEBREW) chardal etiopy; (HMONG DAW) zaub ntsuab; (HUNGARIAN) abesszin kel; (ITALIAN) senape; (LUYIA) likabichi[1]; (MACEDONIAN) abisinska repka; (MANINKAKAN, WESTERN) navé; (MONGOLIAN) baitsaa; (POLISH) gorczyca abisyńska, goryczel; (PORTUGUESE) mostarda da Etiópia; (ROMANIAN) muştar; (RUSSIAN) gorchitsa; (SPANISH) mostaza de Abisinia; (SWAHILI) loshuu, figiri; (TAJIKI) khardal; (TURKISH) habeş hardali; (VLAAMS) Ethiopische mosterd, Abessijnse mosterd; (WOLOF) mutard.

30.4. *Brassica juncea* (L.) Czern. var. *capitata* Hort. (EP) leaves; (P) P,C.

(ASSAMESE) soriah; (BURMESE) mon-hyin; (BELARUSAN) gorchitza sareptsaja; (CATALAN-VALENCIAN-BALEAR) mostassa; (CHINESE, MANDARIN) jie qiu gai; (CHINESE, WÚ) jie qiu jie; (CHINESE, YUE) bao xin jie, chao zhou da jie cai; (CZECH) hořčice sareptská, brukev sítinovitá; (DANISH) mosterd; (DUTCH) mosterd; (ENGLISH) capitata mustard; (FARSI, WESTERN) khardal; (FILIPINO) mustasa; (FINNISH) sareptansinappi; (FRENCH) moutarde; (GERMAN) Senf; (GREEK) sferoïdis sinapis; (HEBREW) keruv, hardal; (HMONG-DAW) zaub ntsuab; (HMONG-NJUA) zaub ntsuab; (HUNGARIAN) szareptai mustár; (INDONESIAN) sawi jabung; (ITALIAN) senape; (JAPANESE) ketkyyuu-takana; (LATVIAN) galviņu kāposts; (MACEDONIAN) glavesta nivska repka; (MALAY) sawi, sawi pahit; (MONGOLIAN) zerleg khij; (NORWEGIAN) sennepskål; (POLISH) kapusta sitowa, goryczel sarepska, modra; (PORTUGUESE) mostarda de cabeça; (ROMANIAN) muştar comestibil; (RUSSIAN) gorchitsa sareptsaja; (SERER-SINE) mutard; (SPANISH) mostaza; (TAGALOG) mustasa; (TURKISH) hint hardali; (URDU) sarsoo; (VLAAMS) kropmosterd; (WOLOF) mutard.

30.5. *Brassica juncea* (L.) Czern. var. *crassicaulis* Chen and Yang (EP) stem; (P) C.

(BURMESE) mon-hyin; (CHINESE, MANDARIN) sun zi gai; (CHINESE, WÚ) sun zi jie; (CZECH) hořčice; (DANISH) mosterd; (DUTCH) mosterd; (ENGLISH) bamboo shoot mustard; (FARSI, WESTERN) khardal; (FILIPINO) mustasa; (FRENCH) moutarde; (GREEK) sinapis vlastou indo kalamou; (HEBREW) keruv, hardal; (HMONG-DAW) zaub ntsuab; (HMONG-NJUA) zaub ntsuab; (INDONESIAN) sawi pahit; (ITALIAN) senape; (MACEDONIAN) bambisosteblena repka; (MALAY) sawi; (MONGOLIAN) gaimuu; (NORWEGIAN) sennepskål; (POLISH) gorczyca bambusowa; (PORTUGUESE) mostarda; (ROMANIAN) muştar; (RUSSIAN) gorchitsa; (SERER-SINE) mutard; (SPANISH) mostaza; (TAGALOG) mustasa; (URDU) bans saki sarsoo; (VLAAMS) bamboescheutmosterd; (WOLOF) mutard.

30.6. *Brassica juncea* (L.) Czern. var. *crispifolia* L.H. Bailey (EP) leaves; (P) C.

(BINUKID) mustasa; (BURMESE) mon-hyin; (CATALAN-VALENCIAN-BALEAR) mostassa; (CHINESE, MANDARIN) mi tuo gai; (CHINESE, WÚ) xue li hong, mi tuo jie; (CHINESE, YUE) xue li hong, xue cai; (CZECH) hořčice; (DANISH) mosterd; (DUTCH) mosterd; (ENGLISH) cut leaf mustard, curled mustard; (FARSI, WESTERN) khardal; (FILIPINO) mustasa; (FINNISH) käkäräsinappi; (FRENCH) moutarde; (GERMAN) Senf; (GREEK) katsaroto sinapi; (HEBREW) keruv, hardal; (HINDUSTANI) kaisoi; (HMONG-DAW) zaub ntsuab; (HMONG-NJUA) zaub ntsuab (ITALIAN) senape; (MACEDONIAN) kadrava nivska repka; (MALAY) sawi; (MANINKAKAN, WESTERN) navé; (MONGOLIAN) gaimuu baitsaa; (MORISYEN) bred moutard; (NEPALI) rayo saag; (NORWEGIAN) sennepskål; (POLISH) gorczyca karłowa; (PORTUGUESE) mostarda frisada; (ROMANIAN) muştar; (RUSSIAN) gorchitsa; (SERER-SINE) mutard; (SPANISH) mostaza; (SWAHILI) sukuma wiki; (TAGALOG) mustasa; (TURKISH) hardal; (URDU) kat pati sarsoo; (VLAAMS) krulmosterd; (WOLOF) mutard.

30.7. *Brassica juncea* (L.) Czern. var. *foliosa* L.H. Bailey (EP) leaves; (P) C.

(AFRIKAANS) kleinblaar mosterd; (BURMESE) mon-hyin; (CATALAN-VALENCIAN-BALEAR) mostassa; (CHINESE, MANDARIN) xiao ye gai; (CHINESE, WÚ) xiao ye jie; (CHINESE, YUE) ye yong jie cai, jie cai; (CZECH) hořčice; (DANISH) mosterd; (DUTCH) mosterd; (ENGLISH) small leaf mustard; (FARSI, WESTERN) khardal; (FILIPINO) mustasa; (FINNISH) silosinappi; (FRENCH) moutarde à petites feuilles; (GERMAN) Senf; (GREEK) sinapi me mikra phylla; (HEBREW) keruv, hardal; (HMONG-DAW) zaub ntsuab; (HMONG-NJUA) zaub ntsuab; (ITALIAN) senape; (KRIO) kabej-pla-sas, bush-kabej; (MACEDONIAN) malolisna nivska repka; (MALAY) sawi; (MANINKAKAN, WESTERN) navé; (MONGOLIAN) zerleg khij; (NORWEGIAN) sennepskål; (POLISH) gorczyca drobnolistna; (PORTUGUESE) mostarda de folhas pequenas, (ROMANIAN) muştar; (RUSSIAN) gorchitsa; (SERER-SINE) mutard; (SPANISH) mostaza; (TAGALOG) mustasa; (THEMNE) a-mɔthsəra; (URDU) khurd pati sarsoo; (VLAAMS) smalbladige mosterd; (WELSH) marstard manddail; (WOLOF) mutard.

30.8. *Brassica juncea* (L.) Czern. var. *gemmifera* Lee & Lin (EP) stem and axillary buds; (P) C.

(BURMESE) mon-hyin; (CHINESE, MANDARIN) bao zi gai; (CHINESE, WÚ) bao zi jie; (CZECH) goryczel; (DANISH) mosterd; (DUTCH) mosterd; (ENGLISH) gemmiferous mustard; (FARSI, WESTERN) khardal; (FILIPINO) mustasa; (FRENCH) moutarde gemminifère; (GERMAN) Senf; (HEBREW) keruv, hardal; (HMONG-DAW) zaub ntsuab; (HMONG-NJUA) zaub ntsuab; (HUNGARIAN) bimbóskel; (ITALIAN) senape; (MACEDONIAN) papkasta nivska repka; (MONGOLIAN) gaimuu; (NORWEGIAN) sennepskål; (POLISH) gorczyca pączkowa; (PORTUGUESE) mostarda gemmifera; (ROMANIAN) muştar; (RUSSIAN) gorchitsa; (SERER-SINE) mutard; (SPANISH) mostaza; (TAGALOG) mustasa; (VLAAMS) knopdragende mosterd.

30.9. *Brassica juncea* (L.) Czern. var. *involuta* Yang & Chen (EP) leaves; (P) C,P.

(BURMESE) mon-hyin; (CHINESE, MANDARIN) juan xin gai; (CHINESE, WÚ) juan xin jie; (CZECH) hořčice; (DANISH) mosterd; (DUTCH) mosterd; (ENGLISH) involute mustard; (FARSI, WESTERN) khardal; (FILIPINO) mustasa; (FRENCH) moutarde; (GERMAN) Senf; (GREEK) enestramenon sinapi; (HEBREW) keruv, hardal; (HMONG-DAW) zaub ntsuab; (HMONG-NJUA) zaub ntsuab; (ITALIAN) senape; (MACEDONIAN) svitkana nivska repka; (MALAY) sawi; (MANINKAKAN, WESTERN) navé; (NORWEGIAN) sennepskål; (POLISH) gorczyca inwolucyjna; (PORTUGUESE) mostarda; (ROMANIAN) muştar; (RUSSIAN) gorchitsa; (SERER-SINE) mutard; (SPANISH) mostaza; (TAGALOG) mustasa; (VLAAMS) gekrulde mosterd; (WOLOF) mutard.

30.10. *Brassica juncea* (L.) Czern. var. *latipa* Li (EP) leaves; (P) C.

(BURMESE) mon-hyin; (CHINESE, MANDARIN) kuan bing gai; (CHINESE, WÚ) kuan bing jie; (CZECH) hořčice; (DANISH) mosterd; (DUTCH) mosterd; (ENGLISH) wide petiole mustard; (FARSI, WESTERN) khardal; (FILIPINO) mustasa; (FRENCH) moutarde à large pétiole; (GERMAN) Senf; (GREEK) sinapi me plati mischo; (HEBREW) keruv, hardal; (HMONG-DAW) zaub ntsuab; (HMONG-NJUA) zaub ntsuab; (INDONESIAN) sawi pahit; (ITALIAN) senape; (MACEDONIAN) širokopetelkova nivska repka; (MALAY) sawi; (MANINKAKAN, WESTERN) navé; (NORWEGIAN) sennepskål; (POLISH) gorczyca szerokolistna; (PORTUGUESE) mostarda de pecíolos largos; (ROMANIAN) muştar; (RUSSIAN) gorchitsa; (SERER-SINE) mutard; (SPANISH) mostaza; (TAGALOG) mustasa; (VLAAMS) brede petiool mosterd; (WOLOF) mutard.

30.11. *Brassica juncea* (L.) Czern. var. *leucanthus* Chen & Yang (EP) leaves; (P) C,P.

(AFRIKAANS) witblom mosterd; (BURMESE) mon-hyin; (CHINESE, MANDARIN) bao hua gai; (CHINESE, WÚ) bai hua jie; (CZECH) hořčice; (DANISH) mosterd; (DUTCH) mosterd; (ENGLISH) white flowered mustard; (FARSI, WESTERN) khardal; (FILIPINO) mustasa; (FRENCH) moutarde à fleurs blanches; (GERMAN) Senf; (GREEK) sinapi me lefki anthisi; (HEBREW) keruv, hardal; (HMONG-DAW) zaub ntsuab; (HMONG-NJUA) zaub ntsuab; (ITALIAN) senape; (MACEDONIAN) belocvetna nivska repka; (MALAY) sawi; (MANINKAKAN, WESTERN) navé; (NORWEGIAN) sennepskål; (POLISH) gorczyca białokwiatowa; (PORTUGUESE) mostarda de flores brancas; (ROMANIAN) muştar; (RUSSIAN) gorchitsa; (SERER-SINE) mutard; (SPANISH) mostaza; (TAGALOG) mutasa; (TAJIKI) khardali safed; (VLAAMS) witbloemige mosterd; (WELSH) marstard blodaw gwyn; (WOLOF) mutard.

30.12. *Brassica juncea* (L.) Czern. var. *linearifolia* Sun (EP) leaves; (P) C.

(BURMESE) mon-hyin; (CHINESE, MANDARIN) feng wei gai; (CHINESE, WÚ) feng wei jie; (CZECH) hořčice; (DANISH) mosterd; (DUTCH) mosterd; (ENGLISH) line mustard; (FARSI, WESTERN) khardal; (FILIPINO) mustasa; (FRENCH) moutarde à feuilles linéaires; (GREEK) pytidomeno sinapi; (HEBREW) keruv, hardal; (HMONG-DAW) zaub ntsuab; (HMONG-NJUA) zaub ntsuab; (ITALIAN) senape; (MACEDONIAN) dolgolisna nivska repka; (MALAY) sawi; (MANINKAKAN, WESTERN) navé; (NORWEGIAN) sennepskål; (POLISH) gorczyca liliowa; (PORTUGUESE) mostarda de folhas lineares; (ROMANIAN) muştar; (RUSSIAN) gorchitsa; (SERER-SINE) mutard; (SPANISH) mostaza; (TAGALOG) mustasa; (VLAAMS) lijnmosterd; (WOLOF) mutard.

30.13. *Brassica juncea* (L.) Czern. var. *longepetiolata* Yang & Chen (EP) leaves; (P) C.

(BURMESE) mon-hyin (CHINESE, MANDARIN) chang bing gai; (CHINESE, (CHINESE, WÚ) chang bing jie; ; (CHINESE, YUE) gao jiao jie; (CZECH) hořčice; (DANISH) mosterd; (DUTCH) mosterd; (ENGLISH) long petiole mustard; (FARSI, WESTERN) khardal; (FILIPINO) mustasa; (FRENCH) moutarde à long pétiole; (GREEK) sinapi me makri mischo; (HEBREW) keruv, hardal; (HMONG-DAW) zaub ntsuab; (ITALIAN) senape; (MACEDONIAN) dolgopetalna nivska repka; (MALAY) sawi; (MANINKAKAN, WESTERN) navé; (NORWEGIAN) sennepskål; (POLISH) gorczyca długolistna; (PORTUGUESE) mostarda de peciolos longos; (ROMANIAN) muştar; (RUSSIAN) gorchitsa; (SERER-SINE) mutard; (SPANISH) mostaza; (TAGALOG) mustasa; (VLAAMS) lange petiool mosterd; (WOLOF) mutard.

30.14. *Brassica juncea* (L.) Czern. var. *megarrhiza* N. Tsen & S.N. Lee (EP) root; (P) C.

(BURMESE) mon-hyin; (CHINESE, MANDARIN) dai tou gai; (CHINESE, WÚ) da tou jie; (CHINESE, YUE) gen yong jie cai, da tou cai, chong cai; (ENGLISH) tuberous-rooted mustard; (FRENCH) moutarde tubéreuse; (GREEK) sinapi me kovdylodis rizes; (HEBREW) keruv, hardal; (HMONG-DAW) zaub ntsuab; (ITALIAN) senape; (MACEDONIAN) grudčesto korenov nivska repka; (NORWEGIAN) sennepskål; (POLISH) gorczyca korzeniowa; (PORTUGUESE) mostarda tuberosa; (SPANISH) mostaza tuberosa; (VLAAMS) wortelknolmosterd.

30.15. *Brassica juncea* (L.) Czern. var. *multiceps* N. Tsen & S.N. Lee (EP) leaves; (P) P.

(BURMESE) mon-hyin; (CHINESE, MANDARIN) fen nie gai; (CHINESE, WÚ) fen nie jie; (CZECH) hořčice; (DANISH) mosterd; (DUTCH) mosterd; (ENGLISH) tillered mustard; (FARSI, WESTERN) khardal; (FILIPINO) mustasa; (FRENCH) moutarde; (GERMAN) Senf; (GREEK) kalliergoumenou sinapl; (HEBREW) keruv, hardal; (HMONG-DAW) zaub ntsuab; (HMONG-NJUA) zaub ntsuab; (ITALIAN) senape; (MACEDONIAN) izdankova nivska repka; (MALAY) sawi; (MANINKAKAN, WESTERN) navé; (NORWEGIAN) sennepskål; (POLISH) gorczyca uprawna;

(PORTUGUESE) mostarda de vários lançamentos; (ROMANIAN) muştar; (RUSSIAN) gorchitsa; (SERER-SINE) mutard; (SPANISH) mostaza; (VLAAMS) uitlopende mosterd; (WOLOF) mutard.

30.16. *Brassica juncea* (L.) Czern. var. *mutisecta* L.H. Bailey (EP) leaves; (P) C.

(BURMESE) mon-hyin; (CHINESE, MANDARIN) hua ye gai; (CHINESE, WÚ) hua ye jie; (CHINESE, YUE) hua je jie, ye yong jie cai; (CZECH) hořčice; (DANISH) mosterd; (DUTCH) mosterd; (ENGLISH) flower-like leaf mustard; (FARSI, WESTERN) khardal; (FILIPINO) mustasa; (FRENCH) moutarde; (GERMAN) Senf; (GREEK) sinapi me phyllon jananthos; (HEBREW) keruv, hardal; (HMONG-DAW) zaub ntsuab; (HMONG-NJUA) zaub ntsuab; (ITALIAN) senape; (MACEDONIAN) cvetovidna nivska repka; (MANINKAKAN, WESTERN) navé; (NORWEGIAN) sennepskål; (POLISH) gorczyca kwiatowo liściowa; (PORTUGUESE) mostarda; (ROMANIAN) muştar; (RUSSIAN) gorchitsa; (SERER-SINE) mutard; (SPANISH) mostaza; (VLAAMS) bloemachtige bladmosterd; (WOLOF) mutard.

30.17. *Brassica juncea* (L.) Czern. var. *rugosa* (Roxb.) N. Tsen & S.N. Lee (EP) leaves; (P) C.

(AFRIKAANS) bruinmosterd; (AGTA) mustasi; (ASSAMESE) lai sak; (BADAGA) kaḍugu giḍu; (BALTI) baca; (BENGALI) rai; (BICOLANO, CENTRAL) mustasa; (BINUKID) mustasa; (BODO) besor; (BONTOC CENTRAL) mustasa, amek, baloy, kabkabu; (BOSNIAN) senf; (BURMESE) mon-hyin; (CAJUN FRENCH) moutarde; (CATALAN-VALENCIAN-BALEAR) mostassa; (CEBUANO) mustasa; (CHICKASAW) tohi', tohi' okchamali; (CHINESE, GÀN) jie cai; (CHINESE, MANDARIN) dai ye gai; (CHINESE, MIN BEI) jie cai; (CHINESE, WÚ) da ye jie; (CZECH) hořčice; (DAKOTA) wa'hnakapizi'; (DANISH) mosterd; (DUTCH) Sareptamosterd, Indiase mosterd; (ENGLISH) large leaf mustard, brown mustard, Indian mustard, mustard greens, Chinese mustard, vegetable mustard, Indian red mustard; (ESTONIAN) sarepta kapsasrohi, sarepta sinep; (FARSI, WESTERN) khardal; (FIJIAN) sarso; (FILIPINO) mustasa; (FINNISH) sinappi; (FRENCH) chou faux jonc, moutarde de Sarepta, moutarde jonciforme, moutarde indienne, moutarde de Chine; (GAELIC, MANX) mustart; (GERMAN) Sarepta-Senf, Ruten-Kohl; (GIANGAN) mustasa; (GREEK) sinapl lachanikon; (HARARI) sänābi; (HEBREW) keruv, hardal; (HINDI) rai, rāī; raayi, saron, sar.sõ; (HMONG-DAW) zaub ntsuab; (HMONG-NJUA) zaub ntsuab; (IBANAG) mustasa; (IFUGAO) mustasa; (ILOCANO) mustasa; (INDONESIAN) sawi, sawi hijau, sawi-sawi, sesawi, se-sawi, caisim; (ITALIAN) senape indiana; (JAPANESE) taniku takana; (JAVANESE) sawi, (KANNADA) kempu saasive; (KASHMIRI) sàndij, sarshĕph; (KHMER, CENTRAL) khat naa; (KRIO) bush kabej, mɔstad; (KWÉYÒL) moutad; (LADINO) mostádra; (LAHU) g'aw$^{\vee}$ hpeh ca^{-}; (LAO) kaad khièw; (MACEDONIAN) širokolisna nivska repka; (MALAY) sawi pahit, kai choy; (MALTESE) mustarda; (MANINKAKAN, WESTERN) navé; (MARATHI) mohari, mohiri; (MICMAC) lamŭtald'; (MIZO) antam, kauphêk; (MUNDARI) rāi, māni; (MUSKOGEE) mvstv-lane; (NEPALI) rayo; (NEWAR) kā, pal-kā; (NORWEGIAN) sennepskål, sareptasennep; (NYANJA) mpiru; (OROMO, WEST CENTRAL) shinfa; (PAMPANGAN) mustasa; (POLISH) gorczyca wielkolistna indiańska; (PORTUGUESE) mostarda de folhas, mostarda indiana, mostarda vermelha; (ROMANIAN) muştar, muştar sălbatec; (RUSSIAN) gorchitsa; (SANSKRIT) asuri, ksvaka, rajasarsapa, rajika, sarsapah, siddhaartha, theekshnagandha; (SCOTS) mustart; (SERBO-CROATIAN) gorušica sareptska; (SERER-SINE) mutard; (SIDAMO) lentêta; (SLOVENIAN) kapusta sitinovitá, horčica sitinovitá; (SORBIAN, UPPER) žonop; (SOTHO, NORTHERN) mostara; (SOTHO, SOUTHERN) mosetareta; (SPANISH) mostaza de la China, mostaza de India; (SUNDA) sasawi, jabung; (SWEDISH) sareptä senap; (TAGALOG) mustasa; (TAMIL) kadugu; (TAUSUG) mustasa; (THAI) phakkat-khieo, phakkat-khieopli; (TIBETAN) pɛ̄ɛ̄qaṅ; (TONGA) kanganje; (TSWANA) masetete; (TUMBUKA) kanganje; (TURKISH) hardal; (VIETNAMESE) c[ar]i canh, bastutis; (MACEDONIAN) crn sinap; (MALAYALAM) kaduku; (MANINKAKAN, WESTERN) navé; (MALTESE) mustarda sewda; (MARATHI) kali mohri; (MORISYEN) moutard; (NEPALI) kalo tori; (NORWEGIAN) svartsennep; (ORIYA) rai; (PAMPANGAN) mustasa; (PANJABI, EASTERN) banarasi rayee; (POLISH) czarna, gorczyca czarna; (PORTUGUESE) mostarda negra, mostarda preta; (ROMANIAN) muştar negru; (RUSSIAN) gorchitsa chornaja; (SANSKRIT) rajika, asuri, athitheekshna, jwalanti, jwalatprapha, katuasuri, krimika, krishnasarhapa, krishnika, kshava, kshavaka, kshudhabijanana, kshujjanika, kshutabhijanaka, kshutaka, madhurika, raji, rakthasarshapa, rakthika, sarshapa, theekshagandha; (SERBO-CROATIAN) slačica crna, gorušica crna; (SLOVENIAN) kapusta čierna, horčica čierna; (SPANISH) mostaza negra, jenabe ajenabo; (SWEDISH) svartsenap; (TAGALOG) mustasa; (TAJIKI) siohkaram; (TAMIL) sirukadugu, kadugu; (TAUSUG) mustasa; (TELUGU) nalla avalu, avalu; (TURKISH) eşek hardali, kara hardal; (URDU) kali sarsoo, rai; (VLAAMS) zwarte mosterd; (WELSH) cedw du; (WOLOF) mutard.

30.25. *Brassica oleracea* L. var. *alboglabra* (L.H. Bailey) Musil (EP) young flower stalk and leaves; (P) C.

(AFRIKAANS) Sjinese boerkool; (CHINESE, GÀN) jie lan; (CHINESE, MANDARIN) jie lan; (CHINESE, MIN BEI) ge lan cai; (CHINESE, MIN NAN) kè-nâ-á-chhài, keh-nâ-á-chhài; (CHINESE, WÚ) jie lan; (CHINESE, YUE) jie lan; (DANISH) kailan-kål; (DUTCH) kailan; (ENGLISH) Chinese kale, Chinese broccoli, gailon, kailan; (FRENCH) kailan, broccoli de Chine; (GREEK) mprokol; (HUNGARIAN) kínai kel; (INDONESIAN) kailan; (JAPANESE) kailan; (KHMER, CENTRAL) spéi sâå; (MACEDONIAN) kineski kelj; (MALAY) kailan, kailan cina; (MALTESE) il-kabocci cinizi; (MÒORÉ) chũ; (MORISYEN) kai laan; (NEPALI) kale; (PORTUGUESE) kalian, couve chinesa; (SWEDISH) kinesisk kål; (TAGALOG) gailon; (TAJIKI) chinikaram; (THAI) phakkhana; (TURKISH) cin lahanasi; (URDU) phool ghobi; (VLAAMS) kailan, Chinese broccoli.

30.26. *Brassica oleracea* L. var. *botrytis* L. (EP) immature floral stalk, (P) C,R.

(AFRIKAANS) blomkool; (ALBANIAN, TOSK) lule lakër; (ARABIC) carnapeet, zaher; (ASSAMESE) phul kobi, phool kobi; (BADAGA) ju:go:su; (BALTI) phul gòbi; (BASQUE) azalili; (BELARUSAN) kapusta tsvetnaja; (BEMBÉ) le chou dia bifoulou, chou diri na bifoulou; (BENGALI) fulkapi, ful kapi; (BRETON) kaol-bleuñv; (BULGARIAN) cvetno zele; (BURMESE) gobi-pan; (CATALAN-VALENCIAN-BALEAR) col-i-flor; (CHINESE, GÀN) hua cai; (CHINESE, MANDARIN) hua ye cai, hua yeh ts'ai; (CHINESE, MIN BEI) hua cai; cai hua; (CHINESE, MIN NAN) hoe-chhài; (CHINESE, WÚ) hua ye cai, hau cai; (CHINESE, XIANG) hua cai; (CHINESE, YUE) hua ye cai, cai hua, hua cai, ye cai hua; (CREOLE, HAITIAN FRENCH) chouflè; (CRIOULO, CAPE VERDE) coviflor; (CROATIAN) cvjetaca; (CZECH) květák, karfiol; (DAGAARE, SOUTHERN) chũ; (DANISH) blomkål; (DUTCH) bloemkool; (ENGLISH) cauliflower; (ESTONIAN) lillkapsas; (FARSI, WESTERN) gol kalam; (FIJIAN) phul gobi; (FILIPINO) cauliflower; (FINNISH) kukkakaali; (FRENCH) chou fleur; (GAELIC, IRISH) cóilis; (GAELIC, MANX) caayl blaa; (GAELIC, SCOTTISH) colag, càl-colaig, càl-gruthach; (GALICIAN) colifrol; (GEORGIAN) khvavilovani combosto, calnabi; (GERMAN) Blumenkohl; (GOURMANCHÉMA) soũ-poné; (GOURMANTCHÉ) chũ feron; (GREEK) kounoupidi, vraski e lachanodis, anthokrawvi; (GUJARATI) fulkobi, ful kobi, fulvar; (HAUSA) farin; (HEBREW) kruvit; (HINDI) phool gobhy, phul gobi, phul gobhi, phool gobhi, gobhī; (HINDUSTANI) phul gobhi; (HUNGARIAN) karfiol, virágkel; (ICELANDIC) blomk'al; (INDONESIAN) kembang kol, kubis bunga, kol bunga, blumkol; (ITALIAN) cavolfiore; (JAPANESE) kalifurawaa, hana-yasai; (JOLA-FONYI) essoufleur; (KANNADA) hukosu, kosuguddae, hookosu, cauliflower, gobi; (KAZAKH) kapusta cvetnaja; (KHMER, CENTRAL) phkaa spéi; (KIRGHIZ) gulduu kapusta; (KOREAN) ccotgadubaechu; (KWÉYÒL) chov-fle; (LADINO) karnaba, karnabit; (LAO) kalampii dook; (LATVIAN) ziedu kāposti, ziedkāposti; (LITHUANIAN) žiedinis kopūstas; (LUXENBOURGEOIS) Choufleur, Schufflör; (MACEDONIAN) karfiol, običen karfiol; (MAITHILI) phulkōbī; (MALAY) kubis bunga; (MALAYALAM) cauliflower; (MALTESE) pastarda; (MAORI) kareparaoa; (MARANAO) obar-obar a ripolio; (MARATHI) phul kobi, fulvar; (MIZO) parbawr; (MÒORÉ) chũ; (MORISYEN) lisoufler; (MUNDARI) ba'-kobi; (NAVAJO) ch'il łigaií; (NEPALI) kauli; (NGUNDI) le chou dia bifoulou, chou diri na bifoulou; (NIUE) kāpisi fua; (NORWEGIAN) blomkål; (ORIYA) fula kobi, fule kibi; (PANJABI, EASTERN) phool gobhi, phul gobhi; (PASHTO, NORTHERN) gobay; (POLISH) kalafior; (PORTUGUESE) couve flor; (PULAR) soufleur; (ROMANIAN) conopidă, carfiol; (RUSSIAN) kapusta cvetnaja; (SAMOAN) kapisi; (SERBO-CROATIAN) karfiol; (SLOVAK) kveták, karfiol; (SLOVENIAN) cvetača; (SORBIAN, LOWER) kwětkowy kał; (SORBIAN, UPPER) kwětak; (SOTHO, NORTHERN) kholifolawa; (SOTHO, SOUTHERN) kholifolawa; (SPANISH) coliflor; (STELLINGWERFS) bloemkool; (SUNDA) blumkol, kembang kol; (SWAHILI) kolimaua, namna ya mboga, koli maua, kaliflari; (SWEDISH) blomkål; (TAGALOG) koli plawer; (TAJIKI) gulkaram; (TAMIL) gospoovu; (TELUGU) poogobi, poogbi, cauliflower; (THAI) kalam-dok; (TIBETAN) phūūl qōpi; (TOK PISIN) koli flawa; (TSWANA) kholifolawa; (TURKISH) karnibahar; (TURKMEN) gülli kelem; (UKRAINIAN) cvitna kapusta; (UZBEK) gul caram, gul karam; (VIETNAMESE) s[us]p l[ow], c[ar]i hoa; (VLAAMS) bloemkool; (WELSH) bresychen wen; (WOLOF) su-folor; (XHOSA) ikholiflawa, (YIDDISH) blumenkroyt; (ZULU) ukhalifulawa, ukhlifulawa.

30.27. *Brassica oleracea* L. var. *capitata* L. (EP) leaves; (P) C,R,P.

Dialects of Luyia (Bukusu[1], Itsukha[2])

a. common cabbage

(AFRIKAANS) kopkool, kool; (ALBANIAN, TOSK) krejce, lakër; (ALEUT) perutsiat makpatât; (AMHARIC) yäfäräňg g^{w}ammän; (ARABIC) curumb, crumm; (ARMENIAN) kugamb; (ASSAMESE) bondha kobi, bandha kabi; (AZERBAIJANI, NORTHERN) kələm; (BADAGA) ke:be:ju, go:su; (BALTI) ban gòbi; (BASQUE) azeburu; (BELARUSAN) kapusta belakachannaja; (BEMBA) kabeji; (BENGALI) bandhakapi; (BODO) kobi; (BOSNIAN) kupus; (BRETON) kaol; (BUGHOTU) gnahi; (BULGARIAN) bjalo glavesto zele; (BULU) shu; (BURMESE) gobi-dok; (CAJUN FRENCH) chou; (CATALAN-VALENCIAN-BALEAR) col bajana; (CEBUANO) repolyo; (CHECHEN) koopasta; (CHEROKEE) juganhdena; (CHICKASAW) tohi', tohi' ishkobishto' tohi' ishkobo'; (CHINESE, GÀN) bao cai; (CHINESE, HAKKA) kō-li-ts'ói, pō-li-ts'ói, pāo- ts'ói; (CHINESE, MANDARIN) gan lan, kan lan; (CHINESE, MIN BEI) bao cai, gao li cai; (CHINESE, MIN NAN) ko-lê-chhài; (CHINESE, WÚ) gan lan, bao cai; (CHINESE, XIANG) yang bai cai, juan xin cai; (CHINESE, YUE) ye cai; (COMANCHE) kabitsi; (CREE, PLAINS) otehipak; (CREOLE, HAITIAN FRENCH) chou, chou ponmen; (CRIOULO, CAPE VERDE) rapodjo; (CROATIAN) kupus; (CZECH) zelí, zelí hkavkové, zelí bilé hlavkové; (DAGAARE, SOUTHERN) chũ; (DAKOTA) waȟpetaŋka; (DANISH) hovedkål, hvidkål; (DIGO) kabichi; (DUTCH) sluitkool, witte kool; (ENGLISH) common cabbage, white cabbage, headed cabbage; (ESTONIAN) valge peakapsas, kapsas; (FARSI, EASTERN) karam, kalam; (FARSI, WESTERN) kalam; (FIJIAN) cabbage; (FILIPINO) repolyo; (FINNISH) keräkaali, punakaali, suippokaali, valkokaali; (FRENCH) chou cabus, chou blanc, chou pommé; (GAELIC, IRISH) cabáiste, cabáuste savoy, cabáiste samhraidh; (GAELIC, MANX) cabbash; (GAELIC, SCOTTISH) càl, ceirsle; (GALICIAN) repolo; (GANDA) mboga; (GEORGIAN) combosto, kombósto; (GERMAN) Weißkohl, Weißkraut, Kopfkohl; (GIKUYU) kabici; (GOURMANCHÉMA) soũ; (GOURMANTCHÉ) chũ; (GREEK) aspro lachano, kramvi, mappa; (GUJARATI) kobij; (HARARI) ḥūl, krumba; (HAUSA) kabeji; (HEBREW) keruv, keruv ʿalim, kruv hagina, kruv rosh; (HERERO) kol-i; (HILIGAYNON) repólyo; (HINDI) bandh gobi, band gobi, band gobhī, patagobhi, patgobhy, karam kalla, paangobhi; (HINDUSTANI) karamkalla, katran ki chorí; (HMONG DAW)

WESTERN) ka lamé ghomree; (FIJIAN) kurabi; (FILIPINO) kahlrabi; (FINNISH) kyssäkaali, kaalirapi; (FRENCH) chou rave; (GALICIAN) nabicol; (GEORGIAN) colrabi, khviti; (GERMAN) Kohlrabi; (GREEK) vrassiki (kramvi) e goggylodis, goggylokramvi; (GUJARATI) nolkol, nol kol; (HEBREW) kolrabi, kohlrabi; (HINDI) ganth gobi, ganth gobhi, gaant gobhi; (HUNGARIAN) karalábé; (ICELANDIC) hnuđk'al; (INDONESIAN) kolumbi; (ITALIAN) cavolo rapa, col rabano; (JAPANESE) kohrurabi, kyûkei-kanran; (KANNADA) navilu kosu, gedde kosu, navil kosu; (KAZAKH) kolrabi; (KIRGHIZ) kol'rabi kapustasy; (KOREAN) chulgigadubaechu; (LATVIAN) kolrābji; (LITHUANIAN) kaliaropė; (MACEDONIAN) tulija, crven alabaš; (MALTESE) gidra; (MARATHI) nawalkol, nawal kol; (MÒORÉ) chũ; (MORISYEN) sou-nave; (NEPALI) gyanth gobi; (NORWEGIAN) knutekål, kalrabi; (ORIYA) olkobi, ganthi kobi; (PANJABI, EASTERN) gandh gobi, gandh gobhi; (POLISH) kalarepa; (PORTUGUESE) couve rábano, couve rabão; (ROMANIAN) gulie, gulia, calaraba; (RUSSIAN) kol'rabi, kapusta repnaja; (SERBO-CROATIAN) keleraba; (SINHALA) knolkhol; (SLOVAK) kaleráb; (SLOVENIAN) kolerarbica; (SPANISH) col rábano, colirrábano, col rapano; (STELLINGWERFS) koolrapen; (SWAHILI) kolirabi; (SWEDISH) kålrabbi; (TAJIKI) toshkaram; (TAMIL) nool kohl; (TELUGU) gaddagobi; (THAI) kalam-pom; (TURKISH) alabas; (UKRAINIAN) kol'rabi; (URDU) ganth gobi; (VIETNAMESE) su h[af]o, th[of] l[of]; (VLAAMS) koolrabi; (WELSH) meipen ddeilog; (WOLOF) su-rav.

30.31. *Brassica oleracea* L. var. *italica* Plenck (EP) immature flower stalk; (P) C,R.

(AFRIKAANS) brokkoli; (ASSAMESE) broccoli; (BELARUSAN) kapusta agorodnaja, brokoli; (BULGARIAN) brokoli; (CATALAN-VALENCIAN-BALEAR) bròquil; (CHINESE, MANDARIN) qing hua cai; (CHINESE, WÚ) qing hua cai; (CHINESE, YUE) xi lan hua, qing hua cai, lu cai hua; (CREOLE, HAITIAN FRENCH) bwokoli; (CRIOULO, CAPE VERDE) brocolo; (CROATIAN) brokula; (CZECH) prokolice, brokolice; (DANISH) broccoli; (DUTCH) broccoli; (ENGLISH) broccoli, sprouting broccoli, calabrese; (ESTONIAN) brokoli, asparkapsas; (FILIPINO) broccoli; (FINNISH) parsakaali; (FRENCH) chou brocoli, brocoli, brocoli aspergé; (GAELIC, IRISH) brocaili; (GAELIC, MANX) callish; (GALICIAN) brócolo; (GEORGIAN) brocoli, satatsurisebri combosto; (GERMAN) Brokkoli; (GREEK) brokkolo; (HEBREW) berokoli; (HINDUSTANI) kolí, kiekqism; (HUNGARIAN) brokkoli, spárgakel; (ICELANDIC) brokkoli, spergilk'al; (INDONESIAN) brokoli; (ITALIAN) cavolo broccoli; (JAPANESE) burotkorii, burokkori, Italia-kanran; (LATVIAN) brokoli; (LITHUANIAN) brokolis; (MACEDONIAN) prokula, brokoli; (MALAY) broccoli; (MALTESE) brokkli; (MÒORÉ) chũ; (MORISYEN) brokoli; (NEPALI) brokauli; (NORWEGIAN) brokkoli, broccoli, aspargeskål; (POLISH) brokuł włoski; brokoły, kapusta szparagowa, brokuł; (PORTUGUESE) brócolo, couve brócolo, couver broculo, broccoli; (ROMANIAN) broccoli, brocoli; (RUSSIAN) brokoli, sparevaja kapusta; (SAMOAN) kapisi; (SERBO-CROATIAN) kaulin, prokule; (SLOVAK) kel špargl'ový; (SLOVENIAN) brokoli; (SPANISH) brocoli; (SWAHILI) brokoli; (SWEDISH) broccoli; (TAGALOG) brocoli; (TAJIKI) karami brokkali; (THAI) brokkholi; (TOK PISIN) brokoli; (TURKISH) brokkoli, brokoli, cibes; (VLAAMS) brokoli, broccoli; (WELSH) blodfresychgaeaf; (WOLOF) su-brocoli; (XHOSA) imifuno ethile efana nekholiflawa.

30.32. *Brassica oleracea* L. var. *medullosa* Thell. (EP) leaves; (P) C.

(CHINESE, WÚ) qian tou gan lan; (CZECH) dřeňová kapusta; (DUTCH) markstamkool; (ENGLISH) marrow stem kale; (FINNISH) rehuydinkaali; (FRENCH) chou moëllier; (GERMAN) Markkhohl, Markstammkohl, Kuhkohl; (ITALIAN) cavolo; (MÒORÉ) chũ; (NORWEGIAN) fôrmargkål; (POLISH) kapusta pastewna głąbiasta; (PORTUGUESE) couve de medula; (SPANISH) col de meollo, col medular; (VLAAMS) merg(stam)kool.

30.33. *Brassica oleracea* L. var. *ramosa* DC. (EP) leaves; (P) C.

(CZECH) krmná kapusta, větevna tá kapusta; (DUTCH) eeuwige moes, splijtkool; (ENGLISH) thousand-headed kale; (FRENCH) chou à cent têtes, chou à mille têtes, chou branchu; (GAELIC, IRISH) cál ilcheannach; (GERMAN) Tausendkopfkohl; (POLISH) kapusta tysiągłówkowa; (PORTUGUESE) couve de mil cabeças; (SPANISH) col de mile cabezas; (TAJIKI) kalakaram; (VLAAMS) struikkool, stekkool, eeuwig mose.

30.34. *Brassica oleracea* L. var. *sabauda* L. (EP) leaves; (P) C,R.

(AFRIKAANS) savoykool, krulkool; (ALBANIAN, TOSK) lakär e savojës; (ASSAMESE) bondha kobi; (BELARUSAN) kapusta savoiska; (BULGARIAN) savojsko zele; (CATALAN-VALENCIAN-BALEAR) col borrajenca; (CHINESE, MANDARIN) zhou ye gan lan; (CHINESE, WÚ) zhou ye gan lan; (CREOLE, HAITIAN FRENCH) chou frizé; (CZECH) hlávkavá kapusta, kapusta, kapusta savojska; (DANISH) savoykål; (DUTCH) savooiekool; (ENGLISH) savoy cabbage, headed cabbage, cabbage; (ESTONIAN) kähar peakapsas, savoia kapsas; (FINNISH) kurttukaali, savoijinkaali; (FRENCH) chou de savoie, chou de Milan, chou cabus; (GEORGIAN) savois combosto; (GERMAN) Wirsing, Welschkohl, Wirsingkohl; (GREEK) sgourolachano, lachano savoï; (HEBREW) kruv mesulsal; (HUNGARIAN) kelkáposzta; (ICELANDIC) blöruk'al; (INDONESIAN) kubis babat, kol babat, kubis, kobis, kol; (ITALIAN) cavolo verzotto; (JAPANESE) saboi-kyabetsu; (KHMER, CENTRAL) spéi k'daôb; (KIRGHIZ) savoj kapustasy; (KOREAN) bidangadubaechu; (LAO) kalampii; (LATVIAN) virzinkāposti; (LITHUANIAN) garbanotasis kopūstas; (LUXENBOURGEOIS) Kabes; (MACEDONIAN) kelj-glavest, običen kelj; (MALAY) kubis; (MALTESE) kabocci tar-rebbiegha; (MÒORÉ) chũ; (NORWEGIAN) savoikål, savoykål; (POLISH) kapusta włoska; (PORTUGUESE) couve lombarda, couve de Sabóia; (ROMANIAN) varză creaţă; (RUSSIAN) kapusta savojska, kapusta savojska belokočannaja; (SERBO-CROATIAN) kelj; (SLOVAK) kel hlávkový; (SLOVENIAN) chrovt; (SPANISH) col de milan; (STELLINGWERFS) sevojekool; (SWAHILI) kabichi savoy; (SWEDISH) savojkål; (TAGALOG) repolyo; (THAI) kalam-pli; (TOK PISIN) kapis; (TURKISH) kivircik baş lahana, kivircik lahana; (VIETNAMESE) c[ar]i b[aws]p; (VLAAMS) savooie kool, savooikool; (WELSH) safwy.

30.35. *Brassica oleracea* L. var. *viridis* L. (EP) leaves; (P) C.

Dialects of Luyia (Bukusu[1], Itsukha[2])

(AFRIKAANS) boerkool, beeskool; (ALBANIAN, TOSK) lakër e bardhë; (BASQUE) azalandare; (BELARUSAN) kapusta listovaja; (BULGARIAN) listno zele; (CATALAN-VALENCIAN-BALEAR) col crespa; (CHICKASAW) tohi'chaaha'; (CHINESE, MANDARIN) yu yi gan lan, chuan yeh ts'ai; (CHINESE, WÚ) yu yi gan lan; (CHINESE, YUE) yu yi gan lan; (CREOLE, HAITIAN FRENCH) chou frizé; (CRIOULO, CAPE VERDE) covi; (CROATIAN) raštika, lisnati kelj; (CZECH) krmna kapusta, kapusta zimni, kapusta kaderava, jarmuz, kadeřavá kapusta, kadeřávek; (DANISH) grønkål; (DUTCH) boerenkool; (ENGLISH) kale, tree kale, Scotch kale, borecole, collards; (ESTONIAN) lehtkapsas; (FILIPINO) kale, kolis; (FINNISH) lehtikaali, koristekaali; (FRENCH) chou cavalier, chou fourrager, chou vert, chou frisé; (GAELIC, IRISH) cál catach; (GALICIAN) bertoń, louve; (GERMAN) Grünkohl, Blätterkohl, Federkohl, Futterkohl; (GIKUYU) matharũ; (GREEK) krambe he lachanodes, kramvi z akèfalos; (HEBREW) kruv alim; (HINDI) karam-sag; (HINDUSTANI) ek qism kí gobhí; (HMONG DAW) zaub qhwv; (HUNGARIAN) leveles kel, leveles káposzta, takarmánykáposzta; (ICELANDIC) grænk'al; (INDONESIAN) kailan; (ITALIAN) cavolo riccio, cavolo da foraggio, cavolo cavaliere, cavolo a penna; (JAPANESE) keelu, ryokuyô-kanran; (KAZAKH) kapusta listovaja; (KHMER, CENTRAL) spéi barang; (KIRGHIZ) žalbyraktuu kapusta;

(LATVIAN) ziedu kāposts, lapu kāposti; (LITHUANIAN) lapinis kopūstas; (LOMWE, MALAWI) chikasu; (LUYIA) sukuma[1], lisukuma[2]; (MACEDONIAN) kadrav kelj, lisna zelka; (MÒORÉ) chũ; (MUSKOGEE) 'se-tapho-rakko, esse-tapho-rakko; (NORWEGIAN) bladkål, pyntekål, grønnkål; (POLISH) kapusta ogrodowa, jarmuż, kapousta pastewna; (PORTUGUESE) couve de folhas, couve de desfolhar, couve galega, couve cavaleira, couve manteiga, couve galega frisada, collards; (ROMANIAN) varză furajeră, varză de frunze; (RUSSIAN) karpusta listovaja, kapusta kormovaja, kapusta korovia, kapusta listovaja kurčavaja; (SERBO-CROATIAN) lehtikaali, zelen kupus, raštan; (SLOVAK) kel kadernatý, kel kučeravý; (SLOVENIAN) listnati chrovt; (SPANISH) berza, berza col, col caballar, col gallega, col forragera; (STELLINGWERFS) boerekool, moes, boerestruuk, struupmoes; (SWAHILI) sukuma wiki; (SWEDISH) grönkål; (TAGALOG) kales, kolis; (TAJIKI) kahkaram; (THAI) khana-farang; (TURKISH) yeşil lahana, yaprak lahana, kara lahana, kirmizi lahana, karabas lahanasi; (URDU) karam sag; (VLAAMS) boerenkool, stamkool, peterseliekool; (WELSH) bresych deiliog; (WOLOF) su; (XHOSA) uhlobo oluthile lwekhaphetshu.

30.36. *Brassica rapa* L. subsp. *chinensis* (L.) Hanelt (EP) leaves; (P) C.

(AFRIKAANS) sjinese mosterd, seldery mosterd; (BELARUSAN) repa; (BINUKID) pitsey; (BULGARIAN) kitajsko zele; (CAROLINIAN) nappa; (CHINESE, GÀN) qing cai, pu tou bai cai; (CHINESE, MANDARIN) xiao bai cai, pak choi, xiao hsiao yu ts'ai; (CHINESE, MIN BEI) bai cai, bai cai zai; (CHINESE, MIN NAN) kòa-chhài; (CHINESE, WÚ) xiao bai cai ya zhong, qing cai; (CHINESE, YUE) xiao bai cai, qing cai, pu tong bai cai; (CZECH) čínské zelí; (DANISH) pak-choi; (DUTCH) paksoi; (ENGLISH) pak-choi, Chinese mustard, celery mustard, celery cabbage, Chinese cabbage; (ESTONIAN) hiina kapsas; (FIJIAN) china cabbage; (FILIPINO) Chinese pechay; (FINNISH) paksoi, pinaattikiinankaali; (FRENCH) pak-choi, chou de chine, chou de Shangton; (GERMAN) Pakchoi, Chinakohl; (GREEK) pock-tsoï, kineziko lachano; (HEBREW) keruv sini; (HMONG DAW) zaub ntsuab; (HUNGARIAN) kínai kel; (ICELANDIC) blađk'al; (INDONESIAN) petsai, sawi putih, sesawi putih, pecal pakcoi, paksoi, pecai, sawi ijo; (ITALIAN) pak-choi; (JAPANESE) pakuchoi, tai-si, taisi, shakushi-na; (JAVANESE) sawi putih, bakcai, pecai, kluban; (KHMER, CENTRAL) pechhaay, spéi khiew; (KIRGHIZ) pekin kastusy; (KOREAN) bombaechu; (KWÉYÒL) patchouli; (LAHU) g'aw$^{\vee}$ ca^{-} di$_{\vee}$, g'aw$^{\vee}$ ca^{-} o$_{-}$, g'; (LAO) kaad chiin; (LATVIAN) rāceņi, kīnas kāposti; (LITHUANIAN) kininis bastutis, kino kopūstai; (MACEDONIAN) kineska zelka, celerovidna repka; (MALAY) sawi putih, pak choi; (MORISYEN) pak choy; (NIUE) kāpisi Saina; (NORWEGIAN) stilkkål, kinesisk kål; (NYANJA) tanaposi, mpiru; (POLISH) kapusta chińska, pak-choi; (PORTUGUESE) pak-choi, repolho chines "pakchoi"; (PULAR) nawé; (ROMANIAN) varză chinezească; (RUSSIAN) kapusta kitajskaja; (SAMOAN) kapisi; (SERBO-CROATIAN) kineski kupus; (SERER-SINE) nawwé; (SLOVENIAN) kitajski kapus; (SPANISH) pakchoi, col de china; (SUNDA) pecal, samho; (SWAHILI) kabichi chainizi; (SWEDISH) pak-shoi, kinesisk kal; (TAGALOG) pechay; (THAI) phakkat-khaokwangtung, hong-tae, hong-how; (TOK PISIN) kapis; (TURKISH) kolza lahanasi; (VENDA) mutshaeni; (VIETNAMESE) c[ar]i thi[af], c[ar]i ng[oj]t, c[ar]i b[ej] tr[aws]ng; (VLAAMS) paksoi, pak-choi, krulmostaard; (WOLOF) nawé; (YORUBA) leaf cabbage.

30.37. *Brassica rapa* L. subsp. *chinensis* (L.) Hanelt var. *parachinensis* (L.H. Bailey) Hanelt (EP) leaves; (P) C.

(CHINESE, MANDARIN) cai tai; (CHINESE, WÚ) cai xin; (CHINESE, YUE) cai xin, cai tai; (DUTCH) choy sum; (ENGLISH) mock pak choy, choy sum, caisin, flowering white cabbage; (FRENCH) choy sum; (HMONG DAW) zuab paj; (IFUAGO) at-atang; (INDONESIAN) caisin, sawi hijau, sawi kembang; (IVATAN) pichay; (JAPANESE) saishin; (MALAY) sawi hajau, sawi manis, sawi bunga; (MANOBO) pitsey; (MARANAO) pisay a tao; (MORISYEN) bred tompous; (POLISH) kapusta parachińska; (PORTUGUESE) choy sum; (SPANISH) choy sum; (THAI) phakkwangtung, phakkatkheokwangtung; (VLAAMS) choisum.

30.38. *Brassica rapa* L. subsp. *narinosa* (L.H. Bailey) Hanelt (EP) leaves; (P) C.

(CHINESE, MANDARIN) wu ta cai; (CHINESE, WÚ) paio e cai; (ENGLISH) broad beaked mustard; (JAPANESE) taasai; (MACEDONIAN) širokoklunesta repka; (MANINKAKAN, WESTERN) nave; (VLAAMS) breed gesnavelde mosterd.

30.39. *Brassica rapa* L. subsp. *nipposinica* (L.H. Bailey) Hanelt var. *perviridis* L.H. Bailey (EP) leaves; (P) C.

(AFRIKAANS) spinasie mosterd; (DANISH) spinatsennep; (DUTCH) komatsuna; (ENGLISH) spinach mustard, tendergreen mustard; (FINNISH) vihannessinappi; (FRENCH) moutarde épinard; (GREEK) ispaniko sinapi; (ITALIAN) senape spinacio; (JAPANESE) komatsuna; (LITHUANIAN) žaliasis bastutis; (MACEDONIAN) spanakovidna repka; (MANINKAKAN, WESTERN) navé; (POLISH) gorczyca szpinakowa; (PORTUGUESE) mostarda espinafre, komatsuna; (SPANISH) mostaza espinaca; (VLAAMS) mosterdspinazie, komatsuna; (WOLOF) mutard.

30.40. *Brassica rapa* L. subsp. *pekinensis* (Lour.) Hanelt (EP) leaves; (P) C,P.

(AFRIKAANS) Sjinese kool, pe-tsai; (ASSAMESE) mer lai, China lai; (BULGARIAN) kitajsko zele; (CEBUANO) petsay; (CHINESE, GÀN) da bai cai, huang ya cai, jiao bai; (CHINESE, MANDARIN) da bai cai, pe-tsai; (CHINESE, MIN BEI) bao xin cai, shandong bai; (CHINESE, MIN NAN) pau-sim-pėh-chhài, pėh-chhài; (CHINESE, WÚ) huang ya bei cai, da bei cai ya zhong; (CHINESE, YUE) bai cai, da nai cai, huang ya bai, shao cai; (CROATIAN) kinesko zelje; (CZECH) pikingské zelí, činske zeli; (DANISH) kinakål, Kinesisk kål; (DUTCH) Chinese kool, petsai; (ENGLISH) Chinese cabbage, celery cabbage, pe-tsai, Peking cabbage; (ESTONIAN) hiina kapsas; (FILIPINO) pe-tsai; (FINNISH) salaattikiinankaali, pinaattikunankaali; (FRENCH) chou de Chine, chou de Shangton, pé-tsai, chou chinois; (GEORGIAN) chinuri combosto; (GERMAN) Pekingkohl, Schantungkohl, Chinakohl; (GREEK) kineziko lachano; (HEBREW) keruv sini; (HINDUSTANI) amsoi; (HMONG DAW) zaub qhwv; (HUNGARIAN) kinai kel; (ICELANDIC) kinak'al; (ILOCANO) petsay; (INDONESIAN) petsai, sawi ijo, sawi putih; (ITALIAN) cavolo di Cina, cavolo cinese; (JAPANESE) hakusai; (KAZAKH) kapusta kitajskaja; (KHMER, CENTRAL) pe-chhaay, spéi khiew; (KIRGHIZ) pekin kastusy; (KOREAN) bombaechu; (KWÉYÒL) patchoi; (LAO) kaad chiin; (LATVIAN) kīnas kāposti; (LITHUANIAN) pekino bastutis; (LUO) kabej; (MACEDONIAN) kineska zelka, pekinška zelka; (MALAY) kubis cina, kobis cina, kubis panjang, sawi pahit, sawi putih, petsai; (MALTESE) kabocca ciniza; (MORISYEN) petsay, bred de sinn; (NORWEGIAN) kinakål, kinesisk kål; (POLISH) kapusta pekińska; (PORTUGUESE) couve chinesa, couve da China, repolho chinês, petsai; (ROMANIAN) varză chinezească; (RUSSIAN) kapusta kitajskaja; (SAMOAN) kapisi; (SERBO-CROATIAN) kineski kupus; (SLOVENIAN) kitajski kapus; (SPANISH) col de China, repollo chino; (SWAHILI) kabichi chainizi; (SWEDISH) kinesisk kal; (TAGALOG) petsay, pachay Baguio, wong bok; (THAI) phakkat-khaopli, phakkatkhaoyai; (TIBETAN) pɛ̄ɛ̄tsɛɛ; (TOK PISIN) kapis; (TURKISH) çin lâhanasi, cin lâhanasi; (UZBEK) pekin carami; (VIETNAMESE) c[ar]i b[aws]c kinh; (VLAAMS) krulmostaard, Chinese kool, pe-tsai; (YORUBA) leaf cabbage.

30.41. *Brassica rapa* L. subsp. *rapa* (EP) a. enlarged root, b. leaves, and c. inflorescence; (P) C.

a. root

(AFRIKAANS) raap; (ALBANIAN, TOSK) rrepë; (ALEUT) nakkatannak; (ARABIC) lift; (ASSAMESE) chalgum, salgom; (BELARUSAN) repa; (BENGALI) shalgam; (BLACKFOOT) ksíkka's; (BODO) salgum; (BRETON) irvin; (BULGARIAN) rajapa; (CAJUN FRENCH) navet; (CATALAN-VALENCIAN-BALEAR) nap; (CHEROKEE) dakwsani; (CHICKASAW) tanap; (CHINESE, GÀN) die zi luo bo, bian luo bo; (CHINESE, HAKKA) ts'ói-t'êou, lôp'ĕt; (CHINESE, MANDARIN) wu jing, wu ching; (CHINESE, MIN BEI) pan cai; (CHINESE, MIN NAN) chhài-thâu; (CHINESE, WÚ) da tou cai, pan cai tou, yuan gen; (CHINESE, YUE) wu jing; (CREE, PLAINS) otisihkân, otsekan; (CREOLE, HAITIAN FRENCH) navè; (CRIOULO, CAPE VERDE) nabo; (CROATIAN) bijela repa; (CZECH) vodnice, brukev řepák vodnice, rapa; (DAKOTA) ti'psina; (DANISH) majroe, høstroe; (DUTCH) raap, meiraap, stoppelknol; (ENGLISH) turnip, turnip root; (ESTONIAN) naeris; (FARSI, EASTERN) shalgham; (FARSI, WESTERN) shalgham; (FILIPINO) turnip; (FINNISH) nauris, turnipsi; (FRENCH) navet, navet potager; (GAELIC, IRISH) tornapa, tornapa bán, tornapa dearg; (GAELIC, MANX) napin; (GAELIC, SCOTTISH) snèap; (GERMAN) Mairübe, Speiserübe; (GOURMANCHÉMA) navet; (GOURMANTCHÉ) navé, bompieno-tibina; (GREEK) gongyli, kramvi e rocpys; (GUJARATI) salgham; (HEBREW) lefet, kruv halefet, lephet; (HERERO) ombaruru; (HILIGAYNON) singkamás; (HINDI) salgham, sł.algam, shalga, salgam, shaljum; (HINDUSTANI) shaljam, shalgam, shikhámúl; (HUNGARIAN) tarlórépa, takarmány tarlórépa; (ICELANDIC) næpa; (INDONESIAN) kol rap, rapkol, turnip; (INUKTITUT, WEST GREENLANDIC) nautsîaq; (ITALIAN) rapa, rapa coltivata, rapa di primavera; (JAPANESE) kabu, komatsuna; (KANNADA) turnip; (KASHMIRI) āram; (KATI) s̲h̲alum; (KAZAKH) turneps; (KIRGHIZ) čamgur; (KWÉYÒL) navé; (LADINO) návo; (LAKOTA) tiŋpsila, tiŋpsiŋla skaska; (LATVIAN) rācenīs; (LITHUANIAN) ropė; (LOMWE, MALAWI) tanaposi; (LUXENBOURGEOIS) Rommel; (MACEDONIAN) bela repka, zelkina repka; (MALAY) turnip; (MALAYALAM) seema mullanki; (MALTESE) nevew; (MAORI) keha, pora, toonapi; (MARATHI) salgum; (MICMAC) wĕnjooesŭgŭbŭn; (MIZO) antam bulbawk, tam bul bawk; (MOHAWK) otsihkwa ohtera; (MUNDARI) sālgam; (MUSKOGEE) tvnvpse; (NEPALI) salagam; (NEWAR) palak; (NORWEGIAN) nepe, förnepe; (ORIYA) salgum; (PANJABI, EASTERN) gonglu, shalgam, thipper, shulgam gõglõõ; (PASHTO, NORTHERN) t̲epër, shalghama; (POLISH) rzepa, rzepa turnips, turnips; (PORTUGUESE) nabo; (ROMANI, BALKAN) ràdika; (ROMANIAN) turneps, nap porcesc; (RUSSIAN) turneps; (SANTALI) salgam; (SCOTS) neep; (SERBO-CROATIAN) repa, turnips; (SLOVAK) okrúhlica, vodnica; (SLOVENIAN) poletna repa; (SORBIAN, LOWER) kuwala, rěpa; (SORBIAN, UPPER) rěpa; (SOTHO, NORTHERN) rapa; (SOTHO, SOUTHERN) rapa; (SPANISH) nabo, naba; (STELLINGWERFS) knolle, knollegies, raopen, zaandknollen; (SUNDA) engkol beureum; (SWAHILI) tanipu; (SWEDISH) rova, rofra, turnips; (TELUGU) turnip; (TIBETAN) ñuŋmə; (TSWANA) thenipe; (TURKISH) şalgam, kolza lahanasi; (URDU) salgum salgum; (UZBEK) sholgom; (VLAAMS) raapzaad, raap; (WELSH) erfinen; (XHOSA) ithenephu, iteniphu, irabha; (YIDDISH) brukveh; (ZULU) utheniphu, i(li)theniphu.

b. leaves

(ARABIC) lift; (ASSAMESE) pat chalgum; (BALTI) drum-pa; (CRIOULO, CAPE VERDE) nabica; (DANISH) agerkål; (DUTCH) raapstelen, raapsteeltjes, steeltjes; (ENGLISH) turnip greens, leaf turnip; (FRENCH) raves de feuille; (GEORGIAN) talgami; (GERMAN) Rübenstiel, Rüstiele; (GREEK) rapa lachanon; (ITALIAN) rapa da foglia, rapa senza testa; (KASHMIRI) dēmba-gŏga'ji; (KWÉYÒL) navé; (MAORI) poke; (MUSKOGEE) tvnvpse-lane, tvnvpse-esselane; (POLISH) rzepa liściowa; (PORTUGUESE) nabiças; (SPANISH) nabizas; (SWEDISH) rova.

c. inflorescence

(DUTCH) broccoletto; (ENGLISH) broccoli raab, Italian turnip, turnip broccoli, broccoletto; (FRENCH) brocolis de raves, brocoli navet, cima di rapa; (GREEK) phylla revas, lachanides; (HEBREW) keruv ha-lefet; (ITALIAN) cima di rapa, broccoletti di rapa; (MACEDONIAN) brokoli repka; (POLISH) rzepa włoska; (PORTUGUESE) nabo greleiro, nabo de espigos; (SPANISH) nabo de brotes; (VLAAMS) broccoliraap.

30.42. *Bunias orientalis* L. (EP) leaves; (P) C.

(BELARUSAN) sverbiguz ushodi; (CHINESE, MANDARIN) you guo chi qi; (CHINESE, WÚ) you guo shi ji; (CZECH) rukevník východní; (DANISH) takkeklap; (DUTCH) oosterse bunias; (ENGLISH) hill mustard; (FINNISH) ukonpalko; (FRENCH) bunias oriental, bunier; (GERMAN) hohe Zachenschöte; (HUNGARIAN) szümcső; (ITALIAN) cascellore orientale; (LATVIAN) austrumu dižpērkone; (LITHUANIAN) rytinė engra; (MACEDONIAN) istočna repuška; (MANINKAKAN, WESTERN) navé; (MONGOLIAN) bunias; (NORWEGIAN) russekål; (POLISH) gorczyca szczytowa; (PORTUGUESE) saramago oriental; (SPANISH) jaramago oriental; (SWEDISH) ryssgubbe; (TURKISH) bunias, siğilli lahana; (VLAAMS) grote hardvrucht.

30.43. *Capsella bursa-pastoris* (L.) Medikus (EP) young leaves; (P) C.

(AFRIKAANS) herderstassie, geldbeursie; (AMHARIC) feto; (BADAGA) mojje; (BELARUSAN) strelki zvychainaja; (CHINESE, GÀN) da ye ji cai; (CHINESE, MANDARIN) qi cai; (CHINESE, WÚ) ji cai; (CHINESE, YUE) ling jiao cai; (CZECH) kokoška pastuší tobolka; (DANISH) hyrdetaske; (DUTCH) herderstasje; (ENGLISH) shephard's purse; (ESTONIAN) hiirekõrv; (FINNISH) lutukka; (FRENCH) bourse à berger, bourse à pasteur, capselle bourse à pasteur; (GAELIC, IRISH) lus; (GERMAN) Hirtentäschel, gemeines Hirtentäschelkraut, gemeines Hirtentäschel; (GREEK) kapsella e pimenosakkoï; (HEBREW) yalkut ha-ro'im, yalkut haroim; (HUNGARIAN) pásztortáska; (ICELANDIC) hjartarfi; (ITALIAN) borsa di pastore comune, borsacchina; (JAPANESE) nazuna, naeng-ie; (LATVIAN) parastais plikstiņš; (LITHUANIAN) trikertė žvaginė; (MACEDONIAN) obična ovčarska torbička; (MALTESE) gargir il-gemel; (MONGOLIAN) egel avga; (NEPALI) tori ghans; (NORWEGIAN) gjetertaske; (POLISH) tasznik pospolity; (PORTUGUESE) bolsa de pastor, erva do bom pastor; (QUECHUA, CUSCO) bolsa bolsa; (ROMANIAN) traista-ciobanului; (RUSSIAN) soumochnik, pastushaja sumka, obyknovennaja sumka; (SERBO-CROATIAN) pastirska torbica, obična; (SORBIAN, UPPER) jětrowka; (SPANISH) bolsa de pastor común; (STELLINGWERFS) entepoot, lepeldief, lepelties, leugebas, vork en lepel, lepeldiefien, lepeltiesbloeme; (SWEDISH) lomme; (TAGALOG) erva do bom partor; (TAJIKI) jagh-jagh; (TURKISH) çoban çantasi, kuşkuş ekmeği; (TURKMEN) adaty gyjy-gyjy; (VLAAMS) herderstasje; (WELSH) llysiau.

30.44. *Cardamine pratensis* L. (EP) leaves; (P) C,R.

(CHINESE, MANDARIN) xiao hua sui mi qi; (CHINESE, WÚ) xiao hua sui mi ji; (CZECH) řeřišnice luční; (DANISH) engkarse; (DUTCH) pinksterbloem; (ENGLISH) cuckoo-flower; (ESTONIAN) käokann; (FARSI, WESTERN) tarch koohie; (FINNISH) luhtalitukka; (FRENCH) cardamine des prés, cresson des prés; (GERMAN) Wiesen-Schaumkraut; (GREEK) kardamini e livadiki; (HEBREW) kardamin; (HUNGARIAN) réti kakukktorma; (ICELANDIC) hrafnaklukka; (ITALIAN) billeri dei prati, cardamino dei prati; (LATVIAN) ķērsas; (LITHUANIAN) pievinė kartenė; (MACEDONIAN) livadska penlivka; (MONGOLIAN) nugiin zuuch; (NORWEGIAN) engkarse; (POLISH) rzeżucha łąkowa; (PORTUGUESE) agrião dos prados, cardamina dos prados;

(SPANISH) cardamina, mastuerzo de prado, berro de prado; (SWEDISH) ängsbräsma; (TURKISH) su teresi, çayir teresi, çayir köpük otu; (VLAAMS) pinksterbloem; (WELSH) blodyn y gôg.

30.45. *Crambe maritima* L. (EP) petioles and young leaves; (P) C.

(AFRIKAANS) seekool; (ALBANIAN, TOSK) krambë; (BELARUSAN) morskaja kapusta, katran; (BULGARIAN) divo krajmorsko zele; (CHINESE, MANDARIN) bin cai, hai kan lan; (CHINESE, WÚ) hai gan lan, bin cai; (CZECH) katrán přímořský; (DANISH) strandkål, strandkoal; (DUTCH) zeekool; (ENGLISH) sea kale, crambe; (FINNISH) merikaali; (FRENCH) chou marin, crambe; (GERMAN) Meerkohl, echter Meerkohl; (GREEK) kramui e paralios; (HEBREW) karbba yamit, craba yamit; (HUNGARIAN) tátorján; (ITALIAN) cavolo marino, cavolo di mare; (JAPANESE) hamana; (LATVIAN) jūrmalas krambe; (LITHUANIAN) pajūrinė balža; (MACEDONIAN) morska zelka, morski kelj; (NORWEGIAN) strandkål; (POLISH) kapusta morska, modrak morska; (PORTUGUESE) couve marinha, couve bestarda; (ROMANIAN) hodolean; (RUSSIAN) kapusta morskaja; (SERBO-CROATIAN) krambe; (SLOVAK) katran prímorský; (SLOVENIAN) obmorski kapus; (SPANISH) col marina, col maritima, crambe; (SWEDISH) strandköl; (TAJIKI) qatron; (TURKISH) adi deniz lahanasi; (UKRAINIAN) mors'ka kapusta; (VLAAMS) zeekool; (WELSH) morfreshych.

30.46. *Crambe tatarica* Sebeók (EP) petioles and young leaves, root; (P) C,R.

(CZECH) katrán tatarský; (DANISH) tatarisk strandkål; (ENGLISH) tartar bread-plant, tartarian sea kale; (FRENCH) chou marim de Hongrie, chou tartare, crambe tartare, crambe lacinié; (GERMAN) Meerkohl; (GREEK) kramvi e tatariki; (HUNGARIAN) tátorján; (ITALIAN) crambio di tataria; (POLISH) kapusta tatarska, modrak, katran; (PORTUGUESE) couve marinha da Hungria, couve tártara; (ROMANIAN) varza de mare; (RUSSIAN) kapusta morskaja, katran, krambe; (SPANISH) col marina tartara; (TAJIKI) chaldirghon; (TURKISH) tatar, denizlahanasi; (VLAAMS) tartaar, broodplant.

30.47. *Diplotaxis muralis* (L.) DC. (EP) leaves; (P) C,R.

(BELARUSAN) dvuchradka stsennaja; (CHINESE, WÚ) er xing jie; (CZECH) křez zední; (DANISH) mursennep; (DUTCH) muurzandkool; (ENGLISH) wallrocket; (FINNISH) hietasinappi; (FRENCH) roquette des murailles; (GERMAN) Mauerdoppelsame, Mauerrauke; (GREEK) diplotaxis e epitichios; (HEBREW) turayim, turaim muralis; (HUNGARIAN) fali kányazsázsa; (ITALIAN) rucola murale, ruchetta; (LITHUANIAN) mūrinis šatrenis; (MACEDONIAN) obična dvorepka; (NORWEGIAN) rosettmursennep; (POLISH) rukwiel szpalerowa; (PORTUGUESE) saramago amarelo; (SPANISH) jaramago amarillo, mostacilla; (TURKISH) yillik duvar rokasi, divar diplokakcisi; (VLAAMS) kleine zandkool.

30.48. *Eruca sativa* Miller (EP) leaves; (P) R,C.

(BELARUSAN) indaon pjasanny; (BENGALI) shwetsursha; (CHINESE, MANDARIN) zhi ma cai; (CHINESE, WÚ) zhi ma cai; (CZECH) roketa seta; (DANISH) salatsennep; (DUTCH) rucola, raketsla, gekweekte eruca, tuimeruca; (ENGLISH) roquette salad, rocket salad, garden rocket, roman rocket, arugula; (ESTONIAN) hanerohi; (FINNISH) sinappikaali; (FRENCH) eruca, roquette des jardins, roquette, roquette cultivée; (GERMAN) Garten-Senfrauke, Öl-Senfrauke, Garten-Ruke, Feld-Raukensenf; (GREEK) poka; (HEBREW) ben-ḥardal maẕuy, ben-chardal; (HINDI) taramira, seoha; (HUNGARIAN) borsmustár; (ITALIAN) eruca, rucola comune, ruchetta, ruca; (JAPANESE) roketto salada, kibana-suzushiro; (MACEDONIAN) gradinarska eruka; (MALTESE) il-eruka;

(MONGOLIAN) gunjid; (NORWEGIAN) salatsennep; (PANJABI, EASTERN) assu, jamba, tara, usan; (POLISH) rokietta siewna; (PORTUGUESE) eruca, rúcula, pinchão; (RUSSIAN) indaou, eruka posevnaja; (SANSKRIT) bhutaghna, daradharsha; (SERBO-CROATIAN) riga, usjevna; (SPANISH) roqueta común, eruca; (TAJIKI) indon; (TIGRIGNA) endurkuar; (TURKISH) roka, rokka, izgii; (TURKMEN) ekimlik indew; (URDU) taramira; (VLAAMS) rucola, mosterdkruid, zwaardberik.

30.49. *Erysimum amurense* Kitag. (EP) young plants; (P) C.

(CHINESE, MANDARIN) tang jie; (KOREAN) chombujiggangi; (TAJIKI) giohi shirkhok.

30.50. *Lepidium meyenii* Walp. (EP) root; (P) C.

(CHINESE, MANDARIN) ma ka cai; (ENGLISH) maca; (FRENCH) maca; (NORWEGIAN) karse; (POLISH) pieprzyca; (PORTUGUESE) maca; (SPANISH) maca; (TAJIKI) kanagioh; (URDU) haurf halim; (VLAAMS) maca.

30.51. *Lepidium sativum* L. (EP) leaves; (P) R,C.

(AFRIKAANS) tuinkers, peperkers; (ALBANIAN, TOSK) djegëz e kultivuar; (AMHARIC) feto; (ARABIC) carabu, hab-el-rechad; (ASSAMESE) halim sak; (BASQUE) astotzar; (BELARUSAN) klaponik pasjannaja; (BENGALI) aleveri, halim; (BEROM) làbsúr; (BULGARIAN) kres-salata; (CATALAN-VALENCIAN-BALEAR) morritort; (CHINESE, MANDARIN) jia du xing cai, tu hsing ts'ai; (CHINESE, WÚ) jia du xing cai; (CZECH) řeřicha setá; (DANISH) have-karse, karse; (DUTCH) sterkers, tuinkers; (ENGLISH) garden cress, land cress; (ESTONIAN) kress-salat; (FINNISH) vihanneskrassi; (FRENCH) cresson alénois; (FULFULDE, ADAMAWA) lapsur, takerc foondu, foondu; (GEORGIAN) tsitsmati; (GERMAN) Garten-Kresse; (GREEK) cardamon; (GUJARATI) asalio, halim; (HAUSA) labsûr, làbsûr, lafsur, lausur, laùsúr, laùsúrù, làfusûr; (HEBREW) shaḫalayim tarbutyim, shahalaim tarovteem; (HINDI) pani dhleem, haleem, halim, hurf, chausaur, aliverei, chanasoor, chaunsar, hurf, malavan; (HUNGARIAN) kerti zsázsa; (ICELANDIC) garđperla, karsi; (INDONESIAN) slada darat, alim; (ITALIAN) mastruzzo, masturzio d'orto, lepidio, lepidio ortense, crescione inglese; (JAPANESE) gaaden kuresu, koshōsō; (KANNADA) kurtike, kurthike, kurutige, allibija, alli beeja, alavi beeja, alive, alvi, kuratirugi; (KANURI) làpsûr; (LATVIAN) kressalati; (LITHUANIAN) sėjamoji pipirnė, sėjamoji petražole; (MACEDONIAN) kres salata grbac, gradinarska dobreka; (MALTESE) habberxa; (MARATHI) ahaaleev, asaaliya, aliv, ahliva; (MONGOLIAN) tsanguu; (NEPALI) chamsur; (NORWEGIAN) matkarse, karse; (ORIYA) kresh; (PANJABI, EASTERN) halon, halim, shargundei, tezak; (POLISH) pieprzyca siewna, rzezucha ogrodowa; (PORTUGUESE) mastruço, agrião mouro, agrião do seco, agrião de terra seca, mastrugo; (ROMANIAN) creson, hrenițǎ creson, cresonul de grădina; (RUSSIAN) sadovyj kress, klopovnik posevnoi, kress-salat; (SANSKRIT) candrasura, chandrashura, chandrashoora, ashalika, bhadra, chandrika, charmahanthri, darakrishna, drighabeeja, kalamesha, karavi, nandini, pashumehankaarika, rakthabeeja, raktharaaji, sidhaprayojana, suvasura, vaasapushpa, vaasu preshpa; (SANTALI) hạlim; (SERBO-CROATIAN) kres salata, kres; (SLOVAK) žerucha siata; (SLOVENIAN) draguža; (SPANISH) mastuerzo, mastuerzo hortense, berro comúm, berro de tierra, lepidio; (SUNDA) sasawi, cesin; (SWEDISH) smörgåskrasse, kryddkrasse; (TAJIKI) tar-tszak; (TAMIL) aalivirai, aliverai, aliveraai; (TELUGU) adalavitulu, aadithyalu, adeli, adityalu, adiyalu, aandilee; (TURKISH) bahçe teresi, bahar teresi, acı tere; (TURKMEN) ekimlik ajytere; (UKRAINIAN) kres-salat; (URDU) halim; (VLAAMS) tuinkers; (WELSH) berwr yr ardd.

30.52. *Nasturtium officinale* R. Br. (EP) leaves; (P) R,C.

(AFRIKAANS) waterkers, bronkors, bronkos; (ARABIC) karsun mehi; (BASQUE) berro; (BELARUSAN) kustoun lenavy; (BENGALI) halim; (BRETON) beler; (BULGARIAN) voden kreson; (CAROLINIAN) kkangkkung; (CATALAN-VALENCIAN-BALEAR) morritort d'aygua; (CHEROKEE) ajila; (CHINESE, MANDARIN) dou ban cai, shui han ts'ai; (CHINESE, WÚ) dou ban cai; (CHINESE, YUE) xi yang cai, dou ban cai, liang cai, shui tian jie; (CREOLE, HAITIAN FRENCH) kréson; (CRIOULO, CAPE VERDE) garião-vulgar, agrião-de-água; (CZECH) potočnice lékařská, řeřišnice potoční; (DANISH) brøndkarse; (DUTCH) waterkers; (ENGLISH) watercress, common watercress; (ESTONIAN) ürt-allikkerss, kress, mungalill; (FARSI, WESTERN) shahe abic, tare tezak; (FILIPINO) watercress; (FINNISH) isovesikrassi; (FRENCH) cresson d'eau, cresson de fontaine, cresson officinal; (GAELIC, IRISH) biolar; (GAELIC, MANX) burley; (GALICIAN) agrión, cardama, mestranzo, mestruzo; (GERMAN) Brunnenkresse, echte Brunnenkresse; (GREEK) nastourion, enydrakardamon; (HEBREW) gargir ha-neḫalim, gargir hanechalin; (HINDI) peni sag; (HMONG DAW) zaub dej; (HUNGARIAN) vizitorma, vizizsázsa; (IBALOI) tapsoy; (ICELANDIC) vatnakarsi; (IFUGAO) pakhoy, pakhuy; (INDONESIAN) selada air, seladi air, salada air, seladah, kenci; (ITALIAN) nasturzio acquatico, nasturzio, crescione d'acqua; (JAPANESE) kuresson, oranda-garashi; (JAVANESE) cenil, kenci, jembrak, gejembak; (KASHMIRI) nāga-babürü; (KHMER, CENTRAL) krèh-sông; (KRIO) watakrɛs; (KWÉYÒL) kweson, kouson; (LAO) kaad fàlangx; (LITHUANIAN) paprastasis rėžiukas; (LUXENBOURGEOIS) Burekascht; (MACEDONIAN) obična potočarka; (MAITHILI) kumbhī; (MALAGASY) anandrana; (MALAY) semanggi, selada air; (MALTESE) il-krixxuni; (MAORI) koowhitiwhiti, waata-kirihi; (MICMAC) tāpesāoomakāwāāl; (MORISYEN) bred kresson; (NEPALI) sim saag; (NORWEGIAN) brønnkarse; (PAIUTE, SOUTHERN) pawmuhmp′; (PANJABI, EASTERN) piriya halim; (POLISH) rukiew siewna, rukiew wodna; (PORTUGUESE) agrião, agrião de água, agrião do rio; (QUECHUA, CUSCO) chijchi, occoruro; (ROMANIAN) măcris de baltă; (RUSSIAN) kress vodjanoi, zherouha vodnaja, zherouha aptechnaja; (SAMOAN) kapisi vai; (SCOTS) girse; (SERBO-CROATIAN) grabak, potočarka, perestolisna; (SLOVAK) potočnica; (SORBIAN, UPPER) ropucha, žerchej; (SPANISH) berro, berro de agua, mastuerzo acuático; (SUNDA) salada, cai; (SWAHILI) saladi; (SWEDISH) vattenkrasse; (TOK PISIN) wara karis; (TURKISH) su teresi; (TURKMEN) dermanlyk kenarot; (URDU) nilofar; (VIETNAMESE) c[ar]i soong; (VLAAMS) waterkers; (WELSH) berwr dwr; (ZULU) ukhwaningeni(la).

30.53. *Raphanus sativus* L. Caudatus group (EP) immature seed pods; (P) P,C.

(AFRIKAANS) rotstert radys; (ASSAMESE) mula; (CHINESE, MANDARIN) shu wei luo bo; (CHINESE, WÚ) shu wei lou bo; (CHINESE, YUE) shu wie pu tao; (CZECH) ředkev seté; (ENGLISH) rat tail radish; (FRENCH) radis serpent; (GREEK) repani; (HEBREW) tsnon; (HINDI) sungra, singri, mungra; (HUNGARIAN) kerti retek; (INDONESIAN) lobak ekor; (LATVIAN) rutki; (MACEDONIAN) opaškovidna rotkva; (MARWARI) sogrī; (NEPALI) newari mula; (NORWEGIAN) ormereddik; (PANJABI, EASTERN) shōō̄ōgrä; (POLISH) rzodkiew długa; (PORTUGUESE) rabanete serpente; (TAJIKI) turb; (THAI) phakkhithut; (TURKISH) kirmizi turp; (VLAAMS) rattestaartradijs.

30.54. *Raphanus sativus* L. Dakion group (EP) root; (P) C,R,P.

(ALBANIAN, TOSK) rapanidhe e kultivuar; (ARABIC) fejil; (ASSAMESE) China mula; (BELARUSAN) daikon; (BULU) radi; (BURMESE) mone-la-oo, monla; (CAROLINIAN) daikon; (CHINESE, GÀN) luo bai; (CHINESE, HAKKA) ts'ói t'êou; (CHINESE, MANDARIN) luo bo; (CHINESE, MIN NAN) âng-chhài-thâu; (CHINESE, WÚ) zhong guo luo bo; (CROATIAN) rotkvica; (DAGAARE, SOUTHERN) radi; (DANISH) raeddike, japanræddike, kinaræddike; (DUTCH) ramenas; (ENGLISH) Chinese radish, daikon, oriental winter radish, radish; (FINNISH) retikka, retiisi; (FRENCH) rave, daikon, radis; (GEORGIAN) boloki; (GERMAN) Rettich; (GREEK) kineziki rafanis; (HEBREW) ẕenon; (HINDI) muli; (HINDUSTANI) murai; (HMONG DAW) lo pue, zaub ntug hauv paus; (HUNGARIAN) kínai retek, kerti retek; (ICELANDIC) kinahređka; (ILOCANO) rabanos; (INDONESIAN) lobak, lobak cina, lobak jepang, lobak putih; (ITALIAN) rafano; (JAPANESE) daikon; (JAVANESE) lobak; (KAZAKH) redka; (KHMER, CENTRAL) chhaay thaaw; (LAKOTA) paŋgipepe śaśa, tiŋpaiŋla śaśa; (LAO) kaad khaaw; (LITHUANIAN) ridikas; (LUXENBOURGEOIS) Réidech; (MACEDONIAN) kineska rotkva; (MALAY) lobak, lobak putih; (MALTESE) figel ciniz; (MONGOLIAN) tarimal luuvan; (MÒORÉ) radi; (MORISYEN) rav; (NEPALI) seto mula; (NORWEGIAN) kinareddik, reddik, frokostreddik; (POLISH) rzodkiew chińska; (PORTUGUESE) rábano, rabão, daikon, nabo japones; (ROMANIAN) ridiche de grădină; (RUSSIAN) red'ka; (SERBO-CROATIAN) rotkva; (SLOVAK) red'kev; (SPANISH) rábano; (STELLINGWERFS) rammenas; (SUNDA) lobak; (SWEDISH) rättika, trädgårdsrättika; (TAGALOG) labaños; (TAJIKI) turb; (THAI) phakkat-hua, phakkhithut; (TURKISH) çin turpu; (URDU) muli; (UZBEK) turp; (VIETNAMESE) c[ar]i c[ur]i; (VLAAMS) daikon; (YIDDISH) rehtekh.

30.55. *Raphanus sativus* L. Oleiformis group (EP) leaves; (P) C

(ENGLISH) leaf radish; (INDONESIAN) lobak daun.

30.56. *Raphanus sativus* L. Radicula group (EP) root; (P) R,C.

(AFRIKAANS) radys; (ALBANIAN, TOSK) repkë, rapanidhe e kultivuar; (ALEUT) naggatannarsuk; (AMHARIC) qäyysər; (ARABIC) fejil; (ASSAMESE) mula; (AZERBAIJANI, NORTHERN) q ırmızı turp; (BADAGA) muḷḷangi, muḷḷengi; (BALTI) dul; (BASQUE) errefau; (BELARUSAN) redzika pasjannaja; (BENGALI) muli, mula, mura, muri; (BOSNIAN) rotkvica; (BRETON) irvin-ruz; (BULGARIAN) rjapa; (BULU) radi; (BURMESE) mone-la-oo, monla; (CATALAN-VALENCIAN-BALEAR) rave; (CHINESE, GÀN) luo bai; (CHINESE, MANDARIN) ying tao luo bo, lo po; (CHINESE, MIN BEI) cai tou; (CHINESE, WÚ) lai tu, ying tao luo bo; (CHINESE, YUE) luo bai, lai tu; (CREE, PLAINS) mihkotsehkanis, wapikoskatask; (CREOLE, HAITIAN FRENCH) radi; (CROATIAN) rotkva; (CZECH) ředrvička, ředkev; (DAGAARE, SOUTHERN) radi; (DAKOTA) paŋğiśaśa; (DANISH) radis, raeddike; (DUTCH) radijs, ramenas; (ENGLISH) radish, small radish, garden radish, western radish; (ESTONIAN) rõigas; (FARSI, EASTERN) kaddu; (FARSI, WESTERN) kadu tanbel, torobcheh; (FIJIAN) murai; (FILIPINO) labanos; (FINNISH) retikka, retiisi; (FRENCH) radis, ravonet, rave, petit radis; (GAELIC, MANX) rahgyl; (GAELIC, SCOTTISH) meacan-ruadh; (GEORGIAN) tvis boloki, bolóki; (GERMAN) Garten-Rettich, Rettich, Radieschen, Radies; (GOURMANCHÉMA) radi; (GOURMANTCHÉ) radi; (GREEK) rapanaki, rafanis; (GUJARATI) mulo, muli, mula, mura, muri; (HARARI) fuğūl; (HEBREW) ẕenonit, tsnonit, tznon, tznon hagina; (HILIGAYNON) rabanús; (HINDI) mooli, muli, mula, mura, muri, muraayi; (HINDUSTANI) lakha murai, múlí, turb; (HMONG DAW) zaub ntug hauv paus; (HUNGARIAN) hónapos retek, retek; (IBANAG) labanos; (ICELANDIC) radisa, hređka; (ILOCANO) rabanos; (INDONESIAN) lobak merah, radis, rades, lobak; (INUKTITUT, WEST GREENLANDIC) nautsîaq; (ITALIAN) rafano, ravanello, radice; (ITAWIT) rabanus; (IVATAN) rabanos; (JAPANESE) hatsuka daikon, daikon; (JAVANESE) rades; (KANKANAEY) radis ; (KANNADA) mullangi, bilimoolangi, maguni gedde, moolangi; (KASHMIRI) hāra-kani, kand, mujü; (KHMER, CENTRAL) chhaay thaaw; (KIRGHIZ) turp; (KONKANI) mullo; (KOREAN) muu; (KRIO) radish; (KWÉYÒL) wadi, wav; (LADINO) dumalan, ravano, rávo; (LAHU) g'aw$^{\vee}$ ma hpu; (LAKOTA) paŋgipepe śaśa, tiŋpaiŋla śaśa; (LAO) kaad khaaw; (LATVIAN) redīsi, rutki; (LITHUANIAN) ridikėlis; (LUXENBOURGEOIS) Réidech; (MACEDONIAN) rdokva, obična rotkvička; (MALAY) lobak, lobak kecil; (MALAYALAM) mullanki, mullangi; (MALTESE) figel; (MARANAO) rabanos; (MARATHI) moolaa, muri;

33.2. *Campanula rapunculus* L. (EP) roots and first leaves[s]; (P) R,C.

(BELARUSAN) zvanochak repchaty; (CHINESE, MANDARIN) jie geng cai; (CHINESE, WÚ) feng ling cao; (CHINESE, YUE) jie geng cai; (CZECH) zvonek hlízantý, zvonek řepka; (DANISH) rapunsel-klokke, klokke; (DUTCH) rapunzelklokje; (ENGLISH) rampion; (ESTONIAN) rapuntselkellukas; (FINNISH) kauriinkello; (FRENCH) campanule raiponce, raiponce, rampon, rave suavage, cheveux d'évèque; (GERMAN) Rapunzel-Glockenblume, Rapunzel; (GREEK) kampanoula to goggylio; (HEBREW) pa'amonit, paamonit kipachat; (HUNGARIAN) raponc harangvirág; (ICELANDIC) skriđklukka; (ITALIAN) raperònzolo, raponzo, rapònzolo, campanula commestibile; (JAPANESE) kabura kikyou; (LITHUANIAN) storašaknis katilėlis; (MACEDONIAN) repkasto zvonče; (MONGOLIAN) khonkhontsetseg; (NORWEGIAN) rapunkelklokke; (POLISH) dzwonek rapunkuł; (PORTUGUESE) rapúncio, campainha rabanete, rapôncio, campânula; (ROMANIAN) clopoţei, salată de cimp; (RUSSIAN) rapuntsel, kalakolchik repchatyi; (SPANISH) rapónchigo, rapúnculo, ruiponce, raipóntico; (SWEDISH) rapunkeltlocta; (TAJIKI) momogul; (TURKISH) kazayaği kökü, yaban turpu, büyük köklü, çan çiçeği, mantereş, firavun inciri; (VLAAMS) rapunzelklokje, rampion; (WELSH) clychlys erfin.

33.3. *Codonopsis lanceolata* (Siebold & Zucc.) Trautv. (EP) rhizomes, young plants; (P) R,C.

(BINUKID) lakoron-bolan; (CHINESE, WÚ) yang ru; (ENGLISH) deodeog; (FINNISH) peikonkello; (JAPANESE) tsuru-ninjin, deodeog; (NORWEGIAN) vindelklokke; (ROMANIAN) mangold, sfedădefruuze; (TAJIKI) gandabuy; (TURKISH) sarmaşik, çani; (VLAAMS) deodeog.

34. Capparaceae

34.1. *Capparis spinosa* L. (EP) flower buds; (P) P,C,R.

(ALBANIAN, TOSK) kaparinë; (CHINESE, MANDARIN) ci shan gan; (CHINESE, WÚ) ci shan gan; (CZECH) kapara trnitá; (DANISH) kapers; (DUTCH) kapper-boon, kapperstruik; (ENGLISH) capper, capper bush, caper plant, european capper bush, true caper berry; (FINNISH) kapris; (FRENCH) câprier; (GERMAN) Kappernstrauch; (GREEK) kappari; (GUJARATI) kalo kantharo, kantharo; (HEBREW) ẕalaf koẕani, tsalaph kotsani; (HINDI) baeree, kabra, kabar, ber, kander, kiyaari; (HUNGARIAN) tüskés kápricserje; (ICELANDIC) kapersrunni; (INDONESIAN) kapri, kayu tujuh; (ITALIAN) càppero, càpparo; (KANNADA) maraat moggu, mullukatthari, mullukattari, kathari mullina gida, katrimullinagidda, marattamaggu, sambaar moggu; (LATVIAN) dzeloņainā kapera; (LITHUANIAN) dygliuotasis kaparis; (MALAY) melada; (MALTESE) kappar; (MARATHI) kabar; (NEPALI) bagh mukhwa; (NORWEGIAN) kapers; (PANJABI, EASTERN) kaur, krēr, barar; (POLISH) kapary cierniste; (PORTUGUESE) alcaparra, alcaparreira; (ROMANIAN) capere; (SANSKRIT) himsra, kakadani, kartotti, kaakadani, kartotthi; (SERBO-CROATIAN) kapra; (SLOVAK) kapary; (SPANISH) alcaparra; (SWEDISH) kapris; (TAGALOG) alcaparras; (TAJIKI) kavar; (TELUGU) kokilakshamu, kokilaakshamu; (TURKISH) adi gebreotu, adi kebre, kapari; (TURKMEN) adaty göýül; (URDU) kahar, kabar; (VLAAMS) kapper.

34.2. *Cleome gynandra* L. (EP) leaves, young shoots and fruits; (P) C.

Dialects of Luyia (Bukusu[1], Kisa[2], Tachoni[3], Tiriki[4])

(AFRIKAANS) palmbossie, vingerblaartee, Sjinese klokblom; (AKAN-AKUAPEM) tɛtɛ; (AKAN-ASANTE) tɛtɛ, tɛtɛtɛtɛ; (AKLANON) halaya, hulaya; (BAMANANKAN) naségé; (BAOULÉ) wètè; (CEBUANO) balaya; (CHINESE, MANDARIN) bai hua cai; (CHINESE, WÚ) bai hua cai;

(DANGME) kɛtɛ; tɛtɛ; (DANGME-KROBO) tɛtɛ; (DOONDO) loukaya-louakoutou; (ENGLISH) cat's whiskers, bastard mustard, spider flower, African spiderflower, kaffir cabbage; (ÉWÉ) soboe, agmãgã, awumɛ, sɔbui, sɔgbee, solui, sɔlwi; (FILIPINO) apoi-apoian; (FRENCH) feuilles caya, mozambé, mouzambe, cleome; (FULFULDE) gorbusa, worba, kinaski; (FULFULDE, ADAMAWA) gasiyaahu, kinaski; (GA) kɛtɛ, tɛtɛ; (GERMAN) Senfkapper; (GIKUYU) thagiti, thageti; (HAUSA) tabaday, garsar, gassayaa, diyar ungwa, tabadai, tabadamashe; (HEBREW) baashan, baashan gynandra; (HINDI) hulul; (ILOCANO) manabo, tantandok, tantandok-a-dadakel; (INDONESIAN) mamang, langsana merah, mamam; (ITAWIT) davuravu; (JAVANESE) boboan, bobowan, enceng-enceng, leng-lengan, mamam; (KHMER, CENTRAL) momiënh; (KOONGO) mvoua-vouagna; (LAO) siènz; (LUYIA) tsisaka[2,4], esaka[1], lisaka[3]; (MADURA) bhubhuwan; (MALAY) manam, mamang; (MANINKA) oassi-ni-kuna; (MONGOLIAN) jalguun tsetseg; (MÒORÉ) kelebdo, tiɛlɛbdo; (NEPALI) hur hure; (PORTUGUESE) mussambé cor de rosa, mussambé de cinco folhas; (SERER-SINE) korgona, safoybidum; (SONGHAY) húbéy; (SPANISH) caya, volatín, volantincillo; (SUNDA) mamam; (SWAHILI) mgagani; (TAGALOG) apoyapoyan, apoy-apoyan, bala-balanoyan; (TAJIKI) kaul; (THAI) phak sian, phak sian khao, phak som sian; (TIV) gashia; (TURKISH) örümcek bitkisi, örümcek çiçeği, kleam; (TWI) tɛtɛ, tɛtɛtɛrɛ; (VIETNAMESE) m[af]n m[af]n tr[aws]ng; (VLAAMS) kattesnorharve; (WOLOF) gor bu di daw, nakayorel, n-gor si bidaw; (YORUBA) ekesi-masun, èkùyá, èkùyáko, èkùyalé; (ZARMA) foubey.

34.3. *Platycodon grandiflorus* (Jacq.) A. DC. (EP) leaves; (P) C.

(CHINESE, MANDARIN) jie geng; (CHINESE, WÚ) ju geng; (DANISH) ballonklokke; (ENGLISH) Chinese bellflower, Japanese bellflower, balloonflower; (FINNISH) jalokello; (FRENCH) platycodon à grandes fleurs; (HUNGARIAN) ballonnövény; (ICELANDIC) belgklukka; (JAPANESE) kikyô; (LITHUANIAN) stambiažiedis plačvarpis; (NORWEGIAN) vanlig flatklokke; (POLISH) rozwar wielkokwiatowy; (TURKISH) balon çiçeği, genişçan, muhteşem çan, çan çiçeği; (URDU) kal hal kal; (VLAAMS) Chinees klokje, Japans klokje.

35. Chenopodiaceae

35.1. *Atriplex hortensis* L. (EP) leaves; (P) C.

(AFRIKAANS) melde, hofmelde; (ARABIC) el-hejaz; (ARMENIAN) mochrateluk; (ASSAMESE) pahari paleng, paharipalang, purvi sak; (AZERBIAJANI) sirken; (BELARUSAN) lebada sadovaja; (BENGALI) paharipalang; (BULGARIAN) gradinska loboda; (CATALAN-VALENCIAN-BALEAR) armolls; (CHINESE, MANDARIN) shi yong bin li, shan po ts'ai; (CHINESE, WÚ) shi young bin li, yu qian bo cai; (CZECH) lebeda zahradní; (DANISH) have-melde, melde; (DUTCH) tuinmelde; (ENGLISH) orach, orache, garden orach; (ESTONIAN) malts; (FARSI, WESTERN) esfenaje-roomee; (FINNISH) tarhamaltsa; (FRENCH) arroche, arroche épinard, arroche cultivée, arroche des jardins; (GAELIC, IRISH) eilifleog; (GEORGIAN) tutubo; (GERMAN) Gartenmelde; (GREEK) lapatoi v lita levoudia; (HEBREW) maluaḥ, maluach; (HINDI) korake, surake, paharipalang, chakwat; (HUNGARIAN) kerti laboda; (ICELANDIC) hrimblaðka; (ITALIAN) atriplice; (JAPANESE) yama-hourensou; (KANNADA) chakothana soppu, chakkotane soppu, chakottae soppu, chakkotha; (LATVIAN) balodenes, dārza balodene; (LITHUANIAN) daržinė balandūnė; (MACEDONIAN) pitoma loboda, gradinarska loboda; (MALTESE) bjanka; (MARATHI) chandanbatva; (MONGOLIAN) shornoi; (NORWEGIAN) hagemelde, svinemelde; (PANJABI, EASTERN) pahari palak, korake, suraka; (POLISH) łoboda ogradowa; (PORTUGUESE) armoles, erva armoles; (ROMANIAN) lobodă de grădină; (RUSSIAN) lebeda sadovaja; (SERBO-CROATIAN) loboda, vrtna; (SLOVAK) lebeda záhradná; (SLOVENIAN) vrtna loboda; (SPANISH) armuelle, armuelle de huerta; (SWEDISH) trädgårdsmålla; (TAJIKI) shura; (TELUGU) mattubachali; (TURKISH) koyun sarmaşiği, bahçe karapazisi; (UKRAINIAN) červone loboda; (URDU) pahari palak; (VLAAMS) tuinmelde.

35.2. *Bassia scoparia* (L.) A.J. Scott (EP) tender shoots; (P) C.

(BELARUSAN) proutnjak venichny; (CHINESE, MANDARIN) di fu; (CHINESE, WÚ) di fu; (CHINESE, XIANG) di er zi; (CZECH) bytel metlatý, bytel; (DANISH) sommercypres; (DUTCH) studentenkruid; (ENGLISH) summer cypress, mock cypress, Mexican burningbush, kochia, broom goosefoot; (ESTONIAN) kohhia; (FINNISH) kesäsypressi; (FRENCH) kochie, kochia à balais, ansérine belvédère; (GERMAN) Besenkraut, Besenradmelde; (HEBREW) kokhyat ha-makhbed, kochia; (HUNGARIAN) kerti seprőfű; (ITALIAN) belvedere, granata, scoparia; (JAPANESE) houkigi; (LATVIAN) vasaras eglīte; (LITHUANIAN) šluotelinė kochija; (MACEDONIAN) obična metlica; (MALTESE) kokja; (MONGOLIAN) savaan togtorgono; (NORWEGIAN) kostmelde; (POLISH) mietelnik żakula, mietelnik; (PORTUGUESE) belverde, valverde, mirabela; (ROMANIAN) mături; (RUSSIAN) kochija, proutnjak; (SERBO-CROATIAN) metlica, gajena; (SPANISH) mirabel, albahaca larga, sisallo, barriola; (SWEDISH) sommar cypress; (TAJIKI) jorubak; (TURKISH) yaz servis, bozkirotu, süpürge kahya, süpürge out; (TURKMEN) adaty sübselik; (VLAAMS) studentenkruid.

35.3. *Beta vulgaris* L. Cicla group (EP) leaves; (P) C.

(AFRIKAANS) snybeet, spinasiebeet; (ALBANIAN, TOSK) pazi, salklë; (ARABIC) salk; (ASSAMESE) palang sag; (BASQUE) belete; (BELARUSAN) buraki listavaja; (BENGALI) palang; (BULGARIAN) mangold; (CHINESE, GÀN) jun da, nui pi cai; (CHINESE, MANDARIN) tian cai, t'ien ts'ai; (CHINESE, MIN BEI) nui pi cai, jun da cai; (CHINESE, MIN NAN) ka-boah-chhài; (CHINESE, WÚ) tian cai, ye tian cai, jun da cai, niu pi cai; (CHINESE, XIANG) niu pi cai; (CHINESE, YUE) jun da cai, zhu po cai, biao cai; (CROATIAN) blitva; (CZECH) mangold, cvikla; (DANISH) bladbede, snitbladbede, sølvbede, mangold; (DUTCH) snijbiet; (ENGLISH) chard, Swiss chard, spinach beet, leaf beet; (ESTONIAN) lehtpeet; (FINNISH) lehtimangoldi; (FRENCH) bette, blette, côte de blette, poirée à couper, poirée ordinaire, poirée à carde, arde; (GALICIAN) lampaza; (GEORGIAN) photlovani tsharkhali, mangoldi; (GERMAN) Mangold; (GIKUYU) thibinici; (GREEK) seskoulo; (GUJARATI) palak; (HEBREW) selek matsur; (HINDI) palak; (HUNGARIAN) mángold; (ICELANDIC) strandblaðka, blaðbeðja; (INDONESIAN) bit bodas; (ITALIAN) bieta a foglia, bietola da costa; (JAPANESE) hudansōu, fudansō; (KANNADA) spinak soppu, paalak soppu; (LATVIAN) mangolds; (LITHUANIAN) mangoldas; (MACEDONIAN) lisna mangolds, meko zelje, lisna cveklo; (MALAYALAM) visalacheera; (MALTESE) is-sejq, pitravi tal-werqa; (MARATHI) palak; (MORISYEN) poire; (NEPALI) palungo; (NORWEGIAN) bladbete, mangold, snittbladbete; (ORIYA) mitha palanga saga; (PANJABI, EASTERN) palak, dāssi palak; (POLISH) burak liściowy, mangold, boćwina; (PORTUGUESE) acelga; (ROMANIAN) sfeclă de frunze, sfecla mangold, mangold; (RUSSIAN) svyokla listovaja, svyokla mangold, salat rimski; (SERBO-CROATIAN) mangold; (SLOVAK) cvikla; (SLOVENIAN) mangold, cvikla objčajná; (SPANISH) bleda, acelga; (SWAHILI) spinachi, swisi chadi; (SWEDISH) mangold, bladbete; (TAJIKI) lablabu; (TAMIL) vusavyeley, pasalai keerai; (TELUGU) teegabatchali, dumpabucchale; (TURKISH) pazi; (UKRAINIAN) mangol'd; (URDU) chuchander; (VLAAMS) snijbiet; (YIDDISH) schav.

35.4. *Beta vulgaris* L. Crassa group (EP) root and leaves; (P) C,R.

(AFRIKAANS) beet; (ALBANIAN, TOSK) panxhar tryeze; (ALEUT) nakpatâyâk aupalluktak; (ARABIC) silig; (ARMENIAN) čakndech; (ASSAMESE) beet mula, beet; (AZERBIAJANI) čugundur; (BELARUSAN) buraki zvychainy, burak; (BENGALI) beet; (BOSNIAN) repa; (BRETON) boetrabez ruz; (BULGARIAN) salatno cveklo; (CAJUN FRENCH) betterave; (CATALAN-

VALENCIAN-BALEAR) bleda vermella, remolatxa; (CEBUANO) remolatsa; (CHECHEN) buraak; (CHICKASAW) tanap homma'; (CHINESE, HAKKA) t'iâm ts'ói-t'êou; (CHINESE, MANDARIN) gen tian cai; (CHINESE, WÚ) gen tian cai; (CHINESE, YUE) hong lai tou, gen tian cai; (CREE, PLAINS) mihko otsehkan ocepihk, mihkoskatask; (CREOLE, HAITIAN FRENCH) bètrav; (CRIOULO, CAPE VERDE) beterraba; (CROATIAN) cikla; (CZECH) řepa burák pravá, řepa obecná pravá; (DAKOTA) paŋġišaša; (DANISH) rødbede, guldbede; (DOONDO) lukaya lua nkuyu; (DUTCH) rode biet, kroot; (ENGLISH) beet, beetroot, garden beet, table beet, red beet; (ESTONIAN) peet; (FARSI, WESTERN) chukandar, choghondar; (FILIPINO) beet; (FINNISH) juurikas, punajuuri; (FRENCH) betterave rouge, betterave potagère; (GAELIC, IRISH) biatas milis; (GAELIC, MANX) beetys; (GAELIC, SCOTTISH) biotais; (GEORGIAN) tsharkhali, čarchali; (GERMAN) rote Rübe, rote Beete; (GOURMANTCHÉ) bompieno-tibina; (GREEK) pantzari, teftlo, uok-kinogouli; (GUJARATI) beet; (HEBREW) selek, selek maẕuy, selek adom; (HINDI) chakandar, chukandar, chakwat, palak; (HINDUSTANI) chuqandar; (HUNGARIAN) cukorrépa, cékla; (ICELANDIC) rauđrofa; (INDONESIAN) bit; (ITALIAN) bietola da orta, barba bietola d'insalata; (JAPANESE) biito, shokuyōbīto, kaensai; (JAVANESE) bit; (KANNADA) beet root; (KAZAKH) svekla stolovaja; (KIOWA) gyH-guɑdl-dɑ̄ '-dei; (KIRGHIZ) aš kyzylčasy; (KIRIBATI) bite; (KOONGO) diamba-dambata; (KURDISH, NORTHERN) silq; (KWÉYÒL) betwav; (LADINO) panjár; (LAKOTA) paŋgipepe śaśa, tipŋsiŋla śaśa; (LATVIAN) biete; (LITHUANIAN) runkelis, burokėlis; (LUXENBOURGEOIS) roud Rommel; (MACEDONIAN) cveklo, stočna repa; (MAITHILI) chukandar; (MALAY) avar bit, bit; (MALAYALAM) beet root; (MALTESE) pitravi; (MARATHI) beet; (MICMAC) mĕgwibŭnĕksit; (MIZO) antam bulbawk; (MORISYEN) betrav; (MUSKOGEE) yvlonkv-cate; (NEPALI) chukandar, guliyo mula; (NGUNDI) loukaya loua moukouyou; (NIUE) pitiluti; (NORWEGIAN) matbete, rødbete, gulbete, kvitbete, bete; (ORIYA) beet; (PAMPANGAN) remolatsa; (PANJABI, EASTERN) chakandar, chikandar; (PASHTO, NORTHERN) jaghwandër, c̲hug̲handar, c̲huqandar; (POLISH) burak ćwikłowy; (PORTUGUESE) beterraba hortícola, beterraba de salada, beterraba vermelha, beterraba de mesa, beterraba; (ROMANI, BALKAN) chookoondòori; (ROMANIAN) sfeclă obisnuită, sfeclă roşie; (RUSSIAN) svekla stolovaja; (SANSKRIT) palakya, palanki, raktha gmjna; (SANTALI) palak aṛak; (SCOTS) beetraw; (SERBO-CROATIAN) cvekla; (SINHALA) rathu ala; (SLOVAK) červena repa šalátová; (SLOVENIAN) rdeča pesa, jedlá repa; (SORBIAN, UPPER) čerwjena rěpa; (SOTHO, NORTHERN) pete; (SOTHO, SOUTHERN) pete; (SPANISH) remolacha hortelana, remolacha roja, remolacha de mesa; (STELLINGWERFS) biet; (SUNDA) bit; (SWAHILI) biti ruti; (SWEDISH) rödbeta; (TAGALOG) remolatsa pula; (TELUGU) beetudumpa, (THAI) phakkat-daeng, phakkat-farang; (TSWANA) pete; (TURKISH) pancar, kirmizi pancar; (TURKMEN) adaty şugundyr; (UKRAINIAN) stolovij burjak; (URDU) chukandar; (UZBEK) laviagui, oš lavlagi; (VENDA) bitirùthi; (VIETNAMESE) c[ar]i d[uw][owf]ng, c[ur] d[eef]n; (VLAAMS) rode biet; (WOLOF) bétérav; (XHOSA) ibhitruthi, ibhithi; (YIDDISH) boorik; (ZULU) ibhithrudi, isithombo sebhitrudi; (ZUNI) mo/-shilowa, piz.

35.5. *Chenopodium bonus-henricus* L. (EP) leaves; (P) C.

(ASSAMESE) jilmil, bathua; (BELARUSAN) lebeda dobrova genricha; (CHINESE, MANDARIN) li cai; (CHINESE, WÚ) li cai, hui tiao cai; (CHINESE, YUE) ten cai, zhu mu cai, hui cai; (CZECH) merlík všedobr; (DANISH) stolthenriks-ga'sefod; (DUTCH) brave hendrik; (ENGLISH) Good King Henry, mercury; (FINNISH) hyvänheikinsavikka; (FRENCH) chénopode bon-Henri, épinard saurage, bon-Henri; (GERMAN) guter Heinrich; (GREEK) agrio spanaki, louvoudia; (HEBREW) kaf-ha-avaz, kaf-avaz; (HUNGARIAN) parlagi libatop; (ITALIAN) farinello buono-enrico; (LITHUANIAN) panaminė balanda; (MACEDONIAN) pepelesta loboda; (MONGOLIAN) luuli; (NORWEGIAN) stolthenrik; (POLISH) komosa strzałkowata; (PORTUGUESE) quenopódio bom-henrique, anserina bom-henrique; (ROMANIAN) spanacul ciobanilor; (RUSSIAN) mar dorava genricha; (SPANISH) zurrón, anserina, buen enrique; (SWEDISH) lungrot; (TAJIKI) shirai buginak; (TURKISH) yabani ispanak; (VLAAMS) brave hendrik.

35.6. *Chenopodium quinoa* Willd. subsp. *quinoa* (EP) leaves; (P) C.

(ALBANIAN, TOSK) minuer; (ASSAMESE) jilmil, bathua; (CHINESE, MANDARIN) qi nu ya li; (CHINESE, WÚ) qi nu ya li; (CRIOULO, CAPE VERDE) quinoa; (CZECH) merlík čilský, kinoa; (DANISH) quinoa; (DUTCH) ganzevoet; (ENGLISH) quinoa, white quinoa, goosefoot; (FINNISH) kvinoa, kvinoasavikka; (FRENCH) quinoa, ansérine quinoa, petit riz du Pérou, ansérine du Pérou; (GERMAN) Reismelde, Peruanischer Reisspinat; (GREEK) kockia saromatiki; (HINDI) bathuā; (HUNGARIAN) fehér libatop, kvinoa-cserje; (ITALIAN) quinoa; (MACEDONIAN) loboda, kvinoa; (MIZO) thlai hnah nei chi, zam-zo lam pawl; (MONGOLIAN) luuli; (NORWEGIAN) perumelde; (POLISH) komosa ryżowa; (PORTUGUESE) arroz míudo do Perú, espinafre do Perú, quinoa, quinoa branca; (QUECHUA, CUSCO) quinoa; (ROMANIAN) spanac sălbatic; (RUSSIAN) lebeda kvinoa; (SLOVAK) mrlík čílsky; (SPANISH) quinua; (SWEDISH) quinoa, mjöllmolla; (TURKISH) Peru çeltiği, küçük pirinç; (VLAAMS) Peruaanse spinazie.

35.7. *Salsola komarovii* Iljin. (EP) leaves and young shoots; (P) C.

(CHINESE, MANDARIN) wu chi zhu mao cai; (CHINESE, WÚ) wu chi zhu mao cai; (DANISH) komarov-sodaurt; (DUTCH) Komarov loogkruid; (ENGLISH) barilla plant, Komarov Russian thistle; (FINNISH) otakilokki; (FRENCH) salsola de Komarov; (GREEK) almyridi; (HEBREW) milchit; (HUNGARIAN) ballagófű; (ITALIAN) salsola di Komarov; (JAPANESE) okahijiki; (MONGOLIAN) budargana; (NEPALI) guluf; (NORWEGIAN) sodaurt; (POLISH) solanka komarowa; (PORTUGUESE) salsola de Komarov; (SPANISH) salsola de Komarov; (TAJIKI) shuragioh; (TURKISH) sodaotu, dikenli çöğen; (VLAAMS) okhijiki.

35.8. *Salsola soda* L. (EP) leaves and young shoots; (P) C.

(CHINESE, MANDARIN) su da zhu mao cai; (CHINESE, WÚ) su da zhu mao cai; (DUTCH) soda kruid; (ENGLISH) salsola; (FRENCH) salsola soda, herbe au verre, soude commune; (GERMAN) Soda-Salzkraut; (GREEK) rossiko kirsiam; (HEBREW) milḫit ha-borit, milchit haborit; (HUNGARIAN) sziki ballagófű; (ITALIAN) roscano, bacicci; (MACEDONIAN) solena solnenka; (MALTESE) haxixa tar-rmiet; (MONGOLIAN) budargana; (NORWEGIAN) sodaurt; (POLISH) solanka sodowa; (PORTUGUESE) salsola; (SPANISH) barrilla, peralejo; (TAJIKI) ushnon; (TURKMEN) sodaly şora; (VLAAMS) salsola.

35.9. *Spinacia oleracea* L. (EP) leaves; (P) C,R.

The generic English term "spinach" is often used (e.g. in Africa) for a cross-section of cooked leaf crops.
[1]Itsukha dialect of Luyia.

(AFRIKAANS) spinasie; (ALBANIAN, TOSK) spinap; (ARABIC) isfanahk; (ARMENIAN) spanach; (ASSAMESE) palang sak, paleng, mitha-paleng, palangsag, palang sag; (AZERBAIJANI, NORTHERN) ispanaq; (BADAGA) coppu, soppu; (BASQUE) espinaka; (BELARUSAN) palang, spinat agorodnni; (BENGALI) palang, pinnis; (BOSNIAN) špinat, spanać; (BRETON) pinochez; (BULGARIAN) spanak; (CAJUN FRENCH) épinard; (CATALAN-VALENCIAN-BALEAR) espinac; (CHINESE, GÀN) bo cai; (CHINESE, HAKKA) kôc-ts'ói; (CHINESE, MANDARIN) bo cai, po leng; (CHINESE, MIN BEI) bo lin cai, hong gen cai; (CHINESE, MIN NAN) poe-lêng-á-chhài; (CHINESE, WÚ) bo cai, chi geng cai, jiao cai; (CHINESE, XIANG) che gen cai, hong gen cai;

(CHINESE, YUE) bo cai, jiao cai; (CREE, PLAINS) kistekan ohci nepeya ka mecihk; (CRIOULO, CAPE VERDE) espinafre; (CREOLE, HAITIAN FRENCH) épina; (CROATIAN) spinat; (CZECH) špenát setý, špenát; (DAGAARE, SOUTHERN) ipinar; (DANISH) spinat; (DUTCH) spinazie; (ENGLISH) spinach, garden spinach, spinage; (ESTONIAN) spinat; (FARSI, EASTERN) pâlak; (FARSI, WESTERN) esfenâj; (FILIPINO) spinach; (FINNISH) pinaatti; (FRENCH) êpinard; (GEORGIAN) ispanachi; (GERMAN) Spinat; (GIKUYU) thibinici; (GOURMANTCHÉ) ibardji-djari; (GREEK) spanaka; (GUJARATI) palak; (HAUSA) alayyaho; (HEBREW) tered, tered ha-ginna, tered hagina; (HINDI) paalak, vilayati palak, vilayti palak, isfanaj, palak, paalkashaak, paalaki, pinnis, sag, sagpalak; (HINDUSTANI) ek qism ká ság, pálak; (HMONG DAW) zaub spinach; (HUNGARIAN) spenót, paraj; (ICELANDIC) spinat; (INDONESIAN) spinasi, beyam belanda, peleng, puileng, horenso; (ITALIAN) spinacio, spinaci; (JAPANESE) hourensou, horenso; (KANNADA) spinak soppu, spinaksoppu, spinachsoppu, paalak soppu, spinach; (KASHMIRI) hākh; (KATI) pālak; (KAZAKH) shpinat ogorodnyj, shpiynat; (KIRGHIZ) shpinat; (KOREAN) sigumchi; (KRIO) spinach; (KWÉYÒL) zépina; (LADINO) espináka; (LATVIAN) spināti, zépinat; (LITHUANIAN) daržinis špinatas, špinatai; (LUYIA) isipinachi[1]; (MACEDONIAN) spanak, obična spanak; (MAITHILI) palākī; (MALAY) poh choy, bayam; (MALAYALAM) visalacheera; (MALTESE) spinaci; (MARANAO) sosolon a gadong a kamo; (MARATHI) palak, paalak; (MARINDUQUE) sosolon a gadong a kamo; (MÒORÉ) pinar; (MORISYEN) sok choy; (NAVAJO) waa'; (NEPALI) palungo, palak; (NIUE) sinapi; (NORWEGIAN) spinat; (ORIYA) palaksag, palanga saga, mithapalanga, mitha palanga, mitha; (OROMO, WEST CENTRAL) lima; (PANGASINAN) espínace; (PANJABI, EASTERN) palak, walati pälak, isfanak; (PASHTO, NORTHERN) sâg, pâlëk, pālak; (POLISH) szpinak, szpinak warzywny; (PORTUGUESE) espinafre, espinafre europeu; (ROMANI, BALKAN) spanakoos; (ROMANIAN) spinat, spanacul, spanac; (RUSSIAN) špinat, špinat ogorodnyj; (SANSKRIT) palankya, chhurika, chirtchhada, gramini, gramyavallabha, kshurapathrika, kshurika, madhura, snigdhapathra, supathra, vastukakara; (SERBO-CROATIAN) spanač; (SHONA) sipinachi; (SLOVAK) špenát; (SLOVENIAN) špinača; (SORBIAN, UPPER) spinat; (SPANISH) espinaca; (STELLINGWERFS) spinazie; (SWAHILI) mayugwa, muhugo, viazi, mchicha; (SWEDISH) spenat; (TAGALOG) spinach; (TAJIKI) ismalon; (TAMIL) vusavyeley keerai, vasaiyilaikkeerai, vasayleykiray, vusavyeley, pasalai keerai; (TATAR) shpinat; (TELUGU) teegabatchali, dumpabachhali, dummpabaucchale, matturbachhali, paala koora; (THAI) puaileng; (TIBETAN) pcctsee; (TURKISH) ispanak; (TURKMEN) ysmanak; (UKRAINIAN) špinat; (URDU) palak; (UYGHUR) paläk; (UZBEK) shpinat, ismolok, ismaloq; (VENDA) tshipinìtshi; (VIETNAMESE) rau ba lăng, rau nh[af] ch[uf]a, b[os] x[oo]i; (VLAAMS) spinazie; (WOLOF) mbum; (XHOSA) isipinatshi; (YIDDISH) shpinaht; (ZULU) isipinashi.

35.10. *Suaeda glauca* (Bunge) Bunge (EP) young stems, leaves, plants; (P) C.

(CHINESE, MANDARIN) jian peng; (CHINESE, WÚ) jian peng; (ENGLISH) common seepweed; (FINNISH) kilokki; (FRENCH) soude, soude asperge; (HEBREW) ukkam, ukam; (HUNGARIAN) sóballa; (JAPANESE) matsuna; (MONGOLIAN) budarga; (NORWEGIAN) saftmelde; (POLISH) dziurawiec pospolity; (TAJIKI) shuragioh; (TURKISH) sahil sodasi, sahil suda bitkisi; (VLAAMS) gewoon schorrekruit.

36. Convolvulaceae

36.1. *Convolvulus japonicus* Thunb. (EP) root; (P) C.

(BRETON) troell, gweerez, bezvoud, lavrig; (CHINESE, MANDARIN) chang lie xuan hua; (DANISH) japansk snorle; (DUTCH) Japanse winde; (ENGLISH) rose glorybind, Japanese bindweed; (FINNISH) kierto; (FRENCH) liseron du Japon; (GREEK) kalystegi, iaponiki; (HEBREW) ḥavalbal yappani, charalbal yapani; (ITALIAN) vilucchio giapponese; (JAPANESE) hirugao, kohirugao; (LUO) nero icok; (MACEDONIAN) slačec japonski; (MALTESE) leblieb; (MONGOLIAN) japan sedergene; (NORWEGIAN) vindel; (PORTUGUESE) campainha do Japão; (ROMANIAN) batata èdesburgouya; (SPANISH) boniato del Japón; (TAJIKI) pechak; (TURKISH) sarmaşik, gündüz sefasi; (VLAAMS) hirugao.

36.2. *Ipomoea aquatica* Forsskal (EP) tender shoots and leaves; (P) C.

(AFRIKAANS) watespinasie; (AMHARIC) karia-hareg; (ASSAMESE) kolmow sak; (BAMANANKAN) bafaraka; (BENGALI) kalmisak, kalmi sag; (BICOLANO, CENTRAL) akangkong; (BINUKID) tangkung; (BURMESE) ka-zun-ywet; (BUTUANON) tangkong; (CEBUANO) tangkong; (CHINESE, GÀN) kong xin cai; (CHINESE, MANDARIN) weng cai; (CHINESE, MIN BEI) kong xin cai, teng cai; (CHINESE, MIN NAN) èng-chhài; (CHINESE, WÚ) weng cai, kong xin cai; (CHINESE, XIANG) kong xin cai, teng teng cai; (CHINESE, YUE) weng cai, tong cai, kong xin cai; (CHUWABU) phumde; (CZECH) povijnice vodní; (DANISH) kangkong, bladgrøntsag; (DUTCH) kangkong; (ENGLISH) water spinach, kangkong, swamp morning glory, water convolvulus; (FILIPINO) kangkong; (FRENCH) liseron d'eau, patate aquatique; (FULFULDE, ADAMAWA) delbol; (GA) kɔkɔle biakɔ; (GERMAN) Sumpf-Trichterwinde; (GREEK) ypomia, ydrocharis; (GUJARATI) nalanibhaji; (GUN) tɔwɛlikan; (HANUNOO) tankun; (HAUSA) awarwaro, fùrén gàdú, tarfo, taufau, terfai; (HEBREW) lefufit ha-mayim, lefufit hamaim; (HILIGAYNON) tangkung; (HINDI) kalmisag, karmi, kalmishaak, patuasag, patuasaga, nilkamli, sarrali; (HINDUSTANI) kermi, karemwa; (HMONG DAW) phav npoob; (IBANAG) balangak; (ILOCANO) balangog, galatgat, balangeg, balanged, kangkong, tangkong; (INDONESIAN) kangkung, kankong; (ITALIAN) patate acquatica, villucchio d'acqua; (ITAWIT) balangag; (IZON) èméin fị tùò, tànà nītúó, tọrụbẹ́ịn, tànàïn túó; (JAPANESE) you-sai, en-sai; (JAVANESE) kangkung, kangkung jlidri; (KANNADA) bill hambu, balle soppu, neeru bili gedde hambu, neeru bill gedde hoo, neeru 'hambu; (KHMER, CENTRAL) trâkuön; (LAHU) meu_ nyi⁻ meu$^\vee$ ve$^\wedge$; (LAO) bôngz; (LOMWE, MALAWI) thelele; (LUO) icok pii; (MACEDONIAN) voden spanak; (MALAY) kangkong, kankung; (MALAYALAM) ippomia; (MARATHI) nadishaka, naalichi-bhaaji, nalichi bhaji, naala; (MENDE) kogidi, kpokpo; (MORISYEN) bred sifflet; (NEPALI) kalmi saag, karmaiya saag; (PAMPANGAN) akangkong, aksngkong; (PANGASINAN) kañgkóñg; (PANJABI, EASTERN) ganthian, nali, nari; (POLISH) szpinak wodny; (PORTUGUESE) batata aquática, cancon; (RONGA) chidledelane; (SAMOAN) 'umala vai; (SANSKRIT) kalambi, kadambi, kalaka, kalambika, kalashaka, kechuka, naadi, naadika, naalika, pattashaka, pechu, pechuli, shathaparna, shradhashaka, vishvarochanna; (SENA) chijangumba, pfunde; (SERER-SINE) n-dapul, nof mbal; (SINHALA) kankun; (SONGHAY) tàlhánà; (SPANISH) batatilla acuatica, boniato de agua, camotillo; (SUNDA) kangkung; (TAGALOG) kangkong; (TAJIKI) guli oshiqi pechon; (TAMIL) vellaikeerai, vallai keerai, koilangu, sarkaraeivalli, vallai kodi; (TAUSUG) kangkung, tangkung; (TELUGU) tutikura, thantikada, thootikoora; (THAI) phakbung, phakthotyot; (TOK PISIN) kangkong, kango; (TURKISH) sucul gece sefasi; (URDU) kalmisag; (VIETNAMESE) kang kung, rau mu[oos]ng; (VLAAMS) kankoeng dagoeblad, waterspinazie; (WÁRAY-WÁRAY) tangkong.

36.3. *Ipomoea batatas* (L.) Lam. var. *batatas* (EP) roots and leaves[S]; (P) C.

Dialects of Luyia (Bukusu[1], Marama[2]). A few entries in several of the African languages appear to represent cultivar names.

(AFRIKAANS) patat; (AGTA) kamoti; (AGTA, CASIGURAN DUMAGAT) kamosi; (AKAN-AKUAPEM) abrɔdwemaa, ntɔmmɔ; (AKAN-FANTE) saantom; (ALBANIAN, TOSK) batatë; (AMHARIC) mätatis, bäṭaṭis, səkkʷar dənnəč; (ANUAK) ājwāallā, ājwáallɛ́; (ARABIC) ba ta tah helua, bombe; (ASSAMESE) mitha alu, mitha aloo; (BAATONUM) dankali; (BALANTA-KENTOHE) patate; (BAMANANKAN) kondouba, konduba, patato, toma ulé; (BAOULÉ) alen'da, tingo; (BASQUE) batata; (BASSARI) fútutè, ɔ-pútutè, uridschŏ; (BEMBA) cumbu; (BEMBÉ) mbal' kongo; (BENGALI) mishti alu, lalalu, ranga alu; (BEROM) dànkál; (BÉTÉ) bababeu; (BINUKID) kesila', kamuti; (BLACKFOOT) ómahkataaki, paataak, maatáák; (BODO) tahguna; (BONTOC CENTRAL) tugi, baydan, bayyasi, binaydan, gamoting, kalbooy, obi; (BUDIK) peutètè; (BUHID) amunti mati gas, balingkawak malagsi, balingkawak maraska, gabuk, kamuti, karonsing, lagsi urungan, lumak, mabanwa, magigi, mayamas, medanglis, miling, rabu, salanggi, sarangan, sungot, tabubuan, talise, tarifuan, tifalia, tumantigip, uat sirut; (BULGARIAN) batati; (BULLOM SO) g-bamɛ, g-bami, gbam-dɛ; (BULU) ndua; (BURMESE) ka-zun-oo, myonk-ni; (CAJUN FRENCH) potat douce; (CAROLINIAN) gamwuuti, ghómwuuti; (CATALAN-VALENCIAN-BALEAR) maniato, batata; (CEBUANO) tigsi, kamote; (CHICKASAW) ahi' champoli'; (CHINESE, HAKKA) fān-choû; (CHINESE, MANDARIN) gan shu, shu; (CHINESE, MIN NAN) han-chî; (CHINESE, WÚ) gan sun; (CHINESE, XIANG) hong shu, hong shao; (CHINESE, YUE) fan shu, huong shu; (COMANCHE) kamúutuʔ; (CREOLE, HAITIAN FRENCH) patat; (CRIOULO, CAPE VERDE) batata doci; (CZECH) batáty, sladké brambory; (DAGAARE, SOUTHERN) gnangnour; (DAGBANI) wulejɔ; (DAKOTA) paŋǧitaŋka, mdo sku'ya; (DANGME-KROBO) anãgo, atomo; (DANISH) sød kartoffel, batat; (DIGO) myogwe; (DITAMMARI) yakɛyɛra; (DUTCH) bataat, zoete aardappel; (EDE NAGO) oduku; (EDO) iyan-ebo; (EFIK) biâ ḿbàkárá; (ENGLISH) sweetpotato, sweet potato; (ESTONIAN) maguskartul, bataat; (ÉWÉ) anagote, dzete, anago, anago-te; (FARSI, EASTERN) kachâlu; (FARSI, WESTERN) kachâlu; (FIJIAN) kumala; (FILIPINO) kamote; (FINNISH) bataatti; (FON) wɛli; (FRENCH) patate douce, batate; (FULFULDE, ADAMAWA) kudaku, kudakūji, pēmbere, pēmberēji, duderpurre, kuudaku, laire ngabbu; (GA) atómò, blɔfo-atómò, blɔfoo atomo, atomo; (GADDANG) asilang; (GANDA) lumonde; (GBE, WACI) jete; (GEN) jete; (GERMAN) Batate Süsskartoffel, Süsskartoffel; (GIKUYU) mĩrĩo/gwăcĩ; (GOLA) n-joule; (GONJA) jibelbi; (GOURMANCHÉMA) dayeré, dayessé; (GOURMANTCHÉ) komtouwu, o-kourko; (GREEK) glykopatata; (GUJARATI) shakaria, kanangi, sakaria; (GUN) wueli; (GURO) baàtɛ́n; (HANUNOO) banbid, inasuway, kamuti; (HARARI) biṭāṭis; (HARSŪSI) fendāl; (HAUSA) dankali, bà-fádámeè, ba-lawure, ba-nkuli, daakataá, daŋkálii, dánkálìi, gàmàgàrí, iyaya, kudaku, kugundugu, laáwúr; (HAWAIIAN) ʻuala; (HEBREW) batata, lefufit ha-batata, lefufit habatatah; (HILIGAYNON) kamúti; (HINDI) shakarkandi, sakarkand, shakardand, shakar kand, mitha alu, mitaalu; (HINDUSTANI) i shakarqand, metka peteta; (HMONG DAW) qos liab; (HUNGARIAN) batáta, édesburgonya; (IBALOI) dokto, dukto; (IBANAG) kamosi, patatas, tugi; (IFUGAO) lapni, babain di dupduphug, bagagan an dumaglo, ballagan gibbang, ballagangan mangitit, bugagan an dumaghag, buggapas di matikano, dapnen, haway ad hingyon, kamutit tilit an wakal, kohbuti, lagah, lamtung, lapnay, lapne, lumug an liggod, lumullung, u1dimme, uldino; (IGBO) ndụku, ndukwu, jí nwa nnu, jí-bèkéè, ọtọlị, jí-óyìbó, oi-bèkéè, kúkúǹdùkú; (ILOCANO) kamotit, kamotig, kamote, balangeg; (INDONESIAN) ubi jalar, ubi, keledek, ketela rambat; (ISNAG) kela; (ITALIAN) patata americana, patata dolce, batata; (ITAWIT) ganagan; (IVATAN) wakay; (IZON) bèkè búrù; (JAPANESE) satsuma-imo; (JAVANESE) katela, katela rambat, tela, tela rambat, bodin; (JOLA-FONYI) efatate, batatar; (JULA) osso; (KABIYÉ) aouhiae; (KALAGAN) gubi, luktu; (KANKANAEY) obi; (KANNADA) genasu, sihi genasu, bellagenasu, genasu gedde; (KANURI) dàngálì, kúndùwú, kuwunduwu; (KASHMIRI) shĕkar; (KHMER, CENTRAL) dâmlô:ng chvië; (KIRIBATI) te kumara; (KISSI, NORTHERN) g-banja, k-panja, k-pande-wendẽ; (KONKANI) cheenikanang, kanang, konong; (KONO) bundɛ, yukɛnɛ; (KOREAN) goguma; (KRIO) pɛtɛtɛ, swit pɛtɛtɛ, ɔdɔkun, kukunduku; (KWÉYÒL) patat; (LADINO) patatas dulse; (LAHU) meu̠ nyiˉ, meu̠ nyi; (LAO) man kè:w; (LATVIAN) saldais kartupelis, batāte; (LITHUANIAN) valgomasis batatas; (LOKO) g-bɔgu; (LOMWE, MALAWI) mbatata, kholowa; (LUO) icok; (LUYIA) kambouni[1], ambouni[2]; (MACEDONIAN) batati, sladok kompir; (MAITHILI) alhuā, sakarkan, sekunā; (MALAGASY) vavibomanga; (MALAY) ubi keladi, ubi keladek, keledek; (MALAYALAM) cheeni kizangu, cheenikilanna, kapakalenga, madhura kilanna, vallikilanna; (MALIMBA) ampete, bɔkié, potekai kai, tokpohun, yenkisa, ntada; (MALTESE) patata helwa; (MANDING-MANDINKA) pataati, patatô, mandoroka, tokpohunka, wuse, wusengbe, yenkisa ka; (MANINKAKAN, EASTERN) balawura, bɔkui, gbanka yanka, lofolafan, wusi; (MANINKAKAN, WESTERN) kinkio; (MANOBO) buklad, isila, katila, mundu; (MAORI) kuumara, kuumera, kumara; (MARANAO) mamis a obi, obi, patatas, rapak, samboangan; (MARATHI) ratalu, rathaal, rathali; (MARQUESAN) 'uma'a, kuma'a; (MEHRI) fəndəlōt; (MENDE) njowo; (MIZO) kawlbahra; (MÒORÉ) nayu, nãyũi; (MORISYEN) patat; (MUNDARI) sakarkāndā, sāngā, punḍi sāngā, ārā sāngā; (MUSKOGEE) vhv-cvmpv; (NAVAJO) nímasii łikańí; (NEPALI) sakarkhanda; (NGIZIM) dànkáli; (NGUNDI) mbala za sucari; (NIUE) timala, kumara; (NORWEGIAN) knolldaggry, søtpotet, batat; (NUPE-NUPE-TAKO) dùkú, kàmâ dùkú, kúmúṇdùkú; (NYAKYUSA-NGONDE) imbatata; (NYANJA) mbatata, cimungulu; (ORIYA) kandamula, kandmula; (OROMO, WEST CENTRAL) dinicha miaa; (PALAWANO, CENTRAL) kambang-kambang; (PAMPANGAN) kamuti; (PANGASINAN) kamoti, kamóte; (PANJABI, EASTERN) shakarkandi; (POLISH) batat, patat, kartofel słodki; (PORTUGUESE) batata doce, batata; (POTAWATOMI) wishpupnyak; (PULAAR) pute, wusé, kunduwana; (PULAR) coundou; (RAROTONGAN) kamara; (ROMANIAN) batat, batat edesburgonya, cartof dulce, batatul; (RUSSIAN) batat, patat, sladkii kartofel; (SAMA, CENTRAL) panggi-bahan; (SAMBAL, BOTOLAN) kamoti; (SAMBAL, TINÀ) magbabaukok kamote, patatas; (SAMOAN) 'umala; (SANSKRIT) pindaalu, kandagranthi, pichhila, pindithaka, rakthaalu, romashakandaka, svadukandaka; (SANTALI) sakarkenda; (SENA) mbambaira; (SENOUFO, CEBAARA) bgossoloume; (SERBO-CROATIAN) batat, sladak krompir; (SHONA) bura, dima, mbambaira; (SINHALA) bathala; (SLOVAK) sladké zemiaky, batát; (SLOVENIAN) sladki krompir; (SONGHAY) kúudékà; (SOTHO, NORTHERN) morepa; (SOTHO, SOUTHERN) patata; (SPANISH) batata, boniato, camote, papa dulce, moniato; (SUBANON) gubi; (SUMO-MAYANGNA) pai; (SUNDA) boled, huwi boled, huwi mantang; (SUSU) ouré, uré, torekélé-uré, uré gbuéli, uré firke, wurɛ; (SWAHILI) kiazi, viazi vitamu, kiazi kitamu; (SWEDISH) sötpotatis, batat; (TAGALOG) kamote, kamuti, ubing-kamote; (TAGBANWA) kamote, lidib, kamoteng bagin, kamuti; (TAJIKI) batat, kartoshkai shirin; (TAMIL) sakkareivelleikilangu, sakkareivelleikizhangu, sadrakaraivalli kizhangu, eeni kilangu, seenikilangu, sarkaraivalli, segappuvalli kizhangu kody, vallikilangu; (TAUSUG) pangibagun, panggi bagun; (TELUGU) chelagada, chilagadadumpa, genasu gadda, chilagadam, genasu, mohanamu, thella dumpa; (TEM) anago, toonini; (THAI) man-thet; (THEMNE) ɛ-muna, *a*-bɔko, *a*-muna; (TIRURAY) batatas, ubi; (TIV) a̱tsaka̱, atsaka; (TOK PISIN) kaukau; (TONGA) mbohole; (TONGAN) kumala; (TSWANA) potata; (TULU) kereṅgu, kereng; (TUMBUKA) mtotomera, mbwete; (TURKISH) sarmaşik patatesi, tatli patates; (TWI) nt ɔmmɔ, abɔrɔdwobaa; (URDU) sakarkandi; (VAI) n-jowe; (VENDA) murambo, luṱhariàne, mbudzana; (VIETNAMESE) khoai lang; (VLAAMS) zoete aardappel, bataat; (WAAMA) fɔrɔm w teema; (WAMEY) tata; (WÁRAY-WÁRAY) kamote, kadabáw, kaindáy, kinampáy, kinaraswi, kinarosa, karingkít, katipunan, pinutian; (WOLOF) patas, patat, linene; (XHOSA) ibhatata; (YIDDISH) bahtahteh; (YOM) manur; (YORUBA) oduku, anamo, ànàmọ̀, ẹdumṣi, kúkúndùnkún, òdukún, òdunkún, òdukún fúnfún, òdukún pupa; (ZARMA) kudaku; (ZULU) ubhatata, umhlaza.

37. Crassulaceae

37.1. *Sedum sarmentosum* Burge (EP) leaves; (P) C.

(CHINESE, MANDARIN) qiang wei gen jing tian; (CHINESE, WÚ) qiang wei gen jing tian; (CZECH) rozchodnik; (DUTCH) muurpeper; (ENGLISH) sedum, stonecrop; (ESTONIAN) kukehari; (FINNISH) maksaruoho; (FRENCH) sédum âcre; (GERMAN) scharfer Mauerpfeffer; (GREEK) sedon; (HEBREW) z̤urit, tzurit; (HUNGARIAN) varjúháj; (ITALIAN) borracina acre; (JAPANESE) tsuru-mannen-gusa; (MACEDONIAN) žednik; (MONGOLIAN) mogoin idee; (NORWEGIAN) bergknapp; (POLISH) rozchodnik; (PORTUGUESE) saião acre, pimenta das paredes; (ROMANIAN) larbă grasă; (RUSSIAN) ochitok, jadrenets, (SPANISH) pampajarito, pan de cuco; (TURKISH) sürünücü dam koruğu, dam koruğu; (VLAAMS) sedum.

38. Cucurbitaceae

38.1. *Benincasa hispida* (Thunb.) Cogn. (EP) immature/mature fruit; (P) C, P.

Dialects of Luyia (Bukusu[1], Marama[2]).

(AFRIKAANS) sjinese pampoen; (AGTA) tabuguk; (ASSAMESE) kumura; (BENGALI) chal kumra; (BICOLANO, CENTRAL) rodal; (BURMESE) kyauk-pha-yon; (CEBUANO) tibiayon, kandol, tibiyayon; (CHINESE, GÀN) dong gua; (CHINESE, HAKKA) toūng kouā; (CHINESE, MANDARIN) dong gua; (CHINESE, MIN BEI) bai gua, zheng gua; (CHINESE, WÚ) dong gua, bai gua, zheng gua; (CHINESE, YUE) dong gua; (CRIOULO, CAPE VERDE) melon; (CZECH) beninkasa, tykev vosková; (DANISH) voksagurk; (ENGLISH) wax gourd, white gourd, Chinese preserving melon, ash gourd, wad gourd, white gourd melon, white pumpkin, tallow gourd; (FIJIAN) vago; (FILIPINO) kondol, malinga melon; (FINNISH) vahakurpitsa; (FRENCH) courge à la cire, courge cireuse; (GERMAN) Wachskürbis, weißer Kürbis; (GREEK) peponi; (GUJARATI) kohala, bhuru, kohlu; (HANUNOO) tabagwang, tambalyun; (HINDI) petha, kumhra, pani kumhra, golkaddu, bhathua, kohala, kondha, kudimah, kumra, paetha, phutika; (HMONG DAW) tao-tue, taub twg; (IBANAG) tanghoy, tangkay; (ILOCANO) tangkoy; (INDONESIAN) kundur, beligo, beligu, baligo, baligu, tangkue; (ITALIAN) benincasa; (ITAWIT) tangkoy; (IVATAN) kondol; (JAPANESE) tougan, kamo-uri; (JAVANESE) bligo, baligo; (KANNADA) budugumbla, bagagumbala kayi, boodukumbalakaayi, kooshmaanda; (KHMER, CENTRAL) trâllaach; (KONKANI) cualim, kunolem; (LAHU) a teh hpui shi̠, hpeuᵛ meu¯ hpeu shi̠; (LAO) tônx; (LUYIA) eshimukaa[2], chimuka[1]; (MACEDONIAN) vosočna tikvička; (MALAY) kundur, kundor; (MALAYALAM) kumballanga, elavan, kumbulam, kumpalam, kumpalanna; (MALTESE) bettieh tax-xitwa; (MARATHI) kohala; (MARQUESAN) hue puo; (MORISYEN) tekwa, kalbas sinwa; (NEPALI) kubhindo; (ORIYA) pani kakharu; pazdaba; (PANGASINAN) góndol, gundol; (PANJABI, EASTERN) petha, petha kaddoo, pāthha; (POLISH) melon chiński, zimowy; (PORTUGUESE) benincasa, abóbora de água; (RAROTONGAN) ʻue roro; (SAMOAN) fagu, melo; (SANSKRIT) kooshmanda, kooshmaanda, brihatphala, ghrinavasa, gramyakarkati, karkaru, karkotika, kooshmaandaka, kooshmaandi, kooshpandaha, kumbhanda, kunjaphala, Naagapushphaphala, peethapushpa, pushpaphala, shikhivardhaka, somaka, sthirphala, suphala, timisha, valli phala; (SINHALA) alu puhul; (SPANISH) calabaza branca, calabaza china; (SUBANON) kundul; (SUNDA) baligo, kundur, leyor; (TAGALOG) kundol, kondol; (TAGBANWA) tambulok; (TAMIL) kalyanappushinikkay, pusanikkai, pushini, sambalpushani kaai, sambal poosani, pushani kai, kalyana pusineekai; (TELUGU) budidagummadi, budida-gummadi, boarda goomoodoo, boodidegummadi, pendligummadikaaya; (THAI) fak, faeng, mafaeng; (TONGAN) fangu; (TURKISH) çin mum kabaği, beyaz kabak, kiş kavunu; (URDU) patha; (VIETNAMESE) b[is] dao, b[is] xanh; (VLAAMS) waskalebas, tonggwa; (WÁRAY-WÁRAY) kondól.

38.2. *Citrullus lanatus* (Thunb.) Matsum. & Nakai var. *citroides* (L. H. Bailey) Mansf. (EP) fruit; (P) P.

(AFRIKAANS) sitroen; (ASSAMESE) gorokhia tiyoh; (CHINESE, MANDARIN) xiao xi gua; (CHINESE, MIN NAN) hiu-î; (CHINESE, WÚ) xiao xi gua; (CHINESE, YUE) ju yuan xi gua; (ENGLISH) citron, preserving melon; (ESTONIAN) näsasidrun; (FARSI, WESTERN) toranj; (FINNISH) sukaattisitruuna; (FRENCH) citre, méreville; (GREEK) kitron; (LOMWE, MALAWI) cikho; (MACEDONIAN) citronska lubenica; (MALTESE) citru; (MÒORÉ) pastec; (POLISH) kawon cytrynowy; (SERER-SINE) safu gay nak; (TURKISH) ağaç kavunu; (VLAAMS) citroen.

38.3. *Citrullus lanatus* (Thunb.) Matsum. & Nakai var. *lanatus* (EP) ripe fruit and seeds[S]; (P) R.

[1]Itsukha dialect of Luyia.

(AFRIKAANS) waatlemoen; (AKAN-AKUAPEM) anemuna, ɛfere; (AKAN-FANTE) akate, anamuna, anemuna; (ALBANIAN, TOSK) shalqi, karpuz; (AMHARIC) mähale; (ANUAK) b̄ʌthíik, b̄ʌthíikì; (APACHE, JICARILLA) k'éch'iya̲'éé; (ARABIC) bateekh, belikh, fundi, hadj en-nas; (ARMENIAN) dzmeuk; (ASSAMESE) tormuj, turmuj; (AZERBAIJANI, NORTHERN) qarpɪz; (BALTI) bwaar; (BAMANANKAN) tsara; (BAOULÉ) n'viêlè; (BASQUE) angurri; (BASSARI) inam; (BELARUSAN) kavun; (BENGALI) tarmuj, turbuj, tarbuz; (BEROM) cɛ́s; (BÉTÉ) lilé; (BICOLANO, CENTRAL) sandiya; (BINUKID) degita'; (BLACKFOOT) ómahkotssko'nssimaan; (BODO) turumja; (BOSNIAN) lubenica, karpuza; (BULGARIAN) dinja; (BULLOM SO) bɔhrɔ, yɛntɛ, cham-dɛ, sãhũ-bombom-dɛ, sãŋga-lɛ; (BULU) abok; (BURMESE) pha-ye-thi, pha-rai; (CAJUN FRENCH) melon doux; (CAROLINIAN) saliiya; (CATALAN-VALENCIAN-BALEAR) cindria, síndria; (CEBUANO) melon; (CHECHEN) xorbaz, horbaz; (CHEROKEE) gvgisdi; (CHICKASAW) istokchank; (CHINESE, HAKKA) sī kouā; (CHINESE, MANDARIN) xi gua, hsi kwa; (CHINESE, MIN NAN) si-koe, koa, koe; (CHINESE, WÚ) xi gua; (CHINESE, YUE) xi gua; (COMANCHE) ohapi̱, ohape, puhi bihnáaʔ; (CREOLE, HAITIAN FRENCH) mélon dio, melon dlo; (CRIOULO, CAPE VERDE) melancia; (CROATIAN) lubenice; (CZECH) meloun vodní, lubenice; (DAGAARE, SOUTHERN) pastic; (DAGBANI) neri; (DAKOTA) sa'kayutapi, ṡpaŋ'ṡniyutapi; (DANISH) vandmelon; (DUTCH) watermeloen; (EDO) ògì; (EFIK) íkpán; (ENGLISH) watermelon, water melon; (ESTONIAN) arbuus; (ÉWÉ) atsẽtsẽa; (FARSI, EASTERN) tarboz; (FARSI, WESTERN) handevâne, handevaneh; (FIJIAN) tarbuge; (FILIPINO) watermelon, pakwan; (FINNISH) vesimeloni, arbuusi; (FRENCH) pastèque, melon d'eau; (FULFULDE) sikildjé; (FULFULDE, ADAMAWA) deeni, gunaaru, pari, parīje, cikirre, dende, dene; (GA) wátrɛ̀; (GAELIC, IRISH) mealbhacán; (GAELIC, MANX) melon; (GEORGIAN) sazamtro, sazámthro; (GERMAN) Wassermelone; (GOLA) legiɛ-kwi; (GOURMANCHÉMA) louri nafolo; (GOURMANTCHÉ) bompiène-nouori yamarti; (GREEK) karpouzi, ydropepon; (GUJARATI) tarbuz, karigu; (HANUNOO) sandiya sa raba; (HARSŪSI) beṭṭīx, bettēx; (HAUSA) kankana, guna, àgúshií, gúna; (HEBREW) avatiaḫ, avatiach; (HERERO) ≠khontsamab; (HILIGAYNON) sandíya; (HINDI) tarbuz, mathira, khamuja, turbuj, tarmuj, kalinda, jamaika, tharbooza, halinda, hindwana, karbooj, samanka, tarbooj, tharbooj, tharbooz; (HINDUSTANI) tarbuza, tarbuc; (HMONG DAW) dib liab; (HOPI) kawayvatnga, kuwanyesvakawayvatnga, kyensiskawayvatnga; (HUNGARIAN) görögdinnye; (IBANAG) sandiya; (ICELANDIC) vatnsmelona; (IGBO) egusi, ègúsí, élìlì, nwanru, ogili-irele, nkbuluko, élìlè; (ILOCANO) sandia, sandiya; (INDONESIAN) semangka, keramboja, mendikai, tembikai, mandiké, bĕtek; (ITALIAN) cocomero, anguria, melone d'acqua; (ITAWIT) sandiya; (IVATAN) simon; (JAPANESE) suika, sui kwa; (JAVANESE) semangka, watesan; (JOLA-FONYI) foussaro; (JULA) saradèni; (KANNADA) kallangadi, kallangadi hannu, bachhangaayi, dasamekke, kaalingana balli, kaalenga hannu, kallangadi balli, kallangare; (KANURI) bàmbúsə, fálí gúnà; (KASHMIRI) hĕndawan; (KATI) karbiza; (KAZAKH) arbuz stolovyj, qarbɪz; (KHMER, CENTRAL) 'öö'w llök; (KIOWA) 'ei-sɑhyei-gɑ; (KIRGHIZ) darbyz, darbɪz; (KISSI, NORTHERN) leŋguõ; (KONKANI) bachanga, kaling; (KONO) kɔ-saa, saa; (KOREAN) subak; (KRIO) bara, watamɛlɔn, bara-ɛgusi, e-gbara, ɛgusi; (KURANKO) inabe; (KURDISH, NORTHERN) shuti; (KWANGALI) etanga rouitji; (LADINO) karpùz, karpuze; (LAHU) teh hka⁻ shi̠; (LAO) môô, tèèng môô; (LATVIAN) galda arbūzs, arbūzs; (LUXENBOURGEOIS) Meloun; (LITHUANIAN) tikrasis arbūzas; (LOMWE, MALAWI) mabvende; (LUYIA) imeloni[1]; (MACEDONIAN) lubenica, obična lubenica; (MAITHILI) tārbhuj; (MALAY) tĕmbikai, sikoi, sĕmangka, bĕteka, kĕmĕndikai, mĕndikai, timun tĕmbikai; (MALAYALAM) thannir mathan, mandeki patak, vattakka; (MALTESE) dullieha; (MANDING-MANDINKA) saura, sarakumba; (MANINKAKAN, EASTERN) sara kumba; (MANINKAKAN, WESTERN) sarro; (MANU'A SAMOAN) meleni; (MARANAO) dagita; (MARATHI) kalingad, kalingada, turbuj, tarbuz, tarmuj, tharbooj; (MARINDUQUE) dagita; (MARWARI) matīro, paniphal, lāmpl̠ī; (MENDE) koja, pu-goja; (MICMAC) ĕchkooch; (MIZO) dawnfawh; (MÒORÉ) pastec; (MORISYEN) melon do; (MUSKOGEE) cvstvlē; (NAMA-DAMARA) enuwa; (NAVAJO) ch'ééhjiyáán; (NDONGA) katjama; (NEPALI) kharbuja; (NEWAR) jhadhī; (NICOBARESE, CAR) pan-tē-ka, pan-tëa-ka; (NIUE) meleni; (NORWEGIAN) vannmelon; (NYANJA) cimwela, vwende; (ORIYA) tarubhuja, tarubhuj; (PAIUTE, SOUTHERN) maw´dunts; (PANGASINAN) pakuán; (PANJABI, EASTERN) tarbuz, tarbooz, tandur, tarbuj, hathwänhä, matërä; (PASHTO, NORTHERN) hind wân̠a, hindwāna, tarbūza; (POLISH) kawon jadalny, kawon, arbuz; (PORTUGUESE) melancia; (POTAWATOMI) ashktamo; (PULAAR) budi, dende, wudduru; (PULAR) boudi toubaco; (ROMANI, BALKAN) hurboozoo; (ROMANIAN) pepene verde, pepenele verde, lubeniţă, harbuz; (RUSSIAN) arbouz stolovyj, arbouz, kavoun; (SAMBAL, BOTOLAN) katimon; (SAMOAN) meleni; (SANSKRIT) chayapula, kuttoowombi, alpapramanaka, brihadgold, chayaphala, chelana, chithra, chithraphala, chithravallika, ghrinaphala, godumba, kaalinda, kaalinga, krishnabeeja, lathapanasa, madhuraphala, mansala, mansaphala, meta, mutrala, natamra, raajathinisha, rakthabeeja, setha, shirnavrintha, sukhasha, sukhavasa, suvartula, tharabooja, vritthaphala; (SENOUFO, CEBAARA) tahou; (SERBO-CROATIAN) lubenica; (SERER-SINE) ferèf, béref, ségal; (SHONA) bvembe, gebe, kavi, mwiwa, nwisi, nwiwa, visi; (SLOVAK) vodový melón, melón vodný, dyňa červená; (SLOVENIAN) lubenica; (SOMALI) ḥabḥab; (SONGHAY) kànéỳ; (SOTHO, NORTHERN) legapu; (SOTHO, SOUTHERN) lehapu; (SPANISH) sandia; (SUNDA) samangka; (SWAHILI) tikiti, tikiti maji, tango; (SWEDISH) vattenmelon; (TAGALOG) pakwan; (TAJIKI) hanzal; (TAMIL) tharbuza, palam panna, dharbusini, peechhikaai, pitcha, pullum, vartaikaai; (TATAR) karbız; (TAUSUG) timon; (TELUGU) putcha, putcha kaaya, purcha, kalingarakaya, darbooji; (TEM) kanjinga; (THAI) taeng-mo, taeng-chin, matao; (THEMNE) *ma*-tent; (TOK PISIN) melon; (TSWANA) legapu; (TULU) bachangaayi; (TUMBUKA) cimwamaji; (TURKISH) karpuz, kaun; (TURKMEN) adaty garpyz, garpyz; (UKRAINIAN) kavun stolovij; (URDU) tarbuz; (UYGHUR) tavuz; (UZBEK) tarvuz, tarbuz; (VAI) sa, sara; (VENDA) bvani, l̨ihabu; (VIETNAMESE) d[uw]a h[aa][us], d[uw]ad[or]; (VLAAMS) watermeloen; (WOLOF) xâl, khâl, béref, hal; (XHOSA) ivatala; (YAO) liticiti; (YIDDISH) kahveneh, ahrbooz; (YORUBA) egusi, bàrà, ẹ̀gúsí, ẹ̀gúsí agvè, ẹ̀gúsí bora, egúnsi, ẹ̀gúsí maga; (ZARMA) kanay; (ZULU) ikhabe, u(lu)khabe; (ZUNI) mulu:na.

38.4. *Coccinia grandis* (L.) Voigt (EP) fruit, tender shoots[S] and leaves[S]; (P) C.

[1]Itsukha dialect of Luyia.

(AMHARIC) werk-bemeda; (ASSAMESE) kunduli; (BAMANANKAN) bagéna ka dégé; (BENGALI) telakuch, telacucha, tela kucha; (BURMESE) kin-mone-thi; (CHINESE, MANDARIN) hong gua; (CHINESE, WÚ) hong qua; (DANISH) skariagenagurk; (ENGLISH) ivy gourd, tindora, kovai fruit, small gourd, scarlet gourd; (GIKUYU) kĩnya; (GREEK) koukourvita; (GUJARATI) gilodia, tindora, ghobe, gluru, ghobe, ghuru; (HAUSA) gùrjín-dájìi, gwanduwa; (HINDI) tindora, kanduri, tinda, tendus, kundroo, kunduzi, kunduri, kundru, kanduri-ki-bel, bhimb, gulakaankh, kunali, kundaru, kunderi, kundori, thirkola; (HMONG DAW) taub txaij; (INDONESIAN) labu air; (IZON) ẹmbadákà; (JAPANESE) yasai karasuuri; (JAVANESE) papasan, kemarongan, bolu teke; (KANNADA) thonde kayi, thonde kaayi, tondekai, kaage thonde, konde balli, sihi thonde, thonde balli; (KHMER, CENTRAL) slök baahs; (KONKANI) pendli, tendulem, thendle; (LAHU) hpa^ hkeh^; (LAO) tam ling, tam nin; (LOMWE, MALAWI) cikho; (LUYIA) tsindashe[1]; (MAITHILI) kunḍal; (MALAY) pepasan, papasan; (MALAYALAM) kowva, kovalam, gwel, kova, kwel, thondi; (MALTESE) qara tork; (MARATHI) tondli, bimbi, thondali, tendli, kondvalli, thendli, zidadi; (NEPALI) ban kiri; (ORIYA) kaichi, kakudi; (PANJABI, EASTERN) kanouri, tinda, tendu, tendus; (POLISH) tykwa bluszczowa; (SANSKRIT) bimba, bimbee, bimbika, bimbaka, chharindini, danthachhadaa, danthachhadopama, govhi, jhundikeshi, kaamboja, karmakari, katubimbi, katuka, katutundika, oshthi, oshtopamaphala, piluparni, rakthaphala, ruchiraphala, thikthabimbi, thikthakhya, thikthathundi, thundi, thundika, thundikeri, thundiparyyayaga; (SERER-SINE) safu gaynak; (SONGHAY) lombaria; (SPANISH) pepino cimarrón; (TAMIL) kovai, kovaikai, kavaikai, kovaikaai, thondai; (TELUGU) dondakaya, kaakidonda, donda kaya, bimbika, donda, dondatheege, kaaidonda; (THAI) tamlueng, phaktamlung, phakkhaep; (TULU) manoli; (TURKISH) ağma kabaği; (URDU) tinda, kanduzi; (VIETNAMESE) hoa b[as]t, rau b[as]t; (VLAAMS) klimopkalebas; (WOLOF) yombu mbet, yomb u mbot;.

38.5. *Cucumeropsis mannii* Naudin (EP) fruit and seeds; (P) C,R.

(AJA) guci; (AKAN-AKUAPEM) akatewa, akyèyèa; (AKAN-ASANTE) akatewa; (AKAN-FANTE) akate, akatsewa; (BAOULÉ) n'viele; (BAMILEKE) mpock; (BULLOM SO) yɛntɛ, cham-dɛ, sãhũ-tuntun-dɛ; (BULU) ngon; (CHINESE, MANDARIN) gang guo gua; (CHINESE, WÚ) gang guo gua; (CRIOULO, UPPER GUINEA) tomatom; (EDE NAGO) itoɔ; (EFIK) òkôkòn; (ENGLISH) white-seeded melon, dark egusi; (EWONDO) ngon; (ÉWÉ) abgesi, gusi; (FON) gusi; (FRENCH) ononde, graines d'quonde; (GA) agushi, agúshi; (GBAYA) gala, ngoumbé; (GEN) gusi; (GOLA) legiɛ; (GREEK) koukoumero-pbis, edodimos; (GUN) gusito; (HAUSA) àgúshií; (IGBO) àhụ́, àhụ́ elu, aki, egusi; (IZON) ogúsi; (JAPANESE) egusi-ito; (KANNADA) tondekai; (KISSI, NORTHERN) heŋguo, seŋgo, seŋgo-pɔmbɔ, seŋgu-chielachalo, hiŋgo; (KONO) kõ-sa, sa, sa-nɛsɛ; (KRIO) egusi, wet-egusi, ɛgusi, wet-ɛgusi; (LOKO) koeja; (MALIMBA) mayentɛrɛba; (MANDING-MANDINKA) nantéo, sara; (MANINKAKAN, EASTERN) sara-mɛsɛ, sarɛ; (MENDE) koja, koja-mumu, koja-wawa, fɛŋɛ-goja; (POLISH) melon białonasienny; (PULAAR) mantem, budi; (SENOUFO, CEBAARA) n'gatahou; (SPANISH) calabaza pamué; (SUSU) kɔbɛ, mase; (THEMNE) ma-tent, ma-tent-ma-temne, *an*-tent, *ma*-teint-*ma*-temne; (TWI) akyekyea akatewa, akatoo; (VAI) sa, sara; (VLAAMS) witzadige meloen, egusi-ito; (YORUBA) itoɔ, ito, iroko, ẹ̀gúsí-ìtóò, itóò, ẹgúsí itóò, itórò, ìtórò, itoro.

38.6. *Cucumis anguria* L. var. *anguria* (EP) immature fruit; (P) P,C.

(AFRIKAANS) Wes-Indiese agurkie; (ARABIC) faghoos; (CHINESE, MANDARIN) xi yih du huang gua, hsi yin tu huang kua; (CHINESE, WÚ) xi yin du huang gua; (CREOLE, HAITIAN FRENCH) mélon frans; (CRIOULO, CAPE VERDE) pepino-de-macaco, pepino santeiro; (CZECH) angurie, americká okurka; (DANISH) anguria-agurk; (DUTCH) eivormige komkommer; (ENGLISH) West Indian gherkin, bur gherkin, gooseberry gourd, wild spiny cucumber; (FRENCH) concombre des antilles, cornichon des Antilles, angourie; (GERMAN) Angurien-Gurke, Westindische-Gurke, Pfeffer-Gurke; (HEBREW) kishu, kishue; (HMONG DAW) dib qus; (HUNGARIAN) anguriauborka; (ITALIAN) cetriolo americano; (JAPANESE) gaakin; (NEPALI) khiree; (NIUE) atiu, atiu hega; (POLISH) ogórek antylski; (PORTUGUESE) pepino das Antilhas, cornichão das Antilhas, machiché, maxixé; (RUSSIAN) anguria; (SAMOAN) atiu vao; (SENA) cassango; (SPANISH) pepino de sábana, patilla de golero, pepino de monte; (SRANAN) kundru; (SUSU) fɔtɛnalinyi; (SWEDISH) Vest-Indisk-gurke; (TURKISH) brezilya hiyari; (VLAAMS) West-Indische augurk.

38.7. *Cucumis melo* L. subsp. *melo* var. *cantalupensis* Naudin (EP) fruit; (P) R.

a. cantaloupe group

(AFRIKAANS) spanspek; (ALBANIAN, TOSK) pjepär, pjepër, pjepën; (AMHARIC) yäbahər qulqwal; (ARABIC) cantaloupe, faghoos; (ASSAMESE) bangi; (AZERBAIJANI, NORTHERN) qovun; (BASQUE) meloi; (BENGALI) kharmuj; (BODO) kurmuja; (BRETON) sukrin; (BULGARIAN) păpeš; (BULU) ngon ntagan; (BURMESE) tha-kwa-hmwe; (CAJUN FRENCH) melon français; (CATALAN-VALENCIAN-BALEAR) meloner; (CEBUANO) atimon, katimon; (CHECHEN) shaamali, paasta; (CHEROKEE) gawsvgi; (CHICKASAW) istokchank bila'ma';

(CHINESE, GÀN) xiang gua; (CHINESE, MANDARIN) ying pi tian gua; (CHINESE, MIN BEI) xiang gua; (CHINESE, MIN NAN) hiu-koe, phang-koe, lâi-á-koe, koa, koe; (CHINESE, WÚ) ying pi tian gua; (CHINESE, YUE) chu pi tian gua; (COMANCHE) taibo bihnaa?; (CRIOULO, CAPE VERDE) melon; (CROATIAN) dinja; (CZECH) kantalup, ordrudova skupina, cukroveho meloun; (DAGAARE, SOUTHERN) milon; (DAKOTA) sa'kayutapi, śpaŋ'śniyutapi; (DANISH) melon; (DUTCH) kanteloep, knobbelmeloen; (ENGLISH) cantaloupe; (ESTONIAN) kantaluup; (FARSI, EASTERN) kharbuza; (FARSI, WESTERN) tâlebi, kharboze; (FILIPINO) cantaloupe; (FINNISH) cantaloupe, meloni; (FRENCH) cantaloup, melon cantaloup; (FULFULDE, ADAMAWA) gunaaji; (GAELIC, IRISH) mealbhacán; (GAELIC, MANX) mellyn, asnagh; (GAELIC, SCOTTISH) meal-bhucan; (GALICIAN) meloeiro; (GEORGIAN) nesvi; (GERMAN) Kantaloupe-Melone, Rippen-Melone; (GOURMANCHÉMA) milo; (GOURMANTCHÉ) bompiène-nouori yamarti; (GREEK) peponi, idos peponos; (GUJARATI) sakkar teti; (HANUNOO) katimun, katimun malagti; (HEBREW) melon; (HERERO) etanga; (HINDI) kharbooja, kharbuza; (HINDUSTANI) kankri; (HMONG DAW) dib pag; (HOPI) melooni, patomelooni; (HUNGARIAN) kantalup sárgadinnye, sárgadinnye; (ICELANDIC) kantaloupmelona; (IGBO) egusi; (ILOCANO) itimon, murud; (INDONESIAN) blewak kantelup, blewah, belewa, semangka, belanda; (ITALIAN) melone cantalupo; (JAPANESE) kantarohpu; (JAVANESE) semangka londo, blewek, bluwak, bluwek, blewah, kerahi; (JULA) zrèn; (KANNADA) kekkarike; (KATI) karbiza; (KAZAKH) qawın; (KIRGHIZ) koon; (KOREAN) chamoe; (KURDISH, NORTHERN) kālēk; (LAHU) a⁻ hpeh$^{\vee}$ bvuh$_{\wedge}$; (LATVIAN) melone; (LITHUANIAN) melionas; (LUXENBOURGEOIS) Meloun; (MACEDONIAN) dinja, kantalupska dinja; (MALAY) tembikai wangi; (MALAYALAM) thai kumbalom; (MARANAO) dagita; (MARATHI) kharbuj; (MICMAC) ěchkooch; (MÒORÉ) mèlo; (MORISYEN) melon; (MUSKOGEE) fvmēcv; (NAVAJO) da'neesk'ání dich'ízhígíí; (NORWEGIAN) kantaluppmelon, melon; (ORIYA) khar bhaja ba, kasturi tarabhuja; (PANGASINAN) gundól; (PANJABI, EASTERN) kharbuza, tar; (POLISH) melon kantalup, kantalup, melon; (PORTUGUESE) melão cantalupe; (PULAAR) dende; (ROMANIAN) kantalup, pepene; (RUSSIAN) dynja; (SENOUFO, CEBAARA) kagnon-gberigui; (SERBO-CROATIAN) dinja; (SLOVAK) kanaloupa, melón cukrový, melón žltý; (SLOVENIAN) dinja; (SPANISH) melón cantaloupe; (SUNDA) blewah, hirbis; (SWAHILI) namna ya tikitimaji; (SWEDISH) cantaloupe, melon; (TAGALOG) melon, katimon, milon; (TAGBANWA) melo; (TAJIKI) handalak; (TAMIL) kakkirikaai; (TATAR) kavıın; (TAUSUG) inkug; (TELUGU) karbuja; (TURKISH) van kavunu, cep kavunu, kavunu, kantalup kavunu, kavun; (TURKMEN) gawun, goýun; (UKRAINIAN) dinja zvicajna stolova; (UYGHUR) qoghun; (UZBEK) covun, kovun, qovun; (VIETNAMESE) qua dua hău; (VLAAMS) cantaloupe, meloen; (WOLOF) xâlu tabab; (YIDDISH) deenkah.

b. muskmelon, Persian melon, netted melon, sweet melon

Previously classed as *Cucumis melo* L. Reticulatus group; [1] Indicates a generic term for 'melon' (*C. melo*) used by several African languages.

(AFRIKAANS) spanspek; (ALBANIAN, TOSK) pjepër, pjepën; (ARABIC) faghoos; (ASSAMESE) siral, chiral, bangi; (AZERBAIJANI, NORTHERN) qovun; (BADAGA) kumbala, kumbua; (BALTI) ghoon; (BAMANANKAN) kongo dĩé[1]; (BASQUE) meloi; (BASSARI) inabe; (BENGALI) kharmuj, khurmuj, kakri; (BODO) kurmuja; (BULGARIAN) papeš; (CATALAN-VALENCIAN-BALEAR) meloner, mesc meló; (CHEROKEE) gawsvgi; (CHINESE, HAKKA) hiōng kouā, t'iâm kouā; (CHINESE, MANDARIN) wang wen tian gua, tien kua; (CHINESE, WÚ) wang wen tian gua; (CRIOULO, CAPE VERDE) melon; (CZECH) meloun cukrový; (DANISH) melon; (DUTCH) meloen; (ENGLISH) muskmelon, Persian melon, netted melon, sweet melon; (ESTONIAN) melon; (FARSI, EASTERN) kharbuza; (FARSI, WESTERN) kharbuzeh, kharboze, talebie, kharbozch kotor; (FILIPINO) melon; (FINNISH) verkkomeloni, meloni; (FRENCH) melon brode, melon réticulé, melon; (FULFULDE, ADAMAWA) dende[1], sikirre[1], chikilje[1], cikiire[1], denaare[1], dene bamɗi[1], nyanaare[1]; (GA) wátèrɛ[1], watrɛ̀[1]; (GAELIC, IRISH) mealbhacán; (GAELIC, SCOTTISH) meal-bhucan; (GALICIAN) meloeiro; (GEORGIAN) nesvi; (GERMAN) Zuckermelone, Netz-Melone; (GOURMANTCHÉ) linouari, anouari; (GREEK) peponi, idos peponos; (GUJARATI) sakkar teti, kharbuja, sakkartoti, chibdu; (HARSŪSI) yeḥḥ; (HAUSA) bumbus[1], kankana[1], batanya[1], gulli[1], gunaà[1], gurjií[1], gurji[1], gurji ƙwantal[1], gurli kwantal[1], gwandar beri[1], kuliri[1], kwantal[1], shamman[1]; (HEBREW) melon; (HINDI) kharbuza, kharbooj, kharrbuja, kharbūjā, laalmee; (HINDUSTANI) kharbúza, tarbúz, phút, khurpúza; (HMONG DAW) dib pag; (HOPI) melooni, hápulmeloni, hopimelooni; (HUNGARIAN) sárgadinnye; (ICELANDIC) netmelona; (IGBO) isikiripum[1]; (INDONESIAN) katelup, bhalvengkak, blewah, belewa, semangka, belanda; (ITALIAN) melone retado, melone; (JAPANESE) netto melon, makuwauri; (JAVANESE) semangka londo, blewek, bluwak, bluwek, blewah, kerahi; (KANNADA) chakkotane soppu, kekkarike, budame kayi, kharbooja; (KAZAKH) dynja, qawɪn; (KIRGHIZ) koon; (KONKANI) chibdud, gibud, molao; (KONKOMBA) inabe[1]; (KOREAN) chamoe; (KURDISH, NORTHERN) kālēk; (LATVIAN) melone; (LUXENBOURGEOIS) Meloun; (MACEDONIAN) dinja, persiska dinja; (MAITHILI) kharbhuj, gurmhī, phalēj; (MALAY) tembikai wangi; (MALAYALAM) thai kumbalom; (MANDING-MANDINKA) kurubombonmesergo[1]; (MALTESE) bettieh tax-xibka; (MANU'A SAMOAN) atiu, meleni; (MARATHI) chandanbatva, kharbuja, kharbuj, chibunda; (MEHRI) bəṯṯə́xt, bəṯṯáyx, gawḥ; (MIZO) hmazil; (MÒORÉ) mèlo; (MORISYEN) melon̄ (NAVAJO) da'neesk'ání dich'ízhígíí; (NEPALI) tarbuja; (NORWEGIAN) nettmelon, melon, vintermelon; (NUPE-NUPE-TAKO) pàràgi; (ORIYA) khar bhuja ba, kasturi tatabhuja; (PANJABI, EASTERN) kharbooza, kharbuza; (PASHTO, NORTHERN) khaṯēkay, k̲h̲arbūza, dastambol, k̲h̲atakai, sarda, hanḍyālai; (POLISH) melon siatkowany; (PORTUGUESE) melão, melão reticulado, melão cantalupe; (ROMANIAN) pepene; (RUSSIAN) dynja; (SAMOAN) utiu; (SANSKRIT) kharbuja, kharbooja, madhupaka, kalinga, ervaru, ervaruh, karkati, thrapusa; (SERBO-CROATIAN) dinja; (SERER-SINE) khad[1], nak nãngoy[1]; (SLOVAK) melón žltý, melón cukrový; (SLOVENIAN) dinja; (SOMALI) bartiḥ; (SONGHAY) al bata[1], múné[1]; (SPANISH) melón bordado, melón; (SUNDA) blewah, hirbis; (SWAHILI) tikiti; (SWEDISH) melon; (TAGALOG) katimon, melon; (TAMIL) kakkirikaai, malankody, mulampazham, velapalam, kakkiri kaci, karbujadosa, kattuttumatti, peddadosrai, peddakai; (TATAR) kavɪn; (TELUGU) kharbujadosa, karbooja dosa, putzakova, kharbuja, kasturi tarabuja, vela pandu; (TEM) kaenia[1], kaenja[1]; (TIV) icegher[1]; (TURKISH) ağ kavunu, kavun; (TURKMEN) gawun; (UKRAINIAN) dinja zvicajna stolova; (URDU) kharbuza; (UYGHUR) qoghun; (UZBEK) qovun, kovun; (VIETNAMESE) quả dua hấu; (VLAAMS) muskmelon, meloen; (WOLOF) hal u mbot[1], yambebute[1], yomba mböt[1], yamba[1]; (XHOSA) umxoxozi; (ZULU) isiponspeki.

38.8. *Cucumis melo* L. subsp. *melo* var. *chito* (C. Morren) Naudin (EP) fruit; (P) R.

(ALBANIAN, TOSK) pjepër, pjepën; (CHECHEN) paasta; (CHINESE, MANDARIN) mi gan tian gua; (CHINESE, WÚ) ning meng gua; min gan tian gua; (DUTCH) mango meloen; (ENGLISH) mango melon, garden melon, lemon melon; (FILIPINO) melon; (FINNISH) mangomeloni; (FRENCH) melon mango; (GERMAN) Mango-Melone; (GREEK) peponi, idos peponos; (HEBREW) melon; (HMONG DAW) txiv nkhaus taw; (HUNGARIAN) kerti sárgadinnye, sárgadinnye; (INDONESIAN) melon; (ITALIAN) melone mango; (JAPANESE) shiro-uri; (LATVIAN) deserta melone; (LOMWE, MALAWI) luwimbe; (LUXENBOURGEOIS) Meloun; (MACEDONIAN) gradinarska dinja; (MÒORÉ) mèlo; (NEPALI) ghurmi; (NORWEGIAN) prydmelon, melon; (NYANJA) kayimbe; (POLISH) melon cytrynowy; (PORTUGUESE) melão mango; (PULAR) denné toubaco; (ROMANIAN) pepene; (SERBO-CROATIAN) dinja; (SLOVAK) melón žltý; (SPANISH) melón mango; (SWEDISH) melon; (TAGALOG) melon; (TAJIKI) anbah; (TURKISH) kavun; (UZBEK) covun; (VLAAMS) citroenmeloen; (YAO) mpombe; (YIDDISH) melun.

38.9. *Cucumis melo* L. subsp. *melo* var. *conomon* (Thunb.) Makino (EP) young fruit; (P) P,C,R.

(ALBANIAN, TOSK) pjepër, pjepën; (CHECHEN) paasta; (CHINESE, MANDARIN) yue gua; (CHINESE, WÚ) yue gua, xian gua; (CHINESE, XIANG) cao gua; (CHINESE, YUE) bei gua, yue gua; (CZECH) dyne meloun, cujrovy meloun; (ENGLISH) oriental pickling melon; (FILIPINO) melon; (FINNISH) meloni; (GREEK) peponi, idos peponos; (HEBREW) melon, kishu matok; (HUNGARIAN) sárgadinnye; (ICELANDIC) melona; (JAPANESE) shiro-uri; (LUXENBOURGEOIS) Meloun; (MACEDONIAN) obična dinja; (MANINKAKAN, WESTERN) toubab sarro; (MÒORÉ) mèlo; (NORWEGIAN) grønnsakmelon, melon; (POLISH) melon wschodni; (PORTUGUESE) melão, uri; (ROMANIAN) pepene galben, pepene; (RUSSIAN) dynja; (SLOVAK) melón žltý; (SPANISH) melón manzana, pepino limón; (SWEDISH) melon; (TAGALOG) melon; (URDU) kharbuza; (VLAAMS) oosterse pekelmeloen; (YIDDISH) melun.

38.10. *Cucumis melo* L. subsp. *melo* var. *flexuosus* (L.) Naudin (EP) immature fruits; (P) R,P.

(ALBANIAN, TOSK) pjepër, pjepën; (ARABIC) khiar yapany; (ASSAMESE) Japani tiyoh; (CHINESE, MANDARIN) she tian gua; (CHINESE, WÚ) she tian gua; (ENGLISH) Armonian cucumber, Japanese cucumber, snake melon, serpent melon; (FRENCH) melon serpent, concombre serpent; (GERMAN) Schangen-Melone; (GREEK) peponi, idos peponos; (HMONG DAW) dib nyiupeej; (HUNGARIAN) sárgadinnye; (LATVIAN) līkās melone; (LITHUANIAN) gyvatagurkis; (LUXENBOURGEOIS) Meloun; (MACEDONIAN) japonska dinja; (NEPALI) kakadi; (NORWEGIAN) slangemelon, melon; (PANJABI, EASTERN) tar; (POLISH) melon japoński; (PORTUGUESE) melão serpente, uri; (ROMANIAN) pepene; (SERBO-CROATIAN) dinja; (SLOVAK) melón žltý; (SPANISH) melón serpiente; (SWEDISH) melon; (TAGALOG) melon; (TAJIKI) tarra; (TURKISH) acur, yilan kavunu, hittik, hitti; (VLAAMS) Japanse komkommer; (YIDDISH) melun.

38.11. *Cucumis melo* L. subsp. *melo* var. *inodorus* H. Jacq. (EP) fruit; (P) R.

(AFRIKAANS) winterspanspek; (ALBANIAN, TOSK) pjepär, pjepër, pjepën; (ARABIC) shahed; (BASQUE) meloi; (BULGARIAN) păpeš; (CATALAN-VALENCIAN-BALEAR) meloner; (CHECHEN) paasta; (CHINESE, MANDARIN) song tian gua; (CHINESE, WÚ) dong tian gua; (CROATIAN) dinja; (CZECH) meloun cukrový; (DAKOTA) sa'kayutapi, śpaŋ'śniyutapi; (DANISH) melon; (ENGLISH) winter melon; honeydew melon; casaba melons; (ESTONIAN) mesikaste, lehemesi; (FARSI, EASTERN) kharbuza; (FARSI, WESTERN) kharboze, kharbozeh asaly; (FIJIAN) timo; (FINNISH) meloni; (FRENCH) melon, melon d'hiver; (GAELIC, IRISH) mealbhacán; (GAELIC, MANX) drulght, molley; (GAELIC, SCOTTISH) meal-bhucan; (GALICIAN) meloeiro; (GERMAN) Melone; (GREEK) peponi, idos peponos; (HEBREW) melon; (HINDI) kharbuza; (HMONG DAW) dib pag; (HUNGARIAN) sárgadinnye; (ICELANDIC) hunangsmelona; (IGBO) egusi; (INDONESIAN) bhalvengkak; (ITALIAN) melone; (JAPANESE) wintaa melon; (KATI) karbiza; (KIRGHIZ) koon; (KOREAN) chamoe; (KURDISH, NORTHERN) kālēk; (LATVIAN) melone; (LUXENBOURGEOIS) Meloun; (MACEDONIAN) dinja, bezmirisna dinja; (MALAY) tembikai susu; (MARQUESAN) katiu, 'atiu; (MICMAC) ĕchkooch; (MÒORÉ) mèlo; (NORWEGIAN) vintermelon, melon; (PANGASINAN) gundól; (PANJABI, EASTERN) kharbooza, kharbuza; (POLISH) melon zimowy; (PORTUGUESE) melão, melão amarelo; (ROMANIAN) pepene; (RUSSIAN) dynja; (SAMOAN) 'atiu; (SANTALI) phuṭ, trạbhuj; (SERBO-CROATIAN) dinja; (SLOVAK) melón cukrový, melón žltý; (SLOVENIAN) dinja; (SPANISH) melón; (SWEDISH) melon; (TAGALOG) honeydew; (TONGAN) 'atiu; (TURKISH) kiş kavunu; (UKRAINIAN) dinja zvicajna stolova; (UZBEK) covun, kovun; (VIETNAMESE) qua dua hău; (VLAAMS) honingdauw, casba, meloen; (YIDDISH) melun.

38.12. *Cucumis metuliferus* E. Meyer, ex Naudin (EP) fruits; (P) R.

(AFRIKAANS) rooikomkommer; (BULU) ngon; (CZECH) kiwano; (DAGAARE, SOUTHERN) kũmbié; (DANISH) afrika agurk; (DUTCH) zuiltjesdragende komkommer; (ENGLISH) African horned cucumber; (FINNISH) kivakurkku, kiwano; (FON) gbowunɔ; (FRENCH) concombre africain, concombre porte bornes, kiwano, melon à cornes; (GOURMANTCHÉ) akana; (GREEK) afrikaniko kerato-ides aggouri; (GUN) gbowunɔ; (HAUSA) buùrár zaákiì, noónònkuúraá; (HMONG DAW) dib tuaj pos; (ITALIAN) kiwano; (JAPANESE) kiwano furuutsu; (JOLA-FONYI) sidèldéla; (KUNG-EKOKA) dz'aa!'hau; (KWANGALI) ntende; (LOMWE, MALAWI) mankhaka; (MACEDONIAN) afrikanska mrežesta tikyička; (MÒORÉ) kũba; (NORWEGIAN) hornmelon; (NYANJA) kangamkhwani; (PANJABI, EASTERN) khērä; (POLISH) kiwano, ogórek afrykański; (PORTUGUESE) kiwano, pepino africano, kumbundu, umbundu; (SERER-SINE) dandalub; (SPANISH) kiwano; (SWAHILI) matango; (TURKISH) Afrika hiyari; (VLAAMS) Afrikaanse gehoornde komkommer; (WELSH) ciwcynber cormos; (WOLOF) dândalub.

38.13. *Cucumis sativus* L. (EP) immature fruit; (P) R,P.

a. immature fruit

(AFRIKAANS) komkommer; (ALBANIAN, TOSK) kastravec, trangull; (AMHARIC) qʷasa, yä *f*äräng̃ dubba; (ARABIC) khiar, faggus; (ASSAMESE) tiyoh, tihu; (AZERBAIJANI, NORTHERN) xiyar, yerpənək; (BADAGA) ka:su; (BALANTA-KENTOHE) betbinho; (BALTI) lalu, latu, ladu; (BASQUE) luzker; (BELARUSAN) ogurets; (BEMBA) cibimbi; (BENGALI) shasha, khira; (BINUKID) timun; (BODO) thaihbeng; (BONTOC CENTRAL) kasimum, kasimun; (BOSNIAN) krastavac, krastavica; (BUHID) unduy; (BULGARIAN) krastavica; (BULU) ngon; (BURMESE) tha-kwa, thakhwa; (CAJUN FRENCH) cocomb; (CAROLINIAN) gúúngár, ghúúngar; (CATALAN-VALENCIAN-BALEAR) cogombre; (CEBUANO) kalabaga, pipino; (CHEROKEE) gagama; (CHICKASAW) chauk haloppa', istokchankapa'; (CHINESE, GÀN) huang gua; (CHINESE, HAKKA) nêt-kouā; (CHINESE, MANDARIN) hu gua, huang kua; (CHINESE, MIN BEI) cai gua, ci gua; (CHINESE, MIN NAN) chhì-koe, am-koe, chhài-koe, koa, koe; (CHINESE, XIANG) hung gua; (CHINESE, WÚ) hu gua, wang gua, huang gua; (CHINESE, YUE) haung gua, qing gua, cai gua; (COMANCHE) poko ruhkaʔ, soko rahkaʔmiitsaʔ; (CREE, PLAINS) kohkompaninawak; (CREOLE, HAITIAN FRENCH) konkonm; (CRIOULO, CAPE VERDE) pepino; (CROATIAN) krastavci; (CZECH) okurka, okurka setá, dyně; (DAGAARE, SOUTHERN) boolè; (DAKOTA) sa'kayutapi yuhiħi; (DANISH) agurk, Asie-agurk, drueagurk; (DUTCH) komkommer, augurk; (ENGLISH) cucumber; (ESTONIAN) kurk; (FARSI, EASTERN) bâdrang; (FARSI, WESTERN) khiyâr; (FIJIAN) khira; (FILIPINO) pepino; (FINNISH) kurkku; (FRENCH) concombre; (GA) kokómbà; (GAELIC, MANX) cucowr; (GAELIC, SCOTTISH) cularan; (GALICIAN) colombro; (GEORGIAN) kitri; (GERMAN) Salat-Gurke, Schlangengurke, Schälgurke, Gurke; (GOLA) kipɔsi; (GOURMANTCHÉ) bompiène-kana; (GREEK) angouri; (GUJARATI) khira, kakadi; (HANUNOO) katimun diriit, katimun mariruk, kunduy, kunduy palatay; (HAUSA) kokwamba; (HEBREW) melafefon, kishu ha-ginna; (HERERO) ekungu; (HILIGAYNON) pipínu; (HINDI) kheera, kheeraa, khira, kakrī; (HINDUSTANI) khírá, kakṛí, khiyár; (HMONG-DAW) dib; (HMONG-NJUA) dlib; (HUNGARIAN) uborka; (IBALOI) kasimun; (IBANAG) pepino; (ICELANDIC) gurka, agurka; (INDONESIAN) bonteng ketimun, ketimun, katimun, timun, benteng; mentimun, entimun, tetimun; (INUKTITUT, WEST GREENLANDIC) inalualiussaq; (ITALIAN) cetriolo, cetriolino, cocomero; (ITAWIT) capinu; (JAPANESE) kyuuri, ki uri; (JAVANESE) timun, katimun, ketimun; (JOLA-FONYI) kouconcombre; (JULA) concombrou; (KANKANAEY) kasimun; (KANNADA) savathe layi, sauthe kaayi, maghe kaayi, moge kaayi, sauthe balli; (KANURI) ngùrlí; (KASHMIRI) lörü; (KONKANI) kankri, thause; (KAZAKH) ogurec, qiyar; (KHMER, CENTRAL) trâsâk; (KIOWA) sɑhyei-'ei-goup; (KIRGHIZ) badyran, badɪrang; (KISSI, NORTHERN) kumba; (KOREAN) oi; (KRIO) kɔkumba; (KURDISH, NORTHERN) ārū, k̲h̲ēyār, kālyār, chikek; (KWÉYÒL) konkonm, kòkònm; (LADINO) pepino, pipino, hohómbro; (LAHU) a¯ hpehᵛ shi˷; (LAO) tèèng; (LATVIAN) gurķi; (LITHUANIAN) paprastasis agurkas; (LUXENBOURGEOIS) Kornischong; (MACEDONIAN) krastavica, obična krastavica; (MAITHILI) kẖīrā, kākṛī, sohās; (MALAY) timun; (MALAYALAM) vellarikka, kakkari, mullen belleri; (MALIMBA) kɔkumba; (MALTESE) hjar; (MARANAO) bulutukun a litu, pipino sikwa; (MARATHI) kakdi, khira, thavase, kaakdi, kheera; (MARWARI) kākaḍi, kāchro; (MENDE) kikpɔsi, kɔkumba; (MICMAC) ĕskŭdŭmŭgāwā'; (MIZO) fanghma; (MOHAWK) kanon'onhserakeras; (MÒORÉ) cõcõbr; (MORISYEN) konkom; (MUNDARI) jhingā, tāyār; (MUSKOGEE) kakompv, takompv; (NAVAJO) da'neesk'ání dik'ózhígíí; (NDALI) makaka; (NEPALI) kankro; (NIUE) kukama; (NORWEGIAN) agurk, ogurke, asieagurk, drueagurk; (NUPE-NUPE-TAKO) àgbiàdià, gbiàja; (NYANJA) mankhaka; (ORIYA) kakudi; (PANGASINAN) pepíno; (PANJABI, EASTERN) kheera; (PASHTO, NORTHERN) bâdrang, bādrang, tara, tarai; (POLISH) ogórek, ogórek siewny; (PORTUGUESE) pepino, pepino de salada; (POTAWATOMI) kokobe'; (QUECHUA, CUSCO) worachia; (ROMANIAN) castravete, castravetele; (RUSSIAN) ogurec, ogurets posevnoi, ogurets dlinno plodny, ogurets korotka; (SAMBAL, TINÀ) pipino; (SAMOAN) kukama; (SANSKRIT) thrapusha, bahuphala, kandalu, kantakilatha, kantakiphala, kantalu, koshaphala, peethapushpa, sudhavaasa, susheethala, thrapukarkati, thundilaphala; (SANTALI) tahe̱r; (SERBO-CROATIAN) krastavac; (SHONA) gaka; (SINHALA) pipn̄ña; (SLOVAK) uhorka; (SLOVENIAN) kumara; (SOMALI) ẖiyar; (SORBIAN, UPPER) kórka; (SOTHO, NORTHERN) phara; (SOTHO, SOUTHERN) komokomore; (SPANISH) pepino, cohombro; (STELLINGWERFS) komkommer, augurk, augurkien; (SUNDA) katimun, timun, bonteng; (SWAHILI) tango, mbilimbi, matango; (SWEDISH) gurka; (TAGALOG) pepino, pipino, tabogok; (TAJIKI) bodiring; (TAMIL) vellarikai, vellari, vellarikkai, kakrikai, kakrikaai, khoorapelum, muhivetti, mulluvellari, pipingkaai; (TATAR) kɪyar; (TAUSUG) maras, madas; (TELUGU) dosa, dosakaya, dosekaaya, ujakaayipa; (THAI) taeng-kwa, taeng-ran, taeng-om; (THEMNE) *ma*-kɔkumba; (TIBETAN) qānra; (TIV) ingboko; (TOK PISIN) kukamba; (TSWANA) phare; (TULU) tekkarè, thekkare; (TUMBUKA) uwimbi; (TURKISH) salatalik, hiyar; (TURKMEN) adaty hyýar, hyÿar; (UKRAINIAN) ogirok; (URDU) khira; (UYGHUR) tärkhämäk; (UZBEK) bodring; (VIETNAMESE) d[uw]a chu[ooj]t; (VLAAMS) komkommer, augurk; (WÁRAY-WÁRAY) pepino; (WELSH) ciwcymber; (WOLOF) concombre; (XHOSA) ikomkomire, inkonkomire; (YAO) mangaka, masakasa; (YIDDISH) uhgerkeh; (ZULU) ikhukhamba, khukhamba.

b. very small, immature fruit

(BURMESE) thakhwa; (CEBUANO) kalabaga; (DANISH) agurk; (DUTCH) augurk; (ENGLISH) pickling cucumber, gerkin; (FILIPINO) gerkin; (FRENCH) cornichon, cornichon au vinaigre; (GERMAN) Essiggurke, Gewürzgurke, Einlegegurke; (GREEK) koukoumic, edodimos; (HUNGARIAN) uborka; (ICELANDIC) prugugurka; (INDONESIAN) mentimun; (ITALIAN) cetriolino sotto acete; (JAVANESE) mentimun; (KHMER, CENTRAL) trâsâk; (LADINO) pepiniko, pipinko; (LAO) tèèng; (MALAY) mentimun; (MORISYEN) kornichon; (NEPALI) kankro; (POLISH) ogórek; (PORTUGUESE) pepino para conserva, cornichon; (SPANISH) pepinillo en vinagre; (SUNDA) bonteng; (TAGALOG) pipino; (TOK PISIN) kuikamba; (VIETNAMESE) d[uw]a leo.

38.14. *Cucurbita argyrosperma* C. Huber. (EP) young/mature fruit, and seeds[S]; (P) C,R (seed)

Dialects of Luyia (Bukusu[1], Itsukha[2])

(AFAR) dubba; (AFRIKAANS) pampoen; (AMHARIC) dubba; (APACHE, JICARILLA) náayizee, náayiizéé; (ARABIC) qara sudani; (ASSAMESE) ronga lao, squash; (AZERBAIJANI,

NORTHERN) balqabaq; (BADAGA) kumbaḷa, kumbua, kumba:ke; (BALTI) wan; (BAMILEKE) mfoubou; (BEMBA) cipushi; (BENGALI) saphuril kumra; (BLACKFOOT) sáóppiso; (BODO) jøuat, kahklou; (BOSNIAN) kaša, vošni sok, tikva, bundeva; (BULU) abok; (BURMESE) shwe-pha-yone; (CAJUN FRENCH) juremon; (CAROLINIAN) galamasa, ghalamaasa, ghalabwaasa; (CATALAN-VALENCIAN-BALEAR) carbassa, carabassa; (CHEROKEE) iya; (CHICKASAW) osto, olbi', olbi' lakna', olbi' tohbi, ostahi, olbi' okchamali'; (CHINESE, GÀN) sun gua, bei gua, fan gua; (CHINESE, HAKKA) vông-p'oû, lăc-tsón; (CHINESE, MANDARIN) yü-kwa, mo xi ge nan gua; (CHINESE, MIN BEI) fan gua, jin gua; (CHINESE, MIN NAN) kim-koe; (CHINESE, WÚ) mo xi ge nan gua; (COMANCHE) nakwu̲siʔ, maki̲tsaru, tsaki̲tsetu; (CREOLE, HAITIAN FRENCH) joumou, jiromon, jiwomon; (DAGAARE, SOUTHERN) yor; (DAKOTA) wamnu, wakmu, wagmu; (DANISH) ayote, centnergraeskar; (DUTCH) reuzenkalebas; (ENGLISH) pumpkin, cushaw pumpkin, winter squash; (ESTONIAN) kõrvits; (ÉWÉ) agushi; (FARSI, EASTERN) kaddu; (FARSI, WESTERN) kadu tanbel; (FIJIAN) khadu; (FINNISH) talvisquash; (FRENCH) potiron; (FULFULDE) bombo, waijoré, waygodjé; (FULFULDE, ADAMAWA) foly̲ere, waigōre, baigōje; (GA) skibonte; (GAELIC, MANX) pumpkin; (GBAYA) saa; (GEORGIAN) magarkaniani gogra, khokera; (GERMAN) Riesen-Kürbis; (GOURMANCHÉMA) gan; (GREEK) chimerino kolokynthi; (GUJARATI) pilum kohalum; (HARARI) dubba; (HAUSA) goji, kubewa, kabewa; (HEBREW) dela'at gedola; (HERERO) otjimbakuna; (HINDI) vilayati kaddu, pila kohda, kashiphal, mithakadhu, lāukī; (HINDUSTANI) laddú, laukí, ko̲nhṛá; (HOPI) patnga; (ICELANDIC) vetrargrasler; (IGBO) ụgbọgwụ, ụgụ; (ITALIAN) zucca gigante; (IZON) ondú; (JAPANESE) seiyou-kabocha, kabocha; (KANNADA) gumbala; (KASHMIRI) al, ala-bạrag; (KATI) ālo; (KIRIBATI) te bambin, te bangke, te baukin; (KRIO) pɔ̃kin, pɔnkin; (KWÉYÒL) jòmou, jamou; (LADINO) kalavása, kalavasita; (LATVIAN) lieaugļu ķirbis, ķirbis; (LUO) okono; (LUYIA) lihondo^{2}, liondo^{1}; (MAITHILI) kadīmā; (MAORI) paukena; (MARATHI) lal bhopla; (MICMAC) ĕchkoochkool, ĕchkooaakw; (MIZO) um, mai; (MOHAWK) onon'onhsera; (MÒORÉ) yogre; (MUNDARI) kaduā, kākṛu; (MUSKOGEE) rvfo-tvhvyv, cvse, tvhvyv; (NAVAJO) naayízí łichí'í, naayízí; (NEPALI) pharsi; (NEWAR) jawo, jawo-si, jau-se, phat-si, māku; (NIUE) mōtini; (NUPE-NUPE-TAKO) èbĕ; (OROMO, WEST CENTRAL) bukke; (PALI) alābu, lābu, alāpu, lāpu; (PASHTO, NORTHERN) kadu; (POLISH) dynia pośrednia; (PORTUGUESE) abóbora menina; (POTAWATOMI) ogwîssimaû'n, winbati; (PULAAR) wuuduru; (ROMANI, BALKAN) doodòom; (RUSSIAN) tykva krupnoplodnaja; (SANSKRIT) dadhiphala; (SERER-SINE) o'limb, féet; (SHONA) mbambuni, mbambunu, mbomboni, nhanga, budzi, dikiti, gavu, nyungu; (SIDAMO) bâqula, qônco; (SOMALI) dubbai; (SORBIAN, LOWER) banja; (SORBIAN, UPPER) banja, kirbs; (SOTHO, NORTHERN) lefoli; (SOTHO, SOUTHERN) mokopu; (SPANISH) zapallo; (SWAHILI) mboga, boga; (TAGALOG) kalabasa; (TAGBANWA) kalabasa; (TAMIL) poashani; (TELUGU) gummadi kayi; (TIV) a̲gba̲du̲; (TSWANA) lephutshe; (TULU) kumbuḍa; (TURKISH) bal kabaği, misk kabaği; (TWI) ɛfere; (URDU) villaty kaddu; (VENDA) fhuri, furi, lufhuri, mudzungu, dzungu; (VLAAMS) pompoen; (WÁRAY-WÁRAY) karubasa; (XHOSA) ithanga, iphuzi; (YIDDISH) kirbes, dinee-eh, kahbahk; (ZULU) ithanga, i(li) thanga; (ZUNI) mo:teyalha.

38.15. *Cucurbita ficifolia* Bouché (EP) fruits; (P) C.

(CHINESE, MANDARIN) hei zi nan gua; (CHINESE, WÚ) hei zi nan gua; (CZECH) tykev fíkolistá; (DAGAARE, SOUTHERN) yor; (DANISH) figenblad-græskar; (ENGLISH) Malabar gourd, fig-leaf gourd, black-seeded squash; (FIJIAN) laki; (FILIPINO) Malabar gourd; (FRENCH) gourge de Malabar, courge à feuilles de figuier, courge de Siam, courge à graines noires, melon de Malabar; (GERMAN) Feigenblattkürbis; (GIKUYU) kahũrũra; (GONJA) bulji, kingama; (GREEK) kolokythito fiko-phyllo; (INDONESIAN) waluh, labumerah; (ITALIAN) cocòmero del Malabar, zucca del Malabar; (JAPANESE) kurodane kabocha; (MACEDONIAN) Malabar tikbička; (POLISH) dynia figolistna; (PORTUGUESE) abóbora chila, abóbora gila, gila; (SPANISH) chilacayote, cidra cayote, chirigaita; (TURKISH) siyam kabaği; (VLAAMS) reuzenkalebas, kurbis; (WOLOF) xal u mbot; (ZARMA) laptanda.

38.16. *Cucurbita maxima* Duchesne (EP) mature fruit, and seeds[S]; (P) C.

(AFAR) dubba; (AFRIKAANS) pampoen; (AGTA) kalabisa; (AKAN-AKUAPEM) ɛfrɛ; (AKAN-ASANTE) ɛfere, ɛfre; (AKAN-FANTE) ɛfir; (ALBANIAN, TOSK) shpurdhë; (AMHARIC) dubba; (APACHE, JICARILLA) náayizee, náayiizéé; (ARABIC) qara sudani, arä, garä; (ASSAMESE) squash, ronga lao, kumura; (AZERBAIJANI, NORTHERN) balqabaq; (BADAGA) kumbaḷa, kumbua, kumba:ke; (BAFIA) mezeng, nbamg; (BALTI) wan; (BAMANANKAN) ié; (BAMILEKE) mfoubou; (BASSARI) kaere, katekatego; (BELARUSAN) garbus vjalizki; (BEMBA) cipushi; (BENGALI) kumra, saphuril kumra; (BEROM) bátìrí, ebɛ́; (BINUKID) bebasal; (BLACKFOOT) sáóppiso; (BODO) jøuat, kahklou; (BONTOC CENTRAL) kumbasa, cumbasa, kalobas, kalobasa; (BOSNIAN) tikva, bundeva, kaša, vošni sok; (BUHID) badu, kawbasa; (BULGARIAN) kestenka; (BULU) abok; (CAJUN FRENCH) juremon; (CAROLINIAN) ghalamaasa, galamasa, ghalabwaasa; (CATALAN-VALENCIAN-BALEAR) carabassa rodona, carbassa, carabassa; (CEBUANO) kalabasa, kalabasang-pula; (CHEROKEE) iya; (CHICKASAW) osto, olbi', olbi' lakna', olbi' tohbi, ostahi, olbi' okchamali'; (CHINESE, HAKKA) vông-p'oû, lăc-tsón; (CHINESE, MANDARIN) yin du nan gua, yü kwa; (CHINESE, MIN NAN) kim-koe; (CHINESE, WÚ) yin du nan gua; (COMANCHE) nakwu̱siʔ, maki̱tsaru, tsaki̱tsetu; (CREOLE, HAITIAN FRENCH) joumou, jiromon, jiwomon; (CRIOULO, CAPE VERDE) bobra; (CROATIAN) patison; (CZECH) tykev velkoplodá; (DAGAARE, SOUTHERN) yor; (DAGBANI) ɣoɣli, kaere, kilbundo; (DAKOTA) wamnu, wakmu, wagmu; (DANISH) centnergraeskar; (DIGO) mrenje; (DUTCH) reuzenkalebas, reuzenpompoen; (EDE NAGO) elegbe; (EDO) éyèn; (EFIK) ḿfrì ǹdisè, ḿfrì, ǹdìsè, ǹnàńí; (ENGLISH) winter squash, pumpkin, giant pumpkin, melon pumpkin, red gourd, squash gourd; (ESTONIAN) suureviljalist; (ÉWÉ) agushi; (EWONDO) abok, mendzen, mindzin-mebok, ndzen; (FARSI, EASTERN) kaddu; (FARSI, WESTERN) kadu tanbel; (FIJIAN) khadu; (FILIPINO) kalabasa; (FINNISH) jättikurpitsa, jättiläis kurpitsa; (FON) ayikpɛn; (FRENCH) potiron, grosse courge, giraumon; (FULFULDE) bombo, waijoré, waygodjé; (FULFULDE, ADAMAWA) folyere, waigōre, baigōje, feraare, fol'yere, waygoore; (GA) sakribonte; (GAELIC, MANX) pumpkin; (GAELIC, SCOTTISH) brùth; (GBAYA) saa; (GEORGIAN) didnakopha gogra; (GERMAN) Riesen-kürbis, Zentnerkürbisse; (GIKUYU) irenge; (GONJA) kawurenchu; (GOURMANCHÉMA) gan; (GOURMANTCHÉ) liferi; (GREEK) kolokythi, kolokynthi; (GUJARATI) kohla, pilum kohalum; (GUN) okpɛ; (HANUNOO) labasa, tabagwang; (HARARI) dubba; (HAUSA) goji, kubewa, kabewa, akwato, bàkánùwaá, gámmòn, fátaàkeé, ganwon fatake, gesuma, rugudu, ruguguwa, waáwán-goónaá, ƴar garii, goojíí, kàbaíwaà, kúbeèwaá, kankana, balayal, goòjíí, kabus, kabushi; (HEBREW) dela'at, dela'at haginna, delaat gedola; (HERERO) otjimbakuna; (HINDI) vilayati kaddu, kaddu, lal kumra, mitha kumra, kaddv, kadimah, sitaphal, pila kohda, kashiphal, mithakadhu, lāukī, laal kaddu, meetta kaddu; (HOPI) patnga,; (HUNGARIAN) sütótök, óriástök; (IBALOI) kalumbas; (IBANAG) kalavasa; (ICELANDIC) risagrasker; (IFUGAO) kaluhaba; (IGBO) ụgbọgwụ, ụgụ, ukọrọ, ụ́gbọ̀gụ̀lụ̀, ányụ̀, ákwụ̀kwọ́, ụ́gbọ̀ghọ̀rọ̀, ụ́gbọ̀ghụ̀rụ̀; (ILOCANO) karabasa; (INDONESIAN) labumerah; (ITALIAN) zucca gigante; (IZON) ondú; (JAPANESE) seiyou, seiyou- kabocha; (KANNADA) kumbalakai, seegumbala, kumbala kayi, kumbala kaayi, gumbala, chini kaayi, govekaayi, sakkare kumbala; (KANURI) kàfétò, sàádə̀; (KASHMIRI) al, ala-bạrag; (KATI) ālo; (KAZAKH) tykva krupnoplodnaja; (KIRIBATI) te bambin, te bangke, te baukin; (KISSI, NORTHERN) njalin; (KONKANI) duddi, duddini; (KONO) g-bi; (KOREAN) tokhobak; (KRIO) pɔnkin; (KWÉYÒL) jamou; (LADINO) kalavása, kalavasita; (LATVIAN) lieaugḷu ķirbis, ķirbis, lielauglu kirbji; (LITHUANIAN) didysis moliūgas; (LOMWE, MALAWI) maungu, msuku; (LUO) okono; (MACEDONIAN) tikva pecenka, zimska bela tikva , tikvička, obična bela tikva; (MAITHILI) kadīmā; (MALAGASY) pongy; (MALAY) labu manis; (MALAYALAM) mathanga, mattanga, schakeri schora; (MALIMBA) sendjap; (MALTESE) qargha hamra; (MANINKAKAN, EASTERN) guié; (MAORI) paukena; (MARANAO) bebasal, labogandang, sinambong; (MARATHI) kashi bhopla, lal bhopla, bhomphala; (MENDE) towa, towɛi; (MICMAC) ĕchkoochkool, ĕchkooaakw; (MIZO) um, mai; (MOHAWK) onon'onhsera; (MÒORÉ) yogre; (MORISYEN) ziromon; (MUNDARI) kaduā, kākṛu; (MUSKOGEE) rvfo-tvhvyv, cvse, tvhvyv; (NAVAJO) naayízí łichí'í; (NDALI) tange; (NEPALI) pharsi; (NEWAR) jawo, jawo-si, jau-se, phat-si, māku; (NIUE) mōtini; (NORWEGIAN) vintergresskar, pumpkin, gresskar, kjempegresskar; (NUPE-NUPE-TAKO) èbĕ; (NYAKYUSA-NGONDE) mungu; (NYANJA) dzungu; (ORIYA) balati kakharu; (OROMO, WEST CENTRAL) bukke; (PALI) alābu, lābu, alāpu, lāpu; (PAMPANGAN) kulabasa, kulubasa; (PANGASINAN) kalubasa; (PANJABI, EASTERN) halwa kadu; (PASHTO, NORTHERN) kadu; (POLISH) dynia olbrzymis; (PORTUGUESE) abobora menina, girimú, moranga, moganga; (POTAWATOMI) ogwîssimaû'n, winbati; (PULAAR) wuuduru; (QUECHUA, CUSCO) ayunca; (ROMANI, BALKAN) doodòom; (ROMANIAN) dovleac, dovleac alb; (RUSSIAN) tykva krupnoplodnaja; (SAMBAL, BOTOLAN) kalubaha; (SANSKRIT) dadhiphala, gudayogaphala, dangari, gramya, kooshmaanda, peethakooshmaanda, peethaphala, peethapushpa; (SERBO-CROATIAN) bundeva pečenka, vela tikva; (SERER-SINE) nadio, a dyeng; (SHONA) mbambuni, mbambunu, mbomboni, nhanga, budzi, dikiti, gavu, nyungu; (SIDAMO) bâqula, qônco; (SINHALA) wattakka; (SLOVAK) tekvica velká; (SLOVENIAN) jedilna buča; (SOMALI) dubbai; (SONGHAY) bâg, banga, dombos, nadé, nađé, yomba; (SORBIAN, LOWER) banja; (SORBIAN, UPPER) banja, kirbs; (SPANISH) calabaza grande, calabaza de Castilla, zapallo; (SWAHILI) tikiti, mboga, boga; (SWEDISH) jättepumpa, pumpa; (TAGALOG) kalabasa, kalabsa, bado, kalabasang-bilog, kalabasang -pula, kalubasa; (TAGBANWA) kalabasa, basuk; (TAJIKI) gahvorakadu; (TAMIL) parangikayi, poosani, arasanikaai, pooshini, parangikaai, sirmai pussinkody; (TAUSUG) kabasi, kalabasi; (TELUGU) gummadi, erra gummadi, alugadda, seemagummadi, poashani, gummadi kaaya, thiyya gummadi; (TEM) kaere, kodshŏdo; (THEMNE) *ma*-ali, a-ŋali, *a*-kali; (TIV) àgbàdù, a̱gba̱du̱, furum; (TONGA) tanje; (TULU) kumbuḍa, kembuda, seepe kumbuda; (TUMBUKA) jangu; (TURKISH) meycaci kabaği, su kabaği, bal kabaği, kiş kabaği, helvaci kabagi; (TURKMEN) haýwabjädi; (TWI) ɛfere; (UKRAINIAN) garbuz velikoplìdnij; (URDU) villaty kaddu; (UZBEK) ošķovok; (VENDA) fhuri, furi, lufhuri, mudzungu, dzungu; (VLAAMS) reuzekalebas, kurbis, pompoen, ronde pompoen; (WAAMA) yambi; (WÁRAY-WÁRAY) karubasa; (WOLOF) yomb, nadio, laket; (XHOSA) ithanga, iphuzi; (YAO) msuku; (YIDDISH) kahbahk, kirbes, dinee-eh; (YORUBA) elegbe, elegede, apala, élégédé; (ZULU) i(li)thanga; (ZUNI) mo:teyalha.

38.17. *Cucurbita moschata* Duchesne (EP) young and mature fruit; (P) C.

(AFAR) dubba; (AFRIKAANS) selonspampoen; (AGTA) kalabisa; (APACHE, JICARILLA) náayizee, náayiizéé; (ARABIC) qara sudani; (ASSAMESE) squash; (AZERBAIJANI, NORTHERN) balqabaq; (BADAGA) kumbaḷa, kumbua, kumba:ke; (BAFIA) mezeng, nbamg; (BALTI) wan; (BAMILEKE) mfoubou; (BEMBA) cipushi; (BLACKFOOT) sáóppiso; (BODO) jøuat, kahklou; (BONTOC, CENTRAL) cumbasa, kalobas, kalobasa, kumbasa; (BOSNIAN) kaša, vošni sok, tikva, bundeva; (BUHID) badu; (BULU) abok, ndzeng abog; (BUTUANON) kawbasa; (CAJUN FRENCH) juremon; (CAROLINIAN) ghalamaasa, galamasa, ghalabwaasa; (CEBUANO) kalabasang-pula; (CHEROKEE) iya; (CHICKASAW) olbi', olbi' lakna', olbi' tohbi, ostahi, olbi' okchamali'; (CHINESE, MANDARIN) zhong guo; (CHINESE, WÚ) zhong guo nan gua, jin gua, wo gua, pan gua; (CHINESE, XIANG) nan qua; (CHINESE, YUE) nan gua, pan gua; (CATALAN-VALENCIAN-BALEAR) carbassa, carabassa; (COMANCHE) nakwu̱si?, maki̱tsaru, tsaki̱tsetu; (CRIOULO, CAPE VERDE) abóbora de sequeiro porco; (CROATIAN) bundeva; (CZECH) tykev pižmová; (DAGAARE, SOUTHERN) yor; (DAKOTA) wamnu, wakmu, wagmu; (DANISH) centnergræskar, moskusgræskar; (DUTCH) muskaatpompoen; (ENGLISH) musky pumpkin, musky squash, musky gourd, butternut squash, winter squash; (ÉWÉ) agushi; (EWONDO) abok, mendzen, mindzin-mebok, ndzen; (FARSI, EASTERN) kaddu; (FARSI, WESTERN) kadu tanbel; (FILIPINO) kalabasa;

(FRENCH) courge muscarde, courge musquée, giraumon; (FULFULDE) bombo, waijoré, waygodjé; (FULFULDE, ADAMAWA) folẏere, waigōre, baigōje; (GA) skibonte; (GAELIC, SCOTTISH) brùth; (GBAYA) saa; (GEORGIAN) muskaturi gogra, taphla gogra; (GERMAN) Moschuskürbis, Bisamkürbis; (GONJA) bulji, kingama; (GOURMANCHÉMA) gan; (GOURMANTCHÉ) afera; (GREEK) kolokynthito moschosmo; (HANUNOO) labasa, tabagway; (HARARI) dubba; (HAUSA) goji, kubewa, kabewa; (HINDI) vilayati kaddu, lal kumra, mitha kumra, kadimah, sitaphal, lāukī; (HINDUSTANI) konhra; (HMONG DAW) taub dag, taub; (HOPI) patnga; (HUNGARIAN) pézsmatök; (IBALOI) kalumbas, kalumbasa; (IBANAG) kalayasa; (IFUGAO) kaluhaba; (IGBO) ụgụ; (ILOCANO) karabasa; (INDONESIAN) labu merah, labu ambon, labu parang, waluh, luba, labu; (ISNAG) pinino; (ITALIAN) zucca moscada, zucca muschiata, zucca torta; (IVATAN) kabasa; (JAPANESE) nihon kabocha, kabocha; (JAVANESE) waluh, waloh; (KALINGA, BUTBUT) atong; (KANNADA) kandugumbala; (KATI) ālo; (KIRIBATI) te bambin, te bangke, te baukin; (KRIO) pɔnkin; (LADINO) kalavása, kalavasita; (LATVIAN) parastais ķirbis; (LITHUANIAN) muskusinis moliūgas; (LUO) okono; (MACEDONIAN) zimska muskatna tikya, obična muskatna; (MAITHILI) kadīmā; (MALAY) labu merah; (MALIMBA) sendjap; (MAORI) paukena; (MARANAO) bebasal, labogandang, sinambong; (MICMAC) ĕchkoochkool, ĕchkooaakw; (MIZO) um, mai; (MOHAWK) onon'onhsera; (MÒORÉ) yoɣole; (MUNDARI) kaduā, kākṛu; (MUSKOGEE) tvhvyv, cvse, rvfo-tvhvyv; (NAVAJO) naayízí; (NEWAR) jawo, jawo-si, jau-se, phat-si, māku; (NIUE) mōtini; (NORWEGIAN) vintergresskar, muskatgresskar; (NUPE-NUPE-TAKO) èbĕ; (OROMO, WEST CENTRAL) bukke; (PALI) alābu, lābu, alāpu, lāpu; (PAMPANGAN) kulubasa; (PANGASINAN) kalubasa; (PANJABI, EASTERN) halwa kaddoo, chapanh kadoo; (PASHTO, NORTHERN) kadu; (POLISH) dynia piżmowa; (PORTUGUESE) abóbora almiscarada, abóbora da Guiné, abóbora moscata, abóbora preta; (POTAWATOMI) ogwîssimaû'n, winbati; (PULAAR) wuuduru; (ROMANI, BALKAN) doodòom; (ROMANIAN) ludăi, bostan moscat; (RUSSIAN) tykva krupnoplodnaja; (SAMBAL, BOTOLAN) kalubaha; (SANTALI) kohṇḍa; (SERBO-CROATIAN) borović; (SHONA) mbambuni, mbambunu, mbomboni, nhanga, budzi, dikiti, gavu, nyungu; (SIDAMO) bâqula, qônco; (SINHALA) batana; (SOMALI) dubbai; (SONGHAY) mara kachia; (SORBIAN, LOWER) banja; (SORBIAN, UPPER) banja, kirbs; (SPANISH) calabaza moscada, calabaza almiscarada, calabaza almisclada, calabaza amelonada; (SUNDA) waluh; (SWAHILI) tikiti, mboga, boga, ponda, seta, songa; (TAGALOG) kalabasa, bado, kalabasang-bilog, kalabasang-pula, kalubasa; (TAGBANWA) kalabasa; (TAJIKI) palavkadu; (TAUSUG) kabasi, kalabasi; (TELUGU) potti gummadi; (THEMNE) ma-ali; (TIV) a̱gba̱du̱; (TULU) kumbuḍa; (TURKISH) balkabaği; misk kabaği; (TURKMEN) palowkädi; (TWI) ɛfere; (URDU) data, safad kaddu sitaphal; (VENDA) fhuri, furi, lufhuri, mudzungu, dzungu; (VLAAMS) reuzekalebas, pompoen; (WOLOF) yomb, nadio; (XHOSA) ithanga, iphuzi; (YIDDISH) kahbahk; (ZARMA) laptanda; (ZULU) i(li)thanga.

38.18. *Cucurbita pepo* L. (EP) young and mature fruit; (P) C,R.

(AFAR) dubba; (AFRIKAANS) babamurgskorsie, skorsie, murgpapoen, murg-vangroente; (AKAN-AKUAPEM) akomfem-fere, ɛfere-pa, ɛfrɛ, fere-fita; (AKAN-ASANTE) ɛfrɛ; (AKAN-FANTE) akamfir, ɛfir, ɛfrɛ; (ALBANIAN, TOSK) kungull; (AMHARIC) dubba; (APACHE, JICARILLA) náayizee, náayiizéé; (ARABIC) garra, qara sudani, arä, garä, garä kossa, garä stambuli; (ASSAMESE) squash, ronga lao; (AZERBAIJANI, NORTHERN) balqabaq; (BADAGA) kumba:ke, kumbaḷa, kumbua; (BALTI) wan; (BAMANANKAN) ié; (BAMILEKE) mfoubou; (BASQUE) kalabazin; (BASSARI) ɔ-syẹlis, a-tyẹlis, ɛ-tyẹlis, kaere, katekatego; (BELARUSAN) garbus zvichainy, garbuz; (BEMBA) cipushi; (BEMBÉ) bibeté; (BENGALI) saphuril kumra; (BEROM) yɛ̀ɛ́; (BLACKFOOT) sáóppiso; (BODO) skuash, jøuat, kahklou; (BOSNIAN) kaša, vošni sok, tikva, bundeva; (BRETON) koulourdrenn; (BUDIK) i-dìm, mi-rìm; (BULGARIAN) obiknovena tikva; (BULLOM SO) lam, lamgo, lam-dɛ; (BULU) abok, ngon; (CAJUN FRENCH) si brèam, juremon; (CAROLINIAN) ghalamaasa, galamasa, ghalabwaasa; (CATALAN-VALENCIAN-BALEAR) carabassa, carbassa;

(CHEROKEE) iya; (CHINESE, GÀN) mei zhou nan gua; (CHINESE, MANDARIN) meiguo nan gua, nan kua; (CHINESE, MIN BEI) ba jiao gua; (CHINESE, WÚ) mei guo nan gua, xi hulu, mei zhou nan gua; (CHINESE, YUE) xi hu lu, mei zhou nan gua; (COMANCHE) nakwu̱si?, maki̱tsaru, tsaki̱tsetu; (CREOLE, HAITIAN FRENCH) jiromon, joumou, jiwomon; (CRIOULO, CAPE VERDE) bobrinha; (CROATIAN) tikvica; (CZECH) tykev obecná, tykev turek, dýně, tykev; (DAGAARE, SOUTHERN) yor; (DAGBANI) kaere, kilbundo; (DAKOTA) wamnu, wakmu, wagmu; (DANISH) mandelgræskar, graeskar, centnergraeskar; (DUTCH) courgette, zucchetti, kalebas, patisson; (EDE NAGO) elegbe; (EDO) éyèn; (EFIK) ḿfrì ǹdìsè, ḿfrì, ǹdìsè, ǹnàńí; (ENGLISH) summer squash, courgette, zucchini, marrow, pumpkin, common field pumpkin; (ESTONIAN) kõrvits; (ÉWÉ) agushi; (FARSI, EASTERN) kaddu; (FARSI, WESTERN) kadu tanbel; (FIJIAN) khadu; (FILIPINO) kalabasa; (FINNISH) kesäkurpitsa; (FON) ayikpɛn; (FRENCH) courgette, courge pépon, potiron; (FULFULDE) bombo, waijoré, waygodjé; (FULFULDE, ADAMAWA) folyere, waigōre, baigōje, fol'yere, kucceere; (GA) sakribonte, skibonte; (GAELIC, MANX) pumpkin; (GALICIAN) cabaceira; (GBAYA) saa; (GEORGIAN) kabaki, patissoni; (GERMAN) Garten-Kürbis, Zucchini, Chiromon-Kürbis, Riesen-Kürbis; (GOLA) die; (GOURMANCHÉMA) pikampob, gan; (GOURMANTCHÉ) bompiène-nouara; (GREEK) kolkythaki, kolokythi; (GUJARATI) pilum kohalum; (GUN) okpɛ; (HARARI) dubba; (HAUSA) goji, kubewa, kabewa, akwato, bàkánúwaà, balaya, gamon fatake, ganwon fatake, gesuma, goòjíí, kabus, kabushi, rugudu, ruguguwa, wawan gona, garii; (HEBREW) dela'at ha-sade, delaat hasade; (HILIGAYNON) kalabása; (HINDI) chappan kaddu, kadimah, kumra, safedkaddu, vilayati kaddu, pila kohda, kashiphal, mithakadhu, lāukī; konda, kumra, safedkaddu; (HMONG DAW) taub, taub dag; (HOPI) patnga; (HUNGARIAN) tök, spárgatök, közönséges tök, főzeléktök, úritök, tojás alakú főzeléktök; (ICELANDIC) kurbitur, dvergbitur, grasker; (IGBO) ụgbọgwụ, ụgụ, ukọrọ, ụ́gbọ̀gụ̀lụ̀, ányụ̄, ụ́gbọghụ̀lụ̀, ụ́gbọ̀gọ̀rọ̀; (INDONESIAN) luba, pucuk labu, labu merah; (ITALIAN) zucca, zucchini, zucchino, zuccheti, zuccheta, zucca gigante; (IZON) udẹmụdẹ́, utí; (JAPANESE) pepo-kabocha, seiyou-kabocha; (KANNADA) gumbala, boodugumbala; (KANURI) fámfam, fánnâ díwì, kàfétò, sàádə̀; (KASHMIRI) al, ala-bạrag; (KATI) ālo; (KAZAKH) tykva tverdokoraja; (KIRGHIZ) aškabak; (KIRIBATI) te bambin, te bangke, te baukin; (KISSI, NORTHERN) chalẽ; (KONKANI) bobrini, bobro, concondudi, kunvale; (KONKOMBA) dedschaelbɔne; (KONO) g-bi, g-bili, g-biya, korankɔ-c-bi; (KOREAN) anzunhobak; (KRIO) skwash, pɔnkin, põkin; (KWÉYÒL) jòmou; (LADINO) kalavása, kalavasita; (LAHU) hpeu$^{\vee}$ meu^{-} shi$_{-}$, hpui$^{\vee}$ mui^{-} shi$_{-}$; (LATVIAN) parastais ķirbis, lieaugļu ķirbis, ķirbis; (LITHUANIAN) paprastasis moliūgas; (LOKO) kuwe, tea, yeŋgelesi; (LOMWE, MALAWI) maungu; (LUO) okono; (MACEDONIAN) tikvički, tikva letna obična tikvička, zimska obična, obična tikva; (MAITHILI) kadīmā; (MALAGASY) tsirebika; (MALAY) labu air; (MALAYALAM) kumbalam; (MALIMBA) kutu; (MALTESE) gurgett, qara hamra; (MANDING-MANDINKA) miraŋ, gbili, jee, ye; (MANINKAKAN, EASTERN) guié, g-bili, gbulɛ ye yee, ye-gbɛ, ye-kumba, ye-nyɛ̃; (MANKANYA) umbôgre; (MAORI) paukena; (MARATHI) lal bhopla, kohala; (MENDE) blamange, koja-lowa, sɛgbula-lowa, towa, towa-wawa; (MICMAC) ĕchkoochkool, ĕchkooaakw; (MIZO) um, mai; (MOHAWK) onon'onhsera; (MÒORÉ) yogré; (MORISYEN) patisson, kourzet; (MUNDARI) kaduā, kākṛu; (MUSKOGEE) tvhvyv, cvse; (NAVAJO) naayízí, naayízí łichí'í; (NEPALI) pharsi; (NEWAR) jawo, jawo-si, jau-se, phat-si, māku; (NGIZIM) gàaɓíyǎw; (NGUNDI) bibeté; (NIUE) mōtini; (NORWEGIAN) sommergresskar, mandelgresskar, summersquash, gresskar, pumpkin; (NUPE-NUPE-TAKO) èbě; (OROMO, WEST CENTRAL) bukke; (PALI) alābu, lābu, alāpu, lāpu; (PANGASINAN) kalobása; (PANJABI, EASTERN) halwa-kaddoo, kasookola; (PASHTO, NORTHERN) kadu; (POLISH) dynia zwyczajna, cukinia-zucchino, kabaczek; (PORTUGUESE) aboborinha, aboborita, courgette, abóbora porqueira; (POTAWATOMI) ogwîssimaû'n, winbati; (PULAAR) budi, búbi, bodi, wuuduru; (QUECHUA, CUSCO) chapayu; (ROMANI, BALKAN) dodòom; (ROMANIAN) dovleac comun, bostan; (RUSSIAN) tykva tverdokoraja, tykva krupnoplodnaja; (SANSKRIT) dadhiphala, kurkaru, karkaru, kooshmaandi; (SERBO-CROATIAN) tikova, tikva, bela bundeva, tikvice; (SERER-SINE) a dyeng; (SHONA) mbambuni, mbambunu, mbomboni, nhanga, budzi, dikiti, gavu, nyungu; (SIDAMO) bâqula, qônco; (SLOVAK) tekvica; (SLOVENIAN) bučke; (SOMALI) dubbai; (SONGHAY) labtenda, léptándà; (SORBIAN, LOWER) banja; (SORBIAN, UPPER) banja, kirbs; (SPANISH) calabacin, zapallito, zapallo, calabaza común, calabaza de San Juan; (STELLINGWERFS) klabassen, pompoen, pronkmeloen; (SWAHILI) tikiti, mboga, boga, ponda, seta, songa; (SWEDISH) courgette, pumpa, matpumpa; (TAGALOG) kalabasa; (TAGBANWA) kalabasa; (TAJIKI) oshkadu; (TAMIL) suraikayi, poashani, poosanikaai, suraikaai, kalyaanapukuni; (TELUGU) gummadi kayi, boodadegummadi, boodigummadi, pottigummadi; (THEMNE) a-ŋali, a-kali, *a*-kali-*a*-bana, *a*-ŋali, *a*-yiŋkəres, (TIV) a̱gba̱du̱, furu̱m, àgbàdù, furum, icegher-nyian, icegher-pupuu; (TULU) kumbuḍa, karukumbuda; (TURKISH) sakiz kabaği, girit kabaği, adi kabak, yag kabaği, bal kabaği, kabak; (TURKMEN) daşkädi; (TWI) ɛfere; (UKRAINIAN) garbuz tverdokorij; (UZBEK) cullcha-covokk, sapcha-kovock, ķovoķča covock; (VAI) k-po, pori; (VENDA) fhuri, furi, lufhuri, mudzungu, dzungu; (VIETNAMESE) qua bi; (VLAAMS) courgette, spaghettipompoen, patisson, pompoen; (WAMEY) i-tyirir; (WOLOF) yomb, nadio, banga, dôbos, dombos, nadé yomba, laket; (XHOSA) ithanga, iphuzi; (YIDDISH) kahbahk, kirbes, dince-eh; (YORUBA) elegbe, elegede, apala, élégédé; (ZULU) i(li) thanga; (ZUNI) me:-mo.

38.19. *Cyclanthera pedata* (L.) Schrader (EP) immature fruit; (P) C.

(ASSAMESE) bon tiyoh; (CHINESE, MANDARIN) sai huang gua; (CHINESE, WÚ) sai huang gua; (ENGLISH) cyclanthera, pepnio, wild cucumber, korilla, achoccha; (FRENCH) concombre grimpant; (MACEDONIAN) obična lejka; (POLISH) dziki, caihua; (PORTUGUESE) taiuiá de cipó, taimiá de cipó; (QUECHUA, CUSCO) achoccha; (SPANISH) pepino de comer, pepino de rellenar, caigua; (TURKISH) siklan tera; (VLAAMS) wilde komkommer, caibus; (WELSH) ciwcymber gwyllt.

38.20. *Lagenaria siceraria* (Molina) Standley (EP) immature fruit, tender shoots[S] and leaves[S]; (P) C.

Dialects of Luyia (Bukusu[1], Tachoni[2]).

(AFAR) xala; (AFRIKAANS) maranka, kalbas, kababas, pypsteelboon, popkalabas, bierkalbas, marakka; (AJA) etle; (AKAN-AKUAPEM) adenkum; (ALBANIAN, TOSK) susak; (AMHARIC) qəl; (ANUAK) kὲɛnɔ̀, kéenè; (ASSAMESE) jati lao, jatilas, pani lao, lau, bogalau; (BAATONUM) kaaru; (BALANTA-KENTOHE) fôóti; (BALTI) xòs-sor; (BAMANANKAN) bara, fié, filé, file; (BAOULÉ) awa, toa yamman; (BASSARI) a-ngɔ̀ẁ; (BEMBA) mungu; (BENGALI) lau, tikta lana, tiktalau, kodu, kodulau; (BÉTÉ) fiepa; (BINUKID) terayag; (BODO) khumbra, udasi; (BONTOC, CENTRAL) payong, tabungaw; (BUDIK) gi-ngóm; (BUHID) baringan, barunggan; (BULLOM SO) sasa, taga, chãsa-lɛ; (BURMESE) bu-thee; (BUTUANON) balantiyong; (CEBUANO) kalubay, kalubai, sikai, kalabasang-puti, sikay; (CHICKASAW) ishtakafa' toba', lokosh, shokshi'; (CHINESE, GÀN) ye kai hua, hu zi; (CHINESE, HAKKA) p'ôu-tz è; (CHINESE, MANDARIN) bian pu; (CHINESE, MIN BEI) bian pu; (CHINESE, MIN NAN) hô˙-lô˙-pû, hô-lô, lô, pû, hô˙; (CHINESE, WÚ) hu zi, ye kai hua, bian pu, hu gua; (CHINESE, XIANG) hu zi, hu gua; (CHINESE, YUE) hu gua, pu gua, bin pu; (CRIOULO, CAPE VERDE) cabaceira; (CRIOULO, UPPER GUINEA) cabaça; (CZECH) lagenárie, kalebasa; (DAGAARE, SOUTHERN) kpmane; (DAGBANI) garle, tschocho; (DAKOTA) wamnuha; (DANGME-KROBO) akpe, kpatu, tɔ; (DANISH) flaskegræskar, kalabas; (DENDI) gaasu; (DIGO) mchiburú; (DITAMMARI) muyaarmu; (DUTCH) fleskalebas, flessepompoen; (EDE NAGO) igba; (EDO) ẹ̀kpérẹ̀, uko ẹ̀kpérẹ, okpan; (EFIK) àyàyá, ìkó; (ENGLISH) bottle gourd, calabash gourd, white-flowered gourd, white pumpkin; (ESTONIAN) pudelkõrvits; (ÉWÉ) etre, aɖaŋga; (FARSI, WESTERN) kadoghalami; (FILIPINO) upo; (FINNISH) pullokurpitsa, kalebassi; (FON) ka; (FRENCH) calebasse, courge bouteille, calebassier, gourde; (FULFULDE) bombo; (FULFULDE, ADAMAWA) gonogono, fãndu, pāli, junguru, gonogonōji, goraru, gulum butu; (GA) tsene, akpaki; (GEN) tre; (GERMAN)

Flaschen-Kürbis, Trompetenkürbis, Kalebassenkürbis; (GIKUYU) rũngũ, kĩnya, mangu; (GOLA) k-pai, k-pali; (GONJA) denken, chefol; (GOURMANCHÉMA) zéan; (GOURMANTCHÉ) kégniéga; (GREEK) kraskentia, koui; (GUJARATI) dudhi, dudio, tumada, kadwitumbade, kadvitumbadi, bopla; (HANUNOO) balak, bandug, tabayag, tabayug, tamuyu, upu; (HARARI) qulu'; (HAUSA) ƙwarya, gora, kumbo, kwacciya, ƙoƙo, masaki, duma, gora, gurāka; (HEBREW) kara; (HERERO) ondjupa; (HILIGAYNON) kalabasang-puti; (HINDI) kaddu, kaddis, lauki, lāukī, tumri, ghia, lauka, daani, aal, kadutumbi, kashiphal, kashiphal, kadavithumbi, lau, thumri, alkaddu, golkaddu, kaddu, kaduthumbari, kaduthumbi, lauaa, lauka, lauki, mithithumbi, thitalau; (HINDUSTANI) lauki; (HMONG DAW) taub hwb; (HOPI) tawiya; (HUNGARIAN) lopótök, nyakas kobaktök; (IGBO) uko, mbubo; (ILOCANO) tabungaw, tabungau; (INDONESIAN) labu panjang, labu panjan, labu air, labu putih, kukuk, labu kendit; (ITALIAN) zucca da vino, zucca da nuoto, zucca bottiglia, cocozza; (IVATAN) tavayay; (JAPANESE) yugao, maru-yugao, hyoutan; (JAVANESE) walub kenti, waluh deleg, kukuk; (JOLA-FONYI) ka tuk; (JULA) flé; (KANKANAEY) tabungaw; (KANNADA) halagumbla, halagumbala, sorekaayi, kashi sore, sore-kayi, soreballi, eesugaayi, eesugaayi balli, eesugumbala, haalu gumbala, kahisore, khaisore, soreballi; (KASHMIRI) ala-bạrag; (KHMER, CENTRAL) khlôôk; (KIOWA) t'ą̄'-gα; (KISSI, NORTHERN) bala, tala; (KONKANI) dudi, dudio, mardudi; (KONO) ba, g-bala, g-bara; (KRIO) kalbas, abɔrɔ; (LADINO) kalavása; (LAHU) a¯ hpo˄ k'o˗, a¯ mo˄ k'o˗, i¯ mo˄ k'o˗; (LAO) namz taux; (LATVIAN) legenārija, turku ķirbis; (LITHUANIAN) paprastasis ilgmoliūgis; (LOKO) kua; (LOMWE, MALAWI) chipanda, mbonda; (LUO) keno; (LUYIA) lurabu[1], kumwendo[2], emuka[1], sesebebe[2]; (MACEDONIAN) aglesta lufa; (MAITHILI) sajmani; (MALAY) labu putin, labu botol; (MALAYALAM) churakka, anapakai, katuchurum, bellashora, garadudi, kaippancura; (MALIMBA) kəbolo, datogo; (MANDING-MANDINKA) kalamaa, miraŋ, udungue, mirandjô-lô; (MANDJAK) pucúo; (MANINKAKAN, WESTERN) fé; (MAORI) hew, hue, wenewene, ramgatora, whaangai, pohaua, kookakoware; (MARANAO) labogandang; (MARATHI) bhopala, dudhya, ran-bhopla, kadubhopla; bhopla, doodhbhoplaa, doodhya; (MENDE) kpula, kpulɔ, tawa, tawa-wawa; (MIZO) um, mai; (MÒORÉ) kãndé, gole, wamde; (MORISYEN) kalbas; (MUNDARI) suku; (MUSKOGEE) fepē; (NDALI) madyaundu; (NEPALI) lauka; (NEWAR) dupoloḍ; (NGIZIM) gádákùwà; fə̀nà, mâbú, áɗiyú; (NIUE) kākaiona; (NORWEGIAN) flaskefrukt, flaskegresskar, kalebass; (NUPE-NUPE-TAKO) evo, bàbò, bingi, ebò, èpuṇ, gbàla, kpasà; (NYAKYUSA-NGONDE) matakale; (NYANJA) mphonda; (ORIYA) lau, tikta lana, kodu; (OROMO, WEST CENTRAL) masakula, bukke; (PALI) tumbī; (PANGASINAN) tabiyayong; (PANJABI, EASTERN) ghia kaddoo, ghiya kaddu, ghia kadu, ghiya, golkaddu, kabulikaddu, keddi, lauki, lōki, tumba; (POLISH) tykwa pospolita, tykwa zwyczajna; (PORTUGUESE) cabaça, colombro, abóbora carneira; (PULAAR) boliiru, córè, faandu, fahándu, hudusuru, lami-córè, ordè, horde, hɔrədɛ, sumalle; (ROMANIAN) tigvă; (RUSSIAN) gorljanka obyknovennaja, boutylochnaja tykva; (SAMBAL, TINÀ) kalubahan maputi; (SAMOAN) fagufagu samoa; (SANSKRIT) katutumbi, alabu, iksvakuh, ikshavaaku, katuthumbi, alaaboo, brihathphala, danthabeeja, katukalabu, katuthikthaka, katuthumbini, kshathriyaveera, labuka, lamba, mahaaphala, nripatmaja, phalini, pindaphala, raajputhri, thikthaka, thikthathumbi, thumba, thumbee, thumbaka; (SANTALI) kạdu, ho̱to̱ṭ'; (SENOUFO, CEBAARA) kpome-ségue; (SERBO-CROATIAN) pivna tikva, tikvić; (SERER-SINE) fed, limb, mbed; (SIDAMO) buqqê; (SLOVAK) tekvica kalabassa, lagenária; (SOMALI) dubbo; (SONGHAY) gáasí, tàndà; (SPANISH) calabaza de peregrinos, calabaza de San Roque, calabaza viñatera, calabaza viñadera, cojombro; (SUBANON) buliyangin; (SUNDA) waluh bodas, waluh panjang, kukuk; (SUSU) lingué; (SWAHILI) vibuyu; (SWEDISH) kalebass; (TAGALOG) upo, opo; (TAJIKI) chilimkadu, obkadu; (TAMIL) shorakkai, sura-kai, sorakkaai, shorakkaai, surai; (TAUSUG) labu; (TELUGU) anapa, anapakai, sorakaya, anapakaaya, anugakaaya, sorakaaya, anugadaya, gubbakaya, kundanuga, nelanuga, alaabuvu, anapachettu, gubbakaaya, kundaanuga, nelaanuga; (TEM) langa, sounounga, sununga, dschola, tjikare; (THAI) namtao, manamtao, khi-luu-saa; (THEMNE) ɛ-pepe, ɛ-kalma, ka-bos, *an*-bos; (TOK PISIN) sel kambang, botol wara; (TULU) kaidu thule; (TUMBUKA) jangu, mswera; (TURKISH) kantar kabaği, testi kabaği, su kabaği; (TWI) kora, kora wa; (URDU) loki, kadugol, tumbari; (UZBEK) tomosha-covock;

(VAI) koŋgoe; (VENDA) mphapha; (VIETNAMESE) b[aa][uf]; (VLAAMS) fleskalebas; (WAAMA) tɔ̃nɔ̃tµ; (WAMEY) i-ntyen, i-ywù; (WÁRAY-WÁRAY) upó; (WOLOF) gamba; (XHOSA) iselwa; (YORUBA) igba; (ZARMA) gaasu, gumbutu; (ZUNI) mo:-k/isi, to.

38.21. *Luffa acutangula* (L.) Roxb. (EP) immature fruit; (P) C.

(AFRIKAANS) loefa; (AKLANON) buyo-buyo; (ARABIC) bamya seeny; (ASSAMESE) jika; (BENGALI) jhinga, sataputi, ghoshalata; (BICOLANO, CENTRAL) patola, tabobog, sekwa; (BINUKID) sikuwa'; (BONTOC CENTRAL) kabatiti; (BURMESE) kw-we-thi; (CEBUANO) sayuka, tabobog, patola, buyo-buyo, saykuwa; (CHINESE, MANDARIN) ling jiao si gua; (CHINESE, MIN BEI) jiao gua; (CHINESE, WÚ) leng jiao si gua; (CHINESE, YUE) you leng si gua, yue si gua; (CZECH) lufa ostrohranná; (DAGAARE, SOUTHERN) yor-kpmane; (DANISH) kantagurk; (ENGLISH) angled loofah, Chinese okra, ridged gourd, ribbed gourd, ribbed luffa; (FILIPINO) patolang Tagalog; (FRENCH) concombre papengaie, papangaye, eponge végétale torcho; (GOURMANCHÉMA) toutou; (GOURMANTCHÉ) ifouamani; (GREEK) loufa, oxygonos; (GUJARATI) turia, ghisoda, sirola; (HANUNOO) patula; (HAUSA) sos-; (HEBREW) lufa; (HINDI) tori, kali tori, jhinga tori, jhingli, turi, arraturai, jhinga turai, jhimani, karvitharu, thorai, bhiranga thorayi, jinga, kaali thoori, sataputithorai, tharoi, thuri; (HINDUSTANI) teroi; (HMONG DAW) soo-qua, xwb kuab, taub ntxauv yias; (HMONG-NJUA) taub ntxuav yag; (ILOCANO) kabatiti; tabau-tabau; (INDONESIAN) gambas, petola, ketola manis, oyong, jingi, ceme, emes; (ITAWIT) kapatitit; (JAPANESE) tokado hechima; (JAVANESE) gambas, kacur, ceme; (KANNADA) heere kayi, heere balli, kirekayi, hirekayi, heera kay, hire balli, hire kaayi; (KHMER, CENTRAL) ronôông chrung; (KONKANI) gonsali, sirgonsolim, sirgousali; (LAO) looy; (LATVIAN) lufa; (LITHUANIAN) briaunotoji plaušenė; (LUO) otigo; (MALAY) petola segi, ketola; (MALAYALAM) peechanga, peecha kam, piccinna, pichenga, puichenggah, djinji, cherupeerum, pichinna, puttalpiram; (MALTESE) bamja ciniża; (MARATHI) dodka, dodali, sataputi, shirola, sirola, dodke; (MENDE) pu-bɔndɔ; (MÒORÉ) nedga; (MORISYEN) pipangay; (NEPALI) pate ghiraunla, ram turai; (NORWEGIAN) kantluffa, luffa; (ORIYA) junhi, janhi; (PANJABI, EASTERN) kalitori, ghēä tōri, räm tōri; (POLISH) trukwa ostrokątna; (PORTUGUESE) lufa riscada, gousalim, bucha de purga; (SANSKRIT) jhongaka, jhingaka, koshaathakee, jalini, dharnargava, kosataki, koshataki, deerghaphala, dhamargowa, dhaaraphala, gramya, jaalinee, karkotaki, kritavedhana, laghukoshathaki, mrudangphal, peetapushpa, raajakoshaathakee, raajimathphala, shathaputhri, sukosha, supushpa, svaduphala; (SANTALI) paroṛ jhiṅgạ; (SINHALA) wetakolu; (SPANISH) calabaza de aristas, dringi; (SUNDA) emes, oyong, kimput; (SWAHILI) dodoki; (TAGALOG) patola, tabobog, patolang-balimbing; (TAJIKI) dastmolkadu; (TAMIL) peerkankai, parkankai, peerakai, pekankai, pikunkai, pirkankai, peekumchedy, peyppeechukku, peyppeerakam, pikunkaai, pirkinkody; (TAUSUG) patula-ba-imbing; (TELUGU) beera koya, beerakaaya, beera, birakaya, burkai, burkaai; (THAI) buap, buap-liam, manoi-liam; (THEMNE) *an*-rɔnthama-a-potho; (TULU) därepīrè; (TURKISH) keskin köşeli lif kabaği; (TWI) akaton; (URDU) kali tori, torai; (VIETNAMESE) m[uw][ows]p kh[is]a, m[uw][ows] p t[aa][uf]; (VLAAMS) Chinese oker, soekwa, gambas; (WÁRAY-WÁRAY) patola.

38.22. *Luffa aegyptiaca* P. Miller (EP) immature fruit and leaves[5]; (P) C.

(AFRIKAANS) vadoekplant, wasvadoekkalbas, skropkalabas, wasvadoek; (AJA) yovocan; (ARABIC) luff; (ASSAMESE) bhul, bhat karela, bhol; (BAATONUM) yaya korobu; (BAOULÉ) flominan; (BEMBÉ) mussassaka; (BENGALI) dhundal, dandhul, hastighasha; (BICOLANO, CENTRAL) patola, kabatiti; (BINUKID) sikuwa'; (BODO) jingka; (BULGARIAN) lufa; (CEBUANO) patola; (CHINESE, GÀN) si gua; (CHINESE, MANDARIN) si gua, ssu-kuo, shui kwa; (CHINESE, MIN BEI) si gua, shui gua; (CHINESE, WÚ) si gua; (CHINESE, YUE) shui gua, pu tong si gua, si gua; (CZECH) lufa válcovatá, lufa; (DAGAARE, SOUTHERN) sapor-kpmane; (DANISH) luffasvamp, frottersvamp; (DENDI) sɛɛsu; (DUTCH) nenwa, luffa; (EDE NAGO) ɛrun; (ENGLISH) smooth loofah, dishcloth gourd, sponge gourd, vegetable sponge; (ESTONIAN) lufa, käsnkõrvits; (FIJIAN) vusovuso; (FILIPINO) patolang kastila; (FINNISH) pesusienikurkku; (FON) asangokan; (FRENCH) loofah, éponge végétale, seviette des pauvres, courge-torchon; (GA) blɔfo katso; (GEN) yovokusa; (GERMAN) Schwammgurke, Schwammkürbis, Luffa; (GOURMANTCHÉ) djouamani; (GREEK) idos kolokynthis; (GUJARATI) galka, galaka, turia, martigonsali; (GUN) kuvit tekan; (HANUNOO) tabuyuk; (HEBREW) lufa gelilit; (HINDI) kali tori, ghiya tori, ghiatorui, ghiyathurayi, nerua, nanua, nenua, gilki, ghiaturi, purula; (HMONG DAW) xwb kuab; (HUNGARIAN) szivacstök, hálóstök; (IFUGAO) anappan oongal, kamatili, kambatiti, kunappan oongal; (ILOCANO) kabatiti, kabatitit-aso, tabaw-tabaw; (INDONESIAN) belustru, blustru, blustra, luffah, oyong lufah, booloostroo; (ITALIAN) luffa, spugna vegetale, zucca da spugne; (ITAWIT) kabantengga; (JAPANESE) hechima, naga ito-uri; (JAVANESE) blustru, blestru, bestru; (JULA) fou; (KANNADA) tuppahirakai, thuppa heera kavi, thuppa heere kayi; thuppeere kaayi, bolaheere, thupari balli, thupri kaayi; (KAZAKH) ljufa; (KHMER, CENTRAL) ronôông muul; (KONKANI) porgonsali, porgousali; (KOONGO) nsania; (KOREAN) susemiori; (KRIO) sapo, sapo-ɔkrɔ; (LAHU) ma⁻ naweh; (LAO) bwàp khôm; (LATVIAN) lufa; (LOMWE, MALAWI) chinkhupule; (LUO) keno; (MACEDONIAN) lufa; (MAITHILI) ghiurā; (MALAY) petola manis, petola balat, ketola manis, petola buntal; (MALAYALAM) chorakka, kattupeechal, kaattu picinna, kattupiccinna, pichinna, puttalpiram; (MANU'A SAMOAN) meleni vao; (MARANAO) sikwa; (MARATHI) gosale, ghosaalen, ghiya turai, ghosali, ghadaghosali, paroshi, turi bhol, paarose; (MENDE) safo, kalo-hũwo; (MÒORÉ) nedga; (MORISYEN) lenwa; (NDALI) makwasaokwasa; (NEPALI) ghiraunla, palo; (NORWEGIAN) svampluffa, luffa; (NYAKYUSA-NGONDE) mantwatwa; (NYANJA) cinkupule, ntembere; (ORIYA) tarada; (PANJABI, EASTERN) ghiatori, ghia tori, ghēä tōri, räm tōri, ghigandoli, ghiturai; (POLISH) trukwa egipska, gąbczak walcowaty; (PORTUGUESE) lufa, esponja vegetal, bucha, bucha dos paulistas, esfregão, fruta dos paulistas; (RAROTONGAN) po'ue, motini; (ROMANIAN) lufă; (RUSSIAN) ljufa, tykva mochalochnaja, luffa, rastiteinaja goubka; (SAMBAL, BOTOLAN) kabatiti; (SAMBAL, TINÀ) kabatiti; (SAMOAN) meleni vao; (SANSKRIT) rajakoshataki, raja kosa taki, dhaamaargava, dirghapatola, dirgha patolika, dhamargava, dhamargavah, koshataki, dhaamagarva, koshtakee, mahaajaalini, mahaakoshaathakee, raaja koshaathakee; (SENOUFO, CEBAARA) tcha tehou; (SERBO-CROATIAN) lufa; (SINHALA) diya labu; (SLOVENIAN) lufa; (SPANISH) calabaza de aristas, esponja, servilleta del pobre; (SUNDA) lopang, oyong, kimput, jaat, emes; (SWAHILI) dodoki; (SWEDISH) luffa; (TAGALOG) loofah, patola, batutang-uwak, patolang-bilog, patolang-ligaw, patolang-uwak, salag-salag, tabobog, tabobok; (TAGBANWA) patola, kabawang; (TAJIKI) dastmolkadui shomi; (TAMIL) muzhkuku peerkai, mozhukupirkankai, mozhuku pirkankai, mozhuku pavakka, pichukku, pikku, guttibeera, nuraipeerku; (TAUSUG) patula-amu; (TELUGU) netibera, netibira, nunibeera, nunebira, guthibira, guttibira, nethi beera, gutti beera; (THAI) buap-hom, buap-klom; (THEMNE) an-ronthama; (TONGAN) mafa'i, fangu 'a kuma; (TULU) bogupeere; (TUMBUKA) metowo; (TURKISH) lif kabağí; (TWI) aborɔfo sapo; (URDU) ghiya tori, turi; (UZBEK) cozonyuvguich; (VIETNAMESE) m[uw][ows]p h[uw][ow]ng; (VLAAMS) sponskalebas; (WÁRAY-WÁRAY) patola-ihalas; (YAO) lidadoci; (YORUBA) ɛrun, erun kaankan.

38.23. *Momordica charantia* L. (EP) immature fruit and young leaves[5]; (P) C.

(AFRIKAANS) balsempeer; (AGTA) maragusu; (AJA) juke asu; (AKAN-ASANTE) nyinya; (AMHARIC) märara qəl; (ASSAMESE) tita kerela; (BEMBÉ) lumbuzi-busi; (BENGALI) kankrol, karala, karela, kerula uchchhe, baramasiya, jethuya, potikakar; (BICOLANO, CENTRAL) paria, amaragoso, marigoso, parya; (BINUKID) kebering, periya'; (BODO) kangkrikola; (BONTOC CENTRAL) palia, paliya; (BUHID) ampalya; (BURMESE) kyet-hni-ga; (CAROLINIAN) atmagoso; (CEBUANO) palia, margaso, margoso, paliya, palya; (CHINESE, GÀN) ku gua; (CHINESE, HAKKA) fòu kouā; (CHINESE, MANDARIN) ku gua; (CHINESE, MIN BEI) jin li zhi, liang gua;

(CHINESE, MIN NAN) khó-koe; (CHINESE, WÚ) ku qua, jin li zhi; (CHINESE, YUE) liang gua, ku gua; (CRIOULO, CAPE VERDE) são caetano; (CZECH) momordika, hořká okurka; (DANISH) karela; (DENDI) baadooma atakluma dunglin; (DUTCH) springkomkommer; (EDE NAGO) ejinrin; (ENGLISH) bitter gourd, bitter melon, balsam pear, bitter cucumber, African cucumber, carilla fruit; (FIJIAN) karaila; (FILIPINO) ampalaya; (FINNISH) karvaskurkku; (FON) nyɛsikɛn; (FRENCH) assorossie, margase, pomme de merveille, momordique a feuilles de vigne, paroka; (FULFULDE) habirou; (GA) nyanyra; (GADDANG) papet; (GBE-VHE) kakle; (GEN) anyianyasan; (GERMAN) Balsamapfel, Wunder-Balsampfel, bitterer Balsamkürbis; (GIKUYU) kareera; (GREEK) idos kolokynthis; (GUJARATI) kantola, karela, karelo, karelu, korela, karate, karli; (HANUNOO) maragusu, puryaw; (HEBREW) le'usit, gedolat peri, leusit; (HILIGAYNON) amargóso; (HINDI) karela, kareli, kaveli, karola, bura kurella; (HINDUSTANI) karaila; (HMONG DAW) jee dee ee-ah, dee ee-ah, dib iab; (HMONG-NJUA) dlib ab; (HUNGARIAN) mormondika; (IBALOI) paldiya; (IBANAG) apape, afafe, ahahe; (IFUGAO) palia, paliya, palya, apafit, papet; (ILOCANO) paria, parya; (INDONESIAN) peria, pare pahit, paria, pare, pepare; (ITALIAN) balsamini lunghi, pomo balsamo, caranza, vite indiana; (JAPANESE) tsuru reishi, niga-uri; (JAVANESE) pare; (JOLA-FONYI) é déhindél; (KANNADA) madihagala, hagal, hagola, hagalaka yi, hagala kayi, kagalakayi, haagala kaayi, haagala, midi kaayi; (KASHMIRI) karēl; (KHMER, CENTRAL) mreah; (KONKANI) karatim, kaaraathe; (KRIO) sapodila; (LAO) haix, s'aix; (MALAY) peria, peria laut, periok; (MALAYALAM) pavakka, paval kaipavalli, pantipavel, pavakkacheti, paval, kaippa, kaippavalli, kappakka, paaval, paavakka chedi, pantipaaval; (MARATHI) kartoli, karla, karala, karle, karli, karate, karale; (MIZO) changkha; (MORISYEN) margoz; (NEPALI) tite karela; (NGUNDI) lubuzi; (NIUE) vine Initia; (NORWEGIAN) balsamfrukt; (ORIYA) kankada, kalara, karena; (PAMPANGAN) apalya, apalia; (PANGASINAN) palyá; (PANJABI, EASTERN) karela, karala, karila; (PASHTO, NORTHERN) tar<u>kh</u>a kāoṛa; (POLISH) balsamka ogórkowata; (PORTUGUESE) balsamina longa, melão de São Caetano; (SAMBAL, BOTOLAN) ungot-ungot, ungot-ungot bali; (SAMBAL, TINÀ) apalyat, ongot-ongot; (SANSKRIT) karavella, karavellah, kaaravella, kaaravellaka, karavellaka, karavellam, sushavi, ambuvallika, brihadvalli, chiripathra, kandakataka, kandura, kantaphalla, kaaraka, kaaravalli, kaaravalli latha, kaaravelli, kathilla, kathillaka, katilla, katillaka, krimighna, patu, peethapushpa, rahavalli, sukanda, sukandaka, sookshmavalli, sushavi, susuvi, toyavalli, ugrakanda, urdhvasita, varivalli, vishakantaki; (SANTALI) kacan arak, karla; (SERER-SINE) bir-bob; (SINHALA) karawila; (SPANISH) balsamina, calabaza africana, cundeamor, estropajo; (SUBANON) puliya, puliya nataw; (SUMO-MAYANGNA) ka, miniklalasni; (SUNDA) paria; (SWAHILI) karela; (TAGALOG) ampalaya, margoso, ampaliya; (TAGBANWA) ampalaya; (TAMIL) palupaakai, pakarkai, pakal, pavakka, pavakkachedi, pavakkayi, pavakkaayi, kombupaavchedy, paaharkkaai, pakaloo, pavakka chedi, tholoo-pany; (TAUSUG) paria, saligun, paliya, parya; (TELUGU) agakarakaya, kaakara kaay, kakarakaya, kakara, metakakara, tellakakara, urakakara, kaakara, mettakaakara, oorakaakara, potikankara, thellakaakara; (THAI) mara, phakha, maha; (TONGAN) vaine 'initia; (TULU) kaňčala, kaňčäla, kanchaala; (TUPURI) nitui; (TURKISH) asma yaprakli mucize elmasi; (TWI) nyinya, nyanya, nikyinikyi, nkonkomi; (URDU) karela; (VIETNAMESE) m[uw][ows]p d[aws]ng, kh[oor] qua; (VLAAMS) bitterkomkommer, bittere komkommer, sopropo; (WÁRAY-WÁRAY) marigosó; (WOLOF) mbermef; (YOM) bɔ̃kpinkɔ̃nɔ̃τμ; (YORUBA) ejinrin.

38.24. *Praecitrullus fistulosus* (Stocks) Pang. (EP) fruit; (P) R.

(ENGLISH) squash melon, round melon; (GREEK) idos kolokynthis; (POLISH) melon gąbczasty; (VLAAMS) pulpmelon.

38.25. *Sechium edule* (Jacq.) Swartz (EP) fruit, tender shoots[s] and leaves[s]; (P) C.

(AFRIKAANS) soe-soe; (ARABIC) chaco; (ASSAMESE) squash; (BICOLANO, CENTRAL) sayote; (BINUKID) seyuti; (BONTOC CENTRAL) tsayote, sayyot; (BULGARIAN) čajot; (BURMESE) gurkha thi, suka thi; (CEBUANO) tsayote, salyote; (CHINESE, MANDARIN) cai yao li, shun jên kua; (CHINESE, WÚ) cai yao li, fa shou gua; (CHINESE, YUE) fu shou gua, he zhang gua; (CREOLE, HAITIAN FRENCH) militon; (CZECH) čajot; (DANISH) kayote, chayote; (DUTCH) groente peer, kajot; (ENGLISH) chayote, mirliton, vegetable pear, cho-cho, chow-chow; (EWONDO) chouchou; (FILIPINO) sayote; (FINNISH) kajottikurpitsa; (FRENCH) chayotte, chayte, chouchou, chouchoutte, christophine; (GERMAN) Chayote; (GIANGAN) tsayote; (HINDUSTANI) kajot; (HMONG DAW) tao tao, taub thaj, taub taj, taub maum; (HMONG-NJUA) taub taaj; (HUNGARIAN) egymagvútök; (IBANAG) tsayote; (IFUGAO) hayuti, hayyuti; (ILOCANO) tsayote; (INDONESIAN) labu siam, labu siyem, labu siem, labu gambas, labu jepang, waluh siyem, waluh jepang; (ITALIAN) chayote, zuccheta africana, saiotta; (ITAWIT) sayoti; (JAPANESE) hayato uri; (JAVANESE) waluh jepang, waluh siyem, jipang; (KANNADA) seeme badane, seeme badane kaayi, chow-chow kaayi; (KHMER, CENTRAL) su-suu; (KRIO) cho-cho; (LAHU) gaw la$_{\wedge}$ hka$^{\vee}$ shi$_{-}$; (LAO) savëëx, nooy th'ai; (MALAY) labu siam; (NEPALI) iskoosh; (NORWEGIAN) kajote; (NYANJA) ngowe; (PAMPANGAN) tsayote; (POLISH) kolczoch jadalny; (PORTUGUESE) chuchú, alcaiota, caiota, pepinela; (RAROTONGAN) christophene, chayote; (RUSSIAN) cajot; (SPANISH) cayote, chayote, chayotera, talote, pipinela, chocho; (SUNDA) waluh gambas, lejet, waluh siyem, gambas; (SWAHILI) chayote; (TAGALOG) sayote, chayote, tsayote, kayote; (TAMIL) seeme kattirikkai, merakaayi, seema-kathirikkaayi; (TAUSUG) tsayote; (THAI) ma-kheua-kreua, taeng-kariang; (TOK PISIN) choko, sako; (TURKISH) dikenli kabak, armut; (VIETNAMESE) xu-xu, trai su, su su; (VLAAMS) chayote; (WÁRAY-WÁRAY) sayoté; (YAO) citungula.

38.26. *Sicana odorifera* (Vell.) Naudin (EP) immature/mature fruit; (P) C.

(CHINESE, MANDARIN) she xiang gua; (CHINESE, WÚ) she xiang gua; (DANISH) casabanana; (ENGLISH) casabanana, curuba; (FRENCH) calebasse zombi, mélocoton; (GREEK) idos peponiou; (PORTUGUESE) coroá, cruá, melão de caboclo; (SPANISH) casabanana, cajuá, cajuba, melocotón, melon de odor; (VLAAMS) paui, curuba.

38.27. *Telfairia occidentalis* Hook. f. (EP) seeds, leaves and tender shoots; (P) C.

(CHINESE, MANDARIN) xi fei li; (CHINESE, WÚ) xi fei li; (CZECH) telfairie; (DAGAARE, SOUTHERN) sũlou; (EDE NAGO) egusi; (ENGLISH) fluted gourd, fluted pumpkin, telfairia; (EWONDO) ekobon; (GOURMANCHÉMA) kolé; (GOURMANTCHÉ) abaga; (GREEK) avlakoto kolokythi; (GUN) aloko; (IGBO) ofi, ụgụ; (KISSI, NORTHERN) lambu-lambo; (KONO) pɔndɔ-koko; (KRIO) ɔroko; (LUO) okono; (MENDE) gɔnugbe; (MÒORÉ) bãré; (POLISH) tykwica zachodnia; (PORTUGUESE) sabina; (SWAHILI) vibuyu; (VLAAMS) gegroefde pompoen; (YORUBA) egusi, iroko.

38.28. *Telfairia pedata* (Sm. ex Sims) Hook (EP) seeds; (P) C.

(AFRIKAANS) oesterneut; (AKAN-AKUAPEM) bɔmmɔfrana, krobonko; (CHINESE, MANDARIN) wen li; (CHINESE, WÚ) wen li; (CZECH) telfairie; (DANGME) krobonko; (DANGME-KROBO) krobonko; (EFIK) ùbụñ; (ENGLISH) oyster nut, kwemi nut, fluted pumpkin; (ÉWÉ) ahinsaw; (FRENCH) bane, châtaigne de l'Inhambane, kouéme; (GA) ansao, krobonko; (GERMAN) Talerkürbis; (GOLA) gɔniŋgbe, gɔnuŋgbe; (IGBO) ụ́gụ̄, óhí; (KISSI, NORTHERN) lambulambo; (KONO) pɔndo-koko; (KRIO) ɔroko; (MENDE) gɔnugbe; (NDALI) matandu; (NORWEGIAN) østersnøtt; (POLISH) tykwica stopowa; (PORTUGUESE) castanha de Inhambane; (SPANISH) kueme; (SWAHILI) kweme; (TSONGA) cuangua; (VLAAMS) oesternoot; (YORUBA) iroko, ìrókò, apiroko, ẹgúsí.

38.29. *Trichosanthes cucumerina* L. (EP) immature fruit, leaves[s] and tender shoots[s]; (P) C.

(AFRIKAANS) pofadderboontjie; (ASSAMESE) dhunduli; (BENGALI) chichinga, chichiuga; (BURMESE) pe-lin-mway; (CEBUANO) sekwa, sekwa, surosandiya; (CHINESE, MANDARIN) wang gua; (CHINESE, WÚ) she gua; (CHINESE, YUE) chang hua lou; (CRIOULO, UPPER GUINEA) camatom; (CZECH) hadí tykev; (DAGAARE, SOUTHERN) koolé; (DANISH) trikosantes, lang tomat; (DUTCH) haarbloem; (ENGLISH) snake gourd; (ÉWÉ) ego; (FIJIAN) lumba loki; (FILIPINO) snake gourd; (FRENCH) courge serpent, serpent végétal, serpent japonais; (GERMAN) Schlangengurke, Schlangen-Haargurke; (GOURMANCHÉMA) pikampolo; (GREEK) to ofioides; (GUJARATI) pandola, padavali; (HANUNOO) buntik diriit, ramudyu; (HEBREW) z̤ayez̤et neh̤ashit; (HINDI) chachera, chichinda, chachinda, chachendha, chic'yada, gularthori, purwul; (HMONG DAW) taub txaij; (ILOCANO) karkarabasa-ti-aso, pakupis, parparya; (INDONESIAN) peria belut, paria belut, paria ular, lindung; (ITALIAN) serpentona, zucchetta cinese; (JAPANESE) hebi uri; (JAVANESE) pare welut, pare pelus, pare jawi; (KALAGAN) sonson di otot; (KANNADA) padavala kayi, padavalaballi, padavalakayi, padavalu kayl, adla balli, adla kaayi, hadla kaayi, padle kaayi, patla; (KONKANI) podolem; (KRIO) snɛk-tamatis, snek tamatis; (LAO) ngoo ngèèwz; (MAITHILI) kaitā; (MALAY) petola ular, ketola ular, timun bengkok; (MALAYALAM) padavalanga, padavalam, potolam; (MARATHI) padwal, padvala, padaaval, padual, tarakaankadee; (MENDE) gbondo-lani; (MORISYEN) patol; (NEPALI) chichinda; (ORIYA) chhachinda, chhachhindara; (PANJABI, EASTERN) chachinda, chachinda karul, galartori, galartori pandol, parrhōl; (POLISH) wężowata japońska; (PORTUGUESE) abóbora serpente, patola; (SANSKRIT) chachinda, chichinda, ahiphala, brihathphala, chichunda, chinakarkatika, deerghaphala, grihakoolaka, potala, shvetharaaji, sudheergha, veshmakula; (SERER-SINE) mété durubab; (SINHALA) pathola; (SPANISH) anguina, calabaza de culebra, culebrina, serpiente vegetal; (SUBANON) tabugok; (SUNDA) paria belut, paria rintit; (SWAHILI) vibuyu; (TAGALOG) pakupis, tabubok, melon-daga, melon-melonan, tabugok, upong-ahas; (TAMIL) pudalangai, pudal, puttla, pudol; (TELUGU) potlakaya, potla kaayi, potlakaaya; lingapotla, polla kaayi, potla, gingaaptal; (THAI) buap ngu, nom phichit, ma noi; (THEMNE) *an*-rɔthma; (TULU) paṭlakāyi, patla kaayi; (TURKISH) yilan kabaği, çin kabaği; (URDU) chi chinda; (VIETNAMESE) d[aa]y na t[aa]y, d[uw]a n[us]i, m[uw][ows]p t[aa]y; (VLAAMS) slangkomkommer; (YORUBA) tòmátò eléjò.

38.30. *Trichosanthes cucumeroides* (Ser.) Maxim. (EP) immature fruit; (P) C.

(ASSAMESE) dhunduli; (CHINESE, WÚ) wang gua; (CHINESE, XIANG) huang qua; (CZECH) čičinda; (DAGAARE, SOUTHERN) koolé; (ENGLISH) Japanese snake gourd; (FINNISH) käärmekurpitsa; (FRENCH) patola, kikenga; (GOURMANCHÉMA) pikampolo; (GREEK) iaponiki ofioides kolohynthis; (HEBREW) z̤ayez̤et neh̤ashit; (HMONG DAW) taub txaij; (JAPANESE) karasuuri; (PORTUGUESE) sabina da Índia, abóbora serpente; (SINHALA) pathola; (SUNDA) areuj tiwuk; (TURKISH) hiyar tüy çiçek; (VIETNAMESE) hoa b[as]t; (VLAAMS) Japanse slangkomkommer; (YORUBA) tomato elejo.

38.31. *Trichosanthes dioica* Roxb. (EP) immature fruit and tender shoots[s]; (P) C.

(ASSAMESE) potol; (BENGALI) patal, potol; (CHINESE, MANDARIN) ye she gua; (CHINESE, WÚ) ye she gua; (CZECH) parval; (ENGLISH) pointed gourd, parval, patol, parvar; (FIJIAN) taroi; (FRENCH) patole; (GOURMANTCHÉ) léiri; (GREEK) myterikolokynthi, koukourita; (GUJARATI) parwar, patal; (HINDI) parwal, parvar, parval, parmal, palwal, palval, parora, patoli;

(ITALIAN) patole; (KANNADA) kadu padavala, kaadu padavala, kabi padavala, kaadupadavala; (MALAYALAM) patolam, kaattu patavalam; (MARATHI) parwar, parval, padval; (NEPALI) parwal; (ORIYA) potala, patal; (PALI) paṭolo; (PANJABI, EASTERN) parmal, parwal; (POLISH) tykwa nakrapiana; (PORTUGUESE) patola; (SANSKRIT) putulika, patola, beejagarbh, kaasbhanjan, karkshachhad, koolak, kulaka, kushtaha, raajiphal; (SPANISH) patole; (TAMIL) kombu, pudalai, peyupadal, kambuyudaalai; (TELUGU) panatapotla, adavi patola, kambupotala, kommu potla, kommupotla; (TURKISH) iki meskenli tüy çiçek; (URDU) parval; (VLAAMS) scherpe komkommer.

39. Euphorbiaceae

39.1. *Cnidoscolus aconitifolius* (Miller) I.M. Johnston (EP) leaves; (P) C.

(ENGLISH) chaya; (FRENCH) manioc batard, ricin batard; (PORTUGUESE) chaya; (SPANISH) chaya, manolo, papayuelo; (VLAAMS) chaya.

39.2. *Cnidoscolus chayamansa* McVaugh (EP) leaves; (P) C.

(ENGLISH) chaya; (FRENCH) chaya; (PORTUGUESE) chaya; (SPANISH) chaya; (VLAAMS) chaya.

39.3. *Codiaeum variegatum* (L.) Blume (EP) young leaves; (P) C.

(AFRIKAANS) kroton; (ASSAMESE) pat-bahar; (BICOLANO, CENTRAL) buenavista; (BUTUANON) palpagayo; (CEBUANO) kalipayan, kalipayang; (CHINESE, WÚ) bian ye mu; (CRIOULO, CAPE VERDE) croton; (CZECH) kroton; (DANISH) kroton; (ENGLISH) croton; (ESTONIAN) krooton; (FARSI, WESTERN) karchak hendi; (FINNISH) ihmepensas, krootoni; (FRENCH) croton; (GERMAN) Croton; (GREEK) kroton; (HANUNOO) daranaw-lanit, luwaliw samparansisku alistun, sanghusip, sanhusip; (HEBREW) kodan meguvan; (HUNGARIAN) kroton; (ICELANDIC) tigurskruo, kroton; (IFUGAO) ulolon an habhabong; (INDONESIAN) kroton, puring; (ITALIAN) croton; (ITAWIT) buenavista; (JAPANESE) hen-yô-boku; (KANNADA) croton gida; (KRIO) flawa-mangro; (MACEDONIAN) zmejska krv; (MALAY) puding mas; (MARANAO) koyo a bolawan; (NORWEGIAN) kroton; (PALAWANO, CENTRAL) orot sabit tinai; (PANGASINAN) benavista; (POLISH) kroton, krocień; (PORTUGUESE) cróton; (SPANISH) croton; (SUBANON) kalipayan; (SWEDISH) kroton; (TAGALOG) San Francisco; (TAUSUG) kila, kilala-puti; (TURKISH) kroton; (URDU) croton; (VLAAMS) croton; (WÁRAY-WÁRAY) kalipayan.

39.4. *Manihot esculenta* Crantz subsp. *esculenta* (EP) roots and leaves[s]; (P) C.

Dialects of Luyia (Bukusu[1], Marama[2]).

(AFRIKAANS) kassawa; (AJA) kutu; (AKAN-ASANTE) bankì, bédé, kóturuwaáánka; (AKPA) òyìrà; (ANDO) agba; (ANUAK) ābāabūr̭̄, ābāabúré; (ANYIN) agba, bundrè; (ASSAMESE) simolu alu, simalalu, simolu aloo; (BAATONUM) tamguma; (BAFUT) kàsávà; (BAGA) ayoka; (BAMANANKAN) banankou, bananku, bandugu, bâniku; (BAOULÉ) agba; (BASSARI) bandárá; (BEMBA) kalundwe; (BEMBÉ) tsaka; (BENGALI) shimul ali; (BEROM) kit cɔ̀kọt, ròkó; (BETE-BENDI) *ú*-lógò, *ú*-lógò fùng, *ú*-pànyá; (BICOLANO, CENTRAL) balang-hoy, balinghoy, kamoteng-kahoy; (BINUKID) binggala; (BODO) tahganda; (BONTOC CENTRAL) pad-padi, akoy, apaya, linoko, padpaddi, padpaddi, padpadli; (BUDIK) bantára; (BUHID) balinguy langfad, bikul, bunting kauy, malimbunga; (BULGARIAN) tapioka; (BULLOM SO) yek, yeke, yoca, iɛge, iɛke, yeg, pu-langei, yɛkɛ-a-potho, yɛkɛ-lɛ; (BULU) mbong, mboñ; (BURMESE) palaw-pinan-upin; (BUTUANON) balang-hoy; (CATALAN-VALENCIAN-BALEAR) mandioca; (CEBUANO) balangai, kamoting-kahoy, kamuting dutung, balinghoy, balangay, balang-hoy, balang-hoy, kamoteng-kahoy; (CHINESE, MANDARIN) shu shu, shên tzu; (CHINESE, MIN NAN) chhīu-chî-hún; (CHINESE, WÚ) shu shu; (CREOLE, HAITIAN FRENCH) manyòk; (CRIOULO, CAPE VERDE) manoioca; (CZECH) maniok, kasáva; (DAGAARE, SOUTHERN) makēnē; (DANGME-KROBO) agbeli, agbeli kani, fɛdɛ; (DANISH) maniok, kassava; (DENDI) loogo; (DIGO) manga; (DUTCH) cassave, maniok; (EDE NAGO) paki; (EDO) ígaí, igari, igari-ikeripia, igari-makole; (EFIK) àkparamkpa iwá, ibo ìwá, ìwá, ìwá únèñé; (ENGLISH) cassava, manioc, yuca, tapioca plant, Brazilian arrowroot; (ESTONIAN) maniokk, kassaava; (ÉWÉ) agbeli; (EWONDO) kpem; (FARSI, WESTERN) kasah; (FIJIAN) tapioca; (FILIPINO) kamoteng kahoy; (FINNISH) maniokki; (FON) fenyɛn; (FRENCH) manioc, cassave; (FULFULDE, ADAMAWA) leggal roogo, roogo, *m*ba, mbai, mbaiji; (GA) duade; (GANDA) muwogo; (GBAYA) souka; (GBE-VHE) agbeli; (GBE, WACI) kute; (GEN) kute; (GERMAN) Maniok, bittre Cassava, Cassavastrauch, Maniokstrauch, Tapiokapflanze; (GIKUYU) mwanga; (GOLA) *g*bada; (GONJA) banche, dua; (GOURMANCHÉMA) madou kou; (GOURMANTCHÉ) tangoũmbo; (GREEK) kassava, maviskon; (GUN) fenyɛn; (GURO) būū; (HANUNOO) apaya, bubugpag itanum, kayu-kayu, malabunga; (HAUSA) rogo, agingira, ganjigaga, gantamau, ƙaraza, dankali; (HEBREW) maniho, tapioka; (HILIGAYNON) kamútingkáhuy, ligbus; (HINDI) tikhoors maravalli, karrapendalamu, tapioca, sakarkanda; (HINDUSTANI) kasaba; (HMONG DAW) qos ntoo; (HUNGARIAN) kasszáva, éti maniok, tápióka; (IBALOI) sangday; (IBANAG) kahoy; (IFUGAO) kahoy, kahoy an dayaat, kahoy-patata; (IGBO) akpu, ijboru, gari, ji apu, akp*ụ* ǹkọ̀lọ̀, akpu awara nọku, àbàchà, akpḥụǹkọ̀rọ̀, jḥakpḥū, ịgbụrụ, jì-apḥ̣ū, ọ̀dọ̀ghàrà, jìg̃bọ̀, iwa; (ILOCANO) balañgeg, kamote ni moro, balangkoy, kahoy, kasaba, balangeg, kamoti ti moro; (INDONESIAN) ubi kayu, singkong, ketela pohon, kaspe, ubi perancis, ubi jendral; (ITALIAN) cassàve, manioca, mandiòca, iucca; (ITAWIT) camoting; (IZON) imidáka; (JAPANESE) imanoka, kyassaba; (JAVANESE) tela pohung, pohung, bodin, kaspe, tela jabut, ketela puhun, telo; (JOLA-FONYI) efafa, essana, ouante, diõko, dõko, ć sana; (JULA) banangou; (KANKANAEY) as; (KANNADA) maragenasu, baragaalada gedde, gidada genasu, kaadu genasu, kalidala gida, sabbakshi; (KANURI) doya, gàlísà, garisa, ngàdàlá; (KHMER, CENTRAL) dâmlô:ng chhë; (KIRIBATI) te tabicka; (KISI, SOUTHERN) djambalen; (KISSI, NORTHERN) yambai-ma purleŋ, yāmbàliŋg, yambalẽ; (KONKOMBA) kam bon yulè; (KONO) tanga, kono-lange; (KRIO) kasada; (KWÉYÒL) kasav, manyòk; (LAHU) suh^ meu_; (LAO) man tônz; (LATVIAN) manioks; (LITHUANIAN) valgomasis manijokas; (LOKO) bátanga, bantaŋga; (LOMWE, MALAWI) chinangwa, mpwani; (LUO) mogo; (LUYIA) omuoko[2], kumuoko[1]; (MACEDONIAN) kasava juka; (MALAGASY) ravitoto; (MALAY) ubi kayu, ubi benggala, ubi belanda; (MALAYALAM) maracheeni; elilakkilannu, marakilannu, marama chinikilannu, maravaelikilannu, pulaakkilanu, sinakkantal, (MALIMBA) ba limba, batánka, babudo, ko fufe, ko pintan, ku pintan ko, kupintaŋ; (MANDING-MANDINKA) banta nyambo, banta nyambo koyo, banta-nyambo, nyambo, banta nhanhambo, banan gu, banan ku, bananku gué, bara bananku, banaŋgu; (MANINKAKAN, WESTERN) banan'gou banankou, banta-niambo; (MANKANYA) kālofi, tuk; (MANOBO) binggala; (MANU'A SAMOAN) ufi maniota, manioka, tapioca; (MARANAO) banggela, kolempang; (MARATHI) prochugaali chine, tapioca; (MENDE) tangei; (MÒORÉ) bã da kũ; (MORISYEN) maniok; (NDALI) kayabo; (NEPALI) simal tarul; (NGIZIM) dáuyâ; (NGUNDI) muntoba; (NIUE) kāpia, maniota; (NORWEGIAN) knollmaniok, maniok, kassava; (NUPE-NUPE-TAKO) rógò, rógò dzúrúgi; (NYAKYUSA-NGONDE) amayabo; (NYANJA) cinangwa; (PALAWANO, CENTRAL) kumbahang; (PAMPANGAN) kamutingdutung, kamoteng-dutong, kamuting-dutung; (PANGASINAN) kamóteng káhoy, káhoy, kamoteng-keyow; (PANJABI, EASTERN) tapioca; (POLISH) maniok jadalny; (PORTUGUESE) mandioca, mandioca brava, maniba; (PULAAR) pulóók, bãntara, nyambe, bántara, cápè, bantāde, bantara; (PULAR) bantara; (QUECHUA, CUSCO) rumu; (RAROTONGAN) maniota; (ROMANIAN) tapiocă, manihot de tapioc; (RONGA) chimangue, fruvere, gangassolo; (RUSSIAN) maniok s'edobnyj; (SAMBAL, BOTOLAN) modos; (SAMBAL, TINÀ) kamoteng-kahoy; (SAMOAN) manioka, tapioca, ufi la'a, ufi maniota; (SANSKRIT) saadahaa, daarukanda, kalpakanda; (SENA) falinya; (SENOUFO, CEBAARA) gberouh; (SERBO-CROATIAN) manioka; (SERER-SINE) kab, mbuloh, khab ou mbulokh, mbuloh gôr; (SHONA) chinangira, mufarinya, mujumbura, musambwarara; (SINHALA) maññokka; (SLOVAK) maniok jedlý; (SPANISH) yuca, tapioca, mandioca, guacamote; (SUMO-MAYANGNA) malai, maley; (SUNDA) huwi dangdeur, huwi jendral, sampeu, kasapen, singkong; (SWAHILI) mhogo, kopa; (SWEDISH) cassava, maniok; (TAGALOG) kamoteng-kahoy, kamoting-kahoy, balinghoy, pangging-kahuy; (TAGBANWA) lidib kayo, malambunga, kamoteng kahoy, malambonga, sanglay kayo; (TAMIL) kuchi kizhangu, aal vallikkizhangu, maravalli kizhangu, aezhilaikkizhangu, maravalli, putaakkizhangu; (TAUSUG) pangi-kahui, kamote-kahoy; (TELUGU) karra pendalamu, karrapendalamu, koyaganasu gadda, maanupendalanu; (TEM) kungola; (THAI) man-sampalang, man-samrong, man-mai; (THEMNE) ɛ-yoka, ɛ-kanda, ɛ-rɔgbɔ, samouya, mayugbe, *a*-yokâ, *a*-yoka abi, *a*-yoka-*a*fera, *a*-yoka-*a*-kono, *a*-yoka-*a*-yim; (TIV) dowase, duàsə, dua̱sc, duaso, ivambèyon, lougou, rogo̱, rougou, vambèyon; (TONGA) cigawo; (TONGAN) manioke; (TSONGA) gangassolo; (TULU) marakereng; (TUMBUKA) mayao; (TURKISH) aci manyok, tapyoka, manyok, manyuka, yuka; (TWI) bankyi; (URDU) tikhoor maravali; (VAI) ɡbasa, dumbai, nyɔrɔ; (VENDA) mutumbula; (VIETNAMESE) san, khoai-mi, s[aws]n, khoai m[if]; (VLAAMS) maniok, cassave, tapioca; (WAMEY) bãntara, bantara, *u*-nkwav *u*-kan; (WÁRAY-WÁRAY) sumaráy, balanghoy, balangay; (WOLOF) puloh, gniambi, nâbi, ñàmbi, pullóóx, nyambe; (XHOSA) imifuno ethile efana nebhatata; (YALA) ikpāleke, ìjikāpā; (YALUNKA) kondek-habi-na, kunda khabi-fore, kundakhabi-fikai, kundakhabi-na; (YAO) inangwa; (YOM) nugumya; (YORUBA) gbaguda, ege, ogége, agugoyo, dereẹ́fe; (ZARMA) roogo; (ZULU) umdumbula.

39.5. *Sauropus androgynus* (L.) Merrill (EP) leaves; (P) C,R.

(CHINESE, MANDARIN) shou gong mu; (CHINESE, WÚ) shou gong mu; (ENGLISH) common sauropus, star gooseberry; (INDONESIAN) katuk; (JAPANESE) ruridama-no-ki; (JAVANESE) babing; (KANNADA) chakraani, chakramuni; (KHMER, CENTRAL) ngub; (LAO) hvaan baanz; (MAKHUWA) garacia, gurue, inchiricano, inchirikano, kwali; (MALAY) cekur manis, chekor manis, chekkurmanis, asin-asin, cangkok manis; (MINANGKABAU) simani; (SPANISH) katuk; (SRANAN) switiwiwiri; (TAGALOG) binahian; (TAMIL) thavasai murungai; (THAI) phak waan ban; (VIETNAMESE) rau ng[os]t, b[oof] ng[os]t; (VLAAMS) chekor manis.

40. Fabaceae (Leguminoseae)*

40.1. *Arachis hypogaea* L. (EP) immature/mature seeds; (P) C,R.

Dialects of Luyia (Bukusu[1], Marama[2]).

(ABUA) àpàpá; (AFRIKAANS) grondboon, grondboontjie, aardboontjie, aapneut; (AGTA) mane; (AKAN-ASANTE) ńkatíe, ńkatsíe; (AKAN-FANTE) nkatsi; (AKPA) òbrònū; (ALBANIAN, TOSK) kikirik; (AMHARIC) läwz, arašid, oččolini, oččoloni; (ARABIC) fool sudanee, l'gerté; (ASSAMESE) badam; (AZERBAIJANI, NORTHERN) araxis, yerf ındığı; (BAATONUM) saa; (BAGA) tálan; (BALANTA-KENTOHE) souou; (BAMANANKAN) lo-séno, menian-tiga, kã-tiga, tigué, tírā, tiga; (BAOULÉ) doru kusassi, ń-gatè, n'gatêh; (BASSARI) *o*-tika, ɔ-*[ɛ-]*tíkà, ɔ-tíkà ɔ-dúndú, asselim; (BELARUSAN) babovja; (BEMBA) lubalala; (BENGALI) chini badam; (BEROM) yáàpá; (BÉTÉ) bai, blê blê; (BETE-BENDI) *kà*-nākəl, *kà*-nəkəlē; (BICOLANO, CENTRAL) mani, lugosing; (BINUKID) lugasi; (BOKYI) ndugekere; (BONTOC CENTRAL) mani; (BOSNIAN) kiki-riki, kikiriki; (BUDIK) *ma*-bɛ́tɛ́r, ɛ-bɛ́tɛ́r, *ba*-bɛ́tɛ́r; (BULLOM SO) kɛntɛ, kɛnti-lɛ, kentir, kɛntir, kɛntir-lɛ, nkantr, ukántsr, ka-kente, kɛnti; (BULU) owôndô; (CAJUN FRENCH) pistache; (CATALAN-VALENCIAN-BALEAR) cacauet; (CEBUANO) mani; (CHICKASAW) yaaknonkaawaa';

(CHINESE, HAKKA) t'í- t'éou, lŏc-fā-sēn; (CHINESE, MANDARIN) hua sheng; (CHINESE, MIN NAN) hoa-seng, hoa-seng-tāu, thô͘-tāu, lȯk-hoa-seng; (CHINESE, WÚ) hua sheng; (CHINESE, XIANG) luo hua sheng, hua sheng guo; (CREE, PLAINS) pâkân, pâkânis; (CREOLE, HAITIAN FRENCH) pistach, amizgél; (CRIOULO, CAPE VERDE) mancarra; (CRIOULO, UPPER GUINEA) mâkara, mãnkara, macara, mancara, mancarra; (CZECH) podzemnice olejná, buraky, burske ořišky, arašidy; (DAGAARE, SOUTHERN) simiyé; (DAGBANI) sámle, sima, suma; (DAKOTA) yaȟuġapihawaŋkadaŋ; (DANGME) gigẽ; (DANGME-KROBO) akate; (DANISH) jordnøt; (DENDI) densi; (DIGO) njugu nyasa; (DITAMMARI) tikomanti; (DUTCH) aardnootje, aardnoot, pinda; (EDE NAGO) ɛkpa; (EDO) isáẹwẹ, isáẹ́wẹwẹ, iságuẹgua, ságuẹguẹ; (EFIK) m̀bàńsang; (ENGLISH) peanut, ground nut, pea nut, chinese almond, earth nut, goober nut, manilla grain, monkey nut; (ESTONIAN) maapähkel, arahhis; (FARSI, EASTERN) bâdâm-e zamini; (FARSI, WESTERN) badam zaminé; (FIJIAN) peanut; (FILIPINO) mani; (FINNISH) maapähkinä; (FON) aziin; (FRENCH) arachide, pois de terre, cacahouette; (FULFULDE) birihu, biriyu; (FULFULDE, ADAMAWA) birigue, sunkaaru, tigage, yiriire, biriiwu; (GA) dkatiɛ, nkati, ŋkátiɛ; (GADDANG) amani; (GAELIC, MANX) thalooin; (GANDA) bingebwa; (GBE-VHE) azi, makate, azi nu dowowowo; (GBE, WACI) azin; (GEN) azin; (GERMAN) Erdnuß; (GIANGAN) mani; (GIKUYU) njũgũ-karanga; (GOLA) *n*gele, *n*gere, ke-gele, lege; (GOURMANCHÉMA) chicarain, á-dsina, a-tina, gúdīa; (GOURMANTCHÉ) acktina; (GREEK) arachida e ypogeios; (GUJARATI) mugphali; (GURO) bɛ̀ɛ́; (HEBREW) boten, egoz adama; (HERERO) !naran, ozosukua; (HILIGAYNON) maní; (HINDI) mungphali, mūg. phalī, vilayati mung, moongphali, vilaaythimung; (HINDUSTANI) pienda; (HMONG DAW) txiv laum huab xeeb; (HOPI) tuva; (HUNGARIAN) földimogyoró; (IBALOI) mene; (IBANAG) mani; (ICELANDIC) hneta, jarđhneta; (ICEVE-MACI) ìgírì, *ò*-télès(h)á; (IFUGAO) mani, lulungngin; (IGBO) apapa, àhụ ekere, aseboko, mgbam ekere, ọ̀kpà, ọ̀pàpa, akpa nwẹkẹlẹgu, àhụekere, àkpa ekere, ọ̀kpa, ọ̀kpa akele, ọ̀kpa ekele, ọ̀kpa nwekele, ọkpa ẹkẹrẹ, ahu ekere; (ILOCANO) mani, kalman; (INDONESIAN) kacang tanah, kacang Jawa, kacang Manila; (ITALIAN) arachide, nocciolina americana; (IZON) apapá, isánguẹguẹ; isele tịn, obori kunẹị; (JAPANESE) ratkasei, nankin-mame; (JAVANESE) kacang brol, kacang brul, kacang brudul, kacang jebrol, kacang pendem, kacang srentul, kacang srintil, kacang tuban; (JOLA-FONYI) badâgot, badãngot, badâgata, badiangot, badyankat, egerda, emangera, é mangera, é gerda, é gertay, jangata, kukuy, mankara; (JULA) tiga; (KALAGAN) kummangi; (KANNADA) nela-gadale, baeru kadale, kadale kaayi, kalle kaayi, nelagadale kaayi, shaenga; (KANURI) gangála, kolche, koldi, koldji, kólji, ngángálà; (KAZAKH) arachis; (KHMER, CENTRAL) sânndaèk dèi; (KISI, SOUTHERN) tion; (KISSI, NORTHERN) tēi, tion, ka, tehũ, tiɔ̃ŋ, kono-niki, ya; (KONKANI) mussombibikan, nelagadle; (KONKOMBA) nama, sema, nasilma, n'silim, silma, s'ma; (KONO) yáála, ya; (KRIO) kpàkɔ́, akpororo, gra(w)mpi, graná, granat, gránɛt; (KURANKO) tige; (KWANGALI) osukwa; (KWÉYÒL) pistach; (LADINO) fistók, fustukes; (LAHU) mi$_{\vee}$ naw$^{\wedge}$ shi$_{-}$; (LAO) thwàx din, thwàx ho; (LATVIAN) zemesrieksts; (LIMBA) fúdēre, kudere, kutank, madɛre; (LITHUANIAN) žemės riešutas; (LOKO) ndsára, *n*jagã, nyo; (LOMWE, MALAWI) m'tedza, manduwi; (LUO) amaido; (LUXENBOURGEOIS) Afenoss; (LUYIA) tsichungu[2], chinjugu[1], chimaito[1]; (MACEDONIAN) obični kikiriki; (MALAY) kacang tanah, kacang China, kacang Jawa, kacang goring; (MALAYALAM) nelakkaadalaka, verkkala; (MALTESE) karawett; (MANDING-MANDINKA) kã tiga, tiga, tio, tiya, tio ba, mantiga, loti-ô, tiô; (MANDINKA) mantiga, ta, tiga, tika, tío, tíra, tiya, to; (MANDJAK) upêtche, upiate, tiga; (MANINKA) tiga, menian tiga; (MANKANYA) õ nkara, ó pets, lik, ulique, uncarra; (MANOBO) Iugasi, lugasin, lugasing; (MARANAO) logasi, logasing, lugasing; (MARATHI) bhui mug, bhooyimoonga, bhooichane, bhooisheng, shaenga daana, vilaaythimug; (MENDE) kono-niki, níkì, níkilì, níklì, nyui; (MIZO) bodom; (MÒORÉ) nanguri, sum-kam, nãguri; (MORISYEN) pistas; (MUSKOGEE) kvnhtvlako, kvmhtvlako, kvmohtvlako, kvnohtvlako; (NAMA-DAMARA) ombudufukwa; (NDONGA) ndongo; (NEPALI) badaam; (NGIZIM) gwádánù; (NIUE) pinati; (NORWEGIAN) peanøtt, jardnøtt; (NUPE-NUPE-TAKO) guzịa; (NYANJA) ntedza, nsawa, masyawala; (OGBIA) ápàpá; (OGBIA-AGHOLO) ápàpá; (PAMPANGAN) mani; (PANGASINAN) maní; (PANJABI, EASTERN) mɔ̃gfəli; (POLISH) orzech ziemny, arachid; (PORTUGUESE) amendoim; (POTAWATOMI) bkanesen; (PULAAR) gerté, daridé, dege, gerte, guérte, tiga, biriyu, gertere, tigagué, tigare, yertere; (QUECHUA, CUSCO) inchic; (ROMANI, BALKAN) fastutaya; (ROMANIAN) arahida, alune americane; (RUSSIAN) arahis, zavilets, zemljanoi oreh, kitaiskii oreh, (SAMA, CENTRAL) batung; (SAMOAN) pinati; (SANGA) aju; (SANSKRIT) buchanaka, bhoochanaka, mandapi, bhoomija, bhooshimbika, bhoostha, rakthabeeja, sneha beejaka, thribeeja; (SANTALI) ciṇạ badam; (SCOTS) puggie nut; (SENOUFO, CEBAARA) masa; (SERER-SINE) arène, arèn, arer o ol, óliak, ópiat, ria ol, tia fana, tya, tiya; (SHONA) nzungu; (SIDAMO) occolône; (SINHALA) ratakaju; (SLOVENIAN) podzemnica olejná; (SONGHAY) matiga, démsì; (SORBIAN, UPPER) worješk, zažne běrny; (SOTHO, NORTHERN) tokomane; (SOTHO, SOUTHERN) letokomane; (SPANISH) aráquida, cacahuete, maní; (SUBANON) mini; (SUMO-MAYANGNA) pinda; (SUNDA) kacang suuk, kacang taneuh, suuk; (SUSU) kansi; (SWAHILI) karanga, njugu; (SWEDISH) jordnöt; (TAGALOG) mani, batung-China; (TAJIKI) chormaghzi zamini; (TAMIL) nelakadalai, nelakkadalai, vaerkkadalai, manilakottai; (TAUSUG) batung-china, baatung, batung-larot, batungsina; (TELUGU) verusenagalu, vaerushanagalu, nela sanagalu; (TEM) áku, akua, keke; (THAI) thua lisong, thua din; (THEMNE) *a*-kan(d), *ma*-kantr, *ma*-kand; (TIBETAN) patam; (TIV) àbṵ̀n, a̱bṵn, abun ahi, ibun ihi, ihi; (TONGA) mbalala; (TSWANA) letonkomane; (TULU) nelakadale; (TUMBUKA) syaba, matewere; (TURKISH) yerfistiği, yer fistiği; (TWI) adaduanan, nkateɛ; (URDU) mung phali; (VAI) gɛnde, gɛ́ndɛri, gɛ́ndɛi, genderi, nyuê; (VENDA) ṅwa-ntshutshulùlu; (VIETNAMESE) dâu phong, lac; (VLAAMS) grondnoot, aardnoot, pinda; (WAMEY) *va*-ngal, *i*-kwɛl; (WOLOF) guèrté, gèrté, aren, gerte; (XHOSA) indongomane; (YALA) èwì; abrinọ̄; (YALUNKA) kansínna, kansina; (YAO) ntesa; (YIDDISH) stashkeh; (YOM) n τrδm; (YORUBA) ɛkpa, egpa, ẹ̀pà gidi, ogódsia, égba(-yo), égba; (ZARMA) damsi, kolonse; (ZULU) itongomane, u(lu)dlubu.

40.2. *Cajanus cajan* (L.) Millsp. (EP) immature pods/leaves[S]; (P) C.

(AFRIKAANS) duiweboon; (AJA) eklwi; (AKLANON) kadyos; (ARABIC) adas asfar; (ASSAMESE) rohor mah; (BAATONUM) aduwa; (BENGALI) arhar, tur, tuver; (BEROM) syìlà làsárá; (BICOLANO, CENTRAL) tabios, tabiyos, tabyos; (BODO) musur, kokleng, rohor; (BONTOC CENTRAL) kidis, kaldis, kardis, keldis; (BUHID) adius, kadyos; (BULLOM SO) sitil; (CEBUANO) kadios, tabios, tabiyos, tabyos; (CHINESE, MANDARIN) shu dou, mu tou; (CHINESE, WÚ) shu dou; (CREOLE, HAITIAN FRENCH) pwa kongo; (CRIOULO, CAPE VERDE) congo, fijon congo; (CRIOULO, UPPER GUINEA) feijão; (CZECH) kajan, hajanus; (DAGBANI) adua; (DANISH) sertebønne, ærtebønne, jordært; (DIGO) mbalazi; (DOONDO) nkassa; (DUTCH) duivenerwt, wandoe; (EDE NAGO) otili; (EDO) ọlẹle, ọrẹle; (ENGLISH) pigeon pea, cajan pea, Congo pea, yellow dhal, cadjan pea, congo bean, dhal, dal, dhol, no-eye pea, red gram; (FIJIAN) dahl; (FILIPINO) pigeon pea, kadios; (FINNISH) kyyhkyherne; (FON) klwekun; (FRENCH) pois d'angola, pois cajan, pois du Congo, pois d'Angole, ambrévade; (GA) atii, blɔfoyɔ, yɔ, yɔ̀ɔ; (GBE-VHE) atiyi; (GBE, WACI) eklwi; (GERMAN) Straucherbse, Strauchbohne, Taubenerbse; (GIKUYU) njũgũ; (GONJA) kichepo, kíchépò, kùdùá, kùkúnchi; (GREEK) mpizeli tou kojkou; (GUJARATI) arhar, tur, tuver, turdal, dangri; (GUN) klwekun; (HANUNOO) kadyis, kadyis piyugu; (HAUSA) aduwa, dan mata, waáákén na Makka, wáákén dànfámíí, wáákén Másàr, wáákén Sàntànbûl, wáákén tàntábàràá, wáákén tuuraawáá, wáákén yán máátáá; (HILIGAYNON) kadyos; (HINDI) adhaki, tuvari, tuvarika, tur, arhar, tuver, arhar, arardal, arhardal, thoor, thuvar; (HINDUSTANI) arhar; (HMONG DAW) taum ntoo; (HUNGARIAN) galambborsó, katjánbab; (IBANAG) kardis; (IFUGAO) bakyug, kudis, kuldih, kuldis, oldih all bakyug an nagalgaltan, oldih an mumpaha, oldih hin inadne; (IGBO) fịọ fịọ, fio-fio, hịọhịọ, shọọ shọọ, vịọvịọ; (ILOCANO) kardis, kaldis, kidis; (INDONESIAN) kacang Bali, kacang iris, kacang gude, kacang kayu, hiris, kachung kadyu, gude; (ITALIAN) caiano, pisello dei tropici, pisello ardoreo; (ITAWIT) kardis; (JAPANESE) juto, ki-mame; (JAVANESE) kacang gude, kacang kayu, gude; (KANKANAEY) kidis; (KANNADA) togare; thogari, thogari bele, thuvara, athaki, baele, byale, dhaal, kari uddu, thogari bele, thogari uddu, thovari, turuku thogari, tuvari; (KHMER, CENTRAL) sândaèk dai, sândaèk kroëb sâ, sândaèk klöng; (KISSI, NORTHERN) yawendo; (KONKANI) tori, thori; (KONO) sɔ-mɛsɛ; (KRIO) koŋgo-binch; (KURANKO) konshõ, kunsɛ; (KWÉYÒL) pwa Angòl; (LAHU) hk'aw$_{\wedge}$ zuh naw$^{\wedge}$; (LAO) thwàx h'ê; (LIMBA) konsoba, koso, konso; (LOKO) konjoi, konyoi, wimbiona, wonjo; (LOMWE, MALAWI) khobwe, epweri; (LUO) apena; (MACEDONIAN) gulapski grašok; (MAITHILI) rāhaṛi; (MALAY) kacang, kacang dal, kacang hiris; (MALAYALAM) thuvaran, thuvara, kakshi, adhaki, aadhaki, kaakshi, thora paeru; (MANDING-MANDINKA) konshõ, kunsɛ; (MARANAO) kayoranti; (MARATHI) thoora, thoori, thoovar, arhar, tur, tuver, turi; (MENDE) biandobwe, binši; (MORISYEN) ambrevad; (NDALI) imbange; (NEPALI) rahar; (NGUNDI) muwandu; (NORWEGIAN) cajanbønne; (NYAKYUSA-NGONDE) imbange; (NYANJA) nandolo; (ORIYA) horodo, kandulo; (PANJABI, EASTERN) arhar, dinger, tohar; (POLISH) nikła indyjska; (PORTUGUESE) ervilha de angola, ervilha do congo, feijão do congo, feijão guandú, guandú, jisongo; (PULAAR) konso-toge; (QUECHUA, CUSCO) puspu-poroto; (RONGA) unhafa, mbowini, imnowini; (RUSSIAN) kajanus; (SANSKRIT) adhaki, tuvari, tuvarika, tuwari, aadhaki, thuvari, kakshi, karvira bhuja, mritana, mritala, mrittalaka, shanapushpika, supya, surashthaja, thuvarika, vrithabeeja; (SANTALI) raheṛ; (SENA) mbwete; (SERBO-CROATIAN) golubiji grašak léca; (SINHALA) thora parippu; (SPANISH) frijol del monte, guisante de Angola, guando; (SUBANON) tagyas; (SUNDA) kacang hiris, hiris; (SUSU) ɛ-konsoi, fote ɛtugi; (SWAHILI) mbaazi; (SWEDISH) kyyhkyherne; (TAGALOG) kadyos, kadios, kalios, kagyos, kagyus, tabios, gablos, kidis, gablos, kagiyos, kalyos; (TAGBANWA) kadyos, kadiyus; (TAMIL) tuvarai, thovaray, thuvarai, adagam, iyavai, paruppu, aadagam, aadagi, karpoo woolando, thovarai; (TELUGU) kandi, kandulu, kandula, adhaki, ettakandulu, aadhaki, errakandulu, kondakandi, peddakandi, peddakondakandi, polukandi, potukandalu, sinnakandi; (TEM) adua, adoua; (THAI) thua rae, thua maetaai, ma hae; (THEMNE) ɛ-konshoŋ, alev a batur; (TIV) alev; (TONGA) itambe za miti; (TULU) togari, togarè, thogari; (TUMBUKA) mbenge, nyandolo; (TURKISH) güvercin bezelyesi, hint güvercin bezelyesi; (URDU) arhar; (VIETNAMESE) cay dau chieu, dau sang, dau thong; (VLAAMS) duivenerwt, cajan; (YALUNKA) konso-tugo-na; (YAO) mbelemende; (YOM) paltiτμ; (YORUBA) otili, òtili, otimi.

40.3. *Canavalia ensiformis* (L.) DC. (EP) immature seed; (P) C.

(AFRIKAANS) jackboon; (AKLANON) lagaylay; (ARABIC) faul; (ASSAMESE) ramtal urahi; (BAATONUM) suwi; (BAMANANKAN) nguo; (BASSARI) *va*-péna, ɛ-pen, ɛ-mbáràx; (BEMBÉ) budi; (BENGALI) makham shim; (BEROM) kwàràk, kwàràk jɛ̀ɛ́jɛ̀ɛ́t; kwàràk pwɛ̀ŋ, kwàràk sùnàŋ; (BONTOC, CENTRAL) kamling; (BUDIK) *ma*-pɛ́l, *ma*-mabarà, *ma*-pɛ́l *ma*-pɛsya; (CHINESE, MANDARIN) ai xing dao dou; (CHINESE, WÚ) ai xing dao dou; (CHINESE, YUE) yang dao dou; (CRIOULO, CAPE VERDE) fava-rica; (CZECH) kanaválie mečovitá; (DAGBANI) kpaliegu; (DANISH) jackbønne; (DENDI) baaliho; (EDE NAGO) popɔndo; (EFIK) ìbábá; (ENGLISH) jack bean, horse bean; (FILIPINO) pataning-dagat; (FON) legba kpakun; (FRENCH) haricot de Madagascar, haricot sabre; (GEN) ayivi; (GERMAN) Jackbohne, Schwertbohne, Madagaskarbohne; (GONJA) kudua, kukunchi; (GREEK) kyamos; (GUJARATI) abvo, abo; (HANUNOO) tablug-batug; (HAUSA) bar ta ta tafi, ɓáránkácíí, ɗan zago, ladiko, roɓaji, wáákén ankwai, wáákén ankwai, wáákén bísáá; (HEBREW) kanavalya; (HINDI) gojiaa sema, bara sem, barasem, sufed kadsumbal, kadsmabu, khadsambal, lalkudsumbal, safedkudsumbal, sema, surasema; (IGBO) ẹkwulo mmịmì; (ILOCANO) badang-badang; (INDONESIAN) kacang parang, kacang prasman, kara pedang, kara kaji; (ITALIAN) canavalia; (JAPANESE) nata-mame, tachi-nata-mame; (JAVANESE) kara bendo, kara pedang, kara haji, kara dodog, kara loke, kara wedung, kacang nyonya, krandang; (JOLA-FONYI) balèngan, buneknena, balêngan; (KANKANAEY) balballiga; (KANNADA) thmate balli, kaadavare, sambe, shambe avare, shimbe balli, shimbi, thamate balli, thumbe kanji; (KRIO) kpokpondo, ɔsbinch; (KURANKO) kukurɔ; (MALAY) kacang parang puteh; (MALAYALAM) vallaringa, baramarecha, kanavaala, kaattuvaalamarakkaya, vaalamara; (MANDING-MANDINKA) tobalô-sosso; (MANDINKA) fâto, tobalõ soso; (MANINKA) fãnto; (MARATHI) abai, ghevda, pandhri abai, pondhri abi, gavara, govari, khadasambala, kisamari; (MENDE) bwepapwe; (NORWEGIAN) jackbønne; (ORIYA) maharada; (PANJABI, EASTERN) bean; (POLISH) kanawalia mieczokształtna; (PORTUGUESE) feijão sabre, feijão de porco, feijao bravo; (PULAAR) pãnté, pãntoy, pãte, pâtowi, pâtoy, fanta, tàgalê; (SANSKRIT) asishimbi, brihathshimbi, khadgashimbi, mahaashimbi, neelashimbika, shimbi, sthoolashimbi; (SANTALI) tihon; (SERER-SINE) ñav durubab; (SPANISH) judia sable, canavalia, judia de puerco, judia de caballo; (SUNDA) kara bedog, kacang mekah, kacang prasman, koas parasman, koas bakol, koas pedang; (TAGALOG) habas, pataning-dagat, pataning-espada; (TAMIL) kaattutthambattan, vella thambattankai, vellai tambattai, vellaitambattai, kodithambattan, kolhiyavarai, kottuthambattam, peyavarai, sivapputthambattai, thamattan kodi, thambattai, valavarai, vellaitthambattai; (TELUGU) chamba kaasya, tammatikaya, vella tamma, vellatamma, adavithamma, errathamma, kaaranuputhige, karuthamma, shimbi, thamba, thamma; (TIV) awon du; (TURKISH) kiliç fasulyesi, Madagaskar fasulyesi, bakla; (URDU) bara sem; (VLAAMS) jackboon; (WAMEY) *va*-pèna; (WÁRAY-WÁRAY) magtambokan; (WOLOF) sèb u tubab; (YORUBA) popɔndo, papanla, pòpondó, poponla.

40.4. *Canavalia gladiata* (Jacq.) DC. (EP) immature seed. (P) C.

(AFRIKAANS) swaardboon, sewejaarboontjie, gifboontjie; (ASSAMESE) kotari dobua mah, ramtal urahi; (BENGALI) makhan shim; (CEBUANO) magtambokau, magtabokaw; (CHINESE, GÀN) dao dou; (CHINESE, MANDARIN) dao dou; (CHINESE, MIN BEI) dao dou; (CHINESE, WÚ) dao dou; (CHINESE, XIANG) dao ba dou; (CHINESE, YUE) dao dou; (CZECH) kanaválie; (DANISH) sværdbønne; (ENGLISH) sword bean, horse bean; (FILIPINO) habas; (FRENCH) pois sabre, pois sabre rouge, dolic en sabre, haricot sabre, haricot rouge; (GERMAN) Schwertbohne; (GUJARATI) abva, abo; (HANUNOO) buray dipay; (HEBREW) kannavlya seyfanit; (HINDI) bara sem, barasem, lal kadsumbal; (HMONG DAW) taum ntaj; (INDONESIAN) kara pedang, kacang parasman, koas bakol; (ITALIAN) canavalia; (JAPANESE) nata-mame; (KANNADA) thmate balli, shembi avare, shembiavari tumbeki; (KHMER, CENTRAL) tioeuhs; (LAO) (khùa) ‘khao ‘khièo; (MALAY) kacang parang, kacang polong; (MALAYALAM) vallaringa, baramareha; (MARATHI) abai, ghevda; (NEPALI) taraware simi; (NORWEGIAN) sverdbønne; (ORIYA) maharada; (PANJABI, EASTERN) bean; (POLISH) kanawalia szablasta; (PORTUGUESE) feijão sabre, fava contra; (SANSKRIT) asishimbi; (SPANISH) haba de burro, judia sable; (TAGALOG) habas, pataning Espanya; (TAMIL) vella thambattankai, segapu thambattai, segaputham battai; (TELUGU) thamma kaaya, tammatikaya, yerra tamma; (THAI) thua-phra; (URDU) talvari sem, bara sem; (VIETNAMESE) d[aaj]u r[uwj]a; (VLAAMS) zwaardboon.

40.5. *Cicer arietinum* L. (EP) seeds; (P) C.

(AFAR) sumbura; (ALBANIAN, TOSK) qiqër; (AMHARIC) šəmbəra, šənbəra, šumbəra, dubbe; (ARABIC) hummus; (ASSAMESE) boot mah; (BADAGA) kudure kaḍale, caṇagi; (BASQUE) garbantzu; (BELARUSAN) nont; (BENGALI) chola, but, butkalai; (BODO) sana; (BULGARIAN) nachut; (CATALAN-VALENCIAN-BALEAR) cigro; (CHINESE, MANDARIN) ji er dou, ying tsui tou; (CHINESE, WÚ) ji er dou; (CREOLE, HAITIAN FRENCH) garbans, gabans; (CRIOULO, CAPE VERDE) gron di bico; (CZECH) cizrna beraní; (DANISH) kikerært, kikært; (DUTCH) kikkererwt, keker, spaanse erwt; (ENGLISH) garbanzo bean, chick pea, Bengal gram, common gram; (FARSI, EASTERN) nakhod; (FARSI, WESTERN) nokhod, nokhod chi; (FIJIAN) chana; (FILIPINO) garbansos; (FINNISH) kohviherne; (FRENCH) pois-chiche; (GAELIC, MANX) pishyr chirkey; (GERMAN) Kichererbse; (GREEK) erevinthos, pevithi; (GUJARATI) chana, chania; (HARARI) šumbura, ‘atär; (HEBREW) ḫimẓa tarbutit, humus, chimtza tarbutit, himzatarbutit;

(HINDI) chana, boot, but, chunna, chanaka, chola, rahilaa, rohilaa; (HUNGARIAN) csicseriborsó, bagolyborsó; (INDONESIAN) kacang Arab, kacang kuda; (ITALIAN) cece, garbanzo; (JAPANESE) hiyokomame; (KANNADA) kadale, chana, karikadale, kempukadale, kudoly, channa, hasi kadale, kadale kaalu, kale; (KASHMIRI) cana; (KAZAKH) nut; (KONKANI) chono, chano; (KURDISH, NORTHERN) nok; (LADINO) bilibíz; (LATVIAN) aunazirni; (LITHUANIAN) sėjamasis avinžirnis; (MACEDONIAN) naut, običen slanutok; (MALAY) kacang kuda; (MALAYALAM) kadale-kaadi; (MALTESE) cicri; (MARATHI) harbara, chano, harbharaa; (MORISYEN) gram; (NEPALI) kauli chana; (NEWAR) canā, chohalā; (NORWEGIAN) kikert, bukkeert; (PALI) caṇako, kalāyo; (PANGASINAN) garbánzos; (PANJABI, EASTERN) channa, chola; (PASHTO, NORTHERN) c̲h̲an̲r̲a, nakhwëd; (POLISH) ciecierzyca pospolita, cieciorka; (PORTUGUESE) grão de bico, gravanço, grao; (ROMANIAN) năut; (RUSSIAN) nout, touretskii garoh, baranii garoh, garohovnik, nout obyknovennyji; (SANSKRIT) chanaka, canakah, canaka, balabhaishajya, balabhojya, harimantha, jeevana, kanchuki, krishnachanchuka, sakalapriya, sugandha, vaajibhakshya, vaajimantha; (SANTALI) buṭ; (SERBO-CROATIAN) leblebija, naut; (SIDAMO) shumburà; (SINHALA) kaddla parippu; (SLOVAK) cícer baraní; (SLOVENIAN) čičerika; (SPANISH) garbanzo; (SUNDA) kacang arab; (SWAHILI) dengu; (SWEDISH) kikärt; (TAGALOG) garbanzo, garbansos; (TAJIKI) nakhud; (TAMIL) kadalai, chunda kadalai, uppukkadalai; (TELUGU) sanagalu, sanagulu, chanagalu, harimandhakam, harimandhakamu, jennaga, senaga, shenagalu; (THAI) thua hua chang; (TULU) kadale, kaḍalè; (TURKISH) nohut; (TURKMEN) adatl nohut, nochut; (URDU) chana, but; (VLAAMS) kekererwt, spaanse erwt, sissererwten; (YIDDISH) ahrbis.

40.6. *Cyamopsis tetragonoloba* (L.) Taubert (EP) immature pods and seeds; (P) C.

(AFRIKAANS) guarboon; (ASSAMESE) thupi mah, thupiurahi, thupi urahi; (BENGALI) guar; (CHINESE, MANDARIN) sui dou; (CHINESE, WÚ) sui dou; (DANISH) guar, guarplante, indisk nytteplant; (DUTCH) guar; (ENGLISH) cluster bean, guar; (FIJIAN) guar; (FRENCH) guar; (GERMAN) Guarpflanze, Cyamopsis; (GUJARATI) guwar, gavar; (HINDI) gowar, guar, guvar, ramphali, darhari, gawarphali; (INDONESIAN) tarum; (ITALIAN) guar, fagioli agrappolo; (KANNADA) gori kayi, gorikayi, chavuli kaayi, gavaara kaayi, gori kaayi, govardhana kaayi; (MACEDONIAN) grozdest bob; (MALAY) kacang siam; (MALAYALAM) kothavara; (MARATHI) gowar, guwar, bavachi, govaari, gavaara; (NEPALI) guar simi; (NORWEGIAN) guar; (ORIYA) gunar chhuin, guvar chhuim; (PANJABI, EASTERN) guara, kulti, guar, kuwara; (POLISH) fasola groniasta; (PORTUGUESE) guar; (SANSKRIT) bakuchi, dridhabija, gorani, dridhabeeja, gorakshaphalini, nishandyaghni, sushaka, vakrashimbi; (SPANISH) guar; (TAMIL) kothaveray, kothavarnkai, seeniavaraikai; (TELUGU) gorchikudu, goruchikkudu, goru chikkudu; (TULU) chowlikodu; (URDU) guar; (VLAAMS) trosboon.

40.7. *Flemingia vestita* Benth. ex Bak. (EP) tubers; (P) C.

(ASSAMESE) makhioti; (CHINESE, MANDARIN) bei mao qian jin ba; (CHINESE, WÚ) bei mao qian jin ba; (DANISH) flemingia; (ENGLISH) flemingia; (FRENCH) flemingia; (INDONESIAN) otok-otok, lapa-lapa; (POLISH) flemingia; (VLAAMS) flemingia.

40.8. *Glycine max* (L.) Merr. (EP) immature and sprouted seeds; (P) C.

[1]Itsukha dialect of Luyia.

(AFRIKAANS) sojaboon, soyaboontjie, sojaboontjie; (ALBANIAN, TOSK) sojë; (ARABIC) ful suyah; (ASSAMESE) soya bean, garo mah; (BASQUE) soia; (BELARUSAN) soja vjalikaja; (BENGALI) gari kalai; (BULGARIAN) soja; (BULU) soja; (BURMESE) lasi, pengapi, peryatpym;

(CHINESE, GÀN) mao dou; (CHINESE, MANDARIN) mao dou, hei tou; (CHINESE, MIN BEI) mao dou; (CHINESE, MIN NAN) n̂g-tāu, tōa-tāu, o˙-tāu, tāu; (CHINESE, WÚ) cai young da dou, mao dou; (CHINESE, XIANG) qing dou; (CHINESE, YUE) mao dou, cai young da dou; (CREOLE, HAITIAN FRENCH) soja; (CRIOULO, CAPE VERDE) soja; (CZECH) sója luštinatá, sója; (DANISH) sojabønne; (DUTCH) sojaboon; (ENGLISH) soybean, soy bean, soya, soya bean; (ESTONIAN) soyauba; (FILIPINO) utaw; (FINNISH) soijapapa; (FRENCH) soia, soja, pois oléagineux de Chine; (GAELIC, IRISH) pónaire shoighe; (GERMAN) Sojabohne; (GIKUYU) soya; (GREEK) sogia, kyamos sogias; (GUJARATI) soyabin; (HEBREW) soya; (HINDI) bhat, bhatwar, bhetmas, ramkurthi; (HINDUSTANI) bhatwas; (HMONG DAW) taum panv; (HUNGARIAN) szójabab; (ICELANDIC) sojabaun; (INDONESIAN) kedelai, kacang kedelai, kedele, rĕtak mĕdjong, kacang Jepun, kacang bulu; (ITALIAN) soia, soya; (JAPANESE) daizu, edamame; (JAVANESE) dele, dekeman, demekan, gadele, kedele, kedungsul, dangsul; (KANNADA) soya bean, soya beeja; (KAZAKH) soja; (KHMER, CENTRAL) sândaèk sieng, sândaèk an gen sar; (KOREAN) ccongnamul; (LAHU) naw$^{\wedge}$ hk'eh$^{\vee}$; (LAO) thwàx khôn, thwàx tê; (LATVIAN) sojas, soja; (LITHUANIAN) gauruotoji soja; (LOMWE, MALAWI) khobwe; (LUYIA) tsisoya[1]; (MACEDONIAN) soja, obična soja; (MALAY) kacang soya, kacang bulu rimau, kacang Jepun; (MALAYALAM) soybean; (MARATHI) soybin; (MÒORÉ) soza kolgo; (MORISYEN) soza; (NEPALI) bhatmas; (NORWEGIAN) soya, soja; (ORIYA) soyabin; (PANJABI, EASTERN) soyabean, sörä bēn; (POLISH) soja zwyczajna, soja; (PORTUGUESE) soja; (ROMANIAN) soia de cultură, soia; (RUSSIAN) soja; (SANTALI) disom ho̱ṟe̱c', poṇḍ disom ho̱ṟe̱c'; (SERBO-CROATIAN) soja; (SINHALA) soya bonchi; (SLOVAK) sója; (SLOVENIAN) soja; (SPANISH) soja, soya; (SUNDA) kedele, kacang bulu, kacang jepun, kacang kedele; (SWAHILI) maharagwe soya, njegere; (SWEDISH) soja böna; (TAGALOG) utaw, balatong; (TAMIL) soya payaru; (TELUGU) soya chikkudu; (THAI) thua lueang, thua phra lueang, thua rae; (TIBETAN) ţenser; (TURKISH) soya, soya fasülyesi; (URDU) soybean, bhat; (VIETNAMESE) dâu tuong, dâu nành, quantan; (VLAAMS) sojaboon; (WELSH) ffa soy; (YIDDISH) soyeh; (ZULU) uhlobo lukabhontshisi.

40.9. *Lablab purpureus* (L.) Sweet (EP) immature seeds, (P) C.

Dialects of Luyia (Bukusu[1], Maragoli[2]).

(AFRIKAANS) lablabboon, lablab-boontjie; (AKLANON) bataw; (ALBANIAN, TOSK) dolikë; (ANYIN) guangono abrua; (ARABIC) lubish; (ASSAMESE) desi urahi; (BADAGA) hittili avare, avare, ma:navare; (BENGALI) deshi shim, shim; (BICOLANO, CENTRAL) batau, bataw, bat-aw; (BINUKID) biyarey; (BONTOC CENTRAL) itab; (BUHID) taptu; (BULGARIAN) lestovička; (BURMESE) pe-gyi; (CEBUANO) baglau, batau, baglaw; (CHINESE, MANDARIN) que dou, pai pien tou; (CHINESE, WÚ) que dou, e mei dou, mei dou; (CHINESE, YUE) bian dou, mian tou; (CRIOULO, CAPE VERDE) feijão-pedra; (CZECH) lablab purpurový, dolichos lablab; (DANISH) hjelmbønne; (DUTCH) slingerboon, lablab; (EDE NAGO) awujɛ; (ENGLISH) hyacinth bean, Indian bean, lubia bean, lablab, lab-lab bean; (ESTONIAN) punauba; (FARSI, WESTERN) lubia; (FINNISH) hyasinttipapu; (FRENCH) dolique d'egypte, pois indien; (GADDANG) bitabit; (GERMAN) Helmbohne, Faselbohne; (GIKUYU) njahĩ; (GREEK) yakimthos; (GUJARATI) wal, val; (GUN) akla; (HANUNOO) buray; (HAUSA) wáákén baibààyíí, wáákén dànfàmíí, wáákén dàrníí, wáákén gwààrí, wáákén Sàntànbûl, wáákén tàntábàráá, wáákén tuuraawaa, wáákén yán-máátáá, wáákén Màsàr; (HEBREW) lavlav; (HINDI) sem, baloor, bhata vaasu, bhatavans, betharasu, borboti, kaalalobia, katajang, lobia, lobiya, makhansim, raajashimbi, valm, wall; (HUNGARIAN) lablab; (IBANAG) bassaw, patani; (IFUGAO) ita, itab, itab an gayyang; (ILOCANO) parda, parda atap; (INDONESIAN) komak, kara-kara, kacang kara, kacang bado, kacang biduk, kekara; kara uceng, mochakotta, roay; (ITALIAN) dolico, fagiolo lablab; (ITAWIT) bassaw; (JAPANESE) fiji mame; (JAVANESE) kara, kara wedus, kara putih, kara andong, kara krupuk, kara lunjar, kara uceng, kekara; (KANKANAEY) itab, payda; (KANNADA) avare balli, avare, avare baele, avare kaalu, avarekaayi, bili maniavare, holadavare, maniavare; (KOREAN) ccachicong; (LAHU) aw- k'a_ naw^ pa_ shi_, naw^ pa_ shi_; (LATVIAN) vitenu lobija; (LOMWE, MALAWI) kukusu; (LUYIA) njawu[1], sikandakanda[2]; (MAITHILI) bōṛā, sīm; (MAKHUWA) cungudze; (MALAY) kacang kara, kara-kara, kekara; (MALAYALAM) ramacha; (MALTESE) giacinti; (MANDING-MANDINKA) nalvo; (MARATHI) anvare, adavebaala, pandharepavate, parvate, paote, tambadepavade, val, wal, pavta; (MORISYEN) zantak; (NAMA-DAMARA) !hureb; (NEPALI) bakula; (NORWEGIAN) hjelmbønne; (NYAKYUSA-NGONDE) mankhamba; (NYANJA) mkhungudzu; (ORIYA) jhata simba; (PANGASINAN) baktaw, pagda; (PANJABI, EASTERN) lobia desi; (POLISH) fasolnik egipski, wsiega pospolita; (PORTUGUESE) feijão cutelinho, feijão lab-lab, lablab; (ROMANIAN) fasole turcească; (RONGA) chicharo, chicharus; (RUSSIAN) lobija; (SAMBAL, TINÀ) patani; (SANSKRIT) nishpava, ncspava, raajashimbi, shimbi, shvethashimbika, simbi, vallaka; (SERBO-CROATIAN) lastima usta; (SLOVAK) dolichos; (SPANISH) frijol del antibo, dólico lablab, dólico gigante; (SUNDA) kacang jeriji, kacang peda, roway krupuk, roway peda, roway bedog, roway katopes; (SWAHILI) ngwara, fiwi; (SWEDISH) lablab, lablabpapu; (TAGALOG) batau, bataw, bulay, sibatsi, itab, pardá, baray malagti, patan, isibatsi; (TAMIL) mochchai, avarai, avarai-motchai, thattaam payiru; (TAUSUG) bataw; (TELUGU) anapa, chikkudu, adavichikkudu, alsanda, anapachikkudu, annapa, anumulu, boberulu, thellachikkudu, thellachikurkai; (THAI) thua paep; (TONGAN) pini'ae puaka; (TSONGA) tingolocotse, tingolocotze; (TULU) abarè; (TUMBUKA) mambamba; (TURKISH) misir börülcesi; (URDU) lobia, sem; (UZBEK) loviya; (VIETNAMESE) dau-vang, dâu van; (VLAAMS) Indische boon, lubiaboon; (WAAMA) nɔɔti yani; (YAO) ngwingwisu; (YORUBA) awujɛ, awúję.

40.10. *Lathyrus sativus* L. (EP) immature pods/seed; (P) C.

(ALBANIAN, TOSK) buxhak, vingëz e koltivuar; (ARABIC) jilbana; (ASSAMESE) khesari mah, kola mah, khesari, teora; (BASQUE) aizcoles; (BELARUSAN) china kloubnenosnaja; (BENGALI) khesari; (BULGARIAN) posevna sekirče; (BURMESE) pèsali; (CATALAN-VALENCIAN-BALEAR) guixa; (CHINESE, MANDARIN) ts'ao wan tou; (CHINESE, WÚ) xiang wan dou; (CZECH) hrachor setý; (DANISH) græs-fladbælg, fladbælg; (DUTCH) zaailathyrus; (ENGLISH) grass pea, chickling pea, chickling vetch, white vetch; (ESTONIAN) põld-seahernes; (FINNISH) peltonätkemä; (FRENCH) gesse blanche, gesse cultivée, pois carré; (GALICIAN) chicharo de raposo; (GERMAN) Saat-Platterbse, Deutsche-Kicher; (GREEK) lathuri; (HEBREW) tofaḥ tarbuti, tophach; (HINDI) khesari, latri, tiuri, teora, kassar, kass'ur, kasari, kassur, lathri; (HUNGARIAN) szegletes lednek, szeges lednek, vetési lednek, szeges borsó; (ITALIAN) cicerchia, robiglio; (JAPANESE) garasy mame; (KANNADA) chapparada avare, kaesari baele, kaesari thogari, visha thogari; (KAZAKH) china posevnaja; (LATVIAN) sejas dedestina; (LITHUANIAN) sėjamasis pelėžirnis; (MACEDONIAN) običen graor; (MALTESE) fevetta-ciccardi; (MARATHI) laakh; (MONGOLIAN) tarimal tömördöö; (NEPALI) khesari; (NORWEGIAN) fôrskolm, vikke; (ORIYA) khesra; (PANJABI, EASTERN) kisari, chural, karas, karil, matri, dōōkən; (POLISH) lędźwian siewny; (PORTUGUESE) chícharos, chícharos comum, ervilha de cheiro; (ROMANIAN) lintea pratulnide de cultura, orăstică, măzăroi; (RUSSIAN) čina posevnaja; (SANSKRIT) thriputa, kalaaya, khandikaa, lanka, sandika; (SANTALI) kesari; (SERBO-CROATIAN) grahor poljski, grahor, sastrica; (SLOVAK) hrachor siaty; (SPANISH) arvejo, almorta, muela, tito, guija; (SWEDISH) plattvial, åkervial; (TAJIKI) sidirgh; (TSONGA) chicharos; (TURKISH) adi mürdümük, mürdümük, ispanyol bezelyesi, culbant, külür, çayir bezelyesi; fiğ; (URDU) jnugli matar; (VLAAMS) graserwt.

40.11. *Lathyrus tuberosus* L. (EP) tuber; (P) C.

(BEMBA) lutoyo; (CHINESE, GÀN) lian shu, di gua; (CHINESE, MANDARIN) wan dou shu; (CHINESE, MIN BEI) di gua, bai di gua, he ge; (CHINESE, WÚ) bai yen dou, yu dou, wan dou shu; (CHINESE, YUE) mian dou, he bao dou; (CZECH) hrachor hlíznatý; (DANISH) knoldfladbælg; (DUTCH) aardaker; (ENGLISH) earth chestnut, groundnut, earth-nut pea; (FINNISH) maapähkinä; (FRENCH) gesse tubéreuse, anette, châtaigne de terre; (GERMAN) Knollen-Platterbse; (GREEK) lathouros o kordy-lodis; (HEBREW) tofaḥ ha-peka'ot; (HUNGARIAN) mogyorós lednek; (ITALIAN) cicerchia tuberosa, ghianda di terra, tartufo di prato; (KAZAKH) china klubnevaja; (LATVIAN) bumbuļu dedestiņa; (LITHUANIAN) gumbinis pelėžirnis; (MACEDONIAN) orešesest graor; (MONGOLIAN) bultsuut tömördöö; (NORWEGIAN) jordskolm; (POLISH) groszek ziewny, groszek bulwiasty; (PORTUGUESE) chícharo tuberoso; (ROMANIAN) oreşnita tuberculifera; (RUSSIAN) china kloubnenosnaja; (SERBO-CROATIAN) grahor, gomoljasti; (SPANISH) arveja tuberosa, guija, cicércula tuberosa; (TURKISH) yumrulu mürdümük (TURKMEN) klubenli eşekmäş; (VLAAMS) aardaker.

40.12. *Lens culinaris* Medikus (EP) immature pods, sprouted seeds; (P) C,R.

(AFAR) qadas, missiri; (AFRIKAANS) lensie; (ALBANIAN, TOSK) thierëzë; (AMHARIC) massər; (ARABIC) adas; (ASSAMESE) mosur mah, masurmoha; (AZERBAIJANI, NORTHERN) mərci; (BADAGA) be; (BALTI) sran-ma, sran-zhung; (BASQUE) tilista; (BELARUSAN) chechevista pasjanny; (BEMBA) ilanda; (BENGALI) masur, masser, masuri; (BULGARIAN) lešča; (CATALAN-VALENCIAN-BALEAR) llentia; (CHINESE, HAKKA) pièn t'éou; (CHINESE, MANDARIN) jin mai wan, hsiao pien ton; (CHINESE, MIN NAN) lėk-tāu; (CHINESE, WÚ) jin mai wan, bin dou; (CHINESE, XIANG) mai wan dou; (CREOLE, HAITIAN FRENCH) pwa pijon; (CZECH) čočka jedlá, čočka kuchyňská; (DANISH) linse; (DUTCH) linze; (ENGLISH) lentil; (ESTONIAN) lääts; (FARSI, EASTERN) dâl nakhod, dâl; (FARSI, WESTERN) adas; (FILIPINO) lentil; (FINNISH) linssi, kylvövirvilä; (FRENCH) lentille; (GAELIC, MANX) pishyr lughag; (GAELIC, SCOTTISH) pea sair; (GALICIAN) lentella; (GERMAN) Linse; (GREEK) phake, faki; (GUJARATI) masur, mascur, masser, masuri; (HARARI) missir; (HEBREW) 'adasha, 'adash tarbutit, adasha tarbutit, adasha; (HINDI) musur, mascur, masser, masuri, masur, masoor, mūg, masūr; (HINDUSTANI) masúr, mothí, moth; (HUNGARIAN) lencse, főzeléklencse; (ICELANDIC) gular baunir, linsubaunir; (INDONESIAN) kacang adas, miju; (INUKTITUT, WEST GREENLANDIC) qáquaussaq; (ITALIAN) lenticchia; (JAPANESE) hiramame, renzumame; (KANNADA) massur, chanangi, channangi baele, lentil, masoor baele; (KAZAKH) zhasımıq; (KHMER, CENTRAL) lânti; (KIRGHIZ) jasımık; (KIRIBATI) rentira; (KURDISH, NORTHERN) nîsik; (KWÉYÒL) lantil, pwa; (LADINO) lentèjas; (LAHU) ja$_{\wedge}$ meu$^{\vee}$ naw$^{\wedge}$, ka$_{-}$ la$_{-}$ beh$^{\vee}$; (LATVIAN) rupjseklu lecas, lēcas; (LITHUANIAN) valgomasis lęšis; (MACEDONIAN) leka socivo, obična leka; (MAITHILI) masurī, ghāghrā, dāli, dāil, māhamūṅ; (MALAY) kacang dal; (MALTESE) l-ghazz; (MARATHI) masur, masser, masuri, masoori; (MIZO) be hring chi khat, thali chi khat, dailuah sen chi; (MORISYEN) lanti; (NEPALI) musuro; (NGUNDI) lentiye; (NORWEGIAN) linse; (OROMO, WEST CENTRAL) misira; (PALI) masuro; (PANJABI, EASTERN) masur, malka masur, musri; (PASHTO, NORTHERN) dâl, mëshëng, nësk, mot, mahe; (POLISH) soczewica jadalna, soczewica; (PORTUGUESE) lentilha; (ROMANI, BALKAN) lìnta; (ROMANIAN) linte, linte comestibilă; (RUSSIAN) chechevista sedovnja, čečevica; (SANSKRIT) masoora, gabholika, gurubeeja, halasaka, kalyaanabeeja, mangaliya, masooree, masooraka, masoorika, prithubeejaka, ragadali, sura, thamboolaraaga, vrihikanchana; (SERBO-CROATIAN) sočivo; (SIDAMO) missirà; (SINHALA) rathu parippu; (SLOVAK) šošovica; (SLOVENIAN) leča; (SORBIAN, UPPER) čóčk, čóčka; (SPANISH) lenteja; (SUNDA) kacang adas; (SWAHILI) dengu, adesi, lentil; (SWEDISH) lins; (TAJIKI) nask, adas; (TAMIL) misurpurpu; (TATAR) yasmık; (TELUGU) chirisanagalu; (THAI) thua daeng; (TIBETAN) ţēma; (TURKISH) mercimek; (TURKMEN) merjimek; (URDU) masur, adas; (UYGHUR) yesimuq; (UZBEK) adas, yasmiq; (VIETNAMESE) giöng dâu bièn; (VLAAMS) linze; (XHOSA) ilentile, uhlobo oluthile lwesityalo esifana; (YIDDISH) lindz; (ZULU) udali, ulensinsi.

40.13. *Lotus tetragonolobus* L. (EP) immature pods; (P) C.

(AFRIKAANS) gevlerkte ert; (ASSAMESE) kontal urohi; (CHINESE, MANDARIN) duan ce yyi dou; (CHINESE, WÚ) duan jia yi dou; (CZECH) ledenec nachový; (DANISH) aspargesært; (DUTCH) asperge-erwt; (ENGLISH) asparagus pea, winged pea; (ESTONIAN) aspar, spargel; (FIJIAN) sam; (FINNISH) parsaherne; (FRENCH) pois asperge, lotier rouge, lotier cultivé; (GERMAN) rote Spargelbohne, Spargelerbse, Fluegelerbse; (GREEK) tetrocgonolovos oorphyros; (HEBREW) arba kenafot; (HUNGARIAN) bársony kerep; (ITALIAN) veccia pisello, lotto rosso, pisello asparagio; (LITHUANIAN) raudonasis siensparnis; (MACEDONIAN) krilobob, krilest grav; (MALTESE) figgila hamra; (NEPALI) pate simi; (NORWEGIAN) vingeert; (POLISH) głabigroszek szkarłatny; (PORTUGUESE) loto vermelho; (RUSSIAN) tetragonolobus purpurnyj; (SERBO-CROATIAN) grašak špargar; (SLOVENIAN) škrlatną nokota; (SPANISH) bocha cultivada, guisantillo rojo; (SWAHILI) asiparagasi; (SWEDISH) sparrisärt; (TAGALOG) sequidillas; (TURKISH) kirmizi lüfer çiçeği, kuşkonmaz; (VLAAMS) asperge-erwt.

40.14. *Lupinus* spp.[8] (EP) seeds; (P) C.

(e.g. *L. albus, L. angustifolius, L. mutabilis, L. termis*)

(AFRIKAANS) lupien, wolfsboontjie; (ALBANIAN, TOSK) lupin i bardhë; (ASSAMESE) lupin; (BELARUSAN) lubin; (BENGALI) turmuz; (BLACKFOOT) aisattsikohtako; (CHINESE, MANDARIN) bai hua yu shan dou; (CHINESE, WÚ) bai hua yu shan dou; (CZECH) viči bob, lupina; (DANISH) lupin; (DUTCH) lupine; (ENGLISH) lupin; (ESTONIAN) lupiin, hundiuba; (FARSI, WESTERN) baghlayí-mesree; (FINNISH) lupiini; (FRENCH) lupin, pois de loup; (GAELIC, MANX) lupeen; (GALICIAN) cardea; (GERMAN) Lupine, Wolfsbohne; (GREEK) loupimos o lefkos; (HEBREW) turmus; (HINDI) turmas; (HUNGARIAN) csillagfürt, fehér csillagfürt; (ICELANDIC) lupina; (ITALIAN) lupino; (LADINO) atramùz; (LATVIAN) baltā lupīna; (LITHUANIAN) lubinas; (MACEDONIAN) lupina; (NORWEGIAN) lupin, hvit lupin; (PANJABI, EASTERN) turmuz, zurmish; (POLISH) łubin; (PORTUGUESE) tremoço, tremoceiro; (QUECHUA, CUSCO) chocho, tarwi; (ROMANIAN) lupin, lupin alb; (RUSSIAN) lupin, loupin, volchii bob; (SERBO-CROATIAN) lupina bela; (SLOVAK) vlcí bôb biely; (SPANISH) altramuz, lupino; (STELLINGWERFS) lupine, lupiene; (SWEDISH) lupin, vitlupin; (TSONGA) tremoco; (TURKISH) lüpen, aci bakla, yahudi baklasi; (URDU) tarmus; (VLAAMS) lupine; (WELSH) bys y blai`dd.

40.15. *Macrotyloma geocarpum* (Harms) Marechal & Baudet. (EP) seed; (P) C.

(BAATONUM) suwi; (BAMANANKAN) dougou folo, dugu folo, ntigi folo, bendi, bindi, dugus falo; (BASSARI) *a*-nyanax, issilimpa; (BUDIK) *ma*-kánda; (DAGBANI) sempi; (DANISH) hausaknold, hausa jordnød; (EDE NAGO) nadu; (ENGLISH) Kersting's groundnut, hausa groundnut; (ÉWÉ) azi; (FON) doyi; (FRENCH) haricot de behanzin, haricot royal, pararu; (GEN) doyi; (GERMAN) Kouarourou; (GOURMANCHÉMA) youlou-sunmõ; (GOURMANTCHÉ) adja ã tina; (GREEK) arachis e ypogeios; (HAUSA) konarourou, kwààrúúrú; (IGBO) àkịdī àlà, ahụ ekere; (KONKOMBA) tumpiɛla; (LOMWE, MALAWI) m'tedza; (MANDINKA) tia kolèñ, tia kolõ; (MÒORÉ) dieguem tenguéré; (PULAAR) binde; (TAMIL) kollu; (TEM) kandɛla; (TURKISH) yer fistiği; (VLAAMS) hausa aardnoot; (WAMEY) *i*-kwol *i*-dyag; (YORUBA) nadu.

---

[8] Many of the *Lupinus* species contain toxic alkaloids that must be removed before consumption.

40.16. *Macrotyloma uniflorum* (Lam.) Verdc. (EP) seed; (P) C.

(BADAGA) koḷḷu; (BENGALI) kurtikalai; (BURMESE) pè-bi-zât; (CHINESE, MANDARIN) chang geng pi dou; (CHINESE, WÚ) chang geng pi dou; (DANISH) kalthi; (ENGLISH) horse gram; (FRENCH) grain de cheval, hurali; (GERMAN) Pferdekorn; (GUJARATI) kalathi, kvlit; (HINDI) kulthi; (INDONESIAN) kekara, kacang karas; (ITALIAN) dolico cavallino; (KANNADA) hurali; (KAZAKH) jonyshka, lucerna posevnaja; (MARATHI) kulith, kulthi; (SANSKRIT) kulaththa; (SUNDA) kacang karas, rarawea; (TELUGU) ulavalu; (TULU) kuḍụ; (URDU) kulthi; (VLAAMS) paardekeker.

40.17. *Medicago sativa* L. subsp. *sativa* (EP) leaves, young shoots, and (and several closely related species) sprouted seeds; (P) R,C.

(AFRIKAANS) alfalfa, lusern, klawer, lusering, makklawer; (ALBANIAN, TOSK) jonxhë e kultivuar; (ASSAMESE) lucerne ghah; (BELARUSAN) lutsena pasjannaja; (BENGALI) lecern; (CHINESE, MANDARIN) zi hua mu xu; (CHINESE, WÚ) zi hua nu xu; (CHINESE, YUE) mu xu; (CREOLE, HAITIAN FRENCH) alfalfa; (CRIOULO, CAPE VERDE) anafe, alfalfa; (CZECH) tolice vojtěşka, tolice setá, vojtěška, lucina; (DANISH) lucerne; (DUTCH) luzerne, lucerne; (ENGLISH) alfalfa, alfalfa sprouts, lucerne; (ESTONIAN) lutsern; (FARSI, WESTERN) younjeh; (FINNISH) sinimailanen; (FRENCH) luzerne; (GAELIC, IRISH) alfalfa; (GAELIC, MANX) alfalfey; (GERMAN) Saat-Luzerne; (GREEK) midiki, triphyli; (GUJARATI) gadab rajko, vilayti-ghas; (HEBREW) aspeset tarbutit, aspeset; (HINDI) wilayti-gawuth, lasunghas; (HMONG DAW) zaub nyuj; (HOPI) kawàytusaqa; (HUNGARIAN) lucerna, takarmány lucerna; (ICELANDIC) refasm'ari; (INDONESIAN) alfalfa; (ITALIAN) erba medica, erba Spagna; (JAPANESE) alu falufa, murasaki-uma-goyashi; (KANNADA) vilayati-hullu, kudure masale; (KIOWA) 'H'-sαhyei-gα; (LATVIAN) sējas lucerna; (LITHUANIAN) mėlynžiedė liucerna; (MACEDONIAN) obična lucerka; (MALAY) alfalfa; (MALAYALAM) lucerne; (MARATHI) lasun ghas, vilay-ati-gavat, vilaiti ghavat; (MONGOLIAN) tarimal tsargas; (NEPALI) alfalfa; (NORWEGIAN) blåluserne, luserne; (ORIYA) lusarna; (PANJABI, EASTERN) lusan; (POLISH) lucerna siewna, alfalfa; (PORTUGUESE) luzerna, alfalfa; (ROMANIAN) lucerna comună, lucernă; (RONGA) luzerna; (RUSSIAN) lutserna posevnaja, lutserna obyknovennaja, lutsarna sinjaja; (SERBO-CROATIAN) lucerka sijata; (SLOVAK) lucerna siata; (SPANISH) alfalfa, mielga; (SWAHILI) luseni; (SWEDISH) alfalfa, blålucern; (TAJIKI) junuchqa, beda; (TAMIL) kuthiraimasal; (TELUGU) garam masal; (TURKISH) adi yonca, yonca; (TURKMEN) ekimlik ýorunja; (URDU) losan; (VLAAMS) luzerne; (XHOSA) ilusini; (ZUNI) ja.

40.18. *Mucuna pruriens* (L.) DC. var. *utilis* (Wall, ex Wight) Baker, ex Burck. (and several other species, e.g. *M. pachylobia*) (EP) seeds; (P) C.

(AFRIKAANS) fluweelboon; (AJA) ovɛ wiwi; (AKWA) tsokombéle; (ANDO) nduï; (ASSAMESE) makhan urahi; (BAATONUM) tabaru; (BALANTA-KENTOHE) metáftum; (BAOULÉ) n'duouin, duï; (BASSARI) *va*-diegon, *a*-ndiɔxón; (BENGALI) taru kala, tarukala, alkolshi, alkusa, alkushi, bichchoti; (BEROM) ekɔ́s; (BÉTÉ) bagou; (BICOLANO, CENTRAL) hipoi, hipoy, nipoy; (BONTOC, CENTRAL) kamkamiling; (BUDIK) *ɓe*-yowél; (CEBUANO) nipai, kanipay, nipay, nipay; (CHICKASAW) bala' woksho'; (CHINESE, GÀN) lao shu dou, hu gua dou; (CHINESE, MANDARIN) li dou; (CHINESE, MIN BEI) bu gua dou, gou gua dou; (CHINESE, WÚ) li dou; (CHINESE, YUE) hu gua dou; (CRIOULO, CAPE VERDE) ganhoma, feijão-de-bitcho, feijão-de-lagartixa; (CRIOULO, UPPER GUINEA) gañoma, nanoma, nayuma, ganhomá; (DANGME) tʃakatʃaka; (DANISH) fløjlsbønne; (DENDI) gbengbele; (DIGO) ulanga; (DUTCH) jeukboontje; (EDE NAGO) werekpe; (EDO) íghekpe; (ENGLISH) velvet bean, buffalo bean, cowage, cowhage, common cowitch, cat bean; (FILIPINO) pataning espana; (FON) azɔ; (FRENCH) pois velu, pois a gratter, pois pouilleux, pois mascate, mucuna, liane à gratter; (FULFULDE, ADAMAWA) nyanyare gaduru; (GA) aŋleŋle; (GBE-VHE) adza, edza; (GERMAN) Juckbohne, Juckfasel; (GREEK) veloudinos faslolos; (GUJARATI) kavachan, kandchan, kivanch, kavatch, kiwach; (HANUNOO) anipay; (HAUSA) kàkàkà, marara, sansoni; (HINDI) khamch, kaevanch, tohar sem, baikhujani, erka, goncha, kaunch, kiwanch, kajohara, kamaach, kawanch, kivachh, kivanch, konch; (IGBO) àgbàlà, àgbalà, akugba, àgbàlà ufe; (INDONESIAN) kekara juleh, kacang miang, kacang babi, kacang benguk, benguk, kara benguk, kerawai, rawe; (ITALIAN) fagiolo antelmintico, fagiolo de nigriti; (JAPANESE) fûki-mame; (JAVANESE) kara benguk, kara rawe, benguk; (JOLA-FONYI) é mo, ennańey, é gañuma, é ganuna, gaňa; (KANNADA) nasagunni, nasugunni, nasukunni, hasaguni, anipe balli, chinakee beeja, kaadavare, koogari, markata, nayi sonagu balli, naayi sonku, nasagurrni, nasugunni kaayi, nasukoogari, sonagee beeja, turanchi, turuchi avare; (KHMER, CENTRAL) khnhaè; (KONKANI) khavalya valli; (LAHU) hpui^v a¯ le te; (LAO) tam nhè; (LOMWE, MALAWI) kalikonda; (MACEDONIAN) kadifest bob; (MAKHUWA) oyele; (MALAY) kacang babi, kekaras gatal; (MALAYALAM) naicorna, naikorana, naikkuruna, shorivalli, shoriyanam, naaykkurana, naaykorana; (MANDING-MANDINKA) bala gana, balãngaña, mafora, mapura, ñañanako, sa soo, maforá, manforá, nhanhamacô; (MANDINKA) konyurau, korodu; (MANDJAK) bako ñal, búconhoále; (MANKANYA) nañuma, o gan, um-háe; (MARATHI) khajori, bin, kanchkuri, kavacha, kuhili, kaajkuhilee, kuyeli; (MONGOLIAN) nugiin tsargas; (NEPALI) mishri khand; (NGUNDI) matchécha; (NIUE) feteka; (NYANJA) kalongonda; (ORIYA) velvet simba, alokushi, baidhonko, kachu, kaincho, morkotomado; (PAMPANGAN) ipe; (PANGASINAN) lepa; (PANJABI, EASTERN) makhmali sem, gunchgaji, kanaucha, kawanch, knochkari; (POLISH) swierzbiec właściwy, wspinacz właściwy; (PORTUGUESE) pó de mico, olho de burro, café de Mato Grosso; (PULAAR) kossé, ñañdébãg, ngagna, cóssè, ganhaca, nhandebágue, talcódja; (SANSKRIT) aaathmaguptha, kapikachhu, atmagupta, kapikacchu, kapikacchuh, kapikachchhu, kapikachliu, adhyanda, ajada, ajalomi, ajavha, alkushee, arshabi, badari, chanda, durabhigraha, dusparsha, gatrabhanga, guru, jada, kachhumathi, kachhura, kandukaari, kandura, kapiprabha, kapiromaphala, kashiroma, kundali, maharshabhilangali, markati, praavrisha, pravrishayana, rishabhajata, rishabhi, rishyaprokta, romalu, romavalli, sadhyashotha, shimbi, shukapindi, shookashimba, shookashimbika, shukavathi, sugupta, svaguptha, svayamguptha, theekshna, vaanari, vanashukari, varahika, vrishya, vyaaghra, vyanga; (SENA) uri:ri; (SERER-SINE) balãngaña; (SHONA) uliri; (SPANISH) picapica, ojo de zamuro, chiporro; (SUNDA) kowas, kacang kowas, benguk; (SUSU) fɔte ɛtugi; (SWAHILI) upupu; (TAGALOG) nipai, lipay, nipay; (TAMIL) amudari, amudaari, arugrattam, kozhiavarai, punaikkali, punaippidukkan, punaikalei, sugasimbi, naikurni, poonaikkali, atthumbutthar, irishiya, kadigandu, kavi, maaguttam, ponnepoonduku averaikody, poonaippidukkan, poonaiyavarai, sugasimbi; (TELUGU) nugu chikkudu, doolagondi, pilliyadagu, nuguchikkudu, dulagondi, duradagondi, piliadagu kaila, pilliadugu,: dooradagondi, nayikkorana, praavrishayani, theeta kogial, thelladooradagondi, thotadoolagondi; (TEM) pumbela; (THAI) mamui; (TIV) imo; (TONGA) ukazi mudzi; (TULU) nāyisoṇaṅgu̧, nayi songuballi, naayi nonang; (TURKISH) mucuna; (URDU) kavancha; (VIETNAMESE) d[aaj]u m[ef] o r[uwf]ng, m[os]c m[ef]o; (VLAAMS) fluweelboon; (WAMEY) *va*-tiitien, *a*-tyityin; (YAO) tangale; (YOM) tɔ̃m; (YORUBA) werekpe, èsin, èsinfunfun, èsinsin, esise, ewé iná, hérepe, wèpè, wèrepe, yèrepe;

40.19. *Neptunia prostrata* (Lam.) Baill. (EP) leaves and tender shoots; (P) C.

(ASSAMESE) pani lajuki; (CHINESE, MANDARIN) ya cai; (CHINESE, WÚ) ya cai; (DANISH) vandmimose; (ENGLISH) water mimosa; (FRENCH) neptunie potagère; (GERMAN) gemeine Wassermimose; (GREEK) mimoza e ydrocharis; (INDONESIAN) kemon; (JAPANESE) mizu-ojigi-sô; (KANNADA) nidraaganti; (KHMER, CENTRA) kânhchhnaèt; (LAO) ('phak) kas'éèd; (MACEDONIAN) vodna mimoza; (MALAY) akar keman hantu, keman gajah, keman air, tangki; (MALAYALAM) nittithoddavaddi; (POLISH) mimoza wodna; (PORTUGUESE) carurú d'água,

beans; (KISSI, NORTHERN) kawa pompohu, kowõ, kawolo; (KONO) soɛ, sɔɛ, suɛ; (KRIO) kalaba binch, binch; (KURANKO) sosɛ; (LAHU) naw^ tu^v ma; (LATVIAN) limas pupinas; (LIMBA) na-tamba, ya-tamba, koro goro; (LITHUANIAN) lenktasėklės pupelė; (LOKO) alnɔmgo, tɔghɔ; (LUO) oranga; (MACEDONIAN) lima grav; (MALAY) kacang serinding, kacang China, kacang Jawa, kekara kratok; (MANDING-MANDINKA) kissi soso, sɔso, tubabu soso; (MANDINKA) tubabu soso; (MANINKAKAN, WESTERN) toubabou soso; (MARANAO) dila a salaw, kapan, papalak; (MARATHI) double been; (MENDE) tɔwɔ, caka, hɔndɔ, tɔgugui, tɔle, tɔw; (MORISYEN) gro poi; (MUSKOGEE) tvlak-taphv, tvlak-tvptahv, tvlaktapho, tvlak-tvpho, tvlak-taphuce, tvlak-tapho; (NDALI) mankhamba; (NEPALI) simi; (NGUNDI) madeso; (NORWEGIAN) limabønne; (NUPE-NUPE-TAKO) zomgbara; (NYANJA) ntambohodo, makwera; (ORIYA) lamba simba; (PAMPANGAN) bule; (PANJABI, EASTERN) wade rawan, lobia, lobiya; (POLISH) fasola limeńska, fasola półksiezycowata; (PORTUGUESE) feijão de Lima, feijao carolino de Lima; (PULAAR) myɛbɛ, (QUECHUA, CUSCO) huarhui; (ROMANIAN) fasole de lima, fasole de luna; (RUSSIAN) fasol limskaja, fasol' lima, fasol serpovindnaja, lima; (SAMBAL, BOTOLAN) antak, iilo; (SANGA) iṣan-take; (SANSKRIT) akshipeedaka; (SERBO-CROATIAN) lima grah; (SLOVENIAN) lima fižol; (SPANISH) judia de Lima, poroto de Lima, poroto de manteca; (SRANAN) sebijari; (SUBANON) gyulis; (SUNDA) kacang mas, roway minyak, kacang roway; (SUSU) fɔaiswri, toge, togoxunbe, togué; (SWAHILI) maharagwe lima; (SWEDISH) lima böna; (TAGALOG) patani, buringi, butingi, sibatse simaron, patáni, zabache, bulay-patani; (TAGBANWA) kapan; (TAMIL) binusaveraikodi; (TELUGU) double beans; (THAI) thua rachamat; (THEMNE) ɛ-lel, ɛ-lɛkɔbɔtɔ, *a*-lel, *a*-lil, *ma*-panta-*ma*-insal; (TIV) adenger, lele, segha; (TUMBUKA) cimbamba, mandondo, mamberemende; (TURKISH) lima fasulyesi; (TWI) adua, apateram, asa ɛdua, ase dua, ɛdua, ɛdua nton, ɛdua fita; (URDU) sem, lobia; (VAI) sɔ; (VIETNAMESE) dâu ngu; (VLAAMS) rangoonboon, lima boon; (YALUNKA) fara-fara-tugo-na, tugo-na; (YIDDISH) bebil, fahsulia; (YORUBA) awuje, aton, awújẹ, ẹ̀wa, ẹwẹ, ẹwujẹ, kòkondó pòpondó.

40.26. *Phaseolus vulgaris* L. var. *vulgaris* (EP) immature pods and seeds; (P) C.

In many instances the generic term for bean is used for both immature pods and mature seeds.

(AFRIKAANS) boontjie, groenboon, tuinboon, snyboontjie, vanzylsesuikerboontjie; (ALBANIAN, TOSK) fasule, groshë; (ALEUT) taktôjk; (AMHARIC) mekke, adängʷarre; (APACHE, JICARILLA) na'olee, naa'oléé; (ARABIC) fasulyah; (ASSAMESE) motor mah; (AZERBAIJANI, NORTHERN) lobya, paxla, paxlal | bitki, maş, baş, kəllə, təpə, sollə, pul; (BADAGA) aravattavare, gujji avare, majjigavare, tu:ravare, pura:vare; (BALTI) moto, mutu, mamantu; (BAOULÉ) aloa; (BASQUE) babarrun; (BELARUSAN) fasol zvychainaja; (BEMBA) cilemba; (BENGALI) pharash bin; (BÉTÉ) popohoulé; (BINUKID) tilas; (BLACKFOOT) áótooksiinaattsi; (BODO) sobaima; (BONTOC CENTRAL) irula, lubias; (BOSNIAN) pasulj, grah; (BRETON) fav, fav glas; (BULGARIAN) fasul; (BULU) kôn, evele kôn; (BURMESE) pe-taung-shay, bo-sa-pè, pè-bya-galè, pègya(ni); (CAJUN FRENCH) been, fève, haricot; (CAROLINIAN) fiyoolis; (CATALAN-VALENCIAN-BALEAR) fasol, mongetes; (CHECHEN) qoe; (CHEROKEE) tuya; (CHICKASAW) bala' albilha, itti' tiyya', bala' sinti, bala' okchemali', balokchamali'; (CHINESE, GÀN) yun dou, jin dou, si ji dou; (CHINESE, HAKKA) t'éou; (CHINESE, MANDARIN) cai dou, ts'ai tou; (CHINESE, MIN BEI) si ji dou, yun dou; (CHINESE, MIN NAN) bín-tāu, chhài-tāu, âng-tāu, tāu; (CHINESE, WÚ) cai dou, dao dou, yun dou; (CHINESE, XIANG) da wan dou, run dou, si ji dou; (CHINESE, YUE) cai dou, yu dou, si ji dou; (COMANCHE) pihúraa; (CREE, PLAINS) miyîcîmin; (CREOLE, HAITIAN FRENCH) pwa tann, pwa; (CRIOULO, CAPE VERDE) fijon-bonginho; (CROATIAN) grah mahunar, niski; (CZECH) fazol obecný, fazol zahradní; (DAGAARE, SOUTHERN) nassar bēon; (DAKOTA) omnica; (DANISH) bønne, buskbønne, stangbønne, brækbønne, brydbønne, havebønne; (DIGO) maragwe; (DUTCH) slaboon, stokboon, sperzieboon, prinsesseboon, Franse boon; (ENGLISH) green bean, snap bean, French bean, string bean; (ESTONIAN) aeduba, uba; (ÉWÉ) ayi; (FARSI, EASTERN) lubiyâ; (FARSI, WESTERN) lubiyâ; (FIJIAN) bean; (FILIPINO) habichuelas; (FINNISH) tarhapapu; (FRENCH) haricot vert, haricot comun, phaséole; (FULFULDE, ADAMAWA) nyebbere, nyebbe; (GAELIC, IRISH) pónaire fhrancach, pónaire dhuánach, pónaire fhada; (GAELIC, MANX) poanrey; (GAELIC, SCOTTISH) pònair; (GALICIAN) feixó, faba, haba; (GANDA) bigyanjaro; (GEORGIAN) lobio, tsértsvi; (GERMAN) Gartenbohne, Brechbohne, Salatbohne, Prinzeßbohne; (GIKUYU) mboco; (GONJA) chebe; (GOURMANCHÉMA) sunna; (GOURMANTCHÉ) toũm bani; (GREEK) fasoli; (GUJARATI) phanasi; (HARARI) atärä, atär; (HARSŪSI) lōbi; (HAUSA) wāke, ganyer rafi; (HEBREW) sheuit hagima, sheuit haginna, she'u'it ha-ginna; (HERERO) bûiki-i, ekunde; (HINDI) jungli sem, bakla, lobia, frash bean, rajmah, pharasbean; (HINDUSTANI) bonki, lobhiyá, báklá; (HMONG-DAW) taum lag; (HMONG-NJUA) taum laag; (HOPI) pàapu, suqömvàapu, palavàapu, maawiwngwu, wupavàapu, mori; (HUNGARIAN) paszuly, bab, veteménybab; (IBANAG) bitsuwelas; (ICELANDIC) baunir; (IFUGAO) awwat an luyu, awwat an mumbolah, puddila an maluhadan, puddila an mummool; (IGBO) akidi, agwa, aloso, akisi; (INDONESIAN) kacang merah, kacang buncis, buncis, boncis, buncis merah, katjang djogo; (INUKTITUT, EASTERN CANADIAN) nilernait, nilirnaq, pisi; (ITALIAN) fagiolino, fagiolo; (IZON) ọzáụ; (JAPANESE) ingen mame, ingen; (JAVANESE) buncis, koro buncis, koro tonggak, koro lompeh, ucet; (JOLA-FONYI) koussaac; (JULA) sosso; (KANNADA) hurali kayi, tingalavari, thingalavare; (KANURI) ngálò; (KASHMIRI) bākala; (KAZAKH) fasol obyknovennaja, burshaq; (KHMER, CENTRAL) sândaèk barang; (KIOWA) t̂ɑdl-t'ɑn; (KIRGHIZ) buurchak; (KOREAN) gangnamcong; (KRIO) binch; (KURDISH, NORTHERN) fāsolyah; (KWANGALI) ekunde; (KWÉYÒL) pwa, pwa-salad; (LADINO) fijón, avas, fijon verde, fasulyas, fijon amariyo, barbunyas, fijon blanco; (LAHU) naw^ ta^v shi_, naw^ nyi^- shi_; (LAO) thwàx fàlangx; (LATVIAN) parastās pupiņas, pupinas, pupa; (LITHUANIAN) daržinės pupelės, pupelės; (LOMWE, MALAWI) mphwanda; (LUO) oranga; (LUXENBOURGEOIS) Boun; (MACEDONIAN) grav, običen grav; (MALAY) kacang buncis, kacang merah; (MALAYALAM) beans; (MANINKAKAN, WESTERN) sosso; (MALTESE) fazola il-hadra; (MARATHI) shravan gevda, shravanghevda; (MEHRI) dəgərīt, dēgər; (MICMAC) wegoon, papawā'; (MIZO) kawm neia rahchi; (MOHAWK) ohsahèta; (MORISYEN) zarico ver; (MUNDARI) mānāl; (MUSKOGEE) tvlak-lane, tvlak-cvpko, tvlako, tvlak-lvste, tvlak-espokhv, tvlak' encakhv; (NAMA-DAMARA) omakunde; (NAVAJO) naa'olí dooł'izhí, naa'olí; (NDALI) malima; (NDONGA) makunde; (NEPALI) asaare simi; (NGIZIM) ríidùwà, āɾ̃dùwà; (NIUE) pine; (NORWEGIAN) hagebønne, låg hagebønne, høg hagebønne, bønne, skolmebønne; (NUPE-NUPE-TAKO) zongbara, ezo; (NYANJA) mbwanda, cimbamba, kayera; (ORIYA) farasi simba; (OROMO, WEST CENTRAL) bakella; (PAIUTE, SOUTHERN) muhdee´, modee´, muhdee´vees, ungkaw´muhdee´; (PANGASINAN) agáyep, taúsi, bitsuwelas; (PANJABI, EASTERN) loba fasoulia, babri, frãss bēn, frans bean, räjmã; (PASHTO, NORTHERN) fâsuliya, bag̱ri, gung̱ri, lobiyā; (POLISH) fasola zwykła; (PORTUGUESE) feijão, feijão comum, feijão verde, feijão vagem; (POTAWATOMI) mîskodi'ssîmîn, koje's, koje'suk; (PULAAR) ñewre; (PULAR) haricot, niébé; (QUECHUA, CUSCO) porotillo purutu; (ROMANI, BALKAN) bòboos; (ROMANIAN) fasole comună; (RUSSIAN) fasol' obyknovennaja; (SAMBAL, BOTOLAN) kaling, utong; (SAMBAL, TINÀ) bitsuwelas; (SANSKRIT) brihath shimbe, raajamaasha; (SENOUFO, CEBAARA) djagueli; (SERBO-CROATIAN) pasulj; (SERER-SINE) ñaw; (SHONA) bhinzi; (SIDAMO) bâqêla; (SINHALA) bonchi; (SLOVAK) fazul'a záhradná, fazul'a obyčajná; (SLOVENIAN) fižol; (SORBIAN, LOWER) tšuki, šmykace; (SORBIAN, UPPER) buna; (SOTHO, NORTHERN) nawa; (SOTHO, SOUTHERN) nawa; (SPANISH) judia común, judia verde, faséolo; (STELLINGWERFS) bone, grune bone, brune bone, greuneters, droogeters, pronker, pronkbone, schoezole, sniedebone, stoksniedebonen, stamsniede bonen, wiekschellen, hadschellen; (SUMO-MAYANGNA) sinak; (SUNDA) buncis, kacang rompes, kacang kopak, kacang beureum, kacang jogo, kacang tunggak; (SWAHILI) maharagwe, maragwe, choroko, fiwi; (SWEDISH) böna, trädgårdsböna, bruna bönor, störböna; (TAGALOG) habichuelas, bitsuwelas, biringi, butingi, mula; (TAJIKI) lubio; (TAMIL) avarai, seemai avarai, seemai averai kody; (TATAR) fasol'; (TELUGU) bangalore beans; (THAI) thua khaek, thua phum; (TIBETAN) ʈɛma; (TIV) i̱li̱ç̂v; (TONGA) indima; (TSWANA) nawa; (TUMBUKA) ncungu; (TURKISH) fasülye, fasulya, pakla; (TURKMEN) ekilýan noýba; (UKRAINIAN) ovočeva kvasolja; (URDU) lobia; (UYGHUR) purchaq; (UZBEK) loviya; (VENDA) n̪awa, lun̪awa; (VIETNAMESE) dau xanh, dâu ve; (VLAAMS) boon, kookboon, snijboon, sperzieboon, slaboon, princesseboon; (WOLOF) harico (niébé); (XHOSA) imbotyi; (YIDDISH) bebil, fahsulia; (ZARMA) dunguri; (ZULU) ulentshisi, ubhontshisi; (ZUNI) no.

40.27. *Pisum sativum* L. subsp. *sativum* var. *macrocarpon* (EP) immature pods; (P) C,R.

(AFRIKAANS) eetbare peulert; (BURMESE) holantan-pe; (CHINESE, MANDARIN) tian cai wan dou; (CHINESE, MIN NAN) tāu-ngeh; (CHINESE, YUE) tian cai wan dou; (CHINESE, WÚ) tian cai wan dou; (DANISH) sukkerært; (DUTCH) peul; (ENGLISH) edible podded pea, sugar pea, snow pea, Chinese pea; (FILIPINO) chicharo; (FRENCH) pois mange tout, pois sucré; (GERMAN) Zuckererbse; (HMONG DAW) taum mog; (INDONESIAN) kacang; (ITALIAN) pisello mangiatutto, pisello dolce, taccola; (JAPANESE) kinusaya; (MACEDONIAN) krupnoploden grašok; (MALAY) kacang pi; (MALTESE) pixelli fil-mizwett; (MONGOLIAN) tarimal vandui; (MORISYEN) tipoi manz-tou; (NEPALI) sano kerau, matar kosha; (NORWEGIAN) sukkerert; (PANGASINAN) sitaró; (PANJABI, EASTERN) matar; (POLISH) groch konserwowy; (PORTUGUESE) ervilha torta, ervilha de quebrar, ervilha tipo come-tudo, ervilha come-tudo; (ROMANIAN) mazăre cultivată; (SPANISH) guisante tirabeque, guisante come-toto, guisante azucarado; (SWAHILI) kunde, njegere; (URDU) matar; (VLAAMS) eetbare peulerwt.

40.28. *Pisum sativum* L. subsp. *sativum* var. *sativum* (EP) immature seeds, tender shoots[S]; (P) C.

(AFAR) qàtar; (AFRIKAANS) groenert, tuinert, ertjie; (ALBANIAN, TOSK) rrillë, bizele; (AMHARIC) atär; (ARABIC) basella; (ASSAMESE) motor mah; (AZERBAIJANI, NORTHERN) noxud; (BADAGA) kaḍale, kaḷḷe; (BALTI) garaz, pok-sran; (BASQUE) idar; (BELARUSAN) goroh pasjanny; (BEMBA) ntongwe; (BEMBÉ) makata-ta; (BENGALI) matar, bara matar; (BICOLANO, CENTRAL) sitsaro; (BLACKFOOT) oohkominoko; (BODO) bud, motor; (BONTOC CENTRAL) sitsaro; (BOSNIAN) grašak; (BRETON) piz bihan; (BULGARIAN) gradinski grach; (BURMESE) shwe-pe, sadaw-pè; (CATALAN-VALENCIAN-BALEAR) pèsol; (CEBUANO) sitsaro; (CHECHEN) qoe; (CHICKASAW) naahollimbala', bala' lhibowa' bala'; (CHINESE, GÀN) wan dou, mai dou, he lan dou; (CHINESE, HAKKA) hô-lân t'éou; (CHINESE, MANDARIN) ai xing cai dou, wan dou, wan tou; (CHINESE, MIN BEI) he lan dou, jin dou; (CHINESE, MIN NAN) hô͘-lîn-tāu-jîn, hô͘-liân-tāu, hô͘-lîn-tāu, hoan-á-tāu, le̍k-tāu, tāu, tāu-jîn; (CHINESE, WÚ) he lan dou, wan dou; (CHINESE, XIANG) mai dou; (CHINESE, YUE) he lan dou, wan dou; (COMANCHE) toponibihuura?; (CREE, PLAINS) mîcîmin; (CREOLE, HAITIAN FRENCH) pwa frans; (CRIOULO, CAPE VERDE) ervilha; (CROATIAN) grašak; (CZECH) hrách zahradny, hrách setý, hrách setý pravý; (DAKOTA) omuica hmiyaŋyaŋ; (DANISH) havaeært, ært; (DUTCH) erwt, doperwt; (ENGLISH) pea, garden pea; (ESTONIAN) hernes; (FARSI, EASTERN) matar; (FARSI, WESTERN) nokhod sabz, nokhod; (FIJIAN) matar; (FILIPINO) guisantes; (FINNISH) tarhaherne, kylvöherne; (FRENCH) pois potager, petit pois, pois; (GA) yɔɔ; (GAELIC, IRISH) pis; (GAELIC, MANX) pishyr; (GAELIC, SCOTTISH) peasair; (GANDA) kawo; (GEORGIAN) barda; (GERMAN) Erbse, Garten-Erbse; (GIANGAN) sitsaro; (GIKUYU) minji; (GOURMANCHÉMA) nassara-tchine; (GREEK) arakas, mpizeli; (GUJARATI) vatana, watani; (HARARI) antärä; (HAUSA) waken Turawa; (HEBREW) afuna, afuna tarbuti; (HERERO) ergi-i; (HILIGAYNON) gisántes; (HINDI) muttar, matar; (HINDUSTANI) maṭar, kábulí; (HMONG DAW) taum mog; (HUNGARIAN) borsó, zöldborsó veteményborsó; (IBANAG) sitsaro; (ICELANDIC) ertur; (IFUGAO) sitsaro; (ILOCANO) sitsaro; (INDONESIAN) ercis, kacang ercis, kacang kapri, kacang polong, kapri, kacang polong èrtjis; (INUKTITUT, EASTERN CANADIAN) pattak, pisi, pisiujait; (ITALIAN) pisello coltivato, pisello; (JAPANESE) endou, endō;

(JAVANESE) kacang polong, kapri, polong genjah, polong picis, polong priangan; (KANNADA) batani, batagadle; (KANURI) ngálò nàsàrá; (KASHMIRI) hĕm, kara; (KAZAKH) goroch posevnoji, burshaq; (KHMER, CENTRAL) sândaèk muul; (KIOWA) t̂ɑdl-t'ɑn-sɑhyei-gɑ; (KIRGHIZ) buurchak; (KOREAN) wandu; (KRIO) piz; (KWANGALI) ongwinḓi; (KWÉYÒL) pwa; (LADINO) bizèlia, pinzèla; (LAKOTA) omnicagmigmi; (LATVIAN) sējas zirņi, zirnis pukuzirnīši, zirni; (LITHUANIAN) sėjamasis žirnis, daržo žirniai; (LOMWE, MALAWI) nsawawa, kabaifa; (LUO) ngor; (LUXENBOURGEOIS) Irbes, Ierz; (MACEDONIAN) grašok, običen grašok; (MAITHILI) maṭar; (MALAY) kacang wangi, kacang ercis; (MALAYALAM) pattani; (MALTESE) pizelli; (MARATHI) watani, matar; (MICMAC) alawā'; (MIZO) be; (MOHAWK) onekwa; (MONGOLIAN) tarimal vandui; (MORISYEN) tipoi; (MUNDARI) matar; (MUSKOGEE) tvlak-pakkuce; (NAMA-DAMARA) ofukwa; (NAVAJO) naa'ołí nímazí; (NDONGA) edunde; (NEPALI) kerau, matar kosha; (NEWAR) kay-gulī; (NORWEGIAN) hageert, ert, erter, pillert; (ORIYA) matar; (PAIUTE, SOUTHERN) wechee´nupuv; (PAMPANGAN) sitsaro; (PANGASINAN) gisántes, gizantes; (PANJABI, EASTERN) mattar, mater, matar; (PASHTO, NORTHERN) mëṯër, matar, kṛāk; (POLISH) groch, groch zwyczajny, groch zwyczajny ogrodowy; (PORTUGUESE) ervilha, ervilha de debulhar, ervilheira, griséus, ervilha seca; (PULAAR) ñewre; (ROMANIAN) mazăre cultivată, mazăre; (RUSSIAN) goron pasevnoi, goroh agorodnyi, goroh ovoščnoj; (SAMOAN) pi; (SANTALI) mo̱ṯo̱r; (SCOTS) pey; (SERBO-CROATIAN) grašak; (SHONA) ndodzi, pizi; (SLOVAK) hrášok, hrach; (SLOVENIAN) grah; (SORBIAN, LOWER) groch; (SORBIAN, UPPER) hroch, hróšatko; (SOTHO, NORTHERN) erekisi; (SOTHO, SOUTHERN) erekisi; (SPANISH) guisante, guisante común, pésol, arveja; (STELLINGWERFS) aart, atepoelen; (SUNDA) gamet, kacang kapri, kacang polong; (SWAHILI) kunde, njegere, dengu, adesi, mbaazi; (SWEDISH) ärt, grä ärt, spritärt; (TAGALOG) bichuelas, habichuelas, sitsaro, citzaro; (TAJIKI) nakhudi rusi; (TAMIL) pattani; (TATAR) borchak; (TAUSUG) sitsaro; (TELUGU) bataneelu; (TIBETAN) ţēma; (TSWANA) erekisi; (TURKISH) bezelye, sebze bezelyesi; (TURKMEN) gök nohut; (UKRAINIAN) ovočevij goroch; (URDU) matar; (UYGHUR) purchaq; (UZBEK) kok nuchat, no'xat; (VENDA) eregìsi; (VIETNAMESE) dâu hà-lan, dâu hoà lan; (VLAAMS) doperwt, erwt; (WELSH) pys; (XHOSA) ierityisi, iertyisi; (YIDDISH) ahrbis; (ZULU) inhlamuu kaphizi, inhlamvu kaphizi; (ZUNI) no-k/amolliya.

40.29. *Psophocarpus tetragonolobus* (L.) DC. (EP) immature pods, seeds, leaves[S], root[S]; (P) C,R.

(AFRIKAANS) gevlerkte boon; (AGTA) kaboy; (AKLANON) segidilya, sererelya; (BENGALI) chara-koni-sem, lakar sem; (BICOLANO, CENTRAL) parupagulong, segidilya; (BINUKID) keribang; (BONTOC CENTRAL) beyed, buligan; (BUHID) abe; (BURMESE) pe-zaung-ya, pe-saung-ya, hto-pong, ku-bemya; (CEBUANO) sererella, segadilla, kamaluson, karabansos; (CHINESE, GÀN) bian dou; (CHINESE, MANDARIN) yi dou; (CHINESE, MIN BEI) mian dou; (CHINESE, WÚ) yi dou; (CZECH) psofokarpus, praskavec ledencový; (DANISH) goabønne; (ENGLISH) winged bean, Goa bean, asparagus bean, asparagus pea; (FILIPINO) siguidilyas; (FINNISH) goa-papu; (FRENCH) pois ailé, pois carré, haricot ailé; (GADDANG) bulligan; (GERMAN) Goabohne; (HANUNOO) buray dipay, kabay, kabay mabiru, kabay mararas; (HILIGAYNON) balagay; (IBANAG) amale; (IFUGAO) buligan, bulligan an gayyang, bulligan an mumpaha; (ILOCANO) pallang, palag, palam, kalamismis; (INDONESIAN) kecipir, kacang botor, kacang belimbing; (ITAWIT) amalek, amalet; (IVATAN) puhog; (JAPANESE) shikaku mame; (JAVANESE) kecipir, cipir, cicipir, ketypir; (KANKANAEY) buligan; (KANNADA) shambe kayi; (KHMER, CENTRAL) prâpiëy; (LAO) thwàx ph'uu; (MACEDONIAN) krilest bob; (MALAY) kacang botor, kacang kelisa, kacang sirek; (MARANAO) kalamismis; (MORISYEN) poi kare; (NEPALI) pate simi; (NORWEGIAN) goabønne; (PAMPANGAN) sigarillas; (PANGASINAN) gabuy, sigarilyas; (POLISH) łust głąbigroszek; (PORTUGUESE) fava de cavalo, feijão alado; (SAMBAL, TINÀ) sigarilyas; (SINHALA) dhambala; (SPANISH) frijol alado, dólico de Goa; (SUNDA) jaat, roway jaat; (TAGALOG) sigarilyis, sigarillas, sigadillas, sigadilyas, kabey, kalamismis, cigarillas, segidilyas, sigarilyas; (TAMIL) morisuavarai;

(TAUSUG) batung-baimbing; (THAI) thua plu, thua-phu; (TOK PISIN) asbin; (VIETNAMESE) d[aaj]u r[oof]ng, d[aaj]u kh[ees]; (VLAAMS) gevieugelde boon; (WÁRAY-WÁRAY) segidilya.

40.30. *Pueraria montana* (Lour.) Merr. var. *lobata* (Willd.) Maesen & S. M. Almeida, ex Sanjappa & Predeep (EP) roots, leaves[S] and tender shoots[S]; (P) C.

(ALBANIAN, TOSK) puerarie; (ARABIC) kaskt, kaskt yapani; (ASSAMESE) kudzu; (BENGALI) kudzu; (BONTOC, CENTRAL) baai, baay, baay; (CHINESE, MANDARIN) ge teng, ko; (CHINESE, MIN BEI) ten ge; (CHINESE, WÚ) ge teng; (CHINESE, YUE) ten ge; (CREOLE, HAITIAN FRENCH) kudzou; (CZECH) kudzu subtropické, pueraria; (DANISH) kudzibønne; (DUTCH) koedzoe; (ENGLISH) kudzu, tropical kudzu, Japanese arrowroot; (FIJIAN) yaka; (FILIPINO) kudzu; (FRENCH) kudzu, kudzu du Japon, koudzou, vigne japonaise, puéraria; (GERMAN) Kudzu-bohne, Kopoubohne; (GUJARATI) kudzu; (HANUNOO) bagunaw malagti; (HEBREW) puerarya se'jra, pueraria se'ira; (HINDI) dab, kudzu; (HUNGARIAN) fojtóbab; (INDONESIAN) kudzu; (ITALIAN) pueraria, pueraria strisciante; (JAPANESE) kuzu, kudzu; (KANNADA) gamadi hullu; (MACEDONIAN) kudzu; (MANOBO) tahaunon; (MANU'A SAMOAN) va'a, fuefue fai va'a; (MARATHI) kudzu; (NIUE) aka, aka fala, fou gau; (ORIYA) kudajnina; (PANJABI, EASTERN) kudzu; (POLISH) opornik łatkowaty; (PORTUGUESE) kudzu, kudzu tropical, pueraria, puero; (RUSSIAN) kudzu; (SAMOAN) a'a; (SANTALI) jaṅ tirṛạ, patal kohṇḍa, tirṛạ da; (SERBO-CROATIAN) puerarija; (SLOVAK) pueraria; (SPANISH) kudzú común, kudzú tropical, hierba kudzú; (SUNDA) tobi; (SWAHILI) kuzu; (TELUGU) kudju; (THAI) tamyakhrua; (TONGAN) aka, akataha, fue'aepuaka; (TURKISH) pueraria, kaba tüylü Japon sarmaşiği, adi Japon sarmaşiği, kudzu; (VIETNAMESE) cu nang, cu s[aws]n d[aa]y; (VLAAMS) kudzu.

40.31. *Sphenostylis stenocarpa* (Hochst. ex. A. Rich.) Harms. (EP) tubers and seeds[S]; (P) C.

[1]Itsukha dialect of Luyia.

(AFRIKAANS) jamboon; (BASSARI) sessonge; (CHINESE, MANDARIN) suan pan zi; (CHINESE, WÚ) suan pan zi; (CHINESE, XIANG) xiao luo bu, si ji luo bu; (DAGBANI) gundosollo; (DANISH) afrikansk yamsbønne; (EDO) ihiehie, ikhekhe; (ENGLISH) African yam bean; (ÉWÉ) kutreku; (FRENCH) haricot igname, pomme de terre du Batéke, pomme de terre du Mossi, diégemtenguéré; (GBE-VHE) kutreku; (GERMAN) Afrikanisch Knollenbohne; (GOURMANTCHÉ) aan touna; (HAUSA) bidi, girigiri, serikibaba; (IGBO) okpodudu, àkịdị nwa-ngwu, odudu, oka odudu, ọkpa odudu, podu; (IZON) ịyáụ́; (KABIYÉ) tschanguilou; (KONKOMBA) sumpelegu, sumpae légu; (LUYIA) amaruku[1]; (MAKHUWA) n'toco; (MÒORÉ) diéguem tenguéré, noruko; (POLISH) fasola afrykańska; (TEM) kotonosu, kotonosou; (TIV) ahuma; (VLAAMS) Afrikaanse yamboon; (YORUBA) otili, sèsé, ṣẹṣẹ, ṣẹṣẹ erèé.

40.32. *Trigonella foenum-graecum* L. (EP) leaves, tender shoots and immature pods; (P) C.

(AFAR) abqaake, culba; (AMHARIC) abəš; (ASSAMESE) menthi guti; (BADAGA) mette; (BELARUSAN) pazitnik; (BENGALI) methi, methi-shak, methhuka, hoemgreeb; (BURMESE) penantazi; (CHINESE, MANDARIN) hu lu ba; (CHINESE, WÚ) hu lu ba; (CZECH) pískevice, pískavice řecké seno; (DANISH) bukkehorn; (DUTCH) hoornklaver, fenugriek; (ENGLISH) fenugreek, methi, goat's horn, Greek hay-seed; (ESTONIAN) lambalääts; (FIJIAN) mehti; (FINNISH) sarviapila; (FRENCH) trigonelle, foin grec, trigonelle fenugrec, fenugrec, sénegré; (GERMAN) Schabzierger-Klee, Bockshorn-Klee, Griechisch-Heu, Griechischer Bockshornklee; (GREEK) trigonella e frekocho-rios; (GUJARATI) methi, methini, bhaji; (HARARI) ḥilbät; (HEBREW) ḥilbe, gargeranit yevanit; (HINDI) methi, muthi, maythi, maethi, muthi; (HINDUSTANI) menthe, shambalid, hulbá; (HUNGARIAN) görögszéna; (INDONESIAN) klabet, kelabet; (ITALIAN) fieno greco; (JAPANESE) koroha; (JAVANESE) klabet; (KANNADA) menthya, mentesoppu, menk-palle, mente, menthya soppu, menthe; (KONKANI) metthi; (LITHUANIAN) vaistinė ožragė; (MACEDONIAN) grčka rogasta dete lina; (MAITHILI) mēthī; (MALAY) halba, kelabat; (MALAYALAM) venthayom, venthayam, uluva, venthiam, venthaiam; (MALTESE) helba fienu; (MARATHI) methi; (MIZO) an hnah; (MONGOLIAN) chireg; (MORISYEN) meti; (NEPALI) methi; (NEWAR) mithĭ; (NORWEGIAN) bukkehorn; (ORIYA) methi; (PANJABI, EASTERN) methi, methini, methri, methun, mātha, māthi; (PASHTO, NORTHERN) malkhoza; (POLISH) kozieradka pospolita, kozieradka; (PORTUGUESE) feno grego, fenacho, alforvas; (ROMANIAN) molotru; (RUSSIAN) pazhitnik, triganella; (SANSKRIT) methika, methi, chandrika, asumodhagam, pitabija, bahuparni, bahupathrika, deepani, gandhabeeja, gandhaphala, jyothi, kairavi, kunchika, mantha, methini, misrapushpa, munindrika, peethabeeja, vallari, vedhani; (SERBO-CROATIAN) piskavica, mnogosjemena; (SONGHAY) al harba; (SPANISH) alholva, fenogreco, heno griego; (SWAHILI) fenigiriki; (TAMIL) vendayam, ventagam, vendiyam, vethayam; (TELUGU) menti, mentikoora, mentikura, mentulu; menthulu; (TULU) mettè, metthi; (TURKISH) çemen, çemen otu, buyotu; (TURKMEN) sypally üçýaprak, boýdäne; (URDU) methi; (VIETNAMESE) h[oof] l[oo] ba; (VLAAMS) fenegriek.

40.33. *Tylosema esculentum* (Burch.) A. Schreib (EP) immature pods and root; (P) C.

(AFRIKAANS) gemsbokboontjie, gemsbokwortel, maramaboon, tama, tammani, ammani; (ASSAMESE) kotra, ram jokhola; (BULU) kôn; (CHINESE, WÚ) yang ti jia; (DANISH) maramabønne; (ENGLISH) marama bean, gemsbuck bean; (FRENCH) bauhinia; (GERMAN) Elandsbohne; (HEBREW) bohinya neekhelet; (INDONESIAN) kupu-kupu; (ITALIAN) bauhinia; (KUNG-EKOKA) n//ang; (NAMA-DAMARA) //nâus, //nâuhaib; (PANJABI, EASTERN) kachnär; (PORTUGUESE) bauhinia, feijão de Marama; (SPANISH) bauhinia, judia de Marama; (TURKISH) dağ abanozu, havhinya; (URDU) katchnar; (VENDA) gomondo; (VLAAMS) marama boon.

40.34. *Vicia faba* L. (EP) immature seeds; (P) C.

a. broad bean

(AFAR) bagal; (AFRIKAANS) boerboon, perdeboon, boerboontjie, botterboontjie, platte-peters; (ALBANIAN, TOSK) bathë; (AMHARIC) däng^{w}ällo; (ARABIC) ful; (ASSAMESE) urohi, urahi; (BADAGA) eḷḷanavare, eḷḷavare, moravare; (BASQUE) baba; (BELARUSAN) ruskija babi; (BENGALI) baklashim; (BULGARIAN) bakla; (BURMESE) sandusi; (CATALAN-VALENCIAN-BALEAR) faba, faves; (CHINESE, MANDARIN) ye hu lu ba, tsan tou; (CHINESE, MIN NAN) chhân-tāu, chhâm-tāu, hông-tè-tāu; (CHINESE, WÚ) can dou; (CRIOULO, CAPE VERDE) favona; (CROATIAN) bob; (CZECH) bob obecný, bob, bob zahradní; (DANISH) valsk bønne, hestebønne; (DUTCH) grote boon, tuinboon; (ENGLISH) broad bean, horse bean, fava bean; (ESTONIAN) põlduba; (FILIPINO) broad bean, faba bean; (FINNISH) härkäpapu; (FRENCH) fève, fève de marais, fève à gros grain; (GA) akpatramɔ; (GAELIC, IRISH) pónaire leathan; (GALICIAN) faba, faba loba, faballón, faboco, feixaco; (GEORGIAN) tsertsvi; (GERMAN) Pferdebohne, Saubohne, Puffbohne, Großebohne, Dickebohne; (GREEK) koukia chlora, koukki; (HEBREW) pol, ful, pol hagina, bikyat hagina; (HINDI) baakla, bakla, anhuri, kala matar; (HMONG DAW) taum nees; (HUNGARIAN) lóbab, lóbükköny, babó; (ICELANDIC) hestabaunir; (INDONESIAN) kara pedang, kacang babi, kara oncet; (ITALIAN) fava, fava maggiore, fava grossa; (JAPANESE) sora mame, otafuku mame; (JAVANESE) oncet, kara oncet; (KANNADA) kadu huralikayee, kadu huruli kayi, dodda hurali kaayi, kaadu hurali kaayi; (KASHMIRI) bōgala; (KAZAKH) konskij bob; (KOREAN) boracong; (KRIO) brɔd binch, kokondo; (LADINO) fava, ávas frèskas; (LATVIAN) darza pupa, pupas; (LITHUANIAN) pupa, dārzo pupos; (LUO) oranga; (MACEDONIAN) obična bakla; (MALTESE) il-ful; (MARATHI) mothi bin; (MONGOLIAN) gish; (MORISYEN) bokla; (NEPALI) bakula; (NORWEGIAN) bønnevikke, hastebønne, åkervikke; (ORIYA) simba; (PANJABI, EASTERN) raj manh, chas tang, kabli bakla, baklä, mattz-rewari, raj-rawan; (POLISH) bobik, bób; (PORTUGUESE) fava, faveira, feijão fava; (ROMANIAN) bobul, bob; (RUSSIAN) konskij bob, bob kormovoi; (SERBO-CROATIAN) konjsko bob; (SLOVAK) bôb záhradný, bôb obyčajný; (SLOVENIAN) vrtni bob; (SPANISH) haba, hada major, hada de huerta; (STELLINGWERFS) sloffenbonen; (SUNDA) kacang babi; (SWAHILI) choroko, fiwi; (SWEDISH) bondböna, hästböna; (TAJIKI) boqilo; (TELUGU) peddachikkudu; (THAI) thua yang; (TIBETAN) ka̱ṭɛma; (TURKISH) bakla; (URDU) sem, bakla, kala matar, raj-rawam; (VLAAMS) tuinboon, paardeboon.

b. pigeon pea

(DANISH) bønne med amå frø; (DUTCH) duiveboon; (ENGLISH) pigeon bean, tick pea; (FRENCH) féverole, féverole à petites graines, petite fève; (GERMAN) Tautenbohne; (ITALIAN) favino; (PORTUGUESE) fava cavalinha, fava miúda, feverole; (SPANISH) haba baby.

40.35. *Vigna aconitifolia* (Jacq.) Maréchal (EP) immature pods and seeds; (P) C.

(AFRIKAANS) motboon; (ARABIC) fasulia; (BENGALI) kheri; (CHINESE, MANDARIN) e dou; (CHINESE, WÚ) hu dou, tu dou, wo dou, e dou; (CHINESE, YUE) can dou; (CZECH) vigna; (DANISH) måttebønne; (ENGLISH) moth bean, mat bean, dew bean; (FRENCH) haricot papillon; (GERMAN) Mat-Bohne, Mattenbohne; (GUJARATI) mut, math, bhringga, mot, methkalai, moth, mothi, urad; (HEBREW) luvya; (HINDI) moth, bhringga; (KANNADA) madike, thuraka hesaru, madaki; (MALAY) mittikelu; (MARATHI) math, matki, mat, matha; (PANJABI, EASTERN) bhioni, mōr; (POLISH) fasolnik matowy; (SANSKRIT) makushtha, makushthaka, amritha, aranyamudga, bassunta, khandi, krimilaka, kulinaka, madyaka, mapashtha, mayastha, mudjashtaka, nigudhaka, raajamudga, vallimudga, vanamudga, varaka, vasunta; (TAMIL) tulukkapayir, tulkapayir; (TELUGU) kuncumapesalu, kunkumapesalu, minumulu, pijjipesara; (THAI) matpe; (URDU) moth; (VLAAMS) motboon.

40.36. *Vigna angularis* (Willd.) Ohwi & Ohashi (EP) seeds; (P) C.

(AFRIKAANS) adzukiboon; (CHINESE, MANDARIN) hong dou; (CHINESE, WÚ) hong dou; (CZECH) vigna adzuki; (DANISH) adzukibønne; (DUTCH) adzukiboon; (ENGLISH) adzuki bean; (FINNISH) adsukipapu; (FRENCH) haricot à feuilles angulaires, haricot adzuki, pois azuki; (GERMAN) Adzuki-Bohne; (HEBREW) luvya; (HMONG DAW) noob taum; (HUNGARIAN) adzuki-bab; (ITALIAN) fagiolo adzuki; (JAPANESE) azuki; (KANNADA) ghattada uddu; (MACEDONIAN) aglest grav; (NORWEGIAN) adsukibønne; (POLISH) fasolnik adzuki; (PORTUGUESE) feijão adzuqui, feijão azuki; (SPANISH) judia adzuki, poroto arroz; (SWEDISH) adzukibönor; (TURKISH) aszuki fasulyesi; (VLAAMS) adzukiboon.

40.37. *Vigna mungo* (L.) Hepper var. *mungo* (EP) immature pods and seeds; (P) C.

(ARABIC) homos; (ASSAMESE) mati mah; (BADAGA) uddu, uḷundu; (BENGALI) mash-kalai; (CHINESE, MANDARIN) ji dou, hei chi tou; (CHINESE, WÚ) jie dou; (CHINESE, YUE) hu dou; (CZECH) fazol mungo, mungo, vigna mungo; (DANISH) sort mungbønne, urdbønne; (DUTCH) mungboon, oerd, oerdoe; (ENGLISH) black gram, urd; (FIJIAN) urd; (FRENCH) haricot velu, haricot d'Angola, mungo; (GERMAN) Urdbohne, Linsen-Bohne; (GUJARATI) adad, arad; (HEBREW) luvya; (HINDI) urd, urad; dailuah chi khat moong; (HUNGARIAN) mungó bab, urd-bab, indiai bab;

(INDONESIAN) kacang mungo, kacang ijo; (ITALIAN) fagiolo de salsa, fagiolo mungo, fagiolo urd; (JAPANESE) ketsuruazuki; (KANNADA) uddu; (MACEDONIAN) mungo grav; (MALAYALAM) uzhunnu, kaattu ulinu; (MARATHI) udid, maga, moong; (MORISYEN) zamberik noir; (MUNDARI) uṛdi; (NEPALI) maas; (NORWEGIAN) urdbønne; (PALI) muggo; (PANJABI, EASTERN) mãh, urd; (POLISH) fasolnik mungo, fasola mungo; (PORTUGUESE) feijão do Congo, feijão mungo; (RUSSIAN) fasol' mungo; (SANSKRIT) mudga, mudgah, hayaanda, maasaparni, maasha, soopshreshta; (SANTALI) birsaṅ, muṅ, rambṛa; (SINHALA) undu; (SPANISH) frijol mungo, judia de urd; (SWEDISH) suart mungböna, mungoböna; (TAMIL) ulundu, ulundu, payaru; (TELUGU) minumulu, karuminimulu, nallaminumulu, pesalu, pesalu pessara, uddulu; (TULU) urdu; (TURKISH) mungo fasulyesi, cilband; (URDU) urud; (VIETNAMESE) dau den; (VLAAMS) mungoboon.

40.38. *Vigna radiata* (L.) R. Wilczek var. *radiata* (EP) immature pods, sprouted seed, seeds; (P) C.

(AFRIKAANS) mungboon; (ASSAMESE) masur mah; (BASSARI) *a*-tɛ̀mb-tɛ̀mb; (BENGALI) mung; (BINUKID) belatung; (BODO) mugu; (BUDIK) *ga*-tɛmbɛr sɛmbɛr; (BUHID) balatung; (BULGARIAN) papuda; (BURMESE) pè-di-sien; (BUTUANON) monggos; (CEBUANO) mungo, balatong, monggos, munggo, munggos; (CHINESE, MANDARIN) lu dou, lan tou; (CHINESE, WÚ) cai young da dou, lu dou; (CZECH) vigna zlatá; (DANISH) mungbønne; (DIGO) podzo; (DUTCH) tauge, mungboon; (ENGLISH) mung bean, green gram, golden gram; (FIJIAN) mung; (FILIPINO) mungo; (FINNISH) mungopapu; (FRENCH) ambérique, haricot doré, haricot mungo; (GADDANG) balatung; (GERMAN) Mungbohne; (GIKUYU) ndengũ; (GUJARATI) mag; (HANUNOO) balatun, munggu; (HEBREW) luvya; (HINDI) mung, pessara, urad, urid; (HMONG DAW) nong taao, pus sha, noob taum; (HUNGARIAN) mash-bab, aranybab; (IBANAG) balatong; (IFUGAO) mungo, balatong, bahilang, balatung, balatung an iitang; (ILOCANO) mungo, balatong; (INDONESIAN) kacang uci, kacang mungo, rĕtĕk hidjo, kacang hijau, arta ijo; (ITALIAN) fagiolo aureo, fagiolo mungo; (JAPANESE) ryokutou, bundo, yae-nari; (KANKANAEY) amtak; (KANNADA) hesaru, hesaru kaalu; (KASHMIRI) hĕm; (KAZAKH) mas; (KHMER, CENTRAL) sândaèk ba:y; (KOREAN) rockdu; (LAHU) naw^ naw˗; (LAO) thwàx khiêw, thwàx ngo:k, thwàx sadê:k; (LATVIAN) zetainās pupinas; (LOMWE, MALAWI) namurovo; (LUO) oranga; (MACEDONIAN) zelen grav, zlatest grav; (MALAY) kacang hijau; (MALAYALAM) cherupayaru; (MARANAO) monggos, ramsik; (MARATHI) mug, udid; (MORISYEN) zamberik rouz; (NEPALI) mung; (NORWEGIAN) mungbønne; (NYAKYUSA-NGONDE) imposo; (NYANJA) mphodza; (PALI) māso; (PANJABI, EASTERN) mung; (POLISH) fasolnik złoty, fasola złota; (PORTUGUESE) ervilha de Jerusalem, feijão mungo, mungo-verde, feijão da Índia; (PULAAR) layri; (RUSSIAN) mas; (SANSKRIT) maasa, mudga; (SINHALA) mung eta; (SPANISH) judia de mungo, frijol mungo, frijol de oro, poroto mungo; (SWAHILI) choroko; (SWEDISH) mungböna; (TAGALOG) munggo, mungo, mongo, balatong, monggo, monggos, munggos; (TAMIL) pasi-payaru, patcha-payaru; (TAUSUG) batung-munggo; (TELUGU) uthulu, kaaruminumulu, patchapessalu, pachha pesalu; (THAI) thua khieo, thua thong; (TIBETAN) ţɛncan; (TONGA) kakhoma; (TUMBUKA) mposo, kakhoma; (TURKISH) maş, mung fasulyesi, kudüs bezelyesi; (URDU) mung; (VENDA) ṱhumbe; (VIETNAMESE) dâu xanh, dâu chè; (VLAAMS) mungoboon; (YAO) mbweso.

40.39. *Vigna subterranea* (L.) Verdc. (EP) immature/mature seeds; (P) C.

Dialects of Luyia (Bukusu[1], Kisa[2], Maragoli[3], Tachoni[4], Tiriki[5], Wanga[6]).

(AFRIKAANS) jagoboon, kaffergrondboontjie; (BAMANANKAN) tiga ni nguélé, tiga ni nkourou, tiga ouélé; (BASSARI) *i*-vèl; (BUDIK) *e*-bɛ̀r, *ge*-mɛ̀r, *ma*-wɛ̀r; (CHINESE, MANDARIN) ban ba la hua sheng; (CHINESE, WÚ) ban ba la hua sheng; (CRIOULO, UPPER GUINEA) mãnkara di biago;

(CZECH) voandzeja podzemní; (DAGAARE, SOUTHERN) singbũlé; (DANISH) bambarajordnød; (DIGO) njugu mawe, tendegwa mawe; (DUTCH) bambarra aardnootje; (ENGLISH) bambara groundnut, Madagascar groundnut, African peanut; (ÉWÉ) azindgoe; (FRENCH) pois bambara, pois sousterrain, pois de terre, voandzou, arachide de bambarra; (FULFULDE, ADAMAWA) deppu, deppi, mbirīwu, birīji; (GA) akwɛŋ; (GERMAN) Bambara-Erbse, Angola-Erbse, Erderbse; (GONJA) keku, kukwi; (GOURMANCHÉMA) chian; (GOURMANTCHÉ) atimouinna; (HARARI) lōz; (HAUSA) gyanɗa, gujiya, gyaɗa, ƙūlu, ƙwāma; (IGBO) ọkpa; (INDONESIAN) kacang Bogor, kacang bogo, kacang manila, kacang banten; (ITALIAN) pisello di terra; (JAVANESE) kacang banten, kara pendem; (JOLA-FONYI) bamoyu, bamoyon-moyon; (KISI, SOUTHERN) tigan binio; (KRIO) akpároro; (LOMWE, MALAWI) eto; (LUYIA) tsimbande[2,3,5], chimbande[1,4], simbande[6]; (MAKHUWA) djama; (MANDING-MANDINKA) kuru, tiga ba, tiga ni kuélé, tiga ni kuru, yuro blé, tio kolon; (MANDINKA) takolô, tessurom, tiga kolô; (MANINKA) béré ntiga, tiga bélé, tiga kolõ; (MANINKAKAN, WESTERN) tiga bélé; (MÒORÉ) suma; (MORISYEN) pistas malgas; (NDALI) niumu; (NGIZIM) túkúr̃wá; (NYAKYUSA-NGONDE) isugu; (NYANJA) nzama; (POLISH) sorzycha ziemna; (PORTUGUESE) mandubi de Angola, jinguba de Cambambé, mancarra redonda; (PULAAR) guirigue, tiagué dabbii, bambaranke; (SERER-SINE) gadianga; (SHONA) nyimo, nzama; (SPANISH) bambara, cacachuete de bambara, guisante de tierra, mani de Bambarra, mani africano; (SUNDA) kacang bogor, kacang bandung, kacang gengge, kacang jogo, kacang gondolo, kacang manila; (SUSU) kansi tintéré, nkansi, toré gélé; (SWAHILI) njugu mawe, njugu nyasa; (TEM) soué; (TIGRIGNA) fool; (TIV) i̱gbough a̱hi̱; (TONGA) zyama, njuwu; (TUMBUKA) niumu; (TURKISH) yer bezelyesi, Afrika nohudu; (VLAAMS) bambara aardnoot; (WOLOF) gadiaga, gadiãnga, gayaña, gèrté bâbara, gèrté bambara, gèrté pergne; (XHOSA) indongomane; (YAO) njama, sugama; (ZARMA) damsi kukurku.

40.40. *Vigna umbellata* (Thunb.) Ohwi & H. Ohashi (EP) seeds; (P) C.

(AFRIKAANS) rysboon; (ASSAMESE) rice bean; (BINUKID) mangulasi; (BONTOC, CENTRAL) kilkilang, pagsey, pakasay, paksay, tauhuri; (BURMESE) pé-yin; (CHINESE, MANDARIN) mi dou; (CHINESE, WÚ) mi dou; (CZECH) vigna rýžová; (DANISH) risbønne; (ENGLISH) rice bean; (FRENCH) haricot sauvage, haricot de riz; (GERMAN) Reis-Bohne; (HANUNOO) bihanti, marayu, tambasu, tokayan; (HEBREW) luvya; (HINDI) sutri, ghurush; (HMONG DAW) taum nplej; (IFUGAO) lawiging; (ILOCANO) sinay; (INDONESIAN) kacang oci, kacang tolo, kacang uci; (JAPANESE) shima-tsuru-azuki; (JAVANESE) kacang ruji, kacang oci, kacang uci, kacang enjing; (KANKANAEY) kilkilang; (KHMER, CENTRAL) sândaêk ângkât miëhs, sândaêk riech mieh; (LAO) thwâx la:ng tê:k, thwàx sadê:t pa:x, thwàx phi; (MACEDONIAN) orizoviden grav; (MALAY) kacang sepalit; (MALAYALAM) kaattulunnu; (MARANAO) tawhuri; (NEPALI) masyang; sita mas, pau maia; (NORWEGIAN) risbønne; (PANGASINAN) linay; (PANJABI, EASTERN) ghurush, räj mung; (POLISH) fasolnik ryżowy; (PORTUGUESE) feijão arroz; (SANSKRIT) maasa parni; (SPANISH) frijol arroz, frijol mambi, judia arroz; (SUBANON) takuri; (SUNDA) kacang kenyut, kacang oci; (TAGALOG) anipai, kapilan, pagsei, abas, tapilan; (TAUSUG) batung-amu; (THAI) thua daeng, thua pae, ma pae; (TURKISH) pirinç fasulyesi; (URDU) sutri (VIETNAMESE) dâu gao; (VLAAMS) rijstboon; (WÁRAY-WÁRAY) anipay.

40.41. *Vigna unguiculata* (L.) Walp. subsp. *cylindrica* (L.) Van Eselt. ex Verdc. (EP) immature pods and seeds; (P) C.

(CHINESE, GÀN) chang dou jiao, xin dou jiao, dou jiao; (CHINESE, MANDARIN) jiang dou; (CHINESE, MIN BEI) dou jiao, dai dou, cai dou; (CHINESE, WÚ) dai dou, dou jiao, gang dou; (CHINESE, XIANG) dou jiao zi; (CHINESE, YUE) fan jiang dou, shi yong zi li; (CREOLE, HAITIAN FRENCH) pwa inkoni; (DANISH) catjang; (ENGLISH) catjang; (FINNISH) lehmänpapu;

(FRENCH) dolique de Chine, dolique géant, dolique mongette; (GA) yɔɔ; (GERMAN) Catjangbohne; (GIKUYU) thoroko; (GREEK) vigna e kylindiki; (HEBREW) luvya gelilit; (HUNGARIAN) négyszögletű tehénborsó; (INDONESIAN) kacang panjang, kacang merah, kacang peudjit, kacang tunggak; (ITALIAN) vigna cinese, fagginola del occhio; (JAPANESE) hata-sasage; (KHMER, CENTRAL) sândaêk sâ, sândaêk khmau, sândaêk krâhâm; (KUNG-EKOKA) ka´ani; (KWANGALI) gakakurukadi; (LAO) thwàx siênx; (LUO) oranga; (NORWEGIAN) catjangbønne; (PANJABI, EASTERN) rawã; (POLISH) fasolnik cylindryczny; (PORTUGUESE) feijão fradinho; (SANTALI) boḍa ghaṅgra, cihṛi ghaṅgra, ghangra, ghaṅgri; (SPANISH) judia catjang; (SWAHILI) choroko, kunde; (TAJIKI) lubio; (THAI) thua khaao, thua rai, po-thoh-saa; (TULU) auḍe; (TURKISH) catjang börülcesi; (TWI) yɔɔ; (VIETNAMESE) dâu ca, dâu trang, dâu do; (VLAAMS) kitjangboon.

40.42. *Vigna unguiculata* (L.) Walp. subsp. *sesquipedalis* (L.) Verdc. (EP) immature pods and seeds; (P) C.

(AFRIKAANS) jaartsboon; (BICOLANO, CENTRAL) hamtak, hantak; (BONTOC CENTRAL) otong, batong, antak; (BUTUANON) batong; (CEBUANO) hamtak, laston, balatong, batong, hantak; (CHINESE, MANDARIN) chang jiang dou; (CHINESE, WÚ) chang gang dou; (CHINESE, YUE) dou jao, chang jiang dou; (DAGAARE, SOUTHERN) béon; (DANISH) aspargesbønne, meterbønne; (DUTCH) Indiaansche boon, kouseband; (ENGLISH) yard-long bean, long bean, asparagus bean, snake bean; (FILIPINO) sitao, paayap; (FINNISH) lehmänpapu; (FRENCH) haricot asperge, haricot kilomètre, dolique asperge; (GA) yɔɔ; (GADDANG) kawasan; (GERMAN) Spargelbohne, Langbohne; (GREEK) vigna e imitripodios; (HEBREW) luvya arukat tarmil; (HILIGAYNON) latoy; (HINDUSTANI) kosbanti; (HMONG DAW) taao-hla-chao, taum hlab tshos; (IBALOI) durias; (IBANAG) anta, agaya; (IFUGAO) kauat, awwat, awwat an gayyang, awwat an ubbun, awwat an umadukke; (ILOCANO) otong, utong, paayap; (INDONESIAN) kacang panjang, kacang usus, kacang belut, kacang merah, kacang tunggak, kacang cina, kacang lanjaran, kacang tolo, kacang bengkok; (ITALIAN) fagiolo asparagio, fagiolo lungo un metro, fagiolino asparago, fagioletto americano; (ITAWIT) antak; (IVATAN) agayap; (JAPANESE) jûroku-sasage; (JAVANESE) kacang gleyor, kacang lanjaran, kacang usus, kacang panjang, kacang tunggak; (KANKANAEY) batong; (KHMER, CENTRAL) sândaèk troeung; (LUO) oranga; (MALAY) kacang panjang, kacang belut, kacang perut ayam; (MARANAO) gimbor, kaloway, kolawit, kolili; (MONGOLIAN) tumen navchit; (MÒORÉ) bẽnga; (MORISYEN) voem; (NEPALI) tare bodi; (NIUE) pine fua loloa; (NORWEGIAN) meterbønne; (PAMPANGAN) bule, kamanyang; (PANGASINAN) agayep, paa-yap; (PANJABI, EASTERN) rawã, rɔ̃gi; (POLISH) fasolnik długi podwórkowy; (PORTUGUESE) feijão chicote, feijão a metro, feijão frade alfange; (SAMBAL, BOTOLAN) antak; (SAMBAL, TINÀ) kupis; (SINHALA) ma karal; (SPANISH) judia espárrago, judia chinas, dólico espárrago, poroto metro, chancha por metro; (SUNDA) kacang belut, kacang dadap, kacang panjang, kacang turus, kacang ruji, kacang landep, kacang landes, kacang peujit; (SWAHILI) choroko; (SWEDISH) sparrisböna; (TAGALOG) sitaw, sitao, karakala, harakala, kibal, hamtak, banor, batung-lupis, paayap; (TAGBANWA) sitaw, antak, balatong; (TAUSUG) batung hidyaw; (THAI) tua fak yaow, tua Phnom; (TURKISH) kuşkonmaz börülcesi; (TWI) yɔɔ; (VIETNAMESE) dâu dûa, daû giai áo, d[aaj]u d[uwj] a, d[aaj]u gi[ar]i [as]o; (VLAAMS) kouseband; (WÁRAY-WÁRAY) hantak, batong, balatong.

40.43. *Vigna unguiculata* (L.) Walp. subsp. *unguiculata* (EP) immature pods and seeds; (P) C.

Dialects of Luyia (Bukusu[1], Itsukha[2], Marama[3], Tachoni[4]).

(AFRIKAANS) akkerboon, koertjie, kafferboontjie; (AJA) kpawun kwi; (ARABIC) homos shami, adelagan, adlagna; (ASSAMESE) danghodi, lesera mah, lasara mah, urohi, mahor-pat, urhi-mah, urobi-mahor-pat; (BAMANANKAN) soso, doungouri soso, chuo, diontamba, dunguri soso, duri, mamuluku, nébé; (BASSARI) *o*-diod, *o*-jod, ɔ-syòdi, ɛ-tiɔ̀di, kadje; (BENGALI) barbati, ramhikolai; (BEROM) syaàm tuk, syìlà, syìlà jẹ̀ẹ́jèét, syìlà pyẹng, syìlà syìnàng, syìlà syìt; (BOBO MADARÉ, SOUTHERN) dɛ́mèn, nìná; (BUDIK) *mw*-nyí; (BULLOM SO) hɔbdɔɛ; (CHICKASAW) bala' losayyi', bala' ishkin losa'; (CHINESE, MANDARIN) pu tong jiang dou; (CHINESE, WÚ) pu tong gang dou; (CHINESE, YUE) jiang dou; (CREOLE, HAITIAN FRENCH) pwa enkoni; (CRIOULO, CAPE VERDE) feijão-bongolon, bongolon-de-angola, fijon fradi; (CRIOULO, UPPER GUINEA) fidiom mãnkaña, feijão mancanha; (DAGAARE, SOUTHERN) béon; (DAGBANI) toa, tuya, zanzi; (DANGME-KROBO) yɔɔ; (DANISH) vignabønne, foderplante, bælgplante; (DENDI) duguri; (DIGO) kunde; (DITAMMARI) yatura; (DUTCH) vignaboon; (EDE NAGO) ewe; (EDO) ée, ere nofum, ere nubaa; (EFIK) ọ̀kọ́ti; (ENGLISH) black-eyed pea, cowpea, southern pea; (ÉWÉ) aboboe; (FIJIAN) borah; (FINNISH) lehmänpapu; (FON) ayikun, aiku vovo, aiku wagbé lafo, aiku wéwé, doi ku; (FRENCH) haricot à oeil noir, haricot dolique, niébé, dolique chinois; (FULFULDE) hako niébé; (FULFULDE, ADAMAWA) nyebbe, nyebbere; (GA) yɔɔ; (GBE-VHE) ayi, gbenku, kelẽge, kɔnukpekpe, (GERMAN) Kuherbse, Kuhbohne, Augenbohne; (GIKUYU) thoroko, mathoroko; (GOLA) hɔndɔ; (GONJA) chebe, kichebɛ, chèbèɛ; (GOURMANCHÉMA) souna; (GOURMANTCHÉ) atouna; (GREEK) vigna e siniki; (GUJARATI) chorap, chola, choli; (HAUSA) báƙin wáákéé, ba-ude, fárín wáákéé, gayan gayan, ján wáákéé, kereria, kwama, tsereriya, wáákée, wáákén awaki, wáákén bareya, wáákén daamánáá, wáákén damfami, wáákee, wááken gizo, yawa, waké; (HEBREW) luvya tarbutit; (HINDI) lobia, lobya, lobiyaa, lobiyā, chowli, riantshi, rawans, raunsa, ramas, raish, santa; (HUNGARIAN) tehénborsó, kínai tehénborsó; (IGBO) agwa, àgwà, akịdị, igara, kpamaje, akpaka emule; (INDONESIAN) kacang panjang, kacang tunggak; (ITALIAN) fagiolino dall'occhio, veccia; (IZON) ụmgbọ́; (JAPANESE) sasage; (JUNKUN TAKUM) sò; (KANNADA) alasande, tadagunny, todagunny, kursan-pyro, alasandi, alasandige, halasande, kaaraamani, kursana pairu, kursandi, tadagani, tarani kaayi; (KANURI) ligi, ngalo; (KHMER, CENTRAL) sândaêk kâng, sândaêk ângkuy; (KONKOMBA) isanje, isaniye, n'tu; (KONO) sɔ-mɛsɛ; (KORO) bisar; (KRIO) akara binch, blak-ai-binch, watalo-binch; (KURDISH, NORTHERN) lobyā; (LAHU) i¯ cu ka naw^, naw^ i¯ cu ka, a¯ cu ka naw^; (LAO) thwàx do; (LIMBA) batamba, (LOKO) gutɔyɛ; (LOMWE, MALAWI) khobwe; (LUO) ngor; (LUYIA) sikhubi[1], likhuvi[2], esikhubi[4], tsing'oli[3]; (MACEDONIAN) crn bob; (MALAY) kacang putih, kacang bol, kacang toonggak, kacang merah; (MALAYALAM) mambayar, kottapayuru, vellapayaru; (MALIMBA) uaka; (MANDING-MANDINKA) soso; (MANDING-MANINKA) nébé, niébé, niébé golo; (MANDINKA) nébé; (MANINKA) gnébé, niébé, niébé golo, soso, nébé; (MARATHI) chavli, chavali, chaoli, alasunda; (MENDE) hɔndɔ, njui, nyɛkɛnde; (MIZO) be chi khat; (MÒORÉ) bēnga; (MORISYEN) voem; (MUSKOGEE) tvlak-cvpko, torwv-lvstēcv; (NEPALI) bodi; (NORWEGIAN) øyebønne; (NYAKYUSA-NGONDE) inandala; (NYANJA) khobwe, nseula; (ORIYA) baragudi; (PANJABI, EASTERN) lobia, rawã, rɔ̃gi; (POLISH) wspięga chińska, fasolnik chiński; (PORTUGUESE) feijão frade, feijão careto, feijão miúdo, feijão da China, caupi, feijão de corda caupi; (PULAAR) ñébé belli, ñébé limbok, sèb, nhébè limboquê; (ROMANIAN) fasoliţă; (RONGA) nhumba, nhabubo, timbawene, nhemba; (SANGA) iṣu; (SANSKRIT) nishapava, dirghabija, chavala, raajamaasha; (SERBO-CROATIAN) kineski pasulj; (SERER-SINE) ñav, ñav a kob, ñav basa, ñav fasa, onau; (SHONA) nyemba; (SINHALA) cowpea; (SONGHAY) doungouri, dunguri, dúŋgúrì; (SPANISH) caupi, judia de vaca, frijol de vaca, frijol de pinta negra, frijol de ombligo negro, caragilates; (SUSU) togé, konso-toge; (SWAHILI) choroko, kunde; (TAGALOG) poagap, batong, kibal, otong; (TAMIL) karamani, thatapayuru, thatapayaru, thattapayir, caramunnypre, caramunnipayara; (TELUGU) bobberulu, bobbarlu, alusendi, alsandalu, alasandalu, duntupesalu, alasandalu, kaaraamanulu; (TEM) kedsche, sona; (THAI) tua dam; (THEMNE) *ka*-binsh, ɛ-tuɛ, (TIV) alev, arebe; (TSONGA) nhemba; (TULU) latthane; (TUMBUKA) ngunde; (TURKISH) börülce; (TURKMEN) hý taý noýbasy (TWI) yɔɔ, adua, akitereku, ase, ase kɔkɔ, asedua, asekɔkɔ, ɛdua; (URDU) lobya; (VAI) sɔ-kpɔ-mɛsɛ; (VIETNAMESE) dôu den, dôu trang, dôu tua; (VLAAMS) ogenboon, cowpea; (WAAMA) yaniporyi; (WAMEY) *va*-tiur, *va*-tiuri, *u*-suri, *wa*-tyuri; (WOLOF) ñébé, ñébé golo, niébé, seb, söb, sebu, sep; (YALUNKA) mululu thogo-na, seneni thogo-na, thogo hongkure-na, toke luba; (YAO) ngunde; (YOM) tira; (YORUBA) ẹ̀wa, ẹ̀wa ebeji, ẹ̀wa dudu; (ZARMA) doungouri, dunguri.

41. Gnetaceae

41.1. *Gnetum gnemon* L. (EP) leaves, tender shoots and fruits; (P) C.

(BICOLANO, CENTRAL) nabo; (BUHID) bagu; (CEBUANO) bago-sili, banago; (CHINESE, WÚ) xian zheu mai ma teng; (DUTCH) tamme gnetumboom; (ENGLISH) buko; (EWONDO) okok; (FILIPINO) bago; (GREEK) gneton tou gnemou; (HANUNOO) lunud; (HEBREW) gneton; (INDONESIAN) melinjo, belinjo, maninjo, meninjo, karuk; (JAVANESE) so, melinjo, trangkil, bagu, bogor, grintul, maninjo, wit grintul; (LATVIAN) gneti; (LITHUANIAN) tikrasis gnetas; (MAGUINDANAO) ma-gatungal; (MALAY) meninjau, emping melinju, kripik melinju; (MANOBO) kugitis; (PORTUGUESE) genemo; (SPANISH) bago, bulso, genemo; (SUBANON) babayong; (SUNDA) tangkil, tangkil sake, ki tangkil, maninjo; (TAGALOG) gnemo, bago, lamparan; (TAUSUG) kugitas; (VLAAMS) buko.

42. Haloragaceae

42.1. *Myriophyllum aquaticum* (Vell.) Verdc. (EP) shoot tips; (P) C.

(CHINESE, WÚ) za; (DANISH) papegoejerjer; (ENGLISH) parrot's feather; (GREEK) myriophyllo; (HEBREW) elef hehaleh; (INDONESIAN) paris; (ITALIAN) millefoglio d'acqua; (JAVANESE) paris; (POLISH) wywłócznik; (PORTUGUESE) pinheirinha de água; (THAI) sarai-yipun; (TURKISH) bin yaprak; (VIETNAMESE) rong du[oo]i ch[os].

43. Icacinaceae

43.1. *Icacina oliviformis* (Poir.) J. Raynal (EP) tubers; (P) C.

(AKAN-ASANTE) abubu ntɔpe; (BALANTA-KENTOHE) fóè, sóngol; (BASSARI) *a*-nagan, *a*-narham, *a*-nàxàn; (BUDIK) gi-ndél; (CHINESE, WÚ) cha zhu yu; (CRIOULO, UPPER GUINEA) mãnganasa, manganace, manganaz; (DAGBANI) takwara, tankoru; (ENGLISH) false yam; (FON) azɔkwin; (FRENCH) bankanas, basouna pané; (GBE-VHE) miṅa; (IGBO) ututo ogiri; (JOLA-FONYI) bu bambulaf, bu tima, fu timay, furabá, kurabã, u rabã, furabân, butima, futima; (MANDING-BAMBARA) bankanazé; (MANDING-MANDINKA) bankanase, bankanesi, manankaso, manankoso, mankanaso, wamgaso, wangaso, manacossô, mancanassô; (MANDINKA) bâkanazé, mâkanas, mâkanaso; (MANDJAK) be nasia, mãnkanas, unasse; (MANKANYA) be nasin, unasse; (PORTUGUESE) manganaz; (PULAAR) mãnganatié, mãnkanase, silla, bankanase, bankanesi, manankaso, sila, mancanadje, manganadje, silá; (SERER-SINE) 'ba, iba; (VLAAMS) valse yam; (WAMEY) dagan, *a*-ndagana, ndarhan, *a*-ndàxán, *u*-ryakán; (WOLOF) bâkanas, bânkanas, *n*diangam, mâkanas, bananafi, bankanafi, bankanase, bankanesi, manankasa, manankaso, mbahanasa.

44. Lamiaceae (Labiatae)*

44.1. *Lycopus lucidus* Turcz. ex Benth. (EP) rhizomes; (P) C,P.

(CHINESE, MANDARIN) di gua er miao; (CHINESE, WÚ) di gua er miao; (DUTCH) wolfspoot; (ENGLISH) shiny bugleweed; (ESTONIAN) parkhein; (FRENCH) lycope; (GERMAN) Wolfstrapp; (GREEK) lykopous; (HEBREW) kaf ha-zeev, kaf-hazeev; (ITALIAN) erba sega; (JAPANESE) shirone; (MACEDONIAN) sjajna volča noga; (NORWEGIAN) klourt; (POLISH) karbieniec; (PORTUGUESE) menta de lobo; (SPANISH) menta de lobo; (TURKISH) su keneviri, parlak kurtayaği, kurt ayaği; (VLAAMS) shirone.

44.2. *Mentha pulegium* L. (EP) leaves; (P) C.

(AFRIKAANS) ment; (ASSAMESE) paduna; (BELARUSAN) gubakvetkavaja; (BRETON) bent-dour, bent, ment; (CHINESE, MANDARIN) chun e bo he; (CHINESE, WÚ) bo he, chun e bo he; (CHINESE, YUE) qing bo he; (CZECH) máta, máta polej, polej obecná; (DANISH) polej-mynte; (DUTCH) polei; (ENGLISH) European pennyroyal, pennyroyal mint, pudding grass, flea mint; (ESTONIAN) kirbumünt; (FINNISH) puolanminttu; (FRENCH) baume petit, menthe pouliot; (GALICIAN) poego, poenxo, pueixo; (GEORGIAN) ombalo; (GERMAN) Flohkraut, Polei, Poleiminze; (GREEK) idyosmos o glichon; (HEBREW) na'anat ha-kadurim, naanat hakadurim, mentha; (HUNGARIAN) csombor menta; (INDONESIAN) daunpoko; (ITALIAN) menta puleggia, menta salvatica, poleggio, pulezzo; (LITHUANIAN) taškuotoji mėta; (MACEDONIAN) mačkino nane; (MALTESE) plejju; (MONGOLIAN) batrash; (NORWEGIAN) polemynte; (POLISH) mięta pieprzowa, miłta; (PORTUGUESE) poejo; (ROMANIAN) mentă de baltă; (RUSSIAN) mjata bolotnaja, mjata blotnaja; (SERBO-CROATIAN) metvica poljačak; (SPANISH) póleo, póleo común, póleo menta, póleo negro; (TAJIKI) pudinai dashti; (TURKISH) yarpuz, avşanotu, poleu nanesi; (VLAAMS) polei.

44.3. *Mentha spicata* L. (EP) leaves and inflorescence; (P) R.

(AFAR) naqnaq; (AFRIKAANS) kruisement; (ASSAMESE) paduna; (AZERBAIJANI, NORTHERN) nanə, zərbxana, pul çap edilən yer; (BENGALI) pahari pudina, pudina; (BODO) osunda; (BOSNIAN) nana, menta, metvica; (BRETON) bent-liorzhoù, bent, ment; (BURMESE) pu-see-nan; (CHINESE, MANDARIN) liu lan xiang; (CHINESE, WÚ) liu lan xiang; (CHINESE, YUE) xiang hua cao, bo he, liu lan xiang; (CREE, PLAINS) ka tahkeyawepayesik; (CRIOULO, CAPE VERDE) hortelon; (CZECH) máta klasnatá; (DAKOTA) ma'zaskaokaġe; (DANISH) grøn mynte, krusemynte; (DUTCH) groene munt, kruizemunt; (ENGLISH) spearmint, green mint, lamb mint, brown mint, garden mint, mackerel mint; (ESTONIAN) roheline münt; (FARSI, WESTERN) na'nâ; (FINNISH) viherminttu; (FRENCH) baume verte, menthe en épi, menthe de Notre-Dame, menthe crépue, menthe verte; (GAELIC, SCOTTISH) meant; (GALICIAN) amenta, lecimbre, herba boa; (GERMAN) grüne Minze, grüne Rossminze, Krauseminze, Spearmint; (GREEK) idyosmos o prasiyos; (HEBREW) na'ana meshubelet, naanat spicata, mentha; (HINDI) pahari pudina, pahaaripudina, pudina; (HMONG-DAW) pum hub; (HMONG-NJUA) pum hum; (HUNGARIAN) zöld a paramenta; (ICELANDIC) grenmynta; (INDONESIAN) permen; (ITALIAN) mentha verde, mentha gentile, mentha romana, mentha crespa; (KANNADA) bettada pudina; (KAZAKH) zhalbız; (KIRGHIZ) jalbız; (KRIO) mint; (KURDISH, NORTHERN) nēanē; (LITHUANIAN) šaltmėtė; (MACEDONIAN) klasesto nane; (MALTESE) naghnieh; (MARATHI) phhari pudina, pudina; (MICMAC) ŭpsĕskw'; (MIZO) pudina, ẓarbkhāna; (MONGOLIAN) batrash; (MORISYEN) lamant; (NEPALI) babari, pudina; (NORWEGIAN) krusmynte, grønnmynte; (PASHTO, NORTHERN) welanag, nana, taksāl; (POLISH) mięta zielona; (PORTUGUESE) hortelã, hortelã comum, hortelã verde; (PULAAR) manto, naanaa; (SANSKRIT) pudina, poothiha, rochani; (SIDAMO) ha'nassho; (SPANISH) hierba buena, hierba santa, hierba sana mentha, mentha romana, mentha crespa; (SWEDISH) grönmynta; (TAJIKI) pudina hulluj; (TAMIL) pudina; (TATAR) bötnek; (TELUGU) pudina, menthi; (TIV) va̱r inyaregh; (TURKISH) yeşil nane, nane, kivircik nane, rumi nane; (TURKMEN) narpyz; (URDU) pahari, podina; (UYGHUR) yalpuz; (UZBEK) yalpiz; (WELSH) mintys ysbigog; (XHOSA) inxina, uhlobo oluthile lwenxina esetyenziswa ekuqholeni.

44.4. *Ocimum americanum* L. (EP) young leaves; (P) R,C.

(AFRIKAANS) wilde basielkruid; (ASSAMESE) tuloshi; (BAOULÉ) mien; (CHINESE, MANDARIN) hui luo bo; (CHINESE, WÚ) hui luo le; (ENGLISH) hoary basil, American basil; (GERMAN) Herero Buschtee; (GREEK) okimon; (HEBREW) rehan canum; (HERERO) ohauveve; (HINDI) kala tulsi, kaali thulasi, mamri, ajagandha, baabari, banthulasi, thulasi aarjaki; (IGBO) alulu-isi; (INDONESIAN) kemangi, serawung, selasih putih; (JULA) fêfê; (KANNADA) nayi tulasi, naayi thulasi, hola thulasi, raama thulasi; (KRIO) patmajin, patmɛnji; (KUNG-EKOKA) n!angtsu; (KWANGALI) rusivi; (KWANYAMA) endjikaui; (MALAY) selaseh, kemangi, ruku-ruku; (MALAYALAM) kaatturamathulasi; (MARATHI) raana thulasi; (NAMA-DAMARA) /gaubeb, //gammeb; (PORTUGUESE) manjericão branco; (SANSKRIT) ajaka, aarjaka, gambhira, kuthera, gambheera, gandhapanirjjaka, jambeera, kaalmaala, kaalmallika, kathinjara, kshudraarjaka, kshudraparna, kshudrathulasi, mukhaarjaka, surasa, ugragandha; (SPANISH) albahaca velluda; (TAJIKI) sadaraihon; (TAMIL) nayitulasi, naayithulasi, kanjamkorai; (TELUGU) kukka tulasi, kukkathulasi; (THEMNE) ɛ-soro; (THAI) mang-luk, maenglak; (TULU) naayi tholachi; (URDU) kali tulsi; (VIETNAMESE) rau h[us]ng.

44.5. *Ocimum basilicum* L. (EP) leaves; (P) R,C.

(AJA) hanase; (AKAN-ASANTE) solésolé; (AKLANON) kamangi; (ALBANIAN, TOSK) bozilok; (AMHARIC) mätafät, bassobəlla, zäqa, zäqaqəbe; (ANDO) ɛ́mian nvan van; (ARABIC) azeftenni, gamongeï, gumugei, jeftuni, jiftenni; (ASSAMESE) tuloshi; (BAMANANKAN) sukula, chukula, syokola, tiukula; (BAOULÉ) emia, mié, aremangnè; (BASSARI) ɛ-sɛ́ngɔ̀ng; (BENGALI) babuitulsi, debunsha, khubkalam, pashanabeddie; (BEROM) mà'zɛŋ; (BÉTÉ) srabê-klin'nlin; (BOBO MADARÉ, SOUTHERN) sìyɛwǎrò; (BUDIK) ñaramèl, nyaramél, nyaramɛ́l; (BURMESE) pin-sein; (CAROLINIAN) taibwo, teibwo; (CEBUANO) bowak, kaluuy; (CHINESE, MANDARIN) hsiang tsai; (CHINESE, WÚ) mao luo le; (CHINESE, YUE) lo le; (CREE, PLAINS) kîkway katakwastahk kestepowinihk; (CRIOULO, CAPE VERDE) mangirona; (CZECH) bazalka pravá; (DANGME) damie, danue; (DANGME-KROBO) danwe; (DANISH) basilikum; (DUTCH) basilicum, bazielkruid, koningskruid; (EDO) ihiri; (EFIK) amana, ḿfáng; (ENGLISH) sweet basil, common basil; (ESTONIAN) vürtsbasiilik; (ÉWÉ) ahame; (FARSI, WESTERN) reyhan; (FILIPINO) salanoi; (FON) kesu kesu; (FRENCH) basilic, basilic commun; (FULFULDE, ADAMAWA) jambaljoohi, kacukacunga, urngol; (GA) kowe; (GBE, WACI) hanase; (GEN) koklodame; (GEORGIAN) rehani; (GERMAN) Basilikum, Basilienkraut; (GREEK) vasilikos; (GUJARATI) damaro, nasabo, sabza, sabje, takamariya; (GUN) akɔhun; (GURO) sĕrébé; (HANUNOO) buranuy, sagbu; (HAUSA) ɗaiɗóóyàà; ɗàiɗóóyà, ɗàiɗóóyàr, ɗóɗaya, ɗóɗowa, ɗóɗoya, ɗóɗoya ta kare, ɗóɗoya tagida, ɗwai-ɗwaiya, ka fi amara; (HEBREW) rehan matok; (HINDI) babui tulsi, banathulasi, marua, bahari, barbar, gvlal tulsi, kali tulsi, sabzah, kaali thulasi, babayi, babuithulsi, baabul, bahari, barbar, bavare, maala thulasi, mamaree, nivaksho, rihan, sabzah, thukhmerihan (HMONG DAW) txuj lom, zaub, txis theem; (HMONG-NJUA) txis theem; (HUNGARIAN) bazsalikom; (IBANAG) albahaka; (ICELANDIC) basilikum; (IGBO) nchu agwunta, nchu agwụ; (ILOCANO) biday; (INDONESIAN) selasih putih, selaseh, selasih, kemangi, surawung; (ITALIAN) basilico; (ITAWIT) apiang; (JAVANESE) selasih, telasih, kemangi; (JOLA-FONYI) é fulungèy; (JULA) soukonanbourou; (JUNKUN TAKUM) afyinu, shinu; (KANNADA) bhoothulasi, kama kasturi, kaama kasthoori, sajjagida, sajjebiya, ramkasturi, kamagaggare, ramathana champa, sabjaa gida; (KANURI) kabur; (KAZAKH) basilik; (KHMER, CENTRAL) chi neang vong; (KONO) peiŋga, wose-yamba; (KRIO) patmage, patmɛnji; (KURANKO) sogoi; (LAO) phak 'i:x tu:x; (LATVIAN) baziliks; (LOKO) sogho; (MAKHUWA) marrupani; (MALAY) selaseh, selasi, selasi hitam, selasi jantan, kemangi; (MALAYALAM) kattutulasi, pachcha, pachha, tirunitru, raama thulasi, thirunithru; (MALIMBA) fufurufuru; (MANDING-MANDINKA) patmagi; (MANINKA) sukola; (MANINKAKAN, EASTERN) sukola; (MARATHI) marva, sabza, sabja, sabjaa, tukhamariya; (MENDE) pasmanja, pu-gbawo; (MÒORÉ) yulyungu, niuniununga, yulonguga; (MORISYEN) bazilik; (MUNDARI) tulsi; (NEPALI) babari phool; (NIUE) manogi, momili elo; (NORWEGIAN) basilik, basilikum; (ORIYA) dhalatulasi, dhala tulasi, kapur kanti; (PANGASINAN) biray; (PANJABI, EASTERN) furrunj-mushk, furrunjnushk, babri, nigand, tulsi; (POLISH) bazylia pospolita; (PORTUGUESE) basílico, manjericão grande, manjerico da folha grande; (PULAAR) gugumã, ńaramel, ngunguné, sukora; (RAROTONGAN) miri; (ROMANIAN) busuioc; (RONGA) hamabo; (SAMOAN) la'au souga, sauga, la'a sawa, lau sauga; (SANSKRIT) munjariki, arjakah, surasa, varvara, arjaka, barbari, manjariki, ajagandhika, apetaraakhsasi, asurasa, barba, barbara, bisvathulasi, karahi, kharapushpa, pashugandha, surabhi, surasa, thulasidvesha, thungi, varvaree; (SANTALI) bharbhạri, dimbu baha, kạpur mạli baha, mạli baha; (SENOUFO, CEBAARA) coufirigui; (SERBO-CROATIAN) bosiljak; (SERER-SINE) lubalub; (SHONA) bunganye, catandambundo; (SLOVAK) bazalka; (SPANISH) albahaca, alhábega; (SUBANON) bawing sulug, gentud, gluntud; (SUNDA) selasih, solasih, surawung; (SUSU) barikiri; (SWEDISH) basilika; (TAGALOG) balanoy, solasi, albahaka, balansi, kolabtos; (TAGBANWA) antanganan, baing, bawing, baying; (TAJIKI) raihon; (TAMIL) tirnirupachai, thirnupatchi, tirnutpatchi, karpoorathulasi, tirunitru, karpura tulasi, paachhaayi, thirunithru, thirunuthpatchi; (TAUSUG) bawing, ruku-ruku; (TELUGU) bhutulasi, bhoothulasi, rudrajada, vepudupachha, vaepudupachha, vibudipatri, gakarakau, komma gaggiri, sabja, rudrajada, viboodipathri; (THAI) horapha; (THEMNE) *an*-soro; (TIV) kungulaku; (TULU) kāmakastūri; (TURKISH) reyhan, fesleğen, irfan, peslan; (TURKMEN) adaty nazbaýgül; (URDU) rihan, tulsi, janglitulasi; (UZBEK) raikhon; (VIETNAMESE) h[us]ng gi[or]i, [es] tr[aws] ng, [es] t[is]a; (WAMEY) bisial, é sasãg, komeŋa; (WÁRAY-WÁRAY) sangíg; (WOLOF) lepalep bu diigen, ngugum, ngungun, ngumgum; (YORUBA) ofɛn, eferi, efirin wewe, efínrín arùntantan, efínrín ata, efínrín márúgbóṣányán, efínrín wẹ́wẹ́, efínrín-ajá; (ZARMA) koro gabu.

44.6. *Origanum vulgare* L. (EP) flowering plant and inflorescence; (P) R,C.

(AFRIKAANS) marjolein; (BELARUSAN) matserdushka zvychainaja; (BRETON) marjol-ki; (CHINESE, MANDARIN) hua bo he; (CHINESE, WÚ) hua bo he; (CRIOULO, CAPE VERDE) oregon, (CZECH) dobromysl obecná; (DANISH) merian; (DUTCH) marjolein, wilde marjolein; (ENGLISH) marjoram, wild marjoram, oregano; (ESTONIAN) majoraan; (FINNISH) mäkimeirami, oregano; (FRENCH) origan, origan commun, origan vulgare, marjolaine, marjolaine sauvage; (GAELIC, IRISH) oragán; (GALICIAN) aurego, ourego; (GERMAN) Germeiner-Dost, brauner Dosten, wilder Majoran, Oregano; (GREEK) origunon to amaranton; (HEBREW) ezov, azovit, za'tar, ezovit metzua; (HINDI) sathra; (HUNGARIAN) szurokfű; (ICELANDIC) meiran; (INDONESIAN) oregano; (ITALIAN) origano, maggiorana, maggiorana selvatica, regamo, acciughero; (KANNADA) maruga; (KAZAKH) dushytsa; (LADINO) oregano, origáno, rigano; (LATVIAN) parastā raudene; (LITHUANIAN) raudonėlis; (MACEDONIAN) običen origan; (MALTESE) merqtux; (MONGOLIAN) egel khakhuun; (NORWEGIAN) bergkung, kung; (PANJABI, EASTERN) mirzanjosh; (POLISH) lebiodka pospolita; (PORTUGUESE) orégão, ouregão; (ROMANIAN) sovirf, solvirf; (RUSSIAN) doushitsa obyknovennaja; (SERBO-CROATIAN) mravinac, obični; (SPANISH) orégano, orégano común, oriégano, ourego; (SWEDISH) kungsmynta; (TAGALOG) oregano, suganda, torongil de Limon; (TAJIKI) kokoti; (TELUGU) mridu-maruvamu; (TURKISH) adi mercanköşk, güveyi otu, güveyiotu, kekik otu, kekikotu; (VLAAMS) wilde marjolein; (WELSH) mintys y graig.

44.7. *Perilla frutescens* (L.) Britton var. *crispa* (Thunb.) W. Deane (EP) leaves and seeds[5]; (P) C,R.

(AFRIKAANS) perilla, Japanese oliesaad; (ALBANIAN, TOSK) perilë; (ASSAMESE) arim, angami, kenia; (BENGALI) ban tulsi; (CHINESE, MANDARIN) bai su; (CHINESE, WÚ) bai shu, zi su; (CHINESE, YUE) zi su; (CZECH) perila; (DANISH) perilla; (DUTCH) perille; (ENGLISH) perilla, beef-stake plant, perilla, perilla mint, Chinese basil; (FINNISH) kähäräveripeippi; (FRENCH) mélisse, perille, perille de Nankin; (GERMAN) Nessellippe, Perille, gemeine Schwarzenessel; (GREEK) perilla; (HEBREW) reḥan; (HINDI) bhanjira; (HUNGARIAN) feketecsalán; (INDONESIAN) daun shiso; (ITALIAN) perilla; (JAPANESE) shiso; (LAO) chieng, chi:n; (LITHUANIAN) krūminė perilė; (MACEDONIAN) žbunest bosilek; (NEPALI) silam; (NORWEGIAN) buskpurpurnesle, perilla; (POLISH) pachnotka zwyczajna; (PORTUGUESE) perilha, perila, chissô; (ROMANIAN) perila, urzicuțe negre; (SERBO-CROATIAN) crna kopriva; (SPANISH) perilla; (SWEDISH) bladmynta, perilla; (THAI) nga-khi-mon, nga-mon; (TURKISH) kara isirgan, perilla; (URDU) misra mukalab; (VIETNAMESE) t[is]a t[oo], t[uwr] t[oo]; (VLAAMS) shiso.

44.8. *Plectranthus esculentus* N.E. Br. (EP) tubers; (P) C,R,P.

(AFRIKAANS) kafferaartappel; (BASSARI) ɛ-ɓwdy ɛ-ɓanàx, ɛ-ɓwdy ɛ-pɛsyàx; (BEROM) vàát; (BUDIK) ɛ-ɓen *i*-diík, mɛn mwdiík; (CHINESE, MANDARIN) ka fei shu; (CHINESE, WÚ) ka fei shu; (DANISH) kaffirknold, kaffirkartoffa; (DUTCH) kaffiraardappel; (ENGLISH) country potato, Livingstone potato, Kaffir potato, Madagascar potato, hausa potato; (FRENCH) geunaine, plectranthe, pomme de terre de Madagascar; (FULFULDE, ADAMAWA) paiça gué goié, bugumwol, bugumwol gorɗi, dizgoru, bugumji, bugumji gordi, mettawela, samaiwol, samaiji, mettawelāji; (GERMAN) Kaffirkartoffel, Harfenstrauch; (HAUSA) kahuta; (INDONESIAN) kentang Jawa, kentang hitam, kembili, ubi kembili; (JAVANESE) gembili, kumbili jawa, kentang jawa, kentang jembut, kentang iler, kentang ireng, kentang kentul, kentang klici; (JUNKUN TAKUM) suma, sunma; (MALAY) ubi kembili, ubi keling; (MENDE) daugɔ; (MÒORÉ) bayoya; (NORWEGIAN) møllplante; (NYANJA) buye, buye wamthengo; (POLISH) ziemniak wiejski; (PORTUGUESE) jitamba; (SUNDA) huwi kentang, kentang bogor, kumeli; (THAI) man-khinu, man-nu; (THEMNE) ka-bunkɔ; (TIV) agbishe, i̱gbi̱se; (TONGA) njowera, njowe; (TUMBUKA) njowera; (TURKISH) mahmuz çiçeğI; (VLAAMS) Livingstone aardappel; (YAO) nyumbu.

44.9. *Satureja hortensis* L. (EP) leaves and young shoots; (P) C.

(BELARUSAN) chabor; (CHINESE, MANDARIN) mu li bo he; (CHINESE, WÚ) mu li bo he; (CZECH) saturejka zahradní; (DANISH) almindelig sar, sar, bønneurt; (DUTCH) bonenkruid, boonenkruid; (ENGLISH) summer savory, annual savory, savory; (ESTONIAN) piparrohi; (FINNISH) kesäkynteli; (FRENCH) sarriette, sarriette annuelle, sarriette commune, sarriette des jardins, herbe de Saint-Julien, savourée; (GEORGIAN) qondari; (GERMAN) Sommer-Bohnenkraut, Bohenkraut, Weinkraut; (GREEK) satouria; (HEBREW) ẕatra, tzatra hortensis; (HUNGARIAN) csombor, borsfű; (ICELANDIC) sar; (ITALIAN) santoreggia, satureja, coniella; (JAPANESE) kidachi-hakka; (LATVIAN) kalnumētras, ceberiņš; (LITHUANIAN) darželinis dašis; (MACEDONIAN) gradinarska čubric; (NORWEGIAN) hagesar, sar; (POLISH) cząber ogrodowy; (PORTUGUESE) segurelha, alfavaca do campo, remédio do vaquerio; (SPANISH) ajedrea, ajedrea común, ajedrea de jardín, ajedrea blanca, sabouja; (SWEDISH) kyndel; (TURKISH) bakla otu, geyikotu, sater, kekik; (TURKMEN) bag çaberi; (UZBEK) gulraikhon, gulralkhon; (VLAAMS) zomerbonekruid, tuinbonekruid.

44.10. *Solenostemon rotundifolius* (Poiret) J.K. Morton (EP) tubers; (P) C.

(AJA) bitunyakpokpu; (BAMANANKAN) oussou-ni-fing, usu ni fin, usu ni gé, fa-birama, ussu-nifing, ussu-ni-fing, ussu-ni-gue; (BEROM) ekú; (DAGBANI) piaha, piha; (DANISH) hausakartoffel; (DUTCH) hausa aardappel; (EDE NAGO) krodin; (ENGLISH) hausa potato; (FON) yehwlen; (FRENCH) pomme de terre des Hausa; (FULFULDE, ADAMAWA) bugumji debbi, bugumwu, metawela, metabela, noppi bamɗi; (GERMAN) Hausakartoffel; (GOURMANCHÉMA) apara;

51. Onagraceae

51.1. *Oenothera biennis* L. (EP) leaves and tender shoots; (P) R,C.

(AFRIKAANS) nagblom; (BELARUSAN) skripnevaja, aslynnik dvuhgadov; (CHINESE, WÚ) yue jian cao; (CHINESE, YUE) yue jian cao, ye lai xiang; (CZECH) pupalka dvouletá; (DANISH) toårig natlys; (DUTCH) gewone teunisbloem; (ENGLISH) evening primrose; (FINNISH) iltahelokki; (FRENCH) onagre bisannuelle, herbe aux ânes; (GAELIC, SCOTTISH) s(e)òbhrach; (GERMAN) gemeine Nachtkerze; (GREEK) primoula, iranthes, doskraki; (HEBREW) ner ha-layal, ner haleila; (HOPI) poliisi; (HUNGARIAN) parlagi ligetszépe; (ICELANDIC) nattljos; (INDONESIAN) daun panu; (ITALIAN) enàgra comune; (LATVIAN) divgadīgā naktssvece; (LITHUANIAN) dvimetė nakviša; (MACEDONIAN) dvegodišna večernica; (NEPALI) pitambar phool; (NORWEGIAN) rapontika; (POLISH) wiesiołek dwuletni; (PORTUGUESE) onagra, erva dos burros, boa tarde, minuana; (POTAWATOMI) owesa'wanakûk; (SERBO-CROATIAN) pupoljka, dvogodišnj; (SPANISH) onagra, onagra bienal, onagra común, hierba del asno, enotera; (SWEDISH) nattljus; (TURKISH) adi eşekçiçeği eşek çiçeği; (VLAAMS) middelste teunisbloem; (WELSH) melyn yr hwyr.

52. Orobanchaceae

52.1. *Orobanche crenata* Forssk. (EP) shoots; (P) C.

(ASSAMESE) bonia bau; (BRETON) bara-gad; (CHINESE, WÚ) lie dong; (CZECH) záraza, záraza vroubená; (DANISH) gyvelkvæler; (ENGLISH) broomrape; (ESTONIAN) soomukas; (FINNISH) loisio, näive; (FRENCH) orobanche chevelue; (GERMAN) Pferdebohne Sommerwurz, Ackerbohne Sommerwurz; (GREEK) orovagchi; (HEBREW) ʿaleket ḫaruka, aleket harutza; (HUNGARIAN) szádorgó; (ICELANDIC) ertusnikja; (ITALIAN) orobanche della fava; (LITHUANIAN) puošnioji džioveklė; (MACEDONIAN) sina kitka; (MALTESE) budebbus; (MONGOLIAN) guvshaakhai; (NEPALI) jhokara jhar; (NORWEGIAN) snylterot; (POLISH) zaraza; (PORTUGUESE) orobanca, rabo de raposa, erva toira; (ROMANIAN) lupolie verigel; (RUSSIAN) zaraziha; (SPANISH) jopo de las habas, orobanca; (TAJIKI) shumgioh; (TURKISH) canavarotu, bostanbozan, muhteşem canavarotu; (VLAAMS) bremraap; (WELSH) orfanc.

53. Oxalidaceae

53.1. *Oxalis tuberosa* Molina (EP) tubers; (P) C.

(ASSAMESE) tengeshi, sengeri; (CHINESE, MANDARIN) nan mei cu jiang cao; (CHINESE, WÚ) nan mei cu jiang cao; (CZECH) oka, šťavel hliznatý; (DANISH) oka; (DUTCH) knolklaverzuring; (ENGLISH) oca, oka; (FINNISH) käenkaali; (FRENCH) oxalide tubéreuse, truffette acide; (GERMAN) Knollen-Sauerklee, Knolliger Sauerklee; (GREEK) oxalida; (HEBREW) ḫamẕiẕ ha-pekaʿot; hamtzitz; (HUNGARIAN) gumós madársóska; (INDONESIAN) semanggig; (ITALIAN) acetosella tuberosa; (LITHUANIAN) gumbinis kiškiakopūstis; (MACEDONIAN) krtolesta krasatka; (MONGOLIAN) bultsuut isreg; (NEPALI) chari amilo; (NORWEGIAN) knollgaukesyre; (POLISH) szczawik bulwiasty; (PORTUGUESE) oca, azedinha; (QUECHUA, CUSCO) oca; (SPANISH) oca, huisisai, ibia, aleluya tuberosa; (TAJIKI) turshak; (TURKISH) yumrulu ekşi üçgül, tirtil, kuzu kulaği; (VLAAMS) oka.

54. Passifloraceae

54.1. *Passiflora biflora* Lam. (EP) shoots, young leaves, flowers; (P) C,R.

(ASSAMESE) khasia bael; (CHINESE, WÚ) xi fan lian; (CHINESE, YUE) xi fan lian; (CZECH) mučenka dvoukvětá; (DANISH) passionsblomst; (DUTCH) passiebloem; (ENGLISH) passion flower; (ESTONIAN) kannatuslill; (FRENCH) passiflore à deux fleurs, fleurs de la passion; (GERMAN) Passionsblume; (GREEK) passiflora; (HEBREW) sheonit; (HUNGARIAN) golgotavirág; (INDONESIAN) arjuno sosro, ketungkeng, bunga kupu-kupu; (ITALIAN) passiflora; (JOLA-FONYI) kuña ñãngol; (NEPALI) ghadi phool; (POLISH) męczennica; (PORTUGUESE) flor da paixão, chagas; (SERER-SINE) mété a koy; (SPANISH) passionaria, flor de la pasión; (TURKISH) çarkifelek, firildak çiçeği; (URDU) passion flower; (WÁRAY-WÁRAY) mirinda; (WOLOF) maribisab.

55. Pedaliaceae

55.1. *Sesamum radiatum* Schumacker (EP) young shoots; (P) C.

(AKAN-AKUAPEM) torotoro; (BAMANANKAN) nudeku; (BAOULÉ) fiandron; (BASSARI) ɛ-xɔ́yí-kɔ́yí, ngbocho, umpocho; (BUDIK) *nyw*-ywdy; (BULLOM SO) yɛnthɛ-pokan-dɛ, bulkɔ, bukɔ-lɛ; (CHINESE, WÚ) xiang ma; (CHINESE, YUE) zhi ma; (CRIOULO, CAPE VERDE) tres-marias; (CZECH) sezam paprsčitý; (DAGBANI) salu; (DANGME-KROBO) zãpui; (ENGLISH) gogoro; (FULFULDE) gouboudo; (FULFULDE, ADAMAWA) noome baroo; (GA) oturo; (GOLA) doloŋgbo, dombo; (GREEK) sousami; (HAUSA) kárkáshíí, nomen-bàrééwáá, ríídín bàrééwáá; (HUNGARIAN) szezám; (IGBO) ọwalo ani, oke ahịhịa, ódùm, ọrọ oji; (INDONESIAN) wijen; (JOLA-FONYI) béné; (KANURI) máráshi; (KISSI, NORTHERN) ninki, chaŋameyo; (KONKOMBA) baka; (KORO) gbinjɔ̀; (KURANKO) kabul, niŋgi, ninja, niŋke, ninta, ninki; (KRIO) bɛni-lif, bulukɔ; (LOKO) tabulɔgɔ, taminɔghɔ; (MALIMBA) budelɔmo, bulugɔ; (MANDING-MANDINKA) ninkié, ndeku, udeku, benefing-jongo, beno, lɔ-nama, nama, ninkye; (MENDE) tabukɔ, taugɔ, bukɔ, daugɔ; (MÒORÉ) bunda; (PANJABI, EASTERN) til; (POLISH) sezam; (PORTUGUESE) sésamo selvagem, sésamo bravo; (PULAAR) dubulé, laka; (SPANISH) sésamo slavaje; (SUSU) lagai, laxai, salakhoe, sɛlakhonyi, tiliŋgi-na; (SWAHILI) ufuta; (TAJIKI) kunjid, simsim; (TEM) kuseru; (THEMNE) *ka*-buləkɔ, *ka*-bunkɔ, (TIV) ne; (VAI) bukɔ, tabuka; (WAMEY) *a*-mbir; (WOLOF) bene, béné; (YALUNKA) liŋgi-na; (YORUBA) ẹ̀kukù.

56. Phytolaccaceae

56.1. *Phytolacca acinosa* Roxb. (EP) leaves and young shoots;(P) C.

(ASSAMESE) jaiong; (CHINESE, WÚ) yin du shang luo; (CHINESE, YUE) shang lu; (CZECH) líčidlo; (DANISH) almondolig kermesbaer; (ENGLISH) Indian poke; (FRENCH) shanglu; (GREEK) phytolakka; (HINDI) matazor, sarangun; (HUNGARIAN) álkörmös; (ITALIAN) cremesina acinosa; (NEPALI) jaringo; (PANJABI, EASTERN) lubar, rinsag; (URDU) matazor; (VIETNAMESE) th[uw][ow]ng l[uj]c.

56.2. *Phytolacca americana* L. (EP) leaves and young shoots;(P) C.

(AFRIKAANS) karmosymbos; (CHICKASAW) koshibba'; (CHINESE, MANDARIN) yang shang lu; (CHINESE, YUE) mei guo shang lu; (CRIOULO, CAPE VERDE) uva-macaco; (CZECH) líčidlo americké; (DANISH) amerikansk kormesbaer; (DUTCH) Amerikaanse karmozijnbes, westerse karmozijnbes; (ENGLISH) poke salat, poke, pokeweed; (FRENCH) raisin d'Amérique, morelle à grappe; (GERMAN) Amerikanische Kermesbeere, gewohnliche Kermesbeere; (HMONG DAW) kub nyuj; (HUNGARIAN) álkörmös; (ITALIAN) cremesina uva-turca, ura da colorare; (JAPANESE) yôshu-yama-gobô; (POLISH) szkarłatka amerykańska; (PORTUGUESE) fitolaca; (SPANISH) fitolaca, bledo carbonero, hierha carmin decandra; (TURKISH) Amerikan şekerci, boyasi Amerikan üzümü; (VIETNAMESE) th[uw][ow]ng l[uj]c.

56.3. *Phytolacca esculenta* Van Houtte (EP) leaves and young shoots;(P) C.

(CHINESE, MANDARIN) ke si shang lu; (CHINESE, WÚ) shan luo bo; (CHINESE, YUE) shang lu; (CZECH) líčidlo jedlé; (DANISH) almendelig kermesbaer; (ENGLISH) pokeweed; (ESTONIAN) kermesmari; (FRENCH) yama gobo; (ITALIAN) cremesina edule; (JAPANESE) yama gobô.

56.4. *Phytolacca octandra* L. (EP) leaves and young shoots; (P) C.

(CHINESE, YUE) ba rui shang lu; (CZECH) líčidlo; (ENGLISH) inkweed, dyeberry.

57. Plantaginaceae

57.1. *Plantago coronopus* L. var. *sativa* Fiori (EP) leaves; (P) R,C.

(AFRIKAANS) smalblaarplantago; (CHINESE, MANDARIN) chou ji zhuang che qian; (CHINESE, WÚ) chou ji zhuang che qian; (CHINESE, YUE) qian guan cao, che qian cao; (CZECH) jitrocel, jitrocel vraní nožka; (DANISH) fliget vejbred, hærtshørnvejbred; (DUTCH) hertshoornweegbree; (ENGLISH) buckhorn plantain; (FINNISH) liuskaratamo; (FRENCH) corne de cerf; (GAELIC, IRISH) adharca; (GERMAN) Hirschhorn-Salat, Krähenfuß-Wegerich, Schlitz-Wegerich; (GREEK) plantagos o koronopous; (HEBREW) leḥekh shasuʿa, lechech; (ITALIAN) corònopo, corno di cervo; (MACEDONIAN) elenskorogoviden tegavec, pitom tegfavec; (MALTESE) salib l-art; (MONGOLIAN) tavan salaa; (NORWEGIAN) flikkjempe; (POLISH) babka lancetowata, babka; (PORTUGUESE) diabelha, estrela-mar, galapito, guiabelha; (ROMANIAN) patlagină; (RUSSIAN) padarozhnik; (SPANISH) estrellamar, cuerno de ciervo; (TAJIKI) bargi zulf; (TURKISH) geyik boynuzu, karga ayaği, parça yaprakli sinirotu, sinir out; (TURKMEN) elekli atgulak; (VLAAMS) hertshoornweegbree.

58. Polygonaceae

58.1. *Rheum rhabarbarum* L. (EP) petioles; (P) C.

(AFRIKAANS) rabarber; (ALEUT) senartok, sênâluk; (ARABIC) rawend zahar; (ARMENIAN) chandzil; (ASSAMESE) rhubarb; (BELARUSAN) reven; (BLACKFOOT) pokínssomo; (BULGARIAN) kădravolisten reven; (CATALAN-VALENCIAN-BALEAR) ruibarbre; (CHINESE, HAKKA) t'ái vông; (CHINESE, MANDARIN) tang da huang, ta huang; (CHINESE, WÚ) tang da huang; (CREE, PLAINS) sewipak; (CROATIAN) rabarbara; (CZECH) reveň, reveň rebarbora, raveň kadeřavá; (DAKOTA) iyoptaiyeyapizi, caŋȟdoȟuhu opemni ka'ġapi; (DANISH) rabarber; (DUTCH) rabarber; (ENGLISH) rhubarb, pieplant; (ESTONIAN) rabarber, kurdlehine rabarber; (FARSI, EASTERN) rawâsh; (FARSI, WESTERN) reevand, reevase; (FILIPINO) rhubarb; (FINNISH) raparperi; (FRENCH) rhubarbe, rhapontic; (GAELIC, IRISH) biabhóg; (GAELIC, MANX) rubarb; (GAELIC, SCOTTISH) rùbrab; (GEORGIAN) revandi; (GERMAN) Rhabarber, Rhapontic; (GREEK) rion, pion; (HEBREW) rabarber, ribbas ha-ginna; (HINDI) revand chini; (HINDUSTANI) rewand; (HUNGARIAN) rebarbara; (ICELANDIC) rabarbari; (INDONESIAN) rabarber, kalĕmbak; (ITALIAN) rabarbaro, rapontico; (JAPANESE) rubaabu, kara daio; (KASHMIRI) rumach; (KATI) radsâ; (KIRGHIZ) yškyn; (KWÉYÒL) wibab; (LATVIAN) parastais rabarbers, rabarbers; (LITHUANIAN) daržovinis rabarbaras, rabarbarai; (LUXENBOURGEOIS) Rubarb; (MACEDONIAN) reven, običen reven; (MADURA) talembak; (MALAY) rubab, kelembak; (MONGOLIAN) gishuune; (NEPALI) chulthe amilo; (NORWEGIAN) matrabarbra, rabarbra; (PASHTO, NORTHERN) renand, ranāsh; (POLISH) rabarbar ogrodowy, rabarbar, rzewień, rabarbar kedzierzany; (PORTUGUESE) ruibarbo, ruibardo hortense; (ROMANIAN) revent, reventul; (RUSSIAN) reven, reven'ogorodnyj; (SERBO-CROATIAN) rabarbara, rabarber, raved; (SLOVAK) rebarbora; (SLOVENIAN) rabarbara; (SORBIAN, UPPER) rjewjeń; (SPANISH) ruibarbo, rapóntico; (SWEDISH) rabarber; (TAGALOG) rhubarb, rubarbo; (TAJIKI) rebos; (TIBETAN) chūūcuu; (TURKISH) ravent, bahçe işgini; (UKRAINIAN) revìn; (URDU) revand chini; (UZBEK) rovoch, rivož; (VIETNAMESE) cay dai hoang; (VLAAMS) rabarber; (WELSH) rhiwbob; (XHOSA) isityalo segadi ekutyiwa kuso isiqa xa siphekiweyo; (YIDDISH) rahbahrber.

58.2. *Rumex acetosa* L. (EP) leaves; (P) R,C.

(AFRIKAANS) maksuring; (ALBANIAN, TOSK) lepjetë e thartë; (ARABIC) dierb; (ARMENIAN) aveluk; (ASSAMESE) chuka sak; (AZERBIAJANI) avelik; (BADAGA) go:ngu; (BASQUE) andrebelar; (BELARUSAN) shchane zvichainie; (BENGALI) chukapalam; (BRETON) triñchin; (BULGARIAN) grandinski kiselec; (CATALAN-VALENCIAN-BALEAR) agrella; (CHINESE, MANDARIN) suan mo; (CHINESE, WÚ) suan mu; (CZECH) šťovík, šťovík kyselý, kyseláč luční; (DANISH) almindelig syre, syre; (DUTCH) tuinzuring, veldzuring; (ENGLISH) sorrel, garden sorrel; (ESTONIAN) hapu oblikas; (FARSI, WESTERN) torshak; (FINNISH) niittysuolaheinä, tarhasuolaheinä; (FRENCH) grand oseille, rumex oseille, surelle; (GAELIC, IRISH) samadh; (GAELIC, MANX) shughlaig; (GAELIC, SCOTTISH) sealbhag; (GALICIAN) aceda; (GEORGIAN) mdzauna, golo; (GERMAN) Garten-Sauerampfer; (GREEK) lapathon; (HEBREW) ḥumʿat ha-ginna, chumeat hagina; (HINDI) khatta palak, chuka; (HOPI) alasani; (HUNGARIAN) mezei sóska, sóska; (ICELANDIC) tunsura; (INDONESIAN) daoon sooring, daun suring, surengan; (ITALIAN) acetosa, acetosa maggiore; (JAPANESE) sorelu, sukampo, suiba; (KAZAKH) shcavel; (KHMER, CENTRAL) slök mchuu; (KIRGHIZ) kuzu kulak; (KWÉYÒL) lozei; (LATVIAN) parastā skābene, skābene; (LITHUANIAN) valgomoji rūgštynė; (MACEDONIAN) kiselec, običen štavej-kiselec; (MALTESE) agretta; (NEPALI) amilo ghans; (NORWEGIAN) engsyre, syre; (POLISH) szczaw zwyczajny, szczaw; (PORTUGUESE) azedas, azedinha da horta, azeda brava, azedinha; (ROMANIAN) sterie, măcrisul, măcriş; (RUSSIAN) shchavel, shchavelnik, ščvel' morščinistyj, shchavel morščinistyj; (SERBO-CROATIAN) kiselica, kiselica obična, kiseljak; (SLOVAK) šťav, štiav kyslý; (SLOVENIAN) kislica; (SORBIAN, UPPER) kisykał, šćehel; (SPANISH) acedera, acedera común; (SWEDISH) ängssyra, trädgårdssyra; (TAJIKI) shulkha, turshak; (TURKISH) kuzu kulaği, kuzukulaği; (TURKMEN) adaty turşuja; (UKRAINIAN) ščavel; (URDU) khatta palak; (VIETNAMESE) rau chua, rau b[os] x[oo]i, toan th[ar]o; (VLAAMS) zuring, zurkel; (WELSH) suran y cwn; (XHOSA) uhlobo oluthile lomuncwane; (ZULU) isumuncwane.

58.3. *Rumex patientia* L. (EP) leaves; (P) C.

(AFRIKAANS) tongblaar; (ASSAMESE) chuka sak; (BASQUE) lapaitz; (BELARUSAN) shchane shpinatna; (BULGARIAN) gradinski lapad; (CATALAN-VALENCIAN-BALEAR) panadella; (CHINESE, MANDARIN) ba tian suan mo, po tsai suanmo; (CHINESE, WÚ) ba tian suan mu; (CZECH) šťovík zahradní; (DANISH) engelsk spinat; (DUTCH) patientie; (ENGLISH) dock, patience dock; (ESTONIAN) oblikas; (FARSI, WESTERN) revas; (FINNISH) pinaattihierakka; (FRENCH) patience, rumex patience, oseille épinard; (GAELIC, IRISH) copóg; (GAELIC, MANX) cabbag; (GALICIAN) carves; (GERMAN) Garten-Ampfer, Gemuse-Ampfer, Ewiger Spinat; (GREEK) lapatho, siderolopatho; (HUNGARIAN) paréj lorom, angol sóska; (ICELANDIC) sura; (ITALIAN) erba pazienza, pazienza; (JAPANESE) france-suiba; (KWÉYÒL) dòck, basen; (LATVIAN) skābene; (LITHUANIAN) špinatinė rūgštynė, rūgštynis; (MACEDONIAN) štavel, pitom štavej; (NORWEGIAN) hagesyre; (POLISH) szczaw żółty; (PORTUGUESE) labaça, paciencia dos jardins, azeda espinafre, azeda-crespa, língua de vaca; (ROMANIAN) stevia; (RUSSIAN) ščavel' špinatnyj; (SERBO-CROATIAN) zimski špinat, kiselica, pitoma; (SLOVAK) šťav špenátny; (SLOVENIAN) vrtna kislaca; (SPANISH) hierba de la paciencia, acedera espanica; (SWAHILI) kata, katiza, fupisha, punguza; (SWEDISH) engelsk spenat; (TIBETAN) ṭhu̱sam; (TURKISH) bahçe labadasi; (UZBEK) shovul, šovul; (VLAAMS) Engelse winterspinazie; (WELSH) tafol.

58.4. *Rumex scutatus* L. (EP) leaves; (P) C.

(ASSAMESE) chuka sak; (BENGALI) amrula; (BRETON) triñchin-gad; (CHINESE, MANDARIN) yuan ye juan mo; (CHINESE, WÚ) yuan ye suan mu; (CZECH) šťovík štitnatý, kyseláč štítnatý; (DANISH) romersk syre; (DUTCH) spaanse zuring; (ENGLISH) French sorrel, garden sorrel; (FINNISH) suolaheinä; (FRENCH) rumex à écusson, oseille ronde, surelle; (GERMAN) Schild-Ampfer; (GREEK) galliko lapatho; (HEBREW) ḥum'a; (HINDI) ambavati; (HUNGARIAN) kerti sóska; (INDONESIAN) daun suring, surengan; (ITALIAN) acetosa romana, romice scudata; (MACEDONIAN) štitest štavej; (MALTESE) agretta franciza; (POLISH) szczaw tarczolistny; (PORTUGUESE) azeda romana, azedinha; (SPANISH) acedera romana, badolas, romaza escudada; (TURKISH) kalkan labada, eşkileme; (VLAAMS) Franse zuring.

59. Portulacaceae

59.1. *Montia perfoliata* (Donn, ex Willd.) Howell (EP) leaves; (P) R,C.

(ASSAMESE) malbhog khutura; (BELARUSAN) partulakavia; (CHINESE, MANDARIN) chuan ye chun mei cao; (CHINESE, WÚ) chuan ye chun mei cao; (CRIOULO, CAPE VERDE) bordulega; (CZECH) batolka prorosetlá; (DANISH) vinterportulak; (DUTCH) winterpostelein; (ENGLISH) miner's lettuce, winter purslane; (FINNISH) salaattikleitonia, talviportulakka; (FRENCH) pourpier d'hiver, claytone de Cuba, claytonia perfoliée; (GERMAN) Winterportulak, Kuba-Spinat; (GREEK) klaylania ediapirophyllos; (ITALIAN) portulaca d'inverno, spinacio di Cuba; (MACEDONIAN) zimska tučnica; (MALTESE) burdlieka tax-xitwa; (NORWEGIAN) vinterportulakk; (POLISH) portulaka zimowa; (PORTUGUESE) beldroega de Inverno, claitónia de Cuba; (SPANISH) verdolaga de invierno, verdolaga de Cuba; (TURKISH) kiş semizotu, kaynak otu; (VLAAMS) winterpostelein.

59.2. *Portulaca oleracea* L. (EP) leaves; (P) R,C.

(AFRIKAANS) porslein, porselein, postelein, misbredie, bobbejaandruif, purslaan, rooipootjieporslein, varkkos; (AJA) aflatɔkpwi; (AKAN-AKUAPEM) adwera, adworɛ; (AKAN-ASANTE) adida, adwera; (AKAN-FANTE) afare; (AMHARIC) antare; (ARABIC) bighel, arjoubat, l'gardim; (ASSAMESE) malbhog khutura; (BALANTA-KENTOHE) ensámelata, ensámelate; (BAMANANKAN) missidi koumbaré, missi kumbéré, mazabi, missidi kumbaré, san mafé; (BAOULÉ) akpolékankan; (BASQUE) ketorki; (BELARUSAN) partulak agorodnni; (BENGALI) nunia sag, nunia shag, baraloniya; (BICOLANO, CENTRAL) alusiman, ausiman, gulasiman, olasiman, sahikan, sahihan, ulisiman, awsiman, ulasiman; (BONTOC CENTRAL) dupdupil, sedlan; (BULGARIAN) kulturna tučenica; (BULLOM SO) pɔm-*ta*-gbɛŋguɛ, chɔchɔ-bɛl-lɛ; (CATALAN-VALENCIAN-BALEAR) verdolaga; (CEBUANO) bilang-bilang; (CHINESE, GÀN) ma chi xian; (CHINESE, MANDARIN) ma chi xian, ma ch'in hsien; (CHINESE, MIN BEI) zhu mu cai; (CHINESE, WÚ) jiang ban cao, gua zi cai, ma chi xian; (CHINESE, YUE) gua zi cai, ma chi xian; (CRIOULO, CAPE VERDE) bredo-fehea; (CRIOULO, UPPER GUINEA) baldroega; (CZECH) šrucha zelná, portulák; (DAGBANI) zinerigu;

(DANGME-KROBO) nɛrɛƒu; nɛrɛyu; (DANISH) have-portulak, portulak; (DOONDO) makata ma kombo, mindia mi kombo; (DUTCH) postelein; (EDE NAGO) kpakpasan; (EDO) ebé ehófẹn; (EFIK) éférémbákàrà; (ENGLISH) purslane, garden purslane, common purslane, pot purslane; (ESTONIAN) portulak; (EWONDO) elelengue; (FILIPINO) olasiman, gulasiman; (FINNISH) vihannesportulakka; (FON) afla; (FRENCH) pourpier potager, pourpier; (FULFULDE, ADAMAWA) becalhi, endu nagge, teddu tekkunde; (GA) awɔrɔke, awrɔké; (GALICIAN) baldroaga; (GBE-VHE) afla(a), afla(a)-tokpui, aflãtokpui, devio-fe'ama, devioƒeama; (GBE, WACI) aflavi; (GEN) alalokpwe; (GERMAN) Portulak, Rapunzelsalat; (GOLA) saku; (GREEK) andrachni e lachanira, andrakia; (GUJARATI) ghol, loni, mhotiluni; (GUN) nɛkan flogbohaya; (HANUNOO) alusiman; (HAUSA) babbajibji, dááɓùrín sáánìyáá, fàsa ƙaba, fàsà kùmrúrí, gure, hálsèn rààgóó, hálsèn sàànìyáá, hálsèn túnkìyáá, rùb-dà-túkúnyá, sárkín jíbjíí; (HEBREW) regelat ha-ginna, regelat haginah; (HERERO) ojitandavare; (HINDI) kulfa, khursa, baralaniya, badaa kulfa, baralunia, chhotalunia, kurfa, kurfe-ka-sag, kursa, lonia, lunia, lunikulfa, lunuk, muncha, munya, nonkha, nonkhalunuk; (HINDUSTANI) almonia, ek̩ nobát; (HOPI) pihala; (HUNGARIAN) köver porcsin; (ICELANDIC) portulakk; (IFUGAO) bakbakad, lungum,ba-kakko, bakukku, lungun, ngalub; (ILOCANO) ngalug, ngaluk; (INDONESIAN) poslèn, gelang, krokot, putri india; (ITALIAN) portulaca, porcellana; (ITAWIT) ngalug; (IVATAN) ngayuy; (IZON) oké beri, ị mẹ́ẹ́lẹ́ị́ tuo, obórí mɛ̱ɛ̱lɛ̱í̱; (JAPANESE) tachi suberi hiyu, souberi fiyou; (JAVANESE) krokot, kerayapan; (KANKANAEY) aluloy, balballay, dupduppilaw; (KANNADA) dodda goni soppu, doddagoni soppu, duda-gorai, golisoppu, gonikasa, kirigoli; (KASHMIRI) kuna; (KAZAKH) kara ot, portulak ogorodnyj; (KHMER, CENTRAL) maông dâb phkaa; (KISSI, NORTHERN) kundiapɔnɔ; (KONKANI) gol, golchibagi; (KONO) *g*bɔŋgbɔ-*n*jamba; (KRIO) arata-yes, arata-yace; (KUNG-EKOKA) bobo; (KWANYAMA) oshimhelewene; (LAHU) mvuh$_{\wedge}$ nyi^{-} k'o, mvuh$_{\wedge}$ hpo$_{\wedge}$ teh^{-}, pfuh ti$_{\vee}$ mvuh$_{\wedge}$; (LAO) taa kôngz, biaz; (LATVIAN) portulaks; (LITHUANIAN) paprastoji portulaka, daržovine portulaka; (MACEDONIAN) tučnica, obična tučnica; (MADURA) re-sereyan; (MAITHILI) nōnī; (MALAY) jerami, gelang pasir, segan jantan, rumput beremi; (MALAYALAM) pasali keera, koricchira, korichhira, koluppa; (MALTESE) burdlieka; (MANDING-MANDINKA) mazabi; (MANDJAK) umbintchim; (MANKANYA) belôluga; (MARANAO) biala; (MARATHI) ghol, ghole, bhuigholi, kurfah, mhotighol, pasalai; (MENDE) nyina-wolo, tɔbui, tɔluga, njema, tɔndɔ wuli, tɔɔge; (MONGOLIAN) nogoon gazriin morin shud; (MORISYEN) pourpie; (NEPALI) nundhiki; (NIUE) kamole; (NORWEGIAN) matportulakk, portulakk; (NYAKYUSA-NGONDE) matakogawaoli; (NYANJA) matakoatsanu; (ORIYA) kulpha saga, purunisag; (PANGASINAN) kantataba; (PANJABI, EASTERN) kulfa, lonak; (POLISH) portulaka siewna; (PORTUGUESE) beldroega; (PULAAR) dépè; (QUECHUA, CUSCO) llutu-llutu, llutuyuyu; (RAROTONGAN) pokea; (ROMANIAN) grăsită; (RUSSIAN) portulak ogorodnyj; (SAMBAL, BOTOLAN) taglabag dikot; (SAMOAN) gigie, tamole; (SANSKRIT) brhalloni, ghotika, loni, lonika, lunia, brihathlonika, lona; (SANTALI) moṭa uṛic alañ, uṛic alañ; (SERBO-CROATIAN) portulak, portulac, povaljan; (SERER-SINE) taf a nuk, tiaf a nuk; (SLOVAK) portulaka kapustna siata; (SLOVENIAN) tolščak; (SONGHAY) san mafé; (SPANISH) verdolaga; (STELLINGWERFS) posselein, postelein; (SUBANON) glunsiman; (SUMO-MAYANGNA) dislah pula; (SUNDA) gelang, poslen; (SUSU) bɔrɔbɔrɛ, tiridinkitinyi; (SWAHILI) tako la hassan; (SWEDISH) trädgärdsportlak, portlak; (TAGALOG) kolasiman, golasiman, olasiman, sahihan, makabling, sahikan, ulasiman, ulisiman; (TAGBANWA) aluyu; (TAJIKI) chilmangushtak, khurfa; (TAMIL) pasalai keerai, karikeerai, karikkeerai, parpukire, parukire, paruppukiray, pullikirai, pulikkeerai, parupukkeerai, pasalaikkeerai, passalakkeerai, vazhukkaikeerai; (TELUGU) pavila, pappukoor aaku, ganga-pavilikura, pappukura, peddapavila kura, peddapayilikura, boddupavilikoora, ganga paavilikoora, pappukoora, pedda pavilikoora; (THAI) phakbia-yai, phakbia-dokluang; (THEMNE) *a*-boro, ɛ-tana, *an*-ləbin-ləbin, *ma*-kil-*ma*-a-rɔ̃; (TIRURAY) lensimon; (TONGA) matakaoali; (TONGAN) tamole; (TULU) golipadpe; (TUMBUKA) kokwa; (TURKISH) semizotu; (TURKMEN) adaty semzek; (TWI) adwera, adwere; (UKRAINIAN) portulak; (URDU) kulfa, khurfah; (VAI) kɔnjɛ-folo; (VIETNAMESE) rau sam; (VLAAMS) postelein, zomerpostelein; (WÁRAY-WÁRAY) olasiman; (WOLOF) tâg i mpitöx, tang i mpiteurh, tanguipeter, tãnk i mpétah; (YAO) mataga atsanu; (YORUBA) awá yékú, esan ọmọde, pàpásan, kpakpasan, papassan.

59.3. *Talinum fruticosum* (L.) Juss. (EP) leaves; (P) C.

(AJA) glazwi; (ARABIC) filfil ahmar; (ASSAMESE) peerali paleng, asomiya paleng; (BEMBÉ) pina; (BULU) elok-sup; (CHINESE, MANDARIN) jia ren shen; (CHINESE, WÚ) jia ren shen; (CHINESE, YUE) tu ren shen; (DANGME) amlɔʃi-pui, ŋmlɔŋmlo, amlɔtʃi-pui; (DENDI) gɔrɔnɔmmafɛ; (DUTCH) Surinaamse postelein; (EDO) adodoro, ebọ́dọ̄ndọ̄n; (EFIK) ḿmɔ̀ngḿmɔ̀ng íkɔ̀ng; (ENGLISH) waterleaf, Lagos bologi, Suraim spinach, tilinum, Surinam purslane, sweetheart; (ÉWÉ) yevugboma; (EWONDO) elok-sup; (FON) aglaswe; (FRENCH) grand pourprier, pourprier droit, grassé; (GA) shwee, awrɔke, blɔfo/wie, bɔkɔbɔkɔ, shwie; (GBE, WACI) yovogboma; (GEN) yovogboma; (GERMAN) Talinum, Javaspinat, Brautpflanze; (GUN) aglaswe; (HEBREW) talyon meshulash; (IFUGAO) palawel; (IGBO) gbọlọgi, ofe-bèkeè, ntì̩ ōke; (INDONESIAN) gelang, krokot belanda; (ITALIAN) talia; (IZON) obóríị́mẹ́ẹ̀lẹ́ị́ị́; (JAVANESE) poslen; (JULA) nangobourou; (KOONGO) epinari; (KRIO) gambain bologi, legɔsbɔlɔgi; (MARANAO) biala, karaparapa a sina; (MENDE) kpame, kpaome; (NGUNDI) kipinari; (POLISH) liściowiec wodny; (PORTUGUESE) beldroega grande; (SENOUFO, CEBAARA) damsa fala; (SPANISH) espinaca de Filipinas, espinaca de Java, verdolaga de playa; (SUBANON) galaghati, galigbati; (TAGALOG) talinum, talilong; (TAGBANWA) saluyot; (TAMIL) pasali; (THAI) som-kaoli, som-khon, som-chin; (THEMNE) *ma*-kilmoron (TIV) ashwe; (TOK PISIN) kumu manus; (TURKISH) Surinam semizotu, büyük semizotu, semiz otu; (TWI) nkok ɔdwe, boroboro, busummuru adwerɛ, ɛ-fan; (VIETNAMESE) th[oor] nh[aa]n s[aa]m; (VLAAMS) Surinaamse postelein, posren; (WÁRAY-WÁRAY) sidá-sidâ; (YOM) korforðm; (YORUBA) gbure, gbure funfun.

59.4. *Talinum paniculatum* (Jacq.) Gaertner (EP) young shoots; (P) C.

(ASSAMESE) peerali paleng, noniya; (BENGALI) baraloniya; (CHINESE, MANDARIN) tu jen san; (CHINESE, WÚ) tu ren shen; (CHINESE, YUE) yuan zhui hjua tu ren shen; (DANISH) talinum; (ENGLISH) flameflower; (FRENCH) pourpier, pourpier de la plage; (GUJARATI) moti loni, ghol; (HINDI) khursa, baralaniya, kulfa; (KANNADA) dooddagooni soppu; (KRIO) lɛgɔs-bɔlɔgi, (MARATHI) mhotighol, bhvigholi, kurfah; (ORIYA) purunisag; (PANJABI, EASTERN) lonak, kundar; (PORTUGUESE) beldroega das areias, beldroega de praia; (QUECHUA, CUSCO) cuchiyuyu; (SANSKRIT) brihalloni, lonika, lonamla; (SPANISH) portulaca de playa; (TAGALOG) talinum; (TAMIL) karikeerai, paruppukiray; (YORUBA) gbure pupa.

60. Resedaceae

60.1. *Reseda odorata* L. (EP) leaves and flowers; (P) C.

(AFRIKAANS) minjonet, reseda; (BELARUSAN) resedovia, reseda pahuchaja; (CHINESE, WÚ) mu xi cao; (CHINESE, YUE) mu xi cao; (CZECH) rezeda vonná; (DANISH) have-reseda, reseda; (DUTCH) reseda; (ENGLISH) mignonette; (ESTONIAN) reseeda; (FARSI, WESTERN) esparak; (FINNISH) tuoksureseda; (FRENCH) herbe d'amour, mignonette, réséda odorante; (GERMAN) Garten-Reseda; (GREEK) pezeda o evosmos, pezenta; (HEBREW) rikhpa reḥanit, richpah; (HUNGARIAN) szagos rezeda; (ICELANDIC) ilmkollur; (ITALIAN) reseda, amorino, amoretti d'Egitto, miglionet; (LATVIAN) smaržīgā rezēda; (LITHUANIAN) kvapioji rezeta; (MACEDONIAN) mirizliva rezeda; (MALTESE) denb il-haruf; (NORWEGIAN) hagereseda; (POLISH) rezeda wonna, rezeda ogrodowa; (PORTUGUESE) reseda, reseda de cheiro, erva de amor, minhonete; (SPANISH) réseda, réseda de odor, miñoneta; (SWEDISH) reseda; (TAJIKI) asirak, guli asirak; (TURKISH) kokulu muhabbet çiçeği, muhabbet çiçeği; (URDU) khazam; (VLAAMS) reseda; (WELSH) perllys.

61. Saururaceae

61.1. *Houttuynia cordata* Thunb. (EP) leaves; (P) R,C.

(ASSAMESE) masundari; (CHINESE, MANDARIN) ji cai; (CHINESE, WÚ) ji cai; (DANISH) kamaeleonblad; (ENGLISH) saururis, tsi; (FRENCH) hottuynie, houttuynie en coeur; (GERMAN) Chamäloonblatt; (INDONESIAN) jukut hanyir; (JAPANESE) dokudami; (KHMER, CENTRAL) chii yaab kaa; (LAO) kh'aaw th'oong; (NEPALI) gane; (THAI) phlu-kae, phakkhaothong; (VIETNAMESE) di[ees]p c[as], gi[aas]p c[as], l[as] gi[aas]p; (VLAAMS) saururis.

62. Solanaceae

62.1. *Capsicum annuum* L. var. *annuum* (EP) mature fruit.; (P) R,C.

[1]Bukusu dialect of Luyia.

(AFRIKAANS) soetrissie; (AGTA) ladi, lara; (ALBANIAN, TOSK) spec, pipere; (ARABIC) filfil; (ARMENIAN) pchpech pačokavor; (ASSAMESE) mitha jolokia, capsicum; (AZERBIAJANI) kyrmyzy istpot; (BAATONUM) yɛku boo; (BADAGA) kunna ma:su, masuka:i, ma:su, ma:cu; (BAOULÉ) sralo; (BASQUE) biper; (BELARUSAN) perets; (BÉTÉ) gbofroto yibê; (BICOLANO, CENTRAL) sili, lada; (BINUKID) binenisil, kegu'ku'; (BLACKFOOT) pisstaahkaipokoo; (BONTOC CENTRAL) sili, kalu-bengan, kalubsengan, kalubsengan; (BOSNIAN) crvena paparika, aleva paprika, paprika; (BULGARIAN) sladak piper; (BULU) beta ôndôndô; (CAJUN FRENCH) piment doux; (CATALAN-VALENCIAN-BALEAR) pebre, pebrot; (CEBUANO) sili, kagoko; (CHECHEN) miarza burch, k'oa yotsu burch; (CHICKASAW) hammahomma' champoli', hammahomma' lakna'; hammahomma' okchamali'; (CHINESE, HAKKA) foû-tsiāo; (CHINESE, MANDARIN) tian jiao, tse tin tsiu; (CHINESE, MIN NAN) hiam-chio-á; (CHINESE, WÚ) tain jiao; (CHINESE, YUE) tian jiao, yuan jiao, deng long jiao; (COMANCHE) tuku mahyaʔ; (CREE, PLAINS) papeskomina; (CREOLE, HAITIAN FRENCH) piman dous, piman dou; (CRIOULO, CAPE VERDE) pimenton; (CZECH) sladká paprika, paprika roční; (DAGAARE, SOUTHERN) siman bēzē; (DAKOTA) sucikcikana seca, yamnumnuġapi su; (DANISH) paprika, peberfrugt; (DUTCH) paprika; (EDE NAGO) ata jijɛ; (ENGLISH) bell pepper, green pepper, sweet pepper, pimento pepper, capsicum pepper, bull nose pepper; (ESTONIAN) piment, paprika; (ÉWÉ) yevutadi; (FARSI, EASTERN) morch; (FARSI, WESTERN) felfel dolmeyi, felfel; (FILIPINO) siling-bilog; (FINNISH) vihannespaprika, paprikat; (FON) yovo takin; (FRENCH) poivre d'Espagne, piment d'Espagne, paprica, poivron, piment doux; (FULFULDE, ADAMAWA) chitta, chittāje, chitta kolong; (GA) tɔŋŋgɔ; (GADDANG) palut; (GAELIC, IRISH) piobar glas; (GALICIAN) pemento, pimiento; (GBE, WACI) gboyoboli; (GEN) yebese; (GEORGIAN) cicaka; (GERMAN) Gemüse-Paprika, Paprika, Gewürzpaprika; (GIANGAN) sili; (GIKUYU) biribiri; (GOURMANTCHÉ) diakari-gberi; (GREEK) piperia, prasini piperia, pimenta e pharmakestiki; (GUN) vafo daho; (HANUNOO) lada balaynun; (HARSŪSI) bezār, felfēl; (HAUSA) tattāsai, tasshi; (HEBREW) pilpelet ha-ginna, pilpel; (HILIGAYNON) katúmbal; (HINDI) lalmirich, simla mirch; (HMONG DAW) kua tvob phaw; (HOPI) tsiili; (HUNGARIAN) paprika, étkezési paprika, paradicsompaprika; (IBANAG) sili; (ICELANDIC) paprika; (IFUGAO) sili; (IGBO) ose-oyibo; (ILOCANO) sili, gumpapeng; (INDONESIAN) cabai manis, cabe merah, cabai besar, lombok; cabe besar, cabai, cili, lada merah; (ITALIAN) peperone, peperone dolce, peperoncino; (IZON) ịgịná; (JAPANESE) piiman, shishi-tôgarashi, tōgarashi; (JAVANESE) lombok, mengkreng, cabe; (JOLA-FONYI) etam bakaani; (JULA) frotto bani; (KANNADA) donne menasina kaayi;

(KAZAKH) perec sladkij; (KIRGHIZ) kalempir; (KOREAN) gochu; (KRIO) bell-pɛpɛ, bigpɛpɛ, gadin pɛpɛ, swit pɛpɛ; (KWÉYÒL) pwev piman, pimando; (LADINO) pimyènta, bibèr, pimienton, pimyenton; (LAHU) a⁻ hpe^ chaweh; (LATVIAN) paprika, pipari; (LITHUANIAN) vienmetė paprika, ankštpipiriai; (LOMWE, MALAWI) tsabola; (LUO) kamu laru; (LUXENBOURGEOIS) Pfeffer; (LUYIA) biribiri[1]; (MACEDONIAN) piperka, širokoplodna piperka; (MALAGASY) felfelzina; (MALAY) cili manis, kada manis, lada besar; (MALTESE) bzar ahdar; (MANINKAKAN, WESTERN) kano; (MARANAO) kagoko; (MEHRI) bəzār, bəsbēs; (MENDE) pujɛ, pujɛ -wawa; (MICMAC) dăpesāwāāl; (MOHAWK) tyotskara'kowa; (MÒORÉ) kíparé; (MORISYEN) poivron, pima salad; (MUSKOGEE) homv-lane; (NEPALI) bhede khursani; (NEWAR) kālālā; (NGIZIM) shìitá, cìitá; (NIUE) pepa; (NORWEGIAN) grønnsakpaprika, paprika; (NYANJA) tsobola, mpiripiri; (OROMO, WEST CENTRAL) kundo berberi, mimita, dabusi; (PALI) ibhapipphalí, mágadhí; (PAMPANGAN) larang-bilog, sili; (PANGASINAN) pamínta, síli; (PANJABI, EASTERN) mirch, shionlä mirch; (PASHTO, NORTHERN) mrëch, mrich; (POLISH) papryka raczna wieloowocowa, paprika sladka; (PORTUGUESE) pimento, pimentão, paprika; (PULAR) kiamakou; (ROMANIAN) ardei, ardei gras, ardei dulce, ardei de salată; (RUSSIAN) perets sladkii, perets salatnyi, perec sladkij; (SAMBAL, BOTOLAN) lada-bali; (SAMOAN) polo papalagi, polo ula; (SENOUFO, CEBAARA) sikalikpolougui; (SERBO-CROATIAN) paprika; (SINHALA) malu miris; (SLOVAK) paprika zeleninová, paprika ročná; (SLOVENIAN) paprika; (SOMALI) filfil; (SORBIAN, LOWER) pjepjer; (SORBIAN, UPPER) paprika; (SOTHO, NORTHERN) pherefere yetala; (SOTHO, SOUTHERN) pepere e tala; (SPANISH) pimiento, pimiento dulce, pimiento morron, paprica, chile dulce; (SUNDA) cabe, lombok, sabrang; (SWAHILI) pilipili kichaa, pilipli tamu, chakula, hoho; (SWEDISH) paprika, spansk peppar; (TAGALOG) sili, siling-bilog; (TAJIKI) qalamfur; (TAUSUG) sili, larajanggay; (THEMNE) ta-gbeŋgbe-ta-potho, ta-gbeŋgbe-ta-wir, ta-gbeŋgbe-ta-bana; (TIV) mkem; (TSWANA) pherefere; (TURKISH) biber, dolmalik biber; (TURKMEN) adaty burç, perec; (TWI) ntomkom; (UKRAINIAN) perec'; (URDU) lalmirich; (UZBEK) shirin calampir, achuk calampir, aččik kalampir; (VIETNAMESE) ot; (VLAAMS) paprika, peper; (WAAMA) saabɔtɔ yɛɛku; (WÁRAY-WÁRAY) hulabtog; (WELSH) pupur poeth; (WOLOF) kani salat; (XHOSA) igrin pepa; (YAO) sabola; (YORUBA) ata jijɛ, tatase, atarodo; (ZULU) upelepele oluhlaza.

62.2. *Capsicum annuum* L. var. *longum* Bailey (EP) mature fruit; (P) C.

[1]Bukusu dialect of Luyia.

(AFRIKAANS) brandrissie; (AKAN-ASANTE) makɔ huam; (ALEUT) ônarnartok aupalângajok; (ANDO) ahalimaku, kokuamu, maku; (ASSAMESE) jolokia; (BADAGA) kunna ma:su, masuka:i, ma:su, ma:cu; (BASSARI) kantoge, kiegeman; (BENGALI) lankamorich, lalmorich, lalmirchi; (BICOLANO, CENTRAL) siling-labuyo; (BODO) fanlu; (BONTOC CENTRAL) siling-labuyo; (BULLOM SO) kɛfɛ-bom-dɛ, kɛfɛ-bum; (BULU) beta ôndôndô; (BURMESE) nga-yoke-thi; (CAROLINIAN) amweigh, amwiig, mwiigh; (CATALAN-VALENCIAN-BALEAR) pebre, pebrot; (CEBUANO) siling-labuyo; (CHECHEN) c'ie~ burch; (CHICKASAW) hammahomma', hammahomma' homi'; (CHINESE, GÀN) la jiao; (CHINESE, HAKKA) lăt-tsiāo; (CHINESE, MANDARIN) chang la jiao; (CHINESE, MIN BEI) fan jiao; (CHINESE, MIN NAN) hoan-kiu-á; (CHINESE, WÚ) chang la jiao; (CHINESE, XIANG) jiao zi; (CHINESE, YUE) niu jiao jiao, jian jiao, fan jiao; (COMANCHE) tsiira?; (CREOLE, HAITIAN FRENCH) piman pike, piman bouk, piman zwézo; (CRIOULO, CAPE VERDE) malagueta; (CROATIAN) paprika; (CZECH) paprika čili, kayenský pepř; (DAGAARE, SOUTHERN) simane; (DAGBANI) bɔlumbo; (DANGME-KROBO) toko-kwada; (DANISH) spansk peber; (DUTCH) spaanse peper; (EFIK) aman- ǹtùèn, éfọ̀k ǹtókòn, ǹtókòn ókpó; (ENGLISH) cayenne pepper, chili pepper, hot pepper, chili; (ESTONIAN) kirbik, paprika, punapipar; (ÉWÉ) atadi; (FARSI, EASTERN) morch; (FARSI, WESTERN) filfil-i-surkh, felfel, felfel germez; (FIJIAN) miracha; (FINNISH) maustepaprika; (FRENCH) poivre de

Cayenne, piment de Cayenne, piment d'oiseau; (FULFULDE, ADAMAWA) citta kolong; (GA) akwelewaobii, kpakpoʃitɔ; (GAELIC, IRISH) piobar dearg; (GAELIC, MANX) pibbyr caiaan; (GANDA) kamulari; (GBE-VHE) titikukɔɖɛ; (GEORGIAN) tsitsaca; (GERMAN) Cayenne Pfeffer, Chilli Pfeffer, Vogel Pfeffer, Spanischer Pfeffer, Gewürzpaprika; (GIANGAN) siling-labuyo; (GIKUYU) biribiri; (GOLA) kidɛ-gbɛgbɛ; (GOURMANTCHÉ) diakari-gberi; (GREEK) piperia e makrdkarpos; (GUJARATI) lelmirchi; (HANUNOO) lada balaynun; (HARARI) qondo; (HAUSA) barkono, tankwa, yāji, tàttàasay; (HEBREW) pilpelet, pilpel ḥarif, pilpel cayenne; (HINDI) lalmirich, lalmirach, laalmirich, gachmarich, gachmicichi, gachmirich, mirch, mirach; (HINDUSTANI) gol mirch; (HMONG-DAW) kua txob; (HMONG-NJUA) hov txob nplej; (HOPI) tsiili; (HUNGARIAN) etkezési paprika, fűszer paprika; (IBANAG) siling-labuyo; (ICELANDIC) chilepipar; (IFUGAO) siling-labuyo; (IGBO) ose nsukka, osè nkwu, osè nwañkpi, ose utọrọ; (ILOCANO) siling-labuyo; (INDONESIAN) cabai besar, lombok; (ITALIAN) pepe di Cayenna, peperone rabbioso, peperoncino piccante; (ITAWIT) sili macopa; (JAPANESE) fushimi-tougarashi; (JOLA-FONYI) emama kaani; (JUNKUN TAKUM) shitapú; (KANNADA) mesina kai, menasina kaayi, kempu mensau, byadagi menasina kaayi, khaarada menasinakaayi; (KANURI) tàttàshé; (KASHMIRI) martsa - gõsôñu; (KIOWA) tseidlei; (KIRIBATI) te beneka; (KISSI, NORTHERN) kisã-chuindõ; (KONKANI) putaque pariu; (KONKOMBA) ekandjimau, ikâgju, ikàndio, katetu; (KONO) putu-wa; (KORO) tàtaci; (KRIO) chɛripɛpɛ, kabala pɛpɛ; (KWÉYÒL) piman kayenn, piman-sézòné; (LATVIAN) sarkanie pipari; (LIMBA) makontho; (LOKO) njesugɛ; (LOMWE, MALAWI) tsabola; (LUO) kamu laru; (LUXENBOURGEOIS) Pfeffer; (LUYIA) biribiri[1]; (MACEDONIAN) dolgoplodna piperka; (MALAY) cili, cili besar; (MALAYALAM) upperi-parnki, perangimuluk, chalie, lodachina; (MALTESE) bzar ahmar; (MANDING-MANINKA) kien furtu, sinsi furtu; (MANINKAKAN, WESTERN) kani dimba; (MARATHI) mirsinga, mirchi, laal mirchee, mirchee; (MENDE) pujɛ-wawa; (MICMAC) dăpesāwāāl mĕgwāāgŭl; (MIZO) hmarcha; (MÒORÉ) kíparé; (MORISYEN) tipima; (MUNDARI) marchi; (MUSKOGEE) homv-cate; (NEPALI) machha khursani; (NEWAR) pippali; (NIUE) polo magiho; (NORWEGIAN) spanskepepper, krydderpaprika, paprika; (NYANJA) tsobola, mpiripiri; (OROMO, WEST CENTRAL) berberi, mimita; (PAMPANGAN) siling-labuyo; (PANGASINAN) pamínta, síli; (PANJABI, EASTERN) läl mirch; (PASHTO, NORTHERN) srë mrëch, sara mrich̲; (POLISH) papryka roczna; (PORTUGUESE) malagueta, jindungo, pimento de Caiena, pimenta vermelha, pimentão de Caiena, pimenta malagueta; (POTAWATOMI) wasgagIn; (PULAAR) nyamaku-suokuli; (PULAR) kani salade; (QUECHUA, CUSCO) misqui uchu; (RUSSIAN) perec odnoletniy; (SAMOAN) pepa; (SANGA) itace; (SANSKRIT) marichiphalam, katuveera, katu vira, bruhi, branmaricha, raktha maricha, lanka, marichi phalam, peetthakarini, theekshna; (SERER-SINE) kani tubab, kanii; (SIDAMO) barbarè; (SINHALA) awu miris; (SPANISH) chile, guidnilla, pimiento picante; (SUSU) sikuli *g*béngbé, yaraféli *g*béngbé, sekule-gbɛŋgbɛ; (SWAHILI) pilipili, pilipili hoho, pilipili kali; (SWEDISH) cayenne peppar; (TAGALOG) siling-labuyo, sayul lara, siling-haba; (TAMIL) miligay, malagay, usimilagaay; (TAUSUG) siling-labuyo; (TELUGU) mirapa kaya, mirapa kaayi, mirapakaya, galakonda, sudmirapakaaya; (TEM) tschangei-kubong; (THEMNE) *ta*-gbɛŋgbɛ-*ta*-bana, *ta*-gbɛŋgbɛ-*ta*-bɔli, *ta*-gbɛŋgbɛ-*ta*-but, *ta*-gbɛŋgbɛ-*ta*-potho; (TIBETAN) sīpɛɛn; (TULU) muńci, mumbe; (TURKISH) sivri biber, kirmizi biber; (TWI) heneba, nsafea, makɔ huam; (URDU) lalmirich; (VAI) kefe-baba; (VLAAMS) cayenne peper, spaanse peper; (WÁRAY-WÁRAY) hulabtog; (WELSH) puper glâs; (WOLOF) kani xèn, kani tubab; (XHOSA) itshilisi; (YALUNKA) tala-suruntu-na; (YAO) sabola; (YORUBA) tatase, atarodo, ata, ata afunjalo, ata didun, ata funfun, ata sombo, ata tatase; (ZARMA) tonko; (ZULU) upelepele obomuu; (ZUNI) k/ola.

62.3. *Capsicum baccatum* L. var. *baccatum* (EP) fruit; (P) C.

[1]Bukusu dialect of Luyia.

(BULU) mone ôndôndô; (CHINESE, YUE) la jiao; (CREOLE, HAITIAN FRENCH) piman zwézo; (CRIOULO, CAPE VERDE) pimento piripiri; (ENGLISH) small pepper; (FINNISH) cayennepippuri; (FRENCH) piment d'oiseau, piment zouézeau, piment rouge; (GIKUYU) biribiri; (GOURMANTCHÉ) kambi; (HEBREW) pilpel katan; (HMONG DAW) kua txob; (HUNGARIAN) bagyópaprika; (LAHU) bui$^{\vee}$ nyi ceh$_{\vee}$; (LOMWE, MALAWI) tsabola; (LUO) kamu laru; (LUYIA) biribiri[1]; (MACEDONIAN) topčestoplodna piperka; (NEPALI) jire khursai; (POLISH) pieprz turecki; (PORTUGUESE) pimento cereja, pimento comarim, cumarim; (SPANISH) pimiento loco, pimiento silvestre; (URDU) khurdmirich, chotimirich.

62.4. *Capsicum chinense* Jacq. (EP) fruit; (P) C.

[1]Bukusu dialect of Luyia.

(BRETON) pebr; (ENGLISH) bonnet pepper; (FRENCH) poivre de Chine, piment antillais; (GIKUYU) biribiri; (ITALIAN) peperone cinese; (LUO) kamu laru; (LUYIA) biribiri[1]; (MACEDONIAN) kineska piperka; (MORISYEN) pima kari; (POLISH) papryka chińska; (PORTUGUESE) pimento chinês; (SPANISH) ají, pimiento chinese; (SUMO-MAYANGNA) angmak, anmak; (SWAHILI) pilipli mbuzi.

62.5. *Capsicum frutescens* L. (EP) fruit; (P) P,C.

[1]Bukusu dialect of Luyia.

(ABUA) óbìạn; (AKAN-ASANTE) makɔ nsatea, makɔ ohenba nsatea; (ANDO) kpèsè kpèsè; (ASSAMESE) thupi jolokia, jalakiya; (BAATONUM) yɛku yaki; (BAFUT) iitá; (BALANTA-KENTOHE) sinmeti; (BAMANANKAN) forôto, furtu; (BAOULÉ) kpèsè-kpèsè, atjémankou, mankou n'dinn'din, mankouflolô; (BASSARI) *a*-ngin, *on*-gaine, *on*-grendt, jeginga; (BENGALI) lanka, marich; (BÉTÉ) deklé yibê, yibê; (BETE-BENDI) *ù*-gàgə̀, *ù*-gùgò; (BICOLANO, CENTRAL) siling-labuyo, pasitis, rimo-rimo, sanggariya; (BINUKID) tidek; (BLAAN, SARANGANI) katumbal; (BOBO MADARÉ, SOUTHERN) kapononon; (BOKYI) okre; (BONTOC CENTRAL) siling-labuyo; (BUDIK) kaní-kaní; (BUHID) lada, lada-buid; (BULLOM SO) kɛfɛ-tuntun-dɛ; (BUTUANON) atsal, katumba; (CAROLINIAN) amweigh, amwiig, mwiigh; (CATALAN-VALENCIAN-BALEAR) pebre, pebrot; (CEBUANO) siling-labuyo, gatumbal, katumbal, kitikot, kulikot, lada; (CHINESE, MANDARIN) xiao mi jiao, laat tsiu; (CHINESE, WÚ) xiao mi jiao; (CHINESE, XIANG) chao tian jiao; (CHINESE, YUE) la jiao; (CREOLE, HAITIAN FRENCH) piman zwézo; (CRIOULO, UPPER GUINEA) malaguèta; (CZECH) paprika křovitá; (DAGBANI) kiriga; (DANGME-KROBO) kwada yowi; (DANISH) cayennepeber, chili peber; (DENDI) nyɛɛku; (DITAMMARI) mukan bammu; (DUTCH) chillipeper; (ENGLISH) tabasco pepper; (FILIPINO) siling labuyo; (FINNISH) pensaspaprika, chilipippuri; (FON) atakin; (FRENCH) piment, piment d'oiseaux, pilli, tabasco pepper; (FULFULDE, ADAMAWA) citta boɗe, tasije tokotje; (GA) akwele wabii, kpakposhito; (GAELIC, IRISH) piobar dearg; (GANDA) muwogo; (GBE-VHE) dzakpɔbɔ-tukui; (GERMAN) Tabasco Pfeffer; (GIANGAN) siling-labuyo; (GIKUYU) biribiri; (GOLA) kidɛ-*n*jɛɛ; (GOURMANCHÉMA) yamiai, kânbù; (GOURMANTCHÉ) diakari-mouinni; (GREEK) piperia e thamnodis; (GUJARATI) marcha; (GUN) vafo; (HANUNOO) lada tirindukun-tigbaya; (HAUSA) barkono, bàrkònoo; (HEBREW) pilpelet, pilpel tobasco; (HILIGAYNON) udto-udto; (HINDI) lalmirich, gachmirich, laalmircha, laan-ka-mirchi; (HINDUSTANI) mircha; (HMONG DAW) kus tsao may, kua txob; (HMONG-NJUA) kua tsao, hov txob nplej; (HUNGARIAN) cserjepaprika; (IBANAG) siling-labuyo, ami-ling; (ICEVE-MACI) ànyìfè, igál ìhól; (IFUGAO) siling-labuyo, paktiw, paktiw di inadne, paktiw di malikanu, paktiw kuem pakawin, paktiw palabok, silit-diyablo; (IGBO) osè bèkeè, osē m̃kpi, osè mkpele, osè nkere, ose-mkpili; (ILOCANO) siling-labuyo; (INDONESIAN) cabai rawit, cabe rawit, cabe setan; (ITALIAN) tabasco, peperone ornamentale; (ITAWIT) sili amiling; (JAPANESE) tougara-shi; (JAVANESE) lombok jemprit, lombok rawit, lombok jempling; (JOLA-FONYI) ebaba kani; (KANNADA) menasina kayi, holada menasina kaayi, parangi menasina kaayi; (KANURI) njitta wùjìwújí; (KIOWA) tseidlei; (KIRIBATI) te beneka; (KISSI, NORTHERN) kisã-nyamɔ̃, kisã-pɔbɔ; (KONKANI) mirsang; (KONO) gbɛŋgbɛ-mɛsɛ, putu- mɛsɛ; (KORO) shi; (KRIO) pɛpɛ, bɔd pɛpɛ; (LAHU) a¯ hpe‸ fa^ hk’eh˅, a¯ hpe‸ tcuh¯–eh¯; (LOKO) sugɛ-ŋɛŋɛɛgɛ; (LOMWE, MALAWI) tsabola; (LUO) kamu laru; (LUXENBOURGEOIS) Peffer; (LUYIA) biribiri[1]; (MAGUINDANAO) kasira; (MALAY) cili api, cili padi, lada padi; (MALAYALAM) mulaku, chabai, chabelombok, kapa molaga, kappalmelaka, ladamera, ladamerachina, ladumira, vallia capo molago; (MALIMBA) kukbɛŋbɛ, magbɛŋgbɛ; (MANDING-MANINKA) bakoro fortu, furtu-missé, *k*-pɛŋgbɛ; (MANDJAK) idúu; (MANINKA) forôto, furtu; (MANKANYA) edálê; (MANU’A SAMOAN) polo papalagi, polo uli; (MARANAO) katombar, loya tidek; (MARATHI) mirchi; (MENDE) pujɛ-mumu, pujɛ; (MÒORÉ) kipare, kiparin-moaga; (MORISYEN) gro pima; (NEPALI) jyanmara khursai; (NIUE) polo miti, polo magiho; (NORWEGIAN) tabascopaprika; (NYANJA) tsobola, mpiripiri; (OGBIA) ẹ̀gàrà; (OGBIA-AGHOLO) ìdũ; (ORIYA) lanka; (PAMPANGAN) lara, larang-inanis, siling-labuyo; (PANGASINAN) silingbalang; (PANJABI, EASTERN) lal mirch, mari mirch; (POLISH) papryka wieloletnia; (PORTUGUESE) tabasco, pimenteira, pimentinha; (POTAWATOMI) wasgagIn; (PULAAR) guilè, nhamaco; (RAROTONGAN) oporo; (RONGA) piripiri, socoti; (RUSSIAN) perec mnogoletniy; (SAMBAL, BOTOLAN) lada-balang, tangal; (SAMBAL, TINÀ) lara tangal, malisang balang; (SAMOAN) polo; (SANGA) iṣata; (SANSKRIT) ajadakumaricha, brahumaricha, katuveera, marichiphala, rakthamaricha, ujivala, theekshna, theevrashakti; (SENOUFO, CEBAARA) sikalignirini, sikaliwalih; (SERER-SINE) malgaté; (SINHALA) kochchi; (SPANISH) tabasco, chile, guindilla; (SUBANON) katumbal; (SUMO-MAYANGNA) angmak, anmak; (SUNDA) cabe rawit, cabe cengek, cengek, ceplik; (SUSU) gbéngbé gonkori, *g*bɛŋgbɛ-kuri-kuri, *g*bɛŋgbɛ-xunxuri; (SWAHILI) pilipili kichaa, pilipili; (SWEDISH) chilli spanskpeppar; (TAGALOG) siling labuyo, lara, pasitis, sili-diyablo, siling-bundok, siling-diyablo, siling-maanghang, silingpalay, siling-pasiti; (TAMIL) milakai, milagaai, mullagai; (TAUSUG) siling-labuyo, lada, lara manok; (TELUGU) mirapakaya, mirapakaaya, aggi mirapa, golakanda miraph, konda mirapa, seemamirapakaaya, sudimirapakaaya; (TEM) tschangei; (THAI) phrik khee nuu; (THEMNE) *ta*-gbɛŋgbɛ-*ta*-lɔl-*ta*-lɔl, *ta*-gbɛŋgbɛ-*ta*-temne; (TONGAN) polo, polo fifisi; (TSONGA) sowori; (TURKISH) süs biberi, vkeçi biberi, kuş biberi; (TWI) makɔ nsatea, makɔ ohenba nsatea, faa po moko; (URDU) mirich; (VAI) kefe-pɔ-mɛsɛ; (VIETNAMESE) ot bi; (VLAAMS) cayenne peper; (WAAMA) yɛɛku; (WAMEY) *u*-wong; (WÁRAY-WÁRAY) haráng; (WOLOF) kani buseu; (YALUNKA) *g*bɛŋgbeya-na; (YAO) sabola; (YOM) nyɔkɥn; (YORUBA) atawewe, sombo, ata ẹyẹ, ata gbasẹjọ, ata korukọ, ata nlá, ata pupa, ata sanbaye, ata sísèbè, ata wẹ̀wẹ̀; (ZARMA) tonko.

62.6. *Capsicum pubescens* Ruiz & Pavon (EP) fruit; (P) C.

(DANISH) rocoto; (DUTCH) rocoto; (ENGLISH) rocoto, apple chilli, chilli manzana; (FRENCH) rocoto; (GERMAN) Rocoto, Caraibcher Pfeffer; (ITALIAN) rocoto; (JAVANESE) cabe gondol, cabe bendot, cabe Dieng; (POLISH) pieprz czerwony; (PORTUGUESE) pimento das Caraíbas, rocoto; (QUECHUA, CUSCO) pucuna uchu; (SPANISH) rocoto, chile manzana, pimiento bravo, pimiento caribe, ají; (VLAAMS) rocoto.

62.7. *Cyphomandra betacea* (Cav.) Sendtner (EP) ripe fruit; (P) C.

(AFRIKAANS) boomtamatie; (ASSAMESE) goch bilahi; (CHINESE, MANDARIN) shu fan qie; (CHINESE, WÚ) shu fan qie; (CHINESE, YUE) shu fan qie; (CZECH) rajčenka; (DANISH) trætomat; (DUTCH) boom tomatie, struiktomaat; (ENGLISH) tree tomato, tamarillo; (FINNISH) puutomaatti; (FRENCH) arbe à tomates, tomate arbustive, tomateiro en arbre, tomate en arbre, tamarillo; (GERMAN)

Tomatobaum, Baumtomate, Baumtomatenstrauch; (GREEK) tomata dendrodis; (HUNGARIAN) paradicsomfa; (INDONESIAN) terong belanda, terong menen, terong betanda, terong perus; (ITALIAN) pomodoro arboreo, albero del pomodoro, pionandra, solano bietolone; (JAPANESE) shihomandora; (MALAY) pokok tomato; (MALTESE) sigra tat-tadam; (NEPALI) tyamatar, rukh tamater; (NORWEGIAN) tamarillo; (POLISH) pomidor drzewiasty; (PORTUGUESE) tomate de árvore, tomate chimango, tomateiro arbóreo, tomateiro arbisto; (SPANISH) árbol tomate, tomate arbóreo, palo de tomate, tomate arbol; (SUNDA) terong menen; (SWEDISH) trädtomat; (TAMIL) marathhakkali; (THAI) makhua-thetton; (TURKISH) ağaç domatesi; (VLAAMS) boomtomaat.

62.8. *Lycium chinense* Miller (EP) leaves; (P) C.

(CHINESE, MANDARIN) gou qi; (CHINESE, WÚ) di qu pi, gou qi tou, gou qie; (CHINESE, YUE) gou qi; (DANISH) bredbladet bukketorn; (DUTCH) Chinese boksdoorn; (ENGLISH) Chinese wolfberry, Chinese matrimony vine, Chinese boxthorn, box thorn; (ESTONIAN) taralông; (FRENCH) lyciet de la Chine, lyciet, kaoki; (GERMAN) Chinesischer Bocksdorn; (HEBREW) atad sini, atad seenee; (HUNGARIAN) kínai ördögcérna; (INDONESIAN) daun koki; (ITALIAN) spina santa cinese; (JAPANESE) kuko; (LITHUANIAN) kininis ožerškis; (MACEDONIAN) kineski licium; (MALAY) kau kei, kaukichai, kaukichoy, kei-chi; (MONGOLIAN) makhirs, chonon kharmag; (NORWEGIAN) bukketorn; (POLISH) kolcowój chiński; (PORTUGUESE) cambroeira da China; (SPANISH) cambronera de la China; (SWEDISH) kinesiskt bocktörne; (TAJIKI) chingil; (THAI) kaokichai, kaochichai; (TURKISH) çin şeytan ipliği, çin teneke dikeni, çin tenekedikeni; (URDU) pagandi; (VIETNAMESE) c[aa]u kh[owr]i, kh[owr]i t[uwr], c[aa]u k[yr] t[uwr]; (VLAAMS) Chinese wolfbes.

62.9. *Lycopersicon esculentum* var. *esculentum* P. Miller (EP) ripe fruit; (P) R,C.

[1]Bukusu dialect of Luyia.

(AFAR) kamadooro, tamaàtim; (AFRIKAANS) tamatie; (AKAN-ASANTE) tomatese; (AKAN-FANTE) amɛ osɛbɛ, nsusowa, nturuba, tomates; (AKPA) tòmátò; (ALBANIAN, TOSK) domate; (AMHARIC) tomatim; timatim, təmatəm; (ANYIN) ndrowa-kokre; (ARABIC) tamatum; (ARMENIAN) tomat; (ASSAMESE) bilahi; (AZERBIAJANI) pomidor; (BADAGA) domba; (BAFUT) tòmátò; (BAOULÉ) tomati, tɔmati, gyasala-ndrowa; (BASQUE) tomate; (BASSARI) ɛ-páɗy ɛnd *a*-ɓịmɛn, dikambil, dikamire, kantoge; (BELARUSAN) pamidory; (BEMBA) matimati, tomatoshi; (BEMBÉ) matoumate; (BENGALI) bilati begun, tamatar, vilayithi, baingan; (BEROM) tùmatùr; (BÉTÉ) tomaté; (BETE-BENDI) *ì*-tìtìn ì-bwá; (BICOLANO, CENTRAL) kamatis; (BINUKID) kemantis; (BLACKFOOT) kiníí; (BODO) fantao gøkhøi, lambehra; (BONTOC CENTRAL) ang-angseg; (BOSNIAN) paradajz, rajčica; (BRETON) tomatez; (BUDIK) *ga*-mɛntɛ́ŋ; (BULGARIAN) domat; (BULLOM SO) kipɔse, (BULU) ngôtô, ongoto; (BURMESE) kha-yan-gyin; (CAJUN FRENCH) tomate; (CAROLINIAN) tumóótis, tomwatis; (CATALAN-VALENCIAN-BALEAR) tomatera, tomàquet, tomate; (CEBUANO) kamatis, tamatis; (CHECHEN) pamdor, pomidor; (CHEROKEE) tamahli; (CHICKASAW) akanka' inchota̲'to', akankimimpa', akankintako̲loshi', akankishniha', chaka̲'to'; (CHINESE, GÀN) xi hong shi; (CHINESE, HAKKA) tomáto, fān-k'iô; (CHINESE, MANDARIN) fan qie, fan ch'ieh; (CHINESE, MIN BEI) xi hong shi, fan shi; (CHINESE, MIN NAN) thô-má-toh, kam-á-bit, chhàu-khī-á; (CHINESE, WÚ) fan qie, xi hong shi, yang shi zi; (CHINESE, XIANG) xi hong shi, yang shi zi, xi; (CHINESE, YUE) fan qie, xi hong shi, ju zai; (COMANCHE) kwasinabooʔa tuahkapu; (CREE, PLAINS) okiniy; (CREOLE, HAITIAN FRENCH) tomat; (CRIOULO, CAPE VERDE) tomati, tomateiro, tomatinho; (CRIOULO, UPPER GUINEA) camate; (CROATIAN) rajcica; (CZECH) rajče, rajče jedlé; (DAGAARE, SOUTHERN) tomati; (DAGBANI) kanto; (DAKOTA) waskuyecaśaśa; (DANGME-KROBO) dsukuwiafa, tomatə; (DANISH) tomat; (DENDI) tìmátì, tìmátãì, tìmátò; (DIGO) mtamata; (DUTCH) tomaat; (EDO) ekhu, ẹkhuẹ; (ENGLISH) tomato, love apple; (ESTONIAN) tomat; (ÉWÉ) tomatos; (EWONDO) ngoro; (FARSI, EASTERN) bâdenjân-e rumi; (FARSI, WESTERN) gôje farangi, gowje farangee; (FIJIAN) tamatar; (FILIPINO) kamatis; (FINNISH) tomaatit; (FON) timati; (FRENCH) tomate, pomme d'amour; (FULFULDE, ADAMAWA) hwitare, kwikedje, kwitaje boɗeyi, kuitāje boḍēje; (GA) ameo, amɛ osɛbɛ; (GADDANG) kamasit; (GAELIC, IRISH) tráta; (GAELIC, MANX) tomato; (GAELIC, SCOTTISH) tomàto; (GALICIAN) tomateiro, tomate; (GANDA) nyanya; (GBE-VHE) yevu-te, agbitsa, tomato; (GEORGIAN) pamidóri; (GERMAN) Tomate; (GIANGAN) kamatis; (GIKUYU) nyanya; (GOLA) bɔŋgɛ, tambatas; (GONJA) kamantosi, ikpĩa, kàmántosi, kàtèré; (GOURMANCHÉMA) nassar timon; (GOURMANTCHÉ) tomari; (GREEK) tomate, lykopersikon; (GUJARATI) vilaiti wagan, vilayithi vengan, tameta, vilayati wagan; (HARARI) timātim; (HAUSA) gauta, tumātir, yālo, átùmáátùr, gaútán tùùrààwáá, tumáátur, gaútáá, gaútán ɓéllò; (HEBREW) agvaniya ha-ginna, agvaniyat hagime, agvaniya, agvania; (HERERO) tamatis, otamatina; (HILIGAYNON) kamátis; (HINDI) tamater, tamatar, tamaatar, vilayithi baingan, vilaythibaigun, baingan, vilaaythi bengar; (HINDUSTANI) tomati; (HOPI) tomaati; (HUNGARIAN) paradiscom; (IBANAG) kamatis, kamati; (ICELANDIC) tomatur; (ICEVE-MACI) *à*-nyèkendé ú-genyi; (IFUGAO) umli, kamati; (IGBO) àfụfã, afufa-bekee, tòmatò, tumati; (ILOCANO) butinggan; (INDONESIAN) tomat, reteng, rangam; (ITALIAN) pomodoro, pomidori; (ITAWIT) kamasit; (IVATAN) kamalo-tajit kamatis; (JAPANESE) tomato; (JAVANESE) tomat, blandotan, kermate, ranti bali, ranti gendel, ranti kenong, terong sabrang; (JOLA-FONYI) khoutoumate, bu fimban, ku finbarako, fu fimban, komètan; (JULA) tomati; (KANKANAEY) gubgubbaw; (KANNADA) chapperbhende, chapperbadnekai, tomato, tomato hannu, goode hannu, goode kattu, gonda hannu, kadagolu, kempu chappara badane, thakkaali; (KANURI) tomátum, kə̀màttə̀n; (KASHMIRI) bŏṭa; (KAZAKH) pomidor; (KHMER, CENTRAL) peeng pâh; (KIOWA) k̂α'; (KIRGHIZ) pomidor, tomat; (KIRIBATI) te tomato; (KISSI, NORTHERN) tamasisio; (KONKANI) tambuta; (KONKOMBA) dikambil, dikamire, kantoge, katetu, dikàbil, dikambil, kartung, katetu; (KONO) kibɔnk, mantamba, tamantasi; (KOREAN) tomato; (KRIO) tamatis; (KURDISH, NORTHERN) temate, tēmātah; (KWANGALI) otjiamate; (KWÉYÒL) tanmadòz, ponmd'amov, tamados; (LADINO) tomát; (LAHU) paw le$_{\vee}$ shi$_{\text{-}}$, hkui$^{\vee}$ shi$_{\text{-}}$ nyi$^{\text{-}}$, ma$^{\text{-}}$ hkui$^{\text{-}}$ shon$_{\text{-}}$; (LAKOTA) uŋjiŋjiŋtka; (LAO) khüa sômz; (LATVIAN) tomāti, tomāts; (LIMBA) hubendin, *hu*tamba, tamasis; (LITHUANIAN) pomidoras, pomidorai; (LOKO) tambasisi; (LOMWE, MALAWI) phwetekele; (LUO) nyanya; (LUXENBOURGEOIS) Tomat; (LUYIA) chinyanya[1]; (MACEDONIAN) domat, običen domat; (MAITHILI) ṭamāṭar; (MALAY) tomato; (MALAYALAM) thakkali, paethakkaali; (MALTESE) tadam; (MANDING-BAMBARA) *n*goyo, chéraké n'koyo, dani, tamati, tomati; (MANDING-MANDINKA) mentengo, mantêm-ô; (MANDINKA) métégo; (MANDJAK) cálamate; (MANINKAKAN, WESTERN) tamaté; (MANKANYA) encámato; (MARANAO) kamantis; (MARATHI) welwangi, tambati, vel vangi, bailwangi, tambuta, goot baigun, belavaangi, bhedra, kharooja vaange, thaambeta; (MENDE) kibɔŋgi, namumbɔ; (MIZO) tomato; (MOHAWK) wahyakhahon; (MÒORÉ) tomato, kumba, kumbre; (MORISYEN) pom damour, tomat; (MUNDARI) jalbhātā, bilaiti; (MUSKOGEE) tomatv; (NAMA-DAMARA) edemate; (NAVAJO) ch'il łichí'í; (NDONGA) sidamate; (NEPALI) golbhenda; (NGUNDI) tumatu; (NIUE) tomato; (NORWEGIAN) tomat, busktomat, bifftomat; (NUPE-NUPE-TAKO) yengi; (NYAKYUSA-NGONDE) mapuno; (NYANJA) mapwetekere, matimati; (ORIYA) bilati baign, bapatala ghant; (PAMPANGAN) kamatis; (PANGASINAN) kamátis; (PANJABI, EASTERN) tamatar, tamattar; (PASHTO, NORTHERN) srë bânjan, rumi bânjân; (POLISH) pomidor; (PORTUGUESE) tomate, tomateiro; (PULAAR) mandae, mentou, méntuŋeere, tamatere, mantêm, matamba; (PULAR) tomaté, murtou; (ROMANIAN) pătlăgea rosie, pătlăgele roşii, tomata rosie, tomatele; (RUSSIAN) pomidor, pomidory, tomat, tomaty; (SAMBAL, BOTOLAN) kamati; (SANSKRIT) raktamaci, raktavardhak, ramphala, shak-stmeshdha, rakthavrunthaak, videshivrunthaak; (SENOUFO, CEBAARA) tomtihou; (SERBO-CROATIAN) patlidžan, paradajz, rajčica; (SERER-SINE) meté; (SHONA) domasi, buno, domase, guzungu; (SIDAMO) tumâtume; (SINHALA) thakkali; (SLOVAK) rajčiak; (SLOVENIAN) paradižnik; (SONGHAY) tòmáatì; (SORBIAN, LOWER) tomata; (SORBIAN, UPPER) tomata; (SOTHO, NORTHERN) tamati; (SOTHO, SOUTHERN) tamati; (SPANISH) tomate; (STELLINGWERFS) tomate, temate, tomaat; (SUMO-MAYANGNA) paumak; (SUNDA) tomat, leunca komir, kemir; (SUSU) tamasis; (SWAHILI) nyanya, mnyanya, mtunguja, nyanya tunguja; (SWEDISH) tomat; (TAGALOG) kamatis, kamatis-bundok; (TAGBANWA) kamatis; (TAJIKI) phatilajon; (TAMIL) takkali, seemay tekkali, thakkaali, aerumai thakkaali, sheemaimanathakkaali; (TATAR) pomidor; (TAUSUG) kamatis; (TELUGU) tomato, ramamulaka, seema vankaya, thakkaali pandu, karpoora vankaayi; (TEM) kaela, kɛla; (THAI) makhua-thet, makhua-som; (THEMNE) *a*-tamba, *a*-thamba, ma-thamba; (TIBETAN) ţūmaaţo; (TIV) ingishing, mgisim, mugishim ma batur, ingishigh; (TOK PISIN) tamato; (TSWANA) tamati; (TUMBUKA) mapuno; (TURKISH) domates, firek; (TURKMEN) adaty pomidor, pomidor; (TWI) tomato, nsusuwa, tomatese; (UKRAINIAN) pomídor; (URDU) tamatar; (UYGHUR) pämidur; (UZBEK) pamildory, pamidori, pomidor; (VAI) tambatasi; (VENDA) tamatìsi; (VIETNAMESE) cà-tô-mách, c[af] chua; (VLAAMS) tomaat; (WÁRAY-WÁRAY) kamatis; (WELSH) tomato; (WOLOF) tamaté, kamaté, menté-menté, menté, tamaate, tomaat, tomaté, tomate; (XHOSA) itumata, itumato; (YALA) ìtòmátò; (YIDDISH) pahmidor; (YORUBA) tomati; (ZULU) utamatisi, utumatisi(la); (ZUNI) k/e:zitokka, tome:ta.

62.10. *Lycopersicon pimpinellifolium* (L.) P. Miller (EP) ripe fruit; (P) R,C.

(ASSAMESE) kon bilahi; (BULU) ngôtô; (CHINESE, MANDARIN) cu li fan qie; (CHINESE, WÚ) cu li fan qie; (CHINESE, YUE) cu li fan qie; (DAGAARE, SOUTHERN) tomati; (DANISH) kirsebærtomat; (DUTCH) cherry tomaat; (ENGLISH) currant tomato, cherry tomato; (FINNISH) marjatomaatti; (FRENCH) tomate cerise, tomate cherry, tomate groseille; (GERMAN) Johannisbeertomate, Kirschtomate; (GREEK) tomata e pimpinelophyllos; (HEBREW) agvaniya kitnat peri, agvania; (INDONESIAN) ranti, rambai; (ITALIAN) pomodoro da serbo, pomodoro ciliegia, pomidori salvatico; (KANNADA) kaadu chappara badane, kaadu thamate gida, thippe tomato hannu; (LOMWE, MALAWI) phwetekele; (LUO) nyanya; (MACEDONIAN) div domat; (MÒORÉ) tomato; (NEPALI) golbhenda; (NORWEGIAN) ripstomat; (PANJABI, EASTERN) tamätarē; (POLISH) pomidor porzeczkowy; (PORTUGUESE) tomate cereja, tomate selvagem; (RUSSIAN) smorodinovidniy tomat; (SERER-SINE) mété; (SPANISH) tomate cereza, tomate cimarrón, tomate salvage, tomate cherry, tomatillo; (SWAHILI) tunguja; (SWEDISH) tomat; (VLAAMS) rode bestomaat.

62.11. *Physalis alkekengi* L. (EP) ripe fruit; (P) R,C.

(ARABIC) tamatum el-sholak; (BELARUSAN) fisalis zvychainy; (BRETON) louzaouenn-ar-maen, alkañjez; (CHINESE, MANDARIN) suan jiang; (CHINESE, WÚ) suan jiang; (CZECH) mochyně židorská třešeň; (DANISH) jødekirsebær; (DUTCH) lampionplant, jodenkers; (ENGLISH) strawberry tomato, Chinese lantern plant; (ESTONIAN) harilik füüsal; (FINNISH) lyhtykoiso, juutalaiskirsikka; (FRENCH) coqueret alkékenge, amour en cage; (GERMAN) gemeine Blasenkirsche, Judenkirsche, Lampionpflanze, Lanternenpflanze; (GREEK) fysalis e alkekenslos; (HUNGARIAN) zsidócseresznye; (ITALIAN) fisalis, acchechengio, alchechengi, coralli palloncini; (JAPANESE) shokuyou-houzuki; (KAZAKH) ogorodnaja vichnja; (LITHUANIAN) paprastoji dumlainė; (MACEDONIAN) obična meurka; (NORWEGIAN) japanlykt; (POLISH) miechunka rozdęta; (PORTUGUESE) fisalis, alquequenje, lanterna chinesa, lanterninha chinesa, cassussus, tomatinho de capuz, cerejas de judeu; (ROMANIAN) fizalis; (RUSSIAN) fizalis obyknovennyi, fyosja vishnja obyknovennaja; (SPANISH) tomate inglés, halicábalo, alquequenje, vejiga de perro; (TAJIKI) pardaarusak, padresak; (TURKISH) yahudi kirazi, fener çiçeği; (TURKMEN) adaty fizalis; (URDU) kaknag; (VLAAMS) lampionplant.

62.12. *Physalis peruviana* L. (EP) ripe fruit; (P) R,C.

(AFRIKAANS) appelliefie, kapseklapbes, makappelliefie, geelappelliefie, pampelmoertjie, kaapseleeubekkie, klapbessie, klapbes; (AMHARIC) awwət, awt; (ARABIC) enab el-deep; (ASSAMESE) kopal phuta, kopalphutiye; (BADAGA) poṭṭalu; (BENGALI) tepari, tipariya; (BONTOC, CENTRAL) bayas, kamatis, kamatis tuntog, pulet; (BULGARIAN) peruvianski fizalis; (CHINESE, MANDARIN) xiao guo suan jiang, pi lu suan chiang; (CHINESE, WÚ) xiao quo suan jiang; (CZECH) mochyně peruánská; (DANISH) ananaskirsebær; (DUTCH) Peruaanse lampionplant, kaapse kruisbes; (ENGLISH) cape gooseberry, Peruvian groundcherry, winter cherry; (ESTONIAN) peruu füüsal; (FINNISH) ananaskirsikka, kapinkarviainen; (FRENCH) coqueret du Pérou, alkékenge jaune; (FULFULDE, ADAMAWA) akubembem; (GERMAN) Peruanische Blasenkirsche, Kapstachelbeere, Kapsche Stachelbeere, Andenbeere, Erdkirsche, Physalie; (GREEK) fysalis e perouviani; (GUJARATI) moti popti; (HEBREW) bu'an neekhal; (HINDI) rasbhari, tipari, daeshi kaaknaja, macao, tepariya, thulaathipathi, timpari, tinkari; (HUNGARIAN) ananász cseresznye, ehetö zsidócseresznye; (ICELANDIC) blæjaber; (IFUGAO) kamatit dl hapon; (INDONESIAN) ceplukan, cecendet badak; (ITALIAN) alchechengi peruviano, alchechengi giallo, pomodoro fragola; (JAPANESE) shima houzuki, ke-houzuki; (KANNADA) dodda bedde gida, bndoola, budde hannu, dodda bondoola gida, dodda budde gida, dodda guppatte gida, gudde hannu; (KONKANI) chirput, chiriputla; (LATVIAN) arabu fizale; (MALAYALAM) mottaam puli, nittaayinduna; (MALTESE) tuffieh tal-fanali; (MARATHI) phopti, chirbot, thanmori; (NEPALI) jangali mewa; (NIUE) manini, manini fua lalahi, manini palagi; (NORWEGIAN) barbadoslykt; (NYANJA) jam, jamu; (ORIYA) tipari; (OROMO, WEST CENTRAL) renchi; (PANJABI, EASTERN) rasbhary; (POLISH) miechunka peruwiańska; (PORTUGUESE) alquequenge do Perú, camapú bexiga, alkekengi; (QUECHUA, CUSCO) ahuaimanto; (ROMANIAN) păpălăul; (RUSSIAN) peryanskaja vishnja, peruanskij fizalis; (SANSKRIT) avagutha, kunthali, parpotika, tankari, tankasi; (SLOVAK) machovka peruánská; (SPANISH) capulí, alquequenge del Perú; (SWEDISH) gul judekörs; (TAGALOG) lobo-lobohan; (TAMIL) tholthakkali, pottipallam, perungunni; (TELUGU) nelausiri, budama, buddabusara, busarakaaya, paambudda; (THAI) gusboeri; (TONGAN) ku'usi; (TULU) bondoli; (TURKISH) kap bektaşi üzümü, yenen yahudi kirazi, sarhoş otu; (UZBEK) fizalis; (VLAAMS) jodenkers, golden berry.

62.13. *Physalis philadelphica* Lam. (EP) unripe fruit; (P) R,C.

(AFRIKAANS) doptamatie; (ARABIC) tomatin, bandora; (CHINESE, MANDARIN) da suan jiang; (CHINESE, WÚ) da suan jiang; (CZECH) mochyně; (DANISH) tomatillo; (ENGLISH) tomatillo, husk tomato; (FINNISH) tomatillo; (FRENCH) tomatillo; (GERMAN) Blasenkirsche; (GREEK) fysalis e ixokarpos; (HOPI) kyeeve; (HUNGARIAN) tomatilló; (INDONESIAN) ceplukan; (JAPANESE) ohbudou houzuki; (NORWEGIAN) meksikolykt; (POLISH) pomidor skórzasty; (PORTUGUESE) camapú; (RUSSIAN) fisalis kleikoplodniy; (SPANISH) tomate de cáscara, tomatillo; (TURKISH) çilek domatesi; (VLAAMS) tomatillo.

62.14. *Solanum aethiopicum* L.

a. fruit and leaves (EP) fruit and leaves; (P) C.

(AJA) yɛhwɛshi; (ARABIC) el-tofah el-dahabi; (BAMANANKAN) goyo; (BEMBÉ) mantsa; (BULU) zom, zom amère, zom passu; (CHINESE, YUE) ai sai e bi ya qie; (DANISH) osun; (DENDI) feibii; (DUTCH) peper van Guinée; (EDE NAGO) odu; (ENGLISH) golden apple, mock tomato; (EWONDO) zom, zom amère, zom passu; (FINNISH) otaheiteomena; (FON) gboma; (FRENCH) tomate amère, pomme d'amour; (GBAYA) zom, zom amère, zom passu; (GREEK) gkolden (kitrina) mila; (JOLA-FONYI) khoudia guita; (MANINKAKAN, WESTERN) diato; (NGUNDI) musoso; (POLISH) psianka etiopska; (PORTUGUESE) tunga; (PULAR) diakhatou, khoulougné; (SERER-SINE) diakhatu; (SWAHILI) ngogwe; (TAJIKI) boimjon; (TURKISH) golden elma; (VLAAMS) gouden appel; (WAAMA) kotoroku; (WOLOF) jaxatu, diakhatou; (YOM) karbδημτμ; (YORUBA) odu, osum.

b. young shoots (EP) young shoots; (P) C.

(Formerly *Solanum gilo* Raddi but now considered a member of *S. aethiopicum*)

(BAFIA) susuo; (BAMILEKE) gedjo, nguejo, nsube, sebo; (BEMBÉ) bintsoukouri; (BULU) zom, zong; (ENGLISH) gilo; (EWONDO) zom, zong; (FRENCH) grande morelle, gilo, aubergine africaine; (FULFULDE) gouita, quitadji; (GBAYA) ntono; (KOONGO) bintsoukoulou; (MONGOLIAN) chesentser; (POLISH) psianka gilo; (PORTUGUESE) giló, jiló, jiloeiro; (SPANISH) gilo; (SWAHILI) nyanya chungu, ngogwe; (VLAAMS) gilo, jilo; (YORUBA) igbagba.

c. immature fruit (EP) immature fruit; (P) C.

(Formerly *Solanum integrifolium* Poiret but now considered a member of *S. aethiopicum*)

(ASSAMESE) bhekuri; (BULU) zong, zoñ; (CHINESE, MANDARIN) hong qie; (CHINESE, WÚ) hong qie; (ENGLISH) scarlet eggplant, tomato eggplant; (FRENCH) aubergine moucheté, aubergine tomate; (ITALIAN) melanzana scarlata; (JAPANESE) hira nasu; (JOLA-FONYI) koubatagcé; (LUO) bili nyanya; (MONGOLIAN) eldev navchit chesentser; (MÒORÉ) kumba; (POLISH) oberżyna szkarłatna; (PORTUGUESE) beringela escarlate; (RUSSIAN) baklazan kitaiskiy dekorativniy; (SERER-SINE) batansé; (SPANISH) berenjena escarlata; (VLAAMS) scharlaken aubergine; (WOLOF) batansé.

62.15. *Solanum americanum* Miller (EP) tender shoots, leaves, unripe fruit; (P) R,C.

(ANDO) fô; (ARABIC) timitach eddib; (BAOULÉ) fue; (BULLOM SO) kemba-lɛ; (CEBUANO) bolagtab; (ENGLISH) American black nightshade, caribnettle, glossy nightshade, black nightshade, common nightshade, garden nightshade, hound's berry, morelle, petty morel; (EWONDO) zom douce, ossan; (FRENCH) morelle, herbe à calalou; (GOLA) kimba; (GREEK) solanon to melan; (HAUSA) gaútán kádii; (HEBREW) solanum amerikani; (HINDI) gurkamai, makoya, kabaiya, makoi; (HMONG DAW) zaub iab; (IGBO) añàrà, awa, ipere, ugunakpe, fide; (INDONESIAN) ranti, rampai, loncak, terong hitam, teroung hitam, kampai; (IZON) ịkpịkpị́, dụ́wẹ́ị́ ịkpịkpị́; (JAVANESE) ranti, dara gowok; (KANNADA) kari kaachi gida, kempu kaachi gida, kaage hannina gida, kaaka maachae, kaaka munchee, kaake hannu, kaaki gida, kaaki hannu, kanka pitle, kari kaachi; (KANURI) àljáng; (KISSI, NORTHERN) sahéyo, chayo; (KONO) bundu kemba, kemba; (KRIO) ɛfɔ-odu; (KURANKO) gbirəkunɛ; (LAO) kh'èèngz namz; (LOKO) kemba, kimba; (MALAY) ranti, terong meranti, terong perat; (MANDING-BAMBARA) bogolo, fassa kuna; (MANDING-MANDINKA) sibito jambo, suludjatô, bassa béné, bassia béné, fassa; (MAORI) raupeti, remuroa, pooporo, poroporo, poroporo raupeti; (MARATHI) lalgooka batooli, kaamoni; (MENDE) kemba, kimba; (POLISH) psianka amerykańska; (PORTUGUESE) maria pretinha, erva moura; (PULAAR) djagato-fórô; (QUECHUA, CUSCO) pilliyuyu; (SANSKRIT) bahuphala, bahuthiktha, dhvankshamaachi, ghanahana, guchhaphala, jaghenephala, kaakaahva, kaakamaachi, kaakamaatha, kaakini, katuphala, kushthaaghni, rasaayani, sarvathiktha, sundari, svadupaka, thikthika, vaayasi, vayasavha; (SPANISH) hierba mora, yerba mora; (SUNDA) leunca, leunca badak, leunca manuk, lenuca phit; (SUSU) sɔkɛ; (SWAHILI) mnavu; (TAGALOG) anti; (TAMIL) manathakkaalli, kaasithazhai, manithakkaali, milaguthakkaalii, sirungunni; (TELUGU) gajuchettu, kaachi, kaakamaachi, kaamanchi, kanchipoondu, kasaka; (THAI) mawaeng-nok, ya-tomtok; (TIV) humbughl; (TOK PISIN) karakap; (TULU) kaamanchi, kaachiponda, kakai thappu, kakke thappu; (VAI) kemba; (VIETNAMESE) lu lu d[uwj]c, th[uf] l[uf] d[uwj]c; (YALUNKA) khɔlɛkɔlɛ̃dɛn-na; (YORUBA) ègùnmọ̀, òdú, osun.

62.16. *Solanum ferox* L. (EP) mature fruit; (P) R,C.

(AKLANON) tagatum; (ASSAMESE) bhekuri, kotahi bengena; (BENGALI) ram begun; (BICOLANO, CENTRAL) talong-ayam; (BUHID) tagutung; (BURMESE) sinkade, tarabi; (CEBUANO) tagatum, tagutong, tugutong; (CHINESE, MANDARIN) mao qie; (CHINESE, WÚ) mao qie; (ENGLISH) hairy-fruited eggplant, wild eggplant, wild pea eggplant; (FRENCH) aubergine de Siam à fruit hirsute; (HANUNOO) bayun tinaw mabiru; (IBANAG) basula; (ILOCANO) balbalusangi; (INDONESIAN) terong perat, terong asam; (IVATAN) kamadaka; (KANNADA) haladigulla anesundegida; (LAO) khüa khôn, khüa puux; (MALAY) terong iban, terong asam, terong dayak; (MARANAO) kogong; (MONGOLIAN) chesentser; (PORTUGUESE) beringeleira do mato; (SPANISH) berenjena espinuda, berenjena silvestre; (SUBANON) kabul; (TAGALOG) tabanburo, talong-buru, talong-gubat, talong-talong, tarambulo; (TAGBANWA) tuggong; (TAMIL) molakkai, anachundai; (TAUSUG) dagutung; (TELUGU) mulakkayi; (THAI) ma-uk, mapu; (VIETNAMESE) c[af] bung, cà bung, cà du; (WÁRAY-WÁRAY) tarongtarong.

62.17. *Solanum incanum* L. (EP) unripe fruit; (P) R,C.

(AJA) agbo; (ASSAMESE) tita bhekuri; (BAATONUM) sambiru; (BULU) zong, zoñ; (CHINESE, YUE) dong hai hong; (DIGO) mtungudza; (EDE NAGO) igba aja; (ENGLISH) garden egg; (FON) agbonekwin; (GEN) egbo; (GIKUYU) mũtongu; (PANJABI, EASTERN) bɛ̃gan, batäɔ̃, watäɔ̃; (PORTUGUESE) tomateiro bravo; (SENA) nrolando, mandolando; (SHONA) ndunduludwa, mutenderugua, mtido; (SPANISH) berenjena silvestre; (SWAHILI) ndulele, tura; (TSONGA) tumane, inthuma, ntuma; (URDU) tabag; (WAAMA) saatɛɛka; (YORUBA) igba aja, ikan.

62.18. *Solanum macrocarpon* L. (EP) leaves and fruit; (P) C.

(AKAN-ASANTE) komprinya dua; (ARABIC) bazengan africy; (BEMBÉ) ngabi; (BULLOM SO) gbag; (BULU) olele; (CHINESE, MANDARIN) fei zhou qie; (CHINESE, WÚ) fei zhou qie; (DAGAARE, SOUTHERN) kumbié; (DAGBANI) ga hiɔ, kambayra; (DANGME-KROBO) ga hiɔ; (DANISH) afrikansk ægplante; (DIGO) mtungudza; (DUTCH) Afrikaanse aubergine, antroewa; (ENGLISH) African eggplant, native eggplant, gboma eggplant; (ÉWÉ) agbitsa, ete; (EWONDO) étoué zom; (FRENCH) aubergine indigéne, aubergine gboma, fausse tomate, grosse anghive, petite bringelle; (FULFULDE, ADAMAWA) hwitare bagarmi, garko, gaútán bacarmi, gwargo, jau dari; (GA) atoropo, atrokoba, kva-fyé, sɛbɛ baa; (GANDA) biringanya; (GBAYA) wa ngagu; (GERMAN) Afrikanische Eirpflanze; (GOLA) peili, peli; (GOURMANCHÉMA) lélé timion; (GOURMANTCHÉ) akana; (HAUSA) gorko; (HINDUSTANI) antruwa; (HMONG DAW) lws africa; (IGBO) anara, àfufà, ayàrà, ipere, ofe, ofe ofe, ụfụfà; (INDONESIAN) terong glatik, terong engkol, terong kelapa, terong cina; (ITALIAN) melanzana africana; (KANURI) gargo, gorgo, gorko; (KISI, SOUTHERN) peri-ouo; (KISSI, NORTHERN) peri-uo, pelo; (KONO) pii; (KURANKO) bussu; (LIMBA) gbete; (LOKO) peli; (LOMWE, MALAWI) mabilinganya; (LUO) bili nyanya; (MALAY) terong santan, terong rapoh; (MANDING-MANDINKA) busi, busu; (MANDING-MANINKA) gbussu; (MANINKAKAN, WESTERN) gboussou; (MENDE) gbɔlɔ, peila, peili, tindigbɔlɔ; (MONGOLIAN) tom jimst chesentser; (MÒORÉ) kũmba; (MORISYEN) angiv; (POLISH) oberżyna afrykańska; (PORTUGUESE) beringela africana; (PULAAR) jakato-brurɛ; (SPANISH) berenjena africana; (SUNDA) terong engkol, terong gayung, terong kalapa, terong kopek; (SUSU) bosui, busiburaxe;

(SWAHILI) ngogwe; (THEMNE) *am*-bethe; (TWI) aturopo, atoropo; (VLAAMS) Afrikaanse aubergine, antroewa; (YALUNKA) bosuna; (YORUBA) ẹfoọ̀sùn, ìgbágba, ìgbá-ọ̀sùn, ọ̀sùn.

62.19. *Solanum melongena* L. (EP) immature fruit; (P) C.

(AFRIKAANS) eiervrug, brinjal; (AGTA, CASIGURAN DUMAGAT) tahung, talon; (AKAN-ASANTE) atoropo, guan hwowa, nyadoa, wabem; (AKAN-FANTE) *m*bamba, *n*gyrehye, *n*suprudu, *n*susuwa, *n*trɔba, *n*tropo; (ALBANIAN, TOSK) patlixhan, melzhane; (AMHARIC) bazinğän, bädrăğan; (ANUAK) ōcɔ̀ɔ jùr, ōcóog júr; (ARABIC) badhingan; (ASSAMESE) bengena; jati bengani; (ATTA, PAMPLONA) tarung; (AZERBAIJANI, NORTHERN) badımcan; (BADAGA) badana kae; (BAFUT) àngə̀rɛ; (BAGA) *ta*-kobo; (BAOULÉ) níguema; (BELARUSAN) bakladzan; (BENGALI) begun, kuli begun, bartaku, mahoti hinpoli; (BEROM) pyàál; (BÉTÉ) fré; (BICOLANO, CENTRAL) talong; (BINUKID) tegudtung; (BLAAN, SARANGANI) sgatong, skatong; (BODO) fantao; (BOKYI) otcicen; (BONTOC CENTRAL) talong; (BUHID) bringhinas; (BULGARIAN) patladžan; (BULLOM SO) *g*bɔkhɔ, koso, kpɔlɔ, bak-lɛ, gbɔ-lɛ; (BURMESE) kha-yan-thi, kayan; (BUTUANON) bringhenas; (CAJUN FRENCH) brèam; (CATALAN-VALENCIAN-BALEAR) albergínia, alberginiera; (CEBUANO) bringhinas, aong; (CHINESE, GÀN) qie zi; (CHINESE, HAKKA) k'iô-tzè; (CHINESE, MANDARIN) qie zi, chieh tse, ch'ieh; (CHINESE, MIN BEI) zi cai, hong cai; (CHINESE, MIN NAN) âng-phôe-chhài; (CHINESE, WÚ) qie zi, luo su; (CHINESE, XIANG) qie zi, luo su; (CHINESE, YUE) ai gua, qie gua; (CREOLE, HAITIAN FRENCH) obejinn, obèjin, berejenn; (CRIOULO, CAPE VERDE) bringela; (CROATIAN) patlidzan; (CZECH) lilek vejcoplodý, baklažán; (DAGAARE, SOUTHERN) boolē; (DANGME) agbitʃa; (DANISH) ægplante, ægfrugt, aubergine; (DIGO) bungulia; (DUTCH) aubergine, eierplant; (ENGLISH) eggplant, aubergine, melongene, brinjal, Chinese brinjal, Jew's apple, purple brinjal; (ESTONIAN) baklažaan; (FARSI, EASTERN) bâdenjân syâh, banjân; (FARSI, WESTERN) badangan, bâdemjân; (FIJIAN) baigan; (FILIPINO) talong; (FINNISH) munakoiso; (FRENCH) aubergine, melongène, bringelle; (GA) sɛbɛ, atorókpò, ekpu; (GADDANG) balansina; (GALICIAN) berinxela; (GBE-VHE) agbitsa, gbo; (GEORGIAN) badrigani, badrijáni; (GERMAN) Eierfrucht, Aubergine; (GIKUYU) mbiringanya; (GOLA) pambo, *k*paŋbɔi-*g*begbe, *k*paŋbɔi-gɔni, *k*paŋgbo; (GONJA) ekpu, kàtàré, kàtèré; (GOURMANTCHÉ) boumpiéne-kana; (GREEK) melitzana; (GUJARATI) vengan, ringni, vengni, vantak; (HANUNOO) talun, talun balubun; (HAUSA) gáutáá, gáutan-kúúrúá, gauta; (HEBREW) ḥaẓil, chatzil, solanum he-ḥaẓil; (HILIGAYNON) talúng; (HINDI) brinjal, bangain, bhanta, baingan, bainyan, brinjal, badanjan, bhengana, baigun, bangan; (HINDUSTANI) baygan, baiṉgan, bháṉtá; (HMONG DAW) tsi lu, lws kub tshis; (HUNGARIAN) tojásgyümölcs, törökparadicsum, padlizsán, tojascsucsor; (IBANAG) birenghenas, brenghinas; (ICELANDIC) eggaldin; (IGBO) eggplant, àfụfà oyìbo; (ILOCANO) tarong; (INDONESIAN) terong, encung; (ITALIAN) melanzana, petonciano; (ITAWIT) barangkinit; (IVATAN) vahosa; (IZON) isụsụ́; (JAPANESE) nasu; (JAVANESE) terong, encung; (JULA) goyô; (JUNKUN TAKUM) fye; (KALAGAN) balanhinu, belantinu, tadung; (KANNADA) badane kayi, badane kaayi, dodda badane, badane, bhantaki, gulla, kaachi; (KANURI) gorwò, yáló; (KASHMIRI) buntakī, wãgun; (KAZAKH) baklazhan; (KHMER, CENTRAL) trâb vèèng, trâb put lonhoong; (KIRGHIZ) baklajan, baklažan; (KISSI, NORTHERN) *k*pɔnigojiyo; (KONKANI) vaayingana, vaingim; (KONO) *k*pɔni-yalo, *k*pɔni-yawo; (KOREAN) gasi; (KORO) l'l'; (KRIO) jubɔks, skwash, jakato, bita-bol, jublɔks, gadin ɛg, kɔbɔkɔbɔ; (KWÉYÒL) obédjin, bélanjén, bélanjé; (LADINO) berenjèna, merenjèna, merendjena, mirindjena; (LAHU) ma⁻ hkui⁻ shi ̠; (LAO) khüa poom, khüa hlèèz, khüa ham maaz; (LATVIAN) baklažāns; (LIMBA) kɔbɔkɔbɔ; (LITHUANIAN) baklažanai; (LOKO) wondikoho; (LOMWE, MALAWI) mabilinganya; (LUO) bili nyanya; (MACEDONIAN) patlidžan, običen crn patlidža; (MAGUINDANAO) talum; (MAITHILI) bhāṭā; (MALAY) terong; (MALAYALAM) vazhuthana, nazhuthana, hingula, katthiri, mulakutakkaali, nelavalutina, valutina; (MALTESE) bringiel; (MANINKAKAN, WESTERN) batansé; (MANOBO) bulilinas, sigutung, sihutung, talom, taung, tigudtung; (MARANAO) tagotong, tagotung; (MARATHI) yangi, vange, baange; (MENDE) kpɔni-kojo, bobo-kojo, bobo-gojo, gbɔlɔ, kojo, kɔbɔkɔbɔ; (MINANGKABAU) tiung; (MIZO) bawkbawn; (MÒORÉ) bõda; (MORISYEN) brinzel; (MUNDARI) bengaṛ; (NEPALI) bhanta; (NEWAR) bhantā; (NGUNDI) bitsounkoulou; (NIUE) loku moka fāpogi; (NORWEGIAN) eggsøtvier, auberginer, eggplante, eggfrukt; (ORIYA) baingun, baigan; (PALI) vātiṅgano, vāttākī; (PAMPANGAN) balasenas; (PANGASINAN) talón, talon; (PANJABI, EASTERN) bengan, bataun, bhaingon, baingan, batoon, vataun; (PASHTO, NORTHERN) tor bânjân; (POLISH) oberżyna, gruszka miłosna, bakłażan, bakłażany; (PORTUGUESE) beringela; (PULAAR) jakato; (PULAR) batansé; (ROMANI, BALKAN) patlazhùy; (ROMANIAN) pătlăgea vinătă, vinătă, pătlăgele vinete; (RUSSIAN) baklažhan, baklažhan obycnovennyi, demjanka; (SAMBAL, BOTOLAN) talon, talom; (SAMA, CENTRAL) talum; (SAMBAL, TINÀ) talom; (SANGA) ikwariya, iyalọ; (SANSKRIT) vartaku vatigama, vatigana, bhantaki, vartaku peetaphala, brihathi, vrinthaak, angana, bera, bhantaaki, chithraphala, deerghavarthaki, dushpra dharshini, hinguli, kantakini, kantalu, kantapathrika, kantavarthaki, katphala, kuli, mahaahrihati, mahathi, mahothika, mansalaphala, mishravarnaphala, neelakantaka, neelaphala, neelavriksha, nidralu, nripapriyaphala, peethaphala, raajakooshmaanda, rakthphala, shakabilva, shakabilvaka, shakashreshtha, sinhi, vaarthaaki, vangana, vartha, varthaka, vritthaphala; (SANTALI) beṅgaṛ, bḛndḛ beṅgaṛ; (SENOUFO, CEBAARA) gôyoho; (SERBO-CROATIAN) patlidžan; (SHONA) bwarupwa, mharupwa, mwarupwa, ndyarushwa; (SINHALA) wambotu; (SLOVAK) l'ulok vojcoplodný, l'ul'ok vajcoplodý; (SLOVENIAN) melancana; (SONGHAY) ǹtàwrá; (SPANISH) berenjena, berengena; (SUBANON) tilung; (SUNDA) terong, conkrom, encung; (SUSU) kobokobo, kɔbɔkɔbɔ; (SWAHILI) biringinya, bilinganya; (SWEDISH) aubergin, äggfrukt, äggplanta; (TAGALOG) talonga, talong, tarung; (TAGBANWA) tagogtog, talong, taguntung, tarung; (TAJIKI) boimjon, batirjon; (TAMIL) kathirikai, kathhiri, katharikai, vankaya, achatti, kaaliliyam, kaarkkatthari, kathiri, koyambattai, milaguttakkali, mullukkatthari, pandaagi, piragadi, valudalai, valudunai, vangam, vaartthaagam; (TATAR) baklazhan; (TAUSUG) tolung, talum; (TELUGU) chirivanga, chiruvanga, vangachettu, vanga chettu, vankaya, vankaaya, hingudi, eruvanga, mettavanga, muluka, nidupuvanga, nirettuvanga, niruvanga, vangamu, (THAI) makhua-yao, makhua-chan, makhua-khao; (THEMNE) *a*-kobokobo, *ma*- kɔbɔkɔbɔ, ma-bela; (TULU) badane; (TURKISH) patlican, baldircan; (TURKMEN) badamjan, badymjan; (TWI) ntɔrwa, ntɔrowa, aturopo, atoropo, nniadowa, nnua hwɛ, nnuadewa; (UKRAINIAN) baklažan; (URDU) bangan; (UYGHUR) chäyzä; (UZBEK) bucklujon, baklažon, baqlajon; (VAI) fɔkosie; (VIETNAMESE) ca, c[af] t[is]m, c[af] b[as]t; (VLAAMS) eiervrucht, aubergine; (WÁRAY-WÁRAY) taróng, taron; (WELSH) ŵy planhigyn; (WOLOF) batansé, diakhatu; (YAO) mpilingana; (YIDDISH) pahtledzáhn; (YORUBA) ìgbá-òyibo.

62.20. *Solanum muricatum* Aiton (EP) ripe fruit; (P) R.

(AFRIKAANS) pepino, spanspekboompie; (CHINESE, MANDARIN) xiang gua qie; (CHINESE, WÚ) xiang qua qie; (CZECH) pepino; (DANISH) pepino, melonpære; (DUTCH) pepino, appelmeloen, meloenpeer; (ENGLISH) pepino, melon pear, sweet pepino; (FINNISH) pepino; (FRENCH) melon poire, poire melon, pépino; (GERMAN) Pepino, Melonenbirne; (ITALIAN) pepino, pera melone; (JAPANESE) pepiino; (MALAY) melon pear; (NORWEGIAN) pepinosøtvier; (POLISH) pepino, gruszka melonowa; (PORTUGUESE) pepino doce, pera melão, tomateiro francês; (QUECHUA, CUSCO) kachan; (RUSSIAN) pepino; (SPANISH) pepino, pera mélon, pepino dulce, manguena; (TURKISH) pepino; (VLAAMS) pepino, meloenpeer.

62.21. *Solanum nigrum* L. (EP) mature fruit, leaves and tender shoots; (P) C.

(AFRIKAANS) nastergal, galbessie, nasgal, nasgalbossie, wildenastagal; (AJA) gbohi; (AMHARIC) awt; (ASSAMESE) pokmow, pichkati; (BADAGA) ga:ke; (BELARUSAN) paslen cherni; (BEMBÉ) moussosso; (BENGALI) gurkamai, kakmachi, tulidun; (BICOLANO, CENTRAL) lagkakum, lubi-

lubi; (BINUKID) ketuglew, muti; (BONTOC, CENTRAL) amti, anti; (BRETON) froud, sanab; (BUHID) samaray, sanaray; (CEBUANO) bolagtab, hulablub, kamatis-manok, kuru-kamatis, lubi-lubi, malasili; (CHINESE, MANDARIN) long kui; (CHINESE, WÚ) ma ti cai, ye hai jiao; (CHINESE, YUE) long kui, dong han cai; (CHUWABU) mai-mati-mati; (CRIOULO, CAPE VERDE) santa-maria, malagueta-de-galinha, uva-de-santa maria; (CZECH) lilek černý, lilek; (DANISH) sort natskygge; (DIGO) mnavu; (DUTCH) zwarte nachtschade, agoema; (ENGLISH) wonderberry, black nightshade, common nightshade, poisonberry; (ESTONIAN) must maavits; (FILIPINO) konti, black nightshade; (FINNISH) mustakoiso; (FON) gboma; (FRENCH) morelle noire, tue-chien; (GALICIAN) herba moura; (GBE, WACI) asukusɛ; (GEN) gbonyamɛ; (GERMAN) Nachtschatten, schwarze Nachtschatten, schwarze Teufelskirsche, falsche Blaubeere; (GIKUYU) managu; (GONJA) katare, katere; (GREEK) agriotomatia, styfnos; (GUJARATI) piludi; (HANUNOO) dutun ilamnu, nunti; (HEBREW) solanum shaḫor, ʿinvey shuʿal; (HINDI) makoi, gurkamai, kabaiya; (HINDUSTANI) makoiya; (HMONG-DAW) zaub iab; (HMONG-NJUA) zaub ab; (HUNGARIAN) fekete csúcsor; (IBALOI) natang-ni-aso, nateng; (IFUGAO) amti, amti-iitang; (INDONESIAN) leunca; (ISNAG) bakohan; (ITALIAN) solano nero, erba morella, morella minore, ballerina, pomidorella; (IVATAN) malanateng, nateng; (JAPANESE) inu-hôzuki; (KALAGAN) amti; (KANKANAEY) amti; (KANNADA) karikaachi gida; (KASHMIRI) makō; (KAZAKH) paslen chernyj; (KISI, SOUTHERN) sahéyo; (KWANYAMA) ndemukolenghula; (LATVIAN) melaā naktene; (LITHUANIAN) juodoji kiauliauogė; (MACEDONIAN) crn zrnec; (MALAY) terong telunjuk; (MALAYALAM) thulasi, manatakkali; (MALTESE) gheneb id-dib; (MANINKAKAN) fassa; (MANU'A SAMOAN) magalogalo, magalo, polo vao; (MAORI) raupeti, remuroa, pooporo, poroporo, poroporo raupeti; (MARANAO) moti; (MARATHI) ghati, kakmachi, kamoni, laghukavali, meko; (MONGOLIAN) khar chesentser; (MÒORÉ) ludo, nõraog-kubre; (MORISYEN) bred martin; (NDALI) inafu; (NEPALI) thulo bihin; (NGUNDI) mantsa; (NIUE) polo kai, polo fua; (NORWEGIAN) svartsøtvier; (NYANJA) mnadzi; (PANJABI, EASTERN) mako, kambei, kachmach, riaungi, mkōō; (POLISH) psianka czarna, psianka; (PORTUGUESE) erva moura, erva moira, maria pretinha; (POTAWATOMI) acib'nimic; (RAROTONGAN) poro; (ROMANIAN) solanum; (RONGA) mussowa, chiauenequene; (RUSSIAN) paslen cherniy, paslyon; (SAMBAL, BOTOLAN) unti; (SAMOAN) māgalo, polo vao; (SANSKRIT) kakamachi, kakamaci; (SENA) nadzi; (SERBO-CROATIAN) pomoćnica, crna; (SPANISH) hierba mora, solano negro, tomatitos, tomatillos del diablo; (STELLINGWERFS) hondebeien, nachtschade, nachtschaduw, nachtschaede; (SUBANON) sili-sili; (SWAHILI) mnavu; (SWEDISH) nattskatta; (TAGALOG) kanti, anti, kama-kamatisan, kamatis-kamatisan, konti, kunti, lubi-lubi, onti; (TAJIKI) anguri sagak; (TAMIL) munatakali, manathakkali, manattakkali; (TAUSUG) muti; (TELUGU) kachchipundu, kachi, kamanchi, kaman chichettu, kasaka; (TIRURAY) kawat; (TONGAN) polo kai; (TUMBUKA) msaka, musaka; (TURKISH) köpeküzümü, karayaban yasemini, köpek memesi; (TURKMEN) adaty itüzüm; (URDU) makoya; (VLAAMS) zwarte nachtschade; (WAAMA) kancenfa; (WÁRAY-WÁRAY) mala-sili; (YAO) mnesi; (YORUBA) odu, ogunmo.

62.22. *Solanum quitoense* Lam. (EP) ripe fruit; (P) R.

(CZECH) narančila; (DANISH) naranjilla; (ENGLISH) naranjillo, naranjilla, lulo; (FINNISH) naranjilla; (FRENCH) narangille, orange de Quito; (NORWEGIAN) lulosøtvier; (POLISH) psianka lulo; (VLAAMS) naranjilla, lulo.

62.23. *Solanum torvum* Swartz (EP) tender shoots and immature fruits; (P) C.

(AKAN-ASANTE) kwaonsusua; (AKLANON) taogotan, titina; (ASSAMESE) hati bhekuri, bhit-tita, hathibhekuri; (BAOULÉ) yaokan n'droa; (BASSARI) ɛ-pády̆, ɛ-níiyàn, ɛ-pády̆ ɛ-tyɛ́n; (BENGALI) gota begun, gothbegun, titbaigun; (BÉTÉ) gbi fré; (BINUKID) bokul; (BONTOC CENTRAL) balbalusa, batuang; (BUDIK) *ga*-pèdy̆, ɛ̀-pedy̆; (CEBUANO) taogotan; (CHINESE, MANDARIN) shui qie; (CHINESE, WÚ) shui qie; (DANISH) ærte-natskygge; (EDO) omgbabelara; (ENGLISH) pea eggplant, terongan berries, devil's fig, plate brush; (FRENCH) morelle diable, belangere batarde, z'amorétte, aubergine du diable; (GBE-VHE) susruba; (GREEK) moura; (HILIGAYNON) titina; (HMONG DAW) txiv pos; (IFUGAO) balbaluha; (INDONESIAN) takokak, pokak, terong susu, terong pipit, tekokak, rimang; (IZON) omgbómịnịbɔ́ pún kọrịghẹ́; (JAPANESE) suzume-nasubi; (JAVANESE) terong belut, poka, coko wana, terong pipit, cepoka, cong belut; (JULA) gacoun 'kouêssê; (KANNADA) kadusunde, kadusonde, sonde gida, sondaegida, sundakkai, sundakkayi, sonde, bovasonde, chunde, kaadu badane, kaadu kadane, kaadu kallate, kaadu kallunte, kaadu sonde, karashe; (KONO) kɔlau; (LAHU) ma¯ heh lo¯; (LAO) kh'èèngz faaz, kh'èèngz saph'au; (LOKO) banyenye; (LUO) bili nyanya; (MALAY) terong pipit, terong pipit putch, terong rembang; (MALAYALAM) chunta, kattuchunta, kaattuchunta, putthari chunta; (MANDING-MANDINKA) kɔle-yarɛ; (MARANAO) pariya laot; (MARATHI) doralee, dhoravaange; (MENDE) heva-gojo, hɛva-gɔjɔ, kɔli-gojo, kɔli-gɔjɔ; (MINANGKABAU) terong pipit; (MORISYEN) angiv maron; (POLISH) oberżyna grochowa; (PORTUGUESE) jurubera; (SANSKRIT) brhati, brihathi, gosthabartaku; (SPANISH) berengena cimarrona; (SUBANON) gambol, sebulnusa; (SUNDA) takokak; (TAGALOG) takapasin, talongon, talong-talongan, tandang-aso; (TAMIL) sudaikai, sundaikaai, sundai, malaichundai, kottukkatari, kotthu katthari; (TAUSUG) dagutung, talun-kiya; (TELUGU) chundaikai, kondavuste, kottuvastu, ustekaya, usti; (THAI) makhuea phuang, makhua-phuang, makhwaeng; (THEMNE) *ma*-gboli-*ma*-bana; (TULU) kaattukallante; (TURKISH) üzümsü bitkiler; (TWI) susuruba, samanntɔrewa, saman nsusuwa; (VIETNAMESE) c[af] d[aj]i hoa tr[aws]ng, c[af] hoang; (VLAAMS) erwtaubergine; (WAMEY) *u*-fady; (WÁRAY-WÁRAY) tarampay, tarong-tarong; (YORUBA) ekẹ-ko-sẹ, igbáyinrin ẹlẹgun, ijẹgunajẹ, ikanyanrin, kanyanrin, ṣokudalẹ.

62.24. *Solanum tuberosum* L. (and several related species) (EP) tuber; (P) C.

(AFAR) bataata, dinnit; (AFRIKAANS) aartappel, ertappel; (ALBANIAN, TOSK) patate; (ALEUT) kotopalak; (AMHARIC) dənnəčč, bätata; (ANUAK) ājwāallā, ājwáallɛ́; (APACHE, JICARILLA) nigokats'i; (ARABIC) batates; (ASSAMESE) alu, alooguti; (AZERBAIJANI, NORTHERN) kartof; (BADAGA) ga:su; (BALTI) alu; (BAMILEKE) njó' ndak; (BASQUE) patata; (BELARUSAN) bulba; (BEMBA) cilashi; (BENGALI) alu; (BEROM) dànkál làsàrá; (BICOLANO, CENTRAL) patatas; (BINUKID) petatas; (BLACKFOOT) maatáák, maatáákkistsi, paataak; (BODO) alu; (BONTOC CENTRAL) patatas, batatas; (BOSNIAN) krompir, krumpir; (BRETON) avaloû-douar; (BULGARIAN) kartof; (BULU) ndua ntangan; (BURMESE) ah-lu; (CAJUN FRENCH) pomme de terre; (CAROLINIAN) gamwuuti, bwotaatas; (CATALAN-VALENCIAN-BALEAR) patata; (CEBUANO) patatas; (CHECHEN) kartol, kartog; (CHEROKEE) nuna; (CHICKASAW) ahi', ahi' lhibowa'; (CHINESE, GÀN) yang yu; (CHINESE, HAKKA) choû (ŷt-t'iaô), mā-lâng-choû; (CHINESE, MANDARIN) ma ling shu; (CHINESE, MIN BEI) fan zai shu; (CHINESE, MIN NAN) chî, chû, má-lêng-chî; (CHINESE, WÚ) yan yu, di dan, tu dou, ma ling shu; (CHINESE, XIANG) yang yu zi, did an, tu dou; (CHINESE, YUE) tu dou, ma ling shu, shu zai, he lan shu, di gua; (COMANCHE) paapasi̱, suhu tsitsina?; (CREE, PLAINS) askipwâw; (CREOLE, HAITIAN FRENCH) ponmtè, ponmdetè, pomme dé tè; (CRIOULO, CAPE VERDE) batata inglesa; (CROATIAN) krumpir; (CZECH) brambor, obechý, lilek brambor; (DAKOTA) mdo, bdo, blo; (DANISH) kartoffel; (DIGO) viazi; (DUTCH) aardappel; (ENGLISH) potato; (ESTONIAN) kartul; (FARSI, EASTERN) kachâlu; (FARSI, WESTERN) sib zamini, sibé zamine; (FIJIAN) alu; (FILIPINO) patatas; (FINNISH) peruna; (FRENCH) pomme de terre, patata; (GAELIC, IRISH) práta, prátaí lutaha, prátaí céadfhómhair; (GAELIC, MANX) praase; (GAELIC, SCOTTISH) buntàta; (GANDA) lumonde; (GEORGIAN) cartophili, kartopfíli; (GERMAN) Kartoffel; (GIANGAN) patatas; (GIKUYU) waru; (GOURMANCHÉMA) nassar pièrain; (GOURMANTCHÉ) bompiéne-para; (GREEK) patata; (GUJARATI) batata, aloo, deengli, kando aloo; (HARARI) dinničča; dukuma, kūdaku, lāwur, dankalin Turawa; (HAUSA) dangkálin tuuraawaá; (HEBREW) tappuah adama, tappud, solanüm ha-peka'ot; (HERERO) laian, otjihakautu; (HINDI) alu, salooalu, aaloo; (HINDUSTANI) álú; (HMONG DAW) qos yaj ywm; (HOPI) tumna; (HUNGARIAN) burgonya, krumpli, pityóka; (IBANAG) patatas; (ICELANDIC) kartafla; (IFUGAO) patatas; (IGBO) potato; (ILOCANO) patatas; (INDONESIAN) kentang, gantang, ubi kentang, kembili welanda; (INUKTITUT, EASTERN CANADIAN) manniujaq, patiitasi; (INUKTITUT, WEST GREENLANDIC) igssûssq, nautsîaq; (ITALIAN) pomi di terra, patata; (JAPANESE) jagaimo; (JAVANESE) kentang, kentang walanda, uwi kentang; (JOLA-FONYI) epombitère; (KANNADA) alu gedde, batate, bataate, alu-gidde, alugedde, aaloogedde, ooralu gedde; (KANURI) dàngáli nàsàrá; (KASHMIRI) ālū, ôluh; (KAZAKH) kartofel, kartop; (KHMER, CENTRAL) dâmlô:ng barang; (KIOWA) 'HepiH-'ei-gα; (KIRGHIZ) kartöshkö; (KIRIBATI) te boteto, te poteto; (KONKANI) battato; (KRIO) airish-pɛtɛtɛ, Ayrish-pɛtɛtɛ; (KOREAN) gamsa; (KURDISH, NORTHERN) petate; (KWANGALI) otjihakautu; (KWÉYÒL) ponm, ponm-tè; (LADINO) patátas; (LAHU) a lu¯ shi˛, ya^ yi¯ shi˛; (LAKOTA) blo; (LAO) man fàlangx; (LATVIAN) kartupeļi; (LITHUANIAN) bulvė, ankstyvosios bulves; (LOMWE, MALAWI) mbatata, kholowa a cizungu; (LUO) icok munu; (LUXENBOURGEOIS) Gromper; (MACEDONIAN) kompir, običen kompir; (MAITHILI) ālū; (MALAGASY) ravin'ovy; (MALAY) ubi kentang; (MALAYALAM) urala kizangu, urulakizhangu; (MALTESE) patata; (MANINKAKAN, WESTERN) pombitère; (MAORI) riiwai, hiiwai, huiwaiwaka, kapana, kaapana, mahetau, parareka, parate, taewa, taewha, taiawa, taiwha, ropi, riiwai, popoia; (MARANAO) kapatas; (MARATHI) batata, bataata; (MEHRI) bəḍəḥáyn, bəṭāṭ; (MICMAC) tāpătăt'; (MIZO) alu; (MOHAWK) ohnennàta; (MONGOLIAN) bultsuut chesentser; (MÒORÉ) pom deteere; (MORISYEN) pom de ter; (MUNDARI) ālu; (MUSKOGEE) vhv; (NAMA-DAMARA) oihakautu; (NAVAJO) nímasii; (NDALI) ndofanya; (NDONGA) katafura; (NEPALI) aalu; (NGUNDI) bombreterre; (NIUE) pateta; (NORWEGIAN) potet; (NYAKYUSA-NGONDE) katofeniya; (NYANJA) mbatata wa cizungu, matetesi, kacewere; (ORIYA) bilati alu; (OROMO, WEST CENTRAL) dinicha; (PAIUTE, SOUTHERN) tuh´dus; (PAMPANGAN) papas, patatas; (PANGASINAN) patátas, batatas; (PANJABI, EASTERN) aalu, alu; (PASHTO, NORTHERN) âlu, ālū, kachâlu; (POLISH) ziemniak, ziemniak uprawny; (PORTUGUESE) batata, batateira; (POTAWATOMI) pin, pinyak; (PULAAR) pómpiteer; (PULAR) pompiterry; (QUECHUA, CUSCO) papa; (ROMANI, BALKAN) mùrgili, kòmpiri; (ROMANIAN) cartof; (RUSSIAN) kartofel; (SAMOAN) pateta; (SANSKRIT) golakandah, aalooka, pindaaloo; (SANTALI) ạlu; (SCOTS) tattie; (SENOUFO, CEBAARA) lorgyirike; (SERBO-CROATIAN) krompir (SERER-SINE) pombitère; (SHONA) mbatatisi, dapiri, dapura, gwiri; (SIDAMO) dinnìcca; (SINHALA) arthapal; (SLOVAK) zemiaky; (SLOVENIAN) krompir; (SOMALI) baɗaɗɗo; (SORBIAN, LOWER) kulka; (SORBIAN, UPPER) běrna, kulka, nepl; (SOTHO, NORTHERN) letapola; (SOTHO, SOUTHERN) tapole; (SPANISH) patata, papa; (STELLINGWERFS) eerappel, eerpel; (SUMO-MAYANGNA) pai; (SUNDA) huwi kumeli, huwi walanda, kentang, keumeli; (SWAHILI) kiazi, viazi muiringo, ulaya, kiazi cha kisungu; (SWEDISH) potatis; (TAGALOG) patatas, papas; (TAJIKI) kacholu; (TAMIL) uralakilangu, uruzhai kilangu, urulaikizhangu; (TATAR) bärängge; (TAUSUG) patatas; (TELUGU) aalu gaddalu, bangaladumpa, bengaldumpa, urlagadda, uralagadda; (THAI) man-farang, man-alu; (TIBETAN) shɔ̱ɔqɔ̀ɔ; (TOK PISIN) poteto; (TONGA) mbambaira, mciwile; (TSWANA) lekwele; (TULU) pataate; (TUMBUKA) mbambaira, mciwile; (TURKISH) patates, kümpür; (TURKMEN) ýeralma, kartoşka; (URDU) alo; (UYGHUR) yangyu; (UZBEK) kartoshka; (VENDA) ḓabùla; (VIETNAMESE) khoai tây, khoai t[aa]y; (VLAAMS) aardappel; (WÁRAY-WÁRAY) patatas; (WELSH) taten; (WOLOF) pombitère; (XHOSA) itapile; (YIDDISH) kahrtofl, bulbeh; (ZARMA) pompiter; (ZULU) izambane, i(li)zambane; (ZUNI) k/api-mo.

62.25. *Solanum violaceum* Ortega (EP) mature/immature fruit; (P) R,C.

(ENGLISH) Indian nightshade, Asian nightshade; (LAO) kh'èèngz khôm; (MALAY) terong pipit; (SANSKRIT) brhati, simhi; (SINHALA) tibbatu; (TAGALOG) talong-na-puti, talong-pipit, talong na puti; (THAI) mawaeng, mawaeng-ton, ma-khwaeng-dam; (VIETNAMESE) c[af] d[aj]i hoa t[is] m, c[af] hoang, cà dai hoa tim, cà hoang, cà ân.

63. Tiliaceae

63.1. *Corchorus olitorius* L. (EP) leaves and tender shoots; (P) C,R.

(AFRIKAANS) indiese jute; (ALBANIAN, TOSK) korkore; (AMHARIC) mulukəye; (ARABIC) bamia; (ASSAMESE) mitha mora; (BAMILEKE) nkelan; (BAOULÉ) klouala; (BASSARI) ɛ-syɔ́ɓò; (BELARUSAN) raguinikavaja; (BEMBÉ) dongo dongo dia makya, dong dong dia makaya; (BENGALI) banpat, bhungipat, phunjipat, koshta, lalitapat, nalitapat; (BICOLANO, CENTRAL) sakuyot; (BOBO MADARÉ, SOUTHERN) dongo(ne); (BODO) narjøi; (BUDIK) syándúŋ gokə̀r; (BULLOM SO) *n*geŋge, kriŋ krin, kriŋkrin-dɛ; (CEBUANO) tagabang, tugabang, tukog-banog; (CHINESE, MANDARIN) chang shuo huang ma; (CHINESE, WÚ) chang shuo huang ma; (CZECH) jutovník zelinný; (DAGAARE, SOUTHERN) bior; (DAGBANI) bundu; (DANISH) jute; (DIGO) phombo; (DUTCH) jute, crain-crain; (EFIK) ètínyūng; (ENGLISH) Jew's marrow, bush okra, nalta jute, red jute, tossa jute; (EWONDO) kelen kelen, tege; (FILIPINO) saluyot; (FINNISH) juutti; (FRENCH) corette potagère, coréte, feuilles lalo, jute à capsules cylindriques, méloukhia; (FULFULDE) lalo; (FULFULDE, ADAMAWA) fako, fakuho, laalo, laalomemel; (GA) otoro, oturo; (GBAYA) gbolo; (GBE-VHE) sigli, siŋgli, singli; (GERMAN) Jutepflanze, Toassa-Jute, Jute-Wolle, Nalta-Jute; (GOLA) mahei; (GOURMANCHÉMA) pepanan; (GOURMANTCHÉ) ibardji; (GREEK) mouchlia, korchoros o lachanodis; (GUJARATI) chhunchho; (HANUNOO) tagabang diririt marara, tagabang talun; (HAUSA) lallo, fullane, láálòò, malafia, tùngúrnùwáá, tùrgúnùùwáá, tururuwa; (HEBREW) malukhiya, malukhiya neekhelet; (HINDI) koshta, pata, badi joot, banapaata; (HUNGARIAN) juta; (IBANAG) salayo, saluyu; (IGBO) keren-kere, ahị harā, kẹlarikele, aḧị ḧaṛā, agheregha, ahị hịarā, ahị hịrā; (ILOCANO) sakuyot, saluyot nga bulang, saluyut; (INDONESIAN) jengotan; (ITALIAN) cibo degli ebrei, còrcoro siliquoso, còrcoro ortense, còrcora, jute malva di giudèi, spinacio africano; (JAPANESE) moroheyia, Taiwan-tsunaso; (JULA) kdlalââ; (KANNADA) kaadu senabu, dodda chunchala gida; (KANURI) darraba, gamiaino, gamzaino; (KISI, SOUTHERN) guio, dyo; (KISSI, NORTHERN) dyo, guio, yɔ, yɔ-*n*gengeyo; (KONO) kɔkɔɛ; (KOONGO) dongo dongo dia makya, dong dong dia makaya; (KRIO) kreŋkre, krinkrin, kreŋkreŋ; (KURANKO) séméné, sobon, kilinkilà, kirinkaraŋ; (LAHU) meh (ceh_V); (LIMBA) kereŋkere; (LITHUANIAN) ilgavaisis džiutas; (LOKO) *n*geŋge, ngohai; (LOMWE, MALAWI) phinyu; (LUO) otigo; (MAGUINDANAO) taka, taka-ma-gindanaw, yaka; (MALTESE) bamja; (MANDING-MANDINKA) alo, buror buror, kolo-oro, kereŋ-kereŋ, kirinkilà, salakunda, sɔbɔlɔŋ; (MANINKAKAN, WESTERN) kolo-oro; (MARATHI) chunch, motichunch, mote chunch; (MENDE) *n*geŋge, *n*geŋgle, *n*geŋgle-gbogboli, *n*geŋgle-gowuli, *n*gingere, gringri, yeiŋgei, yeŋgei; (MÒORÉ) bulvaka nafouti, bũlvãka, berga; (NEPALI) sanpat; (NORWEGIAN) jute; (NYANJA) cilenzi, denje; (ORIYA) jhoto, joto, kaunria; (PAMPANGAN) sakuyot; (PANJABI, EASTERN) banphal; (POLISH) juta warzywna, juta kolorowa; (PORTUGUESE) juta, malva dos judeus, carurú da Baía; (PULAAR) sɔrɛŋ; (ROMANIAN) iută, corchorus; (SAMBAL, TINÀ) pasaw; (SANSKRIT) brihachanchu, brihathchanchu, singgika, patta, chanchu, deerghapathri, divyagandha, kalasa, mahaachanchu, nadika, nalika, sing ghika, sthoola chanchu, suchanchuka, sukanaasaka, vishaari; (SENA) ntonono; (SENOUFO, CEBAARA) kavonnigui; (SERBO-CROATIAN) juta, indiski lan; (SERER-SINE) akud, bot a nay, kud; (SLOVAK) juta; (SONGHAY) fákú, mìtìgà; (SPANISH) yute, yute de hojas comestibles, yute de fruto alargado; (SUBANON) tagabang; (SUSU) sórrè, shure, sore, sorofikhe; (SWAHILI) mlenda; (SWEDISH) jute; (TAGALOG) saluyot, pasau, pasaw, pasaw-haba; (TAMIL) peratti, perum ponnaku poondu, sanal; (TELUGU) parintakura, parinta koora; (THEMNE) *an*-kiŋ-kiri, *an*-kiriŋ- kiriŋ; (TIV) atyever; (TONGA) kankhumba; (TUMBUKA) celewa; (TUPURI) nitchani; (TURKISH) jüt, yahudi ebegümeci, mühliye; (TWI) oturo; (URDU) patsun; (VAI) bɛtu; (VLAAMS) struikokra, jute;

(WOLOF) *m*bali, kren kren, kering kering; (YALUNKA) suro-na; (YORUBA) ewudu, abo jaja, ayo, du, du ga-nbe, ẹ́yọ, oọ́yọ́, ọyọ́yọ́, yọ́yọ́; (ZARMA) fakou.

64. Trapaceae

64.1. *Trapa bicornis* Osbeck (EP) seeds; (P) C.

(ASSAMESE) pani shingori; (CHINESE, GÀN) ling jiao; (CHINESE, HAKKA) chòui-lĭt; (CHINESE, MANDARIN) er jiao wan ling; (CHINESE, MIN BEI) ling jiao; (CHINESE, MIN NAN) bé-chî, lêng-kak; (CHINESE, XIANG) ling jiao; (ENGLISH) Asian water chestnut, horn chestnut, horn-nut; (FINNISH) vesipähkinä; (FRENCH) châtaigne d'eau, châtaigne d'eau à deux cornes, châtaigne d'eau à fruit noir, macre; (GERMAN) Wassernuss, Chinesische Wassernuß; (GREEK) kastanon to ydronhares; (HUNGARIAN) sulyom; (ITALIAN) castagna d'acqua; (MACEDONIAN) dvorogčesta oreška; (NEPALI) singara; (POLISH) orzech wodny; (PORTUGUESE) castanha de água; (RUSSIAN) Vodianoi orekh dvurogii; (SPANISH) castaña de água, castaña de agua; (TAGALOG) apulid; (VLAAMS) waternoot.

64.2. *Trapa natans* L. var. *bispinosa* (Roxb.) Makino (EP) seed; (P) C.

(ASSAMESE) shingori, pani shingori; (CHINESE, WÚ) si jiao ling; (ENGLISH) singhara nut, singhara water chestnut, caltrop; (FRENCH) noix de singhara; (GUJARATI) shnghara; (HINDI) singhara, singhaada; (ITALIAN) castagna d'ácqua singara; (KANNADA) singhara, singaara beeja, singaade, singhaada balli; (KASHMIRI) gôru; (MAITHILI) siṅhāṛ; (MARATHI) singhara; (NEPALI) singara; (PANJABI, EASTERN) sēǣgärṛha; (PORTUGUESE) castanha de água singara; (SANSKRIT) shringata, sringataka; (SPANISH) castagna de água singara; (TAMIL) singhara, mullikaai, singhaara, pannimonthan kizhangu, sinvaara; (URDU) singhara; (VLAAMS) waternoot.

64.3. *Trapa natans* L. var. *natans* (EP) seed; (P) C.

(AFRIKAANS) waterkastaiing; (BELARUSAN) vadzjany arech plavanchi; (BENGALI) paniphal; (BODO) sugreng; (CHINESE, MANDARIN) si jiao ling; (CHINESE, WÚ) ling jiao, yuan jiao ling; (CHINESE, YUE) ou ling, ling jiao; (CZECH) kotvice plovoucí dvouostná, kotvice plovoucí pravá, lichořeřišnice větší; (DANISH) hornnød; (DUTCH) waternoot; (ENGLISH) water chestnut, water caltrop; (ESTONIAN) vesipähkel, mungalill; (FINNISH) vesipähkinä; (FRENCH) noix aquatique, mâcre nageante, châtaigne d'eau, saligot, macre; (FULFULDE, ADAMAWA) biriji ndiyam; (GERMAN) schwimmende Wassernuss, Wassernuss; (GREEK) trivoles o cherseos; (HAUSA) ɓigira, gújíyár rúwá, gyàɗár rúwá, ƙwaliliya, kwankalita, kwankwarita, kwankwarito, sankalwa; (HUNGARIAN) sulyom; (ICELANDIC) vatnshneta; (ITALIAN) tribolo acquatico; (JAPANESE) hishi; (JAVANESE) lengkat, lengkong, salaikat; (KANNADA) mullu kombu beeja, mullu kombu balli, neeru acrotu; (KAZAKH) shylim; (KHMER, CENTRAL) krachap; (KWANYAMA) yimbidira; (LATVIAN) ezerrieksts; (LITHUANIAN) plūduriuojantysis agaras; (MACEDONIAN) vodna oreška; (MALAYALAM) karimpolam, ponaver tsjeraua; (MARATHI) shingada, singhaada; (MIZO) thei kawr sak; (POLISH) kotewka orzech wodny; (PORTUGUESE) falsa castanha de água; (SANSKRIT) jalaphala, jalakantaka, trikonaphalam, thrikonaphala, jalashaya, jalasoochi, jalavalli, ksheerashukla, sanghatica, shapaka, shringakanda, shringamoola, shringaruha, shringata, shringataka, shukladugda, thrika, thrikota, varikantaka, varikubshaka, vishani; (SPANISH) abrojo de água, tribulo acuático; (SWEDISH) sjönöt; (TAGALOG) apulid, sungay-kalabaw; (TELUGU) kubyakama, kubjakam, kubyakam, pandi gadda, parike gadda; (THAI) kra chap, khao khwaai, ma ngaeng; (TURKISH) su cevizi, su kestanesi, göl kestanesi, boğa diken, çoban kalkitan; (VIETNAMESE) c[ur] [aas]u, [aas]u n[uw][ows]c; (VLAAMS) waternoot.

65. Tropaeolaceae

65.1. *Tropaeolum majus* L. (EP) leaves, flowers[S]; (P) R.

(AFRIKAANS) kappertjie, kapkappertjie; (ALBANIAN, TOSK) tropeolë; (ASSAMESE) nasturtium; (CHINESE, MANDARIN) jin lian hua; (CHINESE, WÚ) jin lian hua; (CHINESE, YUE) zao jin lian; (CRIOULO, CAPE VERDE) chagas; (CZECH) lichořešnice větši; (DANISH) blomsterkarse, tropæolum; (DUTCH) grote Oostindische kers; (ENGLISH) nasturtium, Indian cress; (FARSI, WESTERN) golé-ladan; (FILIPINO) nasturtium; (FINNISH) köynnöskrassi; (FRENCH) cresson d'Inde, capucine; (GAELIC, MANX) nvs ly sfyrryn; (GALICIAN) flor do sangue; (GERMAN) große Kapuzinerkresse; (GREEK) nastourtion; (HEBREW) kova ha-nazir, nasturzia, tropeolum gadul; (HUNGARIAN) sarkantyúka; (ICELANDIC) skjaldfletta; (ITALIAN) nasturzio comune, crescione del Perù; (JAPANESE) nasutaa chuumu, nôzen-haren; (LITHUANIAN) didžioji nasturtė; (MACEDONIAN) golema latanka; (MALTESE) kabucinella; (MORISYEN) kapousinn; (NORWEGIAN) vanlig blomkarse, blomkarse; (POLISH) nasturcja wielka, nasturcia większa; (PORTUGUESE) chagas, mastruço do Perú, flor de sangue, capuchinha grande; (QUECHUA, CUSCO) ticsau; (ROMANIAN) condurul doamnei mare, condurul doamnei; (RUSSIAN) nasturtsija balshaja sadovaja, kapustin bolshoi; (SERBO-CROATIAN) dragoljub; (SLOVAK) kapucínka; (SPANISH) nasturcio, mastranzo de las Indias, llagas de Cristo; (STELLINGWERFS) kiewietskop; (SWEDISH) indian krasse, källkrasse; (TURKISH) büyük latin çiçeği, hint teresi, frenk teresi; (VLAAMS) Oostindische kers.

65.2. *Tropaeolum tuberosum* Ruiz & Pavon (EP) tubers; (P) C.

(AFRIKAANS) bolkappertjie; (CHINESE, MANDARIN) kuai jing jin lzan hua; (CHINESE, WÚ) kuai jing jin lian hua; (CHINESE, YUE) kuai jing zao jin lian; (DANISH) anu; (DUTCH) knol Oostindische kers; (ENGLISH) tuberous nasturtium; (FRENCH) capucine tubéreuse; (GERMAN) Knollen-Kapuzinerkresse; (GREEK) nastourion; (ITALIAN) nasturzio tuberoso, cappuccina; (LITHUANIAN) gumbinė nasturtė; (MACEDONIAN) krtolesta batanka; (POLISH) nasturcja bulwiasta; (PORTUGUESE) capuchinha tuberosa; (QUECHUA, CUSCO) mashua, añu; (SPANISH) mashua, ysaño, anu, capuchina tuberosa, isaña; (TURKISH) yumrulu latin çiçeği; (VLAAMS) ysano; (WELSH) meri a mari.

66. Urticaceae

66.1. *Pilea glaberrima* (Blume) Blume (EP) leaves; (P) R,C.

(ASSAMESE) pilea; (CHINESE, WÚ) leng shui hua; (DANISH) pilea; (DUTCH) pilea; (ENGLISH) pilea; (FRENCH) pilea; (GERMAN) Pilea; (INDONESIAN) poh-pohan; (ITALIAN) pilea; (JAVANESE) poh-pohan; (PORTUGUESE) pílea; (SPANISH) pilea; (TURKISH) topçu çiçeği.

66.2. *Pilea melastomoides* (Poir.) Wedd. (EP) leaves; (P) R,C.

(ASSAMESE) pilea; (CHINESE, WÚ) shan mai leng shui hua; (DANISH) pilea; (DUTCH) pilea; (ENGLISH) pilea; (FRENCH) pilea; (GERMAN) Pilea; (GREEK) pilea; (INDONESIAN) poh-pohan, bungbrun; (JAVANESE) poh-pohan; (PORTUGUESE) pílea; (SPANISH) pilea;

66.3. *Urtica dioica* L. (EP) leaves; (P) C.

[1]Itsukha dialect of Luyia.

Chapter 2. Common names of the cultivated vegetable crops of the world, listed alphabetically

a cu (17.1. AKHA)
aˉ cu ka nawˆ (40.43. LAHU)
a dyeng (38.16. SERER-SINE)
a dyeng (38.18. SERER-SINE)
aˉ hpa˰ meuˍ (15.2. LAHU)
aˉ hpe˰ chaweh (62.1. LAHU)
aˉ hpe˰ faˆ hk’ehˇ (62.5. LAHU)
aˉ hpe˰ g’awˇ (23.1. LAHU)
aˉ hpe˰ g’awˇ (23.2. LAHU)
aˉ hpe˰ tcuhˉ–ehˉ (62.5. LAHU)
aˉ hpehˇ bvuh˰ (38.7a. LAHU)
aˉ hpehˇ shiˍ (38.13a. LAHU)
aˉ hpo˰ k’oˍ (38.20. LAHU)
a luˉ shiˍ (62.24. LAHU)
aˉ mo˰ k’oˍ (38.20. LAHU)
aˉ pawˇ hpui ma (16.1a. LAHU)
a teh hpui shiˍ (38.1. LAHU)
a’a (40.30. SAMOAN)
aaane manjal (20.4. TULU)
aaathmaguptha (40.18. SANSKRIT)
aachi chohku (27.16. JAPANESE)
aachilkkilangu (11.1. MALAYALAM)
aadagam (40.2. TAMIL)
aadagi (40.2. TAMIL)
aadhaki (40.2. MALAYALAM)
aadhaki (40.2. SANSKRIT)
aadhaki (40.2. TELUGU)
aadi (20.6. FIJIAN)
aadithyalu (30.51. TELUGU)
aakaaʔ (16.1a. COMANCHE)
aal (38.20. HINDI)
aal vallikkizhangu (39.4. TAMIL)
ààlàoko (11.7. YORUBA)
aalivirai (30.51. TAMIL)
aaloo (8.4. HINDI)
aaloo (62.24. MARATHI)
aaloogedde (62.24. KANNADA)
aalooka (62.24. SANSKRIT)
aalooki (8.4. SANSKRIT)
aalu (62.24. NEPALI)
aalu (62.24. PANJABI, EASTERN)
aalu gaddalu (62.24. TELUGU)
aamaada (20.4. HINDI)
aamba haldi (20.4. KONKANI)
aamlaardraka (11.6. SANSKRIT)
aan touna (40.31. GOURMANTCHÉ)
aanandam (20.1. TAMIL)
aandilee (30.51. TELUGU)
aane hullu (17.3. KANNADA)
aankhle jhar (1.1. NEPALI)
aanteelui (7.2. AFRIKAANS)
aapneut (40.1. AFRIKAANS)
aara root (15.2. MARATHI)
aardaker (40.11. DUTCH)
aardaker (40.11. VLAAMS)
aardamandel (10.1. VLAAMS)
aardappel (62.24. DUTCH)
aardappel (62.24. VLAAMS)
aardappelboon (40.22. VLAAMS)
aardappelyam (11.2. VLAAMS)
aardappelyam (11.6. VLAAMS)
aardboontjie (40.1. AFRIKAANS)
aardmandel (10.1. DUTCH)
aardnoot (40.1. DUTCH)
aardnoot (40.1. VLAAMS)
aardnootje (40.1. DUTCH)
aardpeer (27.26. DUTCH)
aardpeer (27.26. VLAAMS)
aardraka (20.6. KANNADA)
aardrakah (20.6. SANSKRIT)
aardrakamu (20.6. TELUGU)
aardrashaka (20.6. SANSKRIT)
aarjaka (44.4. SANSKRIT)
aart (40.28. STELLINGWERFS)
aartappel (62.24. AFRIKAANS)
aartappeljam (11.2. AFRIKAANS)
aarumaatha thandu (23.3. TAMIL)
aarumatha-thandu (23.3. TAMIL)
aashukachu (8.4. SANSKRIT)
[aas]u n[uw][ows]c (64.3. VIETNAMESE)
aba (8.1. IBANAG)
aba (8.4. IBALOI)
aba (8.4. ILOCANO)
aba-aba (8.1. BONTOC, CENTRAL)
ābāabūrˉ˰ (39.4. ANUAK)
ābāabúré (39.4. ANUAK)
a-babani (16.1a. BAGA)
ababika (24.9. SANSKRIT)
àbàchà (39.4. IGBO)
abaga (38.27. GOURMANTCHÉ)
abai (40.3. MARATHI)
abai (40.4. MARATHI)

aburobia (9.1. TWI)
aburuma (7.3. SANGA)
aburuma nẹẹra (7.10. SANGA)
abva (40.4. GUJARATI)
abvo (40.3. GUJARATI)
a-bwɔ (27.14. THEMNE)
ābʌ́y (17.7. ANUAK)
Abyssinian mustard (30.3. ENGLISH)
açafrão da Índia (20.2. PORTUGUESE)
açafron (20.2. CRIOULO, UPPER GUINEA)
acanthopanax espinoso (26.4. SPANISH)
acchechengio (62.11. ITALIAN)
aččik kalampir (62.1. UZBEK)
acciughero (44.6. ITALIAN)
aceda (58.2. GALICIAN)
acedera (58.2. SPANISH)
acedera común (58.2. SPANISH)
acedera espanica (58.3. SPANISH)
acedera romana (58.4. SPANISH)
aceitilla (27.8. SPANISH)
acelga (35.3. PORTUGUESE)
acelga (35.3. SPANISH)
acetosa (58.2. ITALIAN)
acetosa maggiore (58.2. ITALIAN)
acetosa romana (58.4. ITALIAN)
acetosella tuberosa (53.1. ITALIAN)
ach (24.4. BRETON)
àchàlà (17.3. IGBO)
àchàrà (17.3. IGBO)
àchàrà mili (17.3. IGBO)
àchàrà mmako (17.5. IGBO)
acharbondi (27.2. KONKANI)
acharbondi (27.2. MARATHI)
achatti (62.19. TAMIL)
achcaña (27.42. QUECHUA, CUSCO)
achera (9.1. SPANISH)
achicoria (27.9. QUECHUA, CUSCO)
achicoria (27.45. QUECHUA, CUSCO)
achicoria de bruselas (27.10a. SPANISH)
achicoria de hoja roja (27.10b. SPANISH)
achicoria de hojas verdes (27.10a. SPANISH)
achicoria de raíz (27.10a. SPANISH)
achicoria silvestre (27.10a. SPANISH)
achipa (40.21. SPANISH)
achira (9.1. FRENCH)
Achira (9.1. GERMAN)
achira (9.1. SPANISH)
achis (23.3. QUECHUA, CUSCO)
achita (23.3. QUECHUA, CUSCO)
achoccha (38.19. ENGLISH)
achoccha (38.19. QUECHUA, CUSCO)
achuk calampir (62.1. UZBEK)
aci bakla (40.14. TURKISH)
aci manyok (39.4. TURKISH)
aci marul (27.9. TURKISH)
aci ot (13.1. TURKISH)
aci tere (30.51. TURKISH)
aci yaprak (27.24. TURKISH)
acib'nimic (62.21. POTAWATOMI)
Ackerbohne Sommerwurz (52.1. GERMAN)
Ackerknoblauch (7.1. GERMAN)
Ackersalat (67.2. GERMAN)
Acker-Schachtelhalm (1.1. GERMAN)
acktina (40.1. GOURMANTCHÉ)
acsalapu (27.35. HUNGARIAN)
açucena de Trinidade (12.1. PORTUGUESE)
açucenas (13.4. PORTUGUESE)
acur (38.10. TURKISH)
acut (7.10. ITAWIT)
ād (20.6. MAITHILI)
adá (17.3. GBE-VHE)
ada (20.6. ASSAMESE)
ada (20.6. BENGALI)
ada (20.6. HINDI)
ada (20.6. ORIYA)
ada (20.6. SANTALI)
adabau lambho (23.3. GUJARATI)
adabav (23.3. GUJARATI)
adabri (16.1a. EFIK)
adad (40.37. GUJARATI)
adaduanan (40.1. TWI)
adagam (40.2. TAMIL)
adaí (17.3. GA)
adairdumpa (11.2. TELUGU)
adalavitulu (30.51. TELUGU)
adalut (9.1. BURMESE)
adâma (32.1. SIDAMO)
ad-ammã-wo-ba (11.1. AKAN-AKUAPEM)
adaŋga (38.20. ÉWÉ)
adäng^{w}arre (40.26. AMHARIC)
adarak (20.2. HINDI)
adas (24.13. INDONESIAN)
adas (24.13. JAVANESE)
adas (40.12. ARABIC)
adas (40.12. FARSI, WESTERN)
adas (40.12. TAJIKI)
adas (40.12. URDU)
adas (40.12. UZBEK)
adas asfar (40.2. ARABIC)
adas landi (24.13. JAVANESE)

adas londo (24.13. JAVANESE)
adas pedas (24.13. INDONESIAN)
adas pedas (24.13. MALAY)
ʽadash tarbutit (40.12. HEBREW)
ʽadasha (40.12. HEBREW)
adasha (40.12. HEBREW)
adasha tarbutit (40.12. HEBREW)
adatl nohut (40.5. TURKMEN)
adaty arpabadýan (24.13. TURKMEN)
adaty badyýan (24.19. TURKMEN)
adaty burç (62.1. TURKMEN)
adaty fizalis (62.11. TURKMEN)
adaty garpyz (38.3. TURKMEN)
adaty göýül (34.1. TURKMEN)
adaty gyjy-gyjy (30.43. TURKMEN)
adaty hren (30.1. TURKMEN)
adaty hyýar (38.13a. TURKMEN)
adaty itüzüm (62.21. TURKMEN)
adaty käşir (24.11. TURKMEN)
adaty mekgejöwen (17.7. TURKMEN)
adaty nazbaýgül (44.5. TURKMEN)
adaty pomidor (62.9. TURKMEN)
adaty semzek (59.2. TURKMEN)
adaty şogan (7.2. TURKMEN)
adaty sübselik (35.2. TURKMEN)
adaty şugundyr (35.4. TURKMEN)
adaty syşratgy (27.10a. TURKMEN)
adaty turşuja (58.2. TURKMEN)
adavebaala (40.9. MARATHI)
adavi patola (38.31. TELUGU)
adavichikkudu (40.9. TELUGU)
adavidumpathige (11.6. TELUGU)
adavidumpatige (11.2. TELUGU)
adavipasuvu (20.4. TELUGU)
adavithamma (40.3. TELUGU)
adaviyatagathige (11.6. TELUGU)
addakaṁ (20.6. PALI)
ade (11.7. AKAN-AKUAPEM)
ade (11.7. DANGME-KROBO)
ade (11.7. ÉWÉ)
adekute (11.4. ÉWÉ)
Adelaarsvaren (2.1. DUTCH)
adelaarsvaren (2.1. VLAAMS)
adelaarsvaring (2.1. AFRIKAANS)
adelagan (40.43. ARABIC)
adeli (30.51. TELUGU)
adenger (40.25. TIV)
adenkum (38.20. AKAN-AKUAPEM)
adesi (40.12. SWAHILI)
adesi (40.28. SWAHILI)
adhaki (40.2. HINDI)
adhaki (40.2. MALAYALAM)
adhaki (40.2. SANSKRIT)
adhaki (40.2. TELUGU)
adharanga (23.3. HINDI)
adharca (57.1. GAELIC, IRISH)
adhas (24.13. INDONESIAN)
adhe̱ (20.6. SANTALI)
adhyanda (40.18. SANSKRIT)
ādi (20.6. MUNDARI)
adi aci marul (27.10a. TURKISH)
adi aslanağzi (27.45. TURKISH)
adi deniz lahanasi (30.45. TURKISH)
adi eşek marulu (27.42. TURKISH)
adi eşekçiçeği eşek çiçeği (51.1. TURKISH)
adi gebreotu (34.1. TURKISH)
adi Japon sarmaşği (40.30. TURKISH)
adi kabak (38.18. TURKISH)
adi kaplan çiçeği (12.1. TURKISH)
adi karakafesotu (29.2. TURKISH)
adi kebre (34.1. TURKISH)
adi kuzu gevreği (67.2. TURKISH)
adi mercanköşk (44.6. TURKISH)
adi mürdümük (40.10. TURKISH)
adi neptunia (40.19. TURKISH)
adi su oku (6.2. TURKISH)
adi suoku (6.2. TURKISH)
adi yonca (40.17. TURKISH)
adida (59.2. AKAN-ASANTE)
a-dif-wir (27.24. THEMNE)
aditi (23.2. GEN)
adityalu (30.51. TELUGU)
adius (40.2. BUHID)
adiyalu (30.51. TELUGU)
ádíyú (38.20. NGIZIM)
adja ã tina (40.15. GOURMANTCHÉ)
adja-bouinga (11.3. GOURMANTCHÉ)
adjaka (23.3. FULFULDE)
adja-noũga (11.4. GOURMANTCHÉ)
adjo (7.10. CRIOULO, CAPE VERDE)
adjongô (27.8. BAOULÉ)
adjugo somadrewe (11.1. AKAN-AKUAPEM)
adla balli (38.29. KANNADA)
adla kaayi (38.29. KANNADA)
adlagna (40.43. ARABIC)
Adlerfarn (2.1. GERMAN)
adodoro (59.3. EDO)
adodowe (24.7. FON)
adong shinga (7.10. JUNKUN TAKUM)
adoua (40.2. TEM)

adrak (20.2. PASHTO, NORTHERN)
adrak (20.5. URDU)
adrak (20.6. HINDI)
adrak (20.6. HINDUSTANI)
ādrak (20.6. MUNDARI)
adrak (20.6. PANJABI, EASTERN)
adrak (20.6. URDU)
adraka (20.6. SANSKRIT)
adrakh (20.6. HINDI)
adrakh (20.6. KASHMIRI)
á-dsina (40.1. GOURMANCHÉMA)
adsukibønne (40.36. NORWEGIAN)
adsukipapu (40.36. FINNISH)
adtad (23.3. KANKANAEY)
adụ (11.2. IGBO)
ádu (11.2. IGBO)
ádụ̀ (11.4. IGBO)
adu (20.6. GUJARATI)
adua (40.2. DAGBANI)
adua (40.2. TEM)
adua (40.25. TWI)
adua (40.43. TWI)
aduwa (20.6. NEPALI)
aduwa (40.2. BAATONUM)
aduwa (40.2. HAUSA)
adwera (59.2. AKAN-AKUAPEM)
adwera (59.2. AKAN-ASANTE)
adwera (59.2. TWI)
adwere (59.2. TWI)
adworɛ (59.2. AKAN-AKUAPEM)
adza (40.18. GBE-VHE)
adzigo (11.1. ÉWÉ)
adzrɔkpi (27.8. ÉWÉ)
adzugo (11.1. ÉWÉ)
adzugo bolobolo (11.1. ÉWÉ)
adzugo dzedze (11.1. ÉWÉ)
adzugo kpuka (11.1. ÉWÉ)
adzugo nkani (11.1. ÉWÉ)
adzuki bean (40.36. ENGLISH)
adzuki-bab (40.36. HUNGARIAN)
Adzuki-Bohne (40.36. GERMAN)
adzukibønne (40.36. DANISH)
adzukibönor (40.36. SWEDISH)
adzukiboon (40.36. AFRIKAANS)
adzukiboon (40.36. DUTCH)
adzukiboon (40.36. VLAAMS)
adžygovuk (27.45. AZERBAIJANI, NORTHERN)
aed-mustjuur (27.38. ESTONIAN)
aed-piimjuur (27.46. ESTONIAN)
aeduba (40.26. ESTONIAN)
ægfrugt (62.19. DANISH)
ægplante (62.19. DANISH)
aelaaparni (20.1. SANSKRIT)
aele kosu (30.27a. KANNADA)
aelika chaevi koora (24.15. TELUGU)
aenna soora (11.6. KANNADA)
aerial yam (11.2. ENGLISH)
ært (40.28. DANISH)
ærtebønne (40.2. DANISH)
ærte-natskygge (62.23. DANISH)
aerumai thakkaali (62.9. TAMIL)
Aethiopischer Senf (30.3. GERMAN)
ætififill (27.26. ICELANDIC)
ætihvönn (24.1. ICELANDIC)
ætipistill (27.16. ICELANDIC)
aezhilaikkizhangu (39.4. TAMIL)
afafe (38.23. IBANAG)
afare (59.2. AKAN-FANTE)
afaseɛ (11.1. TWI)
afaseo (11.1. GA)
afaséo (11.1. GA)
afasew (11.1. AKAN-AKUAPEM)
afasew (11.1. TWI)
afasew anamasu (11.1. AKAN-AKUAPEM)
afasew apuka (11.1. AKAN-AKUAPEM)
afasew ɔdepa (11.1. AKAN-AKUAPEM)
afasew kani (11.1. AKAN-AKUAPEM)
afasew pa (11.1. AKAN-AKUAPEM)
afasi (11.1. ÉWÉ)
afasó (11.1. GA)
afasse (11.1. KABIYÉ)
a-fɔngwón (11.2. BASSARI)
Afenoss (40.1. LUXENBOURGEOIS)
afera (38.17. GOURMANTCHÉ)
affoo (11.3. SPANISH)
afia edidia (11.4. EFIK)
àfia èdidìà iwá (11.4. EFIK)
àfia iwá (11.4. EFIK)
afia oko (11.7. EFIK)
àfia òkò (11.7. EFIK)
afiou (24.23. PORTUGUESE)
afla (59.2. FON)
afla(a) (59.2. GBE-VHE)
afla(a)-tokpui (59.2. GBE-VHE)
aflatɔkpwi (59.2. AJA)
aflãtokpui (59.2. GBE-VHE)
aflavi (59.2. GBE, WACI)
African cucumber (38.23. ENGLISH)
African eggplant (62.18. ENGLISH)

ahabía (9.1. GA)
ahabia (9.1. TWI)
ahahe (38.23. IBANAG)
ahal ha-gevishim (22.1. HEBREW)
ahal hagvishim (22.1. HEBREW)
ahalimaku (62.2. ANDO)
ahame (44.5. ÉWÉ)
ahan kui (30.58. CHINESE, MANDARIN)
ahi (17.7. NIUE)
ahi' (62.24. CHICKASAW)
ahi' champoli' (36.3. CHICKASAW)
ahị harā (63.1. IGBO)
aĥị ĥaṛā (63.1. IGBO)
ahị hịarā (63.1. IGBO)
ahị hịrā (63.1. IGBO)
ahi' lhibowa' (62.24. CHICKASAW)
ahi taina (17.7. NIUE)
ahichhathra (24.13. SANSKRIT)
ahinsaw (38.28. ÉWÉ)
ahipa (40.20. ENGLISH)
ahipa (40.20. PORTUGUESE)
ahipa (40.20. SPANISH)
ahiphala (38.29. SANSKRIT)
ahleu (17.5. DANGME-KROBO)
ahliva (30.51. MARATHI)
ah-lu (62.24. BURMESE)
ahos (7.10. CEBUANO)
áhos (7.10. HILIGAYNON)
ahos (7.10. ILOCANO)
ahos (7.9a. WÁRAY-WÁRAY)
ahrana (17.5. ANDO)
ahrbis (40.5. YIDDISH)
ahrbis (40.28. YIDDISH)
ahrbooz (38.3. YIDDISH)
àhụ́ (38.5. IGBO)
ahu ekere (40.1. IGBO)
àhụ ekere (40.1. IGBO)
ahụ ekere (40.15. IGBO)
àhụ́ elu (38.5. IGBO)
ahuaimanto (62.12. QUECHUA, CUSCO)
àhụekere (40.1. IGBO)
ahuma (40.31. TIV)
ahus (7.10. BINUKID)
ahus (7.10. IBANAG)
ahwedie (17.5. TWI)
ahwere (17.5. AKAN-FANTE)
ahwereɛ (17.5. AKAN-ASANTE)
ahwerew (17.5. TWI)
ai (27.7. CHINESE, MANDARIN)
ai (27.7. CHINESE, WÚ)
ai (7.10. ROMANIAN)
ai (7.3. VLAAMS)
ai cai (27.7. CHINESE, MIN BEI)
ai gua (62.19. CHINESE, YUE)
ai sai e bi ya qie (62.14a. CHINESE, YUE)
ai xing cai dou (40.28. CHINESE, MANDARIN)
ai xing dao dou (40.3. CHINESE, MANDARIN)
ai xing dao dou (40.3. CHINESE, WÚ)
ai ye (27.7. CHINESE, XIANG)
aibika (45.2. TOK PISIN)
aibike (45.2. VLAAMS)
aigyptiako kremmydi (7.9a. GREEK)
aiku vovo (40.43. FON)
aiku wagbé lafo (40.43. FON)
aiku wéwé (40.43. FON)
ail (7.10. FRENCH)
ail blanc (7.10. FRENCH)
ail chinoise (7.13. FRENCH)
ail civette de Chine (7.13. FRENCH)
ail de cerf (7.14. FRENCH)
ail d'orient (7.1. FRENCH)
ail du Japon (7.7. FRENCH)
ail gros (7.1. FRENCH)
ail ordinaire (7.10. FRENCH)
ail rocambole (7.12. FRENCH)
ail stérile (7.2. FRENCH)
ail-ciboule (7.5. FRENCH)
ailli (7.10. BEMBÉ)
ailli (7.10. KOONGO)
ainbe (21.2. TOK PISIN)
aioskorea ptery-gioforos (11.1. GREEK)
aipo (24.4. CRIOULO, CAPE VERDE)
aipo (24.4. PORTUGUESE)
aipo (24.5. PORTUGUESE)
aipo da agua (24.4. PORTUGUESE)
aipo de cabeça (24.5. PORTUGUESE)
aipo de raíz (24.5. PORTUGUESE)
aipo de talos (24.4. PORTUGUESE)
aipo nabo (24.5. PORTUGUESE)
airish-pɛtɛtɛ (62.24. KRIO)
aisattsikohtako (40.14. BLACKFOOT)
aishwe (45.4. TIV)
aizcoles (40.10. BASQUE)
ajada (40.18. SANSKRIT)
ajadakumaricha (62.5. SANSKRIT)
ajagandha (44.4. HINDI)
ajagandhika (44.5. SANSKRIT)
ajaka (44.4. SANSKRIT)

ajalomi (40.18. SANSKRIT)
ajavha (40.18. SANSKRIT)
ajɔ (11.7. GONJA)
ajedrea (44.9. SPANISH)
ajedrea blanca (44.9. SPANISH)
ajedrea común (44.9. SPANISH)
ajedrea de jardín (44.9. SPANISH)
ajefowo (23.4. YORUBA)
ajenjo (27.7. SPANISH)
ají (62.4. SPANISH)
ají (62.6. SPANISH)
ájíē (11.2. TIV)
ajig (27.8. MONGOLIAN)
ajila (30.52. CHEROKEE)
ajipa (40.21. SPANISH)
ajmod (24.4. HINDI)
ajmod (24.4. URDU)
ajmod (24.19. HINDUSTANI)
ajmoda (24.4. SANSKRIT)
ajmud (24.4. HINDI)
ajmud (24.5. HINDI)
ájo (7.10. LADINO)
ajo (7.10. SPANISH)
ajo común (7.10. SPANISH)
ajo del Japon (7.7. SPANISH)
ajo kiras (7.8. SPANISH)
ajo porro (7.8. SPANISH)
ajo ramano (7.12. SPANISH)
ajôé (16.1b. BULU)
aju (40.1. SANGA)
ajuin (7.3. VLAAMS)
ājwāallā (36.3. ANUAK)
ājwāallā (62.24. ANUAK)
ājwáallɛ́ (36.3. ANUAK)
ājwáallɛ́ (62.24. ANUAK)
ajwáyan khurásání (24.4. HINDUSTANI)
ạk (17.5. SANTALI)
ak kočanduu kapusta (30.27a. KIRGHIZ)
aká (17.7. IZON)
aka (40.30. NIUE)
aka (40.30. TONGAN)
aka' aapaka̱li' (50.2. CHICKASAW)
aka fala (40.30. NIUE)
akaï (11.2. AKAN-ASANTE)
akakadur (20.6. AKAN-ASANTE)
a-kali (38.16. THEMNE)
a-kali (38.18. THEMNE)
a-kali-*a*-bana (38.18. THEMNE)
akalkar (27.2. SANSKRIT)
akam (11.2. AKAN-AKUAPEM)
akam (11.2. AKAN-ASANTE)
akam (11.2. DANGME-KROBO)
akam moto (11.2. AKAN-AKUAPEM)
akamfir (38.18. AKAN-FANTE)
akan (11.2. TWI)
a-kan(d) (40.1. THEMNE)
akana (38.12. GOURMANTCHÉ)
akana (62.18. GOURMANTCHÉ)
akana maigbo (19.1. YORUBA)
åkande (50.2. DANISH)
akangkong (36.2. BICOLANO, CENTRAL)
akangkong (36.2. PAMPANGAN)
akanka' inchota̱'to' (62.9. CHICKASAW)
akankimimpa' (62.9. CHICKASAW)
akankintako̱loshi' (62.9. CHICKASAW)
akankishniha' (62.9. CHICKASAW)
akanmmto (11.2. TWI)
akantusz (21.1. HUNGARIAN)
akar keman hantu (40.19. MALAY)
akar parsi (13.2. INDONESIAN)
akara binch (40.43. KRIO)
akasi (8.4. IGBO)
àkàsì (8.4. IGBO)
àkàsị óyìbó (8.9. IGBO)
akasi-oyibo (8.9. IGBO)
akataha (40.30. TONGAN)
akate (38.3. AKAN-FANTE)
akate (38.5. AKAN-FANTE)
akate (40.1. DANGME-KROBO)
akatewa (38.5. AKAN-AKUAPEM)
akatewa (38.5. AKAN-ASANTE)
akaton (38.21. TWI)
akatoo (38.5. TWI)
akatsewa (38.5. AKAN-FANTE)
akɔdu (16.1a. GBE-VHE)
akɔdua-zẽ (16.1a. GBE-VHE)
akɔdukɔkɔ (16.1a. GBE-VHE)
akɔdukpe (16.1a. GBE-VHE)
akɔhun (44.5. GUN)
a-kɔlma-ka-tar (10.2. THEMNE)
a-kɔtɔ (45.1. THEMNE)
akɔtrɔmɔ (40.25. DANGME-KROBO)
akeka (20.6. TWI)
akekaduru (20.6. TWI)
aken zaya (11.1. BAOULÉ)
akerkål (30.22. NORWEGIAN)
åkersnelle (1.1. NORWEGIAN)
åkervial (40.10. SWEDISH)
åkervikke (40.34a. NORWEGIAN)
akẹsan (27.8. YORUBA)

ala-bạrag (38.16. KASHMIRI)
ala-bạrag (38.18. KASHMIRI)
ala-bạrag (38.20. KASHMIRI)
alabas (30.30. TURKISH)
à-làbásyà (7.3. BETE-BENDI)
alabu (16.1a. SANSKRIT)
alābu (38.14. PALI)
alābu (38.16. PALI)
alābu (38.17. PALI)
alābu (38.18. PALI)
alabu (38.20. SANSKRIT)
a̲la̲busa (7.3. TIV)
aladẹ (11.7. YORUBA)
alafi garmi (27.8. TAJIKI)
alafi mushkhor (27.5. TAJIKI)
alafolo (11.7. IGBO)
alakowak (17.5. BONTOC, CENTRAL)
al-allobo (24.15. KALAGAN)
alalokpwe (59.2. GEN)
alam-am (2.1. IBALOI)
alambusa (40.19. SANSKRIT)
alamoa (11.1. DANGME-KROBO)
alamoa gaga (11.1. DANGME-KROBO)
alamoa gu (11.1. DANGME-KROBO)
alamoa kani (11.1. DANGME-KROBO)
alamoa ku (11.1. DANGME-KROBO)
alamoa puka (11.1. DANGME-KROBO)
alamoa sale (11.1. DANGME-KROBO)
alamoa tun (11.1. DANGME-KROBO)
alamoa tun gaga (11.1. DANGME-KROBO)
alapolo (27.14. IGBO)
alāpu (38.14. PALI)
alāpu (38.16. PALI)
alāpu (38.17. PALI)
alāpu (38.18. PALI)
alari (23.4. YORUBA)
alari funfun (23.4. YORUBA)
alasandalu (40.43. TELUGU)
alasande (40.43. KANNADA)
alasandi (40.43. KANNADA)
alasandige (40.43. KANNADA)
alasani (58.2. HOPI)
alasunda (40.43. MARATHI)
alavi (8.4. KANNADA)
alavi beeja (30.51. KANNADA)
alavu (8.4. GUJARATI)
alaw (13.4. WELSH)
alawā' (40.28. MICMAC)
alayafou (23.3. HAUSA)
alayon (23.3. IFUGAO)
alayyaho (35.9. HAUSA)
alayyon (23.3. IFUGAO)
alayyon an mumbola (23.3. IFUGAO)
alayyon di tindalo (23.3. IFUGAO)
alba (7.3. MÒORÉ)
albacce (7.3. FULFULDE, ADAMAWA)
albahaca (44.5. SPANISH)
albahaca larga (35.2. SPANISH)
albahaca velluda (44.4. SPANISH)
albahaka (44.5. IBANAG)
albahaka (44.5. TAGALOG)
albaharé (7.2. MÒORÉ)
àlbásà (7.3. BEROM)
albasa (7.3. DENDI)
albasa (7.3. FULFULDE, ADAMAWA)
albasa (7.3. HAUSA)
albása (7.3. HAUSA)
àlbásàà (7.2. HAUSA)
àlbásàà (7.3. HAUSA)
àlbásàà gudààjíí (7.3. HAUSA)
àlbásàà mai-gora (7.2. HAUSA)
àlbásàà mai-kara (7.2. HAUSA)
àlbásàà-mai-kara (7.8. HAUSA)
àlbásàà mai-wuya (7.2. HAUSA)
albasal (7.3. FULFULDE, ADAMAWA)
albasan (7.5. ZARMA)
àlbásâr (7.3. NGIZIM)
albatje dadi (7.10. FULFULDE, ADAMAWA)
albatje darri (7.10. FULFULDE, ADAMAWA)
albergínia (62.19. CATALAN-VALENCIAN-BALEAR)
alberginiera (62.19. CATALAN-VALENCIAN-BALEAR)
albero del pomodoro (62.7. ITALIAN)
àlbésel (7.2. SONGHAY)
albugbati (28.1. TAGALOG)
alcachofa (27.16. SPANISH)
alcachofero (27.16. SPANISH)
alcachofra (27.16. GALICIAN)
alcachofra (27.16. PORTUGUESE)
alcachofra brava (27.15. PORTUGUESE)
alcachofra chinesa (44.11. PORTUGUESE)
alcachofra tuberosa (44.11. SPANISH)
alcaiota (38.25. PORTUGUESE)
alcaparra (34.1. PORTUGUESE)
alcaparra (34.1. SPANISH)
alcaparras (34.1. TAGALOG)
alcaparreira (34.1. PORTUGUESE)
alchechengi (62.11. ITALIAN)
alchechengi giallo (62.12. ITALIAN)

alchechengi peruviano (62.12. ITALIAN)
alcocaz (8.4. SPANISH)
ale (20.6. MARATHI)
alebassa (7.2. TEM)
alechuga (27.31. LADINO)
alechuga (27.32. LADINO)
aleefɔ (23.3. DENDI)
aleifu buter (23.3. TIV)
ʿaleket ḥaruka (52.1. HEBREW)
aleket harutza (52.1. HEBREW)
a-lel (40.25. THEMNE)
aleluya tuberosa (53.1. SPANISH)
alem (20.6. KONKANI)
alen (20.6. MARATHI)
alen'da (36.3. BAOULÉ)
alev (40.2. TIV)
alev (40.43. TIV)
alev a batur (40.2. THEMNE)
aleva paprika (62.1. BOSNIAN)
aleveri (30.51. BENGALI)
alface (27.31. CRIOULO, CAPE VERDE)
alface (27.31. GALICIAN)
alface (27.31. PORTUGUESE)
alface (27.32. CRIOULO, CAPE VERDE)
alface da India (27.29. PORTUGUESE)
alface de Alger (27.21. PORTUGUESE)
alface de Alger (67.1. PORTUGUESE)
alface de cordeíro (67.1. PORTUGUESE)
alface de cordeiro (67.2. PORTUGUESE)
alface de cortar (27.32. PORTUGUESE)
alface de repolho (27.31. PORTUGUESE)
alface do campo (67.2. PORTUGUESE)
alface espargo (27.30. PORTUGUESE)
alface francesa (27.31. PORTUGUESE)
alface orelha de mula (27.32. PORTUGUESE)
alface repolhuda (27.31. PORTUGUESE)
alface romana (27.32. PORTUGUESE)
alfalfa (40.17. AFRIKAANS)
alfalfa (40.17. CREOLE, HAITIAN FRENCH)
alfalfa (40.17. CRIOULO, CAPE VERDE)
alfalfa (40.17. ENGLISH)
alfalfa (40.17. GAELIC, IRISH)
alfalfa (40.17. INDONESIAN)
alfalfa (40.17. MALAY)
alfalfa (40.17. NEPALI)
alfalfa (40.17. POLISH)
alfalfa (40.17. PORTUGUESE)
alfalfa (40.17. SPANISH)
alfalfa (40.17. SWEDISH)
alfalfa sprouts (40.17. ENGLISH)
alfalfey (40.17. GAELIC, MANX)
alfavaca do campo (44.9. PORTUGUESE)
alforvas (40.32. PORTUGUESE)
alfuyuan (11.1. BUHID)
algazul (22.1. SPANISH)
Algerisches Baldrian (27.21. GERMAN)
algiersalat (27.21. DANISH)
alhábega (44.5. SPANISH)
alho (7.10. PORTUGUESE)
alho chinês (7.13. PORTUGUESE)
alho comum (7.10. PORTUGUESE)
alho do Japão (7.7. PORTUGUESE)
alho francês (7.8. PORTUGUESE)
alho gigante (7.1. PORTUGUESE)
alho gigante (7.12. PORTUGUESE)
alho grosso de Espanha (7.12. PORTUGUESE)
alho porro (7.8. PORTUGUESE)
alho porro (7.10. PORTUGUESE)
alho porro bravo (7.1. PORTUGUESE)
alho rocambole (7.12. PORTUGUESE)
alho vermelho (7.12. PORTUGUESE)
alholva (40.32. SPANISH)
alhuā (36.3. MAITHILI)
alia (20.6. INDONESIAN)
àlibácá (7.3. HAUSA)
alibanos (30.56. PANGASINAN)
alibosa (7.3. KRIO)
aliè manda (16.1a. BAOULÉ)
a-lil (40.25. THEMNE)
alim (30.51. INDONESIAN)
alimpuyang (20.4. CEBUANO)
alimpuyas (20.4. CEBUANO)
aliv (30.51. MARATHI)
alive (30.51. KANNADA)
aliveraai (30.51. TAMIL)
aliverai (30.51. TAMIL)
aliverei (30.51. HINDI)
àljáng (62.15. KANURI)
aljmoda; (24.4. MARATHI)
alkaddu (38.20. HINDI)
àlkáfún (24.13. SONGHAY)
alkañjez (62.11. BRETON)
alkatxofa (27.16. BASQUE)
alkékenge jaune (62.12. FRENCH)
alkekengi (62.12. PORTUGUESE)
alkolshi (40.18. BENGALI)
álkörmös (56.1. HUNGARIAN)
álkörmös (56.2. HUNGARIAN)
alkusa (40.18. BENGALI)
alkushee (40.18. SANSKRIT)

alkushi (40.18. BENGALI)
all (7.10. CATALAN-VALENCIAN-BALEAR)
all porret (7.1. CATALAN-VALENCIAN-BALEAR)
alla (20.6. KANNADA)
alla-bachhalla (28.1. TELUGU)
allaka (24.9. SANSKRIT)
allalaya (23.6. ITAWIT)
allam (20.6. TAMIL)
allam (20.6. TELUGU)
allamu (20.6. TELUGU)
allay (23.3. GADDANG)
alle (20.6. KONKANI)
alléluia (15.1. FRENCH)
alleluia (15.1. SPANISH)
allemandsharnisk (7.14. DANISH)
alli (50.2. TAMIL)
alli beeja (30.51. KANNADA)
alli kaada (50.2. TELUGU)
allibija (30.51. KANNADA)
allibija (30.58. KANNADA)
alligator weed (23.1. ENGLISH)
allion to kollon (7.9b. GREEK)
allion to nikiforo (7.14. GREEK)
allion to sholnoproeso (7.11. GREEK)
allion to skorodoprolso (7.12. GREEK)
allitamara (50.2. TELUGU)
alli-tamarai (50.2. TAMIL)
allo (7.10. GALICIAN)
allu-batchhalla (28.1. TELUGU)
almakdoonis zu el-gzoor (24.8. ARABIC)
almeirão (27.10a. PORTUGUESE)
almendelig kermesbaer (56.3. DANISH)
almindelig humle (46.1. DANISH)
almindelig sar (44.9. DANISH)
almindelig syre (58.2. DANISH)
almirón (27.10a. SPANISH)
almirón de España (27.36. SPANISH)
almondolig kermesbaer (56.1. DANISH)
almonia (59.2. HINDUSTANI)
almorta (40.10. SPANISH)
almyridi (35.7. GREEK)
alnɔmgo (40.25. LOKO)
àlọ̄ (11.3. YORUBA)
alo (11.7. YORUBA)
ālo (38.14. KATI)
ālo (38.16. KATI)
ālo (38.17. KATI)
ālo (38.18. KATI)
alo (62.24. URDU)
alo (63.1. MANDING-MANDINKA)
aloa (40.26. BAOULÉ)
alocasia (8.1. ENGLISH)
alocasia (8.1. HEBREW)
alodan (9.1. GUN)
alogbati (28.1. ILOCANO)
alokasye (8.1. TURKISH)
alokázie (8.1. CZECH)
alokazja (8.1. POLISH)
aloko (38.27. GUN)
alokushi (40.18. ORIYA)
àlọ̀lọ (11.7. YORUBA)
aloma (27.24. AJA)
aloma (27.24. ÉWÉ)
aloma (27.24. GBE, WACI)
aloma (27.24. TEM)
aloma kulu (27.24. TEM)
alomangi (20.6. SAMBAL, BOTOLAN)
alongoa (27.8. BAOULÉ)
alongoĩ (27.8. BAOULÉ)
a-lontho (45.1. THEMNE)
aloo (62.24. GUJARATI)
aloo yam (11.6. URDU)
alooguti (62.24. ASSAMESE)
alooka (8.4. SANSKRIT)
aloró (15.2. WÁRAY-WÁRAY)
aloso (40.26. IGBO)
alpạluạ (40.23. SANTALI)
alpamaareesha (23.3. SANSKRIT)
alpamarisha (23.3. SANSKRIT)
alpapramanaka (38.3. SANSKRIT)
àlpása (7.3. KORO)
alpinia (20.1. HEBREW)
alpinia galgant wielki (20.1. POLISH)
alquequenge del Perú (62.12. SPANISH)
alquequenge do Perú (62.12. PORTUGUESE)
alquequenje (62.11. PORTUGUESE)
alquequenje (62.11. SPANISH)
alsanda (40.9. TELUGU)
alsandalu (40.43. TELUGU)
alsem (27.7. DUTCH)
alt tohum (24.17. TURKISH)
alternantera (23.1. ITALIAN)
alternantera (23.1. PORTUGUESE)
alternantera (23.1. SPANISH)
alternantera sesil (23.2. SPANISH)
alternantera séssil (23.2. PORTUGUESE)
alternantera sessile (23.2. ITALIAN)
alternanthère (23.1. FRENCH)
alternanthère sessile (23.2. FRENCH)

altramuz (40.14. SPANISH)
alu (8.1. MARATHI)
alu (8.4. MARATHI)
alu (62.24. ASSAMESE)
alu (62.24. BALTI)
alu (62.24. BENGALI)
alu (62.24. BODO)
alu (62.24. FIJIAN)
alu (62.24. HINDI)
álú (62.24. HINDUSTANI)
ālū (62.24. KASHMIRI)
ālū (62.24. MAITHILI)
alu (62.24. MIZO)
ālu (62.24. MUNDARI)
alu (62.24. PANJABI, EASTERN)
âlu (62.24. PASHTO, NORTHERN)
ālū (62.24. PASHTO, NORTHERN)
ạlu (62.24. SANTALI)
alu falufa (40.17. JAPANESE)
alu gedde (62.24. KANNADA)
alu puhul (38.1. SINHALA)
alua ualé (11.1. BAOULÉ)
aluba (27.24. EDE NAGO)
aluba (27.24. YORUBA)
alubarha (7.3. EDO)
àlùbăsa (7.3. NUPE-NUPE-TAKO)
alubati (28.1. PAMPANGAN)
alubosa (7.2. EDE NAGO)
alubosa (7.2. YORUBA)
àlubọsà (7.3. YORUBA)
àlubọ́sà (7.10. YORUBA)
àlubọsà eléwé (7.2. YORUBA)
alubosa gambari (7.3. EDE NAGO)
àlubọsà gambari (7.3. YORUBA)
àlubọsà keta (7.3. YORUBA)
àlubọsà onịṣu (7.2. YORUBA)
alufa (8.4. BUHID)
alugadda (38.16. TELUGU)
alugbati (28.1. BUHID)
alugbati (28.1. CEBUANO)
alugbati (28.1. FILIPINO)
alugbati (28.1. TAGALOG)
alugbati (28.1. WÁRAY-WÁRAY)
alugedde (62.24. KANNADA)
alu-gidde (62.24. KANNADA)
alukamu (8.3. TELUGU)
alukha (11.1. BUHID)
aluloy (59.2. KANKANAEY)
alulu-isi (44.4. IGBO)
aluma (23.3. AMHARIC)
alumpan (20.2. HANUNOO)
alumpan budak (20.4. HANUNOO)
alune americane (40.1. ROMANIAN)
alupahi (8.3. BICOLANO, CENTRAL)
alusendi (40.43. TELUGU)
alusiis (24.17. BONTOC, CENTRAL)
alusiis (24.17. KANKANAEY)
alusiman (59.2. BICOLANO, CENTRAL)
alusiman (59.2. HANUNOO)
alusis (24.17. BONTOC CENTRAL)
aluyu (59.2. TAGBANWA)
alvi (8.4. KANNADA)
alvi (30.51. GUJARATI)
alzangabeel (20.6. ARABIC)
am (17.7. SUMO-MAYANGNA)
ama (17.7. SUMO-MAYANGNA)
ama (23.3. GBE-VHE)
ámàawà (17.5. NGIZIM)
amadaa (16.1a. GA)
amadumbi (8.4. AFRIKAANS)
amaesoppu (23.4. KANNADA)
amaido (40.1. LUO)
amãkãni (8.9. DANGME-KROBO)
amale (40.29. IBANAG)
amalek (40.29. ITAWIT)
amalet (40.29. ITAWIT)
amalore (5.1. KABIYÉ)
aman- ǹtùèn (62.2. EFIK)
amana (44.5. EFIK)
a-manawa (16.1a. THEMNE)
amandan (16.1a. GEN)
amande de terre (10.1. FRENCH)
amani (40.1. GADDANG)
amankani (8.9. AKAN-AKUAPEM)
amankani (8.9. AKAN-ASANTE)
amankani (8.9. DANGME)
amankani (8.9. ÉWÉ)
amankani (8.9. GA)
amaŋkani (8.9. GA)
amankani (8.9. THEMNE)
amankani (8.9. TWI)
amankani antwibo (8.9. AKAN-AKUAPEM)
amankani fita (8.9. AKAN-AKUAPEM)
amankani fufu (8.9. AKAN-AKUAPEM)
amankani hiɔ (8.9. DANGME-KROBO)
amankani kokoo (8.9. AKAN-AKUAPEM)
amankani pa (8.9. AKAN-AKUAPEM)
amankani tchu (8.9. DANGME-KROBO)
amankanidzẽ (8.9. DANGME-KROBO)
amara (19.1. TIV)

amaragoso (38.23. BICOLANO, CENTRAL)
amaragoso (38.23. TAGALOG)
amarant (23.3. DANISH)
amarant (23.3. DUTCH)
Amarant (23.3. GERMAN)
amarant (23.3. NORWEGIAN)
amarant (23.3. SWEDISH)
amarant (23.3. TAJIKI)
amarant (23.3. TURKISH)
amarant (23.3. VLAAMS)
amarante (23.3. FRENCH)
amarante blanche (23.3. FRENCH)
amarante blette (23.3. FRENCH)
amarante épineuse (23.3. FRENCH)
amarante sauvag (23.3. FRENCH)
amarante verte (23.3. FRENCH)
amaranthe (23.3. FRENCH)
amaranthos (23.3. GREEK)
amaranthus (23.3. AFRIKAANS)
amaranthus (23.3. ENGLISH)
amaranti (23.3. LATVIAN)
amaranto (23.3. ITALIAN)
amaranto branco (23.4. PORTUGUESE)
amarantos (23.3. PORTUGUESE)
amarantos (23.3. SPANISH)
amargón (27.10a. SPANISH)
amargóso (38.23. HILIGAYNON)
amaruku (40.31. LUYIA)
a-mas (11.3. THEMNE)
amas (16.1a. BINUKID)
amas (16.1a. MANOBO)
amas (16.1a. MARANAO)
amashindi (17.6. LUYIA)
amatayia kokhumwalo (17.8. LUYIA)
amavivɛ (27.24. FON)
amayabo (39.4. NYAKYUSA-NGONDE)
ambal (48.1a. TAMIL)
ambanam (16.1a. TAMIL)
ambaṛo (45.4. SANTALI)
ambasthaki (45.4. SANSKRIT)
ambavati (58.4. HINDI)
àmbɔ́ (16.1a. BAFUT)
ambe haldi (20.4. KANNADA)
ambérique (40.38. FRENCH)
am-bɛroŋ (8.9. THEMNE)
am-bethe (62.18. THEMNE)
ambhoja (48.1a. SANSKRIT)
ambhoruha (48.1a. SANSKRIT)
a-mbir (20.6. WAMEY)
a-mbir (55.1. WAMEY)
ambo (17.7. AFAR)
am-boŋk (11.2. THEMNE)
àmbótsu (16.1a. BAFUT)
ambouni (36.3. LUYIA)
ambouo-babié (23.3. TEKE)
ambrevad (40.2. MORISYEN)
ambrévade (40.2. FRENCH)
ambuj (48.1a. HINDI)
ambuja (48.1a. SANSKRIT)
ambujanma (48.1a. SANSKRIT)
ambupadma (48.1a. SANSKRIT)
ambur (27.19. MINANGKABAU)
amburoha (48.1a. SANSKRIT)
ambusaara (16.1a. SANSKRIT)
ambuvallika (38.23. SANSKRIT)
a-mɔthsəra (30.7. THEMNE)
amɛ osɛbɛ (62.9. AKAN-FANTE)
amɛ osɛbɛ (62.9. GA)
amek (30.17. BONTOC, CENTRAL)
amendoim (40.1. PORTUGUESE)
amenta (44.3. GALICIAN)
ameo (62.9. GA)
American basil (44.4. ENGLISH)
American black nightshade (62.15. ENGLISH)
america-neri (45.1. JAPANESE)
americká okurka (38.6. CZECH)
Amerikaanse karmozijnbes (56.2. DUTCH)
amerikabōfū (24.18. JAPANESE)
Amerikan şekerci (56.2. TURKISH)
Amerikanische Kermesbeere (56.2. GERMAN)
Amerikanische Winterkresse (30.2. GERMAN)
amerikansk kormesbaer (56.2. DANISH)
amfut (7.10. GADDANG)
amiling (62.5. IBANAG)
amilo ghans (58.2. NEPALI)
amisambɛɛ (7.3. YOM)
amittukaaq (24.11. INUKTITUT, EASTERN CANADIAN)
amizgél (40.1. CREOLE, HAITIAN FRENCH)
am-koe (30.13. CHINESE, MIN NAN)
amlana (48.1a. SANSKRIT)
amlardraka (11.2. SANSKRIT)
amlɔƒi-pui (59.3. DANGME)
amlɔtʃi-pui (59.3. DANGME)
amle malagsi (8.4. BUHID)
amle malayong (8.4. BUHID)
ammani (40.33. AFRIKAANS)
amnon vetamar (68.1. HEBREW)
amo (10.1. KABIYÉ)
amo (10.1. TEM)

apio blanco (24.4. SPANISH)
apio de tallo (24.4. SPANISH)
apio hortense (24.4. PORTUGUESE)
apio nabo (24.5. SPANISH)
apiroko (38.28. YORUBA)
apirrábano (24.5. SPANISH)
apisang (17.5. BONTOC, CENTRAL)
apoi-apoian (34.2. FILIPINO)
apoler (10.2. PANGASINAN)
apon (8.3. TAGALOG)
ápon (8.3. TAGALOG)
apoyapoyan (34.2. TAGALOG)
apoy-apoyan (34.2. TAGALOG)
appaku (5.1. IFUGAO)
appaku (5.1. KANKANAEY)
appamāriso (23.3. PALI)
appelliefie (62.12. AFRIKAANS)
appelmeloen (62.20. DUTCH)
apple chilli (62.6. ENGLISH)
ăpsāāsisk (7.3. MICMAC)
apteegitill (24.12. ESTONIAN)
apteegitill (24.13. ESTONIAN)
apugin (16.1a. ILOCANO)
apuk-apuk madapug (27.14. HANUNOO)
apuk-apuk taga-bas (27.18. HANUNOO)
apuk-ili mabiru (27.14. HANUNOO)
apular (10.2. PANGASINAN)
apulid (10.2. BICOLANO, CENTRAL)
apulid (10.2. FILIPINO)
apulid (10.2. TAGALOG)
apúlid (10.2. TAGALOG)
apulid (64.1. TAGALOG)
apulid (64.3. TAGALOG)
apulid-tsina (10.2. TAGALOG)
apung-apong (8.3. TAGALOG)
apyo (24.4. LADINO)
aqàwtumna (27.26. HOPI)
arä (38.16. ARABIC)
arä (38.18. ARABIC)
ārā sāngā (36.3. MUNDARI)
araalia (26.2. ESTONIAN)
aräb əxi (17.7. HARARI)
arabu fizale (62.12. LATVIAN)
arachid (40.1. POLISH)
arachida e ypogeios (40.1. GREEK)
arachide (40.1. FRENCH)
arachide (40.1. ITALIAN)
arachide de bambarra (40.39. FRENCH)
arachis (40.1. KAZAKH)
arachis e ypogeios (40.15. GREEK)
arad (40.37. GUJARATI)
arahhis (40.1. ESTONIAN)
arahida (40.1. ROMANIAN)
arahis (40.1. RUSSIAN)
arai keerai (23.3. TAMIL)
arairut (15.2. INDONESIAN)
aṛaḱ (23.3. SANTALI)
aṛaḱ ḱudrum (45.4. SANTALI)
arakas (40.28. GREEK)
àràkké dalimi (17.5. HAUSA)
aralia (26.2. NORWEGIAN)
aralia (26.2. PORTUGUESE)
aralia (26.2. SPANISH)
aralia (26.3. PORTUGUESE)
aralia (26.3. SPANISH)
aralia cordiforme (26.2. FRENCH)
aralia cordiforme (26.2. PORTUGUESE)
aralia elevata (26.3. ITALIAN)
aralia elevée (26.3. FRENCH)
aralia japonesa (26.3. PORTUGUESE)
aralia sercowata (26.2. POLISH)
arálie srdčitá (26.2. CZECH)
arálie vyšší (26.3. CZECH)
āram (30.41a. KASHMIRI)
arambai (16.1a. TAMIL)
aranyamudga (40.35. SANSKRIT)
aranybab (40.38. HUNGARIAN)
aranybojtorján (27.36. HUNGARIAN)
araptaşaği (13.5. TURKISH)
aráquida (40.1. SPANISH)
ararang (17.1. BUHID)
arardal (40.2. HINDI)
araro (15.4. TAGALOG)
araroet (15.2. AFRIKAANS)
ararot (9.1. TURKISH)
arārot (15.2. HINDI)
ararot (15.2. TURKISH)
araru' (15.2. FILIPINO)
araru (15.4. IBANAG)
araru (15.4. ITAWIT)
ararut (15.2. BENGALI)
ararut (15.2. GUJARATI)
ararut (15.2. KRIO)
ararut (15.2. MALAY)
araruta (15.2. PORTUGUESE)
araruta (15.2. QUECHUA, CUSCO)
araruta do Tahiti (19.1. PORTUGUESE)
araruttukkilangu (15.2. TAMIL)
àrás (45.4. NGIZIM)
arasanikaai (38.16. TAMIL)

arašid (40.1. AMHARIC)
arašidy (40.1. CZECH)
arata-yace (59.2. KRIO)
arata-yes (59.2. KRIO)
aratta (20.1. MALAYALAM)
aratthai (20.1. TAMIL)
aratthi (16.1a. TELUGU)
araut (15.2. INDONESIAN)
aravattavare (40.26. BADAGA)
aravindam (48.1a. MALAYALAM)
araxis (40.1. AZERBAIJANI, NORTHERN)
arayakkommai (16.1a. TAMIL)
arba kenafot (40.13. HEBREW)
arbaruttam (16.1a. TAMIL)
arbe à tomates (62.7. FRENCH)
arbhar milis (17.7. GAELIC, IRISH)
arbi (8.4. PANJABI, EASTERN)
árbol tomate (62.7. SPANISH)
arboloco (27.39. SPANISH)
arbouz (38.3. RUSSIAN)
arbouz stolovyj (38.3. RUSSIAN)
arbusto del diablo (26.4. SPANISH)
arbuus (38.3. ESTONIAN)
arbuusi (38.3. FINNISH)
arbuz (38.3. POLISH)
arbuz stolovyj (38.3. KAZAKH)
arbūzs (38.3. LATVIAN)
arcangelica (24.1. ITALIAN)
archaggeliki (24.1. GREEK)
arcydzięgiel (24.1. POLISH)
arcydzięgiel japoński (24.2. POLISH)
arde (35.3. FRENCH)
ardei (62.1. ROMANIAN)
ardei de salată (62.1. ROMANIAN)
ardei dulce (62.1. ROMANIAN)
ardei gras (62.1. ROMANIAN)
ardubam (20.1. TAMIL)
ář̃dùwà (40.26. NGIZIM)
arebe (40.43. TIV)
aremangnè (44.5. BAOULÉ)
a-rɛŋ (8.4. THEMNE)
arèn (40.1. SERER-SINE)
aren (40.1. WOLOF)
arène (40.1. SERER-SINE)
arer o ol (40.1. SERER-SINE)
arerut (15.2. INDONESIAN)
aresigam (16.1a. TAMIL)
aretegalu (11.2. TELUGU)
arethegalu (11.6. TELUGU)
areuj tiwuk (38.30. SUNDA)
argui (44.11. MONGOLIAN)
arhar (40.2. BENGALI)
arhar (40.2. GUJARATI)
arhar (40.2. HINDI)
arhar (40.2. HINDUSTANI)
arhar (40.2. MARATHI)
arhar (40.2. PANJABI, EASTERN)
arhar (40.2. URDU)
arhardal (40.2. HINDI)
aribukbuk (11.2. ILOCANO)
aricoma (27.39. SPANISH)
árìgízà (10.1. HAUSA)
arim (44.7. ASSAMESE)
arishina (20.2. KANNADA)
arishtha (7.10. SANSKRIT)
arisina (20.2. KANNADA)
arithige (11.6. TELUGU)
ariti chettu (16.1a. TELUGU)
arjaka (44.5. SANSKRIT)
arjakah (44.5. SANSKRIT)
arjoubat (59.2. ARABIC)
arjuno sosro (54.1. INDONESIAN)
arljak ohyknovennyi (2.1. RUSSIAN)
arljak zvanchine (2.1. BELARUSAN)
armoise (27.7. FRENCH)
armoles (35.1. PORTUGUESE)
armolls (35.1. CATALAN-VALENCIAN-BALEAR)
armon maim (10.2. HEBREW)
Armonian cucumber (38.10. ENGLISH)
armorácia (30.1. PORTUGUESE)
armoracie raifort (30.1. FRENCH)
armuelle (35.1. SPANISH)
armuelle de huerta (35.1. SPANISH)
armut (38.25. TURKISH)
arnarburkni (2.1. ICELANDIC)
arngalāje (7.10. FULFULDE, ADAMAWA)
arngalāre (7.10. FULFULDE, ADAMAWA)
arngalore (7.10. FULFULDE, ADAMAWA)
aro di Egitto (8.4. ITALIAN)
arogbati (28.1. BICOLANO, CENTRAL)
arokōch (8.4. MAITHILI)
aronille (8.4. FRENCH)
aroru (15.2. TAGALOG)
aroru (15.4. TAGALOG)
aroruut (15.2. ESTONIAN)
arouil kari (8.9. MORISYEN)
arpagic (7.11. ROMANIAN)
arracacha (24.6. ENGLISH)
arracacha (24.6. FRENCH)

arracacha (24.6. SPANISH)
arracacha (24.6. VLAAMS)
arracacha mandioquinha (24.6. PORTUGUESE)
arracacharot (24.6. NORWEGIAN)
arragon (27.6. ENGLISH)
Arrakatascha (24.6. GERMAN)
arrarot (19.1. URDU)
arrarouti anatolir kon indion (19.1. GREEK)
arraturai (38.21. HINDI)
arroche (35.1. FRENCH)
arroche cultivée (35.1. FRENCH)
arroche des jardins (35.1. FRENCH)
arroche épinard (35.1. FRENCH)
arrou-rout (15.2. MORISYEN)
arrow rot (15.2. SWEDISH)
arrowhead (6.1. ENGLISH)
arrowhead (6.2. ENGLISH)
arrow-leaved monochoria (18.1. ENGLISH)
arrowroot (15.2. ENGLISH)
arrowroot de Tahiti (19.1. FRENCH)
arrowroot des Antilles (15.2. FRENCH)
arrowroot polynésien (19.1. FRENCH)
arrow-weed (6.1. ENGLISH)
arroz bravo (17.8. PORTUGUESE)
arroz míudo do Perú (35.6. PORTUGUESE)
arroz silvestre (17.8. SPANISH)
arrurruz arundinácc (15.2. SPANISH)
arsaghna (8.3. SANSKRIT)
arshabi (40.18. SANSKRIT)
aršlama (7.5. SERBO-CROATIAN)
arslandisi (27.45. TURKISH)
ārstniecības gurķene (29.1. LATVIAN)
ārstniecības ingvēras (20.6. LATVIAN)
arstniecības pienene (27.45. LATVIAN)
ārstniecības tauksakne (29.2. LATVIAN)
ärt (40.28. SWEDISH)
arta ijo (40.38. INDONESIAN)
artemísia (27.7. PORTUGUESE)
artemisia e drakontios (27.6. GREEK)
arthapal (62.24. SINHALA)
artichaut (27.16. FRENCH)
artichaut commun (27.16. FRENCH)
artichaut de Jérusalem (27.26. FRENCH)
articho (27.16. LADINO)
artichoke (27.16. ENGLISH)
artichoke gedde (27.26. KANNADA)
artichuk (27.16. YIDDISH)
articiocche (27.16. ITALIAN)
artičoka (27.16. BOSNIAN)
articoka (27.16. CROATIAN)
artičoka (27.16. SERBO-CROATIAN)
artičoka (27.16. SLOVAK)
artičoka (27.16. SLOVENIAN)
articokks (27.26. MALTESE)
árticsóka (27.16. HUNGARIAN)
Artischoke (27.16. GERMAN)
artishok (27.16. HEBREW)
artišhok (27.16. RUSSIAN)
artishok yerushalmy (27.26. HEBREW)
artisjok (27.16. AFRIKAANS)
artisjok (27.16. DUTCH)
artisjok (27.16. VLAAMS)
artiskok (27.16. DANISH)
artiskokk (27.16. NORWEGIAN)
artiso (27.16. MORISYEN)
artišok (27.16. BULGARIAN)
artišok (27.16. UKRAINIAN)
artišoka (27.16. SERBO-CROATIAN)
artišokk (27.16. ESTONIAN)
artisokka (27.16. FINNISH)
artišoks (27.16. LATVIAN)
artisyok (27.16. INDONESIAN)
artisyok jahudi (27.26. INDONESIAN)
artthiragam (20.6. TAMIL)
artyčok (27.16. CZECH)
artyčok zeleninový (27.16. CZECH)
artyszok (27.16. POLISH)
ārū (38.13a. KURDISH, NORTHERN)
aru-aru (15.2. TAGALOG)
aru-aru (15.4. TAGBANWA)
árùbásā (7.3. AKPA)
arubī̃ (8.4. MAITHILI)
arugrattam (40.18. TAMIL)
arugula (30.48. ENGLISH)
arui (8.4. HINDUSTANI)
aruna (20.1. SANSKRIT)
arunfanfan (11.1. YORUBA)
aruru (15.4. TAGALOG)
arurruttukizhangu (15.2. TAMIL)
arus (15.2. JAVANESE)
aruuru (15.2. CAROLINIAN)
árvácska (68.1. HUNGARIAN)
arveja (40.28. SPANISH)
arveja tuberosa (40.11. SPANISH)
arvejo (40.10. SPANISH)
arvi (8.4. HINDI)
arvi (8.4. PANJABI, EASTERN)
arvinda (48.1a. SANSKRIT)
as (39.4. KANKANAEY)

aš kok (24.19. KIRGHIZ)
aš kyzylčasy (35.4. KIRGHIZ)
asa ɛdua (40.25. TWI)
asaaliya (30.51. MARATHI)
asaare simi (40.26. NEPALI)
asalio (30.51. GUJARATI)
asam susur (45.4. MALAY)
asangokan (38.22. FON)
asatsuki (7.11. JAPANESE)
asawa' bagwûnik (27.45. POTAWATOMI)
asbin (40.29. TOK PISIN)
[as]c ti s[oo] (27.16. VIETNAMESE)
ascalonia (7.2. SPANISH)
ascalunia (7.2. CATALAN-VALENCIAN-BALEAR)
asɔvle (8.4. ÉWÉ)
ase (40.43. TWI)
ase (44.10. TEM)
ase dua (40.25. TWI)
ase kɔkɔ (40.43. TWI)
aseboko (40.1. IGBO)
asedua (40.43. TWI)
asekɔkɔ (40.43. TWI)
ase-kougbaeta (44.10. KABIYÉ)
a-senduku (23.3. THEMNE)
asete alamoa (11.1. DANGME-KROBO)
ash gourd (38.1. ENGLISH)
ashalika (30.51. SANSKRIT)
ashitaba (24.2. DANISH)
ashitaba (24.2. JAPANESE)
ashköki (24.19. UYGHUR)
ashktamo (38.3. POTAWATOMI)
ashoghna (8.3. SANSKRIT)
āshom (17.7. TIBETAN)
ashukachu (8.4. HINDI)
ashwe (45.4. TIV)
ashwe (59.3. TIV)
Asian nightshade (62.25. ENGLISH)
Asian water chestnut (64.1. ENGLISH)
asiatic pennywort (24.7. ENGLISH)
Asiatic yam (11.1. ENGLISH)
Asiatic yam (11.5. ENGLISH)
asiatisk centella (24.7. DANISH)
asibe (23.3. AKAN-AKUAPEM)
asiê ounsa (20.6. BAOULÉ)
Asie-agurk (38.13a. DANISH)
asieagurk (38.13a. NORWEGIAN)
asilang (36.3. GADDANG)
asin-asin (39.5. MALAY)
asiparagasi (40.13. SWAHILI)
asirak (60.1. TAJIKI)
*a*sïsáng (17.5. BAFUT)
asishimbi (40.3. SANSKRIT)
asishimbi (40.4. SANSKRIT)
asithothpala (50.2. SANSKRIT)
aškabak (38.18. KIRGHIZ)
askalon (7.2. TURKISH)
askipwâw (62.24. CREE, PLAINS)
askol debrin (27.16. BRETON)
aslottahagyma (7.2. HUNGARIAN)
aslrai (30.24. HINDI)
aslynnik dvuhgadov (51.1. BELARUSAN)
asma yaprakli mucize elmasi (38.23. TURKISH)
asmacius (24.3. ROMANIAN)
asnagh (38.7a. GAELIC, MANX)
aso bayerɛ (11.7. AKAN-AKUAPEM)
asogam (16.1a. TAMIL)
asokolo (11.7. IGBO)
asomiya paleng (59.3. ASSAMESE)
asomiya piyaj (7.2. ASSAMESE)
asot agorodnaja (27.42. BELARUSAN)
aspar (13.2. ESTONIAN)
aspar (40.13. ESTONIAN)
asparag (13.2. GAELIC, MANX)
asparag mehudad (13.1. HEBREW)
asparag refui (13.2. HEBREW)
asparag refuyi (13.2. HEBREW)
asparagasi (13.2. SWAHILI)
asparageus (13.2. SUNDA)
asparago (13.2. ITALIAN)
asparago (13.2. TAGALOG)
asparago delle Cabo (25.1. ITALIAN)
asparago pungente (13.1. ITALIAN)
asparagos (13.2. MARANAO)
asparagus (13.2. ENGLISH)
asparagus (13.2. FILIPINO)
asparagus (13.2. INDONESIAN)
asparagus (13.2. MALAY)
asparagus (13.2. TAGALOG)
asparagus (13.2. URDU)
asparagus bean (40.29. ENGLISH)
asparagus bean (40.42. ENGLISH)
asparagus lettuce (27.30. ENGLISH)
asparagus pea (40.13. ENGLISH)
asparagus pea (40.29. ENGLISH)
asparagus refuee (13.2. HEBREW)
asparges (13.1. NORWEGIAN)
asparges (13.2. DANISH)
asparges (13.2. NORWEGIAN)

aspargesært (40.13. DANISH)
aspargesbønne (40.42. DANISH)
aspargeskål (30.31. NORWEGIAN)
aspargo (13.2. PORTUGUESE)
aspargus barry (13.1. ARABIC)
asparkapsas (30.31. ESTONIAN)
aspèj (13.2. CREOLE, HAITIAN FRENCH)
aspèl (13.2. CREOLE, HAITIAN FRENCH)
asperge (13.2. DUTCH)
asperge (13.2. FRENCH)
asperge (13.2. VLAAMS)
asperge du Cap (25.1. FRENCH)
asperge sauvage (13.1. FRENCH)
aspergesalat (27.30. DANISH)
aspergesalat (27.30. DUTCH)
aspergez (13.2. BRETON)
aspersie (13.2. AFRIKAANS)
asperz (13.2. MORISYEN)
asperža (13.2. BULGARIAN)
aspeset (40.17. HEBREW)
aspeset tarbutit (40.17. HEBREW)
aspro lachano (30.27a. GREEK)
asselim (40.1. BASSARI)
assenzio (27.7. ITALIAN)
assorossie (38.23. FRENCH)
assu (30.48. PANJABI, EASTERN)
aster (27.17. ASSAMESE)
aster (27.17. DANISH)
aster (27.17. DUTCH)
aster (27.17. ENGLISH)
aster (27.17. ESTONIAN)
aster (27.17. FRENCH)
Aster (27.17. GERMAN)
aster (27.17. HEBREW)
aster (27.17. PORTUGUESE)
aster (27.17. SPANISH)
aster chiński (27.17. POLISH)
aster jomena (27.28. POLISH)
aster saraypati (27.17. TURKISH)
asteres (27.17. LATVIAN)
asters (27.17. NORWEGIAN)
astir (27.17. GREEK)
astotzar (30.51. BASQUE)
astro (27.17. ITALIAN)
asukar (17.5. WÁRAY-WÁRAY)
asuku (11.7. IGBO)
asukusɛ (62.21. GBE, WACI)
asumodhagam (40.32. SANSKRIT)
asuparagasu (13.2. JAPANESE)
asurasa (44.5. SANSKRIT)
asuri (30.17. SANSKRIT)
asuri (30.24. SANSKRIT)
asyapathra (48.1a. SANSKRIT)
aszat (27.11. HUNGARIAN)
aszuki fasulyesi (40.36. TURKISH)
at kuyruğu (1.1. TURKISH)
at turpu (30.1. TURKISH)
ata (62.2. YORUBA)
ata afunjalo (62.2. YORUBA)
ata didun (62.2. YORUBA)
ata ẹyẹ (62.5. YORUBA)
ata funfun (62.2. YORUBA)
ata gbasẹjọ (62.5. YORUBA)
ata jijɛ (62.1. EDE NAGO)
ata jijɛ (62.1. YORUBA)
ata korukọ (62.5. YORUBA)
àta ḿkpọ̀ñ (8.4. EFIK)
àta ḿkpọ̀ñ mbàkárá (8.9. EFIK)
ata nlá (62.5. YORUBA)
ata pupa (62.5. YORUBA)
ata rrer (20.6. TEM)
ata sanbaye (62.5. YORUBA)
ata sísèbè (62.5. YORUBA)
ata sombo (62.2. YORUBA)
ata tatase (62.2. YORUBA)
ata wẹ̀wẹ̀ (62.5. YORUBA)
ataco-casha (23.3. QUECHUA, CUSCO)
atad seenee (62.8. HEBREW)
atad sini (62.8. HEBREW)
atadi (62.2. ÉWÉ)
atadjô (10.1. BAOULÉ)
atadwe (10.1. AKAN-AKUAPEM)
atadwe awehea (10.1. AKAN-AKUAPEM)
atadwe fufuu (10.1. AKAN-AKUAPEM)
atadwe hene (10.1. AKAN-AKUAPEM)
atadwe tsintsin (10.1. AKAN-AKUAPEM)
atadwe tuntum (10.1. AKAN-AKUAPEM)
atajwe (10.1. TWI)
atakin (62.5. FON)
atalẹ̀ (20.6. YORUBA)
a-tamba (62.9. THEMNE)
atang (8.4. ITAWIT)
ataŋme (10.1. GA)
'atär (40.5. HARARI)
atär (40.26. HARARI)
atär (40.28. AMHARIC)
atärä (40.26. HARARI)
atarodo (62.1. YORUBA)
atarodo (62.2. YORUBA)
atarraga (20.6. TEM)

at-atang (30.37. IFUGAO)
atawewe (62.5. YORUBA)
atbo (17.5. PAMPANGAN)
atbu (17.5. PAMPANGAN)
a-tɛ̀mb-tɛ̀mb (40.38. BASSARI)
atepoelen (40.28. STELLINGWERFS)
ateş fasulyesi (40.24. TURKISH)
athaki (40.2. KANNADA)
a-thamba (62.9. THEMNE)
athitheekshna (30.24. SANSKRIT)
athi-tu (17.5. NEWAR)
athyaga (11.6. TELUGU)
atii (40.2. GA)
a-tiíti *ɓi* yil (19.1. BASSARI)
atimesia (27.6. SWAHILI)
atimon (38.7. CEBUANO)
atimouinna (40.39. GOURMANTCHÉ)
a-tina (40.1. GOURMANCHÉMA)
atingkamas (40.21. KANKANAEY)
atiu (38.6. NIUE)
atiu (38.7b. MANU'A SAMOAN)
'atiu (38.11. MARQUESAN)
'atiu (38.11. SAMOAN)
'atiu (38.11. TONGAN)
atiu hega (38.6. NIUE)
atiu vao (38.6. SAMOAN)
atiyi (40.2. GBE-VHE)
atjémankou (62.5. BAOULÉ)
atmagoso (38.23. CAROLINIAN)
atmagupta (40.18. SANSKRIT)
ato̱falaa' (7.3. CHICKASAW)
ato̱falaa' himona' (7.2. CHICKASAW)
ato̱falla'a' (7.3. CHICKASAW)
ato̱falla'a' homma' (7.3. CHICKASAW)
ato̱falla'a' lakna' (7.3. CHICKASAW)
ato̱falla'a' tohbi' (7.3. CHICKASAW)
ato̱falla'ha'a (7.3. CHICKASAW)
atọka (11.4. IGBO)
atomo (36.3. DANGME-KROBO)
atomo (36.3. GA)
atómò (36.3. GA)
aton (40.25. YORUBA)
atona (45.4. LOKO)
atong (38.17. KALINGA, BUTBUT)
atorókpò (62.19. GA)
atoropo (62.18. GA)
atoropo (62.18. TWI)
atoropo (62.19. AKAN-ASANTE)
atoropo (62.19. TWI)
atouna (40.43. GOURMANTCHÉ)
atowa (17.7. ZUNI)
atramùz (40.14. LADINO)
atriplice (35.1. ITALIAN)
atrokoba (62.18. GA)
atsaka (36.3. TIV)
a̱tsaka (36.3. TIV)
atsal (62.5. BUTUANON)
atsẽtsẽa (38.3. ÉWÉ)
atthiradam (20.6. TAMIL)
atthumam (20.1. TAMIL)
atthumbutthar (40.18. TAMIL)
atti (11.1. YORUBA)
atu (8.4. BULU)
atujapi (20.6. MALAY)
atuklwi (9.1. GEN)
átùmáátùr (62.9. HAUSA)
atung gula (7.2. LUO)
atung gula (7.3. LUO)
atung gula (7.5. LUO)
atung gula (7.10. LUO)
atung gula (7.11. LUO)
atur (45.1. TIV)
aturopo (62.18. TWI)
aturopo (62.19. TWI)
atuu̯r (45.1. TIV)
a-tyẹlis (38.18. BASSARI)
atyever (63.1. TIV)
a-tyityin (40.18. WAMEY)
a-tywnɛ́ngá (45.1. BASSARI)
'au tasi (23.3. SAMOAN)
au zuri (13.3. SWAHILI)
aua (8.4. ILOCANO)
aubergin (62.19. SWEDISH)
aubergine (62.19. DANISH)
aubergine (62.19. DUTCH)
aubergine (62.19. ENGLISH)
aubergine (62.19. FRENCH)
Aubergine (62.19. GERMAN)
aubergine (62.19. VLAAMS)
aubergine africaine (62.14b. FRENCH)
aubergine de Siam à fruit hirsute (62.16. FRENCH)
aubergine du diable (62.23. FRENCH)
aubergine gboma (62.18. FRENCH)
aubergine indigéne (62.18. FRENCH)
aubergine moucheté (62.14c. FRENCH)
aubergine tomate (62.14c. FRENCH)
auberginer (62.19. NORWEGIAN)
auḍe (40.41. TULU)
Augenbohne (40.43. GERMAN)

augstā aralija (26.3. LATVIAN)
augurk (38.13a. DUTCH)
augurk (38.13a. STELLINGWERFS)
augurk (38.13a. VLAAMS)
augurkien (38.13a. STELLINGWERFS)
aula (8.3. HINDI)
aunazirni (40.5. LATVIAN)
aupaluktârsuk (24.11. ALEUT)
aurego (44.6. GALICIAN)
auri (8.4. HINDI)
ausiman (59.2. BICOLANO, CENTRAL)
Australischer Wein (69.1. GERMAN)
austrumu dižpērkone (30.42. LATVIAN)
autasi (23.3. MANU'A SAMOAN)
ava galangal (20.1. ENGLISH)
'ava pui (20.6. SAMOAN)
avagutha (62.12. SANSKRIT)
avakapushpi (24.13. SANSKRIT)
avaloû-douar (62.24. BRETON)
avalu (30.24. TELUGU)
avar bit (35.4. MALAY)
avarai (40.9. TAMIL)
avarai (40.26. TAMIL)
avarai-motchai (40.9. TAMIL)
avare (40.9. BADAGA)
avare (40.9. KANNADA)
avare baele (40.9. KANNADA)
avare balli (40.9. KANNADA)
avare kaalu (40.9. KANNADA)
avarekaayi (40.9. KANNADA)
avas (40.26. LADINO)
ávas frèskas (40.34a. LADINO)
avatengatige (11.2. TELUGU)
avathengathige (11.6. TELUGU)
avatiach (38.3. HEBREW)
avatiah (38.3. HEBREW)
avaw´tuh kaw´menuv (27.5. PAIUTE, SOUTHERN)
avelik (58.2. AZERBAIJANI, NORTHERN)
aveluk (58.2. ARMENIAN)
avenya (27.24. ÉWÉ)
avete (11.3. ÉWÉ)
avigedde (8.4. KANNADA)
avlakoto kolokythi (38.27. GREEK)
avois (8.4. HINDI)
avşanotu (44.2. TURKISH)
avunvɔ (23.4. GEN)
avyssiniakos sinapis (30.3. GREEK)
aw¯ k'a_ naw^ pa_ shi_ (40.9. LAHU)
awa (8.4. ILOCANO)
awa (38.20. BAOULÉ)
awa (62.15. IGBO)
awá yékú (59.2. YORUBA)
awâri (17.7. FARSI, EASTERN)
awarwaro (36.2. HAUSA)
a-wata-yams (11.1. THEMNE)
awayan an pallotan (17.1. IFUGAO)
awɔrɔke (59.2. GA)
awi (11.1. MARANAO)
awi a kakowak (11.1. MARANAO)
awi bitung (17.2. SUNDA)
awi iwung (17.2. SUNDA)
awo (17.3. TIV)
awon du (40.3. TIV)
awonẽwẽ (27.24. AKAN-AKUAPEM)
awonywẽne (27.24. AKAN-AKUAPEM)
awrɔké (59.2. GA)
awrɔke (59.3. GA)
awsiman (59.2. BICOLANO, CENTRAL)
awt (62.12. AMHARIC)
awt (62.21. AMHARIC)
awu miris (62.2. SINHALA)
awudo (11.7. IGBO)
awujɛ (40.9. EDE NAGO)
awujɛ (40.9. YORUBA)
awújẹ (40.9. YORUBA)
awuje (40.25. YORUBA)
awújẹ (40.25. YORUBA)
awumɛ (34.2. ÉWÉ)
awure (11.7. YORUBA)
awusa pɛtɛtɛ (44.10. KRIO)
awusaŋaei (23.3. GA)
awwat (40.42. IFUGAO)
awwat an gayyang (40.42. IFUGAO)
awwat an luyu (40.26. IFUGAO)
awwat an mumbolah (40.26. IFUGAO)
awwat an ubbun (40.42. IFUGAO)
awwat an umadukke (40.42. IFUGAO)
awwət (62.12. AMHARIC)
aya (7.10. GA)
aya (23.3. SAMBAL, BOTOLAN)
aya mantug (23.3. SAMBAL, BOTOLAN)
ayáá (10.1. HAUSA)
áyaá (10.1. HAUSA)
áyaá rigiza (10.1. HAUSA)
àyaà-áyaà (10.1. HAUSA)
ayaare (10.1. FULFULDE, ADAMAWA)
ayaba (16.1a. HAUSA)
àyàbà (16.1a. HAUSA)
ayaba (16.1a. KONKOMBA)

ayaba (16.1a. SANGA)
a̲yaba (16.1a. TIV)
a̲ya̲ba (16.1a. TIV)
àyàbaasì (7.3. IGBO)
ayabaje (16.1a. FULFULDE, ADAMAWA)
àyàbàr dáaji (16.1a. HAUSA)
àyàbàr tuùraàwaá (16.1a. HAUSA)
aya-kabayo (23.3. SAMBAL, BOTOLAN)
a-yams-*a*-fɛra (11.7. THEMNE)
ayantoto (23.3. PAMPANGAN)
ayàrà (62.18. IGBO)
àyàwà (16.1a. KANURI)
àyàyá (38.20. EFIK)
aye (7.10. MÒORÉ)
ayebir (11.7. AKAN-AKUAPEM)
ayebir (11.7. AKAN-FANTE)
ayi (40.26. ÉWÉ)
ayi (40.43. GBE-VHE)
ayikpɛn (38.16. FON)
ayikpɛn (38.18. FON)
ayikun (40.43. FON)
ayin (11.7. YORUBA)
a-yiŋkəres (38.18. THEMNE)
ayivi (40.3. GEN)
aynatoto (23.3. PAMPANGAN)
áyo (7.3. ABUA)
ayo (7.3. IGBO)
ayo (7.10. ÉWÉ)
ayo (7.10. FON)
ayo (7.10. GA)
ayo (7.10. GBE, WACI)
ayo (7.10. GEN)
ayo (7.10. GUN)
àyo (7.10. IGBO)
ayo (63.1. YORUBA)
ayoka (39.4. BAGA)
a-yokâ (39.4. THEMNE)
a-yoka abi (39.4. THEMNE)
a-yoka-*a*fera (39.4. THEMNE)
a-yoka-*a*-kono (39.4. THEMNE)
a-yoka-*a*-yim (39.4. THEMNE)
ayomasa (7.3. FON)
ayón (7.10. IGBO)
ayón (7.10. YORUBA)
ayote (38.14. DANISH)
ayoto (27.33. ÉWÉ)
Ayrish-pɛtɛtɛ (62.24. KRIO)
ayu (7.10. EDE NAGO)
ayu (7.10. YORUBA)
ayun (7.3. SRANAN)
ayún (7.10. IGBO)
ayunca (38.16. QUECHUA, CUSCO)
ayuskus (8.4. BUHID)
azáfran (17.5. SUMO-MAYANGNA)
azáfran (20.2. SUMO-MAYANGNA)
azafrán de la India (20.2. SPANISH)
azalandare (30.35. BASQUE)
azalili (30.26. BASQUE)
azargana (27.11. MONGOLIAN)
azɔ (40.18. FON)
azɔkwin (43.1. FON)
azeburu (30.27a. BASQUE)
azeda brava (58.2. PORTUGUESE)
azeda espinafre (58.3. PORTUGUESE)
azeda romana (58.4. PORTUGUESE)
azeda-crespa (58.3. PORTUGUESE)
azedas (58.2. PORTUGUESE)
azedinha (53.1. PORTUGUESE)
azedinha (58.2. PORTUGUESE)
azedinha (58.4. PORTUGUESE)
azedinha da horta (58.2. PORTUGUESE)
azeftenni (44.5. ARABIC)
azi (40.1. GBE-VHE)
azi (40.15. ÉWÉ)
azi nu dowowowo (40.1. GBE-VHE)
aziin (40.1. FON)
azin (40.1. GBE, WACI)
azin (40.1. GEN)
azindgoe (40.39. ÉWÉ)
azjatycka pępownica (24.7. POLISH)
azovit (44.6. HEBREW)
azucena (13.3. SPANISH)
azucena (13.4. SPANISH)
azuki (40.36. JAPANESE)
ba (38.20. KONO)
'ba (43.1. SERER-SINE)
ba biloun (8.4. TOK PISIN)
ba cai (27.15. VIETNAMESE)
bã da kũ (39.4. MÒORÉ)
ba jiao gua (38.18. CHINESE, MIN BEI)
ba jiao ke (16.1a. CHINESE, WÚ)
ba'-kobi (30.26. MUNDARI)
ba lal saga (23.3. ORIYA)
ba limba (39.4. MALIMBA)
ba luo men shen (27.46. CHINESE, WÚ)
ba ra monjin (27.46. JAPANESE)
ba rui shang lu (56.4. CHINESE, YUE)
ba ta tah helua (36.3. ARABIC)
ba tian suan mo (58.3. CHINESE, MANDARIN)

banjhā (30.27a. MAITHILI)
bankanafi (43.1. WOLOF)
bankanas (43.1. FRENCH)
bânkanas (43.1. WOLOF)
bankanase (43.1. MANDING-MANDINKA)
bankanase (43.1. PULAAR)
bankanase (43.1. WOLOF)
bankanazé (43.1. MANDING-BAMBARA)
bankanesi (43.1. MANDING-MANDINKA)
bankanesi (43.1. PULAAR)
bankanesi (43.1. WOLOF)
bankì (39.4. AKAN-ASANTE)
ba-nkuli (36.3. HAUSA)
bankyi (39.4. TWI)
bannann (16.1a. KWÉYÒL)
bannann fran (16.1a. CREOLE, HAITIAN FRENCH)
banor (40.42. TAGALOG)
banpat (63.1. BENGALI)
banphal (63.1. PANJABI, EASTERN)
bans (17.1. BENGALI)
bans (17.1. HINDI)
báṉs (17.1. HINDUSTANI)
bāns (17.1. HINDUSTANI)
bāns (17.2. HINDI)
báṉs (17.2. HINDUSTANI)
bāns (17.4. HINDI)
báṉs (17.4. HINDUSTANI)
bans ki shakhan (17.1. URDU)
bans ki shakhan (17.2. URDU)
bans ki shakhan (17.4. URDU)
bans saki sarsoo (30.5. URDU)
bans tama (17.1. NEPALI)
bans tama (17.2. NEPALI)
bansa (17.1. URDU)
bānsdo (17.1. MARWARI)
bānsdo (17.2. MARWARI)
bānsdo (17.4. MARWARI)
bãnta (17.5. WOLOF)
banta nhanhambo (39.4. MANDING-MANDINKA)
banta nyambo (39.4. MANDING-MANDINKA)
banta nyambo koyo (39.4. MANDING-MANDINKA)
bantāde (39.4. PULAAR)
bantaŋga (39.4. LOKO)
banta-niambo (39.4. MANINKAKAN, WESTERN)
banta-nyambo (39.4. MANDING-MANDINKA)
bantára (39.4. BUDIK)
bantara (39.4. PULAAR)
bántara (39.4. PULAAR)
bãntara (39.4. PULAAR)
bantara (39.4. PULAR)
bantara (39.4. WAMEY)
bãntara (39.4. WAMEY)
bantara bururé (27.24. PULAAR)
banthulasi (44.4. HINDI)
bantu u sukar (17.5. WOLOF)
banyenye (62.23. LOKO)
bao cai (30.27a. CHINESE, GÀN)
bao cai (30.27a. CHINESE, MIN BEI)
bao cai (30.27a. CHINESE, WÚ)
bao gu (17.7. CHINESE, XIANG)
bao hua gai (30.11. CHINESE, MANDARIN)
bao ta cai (44.11. CHINESE, WÚ)
bao xin cai (30.40. CHINESE, MIN BEI)
bao xin jie (30.4. CHINESE, YUE)
bao zi gai (30.8. CHINESE, MANDARIN)
bao zi gan lan (30.29. CHINESE, MANDARIN)
bao zi gan lan (30.29. CHINESE, WÚ)
bao zi jie (30.8. CHINESE, WÚ)
baoan (23.3. BONTOC CENTRAL)
bapatala ghant (62.9. ORIYA)
bâqêla (40.26. SIDAMO)
baqlajon (62.19. UZBEK)
bäqqollo (17.7. AMHARIC)
bâqula (38.14. SIDAMO)
bâqula (38.16. SIDAMO)
bâqula (38.17. SIDAMO)
bâqula (38.18. SIDAMO)
bar ta ta tafi (40.3. HAUSA)
bara (19.1. WOLOF)
bara (38.3. KRIO)
bàrà (38.3. YORUBA)
bara (38.20. BAMANANKAN)
bara bananku (39.4. MANDING-MANDINKA)
bara lasan (7.12. URDU)
bara matar (40.28. BENGALI)
bara sem (40.3. HINDI)
bara sem (40.3. URDU)
bara sem (40.4. HINDI)
bara sem (40.4. URDU)
bara sir wala lasan (7.1. URDU)
barãʔ (17.1. LIMBA)
baraboi (24.8. ROMANIAN)

barãde (16.1a. MÒORÉ)
bara-ɛgusi (38.3. KRIO)
baragaalada gedde (39.4. KANNADA)
bara-gad (52.1. BRETON)
baragudi (40.43. ORIYA)
barah murai (30.1. FIJIAN)
barahi kund (11.6. URDU)
barajaur (17.7. PANJABI, EASTERN)
barajowar (17.7. HINDI)
barajuar (17.7. HINDI)
barak (20.4. TAGALOG)
barakatz (7.10. BASQUE)
barakue (45.2. BUHID)
barakulanjan (20.1. HINDI)
barakulanjar (20.1. BENGALI)
barakulanjar (20.1. HINDI)
baralaniya (59.2. HINDI)
baralaniya (59.4. HINDI)
baraloniya (59.2. BENGALI)
baraloniya (59.4. BENGALI)
baralunia (59.2. HINDI)
baramarecha (40.3. MALAYALAM)
baramareha (40.4. MALAYALAM)
baramasiya (38.23. BENGALI)
baran (27.45. HINDI)
baran (27.45. PANJABI, EASTERN)
bärängge (62.24. TATAR)
barangkinit (62.19. ITAWIT)
barangse (11.1. BUHID)
baranii garoh (40.5. RUSSIAN)
ɓáránkácíí (40.3. HAUSA)
barar (34.1. PANJABI, EASTERN)
barasem (40.3. HINDI)
barasem (40.4. HINDI)
barata (14.1. PORTUGUESE)
barau (27.45. HINDI)
baray malagti (40.9. TAGALOG)
barba (44.5. SANSKRIT)
barba bietola d'insalata (35.4. ITALIAN)
barba cabreira (27.46. GALICIAN)
barba cabruna (27.47. SPANISH)
barba caprei (27.46. ROMANIAN)
barba caprel (27.47. ROMANIAN)
barba de bode (27.46. PORTUGUESE)
barba de bode (27.47. PORTUGUESE)
barba de cabra (27.46. CATALAN-VALENCIAN-BALEAR)
barba di becco (27.46. ITALIAN)
barba di becco nera (27.38. ITALIAN)
barba di bêcco (27.47. ITALIAN)
barba di prete (27.47. ITALIAN)
barba silvuna (6.2. ITALIAN)
barbadoslykt (62.12. NORWEGIAN)
barbafonte (30.1. ITALIAN)
barbar (44.5. HINDI)
barbara (44.5. SANSKRIT)
barbarè (62.2. SIDAMO)
barbarée printanière (30.2. FRENCH)
barbari (44.5. SANSKRIT)
barbati (40.43. BENGALI)
barbunyas (40.26. LADINO)
barda (40.28. GEORGIAN)
bardan (27.5. GALICIAN)
bardana (27.5. PORTUGUESE)
bardana maggiore (27.5. ITALIAN)
bardana maior (27.5. PORTUGUESE)
bardana mayor (27.5. SPANISH)
bardane (27.5. FRENCH)
bardane géante (27.5. FRENCH)
bardane majeure (27.5. FRENCH)
bare (8.4. SUSU)
bārè (16.1a. TULU)
bãré (38.27. MÒORÉ)
bargi zulf (57.1. TAJIKI)
barhang (16.1a. FARSI, WESTERN)
bari (23.3. BEMBÉ)
bari (23.3. KOONGO)
bari (8.4. SUSU)
bari ajmod (24.4. HINDI)
baribaran (11.5. BICOLANO, CENTRAL)
barida souwi (16.1a. SENOUFO, CEBAARA)
barikiri (44.5. SUSU)
barilla plant (35.7. ENGLISH)
baringan (38.20. BUHID)
barisaunf (24.13. HINDI)
barkono (62.2. HAUSA)
barkono (62.5. HAUSA)
bàrkònoo (62.5. HAUSA)
barmaasàlla (17.7. AFAR)
ɓarnar (40.25. HAUSA)
baroko (23.4. SUNDA)
barr (17.7. MEHRI)
barrilla (35.8. SPANISH)
barriola (35.2. SPANISH)
bársony kerep (40.13. HUNGARIAN)
bartaku (62.19. BENGALI)
bartang (16.1a. FARSI, WESTERN)
bartiẖ (38.7b. SOMALI)
barunggan (38.20. BUHID)
bas (17.1. FIJIAN)

baş (40.26. AZERBAIJANI, NORTHERN)
baş lahana (30.27a. TURKISH)
baş salata (27.31. TURKISH)
bàsal (7.3. AFAR)
basal (7.3. HINDUSTANI)
basal (7.3. SOMALI)
basal el-maa (25.1. ARABIC)
basal il-hniezer (13.5. MALTESE)
basal tal-qsari (7.11. MALTESE)
basala (28.1. TAMIL)
basale (28.1. KANNADA)
basalè (28.1. TULU)
basale soppu (28.1. KANNADA)
basap (45.4. SERER-SINE)
basap (45.4. WOLOF)
basela (28.1. INDONESIAN)
basela (28.1. PORTUGUESE)
basella (28.1. DUTCH)
basella (28.1. ITALIAN)
basella (28.1. NORWEGIAN)
basella (28.1. VLAAMS)
basella (40.28. ARABIC)
basella bianca (28.1. ITALIAN)
basellakkeera (28.1. MALAYALAM)
baselle (28.1. FRENCH)
baselle blanche (28.1. FRENCH)
basen (58.3. KWÉYÒL)
baseng (20.6. ILOCANO)
bashpaka (23.3. SANSKRIT)
basilic (44.5. FRENCH)
basilic commun (44.5. FRENCH)
basilico (44.5. ITALIAN)
basílico (44.5. PORTUGUESE)
basilicum (44.5. DUTCH)
Basilienkraut (44.5. GERMAN)
basilik (44.5. KAZAKH)
basilik (44.5. NORWEGIAN)
basilika (44.5. SWEDISH)
basilikum (44.5. DANISH)
Basilikum (44.5. GERMAN)
basilikum (44.5. ICELANDIC)
basilikum (44.5. NORWEGIAN)
basing (20.6. ILOCANO)
basini bans (17.1. BENGALI)
baskodjé (45.1. FULFULDE)
baskōje (45.1. FULFULDE, ADAMAWA)
baskore (45.1. FULFULDE, ADAMAWA)
baslahana (30.27a. TURKISH)
basor (8.4. BODO)
basouna pané (43.1. FRENCH)
bassa béné (62.15. MANDING-MANDINKA)
bassal (7.3. ARABIC)
bassal el shallut (7.2. ARABIC)
bassal el-ankudy (7.5. ARABIC)
bassal el-masry (7.9a. ARABIC)
bassal el-mustatere (7.5. ARABIC)
bassal el-shifee (7.11. ARABIC)
bassal zu-el-gazoor (7.14. ARABIC)
bassaleé (7.9a. PULAR)
bassap (45.4. SERER-SINE)
bassap a kob (45.3. SERER-SINE)
bassar (20.2. PANJABI, EASTERN)
başsarimsak (7.11. TURKISH)
bassaw (40.9. IBANAG)
bassaw (40.9. ITAWIT)
bassia béné (62.15. MANDING-MANDINKA)
bassobəlla (44.5. AMHARIC)
bassunta (40.35. SANSKRIT)
bastaard smeerwortel (29.3. VLAAMS)
Bastard Amarant (23.3. GERMAN)
bastard mustard (34.2. ENGLISH)
bastun talkanella (3.1. MALTESE)
basuk (38.16. TAGBANWA)
basula (62.16. IBANAG)
basun (11.5. HANUNOO)
bat (16.1a. GADDANG)
bat (16.1a. ITAWIT)
ɓata wandon buzu (40.25. HAUSA)
bataat (36.3. DUTCH)
bataat (36.3. ESTONIAN)
bataat (36.3. VLAAMS)
bataata (62.24. AFAR)
bataata (62.24. MARATHI)
bataate (62.24. KANNADA)
bataatti (36.3. FINNISH)
batäɔ̃ (62.17. PANJABI, EASTERN)
batagadle (40.28. KANNADA)
batag-batag (9.1. BICOLANO, CENTRAL)
batak seven (14.1. TURKISH)
batakan (17.1. CEBUANO)
batamba (40.43. LIMBA)
batana (38.17. SINHALA)
bataneelu (40.28. TELUGU)
bátanga (39.4. LOKO)
batani (40.28. KANNADA)
batánka (39.4. MALIMBA)
batansé (62.14c. SERER-SINE)
batansé (62.14c. WOLOF)
batansé (62.19. MANINKAKAN, WESTERN)
batansé (62.19. PULAR)

batansé (62.19. WOLOF)
batanya (38.7b. HAUSA)
batat (8.3. TAJIKI)
batat (11.1. CZECH)
batat (11.1. RUSSIAN)
batat (36.3. DANISH)
batat (36.3. NORWEGIAN)
batat (36.3. POLISH)
batat (36.3. ROMANIAN)
batat (36.3. RUSSIAN)
batat (36.3. SERBO-CROATIAN)
batát (36.3. SLOVAK)
batat (36.3. SWEDISH)
batat (36.3. TAJIKI)
batat edesburgonya (36.3. ROMANIAN)
batată (11.1. ROMANIAN)
batata (36.3. BASQUE)
batata (36.3. CATALAN-VALENCIAN-BALEAR)
batata (36.3. HEBREW)
batáta (36.3. HUNGARIAN)
batata (36.3. ITALIAN)
batata (36.3. PORTUGUESE)
batata (36.3. SPANISH)
bätata (62.24. AMHARIC)
batata (62.24. GUJARATI)
batata (62.24. MARATHI)
batata (62.24. PORTUGUESE)
batata aquática (36.2. PORTUGUESE)
batata baronesa (24.8. PORTUGUESE)
batata beda (11.7. ARABIC)
batata de Madagascar (44.10. PORTUGUESE)
batata de telinga (8.3. PORTUGUESE)
batata doce (36.3. PORTUGUESE)
batata doci (36.3. CRIOULO, CAPE VERDE)
batata èdesburgouya (36.1. ROMANIAN)
batata el-feel (8.3. ARABIC)
batata inglesa (62.24. CRIOULO, CAPE VERDE)
batata maee (11.1. ARABIC)
batata mura (11.4. ARABIC)
batata rcno (17.8. RONGA)
batata safraa (11.3. ARABIC)
batata seeny (11.6. ARABIC)
batata swdaa (11.3. ARABIC)
batatar (36.3. JOLA-FONYI)
batatas (36.3. TIRURAY)
batatas (62.24. BONTOC, CENTRAL)
batatas (62.24. PANGASINAN)
batatë (36.3. ALBANIAN, TOSK)
batate (36.3. FRENCH)
batāte (36.3. LATVIAN)
batate (62.24. KANNADA)
Batate Süsskartoffel (36.3. GERMAN)
batateira (62.24. PORTUGUESE)
batates (62.24. ARABIC)
batati (36.3. BULGARIAN)
batati (36.3. MACEDONIAN)
batatilla acuatica (36.2. SPANISH)
batatinė dioskorėja (11.6. LITHUANIAN)
bäṭaṭis (36.3. AMHARIC)
batātu dioskoreja (11.6. LATVIAN)
batatul (36.3. ROMANIAN)
batáty (36.3. CZECH)
batau (40.9. BICOLANO, CENTRAL)
batau (40.9. CEBUANO)
batau (40.9. TAGALOG)
bataun (62.19. PANJABI, EASTERN)
bataw (40.9. AKLANON)
bataw (40.9. BICOLANO, CENTRAL)
bat-aw (40.9. BICOLANO, CENTRAL)
bataw (40.9. TAGALOG)
bataw (40.9. TAUSUG)
bateekh (38.3. ARABIC)
batei (16.1a. BÉTÉ)
bathala (36.3. SINHALA)
bathë (40.34a. ALBANIAN, TOSK)
bathu (23.3. BENGALI)
bathua (35.5. ASSAMESE)
bathua (35.6. ASSAMESE)
bathuā (35.6. HINDI)
batila (16.1a. TELUGU)
bátìrí (38.16. BEROM)
batirjon (62.19. TAJIKI)
batlingan (8.4. SAMBAL, BOTOLAN)
batminari (24.5. KOREAN)
batolka prorosetlá (59.1. CZECH)
batong (40.42. BONTOC CENTRAL)
batong (40.42. BUTUANON)
batong (40.42. CEBUANO)
batong (40.42. KANKANAEY)
batong (40.42. WÁRAY-WÁRAY)
batong (40.43. TAGALOG)
batoon (62.19. PANJABI, EASTERN)
batrash (44.2. MONGOLIAN)
batrash (44.3. MONGOLIAN)
batsala (28.1. TELUGU)
battato (62.24. KONKANI)
batuang (62.23. BONTOC, CENTRAL)
batudan (27.29. BONTOC CENTRAL)

batun pijazi (7.5. KIRGHIZ)
batung (40.1. SAMA, CENTRAL)
batung hidyaw (40.42. TAUSUG)
batung-amu (40.40. TAUSUG)
batung-baimbing (40.29. TAUSUG)
batung-China (40.1. TAGALOG)
batung-china (40.1. TAUSUG)
batung-larot (40.1. TAUSUG)
batung-lupis (40.42. TAGALOG)
batung-munggo (40.38. TAUSUG)
batungsina (40.1. TAUSUG)
batūnsīpoli (7.5. LATVIAN)
batutang-uwak (38.22. TAGALOG)
batyay (11.2. HANUNOO)
batzal ha-ginna (7.3. HEBREW)
batzal mitzree (7.9a. HEBREW)
batzal yapani (7.5. HEBREW)
baudan (23.3. IBALOI)
ba-ude (40.43. HAUSA)
baugan (11.5. SAMBAL, BOTOLAN)
baugin (17.1. PAMPANGAN)
bauhinia (40.33. FRENCH)
bauhinia (40.33. ITALIAN)
bauhinia (40.33. PORTUGUESE)
bauhinia (40.33. SPANISH)
baume petit (44.2. FRENCH)
baume verte (44.3. FRENCH)
Baumtomate (62.7. GERMAN)
Baumtomatenstrauch (62.7. GERMAN)
bauna neai (50.2. LAO)
baunir (40.26. ICELANDIC)
bavachi (40.6. MARATHI)
bavang (7.3. MALAYALAM)
bavare (44.5. HINDI)
bawan (23.3. BONTOC, CENTRAL)
bawan (23.3. KANKANAEY)
bawan malagti (7.10. HANUNOO)
bawang (7.3. MALAY)
bawang (7.10. BICOLANO, CENTRAL)
bawang (7.10. BONTOC CENTRAL)
bawang (7.10. FILIPINO)
bawang (7.10. GIANGAN)
bawang (7.10. IBANAG)
bawang (7.10. IFUGAO)
bawang (7.10. ILOCANO)
bawang (7.10. JAVANESE)
báwañg (7.10. PANGASINAN)
bawang (7.10. PAMPANGAN)
bawang (7.10. TAGALOG)
bawang (7.10. TAUSUG)
bawang abang mirah (7.2. INDONESIAN)
bawang acar (7.2. INDONESIAN)
bawang acar (7.2. SUNDA)
bawang bakung (7.5. INDONESIAN)
bawang bakung (7.5. SUNDA)
bawang bali (7.8. SUNDA)
bawang benggala (7.5. INDONESIAN)
bawang besar (7.3. INDONESIAN)
bawang besar (7.3. MALAY)
bawang beureum (7.2. SUNDA)
bawang bodas (7.10. SUNDA)
bawang Bombay (7.3. INDONESIAN)
bawang Bombay (7.3. SUNDA)
bawang cina (7.5. SUNDA)
bawang daun (7.5. INDONESIAN)
bawang daun (7.5. SUNDA)
bawang daven (7.5. INDONESIAN)
bawang ganda (7.4. INDONESIAN)
bawang ganda (7.13. SUNDA)
bawang kecil (7.2. MALAY)
bawang kucai (7.13. INDONESIAN)
bawang kucai (7.13. SUNDA)
bawang lampoeng (7.3. INDONESIAN)
bawang langkio (7.11. INDONESIAN)
bawang manis (7.3. INDONESIAN)
bawang merah (7.2. INDONESIAN)
bawang merah (7.2. JAVANESE)
bawang merah (7.2. MALAY)
bawang merah (7.3. INDONESIAN)
bawang oncang (7.5. JAVANESE)
bawang parai (7.8. SUNDA)
bawang péré (7.8. INDONESIAN)
bawang pere (7.8. SUNDA)
bawang pire (7.8. SUNDA)
bawang pre (7.8. JAVANESE)
bawang prei (7.1. INDONESIAN)
bawang prei (7.8. INDONESIAN)
bawang prey (7.13. INDONESIAN)
bawang putih (7.9a. INDONESIAN)
bawang putih (7.10. INDONESIAN)
bawang putih (7.10. MALAY)
bawang putih (7.10. SUNDA)
bawang sayuran (7.8. INDONESIAN)
bawang sayuran (7.8. JAVANESE)
bawang semprong (7.5. INDONESIAN)
bawang sop (7.5. INDONESIAN)
bawang sop (7.5. SUNDA)
bawang-pula (7.2. TAGALOG)
bawang-puti (7.10. TAGALOG)
bawany (7.10. TAGALOG)

bawing (44.5. TAGBANWA)
bawing (44.5. TAUSUG)
bawing sulug (44.5. SUBANON)
bawkbawn (62.19. MIZO)
bawrh-saiabe (45.1. MIZO)
bawugan (11.5. SAMBAL, BOTOLAN)
bay (15.4. IVATAN)
bayaduri (23.3. INDONESIAN)
bayag-kabayo (11.2. TAGALOG)
bayag-toro (11.2. TAGALOG)
bayam (23.3. INDONESIAN)
bayam (23.3. MALAY)
bayam (35.9. MALAY)
bayam cabut (23.3. INDONESIAN)
bayam daun (23.4. INDONESIAN)
bayam duri (23.3. INDONESIAN)
bayam ekor belanda (23.4. INDONESIAN)
bayam gelatik (23.3. INDONESIAN)
bayam hijau (23.3. MALAY)
bayam hutan (23.3. INDONESIAN)
bayam itik (23.3. INDONESIAN)
bayam kakap (23.3. INDONESIAN)
bayam liar (23.3. INDONESIAN)
bayam merah (23.3. INDONESIAN)
bayam merah (23.3. MALAY)
bayam monyet (23.3. INDONESIAN)
bayam new zelan (22.2. INDONESIAN)
bayam pasir (23.2. MALAY)
bayam petik (23.4. INDONESIAN)
bayam putih (23.2. INDONESIAN)
bayam putih (23.3. INDONESIAN)
bayam selasih (23.3. INDONESIAN)
bayam tanah (23.2. MALAY)
bayam tur (23.3. INDONESIAN)
bayambang (23.3. TAGALOG)
bayas (62.12. BONTOC, CENTRAL)
baydan (36.3. BONTOC, CENTRAL)
bayem (23.3. INDONESIAN)
bayem abang (23.3. JAVANESE)
bayem abrit (23.3. JAVANESE)
bayem biludu (23.4. SUNDA)
bayem bodas (23.3. SUNDA)
bayem cenggeng (23.4. JAVANESE)
bayem cucuk (23.3. SUNDA)
bayem dempo (23.3. JAVANESE)
bayem eri (23.3. JAVANESE)
bayem hayam (23.3. SUNDA)
bayem itik (23.3. JAVANESE)
bayem kakap (23.3. JAVANESE)
bayem keyong (23.3. JAVANESE)
bayem kremah (23.2. JAVANESE)
bayem lemah (23.3. JAVANESE)
bayem lemah (23.3. SUNDA)
bayem menir (23.3. JAVANESE)
bayem raja (23.3. JAVANESE)
bayem ringgit (23.3. JAVANESE)
bayem salasi (23.3. SUNDA)
bayem sapi (23.3. JAVANESE)
bayem sekul (23.3. JAVANESE)
bayem siti (23.3. JAVANESE)
bayem tai sapi (23.3. SUNDA)
bayem tanah (23.3. SUNDA)
baygan (62.19. HINDUSTANI)
baying (44.5. TAGBANWA)
bayir turpu (30.1. TURKISH)
bayoya (44.8. MÒORÉ)
bayulo (11.2. MANDING-MANDINKA)
bayun tinaw mabiru (62.16. HANUNOO)
bayyasi (36.3. BONTOC, CENTRAL)
baz̤al (7.9a. HEBREW)
bazalka (44.5. SLOVAK)
bazalka pravá (44.5. CZECH)
bazela biała (28.1. POLISH)
bazela bílá (28.1. CZECH)
bazengan africy (62.18. ARABIC)
bazielkruid (44.5. DUTCH)
bazilik (44.5. MORISYEN)
baziliks (44.5. LATVIAN)
bazinğän (62.19. AMHARIC)
bazsalikom (44.5. HUNGARIAN)
bazvi pati khoodro poda (18.2. URDU)
bazylia pospolita (44.5. POLISH)
bɔbɔɛ (23.3. KONO)
bɔbɔ-wa (23.3. KONO)
bɔɔ (17.1. MANDING-MANDINKA)
bɔd pɛpɛ (62.5. KRIO)
bɔhɔri (9.1. SUSU)
bɔhrɔ (38.3. BULLOM SO)
bɔkɔbɔkɔ (59.3. GA)
bɔkié (36.3. MALIMBA)
bɔkui (36.3. MANINKAKAN, EASTERN)
bɔlɔgi (27.14. KRIO)
bɔlumbo (62.2. DAGBANI)
bɔmmɔfrana (38.28. AKAN-AKUAPEM)
bɔndɔ (45.1. BULLOM SO)
bɔndɔ (45.1. GOLA)
bɔndɔ (45.1. MENDE)
bɔndɔ-nande (45.1. MENDE)
bɔnde (45.2. KURANKO)
bɔŋgɛ (62.9. GOLA)

bɔrɔ̃ (23.3. MANDING-MANDINKA)
bɔrɔ̃ (23.3. MANINKAKAN, EASTERN)
bɔrɔbɔrɛ (59.2. SUSU)
bɔrɔ̃-m-bɛbɛ (23.3. MANDING-MANDINKA)
bɔrɔ̃-m-bɛdɛ (23.3. MANINKAKAN, EASTERN)
bɔrəbɔrɛ (23.3. MANDING-MANDINKA)
bɔrəbɔrɛ (23.3. MANINKAKAN, EASTERN)
bɔren ɔre-ba (23.3. MALIMBA)
bdo (62.24. DAKOTA)
be (40.12. BADAGA)
be (40.28. MIZO)
be chi khat (40.43. MIZO)
be hring chi khat (40.12. MIZO)
be nasia (43.1. MANDJAK)
be nasin (43.1. MANKANYA)
bea-bea (6.1. INDONESIAN)
beakogan (32.1. BULLOM SO)
bean (40.3. PANJABI, EASTERN)
bean (40.4. PANJABI, EASTERN)
bean (40.26. FIJIAN)
beans (40.26. MALAYALAM)
beàrnan-brìde (27.45. GAELIC, SCOTTISH)
bebasal (38.16. BINUKID)
bebasal (38.16. MARANAO)
bebasal (38.17. MARANAO)
bebil (40.24. YIDDISH)
bebil (40.25. YIDDISH)
bebil (40.26. YIDDISH)
becalhi (59.2. FULFULDE, ADAMAWA)
bé-chî (64.1. CHINESE, MIN NAN)
bed (11.2. BUHID)
beda (40.17. TAJIKI)
bédé (39.4. AKAN-ASANTE)
bəḍəḥáyn (62.24. MEHRI)
bedrennej (24.22. TAJIKI)
bɛ̀ɛ́ (40.1. GURO)
bɛɛ kumpeeyu (23.3. YOM)
beebread (29.1. ENGLISH)
beef-stake plant (44.7. ENGLISH)
beejagarbh (38.31. SANSKRIT)
been (40.26. CAJUN FRENCH)
beera (38.21. TELUGU)
beera koya (38.21. TELUGU)
beerakaaya (38.21. TELUGU)
beeskool (30.35. AFRIKAANS)
beet (35.4. AFRIKAANS)
beet (35.4. ASSAMESE)
beet (35.4. BENGALI)
beet (35.4. ENGLISH)
beet (35.4. FILIPINO)
beet (35.4. GUJARATI)
beet (35.4. MARATHI)
beet (35.4. ORIYA)
beet mula (35.4. ASSAMESE)
beet root (35.4. KANNADA)
beet root (35.4. MALAYALAM)
beetraw (35.4. SCOTS)
beetroot (35.4. ENGLISH)
beetudumpa (35.4. TELUGU)
beetys (35.4. GAELIC, MANX)
beetzan (23.1. HEBREW)
beetzan machseef (23.2. HEBREW)
bẽgan (62.17. PANJABI, EASTERN)
beggar tick (27.8. ENGLISH)
begun (62.19. BENGALI)
behar (17.1. BENGALI)
behurbans (17.1. BENGALI)
bei gai (30.57. CHINESE, YUE)
bei gua (38.9. CHINESE, YUE)
bei gua (38.14. CHINESE, GÀN)
bei jie (30.57. CHINESE, YUE)
bei mao qian jin ba (40.7. CHINESE, MANDARIN)
bei mao qian jin ba (40.7. CHINESE, WÚ)
bei sha shen (24.14. CHINESE, WÚ)
beiba (16.1a. IZON)
Beifuß (27.7. GERMAN)
beil zelje (30.27a. SLOVENIAN)
bekaparnamu (24.7. TELUGU)
beké abangha (16.1a. IZON)
bèkè búrù (36.3. IZON)
bèké òdù (8.9. IZON)
bekkolo (17.7. OROMO, WEST CENTRAL)
bel (11.2. BUHID)
bel (11.5. BUHID)
bel koren livadska kozja brada (27.47. MACEDONIAN)
bel sinap (30.57. MACEDONIAN)
bela bundeva (38.18. SERBO-CROATIAN)
bela repa (30.22. SERBO-CROATIAN)
bela repka (30.22. MACEDONIAN)
bela repka (30.41a. MACEDONIAN)
bela zelka (30.27a. MACEDONIAN)
belanda (38.7a. INDONESIAN)
belanda (38.7b. INDONESIAN)
belangere batarde (62.23. FRENCH)
bélanjé (62.19. KWÉYÒL)
bélanjén (62.19. KWÉYÒL)
belantinu (62.19. KALAGAN)

belatung (40.38. BINUKID)
belavaangi (62.9. MARATHI)
beldroega (59.2. PORTUGUESE)
beldroega das areias (59.4. PORTUGUESE)
beldroega de Inverno (59.1. PORTUGUESE)
beldroega de praia (59.4. PORTUGUESE)
beldroega grande (59.3. PORTUGUESE)
bele sasive (30.57. KANNADA)
belembe (8.8. DANISH)
belembe (8.8. FRENCH)
belembe (8.8. PORTUGUESE)
belembe (8.8. SPANISH)
belembe (8.8. VLAAMS)
beler (30.52. BRETON)
belesagan (23.3. BINUKID)
belete (35.3. BASQUE)
belewa (38.7a. INDONESIAN)
belewa (38.7b. INDONESIAN)
belgklukka (34.3. ICELANDIC)
beli koren (27.46. SLOVENIAN)
beligo (38.1. INDONESIAN)
beligu (38.1. INDONESIAN)
belikh (38.3. ARABIC)
belinjo (41.1. INDONESIAN)
bell pepper (62.1. ENGLISH)
bella di notte espansa (49.1. ITALIAN)
bellagenasu (36.3. KANNADA)
bellaraai (11.6. KANNADA)
bellarai (11.2. KANNADA)
bellashora (38.20. MALAYALAM)
beḷḷe benguve (7.10. BADAGA)
bellede nuit (49.1. FRENCH)
bell-pɛpɛ (62.1. KRIO)
bellulli belluli (7.10. KANNADA)
belocvetna nivska repka (30.11. MACEDONIAN)
bělohořčice setá (30.57. CZECH)
belôluga (59.2. MANKANYA)
belušna zelena (24.4. SLOVENIAN)
belustru (38.22. INDONESIAN)
belvedere (35.2. ITALIAN)
belverde (35.2. PORTUGUESE)
bemthaamara (48.1a. MALAYALAM)
benarasi rai (30.24. HINDI)
benavista (39.3. PANGASINAN)
ben-chardal (30.48. HEBREW)
benda (45.1. TELUGU)
benda kaya (45.1. KANNADA)
bendai (45.1. MALAYALAM)
bendai kai (45.1. TAMIL)
bendakainaru (45.1. KANNADA)
bendakaya (45.1. TELUGU)
beṇḍe (45.1. BADAGA)
bende (45.1. KANNADA)
beṇḍè (45.1. TULU)
be̱nde̱ beṅgaṛ (62.19. SANTALI)
bende kaayi (45.1. KANNADA)
bende kaayi gida (45.1. KANNADA)
bende kayi (45.1. KANNADA)
bende naaru (45.1. KANNADA)
bendelløk (7.12. NORWEGIAN)
bend̲ëy (45.1. PASHTO, NORTHERN)
bendi (40.15. BAMANANKAN)
bendi (45.1. MALAY)
bendi (45.1. TELUGU)
bendikay (45.1. HINDI)
bendo (17.1. ORIYA)
bendo (45.1. KONKANI)
béné (55.1. JOLA-FONYI)
bene (55.1. WOLOF)
béné (55.1. WOLOF)
benefing-jongo (55.1. MANDING-MANDINKA)
benekli altin dikeni (27.37. TURKISH)
bēnga (40.42. MÒORÉ)
bēnga (40.43. MÒORÉ)
Bengal gram (40.5. ENGLISH)
bengaldumpa (62.24. TELUGU)
bengan (62.19. PANJABI, EASTERN)
bengar (62.19. MUNDARI)
beṅgaṛ (62.19. SANTALI)
bengena (62.19. ASSAMESE)
bengkuang (40.21. INDONESIAN)
bengkuang (40.22. INDONESIAN)
bengkuwang (40.21. JAVANESE)
bengkuwang (40.21. MALAY)
be̱ṅgo̱ nāṛĩ (11.2. SANTALI)
bengue (7.3. BADAGA)
benguk (40.18. INDONESIAN)
benguk (40.18. JAVANESE)
benguk (40.18. SUNDA)
benguve (7.3. BADAGA)
ɓẹ-ngwaf u-naɗ (11.2. WAMEY)
bengwil (17.1. BONTOC, CENTRAL)
ben-ḥardal maẕuy (30.48. HEBREW)
benibana-borogiku (27.14. JAPANESE)
benibanaingen (40.24. JAPANESE)
bɛni-lif (55.1. KRIO)
benincasa (38.1. ITALIAN)
benincasa (38.1. PORTUGUESE)

beninkasa (38.1. CZECH)
beno (55.1. MANDING-MANDINKA)
bent (44.2. BRETON)
bent (44.3. BRETON)
bent-dour (44.2. BRETON)
benteng (38.13a. INDONESIAN)
bɛntisɔk-lɛ (23.3. BULLOM SO)
bent-liorzhoù (44.3. BRETON)
bentul (8.4. INDONESIAN)
bentul (8.9. INDONESIAN)
bentul (8.9. JAVANESE)
bentul (8.9. SUNDA)
béon (40.42. DAGAARE, SOUTHERN)
béon (40.43. DAGAARE, SOUTHERN)
bər (17.7. MEHRI)
ber (34.1. HINDI)
bér palmolistý (17.6. CZECH)
bera (45.1. HAUSA)
bera (62.19. SANSKRIT)
berambang (7.2. INDONESIAN)
berberi (62.2. OROMO, WEST CENTRAL)
béré ntiga (40.39. MANINKA)
berebe-ghịsa (16.1a. IZON)
béref (38.3. SERER-SINE)
béref (38.3. WOLOF)
berejenn (62.19. CREOLE, HAITIAN FRENCH)
berengena (62.19. SPANISH)
berengena cimarrona (62.23. SPANISH)
berenjèna (62.19. LADINO)
berenjena (62.19. SPANISH)
berenjena africana (62.18. SPANISH)
berenjena escarlata (62.14c. SPANISH)
berenjena espinuda (62.16. SPANISH)
berenjena silvestre (62.16. SPANISH)
berenjena silvestre (62.17. SPANISH)
berga (63.1. MÒORÉ)
Bergaralie (26.2. GERMAN)
bergknapp (37.1. NORWEGIAN)
bergkung (44.6. NORWEGIAN)
beribă (16.1a. IZON)
beringela (62.19. PORTUGUESE)
beringela africana (62.18. PORTUGUESE)
beringela escarlate (62.14c. PORTUGUESE)
beringeleira do mato (62.16. PORTUGUESE)
berinxela (62.19. GALICIAN)
berle à sucre (24.23. FRENCH)
Bermuda arrowroot (15.2. ENGLISH)
Bermudinė maranta (15.2. LITHUANIAN)
běrna (62.24. SORBIAN, UPPER)
bernagie (29.1. AFRIKAANS)
bernagie (29.1. DUTCH)
bernagie (29.1. VLAAMS)
berokoli (30.31. HEBREW)
berolu (15.2. MALAY)
berr (17.7. HARSŪSI)
berralas (29.1. GALICIAN)
berro (30.52. BASQUE)
berro (30.52. SPANISH)
berro comúm (30.51. SPANISH)
berro de agua (30.52. SPANISH)
berro de Pará (27.2. SPANISH)
berro de prado (30.44. SPANISH)
berro de tierra (30.2. SPANISH)
berro de tierra (30.51. SPANISH)
bertalha (28.1. PORTUGUESE)
bertoń (30.35. GALICIAN)
berw'r mynydd (30.2. WELSH)
berwr dwr (30.52. WELSH)
berwr yr ardd (30.51. WELSH)
berza (30.35. SPANISH)
berza col (30.35. SPANISH)
bəṣāl (7.3. MEHRI)
besar (7.3. MALAY)
besar (20.2. NEPALI)
besbes (24.13. ARABIC)
bəsbēs (62.1. MEHRI)
bəṣəlēt (7.3. MEHRI)
Besenkraut (35.2. GERMAN)
Besenradmelde (35.2. GERMAN)
besor (30.17. BODO)
bestru (38.22. JAVANESE)
besusa (40.21. INDONESIAN)
besusu (40.21. JAVANESE)
beta ôndôndô (62.1. BULU)
beta ôndôndô (62.2. BULU)
bêtai bêtai (11.1. BAOULÉ)
bəṭāṭ (62.24. MEHRI)
betbinho (38.13a. BALANTA-KENTOHE)
bete (35.4. NORWEGIAN)
bété bété (11.1. BAOULÉ)
bété bété alengbe (11.1. BAOULÉ)
bété bété kpa (11.1. BAOULÉ)
bětek (38.3. INDONESIAN)
běteka (38.3. MALAY)
betel yam (11.6. ENGLISH)
bétérav (35.4. WOLOF)
beterraba (35.4. CRIOULO, CAPE VERDE)
beterraba (35.4. PORTUGUESE)
beterraba de mesa (35.4. PORTUGUESE)

beterraba de salada (35.4. PORTUGUESE)
beterraba hortícola (35.4. PORTUGUESE)
beterraba vermelha (35.4. PORTUGUESE)
betharasu (40.9. HINDI)
bètrav (35.4. CREOLE, HAITIAN FRENCH)
betrav (35.4. MORISYEN)
bettada pudina (44.3. KANNADA)
bəṭṭáyx (38.7b. MEHRI)
bette (35.3. FRENCH)
betterave (35.4. CAJUN FRENCH)
betterave potagère (35.4. FRENCH)
betterave rouge (35.4. FRENCH)
bettēx (38.3. HARSŪSI)
bəṭṭə́xt (38.7b. MEHRI)
bettieh tax-xibka (38.7b. MALTESE)
bettieh tax-xitwa (38.1. MALTESE)
beṭṭīx (38.3. HARSŪSI)
bɛtu (63.1. VAI)
betwav (35.4. KWÉYÒL)
beuaz baslahana (30.27a. TURKISH)
beubeureuman (69.1. JAVANESE)
beudobaunso (17.1. ORIYA)
beyam belanda (35.9. INDONESIAN)
beyaz hardal (30.57. TURKISH)
beyaz kabak (38.1. TURKISH)
beyed (40.29. BONTOC CENTRAL)
ɓe-yowél (40.18. BUDIK)
bezār (62.1. HARSŪSI)
bəzār (62.1. MEHRI)
bezelye (40.28. TURKISH)
bezmirisna dinja (38.11. MACEDONIAN)
bezvoud (36.1. BRETON)
bgossoloume (36.3. SENOUFO, CEBAARA)
bhā (48.1a. PANJABI, EASTERN)
bhāæ (48.1a. PANJABI, EASTERN)
bhaanuphala (16.1a. SANSKRIT)
bhadra (30.51. SANSKRIT)
bhadramustha (10.1. SANSKRIT)
bhaingon (62.19. PANJABI, EASTERN)
bhaji (23.3. MARATHI)
bhaji (40.32. GUJARATI)
bhalvengkak (38.7b. INDONESIAN)
bhalvengkak (38.11. INDONESIAN)
bhanana (16.1a. SHONA)
bhandira (23.3. SANSKRIT)
bhanjira (44.7. HINDI)
bhanta (62.19. HINDI)
bháṉtá (62.19. HINDUSTANI)
bhanta (62.19. NEPALI)
bhantā (62.19. NEWAR)
bhantaaki (62.19. SANSKRIT)
bhantaki (62.19. KANNADA)
bhantaki (62.19. SANSKRIT)
bharakalijan (20.1. HINDI)
bharbhạri (44.5. SANTALI)
bhat (40.8. HINDI)
bhat (40.8. URDU)
bhat karela (38.22. ASSAMESE)
bhāṭā (62.19. MAITHILI)
bhata vaasu (40.9. HINDI)
bhatavans (40.9. HINDI)
bhathua (38.1. HINDI)
bhathur (27.45. MARATHI)
bhatmas (40.8. NEPALI)
bhatwar (40.8. HINDI)
bhatwas (40.8. HINDUSTANI)
bhede khursani (62.1. NEPALI)
bhedra (62.9. MARATHI)
bhekaparni (24.7. SANSKRIT)
bheki (24.7. SANSKRIT)
bhekuri (62.14c. ASSAMESE)
bhekuri (62.16. ASSAMESE)
bhenda (45.1. MARATHI)
bhenda (45.1. SANSKRIT)
bhende (45.1. KANNADA)
bhendi (45.1. ASSAMESE)
bhendi (45.1. MARATHI)
bhendi (45.1. ORIYA)
bhengana (62.19. HINDI)
bhenght (50.2. HINDI)
bhet phul (50.2. ASSAMESE)
bhetmas (40.8. HINDI)
bhida (45.1. GUJARATI)
bhimb (38.4. HINDI)
bhinda (45.1. GUJARATI)
bhinda (45.1. SANSKRIT)
bhindatika (45.1. SANSKRIT)
bhindatori (45.1. PANJABI, EASTERN)
bhindi (45.1. FIJIAN)
bhindi (45.1. HINDI)
bhindi (45.1. HINDUSTANI)
bhindi (45.1. NEPALI)
bhindi (45.1. PANJABI, EASTERN)
bhindi (45.1. URDU)
bhindu (45.1. GUJARATI)
bhinzi (40.26. SHONA)
bhioni (40.35. PANJABI, EASTERN)
bhiranga thorayi (38.21. HINDI)
bhiringi saag (23.2. NEPALI)
bhiruvolikanda (11.2. HINDI)

bhit-tita (62.23. ASSAMESE)
bhol (38.22. ASSAMESE)
bhomphala (38.16. MARATHI)
bhoochanaka (40.1. SANSKRIT)
bhooichane (40.1. MARATHI)
bhooisheng (40.1. MARATHI)
bhoomija (40.1. SANSKRIT)
bhoomikakshara (30.56. SANSKRIT)
bhooshimbika (40.1. SANSKRIT)
bhoostha (40.1. SANSKRIT)
bhoothulasi (44.5. KANNADA)
bhoothulasi (44.5. TELUGU)
bhoottah (17.7. HINDI)
bhooyimoonga (40.1. MARATHI)
bhopala (38.20. MARATHI)
bhopla (38.20. MARATHI)
bhringga (40.35. GUJARATI)
bhringga (40.35. HINDI)
bhubhuwan (34.2. MADURA)
bhui mug (40.1. MARATHI)
bhuigholi (59.2. MARATHI)
bhul (38.22. ASSAMESE)
bhungipat (63.1. BENGALI)
bhuripushpa (24.13. SANSKRIT)
bhuru (38.1. GUJARATI)
bhutaghna (30.48. SANSKRIT)
bhutagna (7.10. SANSKRIT)
bhutta (17.7. BENGALI)
bhutta (17.7. HINDI)
bhuṭṭá (17.7. HINDUSTANI)
bhutulasi (44.5. TELUGU)
bhvigholi (59.4. MARATHI)
bi ji (10.2. CHINESE, WÚ)
bi qi (10.2. CHINESE, GÀN)
bi qi (10.2. CHINESE, MANDARIN)
bia (8.1. ITAWIT)
biâ ḿbàkárá (36.3. EFIK)
biabhóg (58.1. GAELIC, IRISH)
bia-bia (6.2. INDONESIAN)
bia-bia (18.1. INDONESIAN)
biala (59.2 MARANAO)
biala (59.3. MARANAO)
bian dou (40.9. CHINESE, YUE)
bian dou (40.29. CHINESE, GÀN)
bian fu chi (27.5. CHINESE, WÚ)
bian luo bo (30.41a. CHINESE, GÀN)
bian pu (38.20. CHINESE, MANDARIN)
bian pu (38.20. CHINESE, MIN BEI)
bian pu (38.20. CHINESE, WÚ)
bian ye mu (39.3. CHINESE, WÚ)
biandobwe (40.2. MENDE)
biao cai (35.3. CHINESE, YUE)
biatas milis (35.4. GAELIC, IRISH)
biaz (59.2. LAO)
bibèr (62.1. LADINO)
biber (62.1. TURKISH)
bibeté (38.18. BEMBÉ)
bibeté (38.18. KOONGO)
bichchoti (40.18. BENGALI)
bichhu boot (66.3. URDU)
bichhu booti (66.3. HINDI)
bichuelas (40.28. TAGALOG)
bidangadubaechu (30.34. KOREAN)
biday (44.5. ILOCANO)
bident (27.8. FRENCH)
bident pileux (27.8. FRENCH)
bidi (40.31. HAUSA)
bidii (20.2. SWAHILI)
bidil (17.1. BADAGA)
bidilu (17.1. BADAGA)
bidungalu (17.1. KANNADA)
biɛfo hiɛ (11.5. DANGME-KROBO)
bierkalbas (38.20. AFRIKAANS)
bierpolle (46.1. STELLINGWERFS)
bieslook (7.5. VLAAMS)
bieslook (7.11. DUTCH)
biet (35.4. STELLINGWERFS)
bieta a foglia (35.3. ITALIAN)
biete (35.4. LATVIAN)
bietola da costa (35.3. ITALIAN)
bietola da orta (35.4. ITALIAN)
bifftomat (62.9. NORWEGIAN)
biga (8.1. CEBUANO)
biga (8.1. FILIPINO)
biga (8.1. ILOCANO)
biga (8.1. PAMPANGAN)
biga (8.1. TAGALOG)
biga-bigaan (18.2. TAGALOG)
biget (16.1a. AGTA)
bighel (59.2. ARABIC)
ɓigira (64.3. HAUSA)
bigpɛpɛ (62.1. KRIO)
bigyanjaro (40.26. GANDA)
bihanti (40.40. HANUNOO)
biito (35.4. JAPANESE)
bijela repa (30.41a. CROATIAN)
bijeli (23.3. SERBO-CROATIAN)
bijeli korijen (27.46. SERBO-CROATIAN)
bijeli kupus (30.27a. SERBO-CROATIAN)
bijeli luk (7.10. BOSNIAN)

bikul (39.4. BUHID)
bikyat hagina (40.34a. HEBREW)
bilaan aba (8.4. IFUGAO)
bila-bila (8.1. KANKANAEY)
bilagut (18.2. ILOCANO)
bilahi (62.9. ASSAMESE)
bilaiti (62.9. MUNDARI)
bilan mumpaha (8.1. IFUGAO)
bilan ngalngalitan (8.10. IFUGAO)
bilanamanut (23.4. MAGUINDANAO)
bilang-bilang (59.2 CEBUANO)
bilati alu (62.24. ORIYA)
bilati baign (62.9. ORIYA)
bilati begun (62.9. BENGALI)
bilbila (8.1. BONTOC, CENTRAL)
bili basale (28.1. KANNADA)
bili maniavare (40.9. KANNADA)
bili nyanya (62.14c. LUO)
bili nyanya (62.18. LUO)
bili nyanya (62.19. LUO)
bili nyanya (62.23. LUO)
bilibíz (40.5. LADINO)
bilimoolangi (30.56. KANNADA)
bilinganya (62.19. SWAHILI)
bill hambu (36.2. KANNADA)
billeri dei prati (30.44. ITALIAN)
bìlogolova kapusta (30.27a. UKRAINIAN)
bilvamoola (11.2. SANSKRIT)
bílý jam (11.1. CZECH)
bimba (38.4. SANSKRIT)
bimbaka (38.4. SANSKRIT)
bimbee (38.4. SANSKRIT)
bimbi (38.4. MARATHI)
bimbika (38.4. SANSKRIT)
bimbika (38.4. TELUGU)
bimbóskel (30.8. HUNGARIAN)
bimbóskel (30.29. HUNGARIAN)
bin (40.18. MARATHI)
bin cai (30.45. CHINESE, MANDARIN)
bin cai (30.45. CHINESE, WÚ)
bin dou (40.12. CHINESE, WÚ)
bin pu (38.20. CHINESE, YUE)
bin yaprak (42.1. TURKISH)
binagyang (8.6. BINUKID)
binagyang (8.7. MANOBO)
binahian (39.5. TAGALOG)
binaydan (36.3. BONTOC, CENTRAL)
binch (40.25. KRIO)
binch (40.26. KRIO)
binda (45.1. GUJARATI)
binda (45.1. SWAHILI)
binde (40.15. PULAAR)
Bindersalat (27.32. GERMAN)
bindi (40.15. BAMANANKAN)
bindsalat (27.32. DANISH)
Bindsalat (27.32. GERMAN)
bindsalat (27.32. NORWEGIAN)
bindsallat (27.32. SWEDISH)
bindui (7.5. AFRIKAANS)
binenisil (62.1. BINUKID)
bing hau (22.1. CHINESE, WÚ)
bing hau (22.1. CHINESE, MANDARIN)
bing ye ri shong hua (22.1. CHINESE, YUE)
bingebwa (40.1. GANDA)
binggala (39.4. BINUKID)
binggala (39.4. MANOBO)
bingi (38.20. NUPE-NUPE-TAKO)
binši (40.2. MENDE)
bín-tāu (40.26. CHINESE, MIN NAN)
bintsoukoulou (62.14b. KOONGO)
bintsoukouri (62.14b. BEMBÉ)
binusaveraikodi (40.25. TAMIL)
binzari (20.2. SWAHILI)
biolar (30.52. GAELIC, IRISH)
bior (63.1. DAGAARE, SOUTHERN)
biotais (35.4. GAELIC, SCOTTISH)
biper (62.1. BASQUE)
bir (45.4. DAGAARE, SOUTHERN)
bir alpạlua (40.23. SANTALI)
bir khil köktat (27.31. UYGHUR)
bir khil köktat (27.32. UYGHUR)
bira (8.1. ILOCANO)
bira (8.1. INDONESIAN)
bira (8.1. SUNDA)
birah (8.1. INDONESIAN)
birah hitam (8.10. MALAY)
birah hutan (8.7. MALAY)
birah keladi (8.4. MALAY)
birah nĕgĕri (8.1. MALAY)
birakaya (38.21. TELUGU)
birali kand (40.22. NEPALI)
biray (44.5. PANGASINAN)
bir-bob (38.23. SERER-SINE)
birdo (23.3. MARANAO)
birenghenas (62.19. IBANAG)
biri (27.2. FILIPINO)
biri (27.4. BONTOC CENTRAL)
biribiri (62.1. GIKUYU)
biribiri (62.1. LUYIA)
biribiri (62.2. GIKUYU)

biribiri (62.2. LUYIA)
biribiri (62.3. GIKUYU)
biribiri (62.3. LUYIA)
biribiri (62.4. GIKUYU)
biribiri (62.4. LUYIA)
biribiri (62.5. GIKUYU)
biribiri (62.5. LUYIA)
birigue (40.1. FULFULDE, ADAMAWA)
birihu (40.1. FULFULDE)
biriiwu (40.1. FULFULDE, ADAMAWA)
birīji (40.39. FULFULDE, ADAMAWA)
biriji ndiyam (64.3. FULFULDE, ADAMAWA)
biringanya (62.18. GANDA)
biringi (40.26. TAGALOG)
biringinya (62.19. SWAHILI)
biriyu (40.1. FULFULDE)
biriyu (40.1. PULAAR)
birsan (40.37. SANTALI)
birú manso (9.1. PORTUGUESE)
b[is] dao (38.1. VIETNAMESE)
b[is] xanh (38.1. VIETNAMESE)
bisab (45.1. WOLOF)
bisab (45.4. WOLOF)
bisakusuma (48.1a. SANSKRIT)
Bisamkürbis (38.17. GERMAN)
bisap (45.4. WOLOF)
bisaprasuna (48.1a. SANSKRIT)
bisar (40.43. KORO)
bisbas (24.12. ARABIC)
biser luk (7.1. SERBO-CROATIAN)
biseren luk (7.1. BULGARIAN)
bisial (44.5. WAMEY)
bissab (45.1. WOLOF)
bissab (45.3. WOLOF)
bissab (45.4. WOLOF)
bissab u ala (45.2. WOLOF)
bissap (45.4. FRENCH)
bisvathulasi (44.5. SANSKRIT)
bit (35.4. INDONESIAN)
bit (35.4. JAVANESE)
bit (35.4. MALAY)
bit (35.4. SUNDA)
bit bodas (35.3. INDONESIAN)
bitabit (40.9. GADDANG)
bita-bol (62.19. KRIO)
bitalé (27.24. BAMILEKE)
bita-lif (27.24. KRIO)
biṭāṭis (36.3. HARARI)
bite (35.4. KIRIBATI)
bitetam (45.1. BULU)
bitetam (45.1. EWONDO)
biti ruti (35.4. SWAHILI)
bitirùthi (35.4. VENDA)
bito (45.4. MÒORÉ)
bito-an (17.7. MARANAO)
bitsounkoulou (62.18. KOONGO)
bitsuwelas (40.26. IBANAG)
bitsuwelas (40.26. PANGASINAN)
bitsuwelas (40.26. SAMBAL, TINÀ)
bitsuwelas (40.26. TAGALOG)
bittas (27.24. KRIO)
bitter cucumber (38.23. ENGLISH)
bitter gourd (38.23. ENGLISH)
bitter jam (11.4. AFRIKAANS)
bitter leaf (27.24. ENGLISH)
bitter melon (38.23. ENGLISH)
bitter yam (11.4. ENGLISH)
bitterblaar (27.24. AFRIKAANS)
bitterblad (27.24. DUTCH)
bitterblad (27.24. VLAAMS)
bittere komkommer (38.23. VLAAMS)
bittere yam (11.4. VLAAMS)
bitterer Balsamkürbis (38.23. GERMAN)
bitterkomkommer (38.23. VLAAMS)
bitterwort (27.45. ENGLISH)
bittre Cassava (39.4. GERMAN)
bitung (17.2. SUNDA)
bitungulu (7.3. LUYIA)
bitunyakpokpu (44.10. AJA)
bitxaleka (30.1. BASQUE)
bitz̤öni matok (10.2. HEBREW)
biyarey (40.9. BINUKID)
biyarey (40.25. BINUKID)
bizele (40.28. ALBANIAN, TOSK)
bizèlia (40.28. LADINO)
bjalo glavesto zele (30.27a. BULGARIAN)
bjanka (35.1. MALTESE)
bkanesen (40.1. POTAWATOMI)
bʌ̄ʌlʌ̄ (16.1a. ANUAK)
blaarslaai (27.32. AFRIKAANS)
black gram (40.37. ENGLISH)
black jack (27.8. ENGLISH)
black mustard (30.24. ENGLISH)
black nightshade (62.15. ENGLISH)
black nightshade (62.21. ENGLISH)
black nightshade (62.21. FILIPINO)
black salsify (27.38. ENGLISH)
black-eyed pea (40.43. ENGLISH)
black-seeded squash (38.15. ENGLISH)
bladbede (35.3. DANISH)

bladbedja (35.3. ICELANDIC)
bladbete (35.3. NORWEGIAN)
bladbete (35.3. SWEDISH)
bladgrøntsag (36.2. DANISH)
bladkål (30.35. NORWEGIAN)
bladk'al (30.36. ICELANDIC)
bladkool (30.21. VLAAMS)
bladmynta (44.7. SWEDISH)
bladpersille (24.19. NORWEGIAN)
bladpeterselie (24.19. VLAAMS)
bladsalat (27.32. ICELANDIC)
bladsalatfifill (27.10a. ICELANDIC)
bladsallat (27.32. SWEDISH)
bladselleri (24.4. DANISH)
bladtetragonia (22.2. NORWEGIAN)
bladylaukur (7.8. ICELANDIC)
blæjaber (62.12. ICELANDIC)
blafo (17.7. ÉWÉ)
blak-ai-binch (40.43. KRIO)
blakinė kalendra (24.9. LITHUANIAN)
blålucern (40.17. SWEDISH)
blåluserne (40.17. NORWEGIAN)
blamange (38.18. MENDE)
blanched celery (24.4. ENGLISH)
blandotan (62.9. JAVANESE)
Blasenkirsche (62.13. GERMAN)
blat de moro (17.7. CATALAN-VALENCIAN-BALEAR)
Blätterkohl (30.35. GERMAN)
Blattsalat (27.32. GERMAN)
blauwe druifjes (13.5. DUTCH)
blauwe kool (30.27b. STELLINGWERFS)
blauwe morgenster (27.46. DUTCH)
blɔfo katso (38.22. GA)
blɔfo kodu (16.1a. DANGME-KROBO)
blɔfo sabola (7.3. GA)
blɔfo sabola (7.9a. GA)
blɔfo-atómò (36.3. GA)
blɔfoƒwie (59.3. GA)
blɔfoo atomo (36.3. GA)
blɔfoyɔ (40.2. GA)
bʌ́ʌlé (16.1a. ANUAK)
blê blê (40.1. BÉTÉ)
bleachtán (27.42. GAELIC, IRISH)
bleda (35.3. SPANISH)
bleda vermella (35.4. CATALAN-VALENCIAN-BALEAR)
bledo (23.3. GALICIAN)
bledo carbonero (56.2. SPANISH)
bledos blancos (23.3. SPANISH)
bledselleri (24.4. DANISH)
bleekselderij (24.4. DUTCH)
bleekselderij (24.4. VLAAMS)
blɛfo-tobi (9.1. DANGME)
blɛfo-tobi (9.1. DANGME-KROBO)
Bleichsellerie (24.4. GERMAN)
bleifarbiger Amarant (23.3. GERMAN)
blekselleri (24.4. SWEDISH)
blenggi (27.41. INDONESIAN)
blestru (38.22. JAVANESE)
blette (35.3. FRENCH)
blewah (38.7a. INDONESIAN)
blewah (38.7a. JAVANESE)
blewah (38.7a. SUNDA)
blewah (38.7b. INDONESIAN)
blewah (38.7b. JAVANESE)
blewah (38.7b. SUNDA)
blewak kantelup (38.7a. INDONESIAN)
blewek (38.7a. JAVANESE)
blewek (38.7b. JAVANESE)
bligo (38.1. JAVANESE)
bliosán (27.16. GAELIC, IRISH)
blitva (35.3. CROATIAN)
blo (62.24. DAKOTA)
blo (62.24. LAKOTA)
blodfresychgaeaf (30.31. WELSH)
blodyn y gôg (30.44. WELSH)
bloemachtige bladmosterd (30.16. VLAAMS)
bloemkool (30.26. DUTCH)
bloemkool (30.26. STELLINGWERFS)
bloemkool (30.26. VLAAMS)
bloesum banaan (16.1b. DUTCH)
blofouè douo (11.5. BAOULÉ)
blomkål (30.26. DANISH)
blomk'al (30.26. ICELANDIC)
blomkål (30.26. NORWEGIAN)
blomkål (30.26. SWEDISH)
blomkarse (65.1. NORWEGIAN)
blomkool (30.26. AFRIKAANS)
blomst banan (16.1b. DANISH)
blomsterböna (40.24. SWEDISH)
blomsterkarse (65.1. DANISH)
blöruk'al (30.34. ICELANDIC)
blow ball flower (27.45. ENGLISH)
blue lotus of India (50.2. ENGLISH)
blue taro (8.10. ENGLISH)
Blumenkohl (30.26. GERMAN)
blumenkroyt (30.26. YIDDISH)
blumkol (30.26. INDONESIAN)
blumkol (30.26. SUNDA)

budarga (35.10. MONGOLIAN)
budargana (35.7. MONGOLIAN)
budargana (35.8. MONGOLIAN)
buddabusara (62.12. TELUGU)
budde hannu (62.12. KANNADA)
budebbus (52.1. MALTESE)
budelɔmo (55.1. MALIMBA)
budi (38.3. PULAAR)
budi (38.5. PULAAR)
budi (38.18. PULAAR)
budi (40.3. BEMBÉ)
budibaale (16.1a. KANNADA)
budidagummadi (38.1. TELUGU)
budida-gummadi (38.1. TELUGU)
budugumbla (38.1. KANNADA)
budzi (38.14. SHONA)
budzi (38.16. SHONA)
budzi (38.17. SHONA)
budzi (38.18. SHONA)
buen enrique (35.5. SPANISH)
buenavista (39.3. BICOLANO, CENTRAL)
buenavista (39.3. ITAWIT)
buffalo bean (40.18. ENGLISH)
buffalo spinach (27.19. ENGLISH)
buga (23.3. LUO)
bugagan an dumaghag (36.3. IFUGAO)
buggapas di matikano (36.3. IFUGAO)
bugraya (11.1. BUHID)
bugsi (11.1. BUHID)
bugumji (44.8. FULFULDE, ADAMAWA)
bugumji debbi (44.10. FULFULDE, ADAMAWA)
bugumji gordi (44.8. FULFULDE, ADAMAWA)
bugumwol (44.8. FULFULDE, ADAMAWA)
bugumwol gordi (44.8. FULFULDE, ADAMAWA)
bugumwu (44.10. FULFULDE, ADAMAWA)
bui^{V} nyi ceh$_{V}$ (62.3. LAHU)
bûiki-i (40.26. HERERO)
buinneag Bhruisealach (30.29. GAELIC, SCOTTISH)
buisson du diable (26.4. FRENCH)
b[uj] b[aws]p (45.1. VIETNAMESE)
bujad (2.1. SERBO-CROATIAN)
b[uj]p gi[aas]m (45.4. VIETNAMESE)
buk (8.3. THAI)
bukawe (17.2. TAGALOG)
bukɔ (55.1. MENDE)
bukɔ (55.1. VAI)
bukɔ-lɛ (55.1. BULLOM SO)
bukke (38.14. OROMO, WEST CENTRAL)
bukke (38.16. OROMO, WEST CENTRAL)
bukke (38.17. OROMO, WEST CENTRAL)
bukke (38.18. OROMO, WEST CENTRAL)
bukke (38.20. OROMO, WEST CENTRAL)
bukkeert (40.5. NORWEGIAN)
bukkehorn (40.32. DANISH)
bukkehorn (40.32. NORWEGIAN)
bukketorn (62.8. NORWEGIAN)
buklad (36.3. MANOBO)
buko (41.1. ENGLISH)
buko (41.1. VLAAMS)
buk-ro (19.1. THAI)
bukyot (11.2. SAMBAL, BOTOLAN)
bulak-manok (27.14. TAGALOG)
bulay (40.9. TAGALOG)
bulay-patani (40.25. TAGALOG)
bulb onion (7.3. ENGLISH)
bulba (62.24. BELARUSAN)
bulb-bearing yam (11.2. ENGLISH)
bulbeh (62.24. YIDDISH)
bulbil-bearing yam (11.2. ENGLISH)
bule (40.25. PAMPANGAN)
bule (40.42. PAMPANGAN)
buley (11.5. BINUKID)
bulig (16.1a. BINUKID)
bulig (16.1a. MANOBO)
buligan (40.29. BONTOC CENTRAL)
buligan (40.29. IFUGAO)
buligan (40.29. KANKANAEY)
bulilinas (62.19. MANOBO)
buliyangin (38.20. SUBANON)
buliyas (7.3. IVATAN)
bulji (38.15. GONJA)
bulji (38.17. GONJA)
bulkɔ (55.1. BULLOM SO)
bull nose pepper (62.1. ENGLISH)
bulligan (40.29. GADDANG)
bulligan an gayyang (40.29. IFUGAO)
bulligan an mumpaha (40.29. IFUGAO)
bulmuj (24.11. KASHMIRI)
buloh batu (17.2. MALAY)
buloh beting (17.2. MALAY)
buloh betong (17.2. MALAY)
buloh duri (17.1. MALAY)
buloh kuning (17.1. MALAY)
buloh minyak (17.1. MALAY)
buloh panching (17.2. MALAY)
buloh sikai (17.1. MALAY)

burkoli (30.22. SUNDA)
burley (30.52. GAELIC, MANX)
burnotis (23.3. LITHUANIAN)
burokėlis (35.4. LITHUANIAN)
burokkori (30.31. JAPANESE)
buror buror (63.1. MANDING-MANDINKA)
burotkorii (30.31. JAPANESE)
burshaq (40.26. KAZAKH)
burshaq (40.28. KAZAKH)
burske ořišky (40.1. CZECH)
Bürüksel lahanasi (30.29. TURKISH)
burut (11.5. HANUNOO)
buruy (8.2. HANUNOO)
busag (16.1a. TAUSUG)
busanto (45.4. MALIMBA)
busarakaaya (62.12. TELUGU)
bušay (50.2. AMHARIC)
Buschmannspfeil (27.8. GERMAN)
bush kabej (30.17. KRIO)
bush okra (63.1. ENGLISH)
bush-kabej (30.7. KRIO)
bush-shuga-kɛn (17.3. KRIO)
busi (62.18. MANDING-MANDINKA)
busiburaxe (62.18. SUSU)
buskamarant (23.3. NORWEGIAN)
buskbønne (40.26. DANISH)
buskpurpurnesle (44.7. NORWEGIAN)
busktomat (62.9. NORWEGIAN)
buslig (10.2. BICOLANO, CENTRAL)
buslig (10.2. CEBUANO)
buslig (10.2. ILOCANO)
buslig (10.2. TAGALOG)
bussu (62.18. KURANKO)
bustān afrōz (23.3. KASHMIRI)
busthirse (17.6. NORWEGIAN)
busu (62.18. MANDING-MANDINKA)
busu bulé (11.2. BALANTA-KENTOHE)
busuioc (44.5. ROMANIAN)
busummuru adwerɛ (59.3. TWI)
but (40.5. BENGALI)
but (40.5. HINDI)
buṭ (40.5. SANTALI)
but (40.5. URDU)
buta (17.7. MARATHI)
buta (17.7. ORIYA)
buth sarana (9.1. SINHALA)
bu-thee (38.20. BURMESE)
butig (8.4. CEBUANO)
butima (43.1. JOLA-FONYI)
butinggan (62.9. ILOCANO)
butingi (40.25. TAGALOG)
butingi (40.26. TAGALOG)
butkalai (40.5. BENGALI)
buto sa a-i (17.7. MARANAO)
butsarana (9.1. BURMESE)
butter bean (40.25. ENGLISH)
butterbur (27.35. ENGLISH)
butterhead lettuce (27.31. ENGLISH)
butternut squash (38.17. ENGLISH)
Buttersalat (27.31. GERMAN)
butuan (11.2. TAGBANWA)
butungulu (7.3. GANDA)
būū (39.4. GURO)
buùrár zaákiì (38.12. HAUSA)
buurcha (40.28. KIRGHIZ)
buurchak (40.26. KIRGHIZ)
buxhak (40.10. ALBANIAN, TOSK)
buyah (7.5. IFUGAO)
buyas (7.5. IFUGAO)
buye (44.8. NYANJA)
buye wamthengo (44.8. NYANJA)
buyo-buyo (38.21. AKLANON)
buyo-buyo (38.21. CEBUANO)
buyotu (40.32. TURKISH)
büyük dulavratotu (27.5. TURKISH)
büyük galanga (20.1. TURKISH)
büyük isirganotu (66.3. TURKISH)
büyük köklü (33.2. TURKISH)
büyük latin çiçeği (65.1. TURKISH)
büyük semizotu (59.3. TURKISH)
büyük yam (11.1. TURKISH)
buz çiçeği (22.1. TURKISH)
buz otu (22.1. TURKISH)
bvani (38.3. VENDA)
bvembe (38.3. SHONA)
bvuh^{-} meu$^{\vee}$ ceh$_{\vee}$ (8.3. LAHU)
bwà (48.1a. LAO)
bwà khiiz bèèz (50.2. LAO)
bwaar (38.3. BALTI)
bwaay (17.1. CAROLINIAN)
bwaay (17.2. CAROLINIAN)
bwaay (17.4. CAROLINIAN)
bwàp khôm (38.22. LAO)
bwarupwav (62.19. SHONA)
bwàx fàlangx (7.3. LAO)
bwepapwe (40.3. MENDE)
bwokoli (30.31. CREOLE, HAITIAN FRENCH)
bwotaatas (62.24. KHMER, CENTRAL)
bwula (8.6. CAROLINIAN)

byadagi menasina kaayi (62.2. KANNADA)
byale (40.2. KANNADA)
bylica głupich (27.6. POLISH)
bys y blai'dd (40.14. WELSH)
bytel (35.2. CZECH)
bytel metlatý (35.2. CZECH)
bzar ahdar (62.1. MALTESE)
bzar ahmar (62.2. MALTESE)
bδkpinkδnδτμ (38.23. YOM)
ca (62.19. VIETNAMESE)
cà ân (62.25. VIETNAMESE)
cà bung (62.16. VIETNAMESE)
cà dai hoa tim (62.25. VIETNAMESE)
cà du (62.16. VIETNAMESE)
cà hoang (62.25. VIETNAMESE)
c[aa]u k[yr] t[uwr] (62.8. VIETNAMESE)
c[aa]u kh[owr]i (62.8. VIETNAMESE)
caayl (30.23. GAELIC, MANX)
caayl blaa (30.26. GAELIC, MANX)
cabaça (38.20. CRIOULO, UPPER GUINEA)
cabaça (38.20. PORTUGUESE)
cabaceira (38.18. GALICIAN)
cabaceira (38.20. CRIOULO, CAPE VERDE)
cabai (62.1. INDONESIAN)
cabai besar (62.1. INDONESIAN)
cabai besar (62.2. INDONESIAN)
cabai manis (62.1. INDONESIAN)
cabai rawit (62.5. INDONESIAN)
cabáiste (30.27a. GAELIC, IRISH)
cabáiste samhraidh (30.27a. GAELIC, IRISH)
cabakan (23.3. NEWAR)
cabáuste savoy (30.27a. GAELIC, IRISH)
cabbag (58.3. GAELIC, MANX)
cabbage (30.27a. FIJIAN)
cabbage (30.27b. ENGLISH)
cabbage (30.28. ENGLISH)
cabbage (30.34. ENGLISH)
cabbash (30.27a. GAELIC, MANX)
cabbitch (30.27a. SCOTS)
cabe (62.1. JAVANESE)
cabe (62.1. SUNDA)
cabe bendot (62.6. JAVANESE)
cabe besar (62.1. INDONESIAN)
cabe cengek (62.5. SUNDA)
cabe Dieng (62.6. JAVANESE)
cabe gondol (62.6. JAVANESE)
cabe merah (62.1. INDONESIAN)
cabe rawit (62.5. INDONESIAN)
cabe rawit (62.5. SUNDA)
cabe setan (62.5. INDONESIAN)
cabezas de negrito (10.2. SPANISH)
cabola (7.3. CRIOULO, CAPE VERDE)
cabolinha (7.5. CRIOULO, CAPE VERDE)
cacachuete de bambara (40.39. SPANISH)
cacahouette (40.1. FRENCH)
cacahuete (40.1. SPANISH)
cacauet (40.1. CATALAN-VALENCIAN-BALEAR)
cacomite (12.1. FRENCH)
cactus (32.1. CATALAN-VALENCIAN-BALEAR)
cadillo negro (27.8. SPANISH)
cadjan pea (40.2. ENGLISH)
c[af] b[as]t (62.19. VIETNAMESE)
c[af] bung (62.16. VIETNAMESE)
c[af] chua (62.9. VIETNAMESE)
c[af] d[aj]i hoa t[is]m (62.25. VIETNAMESE)
c[af] d[aj]i hoa tr[aws]ng (62.23. VIETNAMESE)
c[af] hoang (62.23. VIETNAMESE)
c[af] hoang (62.25. VIETNAMESE)
c[af] r[oos]t (24.11. VIETNAMESE)
c[af] t[is]m (62.19. VIETNAMESE)
café de Mato Grosso (40.18. PORTUGUESE)
cago (20.2. FIJIAN)
cahctus bahr (32.1. YIDDISH)
cahou (7.9a. KABIYÉ)
cahoun (7.3. KABIYÉ)
cai (30.52. SUNDA)
cai cong (7.14. CHINESE, MIN BEI)
cai dou (40.26. CHINESE, MANDARIN)
cai dou (40.26. CHINESE, WÚ)
cai dou (40.26. CHINESE, YUE)
cai dou (40.41. CHINESE, MIN BEI)
cai gua (38.13a. CHINESE, MIN BEI)
cai gua (38.13a. CHINESE, YUE)
cai hua (30.26. CHINESE, MIN BEI)
cai hua (30.26. CHINESE, YUE)
cai jue (5.1. CHINESE, MANDARIN)
cai tai (30.20. CHINESE, YUE)
cai tai (30.37. CHINESE, MANDARIN)
cai tai (30.37. CHINESE, YUE)
cai tou (30.56. CHINESE, MIN BEI)
cai xin (30.37. CHINESE, WÚ)
cai xin (30.37. CHINESE, YUE)
cai yao li (38.25. CHINESE, MANDARIN)
cai yao li (38.25. CHINESE, WÚ)
cai young da dou (40.8. CHINESE, WÚ)
cai young da dou (40.38. CHINESE, WÚ)
cai young da dou (40.8. CHINESE, YUE)

caiano (40.2. ITALIAN)
caibus (38.19. VLAAMS)
caigua (38.19. SPANISH)
caihua (38.19. POLISH)
cainneann (7.8. GAELIC, IRISH)
cainneann (7.8. GAELIC, SCOTTISH)
caiota (38.25. PORTUGUESE)
cairussú (24.7. PORTUGUESE)
caisearbhán (27.45. GAELIC, IRISH)
caisim (30.17. INDONESIAN)
caisin (30.37. ENGLISH)
caisin (30.37. INDONESIAN)
cajan (40.2. VLAAMS)
cajan pea (40.2. ENGLISH)
cajanbønne (40.2. NORWEGIAN)
čajot (38.25. BULGARIAN)
čajot (38.25. CZECH)
cajot (38.25. RUSSIAN)
cajuá (38.26. SPANISH)
cajuba (38.26. SPANISH)
caka (40.25. MENDE)
čakanka obyčajná (27.10a. SLOVAK)
čakanka šalatná (27.10a. SLOVAK)
čakanka štrbák (27.9. SLOVAK)
čakndech (35.4. ARMENIAN)
càl (30.27a. GAELIC, SCOTTISH)
cál catach (30.35. GAELIC, IRISH)
cál ilcheannach (30.33. GAELIC, IRISH)
calabacin (38.18. SPANISH)
calabash gourd (38.20. ENGLISH)
calabaza africana (38.23. SPANISH)
calabaza almiscarada (38.17. SPANISH)
calabaza almisclada (38.17. SPANISH)
calabaza amelonada (38.17. SPANISH)
calabaza branca (38.1. SPANISH)
calabaza china (38.1. SPANISH)
calabaza común (38.18. SPANISH)
calabaza de aristas (38.21. SPANISH)
calabaza de aristas (38.22. SPANISH)
calabaza de Castilla (38.16. SPANISH)
calabaza de culebra (38.29. SPANISH)
calabaza de peregrinos (38.20. SPANISH)
calabaza de San Juan (38.18. SPANISH)
calabaza de San Roque (38.20. SPANISH)
calabaza grande (38.16. SPANISH)
calabaza moscada (38.17. SPANISH)
calabaza pamué (38.5. SPANISH)
calabaza viñadera (38.20. SPANISH)
calabaza viñatera (38.20. SPANISH)
calabrese (30.31. ENGLISH)
calalou (8.8. ENGLISH)
calalou (8.8. FRENCH)
calalou (8.8. VLAAMS)
calalu (8.8. PORTUGUESE)
calalú (8.8. SPANISH)
cálamate (62.9. MANDJAK)
calaraba (30.30. ROMANIAN)
calathea (15.1. HEBREW)
càl-colaig (30.26. GAELIC, SCOTTISH)
calebasse (38.20. FRENCH)
calebasse zombi (38.26. FRENCH)
calebassier (38.20. FRENCH)
càl-gruthach (30.26. GAELIC, SCOTTISH)
çali fasulyesi (40.23. TURKISH)
cali fasulyesi (40.24. TURKISH)
calingan rambat (24.7. JAVANESE)
callish (30.31. GAELIC, MANX)
calnabi (30.26. GEORGIAN)
caltrop (64.2. ENGLISH)
çam otu (1.1. TURKISH)
camapú (62.13. PORTUGUESE)
camapú bexiga (62.12. PORTUGUESE)
camate (62.9. CRIOULO, UPPER GUINEA)
camatom (38.29. CRIOULO, UPPER GUINEA)
camawak (27.42. SUNDA)
cambroeira da China (62.8. PORTUGUESE)
cambronera de la China (62.8. SPANISH)
čamgur (30.41a. KIRGHIZ)
camó dandandim (11.2. MANDING-MANDINKA)
camote (36.3. SPANISH)
camotillo (36.2. SPANISH)
camoting (39.4. ITAWIT)
campainha do Japão (36.1. PORTUGUESE)
campainha rabanete (33.2. PORTUGUESE)
campânula (33.2. PORTUGUESE)
campanula commestibile (33.2. ITALIAN)
campanule raiponce (33.2. FRENCH)
çan çiçeği (33.2. TURKISH)
çan çiçeği (34.3. TURKISH)
can dou (40.34a. CHINESE, WÚ)
can dou (40.35. CHINESE, YUE)
can koko (20.2. FON)
can shu (11.1. CHINESE, YUE)
cana (9.1. ROMANIAN)
cana (17.5. CRIOULO, CAPE VERDE)
cana (40.5. KASHMIRI)
canā (40.5. NEWAR)
caña comestible (9.1. SPANISH)

cana comestível (9.1. PORTUGUESE)
cana da Índia (9.1. PORTUGUESE)
cana de açúcar (17.5. PORTUGUESE)
caña de azúcar (17.5. SPANISH)
caña de las Indias (9.1. SPANISH)
cana-de-india (9.1. CRIOULO, CAPE VERDE)
caṇagi (40.5. BADAGA)
canaka (40.5. SANSKRIT)
canakah (40.5. SANSKRIT)
caṇako (40.5. PALI)
cañamiel (17.5. SPANISH)
cañamo de Guinea (45.4. SPANISH)
canavalia (40.3. ITALIAN)
canavalia (40.3. SPANISH)
canavalia (40.4. ITALIAN)
canavarotu (52.1. TURKISH)
cancon (36.2. PORTUGUESE)
candrasura (30.51. SANSKRIT)
cangarinha (27.36. PORTUGUESE)
cangkok manis (39.5. MALAY)
canhamo do Decão (45.4. PORTUGUESE)
caŋhaŋpi hu (17.5. DAKOTA)
caŋḣdoḣuhu opemni ka'ġapi (58.1. DAKOTA)
çani (33.3. TURKISH)
canja (45.1. CRIOULO, UPPER GUINEA)
canker (27.45. ENGLISH)
cankerwort (27.45. ENGLISH)
canna (9.1. FILIPINO)
canna (9.1. FRENCH)
canna (9.1. NORWEGIAN)
canna commestible (9.1. ITALIAN)
canna da zucchero (17.5. ITALIAN)
canna d'Inde (9.1. FRENCH)
canna d'India (9.1. ITALIAN)
canna dolce (9.1. ITALIAN)
canna edule (9.1. ITALIAN)
canna hodi (9.1. HEBREW)
canne à sucre (17.5. FRENCH)
canonga (27.8. NYANJA)
canonges (67.2. CATALAN-VALENCIAN-BALEAR)
cantaloup (38.7a. FRENCH)
cantaloupe (38.7a. ARABIC)
cantaloupe (38.7a. ENGLISH)
cantaloupe (38.7a. FILIPINO)
cantaloupe (38.7a. FINNISH)
cantaloupe (38.7a. SWEDISH)
cantaloupe (38.7a. VLAAMS)
cao di bo lou men shen (27.47. CHINESE, MANDARIN)
cao di bo luo men shen (27.47. CHINESE, WÚ)
cao gua (38.9. CHINESE, XIANG)
cao shi can (44.11. CHINESE, MANDARIN)
cao shi can (44.11. CHINESE, MIN BEI)
cao shi can (44.11. CHINESE, WÚ)
cao shi can (44.11. CHINESE, YUE)
cápè (39.4. PULAAR)
cape asparagus (25.1. ENGLISH)
cape gooseberry (62.12. ENGLISH)
cape pondweed (25.1. ENGLISH)
caper plant (34.1. ENGLISH)
capere (34.1. ROMANIAN)
capim da Rodésia (17.3. PORTUGUESE)
capim elefante (17.3. PORTUGUESE)
capim napier (17.3. PORTUGUESE)
capinu (38.13. ITAWIT)
capitata mustard (30.4. ENGLISH)
càpparo (34.1. ITALIAN)
cappata (23.5. ZARMA)
capper (34.1. ENGLISH)
capper bush (34.1. ENGLISH)
càppero (34.1. ITALIAN)
cappuccina (65.2. ITALIAN)
câprier (34.1. FRENCH)
căprişor (10.1. ROMANIAN)
căprişor de mîncare (10.1. ROMANIAN)
capselle bourse à pasteur (30.43. FRENCH)
capsicum (62.1. ASSAMESE)
capsicum pepper (62.1. ENGLISH)
capuchina tuberosa (65.2. SPANISH)
capuchinha grande (65.1. PORTUGUESE)
capuchinha tuberosa (65.2. PORTUGUESE)
capucine (65.1. FRENCH)
capucine tubéreuse (65.2. FRENCH)
capulí (62.12. SPANISH)
cará (11.1. PORTUGUESE)
cará aereo (11.2. PORTUGUESE)
cará amarelo (11.6. PORTUGUESE)
cará branco (11.1. PORTUGUESE)
cará branco (11.1. SPANISH)
cará da terra (11.1. PORTUGUESE)
cará de Angola (11.1. PORTUGUESE)
cará de sapateiro (11.2. PORTUGUESE)
cará do Pará (11.3. PORTUGUESE)
cará doce (11.8. PORTUGUESE)
cará maco (15.2. SPANISH)
cará mimoso (11.8. PORTUGUESE)
cará tayá (8.1. SPANISH)

carabassa (38.14. CATALAN-VALENCIAN-BALEAR)
carabassa (38.16. CATALAN-VALENCIAN-BALEAR)
carabassa (38.17. CATALAN-VALENCIAN-BALEAR)
carabassa (38.18. CATALAN-VALENCIAN-BALEAR)
carabassa rodona (38.16. CATALAN-VALENCIAN-BALEAR)
carabu (30.51. ARABIC)
caragilates (40.43. SPANISH)
Caraibcher Pfeffer (62.6. GERMAN)
čaramekə (11.8. PUMÉ)
caramunnipayara (40.43. TAMIL)
caramunnypre (40.43. TAMIL)
caranza (38.23. ITALIAN)
carbassa (38.14. CATALAN-VALENCIAN-BALEAR)
carbassa (38.16. CATALAN-VALENCIAN-BALEAR)
carbassa (38.17. CATALAN-VALENCIAN-BALEAR)
carbassa (38.18. CATALAN-VALENCIAN-BALEAR)
carcade (45.4. ITALIAN)
čarchali (35.4. GEORGIAN)
carciofo (27.15. ITALIAN)
carciofo (27.16. ITALIAN)
card bort (27.36. CATALAN-VALENCIAN-BALEAR)
cardama (30.52. GALICIAN)
cardamina (30.44. SPANISH)
cardamina dos prados (30.44. PORTUGUESE)
cardamine des prés (30.44. FRENCH)
cardamino dei prati (30.44. ITALIAN)
cardamon (30.51. GREEK)
cardea (40.14. GALICIAN)
cardeda (27.42. ITALIAN)
cardillo (27.36. SPANISH)
cardo (27.15. CRIOULO, CAPE VERDE)
cardo branco (27.37. PORTUGUESE)
cardo coalhador (27.15. PORTUGUESE)
cardo comestible (27.15. SPANISH)
cardo de huerta (27.15. SPANISH)
cardo de ouro (27.36. PORTUGUESE)
cardo do coalho (27.15. PORTUGUESE)
cardo hortense (27.15. PORTUGUESE)
cardo scòlimo (27.36. ITALIAN)
cardon (27.15. FRENCH)
cardon (27.15. NORWEGIAN)
cardon d'Espagne (27.36. FRENCH)
cardoncello salvatico (27.36. ITALIAN)
cardone (27.15. ITALIAN)
cardonul (27.15. ROMANIAN)
cardoon (27.15. ENGLISH)
carduille (27.36. FRENCH)
careless weed (23.3. ENGLISH)
carfiol (30.26. ROMANIAN)
c[ar]i b[aws]c kinh (30.40. VIETNAMESE)
c[ar]i b[aws]p (30.27a. VIETNAMESE)
c[ar]i b[aws]p (30.27b. VIETNAMESE)
c[ar]i b[aws]p (30.34. VIETNAMESE)
c[ar]i b[ej] tr[aws]ng (30.36. VIETNAMESE)
c[ar]i b[ej] xanh (30.17. VIETNAMESE)
c[ar]i c[ur]i (30.54. VIETNAMESE)
c[ar]i c[ur]i (30.56. VIETNAMESE)
c[ar]i c[us]c (27.23. VIETNAMESE)
c[ar]i canh (30.17. VIETNAMESE)
c[ar]i d[uw][owf]ng (35.4. VIETNAMESE)
c[ar]i hoa (30.26. VIETNAMESE)
c[ar]i ng[oj]t (30.36. VIETNAMESE)
c[ar]i soong (30.52. VIETNAMESE)
c[ar]i thi[af] (30.36. VIETNAMESE)
caribnettle (62.15. ENGLISH)
cariéad (24.11. GAELIC, IRISH)
carilla fruit (38.23. ENGLISH)
caringan (27.8. INDONESIAN)
carisso-da-guiné (17.1. CRIOULO, CAPE VERDE)
çarkifelek (54.1. TURKISH)
carmantina (21.1. ITALIAN)
carnapeet (30.26. ARABIC)
carot (24.11. MÒORÉ)
carot (24.11. SERER-SINE)
carota (24.11. ITALIAN)
carota rossa (24.18. ITALIAN)
caroti (24.11. BAOULÉ)
caroti (24.11. GOURMANTCHÉ)
caroti (24.11. JULA)
carotte (24.11. CAJUN FRENCH)
carotte (24.11. FRENCH)
carpas (24.5. HEBREW)
carradje (24.11. GAELIC, MANX)
carrapicho grande (27.5. PORTUGUESE)
carrot (24.11. ENGLISH)
carrot (24.11. FILIPINO)
carrot (24.11. MALAYALAM)
carrot (24.11. WOLOF)
carroti (24.11. DAGAARE, SOUTHERN)

carroti (24.11. GOURMANCHÉMA)
cart comu (27.15. CATALAN-VALENCIAN-BALEAR)
cartof (62.24. ROMANIAN)
cartof brambor (11.1. ROMANIAN)
cartof dulce (36.3. ROMANIAN)
cartophili (62.24. GEORGIAN)
carurú (23.3. PORTUGUESE)
caruru azedo (45.4. PORTUGUESE)
carurú da Baía (63.1. PORTUGUESE)
carurú d'água (40.19. PORTUGUESE)
carves (58.3. GALICIAN)
carxofa (27.16. CATALAN-VALENCIAN-BALEAR)
casaba melons (38.11. ENGLISH)
casabanana (38.26. DANISH)
casabanana (38.26. ENGLISH)
casabanana (38.26. SPANISH)
casba (38.11. VLAAMS)
cascellore orientale (30.42. ITALIAN)
cashnich (24.9. UZBEK)
cassango (38.6. SENA)
cassava (39.4. ENGLISH)
cassava (39.4. SWEDISH)
Cassavastrauch (39.4. GERMAN)
cassave (39.4. DUTCH)
cassave (39.4. FRENCH)
cassàve (39.4. ITALIAN)
cassave (39.4. VLAAMS)
cassussus (62.11. PORTUGUESE)
castagna d'acqua (64.1. ITALIAN)
castagna d'ácqua singara (64.2. ITALIAN)
castagna de água singara (64.2. SPANISH)
castaña de água (64.1. SPANISH)
Castaña de agua (64.1. SPANISH)
castanha de água (64.1. PORTUGUESE)
castanha de água singara (64.2. PORTUGUESE)
castanha de Inhambane (38.28. PORTUGUESE)
castanheiro de água (10.2. PORTUGUESE)
castawwydden dwr (10.2. WELSH)
castravete (38.13a. ROMANIAN)
castravetele (38.13a. ROMANIAN)
cat bean (40.18. ENGLISH)
catandambundo (44.5. SHONA)
catinhá-ô (7.2. MANDING-MANDINKA)
catjang (40.41. DANISH)
catjang (40.41. ENGLISH)
catjang börülcesi (40.41. TURKISH)
Catjangbohne (40.41. GERMAN)
catjangbønne (40.41. NORWEGIAN)
catoco (11.2. MANKANYA)
cà-tô-mách (62.9. VIETNAMESE)
cat's whiskers (34.2. ENGLISH)
cau galek (16.1a. SUNDA)
cau tanduk (16.1a. SUNDA)
cauda de cavalo (1.1. PORTUGUESE)
cauliflower (30.26. ENGLISH)
cauliflower (30.26. FILIPINO)
cauliflower (30.26. KANNADA)
cauliflower (30.26. MALAYALAM)
cauliflower (30.26. TELUGU)
caupi (40.43. PORTUGUESE)
caupi (40.43. SPANISH)
cauraiya (23.3. HINDUSTANI)
cavalinha (1.1. PORTUGUESE)
cavalinha dos campos (1.1. PORTUGUESE)
cavolfiore (30.26. ITALIAN)
cavolo (30.32. ITALIAN)
cavolo a penna (30.35. ITALIAN)
cavolo bianco (30.27a. ITALIAN)
cavolo broccoli (30.31. ITALIAN)
cavolo cappuccio (30.27a. ITALIAN)
cavolo cavaliere (30.35. ITALIAN)
cavolo cinese (30.40. ITALIAN)
cavolo da foraggio (30.35. ITALIAN)
cavolo di Bruxelles (30.29. ITALIAN)
cavolo di Cina (30.40. ITALIAN)
cavolo di mare (30.45. ITALIAN)
cavolo marino (30.45. ITALIAN)
cavolo navone (30.21. ITALIAN)
cavolo rapa (30.30. ITALIAN)
cavolo riccio (30.35. ITALIAN)
cavolo rossa (30.27b. ITALIAN)
cavolo senape-nera (30.24. ITALIAN)
cavolo tronchuda (30.28. ITALIAN)
cavolo verzotto (30.34. ITALIAN)
cay dai hoang (58.1. VIETNAMESE)
cay dau chieu (40.2. VIETNAMESE)
caya (34.2. SPANISH)
cayenne peper (62.2. VLAAMS)
cayenne peper (62.5. VLAAMS)
cayenne peppar (62.2. SWEDISH)
cayenne pepper (62.2. ENGLISH)
Cayenne Pfeffer (62.2. GERMAN)
Cayenne Yamswurzel (11.3. GERMAN)
cayennepeber (62.5. DANISH)
cayennepippuri (62.3. FINNISH)
çayir bezelyesi (40.10. TURKISH)

chikandar (35.4. PANJABI, EASTERN)
chikasu (30.35. LOMWE, MALAWI)
chikek (38.13a. KURDISH, NORTHERN)
chikesoppu (22.2. KANNADA)
chikilje (38.7b. FULFULDE, ADAMAWA)
chikkeere soppu (23.3. KANNADA)
chikkere soppu (23.3. KANNADA)
chikkudu (40.9. TELUGU)
ch'il łichí'í (62.9. NAVAJO)
ch'il łigaaí (27.31. NAVAJO)
ch'il łigaaí (27.32. NAVAJO)
ch'il łigaií (30.26. NAVAJO)
ch'il łigaií (30.27a. NAVAJO)
chilacayote (38.15. SPANISH)
chilagadadumpa (36.3. TELUGU)
chilagadam (36.3. TELUGU)
chilaka thota koora (23.3. TELUGU)
chilazi (8.3. LOMWE, MALAWI)
chilazi (11.1. LOMWE, MALAWI)
chilazi (11.2. LOMWE, MALAWI)
chilazi (11.3. LOMWE, MALAWI)
chilazi (11.7. LOMWE, MALAWI)
chilbughum (1.1. TAJIKI)
chilchil (23.4. PANJABI, EASTERN)
chile (62.2. SPANISH)
chile (62.5. SPANISH)
chile dulce (62.1. SPANISH)
chile manzana (62.6. SPANISH)
chilepipar (62.2. ICELANDIC)
chili (62.2. ENGLISH)
chili peber (62.5. DANISH)
chili pepper (62.2. ENGLISH)
chilikaesoppu (23.3. KANNADA)
chilikeerae soppu (23.3. KANNADA)
chilikesoppu (23.3. KANNADA)
chilikiraesoppu (23.3. KANNADA)
chilimkadu (38.20. TAJIKI)
chilipippuri (62.5. FINNISH)
chilli manzana (62.6. ENGLISH)
Chilli Pfeffer (62.2. GERMAN)
chilli spanskpeppar (62.5. SWEDISH)
chillipeper (62.5. DUTCH)
chilmangushtak (59.2. TAJIKI)
chimaito (40.1. LUYIA)
chimangue (39.4. RONGA)
chimapaya (27.4. ENGLISH)
chimbande (40.39. LUYIA)
chimerino kolokynthi (38.14. GREEK)
chimeu dulce (24.13. ROMANIAN)
chimtza tarbutit (40.5. HEBREW)
chimuka (38.1. LUYIA)
ch'in (24.5. CHINESE, MANDARIN)
chi:n (44.7. LAO)
ch'in ts'ai (24.4. CHINESE, MANDARIN)
china cabbage (30.36. FIJIAN)
china kloubnenosnaja (40.10. BELARUSAN)
china kloubnenosnaja (40.11. RUSSIAN)
china klubnevaja (40.11. KAZAKH)
China lai (30.40. ASSAMESE)
China leek (7.13. ASSAMESE)
China mula (30.54. ASSAMESE)
china posevnaja (40.10. KAZAKH)
chinakarkatika (38.29. SANSKRIT)
chinakee beeja (40.18. KANNADA)
Chinakohl (30.36. GERMAN)
Chinakohl (30.40. GERMAN)
Chinalauch (7.13. GERMAN)
chinangira (39.4. SHONA)
chinangwa (39.4. LOMWE, MALAWI)
chin-baung (45.4. BURMESE)
chinchatakam (20.6. MALAYALAM)
chin-chên-ts'ao (13.3. CHINESE, MANDARIN)
chinchiver (20.6. MALAYALAM)
chinchon (27.32. FRENCH)
Chinees klokje (34.3. VLAAMS)
chinese almond (40.1. ENGLISH)
Chinese arrowhead (6.1. ENGLISH)
Chinese artichoke (44.11. ENGLISH)
Chinese basil (44.7. ENGLISH)
Chinese bellflower (34.3. ENGLISH)
Chinese bieslook (7.13. VLAAMS)
Chinese boksdoorn (62.8. DUTCH)
Chinese boxthorn (62.8. ENGLISH)
Chinese brinjal (62.19. ENGLISH)
Chinese broccoli (30.25. ENGLISH)
Chinese broccoli (30.25. VLAAMS)
Chinese cabbage (30.36. ENGLISH)
Chinese cabbage (30.40. ENGLISH)
Chinese chive (7.13. ENGLISH)
Chinese kale (30.25. ENGLISH)
Chinese kool (30.40. VLAAMS)
Chinese lantern plant (62.11. ENGLISH)
Chinese leek (7.13. ENGLISH)
Chinese matrimony vine (62.8. ENGLISH)
Chinese mustard (30.17. ENGLISH)
Chinese mustard (30.36. ENGLISH)
Chinese oker (38.21. VLAAMS)
Chinese okra (38.21. ENGLISH)
Chinese pea (40.27. ENGLISH)

Chinese pechay (30.36. FILIPINO)
Chinese preserving melon (38.1. ENGLISH)
Chinese radish (30.54. ENGLISH)
Chinese spinach (23.3. ENGLISH)
Chinese water chestnut (10.2. ENGLISH)
Chinese water-lily (48.1a. ENGLISH)
Chinese wolfberry (62.8. ENGLISH)
Chinese wolfbes (62.8. VLAAMS)
Chinese yam (11.5. ENGLISH)
Chinese yam (11.6. ENGLISH)
Chinesekool (30.40. DUTCH)
Chinesische Wassernuß (64.1. GERMAN)
ChinesischeYamswurze (11.6. GERMAN)
Chinesischer Bocksdorn (62.8. GERMAN)
chingil (62.8. TAJIKI)
chingsäy (24.4. UYGHUR)
chin-hgüan-ts'ao (13.3. CHINESE, MANDARIN)
chini badam (40.1. BENGALI)
chini kaayi (38.16. KANNADA)
chini sir naza (6.1. URDU)
chini yam (11.1. URDU)
chinikaram (30.25. TAJIKI)
chinjugu (40.1. LUYIA)
chinkhupule (38.22. LOMWE, MALAWI)
chinólyio (24.13. LADINO)
chīntshεε (24.4. TIBETAN)
chinuri combosto (30.40. GEORGIAN)
chinyanya (62.9. LUYIA)
chio (16.1a. CHINESE, MIN NAN)
chipanda (38.20. LOMWE, MALAWI)
chipanza (27.24. SHONA)
chipari-aalu (11.1. MARATHI)
chiporro (40.18. SPANISH)
chiral (38.7b. ASSAMESE)
chirbot (62.12. MARATHI)
chireg (40.32. MONGOLIAN)
chirigaita (38.15. SPANISH)
chiripathra (38.23. SANSKRIT)
chiriputla (62.12. KONKANI)
chirisanagalu (40.12. TELUGU)
chirivanga (62.19. TELUGU)
chirivia (24.18. PORTUGUESE)
chirivía (24.18. SPANISH)
chirivía de azúcar (24.23. SPANISH)
Chiromon-Kürbis (38.18. GERMAN)
chirput (62.12. KONKANI)
chirtchhada (35.9. SANSKRIT)
chiruvanga (62.19. TELUGU)
chisaca (27.2. SPANISH)
chiso (10.1. HAUSA)
chisokono (27.8. TUMBUKA)
chissô (44.7. PORTUGUESE)
chithra (38.3. SANSKRIT)
chithraphala (38.3. SANSKRIT)
chithraphala (62.19. SANSKRIT)
chithravallika (38.3. SANSKRIT)
chitlanje ghams (27.22. NEPALI)
chitta (62.1. FULFULDE, ADAMAWA)
chitta kolong (62.1. FULFULDE, ADAMAWA)
chittāje (62.1. FULFULDE, ADAMAWA)
chitti saronh (30.57. PANJABI, EASTERN)
chitunguu madzi (7.3. DIGO)
chitunguu saumu (7.10. DIGO)
ch'iu k'uei (45.1. CHINESE, MANDARIN)
chive (7.11. ASSAMESE)
chive (7.11. ENGLISH)
chives (7.11. TAGALOG)
chiwere tsanga (17.4. SHONA)
chizambana (17.8. RONGA)
chizran (17.1. HEBREW)
chizran (17.2. HEBREW)
chizran (17.4. HEBREW)
chmel (46.1. KAZAKH)
chmel' (46.1. SLOVAK)
chmel obecny (46.1. CZECH)
chmel otáčivý pazochy (46.1. CZECH)
chmiel zwyczajny (46.1. POLISH)
choach akod (27.37. HEBREW)
choach spharadi (27.36. HEBREW)
cho-cho (38.25. ENGLISH)
cho-cho (38.25. KRIO)
chocho (38.25. SPANISH)
chocho (40.14. QUECHUA, CUSCO)
choghondar (35.4. FARSI, WESTERN)
chohalā (40.5. NEWAR)
choisum (30.37. VLAAMS)
choke gohbi (30.29. URDU)
choke kobee (30.29. HINDI)
choko (38.25. TOK PISIN)
chola (40.5. BENGALI)
chola (40.5. HINDI)
chola (40.5. PANJABI, EASTERN)
chola (40.43. GUJARATI)
cholai (23.3. BENGALI)
cholai (23.3. GUJARATI)
cholai (23.3. HINDI)
cholam (17.7. MALAYALAM)
choli (23.3. HINDI)
choli (40.43. GUJARATI)

chombujiggangi (30.49. KOREAN)
chon yai (2.1. THAI)
chong cai (30.1. CHINESE, YUE)
chong cai (30.14. CHINESE, YUE)
chono (40.5. KONKANI)
chonon kharmag (62.8. MONGOLIAN)
chookoondòori (35.4. ROMANI, BALKAN)
choorana (8.3. KANNADA)
choorna gedde (8.3. KANNADA)
chop suey green (27.23. ENGLISH)
chorakka (38.22. MALAYALAM)
chorap (40.43. GUJARATI)
chormaghzi zamini (40.1. TAJIKI)
chorogi (44.11. JAPANESE)
choroko (40.26. SWAHILI)
choroko (40.34a. SWAHILI)
choroko (40.38. SWAHILI)
choroko (40.41. SWAHILI)
choroko (40.42. SWAHILI)
choroko (40.43. SWAHILI)
chorom-samit (24.12. ARMENIAN)
chosen azami (27.16. JAPANESE)
chota kanwal (50.2. PANJABI, EASTERN)
chotimirich (62.3. URDU)
chotti chaulai (23.3. HINDI)
chou (8.9. KWÉYÒL)
chou (30.27a. CAJUN FRENCH)
chou (30.27a. CREOLE, HAITIAN FRENCH)
chou à cent têtes (30.33. FRENCH)
chou à faucher (30.23. FRENCH)
chou à grosses côtes (30.28. FRENCH)
chou à mille têtes (30.33. FRENCH)
chou blanc (30.27a. FRENCH)
chou branchu (30.33. FRENCH)
chou brocoli (30.31. FRENCH)
chou cabus (30.27a. FRENCH)
chou cabus (30.27b. FRENCH)
chou cabus (30.34. FRENCH)
chou caraibe (8.9. FRENCH)
chou cavalier (30.35. FRENCH)
chou chinois (30.40. FRENCH)
chou de Chine (30.36. FRENCH)
chou de Chine (30.40. FRENCH)
chou de Milan (30.34. FRENCH)
chou de savoie (30.34. FRENCH)
chou de Shangton (30.36. FRENCH)
chou de Shangton (30.40. FRENCH)
chou de Sibérie (30.23. FRENCH)
chou diri na bifoulou (30.26. BEMBÉ)
chou diri na bifoulou (30.26. KOONGO)
chou faux jone (30.17. FRENCH)
chou fleur (30.26. FRENCH)
chou fourrager (30.35. FRENCH)
chou frisé (30.35. FRENCH)
chou frizé (30.34. CREOLE, HAITIAN FRENCH)
chou frizé (30.35. CREOLE, HAITIAN FRENCH)
chou ji zhuang che qian (57.1. CHINESE, MANDARIN)
chou ji zhuang che qian (57.1. CHINESE, WÚ)
chou marim de Hongrie (30.46. FRENCH)
chou marin (30.45. FRENCH)
chou moëllier (30.32. FRENCH)
chou navet (30.21. FRENCH)
chou pommé (30.27a. FRENCH)
chou ponmen (30.27a. CREOLE, HAITIAN FRENCH)
chou rav (30.30. CRIOULO, CAPE VERDE)
chou rave (30.30. FRENCH)
chou rouge (30.27b. FRENCH)
chou tartare (30.46. FRENCH)
chou tronchuda (30.28. FRENCH)
chou vert (30.35. FRENCH)
chou yeh wo chü (27.32. CHINESE, MANDARIN)
choû (ŷt-t'iaô) (62.24. CHINESE, HAKKA)
chouchou (38.25. EWONDO)
chouchou (38.25. FRENCH)
chouchoutte (38.25. FRENCH)
chouflè (30.26. CREOLE, HAITIAN FRENCH)
Choufleur (30.26. LUXENBOURGEOIS)
chòui-lĭt (64.1. CHINESE, HAKKA)
choulai (23.3. URDU)
chounavet (30.22. FRENCH)
chou-ponm (30.27a. KWÉYÒL)
choux de Bruxelles (30.29. FRENCH)
chov-fle (30.26. KWÉYÒL)
chow-chow (38.25. ENGLISH)
chow-chow kaayi (38.25. KANNADA)
chowlaayi (23.3. HINDI)
chowli (40.43. HINDI)
chowlikodu (40.6. TULU)
choy sum (30.37. DUTCH)
choy sum (30.37. ENGLISH)
choy sum (30.37. FRENCH)
choy sum (30.37. PORTUGUESE)
choy sum (30.37. SPANISH)
choya (45.4. SWAHILI)

chrach (18.1. KHMER, CENTRAL)
chrach (18.2. KHMER, CENTRAL)
chrâlong (28.1. KHMER, CENTRAL)
chren (30.1. KAZAKH)
chren (30.1. KIRGHIZ)
chren (30.1. RUSSIAN)
chren (30.1. SLOVAK)
chrĕn (30.1. SORBIAN, UPPER)
chřest (13.2. CZECH)
chřest lekařsky (13.2. CZECH)
chřest obecný (13.2. CZECH)
chrìn (30.1. UKRAINIAN)
christophene (38.25. RAROTONGAN)
christophine (38.25. FRENCH)
chrizantema uvenchanaja (27.23. BELARUSAN)
chrjan (30.1. BULGARIAN)
chrovt (30.34. SLOVENIAN)
chrysant (27.23. DUTCH)
chrysant (27.23. VLAAMS)
chrysanthème (27.23. FRENCH)
Chrysantheme (27.23. GERMAN)
chrysanthème des jardins (27.23. FRENCH)
chrysanthemo to stefanoto (27.23. GREEK)
chrysanthemum (27.23. FRENCH)
chryzantéma (27.23. SLOVAK)
chrzan (30.1. POLISH)
chrzan pospolity (30.1. POLISH)
chû (11.1. CHINESE, MIN NAN)
chû (11.2. CHINESE, MIN NAN)
chû (11.3. CHINESE, MIN NAN)
chû (11.4. CHINESE, MIN NAN)
chû (11.5. CHINESE, MIN NAN)
chû (11.6. CHINESE, MIN NAN)
chû (11.7. CHINESE, MIN NAN)
chû (11.8. CHINESE, MIN NAN)
chũ (30.27a. GOURMANTCHÉ)
chũ (30.25. MÒORÉ)
chũ (30.26. DAGAARE, SOUTHERN)
chũ (30.26. MÒORÉ)
chũ (30.27a. DAGAARE, SOUTHERN)
chũ (30.27a. MÒORÉ)
chũ (30.28. MÒORÉ)
chũ (30.29. MÒORÉ)
chũ (30.30. MÒORÉ)
chũ (30.31. MÒORÉ)
chũ (30.32. MÒORÉ)
chũ (30.34. MÒORÉ)
chũ (30.35. MÒORÉ)
chû (62.24. CHINESE, MIN NAN)
chü chi (27.16. CHINESE, MANDARIN)
chũ feron (30.26. GOURMANTCHÉ)
chu kui (24.17. CHINESE, YUE)
chu^{V} pi (20.6. LAHU)
chu pi tian gua (38.7a. CHINESE, YUE)
chua (23.3. HINDI)
chuamarasa (23.3. BENGALI)
chuamarsa (23.3. HINDI)
chüan hsin wo chü (27.31. CHINESE, MANDARIN)
chuan ye chun mei cao (59.1. CHINESE, MANDARIN)
chuan ye chun mei cao (59.1. CHINESE, WÚ)
chuan yeh ts'ai (30.35. CHINESE, MANDARIN)
chuchander (35.3. URDU)
chuchú (38.25. PORTUGUESE)
chufa (10.1. ENGLISH)
chufa (10.1. FRENCH)
chufa (10.1. PORTUGUESE)
chufa (10.1. QUECHUA, CUSCO)
chufa (10.1. SPANISH)
c̲hug̲handar (35.4. PASHTO, NORTHERN)
chui (27.17. VLAAMS)
chuka (23.3. BENGALI)
chuka (58.2. HINDI)
chuka sak (58.2. ASSAMESE)
chuka sak (58.3. ASSAMESE)
chuka sak (58.4. ASSAMESE)
chukandar (35.4. FARSI, WESTERN)
chukandar (35.4. HINDI)
chukandar (35.4. MAITHILI)
chukandar (35.4. NEPALI)
chukandar (35.4. URDU)
chukapalam (58.2. BENGALI)
chukar (45.4. BENGALI)
chukiar (45.4. ASSAMESE)
chukku (20.6. MALAYALAM)
chuko (23.3. BENGALI)
chuko (23.3. GUJARATI)
chukula (44.5. BAMANANKAN)
chulai (23.3. PANJABI, EASTERN)
chulgigadubaechu (30.30. KOREAN)
chulthe amilo (58.1. NEPALI)
chumbera (32.1. SPANISH)
chumeat hagina (58.2. HEBREW)
chumli sag (23.3. HINDI)
chun cai (31.1. CHINESE, MANDARIN)
chun cai (31.1. CHINESE, MIN BEI)
chun cai (31.1. CHINESE, WÚ)

cisosoci (27.8. YAO)
cisso (69.1. PORTUGUESE)
cissus (69.1. DUTCH)
cissus (69.1. ITALIAN)
čistec (44.11. MACEDONIAN)
čistec hlízantý (44.11. CZECH)
citambel (50.2. MALAYALAM)
citre (38.2. FRENCH)
citroen (38.2. VLAAMS)
citroenmeloen (38.8. VLAAMS)
citron (38.2. ENGLISH)
citronska lubenica (38.2. MACEDONIAN)
citru (38.2. MALTESE)
citta (20.6. HAUSA)
citta (20.6. ZARMA)
citta boɗe (62.5. FULFULDE, ADAMAWA)
citta kolong (62.2. FULFULDE, ADAMAWA)
cìttar àhò (20.6. HAUSA)
cìttár àhóó (20.6. HAUSA)
cìttár-fo (20.6. HAUSA)
cìttár-jábà (20.6. HAUSA)
cìttár-mai-ƙwaya (20.6. HAUSA)
cìttár-mai-yáátsàà (20.6. HAUSA)
citungula (38.25. YAO)
citvor (20.4. HUNGARIAN)
citzaro (40.28. TAGALOG)
cive (7.5. FRENCH)
cive chino (7.13. SPANISH)
cives (7.11. HINDI)
civette (7.11. FRENCH)
ciwcymber (38.13a. WELSH)
ciwcymber gwyllt (38.19. WELSH)
ciwcynber cormos (38.12. WELSH)
ciyao (11.1. TONGA)
ciyao (11.1. TUMBUKA)
ciyao (11.2. TONGA)
ciyao (11.2. TUMBUKA)
ciyao (11.3. TONGA)
ciyao (11.3. TUMBUKA)
ciyao (11.4. TONGA)
ciyao (11.4. TUMBUKA)
ciyao (11.5. TONGA)
ciyao (11.5. TUMBUKA)
ciyao (11.6. TONGA)
ciyao (11.6. TUMBUKA)
ciyao (11.7. TONGA)
ciyao (11.7. TUMBUKA)
ciyao (11.8. TONGA)
ciyao (11.8. TUMBUKA)
cizlagan (66.3. TURKISH)
cizo (10.1. HAUSA)
cizrna beraní (40.5. CZECH)
ɔkafo (11.4. GA)
ɔ́ká̰ị (16.1a. OGBIA)
ɔ́kạí (16.1a. OGBIA-AGHOLO)
ɔkamfo (11.4. GA)
ɔkani (11.3. GA)
ɔkpɔlɔ-bid (9.1. KRIO)
ɔkrɔ (45.1. KRIO)
ɔ́kụ̀rụ̀ (45.1. OGBIA)
claitónia de Cuba (59.1. PORTUGUESE)
clavel chino (27.18. SPANISH)
claytone de Cuba (59.1. FRENCH)
claytonia perfoliée (59.1. FRENCH)
cleome (34.2. FRENCH)
ɔ-léwùrè (8.4. BASSARI)
clopoţei (33.2. ROMANIAN)
cluster bean (40.6. ENGLISH)
cluster yam (11.4. ENGLISH)
clychlys erfin (33.2. WELSH)
ɔmɔkç (19.1. GA)
ɔ-ŋáɓ (11.3. BASSARI)
ɔ-ngànɛ́r (9.1. BASSARI)
ɔ-ngị́n ɔnd *ɓi*- ɗyàr (20.6. BASSARI)
coastal glehnia (24.14. ENGLISH)
çoban çantasi (30.43. TURKISH)
çoban kalkitan (64.3. TURKISH)
cobo (17.8. ENGLISH)
čóčk (40.12. SORBIAN, UPPER)
čóčka (40.12. SORBIAN, UPPER)
čočka jedlá (40.12. CZECH)
čočka kuchyňská (40.12. CZECH)
cock's comb (23.4. ENGLISH)
cockscomb (23.4. ENGLISH)
coco (8.4. BAOULÉ)
cõcõbr (38.13a. MÒORÉ)
cocomb (38.13a. CAJUN FRENCH)
cocomero (38.3. ITALIAN)
cocomero (38.13a. ITALIAN)
cocòmero del Malabar (38.15. ITALIAN)
cocoyam (8.4. ENGLISH)
cocozza (38.20. ITALIAN)
coda cavallina (1.1. ITALIAN)
coendro (24.8. GALICIAN)
coentro (24.9. CRIOULO, CAPE VERDE)
coentro (24.9. PORTUGUESE)
cogollos (8.9. SPANISH)
cogombre (38.13a. CATALAN-VALENCIAN-BALEAR)
cohén (40.22. SPANISH)

cohombro (38.13a. SPANISH)
cóilis (30.26. GAELIC, IRISH)
cojombro (38.20. SPANISH)
cokhy (27.31. UZBEK)
cokhy (27.32. UZBEK)
coko wana (62.23. JAVANESE)
col bajana (30.27a. CATALAN-VALENCIAN-BALEAR)
col borrajenca (30.34. CATALAN-VALENCIAN-BALEAR)
col caballar (30.35. SPANISH)
col crespa (30.35. CATALAN-VALENCIAN-BALEAR)
col de Bruselas (30.29. SPANISH)
col de Brusselles (30.29. CATALAN-VALENCIAN-BALEAR)
col de china (30.36. SPANISH)
col de China (30.40. SPANISH)
col de la Sibéria (30.23. SPANISH)
col de meollo (30.32. SPANISH)
col de milan (30.34. SPANISH)
col de mile cabezas (30.33. SPANISH)
col de pezon gruezo (30.28. SPANISH)
col forragera (30.35. SPANISH)
col gallega (30.35. SPANISH)
col llombarda (30.27b. CATALAN-VALENCIAN-BALEAR)
col marina (30.45. SPANISH)
col marina tartara (30.46. SPANISH)
col maritima (30.45. SPANISH)
col medular (30.32. SPANISH)
col rabano (30.30. ITALIAN)
col rábano (30.30. SPANISH)
col rapano (30.30. SPANISH)
col repollo (30.27a. SPANISH)
col roja (30.27b. SPANISH)
col tronchuda (30.28. SPANISH)
cola de caballo (1.1. SPANISH)
colag (30.26. GAELIC, SCOTTISH)
col-i-flor (30.26. CATALAN-VALENCIAN-BALEAR)
coliflor (30.26. SPANISH)
colifrol (30.26. GALICIAN)
colinabo (30.21. SPANISH)
col-i-nap (30.30. CATALAN-VALENCIAN-BALEAR)
colirrábano (30.30. SPANISH)
colitis (23.3. CEBUANO)
colitis (23.3. TAGALOG)
collards (30.35. ENGLISH)
collards (30.35. PORTUGUESE)
collender (24.9. ENGLISH)
colocasia (8.4. HEBREW)
colocasia (8.4. ITALIAN)
colocasia (8.4. SPANISH)
colocazia (8.4. ROMANIAN)
colombro (38.13a. GALICIAN)
colombro (38.20. PORTUGUESE)
colrabi (30.30. GEORGIAN)
colza (30.21. ENGLISH)
colza (30.21. FRENCH)
colza (30.21. ITALIAN)
colza (30.21. PORTUGUESE)
colza (30.21. SPANISH)
combosto (30.27a. GEORGIAN)
common bamboo (17.1. ENGLISH)
common basil (44.5. ENGLISH)
common cabbage (30.27a. ENGLISH)
common comfrey (29.2. ENGLISH)
common cowitch (40.18. ENGLISH)
common field pumpkin (38.18. ENGLISH)
common ginger (20.6. ENGLISH)
common gram (40.5. ENGLISH)
common hop (46.1. ENGLISH)
common nettle (66.3. ENGLISH)
common nightshade (62.15. ENGLISH)
common nightshade (62.21. ENGLISH)
common onion (7.3. ENGLISH)
common purslane (59.2. ENGLISH)
common sauropus (39.5. ENGLISH)
common seepweed (35.10. ENGLISH)
common tiger flower (12.1. ENGLISH)
common viper's-grass (27.38. ENGLISH)
common watercress (30.52. ENGLISH)
common yam (11.1. ENGLISH)
concombre (38.13a. FRENCH)
concombre (38.13a. WOLOF)
concombre africain (38.12. FRENCH)
concombre des antilles (38.6. FRENCH)
concombre grimpant (38.19. FRENCH)
concombre papengaie (38.21. FRENCH)
concombre porte bornes (38.12. FRENCH)
concombre serpent (38.10. FRENCH)
concombrou (38.13a. JULA)
concondudi (38.18. KONKANI)
condang (19.1. JAVANESE)
condurul doamnei (65.1. ROMANIAN)
condurul doamnei mare (65.1. ROMANIAN)
confrei (29.2. PORTUGUESE)
cong (7.4. CHINESE, WÚ)

cong (7.5. CHINESE, MANDARIN)
cong (7.5. CHINESE, WÚ)
cong (7.9b. CHINESE, WÚ)
cong belut (62.23. JAVANESE)
cong tou (7.3. CHINESE, GÀN)
cong tou (7.3. CHINESE, YUE)
congo (40.2. CRIOULO, CAPE VERDE)
congo bean (40.2. ENGLISH)
Congo pea (40.2. ENGLISH)
congtou (7.3. CHINESE, MANDARIN)
coniella (44.9. ITALIAN)
conkrom (62.19. SUNDA)
conopidă (30.26. ROMANIAN)
conopui (17.1. KONKANI)
consolda (29.3. PORTUGUESE)
consolda maior (29.2. PORTUGUESE)
consolida azzurra (29.3. ITALIAN)
consolida maggiore (29.2. ITALIAN)
consoude d'Upland (29.3. FRENCH)
consoude russe (29.3. FRENCH)
consoude voyageuse (29.3. FRENCH)
consoule officinale (29.2. FRENCH)
constag (24.3. GAELIC, MANX)
consuelda mayor (29.2. SPANISH)
copóg (58.3. GAELIC, IRISH)
coppu (35.9. BADAGA)
coqueirinho-de-jardim (9.1. CRIOULO, CAPE VERDE)
coqueret alkékenge (62.11. FRENCH)
coqueret du Pérou (62.12. FRENCH)
coralli palloncini (62.11. ITALIAN)
corchorus (63.1. ROMANIAN)
còrcora (63.1. ITALIAN)
còrcoro ortense (63.1. ITALIAN)
còrcoro siliquoso (63.1. ITALIAN)
cordyceps (44.11. MALAY)
córè (38.20. PULAAR)
coréte (63.1. FRENCH)
corette potagère (63.1. FRENCH)
coriander (24.9. DUTCH)
coriander (24.9. ENGLISH)
coriander (24.9. SORBIAN, UPPER)
coriandolo (24.9. ITALIAN)
coriandre (24.9. FRENCH)
coriandro (24.9. SPANISH)
coriandru (24.9. ROMANIAN)
corme d'abondance (27.21. FRENCH)
corne de cerf (57.1. FRENCH)
cornichão das Antilhas (38.6. PORTUGUESE)
cornichon (38.13b. FRENCH)
cornichon (38.13b. PORTUGUESE)
cornichon au vinaigre (38.13b. FRENCH)
cornichon des Antilles (38.6. FRENCH)
corno di cervo (57.1. ITALIAN)
cornsalad (67.2. ENGLISH)
cornucopia (27.21. PORTUGUESE)
cornucopia (27.21. SPANISH)
cornucoppia (27.21. ITALIAN)
coroá (38.26. PORTUGUESE)
corònopo (57.1. ITALIAN)
corrydander (24.9. SCOTS)
cosmea caudata (27.12. ITALIAN)
cosmos (27.12. ASSAMESE)
cosmos (27.12. DANISH)
cosmos (27.12. DUTCH)
cosmos (27.12. ENGLISH)
cosmos (27.12. FRENCH)
Cosmos (27.12. GERMAN)
cosmos (27.12. PORTUGUESE)
cosmos (27.12. SPANISH)
cosmos (27.12. TAGALOG)
cóssè (40.18. PULAAR)
costa (16.1a. WÁRAY-WÁRAY)
côte de blette (35.3. FRENCH)
coufirigui (44.5. SENOUFO, CEBAARA)
coundou (36.3. PULAR)
country mallow (45.5. ENGLISH)
country potato (44.8. ENGLISH)
courge à feuilles de figuier (38.15. FRENCH)
courge à graines noires (38.15. FRENCH)
courge à la cire (38.1. FRENCH)
courge bouteille (38.20. FRENCH)
courge cireuse (38.1. FRENCH)
courge de Siam (38.15. FRENCH)
courge muscarde (38.17. FRENCH)
courge musquée (38.17. FRENCH)
courge pépon (38.18. FRENCH)
courge serpent (38.29. FRENCH)
courge-torchon (38.22. FRENCH)
courgette (38.18. DUTCH)
courgette (38.18. ENGLISH)
courgette (38.18. FRENCH)
courgette (38.18. PORTUGUESE)
courgette (38.18. SWEDISH)
courgette (38.18. VLAAMS)
cousuelda azul (29.3. SPANISH)
coutdurul doamnei sodóvaja (67.1. ROMANIAN)
couve bestarda (30.45. PORTUGUESE)
couve brócolo (30.31. PORTUGUESE)

couve cavaleira (30.35. PORTUGUESE)
couve chinesa (30.25. PORTUGUESE)
couve chinesa (30.40. PORTUGUESE)
couve da China (30.40. PORTUGUESE)
couve da Sibéria (30.23. PORTUGUESE)
couve de Bruxelas (30.29. PORTUGUESE)
couve de bruxelles (30.29. PORTUGUESE)
couve de cortar (30.28. PORTUGUESE)
couve de desfolhar (30.35. PORTUGUESE)
couve de folhas (30.35. PORTUGUESE)
couve de medula (30.32. PORTUGUESE)
couve de mil cabeças (30.33. PORTUGUESE)
couve de nervuras (30.28. PORTUGUESE)
couve de Sabóia (30.34. PORTUGUESE)
couve flor (30.26. PORTUGUESE)
couve galega (30.35. PORTUGUESE)
couve galega frisada (30.35. PORTUGUESE)
couve lombarda (30.34. PORTUGUESE)
couve manteiga (30.35. PORTUGUESE)
couve marinha (30.45. PORTUGUESE)
couve marinha da Hungria (30.46. PORTUGUESE)
couve nabiça (30.21. PORTUGUESE)
couve nabo (30.21. PORTUGUESE)
couve nabo (30.22. PORTUGUESE)
couve penca (30.28. PORTUGUESE)
couve rábano (30.30. PORTUGUESE)
couve rabão (30.30. PORTUGUESE)
couve repolho (30.27a. PORTUGUESE)
couve repolho bianca (30.27a. PORTUGUESE)
couve roxa (30.27b. PORTUGUESE)
couve tártara (30.46. PORTUGUESE)
couve tronchuda (30.28. PORTUGUESE)
couver broculo (30.31. PORTUGUESE)
covi (30.35. CRIOULO, CAPE VERDE)
covi bruxelas (30.29. CRIOULO, CAPE VERDE)
coviflor (30.26. CRIOULO, CAPE VERDE)
covi-trenchuda (30.28. CRIOULO, CAPE VERDE)
covun (38.7a. UZBEK)
covun (38.8. UZBEK)
covun (38.11. UZBEK)
cowage (40.18. ENGLISH)
cowek gompel (24.7. JAVANESE)
cowhage (40.18. ENGLISH)
cowpea (40.43. ENGLISH)
cowpea (40.43. SINHALA)
cowpea (40.43. VLAAMS)
cozonyuvguich (38.22. UZBEK)
ɔ-pɔngwón (11.2. BASSARI)
ɔ-pútutè (36.3. BASSARI)
craba yamit (30.45. HEBREW)
craf-y-geifr (7.1. WELSH)
crain-crain (63.1. DUTCH)
crambe (30.45. ENGLISH)
crambe (30.45. FRENCH)
crambe (30.45. SPANISH)
crambe lacinié (30.46. FRENCH)
crambe tartare (30.46. FRENCH)
crambio di tataria (30.46. ITALIAN)
cran (30.1. FRENCH)
crasen el-kharazan (17.1. ARABIC)
creamh (7.10. GAELIC, SCOTTISH)
creamh (13.2. GAELIC, SCOTTISH)
creamh-gàrraidh (7.8. GAELIC, SCOTTISH)
cremesina acinosa (56.1. ITALIAN)
cremesina edule (56.3. ITALIAN)
cremesina uva-turca (56.2. ITALIAN)
cren (30.1. ITALIAN)
crescione d'acqua (30.52. ITALIAN)
crescione del Perù (65.1. ITALIAN)
crescione inglese (30.2. ITALIAN)
crescione inglese (30.51. ITALIAN)
crescionne di Para (27.2. ITALIAN)
creson (30.51. ROMANIAN)
cresonul de grădina (30.51. ROMANIAN)
Cresson (30.2. LUXENBOURGEOIS)
cresson alénois (30.51. FRENCH)
cresson de fontaine (30.52. FRENCH)
cresson de Para (27.2. FRENCH)
cresson de terre (30.2. FRENCH)
cresson d'eau (30.52. FRENCH)
cresson des prés (30.44. FRENCH)
cresson d'Inde (65.1. FRENCH)
cresson du Brésil (27.2. FRENCH)
cresson officinal (30.52. FRENCH)
cresta de gallo (23.4. SPANISH)
cresta di gallo (23.4. ITALIAN)
crête de coq (23.4. FRENCH)
crin (13.4. ROMANIAN)
crisantemo (27.23. ITALIAN)
crisântemo (27.23. PORTUGUESE)
crisantemo (27.23. SPANISH)
crista de galo (23.4. PORTUGUESE)
cristallina (22.1. MALTESE)
crizanteme (27.23. ROMANIAN)
crn bob (40.43. MACEDONIAN)
crn koren (27.38. MACEDONIAN)
crn sinap (30.24. MACEDONIAN)

da shu (11.7. CHINESE, MIN BEI)
da siam (7.10. CHINESE, GÀN)
da suan (7.1. CHINESE, WÚ)
da suan (7.1. CHINESE, YUE)
da suan (7.10. CHINESE, MANDARIN)
da suan (7.10. CHINESE, WÚ)
da suan jiang (62.13. CHINESE, MANDARIN)
da suan jiang (62.13. CHINESE, WÚ)
da tou cai (30.1. CHINESE, YUE)
da tou cai (30.14. CHINESE, YUE)
da tou cai (30.41a. CHINESE, WÚ)
da tou jie (30.14. CHINESE, WÚ)
da tou suan (7.1. CHINESE, MANDARIN)
da tou suan (7.1. CHINESE, WÚ)
dâ volo (45.4. BAMANANKAN)
da wan dou (40.26. CHINESE, XIANG)
da wu feng cao (27.20. CHINESE, MANDARIN)
da wu feng cao (27.20. CHINESE, WÚ)
da xun ma (66.3. CHINESE, WÚ)
da ye ji cai (30.43. CHINESE, GÀN)
da ye jie (30.17. CHINESE, WÚ)
dââ (45.4. JULA)
dááɓùrín sáánìyáá (59.2. HAUSA)
d[aaj]u b[aws]p (45.1. VIETNAMESE)
d[aaj]u d[uwj]a (40.42. VIETNAMESE)
d[aaj]u gi[ar]i [as]o (40.42. VIETNAMESE)
d[aaj]u kh[ees] (40.29. VIETNAMESE)
d[aaj]u m[ef]o r[uwf]ng (40.18. VIETNAMESE)
d[aaj]u r[oof]ng (40.29. VIETNAMESE)
d[aaj]u r[uwj]a (40.4. VIETNAMESE)
daakataá (36.3. HAUSA)
daandelian hoo (27.45. KANNADA)
daani (38.20. HINDI)
daarukanda (39.4. SANSKRIT)
daáwàr kádàà (17.3. HAUSA)
d[aa]y na t[aa]y (38.29. VIETNAMESE)
dab (40.30. HINDI)
dabbale beans (40.25. KANNADA)
dabda ba (11.1. BAMANANKAN)
dàbéré (8.4. BAMANANKAN)
ḓabùla (62.24. VENDA)
dabusi (62.1. OROMO, WEST CENTRAL)
dachinn (8.4. KWÉYÒL)
dâda (11.2. BAMANANKAN)
dâda (11.2. MANINKAKAN, EASTERN)
dada (27.8. KISSI, NORTHERN)
dadakan (11.2. GIANGAN)
dadayem (27.8. FILIPINO)
dadayem (27.8. IVATAN)
dadhiphala (38.14. SANSKRIT)
dadhiphala (38.16. SANSKRIT)
dadhiphala (38.18. SANSKRIT)
daeshavaali pendalam (11.1. TELUGU)
daeshi kaaknaja (62.12. HINDI)
daeva kaanda (19.1. KANNADA)
daevakeli (9.1. MARATHI)
daevakili (9.1. SANSKRIT)
dafaar gbanda (23.2. YOM)
dağ abanozu (40.33. TURKISH)
dagan (43.1. WAMEY)
dagbami (45.4. KURANKO)
dagglarive soppu (23.3. KANNADA)
dagglisoppu (23.3. KANNADA)
dagilja (13.3. ICELANDIC)
dagita (38.3. MARANAO)
dagita (38.3. MARINDUQUE)
dagita (38.7. MARANAO)
daglelics (13.3. VLAAMS)
daglelie (13.3. AFRIKAANS)
daglelie (13.3. DUTCH)
daglilja (13.3. SWEDISH)
daglilje (13.3. DANISH)
daglilje (13.3. NORWEGIAN)
daglisoppu (23.3. KANNADA)
dagmai (8.4. BINUKID)
dagmay (8.4. CEBUANO)
dagutung (62.16. TAUSUG)
dagutung (62.23. TAUSUG)
dahdir (23.3. SERER-SINE)
dahdir gor (23.3. SERER-SINE)
dahl (40.2. FIJIAN)
dai dou (40.41. CHINESE, MIN BEI)
dai dou (40.41. CHINESE, WÚ)
dai tou gai (30.14. CHINESE, MANDARIN)
dai ye gai (30.17. CHINESE, MANDARIN)
ɗáiɗóóyà (44.5. HAUSA)
ɗaiɗóóyàà (44.5. HAUSA)
ɗáiɗóóyàr (44.5. HAUSA)
daiga (8.3. FIJIAN)
daijo (11.1. JAPANESE)
daikon (30.54. BELARUSAN)
daikon (30.54. CAROLINIAN)
daikon (30.54. ENGLISH)
daikon (30.54. FRENCH)
daikon (30.54. JAPANESE)
daikon (30.54. PORTUGUESE)
daikon (30.54. VLAAMS)
daikon (30.56. JAPANESE)

dāil (40.12. MAITHILI)
dail surion (27.24. WELSH)
dailuah chi khat moong (40.37. HINDI)
dailuah sen chi (40.12. MIZO)
daizu (40.8. JAPANESE)
dakasítalikaṁ (50.2. PALI)
dakpan (11.7. DAGBANI)
dakuma (45.4. MANDING-MANDINKA)
dakumi (45.4. MANDING-MANDINKA)
dakuna (27.24. PULAAR)
dakwsani (30.41a. CHEROKEE)
dal (40.2. ENGLISH)
dâl (40.12. FARSI, EASTERN)
dâl (40.12. PASHTO, NORTHERN)
dâl nakhod (40.12. FARSI, EASTERN)
dale-dale (15.1. QUECHUA, CUSCO)
dāli (40.12. MAITHILI)
daling-manuk (23.6. TAUSUG)
dallo mula (30.56. NEPALI)
dalo (8.4. FIJIAN)
dalog (8.4. MARANAO)
dam koruğu (37.1. TURKISH)
damaro (44.5. GUJARATI)
dambho (23.3. GUJARATI)
damie (44.5. DANGME)
damilig (16.1a. IBANAG)
damilik (16.1a. IBANAG)
dâmlô:ng barang (62.24. KHMER, CENTRAL)
dâmlô:ng chhë (39.4. KHMER, CENTRAL)
dâmlô:ng chiem moan (11.1. KHMER, CENTRAL)
dâmlô:ng chvië (36.3. KHMER, CENTRAL)
dâmlô:ng dü:hs (11.2. KHMER, CENTRAL)
dâmlô:ng phluk (11.1. KHMER, CENTRAL)
dâmlô:ng sdâm préi (11.2. KHMER, CENTRAL)
dâmlô:ng sya (11.5. KHMER, CENTRAL)
damsa fala (59.3. SENOUFO, CEBAARA)
damsi (40.1. ZARMA)
damsi kukurku (40.39. ZARMA)
ɗan kanuri (45.1. HAUSA)
ɗan karuna (45.1. HAUSA)
ɗan mata (40.2. HAUSA)
ɗan zago (40.3. HAUSA)
dãna (11.1. GA)
dana (11.2. BAMANANKAN)
dana (11.2. MANINKAKAN, EASTERN)
dãna (11.7. GA)
danadl (66.3. WELSH)
danda (11.2. BAMANANKAN)
danda (11.2. MANDING-MANDINKA)
danda (11.2. MANINKAKAN, WESTERN)
dãnda (11.3. BAMANANKAN)
danda ba (11.1. BAMANANKAN)
dandaalu (11.1. SANSKRIT)
dandalub (38.12. SERER-SINE)
dândalub (38.12. WOLOF)
dan-dan (11.2. MANINKAKAN, EASTERN)
dandaõ (11.2. MANDING-MANDINKA)
dandelion (27.45. ENGLISH)
dandelion (27.45. FILIPINO)
dandhul (38.22. BENGALI)
dané (11.2. SUSU)
da'neesk'ání dich'ízhígíí (38.7a. NAVAJO)
da'neesk'ání dich'ízhígíí (38.7b. NAVAJO)
da'neesk'ání dik'ózhígíí (38.13a. NAVAJO)
dãñĕwali bạrag (24.9. KASHMIRI)
dang qui (24.1. CHINESE, WÚ)
dàngálì (36.3. KANURI)
dàngáli nàsàrá (62.24. KANURI)
däng^{w}ällo (40.34a. AMHARIC)
dangari (38.16. SANSKRIT)
danggu (7.2. IFUGAO)
danggu (7.3. GADDANG)
danghodi (40.43. ASSAMESE)
dangkálin tuuraawaá (62.24. HAUSA)
dangri (40.2. GUJARATI)
dangsul (40.8. JAVANESE)
dani (62.9. MANDING-BAMBARA)
dan-i-noć (68.1. SERBO-CROATIAN)
dànkál (36.3. BEROM)
dànkál làsàrá (62.24. BEROM)
dankali (36.3. BAATONUM)
dankali (36.3. HAUSA)
dànkáli (36.3. NGIZIM)
dankali (39.4. HAUSA)
dánkálii (36.3. HAUSA)
daŋkálii (36.3. HAUSA)
dankalin Turawa (62.24. HARARI)
dant (23.3. MARATHI)
dant y lleew (27.45. WELSH)
danthabeeja (38.20. SANSKRIT)
danthachhadaa (38.4. SANSKRIT)
danthachhadopama (38.4. SANSKRIT)
dantina soppu (23.3. KANNADA)
dantu (23.3. KANNADA)
dantu koora (23.3. TELUGU)
dantu soppu (23.3. KANNADA)
danue (44.5. DANGME)
danwe (44.5. DANGME-KROBO)

danyaalu (24.9. TELUGU)
dao ba dou (40.4. CHINESE, XIANG)
dao dou (40.4. CHINESE, GÀN)
dao dou (40.4. CHINESE, MANDARIN)
dao dou (40.4. CHINESE, MIN BEI)
dao dou (40.4. CHINESE, WÚ)
dao dou (40.4. CHINESE, YUE)
dao dou (40.26. CHINESE, WÚ)
daoon sooring (58.2. INDONESIAN)
daoruang-phama (27.12. THAI)
dăpesāwāāl (62.1. MICMAC)
dăpesāwāāl mĕgwāāgŭl (62.2. MICMAC)
dapiri (62.24. SHONA)
dapnen (36.3. IFUGAO)
dappa genasu (11.1. KANNADA)
dappa naepear hullu (17.3. KANNADA)
dapura (62.24. SHONA)
dara gowok (62.15. JAVANESE)
darachti hanut (24.16. TAJIKI)
daradakandagadda (8.3. TELUGU)
daradharsha (30.48. SANSKRIT)
darakrishna (30.51. SANSKRIT)
daranaw-lanit (39.3. HANUNOO)
darawi (17.3. CEBUANO)
darawi (17.3. TAGALOG)
daray (17.3. CEBUANO)
darbooji (38.3. TELUGU)
darbyz (38.3. KIRGHIZ)
darbız (38.3. KIRGHIZ)
darchin (3.1. FARSI, WESTERN)
darduchhada (24.7. SANSKRIT)
därepīrè (38.21. TULU)
darhari (40.6. HINDI)
daridé (40.1. PULAAR)
darivka (45.1. SANSKRIT)
dark egusi (38.5. ENGLISH)
darraba (63.1. KANURI)
dārza balodene (35.1. LATVIAN)
darza pupa (40.34a. LATVIAN)
dārza sīpols (7.2. LATVIAN)
darželinis dašis (44.9. LITHUANIAN)
daržinė balandūnė (35.1. LITHUANIAN)
daržinė pienė (27.42. LITHUANIAN)
daržinė trūkažolė (27.9. LITHUANIAN)
daržinės pupelės (40.26. LITHUANIAN)
daržinis ibiskas (45.4. LITHUANIAN)
daržinis špinatas (35.9. LITHUANIAN)
dārzo pupos (40.34a. LITHUANIAN)
daržo kukurūzai (17.7. LITHUANIAN)
daržo žirniai (40.28. LITHUANIAN)
daržovine portulaka (59.2. LITHUANIAN)
daržovinis artišokas (27.16. LITHUANIAN)
daržovinis rabarbaras (58.1. LITHUANIAN)
dasamekke (38.3. KANNADA)
dasheen (8.4. ENGLISH)
dashel-flower (27.45. ENGLISH)
dashtir (24.9. TAJIKI)
daşkädi (38.18. TURKMEN)
dasouvigui (23.3. SENOUFO, CEBAARA)
dasouyigui (23.3. SENOUFO, CEBAARA)
das-pĕdas (24.12. INDONESIAN)
dāssi palak (35.3. PANJABI, EASTERN)
dastambol (38.7b. PASHTO, NORTHERN)
dastmolkadu (38.21. TAJIKI)
dastmolkadui shomi (38.22. TAJIKI)
dasun (7.10. INDONESIAN)
dasun merah (7.2. INDONESIAN)
data (38.17. URDU)
datogo (38.20. MALIMBA)
dattakua-thao (69.1. THAI)
dâu ca (40.41. VIETNAMESE)
dâu chè (40.38. VIETNAMESE)
dau den (40.37. VIETNAMESE)
dâu do (40.41. VIETNAMESE)
dâu dûa (40.42. VIETNAMESE)
dâu gao (40.40. VIETNAMESE)
daû giai áo (40.42. VIETNAMESE)
dâu hà-lan (40.28. VIETNAMESE)
dâu hoà lan (40.28. VIETNAMESE)
dâu nành (40.8. VIETNAMESE)
dâu ngu (40.25. VIETNAMESE)
dâu phong (40.1. VIETNAMESE)
dau sang (40.2. VIETNAMESE)
dau thong (40.2. VIETNAMESE)
dâu tuong (40.8. VIETNAMESE)
dâu trang (40.41. VIETNAMESE)
dâu van (40.9. VIETNAMESE)
dâu ve (40.26. VIETNAMESE)
dau xanh (40.26. VIETNAMESE)
dâu xanh (40.38. VIETNAMESE)
daudzposmu sīpoli (7.9a. LATVIAN)
daugɔ (44.8. MENDE)
daugɔ (55.1. MENDE)
daugiaaukštis svoqūnas (7.9a. LITHUANIAN)
daun bawang (7.2. MALAY)
daun bawang (7.5. MALAY)
daun fuki (27.35. INDONESIAN)
daun getang (27.3. INDONESIAN)
daun getang (27.4. INDONESIAN)
daun kaki kuda (24.7. INDONESIAN)

daun kermak (23.2. INDONESIAN)
daun koki (62.8. INDONESIAN)
daun lampeni (27.29. INDONESIAN)
daun lipung (15.2. SUNDA)
daun panjang (27.29. INDONESIAN)
daun panjang (27.29. MALAY)
daun panu (51.1. INDONESIAN)
daun pasli (24.19. MALAY)
daun pegagan (24.15. INDONESIAN)
daun penggaga (24.15. INDONESIAN)
daun rusa (23.2. INDONESIAN)
daun saderi (24.4. MALAY)
daun shiso (44.7. INDONESIAN)
daun sup (24.4. INDONESIAN)
daun suring (58.2. INDONESIAN)
daun suring (58.4. INDONESIAN)
daun tasbeh (9.1. MALAY)
daun tol okeremak (23.2. INDONESIAN)
daun tolod (23.2. SUNDA)
daunpoko (44.2. INDONESIAN)
dau-vang (40.9. VIETNAMESE)
dáuyâ (39.4. NGIZIM)
davung (17.1. BINUKID)
davung (17.2. BINUKID)
davung (17.4. BINUKID)
davuravu (34.2. ITAWIT)
dawahe (40.25. BONTOC, CENTRAL)
dawaso (45.4. BAMANANKAN)
dawe (17.1. FON)
dawnfawh (38.3. MIZO)
dawol (40.25. IFUGAO)
dawol an e-itab (40.25. IFUGAO)
day lily (13.3. ENGLISH)
dayeré (36.3. GOURMANCHÉMA)
dayessé (36.3. GOURMANCHÉMA)
debkoti (23.4. HINDI)
debunsha (44.5. BENGALI)
dedschaelbɔne (38.18. KONKOMBA)
dee ee-ah (38.23. HMONG-DAW)
d[eef]n t[aa]y (22.2. VIETNAMESE)
deeka (11.7. DANGME-KROBO)
deeni (38.3. FULFULDE, ADAMAWA)
deenkah (38.7a. YIDDISH)
deengli (62.24. GUJARATI)
deepani (40.32. SANSKRIT)
deerghakandaka (30.56. SANSKRIT)
deerghamoolaka (30.56. SANSKRIT)
deerghapathra (7.3. SANSKRIT)
deerghapathra (16.1a. SANSKRIT)
deerghapathraka (30.56. SANSKRIT)
deerghapathri (63.1. SANSKRIT)
deerghaphala (38.21. SANSKRIT)
deerghaphala (38.29. SANSKRIT)
deerghavarthaki (62.19. SANSKRIT)
dege (40.1. PULAAR)
dēgər (40.26. MEHRI)
dəgərīt (40.26. MEHRI)
degi (45.2. INDONESIAN)
degita' (38.3. BINUKID)
děhel (24.1. CZECH)
dehi ala (8.4. SINHALA)
dekeman (40.8. JAVANESE)
deklé yibê (62.5. BÉTÉ)
dekute (11.4. ÉWÉ)
dela'at (38.16. HEBREW)
dela'at gedola (38.14. HEBREW)
delaat gedola (38.16. HEBREW)
dela'at haginna (38.16. HEBREW)
dela'at ha-sade (38.18. HEBREW)
delaat hasade (38.18. HEBREW)
delas (8.3. MARANAO)
delbol (36.2. FULFULDE, ADAMAWA)
dele (40.8. JAVANESE)
delgiyu (27.40. JAVANESE)
delgiyu (27.42. JAVANESE)
deling petung (17.2. JAVANESE)
dém pwɛŋ (11.1. BEROM)
dém sunàŋ (11.1. BEROM)
dēmba-gǒga'ji (30.41b. KASHMIRI)
demekan (40.8. JAVANESE)
dɛ́mèn (40.43. BOBO MADARÉ, SOUTHERN)
demjanka (62.19. RUSSIAN)
démsì (40.1. SONGHAY)
denaare (38.7b. FULFULDE, ADAMAWA)
denb id-dib (23.3. MALTESE)
denb il-haruf (60.1. MALTESE)
denb iz-ziemel (1.1. MALTESE)
dende (38.3. FULFULDE, ADAMAWA)
dende (38.3. PULAAR)
dende (38.7a. PULAAR)
dende (38.7b. FULFULDE, ADAMAWA)
dendrokàloomos the platyanthos (17.2. GREEK)
dene (38.3. FULFULDE, ADAMAWA)
dene bamɗi (38.7b. FULFULDE, ADAMAWA)
deng long jiao (62.1. CHINESE, YUE)
dengo danta (23.3. BENGALI)
dengu (40.5. SWAHILI)
dengu (40.12. SWAHILI)

dengu (40.28. SWAHILI)
dengua (23.3. BENGALI)
denivka (13.3. CZECH)
denizlahanasi (30.46. TURKISH)
denje (63.1. NYANJA)
denken (38.20. GONJA)
denné toubaco (38.8. PULAR)
dənnəčč (62.24. AMHARIC)
densi (40.1. DENDI)
dent de lion (27.45. FRENCH)
dent de lleó (27.45. CATALAN-VALENCIAN-BALEAR)
dente de leão (27.45. PORTUGUESE)
dente de leone (27.45. ITALIAN)
dentilion (27.45. ENGLISH)
dentto (23.3. KONKANI)
deo (2.1. PANJABI, EASTERN)
deodeog (33.3. ENGLISH)
deodeog (33.3. JAPANESE)
deodeog (33.3. VLAAMS)
dépè (59.2. PULAAR)
deppi (40.39. FULFULDE, ADAMAWA)
deppu (40.39. FULFULDE, ADAMAWA)
dere otu (24.13. TURKISH)
dereẹ̀fe (39.4. YORUBA)
dermanlyk buraçnik (29.1. TURKMEN)
dermanlyk kenarot (30.52. TURKMEN)
dermanlyk sersepil (13.2. TURKMEN)
dermanlyk tozga (27.45. TURKMEN)
desavala chakrakaeli (16.1a. TELUGU)
deserta melone (38.8. LATVIAN)
deshi shim (40.9. BENGALI)
desi ba khamba alu (11.1. ORIYA)
desi ba khamba alu (11.2. ORIYA)
desi urahi (40.9. ASSAMESE)
desmad (24.18. GAELIC, MANX)
Deutsche-Kicher (40.10. GERMAN)
dev bambu (17.2. TURKISH)
devedikeni (27.11. TURKISH)
devětsil japonský (27.35. CZECH)
devil's fig (62.23. ENGLISH)
devio-fe'ama (59.2. GBE-VHE)
devioƒeama (59.2. GBE-VHE)
dew bean (40.35. ENGLISH)
dewadaru (27.12. JAVANESE)
dhaal (40.2. KANNADA)
dhaamaargava (38.22. SANSKRIT)
dhaamagarva (38.22. SANSKRIT)
dhaaraphala (38.21. SANSKRIT)
dhabalakain (50.2. ORIYA)
dhal (40.2. ENGLISH)
dhala tulasi (44.5. ORIYA)
dhalatulasi (44.5. ORIYA)
dhamargava (38.22. SANSKRIT)
dhamargavah (38.22. SANSKRIT)
dhamargowa (38.21. SANSKRIT)
dhambala (40.29. SINHALA)
dhana (24.9. SANSKRIT)
dhane (24.9. BENGALI)
dhane (24.9. GUJARATI)
dhane (24.9. MARATHI)
dhangiyā (24.9. HINDI)
dhania (24.9. FIJIAN)
dhania (24.9. HINDI)
dhania (24.9. ORIYA)
dhania (24.9. PANJABI, EASTERN)
dhạnia (24.9. SANTALI)
dhanika (24.9. SANSKRIT)
dhaniya (24.9. BENGALI)
dhaniya (24.9. HINDI)
dhaniya (24.9. MIZO)
dhaniya (24.9. NEPALI)
dhaniyaalu (24.9. TELUGU)
dhaniyaka (24.9. SANSKRIT)
dhaniyalu (24.9. TELUGU)
dhanna (24.9. MARATHI)
dhanya (24.9. HINDI)
dhanya (24.9. MARATHI)
dhanya (24.9. SANSKRIT)
dhanya (24.9. URDU)
dhanya beeja (24.9. KANNADA)
dhanyabeeja (24.9. SANSKRIT)
dhanyaka (24.9. SANSKRIT)
dhanyakam (24.9. SANSKRIT)
dharbusini (38.3. TAMIL)
dharnargava (38.21. SANSKRIT)
dhekia (2.1. ASSAMESE)
dhennika (24.9. SANSKRIT)
dhenras (45.1. BENGALI)
dhenrosh (45.1. BENGALI)
dheras (45.1. BENGALI)
dhinmado (23.3. GUJARATI)
dhol (40.2. ENGLISH)
dholi rai (30.57. GUJARATI)
dhonia (24.9. ASSAMESE)
dhoravaange (62.23. MARATHI)
dhulakudi (24.7. HINDI)
dhumala (20.1. SANSKRIT)
dhundal (38.22. BENGALI)
dhunduli (38.29. ASSAMESE)

dhunduli (38.30. ASSAMESE)
dhvankshamaachi (62.15. SANSKRIT)
di can (44.11. CHINESE, WÚ)
di dan (62.24. CHINESE, WÚ)
di er zi (35.2. CHINESE, XIANG)
di fu (35.2. CHINESE, MANDARIN)
di fu (35.2. CHINESE, WÚ)
di gua (40.11. CHINESE, GÀN)
di gua (40.11. CHINESE, MIN BEI)
di gua (40.21. CHINESE, XIANG)
di gua (40.21. CHINESE, YUE)
di gua (62.24. CHINESE, YUE)
di gua er miao (44.1. CHINESE, MANDARIN)
di gua er miao (44.1. CHINESE, WÚ)
di li (10.2. CHINESE, WÚ)
di li (10.2. CHINESE, XIANG)
di qu pi (62.8. CHINESE, WÚ)
di qua (11.1. CHINESE, YUE)
di qua (40.20. CHINESE, WÚ)
dia (17.3. GURO)
diàɓá (7.2. BASSARI)
diàɓá (7.3. BASSARI)
diaba (7.3. MANDING-BAMBARA)
diabelha (57.1. PORTUGUESE)
diabéré (8.4. BAMANANKAN)
diabereen puté (8.9. MANDING-MANDINKA)
diabo (7.9a. MANINKAKAN, WESTERN)
diakari-gberi (62.1. GOURMANTCHÉ)
diakari-gberi (62.2. GOURMANTCHÉ)
diakari-mouinni (62.5. GOURMANTCHÉ)
diakhatou (62.14a. PULAR)
diakhatou (62.14a. WOLOF)
diakhatu (62.14a. SERER-SINE)
diakhatu (62.19. WOLOF)
diamb u sukar (17.5. WOLOF)
diamba-dambata (35.4. KOONGO)
diañ sukar (17.5. SERER-SINE)
dian voli (17.3. GURO)
diao si qiu zhu (17.1. CHINESE, WÚ)
diasan-chiku (17.1. JAPANESE)
diato (62.14a. MANINKAKAN, WESTERN)
dib (38.13a. HMONG-DAW)
dib iab (38.23. HMONG-DAW)
dib liab (38.3. HMONG-DAW)
dib nyiupeej (38.10. HMONG-DAW)
dib pag (38.7a. HMONG-DAW)
dib pag (38.7b. HMONG-DAW)
dib pag (38.11. HMONG-DAW)
dib qus (38.6. HMONG-DAW)
dib tuaj pos (38.12. HMONG-DAW)
dibemre (45.4. TEM)
Dickebohne (40.34a. GERMAN)
did an (62.24. CHINESE, XIANG)
didnakopha gogra (38.16. GEORGIAN)
didysis moliūgas (38.16. LITHUANIAN)
didžioji dilgėlė (66.3. LITHUANIAN)
didžioji nasturtė (65.1. LITHUANIAN)
didžioji varnalėša (27.5. LITHUANIAN)
die (38.18. GOLA)
die zi luo bo (30.41a. CHINESE, GÀN)
diéban (11.2. JOLA-FONYI)
di[ees]p c[as] (61.1. VIETNAMESE)
di[ees]p d[aj]i (27.29. VIETNAMESE)
di[ees]p xo[aw]n (27.10a. VIETNAMESE)
di[ees]p xo[aw]n (27.31. VIETNAMESE)
di[ees]p xo[aw]n (27.32. VIETNAMESE)
diégemtenguéré (40.31. FRENCH)
dieguem tenguéré (40.15. MÒORÉ)
diéguem tenguéré (40.31. MÒORÉ)
diente de léon (27.45. SPANISH)
dienziedis (13.3. LATVIAN)
dierb (58.2. ARABIC)
digbemre (45.4. TEM)
digegbe (11.2. MANINKAKAN, EASTERN)
dihere (11.2. KABIYÉ)
dikàbil (62.9. KONKOMBA)
dikambil (62.9. BASSARI)
dikambil (62.9. KONKOMBA)
dikamire (62.9. BASSARI)
dikamire (62.9. KONKOMBA)
dikenli çöğen (35.7. TURKISH)
dikenli kabak (38.25. TURKISH)
dikiti (38.14. SHONA)
dikiti (38.16. SHONA)
dikiti (38.17. SHONA)
dikiti (38.18. SHONA)
dikondo (16.1a. BEMBÉ)
dikondo (16.1a. KOONGO)
dila (10.1. PANJABI, EASTERN)
dila a manok (23.4. MARANAO)
dila a salaw (40.25. MARANAO)
dilag-dilag (27.4. IFUGAO)
dilang-baka (32.1. FILIPINO)
dilaw (20.2. BUHID)
dilaw (20.2. TAGALOG)
dilaw-pula (20.2. TAGALOG)
dima (36.3. SHONA)
dimariyo (17.7. DITAMMARI)
dimbu baha (44.5. SANTALI)
dimori (10.1. BASSARI)

dock (58.3. ENGLISH)
dòck (58.3. KWÉYÒL)
dŏda-hĕdur (17.7. KASHMIRI)
dodak (27.42. HINDI)
dodali (38.21. MARATHI)
ɗóɗaya (44.5. HAUSA)
dodda badane (62.19. KANNADA)
dodda basale (28.1. KANNADA)
dodda bedde gida (62.12. KANNADA)
dodda bondoola gida (62.12. KANNADA)
dodda budde gida (62.12. KANNADA)
dodda chunchala gida (63.1. KANNADA)
dodda goni soppu (59.2. KANNADA)
dodda guppatte gida (62.12. KANNADA)
dodda honagone (23.1. KANNADA)
dodda hurali kaayi (40.34a. KANNADA)
dodda jeerige (24.13. KANNADA)
dodda jola (17.7. KANNADA)
dodda sompu (24.13. KANNADA)
doddadumprashme (20.1. KANNADA)
doddagoni soppu (59.2. KANNADA)
dŏdijü (27.45. KASHMIRI)
dodka (38.21. MARATHI)
dodke (38.21. MARATHI)
dodoki (38.21. SWAHILI)
dodoki (38.22. SWAHILI)
dodongo (45.1. BEMBÉ)
ɗóɗowa (44.5. HAUSA)
ɗóɗoya (44.5. HAUSA)
ɗóɗoya ta kare (44.5. HAUSA)
ɗóɗoya tagida (44.5. HAUSA)
doggalikura (23.3. TELUGU)
doggoli padpè (23.3. TULU)
dohi kochu (8.8. ASSAMESE)
dohi kochu (8.9. ASSAMESE)
doi ku (40.43. FON)
dõko (39.4. JOLA-FONYI)
dokto (36.3. IBALOI)
dokudami (61.1. JAPANESE)
dokuimo (8.1. JAPANESE)
dolcetta (67.2. ITALIAN)
dolcetta d'Italia (67.1. ITALIAN)
dolchin (3.1. TAJIKI)
dolgolisna nivska repka (30.12. MACEDONIAN)
dolgopetalna nivska repka (30.13. MACEDONIAN)
dolgoplodna piperka (62.2. MACEDONIAN)
dolic en sabre (40.4. FRENCH)
dolichos (40.9. SLOVAK)
dolichos lablab (40.9. CZECH)
dolico (40.9. ITALIAN)
dolico cavallino (40.16. ITALIAN)
dólico de Goa (40.29. SPANISH)
dólico espárrago (40.42. SPANISH)
dólico gigante (40.9. SPANISH)
dólico lablab (40.9. SPANISH)
dolikë (40.9. ALBANIAN, TOSK)
dolique asperge (40.42. FRENCH)
dolique bulbeux (40.21. FRENCH)
dolique bulbeux (40.22. FRENCH)
dolique chinois (40.43. FRENCH)
dolique de Chine (40.41. FRENCH)
dolique d'egypte (40.9. FRENCH)
dolique des Andes (40.20. FRENCH)
dolique géant (40.41. FRENCH)
dolique mongette (40.41. FRENCH)
dolique tubéreux (40.21. FRENCH)
dolique tubéreux (40.22. FRENCH)
dolmalik biber (62.1. TURKISH)
doloŋgbo (55.1. GOLA)
dolycombi (1.1. GREEK)
doman (16.1a. IBALOI)
domar (17.1. INDONESIAN)
domase (62.9. SHONA)
domasi (62.9. SHONA)
domat (62.9. BULGARIAN)
domat (62.9. MACEDONIAN)
domate (62.9. ALBANIAN, TOSK)
domates (62.9. TURKISH)
domba (62.9. BADAGA)
dombo (55.1. GOLA)
dombos (38.16. SONGHAY)
dombos (38.18. WOLOF)
domoda (45.4. BAMANANKAN)
domolnu (27.18. IFUGAO)
donda (38.4. TELUGU)
donda kaya (38.4. TELUGU)
dondakaya (38.4. TELUGU)
dondatheege (38.4. TELUGU)
dong dong dia makaya (63.1. BEMBÉ)
dong dong dia makaya (63.1. KOONGO)
dong feng cai (27.17. CHINESE, MANDARIN)
dong feng cai (27.17. CHINESE, WÚ)
dong gua (38.1. CHINESE, GÀN)
dong gua (38.1. CHINESE, MANDARIN)
dong gua (38.1. CHINESE, WÚ)
dong gua (38.1. CHINESE, YUE)
dong hai hong (62.17. CHINESE, YUE)

drамblinė soruolė (17.3. LITHUANIAN)
drăžkova celina (24.4. BULGARIAN)
dreeka (11.7. ÉWÉ)
dreifarbiges Stiefmütterchen (68.1. GERMAN)
dřeňová kapusta (30.32. CZECH)
drickleurig viooltje (68.1. VLAAMS)
dridhabeeja (40.6. SANSKRIT)
dridhabija (40.6. SANSKRIT)
driekleurig akkerviooltje (68.1. DUTCH)
drighabeeja (30.51. SANSKRIT)
dringi (38.21. SPANISH)
drishopadma (48.1a. SANSKRIT)
drobnjak (7.11. SLOVENIAN)
droogeters (40.26. STELLINGWERFS)
drueagurk (38.13a. DANISH)
drueagurk (38.13a. NORWEGIAN)
druehyacinth (13.5. DANISH)
druifhyacint (13.5. DUTCH)
druight (38.11. GAELIC, MANX)
drum-pa (30.41b. BALTI)
dschola (38.20. TEM)
dsetʃi (27.8. DANGME-KROBO)
dsukuwiafa (62.9. DANGME-KROBO)
du (63.1. YORUBA)
du ga-nbe (63.1. YORUBA)
du huo (26.2. CHINESE, WÚ)
dua (39.4. GONJA)
duade (39.4. GA)
duan ce yyi dou (40.13. CHINESE, MANDARIN)
duan guo hui qin (24.22. CHINESE, MANDARIN)
duan jia yi dou (40.13. CHINESE, WÚ)
dua̱sc (39.4. TIV)
duàsə (39.4. TIV)
duaso (39.4. TIV)
dubba (38.14. AFAR)
dubba (38.14. AMHARIC)
dubba (38.14. HARARI)
dubba (38.16. AFAR)
dubba (38.16. AMHARIC)
dubba (38.16. HARARI)
dubba (38.17. AFAR)
dubba (38.17. HARARI)
dubba (38.18. AFAR)
dubba (38.18. AMHARIC)
dubba (38.18. HARARI)
dubbai (38.14. SOMALI)
dubbai (38.16. SOMALI)
dubbai (38.17. SOMALI)
dubbai (38.18. SOMALI)
dubbe (40.5. AMHARIC)
duɓɓi waandu (19.1. FULFULDE, ADAMAWA)
dubbo (38.20. SOMALI)
dubulé (55.1. PULAAR)
duda-gorai (59.2. KANNADA)
dudak'kahu (27.31. URDU)
duddi (38.16. KONKANI)
duddini (38.16. KONKANI)
duderpurre (36.3. FULFULDE, ADAMAWA)
dudhal (27.45. HINDI)
dudhale (27.45. HINDI)
dudhali (27.45. HINDI)
dudh-batthal (27.45. PANJABI, EASTERN)
dudhe (27.42. NEPALI)
dudheli (27.45. HINDI)
dudhi (27.42. HINDI)
dudhi (38.20. GUJARATI)
dudhya (38.20. MARATHI)
dudi (27.42. URDU)
dudi (38.20. KONKANI)
dudio (38.20. GUJARATI)
dudio (38.20. KONKANI)
dudli (27.45. PANJABI, EASTERN)
dudnik (24.1. RUSSIAN)
dudnik lekavy (24.1. BELARUSAN)
duduku (11.1. YORUBA)
duee (17.1. BUGHOTU)
duee (17.2. BUGHOTU)
duee (17.4. BUGHOTU)
duffin beans (40.25. ENGLISH)
dufo (16.1a. IBANAG)
dug o môn (13.2. SERER-SINE)
dugdhphenee (27.45. SANSKRIT)
dugian (17.1. BICOLANO, CENTRAL)
düğme otu (27.22. TURKISH)
dugu folo (40.15. BAMANANKAN)
dugu-fui (16.1a. SUSU)
dugukoro niamaku (20.6. MANDING-BAMBARA)
dùgùmà nyamugu (20.6. BOBO MADARÉ, SOUTHERN)
duguri (40.43. DENDI)
duguru-nyamuŋhu (20.6. MANDING-MANDINKA)
dugus falo (40.15. BAMANANKAN)
duho (16.1a. IBANAG)
duï (40.18. BAOULÉ)
duilag (16.1a. GAELIC, MANX)

duilleag-bhàite (50.2. GAELIC, SCOTTISH)
duilleog (50.2. GAELIC, IRISH)
duiveboon (40.34b. DUTCH)
duivenerwt (40.2. DUTCH)
duivenerwt (40.2. VLAAMS)
duiweboon (40.2. AFRIKAANS)
duk düa (8.3. LAO)
dukarakanda (11.2. MARATHI)
dukay (11.5. IVATAN)
dukto (36.3. IBALOI)
dùkú (36.3. NUPE-NUPE-TAKO)
dukuma (62.24. HARARI)
dul (30.56. BALTI)
dulagondi (40.18. TELUGU)
dulal (27.45. HINDI)
dulavrat otu (27.5. TURKISH)
dulaw (20.2. CEBUANO)
dulaw (20.2. HILIGAYNON)
dulaw (20.2. WÁRAY-WÁRAY)
duláw (20.2. WÁRAY-WÁRAY)
dullieha (38.3. MALTESE)
dulud (24.7. HANUNOO)
duma (38.20. HAUSA)
dumalan (30.56. LADINO)
dumalum (27.18. IFUGAO)
dumba (17.7. BODO)
dumbai (39.4. VAI)
dumbaraasme (20.1. KANNADA)
dumbarasme (20.1. KANNADA)
dumbe (8.4. SHONA)
dumbraashta (20.1. KANNADA)
dumbu sukar (17.5. WOLOF)
dummpabaucchale (35.9. TELUGU)
dumpabachhali (35.9. TELUGU)
dumpabucchale (35.3. TELUGU)
dumparaashtrakamu (20.1. TELUGU)
dumparaasme (20.1. KANNADA)
dumparasmi (20.1. KANNADA)
dundia (24.9. BODO)
dundu (7.13. NEPALI)
dundu hariya (11.1. DENDI)
dundu kiré (11.4. DENDI)
dundu mbissa (11.2. DENDI)
dundula (28.1. TAGALOG)
dundula (28.1. TAUSUG)
dungari (7.3. GUJARATI)
dungli (7.3. GUJARATI)
dungulpha (7.3. KOREAN)
dunguri (40.26. ZARMA)
dunguri (40.43. SONGHAY)
dúŋgúrì (40.43. SONGHAY)
dunguri (40.43. ZARMA)
dunguri soso (40.43. BAMANANKAN)
duntupesalu (40.43. TELUGU)
dupa (16.1a. ATTA, PAMPLONA)
dupag (16.1a. ATTA, PAMPLONA)
dupat (16.1a. AGTA, CASIGURAN DUMAGAT)
dupdupil (59.2. BONTOC CENTRAL)
dupduppilaw (59.2 KANKANAEY)
dupo (16.1a. IBANAG)
dupoloḍ (38.20. NEWAR)
durabhigraha (40.18. SANSKRIT)
duradagondi (40.18. TELUGU)
durghandh (7.3. SANSKRIT)
duri (40.43. BAMANANKAN)
duri gambas (32.1. INDONESIAN)
duri tiongkok (32.1. INDONESIAN)
duri tiongkok (32.1. JAVANESE)
durias (40.42. IBALOI)
durlabha (20.4. SANSKRIT)
durnamari (8.3. SANSKRIT)
durupnamu (27.27. KOREAN)
du-shen sair (27.8. HEBREW)
dushpra dharshini (62.19. SANSKRIT)
dushytsa (44.6. KAZAKH)
dusparsha (40.18. SANSKRIT)
dutun ilamnun (62.21. HANUNOO)
duu (8.4. GOLA)
duu (8.9. GOLA)
d[uw]a chu[ooj]t (38.13a. VIETNAMESE)
d[uw]a h[aa][us] (38.3. VIETNAMESE)
d[uw]a leo (38.13b. VIETNAMESE)
d[uw]a n[us]i (38.29. VIETNAMESE)
d[uw]ad[or] (38.3. VIETNAMESE)
duwaw (20.2. CEBUANO)
dụ́wẹ́ị́ ịkpịkpị́ (62.15. IZON)
düýpli buten (24.8. TURKMEN)
dvegodišna večernica (51.1. MACEDONIAN)
dvergbitur (38.18. ICELANDIC)
dvgliuotasis artišokas (27.15. LITHUANIAN)
dvimetė nakviša (51.1. LITHUANIAN)
dvispalvė ginūra (27.25. LITHUANIAN)
dvodomna (66.3. SERBO-CROATIAN)
dvogodišnj (51.1. SERBO-CROATIAN)
dvorogčesta oreška (64.1. MACEDONIAN)
dvouzubec chlupatý (27.8. CZECH)
dvozub (27.8. SERBO-CROATIAN)
dvuchradka stsennaja (30.47. BELARUSAN)
ɗwai-ɗwaiya (44.5. HAUSA)

dwarf mallow (45.5. ENGLISH)
dwirantwi (27.8. AKAN-AKUAPEM)
dyaɓá (7.3. BUDIK)
dyaba (7.3. WAMEY)
dyeberry (56.4. ENGLISH)
dygliuotasis kaparis (34.1. LITHUANIAN)
dyňa červená (38.3. SLOVAK)
dyně (38.13a. CZECH)
dýně (38.18. CZECH)
dyne meloun (38.9. CZECH)
dynia figolistna (38.15. POLISH)
dynia olbrzymis (38.16. POLISH)
dynia piżmowa (38.17. POLISH)
dynia pośrednia (38.14. POLISH)
dynia zwyczajna (38.18. POLISH)
dynja (38.7a. RUSSIAN)
dynja (38.7b. KAZAKH)
dynja (38.7b. RUSSIAN)
dynja (38.9. RUSSIAN)
dynja (38.11. RUSSIAN)
dyo (63.1. KISI, SOUTHERN)
dyo (63.1. KISSI, NORTHERN)
dyrka gulrot (24.11. NORWEGIAN)
dza (11.3. ÉWÉ)
dz´aa!´hau (38.12. KUNG-EKOKA)
dzafutu (11.3. ÉWÉ)
dzaglis-satacuri (27.45. GEORGIAN)
dzakpɔbɔ-tukui (62.5. GBE-VHE)
dzeloņainā kapera (34.1. LATVIAN)
dzeltenā dienziede (13.3. LATVIAN)
dze-tɛ (11.7. ÉWÉ)
dzete (36.3. ÉWÉ)
dziki (38.19. POLISH)
dziki ryż (17.8. POLISH)
džindžer (20.6. BOSNIAN)
dzirtetra (24.18. GEORGIAN)
dziurawiec pospolity (35.10. POLISH)
dzivi-dzivi (21.1. ÉWÉ)
dziwaczek (49.1. POLISH)
dziwidło dzwonkowate (8.3. POLISH)
dzmeuk (38.3. ARMENIAN)
dzoboli (11.3. ÉWÉ)
dzogbenyabo (19.1. GBE-VHE)
dzoro (16.1a. SHONA)
džumbir (20.6. BOSNIAN)
dzungu (38.14. VENDA)
dzungu (38.16. NYANJA)
dzungu (38.16. VENDA)
dzungu (38.17. VENDA)
dzungu (38.18. VENDA)
dzwonek rapunkuł (33.2. POLISH)
é badbaday (23.3. JOLA-FONYI)
é borimbor (11.6. JOLA-FONYI)
é déhindél (38.23. JOLA-FONYI)
e dou (40.35. CHINESE, MANDARIN)
e dou (40.35. CHINESE, WÚ)
é fulungèy (44.5. JOLA-FONYI)
é gañuma (40.18. JOLA-FONYI)
é ganuna (40.18. JOLA-FONYI)
é gerda (40.1. JOLA-FONYI)
é gertay (40.1. JOLA-FONYI)
e kama (11.1. JOLA-FONYI)
é kama (11.1. JOLA-FONYI)
é mangera (40.1. JOLA-FONYI)
e mei dou (40.9. CHINESE, WÚ)
é mo (40.18. JOLA-FONYI)
é sana (39.4. JOLA-FONYI)
é sasãg (44.5. WAMEY)
eagle fern (2.1. ENGLISH)
earth almond (10.1. ENGLISH)
earth chestnut (40.11. ENGLISH)
earth nut (40.1. ENGLISH)
earth-nut pea (40.11. ENGLISH)
East Indian arrowroot (19.1. ENGLISH)
East Indian lotus (48.1a. ENGLISH)
èba (17.1. NUPE-NUPE-TAKO)
èba (17.2. NUPE-NUPE-TAKO)
èba (17.4. NUPE-NUPE-TAKO)
eba edi (11.4. EFIK)
eba edi (11.4. IGBO)
ebaba kani (62.5. JOLA-FONYI)
èbáfọ̄ (23.3. EDO)
ebanana (16.1a. NAMA-DAMARA)
ébát-iba (16.1a. EFIK)
ébayibadjay (23.3. JOLA-FONYI)
èbĕ (38.14. NUPE-NUPE-TAKO)
ebέ (38.16. BEROM)
èbĕ (38.16. NUPE-NUPE-TAKO)
èbĕ (38.17. NUPE-NUPE-TAKO)
èbĕ (38.18. NUPE-NUPE-TAKO)
ebé ehófẹn (59.2. EDO)
ebegehi (16.1a. IGBO)
ebegömeci (45.5. TURKISH)
ebegümeci (45.5. TURKISH)
ɛ-ɓɛn (8.4. BUDIK)
ɛ-ɓen *i*-diík (44.8. BUDIK)
e-bὲr (40.39. BUDIK)
èbèsälébō (9.1. EDO)
ɛ-bέtέr (40.1. BUDIK)
ebienlen (23.3. EWONDO)

èbìgè (11.1. EFIK)
èbìgè ǹtán (11.1. EFIK)
èbìgè òsúkpà (11.1. EFIK)
ɛ-bita-lif (27.24. THEMNE)
ebli (17.7. ÉWÉ)
ebli (17.7. GEN)
ebò (38.20. NUPE-NUPE-TAKO)
ebọ́dọ̄ndọ̄n (59.3. EDO)
ebolo (27.14. YORUBA)
ebòlò (27.14. YORUBA)
éborimbor (11.2. JOLA-FONYI)
ebubug endog (11.1. JAVANESE)
ɛ-ɓwdy ɛ-ɓanàx (44.8. BASSARI)
ɛ-ɓwdy ɛ-pɛsyàx (44.8. BASSARI)
eceng gede (18.1. INDONESIAN)
eceng genjer (6.1. SUNDA)
eceng leutik (18.2. INDONESIAN)
eceng padi (18.2. INDONESIAN)
echalot (7.2. CREOLE, HAITIAN FRENCH)
échalote (7.2. FRENCH)
echalott (7.2. CREOLE, HAITIAN FRENCH)
echalotte chinoise (7.4. FRENCH)
ĕchkooaakw (38.14. MICMAC)
ĕchkooaakw (38.16. MICMAC)
ĕchkooaakw (38.17. MICMAC)
ĕchkooaakw (38.18. MICMAC)
ĕchkooch (38.3. MICMAC)
ĕchkooch (38.7a. MICMAC)
ĕchkooch (38.11. MICMAC)
ĕchkoochkool (38.14. MICMAC)
ĕchkoochkool (38.16. MICMAC)
ĕchkoochkool (38.17. MICMAC)
ĕchkoochkool (38.18. MICMAC)
echte Brunnenkresse (30.52. GERMAN)
echte haverwortel (27.46. VLAAMS)
echte Tigerblume (12.1. GERMAN)
echter Meerkohl (30.45. GERMAN)
echtes Pfeilkraut (6.2. GERMAN)
eci (11.1. NUPE-NUPE-TAKO)
eci (11.2. NUPE-NUPE-TAKO)
ecuelle d'eau (24.7. FRENCH)
edálê (62.5. MANKANYA)
ēdamā diskoreja (11.5. LATVIAN)
edamame (40.8. JAPANESE)
edɔ̀ŋ (11.2. BEROM)
eddoe (8.4. ENGLISH)
ede (8.4. IGBO)
édè (8.4. IGBO)
édè áro (8.9. IGBO)
édẹ bèkẹè (8.9. IGBO)
édè éfù (8.4. IGBO)
édè eko (8.9. IGBO)
édè ékò (8.9. IGBO)
édè nkiti (8.4. IGBO)
édè ọ́bà (8.4. IGBO)
édè ọ̄kpọ́ro (8.9. IGBO)
édè ọ́kụ̄ (8.4. IGBO)
édè ōsịlí (8.4. IGBO)
édè óyìbó (8.4. IGBO)
ede-beké (8.9. IGBO)
èdèm ìnày (11.4. EFIK)
edemate (62.9. NAMA-DAMARA)
édesburgonya (36.3. HUNGARIAN)
édeskömény (24.12. HUNGARIAN)
édeskömény (24.13. HUNGARIAN)
ɛ-ɗéwùrè (8.4. BASSARI)
èdiá abañ (11.4. EFIK)
ediaba ou élignon (7.9a. JOLA-FONYI)
edible burdock (27.5. ENGLISH)
edible canna (9.1. ENGLISH)
edible chrysanthemum (27.23. ENGLISH)
edible hibiscus (45.1. ENGLISH)
edible podded pea (40.27. ENGLISH)
èdidià ìwá (11.4. EFIK)
ɛ-dif-wir (27.24. THEMNE)
ed-Indez (17.7. BRETON)
edodimos (38.5. GREEK)
edodimos (38.13b. GREEK)
ed-Turki (17.7. BRETON)
ɛdua (40.25. TWI)
ɛdua (40.43. TWI)
ɛdua fita (40.25. TWI)
ɛdua nton (40.25. TWI)
ẹdumṣi (36.3. YORUBA)
edunde (40.28. NDONGA)
ɛdwew (11.1. AKAN-FANTE)
ɛ-ɗyíɗy (11.4. BUDIK)
edza (40.18. GBE-VHE)
ʃẽ (17.5. GA)
ée (40.43. EDO)
eeni kilangu (36.3. TAMIL)
eerappel (62.24. STELLINGWERFS)
eerpel (62.24. STELLINGWERFS)
eerulli (7.3. KANNADA)
eerulli (7.3. MALAYALAM)
eesugaayi (38.20. KANNADA)
eesugaayi balli (38.20. KANNADA)
eesugumbala (38.20. KANNADA)
eèsún (17.3. YORUBA)
eetbaar bloemriet (9.1. DUTCH)

eetbare krisant (27.23. AFRIKAANS)
eetbare peulert (40.27. AFRIKAANS)
eetbare pulerwt (40.27. VLAAMS)
eeuwig mose (30.33. VLAAMS)
eeuwige moes (30.33. DUTCH)
efafa (39.4. JOLA-FONYI)
efai (23.3. AKAN-ASANTE)
ɛfan (23.3. AKAN-AKUAPEM)
ɛfan (23.3. AKAN-ASANTE)
efan (23.3. TWI)
efan (28.1. TWI)
ɛ-fan (59.3. TWI)
efatate (36.3. JOLA-FONYI)
ɛfɔ-odu (62.15. KRIO)
ɛfere (38.3. AKAN-AKUAPEM)
ɛfere (38.14. TWI)
ɛfere (38.16. AKAN-ASANTE)
ɛfere (38.16. TWI)
ɛfere (38.17. TWI)
ɛfere (38.18. TWI)
éférém̀bákàrà (59.2. EFIK)
ɛfere-pa (38.18. AKAN-AKUAPEM)
eferi (44.5. YORUBA)
èfiáñ (11.3. EFIK)
efínrín arùntantan (44.5. YORUBA)
efínrín ata (44.5. YORUBA)
efínrín márúgbóṣányán (44.5. YORUBA)
efínrín wẹ́wẹ́ (44.5. YORUBA)
efínrín-ajá (44.5. YORUBA)
efio (10.1. AJA)
efio (10.1. ÉWÉ)
ɛfir (38.16. AKAN-FANTE)
ɛfir (38.18. AKAN-FANTE)
efirin wewe (44.5. YORUBA)
ẹ̀fọ èbòlò (27.14. YORUBA)
ẹ̀fọ ebúre (27.14. YORUBA)
ẹfọ tẹ̀tẹ̀ (23.3. YORUBA)
efodu (27.14. YORUBA)
éfọ̀k ǹtókòn (62.2. EFIK)
ẹfoọ̀sùn (62.18. YORUBA)
ɛfrɛ (38.16. AKAN-AKUAPEM)
ɛfre (38.16. AKAN-ASANTE)
ɛfrɛ (38.18. AKAN-AKUAPEM)
ɛfrɛ (38.18. AKAN-ASANTE)
ɛfrɛ (38.18. AKAN-FANTE)
egadagwŏdŭmŭgāwā' (13.2. MICMAC)
egadâkŭnckeech tŭlā' (13.2. MICMAC)
ẹ̀gàrà (62.5. OGBIA)
égba (40.1. YORUBA)
ɛ-gbaŋkaŋ (8.9. THEMNE)
e-gbara (38.3. KRIO)
égba(-yo) (40.1. YORUBA)
egbo (62.17. GEN)
ègbódo (11.1. YORUBA)
ege (39.4. YORUBA)
egéanchio (23.3. GOURMANCHÉMA)
ɛ-gɛ̀b (11.3. BUDIK)
egel avga (30.43. MONGOLIAN)
egel bavran (2.1. MONGOLIAN)
egel khakhuun (44.6. MONGOLIAN)
egerda (40.1. JOLA-FONYI)
eggaldin (62.19. ICELANDIC)
eggfrukt (62.19. NORWEGIAN)
eggplant (62.19. ENGLISH)
eggplant (62.19. IGBO)
eggplante (62.19. NORWEGIAN)
eggsøtvier (62.19. NORWEGIAN)
egipatski kromid (7.9a. MACEDONIAN)
egipatski luk (7.9a. SERBO-CROATIAN)
egipitacoren (17.7. ARMENIAN)
Egiptiese aanteelui (7.9a. AFRIKAANS)
egiptovski luk (7.9a. SLOVENIAN)
ego (38.29. ÉWÉ)
egoz adama (40.1. HEBREW)
egpa (40.1. YORUBA)
eğrelti (2.1. TURKISH)
eğrelti (3.2. TURKISH)
ègùnmo (62.15. YORUBA)
egúnsi (38.3. YORUBA)
egusi (27.14. YORUBA)
egusi (38.3. IGBO)
ègúsí (38.3. IGBO)
ɛgusi (38.3. KRIO)
egusi (38.3. YORUBA)
ẹ̀gúsí (38.3. YORUBA)
egusi (38.5. IGBO)
egusi (38.5. KRIO)
ɛgusi (38.5. KRIO)
egusi (38.7a. IGBO)
egusi (38.11. IGBO)
egusi (38.27. EDE NAGO)
egusi (38.27. YORUBA)
ẹgúsí (38.28. YORUBA)
ẹ̀gúsí agvè (38.3. YORUBA)
ẹ̀gúsí bora (38.3. YORUBA)
ẹgúsí itóò (38.5. YORUBA)
ẹ̀gúsí maga (38.3. YORUBA)
egusi-ito (38.5. JAPANESE)
egusi-ito (38.5. VLAAMS)
ẹ̀gúsí-ìtóò (38.5. YORUBA)

Egyiptomi taro (8.4. HUNGARIAN)
egymagvútök (38.25. HUNGARIAN)
Egyptian bean (48.1b. ENGLISH)
Egyptian lotus (50.2. ENGLISH)
Egyptian onion (7.9a. ENGLISH)
egyptische ui (7.9a. DUTCH)
Egyptische ui (7.9a. VLAAMS)
Egyptisk taro (8.4. SWEDISH)
ehető krizantém (27.23. HUNGARIAN)
ehetö zsidócseresznye (62.12. HUNGARIAN)
ehu (17.5. BUGHOTU)
ehwa (24.11. NDONGA)
Eibisch (45.1. GERMAN)
eie (7.3. SOTHO, NORTHERN)
eie (7.3. SOTHO, SOUTHERN)
eie (7.3. TSWANA)
Eierfrucht (62.19. GERMAN)
eierplant (62.19. DUTCH)
eiervrucht (62.19. VLAAMS)
eiervrug (62.19. AFRIKAANS)
eilifleog (35.1. GAELIC, IRISH)
Einlegegurke (38.13b. GERMAN)
einstape (2.1. NORWEGIAN)
'ei-sɑhyei-gɑ (38.3. KIOWA)
Eisblume (22.1. GERMAN)
ei-t'Hî-dɑ (17.7. KIOWA)
eivormige komkommer (38.6. DUTCH)
èjím (7.3. YALA)
ejinrin (38.23. EDE NAGO)
ejinrin (38.23. YORUBA)
eḳ nobát (59.2. HINDUSTANI)
ek pauda (24.13. HINDUSTANI)
ek qism ká (27.46. HINDUSTANI)
ek qism ká ehhoṭá (27.16. HINDUSTANI)
ek qism ká ság (35.9. HINDUSTANI)
ek qism kí gobhí (30.35. HINDUSTANI)
'eka (20.2. MARQUESAN)
èkà biâ (11.7. EFIK)
ekabé (8.9. BULU)
ε-kalma (38.20. THEMNE)
ekamitsáaʔ (32.1. COMANCHE)
ε-kanda (39.4. THEMNE)
ekandjimau (62.2. KONKOMBA)
ekapaani (24.7. KONKANI)
ekapaani (24.7. MARATHI)
ekɔ́s (40.18. BEROM)
ε-kẹɔkɔrɔ (23.4. THEMNE)
ekefià (11.7. EFIK)
ekẹ-ko-sẹ (62.23. YORUBA)
ekesi-masun (34.2. YORUBA)
ekhu (62.9. EDO)
ẹkhuẹ (62.9. EDO)
ekilýan noýba (40.26. TURKMEN)
ekimlik ajytere (30.51. TURKMEN)
ekimlik indew (30.48. TURKMEN)
ekimlik koriandr (24.9. TURKMEN)
ekimlik turp (30.56. TURKMEN)
ekimlik ýorunja (40.17. TURKMEN)
ekleme (1.1. TURKISH)
eklwi (40.2. AJA)
eklwi (40.2. GBE, WACI)
èkò (11.7. EFIK)
eko (17.3. ÉWÉ)
ekobon (38.27. EWONDO)
ε-koko (8.4. THEMNE)
ε-koko (8.9. THEMNE)
ε-koko-ε-nεthi (8.4. THEMNE)
ε-kol (8.9. THEMNE)
ekon (16.1a. BULU)
ε-konshoŋ (40.2. THEMNE)
ε-konsoi (40.2. SUSU)
ekovi (30.27a. NDONGA)
εkpa (40.1. EDE NAGO)
εkpav (40.1. YORUBA)
ẹkpẹ (11.7. IGBO)
ẹ̀kpére (38.20. EDO)
ekpu (62.19. GA)
ekpu (62.19. GONJA)
ekú (44.10. BEROM)
ẹ̀kukù (55.1. YORUBA)
ekunde (40.26. HERERO)
ekunde (40.26. KWANGALI)
ekungu (38.13a. HERERO)
èkùyá (34.2. YORUBA)
èkùyáko (34.2. YORUBA)
èkùyalé (34.2. YORUBA)
ẹkwulo mmịmị (40.3. IGBO)
eladje (7.2. JOLA-FONYI)
el-adkham (20.1. ARABIC)
elakkibaale (16.1a. KANNADA)
Elandsbohne (40.33. GERMAN)
elaphovoskon (24.18. GREEK)
elavan (38.1. MALAYALAM)
eldev navchit chesentser (62.14c. MONGOLIAN)
ele kosu (30.27a. KANNADA)
ε-lεba (8.4. BULLOM SO)
ε-lεba (8.9. BULLOM SO)
elebaale (16.1a. KANNADA)
elef hehaleh (42.1. HEBREW)

Elefantengras (17.3. GERMAN)
elefantgræs (17.3. DANISH)
elefantgras (17.3. NORWEGIAN)
elefanttiheinä (17.3. FINNISH)
elefantyams (8.3. DANISH)
elegbe (38.16. EDE NAGO)
elegbe (38.16. YORUBA)
elegbe (38.18. EDE NAGO)
elegbe (38.18. YORUBA)
elegede (38.16. YORUBA)
élégédé (38.16. YORUBA)
elegede (38.18. YORUBA)
élégédé (38.18. YORUBA)
ɛ-lɛkɔbɔtɔ (40.25. THEMNE)
elekli atgulak (57.1. TURKMEN)
ɛ-lel (40.25. THEMNE)
elelengue (59.2. EWONDO)
el-emlak (8.1. ARABIC)
elenskorogoviden tegavec (57.1. MACEDONIAN)
eleokramri (30.21. GREEK)
elephant ear (8.1. ENGLISH)
elephant grass (17.3. ENGLISH)
elephant yam (8.3. ENGLISH)
elephant-foot yam (8.2. ENGLISH)
elephantini dioscorea (8.3. GREEK)
ẹlẹshin-másọ (27.8. YORUBA)
eleuterococco (26.4. ITALIAN)
eleuterokokk koliuchii (26.4. RUSSIAN)
eleutherococco espinoso (26.4. SPANISH)
eleutherococoque (26.4. FRENCH)
el-galangal (20.1. ARABIC)
el-hejaz (35.1. ARABIC)
elikevi (27.18. TULU)
elikivi balli (24.7. KANNADA)
elikivi gida (27.18. KANNADA)
elikivi soppu (24.7. KANNADA)
elikivisoppu (27.18. KANNADA)
elilakkilannu (39.4. MALAYALAM)
élìlè (38.3. IGBO)
élìlì (38.3. IGBO)
elili ji okō (16.1a. IGBO)
el-khenshar alkarfy (3.1. ARABIC)
el-khenshar almaee (4.1. ARABIC)
el-klenshar alkhamel (5.1. ARABIC)
el-konbath (1.1. ARABIC)
el-kustnaa (10.2. ARABIC)
eḷḷanavare (40.34a. BADAGA)
eḷḷavare (40.34a. BADAGA)
el-liftia (24.8. ARABIC)
el-maee (10.2. ARABIC)
el-makmalia (14.1. ARABIC)
el-masamy (13.5. ARABIC)
el-naharry (13.3. ARABIC)
eloasson dioscorea (11.5. GREEK)
elok-sup (59.3. BULU)
elok-sup (59.3. EWONDO)
el-qulqass (8.1. ARABIC)
el-safrae (14.1. ARABIC)
el-sawsan (13.3. ARABIC)
el-sawsan (13.4. ARABIC)
el-sawsan (13.5. ARABIC)
el-tofah el-dahabi (62.14a. ARABIC)
eluad (27.8. KANKANAEY)
ẹlubásà (7.3. YORUBA)
elugbati (28.1. BINUKID)
əlukpó (17.5. OGBIA-AGHOLO)
emama kaani (62.2. JOLA-FONYI)
emangera (40.1. JOLA-FONYI)
ẹmbadákà (38.4. IZON)
ɛ-mbáràx (40.3. BASSARI)
emboka (23.3. LUYIA)
èméìn fị̀ tùò (36.2. IZON)
ɛ-mɛn (8.4. BUDIK)
emes (38.21. INDONESIAN)
emes (38.21. SUNDA)
emes (38.22. SUNDA)
emia (44.5. BAOULÉ)
ɛ́mian nvan van (44.5. ANDO)
emiin bagvaakhai (27.45. MONGOLIAN)
emiin khereen nuden (13.2. MONGOLIAN)
emikhonye (17.5. LUYIA)
emilia (27.18. ENGLISH)
emilia (27.18. ITALIAN)
emilia (27.18. PORTUGUESE)
emilia (27.18. SPANISH)
emilie (27.18. FRENCH)
emilie à feuilles de laiteron (27.18. FRENCH)
emilya (27.18. TURKISH)
emina (11.2. EDE NAGO)
ɛmina (11.2. KRIO)
emina (11.2. YORUBA)
ẹmina (11.2. YORUBA)
ẹmina-esi (11.2. YORUBA)
emlik (27.38. AZERBAIJANI, NORTHERN)
emoi (45.1. KONKOMBA)
èmówè (11.4. EDO)
èmọ́wè (11.7. EDO)
emping melinju (41.1. MALAY)
emuka (38.20. LUYIA)

ɛ-muna (36.3. THEMNE)
ʻena (20.2. MARQUESAN)
enab el-deep (62.12. ARABIC)
enaerios (11.3. GREEK)
enàgra comune (51.1. ITALIAN)
enante (24.17. SPANISH)
encámato (62.9. MANKANYA)
enceng-enceng (34.2. JAVANESE)
enciam (27.31. CATALAN-VALENCIAN-BALEAR)
enciam (27.32. CATALAN-VALENCIAN-BALEAR)
encung (62.19. INDONESIAN)
encung (62.19. JAVANESE)
encung (62.19. SUNDA)
endaibu (27.9. JAPANESE)
endāvijas (27.9. LATVIAN)
endewi (27.9. INDONESIAN)
endiba (27.9. FILIPINO)
endiba (27.9. TAGALOG)
endibia (27.9. SPANISH)
endibia acogollada roja (27.10b. SPANISH)
endiivi (27.9. FINNISH)
endiivia (27.9. ESTONIAN)
endiv sallat (27.9. SWEDISH)
endive (27.9. ASSAMESE)
endive (27.9. ENGLISH)
endive (27.9. FILIPINO)
endivez (27.9. BRETON)
endivia (27.9. CATALAN-VALENCIAN-BALEAR)
endívia (27.9. HUNGARIAN)
endivia (27.9. ITALIAN)
endívia (27.9. PORTUGUESE)
endivia (27.9. ROMANIAN)
endívia (27.9. SLOVAK)
endivia (27.9. SPANISH)
endívia sálata (27.9. HUNGARIAN)
endivie (27.9. CZECH)
endivie (27.9. DANISH)
Endivie (27.9. GERMAN)
endivie (27.9. NORWEGIAN)
Endiviensalat (27.9. GERMAN)
endiviesallad (27.9. SWEDISH)
endivij (27.9. RUSSIAN)
endivìj (27.9. UKRAINIAN)
endivija (27.9. CROATIAN)
endivija (27.9. SLOVENIAN)
endivu (27.9. SWAHILI)
endivya (27.9. LADINO)
endjikaui (44.4. KWANYAMA)
endkol siki (30.22. SUNDA)
endō (40.28. JAPANESE)
endou (40.28. JAPANESE)
endu nagge (59.2. FULFULDE, ADAMAWA)
endurkuar (30.48. TIGRIGNA)
endywia (27.9. POLISH)
eŋelfɛl (20.6. GA)
enɛ̀p (23.3. BEROM)
enestramenon sinapi (30.9. GREEK)
enetkheg sharilj (27.7. MONGOLIAN)
èng-chhài (36.2. CHINESE, MIN NAN)
engelenboom (26.3. VLAAMS)
Engelse winterspinazie (58.3. VLAAMS)
engelsk spenat (58.3. SWEDISH)
engelsk spinat (58.3. DANISH)
engelwortel (24.1. AFRIKAANS)
engelwortel (24.1. DUTCH)
Engelwurz (24.1. GERMAN)
enggeitskjegg (27.47. NORWEGIAN)
engifer (20.6. ICELANDIC)
ənginar (27.16. AZERBAIJANI, NORTHERN)
enginar (27.16. TURKISH)
engkarse (30.44. DANISH)
engkarse (30.44. NORWEGIAN)
engkol (30.27a. JAVANESE)
engkol babat (30.22. SUNDA)
engkol beureum (30.41a. SUNDA)
engkol gede (30.27a. SUNDA)
Englischer Senf (30.57. GERMAN)
engôm (11.1. BULU)
engsyre (58.2. NORWEGIAN)
ɛ-níiyàn (62.23. BASSARI)
enit (24.15. KALAGAN)
enjenára (27.16. LADINO)
enjenera (27.16. LADINO)
eŋmɔmi (45.1. GA)
é'ŋmɔmi (45.1. GA)
Ënn (7.3. LUXENBOURGEOIS)
ennańey (40.18. JOLA-FONYI)
enotera (51.1. SPANISH)
en-sai (36.2. JAPANESE)
ensámelata (59.2. BALANTA-KENTOHE)
ensámelate (59.2. BALANTA-KENTOHE)
ənsəlal (24.13. AMHARIC)
ənsolal (24.13. AMHARIC)
entepoot (30.43. STELLINGWERFS)
entimun (38.13a. INDONESIAN)
entoet (16.1a. VLAAMS)
enú (32.1. NUPE-NUPE-TAKO)

enuwa (38.3. NAMA-DAMARA)
enwad (27.8. BONTOC, CENTRAL)
enydrakardamon to pharmakeftikon (30.52. GREEK)
enyewe (23.4. MAKHUWA)
enywe (23.3. MAKHUWA)
ẹ̀pà gidi (40.1. YORUBA)
ɛ-pády (62.23. BASSARI)
ɛ-pády ɛnd a-ɓịmɛn (62.9. BASSARI)
ɛ-pády ɛ-tyɛ́n (62.23. BASSARI)
epana (22.2. GOURMANCHÉMA)
épaympay (10.1. JOLA-FONYI)
ɛ̀-pedy (62.23. BUDIK)
ɛ-pen (40.3. BASSARI)
ɛ-pepe (38.20. THEMNE)
épiaire affine (44.11. FRENCH)
épina (35.9. CREOLE, HAITIAN FRENCH)
epinar (27.19. BULU)
épinard (35.9. CAJUN FRENCH)
êpinard (35.9. FRENCH)
êpinard de la Nouvellle Zêlande (22.2. FRENCH)
épinard de Malabar (28.1. FRENCH)
êpinard d'été (22.2. FRENCH)
épinard du Cong (23.3. FRENCH)
épinard malabar (23.3. FRENCH)
épinard piquant (23.3. FRENCH)
épinard saurage (35.5. FRENCH)
épinard vert (23.3. FRENCH)
epinari (59.3. KOONGO)
ɛ-plantan (16.1a. THEMNE)
ɛ-planti (16.1a. THEMNE)
epombitère (62.24. JOLA-FONYI)
éponge végétale (38.22. FRENCH)
eponge végétale torcho (38.21. FRENCH)
èpuṇ (38.20. NUPE-NUPE-TAKO)
epungu (17.7. NDONGA)
epweri (40.2. LOMWE, MALAWI)
equiseto menor (1.1. SPANISH)
er armydy (27.26. TURKMEN)
er jiao wan ling (64.1. CHINESE, MANDARIN)
er kekju (24.11. AZERBAIJANI, NORTHERN)
er xing jie (30.47. CHINESE, WÚ)
erangayam (7.3. TAMIL)
erba cristallina (22.1. ITALIAN)
erba di napier (17.3. ITALIAN)
erba di Santa Barbara vernale (30.2. ITALIAN)
erba elefantina (17.3. ITALIAN)
erba medica (40.17. ITALIAN)
erba morella (62.21. ITALIAN)
erba pazienza (58.3. ITALIAN)
erba saetta (6.2. ITALIAN)
erba saetta chinese (6.1. ITALIAN)
erba sega (44.1. ITALIAN)
erba Spagna (40.17. ITALIAN)
Erbse (40.28. GERMAN)
ɛ-rɔgbɔ (39.4. THEMNE)
ercis (40.28. INDONESIAN)
ərd (20.2. AMHARIC)
Erdartischecke (27.26. GERMAN)
Erdbirne (27.26. GERMAN)
Erdbirne (27.39. GERMAN)
erdene shish (17.7. MONGOLIAN)
Erderbse (40.39. GERMAN)
Erdkirsche (62.12. GERMAN)
Erdmandel (10.1. GERMAN)
Erdnuß (40.1. GERMAN)
ere nofum (40.43. EDO)
ere nubaa (40.43. EDO)
eregìsi (40.28. VENDA)
erekisi (40.28. SOTHO, NORTHERN)
erekisi (40.28. SOTHO, SOUTHERN)
erekisi (40.28. TSWANA)
eresinge (11.2. BADAGA)
eresingi (11.2. BADAGA)
erevinthos (40.5. GREEK)
erfin gwylt (30.21. WELSH)
erfinen (30.41a. WELSH)
erfinew (30.22. WELSH)
ergi-i (40.28. HERERO)
erguvani zencidarisi (17.3. TURKISH)
erka (40.18. HINDI)
erkenci kis teresi (30.2. TURKISH)
eromo (17.3. BEROM)
erra doggali koora (23.3. TELUGU)
erra gummadi (38.16. TELUGU)
erra ulligadda (7.3. TELUGU)
erraalubachhali (28.1. TELUGU)
errachakrakaeli (16.1a. TELUGU)
erragaddalu (7.3. TELUGU)
erragomguru (45.4. TELUGU)
erragon kaaya (45.4. TELUGU)
erragongaka (45.4. TELUGU)
erragonguru (45.4. TELUGU)
errakandulu (40.2. TELUGU)
erramullugoranta (23.3. TELUGU)
errarati (16.1a. TELUGU)
erra-tamara (48.1a. TELUGU)
errathaamara (48.1a. TELUGU)

errathamma (40.3. TELUGU)
erratoakura (23.3. TELUGU)
erratotakura (23.3. TELUGU)
errefau (30.56. BASQUE)
ert (40.28. NORWEGIAN)
ertappel (62.24. AFRIKAANS)
erter (40.28. NORWEGIAN)
ertjie (40.28. AFRIKAANS)
ertur (40.28. ICELANDIC)
ertusnikja (52.1. ICELANDIC)
eruca (30.48. FRENCH)
eruca (30.48. ITALIAN)
eruca (30.48. PORTUGUESE)
eruca (30.48. SPANISH)
eruka posevnaja (30.48. RUSSIAN)
ɛrun (38.22. EDE NAGO)
ɛrun (38.22. YORUBA)
erun kaankan (38.22. YORUBA)
ẹrũnshà (10.1. YORUBA)
eruvanga (62.19. TELUGU)
erva armoles (35.1. PORTUGUESE)
erva da moda (27.22. PORTUGUESE)
erva de amor (60.1. PORTUGUESE)
erva de orvalho (22.1. PORTUGUESE)
erva do bom partor (30.43. TAGALOG)
erva do bom pastor (30.43. PORTUGUESE)
erva do Espírito Santo (24.1. PORTUGUESE)
erva do fígado (27.18. PORTUGUESE)
erva doce (24.13. PORTUGUESE)
erva dos burros (51.1. PORTUGUESE)
erva dos tinhosos (27.5. PORTUGUESE)
erva frecha (6.2. PORTUGUESE)
erva frecha chinesa (6.1. PORTUGUESE)
erva maura (17.5. RONGA)
erva moira (62.21. PORTUGUESE)
erva moura (62.15. PORTUGUESE)
erva moura (62.21. PORTUGUESE)
erva toira (52.1. PORTUGUESE)
erva-doce (24.13. CRIOULO, CAPE VERDE)
ervaru (38.7b. SANSKRIT)
ervaruh (38.7b. SANSKRIT)
ervilha (40.28. CRIOULO, CAPE VERDE)
ervilha (40.28. PORTUGUESE)
ervilha come-tudo (40.27. PORTUGUESE)
ervilha de angola (40.2. PORTUGUESE)
ervilha de cheiro (40.10. PORTUGUESE)
ervilha de debulhar (40.28. PORTUGUESE)
ervilha de Jerusalem (40.38. PORTUGUESE)
ervilha de quebrar (40.27. PORTUGUESE)
ervilha do congo (40.2. PORTUGUESE)
ervilha seca (40.28. PORTUGUESE)
ervilha tipo come-tudo (40.27. PORTUGUESE)
ervilha torta (40.27. PORTUGUESE)
ervilheira (40.28. PORTUGUESE)
erwt (40.28. DUTCH)
erwt (40.28. VLAAMS)
erwtaubergine (62.23. VLAAMS)
[es] t[is]a (44.5. VIETNAMESE)
[es] tr[aws]ng (44.5. VIETNAMESE)
esaka (34.2. LUYIA)
esalaate (27.33. JOLA-FONYI)
esan (45.3. BULU)
esan ọmọde (59.2. YORUBA)
esang (45.3. BULU)
esan(g) (45.3. EWONDO)
ɛ-santi (16.1a. THEMNE)
escaluña (7.2. SPANISH)
escaravia (24.23. SPANISH)
escaravia tudesca (24.23. SPANISH)
escarchada (22.1. SPANISH)
escarola (27.9. PORTUGUESE)
escarola (27.9. SPANISH)
escarole (27.9. ENGLISH)
Eschlauch (7.2. GERMAN)
escolimo (27.36. SPANISH)
escolimo de Espanha (27.36. PORTUGUESE)
escolimo maculado (27.37. SPANISH)
escólimo malhado (27.37. PORTUGUESE)
escomite (40.23. SPANISH)
escorcioneira (27.38. PORTUGUESE)
escorcomera (27.38. CATALAN-VALENCIAN-BALEAR)
escorzonera (27.38. SPANISH)
escorzonera de España (27.38. SPANISH)
eşek gevreği (27.42. TURKISH)
eşek hardali (30.24. TURKISH)
eşek marulu (27.42. TURKISH)
ɛ-sɛ́ngɔ́ng (44.5. BASSARI)
esfenâj (35.9. FARSI, WESTERN)
esfenaje-roomee (35.1. FARSI, WESTERN)
esfregão (38.22. PORTUGUESE)
eshbal (44.11. HEBREW)
eshimukaa (38.1. LUYIA)
eshitukwo (7.3. LUYIA)
e'shobuk (30.27a. POTAWATOMI)
eshu (20.4. CHINESE, WÚ)
esibe (23.3. AKAN-AKUAPEM)
esikhubi (40.43. LUYIA)
èsin (40.18. YORUBA)
èsinfunfun (40.18. YORUBA)

èsinsin (40.18. YORUBA)
esiri (21.1. YORUBA)
esise (40.18. YORUBA)
eskarola (27.9. BASQUE)
eskarola (27.9. CEBUANO)
eskarola (27.9. PAMPANGAN)
eşkileme (58.4. TURKISH)
ĕskipkodŭmŭgāwā’ (27.32. MICMAC)
ĕskŭdŭmŭgāwā’ (38.13a. MICMAC)
ε-soro (44.4. THEMNE)
esparak (60.1. FARSI, WESTERN)
espargo (13.2. CRIOULO, CAPE VERDE)
espargo (13.2. GALICIAN)
espargo (13.2. PORTUGUESE)
espargo bravo (13.1. PORTUGUESE)
espargo do Cabo (25.1. PORTUGUESE)
espargo silvestre (13.1. PORTUGUESE)
espargueira (13.2. GALICIAN)
esparragèra (13.2. GALICIAN)
espárrago (13.2. SPANISH)
espárrago amarguero (13.1. SPANISH)
espárrago del Cabo (25.1. SPANISH)
espàrrec (13.2. CATALAN-VALENCIAN-BALEAR)
esparrech (13.2. CATALAN-VALENCIAN-BALEAR)
espinac (35.9. CATALAN-VALENCIAN-BALEAR)
espinaca (35.9. SPANISH)
espinaca basela (28.1. SPANISH)
espinaca blanca (28.1. SPANISH)
espinaca de Filipinas (59.3. SPANISH)
espinaca de Java (59.3. SPANISH)
espinaca de Malabar (28.1. SPANISH)
espinaca de Nueva Zelandia (22.2. SPANISH)
espínace (35.9. PANGASINAN)
espinafre (35.9. CRIOULO, CAPE VERDE)
espinafre (35.9. PORTUGUESE)
espinafre da Nova Zelândia (22.2. PORTUGUESE)
espinafre do Perú (35.6. PORTUGUESE)
espinafre europeu (35.9. PORTUGUESE)
espinaka (35.9. BASQUE)
espináka (35.9. LADINO)
esponja (38.22. SPANISH)
esponja vegetal (38.22. PORTUGUESE)
essan (45.4. BULU)
essan (45.4. GBAYA)
essana (39.4. JOLA-FONYI)
essang (45.4. BULU)
essan(g) (45.4. EWONDO)
essang (45.4. GBAYA)
essbare yamswurzel (11.6. DUTCH)
esse-tapho (30.27a. MUSKOGEE)
esse-tapho-rakko (30.35. MUSKOGEE)
Essiggurke (38.13b. GERMAN)
esson (17.3. EWONDO)
essou (30.27a. JOLA-FONYI)
essoufleur (30.26. JOLA-FONYI)
estagon (27.6. BELARUSAN)
estaque (44.11. PORTUGUESE)
estragão (27.6. PORTUGUESE)
estragon (27.6. DANISH)
estragon (27.6. DUTCH)
estragon (27.6. ESTONIAN)
estragon (27.6. FRENCH)
Estragon (27.6. GERMAN)
estragon (27.6. KAZAKH)
estragon (27.6. NORWEGIAN)
estragon (27.6. RUSSIAN)
estragón (27.6. SPANISH)
estragon (27.6. TURKISH)
estragon (27.7. POLISH)
estragone (27.6. ITALIAN)
estrela-mar (57.1. PORTUGUESE)
estrella de mar (27.12. SPANISH)
estrellamar (57.1. SPANISH)
estropajo (38.23. SPANISH)
èsúrú (11.2. YORUBA)
esuru (11.4. KRIO)
èsúru (11.4. YORUBA)
εsuru-yams (11.4. KRIO)
ε-syɔ́ɓò (63.1. BASSARI)
eßbares Blumenrohr (9.1. GERMAN)
èt i buki (19.1. WOLOF)
etage-løg (7.9a. DANISH)
Etagenzwiebel (7.9a. GERMAN)
etam bakaani (62.1. JOLA-FONYI)
ε-tana (59.2. THEMNE)
etanga (38.7b. HERERO)
etanga rouitji (38.3. KWANGALI)
etɔt (11.2. BEROM)
ete (11.1. GEN)
ete (11.3. ÉWÉ)
ete (11.6. ÉWÉ)
ete (62.18. ÉWÉ)
ete (fufute) (11.7. ÉWÉ)
ete (yovote) (11.5. ÉWÉ)
etetsetse (11.2. ÉWÉ)
Ethiopian mustard (30.3. ENGLISH)

Ethiopische mosterd (30.3. VLAAMS)
éti maniok (39.4. HUNGARIAN)
éti palka (10.1. HUNGARIAN)
etiantio (23.4. GOURMANCHÉMA)
ɛ-tiɔ̀di (40.43. BASSARI)
étígí (45.1. EFIK)
ètínyūng (63.1. EFIK)
étkezési paprika (62.1. HUNGARIAN)
etkezési paprika (62.2. HUNGARIAN)
étkezési sárgarépa (24.11. HUNGARIAN)
etle (38.20. AJA)
eto (40.39. LOMWE, MALAWI)
étoué zom (62.18. EWONDO)
etre (38.20. ÉWÉ)
ettagongaka (45.4. TELUGU)
ettagonguru (45.4. TELUGU)
ettakandulu (40.2. TELUGU)
ettakkaaya (16.1a. MALAYALAM)
ettamulugoranta (23.3. TELUGU)
ettavaala (16.1a. MALAYALAM)
ɛ-tuɛ (40.43. THEMNE)
etulùp (27.24. BEROM)
ɛ-tyẹlis (38.18. BASSARI)
european capper bush (34.1. ENGLISH)
European corn-salad (67.2. ENGLISH)
European pennyroyal (44.2. ENGLISH)
euryale (50.1. ENGLISH)
euryale (50.1. FRENCH)
euryale (50.1. URDU)
euryale (50.1. VLAAMS)
evele kôn (40.26. BULU)
evening primrose (51.1. ENGLISH)
evo (38.20. NUPE-NUPE-TAKO)
evronit ha-nesher (2.1. HEBREW)
ẹ̀wa (40.25. YORUBA)
ẹ̀wa (40.43. YORUBA)
ẹ̀wa dudu (40.43. YORUBA)
ẹ̀wa ebeji (40.43. YORUBA)
ɛ-wata-yams (11.1. THEMNE)
ẹwẹ (40.25. YORUBA)
ewe (40.43. EDE NAGO)
ewé iná (40.18. YORUBA)
ewe karo (24.7. EDE NAGO)
ewe karo (24.7. YORUBA)
ewe-kɔdu (16.1a. GBE-VHE)
ewere (20.6. YORUBA)
èwì (40.1. YALA)
ewiger Spinat (58.3. GERMAN)
èwó ùgènyi (17.5. ICEVE-MACI)
ewudu (63.1. YORUBA)
ẹwujẹ (40.25. YORUBA)
ewùrà (11.1. YORUBA)
ewura duduku (11.1. EDE NAGO)
ewura duduku (11.1. YORUBA)
ewùrà funfun (11.1. YORUBA)
ẹwura-esi (11.2. YORUBA)
ewúrò (27.24. YORUBA)
ewúrò jíjẹ (27.24. YORUBA)
ewuro odo (27.44. EDE NAGO)
ewuro odo (27.44. YORUBA)
ewúrò oko (27.24. YORUBA)
ewurodo (27.44. YORUBA)
ɛ-xɔ́yí-kɔ́yí (55.1. BASSARI)
ɛ-yámən (45.4. BASSARI)
ɛ-yams-ɛ-fɛra (11.7. THEMNE)
eyeb (7.3. BOKYI)
éyèn (38.16. EDO)
éyèn (38.18. EDO)
eyinata (27.8. YORUBA)
ə̀yò (7.3. OGBIA)
ə̀yò (7.3. OGBIA-AGHOLO)
ẹ́yọ (63.1. YORUBA)
ɛ-yoka (39.4. THEMNE)
ẹze ọgẹde ji oko (16.1a. IGBO)
ezelsoor (44.11. VLAAMS)
ezerrieksts (64.3. LATVIAN)
ezo (40.26. NUPE-NUPE-TAKO)
ezo ukogi (26.4. JAPANESE)
ezonegi (7.11. JAPANESE)
ezov (44.6. HEBREW)
ezovit metzua (44.6. HEBREW)
ezo-zentei-ka (13.3. JAPANESE)
fa (17.3. KONO)
fa shou gua (38.25. CHINESE, WÚ)
faa po moko (62.5. TWI)
faandu (38.20. PULAAR)
faba (40.26. GALICIAN)
faba (40.34a. CATALAN-VALENCIAN-BALEAR)
faba (40.34a. GALICIAN)
faba bean (40.34a. FILIPINO)
faba loba (40.34a. GALICIAN)
faballón (40.34a. GALICIAN)
fa-birama (44.10. BAMANANKAN)
fabirama (44.10. MANINKA)
faboco (40.34a. GALICIAN)
faeng (38.1. THAI)
fagamanu (9.1. SAMOAN)
fagginola del occhio (40.41. ITALIAN)
faggus (38.13a. ARABIC)

faghoos (38.6. ARABIC)
faghoos (38.7a. ARABIC)
faghoos (38.7b. ARABIC)
fagioletto americano (40.42. ITALIAN)
fagioli agrappolo (40.6. ITALIAN)
fagiolino (40.26. ITALIAN)
fagiolino asparago (40.42. ITALIAN)
fagiolino dall'occhio (40.43. ITALIAN)
fagiolo (40.26. ITALIAN)
fagiolo adzuki (40.36. ITALIAN)
fagiolo antelmintico (40.18. ITALIAN)
fagiolo asparagio (40.42. ITALIAN)
fagiolo aureo (40.38. ITALIAN)
fagiolo coccineo (40.24. ITALIAN)
fagiolo de nigriti (40.18. ITALIAN)
fagiolo de salsa (40.37. ITALIAN)
fagiolo di Lima (40.25. ITALIAN)
fagiolo di Spagna (40.24. ITALIAN)
fagiolo lablab (40.9. ITALIAN)
fagiolo lungo un metro (40.42. ITALIAN)
fagiolo mungo (40.37. ITALIAN)
fagiolo mungo (40.38. ITALIAN)
fagiolo patata (40.21. ITALIAN)
fagiolo patata (40.22. ITALIAN)
fagiolo tepari (40.23. ITALIAN)
fagiolo urd (40.37. ITALIAN)
fagiolone (40.24. ITALIAN)
fagu (38.1. SAMOAN)
fagufagu samoa (38.20. SAMOAN)
fahándu (38.20. PULAAR)
fahsulia (40.24. YIDDISH)
fahsulia (40.25. YIDDISH)
fahsulia (40.26. YIDDISH)
fa'i (16.1a. MANU'A SAMOAN)
fa'i (16.1a. SAMOAN)
fa'i fuga (16.1b. SAMOAN)
fa'i masoa (9.1. SAMOAN)
fa'i papalagi (16.1a. SAMOAN)
fa'i samoa (16.1a. SAMOAN)
fak (38.1. THAI)
faki (40.12. GREEK)
fakli (8.4. BUHID)
fako (63.1. FULFULDE, ADAMAWA)
fakou (63.1. ZARMA)
fákú (63.1. SONGHAY)
fakuho (63.1. FULFULDE, ADAMAWA)
fálí gúnà (38.3. KANURI)
fali kányazsázsa (30.47. HUNGARIAN)
falinya (39.4. SENA)
falo-nyan (20.6. VAI)
falsa castanha de água (64.3. PORTUGUESE)
falsa rosela (45.3. PORTUGUESE)
falsa rosela da Guiné (45.3. PORTUGUESE)
falsa rosella (45.3. SPANISH)
falsa rosella de Guinea (45.3. SPANISH)
falsa serralha (27.18. PORTUGUESE)
falsche Blaubeere (62.21. GERMAN)
false aster (27.28. ENGLISH)
false roselle (45.3. ENGLISH)
false yam (43.1. ENGLISH)
falso taro (8.6. PORTUGUESE)
falso taro (8.6. SPANISH)
fámfam (38.18. KANURI)
famufuéué (23.3. AKAN-ASANTE)
fan (23.3. AKAN-AKUAPEM)
fan ch'ieh (62.9. CHINESE, MANDARIN)
fan ge (40.21. CHINESE, YUE)
fan gua (38.14. CHINESE, GÀN)
fan gua (38.14. CHINESE, MIN BEI)
fan jiang dou (40.41. CHINESE, YUE)
fan jiao (62.2. CHINESE, MIN BEI)
fan jiao (62.2. CHINESE, YUE)
fan qie (62.9. CHINESE, MANDARIN)
fan qie (62.9. CHINESE, WÚ)
fan qie (62.9. CHINESE, YUE)
fan shi (62.9. CHINESE, MIN BEI)
fan shu (36.3. CHINESE, YUE)
fan sui xian (23.3. CHINESE, WÚ)
fan xing (22.2. CHINESE, MANDARIN)
fan xing (22.2. CHINESE, WÚ)
fan xing (22.2. CHINESE, YUE)
fan zai cong (7.3. CHINESE, MIN BEI)
fan zai shu (62.24. CHINESE, MIN BEI)
fanamanu (9.1. SAMOAN)
fãn-choû (36.3. CHINESE, HAKKA)
fãndu (38.20. FULFULDE, ADAMAWA)
fanghma (38.13a. MIZO)
fangu (38.1. TONGAN)
fangu 'a kuma (38.22. TONGAN)
fãn-k'iô (62.9. CHINESE, HAKKA)
fänkål (24.13. SWEDISH)
fankokoré (23.3. AKAN-ASANTE)
fanlu (62.2. BODO)
fánnâ díwì (38.18. KANURI)
fanouilh-mor (24.12. BRETON)
fanta (40.3. PULAAR)
fantao (62.19. BODO)
fantao gøkhøi (62.9. BODO)
fãnto (40.3. MANINKA)
fa-nyɔ-kenɛ (17.5. KONO)

fara-fara-tugo-na (40.25. YALUNKA)
farar áláyyafóo (23.4. HAUSA)
farasi (8.3. ORIYA)
farasi simba (40.26. ORIYA)
fárfara japonesa (27.35. SPANISH)
farfaraccio giapponese (27.35. ITALIAN)
farfúgio (27.20. PORTUGUESE)
farfugio (27.20. SPANISH)
farin (30.26. HAUSA)
fárín wáákéé (40.43. HAUSA)
farinello buono-enrico (35.5. ITALIAN)
farksafog (27.8. HUNGARIAN)
fàsa ƙaba (59.2. HAUSA)
fàsà kùmrúrí (59.2. HAUSA)
fasab (45.4. SERER-SINE)
fasaka (11.3. BAMANANKAN)
Faselbohne (40.9. GERMAN)
faséolo (40.26. SPANISH)
fasol (40.26. CATALAN-VALENCIAN-BALEAR)
fasol' (40.26. TATAR)
fasol krivava (40.24. BELARUSAN)
fasol' lima (40.25. RUSSIAN)
fasol limskaja (40.25. RUSSIAN)
fasol' mungo (40.37. RUSSIAN)
fasol obyknovennaja (40.26. KAZAKH)
fasol' obyknovennaja (40.26. RUSSIAN)
fasol ognennaja (40.24. RUSSIAN)
fasol serpovindnaja (40.25. RUSSIAN)
fasol zvychainaja (40.26. BELARUSAN)
fasola afrykańska (40.31. POLISH)
fasola groniasta (40.6. POLISH)
fasola limeńska (40.25. POLISH)
fasola mungo (40.37. POLISH)
fasola ostrolistna (40.23. POLISH)
fasola półksiezycowata (40.25. POLISH)
fasola wielokwiatowa (40.24. POLISH)
fasola złota (40.38. POLISH)
fasola zwykła (40.26. POLISH)
fasola zwylla (40.24. POLISH)
fasole comună (40.24. ROMANIAN)
fasole comună (40.26. ROMANIAN)
fasole cu frunze ascuţite (40.23. ROMANIAN)
fasole de lima (40.25. ROMANIAN)
fasole de luna (40.25. ROMANIAN)
fasole mare (40.24. ROMANIAN)
fasole turcească (40.9. ROMANIAN)
fasoli (40.24. GREEK)
fasoli (40.26. GREEK)
fasoli tls limas (40.25. GREEK)
fasoli to oxyphyllo (40.23. GREEK)
fasoliţă (40.43. ROMANIAN)
fasolnik adzuki (40.36. POLISH)
fasolnik chiński (40.43. POLISH)
fasolnik cylindryczny (40.41. POLISH)
fasolnik długi podwórkowy (40.42. POLISH)
fasolnik egipski (40.9. POLISH)
fasolnik matowy (40.35. POLISH)
fasolnik mungo (40.37. POLISH)
fasolnik ryżowy (40.40. POLISH)
fasolnik złoty (40.38. POLISH)
fāsolyah (40.26. KURDISH, NORTHERN)
fassa (62.15. MANDING-MANDINKA)
fassa (62.21. MANINKAKAN, WESTERN)
fassa kuna (62.15. MANDING-BAMBARA)
fastutaya (40.1. ROMANI, BALKAN)
fasul (40.26. BULGARIAN)
fasule (40.26. ALBANIAN, TOSK)
fasule gjethempreftë (40.23. ALBANIAN, TOSK)
fasulia (40.35. ARABIC)
fâsuliya (40.26. PASHTO, NORTHERN)
fasulya (40.26. TURKISH)
fasulyah (40.26. ARABIC)
fasulyas (40.26. LADINO)
fasülye (40.26. TURKISH)
fátaàkeé (38.16. HAUSA)
fâto (40.3. MANDINKA)
fā-ts'ói (30.26. CHINESE, HAKKA)
fau lele lili'i (27.18. MANU'A SAMOAN)
faul (40.3. ARABIC)
fausse canne à sucre (17.3. FRENCH)
fausse oseille de Guinée (45.3. FRENCH)
fausse tomate (62.18. FRENCH)
faux-taro (8.6. FRENCH)
fav (40.26. BRETON)
fav glas (40.26. BRETON)
fava (40.34a. ITALIAN)
fava (40.34a. LADINO)
fava (40.34a. PORTUGUESE)
fava bean (40.34a. ENGLISH)
fava cavalinha (40.34b. PORTUGUESE)
fava contra (40.4. PORTUGUESE)
fava de cavalo (40.29. PORTUGUESE)
fava grossa (40.34a. ITALIAN)
fava maggiore (40.34a. ITALIAN)
fava miúda (40.34b. PORTUGUESE)
fava-rica (40.3. CRIOULO, CAPE VERDE)
faveira (40.34a. PORTUGUESE)

feijão sabre (40.4. PORTUGUESE)
feijão tepari (40.23. PORTUGUESE)
feijão vagem (40.26. PORTUGUESE)
feijão verde (40.26. PORTUGUESE)
feijăo-bongolon (40.43. CRIOULO, CAPE VERDE)
feijăo-de-bitcho (40.18. CRIOULO, CAPE VERDE)
feijăo-de-lagartixa (40.18. CRIOULO, CAPE VERDE)
feijăo-pedra (40.9. CRIOULO, CAPE VERDE)
feijoca (40.24. PORTUGUESE)
feixaco (40.34a. GALICIAN)
feixó (40.26. GALICIAN)
fejes káposzta (30.27a. HUNGARIAN)
fejes saláta (27.31. HUNGARIAN)
fejil (30.54. ARABIC)
fejil (30.56. ARABIC)
fekete csúcsor (62.21. HUNGARIAN)
fekete mustár (30.24. HUNGARIAN)
fekete nadálytő (29.2. HUNGARIAN)
feketecsalán (44.7. HUNGARIAN)
feketegyökér (27.38. HUNGARIAN)
felandrio (24.17. SPANISH)
felce aquilina (2.1. ITALIAN)
felci (2.1. MALTESE)
Feld-Raukensenf (30.48. GERMAN)
Feldsalat (67.2. GERMAN)
felfel (62.1. FARSI, WESTERN)
felfēl (62.1. HARSŪSI)
felfel (62.2. FARSI, WESTERN)
felfel dolmeyi (62.1. FARSI, WESTERN)
felfel germez (62.2. FARSI, WESTERN)
felfelzina (62.1. MALAGASY)
fen cong (7.2. CHINESE, WÚ)
fen cong (7.2. CHINESE, YUE)
fen nie cong (7.2. CHINESE, WÚ)
fen nie gai (30.15. CHINESE, MANDARIN)
fen nie jie (30.15. CHINESE, WÚ)
fen nie yang cong (7.2. CHINESE, MANDARIN)
fen zhu zi qi (3.1. CHINESE, MANDARIN)
fen zhu ziqi (3.1. CHINESE, WÚ)
fə̀nà (38.20. NGIZIM)
fenacho (40.32. PORTUGUESE)
Fenchel (24.13. GERMAN)
fenchel'ovoščnoj (24.12. RUSSIAN)
fenchel ovoshcnoj (24.13. KAZAKH)
fendāl (36.3. HARSŪSI)
fəndəlōt (36.3. MEHRI)
fɛŋɛ-goja (38.5. MENDE)
fenegriek (40.32. VLAAMS)
fenel (24.12. SWAHILI)
feneli (24.13. SWAHILI)
fener çiçeği (62.11. TURKISH)
feng chui (24.9. CHINESE, MIN BEI)
feng dou cai (27.35. CHINESE, MANDARIN)
feng dou cai (27.35. CHINESE, WÚ)
feng ling cao (33.2. CHINESE, WÚ)
feng tou tsai (27.35. CHINESE, MANDARIN)
feng wei gai (30.12. CHINESE, MANDARIN)
feng wei jie (30.12. CHINESE, WÚ)
fenhel obyknovennyi (24.13. RUSSIAN)
fenhel zvychainy (24.13. BELARUSAN)
fenicul (24.13. ROMANIAN)
fenicul obecny (24.13. CZECH)
fenícula (24.13. SPANISH)
feniculul de florenta (24.12. ROMANIAN)
fenigiriki (40.32. SWAHILI)
fenikel (24.13. NORWEGIAN)
fenikel (24.13. SLOVAK)
fenkuł (24.13. POLISH)
fenkuł florencki (24.12. POLISH)
fenkuł włoski (24.13. POLISH)
fenkul wloski zwyczyjny (24.12. POLISH)
fennel (24.12. ICELANDIC)
fennel (24.13. ENGLISH)
fennika (24.12. ICELANDIC)
fennikel (24.12. DANISH)
fennikel (24.13. NORWEGIAN)
feno grego (40.32. PORTUGUESE)
fenogreco (40.32. SPANISH)
fenoll (24.12. CATALAN-VALENCIAN-BALEAR)
fenouil (24.13. FRENCH)
fenouil à bulbe (24.12. FRENCH)
fenouil des jardins (24.12. FRENCH)
fenouil doux (24.12. FRENCH)
fenouy (24.12. MORISYEN)
fenugrec (40.32. FRENCH)
fenugreek (40.32. ENGLISH)
fenugriek (40.32. DUTCH)
fenukel záhradný (24.12. SLOVAK)
fenyɛn (39.4. FON)
fenyɛn (39.4. GUN)
fenykl obecný (24.13. CZECH)
fenykl obecný sladký (24.12. CZECH)
fenykl sladký (24.12. CZECH)
fepē (38.20. MUSKOGEE)
feraare (38.16. FULFULDE, ADAMAWA)

ferèf (38.3. SERER-SINE)
fere-fita (38.18. AKAN-AKUAPEM)
ferigă mare (2.1. ROMANIAN)
fesleğen (44.5. TURKISH)
feteka (40.18. NIUE)
fetică de grădină (67.2. ROMANIAN)
fetliti (45.1. GEN)
feto (2.1. CRIOULO, CAPE VERDE)
feto (5.1. PORTUGUESE)
feto (30.43. AMHARIC)
feto (30.51. AMHARIC)
feto aquático (4.1. PORTUGUESE)
feto cinanono (3.1. PORTUGUESE)
feto d'água (4.1. CRIOULO, CAPE VERDE)
feto fêmea (2.1. PORTUGUESE)
feto ordinário (2.1. PORTUGUESE)
fetri (45.1. AJA)
fetri (45.1. ÉWÉ)
fetri (45.1. GBE, VHE)
fetri (45.1. GBE, WACI)
Feuerbohne (40.24. GERMAN)
feuilles caya (34.2. FRENCH)
feuilles lalo (63.1. FRENCH)
fève (40.26. CAJUN FRENCH)
fève (40.34a. FRENCH)
fève à gros grain (40.34a. FRENCH)
fève créole (40.25. FRENCH)
fève de marais (40.34a. FRENCH)
féve d'Egypte (48.1a. FRENCH)
féve hisso (45.1. FON)
fève plate (40.25. CAJUN FRENCH)
féverole (40.34b. FRENCH)
feverole (40.34b. PORTUGUESE)
féverole à petites graines (40.34b. FRENCH)
fevetta-ciccardi (40.10. MALTESE)
fevi (45.1. FON)
ffa dringo (40.24. WELSH)
ffa soy (40.8. WELSH)
ffarwel haf (27.23. WELSH)
ffenigel (24.13. WELSH)
ffenigel florens (24.12. WELSH)
fhuri (38.14. VENDA)
fhuri (38.16. VENDA)
fhuri (38.17. VENDA)
fhuri (38.18. VENDA)
fialka anjutiny glazki (68.1. KAZAKH)
fialka trechcolernaja (68.1. BELARUSAN)
fialka trojfarebná (68.1. SLOVAK)
fialka tryohtsvetnaja (68.1. RUSSIAN)
fiandron (55.1. BAOULÉ)
fico d'India (32.1. ITALIAN)
ficoïde (22.1. FRENCH)
ficoïde cristalline (22.1. FRENCH)
ficoïde glaciale (22.1. FRENCH)
fide (45.1. MANDING-MANDINKA)
fide (62.15. IGBO)
fidiom mãnkaña (40.43. CRIOULO, UPPER GUINEA)
fidjom faba (40.25. CRIOULO, UPPER GUINEA)
fie (10.1. DANGME)
fie (10.1. ÉWÉ)
fié (38.20. BAMANANKAN)
fie ge (10.1. ÉWÉ)
fie yibɔ (10.1. ÉWÉ)
field horsetail (1.1. ENGLISH)
field salad (67.2. ENGLISH)
fieno greco (40.32. ITALIAN)
fiepa (38.20. BÉTÉ)
fiğ (40.10. TURKISH)
figa indyjska (32.1. POLISH)
figel (30.56. MALTESE)
figel ciniz (30.54. MALTESE)
figenblad-græskar (38.15. DANISH)
figenkaktus (32.1. DANISH)
figgila hamra (40.13. MALTESE)
figiri (30.3. SWAHILI)
fig-leaf gourd (38.15. ENGLISH)
figu (16.1a. SHONA)
figueira da Índia (32.1. PORTUGUESE)
figueira-da-india (32.1. CRIOULO, CAPE VERDE)
figuier de barbarie (32.1. FRENCH)
figuier d'Inde (32.1. FRENCH)
Fiji arrowroot (19.1. ENGLISH)
fiji mame (40.9. JAPANESE)
fijn bieslook (7.11. VLAAMS)
fijón (40.26. LADINO)
fijon amariyo (40.26. LADINO)
fijon blanco (40.26. LADINO)
fijon congo (40.2. CRIOULO, CAPE VERDE)
fijon fradi (40.43. CRIOULO, CAPE VERDE)
fijon verde (40.26. LADINO)
fijon-bonginho (40.26. CRIOULO, CAPE VERDE)
fijon-fava (40.25. CRIOULO, CAPE VERDE)
fikenopuntia (32.1. NORWEGIAN)
fikjukaktus (32.1. ICELANDIC)
fil kulaği (8.4. TURKISH)
filalaf (17.3. TAJIKI)

fløjlsbønne (40.18. DANISH)
flominan (38.22. BAOULÉ)
flor da paixão (54.1. PORTUGUESE)
flôr de bananeira (16.1b. PORTUGUESE)
flor de gelo (22.1. PORTUGUESE)
flor de la maravilla (12.1. SPANISH)
flor de la pasión (54.1. SPANISH)
flor de lis (12.1. PORTUGUESE)
flor de plátano (16.1b. SPANISH)
flor de sangue (65.1. PORTUGUESE)
flor do sangue (65.1. GALICIAN)
flor tigre (12.1. PORTUGUESE)
Floransa rezenesi (24.12. TURKISH)
flore banana (16.1b. ITALIAN)
Florence fennel (24.12. ENGLISH)
Florentynse vinkel (24.12. AFRIKAANS)
florinhas de San Francisco (27.17. PORTUGUESE)
flowering white cabbage (30.37. ENGLISH)
flower-like leaf mustard (30.16. ENGLISH)
Fluegelerbse (40.13. GERMAN)
fluted gourd (38.27. ENGLISH)
fluted pumpkin (38.27. ENGLISH)
fluted pumpkin (38.28. ENGLISH)
fluweelboon (40.18. AFRIKAANS)
fluweelboon (40.18. VLAAMS)
flvejlsblad (27.25. DANISH)
fô (62.15. ANDO)
foder-kulsukker (29.3. DANISH)
foderplante (40.43. DANISH)
fodros petrezselyem (24.20. HUNGARIAN)
fóè (43.1. BALANTA-KENTOHE)
fofoŋu (17.5. ÉWÉ)
fogbogui dangarou (11.5. SENOUFO, CEBAARA)
foin grec (40.32. FRENCH)
foinikoulo to glyko (24.12. GREEK)
foinikoulon to koivov (24.13. GREEK)
fojtóbab (40.30. HUNGARIAN)
fokhagyma (7.10. HUNGARIAN)
fokiolo (23.3. DAGAARE, SOUTHERN)
fol'yere (38.16. FULFULDE, ADAMAWA)
fol'yere (38.18. FULFULDE, ADAMAWA)
földimogyoró (40.1. HUNGARIAN)
foleray (45.4. PULAAR)
foléré (45.3. PULAR)
foléré (45.4. PULAAR)
fólerè (45.4. PULAAR)
fólerè ba'di (45.4. PULAAR)
fólerè boleyo (45.4. PULAAR)
folérébadi (45.4. PULAAR)
folha de gelo (22.1. PORTUGUESE)
folléré (45.4. FULFULDE)
follere (45.4. FULFULDE, ADAMAWA)
follere (45.4. PULAAR)
foloñ (23.3. BULU)
folon (23.3. EWONDO)
folong (23.3. BULU)
folẏere (38.14. FULFULDE, ADAMAWA)
folẏere (38.16. FULFULDE, ADAMAWA)
folẏere (38.17. FULFULDE, ADAMAWA)
folẏere (38.18. FULFULDE, ADAMAWA)
fôn (17.7. BULU)
fon'dolo-oungou (23.3. KISSI, NORTHERN)
fondolo-bɛndoŋ (23.3. KISSI, NORTHERN)
fondolo-ungu (23.3. KISSI, NORTHERN)
fondoulo (23.3. KISSI, NORTHERN)
fondoulo saman (23.3. KISSI, NORTHERN)
fondoulo sankoura (23.3. KISSI, NORTHERN)
fondulo (23.3. KISSI, NORTHERN)
fondulo sama (23.3. KISSI, NORTHERN)
fondulo sankura (23.3. KISSI, NORTHERN)
fonfon (17.5. ÉWÉ)
fonfongu (17.5. ÉWÉ)
fönikkel (24.12. NORWEGIAN)
fonoll (24.13. CATALAN-VALENCIAN-BALEAR)
fool (40.39. TIGRIGNA)
fool sudanee (40.1. ARABIC)
foondu (30.51. FULFULDE, ADAMAWA)
foordo (45.3. TUPURI)
foos (7.8. SCOTS)
fôóti (38.20. BALANTA-KENTOHE)
forbicina (27.8. ITALIAN)
forbicina pelosa (27.8. ITALIAN)
fôrmargkål (30.32. NORWEGIAN)
förnepe (30.41a. NORWEGIAN)
forôto (62.5. BAMANANKAN)
forôto (62.5. MANINKA)
fôrskolm (40.10. NORWEGIAN)
fortune-teller (27.45. ENGLISH)
fote ɛtugi (40.2. SUSU)
fotété (23.3. ÉWÉ)
fou (38.22. JULA)
fou gau (40.30. NIUE)
fòu kouā (38.23. CHINESE, HAKKA)
foubey (34.2. ZARMA)
fougere aigle (2.1. FRENCH)
foù-mâc (27.10a. CHINESE, HAKKA)
fôung lô-pʻĕt (24.11. CHINESE, HAKKA)

foûng ts'oi-t'êoi (24.11. CHINESE, HAKKA)
fountain plant (23.3. ENGLISH)
fountotos yakinthos (13.5. GREEK)
foussaro (38.3. JOLA-FONYI)
foû-tsião (62.1. CHINESE, HAKKA)
foxnut (50.1. ENGLISH)
főzeléklencse (40.12. HUNGARIAN)
főzeléktök (38.18. HUNGARIAN)
fragkosikia (32.1. GREEK)
frâken (1.1. SWEDISH)
france-suiba (58.3. JAPANESE)
francia hagyma (7.1. HUNGARIAN)
frans bean (40.26. PANJABI, EASTERN)
Franse boon (40.26. DUTCH)
Franse zuring (58.4. VLAAMS)
franskt estragon (27.6. ICELANDIC)
frantsesporru (13.2. BASQUE)
frash bean (40.26. HINDI)
fré (62.19. BÉTÉ)
French bean (40.26. ENGLISH)
French sorrel (58.4. ENGLISH)
frenk asmasi (69.1. TURKISH)
frenk inciri (32.1. TURKISH)
frenk salatasi (67.2. TURKISH)
frenk teresi (65.1. TURKISH)
friguma (11.2. DAGBANI)
frijol alado (40.29. SPANISH)
frijol arroz (40.40. SPANISH)
frijol de ombligo negro (40.43. SPANISH)
frijol de oro (40.38. SPANISH)
frijol de pinta negra (40.43. SPANISH)
frijol de vaca (40.43. SPANISH)
frijol del antibo (40.9. SPANISH)
frijol del monte (40.2. SPANISH)
frijol mambi (40.40. SPANISH)
frijol mungo (40.37. SPANISH)
frijol mungo (40.38. SPANISH)
frokostreddik (30.54. NORWEGIAN)
frottersvamp (38.22. DANISH)
frotto bani (62.1. JULA)
froud (62.21. BRETON)
Frühlings-Barbarakraut (30.2. GERMAN)
fruta dos paulistas (38.22. PORTUGUESE)
frutto d'ibisco (45.1. ITALIAN)
fruvere (39.4. RONGA)
fteri (2.1. GREEK)
fteri to ithrohari (4.1. GREEK)
fu (17.5. MIZO)
fu fimban (62.9. JOLA-FONYI)
fu gès (45.4. JOLA-FONYI)
fu shou gua (38.25. CHINESE, YUE)
fu timay (43.1. JOLA-FONYI)
fua lele lili'i (27.18. SAMOAN)
Fuchsschwanz (23.3. GERMAN)
fudansō (35.3. JAPANESE)
fúdēre (40.1. LIMBA)
fue (62.15. BAOULÉ)
fue'aepuaka (40.30. TONGAN)
fuefue fai va'a (40.30. MANU'A SAMOAN)
fufurufuru (44.5. MALIMBA)
fügekaktusz (32.1. HUNGARIAN)
fugès (45.4. JOLA-FONYI)
fuğūl (30.56. HARARI)
fuki (27.35. JAPANESE)
fuki (27.35. VLAAMS)
fûki-mame (40.18. JAPANESE)
ful (40.34a. ARABIC)
ful (40.34a. HEBREW)
ful kapi (30.26. BENGALI)
ful kobi (30.26. GUJARATI)
ful suyah (40.8. ARABIC)
fula kobi (30.26. ORIYA)
fule kibi (30.26. ORIYA)
fulkapi (30.26. BENGALI)
fulkobi (30.26. GUJARATI)
fullane (63.1. HAUSA)
fuluy (11.1. BUHID)
fulvar (30.26. GUJARATI)
fulvar (30.26. MARATHI)
funcho (24.13. PORTUGUESE)
funcho aquático (24.17. PORTUGUESE)
funcho de Florença (24.12. PORTUGUESE)
funcho doce (24.12. PORTUGUESE)
funcho oriental (24.17. PORTUGUESE)
fundi (38.3. ARABIC)
funtcho (24.13. CRIOULO, CAPE VERDE)
fupisha (58.3. SWAHILI)
fura (27.26. AFAR)
furabá (43.1. JOLA-FONYI)
furabân (43.1. JOLA-FONYI)
fùrén gàdú (36.2. HAUSA)
furi (38.14. VENDA)
furi (38.16. VENDA)
furi (38.17. VENDA)
furi (38.18. VENDA)
furohrensu-fennelu (24.12. JAPANESE)
furrunj-mushk (44.5. PANJABI, EASTERN)
furrunjnushk (44.5. PANJABI, EASTERN)
fürtös gyöngyike (13.5. HUNGARIAN)
furtu (62.5. BAMANANKAN)

furtu (62.5. MANINKA)
furtu-missé (62.5. MANDING-MANINKA)
furum (38.16. TIV)
furum (38.18. TIV)
furu̱m (38.18. TIV)
fushimi-tougarashi (62.2. JAPANESE)
fusi (16.1a. TONGAN)
fustukes (40.1. LADINO)
fűszer paprika (62.2. HUNGARIAN)
fute giya (20.6. TIRURAY)
futi (16.1a. NIUE)
futima (43.1. JOLA-FONYI)
Futterkohl (30.35. GERMAN)
fútutè (36.3. BASSARI)
fuyu ao (45.5. JAPANESE)
fuyu-garashi (30.2. JAPANESE)
fvmēcv (38.7a. MUSKOGEE)
fye (62.19. JUNKUN TAKUM)
fylostachys (17.4. CZECH)
fymatorrizon selino (24.5. GREEK)
fyosja vishnja obyknovennaja (62.11. RUSSIAN)
fysalis e alkekenslos (62.11. GREEK)
fysalis e ixokarpos (62.13. GREEK)
fysalis e perouviani (62.12. GREEK)
g' (30.36. LAHU)
gâ (45.1. BAMANANKAN)
gã (45.1. BAMANANKAN)
ga hiɔ (62.18. DAGBANI)
ga hiɔ (62.18. DANGME-KROBO)
g'a^ hpfuh tzuh ve^ (23.4. LAHU)
g'a^ na_ tzuh ve^ (23.4. LAHU)
gàaɓíyǎw (38.18. NGIZIM)
gaaden kuresu (30.51. JAPANESE)
gaakin (38.6. JAPANESE)
gaant gobhi (30.30. HINDI)
gáasí (38.20. SONGHAY)
gaasu (10.1. DENDI)
gaasu (38.20. DENDI)
gaasu (38.20. ZARMA)
gaatarin zoomoo (19.1. HAUSA)
gabans (40.5. CREOLE, HAITIAN FRENCH)
gąbczak walcowaty (38.22. POLISH)
gãbdo (7.2. MÒORÉ)
gãbdo (7.3. MÒORÉ)
gabe (8.4. BICOLANO, CENTRAL)
gabholika (40.12. SANSKRIT)
gabi (8.4. BICOLANO, CENTRAL)
gabi (8.4. FILIPINO)
gábi (8.4. HILIGAYNON)
gabi (8.4. TAGALOG)
gabi (8.4. TAGBANWA)
gabi-gabi (18.1. CEBUANO)
gabi-gabihan (18.1. FILIPINO)
gabi-gabihan (18.1. TAGALOG)
gabing San Fernando (8.1. TAGALOG)
gabing uwak (18.2. TAGALOG)
gabing-cebu (8.10. TAGALOG)
gabing-Cebu (8.10. TAGALOG)
gabing-uak (18.2. FILIPINO)
gabiskus sedobnyi (45.1. RUSSIAN)
gablos (40.2. TAGALOG)
gab'nyule (7.3. DAGBANI)
gabuk (36.3. BUHID)
gabutang an mumbola (17.7. IFUGAO)
gabuy (40.29. PANGASINAN)
gachmarich (62.2. HINDI)
gachmicichi (62.2. HINDI)
gachmirich (62.2. HINDI)
gachmirich (62.5. HINDI)
gacoun 'kouêssê (62.23. JULA)
gad ha-sade (24.9. HEBREW)
gad hasadeh (24.9. HEBREW)
gadab rajko (40.17. GUJARATI)
gádákùwà (38.20. NGIZIM)
gaddagobi (30.30. TELUGU)
gaddalu (7.3. TELUGU)
gadde oraga (24.7. KANNADA)
gadele (40.8. JAVANESE)
gadiaga (40.39. WOLOF)
gadianga (40.39. SERER-SINE)
gadiãnga (40.39. WOLOF)
gadin ɛg (62.19. KRIO)
gadin pɛpɛ (62.1. KRIO)
gadinidonda pendalamu (11.1. TELUGU)
gadubaechu (30.27a. KOREAN)
gadung laut (19.1. INDONESIAN)
gadung tikus (19.1. INDONESIAN)
gǣda (7.2. PANJABI, EASTERN)
gagama (38.13a. CHEROKEE)
gagan-gagan (24.7. JAVANESE)
gagatang (27.42. BONTOC CENTRAL)
gagatano (27.42. FILIPINO)
gagattang (27.42. KANKANAEY)
gahala (8.6. SINHALA)
gahilang (17.7. IFUGAO)
gahr (16.1a. HINDUSTANI)
gahvorakadu (38.16. TAJIKI)
gaichaalu (11.2. BENGALI)
gailon (30.25. ENGLISH)

gailon (30.25. TAGALOG)
gaimuu (30.5. MONGOLIAN)
gaimuu (30.8. MONGOLIAN)
gaimuu baitsaa (30.6. MONGOLIAN)
gairleog (7.10. GAELIC, IRISH)
gajar (24.11. BENGALI)
gajar (24.11. FIJIAN)
gajar (24.11. GUJARATI)
gajar (24.11. HINDI)
gájar (24.11. HINDUSTANI)
gājar (24.11. MAITHILI)
gajar (24.11. MARATHI)
gajar (24.11. NEPALI)
gajar (24.11. ORIYA)
gajar (24.11. PANJABI, EASTERN)
gajar (24.11. URDU)
gájar (24.18. HINDUSTANI)
gajaragadda (24.11. TELUGU)
gajarakkilangu (24.11. TAMIL)
gajena (35.2. SERBO-CROATIAN)
gajjar (24.11. PANJABI, EASTERN)
gajjara gedda (24.11. TELUGU)
gajjara kilangu (24.11. TAMIL)
gajjaragedda (24.11. TELUGU)
gajjari (24.11. KANNADA)
gajor (24.11. ASSAMESE)
gajuchettu (62.15. TELUGU)
gajutsu (20.4. JAPANESE)
gaka (38.13a. SHONA)
gakakurukadi (40.41. KWANGALI)
gakarakau (44.5. TELUGU)
ga:ke (62.21. BADAGA)
gako (17.3. BÉTÉ)
gal soğani (7.5. TURKISH)
gala (38.5. GBAYA)
galaghati (59.3. SUBANON)
galaka (38.22. GUJARATI)
galakonda (62.2. TELUGU)
galamaka (15.4. BONTOC, CENTRAL)
galamaka (15.4. KANKANAEY)
galamasa (38.14. CAROLINIAN)
galamasa (38.16. CAROLINIAN)
galamasa (38.17. CAROLINIAN)
galamasa (38.18. CAROLINIAN)
galambbegy saláta (67.2. HUNGARIAN)
galambborsó (40.2. HUNGARIAN)
galanga (20.1. FRENCH)
Galanga (20.1. GERMAN)
galanga (20.1. ITALIAN)
galanga (20.1. SPANISH)
galanga de l'Inde (20.1. FRENCH)
galanga maior (20.1. PORTUGUESE)
galanga majeur (20.1. FRENCH)
galanga mayor (20.1. SPANISH)
galangajuuri (20.1. FINNISH)
galangal (20.1. ENGLISH)
galangawortel (20.1. VLAAMS)
galangsi (27.22. JAVANESE)
galapito (57.1. PORTUGUESE)
galartori (38.29. PANJABI, EASTERN)
galartori pandol (38.29. PANJABI, EASTERN)
galatgat (36.2. ILOCANO)
galayyong an napagatan (23.3. IFUGAO)
galbessie (62.21. AFRIKAANS)
galda arbūzs (38.3. LATVIAN)
galgán velký (20.1. CZECH)
Galgant (20.1. GERMAN)
galiang (8.1. CEBUANO)
galiang (8.7. BICOLANO, CENTRAL)
galigbati (59.3. SUBANON)
galik (7.10. TOK PISIN)
galinggang (27.22. JAVANESE)
galinsoga (27.22. ENGLISH)
galinsoga (27.22. PORTUGUESE)
galinsoga (27.22. SERBO-CROATIAN)
galinsoga (27.22. TAGALOG)
galinsoga à petites fleurs (27.22. FRENCH)
galinsoga comune (27.22. ITALIAN)
galinsoga de flores pequeñas (27.22. SPANISH)
galinsoga glabre (27.22. FRENCH)
galinzoga drobnakvetkava (27.22. BELARUSAN)
gàlísà (39.4. KANURI)
galiwi-giwan (24.17. BONTOC, CENTRAL)
galka (38.22. GUJARATI)
galkidea (27.45. BASQUE)
gallan (8.1. CEBUANO)
gallan (8.7. ENGLISH)
gallan (8.7. SPANISH)
gallan (8.7. VLAAMS)
gallant soldier (27.22. ENGLISH)
galliko lapatho (58.4. GREEK)
gallinella campanulata (67.1. ITALIAN)
gallinette (67.2. ITALIAN)
galviņu kāposts (30.4. LATVIAN)
galviņu kāposts (30.27a. LATVIAN)
galvinsalāti (27.31. LATVIAN)
galyan (8.7. CEBUANO)
galyang (8.1. CEBUANO)
galyang (8.7. BICOLANO, CENTRAL)

galyň ýaprakly kupyr (24.3. TURKMEN)
gamadi hullu (40.30. KANNADA)
gàmàgàrí (36.3. HAUSA)
gamba (38.20. WOLOF)
gamba mané (45.1. MÒORÉ)
gambain bologi (59.3. KRIO)
gàmbàrám (45.1. NGIZIM)
gambas (38.21. INDONESIAN)
gambas (38.21. JAVANESE)
gambas (38.21. VLAAMS)
gambas (38.25. SUNDA)
gambas (45.1. INDONESIAN)
gambheera (44.4. SANSKRIT)
gambhira (44.4. SANSKRIT)
gambo (45.1. KWÉYÒL)
gambol (62.23. SUBANON)
ga-mɛntɛ́ŋ (62.9. BUDIK)
gamet (40.28. SUNDA)
gamet (45.3. INDONESIAN)
gamet walanda (45.4. SUNDA)
gamiaino (63.1. KANURI)
//gammeb (44.4. NAMA-DAMARA)
gámmòn (38.16. HAUSA)
gamon fatake (38.18. HAUSA)
gamongeï (44.5. ARABIC)
gamoting (36.3. BONTOC, CENTRAL)
gamput di gimayyaw (7.13. IFUGAO)
gamsa (62.24. KOREAN)
gamwuuti (36.3. CAROLINIAN)
gamwuuti (62.24. KHMER, CENTRAL)
gamzaino (63.1. KANURI)
gan (38.14. GOURMANCHÉMA)
gan (38.16. GOURMANCHÉMA)
gan (38.17. GOURMANCHÉMA)
gan (38.18. GOURMANCHÉMA)
gan lan (30.27a. CHINESE, MANDARIN)
gan lan (30.27a. CHINESE, WÚ)
gan shu (36.3. CHINESE, MANDARIN)
gan sun (36.3. CHINESE, WÚ)
gan zhe (17.5. CHINESE, WÚ)
gaňa (40.18. JOLA-FONYI)
gana ki pholdari (17.5. URDU)
ganagan (36.3. ITAWIT)
ganahār (23.3. KASHMIRI)
ganana (28.1. PAMPANGAN)
gand gobi (30.29. PANJABI, EASTERN)
ganda (7.3. BINUKID)
ganda (7.3. PANJABI, EASTERN)
ganda (7.10. BINUKID)
ganda (7.13. BINUKID)
ganda (7.13. CEBUANO)
ganda (7.13. TAGBANWA)
ganda (20.4. SAMBAL, TINÀ)
ganda isi (7.11. INDONESIAN)
ganda kawat (7.13. INDONESIAN)
gandabuy (33.3. TAJIKI)
gandana (7.2. HINDI)
gandana (7.2. PANJABI, EASTERN)
gandana (7.8. FARSI, EASTERN)
gandáná (7.8. HINDUSTANI)
gandana (7.8. PASHTO, NORTHERN)
gandana (7.11. FARSI, EASTERN)
gandh gobhi (30.30. PANJABI, EASTERN)
gandh gobi (30.30. PANJABI, EASTERN)
gandhabeeja (40.32. SANSKRIT)
gandhadhika (24.13. SANSKRIT)
gandhamoola (20.1. KANNADA)
gandhamoola (20.1. SANSKRIT)
gandhamoola (45.1. SANSKRIT)
gandhamoolaka (20.4. SANSKRIT)
gandhamula (45.1. SANSKRIT)
gandhan (7.2. PANJABI, EASTERN)
gandhapanirjjaka (44.4. SANSKRIT)
gandhaphala (40.32. SANSKRIT)
gạndhạri aṛak̓ (23.3. SANTALI)
gandhasara (20.4. SANSKRIT)
gandhavaaruni (20.1. SANSKRIT)
gandhmula (45.1. SANSKRIT)
gandola (28.1. INDONESIAN)
gandola (28.1. SUNDA)
gandos (8.4. PAMPANGAN)
gandum (17.7. TAUSUG)
gandus (8.1. PAMPANGAN)
gandus (8.4. PAMPANGAN)
gane (61.1. NEPALI)
ganet (8.4. AGTA)
ganet (8.4. AGTA, CASIGURAN DUMAGAT)
gang dou (40.41. CHINESE, WÚ)
gang guo gua (38.5. CHINESE, MANDARIN)
gang guo gua (38.5. CHINESE, WÚ)
gang zhu (17.4. CHINESE, MANDARIN)
gang zhu shu (17.4. CHINESE, WÚ)
ganga paavilikoora (59.2. TELUGU)
gangachulia (23.4. ORIYA)
gangála (40.1. KANURI)
gàngàmóó (20.2. HAUSA)
gangamu (20.2. AKAN-ASANTE)
gangamuki (20.2. FULFULDE, ADAMAWA)
ganganlubassa (7.3. GUN)
ganga-pavilikura (59.2. TELUGU)

gangassolo (39.4. RONGA)
gangassolo (39.4. TSONGA)
ganggagan (24.7. JAVANESE)
gangnamcong (40.26. KOREAN)
ganhaca (40.18. PULAAR)
ganhar (23.3. HINDI)
ganhoma (40.18. CRIOULO, CAPE VERDE)
ganhomá (40.18. CRIOULO, UPPER GUINEA)
ganjigaga (39.4. HAUSA)
ganjong (9.1. DUTCH)
ganjong (9.1. MALAY)
gannā (17.5. HINDI)
gannä (17.5. PANJABI, EASTERN)
ganna (17.5. PASHTO, NORTHERN)
ganola (28.1. SUNDA)
gañoma (40.18. CRIOULO, UPPER GUINEA)
gantamau (39.4. HAUSA)
gantang (62.24. INDONESIAN)
gante mula (30.22. NEPALI)
ganth gobhi (30.30. HINDI)
ganth gobi (30.30. BENGALI)
ganth gobi (30.30. HINDI)
ganth gobi (30.30. URDU)
ganthi kobi (30.30. ORIYA)
ganthian (36.2. PANJABI, EASTERN)
ganwon fatake (38.16. HAUSA)
ganwon fatake (38.18. HAUSA)
ganyer rafi (40.26. HAUSA)
ganyol (9.1. SUNDA)
ganyong (9.1. INDONESIAN)
ganyong (9.1. JAVANESE)
ganyong (9.1. SUNDA)
ganzebloem (27.23. DUTCH)
ganzevoet (35.6. DUTCH)
gao jiao jie (30.13. CHINESE, YUE)
gao li cai (30.27a. CHINESE, MIN BEI)
gaolgarzaban (29.1. FARSI, WESTERN)
ga-pángòl (11.2. BUDIK)
ga-pèďy (62.23. BUDIK)
garä (38.16. ARABIC)
garä (38.18. ARABIC)
garä kossa (38.18. ARABIC)
garä stambuli (38.18. ARABIC)
garacia (39.5. MAKHUWA)
garadudi (38.20. MALAYALAM)
garam masal (40.17. TELUGU)
garasy mame (40.10. JAPANESE)
garaz (40.28. BALTI)
garbanotasis kopūstas (30.34. LITHUANIAN)
garbans (40.5. CREOLE, HAITIAN FRENCH)
garbansos (40.5. FILIPINO)
garbansos (40.5. TAGALOG)
garbantzu (40.5. BASQUE)
garbanzo (40.5. ITALIAN)
garbanzo (40.5. SPANISH)
garbanzo (40.5. TAGALOG)
garbanzo bean (40.5. ENGLISH)
garbánzos (40.5. PANGASINAN)
garbasilla (17.7. AFAR)
garbus vjalizki (38.16. BELARUSAN)
garbus zvichainy (38.18. BELARUSAN)
garbuz (38.18. BELARUSAN)
garbuz tverdokorij (38.18. UKRAINIAN)
garbuz velikoplìdnij (38.16. UKRAINIAN)
garchisha belaja (30.57. BELARUSAN)
garđakerfill (24.3. ICELANDIC)
garden amaranth (23.3. ENGLISH)
garden beet (35.4. ENGLISH)
garden cress (30.51. ENGLISH)
garden egg (62.17. ENGLISH)
garden endive (27.9. ENGLISH)
garden melon (38.8. ENGLISH)
garden mint (44.3. ENGLISH)
garden myrrh (24.16. ENGLISH)
garden nightshade (62.15. ENGLISH)
garden orach (35.1. ENGLISH)
garden pea (40.28. ENGLISH)
garden purslane (59.2. ENGLISH)
garden radish (30.56. ENGLISH)
garden rocket (30.48. ENGLISH)
garden sorrel (58.2. ENGLISH)
garden sorrel (58.4. ENGLISH)
garden spinach (35.9. ENGLISH)
garđperla (30.51. ICELANDIC)
gardu (27.15. BASQUE)
gargâ bos (32.1. WOLOF)
garga mbosé (32.1. WOLOF)
gargeranit yevanit (40.32. HEBREW)
gargir hanechalin (30.52. HEBREW)
gargir ha-neḫalim (30.52. HEBREW)
gargir il-gemel (30.43. MALTESE)
gargo (62.18. KANURI)
gargy daly (17.7. AZERBAIJANI, NORTHERN)
gari (39.4. IGBO)
gari kalai (40.8. BENGALI)
garião-vulgar (30.52. CRIOULO, CAPE VERDE)
garii (38.18. HAUSA)

garijara (24.11. SANSKRIT)
garisa (39.4. KANURI)
garko (62.18. FULFULDE, ADAMAWA)
garland chrysanthemum (27.23. ENGLISH)
garle (38.20. DAGBANI)
garleid (7.10. GAELIC, MANX)
garlic (7.10. ENGLISH)
garlic (7.10. PANJABI, EASTERN)
garlleg (7.10. WELSH)
garnalenmoe (23.2. DUTCH)
garo mah (40.8. ASSAMESE)
garohovnik (40.5. RUSSIAN)
garpyz (38.3. TURKMEN)
garra (38.18. ARABIC)
garsar (34.2. HAUSA)
Garten-Ampfer (58.3. GERMAN)
Gartenbohne (40.26. GERMAN)
Garten-Erbse (40.28. GERMAN)
Garten-Kerbel (24.3. GERMAN)
Garten-Kerbelkraut (24.3. GERMAN)
Garten-Kress (30.2. GERMAN)
Garten-Kresse (30.51. GERMAN)
Garten-Kürbis (38.18. GERMAN)
Gartenmelde (35.1. GERMAN)
Garten-Reseda (60.1. GERMAN)
Garten-Rettich (30.56. GERMAN)
Garten-Ruke (30.48. GERMAN)
Garten-Sauerampfer (58.2. GERMAN)
Garten-Schwarzwurzel (27.38. GERMAN)
Garten-Senfrauke (30.48. GERMAN)
garut (15.2. INDONESIAN)
garut (15.2. JAVANESE)
gashia (34.2. TIV)
gasi (62.19. KOREAN)
gasiogapi (26.4. KOREAN)
gasiogapinamu (26.4. KOREAN)
gasiyaahu (34.2. FULFULDE, ADAMAWA)
gassayaa (34.2. HAUSA)
ga:su (62.24. BADAGA)
gãṭ (20.2. SANTALI)
ga-tɛmbɛr sɛmbɛr (40.38. BUDIK)
gathalu (11.2. MARATHI)
gatrabhanga (40.18. SANSKRIT)
gaṭṭhá (7.3. HINDUSTANI)
gatudan (27.29. KANKANAEY)
gatumbal (62.5. CEBUANO)
gatunek cibory (10.1. POLISH)
gau (17.1. BUGHOTU)
gau (17.2. BUGHOTU)
gau (17.4. BUGHOTU)
gauai-gauai (6.1. CEBUANO)
gáuai-gáuai (6.2. TAGALOG)
/gaubeb (44.4. NAMA-DAMARA)
gau-gau (19.1. FILIPINO)
gaulchokgguhwa (45.1. KOREAN)
gauratu (17.1. BUGHOTU)
gauratu (17.2. BUGHOTU)
gauratu (17.4. BUGHOTU)
gauri (17.7. FULFULDE, ADAMAWA)
gaurīje (17.7. FULFULDE, ADAMAWA)
gauruotoji soja (40.8. LITHUANIAN)
gauta (62.9. HAUSA)
gauta (62.19. HAUSA)
gáutáá (62.9. HAUSA)
gaútáá (62.19. HAUSA)
gaútán bacarmi (62.18. FULFULDE, ADAMAWA)
gaútán ɓéllò (62.9. HAUSA)
gaútán kádii (62.15. HAUSA)
gaútán tùùrààwáá (62.9. HAUSA)
gáutan-kúúrúá (62.19. HAUSA)
gavaara (40.6. MARATHI)
gavaara kaayi (40.6. KANNADA)
gavar (40.6. GUJARATI)
gavara (40.3. MARATHI)
gavi (8.4. IBANAG)
gavu (38.14. SHONA)
gavu (38.16. SHONA)
gavu (38.17. SHONA)
gavu (38.18. SHONA)
g’aw$^{\vee}$ ca^{-} di$_{\vee}$ (30.36. LAHU)
g’aw$^{\vee}$ ca^{-} o$_{-}$ (30.36. LAHU)
g’aw$^{\vee}$ hk’a$^{\vee}$ bo$_{-}$ (27.18. LAHU)
g’aw$^{\vee}$ hkveh$^{\vee}$ bo$_{-}$ (27.18. LAHU)
g’aw$^{\vee}$ hpeh ca (30.17. LAHU)
gaw la$_{\wedge}$ hka$^{\vee}$ shi$_{-}$ (38.25. LAHU)
g’aw$^{\vee}$ ma hpu (30.56. LAHU)
gawarphali (40.6. HINDI)
gaway (6.2. WÁRAY-WÁRAY)
gaway (8.4. CEBUANO)
gawáy (8.4. WÁRAY-WÁRAY)
gaway-gaway (6.2. WÁRAY-WÁRAY)
gawēje (17.7. FULFULDE, ADAMAWA)
gawgaw (19.1. TAGALOG)
gaw-gaw (19.1. TAGALOG)
gawḥ (38.7b. MEHRI)
gawri ngabbu (17.3. FULFULDE)
gawsvgi (38.7a. CHEROKEE)
gawsvgi (38.7b. CHEROKEE)
gawudɛ (17.5. PULAAR)

gboussou (62.18. MANINKAKAN, WESTERN)
gbowunɔ (38.12. FON)
gbowunɔ (38.12. GUN)
gboyoboli (62.1. GBE, WACI)
gbuhaŋga (23.3. MALIMBA)
gbũhɔ̃ (2.1. MENDE)
gbulɛ ye yee (38.18. MANINKAKAN, EASTERN)
gbure (59.3. YORUBA)
gbure funfun (59.3. YORUBA)
gbure pupa (59.4. YORUBA)
g-busi (8.4. KONO)
gbussu (62.18. MANDING-MANINKA)
gɔniŋgbe (38.28. GOLA)
gɔnugbe (38.27. MENDE)
gɔnugbe (38.28. MENDE)
gɔnuŋgbe (38.28. GOLA)
gɔrɔnɔmmafɛ (59.3. DENDI)
ge cong (7.14. CHINESE, WÚ)
ge lan cai (30.25. CHINESE, MIN BEI)
gē mūz (16.1a. HARARI)
ge teng (40.30. CHINESE, MANDARIN)
ge teng (40.30. CHINESE, WÚ)
gebe (38.3. SHONA)
gébú (7.3. GONJA)
gebu (7.5. GONJA)
gece sefasi (49.1. TURKISH)
gedang sungu (16.1a. JAVANESE)
gʷədäre (8.4. AMHARIC)
gedde kosu (30.30. KANNADA)
gedi (45.2. INDONESIAN)
gedi (45.2. JAVANESE)
gedi (45.2. SUNDA)
gedi (45.2. VLAAMS)
gedjo (62.14b. BAMILEKE)
gedolat peri (38.23. HEBREW)
gedrek (28.1. JAVANESE)
geel fluweelblad (14.1. VLAAMS)
geelappelliefie (62.12. AFRIKAANS)
geeluintjie (10.1. AFRIKAANS)
geelwortel (20.2. DUTCH)
geelwortel (20.2. VLAAMS)
geelwortel (24.11. AFRIKAANS)
geen amaranth (23.3. ENGLISH)
geerakasaale (23.3. KANNADA)
gefleckte Golddistel (27.37. GERMAN)
gegroefde pompoen (38.27. VLAAMS)
geitskjegg (27.46. NORWEGIAN)
gejembak (30.52. JAVANESE)
gekrulde mosterd (30.9. VLAAMS)
gekweekte eruca (30.48. DUTCH)
gelang (59.2. INDONESIAN)
gelang (59.2. SUNDA)
gelang (59.3. INDONESIAN)
gelang pasir (59.2. MALAY)
Gelbwurzel (20.2. GERMAN)
geldbeursie (30.43. AFRIKAANS)
gele morgenster (27.47. DUTCH)
gele morgenster (27.47. VLAAMS)
gele mosterd (30.57. DUTCH)
gele yam (11.3. VLAAMS)
geli-geli (8.7. MALAY)
gelo (8.4. JAVANESE)
geltonoji ciberžolė (20.4. LITHUANIAN)
gember (20.6. DUTCH)
gember (20.6. VLAAMS)
gembili (11.5. INDONESIAN)
gembili (11.5. JAVANESE)
gembili (44.8. JAVANESE)
gembolo (11.1. JAVANESE)
gembolo (11.5. JAVANESE)
gemeine Kuhblume (27.45. GERMAN)
gemeine Nachtkerze (51.1. GERMAN)
gemeine Schwarzenessel (44.7. GERMAN)
gemeine Wassermimose (40.19. GERMAN)
gemeine Wegwarte (27.10a. GERMAN)
gemeine Blasenkirsche (62.11. GERMAN)
gemeiner Beinwell (29.2. GERMAN)
gemeiner Feldsalat (67.2. GERMAN)
gemeiner Hopfen (46.1. GERMAN)
gemeiner Pastinak (24.18. GERMAN)
gemeines Eiskraut (22.1. GERMAN)
gemeines Hirtentäschel (30.43. GERMAN)
gemeines Hirtentäschelkraut (30.43. GERMAN)
gemeines Pfeilkraut (6.1. GERMAN)
ge-mɛ̀r (40.39. BUDIK)
gemerokallis (13.3. TURKMEN)
gemmer (20.6. AFRIKAANS)
gemmiferous mustard (30.8. ENGLISH)
gemsbokboontjie (40.33. AFRIKAANS)
gemsbokwortel (40.33. AFRIKAANS)
gemsbuck bean (40.33. ENGLISH)
Gemüse-Ampfer (58.3. GERMAN)
Gemüse-Paprika (62.1. GERMAN)
gen qin cai (24.5. CHINESE, MANDARIN)
gen qin cai (24.5. CHINESE, WÚ)
gen tian cai (35.4. CHINESE, MANDARIN)
gen tian cai (35.4. CHINESE, WÚ)

gen tian cai (35.4. CHINESE, YUE)
gen xiang qin (24.21. CHINESE, MANDARIN)
gen yong jie cai (30.14. CHINESE, YUE)
genasu (36.3. KANNADA)
genasu (36.3. TELUGU)
genasu gadda (36.3. TELUGU)
genasu gedde (36.3. KANNADA)
genda (7.2. SUBANON)
genda (7.8. MARANAO)
gɛnde (40.1. VAI)
gɛ́ndɛi (40.1. VAI)
genderi (40.1. VAI)
gɛ́ndɛri (40.1. VAI)
gendola (28.1. INDONESIAN)
gendola (28.1. MALAY)
gendolak (28.1. JAVANESE)
gendolan (28.1. SUNDA)
gɛnɛ (27.14. VAI)
genebra (11.2. CRIOULO, UPPER GUINEA)
genemo (41.1. PORTUGUESE)
genemo (41.1. SPANISH)
geng xiang qin (24.21. CHINESE, WÚ)
gengibre (20.5. PORTUGUESE)
gengibre (20.6. CRIOULO, CAPE VERDE)
gengibre (20.6. PORTUGUESE)
gengibre amarelo (20.6. PORTUGUESE)
gengibre das boticas (20.6. PORTUGUESE)
gengibre dourada (20.2. PORTUGUESE)
gengipe (20.6. CRIOULO, UPPER GUINEA)
gen-hārī (23.3. MAITHILI)
genişçan (34.3. TURKISH)
genja (20.6. GOLA)
genjer (14.1. INDONESIAN)
genjerot (28.1. INDONESIAN)
genjerot (28.1. JAVANESE)
genjulak (28.1. JAVANESE)
gēnmēsāmi (17.7. KURDISH, NORTHERN)
genth (11.2. HINDI)
genthi (11.2. HINDI)
genttee (11.2. HINDI)
gentud (44.5. SUBANON)
gerebcsin (27.17. HUNGARIAN)
gerkin (38.13b. ENGLISH)
gerkin (38.13b. FILIPINO)
Germeiner-Dost (44.6. GERMAN)
gerte (40.1. PULAAR)
gerté (40.1. PULAAR)
gerte (40.1. WOLOF)
gèrté (40.1. WOLOF)
gèrté bâbara (40.39. WOLOF)
gèrté bambara (40.39. WOLOF)
gèrté pergne (40.39. WOLOF)
gertere (40.1. PULAAR)
gesse blanche (40.10. FRENCH)
gesse cultivée (40.10. FRENCH)
gesse tubéreuse (40.11. FRENCH)
gesteelde mosterd (30.20. VLAAMS)
gesuma (38.16. HAUSA)
gesuma (38.18. HAUSA)
getang (27.3. MALAY)
getang (27.4. MALAY)
ge-tɛ (11.7. ÉWÉ)
gethi (11.2. HINDI)
getti di bambú (17.1. ITALIAN)
geunaine (44.8. FRENCH)
gevieugelde boon (40.29. VLAAMS)
gevlekte tuindistel (27.37. VLAAMS)
gevlerkte boon (40.29. AFRIKAANS)
gevlerkte ert (40.13. AFRIKAANS)
gewohnliche Kermesbeere (56.2. GERMAN)
gewone melkdistel (27.18. VLAAMS)
gewone melkdistel (27.42. DUTCH)
gewone melkdistel (27.42. VLAAMS)
gewone misbredie (23.3. AFRIKAANS)
gewone paardebleom (27.45. VLAAMS)
gewone smeerwortel (29.2. DUTCH)
gewone smeerwortel (29.2. VLAAMS)
gewone teunisbloem (51.1. DUTCH)
gewone veldsla (67.2. VLAAMS)
gewoon schorrekruit (35.10. VLAAMS)
Gewürzgurke (38.13b. GERMAN)
Gewürzpaprika (62.1. GERMAN)
Gewürzpaprika (62.2. GERMAN)
geyik boynuzu (57.1. TURKISH)
geyik sarmisaği (7.14. TURKISH)
geyikotu (44.9. TURKISH)
gezer (24.11. HEBREW)
gêzer (24.11. KURDISH, NORTHERN)
gezer hagina (24.11. HEBREW)
gezer haginah (24.11. HEBREW)
gezer ha-ginna (24.11. HEBREW)
gezer sahmy hindy (19.1. ARABIC)
gezer sahmy hindy (20.4. ARABIC)
gezwollen stengelmosterd (30.19. VLAAMS)
ghadaghosali (38.22. MARATHI)
ghadi phool (54.1. NEPALI)
ghāghrā (40.12. MAITHILI)
ghalabwaasa (38.14. CAROLINIAN)
ghalabwaasa (38.16. CAROLINIAN)

ghalabwaasa (38.17. CAROLINIAN)
ghalabwaasa (38.18. CAROLINIAN)
ghalamaasa (38.14. CAROLINIAN)
ghalamaasa (38.16. CAROLINIAN)
ghalamaasa (38.17. CAROLINIAN)
ghalamaasa (38.18. CAROLINIAN)
ghallet is-serduq (23.4. MALTESE)
ghanahana (62.15. SANSKRIT)
ghanasvana (23.3. SANSKRIT)
ghangra (40.41. SANTALI)
ghạṅgri (40.41. SANTALI)
ghat jnār (17.7. PASHTO, NORTHERN)
ghati (62.21. MARATHI)
ghattada uddu (40.36. KANNADA)
ghēä tōri (38.21. PANJABI, EASTERN)
ghēä tōri (38.22. PANJABI, EASTERN)
ghemikanda (8.3. TELUGU)
gheneb id-dib (62.21. MALTESE)
ghevda (40.3. MARATHI)
ghevda (40.4. MARATHI)
ghia (38.20. HINDI)
ghia kaddoo (38.20. PANJABI, EASTERN)
ghia kadu (38.20. PANJABI, EASTERN)
ghia tori (38.22. PANJABI, EASTERN)
ghianda di terra (40.11. ITALIAN)
ghiatori (38.22. PANJABI, EASTERN)
ghiatorui (38.22. HINDI)
ghiaturi (38.22. HINDI)
ghigandoli (38.22. PANJABI, EASTERN)
ghimber (20.6. ROMANIAN)
ghimber galben (20.2. ROMANIAN)
ghiraunla (38.22. NEPALI)
ghisoda (38.21. GUJARATI)
ghiturai (38.22. PANJABI, EASTERN)
ghiurā (38.22. MAITHILI)
ghiya (38.20. PANJABI, EASTERN)
ghiya kaddu (38.20. PANJABI, EASTERN)
ghiya tori (38.22. HINDI)
ghiya tori (38.22. URDU)
ghiya turai (38.22. MARATHI)
ghiyathurayi (38.22. HINDI)
ghiyobw (13.3. CAROLINIAN)
ghobe (38.4. GUJARATI)
ghod tapre (24.7. NEPALI)
ghol (59.2. GUJARATI)
ghol (59.2. MARATHI)
ghol (59.4. GUJARATI)
ghole (59.2. MARATHI)
ghómwuuti (36.3. CAROLINIAN)
ghoon (38.7b. BALTI)
ghorghas (8.6. MALTESE)
ghorrai (30.24. HINDI)
ghosaalen (38.22. MARATHI)
ghosali (38.22. MARATHI)
ghosha (24.13. SANSKRIT)
ghoshalata (38.21. BENGALI)
ghoshavathi (24.13. SANSKRIT)
ghotika (59.2. SANSKRIT)
ghoya (8.4. HINDI)
ghrinaphala (38.3. SANSKRIT)
ghrinavasa (38.1. SANSKRIT)
ghrishti (11.2. SANSKRIT)
ghuiya (8.4. HINDI)
ghuiyan (8.4. HINDI)
ghundhun (7.2. BENGALI)
ghurmi (38.8. NEPALI)
ghuru (38.4. GUJARATI)
ghurush (40.40. HINDI)
ghurush (40.40. PANJABI, EASTERN)
ghuschél (20.6. CAROLINIAN)
ghúúngar (38.13a. CAROLINIAN)
ghuya (8.4. HINDI)
ghwiya (8.4. HINDI)
gi[aas]p c[as] (61.1. VIETNAMESE)
giacinti (40.9. MALTESE)
giacinto del pennacchio (13.5. ITALIAN)
giant alocasia (8.1. ENGLISH)
giant bamboo (17.2. ENGLISH)
giant elephant's ear (8.5. ENGLISH)
giant garlic (7.12. ENGLISH)
giant pumpkin (38.16. ENGLISH)
giant swamp taro (8.6. ENGLISH)
giant taro (8.1. ENGLISH)
gibud (38.7b. KONKANI)
gidada genasu (39.4. KANNADA)
gidi (45.2. JAVANESE)
gidi (45.2. SUNDA)
gidjihra (23.2. SANSKRIT)
gidra (30.30. MALTESE)
gi[eef]n t[aa]y (22.2. VIETNAMESE)
giema-vɛlɔ (9.1. MENDE)
gifboontjie (40.4. AFRIKAANS)
gigboi (27.14. LOKO)
gigẽ (40.1. DANGME)
gigie (59.2. SAMOAN)
giginyar birii (19.1. HAUSA)
giglio (13.3. ITALIAN)
giglio (13.4. ITALIAN)
giglio de nilo (48.1a. ITALIAN)
gijimbilla (20.6. OROMO, WEST CENTRAL)

gluvarče (27.45. MACEDONIAN)
glyko kalampoki (17.7. GREEK)
glykopatata (11.2. GREEK)
glykopatata (36.3. GREEK)
gmana (45.1. TEM)
gnahi (30.27a. BUGHOTU)
gnamankou (20.6. BÉTÉ)
gnamankou (20.6. JULA)
gnangnour (36.3. DAGAARE, SOUTHERN)
gnébé (40.43. MANINKA)
gnemo (41.1. TAGALOG)
gneti (41.1. LATVIAN)
gneton (41.1. HEBREW)
gneton tou gnemou (41.1. GREEK)
gniambi (39.4. WOLOF)
gnié (11.1. DAGAARE, SOUTHERN)
gnié (11.2. DAGAARE, SOUTHERN)
gnié (11.3. DAGAARE, SOUTHERN)
gnié (11.4. DAGAARE, SOUTHERN)
gnié (11.5. DAGAARE, SOUTHERN)
gnié (11.6. DAGAARE, SOUTHERN)
gnié (11.7. DAGAARE, SOUTHERN)
gnié (11.8. DAGAARE, SOUTHERN)
gnimikala (17.5. JULA)
gô (giâ)-á chhài (27.32. CHINESE, MIN NAN)
Goa bean (40.29. ENGLISH)
Goabohne (40.29. GERMAN)
goabønne (40.29. DANISH)
goabønne (40.29. NORWEGIAN)
gô-á-chhài (27.32. CHINESE, MIN NAN)
goal halva (23.4. FARSI, WESTERN)
goa-papu (40.29. FINNISH)
goat's horn (40.32. ENGLISH)
goatsbeard (27.47. ENGLISH)
gobay (30.26. PASHTO, NORTHERN)
gobhī (30.26. HINDI)
gobi (30.26. KANNADA)
gobī (30.27a. PASHTO, NORTHERN)
gobi-dok (30.27a. BURMESE)
gobi-pan (30.26. BURMESE)
gobo (27.5. INDONESIAN)
gobou (27.5. JAPANESE)
gobouazami (27.11. ENGLISH)
gobouazami (27.11. JAPANESE)
gobouazami (27.11. VLAAMS)
gobra note (23.3. BENGALI)
gobsilminari (24.19. KOREAN)
goch alu (11.2. ASSAMESE)
goch bilahi (62.7. ASSAMESE)
goch kobi (30.29. ASSAMESE)
gochu (62.1. KOREAN)
godobos (27.19. INDONESIAN)
godobos (27.19. JAVANESE)
godu (7.8. KASHMIRI)
godulbbaegi (27.48. KOREAN)
godumba (38.3. SANSKRIT)
goewerneursboontji (40.25. AFRIKAANS)
goggyli (30.22. GREEK)
goggylokramvi (30.30. GREEK)
gogoro (55.1. ENGLISH)
gogowat (23.4. BONTOC, CENTRAL)
gogu (45.4. TAMIL)
goguma (36.3. KOREAN)
goinjol (17.7. KANNADA)
goirmín (68.1. GAELIC, IRISH)
gôje farangi (62.9. FARSI, WESTERN)
goji (38.14. HAUSA)
goji (38.16. HAUSA)
goji (38.17. HAUSA)
goji (38.18. HAUSA)
gojiaa sema (40.3. HINDI)
gök nohut (40.28. TURKMEN)
gokbilang (17.7. IFUGAO)
gol (59.2. KONKANI)
gol geser (27.17. MONGOLIAN)
gol kalam (30.26. FARSI, WESTERN)
göl kestanesi (64.3. TURKISH)
gol mirch (62.2. HINDUSTANI)
golakanda miraph (62.5. TELUGU)
golakandah (62.24. SANSKRIT)
golasiman (59.2. TAGALOG)
golbhenda (62.9. NEPALI)
golbhenda (62.10. NEPALI)
golchibagi (59.2. KONKANI)
golden apple (62.14a. ENGLISH)
golden bamboo (17.1. ENGLISH)
golden berry (62.12. VLAAMS)
golden elma (62.14a. TURKISH)
golden gram (40.38. ENGLISH)
golden thistle (27.36. ENGLISH)
Goldnarbe (8.9. GERMAN)
Goldwurzel (27.36. GERMAN)
gole (38.20. MÒORÉ)
göleğez (8.4. TURKISH)
golé-ladan (65.1. FARSI, WESTERN)
golem lopušnik (27.5. MACEDONIAN)
golema latanka (65.1. MACEDONIAN)
golemoglavest luk (7.1. MACEDONIAN)
golfo (14.1. PORTUGUESE)
golgotavirág (54.1. HUNGARIAN)

golipadpe (59.2. TULU)
golisoppu (59.2. KANNADA)
golkaddu (38.1. HINDI)
golkaddu (38.20. HINDI)
golkaddu (38.20. PANJABI, EASTERN)
golo (58.2. GEORGIAN)
golubiji grašak léca (40.2. SERBO-CROATIAN)
gomaos (27.14. MARANAO)
gombo (45.1. DANISH)
gombo (45.1. DUTCH)
gombo (45.1. FRENCH)
gombó (45.1. HUNGARIAN)
gombo (45.1. ITALIAN)
gombo (45.1. RUSSIAN)
gombo (45.1. SPANISH)
gombo (45.1. VLAAMS)
Gombro (45.1. GERMAN)
gombro (45.1. ITALIAN)
gombro (45.1. PORTUGUESE)
gombvirág (27.22. HUNGARIAN)
gomchwi (27.34. KOREAN)
gomdhan (17.7. ASSAMESE)
gome neëkhal (10.1. HEBREW)
gomeh (10.1. HEBREW)
gomenna (30.27a. OROMO, WEST CENTRAL)
gomi (23.2. AJA)
gomoljasti (40.11. SERBO-CROATIAN)
gomoljna zelena (24.5. SLOVENIAN)
gomondo (40.33. VENDA)
goncha (40.18. HINDI)
gonda hannu (62.9. KANNADA)
gondol (38.1. PANGASINAN)
gonglu (30.41a. PANJABI, EASTERN)
go:ngu (58.2. BADAGA)
gongyli (30.22. GREEK)
gongyli (30.41a. GREEK)
gonikasa (59.2. KANNADA)
gonogono (38.20. FULFULDE, ADAMAWA)
gonogonōji (38.20. FULFULDE, ADAMAWA)
gonoretti (27.8. GURO)
gonsali (38.21. KONKANI)
goober nut (40.1. ENGLISH)
Good King Henry (35.5. ENGLISH)
goode hannu (62.9. KANNADA)
goode kattu (62.9. KANNADA)
goojare (8.4. FULFULDE)
goojíí (38.16. HAUSA)
goòjíí (38.16. HAUSA)
goòjíí (38.18. HAUSA)
gooseberry gourd (38.6. ENGLISH)
goosefoot (35.6. ENGLISH)
goot baigun (62.9. MARATHI)
gor bu di daw (34.2. WOLOF)
gora (17.1. HAUSA)
gora (17.2. HAUSA)
gora (17.4. HAUSA)
gora (38.20. HAUSA)
goradu (11.1. GUJARATI)
goradu (11.1. MARATHI)
goradu (11.1. MARATHI)
goradu (11.2. GUJARATI)
goradu (11.2. MARATHI)
gorakshaphalini (40.6. SANSKRIT)
gorani (40.6. SANSKRIT)
gororatthaayi (20.1. TAMIL)
goraru (38.20. FULFULDE, ADAMAWA)
gorbusa (34.2. FULFULDE)
gorchikudu (40.6. TELUGU)
gorchitsa (30.3. RUSSIAN)
gorchitsa (30.5. RUSSIAN)
gorchitsa (30.6. RUSSIAN)
gorchitsa (30.7. RUSSIAN)
gorchitsa (30.8. RUSSIAN)
gorchitsa (30.9. RUSSIAN)
gorchitsa (30.10. RUSSIAN)
gorchitsa (30.11. RUSSIAN)
gorchitsa (30.12. RUSSIAN)
gorchitsa (30.13. RUSSIAN)
gorchitsa (30.15. RUSSIAN)
gorchitsa (30.16. RUSSIAN)
gorchitsa (30.17. RUSSIAN)
gorchitsa (30.20. RUSSIAN)
gorchitsa chornaja (30.24. BELARUSAN)
gorchitsa chornaja (30.24. RUSSIAN)
gorchitsa sareptsaja (30.4. RUSSIAN)
gorchitza sareptsaja (30.4. BELARUSAN)
gorčliva endivija (27.10a. MACEDONIAN)
gorczyca abisyńska (30.3. POLISH)
gorczyca bambusowa (30.5. POLISH)
gorczyca biała (30.57. POLISH)
gorczyca białokwiatowa (30.11. POLISH)
gorczyca czarna (30.24. POLISH)
gorczyca długolistna (30.13. POLISH)
gorczyca drobnolistna (30.7. POLISH)
gorczyca inwolucyjna (30.9. POLISH)
gorczyca japońska (30.18. NORWEGIAN)
gorczyca karłowa (30.6. POLISH)
gorczyca korzeniowa (30.14. POLISH)

gorczyca kwiatowo liściowa (30.16. POLISH)
gorczyca liliowa (30.12. POLISH)
gorczyca pączkowa (30.8. POLISH)
gorczyca pęcherzykowata (30.19. POLISH)
gorczyca sarepska (30.20. POLISH)
gorczyca szczytowa (30.42. POLISH)
gorczyca szerokolistna (30.10. POLISH)
gorczyca szpinakowa (30.39. POLISH)
gorczyca uprawna (30.15. POLISH)
gorczyca wielkolistna indiańska (30.17. POLISH)
gorczycznik wiosenny (30.2. POLISH)
gorgo (62.18. KANURI)
gorgon water lily (50.1. ENGLISH)
gori kaayi (40.6. KANNADA)
gori kayi (40.6. KANNADA)
gorikachu (8.4. HINDI)
gorikayi (40.6. KANNADA)
gorko (62.18. HAUSA)
gorko (62.18. KANURI)
gorljanka obyknovennaja (38.20. RUSSIAN)
goroch posevnoji (40.28. KAZAKH)
görögdinnye (38.3. HUNGARIAN)
görögszéna (40.32. HUNGARIAN)
goroh agorodnyi (40.28. RUSSIAN)
goroh ovoščnoj (40.28. RUSSIAN)
goroh pasjanny (40.28. BELARUSAN)
gorokhia tiyoh (38.2. ASSAMESE)
goron pasevnoi (40.28. RUSSIAN)
gortchitsa belaja (30.57. RUSSIAN)
gôru (64.2. KASHMIRI)
goru chikkudu (40.6. TELUGU)
goruchikkuda (40.6. TELUGU)
gorušica bela (30.57. SERBO-CROATIAN)
gorušica crna (30.24. SERBO-CROATIAN)
gorušica sareptska (30.17. SERBO-CROATIAN)
gorwò (62.19. KANURI)
goryczel (30.3. POLISH)
goryczel (30.8. CZECH)
goryczel sarepska (30.4. POLISH)
gosale (38.22. MARATHI)
gospoovu (30.26. TAMIL)
gosthabartaku (62.23. SANSKRIT)
go:su (30.27a. BADAGA)
gota begun (62.23. BENGALI)
gothbegun (62.23. BENGALI)
gotukola (24.7. JAPANESE)
gotukola (24.7. SINHALA)
gou gua dou (40.18. CHINESE, MIN BEI)
gou gua yu (19.1. CHINESE, YUE)
gou qi (62.8. CHINESE, MANDARIN)
gou qi (62.8. CHINESE, YUE)
gou qi tou (62.8. CHINESE, WÚ)
gou qie (62.8. CHINESE, WÚ)
gouboudo (55.1. FULFULDE)
gouden appel (62.14a. VLAAMS)
gouez-irvin (30.1. BRETON)
gougohi (17.7. BÉTÉ)
gouita (62.14b. FULFULDE)
goųm-gɑ-dH-'ei-gɑ (30.27a. KIOWA)
gourde (38.20. FRENCH)
gourge de Malabar (38.15. FRENCH)
gourobi (11.2. ORIYA)
gousalim (38.21. PORTUGUESE)
gouteron (27.5. FRENCH)
gova (30.27a. SINHALA)
govaari (40.6. MARATHI)
govardhana kaayi (40.6. KANNADA)
govari (40.3. MARATHI)
govekaayi (38.16. KANNADA)
gövelez (8.4. TURKISH)
govhi (38.4. SANSKRIT)
govinajola (17.7. KANNADA)
gowar (40.6. HINDI)
gowar (40.6. MARATHI)
gowje farangee (62.9. FARSI, WESTERN)
goyo (62.14a. BAMANANKAN)
goyô (62.19. JULA)
gôyoho (62.19. SENOUFO, CEBAARA)
goýun (38.7a. TURKMEN)
grä ärt (40.28. SWEDISH)
grabak (30.52. SERBO-CROATIAN)
gradinarska čubric (44.9. MACEDONIAN)
gradinarska dinja (38.8. MACEDONIAN)
gradinarska dobreka (30.51. MACEDONIAN)
gradinarska eruka (30.48. MACEDONIAN)
gradinarska loboda (35.1. MACEDONIAN)
gradinarski pelin (27.6. MACEDONIAN)
gradinska loboda (35.1. BULGARIAN)
gradinski grach (40.28. BULGARIAN)
gradinski lapad (58.3. BULGARIAN)
grænk'al (30.35. ICELANDIC)
græs-fladbælg (40.10. DANISH)
graeskar (38.18. DANISH)
grah (40.26. BOSNIAN)
grah (40.28. SLOVENIAN)
grah mahunar (40.24. CROATIAN)
grah mahunar (40.26. CROATIAN)
grahor (40.10. SERBO-CROATIAN)

grahor (40.11. SERBO-CROATIAN)
grahor poljski (40.10. SERBO-CROATIAN)
grain de cheval (40.16. FRENCH)
graines de lotus (48.1b. FRENCH)
graines d'quonde (38.5. FRENCH)
gram (40.5. MORISYEN)
gramanti (10.1. KRIO)
gramantin (10.1. KRIO)
gramini (35.9. SANSKRIT)
gramya (38.16. SANSKRIT)
gramya (38.21. SANSKRIT)
gramyakarkati (38.1. SANSKRIT)
gramyavallabha (35.9. SANSKRIT)
grana (28.1. TAGALOG)
graná (40.1. KRIO)
granat (40.1. KRIO)
granata (35.2. ITALIAN)
grand bambou (17.1. FRENCH)
grand galanga (20.1. FRENCH)
grand oseille (58.2. FRENCH)
grand pourprier (59.3. FRENCH)
grande bardane (27.5. FRENCH)
grande morelle (62.14b. FRENCH)
grande ortie (66.3. FRENCH)
grande tayove (8.1. FRENCH)
grandinski kiselec (58.2. BULGARIAN)
gránɛt (40.1. KRIO)
granoturco da zucchero (17.7. ITALIAN)
granthila (23.3. SANSKRIT)
grao (40.5. PORTUGUESE)
grão de bico (40.5. PORTUGUESE)
grašak (40.28. BOSNIAN)
grašak (40.28. CROATIAN)
grašak (40.28. SERBO-CROATIAN)
grašak špargar (40.13. SERBO-CROATIAN)
graserwt (40.10. VLAAMS)
grăsită (59.2. ROMANIAN)
grasiu (7.11. AFRIKAANS)
grasker (38.18. ICELANDIC)
graslaukur (7.11. ICELANDIC)
grasløk (7.11. NORWEGIAN)
gräslök (7.11. SWEDISH)
graslook (7.12. VLAAMS)
grašok (40.28. MACEDONIAN)
grass pea (40.10. ENGLISH)
grassé (59.3. FRENCH)
grav (40.26. MACEDONIAN)
gravanço (40.5. PORTUGUESE)
grawa (27.24. AMHARIC)
gra(w)mpi (40.1. KRIO)
grčka rogasta dete lina (40.32. MACEDONIAN)
great burdock (27.5. ENGLISH)
greater galangal (20.1. ENGLISH)
greater nettle (66.3. ENGLISH)
greater yam (11.1. ENGLISH)
great-headed garlic (7.1. ENGLISH)
Greek hay-seed (40.32. ENGLISH)
green amaranth (23.3. ENGLISH)
green bean (40.26. ENGLISH)
green gram (40.38. ENGLISH)
green mint (44.3. ENGLISH)
green pepper (62.1. ENGLISH)
grenmynta (44.3. ICELANDIC)
gresløk (7.11. NORWEGIAN)
gresskar (38.16. NORWEGIAN)
gresskar (38.18. NORWEGIAN)
greuneters (40.26. STELLINGWERFS)
greunte (24.4. STELLINGWERFS)
Griechischer Bockshornklee (40.32. GERMAN)
Griechisch-Heu (40.32. GERMAN)
griežtis (30.22. LITHUANIAN)
grihakoolaka (38.29. SANSKRIT)
grin (23.3. KRIO)
gringri (63.1. MENDE)
grins (23.3. KRIO)
grintul (41.1. JAVANESE)
griséus (40.28. PORTUGUESE)
gro pima (62.5. MORISYEN)
gro poi (40.25. MORISYEN)
groch (40.28. POLISH)
groch (40.28. SORBIAN, LOWER)
groch konserwowy (40.27. POLISH)
groch zwyczajny (40.28. POLISH)
groch zwyczajny ogrodowy (40.28. POLISH)
groefloof (27.10a. VLAAMS)
groenboon (40.26. AFRIKAANS)
groene munt (44.3. DUTCH)
groenert (40.28. AFRIKAANS)
groenloof (27.10a. DUTCH)
groente peer (38.25. DUTCH)
groentevenkel (24.13. VLAAMS)
grof bieslook (7.5. DUTCH)
gromak (13.2. SORBIAN, LOWER)
Gromper (62.24. LUXENBOURGEOIS)
gron di bico (40.5. CRIOULO, CAPE VERDE)
grøn mynte (44.3. DANISH)
grondboon (40.1. AFRIKAANS)
grondboontjie (40.1. AFRIKAANS)

grondnoot (40.1. VLAAMS)
grønkål (30.35. DANISH)
grönkål (30.35. SWEDISH)
grönmynta (44.3. SWEDISH)
grønnkål (30.35. NORWEGIAN)
grønnmynte (44.3. NORWEGIAN)
grønnsakmelon (38.9. NORWEGIAN)
grønnsakpaprika (62.1. NORWEGIAN)
grønsakhibisk (45.1. NORWEGIAN)
groshë (40.26. ALBANIAN, TOSK)
grosse anghive (62.18. FRENCH)
große Brennessel (66.3. GERMAN)
grosse courge (38.16. FRENCH)
große Kapuzinerkresse (65.1. GERMAN)
große Klette (27.5. GERMAN)
Großebohne (40.34a. GERMAN)
großer Galgant (20.1. GERMAN)
groszek bulwiasty (40.11. POLISH)
groszek ziewny (40.11. POLISH)
grote boon (40.34a. DUTCH)
grote brandnetel (66.3. DUTCH)
grote brandnetel (66.3. VLAAMS)
grote engelwortel (24.1. VLAAMS)
grote hardvrucht (30.42. VLAAMS)
grote klis (27.5. DUTCH)
grote klit (27.5. VLAAMS)
grote Oostindische kers (65.1. DUTCH)
ground nut (40.1. ENGLISH)
ground-almond (10.1. ENGLISH)
groundnut (40.11. ENGLISH)
grove bieslook (7.9b. VLAAMS)
grozdest bob (40.6. MACEDONIAN)
grudčesto korenov nivska repka (30.14. MACEDONIAN)
grune bone (40.26. STELLINGWERFS)
grüne Minze (44.3. GERMAN)
grüne Rossminze (44.3. GERMAN)
Grünkohl (30.35. GERMAN)
grusha zemljanaja (27.26. RUSSIAN)
gruszka melonowa (62.20. POLISH)
gruszka miłosna (62.19. POLISH)
gruunte (24.4. STELLINGWERFS)
grzebionatka właściwa (23.4. POLISH)
grzybień (50.2. POLISH)
gua (11.1. BAMANANKAN)
gua zi cai (59.2. CHINESE, WÚ)
gua zi cai (59.2. CHINESE, YUE)
guacamote (39.4. SPANISH)
guagara (11.1. BAMANANKAN)
guakuro (27.18. AKAN-ASANTE)
guamuri (24.12. ASSAMESE)
guamuri (24.13. ASSAMESE)
guan (45.1. MANDING-MANDINKA)
guan ba (45.1. MANDING-MANDINKA)
guan hwowa (62.19. AKAN-ASANTE)
guan yin xian (27.25. CHINESE, WÚ)
guando (40.2. SPANISH)
guandú (40.2. PORTUGUESE)
guangâ (45.1. BAMANANKAN)
guangono abrua (40.9. ANYIN)
guar (40.6. BENGALI)
guar (40.6. DANISH)
guar (40.6. DUTCH)
guar (40.6. ENGLISH)
guar (40.6. FIJIAN)
guar (40.6. FRENCH)
guar (40.6. HINDI)
guar (40.6. ITALIAN)
guar (40.6. NORWEGIAN)
guar (40.6. PANJABI, EASTERN)
guar (40.6. PORTUGUESE)
guar (40.6. SPANISH)
guar (40.6. URDU)
guar simi (40.6. NEPALI)
guara (11.1. BAMANANKAN)
guara (40.6. PANJABI, EASTERN)
guarboon (40.6. AFRIKAANS)
Guarpflanze (40.6. GERMAN)
guarplante (40.6. DANISH)
guasca (27.1. ENGLISH)
guasca (27.1. PORTUGUESE)
guasca (27.1. SPANISH)
guasca (27.22. ENGLISH)
guasca (27.22. VLAAMS)
guaza (8.4. HAUSA)
gubakvetkavaja (44.2. BELARUSAN)
gubbakaaya (38.20. TELUGU)
gubbakaya (38.20. TELUGU)
gubgubbaw (62.9. KANKANAEY)
gubi (36.3. KALAGAN)
gubi (36.3. SUBANON)
guchhandathika (16.1a. SANSKRIT)
guchhaphala (16.1a. SANSKRIT)
guchhaphala (62.15. SANSKRIT)
guci (38.5. AJA)
gùdán àlbásàà (7.3. HAUSA)
gudäre (8.4. AMHARIC)
gudayogaphala (38.16. SANSKRIT)
gudde hannu (62.12. KANNADA)
gude (40.2. INDONESIAN)

gude (40.2. JAVANESE)
gúdīa (40.1. GOURMANCHÉMA)
gudu gudu (11.4. FON)
gueï (17.5. BÉTÉ)
guérte (40.1. PULAAR)
guèrté (40.1. WOLOF)
guge (8.4. BASSARI)
gugumã (44.5. PULAAR)
guhayan (11.5. HANUNOO)
gui zi jiang (27.26. CHINESE, WÚ)
guiabelha (57.1. PORTUGUESE)
guidnilla (62.2. SPANISH)
guié (38.16. MANINKAKAN, EASTERN)
guié (38.18. MANINKAKAN, EASTERN)
guija (40.10. SPANISH)
guija (40.11. SPANISH)
guilè (62.5. PULAAR)
guindilla (62.5. SPANISH)
Guinea arrowroot (15.1. ENGLISH)
Guinea yam (11.7. ENGLISH)
Guinea Yamswurzel (11.3. GERMAN)
Guinea Yamswurzel (11.7. GERMAN)
guinfiné (11.2. MANINKAKAN, EASTERN)
guingambó (45.1. SPANISH)
guio (63.1. KISI, SOUTHERN)
guio (63.1. KISSI, NORTHERN)
guirigue (40.39. PULAAR)
guisante (40.28. SPANISH)
guisante azucarado (40.27. SPANISH)
guisante come-toto (40.27. SPANISH)
guisante común (40.28. SPANISH)
guisante de Angola (40.2. SPANISH)
guisante de tierra (40.39. SPANISH)
guisante tirabeque (40.27. SPANISH)
guisantes (40.28. FILIPINO)
guisantillo rojo (40.13. SPANISH)
guissima (45.4. ZARMA)
guite sindé (11.2. YORUBA)
guite sodé (11.2. YORUBA)
guixa (40.10. CATALAN-VALENCIAN-BALEAR)
gujar (24.18. NEPALI)
gujiya (40.39. HAUSA)
gújíyár rúwá (64.3. HAUSA)
gujji avare (40.26. BADAGA)
gujur (24.18. HINDI)
gul caram (30.26. UZBEK)
gul i sausan (13.3. HINDUSTANI)
gul judekörs (62.12. SWEDISH)
gul karam (30.26. UZBEK)
gul sennep (30.57. DANISH)
gulakaankh (38.4. HINDI)
gulalay (17.3. SUBANON)
gulapitan (40.25. ITAWIT)
gulapski grašok (40.2. MACEDONIAN)
gular baunir (40.12. ICELANDIC)
gularthori (38.29. HINDI)
gulasiman (59.2. BICOLANO, CENTRAL)
gulasiman (59.2. FILIPINO)
gulbete (35.4. NORWEGIAN)
guldaudi (27.23. BENGALI)
guldaudi (27.23. HINDI)
guldbede (35.4. DANISH)
guldtidsel (27.36. DANISH)
gulduu kapusta (30.26. KIRGHIZ)
gul-e-daudi (27.23. URDU)
güleis (8.4. TURKISH)
gulerod (24.11. DANISH)
gulesvati (27.23. MARATHI)
gulhand (27.10a. HINDI)
guli asirak (60.1. TAJIKI)
guli dovudi (27.23. TAJIKI)
guli oshiqi pechon (36.2. TAJIKI)
guli qahqah (27.45. TAJIKI)
gulia (30.30. ROMANIAN)
gulie (30.30. ROMANIAN)
gulihatan (40.25. ITAWIT)
gulipatan (40.25. IBANAG)
guliyo mula (35.4. NEPALI)
gulkaram (30.26. TAJIKI)
gulla (62.19. KANNADA)
gulli (38.7b. HAUSA)
gülli kelem (30.26. TURKMEN)
gullpistill (27.36. ICELANDIC)
gulltistel (27.36. NORWEGIAN)
gullu-gullu (9.1. IFUGAO)
gulma moola (20.6. SANSKRIT)
gulraikhon (44.9. UZBEK)
gulralkhon (44.9. UZBEK)
gulrofa (30.22. ICELANDIC)
gulrot (24.11. ICELANDIC)
gulrot (24.11. NORWEGIAN)
gultodjikhuroz (23.3. UZBEK)
guluf (35.7. NEPALI)
gulugkura (23.4. TELUGU)
gulum butu (38.20. FULFULDE, ADAMAWA)
gulurbaale (16.1a. KANNADA)
gum dhan (17.7. ASSAMESE)
gumagbô (45.1. FON)
gumbala (38.14. KANNADA)

gumbala (38.16. KANNADA)
gumbala (38.18. KANNADA)
gumbinė dioskorėja (11.2. LITHUANIAN)
gumbinė nasturtė (65.2. LITHUANIAN)
gumbinis kiškiakopūstis (53.1. LITHUANIAN)
gumbinis pelėžirnis (40.11. LITHUANIAN)
gumbo (45.1. CAJUN FRENCH)
gumbo (45.1. ENGLISH)
gumbuotasis gurgždis (24.8. LITHUANIAN)
gumbuotoji notra (44.11. LITHUANIAN)
gumbutu (38.20. ZARMA)
gumetakui (20.6. ÉWÉ)
gummadi (38.16. TELUGU)
gummadi kaaya (38.16. TELUGU)
gummadi kayi (38.14. TELUGU)
gummadi kayi (38.18. TELUGU)
gumós baraboly (24.8. HUNGARIAN)
gumós kömény (24.13. HUNGARIAN)
gumós madársóska (53.1. HUNGARIAN)
gumós zeller (24.5. HUNGARIAN)
gumpapeng (62.1. ILOCANO)
gumugei (44.5. ARABIC)
guna (38.3. HAUSA)
gúna (38.3. HAUSA)
gunaà (38.7b. HAUSA)
gunaaji (38.7a. FULFULDE, ADAMAWA)
gunaaru (38.3. FULFULDE, ADAMAWA)
gunalganta liti (17.5. IFUGAO)
gunapendalamu (11.1. TELUGU)
gunar chhuin (40.6. ORIYA)
gunchgaji (40.18. PANJABI, EASTERN)
gundól (38.1. PANGASINAN)
gundól (38.7a. PANGASINAN)
gundól (38.11. PANGASINAN)
gundosollo (40.31. DAGBANI)
gündüz sefasi (36.1. TURKISH)
gungri (40.26. PASHTO, NORTHERN)
gungunsale (11.7. DAGBANI)
güngüzeli (13.3. TURKISH)
gunjid (30.48. MONGOLIAN)
gunuukple (11.7. DAGBANI)
gunzi (17.7. SWAHILI)
gurāka (38.20. HAUSA)
gure (59.2. HAUSA)
gurgett (38.18. MALTESE)
gúrhúl (20.6. CAROLINIAN)
gurji (38.7b. HAUSA)
gurji ƙwantal (38.7b. HAUSA)
gurjií (38.7b. HAUSA)
gùrjín-dájìi (38.4. HAUSA)
gurka (38.13a. ICELANDIC)
gurka (38.13a. SWEDISH)
gurkamai (62.15. HINDI)
gurkamai (62.21. BENGALI)
gurkamai (62.21. HINDI)
Gurke (38.13a. GERMAN)
gurkemeie (20.2. NORWEGIAN)
gurkemeje (20.2. DANISH)
Gurkenkraut (29.1. GERMAN)
gurkha thi (38.25. BURMESE)
gurķi (38.13a. LATVIAN)
gurkmeja (20.2. SWEDISH)
gurkört (29.1. SWEDISH)
gurli kwantal (38.7b. HAUSA)
gurmhī (38.7b. MAITHILI)
guro (45.1. HAUSA)
gursami (11.4. HAUSA)
guru (40.18. SANSKRIT)
guru bhaji (23.2. HINDI)
gurubeeja (40.12. SANSKRIT)
gurue (39.5. MAKHUWA)
guruginja (9.1. TELUGU)
gurugu (23.4. TELUGU)
gurùnngu (17.7. VENDA)
gurvalj (10.2. MONGOLIAN)
gurvalj (10.3. MONGOLIAN)
gurvan öngöt nil (68.1. MONGOLIAN)
gusa (17.1. NUPE-NUPE-TAKO)
gusa (17.2. NUPE-NUPE-TAKO)
gusa (17.4. NUPE-NUPE-TAKO)
gušava nivska repka (30.18. MACEDONIAN)
gusboeri (62.12. THAI)
gusi (38.5. ÉWÉ)
gusi (38.5. FON)
gusi (38.5. GEN)
gusito (38.5. GUN)
guta (17.7. MARATHI)
gutɔyɛ (40.43. LOKO)
guter Heinrich (35.5. GERMAN)
guthibira (38.22. TELUGU)
gutti beera (38.22. TELUGU)
guttibeera (38.22. TAMIL)
guttibira (38.22. TELUGU)
gutukola (24.7. VLAAMS)
gúúngár (38.13a. CAROLINIAN)
guvar (40.6. HINDI)
guvar chhuim (40.6. ORIYA)
güvercin bezelyesi (40.2. TURKISH)
güveyi otu (44.6. TURKISH)
güveyiotu (44.6. TURKISH)

guvshaakhai (52.1. MONGOLIAN)
guwar (40.6. GUJARATI)
guwar (40.6. MARATHI)
g[uwf]ng (20.6. VIETNAMESE)
guy[ees]t (2.1. VIETNAMESE)
guẓia (40.1. NUPE-NUPE-TAKO)
gūžinis kopūstas (30.27a. LITHUANIAN)
guzu-guzu (10.1. HAUSA)
guzungu (62.9. SHONA)
gvgisdi (38.3. CHEROKEE)
gvlal tulsi (44.5. HINDI)
gwaázaá (8.4. HAUSA)
gwáázáá (8.4. HAUSA)
gwaàzaá (8.9. HAUSA)
gwaàzaàmaì goòraá (8.9. HAUSA)
gwádánu (40.1. NGIZIM)
gwalagwaji (17.5. HAUSA)
gwandar beri (38.7b. HAUSA)
gwanduwa (38.4. HAUSA)
gwangwama (9.1. HAUSA)
gwargo (62.18. FULFULDE, ADAMAWA)
gwāza (8.4. HAUSA)
gweerez (36.1. BRETON)
gwel (38.4. MALAYALAM)
gwenn (17.7. KWÉYÒL)
gwezēr (24.11. KURDISH, NORTHERN)
gwiri (62.24. SHONA)
gwo bannann (16.1a. CREOLE, HAITIAN FRENCH)
gwo-lòyon (7.3. KWÉYÒL)
gyaɗa (40.39. HAUSA)
gyàɗár rúwá (64.3. HAUSA)
gyanɗa (40.39. HAUSA)
gyanth gobi (30.30. NEPALI)
gyasala-ndrowa (62.9. BAOULÉ)
gyaùrón ràkéé (17.5. HAUSA)
gyeene (7.2. TWI)
gyermekláncfű (27.45. HUNGARIAN)
gyH-guɑdl-dɑ̄ '-dei (35.4. KIOWA)
gyin (20.6. BURMESE)
gyinantwi (27.8. AKAN-AKUAPEM)
gynura (27.25. CZECH)
gynura (27.25. ENGLISH)
gynura (27.25. TURKISH)
gynura à deux couleurs (27.25. FRENCH)
gyo-janinniku (7.14. VLAAMS)
gyökér (20.2. HUNGARIAN)
gyökérpetrezselyem (24.21. HUNGARIAN)
gyömbér (20.5. HUNGARIAN)
gyömbér (20.6. HUNGARIAN)
gyöngyhagyma (7.1. HUNGARIAN)
gyouja ninniku (7.14. JAPANESE)
győzedelmes hagyma (7.14. HUNGARIAN)
gyulis (40.25. SUBANON)
gyu-myit (7.13. BURMESE)
gyvatagurkis (38.10. LITHUANIAN)
gyvelkvæler (52.1. DANISH)
ha pui k'a_ g aw$^{\vee}$(27.18. LAHU)
haa (16.1a. SAMBAL, BOTOLAN)
haa (16.1a. SAMBAL, TINÀ)
haagala (38.23. KANNADA)
haagala kaayi (38.23. KANNADA)
haako ndiyam (23.3. FULFULDE, ADAMAWA)
haako-ndiyam (23.3. FULFULDE, ADAMAWA)
haalu gumbala (38.20. KANNADA)
haarbloem (38.29. DUTCH)
haarfrüchtiger Feldslat (67.1. GERMAN)
haavubaale (16.1a. KANNADA)
haba (16.1a. SAMBAL, TINÀ)
haba (40.26. GALICIAN)
haba (40.34a. SPANISH)
haba baby (40.34b. SPANISH)
haba de burro (40.4. SPANISH)
haba de Egipto (48.1a. SPANISH)
habas (40.3. TAGALOG)
habas (40.4. FILIPINO)
habas (40.4. TAGALOG)
habb ghaziz (10.1. MALTESE)
habberxa (30.51. MALTESE)
habbidiru (17.1. KANNADA)
hab-el-rechad (30.51. ARABIC)
hab-el-samar (10.1. ARABIC)
habeş hardali (30.3. TURKISH)
habet el-baraka (30.24. ARABIC)
ḥabḥab (38.3. SOMALI)
habichuelas (40.26. FILIPINO)
habichuelas (40.26. TAGALOG)
habichuelas (40.28. TAGALOG)
habirou (38.23. FULFULDE)
hada de huerta (40.34a. SPANISH)
hada major (40.34a. SPANISH)
hades (24.13. SUNDA)
hadí mor španielsky (27.38. SLOVAK)
hadi mord španělsky (27.37. CZECH)
hadí mord španělský (27.38. CZECH)
hadí tykev (38.29. CZECH)
hadj en-nas (38.3. ARABIC)
hadla kaayi (38.29. KANNADA)

halon (23.3. TAGALOG)
halon (30.51. PANJABI, EASTERN)
hálóstök (38.22. HUNGARIAN)
hálsèn rààgóó (59.2. HAUSA)
hálsèn sààniyáá (59.2. HAUSA)
hálsèn túnkìyáá (59.2. HAUSA)
halud (20.2. BENGALI)
halunapa (23.3. TAUSUG)
haluyot (45.1. IFUGAO)
halványitó zeller (24.4. HUNGARIAN)
halwa kaddoo (38.17. PANJABI, EASTERN)
halwa kadu (38.16. PANJABI, EASTERN)
halwa-kaddoo (38.18. PANJABI, EASTERN)
halyum (13.2. HINDI)
hama (20.2. IVATAN)
hamabo (44.5. RONGA)
hama-bôfû (24.14. JAPANESE)
hama-bofu (24.14. VLAAMS)
hamana (30.45. JAPANESE)
hambalid (40.32. HINDUSTANI)
hamei (46.1. ROMANIAN)
hamei volubil (46.1. ROMANIAN)
hamel (46.1. TAJIKI)
hammahomma' (62.2. CHICKASAW)
hammahomma' champoli' (62.1. CHICKASAW)
hammahomma' homi' (62.2. CHICKASAW)
hammahomma' lakna' (62.1. CHICKASAW)
hammahomma' okchamali' (62.1. CHICKASAW)
ḫammanit ha-pekaot (27.26. HEBREW)
hampao (11.2. THAI)
hamtak (40.42. BICOLANO, CENTRAL)
hamtak (40.42. CEBUANO)
hamtak (40.42. TAGALOG)
hamtzitz (53.1. HEBREW)
ḫamẓiẓ ha-pekaʿot (53.1. HEBREW)
!hân (10.1. NAMA-DAMARA)
han qin (24.4. CHINESE, GÀN)
han qin (24.4. CHINESE, YUE)
hana (17.7. NIUE)
hanase (44.5. AJA)
hanase (44.5. GBE, WACI)
ha'nassho (44.3. SIDAMO)
hana-yasai (30.26. JAPANESE)
han-chî (36.3. CHINESE, MIN NAN)
hanchote (11.1. OROMO, WEST CENTRAL)
hanchote (11.2. OROMO, WEST CENTRAL)
hand (27.9. KASHMIRI)
handab (27.10a. ARABIC)
handalak (38.7a. TAJIKI)
handalawi (17.3. CEBUANO)
handee gedde (19.1. KANNADA)
handevâne (38.3. FARSI, WESTERN)
handevaneh (38.3. FARSI, WESTERN)
handi gedde (11.2. KANNADA)
hanḍuā-utu (17.1. MUNDARI)
hanḍuā-utu (17.2. MUNDARI)
hanḍuā-utu (17.4. MUNDARI)
handuyong a lilinaw (27.18. SAMBAL, BOTOLAN)
hanḍyālai (38.7b. PASHTO, NORTHERN)
hanekam (23.3. AFRIKAANS)
hanekam (23.4. AFRIKAANS)
hanekam (23.4. DANISH)
hanekam (23.4. DUTCH)
hanekam (23.4. NORWEGIAN)
hanekam (23.4. VLAAMS)
hanerohi (30.48. ESTONIAN)
hangplachon (27.18. THAI)
!han-i (10.1. NAMA-DAMARA)
haníibi (17.7. COMANCHE)
Hanover salad (30.23. ENGLISH)
hanoverska salata (30.23. MACEDONIAN)
hantak (40.42. BICOLANO, CENTRAL)
hantak (40.42. CEBUANO)
hantak (40.42. WÁRAY-WÁRAY)
hãnti (10.1. GOURMANTCHÉ)
hántí (10.1. SONGHAY)
hanut (24.16. TAJIKI)
hanyanisi (7.3. SHONA)
hanzal (38.3. TAJIKI)
hapa (16.1a. SAMBAL, BOTOLAN)
hapey (17.7. MANOBO)
hapu oblikas (58.2. ESTONIAN)
hápulmeloni (38.7b. HOPI)
har ghunusa (69.1. ASSAMESE)
haradi (20.2. MAITHILI)
haraid (20.2. MAITHILI)
harakala (40.42. TAGALOG)
hāra-kani (30.56. KASHMIRI)
harakputk (24.3. ESTONIAN)
haráng (62.5. WÁRAY-WÁRAY)
haras (24.13. TAGALOG)
harbara (40.5. MARATHI)
harbharaa (40.5. MARATHI)
harbi (30.22. BASQUE)
harbuz (38.3. ROMANIAN)
hardal (30.4. HEBREW)
hardal (30.5. HEBREW)

hardal (30.6. HEBREW)
hardal (30.6. TURKISH)
hardal (30.7. HEBREW)
hardal (30.8. HEBREW)
hardal (30.9. HEBREW)
hardal (30.10. HEBREW)
hardal (30.11. HEBREW)
hardal (30.12. HEBREW)
hardal (30.13. HEBREW)
hardal (30.14. HEBREW)
hardal (30.15. HEBREW)
hardal (30.16. HEBREW)
hardal (30.17. HEBREW)
hardal (30.17. TURKISH)
hardal (30.19. HEBREW)
hardal (30.20. HEBREW)
ḫardal lavan (30.57. HEBREW)
ḫardal shaḫor (30.24. HEBREW)
hardi (20.2. HINDUSTANI)
hardufnin sefaradi (27.38. HEBREW)
hardufnin spharadi (27.38. HEBREW)
haredylle (27.42. NORWEGIAN)
hare's lettuce (27.42. ENGLISH)
hare's palace (27.42. ENGLISH)
hare's thistle (27.42. ENGLISH)
håret kortstråle (27.22. DANISH)
hareuga (27.8. INDONESIAN)
Harfenstrauch (44.8. GERMAN)
hargbae (45.1. GBAYA)
harico (niébé) (40.26. WOLOF)
haricot (40.26. CAJUN FRENCH)
haricot (40.26. PULAR)
haricot à feuilles aigües (40.23. FRENCH)
haricot à feuilles angulaires (40.36. FRENCH)
haricot à oeil noir (40.43. FRENCH)
haricot adzuki (40.36. FRENCH)
haricot ailé (40.29. FRENCH)
haricot asperge (40.42. FRENCH)
haricot comun (40.26. FRENCH)
haricot d'Angola (40.37. FRENCH)
haricot de behanzin (40.15. FRENCH)
haricot de Lima (40.25. FRENCH)
haricot de Madagascar (40.3. FRENCH)
haricot de riz (40.40. FRENCH)
haricot d'Espagne (40.24. FRENCH)
haricot dolique (40.43. FRENCH)
haricot doré (40.38. FRENCH)
haricot igname (40.31. FRENCH)
haricot kilomètre (40.42. FRENCH)
haricot moucheté (40.24. FRENCH)
haricot mungo (40.38. FRENCH)
haricot papillon (40.35. FRENCH)
haricot rouge (40.4. FRENCH)
haricot royal (40.15. FRENCH)
haricot sabre (40.3. FRENCH)
haricot sabre (40.4. FRENCH)
haricot sauvage (40.40. FRENCH)
haricot tepary (40.23. FRENCH)
haricot velu (40.37. FRENCH)
haricot vert (40.26. FRENCH)
haridra (20.2. SANSKRIT)
hari-imo (11.5. JAPANESE)
harilik füüsal (62.11. ESTONIAN)
harimandhakam (40.5. TELUGU)
harimandhakamu (40.5. TELUGU)
harimantha (40.5. SANSKRIT)
hariparna (30.56. SANSKRIT)
harivettha (48.1a. SANSKRIT)
härkäpapu (40.34a. FINNISH)
háromszinü árvácska (68.1. HUNGARIAN)
harstit (27.23. HEBREW)
harum (23.3. CEBUANO)
harum (23.3. WÁRAY-WÁRAY)
harut (15.2. SUNDA)
ḫarẕit (27.23. HEBREW)
hasaguni (40.18. KANNADA)
hasdam (5.1. THAI)
hashesha almalak (24.1. ARABIC)
hasi kadale (40.5. KANNADA)
hasi shunti (20.6. KANNADA)
hasisunti (20.6. KANNADA)
hasivka orličí (2.1. CZECH)
hass twil (27.30. MALTESE)
ḫassa (27.31. HEBREW)
ḫassa (27.32. HEBREW)
hassa odit (27.29. HEBREW)
ḫassa tarbutit (27.31. HEBREW)
ḫassa tarbutit (27.32. HEBREW)
hastate-leaved pondweed (18.1. ENGLISH)
hästböna (40.34a. SWEDISH)
hastebønne (40.34a. NORWEGIAN)
hasthi karni (8.1. SANSKRIT)
hasthidantha (30.56. SANSKRIT)
hasthidanthaka (30.56. SANSKRIT)
hasthipijoo (27.26. SANSKRIT)
hastighasha (38.22. BENGALI)
hastivishaani (16.1a. SANSKRIT)
hata-sasage (40.41. JAPANESE)
hatha (66.3. GIKUYU)
hathibhekuri (62.23. ASSAMESE)

hathichak (27.16. BENGALI)
hathichak (27.16. HINDI)
hathichoke (27.16. BENGALI)
hathichoke (27.16. HINDI)
háthí-pích (27.26. HINDUSTANI)
hathipich (27.26. PANJABI, EASTERN)
hathwänhä (38.3. PANJABI, EASTERN)
hati bhekuri (62.23. ASSAMESE)
hati ghass (17.3. URDU)
hati yam (8.3. URDU)
hatichuk (27.26. HINDI)
hatiko (40.25. HOPI)
hatna lilija (50.2. SORBIAN, UPPER)
hato (45.4. LOKO)
hatseson (6.2. HEBREW)
hatsuka daikon (30.56. JAPANESE)
hattichok (27.26. HINDI)
hatzetzon hitzi (6.2. HEBREW)
hatz̤etz̤on tlat-ʿalim (6.1. HEBREW)
hau cai (30.26. CHINESE, WÚ)
haum (23.3. CEBUANO)
haung gua (38.13a. CHINESE, YUE)
haur cucuk (17.1. SUNDA)
haurf halim (30.50. URDU)
hausa aardappel (44.10. DUTCH)
hausa aardappel (44.10. VLAAMS)
hausa aardnoot (40.15. VLAAMS)
hausa groundnut (40.15. ENGLISH)
hausa jordnød (40.15. DANISH)
hausa potato (44.8. ENGLISH)
hausa potato (44.10. ENGLISH)
hausakartoffel (44.10. DANISH)
Hausakartoffel (44.10. GERMAN)
hausaknold (40.15. DANISH)
hava patatesi (11.2. TURKISH)
havaeært (40.28. DANISH)
ḥavalbal yappani (36.1. HEBREW)
havebønne (40.26. DANISH)
haveeja (24.9. KANNADA)
haveej-farangee (24.18. FARSI, WESTERN)
have-karse (30.51. DANISH)
have-kørvel (24.3. DANISH)
have-melde (35.1. DANISH)
have-portulak (59.2. DANISH)
haver rot (27.46. SWEDISH)
have-reseda (60.1. DANISH)
haverwortel (27.46. DUTCH)
haverwortel (27.46. VLAAMS)
have-skorzoner (27.38. DANISH)
havhinya (40.33. TURKISH)
havich (24.11. FARSI, WESTERN)
havlican (20.1. TURKISH)
havrerod (27.46. DANISH)
havrerot (27.46. NORWEGIAN)
havuç (24.11. TURKISH)
haway ad hingyon (36.3. IFUGAO)
hawerwortel (27.46. AFRIKAANS)
hawksbeard velvetplant (27.14. ENGLISH)
haxixa tar-rmiet (35.8. MALTESE)
haya (10.1. HAUSA)
hayaanda (40.37. SANSKRIT)
hayato uri (38.25. JAPANESE)
hayuti (38.25. IFUGAO)
haýwabjädi (38.16. TURKMEN)
hayyuti (38.26. IFUGAO)
hazeret (30.1. HEBREW)
hazereth (30.1. HEBREW)
haz̤ez̤on (6.2. HEBREW)
ḥaz̤il (62.19. HEBREW)
hɔbdɔɛ (40.43. BULLOM SO)
hɔndɔ (40.25. MENDE)
hɔndɔ (40.43. GOLA)
hɔndɔ (40.43. MENDE)
hɔndi (23.3. LOKO)
hɔndi (23.3. MENDE)
hɔndi-gbogboil (23.3. MENDE)
hɔndi-gowuli (23.3. MENDE)
hɔndi-guli (23.3. MENDE)
hɔrədɛ (38.20. PULAAR)
he (7.11. VIETNAMESE)
he (48.1. CHINESE, MANDARIN)
he bao dou (40.11. CHINESE, YUE)
he ge (40.11. CHINESE, MIN BEI)
he hua (48.1. CHINESE, MANDARIN)
he lan dou (40.28. CHINESE, GÀN)
he lan dou (40.28. CHINESE, MIN BEI)
he lan dou (40.28. CHINESE, WÚ)
he lan dou (40.28. CHINESE, YUE)
he lan shu (62.24. CHINESE, YUE)
he zhang gua (38.25. CHINESE, YUE)
head lettuce (27.31. ENGLISH)
headed cabbage (30.27a. ENGLISH)
headed cabbage (30.27b. ENGLISH)
headed cabbage (30.34. ENGLISH)
heartsease (68.1. ENGLISH)
hebbidiru (17.1. KANNADA)
hebbidru (17.1. KANNADA)
hebbu genasu (11.1. KANNADA)
hebi uri (38.29. JAPANESE)
hechima (38.22. JAPANESE)

hedyo huma (27.29. SUNDA)
heera kay (38.21. KANNADA)
heere balli (38.21. KANNADA)
heere kayi (38.21. KANNADA)
heermoes (1.1. DUTCH)
heermoes (1.1. VLAAMS)
heggenasu (11.2. KANNADA)
hegyes levelű bab (40.23. HUNGARIAN)
hei chi tou (40.37. CHINESE, MANDARIN)
hei gai (30.24. CHINESE, MANDARIN)
hei jie (30.24. CHINESE, WÚ)
hei pi bo luo men shen (27.38. CHINESE, MANDARIN)
hei pi bo luo men shen (27.38. CHINESE, WÚ)
hei tou (40.8. CHINESE, MANDARIN)
hei zi nan gua (38.15. CHINESE, MANDARIN)
hei zi nan gua (38.15. CHINESE, WÚ)
h[ej] (7.13. VIETNAMESE)
helachi sak (27.19. ASSAMESE)
helan mumpahag (8.1. IFUGAO)
helba fienu (40.32. MALTESE)
helecho comun (2.1. SPANISH)
Helmbohne (40.9. GERMAN)
helmit (45.5. HEBREW)
ḥelmit (45.5. HEBREW)
helmivalkosipuli (7.12. FINNISH)
helogan (24.4. WELSH)
helogan y dwr (24.17. WELSH)
helta (17.1. SANTALI)
helta (17.2. SANTALI)
helta (17.4. SANTALI)
helv (15.2. MUSKOGEE)
helvaci kabagi (38.16. TURKISH)
hĕm (40.28. KASHMIRI)
hemerocalis (13.3. PORTUGUESE)
hémérocalle (13.3. FRENCH)
hemerocallis (13.3. HEBREW)
hemmugulu (27.2. KANNADA)
hĕndawan (38.3. KASHMIRI)
hene (10.1. AKAN-AKUAPEM)
heneba (62.2. TWI)
hēng-chhài (23.5. CHINESE, MIN NAN)
heŋguo (38.5. KISSI, NORTHERN)
heno griego (40.32. SPANISH)
hen-yô-boku (39.3. JAPANESE)
'HeṕiH-'ei-gɑ (62.24. KIOWA)
herba boa (44.3. GALICIAN)
herba dosanores (27.5. GALICIAN)
herba moura (62.21. GALICIAN)
herbe à calalou (62.15. FRENCH)
herbe a éléphant (17.3. FRENCH)
herbe à la glace (22.1. FRENCH)
herbe a lapin (27.18. FRENCH)
herbe au verre (35.8. FRENCH)
herbe aux ânes (51.1. FRENCH)
herbe aux flèches (15.2. FRENCH)
herbe aux serpents (27.1. FRENCH)
herbe d'amour (60.1. FRENCH)
herbe de Saint-Julien (44.9. FRENCH)
herbe indienne (20.1. FRENCH)
hercai menekşe (68.1. TURKISH)
hərd (20.2. AMHARIC)
herderstasje (30.43. DUTCH)
herderstasje (30.43. VLAAMS)
herderstassie (30.43. AFRIKAANS)
hérepe (40.18. YORUBA)
Herero Buschtee (44.4. GERMAN)
hernes (40.28. ESTONIAN)
hertshoornweegbree (57.1. DUTCH)
hertshoornweegbree (57.1. VLAAMS)
hesaru (40.38. KANNADA)
hesaru kaalu (40.38. KANNADA)
hestabaunir (40.34a. ICELANDIC)
hestahofur (27.35. ICELANDIC)
hestebønne (40.34a. DANISH)
hestekjørvel (24.17. NORWEGIAN)
hɛva-gɔjɔ (62.23. MENDE)
heva-gojo (62.23. MENDE)
hew (38.20. MAORI)
héyé (11.1. KABIYÉ)
hèyè (11.5. KABIYÉ)
ḥəźawīt (27.45. MEHRI)
ḥəźəwēt (27.45. MEHRI)
hhurika (35.9. SANSKRIT)
hiam-chio-á (62.1. CHINESE, MIN NAN)
hibiscus 'ave-gavi'a (45.4. HEBREW)
hibiscus ne'echal (45.1. HEBREW)
hibiscus root (45.2. ENGLISH)
hiɛ (11.7. DANGME-KROBO)
hierba buena (44.3. SPANISH)
hierba de clavo (24.7. SPANISH)
hierba de la paciencia (58.3. SPANISH)
hierba de los canónigos (67.2. SPANISH)
hierba de Santa Bárbara (30.2. SPANISH)
hierba de Trinidad (12.1. SPANISH)
hierba del asno (51.1. SPANISH)
hierba del ciego (66.3. SPANISH)
hierba del Espíritu Santo (24.1. SPANISH)
hierba elefante (17.3. SPANISH)

hierba kudzú (40.30. SPANISH)
hierba mora (62.15. SPANISH)
hierba mora (62.21. SPANISH)
hierba sana mentha (44.3. SPANISH)
hierba santa (44.3. SPANISH)
hierba socialista (27.18. SPANISH)
hierha carmin decandra (56.2. SPANISH)
hietasinappi (30.47. FINNISH)
highlands pitpit (17.6. ENGLISH)
higuera de las Indias (32.1. SPANISH)
hiina kapsas (30.36. ESTONIAN)
hiina kapsas (30.40. ESTONIAN)
hiirekõrv (30.43. ESTONIAN)
hiiwai (62.24. MAORI)
hilan nagalgaletan (8.10. IFUGAO)
ḥilbät (40.32. HARARI)
h̤ilbe (40.32. HEBREW)
hill mustard (30.42. ENGLISH)
hillua (13.2. BENGALI)
himbjer (20.6. SORBIAN, UPPER)
hime-yurine (13.4. JAPANESE)
himsra (34.1. SANSKRIT)
h̤imẓa tarbutit (40.5. HEBREW)
himzatarbutit (40.5. HEBREW)
hind wâṉa (38.3. PASHTO, NORTHERN)
hindiba (30.1. TURKISH)
hinduva (27.10a. HINDI)
hindwana (38.3. HINDI)
hindwāna (38.3. PASHTO, NORTHERN)
hiŋgo (38.5. KISSI, NORTHERN)
hingudi (62.19. TELUGU)
hingula (62.19. MALAYALAM)
hinguli (62.19. SANSKRIT)
hingurala (11.1. SINHALA)
hinkamas (40.21. TAGALOG)
hinojo (24.12. SPANISH)
hinojo común (24.13. SPANISH)
hinojo de agua (24.17. SPANISH)
hinojo de Florencia (24.12. SPANISH)
hint darisi (17.3. TURKISH)
hint güvercin bezelyesi (40.2. TURKISH)
hint hardali (30.4. TURKISH)
hint inciri (32.1. TURKISH)
hint ispanaği (28.1. TURKISH)
hint kamişi (9.1. TURKISH)
hint kamişi (17.4. TURKISH)
hint marulu (27.29. TURKISH)
hint teresi (65.1. TURKISH)
hint yerelmasi (11.5. TURKISH)
hintinciri (32.1. TURKISH)
hintyerelmasi (11.6. TURKISH)
hịọ̀hịo (40.2. IGBO)
hiōng kouā (38.7b. CHINESE, HAKKA)
hipoi (40.18. BICOLANO, CENTRAL)
hipoy (40.18. BICOLANO, CENTRAL)
hira nasu (62.14c. JAPANESE)
hiramame (40.12. JAPANESE)
hirankhari (27.18. URDU)
hirankhuree (27.18. HINDI)
hirankhuri (27.18. HINDI)
hirbis (38.7a. SUNDA)
hirbis (38.7b. SUNDA)
hire balli (38.21. KANNADA)
hire basale (28.1. KANNADA)
hire kaayi (38.21. KANNADA)
hirekayi (38.21. KANNADA)
hirhindilik (24.8. TURKISH)
hiris (40.2. INDONESIAN)
hiris (40.2. SUNDA)
Hirschhorn-Salat (57.1. GERMAN)
Hirtentäschel (30.43. GERMAN)
hirugao (36.1. JAPANESE)
hirugao (36.1. VLAAMS)
hishi (64.3. JAPANESE)
hitti (38.10. TURKISH)
hittik (38.10. TURKISH)
hittili avare (40.9. BADAGA)
hiu-î (38.2. CHINESE, MIN NAN)
hiu-koe (38.7a. CHINESE, MIN NAN)
h̤iyar (38.13a. SOMALI)
hiyar (38.13a. TURKISH)
hiyar tüy çiçek (38.30. TURKISH)
hiyokomame (40.5. JAPANESE)
hiyu (23.3. JAPANESE)
h̤izran (17.1. HEBREW)
hjar (38.13a. MALTESE)
hjartarfi (30.43. ICELANDIC)
hjelmbønne (40.9. DANISH)
hjelmbønne (40.9. NORWEGIAN)
hjolkrona (29.1. ICELANDIC)
hjulkrone (29.1. DANISH)
hk’ aw$_{\wedge}$ zuh naw$^{\wedge}$ (40.2. LAHU)
hkui$^{\vee}$ chu$^{\vee}$ g’w$^{\vee}$ (23.3. LAHU)
hkui$^{\vee}$ shi$_{-}$ nyi^{-} (62.9. LAHU)
hlávkavá kapusta (30.34. CZECH)
hłojčka soloteje (27.31. SORBIAN, UPPER)
hłójčkaty solotej (27.32. SORBIAN, UPPER)
hmaab ntshaa (28.1. HMONG-NJUA)
hmab btsha (28.1. HMONG-NJUA)
hmab dib iab (27.24. HMONG-DAW)

hmab ntsha (28.1. HMONG-DAW)
hmarcha (62.2. MIZO)
hmazil (38.7b. MIZO)
hmel obyknovennyi (46.1. RUSSIAN)
hmel viushchiisja (46.1. RUSSIAN)
hmelj (46.1. SERBO-CROATIAN)
hmukan (7.10. NEWAR)
hmyit (17.1. BURMESE)
hmyit (17.2. BURMESE)
hmyit (17.4. BURMESE)
hneta (40.1. ICELANDIC)
hnuđk'al (30.30. ICELANDIC)
hnudselleri (24.5. ICELANDIC)
hô˙ (38.20. CHINESE, MIN NAN)
ho (44.10. JUNKUN TAKUM)
hô (48.1. CHINESE, MIN NAN)
hô˙-liân-tāu (40.28. CHINESE, MIN NAN)
hô˙-lîn-tāu (40.28. CHINESE, MIN NAN)
hô˙-lîn-tāu-jîn (40.28. CHINESE, MIN NAN)
hô˙-lô˙-pû (38.20. CHINESE, MIN NAN)
hoa b[as]t (38.4. VIETNAMESE)
hoa b[as]t (38.30. VIETNAMESE)
hoa dira (27.16. VIETNAMESE)
hoa sen (48.1a. VIETNAMESE)
ho[af]ng thu (8.10. VIETNAMESE)
ho[af]ng tinh (15.2. VIETNAMESE)
ḥoaḥ akod (27.37. HEBREW)
ḥoaḥ sefaradi (27.36. HEBREW)
hoan-á-tāu (40.28. CHINESE, MIN NAN)
hoan-béh (17.7. CHINESE, MIN NAN)
hoan-kiu-á (62.2. CHINESE, MIN NAN)
hoary basil (44.4. ENGLISH)
hoa-seng (40.1. CHINESE, MIN NAN)
hoa-seng-tāu (40.1. CHINESE, MIN NAN)
hobana (16.1a. LIMBA)
hobbejza (45.5. MALTESE)
hobo (16.1a. SHONA)
hochino (11.1. OROMO, WEST CENTRAL)
hochino (11.2. OROMO, WEST CENTRAL)
Hochland-Beinwell (29.3. GERMAN)
hocys (45.5. WELSH)
hodan (29.1. TURKISH)
hodekål (30.27a. NORWEGIAN)
hodesalat (27.31. NORWEGIAN)
hodolean (30.45. ROMANIAN)
hoe-chhài (30.26. CHINESE, MIN NAN)
hôe-hiu (24.13. CHINESE, MIN NAN)
hoehjang (24.12. KOREAN)
hoemgreeb (40.32. BENGALI)
hoenderuintjie (10.1. AFRIKAANS)
hofmelde (35.1. AFRIKAANS)
höfuđlaukur (7.3. ICELANDIC)
höfuđsalat (27.31. ICELANDIC)
høg hagebønne (40.26. NORWEGIAN)
hohe Zachenschöte (30.42. GERMAN)
ɣoɣli (38.16. DAGBANI)
hoho (62.1. SWAHILI)
hohómbro (38.13a. LADINO)
hoi (11.2. HAWAIIAN)
hoi (11.2. MARQUESAN)
hoi (11.2. NIUE)
hoi (11.2. TONGAN)
hoičo (15.1. PUMÉ)
hoja de buitre (14.1. SPANISH)
hok (17.2. LAO)
hôk (17.2. LAO)
hôk hnhaix (17.2. LAO)
hola thulasi (44.4. KANNADA)
holada menasina kaayi (62.5. KANNADA)
holadavare (40.9. KANNADA)
hô-lân t'éou (40.28. CHINESE, HAKKA)
holantan-pe (40.27. BURMESE)
holdas bab (40.25. HUNGARIAN)
holdbab (40.25. HUNGARIAN)
holgun (13.2. ARABIC)
holo (24.17. IFUGAO)
holo (27.18. IFUGAO)
hô-lô (38.20. CHINESE, MIN NAN)
holo de babalel (27.14. IFUGAO)
holo de gangagang (27.14. IFUGAO)
holo di ameleng (27.14. IFUGAO)
holog (24.17. IFUGAO)
hom (7.2. THAI)
hom (7.3. THAI)
hôm (23.3. LAO)
hombaale (16.1a. KANNADA)
hombrecillo (46.1. SPANISH)
hom-chin (7.5. THAI)
hom-daeng (7.2. THAI)
hom-huayai (7.3. THAI)
hom-lek (7.2. THAI)
homos (40.37. ARABIC)
homos shami (40.43. ARABIC)
hom-paen (7.11. THAI)
hom-paen (7.13. THAI)
hom-paenyuak (7.4. THAI)
hom-prang (7.4. THAI)
hom-tiam (7.10. THAI)
hom-ton (7.5. THAI)
homv-cate (62.2. MUSKOGEE)

homv-lane (62.1. MUSKOGEE)
hom-yai (7.3. THAI)
honaga-ao-geitô (23.3. JAPANESE)
honagone soppu (23.2. KANNADA)
honagonne (23.2. KANNADA)
honagonne soppu (23.2. KANNADA)
hónapos retek (30.56. HUNGARIAN)
hondebeien (62.21. STELLINGWERFS)
honeydew (38.11. TAGALOG)
honeydew melon (38.11. ENGLISH)
hong cai (62.19. CHINESE, MIN BEI)
hong dou (40.36. CHINESE, MANDARIN)
hong dou (40.36. CHINESE, WÚ)
hong dou kou (20.1. CHINESE, MANDARIN)
hong dou kou (20.1. CHINESE, WÚ)
hong gen cai (35.9. CHINESE, MIN BEI)
hong gen cai (35.9. CHINESE, XIANG)
hong gua (38.4. CHINESE, MANDARIN)
hong hua cai dou (40.24. CHINESE, MANDARIN)
hong hua cai dou (40.24. CHINESE, WÚ)
hong lai tou (35.4. CHINESE, YUE)
hong luo bo (24.11. CHINESE, GÀN)
hong luo bo (24.11. CHINESE, MIN BEI)
hong luo bo (24.11. CHINESE, WÚ)
hong luo bo (24.11. CHINESE, XIANG)
hong luo bo (24.11. CHINESE, YUE)
hong luo bu (24.11. CHINESE, WÚ)
hong luo kui (28.1. CHINESE, MANDARIN)
hong luo kui (28.1. CHINESE, WÚ)
hong qie (62.14c. CHINESE, MANDARIN)
hong qie (62.14c. CHINESE, WÚ)
hong qua (38.4. CHINESE, WÚ)
hong shao (36.3. CHINESE, XIANG)
hong shu (36.3. CHINESE, XIANG)
hongdangmuu (24.11. KOREAN)
hong-how (30.36. THAI)
hong-tae (30.36. THAI)
hông-tè-tāu (40.34. CHINESE, MIN NAN)
hoodingala (9.1. KANNADA)
hoodingana (9.1. KANNADA)
h[oof] d[awf]ng hai m[af]u (69.1. VIETNAMESE)
h[oof] l[oo] ba (40.32. VIETNAMESE)
h[oof]ng b[oos]i th[ar]o (27.18. VIETNAMESE)
hookosu (30.26. KANNADA)
hoom bwàx (7.2. LAO)
hoom pèèn (7.11. LAO)
hoornklaver (40.32. DUTCH)
hop (46.1. AFRIKAANS)
hop (46.1. ASSAMESE)
hop (46.1. DUTCH)
hop (46.1. STELLINGWERFS)
hop (46.1. VLAAMS)
hop shoots (46.1. ENGLISH)
hopimelooni (38.7b. HOPI)
hoppu (46.1. JAPANESE)
hops (46.1. ENGLISH)
hops (46.1. NEPALI)
hopspruite (46.1. AFRIKAANS)
hopysen (46.1. WELSH)
hora (8.1. THAI)
horaa (8.1. THAI)
horapha (44.5. THAI)
horbaz (38.3. CHECHEN)
horčica biela (30.57. SLOVENIAN)
horčica čierna (30.24. SLOVENIAN)
horčica sitinovitá (30.17. SLOVENIAN)
hořčice (30.5. CZECH)
hořčice (30.6. CZECH)
hořčice (30.7. CZECH)
hořčice (30.9. CZECH)
hořčice (30.10. CZECH)
hořčice (30.11. CZECH)
hořčice (30.12. CZECH)
hořčice (30.13. CZECH)
hořčice (30.15. CZECH)
hořčice (30.16. CZECH)
hořčice (30.17. CZECH)
hořčice (30.20. CZECH)
hořčice bílá (30.57. CZECH)
hořčice černa (30.24. CZECH)
hořčice černá (30.24. CZECH)
hořčice habaešská (30.3. CZECH)
hořčice sareptská (30.4. CZECH)
horde (38.20. PULAAR)
horenso (35.9. INDONESIAN)
horenso (35.9. JAPANESE)
hořká okurka (38.23. CZECH)
horn chestnut (64.1. ENGLISH)
horn of plenty (27.21. ENGLISH)
hornmelon (38.12. NORWEGIAN)
hornnød (64.3. DANISH)
horn-nut (64.1. ENGLISH)
horodo (40.2. ORIYA)
horozibiği (23.3. TURKISH)
horozibiği (23.4. TURKISH)
horse bean (40.3. ENGLISH)
horse bean (40.4. ENGLISH)

horse bean (40.34a. ENGLISH)
horse gram (40.16. ENGLISH)
horseradish (30.1. ENGLISH)
horsetail (1.1. ENGLISH)
hortelã (44.3. PORTUGUESE)
hortelã comum (44.3. PORTUGUESE)
hortelã verde (44.3. PORTUGUESE)
hortelon (44.3. CRIOULO, CAPE VERDE)
hosoba-kisuge (13.3. JAPANESE)
høstaralia (26.3. NORWEGIAN)
høstaster (27.17. DANISH)
høstroe (30.41a. DANISH)
hot pepper (62.2. ENGLISH)
hotei-chiku (17.4. JAPANESE)
ho̱to̱t' (38.20. SANTALI)
hottuynie (61.1. FRENCH)
houblon commun (46.1. FRENCH)
houblon lupulin (46.1. FRENCH)
houkigi (35.2. JAPANESE)
hound's berry (62.15. ENGLISH)
hourensou (35.9. JAPANESE)
houttuynie en coeur (61.1. FRENCH)
hov txob nplej (62.2. HMONG-NJUA)
hov txob nplej (62.5. HMONG-NJUA)
hova (16.1a. SHONA)
hovedkål (30.27a. DANISH)
hovedsalat (27.31. DANISH)
hpa^ hkeh^ (38.4. LAHU)
hpa‸ li¯ (18.2. LAHU)
hpa‸ naweh_ (24.7. LAHU)
hpeu˅ meu¯ hpeu shi_ (38.1. LAHU)
hpeu˅ meu¯ shi_ (38.18. LAHU)
hpui˅ a¯ le te (40.18. LAHU)
hpui˅ mui¯ shi_ (38.18. LAHU)
hpui˅ shu_ (7.10. LAHU)
hrach (40.28. SLOVAK)
hrách setý (40.28. CZECH)
hrách setý pravý (40.28. CZECH)
hrách zahradny (40.28. CZECH)
hrachor hlíznatý (40.11. CZECH)
hrachor setý (40.10. CZECH)
hrachor siaty (40.10. SLOVAK)
hrafnaklukka (30.44. ICELANDIC)
hrášok (40.28. SLOVAK)
hrean (30.1. ROMANIAN)
hređka (30.56. ICELANDIC)
hren (30.1. CROATIAN)
hren (30.1. SLOVENIAN)
hreniţă creson (30.51. ROMANIAN)
hridyagandha (24.9. SANSKRIT)
hrimblađka (35.1. ICELANDIC)
hrizantema (27.23. SERBO-CROATIAN)
hrizantema (27.23. TURKMEN)
hroch (40.28. SORBIAN, UPPER)
hromak (13.2. SORBIAN, UPPER)
hróšatko (40.28. SORBIAN, UPPER)
'H '-sɑhyei-gɑ (40.17. KIOWA)
hsi hsiang ts'ung (7.11. CHINESE, MANDARIN)
hsi kwa (38.3. CHINESE, MANDARIN)
hsi yin tu huang kua (38.6. CHINESE, MANDARIN)
hsiang tsai (44.5. CHINESE, MANDARIN)
hsiao pien ton (40.12. CHINESE, MANDARIN)
hsüch wei ts'ai (24.3. CHINESE, MANDARIN)
hto-pong (40.29. BURMESE)
hu cong (7.2. CHINESE, YUE)
hu dou (40.35. CHINESE, WÚ)
hu dou (40.37. CHINESE, YUE)
hu gua (38.13a. CHINESE, MANDARIN)
hu gua (38.13a. CHINESE, WÚ)
hu gua (38.20. CHINESE, WÚ)
hu gua (38.20. CHINESE, XIANG)
hu gua (38.20. CHINESE, YUE)
hu gua dou (40.18. CHINESE, GÀN)
hu gua dou (40.18. CHINESE, YUE)
hu hsiang (24.12. CHINESE, MANDARIN)
hu lo po (24.11. CHINESE, MANDARIN)
hu lu ba (40.32. CHINESE, MANDARIN)
hu lu ba (40.32. CHINESE, WÚ)
hu luo bo (24.11. CHINESE, MANDARIN)
hu luo bo (24.11. CHINESE, WÚ)
hu pi hua (12.1. CHINESE, WÚ)
hu sui (24.9. CHINESE, MIN BEI)
hu ts'ung (7.2. CHINESE, MANDARIN)
hu zi (38.20. CHINESE, GÀN)
hu zi (38.20. CHINESE, WÚ)
hu zi (38.20. CHINESE, XIANG)
hua bo he (44.6. CHINESE, MANDARIN)
hua bo he (44.6. CHINESE, WÚ)
hua cai (30.26. CHINESE, GÀN)
hua cai (30.26. CHINESE, MIN BEI)
hua cai (30.26. CHINESE, XIANG)
hua cai (30.26. CHINESE, YUE)
hua je jie (30.16. CHINESE, YUE)
hua sheng (40.1. CHINESE, MANDARIN)
hua sheng (40.1. CHINESE, WÚ)
hua sheng guo (40.1. CHINESE, XIANG)

hua ye cai (30.26. CHINESE, MANDARIN)
hua ye cai (30.26. CHINESE, WÚ)
hua ye cai (30.26. CHINESE, YUE)
hua ye gai (30.16. CHINESE, MANDARIN)
hua ye jie (30.16. CHINESE, WÚ)
hua yeh ts'ai (30.26. CHINESE, MANDARIN)
huai shan (11.6. CHINESE, MIN BEI)
huai shan (11.6. CHINESE, YUE)
huang du (11.2. CHINESE, MANDARIN)
huang du (11.2. CHINESE, WÚ)
huang gua (38.13a. CHINESE, GÀN)
huang gua (38.13a. CHINESE, WÚ)
huang hua cai (13.3. CHINESE, GÀN)
huang hua cai (13.3. CHINESE, XIANG)
huang hua cai (13.3. CHINESE, YUE)
huang hua chi (27.36. CHINESE, MANDARIN)
huang kua (38.13a. CHINESE, MANDARIN)
huang qiu kui (45.1. CHINESE, MANDARIN)
huang qiu kui (45.1. CHINESE, WÚ)
huang qiu kui (45.1. CHINESE, YUE)
huang qua (38.30. CHINESE, XIANG)
huang shu kui (45.2. CHINESE, WÚ)
huang ya bai (30.40. CHINESE, YUE)
huang ya bei cai (30.40. CHINESE, WÚ)
huang ya cai (30.40. CHINESE, GÀN)
huapaekua (40.21. THAI)
huarhui (40.25. QUECHUA, CUSCO)
hüatsint (13.5. ESTONIAN)
hubendin (62.9. LIMBA)
húbéy (34.2. SONGHAY)
hudansōu (35.3. JAPANESE)
hudhër (7.10. ALBANIAN, TOSK)
hudrë (7.10. ALBANIAN, TOSK)
hudusuru (38.20. PULAAR)
hue (38.20. MAORI)
hue puo (38.1. MARQUESAN)
hughër (7.10. ALBANIAN, TOSK)
hui cai (35.5. CHINESE, YUE)
hui luo bo (44.4. CHINESE, MANDARIN)
hui luo le (44.4. CHINESE, WÚ)
hui tiao cai (35.5. CHINESE, WÚ)
huisampilla (24.6. QUECHUA, CUSCO)
huisisai (53.1. SPANISH)
huiwaiwaka (62.24. MAORI)
hukhukin di gayang (8.4. IFUGAO)
hukosu (30.26. KANNADA)
ḥūl (30.27a. HARARI)
hulablub (62.21. CEBUANO)
hulabtog (62.1. WÁRAY-WÁRAY)
hulabtog (62.2. WÁRAY-WÁRAY)
hulankeeriya (15.2. SINHALA)
hulaya (34.2. AKLANON)
hulbá (40.32. HINDUSTANI)
huldimurga (23.4. BENGALI)
hulpa (23.3. TURKMEN)
hulul (34.2. HINDI)
ḥumʿa (58.4. HEBREW)
ḥumʿat ha-ginna (58.2. HEBREW)
humal (46.1. ESTONIAN)
humala (46.1. FINNISH)
humall (46.1. ICELANDIC)
humbughl (62.15. TIV)
humle (46.1. NORWEGIAN)
humle (46.1. SWEDISH)
humleskudd (46.1. NORWEGIAN)
hummus (40.5. ARABIC)
humped yam (11.1. ENGLISH)
humulas (46.1. URDU)
humus (40.5. HEBREW)
hunangsmelona (38.11. ICELANDIC)
hundbo (27.9. TAJIKI)
hundiuba (40.14. ESTONIAN)
hung gua (38.13a. CHINESE, XIANG)
huo jue cai (2.1. CHINESE, MANDARIN)
huong shu (36.3. CHINESE, YUE)
hupi (8.4. TAGALOG)
hur hure (34.2. NEPALI)
hurahuran boka (27.18. HAUSA)
hurali (40.16. FRENCH)
hurali (40.16. KANNADA)
hurali kayi (40.26. KANNADA)
hurboozoo (38.3. ROMANI, BALKAN)
!hureb (40.9. NAMA-DAMARA)
hurf (30.51. HINDI)
hurrieq (66.3. MALTESE)
hùrud (20.2. AFAR)
hurūd (20.2. SOMALI)
Husarenknopf (27.2. GERMAN)
husi (32.1. COMANCHE)
husk tomato (62.13. ENGLISH)
h[us]ng gi[or]i (44.5. VIETNAMESE)
Hussarenknopf (27.42. GERMAN)
*h*utamba (62.9. LIMBA)
huupi̱hnàa (17.5. COMANCHE)
huvudkål (30.27a. SWEDISH)
huvudsallat (27.31. SWEDISH)
huwi (11.1. INDONESIAN)
huwi (11.1. SUNDA)
huwi blicik (11.1. SUNDA)

huwi boled (36.3. SUNDA)
huwi buah (11.2. SUNDA)
huwi butul (11.5. SUNDA)
huwi buwah (11.1. SUNDA)
huwi ceker (11.5. SUNDA)
huwi cucuk (11.5. SUNDA)
huwi dangdeur (39.4. SUNDA)
huwi gandul (11.1. SUNDA)
huwi hiris (40.21. SUNDA)
huwi jahe (11.5. SUNDA)
huwi jendral (39.4. SUNDA)
huwi kamayung (11.5. SUNDA)
huwi kawoyang (11.5. SUNDA)
huwi kentang (44.8. SUNDA)
huwi kumeli (62.24. SUNDA)
huwi landak (11.5. SUNDA)
huwi mantang (36.3. SUNDA)
huwi nagri (11.1. SUNDA)
huwi taropong (11.5. SUNDA)
huwi upas (11.1. SUNDA)
huwi walanda (62.24. SUNDA)
hu[yf]nh tinh (15.2. VIETNAMESE)
huz̲ha (7.10. PASHTO, NORTHERN)
hvaan baanz (39.5. LAO)
hvidkål (30.27a. DANISH)
hvid-løg (7.10. DANISH)
hvit lupin (40.14. NORWEGIAN)
hvitk’al (30.27a. ICELANDIC)
hvitkål (30.27a. NORWEGIAN)
hvitlaukur (7.10. ICELANDIC)
hvömm (24.1. ICELANDIC)
hvosch (1.1. KAZAKH)
hvost poljave (1.1. BELARUSAN)
hwà ‘i: mu:z (11.2. LAO)
hwà katha:d (11.5. LAO)
hwaechk’a (17.7. CHECHEN)
hwediɛ (17.3. AKAN-ASANTE)
hwitare (62.9. FULFULDE, ADAMAWA)
hwitare bagarmi (62.18. FULFULDE, ADAMAWA)
hý taý noýbasy (40.43. TURKMEN)
hyacinth bean (40.9. ENGLISH)
hyasinttipapu (40.9. FINNISH)
hydrocotyle (24.15. ENGLISH)
hydrocotyle asiatique (24.7. FRENCH)
hydropire (19.1. FRENCH)
hyin-nu-nwe (23.3. BURMESE)
hyoutan (38.20. JAPANESE)
hyrdetaske (30.43. DANISH)
hyvänheikinsavikka (35.5. FINNISH)
hyÿar (38.13a. TURKMEN)
i¯ cu ka naw^ (40.43. LAHU)
ị mẹ́ẹ̣́lẹ́ị́ tuo (59.2. IZON)
i¯ mo^ k’o_ (38.20. LAHU)
i n’gol (17.1. SERER-SINE)
i shakarqand (36.3. HINDUSTANI)
iabéré-koko (8.4. PULAAR)
iaponiki (36.1. GREEK)
iaponiki ofioides kolohynthis (38.30. GREEK)
iba (43.1. SERER-SINE)
ìbábá (40.3. EFIK)
ibara-basu (50.1. JAPANESE)
ibardji (63.1. GOURMANTCHÉ)
ibardji-djari (35.9. GOURMANTCHÉ)
iberou gni (11.3. BÉTÉ)
ibhanana (16.1a. XHOSA)
ibhapipphalí (62.1. PALI)
ibhatata (36.3. XHOSA)
ibhithi (35.4. XHOSA)
ibhithrudi (35.4. ZULU)
ibhitruthi (35.4. XHOSA)
ibia (53.1. SPANISH)
ibiš (45.1. SLOVAK)
ibiš jedlý (45.1. SLOVAK)
ibisco (45.1. ITALIAN)
ibisco da fibra (45.4. ITALIAN)
ibišek (45.1. CZECH)
ibišek jedlý (45.1. CZECH)
ibo ìwá (39.4. EFIK)
ibourdi (16.1a. GOURMANTCHÉ)
ibun ihi (40.1. TIV)
ice plant (22.1. ENGLISH)
icegher (38.7b. TIV)
icegher-nyian (38.18. TIV)
icegher-pupuu (38.18. TIV)
icok (36.3. LUO)
icok munu (62.24. LUO)
icok pii (36.2. LUO)
idabongmale (10.1. KONKOMBA)
idabunyal amo (10.1. KONKOMBA)
idar (40.28. BASQUE)
idaya (7.2. BONTOC, CENTRAL)
idčin benul (20.6. KONKOMBA)
idídùòt ḿkpọ̀n̈ mbàkárá (8.9. EFIK)
i-dìm (38.18. BUDIK)
ìdò (9.1. YORUBA)
idos antidiou (27.9. GREEK)
idos kolokynthis (38.22. GREEK)
idos kolokynthis (38.23. GREEK)
idos kolokynthis (38.24. GREEK)

idos mpanana (16.1a. GREEK)
idos peponiou (38.26. GREEK)
idos peponos (38.7a. GREEK)
idos peponos (38.7b. GREEK)
idos peponos (38.8. GREEK)
idos peponos (38.9. GREEK)
idos peponos (38.10. GREEK)
idos peponos (38.11. GREEK)
idree (27.9. MONGOLIAN)
idree (27.10a. MONGOLIAN)
idù (45.1. YALA)
ìdŭ (62.5. OGBIA-AGHOLO)
idúu (62.5. MANDJAK)
idyosmos o glichon (44.2. GREEK)
idyosmos o prasiyos (44.3. GREEK)
ié (38.16. BAMANANKAN)
ié (38.18. BAMANANKAN)
iɛge (39.4. BULLOM SO)
iɛke (39.4. BULLOM SO)
ierityisi (40.28. XHOSA)
iertyisi (40.28. XHOSA)
ifirombe (17.7. NYAKYUSA-NGONDE)
ifouamani (38.21. GOURMANTCHÉ)
ifteri otu (2.1. TURKISH)
ifunamni (45.2. GOURMANTCHÉ)
ígaí (39.4. EDO)
igál ìhól (62.5. ICEVE-MACI)
ìgángán (11.3. YORUBA)
ìgángán alo (11.3. YORUBA)
igara (40.43. IGBO)
igari (39.4. EDO)
igari-ikeripia (39.4. EDO)
igari-makole (39.4. EDO)
i̱ga̱so alôm lôhô (13.3. TIV)
igba (38.20. EDE NAGO)
igba (38.20. YORUBA)
igba aja (62.17. EDE NAGO)
igba aja (62.17. YORUBA)
igbagba (62.14b. YORUBA)
ìgbágba (62.18. YORUBA)
igbáŋ (17.5. BULLOM SO)
ìgbá-ọ̀sùn (62.18. YORUBA)
ìgbá-òyibo (62.19. YORUBA)
igbáyinrin ẹlẹgun (62.23. YORUBA)
ígbẹ̄ñbe (45.1. AKPA)
i̱gbi̱se (44.8. TIV)
igbon ịkịrai (17.1. IZON)
i̱gbough a̱hi (40.39. TIV)
ịgbụrụ (39.4. IGBO)
íghekpe (40.18. EDO)
igi (17.7. BINUKID)
igi (17.7. BONTOC, CENTRAL)
ígìerù (11.7. EDO)
ịgịná (62.1. IZON)
iginsumpa (23.3. WÁRAY-WÁRAY)
ígìorùa (11.1. EDO)
ìgírì (40.1. ICEVE-MACI)
igname (11.1. ITALIAN)
igname (11.1. VLAAMS)
igname (11.2. ITALIAN)
igname (11.3. ITALIAN)
igname (11.5. ITALIAN)
igname (11.7. ITALIAN)
igname ailée (11.1. FRENCH)
igname blanche (11.7. FRENCH)
igname bulbifere (11.2. FRENCH)
igname couche-couche (11.8. FRENCH)
igname couche-couche (11.8. ITALIAN)
igname Chinese (11.6. ITALIAN)
igname de Chine (11.1. FRENCH)
igname de Chine (11.5. FRENCH)
igname de Chine (11.6. FRENCH)
igname de la Guinée (11.3. FRENCH)
igname de la Guinée (11.7. FRENCH)
igname de la Martinique (11.8. FRENCH)
igname des blancs (11.5. FRENCH)
igname douce (11.8. FRENCH)
igname jaune (11.3. FRENCH)
igname pas possible (11.5. FRENCH)
igname patate (11.2. FRENCH)
igname patate (11.5. FRENCH)
igname pays-nègre (11.3. FRENCH)
igname pousse en l'air (11.2. FRENCH)
igname salvatico (11.4. ITALIAN)
igname sauvage (11.4. FRENCH)
igoũo kambi (20.6. GOURMANTCHÉ)
igrin pepa (62.1. XHOSA)
igssûssq (62.24. INUKTITUT, WEST GREENLANDIC)
igyór (23.4. TIV)
ihi (40.1. TIV)
ihiehie (40.31. EDO)
ihiri (44.5. EDO)
ihmekukka (49.1. FINNISH)
ihmepensas (39.3. FINNISH)
ihname (8.4. CRIOULO, CAPE VERDE)
ìhọ̀họ̀ (10.1. TIV)
'ii hin (18.1. LAO)
iilo (40.25.SAMBAL, BOTOLAN)
iioningdauw (38.11. VLAAMS)

Italiaanse veldsla (67.1. VLAAMS)
Italiaanse venkel (24.12. VLAAMS)
Italia-kanran (30.31. JAPANESE)
Italian corn-salad (67.1. ENGLISH)
Italian parsley (24.20. ENGLISH)
Italian turnip (30.41c. ENGLISH)
Italienisch Petersilie (24.20. GERMAN)
Italienisches Rapünzchen (67.1. GERMAN)
italijanska motovilka (67.1. MACEDONIAN)
italijanski magdonos (24.20. MACEDONIAN)
italiko petroselino (24.20. GREEK)
italyan maydanozu (24.20. TURKISH)
itambe za miti (40.2. TONGA)
ì-tān à-kpə̄ (40.25. BETE-BENDI)
itania (24.9. LUYIA)
itapile (62.24. XHOSA)
iteniphu (30.41a. XHOSA)
ithanga (38.14. XHOSA)
ithanga (38.14. ZULU)
ithanga (38.16. XHOSA)
ithanga (38.17. XHOSA)
ithanga (38.18. XHOSA)
ithenephu (30.41a. XHOSA)
itikna (17.1. SUMO-MAYANGNA)
itimon (38.7. ILOCANO)
iting-iting (23.3. CEBUANO)
ì-tìtìn ì-bwá (62.9. BETE-BENDI)
ito (38.5. YORUBA)
itoɔ (38.5. EDE NAGO)
itoɔ (38.5. YORUBA)
ìtòmátò (62.9. YALA)
itongomane (40.1. ZULU)
itóò (38.5. YORUBA)
itoro (38.5. YORUBA)
itórò (38.5. YORUBA)
ìtórò (38.5. YORUBA)
itshikili (27.10a. XHOSA)
itshilisi (62.2. XHOSA)
itswede (7.3. XHOSA)
itti' tiyya' (7.2. CHICKASAW)
itti' tiyya' (40.24. CHICKASAW)
itti' tiyya' (40.26. CHICKASAW)
itumata (62.9. XHOSA)
itumato (62.9. XHOSA)
i-tyirir (38.18. WAMEY)
ityuna (27.24. TIV)
iucca (39.4. ITALIAN)
iû-chhang (7.3. CHINESE, MIN NAN)
iută (63.1. ROMANIAN)
ivambèyon (39.4. TIV)
ivatala (38.3. XHOSA)
i-vèl (40.39. BASSARI)
ivigaujaq (30.27a. INUKTITUT, EASTERN CANADIAN)
ivisdos o edodimos (45.1. GREEK)
iviskos o avelmóschos (45.2. GREEK)
ivy gourd (38.4. ENGLISH)
ìwá (39.4. EFIK)
iwa (39.4. IGBO)
ìwá únèñé (39.4. EFIK)
íxiavo (45.1. EDO)
ix-xalloti (7.2. MALTESE)
iya (38.14. CHEROKEE)
iya (38.16. CHEROKEE)
iya (38.17. CHEROKEE)
iya (38.18. CHEROKEE)
iyabasi (7.4. HUNGARIAN)
iyabiye (16.1a. KABIYÉ)
iyaġezizi (16.1a. DAKOTA)
iyalọ (62.19. SANGA)
iyan-ebo (36.3. EDO)
ịyáụ́ (40.31. IZON)
iyavai (40.2. TAMIL)
iyaya (36.3. HAUSA)
iyeregh (17.5. TIV)
íyogh (11.3. TIV)
íyòkhó (8.4. EDO)
íyòkhó àkàrá (8.9. EDO)
iyoptaiyeyapizi (58.1. DAKOTA)
ịyọ́rọ̀bùrú (11.1. IZON)
iyou̯gh (11.1. TIV)
iyou̯gh (11.2. TIV)
i-ywù (38.20. WAMEY)
iy-zélandi spenót (22.2. HUNGARIAN)
izambane (62.24. ZULU)
izdankova nivska repka (30.15. MACEDONIAN)
izgii (30.48. TURKISH)
ịzọ́n úsí (17.3. IZON)
ja (40.17. ZUNI)
ja'faree (24.19. FARSI, WESTERN)
ja$_{\wedge}$ meu$^{\vee}$ naw$^{\wedge}$ (40.12. LAHU)
jaalinee (38.21. SANSKRIT)
jääpäivikki (22.1. FINNISH)
jaartsboon (40.42. AFRIKAANS)
jaat (38.22. SUNDA)
jaat (40.29. SUNDA)
jabbo (7.3. MANDING-MANDINKA)
jabéré (8.4. BAMANANKAN)
jabɛrɛ (8.9. PULAAR)

jabéré-koko (8.4. PULAAR)
jabɛrɛ-koko (8.9. PULAAR)
jabero (8.4. MANDING-MANDINKA)
jabô (7.2. MANDING-MANDINKA)
jabung (30.17. SUNDA)
jacatupé das Antilhas (40.22. PORTUGUESE)
jacinthe musquée (13.5. FRENCH)
jacinto bravo (13.5. PORTUGUESE)
jacinto das searas (13.5. PORTUGUESE)
jacinto racimoso (13.5. SPANISH)
jack bean (40.3. ENGLISH)
Jackbohne (40.3. GERMAN)
jackbønne (40.3. DANISH)
jackbønne (40.3. NORWEGIAN)
jackboon (40.3. AFRIKAANS)
jackboon (40.3. VLAAMS)
jacón (27.39. SPANISH)
jada (40.18. SANSKRIT)
jadrenets (37.1. RUSSIAN)
jae (20.6. INDONESIAN)
jae (20.6. JAVANESE)
jafari (24.19. TAJIKI)
jafari farangi (24.3. FARSI, WESTERN)
jafari itolioi (24.20. TAJIKI)
jagaimo (62.24. JAPANESE)
jaghenephala (62.15. SANSKRIT)
jagh-jagh (30.43. TAJIKI)
jaghwandër (35.4. PASHTO, NORTHERN)
jagoboon (40.39. AFRIKAANS)
jagong (17.7. MALAY)
jagong (17.7. SUNDA)
jagung (17.7. INDONESIAN)
jagung (17.7. JAVANESE)
jagung manis (17.7. INDONESIAN)
jagung manis (18.2. INDONESIAN)
jahe (20.6. INDONESIAN)
jahe (20.6. JAVANESE)
jahe (20.6. SUNDA)
jahi (20.6. MALAY)
jahubanaan (16.1a. ESTONIAN)
jai shan qie (45.3. CHINESE, WÚ)
jaibereh (8.4. PULAAR)
jạil (17.1. SANTALI)
jạil (17.2. SANTALI)
jạil (17.4. SANTALI)
jaiong (56.1. ASSAMESE)
jajang betung (17.2. JAVANESE)
jakato (62.19. KRIO)
jakato (62.19. PULAAR)
jakato-brurɛ (62.18. PULAAR)
jakbicharu (13.2. KOREAN)
jakon (27.39. CZECH)
jal jimbua (23.2. HINDI)
jala jambu (23.1. NEPALI)
jalajambo (23.2. GUJARATI)
jalakantaka (64.3. SANSKRIT)
jalakiya (62.5. ASSAMESE)
jalaphala (64.3. SANSKRIT)
jalashaya (64.3. SANSKRIT)
jalasoochi (64.3. SANSKRIT)
jalatiin songino (7.14. MONGOLIAN)
jalavalli (64.3. SANSKRIT)
jalbhātā (62.9. MUNDARI)
jalbɪz (44.3. KIRGHIZ)
jalguun tsetseg (34.2. MONGOLIAN)
jalini (38.21. SANSKRIT)
jalodbhava (48.1a. SANSKRIT)
jalokello (34.3. FINNISH)
ǰaḷu (20.2. TULU)
jam (11.1. AFRIKAANS)
jam (11.2. AFRIKAANS)
jam (62.12. NYANJA)
jam čibulkatý (11.2. CZECH)
jam čínský (11.6. CZECH)
jamaan sakhal (27.46. MONGOLIAN)
Jamaican sorrel (45.4. ENGLISH)
jamaii ya yungiyungi (13.3. SWAHILI)
jamaika (38.3. HINDI)
jamaikahibisk (45.4. NORWEGIAN)
jamaku (28.1. FON)
jamba (30.48. PANJABI, EASTERN)
jamba (45.5. MONGOLIAN)
jambaljoohi (44.5. FULFULDE, ADAMAWA)
jambeera (44.4. SANSKRIT)
jamboon (40.20. AFRIKAANS)
jamboon (40.21. AFRIKAANS)
jamboon (40.31. AFRIKAANS)
jambu (27.2. PORTUGUESE)
jạmi (23.3. SANTALI)
jamou (38.14. KWÉYÒL)
jamou (38.16. KWÉYÒL)
jams (11.1. SERBO-CROATIAN)
jams (11.1. SLOVAK)
jams (11.1. SWEDISH)
jams (11.2. SERBO-CROATIAN)
jams (11.2. SLOVAK)
jams (11.2. SWEDISH)
jams (11.3. SERBO-CROATIAN)
jams (11.3. SLOVAK)
jams (11.3. SWEDISH)

jams (11.4. SERBO-CROATIAN)
jams (11.4. SLOVAK)
jams (11.4. SWEDISH)
jams (11.5. SERBO-CROATIAN)
jams (11.5. SLOVAK)
jams (11.5. SWEDISH)
jams (11.6. SERBO-CROATIAN)
jams (11.6. SLOVAK)
jams (11.6. SWEDISH)
jams (11.7. SERBO-CROATIAN)
jams (11.7. SLOVAK)
jams (11.7. SWEDISH)
jams (11.8. SERBO-CROATIAN)
jams (11.8. SLOVAK)
jams (11.8. SWEDISH)
jamsbønne (40.21. NORWEGIAN)
jamsi (11.1. SRANAN)
jamss (11.1. ESTONIAN)
jamss (11.2. ESTONIAN)
jamss (11.3. ESTONIAN)
jamss (11.4. ESTONIAN)
jamss (11.5. ESTONIAN)
jamss (11.6. ESTONIAN)
jamss (11.7. ESTONIAN)
jamss (11.8. ESTONIAN)
jamssi (11.6. FINNISH)
jamssipapu (40.20. FINNISH)
jamszgyökér (11.1. HUNGARIAN)
jamszgyökér (11.2. HUNGARIAN)
jamszgyökér (11.3. HUNGARIAN)
jamszgyökér (11.4. HUNGARIAN)
jamszgyökér (11.5. HUNGARIAN)
jámszgyökér (11.6. HUNGARIAN)
jamszgyökér (11.7. HUNGARIAN)
jamszgyökér (11.8. HUNGARIAN)
jamu (62.12. NYANJA)
jàmúk (45.4. BEROM)
jaṅ tirṛạ (40.30. SANTALI)
ján wáákéé (40.43. HAUSA)
jāṇa paparde (2.1. LATVIAN)
janapriya (24.9. SANSKRIT)
janar (17.7. BENGALI)
jandae (33.1. KOREAN)
jangali bhendi (45.2. MARATHI)
jangali mewa (62.12. NEPALI)
jangata (40.1. JOLA-FONYI)
janggel (17.7. JAVANESE)
jangli bhendi (45.2. MARATHI)
jangli ghuya (8.4. HINDI)
jangli kaasani kanpool (27.45. HINDI)
jangli suran (8.3. HINDI)
jangli suran (8.3. URDU)
janglitulasi (44.5. URDU)
jangu (38.16. TUMBUKA)
jangu (38.20. TUMBUKA)
janhi (38.21. ORIYA)
jani chungu (27.24. SWAHILI)
jani chungu (27.44. SWAHILI)
janjiwêlo (20.6. SIDAMO)
janka (24.12. SWEDISH)
janli salad (27.33. URDU)
jantung pisang (16.1b. INDONESIAN)
jantung pisang (16.1b. MALAY)
janumarak (23.3. HINDI)
japan sedergene (36.1. MONGOLIAN)
Japanaamse winde (36.1. DUTCH)
Japanese angelica (24.2. ENGLISH)
Japanese angelica (26.3. ENGLISH)
Japanese aralia (26.3. ENGLISH)
Japanese arrowroot (40.30. ENGLISH)
Japanese artichoke (44.11. ENGLISH)
Japanese aster (27.28. ENGLISH)
Japanese bellflower (34.3. ENGLISH)
Japanese bindweed (36.1. ENGLISH)
Japanese bunching onion (7.5. ENGLISH)
Japanese butterbur (27.35. ENGLISH)
Japanese cucumber (38.10. ENGLISH)
Japanese engelwortel (24.2. AFRIKAANS)
Japanese farfugium (27.20. ENGLISH)
Japanese flowering fern (3.2. ENGLISH)
Japanese garlic (7.7. ENGLISH)
Japanese greens (27.23. ENGLISH)
Japanese hornwort (24.10. ENGLISH)
Japanese horseradish (30.58. ENGLISH)
Japanese oliesaad (44.7. AFRIKAANS)
Japanese snake gourd (38.30. ENGLISH)
Japanese wild ginger (20.5. ENGLISH)
Japanese wilde gemmer (20.4. AFRIKAANS)
japanesi toriman (27.8. SRANAN)
Japani ada (20.5. ASSAMESE)
Japani angelica (24.2. ASSAMESE)
japani lasan (7.7. URDU)
japani phol wala fern (3.2. URDU)
Japani tiyoh (38.10. ASSAMESE)
Japaningwer (20.5. GERMAN)
japaninruttojuuri (27.35. FINNISH)
Japanische Aralie (26.3. GERMAN)
Japanische Kartoffel (44.11. GERMAN)
japanische Pestwurz (27.35. GERMAN)
Japanischer Angelicabum (26.3. GERMAN)

Japanknollen (44.11. GERMAN)
japanlykt (62.11. NORWEGIAN)
japanpestrot (27.35. NORWEGIAN)
japanræddike (30.54. DANISH)
Japans hoefblad (27.35. DUTCH)
Japans hoornkruid (24.10. VLAAMS)
Japans klokje (34.3. VLAAMS)
Japansc bosui (7.5. VLAAMS)
Japanse andoorn (44.11. DUTCH)
Japanse andoorn (44.11. VLAAMS)
Japanse bloeiende varen (3.2. VLAAMS)
Japanse engelwortel (24.2. VLAAMS)
Japanse komkommer (38.10. VLAAMS)
Japanse slangkomkommer (38.30. VLAAMS)
Japanse wilde gember (20.5. VLAAMS)
japansk burre (27.5. DANISH)
japansk foelfod (27.20. DANISH)
japansk hestehov (27.35. DANISH)
japansk snorle (36.1. DANISH)
Japon aralyasi (26.2. TURKISH)
Japon çiçekli (3.2. TURKISH)
Japon enginari (44.11. TURKISH)
Japon öksürükotu (27.35. TURKISH)
Japon sarmisaği (7.7. TURKISH)
Japoninis imbieras (20.5. LITHUANIAN)
japonska angelika (24.2. MACEDONIAN)
japonska artičoka (44.11. MACEDONIAN)
japonska dinja (38.10. MACEDONIAN)
japonska vasabija (30.58. MACEDONIAN)
japonski lopuš (27.35. MACEDONIAN)
jar áyaá (10.1. HAUSA)
jara' (17.1. SOMALI)
jara' (17.2. SOMALI)
jara' (17.4. SOMALI)
jaramago amarillo (30.47. SPANISH)
jaramago oriental (30.42. SPANISH)
jararaca mirim (8.3. PORTUGUESE)
jarđhneta (40.1. ICELANDIC)
jardnøtt (40.1. NORWEGIAN)
jaringan (27.8. INDONESIAN)
jaringo (56.1. NEPALI)
jarmuz (30.35. CZECH)
jarmuż (30.35. POLISH)
jasımık (40.12. KIRGHIZ)
jatala (20.4. SANSKRIT)
jati bengani (62.19. ASSAMESE)
jati lao (38.20. ASSAMESE)
jatilas (38.20. ASSAMESE)
jättepumpa (38.16. SWEDISH)
jättialokasia (8.1. FINNISH)
jättikurpitsa (38.16. FINNISH)
jättiläis kurpitsa (38.16. FINNISH)
jau dari (62.18. FULFULDE, ADAMAWA)
jauhobanaani (16.1a. FINNISH)
jaunzelandes spināti (22.2. LATVIAN)
jaunzēlandes spināts (22.2. LATVIAN)
jau-se (38.14. NEWAR)
jau-se (38.16. NEWAR)
jau-se (38.17. NEWAR)
jau-se (38.18. NEWAR)
javanski vinocvet (24.17. MACEDONIAN)
javâri (17.7. FARSI, WESTERN)
Javaspinat (59.3. GERMAN)
javshan (27.7. TAJIKI)
jawār (17.7. KATI)
jawdra (17.7. HINDI)
jawer hayam (23.4. SUNDA)
jawer kotok (23.4. SUNDA)
jawo (38.14. NEWAR)
jawo (38.16. NEWAR)
jawo (38.17. NEWAR)
jawo (38.18. NEWAR)
jawo-si (38.14. NEWAR)
jawo-si (38.16. NEWAR)
jawo-si (38.17. NEWAR)
jawo-si (38.18. NEWAR)
jaxatu (62.14a. WOLOF)
jazar (24.11. ARABIC)
jebubug basu (11.1. JAVANESE)
jebubug basu (11.2. JAVANESE)
jedilna buča (38.16. SLOVENIAN)
jedlá repa (35.4. SLOVENIAN)
jee (38.18. MANDING-MANDINKA)
jee dee ee-ah (38.23. HMONG-DAW)
jeevana (40.5. SANSKRIT)
jeftuni (44.5. ARABIC)
jeginga (62.5. BASSARI)
jelarut (15.2. JAVANESE)
jelegboi (16.1a. MENDE)
jembrak (30.52. JAVANESE)
jɛmiŋga (16.1a. MENDE)
jenabe ajenabo (30.24. SPANISH)
jengger ayam (23.4. INDONESIAN)
jengger ayam (23.4. JAVANESE)
jengger rumpuk (23.4. JAVANESE)
jengibre (20.5. SPANISH)
jengibre (20.6. SPANISH)
jengibre oficinal (20.6. SPANISH)
jengotan (63.1. INDONESIAN)
jenjamb (20.6. KWÉYÒL)

jenjanm (20.6. CREOLE, HAITIAN FRENCH)
jenjanm (20.6. KWÉYÒL)
jenjibre (20.6. LADINO)
jennaga (40.5. TELUGU)
jenubaale (16.1a. KANNADA)
jeow (11.7. DANGME-KROBO)
jerami (59.2. MALAY)
Jerusalem artichoke (27.26. ASSAMESE)
Jerusalem artichoke (27.26. ENGLISH)
Jerusalem artisjok (27.26. AFRIKAANS)
jete (36.3. GBE, WACI)
jete (36.3. GEN)
jethuya (38.23. BENGALI)
jĕtrowka (30.43. SORBIAN, UPPER)
jeukboontje (40.18. DUTCH)
jewar (50.1. PANJABI, EASTERN)
Jew's apple (62.19. ENGLISH)
Jew's marrow (63.1. ENGLISH)
jhadhī (38.3. NEWAR)
jh̃akph̃ū (39.4. IGBO)
jhasan (30.56. NEWAR)
jhata simba (40.9. ORIYA)
jhimani (38.21. HINDI)
jhingā (38.13a. MUNDARI)
jhinga (38.21. BENGALI)
jhinga tori (38.21. HINDI)
jhinga turai (38.21. HINDI)
jhingaka (38.21. SANSKRIT)
jhingli (38.21. HINDI)
jhokara jhar (52.1. NEPALI)
jhongaka (38.21. SANSKRIT)
jhoto (63.1. ORIYA)
jhundikeshi (38.4. SANSKRIT)
ji (11.1. IGBO)
ji (11.2. IGBO)
ji (11.5. IGBO)
jí àbàlà (11.1. IGBO)
jí àbànà (11.1. IGBO)
jí àbí (11.7. IGBO)
jí àgà (11.7. IGBO)
jí alafolo alẹfulu (11.7. IGBO)
ji apu (39.4. IGBO)
jí awudo (11.7. IGBO)
jí àyọ̀bè (11.7. IGBO)
ji cai (30.43. CHINESE, WÚ)
ji cai (61.1. CHINESE, MANDARIN)
ji cai (61.1. CHINESE, WÚ)
ji dou (40.37. CHINESE, MANDARIN)
ji er dou (40.5. CHINESE, MANDARIN)
ji er dou (40.5. CHINESE, WÚ)
ji guan hua (23.4. CHINESE, MANDARIN)
ji guang hua (23.4. CHINESE, YUE)
jí íkē (11.7. IGBO)
jí m̀bàlà àdàkà (11.1. IGBO)
jí m̀bàlà m̀bàlà (11.1. IGBO)
jí m̀bàlà m̀me (11.1. IGBO)
jí nwa nnu (36.3. IGBO)
jí ọ́chá (11.7. IGBO)
ji okō (16.1a. IGBO)
jí òkò jí-akero (11.7. IGBO)
ji quan hua (23.4. CHINESE, WÚ)
ji xue cao (24.7. CHINESE, MANDARIN)
ji xue cao (24.7. CHINESE, WÚ)
jia du xing cai (30.51. CHINESE, MANDARIN)
jia du xing cai (30.51. CHINESE, WÚ)
jia ren shen (59.3. CHINESE, MANDARIN)
jia ren shen (59.3. CHINESE, WÚ)
jia shan yao (11.6. CHINESE, MIN BEI)
jia shan yao (11.6. CHINESE, YUE)
jia yu (9.1. CHINESE, YUE)
jian jiao (62.2. CHINESE, YUE)
jian peng (35.10. CHINESE, MANDARIN)
jian peng (35.10. CHINESE, WÚ)
jian ye cai dou (40.23. CHINESE, WÚ)
jian ye cao dou (40.23. CHINESE, MANDARIN)
jian ye yu jiu hua (18.1. CHINESE, WÚ)
jiang (20.6. CHINESE, MANDARIN)
jiang (20.6. CHINESE, WÚ)
jiang ban cao (59.2. CHINESE, WÚ)
jiang dou (40.41. CHINESE, MANDARIN)
jiang dou (40.43. CHINESE, YUE)
jiang huang (20.2. CHINESE, MANDARIN)
jiang huang (20.2. CHINESE, WÚ)
jiang huang (20.3. CHINESE, WÚ)
jiao bai (17.8. CHINESE, MANDARIN)
jiao bai (17.8. CHINESE, WÚ)
jiao bai (30.40. CHINESE, GÀN)
jiao ban shu (11.1. CHINESE, XIANG)
jiao cai (35.9. CHINESE, WÚ)
jiao cai (35.9. CHINESE, YUE)
jiao gua (38.21. CHINESE, MIN BEI)
jiao yu (9.1. CHINESE, WÚ)
jiao zi (62.2. CHINESE, XIANG)
jì-apḥụ̄ (39.4. IGBO)
jí-bèkéè (36.3. IGBO)
jibelbi (36.3. GONJA)
jicama (40.21. ENGLISH)
jicama (40.21. FRENCH)

jicama (40.21. PORTUGUESE)
jícama (40.21. SPANISH)
jie cai (30.7. CHINESE, YUE)
jie cai (30.17. CHINESE, GÀN)
jie cai (30.17. CHINESE, MIN BEI)
jie cai tou (30.30. CHINESE, MIN BEI)
jie dou (40.37. CHINESE, WÚ)
jie geng (34.3. CHINESE, MANDARIN)
jie geng cai (33.2. CHINESE, MANDARIN)
jie geng cai (33.2. CHINESE, YUE)
jie lan (30.25. CHINESE, GÀN)
jie lan (30.25. CHINESE, MANDARIN)
jie lan (30.25. CHINESE, WÚ)
jie lan (30.25. CHINESE, YUE)
jie lan tou (30.30. CHINESE, YUE)
jie qiu gai (30.4. CHINESE, MANDARIN)
jie qiu jie (30.4. CHINESE, WÚ)
jie qiu wo ju (27.31. CHINESE, WÚ)
jie qiu wo ju (27.32. CHINESE, YUE)
jiftenni (44.5. ARABIC)
jīgbọ̀ (39.4. IGBO)
jigboi (27.14. LOKO)
jigor (17.7. KATI)
jii jii (27.10a. CHINESE, WÚ)
jiji (10.1. HAUSA)
jijimbilla (20.6. OROMO, WEST CENTRAL)
jika (38.21. ASSAMESE)
jikin mutun (11.1. HAUSA)
jilbana (40.10. ARABIC)
jilmil (35.5. ASSAMESE)
jilmil (35.6. ASSAMESE)
jiló (62.14b. PORTUGUESE)
jilo (62.14b. VLAAMS)
jiloeiro (62.14b. PORTUGUESE)
ji-mbo (11.1. IGBO)
jimikand (8.3. HINDI)
jimikand (8.3. PANJABI, EASTERN)
jimmy kand (8.3. HINDI)
jin cai (7.13. CHINESE, GÀN)
jin dou (40.26. CHINESE, GÀN)
jin dou (40.28. CHINESE, MIN BEI)
jin gua (38.14. CHINESE, MIN BEI)
jin gua (38.17. CHINESE, WÚ)
jin li zhi (38.23. CHINESE, MIN BEI)
jin li zhi (38.23. CHINESE, WÚ)
jin lian hua (65.1. CHINESE, MANDARIN)
jin lian hua (65.1. CHINESE, WÚ)
jin mai wan (40.12. CHINESE, MANDARIN)
jin mai wan (40.12. CHINESE, WÚ)
jin niu kou (27.3. CHINESE, WÚ)
jin sun (24.11. CHINESE, MIN BEI)
jin sun (24.11. CHINESE, YUE)
jin zhen cai (13.3. CHINESE, MANDARIN)
jin zhen cai (13.3. CHINESE, MIN BEI)
jin zhen cai (13.3. CHINESE, WÚ)
jin zhen cai (13.3. CHINESE, YUE)
jindungo (62.2. PORTUGUESE)
jing lan (30.30. CHINESE, GÀN)
jing liu gai (30.19. CHINESE, MANDARIN)
jing liu jie (30.19. CHINESE, WÚ)
jinga (38.21. HINDI)
jingga (28.1. JAVANESE)
jingi (38.21. INDONESIAN)
jingka (38.22. BODO)
jinguba de Cambambé (40.39. PORTUGUESE)
jinjà (20.6. IGBO)
jinja (20.6. KRIO)
jinjanm (20.6. CREOLE, HAITIAN FRENCH)
jinjer (20.6. WOLOF)
jinjibrè (20.6. LADINO)
jinjir paku rawan (14.1. MALAY)
jinshar (20.6. GAELIC, MANX)
ji-ocha (11.7. IGBO)
ji-okwu (11.3. IGBO)
jí-óyìbó (36.3. IGBO)
jipang (38.25. JAVANESE)
jiquimilla (27.39. SPANISH)
jire khursai (62.3. NEPALI)
jiri saag (27.29. NEPALI)
jiromon (38.14. CREOLE, HAITIAN FRENCH)
jiromon (38.16. CREOLE, HAITIAN FRENCH)
jiromon (38.18. CREOLE, HAITIAN FRENCH)
jisongo (40.2. PORTUGUESE)
jitamba (44.8. PORTUGUESE)
jite (11.2. GEN)
jite sɛnde (11.2. FON)
jitrocel (57.1. CZECH)
jitrocel vraní nožka (57.1. CZECH)
jiu (7.13. CHINESE, YUE)
jiu cai (7.13. CHINESE, MANDARIN)
jiu cai (7.13. CHINESE, MIN BEI)
jiu cai (7.13. CHINESE, WÚ)
jiu cai (7.13. CHINESE, YUE)
jiu cong (7.8. CHINESE, MANDARIN)
jiu cong (7.8. CHINESE, WÚ)
jiwomon (38.14. CREOLE, HAITIAN FRENCH)

jiwomon (38.16. CREOLE, HAITIAN FRENCH)
jiwomon (38.18. CREOLE, HAITIAN FRENCH)
jnār (17.7. PASHTO, NORTHERN)
jnugli matar (40.10. URDU)
jo: (17.7. BADAGA)
jocong (27.3. INDONESIAN)
jocong (27.4. INDONESIAN)
jødekirsebær (62.11. DANISH)
jodenkers (62.11. DUTCH)
jõenupp (50.2. ESTONIAN)
Johannisbeertomate (62.10. GERMAN)
Johnny-jump-up (68.1. ENGLISH)
jokotowo (8.9. JAVANESE)
jola (17.7. KANNADA)
jola (17.7. TULU)
jolokia (62.2. ASSAMESE)
jo:lu (17.7. BADAGA)
jolu (17.7. KONKANI)
jombang (27.18. INDONESIAN)
jombang (27.42. INDONESIAN)
jombang (27.45. INDONESIAN)
jombang (27.45. JAVANESE)
jòmou (38.14. KWÉYÒL)
jòmou (38.18. KWÉYÒL)
jonar (17.7. BENGALI)
joṇḍra (17.7. SANTALI)
jonge (27.18. SUNDA)
jonghe (27.18. SUNDA)
jonṛa (17.7. MUNDARI)
jonxhë e kultivuar (40.17. ALBANIAN, TOSK)
jonyshka (40.16. KAZAKH)
jopo de las habas (52.1. SPANISH)
jordært (40.2. DANISH)
jordärtskocka (27.26. SWEDISH)
jordmandel (10.1. DANISH)
jordnøt (40.1. DANISH)
jordnöt (40.1. SWEDISH)
jordskocka (27.26. SWEDISH)
jordskok (27.26. DANISH)
jordskokk (27.26. NORWEGIAN)
jordskolm (40.11. NORWEGIAN)
jorubak (35.2. TAJIKI)
Joseph's coat (23.3. ENGLISH)
jotang (27.2. INDONESIAN)
jotang (27.3. INDONESIAN)
jotang (27.4. INDONESIAN)
joto (63.1. ORIYA)
jou yeh jue chuang (21.1. CHINESE, MANDARIN)
jøuat (38.14. BODO)
jøuat (38.16. BODO)
jøuat (38.17. BODO)
jøuat (38.18. BODO)
joumou (38.14. CREOLE, HAITIAN FRENCH)
joumou (38.16. CREOLE, HAITIAN FRENCH)
joumou (38.18. CREOLE, HAITIAN FRENCH)
ju geng (34.3. CHINESE, WÚ)
ju hua (27.9. CHINESE, YUE)
ju hua (27.23. CHINESE, WÚ)
ju mai cai (27.9. CHINESE, WÚ)
ju niu bang (27.11. CHINESE, MANDARIN)
ju niu bang (27.11. CHINESE, WÚ)
ju qu (27.11. CHINESE, YUE)
ju ruo shu (19.1. CHINESE, MANDARIN)
ju ruo shu (19.1. CHINESE, WÚ)
ju yu (27.26. CHINESE, MANDARIN)
ju yu (27.26. CHINESE, WÚ)
ju yu (27.28. CHINESE, YUE)
ju yuan xi gua (38.2. CHINESE, YUE)
ju zai (62.9. CHINESE, YUE)
juacallicma (68.1. QUECHUA, CUSCO)
juan xin cai (30.27a. CHINESE, XIANG)
juan xin gai (30.9. CHINESE, MANDARIN)
juan xin jie (30.9. CHINESE, WÚ)
juan xin wo ju (27.32. CHINESE, YUE)
juár (17.7. HINDUSTANI)
juar (17.7. PANJABI, EASTERN)
jubɔks (62.19. KRIO)
jublɔks (62.19. KRIO)
Juckbohne (40.18. GERMAN)
Juckfasel (40.18. GERMAN)
Judenkirsche (62.11. GERMAN)
judia adzuki (40.36. SPANISH)
judia arroz (40.40. SPANISH)
judia batata (40.21. SPANISH)
judia catjang (40.41. SPANISH)
judia común (40.26. SPANISH)
judia chinas (40.42. SPANISH)
judia de caballo (40.3. SPANISH)
judia de España (40.24. SPANISH)
judia de Lima (40.25. SPANISH)
judia de Marama (40.33. SPANISH)
judia de mungo (40.38. SPANISH)
judia de puerco (40.3. SPANISH)

judia de urd (40.37. SPANISH)
judia de vaca (40.43. SPANISH)
judia escarlata (40.24. SPANISH)
judia espárrago (40.42. SPANISH)
judia pinta (40.24. SPANISH)
judia roja (40.24. SPANISH)
judia sable (40.3. SPANISH)
judia sable (40.4. SPANISH)
judia teparí (40.23. SPANISH)
judia verde (40.26. SPANISH)
jue (2.1. CHINESE, GÀN)
jue (2.1. CHINESE, WÚ)
jue cai (2.1. CHINESE, MIN BEI)
jue cai (2.1. CHINESE, YUE)
jue chuang (21.1. CHINESE, WÚ)
juganhdena (30.27a. CHEROKEE)
jügörü (17.7. KIRGHIZ)
ju:go:su (30.26. BADAGA)
juhudi (20.2. SWAHILI)
juke asu (38.23. AJA)
jukut gajah (17.3. SUNDA)
jukut hanyir (61.1. INDONESIAN)
jukut kriminil (23.1. INDONESIAN)
julivert (24.19. CATALAN-VALENCIAN-BALEAR)
jumki hoo (27.18. KANNADA)
jun da (35.3. CHINESE, GÀN)
jun da cai (35.3. CHINESE, MIN BEI)
jun da cai (35.3. CHINESE, WÚ)
jun da cai (35.3. CHINESE, YUE)
junça doce (10.1. PORTUGUESE)
juncinha (10.1. PORTUGUESE)
jungli sem (40.26. HINDI)
junguru (38.20. FULFULDE, ADAMAWA)
junhi (38.21. ORIYA)
junri (17.7. HINDI)
junsai (31.1. JAPANESE)
junsia (31.1. VLAAMS)
junuchqa (40.17. TAJIKI)
juodasis bastutis (30.24. LITHUANIAN)
juodoji kiauliauogė (62.21. LITHUANIAN)
juqueri-manso (40.19. PORTUGUESE)
juremon (38.14. CAJUN FRENCH)
juremon (38.16. CAJUN FRENCH)
juremon (38.17. CAJUN FRENCH)
juremon (38.18. CAJUN FRENCH)
jūrmalas krambe (30.45. LATVIAN)
jûroku-sasage (40.42. JAPANESE)
jurubera (62.23. PORTUGUESE)
justícia (21.1. PORTUGUESE)
jüt (63.1. TURKISH)
juta (63.1. HUNGARIAN)
juta (63.1. PORTUGUESE)
juta (63.1. SERBO-CROATIAN)
juta (63.1. SLOVAK)
juta kolorowa (63.1. POLISH)
juta warzywna (63.1. POLISH)
jute (63.1. DANISH)
jute (63.1. DUTCH)
jute (63.1. NORWEGIAN)
jute (63.1. SWEDISH)
jute (63.1. VLAAMS)
jute à capsules cylindriques (63.1. FRENCH)
jute malva di giudèi (63.1. ITALIAN)
Jutepflanze (63.1. GERMAN)
Jute-Wolle (63.1. GERMAN)
juto (40.2. JAPANESE)
jutovník zelinný (63.1. CZECH)
juurikas (35.4. FINNISH)
juuripersilja (24.21. FINNISH)
juurseller (24.5. ESTONIAN)
juutalaiskirsikka (62.11. FINNISH)
juutti (63.1. FINNISH)
juvorimakka (17.7. TAJIKI)
juwur (50.1. KASHMIRI)
jwalanti (30.24. SANSKRIT)
jwalatprapha (30.24. SANSKRIT)
jwano (24.5. NEPALI)
jwɛ (8.4. BIROM)
jwɛ-pwɛŋ (8.4. BIROM)
jwɛ-sunàŋ (8.4. BIROM)
jyabtibg (23.3. ILOCANO)
jyanmara khursai (62.5. NEPALI)
jyothi (40.32. SANSKRIT)
ka (11.2. BAOULÉ)
kā (30.17. NEWAR)
ka (38.20. FON)
ka (38.23. SUMO-MAYANGNA)
ka (40.1. KISSI, NORTHERN)
k̂α' (62.9. KIOWA)
ka fei shu (44.8. CHINESE, MANDARIN)
ka fei shu (44.8. CHINESE, WÚ)
ka fi amara (44.5. HAUSA)
ka kéna (19.1. SERER-SINE)
ka̠ la̠ beh˅ (40.12. LAHU)
ka˅ la˅ g'aw˅ ma (24.11. LAHU)
ka lamé ghomree (30.30. FARSI, WESTERN)
ka tahkeyawepayesik (44.3. CREE, PLAINS)
kã tiga (40.1. MANDING-MANDINKA)
ka tuk (38.20. JOLA-FONYI)

kaaru (38.20. BAATONUM)
kaarubachhali (28.1. TELUGU)
kaaruminumulu (40.38. TELUGU)
kaasaalu (8.1. SANSKRIT)
kaasani (27.9. SANSKRIT)
kaasani (27.10a. HINDI)
kaasani (27.10a. SANSKRIT)
kaasbhanjan (38.31. SANSKRIT)
kaashini (27.9. TAMIL)
kaashini (27.10a. TAMIL)
kaashthhila (16.1a. MALAYALAM)
kaasini gida (27.9. KANNADA)
kaasithazhai (62.15. TAMIL)
kaasni (27.10a. TELUGU)
kaat kadalekka (45.5. MALAYALAM)
kaattu kaachil (11.2. MALAYALAM)
kaattu kadale (45.5. TULU)
kaattu karunai (19.1. TAMIL)
kaattu patavalam (38.31. MALAYALAM)
kaattu picinna (38.22. MALAYALAM)
kaattu seeragam (24.13. TAMIL)
kaattu siru keerai (23.3. TAMIL)
kaattu ulinu (40.37. MALAYALAM)
kaattubaala (9.1. MALAYALAM)
kaattuchunta (62.23. MALAYALAM)
kaattukallante (62.23. TULU)
kaattukkizhangu (11.2. TAMIL)
kaattulunnu (40.40. MALAYALAM)
kaattumannaal (20.4. MALAYALAM)
kaatturamathulasi (44.4. MALAYALAM)
kaattutthambattan (40.3. TAMIL)
kaattuvaala (9.1. MALAYALAM)
kaattuvaalamarakkaya (40.3. MALAYALAM)
kaav ywj (8.4. HMONG-NJUA)
kaavattu (11.1. MALAYALAM)
kaaw (17.7. KALINGA, BUTBUT)
kaawno (17.1. CEBUANO)
kab (39.4. SERER-SINE)
kab tsib (17.5. HMONG-DAW)
kaba (17.7. JULA)
kàba (17.7. NUPE-NUPE-TAKO)
kaba tüylü Japon sarmaşiği (40.30. TURKISH)
kababas (38.20. AFRIKAANS)
kabaczek (38.18. POLISH)
kabaifa (40.28. LOMWE, MALAWI)
kàbaíwaà (38.16. HAUSA)
kabaiya (62.15. HINDI)
kabaiya (62.21. HINDI)
kabak (22.2. INDONESIAN)
kabak (22.2. MALAY)
kabak (22.2. SUNDA)
kabak (38.18. TURKISH)
kabaki (38.18. GEORGIAN)
kabala pɛpɛ (62.2. KRIO)
kabaloan (17.1. BICOLANO, CENTRAL)
kabantengga (38.22. ITAWIT)
kabar (34.1. HINDI)
kabar (34.1. MARATHI)
kabar (34.1. URDU)
kabasa (38.17. IVATAN)
kabasi (38.16. TAUSUG)
kabasi (38.17. TAUSUG)
kabata (27.8. TONGA)
kabata (27.8. TUMBUKA)
kabatiti (38.21. BONTOC CENTRAL)
kabatiti (38.21. ILOCANO)
kabatiti (38.22. BONTOC, CENTRAL)
kabatiti (38.22. ILOCANO)
kabatiti (38.22. SAMBAL, BOTOLAN)
kabatiti (38.22. SAMBAL, TINÀ)
kabatitit-aso (38.22. ILOCANO)
kabatsi (40.25. HANUNOO)
kabawang (38.22. TAGBANWA)
kabay (40.29. HANUNOO)
kabay mabiru (40.29. HANUNOO)
kabay mararas (40.29. HANUNOO)
kabbu (17.5. BADAGA)
kabe (8.1. KIRIBATI)
kabej (30.27a. KRIO)
kabej (30.27a. LUO)
kabej (30.40. LUO)
kabeji (30.27a. BEMBA)
kabeji (30.27a. HAUSA)
kabej-pla-sas (30.7. KRIO)
Kabes (30.27a. LUXENBOURGEOIS)
Kabes (30.34. LUXENBOURGEOIS)
käbestä (30.27a. TATAR)
kabewa (38.14. HAUSA)
kabewa (38.16. HAUSA)
kabewa (38.17. HAUSA)
kabewa (38.18. HAUSA)
kabey (40.29. TAGALOG)
kabheji (30.27a. SHONA)
kabi padavala (38.31. KANNADA)
kabichi (30.27a. DIGO)
kabichi (30.27a. LUYIA)
kabichi (30.27a. SHONA)
kabichi (30.27a. SWAHILI)
kabichi chainizi (30.36. SWAHILI)
kabichi chainizi (30.40. SWAHILI)

kabichi savoy (30.34. SWAHILI)
kabici (30.27a. GIKUYU)
kabici (30.27a. NYANJA)
ka-binsh (40.43. THEMNE)
kabishi (30.27a. SWAHILI)
kabitsi̱ (30.27a. COMANCHE)
kabkabatiti (28.1. PANGASINAN)
kabkabu (30.17. BONTOC, CENTRAL)
kabli bakla (40.34a. PANJABI, EASTERN)
ka-boah-chhài (35.3. CHINESE, MIN NAN)
kabocca (30.27a. MALTESE)
kabocca ciniza (30.40. MALTESE)
kabocca taz-zokk (30.29. MALTESE)
kabocci tar-rebbiegha (30.34. MALTESE)
kabocha (38.14. JAPANESE)
kabocha (38.17. JAPANESE)
ka-bonthila (23.3. THEMNE)
ka-bos (38.20. THEMNE)
kaboy (40.29. AGTA)
kabra (34.1. HINDI)
kabu (30.41a. JAPANESE)
kabucinella (65.1. MALTESE)
kabugawan (17.1. BICOLANO, CENTRAL)
kabukanran (30.22. JAPANESE)
kabuki (8.3. LAO)
kabul (55.1. KURANKO)
kabul (62.16. SUBANON)
ka-buləkɔ (55.1. THEMNE)
kábulí (40.28. HINDUSTANI)
kabulikaddu (38.20. PANJABI, EASTERN)
ka-bunkɔ (44.8. THEMNE)
ka-bunkɔ (55.1. THEMNE)
kabur (44.5. KANURI)
kabura kikyou (33.2. JAPANESE)
kabus (38.16. HAUSA)
kabus (38.18. HAUSA)
kabushi (38.16. HAUSA)
kabushi (38.18. HAUSA)
kacan arak̓ (38.23. SANTALI)
kacang (40.2. MALAY)
kacang (40.27. INDONESIAN)
kacang adas (40.12. INDONESIAN)
kacang adas (40.12. SUNDA)
kacang Arab (40.5. INDONESIAN)
kacang arab (40.5. SUNDA)
kacang babi (40.18. INDONESIAN)
kacang babi (40.18. MALAY)
kacang babi (40.34a. INDONESIAN)
kacang babi (40.34a. SUNDA)
kacang bado (40.9. INDONESIAN)
kacang Bali (40.2. INDONESIAN)
kacang bandung (40.39. SUNDA)
kacang banten (40.39. INDONESIAN)
kacang banten (40.39. JAVANESE)
kacang belimbing (40.29. INDONESIAN)
kacang belut (40.42. INDONESIAN)
kacang belut (40.42. MALAY)
kacang belut (40.42. SUNDA)
kacang bendi (45.1. MALAY)
kacang bengkok (40.42. INDONESIAN)
kacang benguk (40.18. INDONESIAN)
kacang beureum (40.26. SUNDA)
kacang biduk (40.9. INDONESIAN)
kacang bogo (40.39. INDONESIAN)
kacang Bogor (40.39. INDONESIAN)
kacang bogor (40.39. SUNDA)
kacang bol (40.43. MALAY)
kacang botor (40.29. INDONESIAN)
kacang botor (40.29. MALAY)
kacang brol (40.1. JAVANESE)
kacang brudul (40.1. JAVANESE)
kacang brul (40.1. JAVANESE)
kacang bulu (40.8. INDONESIAN)
kacang bulu (40.8. SUNDA)
kacang bulu rimau (40.8. MALAY)
kacang buncis (40.26. INDONESIAN)
kacang buncis (40.26. MALAY)
kacang cina (40.42. INDONESIAN)
kacang China (40.1. MALAY)
kacang China (40.25. MALAY)
kacang dadap (40.42. SUNDA)
kacang dal (40.2. MALAY)
kacang dal (40.12. MALAY)
kacang enjing (40.40. JAVANESE)
kacang ercis (40.28. INDONESIAN)
kacang ercis (40.28. MALAY)
kacang gengge (40.39. SUNDA)
kacang gleyor (40.42. JAVANESE)
kacang gondolo (40.39. SUNDA)
kacang goring (40.1. MALAY)
kacang gude (40.2. INDONESIAN)
kacang gude (40.2. JAVANESE)
kacang hijau (40.38. INDONESIAN)
kacang hijau (40.38. MALAY)
kacang hiris (40.2. MALAY)
kacang hiris (40.2. SUNDA)
kacang ijo (40.37. INDONESIAN)
kacang iris (40.2. INDONESIAN)
kacang Jawa (40.1. INDONESIAN)
kacang Jawa (40.1. MALAY)

kacang Jawa (40.25. MALAY)
kacang jebrol (40.1. JAVANESE)
kacang Jepun (40.8. INDONESIAN)
kacang Jepun (40.8. MALAY)
kacang jepun (40.8. SUNDA)
kacang jeriji (40.9. SUNDA)
kacang jogo (40.26. SUNDA)
kacang jogo (40.39. SUNDA)
kacang kapri (40.28. INDONESIAN)
kacang kapri (40.28. SUNDA)
kacang kara (40.9. INDONESIAN)
kacang kara (40.9. MALAY)
kacang karas (40.16. INDONESIAN)
kacang karas (40.16. SUNDA)
kacang kayu (40.2. INDONESIAN)
kacang kayu (40.2. JAVANESE)
kacang kedelai (40.8. INDONESIAN)
kacang kedele (40.8. SUNDA)
kacang kelisa (40.29. MALAY)
kacang kenyut (40.40. SUNDA)
kacang kopak (40.26. SUNDA)
kacang kowas (40.18. SUNDA)
kacang kuda (40.5. INDONESIAN)
kacang kuda (40.5. MALAY)
kacang landep (40.42. SUNDA)
kacang landes (40.42. SUNDA)
kacang lanjaran (40.42. INDONESIAN)
kacang lanjaran (40.42. JAVANESE)
kacang lendir (45.1. MALAY)
kacang Manila (40.1. INDONESIAN)
kacang manila (40.39. INDONESIAN)
kacang manila (40.39. SUNDA)
kacang mas (40.25. INDONESIAN)
kacang mas (40.25. SUNDA)
kacang mekah (40.3. SUNDA)
kacang merah (40.26. INDONESIAN)
kacang merah (40.26. MALAY)
kacang merah (40.41. INDONESIAN)
kacang merah (40.42. INDONESIAN)
kacang merah (40.43. MALAY)
kacang miang (40.18. INDONESIAN)
kacang mungo (40.37. INDONESIAN)
kacang mungo (40.38. INDONESIAN)
kacang nyonya (40.3. JAVANESE)
kacang oci (40.40. INDONESIAN)
kacang oci (40.40. JAVANESE)
kacang oci (40.40. SUNDA)
kacang panjang (40.41. INDONESIAN)
kacang panjang (40.42. INDONESIAN)
kacang panjang (40.42. JAVANESE)
kacang panjang (40.42. MALAY)
kacang panjang (40.42. SUNDA)
kacang panjang (40.43. INDONESIAN)
kacang parang (40.3. INDONESIAN)
kacang parang (40.4. MALAY)
kacang parang puteh (40.3. MALAY)
kacang parasman (40.4. INDONESIAN)
kacang peda (40.9. SUNDA)
kacang pendem (40.1. JAVANESE)
kacang perut ayam (40.42. MALAY)
kacang peudjit (40.41. INDONESIAN)
kacang peujit (40.42. SUNDA)
kacang pi (40.27. MALAY)
kacang polong (40.4. MALAY)
kacang polong (40.28. INDONESIAN)
kacang polong (40.28. JAVANESE)
kacang polong (40.28. SUNDA)
kacang polong èrtjis (40.28. INDONESIAN)
kacang prasman (40.3. INDONESIAN)
kacang prasman (40.3. SUNDA)
kacang putih (40.43. MALAY)
kacang rompes (40.26. SUNDA)
kacang roway (40.25. SUNDA)
kacang ruji (40.40. JAVANESE)
kacang ruji (40.42. SUNDA)
kacang sepalit (40.40. MALAY)
kacang serinding (40.25. MALAY)
kacang siam (40.6. MALAY)
kacang sirek (40.29. MALAY)
kacang soya (40.8. MALAY)
kacang srentul (40.1. JAVANESE)
kacang srintil (40.1. JAVANESE)
kacang suuk (40.1. SUNDA)
kacang tanah (40.1. INDONESIAN)
kacang tanah (40.1. MALAY)
kacang taneuh (40.1. SUNDA)
kacang tolo (40.40. INDONESIAN)
kacang tolo (40.42. INDONESIAN)
kacang toonggak (40.43. MALAY)
kacang tuban (40.1. JAVANESE)
kacang tunggak (40.26. SUNDA)
kacang tunggak (40.41. INDONESIAN)
kacang tunggak (40.42. INDONESIAN)
kacang tunggak (40.42. JAVANESE)
kacang tunggak (40.43. INDONESIAN)
kacang turus (40.42. SUNDA)
kacang uci (40.24. INDONESIAN)
kacang uci (40.38. INDONESIAN)
kacang uci (40.40. INDONESIAN)
kacang uci (40.40. JAVANESE)

kacang usus (40.42. INDONESIAN)
kacang usus (40.42. JAVANESE)
kacang wangi (40.28. MALAY)
kaččūra (20.2. TULU)
kacewere (62.24. NYANJA)
kāch (24.11. KASHMIRI)
kachaay (7.13. KHMER, CENTRAL)
kachalu (8.4. HINDI)
kachâlu (36.3. FARSI, EASTERN)
kachâlu (36.3. FARSI, WESTERN)
kachâlu (62.24. FARSI, EASTERN)
kachâlu (62.24. PASHTO, NORTHERN)
kachan (62.20. QUECHUA, CUSCO)
kachari (20.4. MARATHI)
kachchi (8.4. KANNADA)
kachchipundu (62.21. TELUGU)
kachhi (8.4. SANSKRIT)
kachholam (20.4. MALAYALAM)
kachhu (8.4. SANSKRIT)
kachhumathi (40.18. SANSKRIT)
kachhura (40.18. SANSKRIT)
kachhurikizhanna (20.4. MALAYALAM)
kachi (62.21. TELUGU)
kachil (8.3. MALAYALAM)
kachil (11.1. MALAYALAM)
kachil (11.2. MALAYALAM)
kachkula (16.1a. HINDI)
kachmach (62.21. PANJABI, EASTERN)
kachnär (40.33. PANJABI, EASTERN)
kacholu (62.24. TAJIKI)
kachoora (20.4. HINDI)
kachoora (20.4. SANSKRIT)
kachora (20.4. KANNADA)
kachora (20.4. MARATHI)
kachora (20.4. TAMIL)
kachoram (20.4. TELUGU)
kachoramu (20.1. TELUGU)
kāchro (38.13a. MARWARI)
kachu (8.4. BENGALI)
kachu (8.4. HINDI)
kachu (8.6. SANSKRIT)
kachu (40.18. ORIYA)
kachul (8.3. TAMIL)
kachung kadyu (40.2. INDONESIAN)
kachura (20.4. BENGALI)
kachura (20.4. GUJARATI)
kachura (20.4. HINDI)
kachura (20.4. KANNADA)
kachura (20.4. MARATHI)
kachwi (8.4. SANSKRIT)
kacondang (19.1. SUNDA)
kacukacunga (44.5. FULFULDE, ADAMAWA)
kacunda (19.1. INDONESIAN)
kacur (38.21. JAVANESE)
kada manis (62.1. MALAY)
kadabáw (36.3. WÁRAY-WÁRAY)
kadagolu (62.9. KANNADA)
kadal (16.1a. MUNDARI)
kadalai (40.5. TAMIL)
kadalamu (16.1a. TELUGU)
kadale (40.5. KANNADA)
kadale (40.5. TULU)
kaḍalè (40.5. TULU)
kaḍale (40.28. BADAGA)
kadale kaalu (40.5. KANNADA)
kadale kaayi (40.1. KANNADA)
kadale-kaadi (40.5. MALAYALAM)
kadalí (16.1a. HINDUSTANI)
kadali (16.1a. KANNADA)
kadali (16.1a. MARATHI)
kadalī (16.1a. PALI)
kadali (16.1a. SANSKRIT)
kadali (16.1a. TAMIL)
kadali (16.1a. TELUGU)
kadalīkā (16.1a. PALI)
kadambi (36.2. SANSKRIT)
kadan (13.5. HEBREW)
kadavithumbi (38.20. HINDI)
kadayuhan (23.6. TAGALOG)
kadayum-dagi (27.12. SAMBAL, BOTOLAN)
kaddis (38.20. HINDI)
kaddla parippu (40.5. SINHALA)
kaddu (30.56. FARSI, EASTERN)
kaddu (38.14. FARSI, EASTERN)
kaddu (38.16. FARSI, EASTERN)
kaddu (38.16. HINDI)
kaddu (38.17. FARSI, EASTERN)
kaddu (38.18. FARSI, EASTERN)
kaddu (38.20. HINDI)
kâdé (45.1. PULAAR)
kadeřavá kapusta (30.35. CZECH)
kadeřávek (30.35. CZECH)
kâdia (45.1. WOLOF)
kadifest bob (40.18. MACEDONIAN)
kadigandu (40.18. TAMIL)
kadīmā (38.14. MAITHILI)
kadīmā (38.16. MAITHILI)
kadīmā (38.17. MAITHILI)
kadīmā (38.18. MAITHILI)
kadimah (38.16. HINDI)

kapononon (62.5. BOBO MADARÉ, SOUTHERN)
kāpost (30.27a. LATVIAN)
kapotavankaa (24.7. SANSKRIT)
kapousinn (65.1. MORISYEN)
kapousta pastewna (30.35. POLISH)
kappai keerai (23.3. TAMIL)
kappakka (38.23. MALAYALAM)
kappalmelaka (62.5. MALAYALAM)
kappar (34.1. MALTESE)
kappari (34.1. GREEK)
kappe rotti gida (45.5. KANNADA)
kapper (34.1. VLAAMS)
kapper-boon (34.1. DUTCH)
Kappernstrauch (34.1. GERMAN)
kapperstruik (34.1. DUTCH)
kappertjie (65.1. AFRIKAANS)
kappu saasive (30.24. KANNADA)
kapra (34.1. SERBO-CROATIAN)
kapri (34.1. INDONESIAN)
kapri (40.28. INDONESIAN)
kapri (40.28. JAVANESE)
kapris (34.1. FINNISH)
kapris (34.1. SWEDISH)
kapsas (30.27a. ESTONIAN)
Kapsche Stachelbeere (62.12. GERMAN)
kapseklapbes (62.12. AFRIKAANS)
kapsella e pimenosakkoï (30.43. GREEK)
Kapstachelbeere (62.12. GERMAN)
kapucínka (65.1. SLOVAK)
kapukapu (24.7. RAROTONGAN)
kapuninga (27.8. NDALI)
kapur kanti (44.5. ORIYA)
kạpur mạli baha (44.5. SANTALI)
ka-pura (50.2. THEMNE)
kapusta (30.27a. KAZAKH)
kapusta (30.27a. KIRGHIZ)
kapusta (30.34. CZECH)
kapusta agorodnaja (30.31. BELARUSAN)
kapusta belakachannaja (30.27a. BELARUSAN)
kapusta belokočannaja (30.27a. RUSSIAN)
kapusta belokochannaja (30.27a. KAZAKH)
kapusta brjusselskaja (30.29. KAZAKH)
kapusta Brukselska (30.29. POLISH)
kapusta brussel'skaja (30.29. RUSSIAN)
kapusta bryusselskaja (30.29. BELARUSAN)
kapusta chińska (30.36. POLISH)
kapusta čierna (30.24. SLOVENIAN)
kapusta cvetnaja (30.26. KAZAKH)
kapusta cvetnaja (30.26. RUSSIAN)
kapusta głowiasta (30.27a. POLISH)
kapusta głowiasta biala (30.27a. POLISH)
kapusta głowiasta czerwona (30.27b. POLISH)
kapusta hlávková (30.27a. SLOVAK)
kapusta hlávková biela (30.27a. SLOVAK)
kapusta kachannaja (30.27a. RUSSIAN)
kapusta kaderava (30.35. CZECH)
kapusta kitajskaja (30.36. RUSSIAN)
kapusta kitajskaja (30.40. KAZAKH)
kapusta kitajskaja (30.40. RUSSIAN)
kapusta kormovaja (30.35. RUSSIAN)
kapusta korovia (30.35. RUSSIAN)
kapusta listovaja (30.35. BELARUSAN)
kapusta listovaja (30.35. KAZAKH)
kapusta listovaja kurčavaja (30.35. RUSSIAN)
kapusta morska (30.45. POLISH)
kapusta morskaja (30.45. RUSSIAN)
kapusta morskaja (30.46. ROMANIAN)
kapusta ogrodowa (30.35. POLISH)
kapusta parachińska (30.37. POLISH)
kapusta pastewna głąbiasta (30.32. POLISH)
kapusta pekińska (30.40. POLISH)
kapusta portugalska (30.28. POLISH)
kapusta repka (30.21. SLOVENIAN)
kapusta repnaja (30.30. RUSSIAN)
kapusta ružičková (30.29. CZECH)
kapusta savoiska (30.34. BELARUSAN)
kapusta savojska (30.34. CZECH)
kapusta savojska (30.34. RUSSIAN)
kapusta savojska belokočannaja (30.34. RUSSIAN)
kapusta sitinovitá (30.17. SLOVENIAN)
kapusta sitowa (30.4. POLISH)
kapusta syberyjska (30.23. POLISH)
kapusta szparagowa (30.31. POLISH)
kapusta tatarska (30.46. POLISH)
kapusta tsvetnaja (30.26. BELARUSAN)
kapusta tysiągłówkowa (30.33. POLISH)
kapusta włoska (30.34. POLISH)
kapusta zimni (30.35. CZECH)
kapustin bolshoi (65.1. RUSSIAN)
kapustny (30.1. BELARUSAN)
kara (38.20. HEBREW)
kara (40.9. JAVANESE)
kara (40.25. INDONESIAN)
kara (40.25. JAVANESE)
kara (40.28. KASHMIRI)
kara andong (40.9. JAVANESE)
kara bedog (40.3. SUNDA)

kara bendo (40.3. JAVANESE)
kara benguk (40.18. INDONESIAN)
kara benguk (40.18. JAVANESE)
kara çavdar (20.4. TURKISH)
kara daio (58.1. JAPANESE)
kara dodog (40.3. JAVANESE)
kara haji (40.3. JAVANESE)
kara hardal (30.24. TURKISH)
kara hindiba (27.45. TURKISH)
kara isirgan (44.7. TURKISH)
kara kaji (40.3. INDONESIAN)
kara kimyon (24.9. TURKISH)
kara krupuk (40.9. JAVANESE)
kara lahana (30.35. TURKISH)
kara loke (40.3. JAVANESE)
kara lunjar (40.9. JAVANESE)
kara oncet (40.34a. INDONESIAN)
kara oncet (40.34a. JAVANESE)
kara ot (59.2. KAZAKH)
kara pedang (40.3. INDONESIAN)
kara pedang (40.3. JAVANESE)
kara pedang (40.4. INDONESIAN)
kara pedang (40.34a. INDONESIAN)
kara pendem (40.39. JAVANESE)
kara putih (40.9. JAVANESE)
kara rawe (40.18. JAVANESE)
kara uceng (40.9. INDONESIAN)
kara uceng (40.9. JAVANESE)
kara wedung (40.3. JAVANESE)
kara wedus (40.9. JAVANESE)
kara yemlik otu (27.38. TURKISH)
karabansos (40.29. CEBUANO)
karabas lahanasi (30.35. TURKISH)
karabasa (38.16. ILOCANO)
karabasa (38.17. ILOCANO)
karaboş (44.11. TURKISH)
karachunai (19.1. TAMIL)
karaf (24.4. FARSI, WESTERN)
karafs (24.4. ARABIC)
karafs (24.4. TAJIKI)
karafs (24.5. TAJIKI)
karafs i bos-táni (24.19. HINDUSTANI)
karahana-sô (46.1. JAPANESE)
karahi (44.5. SANSKRIT)
karaib (8.9. CREOLE, HAITIAN FRENCH)
karaila (38.23. FIJIAN)
karaila (38.23. HINDUSTANI)
karakafes (29.2. TURKISH)
karakala (40.42. TAGALOG)
karakap (62.15. TOK PISIN)
kara-kara (40.9. INDONESIAN)
kara-kara (40.9. MALAY)
karakarunai (8.3. TAMIL)
karala (38.23. BENGALI)
karala (38.23. MARATHI)
karala (38.23. PANJABI, EASTERN)
karalábé (30.30. HUNGARIAN)
karale (38.23. MARATHI)
karam (30.23. TAJIKI)
karam (30.27a. FARSI, EASTERN)
karam (30.27a. PASHTO, NORTHERN)
karam (30.27a. TAJIKI)
karam (30.27a. UZBEK)
karam (30.29. TAJIKI)
karam kalla (30.27a. HINDI)
karam sag (30.35. URDU)
karamani (40.43. TAMIL)
karamba (11.2. JOLA-FONYI)
karami brokkali (30.31. TAJIKI)
karamkala (30.27a. PASHTO, NORTHERN)
karamkalla (30.27a. HINDUSTANI)
karam-sag (30.35. HINDI)
kárán àlbásàà (7.3. FULFULDE, ADAMAWA)
kárán sárkíí (17.5. HAUSA)
karanai-kilangu (8.3. TAMIL)
karanaippala (8.3. TAMIL)
karando (11.2. KONKANI)
karanga (40.1. SWAHILI)
karaparapa a sina (59.3. MARANAO)
karaparna (45.1. SANSKRIT)
karas (24.11. HAUSA)
káràs (24.11. KANURI)
karas (40.10. PANJABI, EASTERN)
kàrásə́ (45.4. KANURI)
karashe (62.23. KANNADA)
karasu (45.4. KANURI)
karasuuri (38.30. JAPANESE)
karate (38.23. GUJARATI)
karate (38.23. MARATHI)
karate (50.2. SUNDA)
karati (24.11. GIKUYU)
karatim (38.23. KONKANI)
ka:raṭu (24.11. BADAGA)
karaturp (30.1. TURKISH)
karavella (38.23. SANSKRIT)
karavellah (38.23. SANSKRIT)
karavellaka (38.23. SANSKRIT)
karavellam (38.23. SANSKRIT)
karavi (24.13. SANSKRIT)
karavi (30.51. SANSKRIT)

kassava (39.4. NORWEGIAN)
kassawa (39.4. AFRIKAANS)
kassinaeris (45.5. ESTONIAN)
kassur (40.10. HINDI)
kass'ur (40.10. HINDI)
kasszáva (39.4. HUNGARIAN)
ka-sta (17.3. THEMNE)
kastano idrohars (10.2. GREEK)
kastanon to ydronhares (64.1. GREEK)
kasthoori-manjil (20.4. TAMIL)
kasthooripasupu (20.4. TELUGU)
kastravec (38.13a. ALBANIAN, TOSK)
kasturi tarabhuja (38.7a. ORIYA)
kasturi tarabuja (38.7b. TELUGU)
kasturi tatabhuja (38.7b. ORIYA)
ka:su (38.13a. BADAGA)
kasudu (11.1. BUHID)
ka-suka-ken (17.5. THEMNE)
ka-sul (17.1. THEMNE)
kasumba giya (20.6. TIRURAY)
kat pati sarsoo (30.6. URDU)
kata (58.3. SWAHILI)
katabas sa palaw (27.18. MARANAO)
katabataba (48.1. MARANAO)
katafura (62.24. NDONGA)
kataili-chaulai (23.3. HINDI)
katajang (40.9. HINDI)
katakata (7.3. FULFULDE, ADAMAWA)
kaṭāl (17.5. NEWAR)
katalu (11.1. ASSAMESE)
katang bans (17.1. HINDI)
kàtàré (62.19. GONJA)
katare (62.21. GONJA)
katausi (17.1. BENGALI)
katavandai (45.1. HINDI)
katchnar (40.33. URDU)
katchu (8.4. SANSKRIT)
kate (11.7. DANGME-KROBO)
katebek (23.3. MARANAO)
katekatego (38.16. BASSARI)
katekatego (38.18. BASSARI)
katela (36.3. JAVANESE)
katela pulak (22.2. HINDI)
katela pulak (22.2. URDU)
katela rambat (36.3. JAVANESE)
katelup (38.7b. INDONESIAN)
kaṯɛma (40.34a. TIBETAN)
katepan (24.15. JAVANESE)
kàtèré (62.9. GONJA)
kàtèré (62.19. GONJA)
katere (62.21. GONJA)
katetu (62.2. KONKOMBA)
katetu (62.9. KONKOMBA)
kath alu (11.1. ASSAMESE)
kath 'ièm (7.10. LAO)
kathalu (11.1. ASSAMESE)
kathalu (11.2. ASSAMESE)
kathari mullina gida (34.1. KANNADA)
katharikai (62.19. TAMIL)
kathhiri (62.19. TAMIL)
ka-thi (17.5. KAREN)
kathilla (38.23. SANSKRIT)
kathillaka (38.23. SANSKRIT)
kathinjara (44.4. SANSKRIT)
kathiri (62.19. TAMIL)
kathirikai (62.19. TAMIL)
katifa orf el-deek (23.4. ARABIC)
katifes (27.14. GREEK)
kã-tiga (40.1. BAMANANKAN)
katila (36.3. MANOBO)
katilla (38.23. SANSKRIT)
katillaka (38.23. SANSKRIT)
katimon (38.3. SAMBAL, BOTOLAN)
katimon (38.7. CEBUANO)
katimon (38.7a. TAGALOG)
katimon (38.7b. TAGALOG)
katimun (38.7. HANUNOO)
katimun (38.13a. INDONESIAN)
katimun (38.13a. JAVANESE)
katimun (38.13a. SUNDA)
katimun diriit (38.13. HANUNOO)
katimun malagti (38.7. HANUNOO)
katimun mariruk (38.13. HANUNOO)
katipunan (36.3. WÁRAY-WÁRAY)
katiu (38.11. MARQUESAN)
katiza (58.3. SWAHILI)
katjama (38.3. NDONGA)
katjánbab (40.2. HUNGARIAN)
katjang djogo (40.26. INDONESIAN)
katofeniya (62.24. NYAKYUSA-NGONDE)
katombar (62.5. MARANAO)
katphala (62.19. SANSKRIT)
katran (30.45. BELARUSAN)
katran (30.46. POLISH)
katran (30.46. ROMANIAN)
katran ki chorí (30.27a. HINDUSTANI)
katrán přímořský (30.45. CZECH)
katran prímorský (30.45. SLOVAK)
katrán tatarský (30.46. CZECH)
katrimullinagidda (34.1. KANNADA)

kerako phool (16.1b. NEPALI)
keramboja (38.3. INDONESIAN)
keräsalaatti (27.31. FINNISH)
keratok (40.25. INDONESIAN)
kerau (40.28. NEPALI)
kerawai (40.18. INDONESIAN)
kerayapan (59.2. JAVANESE)
Kerbeirübe (24.8. GERMAN)
kerblík třebule (24.3. CZECH)
keremah (23.2. JAVANESE)
keremek (23.2. INDONESIAN)
keremek (23.2. SUNDA)
keremi (23.2. JAVANESE)
kereng (36.3. TULU)
kereṅgu (36.3. TULU)
keren-kere (63.1. IGBO)
kereŋkere (63.1. LIMBA)
kereŋ-kereŋ (63.1. MANDING-MANDINKA)
kereria (40.43. HAUSA)
kərəviz (24.4. AZERBAIJANI, NORTHERN)
kereviz (24.4. TURKISH)
kereviz (24.5. TURKISH)
keribang (40.29. BINUKID)
kering (17.1. BINUKID)
kering kering (63.1. WOLOF)
kerisan (27.23. SUNDA)
kermate (62.9. JAVANESE)
kermesmari (56.3. ESTONIAN)
kermi (36.2. HINDUSTANI)
kerok batok (24.7. JAVANESE)
kerotsi (24.11. SHONA)
ķērsas (30.44. LATVIAN)
Kersting's groundnut (40.15. ENGLISH)
kerti koriander (24.9. HUNGARIAN)
kerti laboda (35.1. HUNGARIAN)
kerti retek (30.53. HUNGARIAN)
kerti retek (30.54. HUNGARIAN)
kerti sárgadinnye (38.8. HUNGARIAN)
kerti seprőfű (35.2. HUNGARIAN)
kerti sóska (58.4. HUNGARIAN)
kerti zsázsa (30.51. HUNGARIAN)
kerula uchchhe (38.23. BENGALI)
keruv (30.4. HEBREW)
keruv (30.5. HEBREW)
keruv (30.6. HEBREW)
keruv (30.7. HEBREW)
keruv (30.8. HEBREW)
keruv (30.9. HEBREW)
keruv (30.10. HEBREW)
keruv (30.11. HEBREW)
keruv (30.12. HEBREW)
keruv (30.13. HEBREW)
keruv (30.14. HEBREW)
keruv (30.15. HEBREW)
keruv (30.16. HEBREW)
keruv (30.17. HEBREW)
keruv (30.19. HEBREW)
keruv (30.20. HEBREW)
keruv (30.27a. HEBREW)
keruv ʿalim (30.27a. HEBREW)
keruv ha-lefet (30.41c. HEBREW)
keruv ha-nafus (30.21. HEBREW)
keruv niẓanim (30.29. HEBREW)
keruv shaḥor (30.24. HEBREW)
keruv sini (30.36. HEBREW)
keruv sini (30.40. HEBREW)
kervel (24.3. BULGARIAN)
kervel (24.3. DUTCH)
kervel' (24.3. RUSSIAN)
kervel (24.3. VLAAMS)
kervel' korneplodnyj (24.8. RUSSIAN)
kerwil (24.3. AFRIKAANS)
kesäkurpitsa (38.18. FINNISH)
kesäkynteli (44.9. FINNISH)
kesari (40.10. SANTALI)
kesäsypressi (35.2. FINNISH)
kesave dantu (8.4. KANNADA)
kesavina gedde (8.4. KANNADA)
kesavu (8.4. KANNADA)
kesel muwa (16.1b. SINHALA)
kesharee (27.9. NEPALI)
kẹshin (27.8. YORUBA)
keshro (50.2. BODO)
keshutanit meẓuya (46.1. HEBREW)
kesila' (36.3. BINUKID)
keskesteng (27.42. BONTOC, CENTRAL)
keskin köşeli lif kabaği (38.21. TURKISH)
kestenka (38.16. BULGARIAN)
kesu kesu (44.5. FON)
kɛtɛ (34.2. DANGME)
kɛtɛ (34.2. GA)
kete kete (27.8. YORUBA)
kətəkwakyi (16.1a. KRIO)
ketela pohon (39.4. INDONESIAN)
ketela puhun (39.4. JAVANESE)
ketela rambat (36.3. INDONESIAN)
ketimun (38.13a. INDONESIAN)
ketimun (38.13a. JAVANESE)
ketkyyuu-takana (30.4. JAPANESE)
ketmia (45.2. POLISH)

ketmia jadalna (45.1. POLISH)
ketmia szczawiowa (45.4. POLISH)
ketmie (45.1. VLAAMS)
ketmie (45.4. FRENCH)
ketmie à feuilles de manioc (45.2. FRENCH)
ketmie comestible (45.1. FRENCH)
ketmierose (45.4. FRENCH)
ketola (38.21. MALAY)
ketola manis (38.21. INDONESIAN)
ketola manis (38.22. MALAY)
ketola ular (38.29. MALAY)
keto-orvokki (68.1. FINNISH)
ketorki (59.2. BASQUE)
ketsuruazuki (40.37. JAPANESE)
ketuglew (62.21. BINUKID)
ketumbar (24.9. INDONESIAN)
ketumbar (24.9. MALAY)
ketumbit jantan (27.18. INDONESIAN)
ketumbit jantan (27.18. MALAY)
ketungkeng (54.1. INDONESIAN)
keturi (20.4. ASSAMESE)
ketypir (40.29. JAVANESE)
keumeli (62.24. SUNDA)
kɛvɛ (17.1. PULAAR)
kevetsi (30.27a. HOPI)
kewal (17.1. FULFULDE, ADAMAWA)
kewal (17.1. PULAAR)
kewal (17.2. FULFULDE, ADAMAWA)
kewal (17.2. PULAAR)
kewal (17.4. FULFULDE, ADAMAWA)
kewal (17.4. PULAAR)
kewats (24.11. ZUNI)
kewayan (17.1. MANOBO)
kewe (17.1. FULFULDE, ADAMAWA)
kewe (17.2. FULFULDE, ADAMAWA)
kewe (17.4. FULFULDE, ADAMAWA)
kewi (6.1. ARABIC)
kəyoro (11.2. MALIMBA)
k/e:zitokka (62.9. ZUNI)
kha (20.1. THAI)
kh'aad (27.3. LAO)
kh'aad (27.4. LAO)
khaara genasu (20.6. KANNADA)
khaarada menasinakaayi (62.2. KANNADA)
kh'aaw th'oong (61.1. LAO)
kh̃áɓ (11.3. BASSARI)
khab ou mbulokh (39.4. SERER-SINE)
khabetjhe (30.27a. SOTHO, SOUTHERN)
khabetšhe (30.27a. SOTHO, NORTHERN)
khabetshe (30.27a. TSWANA)
khabi (11.7. SUSU)
khabi-gboueli (11.1. SUSU)
khabi-gbueli (11.1. SUSU)
khad (38.7b. SERER-SINE)
khadasambala (40.3. MARATHI)
khadgashimbi (40.3. SANSKRIT)
khadsambal (40.3. HINDI)
khadu (38.14. FIJIAN)
khadu (38.16. FIJIAN)
khadu (38.18. FIJIAN)
khaerot (24.11. THAI)
khaisore (38.20. KANNADA)
khajori (40.18. MARATHI)
khakhiat (6.1. THAI)
khakhiat (18.2. THAI)
khakhvi (7.3. GEORGIAN)
khâl (38.3. WOLOF)
khal khurmi (24.16. URDU)
khalanj (20.1. URDU)
khaloti (24.11. ZULU)
khamalu (11.1. BENGALI)
khamalu (11.1. HINDI)
khamch (40.18. HINDI)
kham-daeng (69.1. THAI)
khamhae (27.12. THAI)
khamhāru (11.1. MAITHILI)
khamin (20.2. THAI)
khamin-chan (20.2. THAI)
khamin-kaeng (20.2. THAI, NORTHERN)
khamin-khao (20.3. THAI)
khamin-khun (20.4. THAI)
khamin-oi (20.4. THAI)
khamuja (38.3. HINDI)
khana ka fern (5.1. URDU)
khana wala canna (9.1. URDU)
khana-farang (30.35. THAI)
khandi (40.35. SANSKRIT)
khandia a kob (45.2. SERER-SINE)
khandikaa (40.10. SANSKRIT)
khanphal (11.1. MARATHI)
khao khwaai (64.3. THAI)
khar bhaja ba (38.7a. ORIYA)
khar bhuja ba (38.7b. ORIYA)
khar chesentser (62.21. MONGOLIAN)
kharapushpa (44.5. SANSKRIT)
kharazan el-maa (17.8. ARABIC)
kharbhuj (38.7b. MAITHILI)
kharbooj (38.7b. HINDI)
kharbooja (38.7a. HINDI)
kharbooja (38.7b. KANNADA)

kharbooja (38.7b. SANSKRIT)
kharbooza (38.7b. PANJABI, EASTERN)
kharbooza (38.11. PANJABI, EASTERN)
kharbozch kotor (38.7b. FARSI, WESTERN)
kharboze (38.7a. FARSI, WESTERN)
kharboze (38.7b. FARSI, WESTERN)
kharboze (38.11. FARSI, WESTERN)
kharbozeh asaly (38.11. FARSI, WESTERN)
kharbuj (38.7a. MARATHI)
kharbuj (38.7b. MARATHI)
kharbuja (38.3. NEPALI)
kharbuja (38.7b. GUJARATI)
kharbūjā (38.7b. HINDI)
kharbuja (38.7b. MARATHI)
kharbuja (38.7b. SANSKRIT)
kharbuja (38.7b. TELUGU)
kharbujadosa (38.7b. TELUGU)
kharbuza (38.7a. FARSI, EASTERN)
kharbuza (38.7a. HINDI)
kharbuza (38.7a. PANJABI, EASTERN)
kharbuza (38.7b. FARSI, EASTERN)
kharbuza (38.7b. HINDI)
kharbúza (38.7b. HINDUSTANI)
kharbuza (38.7b. PANJABI, EASTERN)
k̲h̲arbūza (38.7b. PASHTO, NORTHERN)
kharbuza (38.7b. URDU)
kharbuza (38.9. URDU)
kharbuza (38.11. FARSI, EASTERN)
kharbuza (38.11. HINDI)
kharbuza (38.11. PANJABI, EASTERN)
kharbuzeh (38.7b. FARSI, WESTERN)
khardal (30.3. FARSI, WESTERN)
khardal (30.3. TAJIKI)
khardal (30.4. FARSI, WESTERN)
khardal (30.5. FARSI, WESTERN)
khardal (30.6. FARSI, WESTERN)
khardal (30.7. FARSI, WESTERN)
khardal (30.8. FARSI, WESTERN)
khardal (30.9. FARSI, WESTERN)
khardal (30.10. FARSI, WESTERN)
khardal (30.11. FARSI, WESTERN)
khardal (30.12. FARSI, WESTERN)
khardal (30.13. FARSI, WESTERN)
khardal (30.15. FARSI, WESTERN)
khardal (30.16. FARSI, WESTERN)
khardal (30.17. FARSI, WESTERN)
khardal (30.19. FARSI, WESTERN)
khardal (30.20. FARSI, WESTERN)
khardali safed (30.11. TAJIKI)
kharmuj (38.7a. BENGALI)
kharmuj (38.7b. BENGALI)
!kharo-i (7.3. HERERO)
kharooja vaange (62.9. MARATHI)
kharrbuja (38.7b. HINDI)
khartoi (30.1. UZBEK)
khashilj (24.3. MONGOLIAN)
khasia bael (54.1. ASSAMESE)
khass el-kalab (27.38. ARABIC)
khass hindy (27.29. ARABIC)
khass saki (27.30. ARABIC)
khass waraxi (27.32. ARABIC)
khat naa (30.17. KHMER, CENTRAL)
k̲h̲atakai (38.7b. PASHTO, NORTHERN)
khat̲ĕkay (38.7b. PASHTO, NORTHERN)
khatta palak (58.2. HINDI)
khatta palak (58.2. URDU)
khavalya valli (40.18. KONKANI)
khavhishi (30.27a. VENDA)
khaviskhana (27.38. MONGOLIAN)
kha:x ta: dé:ng (20.1. LAO)
kha-yan-gyin (62.9. BURMESE)
kha-yan-thi (62.19. BURMESE)
khazam (60.1. URDU)
khɔlɛkɔlɛ̃dɛn-na (62.15. YALUNKA)
kh’èèngz faaz (62.23. LAO)
kh’èèngz khôm (62.25. LAO)
kh’èèngz namz (62.15. LAO)
kh’èèngz saph’au (62.23. LAO)
kheera (38.13a. HINDI)
kheera (38.13a. MARATHI)
kheera (38.13a. PANJABI, EASTERN)
kheeraa (38.13a. HINDI)
khemun-yi (17.5. SUSU)
khērä (38.12. PANJABI, EASTERN)
kheri (40.35. BENGALI)
kherò the (24.11. VENDA)
khesari (40.10. ASSAMESE)
khesari (40.10. BENGALI)
khesari (40.10. HINDI)
khesari (40.10. NEPALI)
khesari mah (40.10. ASSAMESE)
khesra (40.10. ORIYA)
k̲h̲ēyār (38.13a. KURDISH, NORTHERN)
khiar (38.13a. ARABIC)
khiar yapany (38.10. ARABIC)
khi-luu-saa (38.20. THAI)
khîn-chhài (24.4. CHINESE, MIN NAN)
khi:ng (20.6. LAO)
khing (20.6. THAI)
khing-daeng (20.6. THAI)

khiorak (29.1. TAJIKI)
khira (38.13a. BENGALI)
khira (38.13a. FIJIAN)
khira (38.13a. GUJARATI)
khira (38.13a. HINDI)
khírá (38.13a. HINDUSTANI)
kḥīrā (38.13a. MAITHILI)
khira (38.13a. MARATHI)
khira (38.13a. URDU)
khiree (38.6. NEPALI)
khiyâr (38.13a. FARSI, WESTERN)
khiyár (38.13a. HINDUSTANI)
khlôôk (38.20. KHMER, CENTRAL)
khminz khünz (20.2. LAO)
khminz khünz (20.4. LAO)
khnhaè (40.18. KHMER, CENTRAL)
khnhei (20.6. KHMER, CENTRAL)
khnhei phlung (20.6. KHMER, CENTRAL)
khó-koe (38.23. CHINESE, MIN NAN)
khoai ch[oof]i (11.2. VIETNAMESE)
khoai d[as]i (11.2. VIETNAMESE)
khoai dao (9.1. VIETNAMESE)
khoai lang (36.3. VIETNAMESE)
khoai m[if] (39.4. VIETNAMESE)
khoai n[uw]a (8.2. VIETNAMESE)
khoai n[uw]a (8.3. VIETNAMESE)
khoai n[uw][ows]c (8.4. VIETNAMESE)
khoai s[as]p (8.1. VIETNAMESE)
khoai s[as]p (8.10. VIETNAMESE)
khoai s[oj] (8.4. VIETNAMESE)
khoai t[aa]y (62.24. VIETNAMESE)
khoai tây (62.24. VIETNAMESE)
khoai t[uwf] (11.5. VIETNAMESE)
khoai v[aj]c (11.1. VIETNAMESE)
khoai-mi (39.4. VIETNAMESE)
khoai-mi-tinh (15.2. VIETNAMESE)
khoai-mon (8.4. VIETNAMESE)
khobwe (40.2. LOMWE, MALAWI)
khobwe (40.8. LOMWE, MALAWI)
khobwe (40.43. LOMWE, MALAWI)
khobwe (40.43. NYANJA)
khojor gert khalgai (66.3. MONGOLIAN)
khokera (38.14. GEORGIAN)
kholifolawa (30.26. SOTHO, NORTHERN)
kholifolawa (30.26. SOTHO, SOUTHERN)
kholifolawa (30.26. TSWANA)
kholowa (36.3. LOMWE, MALAWI)
kholowa a cizungu (62.24. LOMWE, MALAWI)
khonkhontsetseg (33.2. MONGOLIAN)
khonog budaa (17.6. MONGOLIAN)
khonphir (24.9. GUJARATI)
≠khontsamab (38.3. HERERO)
kh[oor] qua (38.23. VIETNAMESE)
khoorapelum (38.13a. TAMIL)
khoosjan (20.1. URDU)
khotbir (24.9. MARATHI)
khothmir (24.9. MARATHI)
khotmir (24.9. MARATHI)
khoudia guita (62.14a. JOLA-FONYI)
khoulougné (62.14a. PULAR)
khoutoumate (62.9. JOLA-FONYI)
kh[owr]i t[uwr] (62.8. VIETNAMESE)
khrayn (30.1. YIDDISH)
khtüm barang (7.3. KHMER, CENTRAL)
khtüm khchâl (7.8. KHMER, CENTRAL)
khtüm krâhââm (7.2. KHMER, CENTRAL)
khtüm sââ (7.10. KHMER, CENTRAL)
khtüm sânlök (7.5. KHMER, CENTRAL)
khüa ham maaz (62.19. LAO)
khüa hlèèz (62.19. LAO)
(khùa) ‘khao ‘khièo (40.4. LAO)
khüa khôn (62.16. LAO)
khüa ngwàng (45.1. LAO)
khüa poom (62.19. LAO)
khüa puux (62.16. LAO)
khüa sômz (62.9. LAO)
khubasi (45.5. HINDI)
khubasi (45.5. KANNADA)
khubasi (45.5. URDU)
khubkalam (44.5. BENGALI)
khuduugiin shivlee (1.1. MONGOLIAN)
khukhamba (38.13a. ZULU)
khulakhudi (24.7. HINDI)
khumbra (38.20. BODO)
khumkhell (7.12. MONGOLIAN)
khunchai (24.4. THAI)
khûn-chhài (24.4. CHINESE, MIN NAN)
khuräsäni ajwain (24.4. PANJABI, EASTERN)
khurásáni ajwáin (24.19. HINDUSTANI)
khurd pati sarsoo (30.7. URDU)
khurdmirich (62.3. URDU)
khurfa (59.2. TAJIKI)
khurfah (59.2. URDU)
khurmuj (38.7b. BENGALI)
khurpúza (38.7b. HINDUSTANI)
khursa (59.2. HINDI)
khursa (59.4. HINDI)
khutura sak (23.3. ASSAMESE)
khvavilovani combosto (30.26. GEORGIAN)

kí-mbə̄ (16.1a. BETE-BENDI)
kimbombolo (28.1. BEMBÉ)
kim-koe (38.14. CHINESE, MIN NAN)
kim-koe (38.16. CHINESE, MIN NAN)
kimpoi (8.4. BINUKID)
kimpoy (8.4. CEBUANO)
kimpul (8.9. INDONESIAN)
kimpul (8.9. JAVANESE)
kimpul (8.9. SUNDA)
kimpul (8.10. INDONESIAN)
kimpul (8.10. JAVANESE)
kimpul (8.10. SUNDA)
kimput (38.21. SUNDA)
kimput (38.22. SUNDA)
kinagrasløk (7.13. NORWEGIAN)
kinahređka (30.54. ICELANDIC)
kínai kel (30.25. HUNGARIAN)
kínai kel (30.36. HUNGARIAN)
kinai kel (30.40. HUNGARIAN)
kínai ördögcérna (62.8. HUNGARIAN)
kínai retek (30.54. HUNGARIAN)
kínai tehénborsó (40.43. HUNGARIAN)
kinakål (30.40. DANISH)
kinak'al (30.40. ICELANDIC)
kinakål (30.40. NORWEGIAN)
kina-løg (7.13. DANISH)
kinamboi (20.2. CEBUANO)
kinamboy (20.2. CEBUANO)
kinampai (11.1. CEBUANO)
kinampay (11.1. CEBUANO)
kinampáy (36.3. WÁRAY-WÁRAY)
kinaræddike (30.54. DANISH)
kinaraswi (36.3. WÁRAY-WÁRAY)
kinareddik (30.54. NORWEGIAN)
kinarosa (36.3. WÁRAY-WÁRAY)
kīnas kāposti (30.36. LATVIAN)
kīnas kāposti (30.40. LATVIAN)
kinaski (34.2. FULFULDE)
kinaski (34.2. FULFULDE, ADAMAWA)
kinaskok (44.11. DANISH)
kinavandkastanie (10.2. DANISH)
kinayams (11.6. NORWEGIAN)
kinchai (24.4. FILIPINO)
kinchai (24.4. TAGALOG)
kínchay (24.4. PANGASINAN)
kin-chio (16.1a. CHINESE, MIN NAN)
kindayohan (23.6. TAGALOG)
kindzi (24.9. GEORGIAN)
kinesisk kål (30.25. SWEDISH)
kinesisk kål (30.36. NORWEGIAN)
kinesisk kal (30.36. SWEDISH)
Kinesisk kål (30.40. DANISH)
kinesisk kål (30.40. NORWEGIAN)
kinesisk kal (30.40. SWEDISH)
kinesisk lök (7.13. SWEDISH)
kinesiskt bocktörne (62.8. SWEDISH)
kineska piperka (62.4. MACEDONIAN)
kineska rotkva (30.54. MACEDONIAN)
kineska zelka (30.36. MACEDONIAN)
kineska zelka (30.40. MACEDONIAN)
kineski kelj (30.25. MACEDONIAN)
kineski kupus (30.36. SERBO-CROATIAN)
kineski kupus (30.40. SERBO-CROATIAN)
kineski licium (62.8. MACEDONIAN)
kineski luk (7.13. MACEDONIAN)
kineski pasulj (40.43. SERBO-CROATIAN)
kinesko zelje (30.40. CROATIAN)
kineziki rafanis (30.54. GREEK)
kineziko lachano (30.36. GREEK)
kineziko lachano (30.40. GREEK)
kineziko skordo (7.7. GREEK)
kingama (38.15. GONJA)
kingama (38.17. GONJA)
kiníí (62.9. BLACKFOOT)
kininis bastutis (30.36. LITHUANIAN)
kininis ožerškis (62.8. LITHUANIAN)
kinintsai (24.4. TAGALOG)
kinintsay (24.4. TAGALOG)
kinkio (36.3. MANINKAKAN, WESTERN)
kin-mone-thi (38.4. BURMESE)
kino kopūstai (30.36. LITHUANIAN)
kinoa (35.6. CZECH)
kinras (27.15. HEBREW)
kinres (27.15. HEBREW)
kinres tarbuti (27.16. HEBREW)
kinsa (24.9. KAZAKH)
kinsay (24.4. CEBUANO)
kinsay (24.4. PAMPANGAN)
kinsay (24.4. TAGALOG)
kinse (24.4. PAMPANGAN)
kinshinsai (13.3. JAPANESE)
kíntsay (24.4. HILIGAYNON)
kintsay (24.4. SAMBAL, TINÀ)
kintsay (24.4. TAGALOG)
kintse (24.4. PAMPANGAN)
kinusaya (40.27. JAPANESE)
kînya (38.4. GIKUYU)
kĩnya (38.20. GIKUYU)
kiogo (10.1. MÒORÉ)
kiok (27.23. CHINESE, MIN NAN)

kistikanis (24.19. CREE, PLAINS)
kisykał (58.2. SORBIAN, UPPER)
kit cɔkọt (39.4. BEROM)
kit jey (11.1. BEROM)
kit kácik (11.1. BEROM)
kit ryáŋ (11.7. BEROM)
kitaiskii oreh (40.1. RUSSIAN)
kitajski kapus (30.36. SLOVENIAN)
kitajski kapus (30.40. SLOVENIAN)
kitajsko zele (30.36. BULGARIAN)
kitajsko zele (30.40. BULGARIAN)
kitikot (62.5. CEBUANO)
kitjangboon (40.41. VLAAMS)
kitri (38.13a. GEORGIAN)
kitrini (11.3. GREEK)
kitron (38.2. GREEK)
kitse-enelas (27.47. ESTONIAN)
kitunguu (7.3. SWAHILI)
kitunguu maji (7.3. SWAHILI)
kitunguu saumu (7.10. SWAHILI)
kiu (20.6. CHINESE, MIN NAN)
kivachh (40.18. HINDI)
kivakurkku (38.12. FINNISH)
kivanch (40.18. GUJARATI)
kivanch (40.18. HINDI)
kivircik (27.31. TURKISH)
kivircik baş lahana (30.34. TURKISH)
kivircik lahana (30.34. TURKISH)
kivircik nane (44.3. TURKISH)
kiwach (40.18. GUJARATI)
kiwanch (40.18. HINDI)
kiwano (38.12. CZECH)
kiwano (38.12. FINNISH)
kiwano (38.12. FRENCH)
kiwano (38.12. ITALIAN)
kiwano (38.12. POLISH)
kiwano (38.12. PORTUGUESE)
kiwano (38.12. SPANISH)
kiwano furuutsu (38.12. JAPANESE)
kiwingan (9.1. IFUGAO)
kiyaari (34.1. HINDI)
kiyáárots (24.11. CAROLINIAN)
kizhangu (11.1. TAMIL)
kizhangu (11.2. TAMIL)
kizhanna (8.3. MALAYALAM)
kizil-bosh caram (30.27b. UZBEK)
kizilot (2.1. TURKISH)
kjeks (24.8. NORWEGIAN)
kjempegresskar (38.16. NORWEGIAN)
kjempeyams (11.1. NORWEGIAN)
kjørvel (24.3. NORWEGIAN)
kjorvel roe (24.8. DANISH)
kkangkkung (30.52. CAROLINIAN)
kkeyzaran (17.1. FARSI, WESTERN)
kkeyzaran (17.2. FARSI, WESTERN)
kkeyzaran (17.4. FARSI, WESTERN)
klabassen (38.18. STELLINGWERFS)
klabet (40.32. INDONESIAN)
klabet (40.32. JAVANESE)
klabishi (30.27a. ZULU)
klapbes (62.12. AFRIKAANS)
klapbessie (62.12. AFRIKAANS)
klaponik pasjannaja (30.51. BELARUSAN)
klasesto nane (44.3. MACEDONIAN)
klawer (40.17. AFRIKAANS)
klaylania ediapirophyllos (59.1. GREEK)
kleam (34.2. TURKISH)
kłębian fasolowy (40.20. POLISH)
kłębian kątowaty (40.21. POLISH)
kłębian ziemniaczany (40.22. POLISH)
klein knopkruid (27.22. DUTCH)
kleinblaar mosterd (30.7. AFRIKAANS)
kleinbluetige Malve (45.5. GERMAN)
kleinblütiges Franzosenkraut (27.22. GERMAN)
kleinblütiges Knopfkraut (27.22. GERMAN)
kleinc yam (11.5. VLAAMS)
kleine bieslook (7.11. VLAAMS)
kleine zandkool (30.47. VLAAMS)
kleśnica (8.4. POLISH)
klewu (11.7. ÉWÉ)
kli (8.7. THAI)
klimopkalebas (38.4. VLAAMS)
klinglin (11.3. BAOULÉ)
kliswortel (27.5. VLAAMS)
klô afan (23.3. BAOULÉ)
kloelfting (1.1. ICELANDIC)
klokke (33.2. DANISH)
klopovnik posevnoi (30.51. RUSSIAN)
klouala (63.1. BAOULÉ)
klourt (44.1. NORWEGIAN)
kluai (16.1a. THAI)
kluban (30.36. JAVANESE)
klubenli eşekmäş (40.11. TURKMEN)
klubenoviden kervel (24.8. BULGARIAN)
klwekun (40.2. FON)
klwekun (40.2. GUN)
knapsakkêrel (27.8. AFRIKAANS)
knapsakkerwel (27.8. AFRIKAANS)
knapsekêrel (27.8. AFRIKAANS)

kne sukar (17.5. HEBREW)
knidi (66.3. GREEK)
knidousa (66.3. GREEK)
knobbelmeloen (38.7a. DUTCH)
Knoblauch (7.10. GERMAN)
knochkari (40.18. PANJABI, EASTERN)
knof Oostindische kers (65.2. DUTCH)
knoffel (7.10. AFRIKAANS)
knoflook (7.10. DUTCH)
knoflook (7.10. VLAAMS)
knolartisjok (27.26. AFRIKAANS)
knoldet hulsvøb (24.8. DANISH)
knoldfennikel (24.13. DANISH)
knold-fladbælg (40.11. DANISH)
knoldselleri (24.5. DANISH)
knölflocka (24.8. SWEDISH)
knolkervel (24.8. VLAAMS)
knolkhol (30.30. SINHALA)
knolklaverzuring (53.1. DUTCH)
knolldaggry (36.3. NORWEGIAN)
knolle (30.41a. STELLINGWERFS)
knollegies (30.41a. STELLINGWERFS)
Knollen Kerbel (24.8. GERMAN)
Knollenbaselle (28.2. GERMAN)
Knollen-Bohne (40.21. GERMAN)
Knollen-Bohne (40.22. GERMAN)
Knollenfenchel (24.12. GERMAN)
Knollen-Kapuzinerkresse (65.2. GERMAN)
Knollen-Petersillie (24.21. GERMAN)
Knollen-Platterbse (40.11. GERMAN)
Knollen-Sauerklee (53.1. GERMAN)
Knollen-Sellerie (24.5. GERMAN)
Knollen-Yam (11.2. GERMAN)
Knollen-Ziest (44.11. GERMAN)
knollfennikel (24.13. NORWEGIAN)
knollgaukesyre (53.1. NORWEGIAN)
Knollige-Bohne (40.21. GERMAN)
knolliger Kälberkropf (24.8. GERMAN)
knolliger Sauerklee (53.1. GERMAN)
knolljordskokk (27.26. NORWEGIAN)
knoll-kervel (24.8. DUTCH)
knollkjeks (24.8. NORWEGIAN)
knollmaniok (39.4. NORWEGIAN)
knollselleri (24.5. NORWEGIAN)
knollsivaks (10.2. NORWEGIAN)
knollsolsikke (27.26. NORWEGIAN)
knollsvinerot (44.11. NORWEGIAN)
knolpeterselie (24.21. DUTCH)
knolraap (30.30. AFRIKAANS)
knolribzand (24.8. DUTCH)
knolselderij (24.5. DUTCH)
knolselderij (24.5. VLAAMS)
knolseldery (24.5. AFRIKAANS)
knolvenkel (24.12. DUTCH)
knolvenkel (24.13. VLAAMS)
knopdragende mosterd (30.8. VLAAMS)
knopkruid (27.22. AFRIKAANS)
knopkruid (27.22. VLAAMS)
knotroot (44.11. ENGLISH)
knubel (7.10. YIDDISH)
knudekål (30.30. DANISH)
Knueweléck (7.10. LUXENBOURGEOIS)
knutekål (30.30. NORWEGIAN)
ko (40.30. CHINESE, MANDARIN)
ko fufe (39.4. MALIMBA)
ko pintan (39.4. MALIMBA)
koa (30.13. CHINESE, MIN NAN)
koa (38.3. CHINESE, MIN NAN)
koa (38.7. CHINESE, MIN NAN)
kòa-chhài (30.36. CHINESE, MIN NAN)
koas bakol (40.3. SUNDA)
koas bakol (40.4. INDONESIAN)
koas parasman (40.3. SUNDA)
koas pedang (40.3. SUNDA)
kobi (30.27a. BODO)
kobi (30.27a. MARATHI)
kobi (30.27a. MUNDARI)
kobi (30.27a. SOMALI)
kobij (30.27a. GUJARATI)
kobis (30.27a. INDONESIAN)
kobis (30.27b. INDONESIAN)
kobis (30.34. INDONESIAN)
kobis cina (30.40. MALAY)
kobokobo (62.19. SUSU)
kobołk (7.10. SORBIAN, UPPER)
Kochbanane (16.1a. GERMAN)
kochchi (62.5. SINHALA)
kochia (35.2. ENGLISH)
kochia (35.2. HEBREW)
kochia à balais (35.2. FRENCH)
kochie (35.2. FRENCH)
kochija (35.2. RUSSIAN)
kochliaris e armorakia (30.1. GREEK)
kocho (16.1a. OROMO, WEST CENTRAL)
Kochsalat (27.32. GERMAN)
kochu (8.4. ASSAMESE)
kociandr setý (24.9. CZECH)
kockia saromatiki (35.6. GREEK)
kocnjit (20.2. VLAAMS)
kôc-ts'ói (35.9. CHINESE, HAKKA)

kodagan (24.7. MALAYALAM)
kodan meguvan (39.3. HEBREW)
kodangal (24.7. MALAYALAM)
kodatharam (20.6. TAMIL)
kodijuttutotakura (23.4. TELUGU)
kodikilangu (11.2. TAMIL)
kodithambattan (40.3. TAMIL)
kodshŏdo (38.16. TEM)
kodu (16.1a. DANGME-KROBO)
kodu (38.20. BENGALI)
kodu (38.20. ORIYA)
kodulau (38.20. BENGALI)
kodupai (23.2. TAMIL)
koe (17.1. RAROTONGAN)
koe (30.13. CHINESE, MIN NAN)
koe (38.3. CHINESE, MIN NAN)
koe (38.7. CHINESE, MIN NAN)
koedzoe (40.30. DUTCH)
koeja (38.5. LOKO)
koendoro (24.9. JAPANESE)
koertjie (40.43. AFRIKAANS)
koe-yaba (7.3. THEMNE)
kofe mindalnyi (10.1. RUSSIAN)
kogidi (36.2. MENDE)
kogo (8.9. MALIMBA)
kogong (62.16. MARANAO)
kohala (38.1. GUJARATI)
kohala (38.1. HINDI)
kohala (38.1. MARATHI)
kohala (38.18. MARATHI)
kohbuti (36.3. IFUGAO)
koheriki (27.8. MAORI)
kohhia (35.2. ESTONIAN)
kohirugao (36.1. JAPANESE)
kohkompaninawak (38.13a. CREE, PLAINS)
kohla (38.16. GUJARATI)
Kohl-Gänsedistel (27.42. GERMAN)
kohlrabi (30.30. ENGLISH)
Kohlrabi (30.30. GERMAN)
kohlrabi (30.30. HEBREW)
Kohlraps (30.21. GERMAN)
Kohlrübe (30.21. GERMAN)
kohlrübe (30.22. GERMAN)
kohlu (38.1. GUJARATI)
kohn furii (29.2. JAPANESE)
kohn salada (67.2. JAPANESE)
kohṇḍa (38.17. SANTALI)
kohoutí hřeben (23.4. CZECH)
kohrurabi (30.30. JAPANESE)
kohu (27.29. TAJIKI)
kohviherne (40.5. FINNISH)
koi (50.2. HINDI)
koilangu (36.2. TAMIL)
koja (38.3. MENDE)
koja (38.5. MENDE)
koja-lowa (38.18. MENDE)
koja-mumu (38.5. MENDE)
koja-wawa (38.5. MENDE)
koje's (40.26. POTAWATOMI)
koje'suk (40.26. POTAWATOMI)
kojo (62.19. MENDE)
kok (8.4. YORUBA)
kök hindiba (27.10a. TURKISH)
kok karn pu (4.1. LAO)
kök kereviz (24.5. TURKISH)
kök maydanoz (24.21. TURKISH)
kok nuchat (40.28. UZBEK)
koka (50.2. HINDI)
kokan (23.4. HINDI)
kokebanan (16.1a. NORWEGIAN)
kokhyat ha-makhbed (35.2. HEBREW)
koki chisha (27.32. JAPANESE)
kokilaakshamu (34.1. TELUGU)
kokilakshamu (34.1. TELUGU)
kokja (35.2. MALTESE)
kòkkino lachano (30.27b. GREEK)
kokkinophyllom kichorion (27.10b. GREEK)
kokleng (40.2. BODO)
koklodame (44.5. GEN)
koko (8.1. ÉWÉ)
koko (8.4. BÉTÉ)
koko (8.4. ÉWÉ)
koko (8.4. KRIO)
koko (8.4. MENDE)
koko (8.4. YAO)
kókò (8.4. YORUBA)
koko (8.9. KONO)
koko (8.9. KRIO)
koko (8.9. LOKO)
koko (8.9. MENDE)
koko (8.9. SUSU)
ƙoƙo (38.20. HAUSA)
kókò funfun (8.4. YORUBA)
kókò pupa (8.4. YORUBA)
kokobe' (38.13a. POTAWATOMI)
kokoha (27.19. JAVANESE)
kokolimakoe (11.7. ÉWÉ)
kokómbà (38.13a. GA)
kokondo (40.34a. KRIO)
kòkondó pòpondó (40.25. YORUBA)

kòkònm (38.13a. KWÉYÒL)
kokoobu (8.4. WAAMA)
kokori (17.7. GOURMANTCHÉ)
kokoška pastuší tobolka (30.43. CZECH)
kokosyams (8.4. DANISH)
kokoti (44.6. TAJIKI)
kokoyam (8.4. HUNGARIAN)
kokuamu (62.2. ANDO)
koku-lɛ (8.9. BULLOM SO)
kokulu muhabbet çiçeği (60.1. TURKISH)
kokwa (59.2. TUMBUKA)
kokwamba (38.13a. HAUSA)
kokwe sozu (16.1a. FON)
kol (16.1a. ASSAMESE)
kol (30.27a. INDONESIAN)
kol (30.27a. LADINO)
kol (30.27a. SUNDA)
kol (30.27b. INDONESIAN)
kol (30.34. INDONESIAN)
kol babat (30.22. SUNDA)
kol babat (30.34. INDONESIAN)
kol bunga (30.26. INDONESIAN)
kol dil (16.1b. ASSAMESE)
kol puti (30.27a. INDONESIAN)
kol rap (30.41a. INDONESIAN)
k/ola (62.2. ZUNI)
kola mah (40.10. ASSAMESE)
kola soriah (30.24. ASSAMESE)
kolabtos (44.5. TAGALOG)
kolanro (24.9. HOPI)
kolasiman (59.2 TAGALOG)
kolawit (40.42. MARANAO)
kolche (40.1. KANURI)
kolcowój chiński (62.8. POLISH)
kolczoch jadalny (38.25. POLISH)
koldi (40.1. KANURI)
koldji (40.1. KANURI)
kolé (38.27. GOURMANCHÉMA)
ko-lê-chhài (30.27a. CHINESE, MIN NAN)
kolempang (39.4. MARANAO)
kolender siewna (24.9. POLISH)
kolerarbica (30.30. SLOVENIAN)
kolhiyavarai (40.3. TAMIL)
kol-i (30.27a. HERERO)
kolí (30.31. HINDUSTANI)
koli flawa (30.26. TOK PISIN)
koli maua (30.26. SWAHILI)
koli plawer (30.26. TAGALOG)
kolili (40.42. MARANAO)
kolimaua (30.26. SWAHILI)
kol-inji (20.1. MALAYALAM)
kolintasan (9.1. CEBUANO)
kolirabi (30.30. SWAHILI)
kolirlja (20.1. MARATHI)
kolis (30.35. FILIPINO)
kolis (30.35. TAGALOG)
kolitis (23.3. TAGALOG)
kō-li-ts'ói (30.27a. CHINESE, HAKKA)
koljander (24.9. AFRIKAANS)
kólji (40.1. KANURI)
kolkoti (17.7. ZARMA)
kolkythaki (38.18. GREEK)
kolliahdros (24.9. GREEK)
kollitsoharto (17.6. GREEK)
kollokasi (27.26. GREEK)
kollu (40.15. TAMIL)
koḷḷu (40.16. BADAGA)
kolmow sak (36.2. ASSAMESE)
kolokasi (8.4. GREEK)
kolokasia edodimos (8.4. GREEK)
kolokasie (8.4. ALBANIAN, TOSK)
kolokasie (8.4. CZECH)
kolokasya (8.4. TURKISH)
kolokaz (8.4. TURKISH)
kolokazija (8.4. BULGARIAN)
kolokazja (8.4. POLISH)
kolokazja błotna (8.6. POLISH)
kolokazja filipińska (8.7. POLISH)
kolokazja jadalna (8.4. POLISH)
kolokynthi (38.16. GREEK)
kolokynthito moschosmo (38.17. GREEK)
kolokythi (38.16. GREEK)
kolokythi (38.18. GREEK)
kolokythito fiko-phyllo (38.15. GREEK)
kolonse (40.1. ZARMA)
kolo-oro (63.1. MANDING-MANDINKA)
kolo-oro (63.1. MANINKAKAN, WESTERN)
kolrabi (30.30. BULGARIAN)
kolrabi (30.30. HEBREW)
kolrabi (30.30. KAZAKH)
kol'rabi (30.30. RUSSIAN)
kol'rabi (30.30. UKRAINIAN)
kol'rabi kapustasy (30.30. KIRGHIZ)
kolrābji (30.30. LATVIAN)
kolumbi (30.30. INDONESIAN)
kólúo (8.4. GURO)
koluppa (23.2. MALAYALAM)
koluppa (59.2. MALAYALAM)
kolza (30.21. TURKISH)
kolza lahanasi (30.36. TURKISH)

kolza lahanasi (30.41a. TURKISH)
komak (27.29. INDONESIAN)
komak (40.9. INDONESIAN)
Komarov loogkruid (35.7. DUTCH)
Komarov Russian thistle (35.7. ENGLISH)
komarov-sodaurt (35.7. DANISH)
komatsuna (30.39. DUTCH)
komatsuna (30.39. JAPANESE)
komatsuna (30.39. PORTUGUESE)
komatsuna (30.39. VLAAMS)
komatsuna (30.41a. JAPANESE)
kombo' (45.1. CHICKASAW)
kombo' awaalhaali' (45.1. CHICKASAW)
kombósto (30.27a. GEORGIAN)
kombu (38.31. TAMIL)
kombupaavchedy (38.23. TAMIL)
komee' (17.7. PAIUTE, SOUTHERN)
komeŋa (44.5. WAMEY)
komètan (62.9. JOLA-FONYI)
Komfrey (29.3. GERMAN)
komkoma (20.2. INDONESIAN)
komkommer (38.13a. AFRIKAANS)
komkommer (38.13a. DUTCH)
komkommer (38.13a. STELLINGWERFS)
komkommer (38.13a. VLAAMS)
komkommerkruid (29.1. AFRIKAANS)
komkommerkruid (29.1. VLAAMS)
komló (46.1. HUNGARIAN)
komma gaggiri (44.5. TELUGU)
kommanatti (16.1a. TELUGU)
kommaratti (16.1a. TELUGU)
kommu ariti (16.1a. TELUGU)
kommu potla (38.31. TELUGU)
kommupotla (38.31. TELUGU)
komo (23.3. HOPI)
komochi-kanran (30.29. JAPANESE)
komochi-kyabetsu (30.29. JAPANESE)
komokomore (38.13a. SOTHO, SOUTHERN)
komorač (24.13. SERBO-CROATIAN)
komosa ryżowa (35.6. POLISH)
komosa strzałkowata (35.5. POLISH)
kompir (62.24. MACEDONIAN)
kòmpiri (62.24. ROMANI, BALKAN)
komprinya dua (62.18. AKAN-ASANTE)
komtouwu (36.3. GOURMANTCHÉ)
kôn (40.23. BULU)
kôn (40.24. BULU)
kôn (40.25. BULU)
kôn (40.26. BULU)
kôn (40.33. BULU)
kon bilahi (62.10. ASSAMESE)
kona (11.1. KONKANI)
≠kon≠āb (17.5. HERERO)
konaphala (11.2. MARATHI)
konarourou (40.15. HAUSA)
konch (40.18. HINDI)
konda (38.18. HINDI)
konda mirapa (62.5. TELUGU)
kondakandi (40.2. TELUGU)
kondavuste (62.23. TELUGU)
konde (16.1a. GOLA)
konde (16.1a. MENDE)
konde balli (38.4. KANNADA)
kondeke (16.1a. KONO)
kondek-habi-na (39.4. YALUNKA)
kondha (38.1. HINDI)
kondol (38.1. FILIPINO)
kondol (38.1. IVATAN)
kondol (38.1. TAGALOG)
kondól (38.1. WÁRAY-WÁRAY)
kondong (16.1a. FULFULDE, ADAMAWA)
kondongji (16.1a. FULFULDE, ADAMAWA)
kondouba (36.3. BAMANANKAN)
konduba (36.3. BAMANANKAN)
kondvalli (38.4. MARATHI)
kondylodis (24.8. GREEK)
koneg (20.4. BONTOC, CENTRAL)
koneng (20.2. SUNDA)
koneng joho (20.3. SUNDA)
koneng lalab (20.3. SUNDA)
koneng manggah (20.3. SUNDA)
koneng pari (20.3. SUNDA)
koneng tegal (20.4. SUNDA)
koneng temen (20.2. SUNDA)
kong xin cai (36.2. CHINESE, GÀN)
kong xin cai (36.2. CHINESE, MIN BEI)
kong xin cai (36.2. CHINESE, WÚ)
kong xin cai (36.2. CHINESE, XIANG)
kong xin cai (36.2. CHINESE, YUE)
kong xin xian (23.1. CHINESE, YUE)
konggu (20.6. BUGHOTU)
kongo dĭé (38.7b. BAMANANKAN)
koŋgo-binch (40.2. KRIO)
koŋgoe (38.20. VAI)
ko̱nhṛá (38.14. HINDUSTANI)
konhra (38.17. HINDUSTANI)
konik (20.4. ILOCANO)
koningskruid (44.5. DUTCH)
konjac (8.2 ENGLISH)
konjac (8.3. JAPANESE)

kora (38.20. TWI)
kora wa (38.20. TWI)
korabica (30.30. CROATIAN)
korajaa (20.1. SANSKRIT)
korake (35.1. HINDI)
korake (35.1. PANJABI, EASTERN)
korankɔ-c-bi (38.18. KONO)
koranro (24.9. HOPI)
korbjel (24.3. SORBIAN, UPPER)
korchef (27.15. ARABIC)
korchoros o lachanodis (63.1. GREEK)
kore (27.10a. ALBANIAN, TOSK)
korela (38.23. GUJARATI)
korenček (24.11. SLOVENIAN)
koreneva selera (24.12. UKRAINIAN)
korenoplodna celina (24.5. BULGARIAN)
korenorepkast magdonos (24.21. MACEDONIAN)
korenov kereviz (24.5. MACEDONIAN)
korforɔm (59.3. YOM)
korgona (34.2. SERER-SINE)
korianda (24.9. JAPANESE)
koriandër (24.9. ALBANIAN, TOSK)
koriander (24.9. DANISH)
koriander (24.9. ESTONIAN)
Koriander (24.9. GERMAN)
koriander (24.9. HUNGARIAN)
koriander (24.9. NORWEGIAN)
koriander (24.9. SLOVAK)
koriander (24.9. SWEDISH)
koriander (24.9. VLAAMS)
koriandron (24.9. GREEK)
koriandrs (24.9. LATVIAN)
koriandru cultivat (24.9. ROMANIAN)
korianta (24.9. KIRIBATI)
korianteri (24.9. FINNISH)
koricchira (59.2. MALAYALAM)
korichhira (59.2. MALAYALAM)
korijander (24.9. SERBO-CROATIAN)
korilla (38.19. ENGLISH)
korintas sa kalasan (9.1. MARANAO)
koristekaali (30.35. FINNISH)
kórka (38.13a. SORBIAN, UPPER)
kộrkọmbọ (16.1a. TIV)
korkore (63.1. ALBANIAN, TOSK)
korn flor (9.1. MORISYEN)
kornauli (24.4. PANJABI, EASTERN)
kornevoj sel'derej (24.5. RUSSIAN)
kornichon (38.13b. MORISYEN)
Kornischong (38.13a. LUXENBOURGEOIS)
koro buncis (40.26. JAVANESE)
koro gabu (44.5. RMA)
koro goro (40.25. LIMBA)
koro lompeh (40.26. JAVANESE)
koro tonggak (40.26. JAVANESE)
korodu (40.18. MANDINKA)
koroha (40.32. JAPANESE)
kørvel (24.3. DANISH)
körvel (24.3. SWEDISH)
kôrvenôges (66.3. ESTONIAN)
kõrvits (38.14. ESTONIAN)
kõrvits (38.18. ESTONIAN)
kõ-sa (38.5. KONO)
kosala saga (23.3. ORIYA)
kosapuvvu (30.27a. TELUGU)
kosarnik (24.8. SLOVENIAN)
kosataki (38.21. SANSKRIT)
kosbanti (40.42. HINDUSTANI)
koshaathakee (38.21. SANSKRIT)
koshaphala (38.13a. SANSKRIT)
koshataki (38.21. SANSKRIT)
koshataki (38.22. SANSKRIT)
koshibba' (56.2. CHICKASAW)
koshōsō (30.51. JAPANESE)
koshta (63.1. BENGALI)
koshta (63.1. HINDI)
koshtakee (38.22. SANSKRIT)
koshtkulayan (20.1. MARATHI)
koshtkulinjan (20.1. MARATHI)
kosloborodnik lugovoj (27.47. KAZAKH)
kosmatec křištálový (22.1. CZECH)
kosmatec krystalový (22.1. CZECH)
kosmos (27.12. ESTONIAN)
kosmos (27.12. GREEK)
kosni (27.9. TAJIKI)
kösnü otu (10.1. TURKISH)
koso (40.2. LIMBA)
koso (62.19. BULLOM SO)
kosol-kosol (18.1. CEBUANO)
kosoma'si' (7.10. CHICKASAW)
kossa fina (27.24. MANDING-MANDINKA)
kossé (40.18. PULAAR)
kostival lékařský (29.2. CZECH)
kostival uplandský (29.3. CZECH)
kostmelde (35.2. NORWEGIAN)
kosu (30.27a. KANNADA)
kosu gadde (30.27a. KANNADA)
kosuguddae (30.26. KANNADA)
kôta (45.4. MANDING-MANDINKA)
kotahi alu (11.7. ASSAMESE)

kotahi bengena (62.16. ASSAMESE)
kotari dobua mah (40.4. ASSAMESE)
ko-tatami (17.1. BAGA)
kotbor (24.9. KONKANI)
kotbori (24.9. KONKANI)
kotewka orzech wodny (64.3. POLISH)
kothamalli (24.9. MALAYALAM)
kothamalli (24.9. TAMIL)
kothambar (24.9. KONKANI)
kothambari (24.9. TULU)
kothambri (24.9. KANNADA)
kothamir (24.9. GUJARATI)
kothavara (40.6. MALAYALAM)
kothavarnkai (40.6. TAMIL)
kothaveray (40.6. TAMIL)
kothimari (24.9. KANNADA)
kothimber (24.9. MARATHI)
kothimbri (24.9. KANNADA)
kothimiri (24.9. TELUGU)
kothinbir (24.9. MARATHI)
koththamalli (24.9. SINHALA)
kothumbari (24.9. KANNADA)
kotia (45.4. MANDING-MANDINKA)
koto (45.2. THEMNE)
kotok bongkok (19.1. SUNDA)
kotomili (24.9. MORISYEN)
kotonosou (40.31. TEM)
kotonosu (40.31. TEM)
kotopalak (62.24. ALEUT)
kotoroku (62.14a. WAAMA)
kötözoő saláta (27.32. HUNGARIAN)
kötözosaláta (27.32. HUNGARIAN)
kotra (40.33. ASSAMESE)
kottamalli (24.9. BADAGA)
kottambari (24.9. TULU)
kottamberi (24.9. TULU)
kottapayuru (40.43. MALAYALAM)
kotthamalii keerai (24.9. TAMIL)
kotthamalli (24.9. TAMIL)
kotthambari (24.9. KANNADA)
kotthu katthari (62.23. TAMIL)
kotthumalli (24.9. MALAYALAM)
kotthumpalari beeja (24.9. MALAYALAM)
kottukkatari (62.23. TAMIL)
kottuthambattam (40.3. TAMIL)
kottuvastu (62.23. TELUGU)
kóturuwaáánka (39.4. AKAN-ASANTE)
kotvice plovoucí dvouostná (64.3. CZECH)
kotvice plovoucí pravá (64.3. CZECH)
kóu (8.4. TEM)
kou (11.1. JULA)
kou (11.3. JULA)
kou fetini (11.5. JULA)
Kouarourou (40.15. GERMAN)
koubatagcé (62.14c. JOLA-FONYI)
koucarote (24.11. JOLA-FONYI)
kouconcombre (38.13a. JOLA-FONYI)
koudzou (40.30. FRENCH)
kouéme (38.28. FRENCH)
koui (38.20. GREEK)
koukia chlora (40.34a. GREEK)
koukki (40.34a. GREEK)
koukou (11.3. KABIYÉ)
koukoumero-pbis (38.5. GREEK)
koukoumic (38.13b. GREEK)
koukourita (38.31. GREEK)
koukourvita (38.4. GREEK)
koúku (8.4. DAGBANI)
koulourdrenn (38.18. BRETON)
koumba (27.18. KISI, SOUTHERN)
koumbi (27.18. MANINKAKAN, WESTERN)
koumboha (17.7. JOLA-FONYI)
koungou (2.1. KOONGO)
kouniak d'annam (8.3. FRENCH)
kounoupidi (30.26. GREEK)
kouoyu (17.7. GONJA)
kourkoumi hie long (20.2. GREEK)
kourzet (38.18. MORISYEN)
kouseband (40.42. DUTCH)
kouseband (40.42. VLAAMS)
kouson (30.52. KWÉYÒL)
koussaac (40.26. JOLA-FONYI)
koutia (45.4. MANINKAKAN, WESTERN)
kova (38.4. MALAYALAM)
kova ha-nazir (65.1. HEBREW)
kovai (38.4. TAMIL)
kovai fruit (38.4. ENGLISH)
kovaikaai (38.4. TAMIL)
kovaikai (38.4. TAMIL)
kovalam (38.4. MALAYALAM)
kove hittu (15.2. KANNADA)
köver porcsin (59.2. HUNGARIAN)
ķovoķča covock (38.18. UZBEK)
kovun (38.7a. UZBEK)
kovun (38.7b. UZBEK)
kovun (38.11. UZBEK)
kowas (40.18. SUNDA)
kowe (44.5. GA)
kowõ (40.25. KISSI, NORTHERN)
kowòt (24.11. KWÉYÒL)

kowva (38.4. MALAYALAM)
koyaganasu gadda (39.4. TELUGU)
koyambattai (62.19. TAMIL)
koyapa (23.3 . CEBUANO)
koyi (50.2. HINDI)
köynnöskrassi (65.1. FINNISH)
koyo a bolawan (39.3. MARANAO)
koyun sarmaşiği (35.1. TURKISH)
kozalec (27.6. CZECH)
kozengobouya (13.1. KABIYÉ)
kozhiavarai (40.18. TAMIL)
kozí brada fialová (27.46. CZECH)
kozí brada luční (27.47. CZECH)
kozia brada pórolistná (27.46. SLOVAK)
kozibrod lakowy (27.47. POLISH)
kozieradka (40.32. POLISH)
kozieradka pospolita (40.32. POLISH)
kozja brada (27.38. SERBO-CROATIAN)
kozja brada (27.46. SERBO-CROATIAN)
kozja brada (27.47. SERBO-CROATIAN)
kozjak (7.2. SERBO-CROATIAN)
kozłek (27.21. POLISH)
kozłek lekarski (27.21. POLISH)
kozlíček polní (67.2. CZECH)
kozlik (27.21. CZECH)
kozmos (27.12. TURKISH)
közönséges bojtorján (27.5. HUNGARIAN)
közönséges tök (38.18. HUNGARIAN)
*k*pã (17.5. VAI)
k-pai (38.20. GOLA)
kpàkɔ́ (40.1. KRIO)
kpakpasan (59.2. EDE NAGO)
kpakpasan (59.2. YORUBA)
kpakpoʃitɔ (62.2. GA)
kpakposhito (62.5. GA)
k-pali (38.20. GOLA)
kpaliegu (40.3. DAGBANI)
kpamaje (40.43. IGBO)
kpame (59.3. MENDE)
*k*paŋ (17.5. VAI)
kpạṇ fiṇni (45.1. NUPE-NUPE-TAKO)
*k*paŋbɔi-gbegbe (62.19. GOLA)
*k*paŋbɔi-gɔni (62.19. GOLA)
k-pande-wendẽ (36.3. KISSI, NORTHERN)
*k*paŋgbo (62.19. GOLA)
kpange (23.4. MENDE)
k-panja (36.3. KISSI, NORTHERN)
kpạṇmi (45.1. NUPE-NUPE-TAKO)
kpạṇmi enạ guru (45.1. NUPE-NUPE-TAKO)
kpạṇmi gbwọ̀dwộ (45.1. NUPE-NUPE-TAKO)
kpạṇmi ròkò (45.1. NUPE-NUPE-TAKO)
kpànsạ́nạ́kó (17.5. NUPE-NUPE-TAKO)
kpaome (59.3. MENDE)
kpasa (11.3. ÉWÉ)
kpasà (38.20. NUPE-NUPE-TAKO)
kpasajɔ (11.7. DAGBANI)
kpatu (38.20. DANGME-KROBO)
*k*pawo (27.2. MENDE)
kpawun kwi (40.43. AJA)
k-pɔklɔ-lɛ (9.1. BULLOM SO)
*k*pɔlɔ (16.1a. KISSI, NORTHERN)
kpɔlɔ (62.19. BULLOM SO)
*k*pɔnigojiyo (62.19. KISSI, NORTHERN)
kpɔni-kojo (62.19. MENDE)
*k*pɔni-yalo (62.19. KONO)
*k*pɔni-yawo (62.19. KONO)
k-pɔwe (9.1. BULLOM SO)
kpekpèou (8.9. KABIYÉ)
kpelekpelei (8.4. YOM)
kpem (39.4. EWONDO)
kpende-mane-lɛ (16.1a. BULLOM SO)
kpeney (11.7. DAGBANI)
k-pɛŋgbɛ (62.5. MANDING-MANINKA)
kpero (8.4. KABIYÉ)
kpèsè kpèsè (62.5. ANDO)
kpèsè-kpèsè (62.5. BAOULÉ)
kpmane (38.20. DAGAARE, SOUTHERN)
k-po (38.18. VAI)
kpo uman (11.7. EFIK)
k-pohi (8.4. LOKO)
kpoji (8.4. MENDE)
kpokpo (36.2. MENDE)
kpokpondo (40.3. KRIO)
kpome-ségue (38.20. SENOUFO, CEBAARA)
k-posi (8.9. VAI)
kprenga (11.7. DAGBANI)
kpula (38.20. MENDE)
kpulɔ (38.20. MENDE)
kra chap (64.3. THAI)
krabilice hlíznatá (24.8. CZECH)
krabljica (24.8. SERBO-CROATIAN)
krachap (64.3. KHMER, CENTRAL)
krachiap-daeng (45.4. THAI)
krachiapkhieo (45.1. THAI)
krachiap-mon (45.1. THAI)
krachiap-prieo (45.4. THAI)
krach-koko (8.4. KRIO)
Krachsalat (27.31. GERMAN)
kradaat (8.1. THAI)
kradaat dam (8.1. THAI)

kradatdam (8.1. THAI)
kradat-dam (8.10. THAI)
krafes (24.4. MALTESE)
krafs romi (24.4. URDU)
Krähenfuß-Wegerich (57.1. GERMAN)
kṛāk (40.28. PASHTO, NORTHERN)
krak-sivolez (7.5. BRETON)
králik indický (27.23. SLOVAK)
kramanti (10.1. KRIO)
krambë (30.45. ALBANIAN, TOSK)
krambe (30.45. SERBO-CROATIAN)
krambe (30.46. ROMANIAN)
krambe he lachanodes (30.35. GREEK)
kramui e paralios (30.45. GREEK)
kramvi (30.27a. GREEK)
kramvi e eleoforos (30.23. GREEK)
kramvi e ofthalmoforos (30.29. GREEK)
kramvi e rapys (24.21. GREEK)
kramvi e rocpys (30.41a. GREEK)
kramvi e tatariki (30.46. GREEK)
kramvi z akèfalos (30.35. GREEK)
krandang (40.3. JAVANESE)
kranskaasjeskruid (45.5. DUTCH)
krąpiel pierzastodzielna (19.1. POLISH)
krapiva dvoudomnaja (66.3. BELARUSAN)
krapiva dvoudomnaja (66.3. RUSSIAN)
kraskentia (38.20. GREEK)
krassikanankaali (30.2. FINNISH)
krastavac (38.13a. BOSNIAN)
krastavac (38.13a. SERBO-CROATIAN)
krastavci (38.13a. CROATIAN)
krastavica (38.13a. BOSNIAN)
krastavica (38.13a. BULGARIAN)
krastavica (38.13a. MACEDONIAN)
krasulica (24.3. SERBO-CROATIAN)
krasuljica (24.3. SERBO-CROATIAN)
krathiam (7.10. THAI)
krathiam-bai (7.8. THAI)
krathiam-chin (7.4. THAI)
krathiam-ton (7.8. THAI)
kratok (40.25. DUTCH)
kratok (40.25. INDONESIAN)
kratok (40.25. JAVANESE)
kratse (8.4. GA)
kratto (7.6. JAPANESE)
Krauseminze (44.3. GERMAN)
kraykray banana (16.1a. KRIO)
krebuljica (24.3. SLOVENIAN)
krèh-sông (30.52. KHMER, CENTRAL)
krejce (30.27a. ALBANIAN, TOSK)
krele (27.9. ALBANIAN, TOSK)
krema (23.2. INDONESIAN)
kremmy di (7.3. GREEK)
kremmy di to koino (7.3. GREEK)
kremmydi (7.3. GREEK)
kremmydi to askoclonio (7.2. GREEK)
kremmydi to syriggodes (7.5. GREEK)
kren (30.1. ALBANIAN, TOSK)
křen (30.1. CZECH)
kren kren (63.1. WOLOF)
křen selský (30.1. CZECH)
kreŋkre (63.1. KRIO)
kreŋkreŋ (63.1. KRIO)
krēr (34.1. PANJABI, EASTERN)
kres (30.51. SERBO-CROATIAN)
kres salata (30.51. SERBO-CROATIAN)
kres salata grbac (30.51. MACEDONIAN)
kresh (30.51. ORIYA)
kresha (7.6. HEBREW)
kresha (7.8. HEBREW)
kréson (30.52. CREOLE, HAITIAN FRENCH)
kress (30.52. ESTONIAN)
kress vodjanoi (30.52. RUSSIAN)
kres-salat (30.51. UKRAINIAN)
kres-salata (30.51. BULGARIAN)
kressalati (30.51. LATVIAN)
kress-salat (30.51. ESTONIAN)
kress-salat (30.51. RUSSIAN)
křez zední (30.47. CZECH)
krilest bob (40.29. MACEDONIAN)
krilest grav (40.13. MACEDONIAN)
krilobob (40.13. MACEDONIAN)
krimighna (38.23. SANSKRIT)
krimika (30.24. SANSKRIT)
krimilaka (40.35. SANSKRIT)
krimos (13.4. GREEK)
krimos imeras (13.3. GREEK)
krin (13.4. SERBO-CROATIAN)
kriŋ krin (63.1. BULLOM SO)
krinkrin (63.1. KRIO)
kriŋkrin-dɛ (63.1. BULLOM SO)
kripik melinju (41.1. MALAY)
krisan (27.23. INDONESIAN)
krishnabeeja (38.3. SANSKRIT)
krishnachanchuka (40.5. SANSKRIT)
krishnakamal (50.2. MARATHI)
krishnasarhapa (30.24. SANSKRIT)
krishnathaamara (9.1. SANSKRIT)
krishnathaamara (9.1. TELUGU)
krishnika (30.24. SANSKRIT)

kristaljas keskpäevalill (22.1. ESTONIAN)
kristályvirág (22.1. HUNGARIAN)
krištolinė pluoštagėlė (22.1. LITHUANIAN)
kritavedhana (38.21. SANSKRIT)
križani (23.3. SERBO-CROATIAN)
krizantém (27.23. HUNGARIAN)
krizantemë (27.23. ALBANIAN, TOSK)
krizantena (27.23. TURKISH)
krkoška hl'uznatá (24.8. SLOVAK)
krmná kapusta (30.33. CZECH)
krmna kapusta (30.35. CZECH)
kro (17.7. BALTI)
krobonko (38.28. AKAN-AKUAPEM)
krobonko (38.28. DANGME)
krobonko (38.28. DANGME-KROBO)
krobonko (38.28. GA)
krocień (39.3. POLISH)
krodakanya (11.2. SANSKRIT)
krodi (11.2. SANSKRIT)
krodin (44.10. EDE NAGO)
krodin (44.10. YORUBA)
krodyu (44.10. YORUBA)
krokot (23.1. JAVANESE)
krokot (23.2. INDONESIAN)
krokot (59.2. INDONESIAN)
krokot (59.2. JAVANESE)
krokot belanda (59.3. INDONESIAN)
kromid rezanec (7.11. MACEDONIAN)
krompir (62.24. BOSNIAN)
krompir (62.24. SERBO-CROATIAN)
krompir (62.24. SLOVENIAN)
kronärtskocka (27.16. SWEDISH)
kroot (35.4. DUTCH)
krooton (39.3. ESTONIAN)
krootoni (39.3. FINNISH)
kropmosterd (30.4. VLAAMS)
kropsla (27.31. DUTCH)
kropsla (27.31. VLAAMS)
kroton (39.3. AFRIKAANS)
kroton (39.3. CZECH)
kroton (39.3. DANISH)
kroton (39.3. GREEK)
kroton (39.3. HUNGARIAN)
kroton (39.3. ICELANDIC)
kroton (39.3. INDONESIAN)
kroton (39.3. NORWEGIAN)
kroton (39.3. POLISH)
kroton (39.3. SWEDISH)
kroton (39.3. TURKISH)
krtolest sončogled (27.26. MACEDONIAN)
krtolesta batanka (65.2. MACEDONIAN)
krtolesta krasatka (53.1. MACEDONIAN)
krtulesta krasulka (24.8. MACEDONIAN)
kruidige venkel (24.12. VLAAMS)
kruisement (44.3. AFRIKAANS)
kruizemunt (44.3. DUTCH)
krukpeterselie (24.19. VLAAMS)
krulkool (30.34. AFRIKAANS)
krulmostaard (30.36. VLAAMS)
krulmostaard (30.40. VLAAMS)
krulmosterd (30.6. VLAAMS)
krulsla (27.32. VLAAMS)
krumba (30.27a. HARARI)
krūminė perilė (44.7. LITHUANIAN)
krumpir (62.24. BOSNIAN)
krumpir (62.24. CROATIAN)
krumpli (62.24. HUNGARIAN)
krupnoploden grašok (40.27. MACEDONIAN)
krüsanteem (27.23. ESTONIAN)
krusemynte (44.3. DANISH)
krusendivie (27.9. NORWEGIAN)
krusmynte (44.3. NORWEGIAN)
kruspersille (24.19. DANISH)
kruv adom (30.27b. HEBREW)
kruv alim (30.35. HEBREW)
kruv hagina (30.27a. HEBREW)
kruv halefet (30.41a. HEBREW)
kruv mesulsal (30.34. HEBREW)
kruv nitzanim (30.29. HEBREW)
kruv rosh (30.27a. HEBREW)
kruvit (30.26. HEBREW)
kruyt (30.27a. YIDDISH)
krydderfennikel (24.12. DANISH)
krydderfennikel (24.12. NORWEGIAN)
krydderpaprika (62.2. NORWEGIAN)
kryddkrasse (30.51. SWEDISH)
krysantemum (27.23. NORWEGIAN)
krysanthemum (27.23. DANISH)
krysi (27.23. ICELANDIC)
krysztłka lśniąca (22.1. POLISH)
kryzakvety chrjan (30.1. BELARUSAN)
ksantosoma brazylijska (8.8. POLISH)
ksantosoma fiołkowa (8.9. POLISH)
kšěn (30.1. SORBIAN, LOWER)
ksharamoola (30.56. SANSKRIT)
kshathrasambhava (45.1. SANSKRIT)
kshathriyaveera (38.20. SANSKRIT)
kshava (30.24. SANSKRIT)
kshavaka (30.24. SANSKRIT)
ksheerashukla (64.3. SANSKRIT)

kshtykan (66.3. KAZAKH)
kshudhabijanana (30.24. SANSKRIT)
kshudraarjaka (44.4. SANSKRIT)
kshudramustha (10.1. SANSKRIT)
kshudraparna (44.4. SANSKRIT)
kshudrathulasi (44.4. SANSKRIT)
kshujjanika (30.24. SANSKRIT)
kshurapathrika (35.9. SANSKRIT)
kshurika (35.9. SANSKRIT)
kshutabhijanaka (30.24. SANSKRIT)
kshutaka (30.24. SANSKRIT)
kshutit metzua (46.1. HEBREW)
ksíkka's (30.41a. BLACKFOOT)
ksvaka (30.17. SANSKRIT)
kú (7.8. CHINESE, MIN NAN)
kũ (11.4. GOURMANCHÉMA)
kũ (11.5. GOURMANCHÉMA)
kũ (11.5. MÒORÉ)
ku ai (27.7. CHINESE, WÚ)
ku cai (27.9. CHINESE, WÚ)
ku cai (27.42. CHINESE, WÚ)
k'u chu (27.10a. CHINESE, MANDARIN)
ku finbarako (62.9. JOLA-FONYI)
ku ges (45.4. JOLA-FONYI)
ku gua (38.23. CHINESE, GÀN)
ku gua (38.23. CHINESE, MANDARIN)
ku gua (38.23. CHINESE, YUE)
ku jii (27.9. CHINESE, WÚ)
ku ju (27.42. CHINESE, MIN BEI)
ku ju cai (27.42. CHINESE, WÚ)
ku ma cai (27.9. CHINESE, WÚ)
ku ma cai (27.42. CHINESE, WÚ)
ku mai (27.42. CHINESE, YUE)
ku pintan ko (39.4. MALIMBA)
ku qu (27.9. CHINESE, MANDARIN)
ku qu cai (27.42. CHINESE, MANDARIN)
ku qua (38.23. CHINESE, WÚ)
ku shu (11.4. CHINESE, WÚ)
ku ye ban jin ju (27.24. CHINESE, WÚ)
kua (38.20. LOKO)
kua tsao (62.5. HMONG-NJUA)
kua tvob phaw (62.1. HMONG-DAW)
kua txob (62.2. HMONG-DAW)
kua txob (62.3. HMONG-DAW)
kua txob (62.5. HMONG-DAW)
kuai gen luo kui (28.2. CHINESE, YUE)
kuai jing jin lian hua (65.2. CHINESE, WÚ)
kuai jing jin lzan hua (65.2. CHINESE, MANDARIN)
kuai jing zao jin lian (65.2. CHINESE, YUE)
kuan bing gai (30.10. CHINESE, MANDARIN)
kuan bing jie (30.10. CHINESE, WÚ)
kuana (11.7. BAOULÉ)
kuanton (23.3. ILOCANO)
kub nyuj (56.2. HMONG-DAW)
kũba (38.12. MÒORÉ)
kubab (45.4. IFUGAO)
kuban kayu rana (27.29. JAVANESE)
Kuba-Spinat (59.1. GERMAN)
kubbaajee (45.5. HINDI)
kúbeèwaá (38.16. HAUSA)
ku-bemya (40.29. BURMESE)
kubewa (38.14. HAUSA)
kubewa (38.16. HAUSA)
kubewa (38.17. HAUSA)
kubewa (38.18. HAUSA)
kuɓewa (45.1. HAUSA)
kubhindo (38.1. NEPALI)
kubi aṛaḱ (30.27a. SANTALI)
kubis (30.27a. INDONESIAN)
kubis (30.27a. MALAY)
kubis (30.27a. SUNDA)
kubis (30.27b. INDONESIAN)
kubis (30.27b. MALAY)
kubis (30.34. INDONESIAN)
kubis (30.34. MALAY)
kubis babat (30.34. INDONESIAN)
kubis bulat (30.27a. MALAY)
kubis bunga (30.26. INDONESIAN)
kubis bunga (30.26. MALAY)
kubis cina (30.40. MALAY)
kubis ketek (30.29. INDONESIAN)
kubis merah (30.27b. INDONESIAN)
kubis panjang (30.40. MALAY)
kubjakam (64.3. TELUGU)
kuboru (11.1. GONJA)
kubyakam (64.3. TELUGU)
kubyakama (64.3. TELUGU)
kucai (7.11. INDONESIAN)
kucai (7.11. MALAY)
kucai (7.13. INDONESIAN)
kucai (7.13. JAVANESE)
kucceere (38.18. FULFULDE, ADAMAWA)
kucha (45.4. BAMANANKAN)
kuchai (7.8. FILIPINO)
kuchai (7.13. FILIPINO)
kuchai (7.13. MALAY)
Küchenzwiebel (7.3. GERMAN)
kú-chhài (7.8. CHINESE, MIN NAN)

kuchi kizhangu (39.4. TAMIL)
k'u-chu (17.4. CHINESE, MANDARIN)
küçük pirinç (35.6. TURKISH)
küçük taze soğan (7.11. TURKISH)
küçük yaprakli düğmeotu (27.22. TURKISH)
kud (63.1. SERER-SINE)
kudajnina (40.30. ORIYA)
kudaku (36.3. FULFULDE, ADAMAWA)
kudaku (36.3. HAUSA)
kudaku (36.3. ZARMA)
kūdaku (62.24. HARARI)
kudakūji (36.3. FULFULDE, ADAMAWA)
kũdbade (8.4. MÒORÉ)
kũdbari (8.1. DAGAARE, SOUTHERN)
kudere (40.1. LIMBA)
kudimah (38.1. HINDI)
kudis (40.2. IFUGAO)
kudju (40.30. TELUGU)
kudoly (40.5. KANNADA)
kudsi lüt çiçeği (48.1a. TURKISH)
kuḍụ (40.16. TULU)
kùdùá (40.2. GONJA)
kudua (40.3. GONJA)
kudure kaḍale (40.5. BADAGA)
kudure ko:ku (24.7. BADAGA)
kudure masale (40.17. KANNADA)
ƙuduru (7.3. HAUSA)
kudüs bezelyesi (40.38. TURKISH)
kudyapa (23.3. BICOLANO, CENTRAL)
kudyapa (23.3. BUHID)
kudyapa (23.3. CEBUANO)
kudyapa (23.3. HANUNOO)
kudyapa (23.6. CEBUANO)
kudzibønne (40.30. DANISH)
kudzou (40.30. CREOLE, HAITIAN FRENCH)
kudzu (40.30. ASSAMESE)
kudzu (40.30. BENGALI)
kudzu (40.30. ENGLISH)
kudzu (40.30. FILIPINO)
kudzu (40.30. FRENCH)
kudzu (40.30. GUJARATI)
kudzu (40.30. HINDI)
kudzu (40.30. INDONESIAN)
kudzu (40.30. JAPANESE)
kudzu (40.30. MACEDONIAN)
kudzu (40.30. MARATHI)
kudzu (40.30. PANJABI, EASTERN)
kudzu (40.30. PORTUGUESE)
kudzu (40.30. RUSSIAN)
kudzu (40.30. TURKISH)
kudzu (40.30. VLAAMS)
kudzú común (40.30. SPANISH)
kudzu du Japon (40.30. FRENCH)
kudzu subtropické (40.30. CZECH)
kudzu tropical (40.30. PORTUGUESE)
kudzú tropical (40.30. SPANISH)
Kudzu-Bohne (40.30. GERMAN)
kueme (38.28. SPANISH)
kugamb (30.27a. ARMENIAN)
kugitas (41.1. TAUSUG)
kugitis (41.1. MANOBO)
kuglesta artičoka (27.16. MACEDONIAN)
kugundugu (36.3. HAUSA)
Kuhblume (27.45. GERMAN)
Kuhbohne (40.43. GERMAN)
Kuherbse (40.43. GERMAN)
kuhig (20.2. IBALOI)
kuhili (40.18. MARATHI)
Kuhkohl (30.32. GERMAN)
kui gen qin (24.5. CHINESE, YUE)
kuicha (7.13. THAI)
kuichai-baiklom (7.11. THAI)
kuichai-farang (7.11. THAI)
kuifhyacint (13.5. VLAAMS)
kuikamba (38.13b. TOK PISIN)
kuitāje boḍēje (62.9. FULFULDE, ADAMAWA)
kujɔ (11.7. GONJA)
kukab (45.4. IFUGAO)
kukah (45.4. IFUGAO)
kukama (38.13a. NIUE)
kukama (38.13a. SAMOAN)
kukamba (38.13a. TOK PISIN)
kukarkand (11.2. MARATHI)
kukbɛŋbɛ (62.5. MALIMBA)
kukehari (37.1. ESTONIAN)
kuki takana (30.18. JAPANESE)
kukiiw (7.11. KHMER, CENTRAL)
kukka tulasi (44.4. TELUGU)
kukkakaali (30.26. FINNISH)
kukkathulasi (44.4. TELUGU)
kuko (62.8. JAPANESE)
kukonjarja (23.4. FINNISH)
kukorica (17.7. HUNGARIAN)
kukri (17.7. HINDI)
kukri (17.7. PANJABI, EASTERN)
kûku kulga (23.4. PANJABI, EASTERN)
kukuk (38.20. INDONESIAN)
kukuk (38.20. JAVANESE)

kulza (30.21. TURKISH)
kuma (17.5. GOLA)
kuma'a (36.3. MARQUESAN)
kumaarajeeva (23.3. SANSKRIT)
kumada (48.1a. SANSKRIT)
kumak (27.29. INDONESIAN)
kuma-kuma (20.2. INDONESIAN)
kumala (36.3. FIJIAN)
kumala (36.3. TONGAN)
kumara (36.3. MAORI)
kumara (36.3. NIUE)
kumara (38.13a. SLOVENIAN)
kumba (27.14. MANINKAKAN, EASTERN)
kumba (27.18. KISSI, NORTHERN)
kumba (38.13a. KISSI, NORTHERN)
kumba (62.9. MÒORÉ)
kumba (62.14c. MÒORÉ)
kũmba (62.18. MÒORÉ)
kumbahang (39.4. PALAWANO, CENTRAL)
kumba:ke (38.14. BADAGA)
kumba:ke (38.16. BADAGA)
kumba:ke (38.17. BADAGA)
kumba:ke (38.18. BADAGA)
kumbala (38.7b. BADAGA)
kumbaḷa (38.14. BADAGA)
kumbaḷa (38.16. BADAGA)
kumbaḷa (38.17. BADAGA)
kumbaḷa (38.18. BADAGA)
kumbala kaayi (38.16. KANNADA)
kumbala kayi (38.16. KANNADA)
kumbalakai (38.16. KANNADA)
kumbalam (38.18. MALAYALAM)
kumballanga (38.1. MALAYALAM)
kumbasa (38.16. BONTOC CENTRAL)
kumbasa (38.17. BONTOC, CENTRAL)
kumbɔnɛ (27.14. KONO)
kumbɛ (27.14. MANDING-MANDINKA)
kumbhanda (38.1. SANSKRIT)
kumbhī (30.52. MAITHILI)
kumbi (2.1. KONO)
kumbi (27.18. MANINKAKAN, EASTERN)
kũmbié (38.12. DAGAARE, SOUTHERN)
kumbié (62.18. DAGAARE, SOUTHERN)
kumbili (2.1. KONO)
kumbili jawa (44.8. JAVANESE)
kumbo (38.20. HAUSA)
kumbre (62.9. MÒORÉ)
kumbua (38.7b. BADAGA)
kumbua (38.14. BADAGA)
kumbua (38.16. BADAGA)
kumbua (38.17. BADAGA)
kumbua (38.18. BADAGA)
kumbuḍa (38.14. TULU)
kumbuḍa (38.16. TULU)
kumbuḍa (38.17. TULU)
kumbuḍa (38.18. TULU)
kumbulam (38.1. MALAYALAM)
kumbundu (38.12. PORTUGUESE)
kumeli (44.8. SUNDA)
kumeny-ɛ (27.14. BULLOM SO)
kumhra (38.1. HINDI)
kumindelan (27.41. INDONESIAN)
kumis ucing (19.1. SUNDA)
kumiyo (11.7. DANGME-KROBO)
kumkuma (20.2. INDONESIAN)
kummangi (40.1. KALAGAN)
kummejt (49.1. MALTESE)
kumpalam (38.1. MALAYALAM)
kumpalanna (38.1. MALAYALAM)
kümpür (62.24. TURKISH)
kumra (38.1. HINDI)
kumra (38.16. BENGALI)
kumra (38.18. HINDI)
kumtõ (2.1. KISSI, NORTHERN)
kumu manus (59.3. TOK PISIN)
kumud (50.2. MAITHILI)
kumuda (50.2. SANSKRIT)
kúmúṇdùkú (36.3. NUPE-NUPE-TAKO)
kumuoko (39.4. LUYIA)
kumura (38.1. ASSAMESE)
kumura (38.16. ASSAMESE)
kumwendo (38.20. LUYIA)
kumwiiba (17.5. LUYIA)
kuna (59.2. KASHMIRI)
kuña ñãngol (54.1. JOLA-FONYI)
kunali (38.4. HINDI)
ku-ɲambo (11.1. WOLOF)
kunappan oongal (38.22. IFUGAO)
kunati (24.9. SANSKRIT)
kunchika (40.32. SANSKRIT)
kunchur (20.4. MALAY)
kuncumapesalu (40.35. TELUGU)
kunda (8.3. SANSKRIT)
kunda khabi-fore (39.4. YALUNKA)
kundaanuga (38.20. TELUGU)
kundakhabi-fikai (39.4. YALUNKA)
kundakhabi-na (39.4. YALUNKA)
kunḍal (38.4. MAITHILI)
kundali (40.18. SANSKRIT)
kundanuga (38.20. TELUGU)

kundar (59.4. PANJABI, EASTERN)
kundaru (38.4. HINDI)
kunde (40.27. SWAHILI)
kunde (40.28. SWAHILI)
kunde (40.41. SWAHILI)
kunde (40.43. DIGO)
kunde (40.43. SWAHILI)
kunderi (38.4. HINDI)
kundiapɔnɔ (59.2. KISSI, NORTHERN)
kundige jola (17.7. KANNADA)
kundimani (9.1. TAMIL)
kundo berberi (62.1. OROMO, WEST CENTRAL)
kundol (38.1. TAGALOG)
kundor (38.1. MALAY)
kundori (38.4. HINDI)
kundroo (38.4. HINDI)
kundru (38.4. HINDI)
kundru (38.6. SRANAN)
kundul (38.1. SUBANON)
kunduli (38.4. ASSAMESE)
kundur (38.1. INDONESIAN)
kundur (38.1. MALAY)
kundur (38.1. SUNDA)
kunduri (38.4. HINDI)
kunduwana (36.3. PULAAR)
kúndùwú (36.3. KANURI)
kunduy (38.13. HANUNOO)
kunduy palatay (38.13. HANUNOO)
kunduzi (38.4. HINDI)
kunéga (45.1. JOLA-FONYI)
kunégo (45.1. JOLA-FONYI)
kung (44.6. NORWEGIAN)
kungkuma (20.2. INDONESIAN)
kungola (39.4. TEM)
kungsmynta (44.6. SWEDISH)
kungulaku (44.5. TIV)
kungull (38.18. ALBANIAN, TOSK)
kunig (20.2. ILOCANO)
kunig (20.2. ITAWIT)
kunig (20.4. KANKANAEY)
kunik (20.2. IBANAG)
kunik (20.2. ITAWIT)
kuning pari (20.3. INDONESIAN)
kuningan (27.22. JAVANESE)
kunir (20.2. INDONESIAN)
kunir (20.2. JAVANESE)
kunir bentis (20.2. JAVANESE)
kunit (20.2. TAGBANWA)
kunja (48.1a. SANSKRIT)
kunjaphala (38.1. SANSKRIT)
kunjar (30.56. SANSKRIT)
kunjid (55.1. TAJIKI)
kunkumapesalu (40.35. TELUGU)
kunna ma:su (62.1. BADAGA)
kunna ma:su (62.2. BADAGA)
kunnan (16.1a. MALAYALAM)
kunnen jaakii (8.4. HAUSA)
kunolem (38.1. KONKANI)
kunsɛ (40.2. KURANKO)
kunsɛ (40.2. MANDING-MANDINKA)
kunta genasu (11.2. KANNADA)
kunthali (62.12. SANSKRIT)
kunti (62.21. TAGALOG)
kunti genasu (11.1. KANNADA)
kunti genasu (11.2. KANNADA)
kunvale (38.18. KONKANI)
kunyir (20.2. SUNDA)
kunyit (20.2. INDONESIAN)
kunyit (20.2. MALAY)
kupintaŋ (39.4. MALIMBA)
kupis (40.42. SAMBAL, TINÀ)
kuppa manjel (27.2. MALAYALAM)
kuppaikkeerai (23.3. TAMIL)
kúpùk (45.1. BEROM)
kupu-kupu (40.33. INDONESIAN)
kupus (30.27a. BOSNIAN)
kupus (30.27a. CROATIAN)
kupus (30.27a. SERBO-CROATIAN)
kupusna repica (30.22. SERBO-CROATIAN)
kur kum (20.2. ARABIC)
kurabã (43.1. JOLA-FONYI)
kurabi (30.30. FIJIAN)
kurada (23.4. MARATHI)
kurat (7.6. POLISH)
kuratirugi (30.51. KANNADA)
kurawet (24.15. SUNDA)
kurawet galeng (24.7. SUNDA)
kurbis (38.15. VLAAMS)
kurbis (38.16. VLAAMS)
kurbitur (38.18. ICELANDIC)
kurdlehine rabarber (58.1. ESTONIAN)
kurdu (23.4. KONKANI)
kurdu (23.4. MARATHI)
kurduqui bagi (23.4. KONKANI)
kuresson (30.52. JAPANESE)
kurfa (59.2. HINDI)
kurfah (59.2. MARATHI)
kurfah (59.4. MARATHI)
kurfe-ka-sag (59.2. HINDI)

kurgirohi (29.1. ESTONIAN)
kurgum (20.2. HAUSA)
kurilo (13.2. NEPALI)
kurjace woko (17.7. SORBIAN, UPPER)
kurk (38.13a. ESTONIAN)
kurkaru (38.18. SANSKRIT)
kurkku (38.13a. FINNISH)
kurkkuyrtti (29.1. FINNISH)
kurkum (20.2. ESTONIAN)
kurkum (20.2. HEBREW)
kurkum zedoar (20.4. HEBREW)
kurkuma (20.2. AFRIKAANS)
kurkuma (20.2. DANISH)
kurkuma (20.2. HUNGARIAN)
kurkuma (20.2. INDONESIAN)
kurkuma (20.2. SLOVAK)
kurkuma (20.3. HUNGARIAN)
kurkuma (20.4. CZECH)
kurkuma dlouhá (20.2. CZECH)
kurkuma długa (20.2. POLISH)
kurkumë (20.2. ALBANIAN, TOSK)
kurkure jhar (1.1. NEPALI)
kurmuja (38.7a. BODO)
kurmuja (38.7b. BODO)
kuro (27.8. NEPALI)
kuro-chiku (17.4. JAPANESE)
kurodane kabocha (38.15. JAPANESE)
kuro-garashi (30.24. JAPANESE)
kuro-guwai (10.3. JAPANESE)
kuro-guwai (10.3. VLAAMS)
kurrat (7.6. ARABIC)
kurrat (7.6. ENGLISH)
kurrat (7.6. FRENCH)
kurrat (7.6. HUNGARIAN)
kurrat (7.6. PORTUGUESE)
kurrat (7.6. SPANISH)
kurrat (7.6. VLAAMS)
kurrat (7.8. ARABIC)
kurrat (7.8. MALTESE)
kurrat ciniz (7.13. MALTESE)
kurrat seeny (7.13. ARABIC)
kurrat slavagg (7.1. MALTESE)
kursa (59.2. HINDI)
kursana pairu (40.43. KANNADA)
kursandi (40.43. KANNADA)
kursan-pyro (40.43. KANNADA)
kurt ayaği (44.1. TURKISH)
kurthike (30.51. KANNADA)
kurtikalai (40.16. BENGALI)
kurtike (30.51. KANNADA)
kurttukaali (30.34. FINNISH)
kuru (40.39. MANDING-MANDINKA)
kurubomboŋmesergo (38.7b. MANDING-MANDINKA)
kurugh agbishe (44.10. TIV)
kuru-kamatis (62.21. CEBUANO)
kurutige (30.51. KANNADA)
kurutige (30.58. KANNADA)
kuş biberi (62.5. TURKISH)
kuš konmas (13.2. UZBEK)
kus tsao may (62.5. HMONG-DAW)
kusay (7.8. PANGASINAN)
kusbar (24.9. HEBREW)
kusbara (24.9. HEBREW)
kuse (7.8. PAMPANGAN)
kuseru (55.1. TEM)
kusher (17.5. BOBO MADARÉ, SOUTHERN)
kusheshaya (48.1a. SANSKRIT)
kushtaha (38.31. SANSKRIT)
kushthaaghni (62.15. SANSKRIT)
kushthanaashaka (11.2. SANSKRIT)
kusiyāl (17.5. NEWAR)
kuşkonmaz (13.2. TURKISH)
kuşkonmaz (40.13. TURKISH)
kuşkonmaz börülcesi (40.42. TURKISH)
kuşkuş ekmeği (30.43. TURKISH)
kuskusjamssi (11.8. FINNISH)
kusthumburi (24.9. SANSKRIT)
kustoun lenavy (30.52. BELARUSAN)
kustumbri (24.9. SANSKRIT)
kustumburi (24.9. SANSKRIT)
kusu (17.3. GUN)
kut kia (2.1. THAI)
kut kin (2.1. THAI)
kuta (10.2. TONGAN)
kuta (45.4. MANDING-MANDINKA)
kutakam (24.7. MALAYALAM)
kutakut (40.25. BONTOC CENTRAL)
kutank (40.1. LIMBA)
kutapa (48.1a. SANSKRIT)
kute (39.4. GBE, WACI)
kute (39.4. GEN)
kuthera (44.4. SANSKRIT)
kuthi (17.7. PANJABI, EASTERN)
kuthiraimasal (40.17. TAMIL)
kutreku (40.31. ÉWÉ)
kutreku (40.31. GBE-VHE)
kutsai (7.1. TAGALOG)
kutsai (7.8. TAGALOG)
kutsay (7.13. TAGALOG)

kutse (7.13. PAMPANGAN)
kuttakut (40.25. KANKANAEY)
kuttoowombi (38.3. SANSKRIT)
kutu (39.4. AJA)
kutu (38.18. MALIMBA)
kuudaku (36.3. FULFULDE, ADAMAWA)
kúudékà (36.3. SONGHAY)
kuuka̠ (7.3. COMANCHE)
kuukanaruhkaʔ (7.3. COMANCHE)
kuumara (36.3. MAORI)
kuumera (36.3. MAORI)
ku'usi (62.12. TONGAN)
küüslauk (7.10. ESTONIAN)
kuut khue (5.1. THAI)
kūūtsɛɛ (13.2. TIBETAN)
kuvalaya (50.2. SANSKRIT)
kuvamavv (15.2. TAMIL)
kuvay (28.1. IBANAG)
kūvè (20.2. TULU)
kuvehittu (15.2. KANNADA)
kuvit tekan (38.22. GUN)
kuwai (6.1. JAPANESE)
kuwai (6.1. VLAAMS)
kuwai (6.2. JAPANESE)
kuwala (30.41a. SORBIAN, LOWER)
kuwanton (23.3. ILOCANO)
kuwantung (23.3. ILOCANO)
kuwanyesvakawayvatnga (38.3. HOPI)
kuwara (40.6. PANJABI, EASTERN)
kuware-shida (5.1. JAPANESE)
kuwazu-imo (8.1. JAPANESE)
kuwe (38.18. LOKO)
kuwintas-kuwintasan (9.1. TAGALOG)
kuwintáson (9.1. WÁRAY-WÁRAY)
kuwunduwu (36.3. KANURI)
kuyawa (11.2. KONO)
kuyeli (40.18. MARATHI)
kuyi (50.2. HINDI)
kuzbara (24.9. ARABIC)
kuzu (40.30. JAPANESE)
kuzu (40.30. SWAHILI)
kuzu gevreği (67.1. TURKISH)
kuzu kulaği (53.1. TURKISH)
kuzu kulaği (58.2. TURKISH)
kuzu kulak (58.2. KIRGHIZ)
kuzuimo (40.21. JAPANESE)
kuzukulaği (58.2. TURKISH)
kuzu-ukon (15.2. JAPANESE)
kva-fyé (62.18. GA)
kvaka (30.22. SLOVAK)
kvan (24.1. DANISH)
kvann (24.1. NORWEGIAN)
kvapioji garduoklė (24.16. LITHUANIAN)
kvapioji rezeta (60.1. LITHUANIAN)
květák (30.26. CZECH)
kveták (30.26. SLOVAK)
kvinoa (35.6. FINNISH)
kvinoa (35.6. MACEDONIAN)
kvinoa-cserje (35.6. HUNGARIAN)
kvinoasavikka (35.6. FINNISH)
kvitbete (35.4. NORWEGIAN)
kvitløk (7.10. NORWEGIAN)
kvitsennep (30.57. NORWEGIAN)
kvlit (40.16. GUJARATI)
kvmhtvlako (40.1. MUSKOGEE)
kvmohtvlako (40.1. MUSKOGEE)
kvnhtvlako (40.1. MUSKOGEE)
kvnohtvlako (40.1. MUSKOGEE)
kvntapadmu (50.1. ORIYA)
kwàárúúrú (40.15. HAUSA)
kwacciya (38.20. HAUSA)
kwada yowi (62.5. DANGME-KROBO)
kwadu (16.1b. TWI)
kwakier lekarski (29.3. POLISH)
kwali (39.5. MAKHUWA)
ƙwaliliya (64.3. HAUSA)
ƙwāma (40.39. HAUSA)
kwama (40.43. HAUSA)
kwáŋ (11.2. BEROM)
ƙwàn kádàà (7.2. HAUSA)
kwangwaviva (24.13. HOPI)
kwankalita (64.3. HAUSA)
kwankwarita (64.3. HAUSA)
kwankwarito (64.3. HAUSA)
kwantal (38.7b. HAUSA)
kwaonsusua (62.23. AKAN-ASANTE)
kwàràk (40.3. BEROM)
kwàràk jὲέjὲέt (40.3. BEROM)
kwàràk pwὲŋ (40.3. BEROM)
kwàràk sùnàŋ (40.3. BEROM)
ƙwarya (38.20. HAUSA)
kwasinabooʔa tuahkapu (62.9. COMANCHE)
kwawsoo´kwepump (66.3. PAIUTE, SOUTHERN)
kwàyz ké: so:n (9.1. LAO)
kwàyz khauz (16.1a. LAO)
kwàyz ph'uuttha so:n (9.1. LAO)
kwel (38.4. MALAYALAM)
kweme (38.28. SWAHILI)
kwemi nut (38.28. ENGLISH)

kwenti (10.1. TIGRIGNA)
kweson (30.52. KWÉYÒL)
kwětak (30.26. SORBIAN, UPPER)
kwětkowy kał (30.26. SORBIAN, LOWER)
kwikedje (62.9. FULFULDE, ADAMAWA)
kwitaje boɗeyi (62.9. FULFULDE, ADAMAWA)
kw-we-thi (38.21. BURMESE)
kyabaej (30.27a. KANNADA)
kyabetsu (30.27a. JAPANESE)
kyalic (7.10. SUMO-MAYANGNA)
kya-lo-wa (17.2. BURMESE)
kyambama (17.3. HAUSA)
kyamos (40.3. GREEK)
kyamos sogias (40.8. GREEK)
kyana gida (9.1. KANNADA)
kıyar (38.13a. TATAR)
kyassaba (39.4. JAPANESE)
kyauk-pha-yon (38.1. BURMESE)
kyeeve (62.13. HOPI)
kyɛmɛ (16.1a. MANDING-MANDINKA)
kyensiskawayvatnga (38.3. HOPI)
kyet-hni-ga (38.23. BURMESE)
kyet-thun (7.3. BURMESE)
kyet-thun-byu (7.10. BURMESE)
kyet-thun-maik (7.9a. BURMESE)
kyet-thun-u-galay (7.2. BURMESE)
kyet-th-wonini (7.3. BURMESE)
kyévitsi (30.27a. HOPI)
kylvöherne (40.28. FINNISH)
kylvövirvilä (40.12. FINNISH)
kyndel (44.9. SWEDISH)
kyrk muun pijaz (7.2. KIRGHIZ)
kyrmyzy istpot (62.1. AZERBAIJANI, NORTHERN)
kyseláč luční (58.2. CZECH)
kyseláč štítnatý (58.4. CZECH)
kyssäkaali (30.30. FINNISH)
kyûkei-kanran (30.30. JAPANESE)
kyuuri (38.13a. JAPANESE)
kyyhkyherne (40.2. FINNISH)
kyyhkyherne (40.2. SWEDISH)
la (45.1. ZARMA)
la gen (30.1. CHINESE, MANDARIN)
la gen (30.1. CHINESE, WÚ)
la$^{\vee}$ hk' a$^{\vee}$ (11.2. LAHU)
la jiao (62.2. CHINESE, GÀN)
la jiao (62.3. CHINESE, YUE)
la jiao (62.5. CHINESE, YUE)
la zi cao (27.22. CHINESE, WÚ)
la'a sawa (44.5. SAMOAN)
laakh (40.10. MARATHI)
laal kaddu (38.16. HINDI)
laal mirchee (62.2. MARATHI)
laalambaari (45.4. HINDI)
laal-ambaari (45.4. MARATHI)
laalbachlu (28.1. HINDI)
laalkamal (48.1a. HINDI)
laalmee (38.7b. HINDI)
laalmircha (62.5. HINDI)
laalmirich (62.2. HINDI)
laalnatiya (23.3. HINDI)
laalo (63.1. FULFULDE, ADAMAWA)
laalomemel (63.1. FULFULDE, ADAMAWA)
láálòò (63.1. HAUSA)
laalsag (23.3. HINDI)
laana draconit (27.6. HEBREW)
laana indica (27.7. HEBREW)
laanchari (23.2. MARATHI)
laan-ka-mirchi (62.5. HINDI)
laat an aba (8.4. IFUGAO)
laat an banig (8.4. IFUGAO)
laat tsiu (62.5. CHINESE, MANDARIN)
lääts (40.12. ESTONIAN)
la'au souga (44.5. SAMOAN)
laáwúr (36.3. HAUSA)
labaça (58.3. PORTUGUESE)
labaños (30.54. TAGALOG)
labanos (30.56. FILIPINO)
labanos (30.56. IBANAG)
labanos (30.56. TAGALOG)
labas (50.2. TAGALOG)
labasa (38.16. HANUNOO)
labasa (38.17. HANUNOO)
lábàshií (23.3. HAUSA)
labaza (27.5. GALICIAN)
labbo fowru (19.1. FULFULDE, ADAMAWA)
lablab (40.9. DUTCH)
lablab (40.9. ENGLISH)
lablab (40.9. HUNGARIAN)
lablab (40.9. PORTUGUESE)
lablab (40.9. SWEDISH)
lab-lab bean (40.9. ENGLISH)
lablab purpurový (40.9. CZECH)
lablabboon (40.9. AFRIKAANS)
lablab-boontjie (40.9. AFRIKAANS)
lablabpapu (40.9. SWEDISH)
lablabu (35.3. TAJIKI)
labogandang (38.16. MARANAO)
labogandang (38.17. MARANAO)

labogandang (38.20. MARANAO)
labonas (30.56. TAGALOG)
labong (17.1. TAGALOG)
labong (17.2. FILIPINO)
labong (17.2. TAGALOG)
labong (17.4. TAGALOG)
làbsúr (30.51. BEROM)
labsûr (30.51. HAUSA)
làbsûr (30.51. HAUSA)
labtenda (38.18. SONGHAY)
lābu (38.14. PALI)
lābu (38.16. PALI)
labu (38.17. INDONESIAN)
lābu (38.17. PALI)
lābu (38.18. PALI)
labu (38.20. TAUSUG)
labu air (38.4. INDONESIAN)
labu air (38.18. MALAY)
labu air (38.20. INDONESIAN)
labu ambon (38.17. INDONESIAN)
labu botol (38.20. MALAY)
labu gambas (38.25. INDONESIAN)
labu jepang (38.25. INDONESIAN)
labu kendit (38.20. INDONESIAN)
labu manis (38.16. MALAY)
labu merah (38.17. INDONESIAN)
labu merah (38.17. MALAY)
labu merah (38.18. INDONESIAN)
labu panjan (38.20. INDONESIAN)
labu panjang (38.20. INDONESIAN)
labu parang (38.17. INDONESIAN)
labu putih (38.20. INDONESIAN)
labu putin (38.20. MALAY)
labu siam (38.25. INDONESIAN)
labu siam (38.25. MALAY)
labu siem (38.25. INDONESIAN)
labu siyem (38.25. INDONESIAN)
labuka (38.20. SANSKRIT)
labumerah (38.15. INDONESIAN)
labumerah (38.16. INDONESIAN)
lac (40.1. VIETNAMESE)
lachang (17.6. MALAY)
lachanides (30.41c. GREEK)
lachano savoï (30.34. GREEK)
lachano vryxellon (30.29. GREEK)
lăc-tsón (38.14. CHINESE, HAKKA)
lăc-tsón (38.16. CHINESE, HAKKA)
lada (62.1. BICOLANO, CENTRAL)
lada (62.5. BUHID)
lada (62.5. CEBUANO)
lada (62.5. TAUSUG)
lada balaynun (62.1. HANUNOO)
lada balaynun (62.2. HANUNOO)
lada besar (62.1. MALAY)
lada merah (62.1. INDONESIAN)
lada padi (62.5. MALAY)
lada tirindukun-tigbaya (62.5. HANUNOO)
lada-balang (62.5. SAMBAL, BOTOLAN)
lada-bali (62.1. SAMBAL, BOTOLAN)
lada-buid (62.5. BUHID)
ladamera (62.5. MALAYALAM)
ladamerachina (62.5. MALAYALAM)
laddú (38.14. HINDUSTANI)
ladi (30.56. KOONGO)
ladi (62.1. AGTA)
ladiko (40.3. HAUSA)
ladio (7.3. MANINKAKAN, WESTERN)
ladio (7.10. MANINKAKAN, WESTERN)
ladj (7.1. SERER-SINE)
ladj (7.10. SERER-SINE)
ladja (20.6. KALINGA, BUTBUT)
ladji (7.1. PULAR)
ladji (7.10. WOLOF)
ladu (38.13a. BALTI)
ladumira (62.5. MALAYALAM)
ladybell (33.1. ENGLISH)
lady's fingers (45.1. ENGLISH)
læge-kulsukker (29.2. DANISH)
lafoy (45.1. ZARMA)
lafsur (30.51. HAUSA)
làfusûr (30.51. HAUSA)
låg hagebønne (40.26. NORWEGIAN)
lagah (36.3. IFUGAO)
lagai (55.1. SUSU)
lagaylay (40.3. AKLANON)
lagbai (8.4. BINUKID)
lagbang (6.2. BICOLANO, CENTRAL)
lagbay (8.4. CEBUANO)
lagbay (8.4. TAGALOG)
lagenária (38.20. SLOVAK)
lagenárie (38.20. CZECH)
laghukavali (62.21. MARATHI)
laghukoshathaki (38.21. SANSKRIT)
lagiknay (23.6. WÁRAY-WÁRAY)
lagikuway (45.2. TAGALOG)
lagikway baniwayun (45.2. HANUNOO)
lagikway malaguhun (45.2. HANUNOO)
lagkakum (62.21. BICOLANO, CENTRAL)
lagna (30.27a. LADINO)
lagolachano (27.46. GREEK)

lagoli (17.3. GIANGAN)
Lagos bologi (59.3. ENGLISH)
Lagos spinach (23.4. ENGLISH)
lagsi urungan (36.3. BUHID)
lagunilla (23.1. SPANISH)
lahaje (45.1. FULFULDE, ADAMAWA)
lahana (30.27a. TURKISH)
lahano (30.27a. LADINO)
lahanon (30.27a. LADINO)
lahi (30.24. HINDI)
lahong (17.1. FILIPINO)
lahsan (7.10. HINDI)
lahsan (7.10. HINDUSTANI)
lahsan ká jawá (7.10. HINDUSTANI)
lahsun (7.10. HINDI)
lahsun (7.10. MAITHILI)
laï (7.10. DAGAARE, SOUTHERN)
lai (7.10. KWÉYÒL)
laï (7.10. MÒORÉ)
lai sak (30.17. ASSAMESE)
lai tu (30.56. CHINESE, WÚ)
lai tu (30.56. CHINESE, YUE)
lâi-á-koe (38.7. CHINESE, MIN NAN)
laian (62.24. HERERO)
laihfang (16.1a. BODO)
laima biin (40.25. JAPANESE)
laire ngabbu (36.3. FULFULDE, ADAMAWA)
lairén (15.1. SPANISH)
laiškinis česnakas (7.11. LITHUANIAN)
laiteron (27.42. FRENCH)
laitron commun (27.42. FRENCH)
laitron épineux (27.40. FRENCH)
laitue (27.31. FRENCH)
laitue (27.32. FRENCH)
laitue à couper (27.32. FRENCH)
laitue asperge (27.30. FRENCH)
laitue asperge romain (27.30. FRENCH)
laitue de brevis (67.2. FRENCH)
laitue de lierre (27.42. FRENCH)
laitue d'Inde (27.29. FRENCH)
laitue frisée (27.32. FRENCH)
laitue indienne (27.29. FRENCH)
laitue lombard (27.32. FRENCH)
laitue pommée (27.31. FRENCH)
laitue romaine (27.32. FRENCH)
laitue sauvage (27.42. BRETON)
laiya (20.6. IFUGAO)
laiya (20.6. SAMBAL, BOTOLAN)
laiyal (20.6. SAMBAL, TINÀ)
laj (7.10. CREOLE, HAITIAN FRENCH)
laja (20.1. SUNDA)
laji (7.10. WOLOF)
laka (55.1. PULAAR)
lakamas (40.21. PAMPANGAN)
lakamas (40.21. PANGASINAN)
lakär e savojës (30.34. ALBANIAN, TOSK)
lakar sem (40.29. BENGALI)
lakat (8.4. BONTOC, CENTRAL)
lake lay (7.13. MORISYEN)
lake zonion (7.5. MORISYEN)
lake zonion (7.11. MORISYEN)
lakër (30.27a. ALBANIAN, TOSK)
lakër e bardhë (30.35. ALBANIAN, TOSK)
lakër e brukselit (30.29. ALBANIAN, TOSK)
laket (38.16. WOLOF)
laket (38.18. WOLOF)
lakha murai (30.56. HINDUSTANI)
laki (38.15. FIJIAN)
lakoron-bolan (33.3. BINUKID)
lal bhopla (38.14. MARATHI)
lal bhopla (38.16. MARATHI)
lal bhopla (38.18. MARATHI)
lal kadsumbal (40.4. HINDI)
lal khada (23.3. ORIYA)
lal kumra (38.16. HINDI)
lal kumra (38.17. HINDI)
läl mirch (62.2. PANJABI, EASTERN)
lal mirch (62.5. PANJABI, EASTERN)
lal murghaka (23.4. URDU)
lal murghka (23.4. HINDI)
lal sag (23.3. BENGALI)
lal sag (23.3. HINDI)
lal sag (23.3. PANJABI, EASTERN)
lal sag (23.3. URDU)
lal sak (23.3. ASSAMESE)
lal shag (23.3. BENGALI)
lala (11.4. KABIYÉ)
lala gnini (23.3. BÉTÉ)
lalakamal (50.2. MARATHI)
lalalu (36.3. BENGALI)
lal-ambadi (45.4. MARATHI)
lal-ambari (45.4. HINDI)
lalambari (45.4. URDU)
lalgooka batooli (62.15. MARATHI)
l'alia (13.4. SLOVAK)
lalitapat (63.1. BENGALI)
lalkudsumbal (40.3. HINDI)
lallo (63.1. HAUSA)
lalmirach (62.2. HINDI)
lalmirchi (62.2. BENGALI)

lalmirich (62.1. HINDI)
lalmirich (62.1. URDU)
lalmirich (62.2. HINDI)
lalmirich (62.2. URDU)
lalmirich (62.5. HINDI)
lal-mista (45.4. BENGALI)
lalmorich (62.2. BENGALI)
lalmurga (23.4. BENGALI)
lalnatiya (23.3. HINDI)
lalo (45.1. MORISYEN)
lalo (63.1. FULFULDE)
lal-shak (23.3. BENGALI)
lalu (38.13a. BALTI)
lam (38.18. BULLOM SO)
lamanda (45.3. NYANJA)
lamant (44.3. MORISYEN)
lamarudu (17.5. FULFULDE, ADAMAWA)
lamb mint (44.3. ENGLISH)
lamba (38.20. SANSKRIT)
lamba simba (40.25. ORIYA)
lambadi (23.4. GUJARATI)
lambadi (25.1. GUJARATI)
lambalääts (40.32. ESTONIAN)
lambehra (62.9. BODO)
lambho (23.3. GUJARATI)
lamb's lettuce (67.2. ENGLISH)
lambu-lambo (38.27. KISSI, NORTHERN)
lambulambo (38.28. KISSI, NORTHERN)
lambut (23.3. AMHARIC)
lam-dɛ (38.18. BULLOM SO)
la-meun (17.5. NICOBARESE, CAR)
lamgo (38.18. BULLOM SO)
lami-córè (38.20. PULAAR)
lamlampaka (27.18. ILOCANO)
lammuɗe (45.4. FULFULDE, ADAMAWA)
lamopinaatti (22.2. FINNISH)
lampaka (27.43. ILOCANO)
lamparan (41.1. TAGALOG)
lampaza (27.5. GALICIAN)
lampaza (35.3. GALICIAN)
lampazo (27.5. SPANISH)
lampenas (27.29. INDONESIAN)
lampenas (27.29. SUNDA)
lampeni (27.29. SUNDA)
Lampionpflanze (62.11. GERMAN)
lampionplant (62.11. DUTCH)
lampionplant (62.11. VLAAMS)
lāmpḻī (38.3. MARWARI)
lampoyang (20.4. AKLANON)
lampuyang (20.2. AKLANON)
lamtung (36.3. IFUGAO)
lamŭtald' (30.17. MICMAC)
lan tou (40.38. CHINESE, MANDARIN)
l'ana (27.6. ENGLISH)
land cress (30.51. ENGLISH)
lanē-papkv (27.32. MUSKOGEE)
lang tomat (38.29. DANISH)
lang vinterkarse (30.2. NORWEGIAN)
lang wei cao (17.3. CHINESE, WÚ)
langa (38.20. TEM)
Langbohne (40.42. GERMAN)
lang-daeng (69.1. THAI)
lange Kurkuma (20.2. GERMAN)
lange petiool mosterd (30.13. VLAAMS)
langguwas (20.4. ILOCANO)
langkawas (20.1. BICOLANO, CENTRAL)
langkio (7.11. INDONESIAN)
langkio (7.11. SUNDA)
langkoas (27.29. SUNDA)
langkuas (20.1. BICOLANO, CENTRAL)
langkuas (20.1. INDONESIAN)
langkuwas (20.4. ILOCANO)
langsana merah (34.2. INDONESIAN)
langskulpet vinterkarse (30.2. DANISH)
lanka (24.9. BENGALI)
lanka (24.9. ORIYA)
lanka (40.10. SANSKRIT)
lanka (62.2. SANSKRIT)
lanka (62.5. BENGALI)
lanka (62.5. ORIYA)
lankamorich (62.2. BENGALI)
lankau'as (20.1. FILIPINO)
lansblaar (6.2. AFRIKAANS)
lansona (7.2. CEBUANO)
lanterna chinesa (62.11. PORTUGUESE)
Laternenpflanze (62.11. GERMAN)
lanterninha chinesa (62.11. PORTUGUESE)
lânti (40.12. KHMER, CENTRAL)
lanti (40.12. MORISYEN)
lantil (40.12. KWÉYÒL)
lanttu (30.22. FINNISH)
lao shu dou (40.18. CHINESE, GÀN)
laos (20.1. HINDUSTANI)
laos (20.1. JAVANESE)
laos jambe (9.1. INDONESIAN)
laos mekah (9.1. INDONESIAN)
lapa (8.4. SAMBAL, TINÀ)
lapadi (23.4. GUJARATI)
lapadi (25.1. GUJARATI)
lapaitz (58.3. BASQUE)

lapa-lapa (40.7. INDONESIAN)
lapatho (58.3. GREEK)
lapathon (58.2. GREEK)
lapatoi v lita levoudia (35.1. GREEK)
lapinis kopūstas (30.35. LITHUANIAN)
lapnay (36.3. IFUGAO)
lapne (36.3. IFUGAO)
lapni (36.3. IFUGAO)
lappa (27.5. GREEK)
lappa bardana (27.5. ITALIAN)
la-psana (30.57. GREEK)
lapsur (30.51. FULFULDE, ADAMAWA)
làpsûr (30.51. KANURI)
laptanda (38.15. ZARMA)
laptanda (38.17. ZARMA)
lăptucă (27.31. ROMANIAN)
lăptuci de grădină (27.32. ROMANIAN)
lāpu (38.14. PALI)
lāpu (38.16. PALI)
lāpu (38.17. PALI)
lāpu (38.18. PALI)
lapu cigorini (27.10a. LATVIAN)
lapu kāposti (30.35. LATVIAN)
lapu salāti (27.31. LATVIAN)
lapu salāti (27.32. LATVIAN)
lapung (8.4. PALAWANO, CENTRAL)
lar (19.1. SERER-SINE)
lar (19.1. WOLOF)
lara (62.1. AGTA)
lara (62.5. PAMPANGAN)
lara (62.5. TAGALOG)
lara manok (62.5. TAUSUG)
lara tangal (62.5. SAMBAL, TINÀ)
larajanggay (62.1. TAUSUG)
larang-bilog (62.1. PAMPANGAN)
larang-inanis (62.5. PAMPANGAN)
larbă grasă (37.1. ROMANIAN)
larbakɔ (11.7. DAGBANI)
laren (15.1. SPANISH)
large leaf mustard (30.17. ENGLISH)
larin (24.11. HERERO)
larut (15.2. SUNDA)
l[as] gi[aas]p (61.1. VIETNAMESE)
lasa (9.1. IVATAN)
lasan (7.10. GUJARATI)
lasan (7.10. URDU)
lasara mah (40.43. ASSAMESE)
laseena (7.10. SANSKRIT)
lashan (7.10. BENGALI)
lashun (7.10. BENGALI)
lashuna (7.10. SANSKRIT)
lasi (40.8. BURMESE)
laskavec (23.3. CZECH)
láskavec (23.3. SLOVENIAN)
lasoen (7.10. INDONESIAN)
lasona (7.2. CEBUANO)
lasona (7.2. PANGASINAN)
lasona (7.3. ILOCANO)
lasona (7.3. MARANAO)
lasona a tukapun (7.10. MARANAO)
lassan (7.10. PANJABI, EASTERN)
lastima usta (40.9. SERBO-CROATIAN)
laston (40.42. CEBUANO)
lastron (27.18. MORISYEN)
lasun (7.10. MARATHI)
lasun (7.10. NEPALI)
lasun ghas (40.17. MARATHI)
lasuna (7.2. HANUNOO)
lasuna (7.2. ITAWIT)
lasuna (7.10. MARATHI)
lasunas (7.10. MARATHI)
lasunghas (40.17. HINDI)
lathapanasa (38.3. SANSKRIT)
lathouros o kordy-lodis (40.11. GREEK)
lathri (40.10. HINDI)
lathuri (40.10. GREEK)
lato ghans (27.2. NEPALI)
latoy (40.42. HILIGAYNON)
latri (40.10. HINDI)
latte (23.3. NEPALI)
latthane (40.43. TULU)
lăt-tsiāo (62.2. CHINESE, HAKKA)
lattuga a cappuccio (27.31. ITALIAN)
lattuga a testa (27.31. ITALIAN)
lattuga asparago (27.30. ITALIAN)
lattuga cappuccia (27.31. ITALIAN)
lattuga di taglio (27.32. ITALIAN)
lattuga romana (27.32. ITALIAN)
lattughella (27.21. ITALIAN)
latu (38.13a. BALTI)
latunda (16.1a. BICOLANO, CENTRAL)
latundal (16.1a. ILOCANO)
latundan (16.1a. TAGALOG)
latva-artisokka (27.16. FINNISH)
lau (38.20. ASSAMESE)
lau (38.20. BENGALI)
lau (38.20. HINDI)
lau (38.20. ORIYA)
lau sauga (44.5. SAMOAN)
lauaa (38.20. HINDI)

lauas (50.1. FILIPINO)
lauas (50.2. TAGALOG)
lauka (38.20. HINDI)
lauka (38.20. NEPALI)
lāukī (38.14. HINDI)
laukí (38.14. HINDUSTANI)
lāukī (38.16. HINDI)
lāukī (38.17. HINDI)
lāukī (38.18. HINDI)
lauki (38.20. HINDI)
lāukī (38.20. HINDI)
lauki (38.20. HINDUSTANI)
lauki (38.20. PANJABI, EASTERN)
laulau (30.17. YORUBA)
laumalika (16.1a. NIUE)
launea a foglia di tarassaco (27.33. ITALIAN)
lausur (30.51. HAUSA)
laùsúr (30.51. HAUSA)
laùsúrù (30.51. HAUSA)
laviagui (35.4. UZBEK)
lavlav (40.9. HEBREW)
lavrig (36.1. BRETON)
lawag (20.2. SUBANON)
lawas (50.2. TAGALOG)
láwàshì (7.3. FULFULDE, ADAMAWA)
lawasi (7.2. FULFULDE, ADAMAWA)
lawasīji (7.2. FULFULDE, ADAMAWA)
lawiging (40.40. IFUGAO)
lawn pennywort (24.15. ENGLISH)
lāwur (62.24. HARARI)
lawwiging (40.25. IFUGAO)
läwz (40.1. AMHARIC)
laxai (55.1. SUSU)
lay (7.10. CREOLE, HAITIAN FRENCH)
lay (7.10. KWÉYÒL)
lay (7.10. MORISYEN)
laya (20.6. BICOLANO, CENTRAL)
lay-a (20.6. BICOLANO, CENTRAL)
laya (20.6. BONTOC CENTRAL)
laya (20.6. IBANAG)
laya (20.6. ILOCANO)
laya (20.6. PAMPANGAN)
lay-a (20.6. SAMBAL, TINÀ)
layal (20.6. SAMBAL, TINÀ)
layi (7.10. MANDING-BAMBARA)
laylaya (20.4. KANKANAEY)
laylaya (23.6. ISNAG)
layri (40.38. PULAAR)
layya (20.6. IFUGAO)
layya (20.6. ILOCANO)
layya (20.6. SAMBAL, BOTOLAN)
lɔ-nama (55.1. MANDING-MANDINKA)
lɔogihye (20.6. LOKO)
le chou dia bifoulou (30.26. BEMBÉ)
le chou dia bifoulou (30.26. KOONGO)
le chou lua mbuaki (30.27a. KOONGO)
le chou lua moutu (30.29. KOONGO)
le chou lua ndombe na (30.27a. KOONGO)
leadán (27.5. GAELIC, IRISH)
leaf beet (35.3. ENGLISH)
leaf cabbage (30.36. YORUBA)
leaf cabbage (30.40. YORUBA)
leaf celery (24.4. ENGLISH)
leaf lettuce (27.32. ENGLISH)
leaf radish (30.55. ENGLISH)
leaf turnip (30.41b. ENGLISH)
lebasar (7.3. KANURI)
lebeda dobrova genricha (35.5. BELARUSAN)
lebeda kvinoa (35.6. RUSSIAN)
lebada sadovaja (35.1. BELARUSAN)
lebeda sadovaja (35.1. RUSSIAN)
lebeda záhradná (35.1. SLOVAK)
lebeda zahradní (35.1. CZECH)
lebiodka pospolita (44.6. POLISH)
leblebija (40.5. SERBO-CROATIAN)
leblieb (36.1. MALTESE)
leča (40.12. SLOVENIAN)
lēcas (40.12. LATVIAN)
lecern (40.17. BENGALI)
lechech (57.1. HEBREW)
lechuga (27.31. LADINO)
lechuga (27.32. LADINO)
lechuga acogollada (27.31. SPANISH)
lechuga arrepollada (27.31. SPANISH)
lechuga de campo (67.2. SPANISH)
lechuga de hojas (27.32. SPANISH)
lechuga de la India (27.29. SPANISH)
lechuga espárrago (27.30. SPANISH)
lechuga par cortar (27.32. SPANISH)
lechuga romana (27.32. SPANISH)
lechuguilla silvestre (27.42. SPANISH)
lechuguino (27.32. SPANISH)
lecimbre (44.3. GALICIAN)
ledenec nachový (40.13. CZECH)
lędźwian siewny (40.10. POLISH)
leek (7.8. ASSAMESE)
leek (7.8. ENGLISH)
leek (7.8. FILIPINO)
leek (7.8. ILOCANO)
leek (7.8. SINHALA)

leek (7.8. TAGALOG)
lefet (30.41a. HEBREW)
lefki dioacorea (11.7. GREEK)
lefoli (38.14. SOTHO, NORTHERN)
lefufit ha-batata (36.3. HEBREW)
lefufit habatatah (36.3. HEBREW)
lefufit hamaim (36.2. HEBREW)
lefufit ha-mayim (36.2. HEBREW)
legapu (38.3. SOTHO, NORTHERN)
legapu (38.3. TSWANA)
legba kpakun (40.3. FON)
legɔsbɔlɔgi (59.3. KRIO)
lɛgɔs-bɔlɔgi (59.4. KRIO)
lege (40.1. GOLA)
legenārija (38.20. LATVIAN)
legevalurt (29.2. NORWEGIAN)
leggal roogo (39.4. FULFULDE, ADAMAWA)
legiɛ (38.5. GOLA)
legiɛ-kwi (38.3. GOLA)
lehapu (38.3. SOTHO, SOUTHERN)
leḥekh shasuʿa (57.1. HEBREW)
lehemesi (38.11. ESTONIAN)
lehjet il-bodbod (27.46. MALTESE)
lehmänpapu (40.41. FINNISH)
lehmänpapu (40.42. FINNISH)
lehmänpapu (40.43. FINNISH)
lehsoon (7.10. HINDI)
lehsun (7.10. HINDI)
lehtikaali (30.35. FINNISH)
lehtikaali (30.35. SERBO-CROATIAN)
lehtimangoldi (35.3. FINNISH)
lehtisalaatti (27.32. FINNISH)
lehtiselleri (24.4. FINNISH)
lehtkapsas (30.23. ESTONIAN)
lehtkapsas (30.35. ESTONIAN)
lehtpeet (35.3. ESTONIAN)
léiri (38.31. GOURMANTCHÉ)
leitaruga (27.45. GALICIAN)
leiteis (27.31. GAELIC, SCOTTISH)
leiteis (27.32. GAELIC, SCOTTISH)
leitis chabáiste (27.31. GAELIC, IRISH)
leituga (27.42. PORTUGUESE)
lejet (38.25. SUNDA)
leka socivo (40.12. MACEDONIAN)
leker (19.1. INDONESIAN)
leker bamboo (17.2. MORISYEN)
leker bambou (17.1. MORISYEN)
lekir (23.3. KANURI)
lėk-tāu (40.12. CHINESE, MIN NAN)
lėk-tāu (40.28. CHINESE, MIN NAN)
lėk-tek-sún (17.1. CHINESE, MIN NAN)
lėk-tek-sún (17.2. CHINESE, MIN NAN)
lėk-tek-sún (17.4. CHINESE, MIN NAN)
lekwele (62.24. TSWANA)
lele (40.25. TIV)
lélé timion (62.18. GOURMANCHÉMA)
lelie (13.4. AFRIKAANS)
lelie (13.4. DUTCH)
lelija (13.4. LITHUANIAN)
lelle (13.4. VLAAMS)
lellux (27.23. MALTESE)
lelmirchi (62.2. GUJARATI)
leman (17.1. OROMO, WEST CENTRAL)
leman (17.2. OROMO, WEST CENTRAL)
leman (17.4. OROMO, WEST CENTRAL)
lembong nyidra (9.1. JAVANESE)
lemon melon (38.8. ENGLISH)
lempoyan (20.4. MARANAO)
lencse (40.12. HUNGARIAN)
lendebê (23.3. AKWA)
leng jiao si gua (38.21. CHINESE, WÚ)
leng shui hua (66.1. CHINESE, WÚ)
lêng-kak (64.1. CHINESE, MIN NAN)
lengkat (64.3. JAVANESE)
lengkong (64.3. JAVANESE)
lengkuas (20.1. INDONESIAN)
lengkuas (20.1. MALAY)
leng-lengan (34.2. JAVANESE)
leŋguõ (38.3. KISSI, NORTHERN)
lenktasėklės pupelė (40.25. LITHUANIAN)
lensie (40.12. AFRIKAANS)
lensimon (59.2 TIRURAY)
lenteja (40.12. SPANISH)
lentèjas (40.12. LADINO)
lentella (40.12. GALICIAN)
lentêta (30.17. SIDAMO)
lenticchia (40.12. ITALIAN)
lentil (40.12. ENGLISH)
lentil (40.12. FILIPINO)
lentil (40.12. KANNADA)
lentil (40.12. SWAHILI)
lentilha (40.12. PORTUGUESE)
lentille (40.12. FRENCH)
lentiye (40.12. KOONGO)
lenuca phit (62.15. SUNDA)
lenwa (38.22. MORISYEN)
leotia (23.3. ORIYA)
lepa (40.18. PANGASINAN)
lepalep bu diigen (44.5. WOLOF)
lepeldief (30.43. STELLINGWERFS)

lepeldiefien (30.43. STELLINGWERFS)
lepelties (30.43. STELLINGWERFS)
lepeltiesbloeme (30.43. STELLINGWERFS)
lepeng (8.1. MARANAO)
lephet (30.41a. HEBREW)
lephutshe (38.14. TSWANA)
lepidio (30.51. ITALIAN)
lepidio (30.51. SPANISH)
lepidio ortense (30.51. ITALIAN)
lepjetë e thartë (58.2. ALBANIAN, TOSK)
léptándà (38.18. SONGHAY)
lepuu (30.56. TIBETAN)
leqala (17.1. SOTHO, SOUTHERN)
leqala (17.2. SOTHO, SOUTHERN)
leqala (17.4. SOTHO, SOUTHERN)
lëqike (27.9. ALBANIAN, TOSK)
leren (15.1. ENGLISH)
leren (15.1. SPANISH)
lereschia giapponese (24.10. ITALIAN)
leret (15.2. INDONESIAN)
Lerz (40.28. LUXENBOURGEOIS)
lešča (40.12. BULGARIAN)
lesera mah (40.43. ASSAMESE)
lesser yam (11.5. ENGLISH)
lestovička (40.9. BULGARIAN)
lesun (7.10. FIJIAN)
lət (45.5. AMHARIC)
letah hayam (23.2. SUNDA)
letapola (62.24. SOTHO, NORTHERN)
letas (27.32. HAUSA)
letasu (27.31. JAPANESE)
leteng (11.2. MARANAO)
lethìsi (27.32. VENDA)
leti (27.31. CREOLE, HAITIAN FRENCH)
léti (27.31. KWÉYÒL)
leti (27.31. MORISYEN)
léti (27.32. CREOLE, HAITIAN FRENCH)
léti (27.32. KWÉYÒL)
letis (27.31. TOK PISIN)
letis (27.32. TOK PISIN)
lētisi (27.32. NIUE)
letisi (27.32. SHONA)
letokomane (40.1. SOTHO, SOUTHERN)
letondal (16.1a. TAGALOG)
letonkomane (40.1. TSWANA)
letsugas (27.31. FILIPINO)
letsugas (27.31. TAGALOG)
letsugas (27.32. BICOLANO, CENTRAL)
letsugas (27.32. BINUKID)
letsugas (27.32. BONTOC CENTRAL)
letsugas (27.32. CEBUANO)
letsugas (27.32. FILIPINO)
letsugas (27.32. GIANGAN)
letsugas (27.32. IBANAG)
letsugas (27.32. IFUGAO)
letsugas (27.32. ILOCANO)
letsugas (27.32. PAMPANGAN)
letsugas (27.32. TAGALOG)
letsugas (27.32. TAUSUG)
lettuce (27.29. KANNADA)
lettuce (27.29. MALAYALAM)
lettuce (27.31. ENGLISH)
lettuce (27.31. FIJIAN)
lettuce (27.31. KANNADA)
lettuce (27.31. MALAYALAM)
lettuce (27.32. ENGLISH)
lettuce (27.32. FIJIAN)
lettuce (27.32. KANNADA)
lettuce (27.32. MALAYALAM)
lettue (27.32. CAJUN FRENCH)
letus (27.29. BENGALI)
letus (27.29. MARATHI)
letus (27.30. INDONESIAN)
letus (27.31. BENGALI)
letus (27.31. MARATHI)
letus (27.32. BENGALI)
letus (27.32. MARATHI)
letuse (27.29. TELUGU)
letuse (27.31. TELUGU)
letuse (27.32. TELUGU)
letusi (27.32. SWAHILI)
letuz (27.31. BRETON)
letuz (27.32. BRETON)
letysen ddail (27.32. WELSH)
letysen gron (27.31. WELSH)
lɛtyus (27.32. KRIO)
leugebas (30.43. STELLINGWERFS)
leunca (62.15. SUNDA)
leunca (62.21. INDONESIAN)
leunca badak (62.15. SUNDA)
leunca komir (62.9. SUNDA)
leunca manuk (62.15. SUNDA)
leusit (38.23. HEBREW)
le'usit (38.23. HEBREW)
leutia (23.3. ORIYA)
lèv (24.7. JOLA-FONYI)
levant garlic (7.1. ENGLISH)
leveles káposzta (30.35. HUNGARIAN)
leveles kel (30.35. HUNGARIAN)
leveles saláta (27.32. HUNGARIAN)

leyanit (49.1. HEBREW)
leyor (38.1. SUNDA)
lezegenn (27.31. BRETON)
l'gardim (59.2. ARABIC)
l'gerté (40.1. ARABIC)
l-ghazz (40.12. MALTESE)
l-gherq mustarda (30.1. MALTESE)
lhanamat (23.3. MALAYALAM)
lhanamat (23.3. MARATHI)
li cai (35.5. CHINESE, MANDARIN)
li cai (35.5. CHINESE, WÚ)
li dou (40.18. CHINESE, MANDARIN)
li dou (40.18. CHINESE, WÚ)
lia (20.6. INDONESIAN)
li-a tamlan (20.6. HANUNOO)
liân (48.1a. CHINESE, MIN NAN)
lian (48.1a. CHINESE, WÚ)
lian (50.2. CHINESE, MANDARIN)
lian ou (48.1a. CHINESE, YUE)
lian shu (40.11. CHINESE, GÀN)
lian shu (40.21. CHINESE, XIANG)
lian zhi (48.1b. CHINESE, YUE)
lian zi (48.1b. CHINESE, WÚ)
lian zi cao (23.2. CHINESE, WÚ)
liane à gratter (40.18. FRENCH)
liane aux voyageurs (69.1. FRENCH)
liang cai (30.52. CHINESE, YUE)
liang gua (11.1. CHINESE, WÚ)
liang gua (38.23. CHINESE, MIN BEI)
liang gua (38.23. CHINESE, YUE)
liang ma (27.42. CHINESE, GÀN)
liang shan ci zhu (17.2. CHINESE, WÚ)
liân-hoe (48.1. CHINESE, MIN NAN)
liao dong dang qui (26.3. CHINESE, MANDARIN)
liao dong hu mu (26.3. CHINESE, WÚ)
libato (28.1. SPANISH)
libato (28.1. TAGALOG)
libó (27.24. PORTUGUESE)
libó d'água (27.44. PORTUGUESE)
libumbe (27.32. LUYIA)
libumbe (30.27a. LUYIA)
libun (27.18. BICOLANO, CENTRAL)
lichořeřišnice větší (64.3. CZECH)
lichořešnice větši (65.1. CZECH)
líčidlo (56.1. CZECH)
líčidlo (56.4. CZECH)
líčidlo americké (56.2. CZECH)
líčidlo jedlé (56.3. CZECH)
lidadoci (38.22. YAO)
liden katost (45.5. DANISH)
lidib (36.3. TAGBANWA)
lidib kayo (39.4. TAGBANWA)
lidodo (23.3. LUYIA)
lidrus (1.1. STELLINGWERFS)
lidruske (1.1. STELLINGWERFS)
lie dong (52.1. CHINESE, WÚ)
lieaugļu ķirbis (38.14. LATVIAN)
lieaugļu ķirbis (38.16. LATVIAN)
lieaugļu ķirbis (38.18. LATVIAN)
lielā nātre (66.3. LATVIAN)
lielauglu kirbji (38.16. LATVIAN)
lielsakņu alokāzija (8.1. LATVIAN)
liên-fã (50.2. CHINESE, HAKKA)
lif kabaği (38.22. TURKISH)
lif yabas (7.2. KRIO)
liferi (38.16. GOURMANTCHÉ)
lift (30.22. ARABIC)
lift (30.41a. ARABIC)
lift (30.41b. ARABIC)
lif-yabas (7.2. KRIO)
ligayan (11.2. BUHID)
ligbus (39.4. HILIGAYNON)
lighippwáy (17.1. CAROLINIAN)
lighippwáy (17.2. CAROLINIAN)
lighippwáy (17.4. CAROLINIAN)
ligi (40.43. KANURI)
ligippwaay (17.1. CAROLINIAN)
ligippwaay (17.2. CAROLINIAN)
ligippwaay (17.4. CAROLINIAN)
ligyu an palawel (9.1. IFUGAO)
ḽihabu (38.3. VENDA)
lihondo (38.14. LUYIA)
lii sui xian (23.3. CHINESE, WÚ)
liifu letasu (27.32. JAPANESE)
liiki (7.8. JAPANESE)
liilia (13.4. ESTONIAN)
liiriyo (13.3. CAROLINIAN)
lijnmosterd (30.12. VLAAMS)
lik (7.8. MALAY)
lik (40.1. MANKANYA)
likabichi (30.3. LUYIA)
likala (17.5. MANDING-MANDINKA)
līkās melone (38.10. LATVIAN)
likatsila (27.33. LUYIA)
li̱ke (17.5. TIV)
likhuvi (40.43. LUYIA)
likhwana (8.4. LUYIA)
likir (19.1. INDONESIAN)
likka (11.1. CAROLINIAN)

likka (11.2. CAROLINIAN)
likka (11.4. CAROLINIAN)
likka (11.5. CAROLINIAN)
likun (17.3. FON)
lilanaja (13.1. BELARUSAN)
lilé (38.3. BÉTÉ)
lilee (13.4. GAELIC, MANX)
lileeah (13.4. YIDDISH)
lilek (62.21. CZECH)
lilek brambor (62.24. CZECH)
lilek černý (62.21. CZECH)
lilek vejcoplodý (62.19. CZECH)
lili (11.7. BASSARI)
lili (11.7. DAGBANI)
lili' (13.3. CHICKASAW)
lili (13.3. KRIO)
lili (13.4. SAMOAN)
lili (13.4. URDU)
lili (13.4. WELSH)
lilia (13.4. POLISH)
lilia wodna (50.2. POLISH)
liliakhunyu (11.6. LUYIA)
liliawir dydd (13.3. WELSH)
lilie (13.4. CZECH)
Lilie (13.4. GERMAN)
lilija (13.4. LATVIAN)
lilija (13.4. MACEDONIAN)
lilija (13.4. RUSSIAN)
lilija (13.4. SORBIAN, UPPER)
lilinaw naudit (27.18. SAMBAL, BOTOLAN)
liliom (13.4. HUNGARIAN)
liliou (13.4. GREEK)
liliowiec żółty (13.3. POLISH)
lilja (13.4. ICELANDIC)
lilja (13.4. SWEDISH)
liljan (13.4. SERBO-CROATIAN)
lilje (13.4. DANISH)
lilje (13.4. NORWEGIAN)
Liljen (13.4. LUXENBOURGEOIS)
liljor (13.4. SWEDISH)
lillkapsas (30.26. ESTONIAN)
lilophal (50.2. HINDI)
lily (13.4. ASSAMESE)
lily (13.4. ENGLISH)
lima (35.9. OROMO, WEST CENTRAL)
lima (40.25. RUSSIAN)
lima bean (40.25. ENGLISH)
lima böna (40.25. SWEDISH)
lima boon (40.25. VLAAMS)
lima fasulyesi (40.25. TURKISH)
lima fižol (40.25. SLOVENIAN)
lima grah (40.25. SERBO-CROATIAN)
lima grav (40.25. MACEDONIAN)
limabab (40.25. HUNGARIAN)
limabaunir (40.25. ICELANDIC)
lima-boby (40.25. CZECH)
Limabohne (40.25. GERMAN)
lima-bønne (40.25. DANISH)
limabønne (40.25. NORWEGIAN)
limaboon (40.25. AFRIKAANS)
lima-boon (40.25. DUTCH)
limanpapu (40.25. FINNISH)
limas pupinas (40.25. LATVIAN)
limb (38.20. SERER-SINE)
limba mielului medicinala (29.1. ROMANIAN)
limnocares (14.1. PORTUGUESE)
limski fasul (40.25. BULGARIAN)
liṅ (30.56. NEWAR)
lin (48.1a. CHINESE, XIANG)
lin ou (48.1a. CHINESE, GÀN)
lin ou (48.1a. CHINESE, WÚ)
linay (40.40. PANGASINAN)
lindung (38.29. INDONESIAN)
lindz (40.12. YIDDISH)
line mustard (30.12. ENGLISH)
linene (36.3. WOLOF)
lĩng (7.10. GOURMANCHÉMA)
ling jiao (64.1. CHINESE, GÀN)
ling jiao (64.1. CHINESE, MIN BEI)
ling jiao (64.1. CHINESE, XIANG)
ling jiao (64.3. CHINESE, WÚ)
ling jiao (64.3. CHINESE, YUE)
ling jiao cai (30.43. CHINESE, YUE)
ling jiao si gua (38.21. CHINESE, MANDARIN)
lingaling (48.1. IBANAG)
lingapotla (38.29. TELUGU)
liŋgi-na (55.1. YALUNKA)
língua de vaca (58.3. PORTUGUESE)
lingué (38.20. SUSU)
linho de Gombo (45.4. PORTUGUESE)
linikway (45.2. BUHID)
linjal (8.4. JAVANESE)
linjik (8.9. INDONESIAN)
linjik (8.9. JAVANESE)
linkkwum (20.6. MANDING-MANDINKA)
linoko (39.4. BONTOC, CENTRAL)
linouari (38.7b. GOURMANTCHÉ)
lins (40.12. SWEDISH)
linsa (8.4. BICOLANO, CENTRAL)

linse (40.12. DANISH)
Linse (40.12. GERMAN)
linse (40.12. NORWEGIAN)
Linsen-Bohne (40.37. GERMAN)
linssi (40.12. FINNISH)
linsubaunir (40.12. ICELANDIC)
lìnta (40.12. ROMANI, BALKAN)
linte (40.12. ROMANIAN)
linte comestibilă (40.12. ROMANIAN)
lintea pratulnide de cultura (40.10. ROMANIAN)
lintukay (17.3. BUHID)
linyololo (45.3. YAO)
linze (40.12. DUTCH)
linze (40.12. VLAAMS)
liondo (38.14. LUYIA)
liôung-sī-ts'ói (13.2. CHINESE, HAKKA)
lip anian (7.2. TOK PISIN)
lipay (40.18. TAGALOG)
lipeta (11.1. YAO)
lipeta (11.2. YAO)
lipeta (11.3. YAO)
lipeta (11.4. YAO)
lipeta (11.5. YAO)
lipeta (11.6. YAO)
lipeta (11.7. YAO)
lipeta (11.8. YAO)
lippoyris (1.1. GREEK)
liremwa (16.1a. LUYIA)
lírio (13.4. CRIOULO, CAPE VERDE)
lirio (13.4. SPANISH)
lírios (13.4. PORTUGUESE)
lirungu (11.6. LUYIA)
lis (13.4. FRENCH)
lis de tigre (12.1. FRENCH)
lis du Nil (48.1a. FRENCH)
lisa (23.3. KASHMIRI)
lisaka (34.2. LUYIA)
lisama (23.3. MÒORÉ)
lisambo (23.3. MÒORÉ)
liściowiec wodny (59.3. POLISH)
liseron d'eau (36.2. FRENCH)
liseron du Japon (36.1. FRENCH)
lisna cveklo (35.3. MACEDONIAN)
lisna mangolds (35.3. MACEDONIAN)
lisna salata (27.32. MACEDONIAN)
lisna salata (27.32. SERBO-CROATIAN)
lisna zelka (30.35. MACEDONIAN)
lisnati kelj (30.35. CROATIAN)
lisou (30.27a. MORISYEN)
lisou rouz (30.27b. MORISYEN)
lisou-fler (30.26. MORISYEN)
listna salata za rjazane (27.32. BULGARIAN)
listnati chrovt (30.35. SLOVENIAN)
listno zele (30.35. BULGARIAN)
listovoj salat (27.32. KAZAKH)
listovoj salat (27.32. RUSSIAN)
lisukuma (30.35. LUYIA)
litchouga (27.31. LADINO)
litchouga (27.32. LADINO)
liticiti (38.3. YAO)
litore (16.1a. LUYIA)
litoto (23.3. LUYIA)
litsuga (27.32. BICOLANO, CENTRAL)
litsuga (27.32. BINUKID)
litsuga (27.32. BONTOC CENTRAL)
litsuga (27.32. CEBUANO)
litsuga (27.32. GIANGAN)
litsuga (27.32. IBANAG)
litsuga (27.32. IFUGAO)
litsuga (27.32. ILOCANO)
litsuga (27.32. PAMPANGAN)
litsuga (27.32. TAUSUG)
litsugas (27.31. TAGALOG)
litsúgas (27.32. HILIGAYNON)
litsugas (27.32. TAGALOG)
litugu (11.6. LUYIA)
litwor lekarski (24.1. POLISH)
liu lan xiang (44.3. CHINESE, MANDARIN)
liu lan xiang (44.3. CHINESE, WÚ)
liu lan xiang (44.3. CHINESE, YUE)
liu li ju (29.1. CHINESE, WÚ)
liuskaratamo (57.1. FINNISH)
livadna (27.47. SERBO-CROATIAN)
livadska penlivka (30.44. MACEDONIAN)
ḽivhele (17.7. VENDA)
livid pigweed (23.3. ENGLISH)
Livingstone aardappel (44.8. VLAAMS)
Livingstone potato (44.8. ENGLISH)
ljekoviti (27.45. SERBO-CROATIAN)
ljubica (68.1. SERBO-CROATIAN)
ljufa (38.22. KAZAKH)
ljufa (38.22. RUSSIAN)
l'l' (62.19. KORO)
llabaza (27.5. GALICIAN)
llaeth ysgallen (27.42. WELSH)
llagas de Cristo (65.1. SPANISH)
llamón (27.39. SPANISH)
llantén (16.1a. SPANISH)

lombaria (38.4. SONGHAY)
lombo (8.4. YAO)
lombo (8.9. BULU)
lombo (8.9. EWONDO)
lombok (62.1. INDONESIAN)
lombok (62.1. JAVANESE)
lombok (62.1. SUNDA)
lombok (62.2. INDONESIAN)
lombok jempling (62.5. JAVANESE)
lombok jemprit (62.5. JAVANESE)
lombok rawit (62.5. JAVANESE)
lômiêt (20.2. KHMER, CENTRAL)
lomka taya (8.9. HINDUSTANI)
lomme (30.43. SWEDISH)
lomoda (45.4. MANDING-MANDINKA)
lompong (8.1. INDONESIAN)
lona (59.2. SANSKRIT)
lonak (59.2. PANJABI, EASTERN)
lonak (59.4. PANJABI, EASTERN)
lonamla (59.4. SANSKRIT)
loncak (62.15. INDONESIAN)
lone (20.6. BULLOM SO)
long bean (40.42. ENGLISH)
long chuan jue (4.1. CHINESE, YUE)
long hao (27.6. CHINESE, WÚ)
long kui (62.21. CHINESE, MANDARIN)
long kui (62.21. CHINESE, YUE)
long petiole mustard (30.13. ENGLISH)
long zedoary (20.4. ENGLISH)
longiloba (6.2. CEBUANO)
longlong (27.42. KALINGA, BUTBUT)
longolongo'uha (23.3. TONGAN)
longolongo'uha (27.18. TONGAN)
longolongo'uha (27.42. TONGAN)
longroot onion (7.14. ENGLISH)
loni (20.6. BULLOM SO)
loni (59.2. GUJARATI)
loni (59.2. SANSKRIT)
lonia (59.2. HINDI)
lonika (23.2. SANSKRIT)
lonika (59.2. SANSKRIT)
lonika (59.4. SANSKRIT)
lontho-lɛ (45.1. BULLOM SO)
lonyon (7.3. KWÉYÒL)
lonyon (7.11. KWÉYÒL)
loofah (38.22. FRENCH)
loofah (38.22. TAGALOG)
lóóghu (11.1. CAROLINIAN)
lóóghu (11.2. CAROLINIAN)
lóóghu (11.4. CAROLINIAN)
lóóghu (11.5. CAROLINIAN)
loogo (39.4. DENDI)
loothari (27.45. SANSKRIT)
lootos (48.1a. ESTONIAN)
looy (38.21. LAO)
lôp'ĕt (30.41a. CHINESE, HAKKA)
lopang (38.22. SUNDA)
łopian większy (27.5. POLISH)
lopka (11.5. BAOULÉ)
lopoč (50.2. SERBO-CROATIAN)
lopótök (38.20. HUNGARIAN)
lopuch větší (27.5. CZECH)
lopukh (13.2. KAZAKH)
lopush vjaliky (27.5. BELARUSAN)
loqɔɔ pɛ̄ɛ̄tshɛɛ (27.32. TIBETAN)
lorgyirike (62.24. SENOUFO, CEBAARA)
lörü (38.13a. KASHMIRI)
losan (40.17. URDU)
lo-séno (40.1. BAMANANKAN)
loshuu (30.3. SWAHILI)
lô-sún (13.2. CHINESE, MIN NAN)
lotier cultivé (40.13. FRENCH)
lotier rouge (40.13. FRENCH)
loti-ô (40.1. MANDING-MANDINKA)
loto d'Egitto (48.1a. ITALIAN)
loto sagrado (48.1a. SPANISH)
loto vermelho (40.13. PORTUGUESE)
lotos (48.1b. GREEK)
lotos kłączowy (48.1a. POLISH)
lotos orzechodajny (48.1b. POLISH)
lotoss (48.1a. LATVIAN)
lotto rosso (40.13. ITALIAN)
lotus (48.1a. DANISH)
lotus (48.1a. INDONESIAN)
lotus bleu d'Egypte (50.2. FRENCH)
lotus do Egipto (48.1a. PORTUGUESE)
lotus égyiptien (48.1a. FRENCH)
lotus ha-mayim (48.1a. HEBREW)
lotus índico (48.1a. PORTUGUESE)
lotus indien (48.1a. FRENCH)
lotus lily (50.2. ENGLISH)
lotus root (48.1a. ENGLISH)
lotus sacré (48.1a. FRENCH)
lotus seed (48.1b. ENGLISH)
lotusfrø (48.1b. NORWEGIAN)
lotusrøtter (48.1a. NORWEGIAN)
lotussaad (48.1b. AFRIKAANS)
lotuswortel (48.1a. VLAAMS)
lotuswortels (48.1a. AFRIKAANS)
lótusz magvak (48.1b. HUNGARIAN)

lotuszaden (48.1b. VLAAMS)
lótuszgyökér (48.1a. HUNGARIAN)
lou shen kui (45.4. CHINESE, MANDARIN)
lou shen kui (45.4. CHINESE, WÚ)
loufa (38.21. GREEK)
lougou (39.4. TIV)
louk chesnok (7.10. RUSSIAN)
louk parei (7.8. RUSSIAN)
louk repčhatyi (7.3. RUSSIAN)
louk repka (7.3. RUSSIAN)
louk zepchatnyi (7.3. RUSSIAN)
loukaya loua moukouyou (35.4. KOONGO)
loukaya-louakoutou (34.2. DOONDO)
louk-rezanets (7.11. RUSSIAN)
louk-scarada (7.11. RUSSIAN)
loupimos o lefkos (40.14. GREEK)
loupin (40.14. RUSSIAN)
louri nafolo (38.3. GOURMANCHÉMA)
louve (30.35. GALICIAN)
louvoudia (35.5. GREEK)
louzaouenn an troc'h (29.2. BRETON)
louzaouenn ar c'halvez (29.2. BRETON)
louzaouenn-ar-maen (62.11. BRETON)
love apple (62.9. ENGLISH)
love-lies-bleeding (23.3. ENGLISH)
løvetand (27.45. DANISH)
løvetann (27.45. NORWEGIAN)
loviya (40.9. UZBEK)
loviya (40.26. UZBEK)
Löwenzahn (27.45. GERMAN)
loya pagirisun (20.6. MARANAO)
loya tidek (62.5. MARANAO)
loyi (20.6. AGTA)
lōz (40.39. HARARI)
lozei (58.2. KWÉYÒL)
l-sprag (13.2. MALTESE)
lu cai hua (30.31. CHINESE, YUE)
lu dou (40.38. CHINESE, MANDARIN)
lu dou (40.38. CHINESE, WÚ)
lu lu d[uwj]c (62.15. VIETNAMESE)
lu si cai (44.11. CHINESE, GÀN)
lu si cai (44.11. CHINESE, WÚ)
lu sun (13.2. CHINESE, MANDARIN)
lu sun (13.2. CHINESE, WÚ)
lu sun (13.2. CHINESE, YUE)
luba (38.17. INDONESIAN)
luba (38.18. INDONESIAN)
lubalala (40.1. BEMBA)
lubalub (44.5. SERER-SINE)
lubar (56.1. PANJABI, EASTERN)
lùbăsa (7.3. NUPE-NUPE-TAKO)
lubenica (38.3. BOSNIAN)
lubenica (38.3. MACEDONIAN)
lubenica (38.3. SERBO-CROATIAN)
lubenica (38.3. SLOVENIAN)
lubenice (38.3. CROATIAN)
lubenice (38.3. CZECH)
lubeniţă (38.3. ROMANIAN)
lubia (40.9. FARSI, WESTERN)
lubia bean (40.9. ENGLISH)
lubiaboon (40.9. VLAAMS)
lubias (40.26. BONTOC, CENTRAL)
lubi-lubi (62.21. BICOLANO, CENTRAL)
lubi-lubi (62.21. CEBUANO)
lubi-lubi (62.21. TAGALOG)
lubin (40.14. BELARUSAN)
łubin (40.14. POLISH)
lubinas (40.14. LITHUANIAN)
lubingan (8.4. IFUGAO)
lubio (40.26. TAJIKI)
lubio (40.41. TAJIKI)
lubioi surkh (40.24. TAJIKI)
lubish (40.9. ARABIC)
lubiyâ (40.26. FARSI, EASTERN)
lubiyâ (40.26. FARSI, WESTERN)
lubuzi (38.23. KOONGO)
lucerka sijata (40.17. SERBO-CROATIAN)
lucerna (40.17. HUNGARIAN)
lucernă (40.17. ROMANIAN)
lucerna comună (40.17. ROMANIAN)
lucerna posevnaja (40.16. KAZAKH)
lucerna siata (40.17. SLOVAK)
lucerna siewna (40.17. POLISH)
lucerne (40.17. DANISH)
lucerne (40.17. DUTCH)
lucerne (40.17. ENGLISH)
lucerne (40.17. MALAYALAM)
lucerne ghah (40.17. ASSAMESE)
luchúgas (27.32. PANGASINAN)
lucina (40.17. CZECH)
lucka (7.12. SLOVENIAN)
ludăi (38.17. ROMANIAN)
ludju (19.1. TEM)
ludo (62.21. MÒORÉ)
lufa (38.21. HEBREW)
lufa (38.21. LATVIAN)
lufa (38.22. BULGARIAN)
lufa (38.22. CZECH)
lufa (38.22. ESTONIAN)
lufa (38.22. LATVIAN)

lufa (38.22. MACEDONIAN)
lufa (38.22. PORTUGUESE)
lufă (38.22. ROMANIAN)
lufa (38.22. SERBO-CROATIAN)
lufa (38.22. SLOVENIAN)
lufa gelilit (38.22. HEBREW)
lufa ostrohranná (38.21. CZECH)
lufa riscada (38.21. PORTUGUESE)
lufa válcovatá (38.22. CZECH)
luff (38.22. ARABIC)
luffa (38.21. NORWEGIAN)
luffa (38.22. DUTCH)
Luffa (38.22. GERMAN)
luffa (38.22. ITALIAN)
luffa (38.22. NORWEGIAN)
luffa (38.22. RUSSIAN)
luffa (38.22. SWEDISH)
luffah (38.22. INDONESIAN)
luffasvamp (38.22. DANISH)
lufhuri (38.14. VENDA)
lufhuri (38.16. VENDA)
lufhuri (38.17. VENDA)
lufhuri (38.18. VENDA)
luftløk (7.9a. NORWEGIAN)
luftyams (11.2. NORWEGIAN)
lugasi (40.1. BINUKID)
lugasi (40.1. MANOBO)
lugasin (40.1. MANOBO)
lugasing (40.1. MANOBO)
lugasing (40.1. MARANAO)
lugiman (11.1. BUHID)
lugosing (40.1. BICOLANO, CENTRAL)
lugtun di pinading (11.2. BONTOC, CENTRAL)
lugtus puggupug (11.5. IFUGAO)
luguǰ u valo (27.42. WOLOF)
luhtalitukka (30.44. FINNISH)
luk (7.3. BOSNIAN)
luk (7.3. BULGARIAN)
luk (7.3. CROATIAN)
luk (7.3. SERBO-CROATIAN)
luk batun (7.5. KAZAKH)
luk batun (7.5. RUSSIAN)
luk mnogojarusnyj (7.9a. KAZAKH)
luk mnogojarusnyj (7.9a. RUSSIAN)
luk porej (7.8. KAZAKH)
luk porej (7.8. RUSSIAN)
luk repcatyj (7.3. KAZAKH)
luk repčatyj (7.3. RUSSIAN)
luk vinogradnyj (7.1. RUSSIAN)
luk vlasac (7.11. CROATIAN)
lukaya lua nkuyu (35.4. DOONDO)
lukeh (19.1. MALAY)
luko (8.4. SAMBAL, TINÀ)
luk-šalot (7.2. RUSSIAN)
luktu (11.1. IFUGAO)
luktu (36.3. KALAGAN)
luku loonu (7.3. SINHALA)
lukyu (7.2. TAGALOG)
lulasi (17.1. NYAKYUSA-NGONDE)
lule lakër (30.26. ALBANIAN, TOSK)
lule vjeshte (27.23. ALBANIAN, TOSK)
lulo (62.22. ENGLISH)
lulo (62.22. VLAAMS)
l'ul'ok vajcoplodý (62.19. SLOVAK)
l'ulok vojcoplodný (62.19. SLOVAK)
lulosøtvier (62.22. NORWEGIAN)
lulungngin (40.1. IFUGAO)
lumak (36.3. BUHID)
lumanda (45.3. TUMBUKA)
lumba loki (38.29. FIJIAN)
lumbuzi-busi (38.23. BEMBÉ)
lumonde (36.3. GANDA)
lumonde (62.24. GANDA)
lumpuyang (20.2. CEBUANO)
lumug an liggod (36.3. IFUGAO)
lumullung (36.3. IFUGAO)
lun ye sha shen (33.1. CHINESE, MANDARIN)
luṋawa (40.26. VENDA)
lungar (2.1. PANJABI, EASTERN)
lunglung an manema (27.43. IFUGAO)
lungrot (35.5. SWEDISH)
lungum (59.2. IFUGAO)
lungun (8.2. BUHID)
lungun (8.3. BUHID)
lungun (59.2 IFUGAO)
lunia (59.2. HINDI)
lunia (59.2. SANSKRIT)
lunikulfa (59.2. HINDI)
lunud (41.1. HANUNOO)
lunuk (59.2. HINDI)
luo bai (30.54. CHINESE, GÀN)
luo bai (30.56. CHINESE, GÀN)
luo bai (30.56. CHINESE, YUE)
luo bo (30.54. CHINESE, MANDARIN)
luo hua sheng (40.1. CHINESE, XIANG)
luo kui (28.1. CHINESE, YUE)
luo su (62.19. CHINESE, WÚ)
luo su (62.19. CHINESE, XIANG)

lúparo (46.1. GALICIAN)
lupeen (40.14. GAELIC, MANX)
lüpen (40.14. TURKISH)
lupien (40.14. AFRIKAANS)
lupiene (40.14. STELLINGWERFS)
lupiin (40.14. ESTONIAN)
lupiini (40.14. FINNISH)
lupin (40.14. ASSAMESE)
lupin (40.14. DANISH)
lupin (40.14. ENGLISH)
lupin (40.14. FRENCH)
lupin (40.14. NORWEGIAN)
lupin (40.14. ROMANIAN)
lupin (40.14. RUSSIAN)
lupin (40.14. SWEDISH)
lupin alb (40.14. ROMANIAN)
lupin i bardhë (40.14. ALBANIAN, TOSK)
lupina (40.14. CZECH)
lupina (40.14. ICELANDIC)
lupina (40.14. MACEDONIAN)
lupina bela (40.14. SERBO-CROATIAN)
lupine (40.14. DUTCH)
Lupine (40.14. GERMAN)
lupine (40.14. STELLINGWERFS)
lupine (40.14. VLAAMS)
lupino (40.14. ITALIAN)
lupino (40.14. SPANISH)
lupisan (11.5. HILIGAYNON)
lupolie verigel (52.1. ROMANIAN)
lùppolo (46.1. ITALIAN)
lúpulo (46.1. PORTUGUESE)
lúpulo común (46.1. SPANISH)
lurabu (38.20. LUYIA)
luré (19.1. MÒORÉ)
luri (45.1. TUPURI)
lus (30.43. GAELIC, IRISH)
lus an tsiocaire (27.10a. GAELIC, IRISH)
lus ny gfaih (23.3. GAELIC, MANX)
lus y chugyr (27.10a. GAELIC, MANX)
lus yn ainle (24.1. GAELIC, MANX)
lus yn 'eail vaayl (27.17. GAELIC, MANX)
l[us]a mi[ee]u (17.8. VIETNAMESE)
lusan (40.17. PANJABI, EASTERN)
lusarna (40.17. ORIYA)
lusengu (17.1. BEMBA)
lusengu (17.2. BEMBA)
lusengu (17.4. BEMBA)
luseni (40.17. SWAHILI)
lusering (40.17. AFRIKAANS)
lusern (40.17. AFRIKAANS)
luserne (40.17. NORWEGIAN)
lussun (7.10. PANJABI, EASTERN)
łust głąbigroszek (40.29. POLISH)
lusun (7.10. BENGALI)
luṱhariàne (36.3. VENDA)
lutos (48.1a. FILIPINO)
lutoyo (40.11. BEMBA)
lutsarna sinjaja (40.17. RUSSIAN)
lutsena pasjannaja (40.17. BELARUSAN)
lutsern (40.17. ESTONIAN)
lutserna obyknovennaja (40.17. RUSSIAN)
lutserna posevnaja (40.17. RUSSIAN)
luttu (11.5. GADDANG)
luttu (11.5. IBANAG)
luttu (11.5. IFUGAO)
lutukka (30.43. FINNISH)
lutya (8.9. BINUKID)
luuli (35.5. MONGOLIAN)
luuli (35.6. MONGOLIAN)
lu-uya (20.6. TAUSUG)
luvër (46.1. ALBANIAN, TOSK)
luvya (40.35. HEBREW)
luvya (40.36. HEBREW)
luvya (40.37. HEBREW)
luvya (40.38. HEBREW)
luvya (40.40. HEBREW)
luvya arukat tarmil (40.42. HEBREW)
luvya gelilit (40.41. HEBREW)
luvya tarbutit (40.43. HEBREW)
luwaf (17.5. TIRURAY)
luwaliw samparansisku alistun (39.3. HANUNOO)
lùwásàr (7.3. KANURI)
luwimbe (38.8. LOMWE, MALAWI)
luy manuk (23.6. HANUNOO)
luy-a (20.6. BICOLANO, CENTRAL)
luya (20.6. BINUKID)
luy-a (20.6. BUTUANON)
luy-a (20.6. CEBUANO)
luya (20.6. FILIPINO)
lúy'a (20.6. HILIGAYNON)
luya (20.6. TAGALOG)
luya (20.6. TAGBANWA)
luy-a (20.6. WÁRAY-WÁRAY)
luyá (20.6. WÁRAY-WÁRAY)
luya-luyahan (20.4. FILIPINO)
luya-luyahan (20.4. TAGALOG)
luyang dilaw (20.2. FILIPINO)
luyang dilaw (20.2. TAGALOG)
luzerna (40.17. PORTUGUESE)

luzerna (40.17. RONGA)
luzerne (40.17. DUTCH)
luzerne (40.17. FRENCH)
luzerne (40.17. VLAAMS)
luzker (38.13a. BASQUE)
lvs (27.31. GAELIC, MANX)
lvs (27.32. GAELIC, MANX)
lvs ny meeyl (23.4. GAELIC, MANX)
lvs ny minnag (27.45. GAELIC, MANX)
lvs ny moyl mooarey (45.5. GAELIC, MANX)
lvs y chree (24.9. GAELIC, MANX)
lws africa (62.18. HMONG-DAW)
lws kub tshis (62.19. HMONG-DAW)
lyciet (62.8. FRENCH)
lyciet de la Chine (62.8. FRENCH)
lycope (44.1. FRENCH)
lyhtykoiso (62.11. FINNISH)
lykiskos (46.1. GREEK)
lykopersikon (62.9. GREEK)
lykopous (44.1. GREEK)
ma bamboo (17.2. ENGLISH)
ma baw da ma (17.2. LISU)
ma bu (17.2. AKHA)
ma chi xian (59.2. CHINESE, GÀN)
ma chi xian (59.2. CHINESE, MANDARIN)
ma chi xian (59.2. CHINESE, WÚ)
ma chi xian (59.2. CHINESE, YUE)
ma ch'in hsien (59.2. CHINESE, MANDARIN)
ma hae (40.2. THAI)
ma¯ heh lo¯ (62.23. LAHU)
ma¯ hkui¯ shi_ (62.19. LAHU)
ma¯ hkui¯ shon_ (62.9. LAHU)
ma ju (27.31. CHINESE, WÚ)
ma ka cai (30.50. CHINESE, MANDARIN)
ma karal (40.42. SINHALA)
ma- kɔbɔkɔbɔ (62.19. THEMNE)
ma khaw neh (17.2. LISU)
ma lan (27.28. CHINESE, WÚ)
ma lan tou (27.28. CHINESE, WÚ)
ma ling shu (62.24. CHINESE, MANDARIN)
ma ling shu (62.24. CHINESE, WÚ)
ma ling shu (62.24. CHINESE, YUE)
ma lo po (30.1. CHINESE, MANDARIN)
ma¯ naweh (38.22. LAHU)
ma noi (38.29. THAI)
ma ngaeng (64.3. THAI)
ma pae (40.40. THAI)
ma t̲a t̲êssha (11.1. SIDAMO)
ma t̲a t̲êssha (11.2. SIDAMO)
ma ti (10.2. CHINESE, WÚ)
ma ti (10.2. CHINESE, YUE)
ma ti cai (62.21. CHINESE, WÚ)
ma zhu (17.2. CHINESE, MANDARIN)
ma zhu (17.2. CHINESE, WÚ)
ma zhu sun (17.2. CHINESE, YUE)
maa (16.1a. KONO)
maa (16.1a. LOKO)
maa (16.1a. MENDE)
mãade (11.7. DANGME-KROBO)
maadhavi (24.13. SANSKRIT)
maaguttam (40.18. TAMIL)
maahuri (23.2. MAORI)
maakaka (2.1. MAORI)
maakastanja (10.1. FINNISH)
maa-kotɔpɔ-gutugutu (16.1a. LOKO)
maakrassi (30.2. FINNISH)
maala thulasi (44.5. HINDI)
ma-ali (38.16. THEMNE)
ma-ali (38.17. THEMNE)
maande (45.1. MÒORÉ)
maanka (8.1. KANNADA)
maanu (45.1. WAAMA)
maanupendalanu (39.4. TELUGU)
maanδr (45.1. YOM)
maapähkel (40.1. ESTONIAN)
maapähkinä (40.1. FINNISH)
maapähkinä (40.11. FINNISH)
maapirn (27.26. ESTONIAN)
maareesha (23.3. SANSKRIT)
maarsha (23.3. SANSKRIT)
maas (40.37. NEPALI)
maasa (40.38. SANSKRIT)
maasa parni (40.40. SANSKRIT)
maasaparni (40.37. SANSKRIT)
maasha (40.37. SANSKRIT)
maatáák (36.3. BLACKFOOT)
maatáák (62.24. BLACKFOOT)
maatáákkistsi (62.24. BLACKFOOT)
maatha (23.3. MARATHI)
maa-wawa (16.1a. LOKO)
maawiwngwu (40.26. HOPI)
maayaalachivel (28.1. MARATHI)
mabaia (11.4. PULAAR)
mabalbalino (48.1. PANGASINAN)
ma-bana (16.1a. THEMNE)
mabanwa (36.3. BUHID)
ma-bela (62.19. THEMNE)
mabenda (45.1. DIGO)
ma-bɛ́tɛ́r (40.1. BUDIK)
mabilinganya (62.18. LOMWE, MALAWI)

mabilinganya (62.19. LOMWE, MALAWI)
mablüloziz (10.1. TURKISH)
mâbú (38.20. NGIZIM)
mabvende (38.3. LOMWE, MALAWI)
maca (30.50. ENGLISH)
maca (30.50. FRENCH)
maca (30.50. PORTUGUESE)
maca (30.50. SPANISH)
maca (30.50. VLAAMS)
macao (62.12. HINDI)
macara (40.1. CRIOULO, UPPER GUINEA)
maccadjukhory (17.7. UZBEK)
mâche (67.2. FRENCH)
mâche à fruits velus (67.1. FRENCH)
mâche d'Italie (67.1. FRENCH)
machha khursani (62.2. NEPALI)
machhaka (20.6. SANSKRIT)
machiché (38.6. PORTUGUESE)
machiku (17.2. JAPANESE)
machovka peruánská (62.12. SLOVAK)
mackerel mint (44.3. ENGLISH)
mačkino nane (44.2. MACEDONIAN)
macre (64.1. FRENCH)
macre (64.3. FRENCH)
mâcre nageante (64.3. FRENCH)
măcriş (58.2. ROMANIAN)
măcris de baltă (30.52. ROMANIAN)
măcrisul (58.2. ROMANIAN)
mâc-tzè (27.31. CHINESE, HAKKA)
mâc-tzè (27.32. CHINESE, HAKKA)
ma:cu (62.1. BADAGA)
ma:cu (62.2. BADAGA)
macucú (40.22. SPANISH)
mād' (17.1. MUNDARI)
mād' (17.2. MUNDARI)
mād' (17.4. MUNDARI)
mãdãa (16.1a. DANGME-KROBO)
Madagascar bean (40.25. ENGLISH)
Madagascar groundnut (40.39. ENGLISH)
Madagascar potato (44.8. ENGLISH)
Madagaskar fasulyesi (40.3. TURKISH)
Madagaskarbohne (40.3. GERMAN)
madagaszkári bab (40.25. HUNGARIAN)
ma-dake (17.4. JAPANESE)
madaki (40.35. KANNADA)
madanaganti (23.2. TELUGU)
madanous (24.19. ARABIC)
madanous etaly (24.20. ARABIC)
madanous leefty (24.21. ARABIC)
madarangabaale (16.1a. KANNADA)
mädarõigas (30.1. ESTONIAN)
madas (38.13a. TAUSUG)
madɛre (40.1. LIMBA)
madeso (40.25. KOONGO)
madesso (40.25. BEMBÉ)
madhaveshtagrishtika (11.2. SANSKRIT)
madhupaka (38.7b. SANSKRIT)
madhura (24.13. SANSKRIT)
madhura (35.9. SANSKRIT)
madhura kilanna (36.3. MALAYALAM)
madhuraphala (38.3. SANSKRIT)
madhuri (24.13. SANSKRIT)
madhurika (24.13. SANSKRIT)
madhurika (30.24. SANSKRIT)
madhvaluka (11.5. SANSKRIT)
madihagala (38.23. KANNADA)
madike (40.35. KANNADA)
madise soonp (24.12. NEPALI)
madise soonp (24.13. NEPALI)
madou kou (39.4. GOURMANCHÉMA)
maduku (45.4. SANGA)
madushaka (28.1. SANSKRIT)
madyaka (40.35. SANSKRIT)
madyaundu (38.20. NDALI)
madzẽ (23.3. ÉWÉ)
mael (8.1. JAVANESE)
mælkebøtte (27.45. DANISH)
maenglak (44.4. THAI)
maeo (17.7. KONKANI)
mäeroigas (30.1. ESTONIAN)
maethi (40.32. HINDI)
mafaeng (38.1. THAI)
mafafa (8.9. PORTUGUESE)
mafa'i (38.22. TONGAN)
mafaka (17.7. SWAHILI)
m[af]n m[af]n tr[aws]ng (34.2. VIETNAMESE)
mafora (40.18. MANDING-MANDINKA)
maforá (40.18. MANDING-MANDINKA)
mag (40.38. GUJARATI)
maga (40.37. MARATHI)
magadhi (11.2. SANSKRIT)
mágadhí (62.1. PALI)
magae (17.1. PANJABI, EASTERN)
magalay (24.9. MARANAO)
magalo (62.21. MANU'A SAMOAN)
māgalo (62.21. SAMOAN)
magalogalo (62.21. MANU'A SAMOAN)
magang (17.5. SANGA)
magar bans (17.1. PANJABI, EASTERN)
magarás (8.9. PORTUGUESE)

magarkaniani gogra (38.14. GEORGIAN)
ma-gatungal (41.1. MAGUINDANAO)
magbabaukok kamote (36.3. SAMBAL, TINÀ)
magbɛŋgbɛ (62.5. MALIMBA)
ma-gboli-*ma*-bana (62.23. THEMNE)
magdanoz (24.19. BULGARIAN)
magdonosi (24.19. GEORGIAN)
maggiorana (44.6. ITALIAN)
maggiorana selvatica (44.6. ITALIAN)
maghe kaayi (38.13a. KANNADA)
magigi (36.3. BUHID)
magimbi (8.1. SWAHILI)
magimbi (8.4. SWAHILI)
magina (20.2. FULFULDE, ADAMAWA)
magina (20.2. HAUSA)
magloire (23.2. FRENCH)
magonho (16.1b. MAKHUWA)
magtabokaw (40.4. CEBUANO)
magtambokan (40.3. WÁRAY-WÁRAY)
magtambokau (40.4. CEBUANO)
maguni gedde (30.56. KANNADA)
maguskartul (36.3. ESTONIAN)
magyar ánisz (24.13. HUNGARIAN)
magyarparéj (23.4. HUNGARIAN)
mãh (40.37. PANJABI, EASTERN)
maha (38.23. THAI)
mahaabharee (20.1. HINDI)
mahaabhareevacha (20.1. SANSKRIT)
mahaachanchu (63.1. SANSKRIT)
mahaahrihati (62.19. SANSKRIT)
mahaajaalini (38.22. SANSKRIT)
mahaakaaya (17.7. SANSKRIT)
mahaakanda (7.3. SANSKRIT)
mahaakoshaathakee (38.22. SANSKRIT)
mahaapadma (48.1a. SANSKRIT)
mahaaphala (38.20. SANSKRIT)
mahaashimbi (40.3. SANSKRIT)
mahaaveerya (11.2. SANSKRIT)
mahākando (7.10. PALI)
mähale (38.3. AMHARIC)
māhamūṅ (40.12. MAITHILI)
mahaparuvallikkodi (11.6. TAMIL)
maharada (40.3. ORIYA)
maharada (40.4. ORIYA)
mā-haraḍī (20.2. NEWAR)
maharagwe (40.26. SWAHILI)
maharagwe lima (40.25. SWAHILI)
maharagwe soya (40.8. SWAHILI)
maharashtri (27.2. SANSKRIT)
maharshabhilangali (40.18. SANSKRIT)
mahasuda (7.10. SANSKRIT)
mahathi (62.19. SANSKRIT)
mahaushad (20.6. SANSKRIT)
mahaushadamu (20.6. TELUGU)
mahaushadha (11.2. SANSKRIT)
mahaushadhi (24.7. SANSKRIT)
mahe (40.12. PASHTO, NORTHERN)
mahea (17.7. SOTHO, NORTHERN)
mahei (63.1. GOLA)
mahetau (62.24. MAORI)
mahinja (20.6. SANSKRIT)
mahmuz çiçeğI (44.8. TURKISH)
mahoa'a (19.1. TONGAN)
mahor-pat (40.43. ASSAMESE)
mahothapala (48.1a. SANSKRIT)
mahothika (62.19. SANSKRIT)
mahoti hinpoli (62.19. BENGALI)
mahtâmin (17.7. CREE, PLAINS)
mai (38.14. MIZO)
mai (38.16. MIZO)
mai (38.17. MIZO)
mai (38.18. MIZO)
mai (38.20. MIZO)
mai cong (7.14. CHINESE, MIN BEI)
mai dou (40.28. CHINESE, GÀN)
mai dou (40.28. CHINESE, XIANG)
mai hio (17.1. LISU)
mai pao (17.2. LAHU)
mai pok (17.2. LAHU)
mai pok (17.2. SHAN)
mai wan dou (40.12. CHINESE, XIANG)
mai'a (16.1a. HAWAIIAN)
mai-bongyai (17.2. THAI)
maih (17.7. SAMBAL, BOTOLAN)
mai-hok (17.2. SHAN)
mai-hok (17.2. THAI, NORTHERN)
mai-hok-khu (17.2. SHAN)
mâi-i (17.7. HERERO)
mai-lai-lo (17.2. THAI)
mai-mati-mati (17.5. CHUWABU)
mai-mati-mati (17.5. SENA)
mai-mati-mati (62.21. CHUWABU)
maintanos (24.19. GREEK)
mai-nuan (17.2. THAI)
mainzo (17.7. GALICIAN)
mai-pang-puk (17.4. SHAN)
mai-phai (17.1. THAI)
mai-phieo (17.2. THAI, NORTHERN)
mai-po (17.2. THAI)
mai-po (17.2. THAI, NORTHERN)

makhua-phuang (62.23. THAI)
makhua-som (62.9. THAI)
makhua-thet (62.9. THAI)
makhua-thetton (62.7. THAI)
makhua-yao (62.19. THAI)
makhuea phuang (62.23. THAI)
makhwaeng (62.23. THAI)
ma-khwaeng-dam (62.25. THAI)
ma-kil-*ma*-a-rɔ̃ (59.2. THEMNE)
ma-kilmoron (59.3. THEMNE)
mäkimeirami (44.6. FINNISH)
maki̲tsaru (38.14. COMANCHE)
maki̲tsaru (38.16. COMANCHE)
maki̲tsaru (38.17. COMANCHE)
maki̲tsaru (38.18. COMANCHE)
makka (17.7. HINDI)
makka (17.7. MARATHI)
makka (17.7. PULAAR)
makka (17.7. PULAR)
makka cholam (17.7. MALAYALAM)
makka jo:lu (17.7. BADAGA)
makka jonnalu (17.7. TELUGU)
makka sholam (17.7. TAMIL)
makka-cholam (17.7. TAMIL)
makkai (17.7. GUJARATI)
makkai (17.7. KANNADA)
makkajo'xori (17.7. UZBEK)
makkajounalu (17.7. TELUGU)
makkari (17.7. GUJARATI)
makkazonnalu (17.7. TELUGU)
makki (17.7. PANJABI, EASTERN)
makklawer (40.17. AFRIKAANS)
makma madugi (23.3. BUHID)
mak-mak lumak (27.14. BUHID)
makō (62.21. KASHMIRI)
mako (62.21. PANJABI, EASTERN)
makoi (17.7. ASSAMESE)
makoi (62.15. HINDI)
makoi (62.21. HINDI)
makoiya (62.21. HINDUSTANI)
makomo (17.8. JAPANESE)
makomo (19.1. JAPANESE)
makontho (62.2. LIMBA)
makoutou ma nzaou (8.9. BEMBÉ)
makoutou ma nzaou (8.9. KOONGO)
makoya (62.15. HINDI)
makoya (62.21. URDU)
maköyü (17.7. KASHMIRI)
maksaruoho (37.1. FINNISH)
maksuring (58.2. AFRIKAANS)
māku (38.14. NEWAR)
māku (38.16. NEWAR)
māku (38.17. NEWAR)
māku (38.18. NEWAR)
maku (62.2. ANDO)
makunde (40.26. NDONGA)
makushtha (40.35. SANSKRIT)
makushthaka (40.35. SANSKRIT)
makuwauri (38.7b. JAPANESE)
makvinkel (24.13. AFRIKAANS)
makwasaokwasa (38.22. NDALI)
makwera (40.25. NYANJA)
malaakaakaayapendalamu (11.2. TELUGU)
Malabar gourd (38.15. ENGLISH)
Malabar gourd (38.15. FILIPINO)
Malabar nightshade (28.1. ENGLISH)
Malabar spinach (28.1. ENGLISH)
Malabar Spinat (28.1. GERMAN)
Malabar tikbička (38.15. MACEDONIAN)
malabarinpinaatti (28.1. FINNISH)
malabarspinasie (28.1. AFRIKAANS)
malabarspinat (28.1. DANISH)
malabiga (8.1. TAGALOG)
malabunga (39.4. HANUNOO)
malachi (45.5. GREEK)
malafia (63.1. HAUSA)
malagay (62.2. TAMIL)
malagueta (62.2. CRIOULO, CAPE VERDE)
malagueta (62.2. PORTUGUESE)
malaguèta (62.5. CRIOULO, UPPER GUINEA)
malagueta-de-galinha (62.21. CRIOULO, CAPE VERDE)
malai (39.4. SUMO-MAYANGNA)
malaichundai (62.23. TAMIL)
malaikkeerai (23.3. TAMIL)
malakakayapendalamu (11.2. TELUGU)
malambonga (39.4. TAGBANWA)
malambunga (39.4. TAGBANWA)
malan kochi (23.3. HAUSA)
malanateng (62.21. IVATAN)
malanca (8.9. CRIOULO, CAPE VERDE)
malanga (6.2. CREOLE, HAITIAN FRENCH)
malangá (8.1. SPANISH)
malanga (8.4. SPANISH)
malanga amarilla (8.9. SPANISH)
malanga karayib (8.9. CREOLE, HAITIAN FRENCH)
mā-lâng-choû (62.24. CHINESE, HAKKA)
malankody (38.7b. TAMIL)

malan-kotshi (23.3. HAUSA)
malasili (62.21. CEBUANO)
mala-sili (62.21. WÁRAY-WÁRAY)
malavan (30.51. HINDI)
malayan kizhangukkodi (11.6. TAMIL)
malayavacha (20.1. SANSKRIT)
malbhog khutura (59.1. ASSAMESE)
malbhog khutura (59.2. ASSAMESE)
male (11.7. GA)
male bamboo (17.2. ENGLISH)
má-lêng-chî (62.24. CHINESE, MIN NAN)
malentaŋ (32.1. PULAAR)
maley (39.4. SUMO-MAYANGNA)
malgaté (62.5. SERER-SINE)
mali (11.7. DANGME-KROBO)
mạli baha (44.5. SANTALI)
malícia d'água (40.19. PORTUGUESE)
malid-palid (27.14. BICOLANO, CENTRAL)
malima (40.26. NDALI)
malimbunga (39.4. BUHID)
malinga melon (38.1. FILIPINO)
malisang balang (62.5. SAMBAL, TINÀ)
malka masur (40.12. PANJABI, EASTERN)
malkat ha-layla (49.1. HEBREW)
malkhoza (40.32. PASHTO, NORTHERN)
mallow (45.5. ENGLISH)
malolisna nivska repka (30.7. MACEDONIAN)
malontho (45.1. LIMBA)
malts (35.1. ESTONIAN)
malu kesel (16.1a. SINHALA)
malu miris (62.1. SINHALA)
maluach (35.1. HEBREW)
maluah (35.1. HEBREW)
malukhiya (63.1. HEBREW)
malukhiya neekhelet (63.1. HEBREW)
malunggay (30.1. FILIPINO)
malva (45.5. FINNISH)
malva (45.5. PORTUGUESE)
malva (45.5. RUSSIAN)
malva (45.5. SPANISH)
malva (45.5. SWEDISH)
malva comune (45.5. ITALIAN)
malva de flor pequena (45.5. SPANISH)
malva de hoja redonda (45.5. SPANISH)
malva domestica (45.5. ITALIAN)
malva dos judeus (63.1. PORTUGUESE)
malva kruglolistna (45.5. BELARUSAN)
malva piccola (45.5. ITALIAN)
malvavija (45.1. BELARUSAN)
Malve (45.5. GERMAN)
malws (45.5. WELSH)
malý jam (11.5. CZECH)
mályva (45.5. HUNGARIAN)
ma-mabarà (40.3. BUDIK)
mamam (34.2. INDONESIAN)
mamam (34.2. JAVANESE)
mamam (34.2. SUNDA)
mamang (34.2. INDONESIAN)
mamang (34.2. MALAY)
mamanggahan (27.12. SUNDA)
mamantu (40.26. BALTI)
mamarai valai (16.1a. TAMIL)
mamaree (44.5. HINDI)
mambamba (40.9. TUMBUKA)
mambayar (40.43. MALAYALAM)
mamberemende (40.25. TUMBUKA)
mami (11.5. TOK PISIN)
mamis a obi (36.3. MARANAO)
māmoe (23.4. SAMOAN)
mamri (44.4. HINDI)
mamui (40.18. THAI)
mamuluku (40.43. BAMANANKAN)
mamuna ali gone (27.22. SHONA)
man (11.2. AKAN-FANTE)
man fàlangx (62.24. LAO)
man hlièmx (11.1. LAO)
man kachu (8.1. ASSAMESE)
man kani (8.4. BAATONUM)
man kè:w (36.3. LAO)
man 'o:nz (11.5. LAO)
man pauz (11.2. LAO)
man ph'au (40.21. LAO)
man tônz (39.4. LAO)
mana (16.1a. MENDE)
mãnãa (16.1a. DANGME-KROBO)
manabo (34.2. ILOCANO)
manacossô (43.1. MANDING-MANDINKA)
managu (62.21. GIKUYU)
ma-nai (11.3. THEMNE)
manakund (11.2. MARATHI)
mānāl (40.26. MUNDARI)
man-alu (62.24. THAI)
manam (34.2. MALAY)
manamtao (38.20. THAI)
manankasa (43.1. WOLOF)
manankaso (43.1. MANDING-MANDINKA)
manankaso (43.1. PULAAR)
manankaso (43.1. WOLOF)
manankoso (43.1. MANDING-MANDINKA)
manaso tɛɛriya (17.7. WAAMA)

manatakkali (62.21. MALAYALAM)
manathakkaalli (62.15. TAMIL)
manathakkali (62.21. TAMIL)
manattakkali (62.21. TAMIL)
ma:navare (40.9. BADAGA)
mana-wa (16.1a. MENDE)
mancanadje (43.1. PULAAR)
mancanassô (43.1. MANDING-MANDINKA)
mancara (40.1. CRIOULO, UPPER GUINEA)
mancarra (40.1. CRIOULO, CAPE VERDE)
mancarra (40.1. CRIOULO, UPPER GUINEA)
mancarra redonda (40.39. PORTUGUESE)
manchikanda (8.3. TELUGU)
manchi-kunda-guddae (8.3. TELUGU)
man-chuak (11.5. THAI)
manchurian spikenard (26.1. ENGLISH)
Manchurian water rice (17.8. ENGLISH)
Manchurian wild rice (17.8. ENGLISH)
manchurisk aralie (26.3. DANISH)
manda aliê (16.1a. BAOULÉ)
mandae (62.9. PULAAR)
mandamin (17.7. POTAWATOMI)
mandan (16.1a. AJA)
mandan (16.1a. TAMIL)
mandapi (40.1. SANSKRIT)
maṇḍare avare (40.24. BADAGA)
mandeki patak (38.3. MALAYALAM)
mandel-græskar (38.18. DANISH)
mandelgresskar (38.18. NORWEGIAN)
mandër (10.1. ALBANIAN, TOSK)
mandga (17.1. MARATHI)
mandiké (38.3. INDONESIAN)
mandioca (39.4. CATALAN-VALENCIAN-BALEAR)
mandiòca (39.4. ITALIAN)
mandioca (39.4. PORTUGUESE)
mandioca (39.4. SPANISH)
mandioca brava (39.4. PORTUGUESE)
mandioquinha salsa (24.6. PORTUGUESE)
mando (23.3. GOLA)
mandolando (62.17. SENA)
mandondo (40.25. TUMBUKA)
mandooka braahmi (24.7. TELUGU)
mandookaparni (24.7. SANSKRIT)
mandookaparnika (24.7. SANSKRIT)
mandooki (24.7. SANSKRIT)
mandookparni (24.7. HINDI)
mandoroka (36.3. MANDING-MANDINKA)
mandubi de Angola (40.39. PORTUGUESE)
mandukaparni (24.7. SANSKRIT)
mandulakávé (10.1. HUNGARIAN)
manduwi (40.1. LOMWE, MALAWI)
mandyana tubig (23.6. SUBANON)
mane (40.1. AGTA)
månebönne (40.25. DANISH)
måneds reddik (30.56. NORWEGIAN)
manehu (2.1. MAORI)
ma-nəi (11.3. THEMNE)
manëz (10.1. ALBANIAN, TOSK)
man-farang (62.24. THAI)
manforá (40.18. MANDING-MANDINKA)
manga (39.4. DIGO)
mangaka (38.13a. YAO)
ma-ngalàl (9.1. BUDIK)
mangaliya (40.12. SANSKRIT)
mangalya (24.13. SANSKRIT)
manganace (43.1. CRIOULO, UPPER GUINEA)
manganadje (43.1. PULAAR)
mãnganasa (43.1. CRIOULO, UPPER GUINEA)
mãnganatié (43.1. PULAAR)
manganaz (43.1. CRIOULO, UPPER GUINEA)
manganaz (43.1. PORTUGUESE)
mangangaria (69.1. MARINDUQUE)
mangani (8.4. ANDO)
mangani (8.4. JULA)
manganni (8.4. BAOULÉ)
manganouwi (8.4. SENOUFO, CEBAARA)
mangará mirim (8.9. PORTUGUESE)
mangaras (8.9. SPANISH)
mangaride (8.9. FRENCH)
mangarito (8.9. PORTUGUESE)
mangarito (8.9. SPANISH)
manggai (16.1a. MARANAO)
mangi (17.7. IBANAG)
mangina (20.2. FULFULDE, ADAMAWA)
mangina (20.2. HAUSA)
mangir (7.12. MONGOLIAN)
mangirona (44.5. CRIOULO, CAPE VERDE)
mangkọng kọng (17.5. YALA)
mang-luk (44.4. THAI)
mango ginger (20.3. ENGLISH)
mango meloen (38.8. DUTCH)
mango melon (38.8. ENGLISH)
mangobyon (1.1. KALAGAN)
mangold (33.3. ROMANIAN)
mangold (35.3. BULGARIAN)
mangold (35.3. CZECH)

mangold (35.3. DANISH)
Mangold (35.3. GERMAN)
mángold (35.3. HUNGARIAN)
mangold (35.3. NORWEGIAN)
mangold (35.3. POLISH)
mangold (35.3. ROMANIAN)
mangold (35.3. SERBO-CROATIAN)
mangold (35.3. SLOVENIAN)
mangold (35.3. SWEDISH)
mangol'd (35.3. UKRAINIAN)
mangoldas (35.3. LITHUANIAN)
mangoldi (35.3. GEORGIAN)
mangolds (35.3. LATVIAN)
Mango-melone (38.8. GERMAN)
mangomeloni (38.8. FINNISH)
mangu (38.20. GIKUYU)
manguena (62.20. SPANISH)
mangulasi (40.40. BINUKID)
manh tong (17.2. VIETNAMESE)
māni (30.17. MUNDARI)
mani (40.1. BICOLANO, CENTRAL)
mani (40.1. BONTOC CENTRAL)
mani (40.1. CEBUANO)
mani (40.1. FILIPINO)
mani (40.1. GIANGAN)
maní (40.1. HILIGAYNON)
mani (40.1. IBANAG)
mani (40.1. IFUGAO)
mani (40.1. ILOCANO)
mani (40.1. PAMPANGAN)
maní (40.1. PANGASINAN)
maní (40.1. SPANISH)
mani (40.1. TAGALOG)
mani africano (40.39. SPANISH)
mani de Bambarra (40.39. SPANISH)
mani ok (45.2. TURKISH)
maniato (36.3. CATALAN-VALENCIAN-BALEAR)
maniavare (40.9. KANNADA)
maniba (39.4. PORTUGUESE)
maniho (39.4. HEBREW)
manihot de tapioc (39.4. ROMANIAN)
manilakottai (40.1. TAMIL)
manilla grain (40.1. ENGLISH)
manimuni (24.7. ASSAMESE)
manini (62.12. NIUE)
manini fua lalahi (62.12. NIUE)
manini palagi (62.12. NIUE)
maninjo (41.1. INDONESIAN)
maninjo (41.1. JAVANESE)
maninjo (41.1. SUNDA)
manioc (39.4. ENGLISH)
manioc (39.4. FRENCH)
manioc batard (39.1. FRENCH)
manioca (39.4. ITALIAN)
maniok (39.4. CZECH)
maniok (39.4. DANISH)
maniok (39.4. DUTCH)
Maniok (39.4. GERMAN)
maniok (39.4. MORISYEN)
maniok (39.4. NORWEGIAN)
maniok (39.4. SWEDISH)
maniok (39.4. VLAAMS)
maniok jadalny (39.4. POLISH)
maniok jedlý (39.4. SLOVAK)
maniok s'edobnyj (39.4. RUSSIAN)
manioka (39.4. MANU'A SAMOAN)
manioka (39.4. SAMOAN)
manioka (39.4. SERBO-CROATIAN)
manioke (39.4. TONGAN)
Maniok-Eibisch (45.2. GERMAN)
maniokk (39.4. ESTONIAN)
maniokki (39.4. FINNISH)
manioks (39.4. LATVIAN)
Maniokstrauch (39.4. GERMAN)
maniota (39.4. NIUE)
maniota (39.4. RAROTONGAN)
manithakkaali (62.15. TAMIL)
manito oskatask (20.6. CREE, PLAINS)
manjal (20.2. MALAYALAM)
manjal (20.2. TAMIL)
manjano (20.2. SWAHILI)
mańǰapatrè (27.7. TULU)
manjariki (44.5. SANSKRIT)
manjericão branco (44.4. PORTUGUESE)
manjericão grande (44.5. PORTUGUESE)
manjerico da folha grande (44.5. PORTUGUESE)
mańǰipatrè (27.7. TULU)
mankachu (8.1. BENGALI)
man-kaeo (40.21. THAI)
mãnkanas (43.1. MANDJAK)
mãnkanase (43.1. PULAAR)
mankanaso (43.1. MANDING-MANDINKA)
mankanda (8.1. HINDI)
mankanda (8.1. URDU)
mankani (8.4. ÉWÉ)
mankani (8.9. AKAN-AKUAPEM)
mankani (8.9. ÉWÉ)
mankanlin (8.4. GBE, WACI)

marakilannu (39.4. MALAYALAM)
marakka (38.20. AFRIKAANS)
marama bean (40.33. ENGLISH)
marama boon (40.33. VLAAMS)
marama chinikilannu (39.4. MALAYALAM)
maramabønne (40.33. DANISH)
maramaboon (40.33. AFRIKAANS)
mara-mara (27.14. WÁRAY-WÁRAY)
maranka (38.20. AFRIKAANS)
maranta (15.2. AFRIKAANS)
maranta (15.2. DANISH)
maranta (15.2. HEBREW)
maranta (15.2. ITALIAN)
maranta (15.2. NORWEGIAN)
maranta (15.2. PORTUGUESE)
maranta (15.2. SPANISH)
maranta třtinovitá (15.2. CZECH)
maranta trzcinowata (15.2. POLISH)
marantamylon (15.2. GREEK)
marante (15.2. FRENCH)
marante en forme de bambou (15.2. FRENCH)
marantia kalamoidis (9.1. GREEK)
marara (40.18. HAUSA)
märara qəl (38.23. AMHARIC)
märara qətal (27.44. AMHARIC)
maras (38.13a. TAUSUG)
marasa (23.3. HINDI)
marasa kaage (8.1. KANNADA)
marasanige (8.1. KANNADA)
máráshi (55.1. KANURI)
marathhakkali (62.7. TAMIL)
maratho (24.12. GREEK)
marathron (24.13. GREEK)
maratimogga (27.2. TELUGU)
maratiteega (27.2. TELUGU)
maratitige (27.2. TELUGU)
marattamaggu (34.1. KANNADA)
maravaelikilannu (39.4. MALAYALAM)
maravalli (39.4. TAMIL)
maravalli kizhangu (39.4. TAMIL)
maravilla bulb (12.1. ENGLISH)
maravilla knol (12.1. VLAAMS)
marayu (40.40. HANUNOO)
march ysgall (44.11. WELSH)
marcha (62.5. GUJARATI)
marchew ogrodowa (24.11. POLISH)
marchew siewna (24.11. POLISH)
marchew zwyczajna (24.11. POLISH)
marchewnik anyżkowy (24.16. POLISH)
marchi (62.2. MUNDARI)
marchobch (13.2. FARSI, WESTERN)
marchrawn-yr-ardir (1.1. WELSH)
marchrhedyn-y-dwr (4.1. WELSH)
marchwej (24.11. SORBIAN, LOWER)
march-ysgall (27.26. WELSH)
mardudi (38.20. KONKANI)
marek kucmerka (24.23. POLISH)
marerasan (20.6. MANOBO)
margarida (27.23. PORTUGUESE)
margarita (27.23. SPANISH)
margaritka (27.23. MACEDONIAN)
margase (38.23. FRENCH)
margaso (38.23. CEBUANO)
margherita (27.23. ITALIAN)
margoso (38.23. CEBUANO)
margoso (38.23. TAGALOG)
margoz (38.23. MORISYEN)
mari mirch (62.5. PANJABI, EASTERN)
maria pretinha (62.15. PORTUGUESE)
maria pretinha (62.21. PORTUGUESE)
maribisab (54.1. WOLOF)
marich (24.9. BENGALI)
marich (62.5. BENGALI)
marichi phalam (62.2. SANSKRIT)
marichiphala (62.5. SANSKRIT)
marichiphalam (62.2. SANSKRIT)
marid puluni (20.6. SUMO-MAYANGNA)
marigoso (38.23. BICOLANO, CENTRAL)
marigosó (38.23. WÁRAY-WÁRAY)
marĩgũ (16.1a. GIKUYU)
marish (23.3. SANSKRIT)
marisha (23.3. SANSKRIT)
marjatomaatti (62.10. FINNISH)
marjolaine (44.6. FRENCH)
marjolaine sauvage (44.6. FRENCH)
marjolein (44.6. AFRIKAANS)
marjolein (44.6. DUTCH)
marjol-ki (44.6. BRETON)
marjoram (44.6. ENGLISH)
markata (40.18. KANNADA)
markati (40.18. SANSKRIT)
Markkhohl (30.32. GERMAN)
markonik tsymanechnalisny (24.3. BELARUSAN)
markstamkool (30.32. DUTCH)
Markstammkohl (30.32. GERMAN)
marocntia kalamoidis (15.2. GREEK)
marog (23.3. AFRIKAANS)
maror aginot (27.42. HEBREW)
marouli (27.31. GREEK)

marouli (salata) (27.32. GREEK)
marrow (38.18. ENGLISH)
marrow stem kale (30.32. ENGLISH)
marrupani (44.5. MAKHUWA)
mārrutki (30.1. LATVIAN)
marsa (23.3. HINDI)
marsilanana (27.18. TAGALOG)
marstard (30.17. WELSH)
marstard blodaw gwyn (30.11. WELSH)
marstard manddail (30.7. WELSH)
martigonsali (38.22. GUJARATI)
martsa - gŏsôñu (62.2. KASHMIRI)
marua (27.14. MAKHUWA)
marua (44.5. HINDI)
maruba-dokoro (11.2. JAPANESE)
maruga (44.6. KANNADA)
marul (27.31. TURKISH)
marul (27.32. TURKISH)
marula (27.32. MACEDONIAN)
marulă (27.32. ROMANIAN)
marule (27.31. ALBANIAN, TOSK)
marule e kultivuar (27.31. ALBANIAN, TOSK)
marule e kultivuar (27.32. ALBANIAN, TOSK)
marulja (27.32. BULGARIAN)
maruppu (20.6. TAMIL)
marurugi (17.1. BICOLANO, CENTRAL)
mārutki (30.1. LATVIAN)
maru-yugao (38.20. JAPANESE)
marva (44.5. MARATHI)
maş (40.26. AZERBAIJANI, NORTHERN)
mas (40.38. KAZAKH)
mas (40.38. RUSSIAN)
maş (40.38. TURKISH)
masa (40.1. SENOUFO, CEBAARA)
masakasa (38.13a. YAO)
masaki (38.20. HAUSA)
masakula (38.20. OROMO, WEST CENTRAL)
masara (17.7. HAUSA)
masara maciji (19.1. HAUSA)
másár̃mì (17.7. NGIZIM)
masbati (28.1. CEBUANO)
mascur (40.12. GUJARATI)
mascur (40.12. HINDI)
mase (38.5. SUSU)
masetete (30.17. TSWANA)
màshá (45.4. KANURI)
mashahwieqorg (27.5. CHECHEN)
mash-bab (40.38. HUNGARIAN)
mashikanda (8.3. TELUGU)
mash-kalai (40.37. BENGALI)
mashua (65.2. QUECHUA, CUSCO)
mashua (65.2. SPANISH)
masimbi (8.4. TONGA)
maskros (27.45. SWEDISH)
maslačak (27.45. SERBO-CROATIAN)
māso (40.38. PALI)
māsoā (9.1. SAMOAN)
masoa (19.1. MANU'A SAMOAN)
māsoā (19.1. SAMOAN)
masoor (40.12. HINDI)
masoor baele (40.12. KANNADA)
masoora (40.12. SANSKRIT)
masooraka (40.12. SANSKRIT)
masooree (40.12. SANSKRIT)
masoori (40.12. MARATHI)
masoorika (40.12. SANSKRIT)
massa ku (11.1. BAMANANKAN)
massaku (11.1. MANDING-MANDINKA)
massər (40.12. AMHARIC)
masser (40.12. BENGALI)
masser (40.12. GUJARATI)
masser (40.12. HINDI)
masser (40.12. MARATHI)
massur (40.12. KANNADA)
mastranzo de las Indias (65.1. SPANISH)
mastruço (30.51. PORTUGUESE)
mastruço do Perú (65.1. PORTUGUESE)
mastrugo (30.51. PORTUGUESE)
mastruzzo (30.51. ITALIAN)
mastuerzo (30.51. SPANISH)
mastuerzo acuático (30.52. SPANISH)
mastuerzo de prado (30.44. SPANISH)
mastuerzo hortense (30.51. SPANISH)
masturzio d'orto (30.51. ITALIAN)
masturzo de jardin (30.2. SPANISH)
ma:su (62.1. BADAGA)
ma:su (62.2. BADAGA)
masuka:i (62.1. BADAGA)
masuka:i (62.2. BADAGA)
masundari (61.1. ASSAMESE)
masur (40.12. BENGALI)
masur (40.12. GUJARATI)
masur (40.12. HINDI)
masūr (40.12. HINDI)
masúr (40.12. HINDUSTANI)
masur (40.12. MARATHI)
masur (40.12. PANJABI, EASTERN)
masur (40.12. URDU)
masur mah (40.38. ASSAMESE)
masuri (40.12. BENGALI)

matrabarbra (58.1. NORWEGIAN)
matri (40.10. PANJABI, EASTERN)
matserdushka zvychainaja (44.6. BELARUSAN)
matsere (17.7. DIGO)
matsubaudo (13.2. JAPANESE)
matsuna (35.10. JAPANESE)
matsyaksi (23.2. SANSKRIT)
mattanga (38.16. MALAYALAM)
mattar (40.28. PANJABI, EASTERN)
måttebønne (40.35. DANISH)
Mattenbohne (40.35. GERMAN)
mattubachali (35.1. TELUGU)
matturbachhali (35.9. TELUGU)
mattz-rewari (40.34a. PANJABI, EASTERN)
mături (35.2. ROMANIAN)
ma-tyánduŋ (45.4. BUDIK)
mau (17.1. MIZO)
mau (17.2. MIZO)
mau (17.4. MIZO)
Mauerdoppelsame (30.47. GERMAN)
Mauerrauke (30.47. GERMAN)
ma-uk (62.16. THAI)
mauka (49.1. DANISH)
mauka (49.1. ENGLISH)
mauka (49.1. FRENCH)
maungu (38.16. LOMWE, MALAWI)
maungu (38.18. LOMWE, MALAWI)
maúpy (11.3. GREEK)
mauri (24.13. BENGALI)
maurloki (7.11. LATVIAN)
maurloki (7.12. LATVIAN)
mausa anthousa (16.1b. GREEK)
Mausohr (67.2. GERMAN)
maustefenkoli (24.12. FINNISH)
maustekirveli (24.3. FINNISH)
maustekurkuma (20.2. FINNISH)
maustepaprika (62.2. FINNISH)
mau-t'ien-hsing (23.2. CHINESE, MANDARIN)
mauve (45.5. FRENCH)
mauve à feuilles rondes (45.5. FRENCH)
mauve ronde (45.5. FRENCH)
maviskon (39.4. GREEK)
mavro sinapi (30.24. GREEK)
mavuju (16.1a. KANNADA)
mawaeng (62.25. THAI)
mawaeng-nok (62.15. THAI)
mawaeng-ton (62.25. THAI)
mawal (23.3. KASHMIRI)
maw'dunts (38.3. PAIUTE, SOUTHERN)
ma-wɛ̀r (40.39. BUDIK)
m[aw]ng t[aa]y (13.2. VIETNAMESE)
m[ax] th [af]y (10.2. VIETNAMESE)
maxixé (38.6. PORTUGUESE)
may dou (17.7. MORISYEN)
mayalu (28.1. MARATHI)
mayamas (36.3. BUHID)
mayao (39.4. TUMBUKA)
mayastha (40.35. SANSKRIT)
maydanoz (24.19. TURKISH)
mayday gülli galinsoga (27.22. TURKMEN)
mayentɛrɛba (38.5. MALIMBA)
mayer (24.11. YIDDISH)
mayi (17.7. CREOLE, HAITIAN FRENCH)
maythi (40.32. HINDI)
mayugbe (39.4. THEMNE)
mayugwa (35.9. SWAHILI)
mayura-shikha (23.4. SANSKRIT)
mayurshikra (23.4. MARATHI)
mazabi (59.2. BAMANANKAN)
mazabi (59.2. MANDING-MANDINKA)
mazăre (40.28. ROMANIAN)
mazăre cultivată (40.27. ROMANIAN)
mazăre cultivată (40.28. ROMANIAN)
măzăroi (40.10. ROMANIAN)
ma'zaskaokaġe (44.3. DAKOTA)
mà'zɛŋ (44.5. BEROM)
mazoumbèl (8.1. CREOLE, HAITIAN FRENCH)
mazoumbèl (8.4. CREOLE, HAITIAN FRENCH)
mazumbel (8.1. CREOLE, HAITIAN FRENCH)
mazumbel (8.4. CREOLE, HAITIAN FRENCH)
*m*ba (39.4. FULFULDE, ADAMAWA)
mbaazi (40.2. SWAHILI)
mbaazi (40.28. SWAHILI)
*m*bada-nyum (20.6. GOLA)
mbahanasa (43.1. WOLOF)
mbai (39.4. FULFULDE, ADAMAWA)
mbaiji (39.4. FULFULDE, ADAMAWA)
mbal' kongo (36.3. BEMBÉ)
m̀bàlà (11.1. IGBO)
mbala ngun (11.1. BEMBÉ)
mbala za sucari (36.3. KOONGO)
mbalala (40.1. TONGA)
mbala-makole (17.7. VENDA)
mbalazi (40.2. DIGO)

*m*bali (63.1. WOLOF)
*m*bamba (62.19. AKAN-FANTE)
mbambaira (36.3. SENA)
mbambaira (36.3. SHONA)
mbambaira (62.24. TONGA)
mbambaira (62.24. TUMBUKA)
mbambuni (38.14. SHONA)
mbambuni (38.16. SHONA)
mbambuni (38.17. SHONA)
mbambuni (38.18. SHONA)
mbambunu (38.14. SHONA)
mbambunu (38.16. SHONA)
mbambunu (38.17. SHONA)
mbambunu (38.18. SHONA)
mbamia (45.1. SWAHILI)
*m*bana (16.1a. MENDE)
m̀bàńsang (40.1. EFIK)
mbãntu (45.4. MANKANYA)
mbatata (36.3. LOMWE, MALAWI)
mbatata (36.3. NYANJA)
mbatata (62.24. LOMWE, MALAWI)
mbatata wa cizungu (62.24. NYANJA)
mbatatisi (62.24. SHONA)
mbed (38.20. SERER-SINE)
mbelemende (40.2. YAO)
mbembe (17.7. GIKUYU)
mbenge (40.2. TUMBUKA)
mbermef (38.23. WOLOF)
m̀bíit (17.3. EFIK)
mbilimbi (38.13a. SWAHILI)
mbinda (45.1. SWAHILI)
mbiringanya (62.19. GIKUYU)
mbirīwu (40.39. FULFULDE, ADAMAWA)
mbissine (17.6. SERER-SINE)
m'blangblô (17.1. BAOULÉ)
m-bo boni (11.2. LOKO)
m'boa (23.3. RONGA)
mboa (23.3. SHONA)
mboamanga (23.3. SHONA)
mbobo-ikokomo (9.1. AKWA)
mboco (40.26. GIKUYU)
mboga (30.27a. GANDA)
mboga (38.14. SWAHILI)
mboga (38.16. SWAHILI)
mboga (38.17. SWAHILI)
mboga (38.18. SWAHILI)
mboga cia gĩthũngũ (27.31. GIKUYU)
mboga cia gĩthũngũ (27.32. GIKUYU)
mboga kama figili (30.56. SWAHILI)
mbohole (36.3. TONGA)
ḿbọ̀kọ̀k (17.5. EFIK)
ḿbọ́kọ́k ékpò (17.3. EFIK)
mbole (11.2. MENDE)
mbole-gbɔlɔ (11.3. MENDE)
mbole-gbɔlu (11.3. MENDE)
mbole-gowɔ (11.7. MENDE)
mbole-gowe (11.7. MENDE)
mbomboni (38.14. SHONA)
mbomboni (38.16. SHONA)
mbomboni (38.17. SHONA)
mbomboni (38.18. SHONA)
mboñ (39.4. BULU)
mbonda (38.20. LOMWE, MALAWI)
mbong (39.4. BULU)
mbono (27.18. AKWA)
m̀bòró (16.1a. EFIK)
mboum bou gor (23.3. WOLOF)
mboum i keur (23.3. WOLOF)
m'bowa (23.3. RONGA)
*m*bowi-hei (17.3. MENDE)
mbowini (40.2. RONGA)
mbox (17.7. WOLOF)
mbubo (38.20. IGBO)
mbudzana (36.3. VENDA)
mbuloh (39.4. SERER-SINE)
mbuloh gôr (39.4. SERER-SINE)
mbum (35.9. WOLOF)
mbum bu digèn (23.3. WOLOF)
mbum bu gôr (23.1. WOLOF)
mbum bu gor (23.3. WOLOF)
mbum bu levet (23.3. WOLOF)
mbum i gor (23.3. WOLOF)
mbum i kör (23.3. WOLOF)
mbumbi (30.17. YAO)
mbwanda (40.26. NYANJA)
mbwẹde (11.1. IGBO)
mbweso (40.38. YAO)
mbwete (36.3. TUMBUKA)
mbwete (40.2. SENA)
mɔ́ɔ̃́ (45.1. KONKOMBA)
mɔ̃gfəli (40.1. PANJABI, EASTERN)
mchiburu (38.20. DIGO)
mchicha (23.1. SWAHILI)
mchicha (23.2. SWAHILI)
mchichá (23.3. DIGO)
mchicha (23.3. SINHALA)
mchicha (23.3. SWAHILI)
mchicha (23.4. SWAHILI)
mchicha (35.9. SWAHILI)
mchunga (27.29. SWAHILI)

mchunga (27.31. SWAHILI)
mchunga (27.32. SWAHILI)
mchunga (27.33. SWAHILI)
mchunga (27.42. SWAHILI)
mciwile (62.24. TONGA)
mciwile (62.24. TUMBUKA)
mɔkoro (11.2. BAATONUM)
mɔlɔkɔ-yɔ (17.3. MANDING-MANDINKA)
mɔstad (30.17. KRIO)
mdo (62.24. DAKOTA)
mdo sku'ya (36.3. DAKOTA)
mdzauna (58.2. GEORGIAN)
meacan bán (24.18. GAELIC, IRISH)
meacan dearg (24.11. GAELIC, IRISH)
meacan-ruadh (30.56. GAELIC, SCOTTISH)
meadow salsify (27.47. ENGLISH)
mealbhacán (38.3. GAELIC, IRISH)
mealbhacán (38.7a. GAELIC, IRISH)
mealbhacán (38.7b. GAELIC, IRISH)
mealbhacán (38.11. GAELIC, IRISH)
meal-bhucan (38.7a. GAELIC, SCOTTISH)
meal-bhucan (38.7b. GAELIC, SCOTTISH)
meal-bhucan (38.11. GAELIC, SCOTTISH)
meant (44.3. GAELIC, SCOTTISH)
męczennica (54.1. POLISH)
medanglis (36.3. BUHID)
me:de (17.1. BADAGA)
meenaakshi (23.2. SANSKRIT)
Meerkohl (30.45. GERMAN)
Meerkohl (30.46. GERMAN)
meeroore (8.4. FULFULDE)
Meerrettich (30.1. GERMAN)
meetta kaddu (38.16. HINDI)
meghanaad (23.3. SANSKRIT)
meghanada (23.3. SANSKRIT)
mĕgwibŭnĕksit (35.4. MICMAC)
meh (ceh_V) (63.1. LAHU)
meh va bvuh-eh (17.2. LAHU)
Mehlbanane (16.1a. GERMAN)
mého (23.2. KISI, SOUTHERN)
mehti (40.32. FIJIAN)
mei chou fang fêng (24.18. CHINESE, MANDARIN)
mei dou (40.9. CHINESE, WÚ)
mei guo fang feng (24.18. CHINESE, MANDARIN)
mei guo nan gua (38.18. CHINESE, WÚ)
mei guo shang lu (56.2. CHINESE, YUE)
mei qui fang feng (24.18. CHINESE, WÚ)
mei zhou fang feng (24.18. CHINESE, YUE)
mei zhou nan gua (38.18. CHINESE, GÀN)
mei zhou nan gua (38.18. CHINESE, WÚ)
mei zhou nan gua (38.18. CHINESE, YUE)
mei'a (16.1a. MARQUESAN)
meiguo nan gua (38.18. CHINESE, MANDARIN)
meika (16.1a. MARQUESAN)
meiki (16.1a. RAROTONGAN)
meipen ddeilog (30.30. WELSH)
meiraap (30.41a. DUTCH)
meiran (44.6. ICELANDIC)
meis (17.7. CAROLINIAN)
mejdan nwoşy (1.1. TURKMEN)
mekanada (23.3. SANSKRIT)
mekgejöwen (17.7. TURKMEN)
mekke (40.26. AMHARIC)
mekke jola (17.7. KANNADA)
mekkejola (17.7. KANNADA)
meko (62.21. MARATHI)
meko zelje (35.3. MACEDONIAN)
meksikan jola (17.7. KANNADA)
meksikolykt (62.13. NORWEGIAN)
me-kyabetsu (30.29. JAPANESE)
melaā naktene (62.21. LATVIAN)
melada (34.1. MALAY)
melafefon (38.13a. HEBREW)
melancana (62.19. SLOVENIAN)
melancia (38.3. CRIOULO, CAPE VERDE)
melancia (38.3. PORTUGUESE)
melanzana (62.19. ITALIAN)
melanzana africana (62.18. ITALIAN)
melanzana scarlata (62.14c. ITALIAN)
melão (38.11. PORTUGUESE)
melão (38.7b. PORTUGUESE)
melão (38.9. PORTUGUESE)
melão amarelo (38.11. PORTUGUESE)
melão cantalupe (38.7a. PORTUGUESE)
melão cantalupe (38.7b. PORTUGUESE)
melão de caboclo (38.26. PORTUGUESE)
melão de São Caetano (38.23. PORTUGUESE)
melão mango (38.8. PORTUGUESE)
melão reticulado (38.7b. PORTUGUESE)
melão serpente (38.10. PORTUGUESE)
melde (35.1. AFRIKAANS)
melde (35.1. DANISH)
melekotu (24.1. TURKISH)
meleni (38.3. MANU'A SAMOAN)
meleni (38.3. NIUE)
meleni (38.3. SAMOAN)
meleni (38.7b. MANU'A SAMOAN)

melon do (38.3. MORISYEN)
melon doux (38.3. CAJUN FRENCH)
melon français (38.7a. CAJUN FRENCH)
mélon frans (38.6. CREOLE, HAITIAN FRENCH)
melon gąbczasty (38.24. POLISH)
melon japoński (38.10. POLISH)
melon kantalup (38.7a. POLISH)
melon mango (38.8. FRENCH)
melón mango (38.8. SPANISH)
melón manzana (38.9. SPANISH)
melon pear (62.20. ENGLISH)
melon pear (62.20. MALAY)
melon poire (62.20. FRENCH)
melon pumpkin (38.16. ENGLISH)
melon réticulé (38.7b. FRENCH)
melon serpent (38.10. FRENCH)
melón serpiente (38.10. SPANISH)
melon siatkowany (38.7b. POLISH)
melón vodný (38.3. SLOVAK)
melon wschodni (38.9. POLISH)
melon zimowy (38.11. POLISH)
melón žltý (38.7a. SLOVAK)
melón žltý (38.7b. SLOVAK)
melón žltý (38.8. SLOVAK)
melón žltý (38.9. SLOVAK)
melón žltý (38.10. SLOVAK)
melón žltý (38.11. SLOVAK)
melona (38.9. ICELANDIC)
melon-daga (38.29. TAGALOG)
melone (38.7a. LATVIAN)
melone (38.7b. ITALIAN)
melone (38.7b. LATVIAN)
Melone (38.11. GERMAN)
melone (38.11. ITALIAN)
melone (38.11. LATVIAN)
melone cantalupo (38.7a. ITALIAN)
melone d'acqua (38.3. ITALIAN)
melone mango (38.8. ITALIAN)
melone retado (38.7b. ITALIAN)
Melonenbirne (62.20. GERMAN)
meloner (38.7a. CATALAN-VALENCIAN-BALEAR)
meloner (38.7b. CATALAN-VALENCIAN-BALEAR)
meloner (38.11. CATALAN-VALENCIAN-BALEAR)
melongene (62.19. ENGLISH)
melongène (62.19. FRENCH)
meloni (38.7a. FINNISH)
meloni (38.7b. FINNISH)
meloni (38.9. FINNISH)
meloni (38.11. FINNISH)
melon-melonan (38.29. TAGALOG)
melonpære (62.20. DANISH)
melooni (38.7a. HOPI)
melooni (38.7b. HOPI)
méloukhia (63.1. FRENCH)
Meloun (38.3. LUXENBOURGEOIS)
Meloun (38.7a. LUXENBOURGEOIS)
Meloun (38.7b. LUXENBOURGEOIS)
Meloun (38.8. LUXENBOURGEOIS)
Meloun (38.9. LUXENBOURGEOIS)
Meloun (38.10. LUXENBOURGEOIS)
Meloun (38.11. LUXENBOURGEOIS)
meloun cukrový (38.7b. CZECH)
meloun cukrový (38.11. CZECH)
meloun vodní (38.3. CZECH)
melun (38.8. YIDDISH)
melun (38.9. YIDDISH)
melun (38.10. YIDDISH)
melun (38.11. YIDDISH)
melyn yr hwyr (51.1. WELSH)
mėlynžiedė liucerna (40.17. LITHUANIAN)
melzhane (62.19. ALBANIAN, TOSK)
mëm phlông (10.2. KHMER, CENTRAL)
me:-mo (38.18. ZUNI)
memomene (23.3. IGBO)
mɛn mwdiík (44.8. BUDIK)
menasina kaayi (62.2. KANNADA)
menasina kayi (62.5. KANNADA)
mɛnde-hedi (16.1a. MENDE)
mendikai (38.3. INDONESIAN)
mĕndikai (38.3. MALAY)
mendzen (38.16. EWONDO)
mendzen (38.17. EWONDO)
mene (40.1. IBALOI)
menekshe stambolli (68.1. ALBANIAN, TOSK)
mengkreng (62.1. JAVANESE)
mengkuwang (40.21. MALAY)
menian-tiga (40.1. BAMANANKAN)
menian tiga (40.1. MANINKA)
meninjau (41.1. MALAY)
meninjo (41.1. INDONESIAN)
menk-palle (40.32. KANNADA)
ment (44.2. AFRIKAANS)
ment (44.2. BRETON)
ment (44.3. BRETON)
menta (44.3. BOSNIAN)

metet (27.24. BULU)
metet (27.24. EWONDO)
methhuka (40.32. BENGALI)
methi (40.32. BENGALI)
methi (40.32. ENGLISH)
methi (40.32. GUJARATI)
methi (40.32. HINDI)
mēthī (40.32. MAITHILI)
methi (40.32. MARATHI)
methi (40.32. NEPALI)
methi (40.32. ORIYA)
methi (40.32. PANJABI, EASTERN)
methi (40.32. SANSKRIT)
methi (40.32. URDU)
methika (40.32. SANSKRIT)
methini (40.32. GUJARATI)
methini (40.32. PANJABI, EASTERN)
methini (40.32. SANSKRIT)
methi-shak (40.32. BENGALI)
methkalai (40.35. GUJARATI)
methri (40.32. PANJABI, EASTERN)
methun (40.32. PANJABI, EASTERN)
meti (40.32. MORISYEN)
metka peteta (36.3. HINDUSTANI)
metlica (35.2. SERBO-CROATIAN)
metowo (38.22. TUMBUKA)
metta (32.1. ZUNI)
metta thaamara (9.1. TELUGU)
mettakaakara (38.23. TELUGU)
mettavanga (62.19. TELUGU)
mettawela (44.8. FULFULDE, ADAMAWA)
mettawelāji (44.8. FULFULDE, ADAMAWA)
mette (40.32. BADAGA)
mettè (40.32. TULU)
metthi (40.32. KONKANI)
metthi (40.32. TULU)
metvica (44.3. BOSNIAN)
metvica poljačak (44.2. SERBO-CROATIAN)
meu_ ce$^{\vee}$ ba$_{\vee}$ te (11.1. LAHU)
meu_ nyi (36.3. LAHU)
meu_ nyi^{-} (36.3. LAHU)
meu_ nyi^{-} meu$^{\vee}$ ve$^{\wedge}$ (36.2. LAHU)
Mexican burningbush (35.2. ENGLISH)
Mexican yam bean (40.21. ENGLISH)
mexikansk yamsbønne (40.21. DANISH)
meycaci kabaği (38.16. TURKISH)
mezei katáng (27.10a. HUNGARIAN)
mezei sóska (58.2. HUNGARIAN)
mezei zsurló (1.1. HUNGARIAN)
mezeng (38.16. BAFIA)
mezeng (38.17. BAFIA)
mez̤ilot (13.5. HEBREW)
mez̤üyaz̤ot (13.5. HEBREW)
ḿfáng (44.5. EFIK)
mfoubou (38.14. BAMILEKE)
mfoubou (38.16. BAMILEKE)
mfoubou (38.17. BAMILEKE)
mfoubou (38.18. BAMILEKE)
ḿfrì (38.16. EFIK)
ḿfrì (38.18. EFIK)
ḿfrì ǹdisè (38.16. EFIK)
ḿfrì ǹdisè (38.18. EFIK)
mgagani (34.2. SWAHILI)
mgbam ekere (40.1. IGBO)
mgboko (17.5. IGBO)
mgisim (62.9. TIV)
mgomba (16.1a. DIGO)
mgomba (16.1a. SWAHILI)
mharupwa (62.19. SHONA)
mhindi (17.7. SWAHILI)
mhogo (39.4. SWAHILI)
mhotighol (59.2. MARATHI)
mhotighol (59.4. MARATHI)
mhotiluni (59.2. GUJARATI)
mhuvuyu (27.8. SHONA)
mi (17.7. KWÉYÒL)
mi dou (40.40. CHINESE, MANDARIN)
mi dou (40.40. CHINESE, WÚ)
mi gan tian gua (38.8. CHINESE, MANDARIN)
mi guo (19.1. CHINESE, MIN BEI)
mi$_{\vee}$ naw$^{\wedge}$ shi_ (40.1. LAHU)
mi$_{\vee}$ shi (20.2. LAHU)
mi tuo gai (30.6. CHINESE, MANDARIN)
mi tuo jie (30.6. CHINESE, WÚ)
mi xian (23.3. CHINESE, WÚ)
mian dou (40.11. CHINESE, YUE)
mian dou (40.29. CHINESE, MIN BEI)
mian tou (40.9. CHINESE, YUE)
mianzi (17.1. SWAHILI)
mianzi (17.2. SWAHILI)
mianzi (17.4. SWAHILI)
miarza burch (62.1. CHECHEN)
mibaelikand (11.2. MARATHI)
mîcîmin (40.28. CREE, PLAINS)
middelste teunisbloem (51.1. VLAAMS)
midi kaayi (38.23. KANNADA)
midiki (40.17. GREEK)
midjo doci (17.7. CRIOULO, CAPE VERDE)
mié (44.5. BAOULÉ)

miechunka peruwiańska (62.12. POLISH)
miechunka rozdęta (62.11. POLISH)
mielga (40.17. SPANISH)
mielies (17.7. AFRIKAANS)
mien (44.4. BAOULÉ)
mien tou (40.25. CHINESE, MANDARIN)
mierikswortel (30.1. DUTCH)
mierikswortel (30.1. VLAAMS)
mięta pieprzowa (44.2. POLISH)
mięta zielona (44.3. POLISH)
mietelnik (35.2. POLISH)
mietelnik żakula (35.2. POLISH)
migdał ziemny (10.1. POLISH)
miglionet (60.1. ITALIAN)
mignonette (60.1. ENGLISH)
mignonette (60.1. FRENCH)
mi-haraḍī (20.2. NEWAR)
mihko otsehkan ocepihk (35.4. CREE, PLAINS)
mihkoskatask (35.4. CREE, PLAINS)
mihkotsehkanis (30.56. CREE, PLAINS)
miju (40.12. INDONESIAN)
mīkstpienes (27.42. LATVIAN)
milagaai (62.5. TAMIL)
milaguthakkaalii (62.15. TAMIL)
milaguttakkali (62.19. TAMIL)
milakai (62.5. TAMIL)
milchit (35.7. HEBREW)
milchit haborit (35.8. HEBREW)
milḥit ha-borit (35.8. HEBREW)
milho (17.7. CRIOULO, CAPE VERDE)
milho (17.7. PORTUGUESE)
milho doce (17.7. PORTUGUESE)
miligay (62.2. TAMIL)
miling (36.3. BUHID)
militon (38.25. CREOLE, HAITIAN FRENCH)
milk gowan (27.45. ENGLISH)
milk thistle (27.42. ENGLISH)
milk weed (27.42. ENGLISH)
millefoglio d'acqua (42.1. ITALIAN)
millo (17.7. GALICIAN)
millo grosso (17.7. GALICIAN)
milo (38.7a. GOURMANCHÉMA)
milon (38.7a. DAGAARE, SOUTHERN)
milon (38.7a. TAGALOG)
miłowój biały (28.1. POLISH)
miłta (44.2. POLISH)
mimita (62.1. OROMO, WEST CENTRAL)
mimita (62.2. OROMO, WEST CENTRAL)
mimoza e ydrocharis (40.19. GREEK)
mimoza wodna (40.19. POLISH)
min gan tian gua (38.8. CHINESE, WÚ)
mina (27.17. FARSI, WESTERN)
miǹa (43.1. GBE-VHE)
minakmakling (39.4. BONTOC, CENTRAL)
mindal zemljanio (10.1. RUSSIAN)
mindia mi kombo (59.2. DOONDO)
mindzin-mebok (38.16. EWONDO)
mindzin-mebok (38.17. EWONDO)
miner's lettuce (59.1. ENGLISH)
ming he (20.5. CHINESE, WÚ)
ming ri cao (24.2. CHINESE, MANDARIN)
ming ri cao (24.2. CHINESE, WÚ)
minhonete (60.1. PORTUGUESE)
mini (40.1. SUBANON)
miniklalasni (38.23. SUMO-MAYANGNA)
minji (40.28. GIKUYU)
minjonet (60.1. AFRIKAANS)
minkonkolo (27.44. BEMBÉ)
mino (27.12. TAJIKI)
miñoneta (60.1. SPANISH)
mint (44.3. KRIO)
mintys y graig (44.6. WELSH)
mintys ysbigog (44.3. WELSH)
minuana (51.1. PORTUGUESE)
minuer (35.6. ALBANIAN, TOSK)
minumulu (40.35. TELUGU)
minumulu (40.37. TELUGU)
mioga (20.5. JAPANESE)
miogá (20.5. PORTUGUESE)
mioga ginger (20.5. ENGLISH)
mirabel (35.2. SPANISH)
mirabela (35.2. PORTUGUESE)
mirach (62.2. HINDI)
miracha (62.2. FIJIAN)
mirah (69.1. JAVANESE)
mirakelblomst (49.1. NORWEGIAN)
mirampilis (49.1. GREEK)
miraŋ (38.18. MANDING-MANDINKA)
miraŋ (38.20. MANDING-MANDINKA)
mirandjô-lô (38.20. MANDING-MANDINKA)
mirapa kaayi (62.2. TELUGU)
mirapa kaya (62.2. TELUGU)
mirapakaaya (62.5. TELUGU)
mirapakaya (24.9. TELUGU)
mirapakaya (62.2. TELUGU)
mirapakaya (62.5. TELUGU)
mirch (62.1. PANJABI, EASTERN)
mirch (62.2. HINDI)

mircha (62.5. HINDUSTANI)
mirchee (62.2. MARATHI)
mirchi (62.2. MARATHI)
mirchi (62.5. MARATHI)
mirhafű (24.16. HUNGARIAN)
miri (44.5. RAROTONGAN)
mirich (62.5. URDU)
miřík celer kořenový (24.4. CZECH)
mi-rìm (38.18. BUDIK)
mirinda (54.1. WÁRAY-WÁRAY)
mirindjena (62.19. LADINO)
mĩrĩo/gwăcĩ (36.3. GIKUYU)
mirizliva rezeda (60.1. MACEDONIAN)
mirliton (38.25. ENGLISH)
mirnik wonny (24.16. POLISH)
mirre (24.16. AFRIKAANS)
mirride delle Alpi (24.16. ITALIAN)
mirride odorata (24.16. ITALIAN)
mirsang (62.5. KONKANI)
mirsinga (62.2. MARATHI)
mirzanjosh (44.6. PANJABI, EASTERN)
misbredie (59.2. AFRIKAANS)
mischoïdes sinapi (30.20. GREEK)
misër i ëmbël (17.7. ALBANIAN, TOSK)
mishi (24.13. SANSKRIT)
mishravarnaphala (62.19. SANSKRIT)
mishreya (24.13. SANSKRIT)
mishri khand (40.18. NEPALI)
mishti alu (36.3. BENGALI)
misi (17.7. TURKISH)
misimisi (9.1. TONGAN)
misimisi koka'anga (15.2. TONGAN)
misir (17.7. LADINO)
misir börülcesi (40.9. TURKISH)
misir soğani (7.9a. TURKISH)
misira (40.12. OROMO, WEST CENTRAL)
misìri (17.7. ROMANI, BALKAN)
misirski kromid (7.12. MACEDONIAN)
misiwapikwan (27.23. CREE, PLAINS)
misk kabaği (38.14. TURKISH)
misk kabaği (38.17. TURKISH)
misk maydanozu (24.16. TURKISH)
mîskodi'ssîmîn (40.26. POTAWATOMI)
miskotu (45.2. TURKISH)
misksoğani (13.5. TURKISH)
misqui uchu (62.2. QUECHUA, CUSCO)
misra mukalab (44.7. URDU)
misrapushpa (40.32. SANSKRIT)
misri piaz (7.9a. URDU)
missi kumbéré (59.2. BAMANANKAN)
missidi koumbaré (59.2. BAMANANKAN)
missidi kumbaré (59.2. BAMANANKAN)
missinikoumbré (23.2. MANINKAKAN, WESTERN)
missir (40.12. HARARI)
missirà (40.12. SIDAMO)
missiri (40.12. AFAR)
misurpurpu (40.12. TAMIL)
mitaalu (36.3. HINDI)
mitha (35.9. ORIYA)
mitha aloo (36.3. ASSAMESE)
mitha alu (36.3. ASSAMESE)
mitha alu (36.3. HINDI)
mitha jolokia (62.1. ASSAMESE)
mitha kumra (38.16. HINDI)
mitha kumra (38.17. HINDI)
mitha mora (63.1. ASSAMESE)
mitha palanga (35.9. ORIYA)
mitha palanga saga (35.3. ORIYA)
mithakadhu (38.14. HINDI)
mithakadhu (38.16. HINDI)
mithakadhu (38.18. HINDI)
mithapalanga (35.9. ORIYA)
mitha-paleng (35.9. ASSAMESE)
mithĭ (40.32. NEWAR)
mithithumbi (38.20. HINDI)
mìtìgà (63.1. SONGHAY)
mitsamkhala (30.22. GEORGIAN)
mitsuba (24.10. DANISH)
mitsuba (24.10. FRENCH)
mitsuba (24.10. INDONESIAN)
mitsuba (24.10. JAPANESE)
mitsubá (24.10. PORTUGUESE)
mitsuda (24.10. VLAAMS)
mittikelu (40.35. MALAY)
miyîcîmin (40.26. CREE, PLAINS)
mizu-ojigi-sô (40.19. JAPANESE)
mizu-warabi (4.1. JAPANESE)
mjata blotnaja (44.2. RUSSIAN)
mjata bolotnaja (44.2. RUSSIAN)
mjölbanani (16.1a. ICELANDIC)
mjöllmolla (35.6. SWEDISH)
mkem (62.1. TIV)
mkhungudzu (40.9. NYANJA)
mkoo (16.1a. DIGO)
mkŏŏ (62.21. PANJABI, EASTERN)
mlasi (17.1. YAO)
mléč zelinný (27.42. CZECH)
mlecz zwyczajny (27.42. POLISH)
mlenda (63.1. SWAHILI)

ḿmɔ̀ngḿmɔ̀ng íkɔ̀ng (59.3. EFIK)
mmetli (24.11. SOTHO, NORTHERN)
mmetli (24.11. SOTHO, SOUTHERN)
mmetli (24.11. TSWANA)
mmidi (17.7. TSWANA)
mnadzi (62.21. NYANJA)
mnaȟcaȟca (13.3. DAKOTA)
mnavu (62.15. SWAHILI)
mnavu (62.21. DIGO)
mnavu (62.21. SWAHILI)
mnesi (62.21. YAO)
mniszek (27.45. POLISH)
mniszek lekarski (27.45. POLISH)
mniszek pospolity (27.45. POLISH)
mnogocveten fasul (40.24. BULGARIAN)
mnogocveten grav (40.24. MACEDONIAN)
mnogosjemena (40.32. SERBO-CROATIAN)
mnyanya (62.9. SWAHILI)
mo xi ge nan gua (38.14. CHINESE, MANDARIN)
mo xi ge nan gua (38.14. CHINESE, WÚ)
mo yu (19.1. CHINESE, MIN BEI)
mo yu (19.1. CHINESE, XIANG)
mo yu (19.1. CHINESE, YUE)
moa alu (11.5. ASSAMESE)
mo̰c (40.23. SANTALI)
mocha (16.1a. SANSKRIT)
mochakotta (40.9. INDONESIAN)
mochchai (40.9. TAMIL)
mochrateluk (35.1. ARMENIAN)
mochyně (62.13. CZECH)
mochyně peruánská (62.12. CZECH)
mochyně židorská třešeň (62.11. CZECH)
mock cypress (35.2. ENGLISH)
mock pak choy (30.37. ENGLISH)
mock tomato (62.14a. ENGLISH)
moco (16.1a. PALI)
modee‘ (40.26. PAIUTE, SOUTHERN)
modos (39.4. SAMBAL, BOTOLAN)
modra (30.4. POLISH)
modrak (30.46. POLISH)
modrak morska (30.45. POLISH)
modravec chocholatý (13.5. CZECH)
modřenec chocholatý (13.5. CZECH)
mŏĕ (10.1. TEM)
moerastaro (8.6. VLAAMS)
moes (30.35. STELLINGWERFS)
moganga (38.16. PORTUGUESE)
moge kaayi (38.13a. KANNADA)
móghomógh (15.2. CAROLINIAN)
mogo (39.4. LUO)
mogoin idee (37.1. MONGOLIAN)
mógomóg (15.2. CAROLINIAN)
mogyoróhogyma (7.2. HUNGARIAN)
mogyorós lednek (40.11. HUNGARIAN)
mohanamu (36.3. TELUGU)
mohari (30.17. MARATHI)
mohini (28.1. SANSKRIT)
mohiri (30.17. MARATHI)
Möhre (24.11. GERMAN)
møifrai (28.1. BODO)
mojje (30.43. BADAGA)
mo:-k/isi (38.20. ZUNI)
mokka (17.7. ORIYA)
mokka (17.7. TELUGU)
mokkajanna (17.7. TELUGU)
mokka-janna (17.7. TELUGU)
mokopu (38.14. SOTHO, SOUTHERN)
moku (21.2. TOK PISIN)
mokuva (50.2. ASSAMESE)
molakkai (62.16. TAMIL)
molao (38.7b. KONKANI)
moli (30.56. URDU)
molike (17.3. KURANKO)
mollë (27.26. ALBANIAN, TOSK)
mollë dheu (27.26. ALBANIAN, TOSK)
molley (38.11. GAELIC, MANX)
møllplante (44.8. NORWEGIAN)
moloko (17.3. MANDING-MANDINKA)
molo-mustasa (27.14. AKLANON)
molotru (40.32. ROMANIAN)
molsal (27.45. VLAAMS)
molsla (27.45. DUTCH)
molslaai (27.45. AFRIKAANS)
molură (24.13. ROMANIAN)
momɔŋa (16.1a. MENDE)
momiënh (34.2. KHMER, CENTRAL)
momili elo (44.5. NIUE)
momogul (33.2. TAJIKI)
momoqaimoq (27.17. TAJIKI)
momordika (38.23. CZECH)
momordique a feuilles de vigne (38.23. FRENCH)
Mondbohne (40.25. GERMAN)
mo̱ndo (8.4. TIV)
mondokwa - doŵe (17.7. LOMWE, MALAWI)
mone ôndôndô (62.3. BULU)
mone-la-oo (30.54. BURMESE)
mone-la-oo (30.56. BURMESE)
mone-la-oo-ni (24.11. BURMESE)

mongetas patuscas (40.24. CATALAN-VALENCIAN-BALEAR)
mongetes (40.26. CATALAN-VALENCIAN-BALEAR)
monggo (40.38. TAGALOG)
monggos (40.38. BUTUANON)
monggos (40.38. CEBUANO)
monggos (40.38. MARANAO)
monggos (40.38. TAGALOG)
mongo (40.38. TAGALOG)
mongoose plant (20.1. ENGLISH)
mon-hyin (30.4. BURMESE)
mon-hyin (30.5. BURMESE)
mon-hyin (30.6. BURMESE)
mon-hyin (30.7. BURMESE)
mon-hyin (30.8. BURMESE)
mon-hyin (30.9. BURMESE)
mon-hyin (30.10. BURMESE)
mon-hyin (30.11. BURMESE)
mon-hyin (30.12. BURMESE)
mon-hyin (30.13. BURMESE)
mon-hyin (30.14. BURMESE)
mon-hyin (30.15. BURMESE)
mon-hyin (30.16. BURMESE)
mon-hyin (30.17. BURMESE)
mon-hyin (30.18. BURMESE)
mon-hyin (30.19. BURMESE)
mon-hyin (30.20. BURMESE)
moniato (36.3. SPANISH)
monkey nut (40.1. ENGLISH)
monkey plant (20.1. ENGLISH)
monks hood (27.45. ENGLISH)
monla (30.54. BURMESE)
monla (30.56. BURMESE)
monocória (18.1. PORTUGUESE)
monocória (18.2. PORTUGUESE)
môô (38.3. LAO)
m[oof]ng t[ow]i (28.1. VIETNAMESE)
moojelăběch' (13.3. MICMAC)
moolaa (30.56. MARATHI)
moolaja (20.6. SANSKRIT)
moolaka (30.56. SANSKRIT)
moolangi (30.56. KANNADA)
moolavha (30.56. SANSKRIT)
mooli (30.56. HINDI)
mooli (30.56. PANJABI, EASTERN)
m[oo]n n[uw][ows]c (8.4. VIETNAMESE)
moong (40.37. MARATHI)
moongphali (40.1. HINDI)
moorputk (24.18. ESTONIAN)
mooz el-gana (16.1a. ARABIC)
moqqi (7.3. ZUNI)
moqqi: techi (7.10. ZUNI)
mōr (24.11. KASHMIRI)
mōr (40.35. PANJABI, EASTERN)
mora shikha (23.4. GUJARATI)
morado (23.3. MARANAO)
morak (27.11. TAJIKI)
moranga (38.16. PORTUGUESE)
moravare (40.34a. BADAGA)
morch (62.1. FARSI, EASTERN)
morch (62.2. FARSI, EASTERN)
morchej (24.11. SORBIAN, UPPER)
morcov (24.11. ROMANIAN)
morcov comestibil (24.11. ROMANIAN)
morcov comun (24.11. ROMANIAN)
morella minore (62.21. ITALIAN)
morelle (62.15. ENGLISH)
morelle (62.15. FRENCH)
morelle à grappe (56.2. FRENCH)
morelle diable (62.23. FRENCH)
morelle noire (62.21. FRENCH)
morepa (36.3. SOTHO, NORTHERN)
morfreshych (30.45. WELSH)
mori (40.26. HOPI)
mori-azami (27.11. JAPANESE)
morisuavarai (40.29. TAMIL)
morkoos (24.11. ROMANI, BALKAN)
morkos (24.11. LITHUANIAN)
morkotomado (40.18. ORIYA)
morkov (24.11. BULGARIAN)
morkov agorodnaja (24.11. RUSSIAN)
morkov kulturnaja (24.11. KAZAKH)
morkov' kul'turnaja (24.11. RUSSIAN)
morkva (24.11. BELARUSAN)
morkva dzikaja (24.11. BELARUSAN)
mormondika (38.23. HUNGARIAN)
mormuj (24.11. KASHMIRI)
morogo wa seleri (24.4. TSWANA)
moroheyia (63.1. JAPANESE)
moron (23.3. BAMANANKAN)
moron (23.3. MANINKAKAN, EASTERN)
moron (24.11. WELSH)
moron dyé (23.3. BAMANANKAN)
morot (24.11. SWEDISH)
morritort (30.51. CATALAN-VALENCIAN-BALEAR)
morritort d'aygua (30.52. CATALAN-VALENCIAN-BALEAR)
mors'ka kapusta (30.45. UKRAINIAN)

mumbanbou (17.1. BEMBÉ)
mumbanbou (17.1. KOONGO)
mumbe (62.2. TULU)
mûmbu (11.1. BEMBA)
mûmbu (11.2. BEMBA)
mûmbu (11.4. BEMBA)
mûmbu (11.5. BEMBA)
mûmbu (11.6. BEMBA)
mûmbu (11.7. BEMBA)
mûmbuv (11.3. BEMBA)
mumtalokh (27.38. TAJIKI)
muṅ (40.37. SANTALI)
munakoiso (62.19. FINNISH)
munatakali (62.21. TAMIL)
muncha (59.2. HINDI)
muńci (62.2. TULU)
mundi gedde (8.1. KANNADA)
mundigida (8.1. KANNADA)
mundla kura (23.3. TELUGU)
mundlatotakoora (23.3. TELUGU)
mundu (36.3. MANOBO)
mundundindundi (27.44. KOONGO)
múné (38.7b. SONGHAY)
mung (40.38. BENGALI)
mung (40.38. FIJIAN)
mung (40.38. HINDI)
mung (40.38. NEPALI)
mung (40.38. PANJABI, EASTERN)
mung (40.38. URDU)
mung bean (40.38. ENGLISH)
mung eta (40.38. SINHALA)
mung fasulyesi (40.38. TURKISH)
mung phali (40.1. URDU)
mungalill (30.52. ESTONIAN)
mungalill (64.3. ESTONIAN)
Mungbohne (40.38. GERMAN)
mungböna (40.38. SWEDISH)
mungbønne (40.38. DANISH)
mungbønne (40.38. NORWEGIAN)
mungboon (40.37. DUTCH)
mungboon (40.38. DUTCH)
mungboon (40.38. AFRIKAANS)
mùnggọ (16.1a. BAFUT)
munggo (40.38. CEBUANO)
munggo (40.38. TAGALOG)
munggos (40.38. CEBUANO)
munggos (40.38. TAGALOG)
munggu (40.38. HANUNOO)
mungil (17.1. MALAYALAM)
mungil (17.1. TAMIL)
mungo (40.37. CZECH)
mungo (40.37. FRENCH)
mungo (40.38. CEBUANO)
mungo (40.38. FILIPINO)
mungo (40.38. IFUGAO)
mungo (40.38. ILOCANO)
mungo (40.38. TAGALOG)
mungó bab (40.37. HUNGARIAN)
mungo fasulyesi (40.37. TURKISH)
mungo grav (40.37. MACEDONIAN)
mungoböna (40.37. SWEDISH)
mungoboon (40.37. VLAAMS)
mungoboon (40.38. VLAAMS)
mungopapu (40.38. FINNISH)
mungo-verde (40.38. PORTUGUESE)
mungphali (40.1. HINDI)
mungra (30.53. HINDI)
mungu (38.16. NYAKYUSA-NGONDE)
mungu (38.20. BEMBA)
munindrika (40.32. SANSKRIT)
munjaatha (11.5. SANSKRIT)
munjariki (44.5. SANSKRIT)
munkankali (27.44. TEKE)
muntoba (39.4. KOONGO)
muntsié (17.4. KOONGO)
muntun (21.1. WAAMA)
munya (59.2. HINDI)
muomva (16.1a. VENDA)
mura (30.56. BENGALI)
mura (30.56. GUJARATI)
mura (30.56. HINDI)
muraayi (30.56. HINDI)
murai (30.54. HINDUSTANI)
murai (30.56. FIJIAN)
murāi (30.56. MUNDARI)
murambo (36.3. VENDA)
mũrangi (17.1. GIKUYU)
mũrangi (17.2. GIKUYU)
mũrangi (17.4. GIKUYU)
murasaki-uma-goyashi (40.17. JAPANESE)
mürdümük (40.10. TURKISH)
mùrgili (62.24. ROMANI, BALKAN)
murgpapoen (38.18. AFRIKAANS)
murg-vangroente (38.18. AFRIKAANS)
muri (30.56. BENGALI)
muri (30.56. GUJARATI)
muri (30.56. HINDI)
muri (30.56. MARATHI)
mūrinis šatrenis (30.47. LITHUANIAN)
murok (24.11. HUNGARIAN)

mursennep (30.47. DANISH)
murtou (62.9. PULAR)
murud (38.7. ILOCANO)
murulauk (7.11. ESTONIAN)
mururé (4.1. PORTUGUESE)
mururé (40.19. PORTUGUESE)
musa (16.1a. OROMO, WEST CENTRAL)
musaka (62.21. TUMBUKA)
musambwarara (39.4. SHONA)
muscari (13.5. PORTUGUESE)
muscari (13.5. SPANISH)
muscari à toupet (13.5. FRENCH)
muscari comune (13.5. ITALIAN)
mûse (16.1a. SIDAMO)
musengere (17.1. SHONA)
musengere (17.2. SHONA)
musengere (17.4. SHONA)
mushaka tembwe (17.1. VENDA)
mushaka tembwe (17.2. VENDA)
mushaka tembwe (17.4. VENDA)
mushakawali (67.2. URDU)
mushakwali (67.2. HINDI)
mushenjere (17.1. SHONA)
mushenjere (17.2. SHONA)
mushenjere (17.4. SHONA)
mushkhor (27.5. TAJIKI)
musilam valli kilangu (11.5. TAMIL)
musipo (17.5. NDONGA)
muskatgresskar (38.17. NORWEGIAN)
muskaturi gogra (38.17. GEORGIAN)
muskmelon (38.7b. ENGLISH)
muskmelon (38.7b. VLAAMS)
muskusinis moliūgas (38.17. LITHUANIAN)
musky gourd (38.17. ENGLISH)
musky pumpkin (38.17. ENGLISH)
musky squash (38.17. ENGLISH)
musoso (62.14a. KOONGO)
musri (40.12. PANJABI, EASTERN)
mussa (45.4. KOONGO)
mussambé cor de rosa (34.2. PORTUGUESE)
mussambé de cinco folhas (34.2. PORTUGUESE)
mussassaka (38.22. BEMBÉ)
mussombibikan (40.1. KONKANI)
mussowa (62.21. RONGA)
must maavits (62.21. ESTONIAN)
mustajuuri (27.38. FINNISH)
mustakoiso (62.21. FINNISH)
muştar (30.3. ROMANIAN)
muştar (30.5. ROMANIAN)
muştar (30.6. ROMANIAN)
muştar (30.7. ROMANIAN)
muştar (30.8. ROMANIAN)
muştar (30.9. ROMANIAN)
muştar (30.10. ROMANIAN)
muştar (30.11. ROMANIAN)
muştar (30.12. ROMANIAN)
muştar (30.13. ROMANIAN)
muştar (30.15. ROMANIAN)
muştar (30.16. ROMANIAN)
muştar (30.17. ROMANIAN)
muştar (30.20. ROMANIAN)
muştar alb (30.57. ROMANIAN)
muştar comestibil (30.4. ROMANIAN)
muştar negru (30.24. ROMANIAN)
muştar sălbatec (30.17. ROMANIAN)
mustard (30.24. ENGLISH)
mustard greens (30.17. ENGLISH)
mustarda (30.17. MALTESE)
mustarda (30.57. MALTESE)
mustarda sewda (30.24. MALTESE)
mustart (30.17. GAELIC, MANX)
mustart (30.17. SCOTS)
mustasa (30.3. FILIPINO)
mustasa (30.4. FILIPINO)
mustasa (30.4. TAGALOG)
mustasa (30.5. FILIPINO)
mustasa (30.5. TAGALOG)
mustasa (30.6. BINUKID)
mustasa (30.6. FILIPINO)
mustasa (30.6. TAGALOG)
mustasa (30.7. FILIPINO)
mustasa (30.7. TAGALOG)
mustasa (30.8. FILIPINO)
mustasa (30.8. TAGALOG)
mustasa (30.9. FILIPINO)
mustasa (30.9. TAGALOG)
mustasa (30.10. FILIPINO)
mustasa (30.10. TAGALOG)
mustasa (30.11. FILIPINO)
mustasa (30.11. TAGALOG)
mustasa (30.12. FILIPINO)
mustasa (30.12. TAGALOG)
mustasa (30.13. FILIPINO)
mustasa (30.13. TAGALOG)
mustasa (30.15. FILIPINO)
mustasa (30.16. FILIPINO)
mustasa (30.17. BICOLANO, CENTRAL)
mustasa (30.17. BINUKID)
mustasa (30.17. BONTOC CENTRAL)

m[uw][ows]p t[aa]y (38.29. VIETNAMESE)
m[uw][ows]p t[aa]y (45.1. VIETNAMESE)
muxalan (27.45. GALICIAN)
muya muya (6.2. BENGALI)
muyaarmu (38.20. DITAMMARI)
muz (16.1a. AMHARIC)
mūz (16.1a. HARARI)
muz abäba (16.1b. AMHARIC)
mûze (16.1a. SIDAMO)
muzhkuku peerkai (38.22. TAMIL)
mvāc (16.1a. NEWAR)
mvoua-vouagna (34.2. KOONGO)
mvstv-lane (30.17. MUSKOGEE)
mvuh˄ hpo˄ tehˉ (59.2. LAHU)
mvuh˄ nyiˉ k'o (59.2. LAHU)
mvwiko (2.1. DIGO)
mwanga (39.4. GIKUYU)
mwarupwa (62.19. SHONA)
mwiigh (62.2. CAROLINIAN)
mwiigh (62.5. CAROLINIAN)
mwingereza (27.22. SWAHILI)
mwiwa (38.3. SHONA)
mw-nyí (40.43. BUDIK)
myal-ki-bhaji (28.1. HINDI)
myɛbɛ (40.25. PULAAR)
myin-khwa (24.7. BURMESE)
myinwa (17.2. BURMESE)
myogwe (36.3. DIGO)
myonk-ni (36.3. BURMESE)
myouga (20.5. JAPANESE)
myriophyllo (42.1. GREEK)
myrris e eyosmos (24.16. GREEK)
myterikolokynthi (38.31. GREEK)
n τrδm (40.1. YOM)
ŋà (17.3. BULLOM SO)
na muice fiadhaich (13.2. GAELIC, SCOTTISH)
naada' (17.7. APACHE, JICARILLA)
naada̱' (17.7. APACHE, JICARILLA)
naadá̱' (17.7. APACHE, JICARILLA)
naadą́ą́' (17.7. NAVAJO)
naadi (36.2. SANSKRIT)
naadika (36.2. SANSKRIT)
naagajemudu (32.1. TELUGU)
Naagapushphaphala (38.1. SANSKRIT)
naahollimbala' (40.28. CHICKASAW)
naakuli (20.1. SANSKRIT)
naala (36.2. MARATHI)
naalichi-bhaaji (36.2. MARATHI)
naalika (36.2. SANSKRIT)
na'ana meshubelet (44.3. HEBREW)
naanaa (44.3. PULAAR)
na'anat ha-kadurim (44.2. HEBREW)
naanat hakadurim (44.2. HEBREW)
naanat spicata (44.3. HEBREW)
naa'oléé (40.26. APACHE, JICARILLA)
naa'ołí (40.26. NAVAJO)
naa'ołí doołl'izhí (40.26. NAVAJO)
naa'ołí nímazí (40.28. NAVAJO)
naarathega (11.1. TELUGU)
naatabam (16.1a. TELUGU)
naati kanda (8.3. TELUGU)
naató'sipisátssaisski (27.45. BLACKFOOT)
naavu (32.1. HOPI)
naayi hakkarike (27.42. KANNADA)
naayi nonang (40.18. TULU)
naayi sonku (40.18. KANNADA)
naayi tholachi (44.4. TULU)
naayi thulasi (44.4. KANNADA)
náayiizéé (38.14. APACHE, JICARILLA)
náayiizéé (38.16. APACHE, JICARILLA)
náayiizéé (38.17. APACHE, JICARILLA)
náayiizéé (38.18. APACHE, JICARILLA)
naayithulasi (44.4. TAMIL)
náayizee (38.14. APACHE, JICARILLA)
náayizee (38.16. APACHE, JICARILLA)
náayizee (38.17. APACHE, JICARILLA)
náayizee (38.18. APACHE, JICARILLA)
naayízí (38.14. NAVAJO)
naayízí (38.17. NAVAJO)
naayízí (38.18. NAVAJO)
naayízí łichí'í (38.14. NAVAJO)
naayízí łichí'í (38.16. NAVAJO)
naayízí łichí'í (38.18. NAVAJO)
naaykkurana (40.18. MALAYALAM)
naaykorana (40.18. MALAYALAM)
naba (30.41a. SPANISH)
nâbi (39.4. WOLOF)
nabica (30.41b. CRIOULO, CAPE VERDE)
nabiças (30.41b. PORTUGUESE)
nabicol (30.21. SPANISH)
nabicol (30.30. GALICIAN)
nabizas (30.41b. SPANISH)
nabo (30.41a. CRIOULO, CAPE VERDE)
nabo (30.41a. PORTUGUESE)
nabo (30.41a. SPANISH)
nabo (41.1. BICOLANO, CENTRAL)
nabo amarelo (30.22. PORTUGUESE)
nabo da Suécia (30.22. PORTUGUESE)
nabo de brotes (30.41c. SPANISH)

nabo de espigos (30.41c. PORTUGUESE)
nabo de Suecia (30.22. SPANISH)
nabo greleiro (30.41c. PORTUGUESE)
nabo japones (30.54. PORTUGUESE)
nabo sueco (30.22. SPANISH)
näčč šənnurt (7.10. AMHARIC)
nachtschade (62.21. STELLINGWERFS)
nachtschaduw (62.21. STELLINGWERFS)
nachtschaede (62.21. STELLINGWERFS)
Nachtschatten (62.21. GERMAN)
nachtschone (49.1. DUTCH)
nachut (40.5. BULGARIAN)
nadé (38.16. SONGHAY)
naɗé (38.16. SONGHAY)
nadé yomba (38.18. WOLOF)
nadika (63.1. SANSKRIT)
nadio (38.16. SERER-SINE)
nadio (38.16. WOLOF)
nadio (38.17. WOLOF)
nadio (38.18. WOLOF)
nadishaka (36.2. MARATHI)
nadu (40.15. EDE NAGO)
nadu (40.15. YORUBA)
naduenosteblena nivska repka (30.19. MACEDONIAN)
nadzi (62.21. SENA)
naeneyakki gida (23.3. KANNADA)
naeng-ie (30.43. JAPANESE)
næpa (30.41a. ICELANDIC)
naeris (30.41a. ESTONIAN)
nafa-nafa (23.3. PULAAR)
nafus (30.21. HEBREW)
nag phani (32.1. NEPALI)
naga ito-uri (38.22. JAPANESE)
nāga-babürü (30.52. KASHMIRI)
nagaraja (20.6. SANSKRIT)
nagaranshadhi (16.1a. SANSKRIT)
nâgaro (10.1. WOLOF)
nagblom (51.1. AFRIKAANS)
nágdauná (13.2. HINDUSTANI)
nagdour (13.2. HINDI)
nag-down (13.2. HINDI)
naggatannarsuk (30.56. ALEUT)
naghnieh (44.3. MALTESE)
nãguri (40.1. MÒORÉ)
nah (45.1. MANDING-MANDINKA)
naha (20.6. IVATAN)
nahebwe (17.7. LOMWE, MALAWI)
nahlalaf (17.6. TAJIKI)
nahui (23.2. MAORI)
nai obi (17.8. TAJIKI)
naicorna (40.18. MALAYALAM)
naikkuruna (40.18. MALAYALAM)
naikorana (40.18. MALAYALAM)
naikurni (40.18. TAMIL)
näive (52.1. FINNISH)
nak nãngoy (38.7b. SERER-SINE)
nakayorel (34.2. WOLOF)
nakhod (40.5. FARSI, EASTERN)
nakhud (40.5. TAJIKI)
nakhudi rusi (40.28. TAJIKI)
nakhwëd (40.5. PASHTO, NORTHERN)
nakkatannak (30.41a. ALEUT)
nakpatâyâk aupalluktak (35.4. ALEUT)
nakwu̱si? (38.14. COMANCHE)
nakwu̱si? (38.16. COMANCHE)
nakwu̱si? (38.17. COMANCHE)
nakwu̱si? (38.18. COMANCHE)
nal (17.1. PANJABI, EASTERN)
nal (50.2. ASSAMESE)
nal (50.2. BENGALI)
nala (48.1a. SANSKRIT)
nalanibhaji (36.2. GUJARATI)
nalbă (45.5. ROMANIAN)
nali (36.2. PANJABI, EASTERN)
nalichi bhaji (36.2. MARATHI)
nalika (48.1a. SANSKRIT)
nalika (63.1. SANSKRIT)
nalina (48.1a. SANSKRIT)
nalitapat (63.1. BENGALI)
nalla avalu (30.24. TELUGU)
nalla kalava (50.2. TELUGU)
nalladoggali (23.3. TELUGU)
nallaminumulu (40.37. TELUGU)
nallani padmamu (50.2. TELUGU)
nallarati (16.1a. TELUGU)
nalta jute (63.1. ENGLISH)
Nalta-Jute (63.1. GERMAN)
nalvo (40.9. MANDING-MANDINKA)
nama (40.1. KONKOMBA)
nama (55.1. MANDING-MANDINKA)
namassa (16.1a. BAMANANKAN)
namassa (16.1a. MANDING-MANDINKA)
namavalai (16.1a. TAMIL)
ɲam-ba (11.1. WOLOF)
ɲamba ba (11.1. WOLOF)
ɲambi (11.3. MANDING-MANDINKA)
ñàmbi (39.4. WOLOF)
ɲambí-ba (11.1. MANDING-MANDINKA)
ñame amargo (11.4. SPANISH)

ñame amarillo (11.3. SPANISH)
ñame branco (11.1. SPANISH)
ñame congo (11.2. SPANISH)
ñame couche-couche (11.8. SPANISH)
ñame de agua (11.1. SPANISH)
ñame de Guinea (11.3. SPANISH)
ñame de la China (11.6. SPANISH)
ñame de tres hojas (11.4. SPANISH)
ñame del aire (11.2. SPANISH)
ñame grande (11.1. SPANISH)
ñame Guieno blanco (11.7. SPANISH)
ñame mapuey (11.8. SPANISH)
ñame morado (11.8. SPANISH)
ñame negro (11.3. SPANISH)
ñame papa (11.5. SPANISH)
ñame pequeño (11.5. SPANISH)
ñame yampí (11.8. SPANISH)
nami (11.7. TAGALOG)
námijìn gaásàyaá (23.3. HAUSA)
námíjìn-gaásàyaá (23.3. HAUSA)
námíjìn-zaákiíbánzaá (23.3. HAUSA)
námíjìn-zaákií-bánzaá (23.3. HAUSA)
naminya (23.3. LOMWE, MALAWI)
namna ya kitunguu kidogo (7.8. SWAHILI)
namna ya mboga (24.4. SWAHILI)
namna ya mboga (30.26. SWAHILI)
namna ya tikitimaji (38.7a. SWAHILI)
namtao (38.20. THAI)
namulepo (27.8. LOMWE, MALAWI)
namumbɔ (62.9. MENDE)
namurovo (40.38. LOMWE, MALAWI)
namyara (27.8. SENOUFO, CEBAARA)
namz taux (38.20. LAO)
nan gua (38.17. CHINESE, YUE)
nan kua (38.18. CHINESE, MANDARIN)
nan mei cu jiang cao (53.1. CHINESE, MANDARIN)
nan mei cu jiang cao (53.1. CHINESE, WÚ)
nan ou suan (7.1. CHINESE, MANDARIN)
nan ou suan (7.6. CHINESE, WÚ)
nan ou ts'ung (7.8. CHINESE, MANDARIN)
nan qua (38.17. CHINESE, XIANG)
nan sha shen (33.1. CHINESE, MANDARIN)
nana (27.8. MENDE)
nana (44.3. BOSNIAN)
na'nâ (44.3. FARSI, WESTERN)
nana (44.3. PASHTO, NORTHERN)
ñañanako (40.18. MANDING-MANDINKA)
nanborou (28.1. JULA)
nanćwibin (23.3. TWI)
ñañdébãg (40.18. PULAAR)
nandini (30.51. SANSKRIT)
nandolo (40.2. NYANJA)
nanə (44.3. AZERBAIJANI, NORTHERN)
nane (44.3. TURKISH)
n//ang (40.33. KUNG-EKOKA)
nangbanjɛn (23.3. GBE, WACI)
nangobourou (59.3. JULA)
n!angtsu (44.4. KUNG-EKOKA)
nanguri (40.1. MÒORÉ)
nanka (18.2. BENGALI)
nankin-mame (40.1. JAPANESE)
nankyo (20.1. JAPANESE)
nan-nan-pin (24.9. BURMESE)
nanoma (40.18. CRIOULO, UPPER GUINEA)
nansacha (10.1. DAGBANI)
nansaga (10.1. DAGBANI)
nantala (27.22. CHUWABU)
nantéo (38.5. MANDING-MANDINKA)
nantwi bin (23.3. AKAN)
nantwi bin (23.3. AKAN-AKUAPEM)
nanua (38.22. HINDI)
nañuma (40.18. MANKANYA)
nanwin (20.2. BURMESE)
nan'yo-mizu-aoi (18.1. JAPANESE)
na'olee (40.26. APACHE, JICARILLA)
nap (30.22. CATALAN-VALENCIAN-BALEAR)
nap (30.22. ROMANIAN)
nap (30.41a. CATALAN-VALENCIAN-BALEAR)
nap porcesc (30.41a. ROMANIAN)
napat althalg (22.1. ARABIC)
napat orguani (23.3. ARABIC)
napi (11.8. SRANAN)
napi cureceşti (30.22. ROMANIAN)
napi porceşti (27.26. ROMANIAN)
napier (17.3. FRENCH)
napier grass (17.3. ENGLISH)
napin (30.41a. GAELIC, MANX)
nappa (30.36. CAROLINIAN)
naqnaq (44.3. AFAR)
narakachora (20.4. MARATHI)
ñaramèl (44.5. BUDIK)
ńaramel (44.5. PULAAR)
!naran (40.1. HERERO)
narančila (62.22. CZECH)
narangille (62.22. FRENCH)
naranjilla (62.22. DANISH)
naranjilla (62.22. ENGLISH)

ngɔngɔi (17.3. MENDE)
*n*gɔrɔ (45.1. GOLA)
*n*gele (40.1. GOLA)
ngele-gba (32.1. MENDE)
ngele-gonu (32.1. MENDE)
n-gengange (11.2. LOKO)
n-gengange (11.4. LOKO)
*n*geŋge (63.1. BULLOM SO)
*n*geŋge (63.1. LOKO)
*n*geŋge (63.1. MENDE)
*n*geŋgle (63.1. MENDE)
*n*geŋgle-gbogboli (63.1. MENDE)
*n*geŋgle-gowuli (63.1. MENDE)
*n*gere (40.1. GOLA)
nget-pyaw (16.1a. BURMESE)
ngeya (17.7. BONTOC, CENTRAL)
nggarut (15.2. JAVANESE)
ǹggùlùmã (45.1. ANYIN)
ngh[eej] (20.2. VIETNAMESE)
ngh[eej] den (20.4. VIETNAMESE)
*n*gingere (63.1. MENDE)
ngiye ma vɔlɔ (9.1. MENDE)
ngo ban ku (23.4. BAMANANKAN)
ng[of] ta (24.9. VIETNAMESE)
ngogwe (62.14a. SWAHILI)
ngogwe (62.14b. SWAHILI)
ngogwe (62.18. SWAHILI)
ngohai (63.1. LOKO)
ngol a sav (13.1. SERER-SINE)
ngolgol (11.2. WOLOF)
ngon (38.5. BULU)
ngon (38.5. EWONDO)
ngon (38.12. BULU)
ngon (38.13a. BULU)
ngon (38.18. BULU)
ngon ntagan (38.7a. BULU)
ngoo ngèèwz (38.29. LAO)
ngor (40.28. LUO)
ngor (40.43. LUO)
n-gor si bidaw (34.2. WOLOF)
ngoro (62.9. EWONDO)
ngoroba blé (23.3. BAMANANKAN)
ngôtô (62.9. BULU)
ngôtô (62.10. BULU)
ngoumbé (38.5. GBAYA)
ngowe (38.25. NYANJA)
*n*goyo (62.9. MANDING-BAMBARA)
n̂g-tāu (40.8. CHINESE, MIN NAN)
ngub (39.5. KHMER, CENTRAL)
nguejo (62.14b. BAMILEKE)
ngugum (44.5. WOLOF)
ngumgum (44.5. WOLOF)
ngumvu (11.1. KOONGO)
ngunde (40.43. TUMBUKA)
ngunde (40.43. YAO)
ngungun (44.5. WOLOF)
ngunguné (44.5. PULAAR)
nguo (40.3. BAMANANKAN)
ngùrlí (38.13a. KANURI)
nguvu (20.2. SWAHILI)
nguwad (27.8. BONTOC, CENTRAL)
ng[uw]u b[af]ng (27.5. VIETNAMESE)
ngwara (40.9. SWAHILI)
ngwingwisu (40.9. YAO)
*n*gyrehye (62.19. AKAN-FANTE)
ǹgyùsmà (7.10. BEROM)
nhabubo (40.43. RONGA)
nhamaco (62.5. PULAAR)
nhami (8.4. CRIOULO, CAPE VERDE)
n'hampara (23.4. MAKHUWA)
nhandebágue (40.18. PULAAR)
nhanga (38.14. SHONA)
nhanga (38.16. SHONA)
nhanga (38.17. SHONA)
nhanga (38.18. SHONA)
nhanhamacô (40.18. MANDING-MANDINKA)
nhébè limboquê (40.43. PULAAR)
nhemba (40.43. RONGA)
nhemba (40.43. TSONGA)
nhumba (40.43. RONGA)
nhungunira (27.8. SHONA)
nh[ux] c[us]c (27.42. VIETNAMESE)
nh[ux] c[us]c d[oof]ng (27.43. VIETNAMESE)
ni meng suan cao (13.3. CHINESE, WÚ)
niakhuri (24.4. GEORGIAN)
niamaku (20.6. MANDING-BAMBARA)
niamaku (20.6. MANDING-MANINKA)
niamaku-limban (20.6. PULAAR)
nidraaganti (40.19. KANNADA)
nidralu (62.19. SANSKRIT)
nidrayam (40.19. TELUGU)
nidul-lif (27.8. KRIO)
nidupuvanga (62.19. TELUGU)
niébé (40.26. PULAR)
niébé (40.43. FRENCH)
niébé (40.43. MANDING-MANINKA)
niébé (40.43. MANINKA)
niébé (40.43. WOLOF)
niébé golo (40.43. MANDING-MANINKA)

niébé golo (40.43. MANINKA)
ni[ee]ng ni[eex]ng (17.8. VIETNAMESE)
Nieu-Seelandse spinasie (22.2. AFRIKAANS)
Nieuw Zeelandse spinazie (22.2. DUTCH)
Nieuwzeelandse spinazie (22.2. VLAAMS)
niga-gashû (11.2. JAPANESE)
nigand (44.5. PANJABI, EASTERN)
niga-uri (38.23. JAPANESE)
nigokats'i (62.24. APACHE, JICARILLA)
nigoko (9.1. MANDING-MANDINKA)
nigudhaka (40.35. SANSKRIT)
níguema (62.19. BAOULÉ)
nigurio (5.1. NEPALI)
nihon kabocha (38.17. JAPANESE)
niittysuolaheinä (58.2. FINNISH)
níkì (40.1. MENDE)
nikièm-zu-belog (23.4. MÒORÉ)
níkilì (40.1. MENDE)
nikka (11.1. CAROLINIAN)
nikka (11.2. CAROLINIAN)
nikka (11.4. CAROLINIAN)
nikka (11.5. CAROLINIAN)
nikła indyjska (40.2. POLISH)
níklì (40.1. MENDE)
nikyinikyi (38.23. TWI)
nilernait (40.26. INUKTITUT, EASTERN CANADIAN)
nilirnaq (40.26. INUKTITUT, EASTERN CANADIAN)
nilkamli (36.2. HINDI)
nilofar (30.52. URDU)
nilophal (50.2. GUJARATI)
nilufar (48.1a. TAJIKI)
nilufar (48.1b. TAJIKI)
nilufari obi (50.2. TAJIKI)
nilüfer (50.2. TURKISH)
niluvu pendalam (11.1. TELUGU)
nímasii (62.24. NAVAJO)
nímasii łikańí (36.3. NAVAJO)
nìná (40.43. BOBO MADARÉ, SOUTHERN)
ninfea (50.2. ITALIAN)
ninfea (50.2. SPANISH)
ninfea stellata (50.2. ITALIAN)
ning meng gua (38.8. CHINESE, WÚ)
niŋgi (55.1. KURANKO)
níníyà (23.3. IZON)
ninja (55.1. KURANKO)
ninjin (24.11. JAPANESE)
niŋke (55.1. KURANKO)
ninki (55.1. KISSI, NORTHERN)
ninki (55.1. KURANKO)
ninkié (55.1. MANDING-MANDINKA)
ninkye (55.1. MANDING-MANDINKA)
ninlabon (18.2. THAI)
ninniku (7.10. JAPANESE)
ninta (55.1. KURANKO)
niohomi (20.6. SUSU)
nionyn (7.3. WELSH)
nionyn cymru (7.5. WELSH)
nionyn dodwy (7.1. WELSH)
nionyn gwraidd hir (7.14. WELSH)
nionyn prew (7.9a. WELSH)
niori (7.10. GEORGIAN)
nipa (24.18. ICELANDIC)
nipai (40.18. CEBUANO)
nipai (40.18. TAGALOG)
nipay (40.18. CEBUANO)
nipay (40.18. TAGALOG)
nipiya (27.32. CREE, PLAINS)
nipiya ohtihipak (27.31. CREE, PLAINS)
nipoy (40.18. BICOLANO, CENTRAL)
nípúnertôq (27.5. INUKTITUT, WEST GREENLANDIC)
nira (7.13. JAPANESE)
nirá (7.13. PORTUGUESE)
niranegi (7.8. JAPANESE)
nirbatte (11.2. KANNADA)
nirettuvanga (62.19. TELUGU)
nirghandha kusuma (9.1. KANNADA)
nirkancha (18.2. TELUGU)
nirokancha (18.2. TELUGU)
nirulli (7.3. KANNADA)
nirulli (7.3. TELUGU)
niruvanga (62.19. TELUGU)
nisara (16.1a. SANSKRIT)
nisara (24.9. SANSKRIT)
nishandyaghni (40.6. SANSKRIT)
nishapava (40.43. SANSKRIT)
nishi-yomogi (27.7. JAPANESE)
nishpava (40.9. SANSKRIT)
nîsik (40.12. KURDISH, NORTHERN)
nising (45.3. TUPURI)
niski (40.26. CROATIAN)
nitchani (63.1. TUPURI)
nittaayinduna (62.12. MALAYALAM)
nittithoddavaddi (40.19. MALAYALAM)
nitui (38.23. TUPURI)
niu bang (27.5. CHINESE, MANDARIN)
niu bang (27.5. CHINESE, WÚ)
niu bang (27.5. CHINESE, YUE)

oduku (36.3. EDE NAGO)
oduku (36.3. YORUBA)
òdukún (36.3. YORUBA)
òdukún fúnfún (36.3. YORUBA)
òdukún pupa (36.3. YORUBA)
ódùm (55.1. IGBO)
ọ̀dúndún odò (27.18. YORUBA)
odunkun (11.5. YORUBA)
òdunkún (36.3. YORUBA)
ödürü (20.6. KASHMIRI)
oduvanchik lekarstvennyj (27.45. KAZAKH)
oduvančhik lekarstvennyj (27.45. RUSSIAN)
oduvanchik lekavy (27.45. BELARUSAN)
odzahwi (20.6. DANGME-KROBO)
oeil de paon (12.1. FRENCH)
oenanthe (24.17. FRENCH)
oenanthe phellandre (24.17. FRENCH)
oerd (40.37. DUTCH)
oerdoe (40.37. DUTCH)
oesterneut (38.28. AFRIKAANS)
oesternoot (38.28. VLAAMS)
oesterwortel (27.36. AFRIKAANS)
ofe (62.18. IGBO)
ofe fiti (17.1. SAMOAN)
ofe ofe (62.18. IGBO)
ofe papalagi (17.1. SAMOAN)
ofe-bèkeè (59.3. IGBO)
ofɛn (44.5. YORUBA)
ofi (38.27. IGBO)
ofio (10.1. EDE NAGO)
ofio (10.1. YORUBA)
ofio omu (10.1. YORUBA)
ọfọĩ (16.1a. OGBIA)
oftobparast (27.26. TAJIKI)
ofukwa (40.28. NAMA-DAMARA)
ogada (17.3. LUO)
ọgede (16.1a. IGBO)
ọ̀gèdè (16.1a. IGBO)
ọ̀gẹ̀dẹ̀ (16.1a. YORUBA)
ọ̀gèdè ñkịtị (16.1a. IGBO)
ọgede ojoko (16.1a. IGBO)
ogége (39.4. YORUBA)
ogenboon (40.43. VLAAMS)
ọghẹdẹ (16.1a. EDO)
ọghẹdebo (16.1a. EDO)
ọghọdọ́gbọ (17.3. EDO)
ògì (38.3. EDO)
ogili-irele (38.3. IGBO)
ogirok (38.13a. UKRAINIAN)
ognon (7.3. BRETON)
ogódsia (40.1. YORUBA)
ogom (23.3. MANOBO)
ogórecznik lekarski (29.1. POLISH)
ogórek (38.13a. POLISH)
ogórek (38.13b. POLISH)
ogórek afrykański (38.12. POLISH)
ogórek antylski (38.6. POLISH)
ogórek siewny (38.13a. POLISH)
ogorodnaja vichnja (62.11. KAZAKH)
ogu édè (8.4. IGBO)
ogunmo (62.21. YORUBA)
ogurec (38.13a. KAZAKH)
ogurec (38.13a. RUSSIAN)
ogurechnaja trev (29.1. KAZAKH)
ogurets (38.13a. BELARUSAN)
ogurets dlinno plodny (38.13a. RUSSIAN)
ogurets korotka (38.13a. RUSSIAN)
ogurets posevnoi (38.13a. RUSSIAN)
ogurke (38.13a. NORWEGIAN)
ogúsi (38.5. IZON)
ogwîssimaû'n (38.14. POTAWATOMI)
ogwîssimaû'n (38.16. POTAWATOMI)
ogwîssimaû'n (38.17. POTAWATOMI)
ogwîssimaû'n (38.18. POTAWATOMI)
ohahpokopi (24.11. COMANCHE)
ohahtuhkapu (24.11. COMANCHE)
ohape (38.3. COMANCHE)
ohapi̱ (38.3. COMANCHE)
ohauveve (44.4. HERERO)
ohbudou houzuki (62.13. JAPANESE)
ohdake (27.11. FINNISH)
óhí (38.28. IGBO)
ohkuro guai (10.2. JAPANESE)
ohnennàta (62.24. MOHAWK)
ō͘-hôai (8.5. CHINESE, MIN NAN)
ohpikecikana (24.19. CREE, PLAINS)
ohsahèta (40.26. MOHAWK)
oi (11.2. RAROTONGAN)
oi (17.5. THAI)
oi (38.13a. KOREAN)
oi-bèkéè (36.3. IGBO)
oi-daeng (17.5. THAI)
oignon (7.3. CAJUN FRENCH)
oignon (7.3. FRENCH)
oignon catawissa (7.9a. FRENCH)
oignon d'Egypte (7.9a. FRENCH)
oignon d'Hiver (7.9b. FRENCH)
oignon-patate (7.2. FRENCH)
oihakautu (62.24. NAMA-DAMARA)
oije (7.3. STELLINGWERFS)

oinniún (7.3. GAELIC, IRISH)
oje (7.3. STELLINGWERFS)
ojitandavare (59.2. HERERO)
ojō (45.4. IGBO)
ojo de zamuro (40.18. SPANISH)
o-jod (40.43. BASSARI)
ok biçimli altin göbek (8.9. TURKISH)
ok boš karam (30.27a. UZBEK)
ok zira (24.13. UZBEK)
ọka (17.7. IGBO)
oka (53.1. CZECH)
oka (53.1. DANISH)
oka (53.1. ENGLISH)
oka (53.1. VLAAMS)
oka odudu (40.31. IGBO)
oka' nampakali' (50.2. CHICKASAW)
okahijiki (35.7. JAPANESE)
okanakamudesha (24.11. NAMA-DAMARA)
okan-anlo (17.3. GOURMANTCHÉ)
okatjako (10.1. HERERO)
ok-bosh caram (30.27a. UZBEK)
oke ahịhịa (55.1. IGBO)
oké beri (59.2. IZON)
oker (45.1. DUTCH)
oker (45.1. VLAAMS)
o'khandia (45.1. SERER-SINE)
okhijiki (35.7. VLAAMS)
okhrakhushi (24.19. GEORGIAN)
okimon (44.4. GREEK)
ö-ki-nī-ön (11.1. NICOBARESE, CAR)
okiniy (62.9. CREE, PLAINS)
oklo (45.1. MUSKOGEE)
oko (11.7. EFIK)
òkò (11.7. EFIK)
okok (41.1. EWONDO)
òkôkòn (38.5. EFIK)
okono (38.14. LUO)
okono (38.16. LUO)
okono (38.17. LUO)
okono (38.18. LUO)
okono (38.27. LUO)
ọ̀kọ́ti (40.25. EFIK)
ọ̀kọ́ti (40.43. EFIK)
o-kourko (36.3. GOURMANTCHÉ)
ọ̀kpa (40.1. IGBO)
ọ̀kpà (40.1. IGBO)
ọkpa (40.39. IGBO)
ọ̀kpa akele (40.1. IGBO)
ọ̀kpa ekele (40.1. IGBO)
ọkpa ẹkẹrẹ (40.1. IGBO)
ọ̀kpa nwekele (40.1. IGBO)
ọkpa odudu (40.31. IGBO)
okpal (50.2. CHICKASAW)
okpan (38.20. EDO)
okpɛ (38.16. GUN)
okpɛ (38.18. GUN)
okpètè (17.5. IGBO)
ọkpete (17.5. IGBO)
okpo (11.7. EFIK)
ọ̀kpọ́ (11.7. EFIK)
ọ̀kpọ́ úmàn íwā (11.7. EFIK)
òkpòdọ̀ghòn (11.3. EDO)
okpodudu (40.31. IGBO)
okpu (17.5a. IZON)
okpuru (11.7. EFIK)
okpu-utere (23.3. EFIK)
okra (45.1. AFRIKAANS)
okra (45.1. DANISH)
okra (45.1. DUTCH)
okra (45.1. ENGLISH)
okra (45.1. ESTONIAN)
okra (45.1. FILIPINO)
okra (45.1. FINNISH)
okra (45.1. FRENCH)
Okra (45.1. GERMAN)
okra (45.1. HUNGARIAN)
okra (45.1. ICELANDIC)
okra (45.1. ILOCANO)
okra (45.1. INDONESIAN)
okra (45.1. NORWEGIAN)
okra (45.1. SWEDISH)
okra (45.1. TAGALOG)
okra (45.1. TAGBANWA)
okra (45.1. VLAAMS)
okrá (45.1. WÁRAY-WÁRAY)
okrasupp (45.1. ESTONIAN)
ókrɔ (45.1. BAFUT)
okre (62.5. BOKYI)
ọkrọ (45.1. IGBO)
okro (45.1. KRIO)
okrúhlica (30.41a. SLOVAK)
okseøje (27.23. DANISH)
öksürükotu (27.35. TURKISH)
oktaak okpalli' (50.2. CHICKASAW)
Okumo (8.9. GERMAN)
okumo (8.9. NORWEGIAN)
okun (9.1. EDE NAGO)
okun (9.1. YORUBA)
okura (45.1. JAPANESE)
okurka (38.13a. CZECH)

okurka setá (38.13a. CZECH)
òkwè (40.25. IGBO)
okwere (17.5. EDO)
ọkwọrọ-ozo (45.4. IGBO)
ọkwụlụ (45.1. IGBO)
ọkwụrụ (45.1. IGBO)
ol (8.3. ASSAMESE)
ol (8.3. BENGALI)
ol (8.3. SANTALI)
ol kabi (30.30. ASSAMESE)
ol kobi (30.30. ASSAMESE)
ol kochu (8.3. ASSAMESE)
ola suan (7.10. CHINESE, YUE)
olasiman (59.2. BICOLANO, CENTRAL)
olasiman (59.2. FILIPINO)
olasiman (59.2. TAGALOG)
olasiman (59.2. WÁRAY-WÁRAY)
olbi' (38.14. CHICKASAW)
olbi' (38.16. CHICKASAW)
olbi' (38.17. CHICKASAW)
olbi' lakna' (38.14. CHICKASAW)
olbi' lakna' (38.16. CHICKASAW)
olbi' lakna' (38.17. CHICKASAW)
olbi' okchamali' (38.14. CHICKASAW)
olbi' okchamali' (38.16. CHICKASAW)
olbi' okchamali' (38.17. CHICKASAW)
olbi' tohbi (38.14. CHICKASAW)
olbi' tohbi (38.16. CHICKASAW)
olbi' tohbi (38.17. CHICKASAW)
old kapi (30.30. BENGALI)
oldih all bakyug an nagalgaltan (40.2. IFUGAO)
oldih an mumpaha (40.2. IFUGAO)
oldih hin inadne (40.2. IFUGAO)
oldkapi (30.30. BENGALI)
ọlẹle (40.2. EDO)
olele (62.18. BULU)
ʻolena (20.2. HAWAIIAN)
olesh (27.9. HEBREW)
olesh tarbuti (27.10a. HEBREW)
olho de burro (40.18. PORTUGUESE)
óliak (40.1. SERER-SINE)
olifantsgras (17.3. VLAAMS)
olifantsgrass (17.3. DUTCH)
olifantsoor (8.6. AFRIKAANS)
olifantspoot (8.3. VLAAMS)
olifantsqras (17.3. AFRIKAANS)
õlikaalikas (30.21. ESTONIAN)
o'limb (38.14. SERER-SINE)
olkabi (30.30. ASSAMESE)
olkobi (30.30. ORIYA)
olla (8.3. SANSKRIT)
ollay (23.3. ITAWIT)
ollo deboi (27.23. GALICIAN)
olloco (28.2. SPANISH)
olluco (28.2. ENGLISH)
ọlọ̀ pàpà nìràgà (19.1. YORUBA)
ölön (10.1. MONGOLIAN)
olorungbin (23.3. YORUBA)
Öl-Senfrauke (30.48. GERMAN)
ólū (11.7. IGBO)
olua (8.3. ORIYA)
òlúbi (27.24. IGBO)
òlúbì (27.24. IGBO)
òlúbù (27.24. IGBO)
òlúgbì (27.24. IGBO)
òlúgbù (27.24. IGBO)
ôluh (62.24. KASHMIRI)
ómahkataaki (36.3. BLACKFOOT)
ómahkotssko'nssimaan (38.3. BLACKFOOT)
omakunde (40.26. NAMA-DAMARA)
ombalo (44.2. GEORGIAN)
ombaruru (24.11. KWANGALI)
ombaruru (30.41a. HERERO)
ombidi (30.27a. NAMA-DAMARA)
ombudufukwa (40.1. NAMA-DAMARA)
òmèjl (16.1a. IGBO)
omgbabelara (62.23. EDO)
omgbómịnịbɔ́ pún kọrịghẹ́ (62.23. IZON)
ómí (11.7. EDO)
omi (11.7. IGBO)
ómí funfun (11.7. EDO)
omiriva (17.7. KWANGALI)
omnica (40.26. DAKOTA)
omnicagmigmi (40.28. LAKOTA)
omuica hmiyaŋyaŋ (40.28. DAKOTA)
omukaigana (32.1. KWANYAMA)
omung'onye (17.5. LUYIA)
omuoko (39.4. LUYIA)
omwenge wosuuka (17.5. NAMA-DAMARA)
ọna (11.4. IGBO)
ọ̀nà (11.4. IGBO)
onagra (51.1. PORTUGUESE)
onagra (51.1. SPANISH)
onagra bienal (51.1. SPANISH)
onagra común (51.1. SPANISH)
onagre bisannuelle (51.1. FRENCH)
önant (24.17. TURKISH)
ônarnartok aupalângajok (62.2. ALEUT)
onas (17.5. BONTOC, CENTRAL)

onas (17.5. IBALOI)
onas (17.5. IVATAN)
onas (17.5. KANKANAEY)
onau (40.43. SERER-SINE)
oncet (40.34a. JAVANESE)
ondelaga (24.7. KANNADA)
ondjupa (38.20. HERERO)
ondú (38.14. IZON)
ondú (38.16. IZON)
one o'clocks (27.45. ENGLISH)
onekwa (40.28. MOHAWK)
onenhste (17.7. MOHAWK)
onerahtakate (27.32. MOHAWK)
on-gaine (62.5. BASSARI)
ongkol (16.1a. VLAAMS)
ongoto (62.9. BULU)
ongot-ongot (38.23. SAMBAL, TINÀ)
on-grendt (62.5. BASSARI)
ongsoy (24.9. TAGALOG)
ongsoy (24.17. TAGALOG)
òŋgùgù dəńdè (16.1a. KRIO)
ongwinḓi (40.28. KWANGALI)
onhou gni (11.1. BÉTÉ)
onibasu (50.1. JAPANESE)
onion (7.3. ENGLISH)
oni-yurine (13.4. JAPANESE)
onjanga (7.3. HERERO)
onnane (27.16. GAELIC, MANX)
onoko (11.1. IGBO)
ononde (38.5. FRENCH)
onon'onhsera (38.14. MOHAWK)
onon'onhsera (38.16. MOHAWK)
onon'onhsera (38.17. MOHAWK)
onon'onhsera (38.18. MOHAWK)
onot not (27.8. LUO)
onti (62.21. TAGALOG)
ontong (16.1a. VLAAMS)
ọ̀nụ̀ (11.4. IGBO)
ònúbù (27.24. IGBO)
onugbu (27.24. IGBO)
ònúgbù (27.24. IGBO)
onwad (27.8. IFUGAO)
onwad (27.12. IFUGAO)
onwad di binobo (27.8. IFUGAO)
onwad di malikanu (27.12. IFUGAO)
onyan (7.2. SUMO-MAYANGNA)
onyanga (7.3. NAMA-DAMARA)
oohkominoko (40.28. BLACKFOOT)
oorakaakara (38.23. TELUGU)
oorakanda (8.3. TELUGU)
ooralu gedde (62.24. KANNADA)
oort (23.3. TAGALOG)
oosithagarai (27.42. TAMIL)
oosoo (8.4. MANDINKA)
oosterse bunias (30.42. DUTCH)
oosterse pekelmeloen (38.9. VLAAMS)
Oostindische kers (65.1. VLAAMS)
oost-indische pijlwortel (19.1. VLAAMS)
oot (8.4. CAROLINIAN)
'öö'w llök (38.3. KHMER, CENTRAL)
oọ́yọ́ (63.1. YORUBA)
ọ̀pàpa (40.1. IGBO)
opaškovidna rotkva (30.53. MACEDONIAN)
opele (8.1. LUO)
opele (8.4. LUO)
opele (8.8. LUO)
opele (8.9. LUO)
ópiat (40.1. SERER-SINE)
opo (38.20. TAGALOG)
opornik łatkowaty (40.30. POLISH)
oporo (62.5. RAROTONGAN)
opunce (32.1. FRENCH)
opuncia (32.1. MONGOLIAN)
opuncie (32.1. CZECH)
opuncja (32.1. POLISH)
opuntia (32.1. ITALIAN)
Opuntie (32.1. GERMAN)
opunzia (32.1. ITALIAN)
orach (35.1. ENGLISH)
orache (35.1. ENGLISH)
oragán (44.6. GAELIC, IRISH)
oranda-garashi (30.52. JAPANESE)
orandamitsuba (24.5. JAPANESE)
oranga (40.24. LUO)
oranga (40.25. LUO)
oranga (40.26. LUO)
oranga (40.34a. LUO)
oranga (40.38. LUO)
oranga (40.41. LUO)
oranga (40.42. LUO)
orange de Quito (62.22. FRENCH)
orăstică (40.10. ROMANIAN)
oray (23.3. TAGALOG)
ordè (38.20. PULAAR)
ordrudova skupina (38.7a. CZECH)
orecchia d'asino (29.2. ITALIAN)
oregano (44.6. ENGLISH)
oregano (44.6. FINNISH)
Oregano (44.6. GERMAN)
oregano (44.6. INDONESIAN)

oregano (44.6. LADINO)
orégano (44.6. SPANISH)
oregano (44.6. TAGALOG)
orégano común (44.6. SPANISH)
orégão (44.6. PORTUGUESE)
oregon (44.6. CRIOULO, CAPE VERDE)
oreh zemljanoi (10.1. RUSSIAN)
oreille mouton (27.44. FRENCH)
ọrẹle (40.2. EDO)
orelhas de asno (29.2. PORTUGUESE)
orešesest graor (40.11. MACEDONIAN)
oreşnita tuberculifera (40.11. ROMANIAN)
orfanc (52.1. WELSH)
örgön navchit tutarga (17.8. MONGOLIAN)
órí (11.7. EDO)
óriástök (38.16. HUNGARIAN)
oriégano (44.6. SPANISH)
oriental celery (24.17. ENGLISH)
oriental garlic (7.13. ENGLISH)
oriental pickling melon (38.9. ENGLISH)
oriental water fern (4.1. ENGLISH)
oriental winter radish (30.54. ENGLISH)
orientalsk løg (7.1. DANISH)
origan (44.6. FRENCH)
origan commun (44.6. FRENCH)
origan vulgare (44.6. FRENCH)
origano (44.6. ITALIAN)
origáno (44.6. LADINO)
origunon to amaranton (44.6. GREEK)
oříšky lotosu (48.1b. CZECH)
óríwò (27.24. EDO)
orizoviden grav (40.40. MACEDONIAN)
orlica pospolita (2.1. POLISH)
ormarundu (20.1. TAMIL)
ormereddik (30.53. NORWEGIAN)
ornebregne (2.1. DANISH)
oro agogo (32.1. YORUBA)
ọrọ oji (55.1. IGBO)
oro-ablao (11.2. BICOLANO, CENTRAL)
orobanca (52.1. PORTUGUESE)
orobanca (52.1. SPANISH)
orobanche chevelue (52.1. FRENCH)
orobanche della fava (52.1. ITALIAN)
orot sabit tinai (39.3. PALAWANO, CENTRAL)
orovagchi (52.1. GREEK)
oroy (8.3. CEBUANO)
ortage (66.3. FRENCH)
orticone (66.3. ITALIAN)
ortie (66.3. FRENCH)
ortie dioique (66.3. FRENCH)
ortiga (66.3. SPANISH)
ortiga dioique (66.3. ITALIAN)
ortiga grande (66.3. ITALIAN)
ortiga japonesa (44.11. SPANISH)
ortiga maggiore (66.3. ITALIAN)
ortiga mayor (66.3. SPANISH)
ortiq (24.19. AZERBAIJANI, NORTHERN)
örümcek bitkisi (34.2. TURKISH)
örümcek çiçeği (34.2. TURKISH)
oruu rwouitji (17.5. KWANGALI)
örvarblad (6.2. ICELANDIC)
orzech wodny (64.1. POLISH)
orzech ziemny (40.1. POLISH)
oš lavlagi (35.4. UZBEK)
osè bèkeè (62.5. IGBO)
osè mkpele (62.5. IGBO)
osē m̃kpi (62.5. IGBO)
osè nkere (62.5. IGBO)
osè nkwu (62.2. IGBO)
ose nsukka (62.2. IGBO)
osè nwam̃kpi (62.2. IGBO)
ose utọrọ (62.2. IGBO)
oseille de Guinée (45.4. FRENCH)
oseille épinard (58.3. FRENCH)
oseille ronde (58.4. FRENCH)
ose-mkpili (62.5. IGBO)
ose-oyibo (62.1. IGBO)
oshb el-maa almodala (24.17. ARABIC)
osheb el-feel (17.3. ARABIC)
oshimhelewene (59.2. KWANYAMA)
oshkadu (38.18. TAJIKI)
oshthi (38.4. SANSKRIT)
oshtopamaphala (38.4. SANSKRIT)
osi (1.1. ESTONIAN)
osĩ (17.3. IZON)
osibã (11.7. AKAN-AKUAPEM)
oskapi' (17.1. CHICKASAW)
oskapi' (17.2. CHICKASAW)
oskapi' (17.4. CHICKASAW)
oskâtâsk (24.11. CREE, PLAINS)
oškovok (38.16. UZBEK)
ošlejch (7.5. CZECH)
osmonde cannelle (3.1. FRENCH)
osmonde du Japon (3.2. FRENCH
osmounda to kinnamomios (3.1. GREEK)
osmunda (3.1. PORTUGUESE)
osmunda (3.1. SPANISH)
osmunda japonesa (3.2. SPANISH)
osmundo do Japão (3.2. PORTUGUESE)

òsókpó (16.1a. OGBIA-AGHOLO)
ọsoñ-ikpọ̀k èbìgè (11.1. EFIK)
osot ogorodnyj (27.42. KAZAKH)
ossan (62.15. EWONDO)
ossissibole (27.14. AKWA)
ossissiboni (27.14. AKWA)
osso (36.3. JULA)
ostahi (38.14. CHICKASAW)
ostahi (38.16. CHICKASAW)
ostahi (38.17. CHICKASAW)
ostak (27.42. SERBO-CROATIAN)
østersnøtt (38.28. NORWEGIAN)
ostindisk pilrod (19.1. DANISH)
osto (38.14. CHICKASAW)
osto (38.16. CHICKASAW)
ostrożeń (27.11. POLISH)
ostryż długi (20.2. POLISH)
ostryż plamisty (20.4. POLISH)
ostrzeszyn jadalny (15.1. POLISH)
osuban (11.7. DANGME-KROBO)
osukwa (40.1. KWANGALI)
osum (62.14a. YORUBA)
osun (23.4. YORUBA)
osun (62.14a. DANISH)
osun (62.15. YORUBA)
ọ̀sùn (62.18. YORUBA)
osunda (44.3. BODO)
ot (62.1. VIETNAMESE)
ot bi (62.5. VIETNAMESE)
otafuku mame (40.34a. JAPANESE)
otaheiteomena (62.14a. FINNISH)
otahkoottsis (32.1. BLACKFOOT)
otakilokki (35.7. FINNISH)
otamatina (62.9. HERERO)
o-tāu (40.8. CHINESE, MIN NAN)
otcicen (62.19. BOKYI)
otehipak (30.27a. CREE, PLAINS)
ò-télès(h)á (40.1. ICEVE-MACI)
otigo (38.21. LUO)
otigo (45.1. LUO)
otigo (63.1. LUO)
o-tika (40.1. BASSARI)
otili (40.2. EDE NAGO)
otili (40.2. YORUBA)
òtili (40.2. YORUBA)
otili (40.31. YORUBA)
otimi (40.2. YORUBA)
otisihkân (30.41a. CREE, PLAINS)
otjiamate (62.9. KWANGALI)
otjihakautu (62.24. HERERO)
otjihakautu (62.24. KWANGALI)
otjikora (30.27a. KWANGALI)
otjimbakuna (38.14. HERERO)
otjimbakuna (38.16. HERERO)
otjipanana (16.1a. KWANGALI)
otogúúh (8.4. CAROLINIAN)
otok-otok (40.7. INDONESIAN)
ọ̀tọ́lì (36.3. IGBO)
otong (40.42. BONTOC CENTRAL)
otong (40.42. ILOCANO)
otong (40.43. TAGALOG)
otorhall (8.4. CAROLINIAN)
otoro (63.1. GA)
ọtọsi (17.1. IGBO)
otre-kani (11.3. ÉWÉ)
otri-kani (11.3. DANGME-KROBO)
otsekan (30.41a. CREE, PLAINS)
otsihkwa ohtera (30.41a. MOHAWK)
otsinekwar ohtera (24.11. MOHAWK)
otu (45.1. BOKYI)
otuk okpo (11.7. EFIK)
oturo (55.1. GA)
oturo (63.1. GA)
oturo (63.1. TWI)
ou (48.1a. CHINESE, MIN BEI)
ou (48.1a. CHINESE, YUE)
ou ch'in (24.19. CHINESE, MANDARIN)
ou he (48.1a. CHINESE, YUE)
ou ling (64.3. CHINESE, YUE)
ou tong hao (27.23. CHINESE, MANDARIN)
ou tong hao (27.23. CHINESE, WÚ)
ou zhou mo yao (24.16. CHINESE, MANDARIN)
ou zhou mo yao (24.16. CHINESE, WÚ)
oua (17.1. BODO)
oua (17.2. BODO)
oua (17.4. BODO)
ouai (8.4. KISI, SOUTHERN)
ouante (39.4. JOLA-FONYI)
ouélé (17.3. JULA)
ounou ounou-zêpouyiê (45.1. BÉTÉ)
ouré (36.3. SUSU)
ouregão (44.6. PORTUGUESE)
ourego (44.6. GALICIAN)
ourego (44.6. SPANISH)
oussou-ni-fing (44.10. BAMANANKAN)
outuerangama (27.8. HERERO)
oval-leaved monochoria (18.2. ENGLISH)
oval-leaved pondweed (18.2. ENGLISH)
ovɛ wiwi (40.18. AJA)

overdadige fedia (27.21. DUTCH)
ovočeva kvasolja (40.26. UKRAINIAN)
ovočevij goroch (40.28. UKRAINIAN)
owa (8.3. ORIYA)
ọwalo ani (55.1. IGBO)
òwéí bùrú (11.3. IZON)
owesa'wanakûk (51.1. POTAWATOMI)
owôndô (40.1. BULU)
oworo-woro (11.4. AKAN-AKUAPEM)
oworo-woro (11.4. GA)
oxalida (53.1. GREEK)
oxalide tubéreuse (53.1. FRENCH)
oxygonos (38.21. GREEK)
oyaba (7.3. LIMBA)
øyebønne (40.43. NORWEGIAN)
oyele (40.18. MAKHUWA)
òyím (7.3. EFIK)
òyím (7.10. EFIK)
òyím ḿbàkárá (7.3. EFIK)
òyìrà (39.4. AKPA)
oyong (38.21. INDONESIAN)
oyong (38.21. SUNDA)
oyong (38.22. SUNDA)
oyong lufah (38.22. INDONESIAN)
ọyọ́yọ̀ (63.1. YORUBA)
oyster nut (38.28. ENGLISH)
oyster plant (27.46. ENGLISH)
'o:yz (17.5. LAO)
ọzáụ (40.26. IZON)
ozomiriva (17.7. HERERO)
ozosukua (40.1. HERERO)
p słodkie (10.2. POLISH)
pa dou dian zhu (17.1. CHINESE, WÚ)
paachhaayi (44.5. TAMIL)
paaharkkaai (38.23. TAMIL)
paaj qab ib (23.4. HMONG-NJUA)
paala koora (35.9. TELUGU)
paalak (35.9. HINDI)
paalak (35.9. MARATHI)
paalak soppu (35.3. KANNADA)
paalak soppu (35.9. KANNADA)
paalaki (35.9. HINDI)
paalkashaak (35.9. HINDI)
paambudda (62.12. TELUGU)
pa'amonit (33.2. HEBREW)
paamonit kipachat (33.2. HEBREW)
paangobhi (30.27a. HINDI)
paang-puk (17.4. THAI)
paapasi (62.24. COMANCHE)
pàapu (40.26. HOPI)
paardebloem (27.45. DUTCH)
paardeboon (40.34a. VLAAMS)
paardekeker (40.16. VLAAMS)
paarose (38.22. MARATHI)
paarse morgenster (27.46. VLAAMS)
paasaangali (28.1. TAMIL)
paasta (38.7a . CHECHEN)
paasta (38.8. CHECHEN)
paasta (38.9. CHECHEN)
paasta (38.11. CHECHEN)
paasta (38.12. CHECHEN)
paasunitpu (24.18. JAPANESE)
paataak (36.3. BLACKFOOT)
paataak (62.24. BLACKFOOT)
paatala (20.1. SANSKRIT)
paati kanda (8.3. TELUGU)
paavakka chedi (38.23. MALAYALAM)
paaval (38.23. MALAYALAM)
paayap (40.42. FILIPINO)
paayap (40.42. ILOCANO)
paayap (40.42. PANGASINAN)
paayap (40.42. TAGALOG)
paazaa (17.7. YOM)
pacala (11.1. FRENCH)
pacar gandula (28.1. MALAY)
pâc-hăp-fā̃ (13.3. CHINESE, HAKKA)
pachay Baguio (30.40. TAGALOG)
pachcha (44.5. MALAYALAM)
pachha (44.5. MALAYALAM)
pachha pesalu (40.38. TELUGU)
pachhabaale (16.1a. KANNADA)
pachharati (16.1a. TELUGU)
pachilandai (16.1a. TAMIL)
pachnotka zwyczajna (44.7. POLISH)
paciencia dos jardins (58.3. PORTUGUESE)
paciorecznik indyjski (9.1. POLISH)
pacoba (16.1a. PORTUGUESE)
pacova (16.1a. PORTUGUESE)
pacpa yuyo (27.22. QUECHUA, CUSCO)
pacunga (27.8. QUECHUA, CUSCO)
padaaval (38.29. MARATHI)
padagoji (20.1. BURMESE)
padam (48.1a. HINDI)
padam (48.1a. ORIYA)
padarozhnik (57.1. RUSSIAN)
padavala kayi (38.29. KANNADA)
padavalaballi (38.29. KANNADA)
padavalakayi (38.29. KANNADA)
padavalam (38.29. MALAYALAM)
padavalanga (38.29. MALAYALAM)

patátas (62.24. LADINO)
patatas (62.24. PAMPANGAN)
patátas (62.24. PANGASINAN)
patatas (62.24. TAGALOG)
patatas (62.24. TAUSUG)
patatas (62.24. WÁRAY-WÁRAY)
patatas dulse (36.3. LADINO)
patate (36.3. BALANTA-KENTOHE)
patate (62.24. ALBANIAN, TOSK)
patate acquatica (36.2. ITALIAN)
patate aquatique (36.2. FRENCH)
patate cochon (40.21. FRENCH)
patate douce (36.3. FRENCH)
patates (62.24. TURKISH)
patato (36.3. BAMANANKAN)
patatô (36.3. MANDING-MANDINKA)
pat-bahar (39.3. ASSAMESE)
patcha-payaru (40.38. TAMIL)
patchapessalu (40.38. TELUGU)
patchoi (30.40. KWÉYÒL)
patchouli (30.36. KWÉYÒL)
pãte (40.3. PULAAR)
pate ghiraunla (38.21. NEPALI)
pate simi (40.13. NEPALI)
pate simi (40.29. NEPALI)
patenais (24.18. FRENCH)
pateta (62.24. NIUE)
pateta (62.24. SAMOAN)
patgobhy (30.27a. HINDI)
patha (38.1. URDU)
pathardi (27.45. GUJARATI)
pãthha (38.1. PANJABI, EASTERN)
pathoja (48.1a. SANSKRIT)
pathola (38.29. SINHALA)
pathola (38.30. SINHALA)
pathyashaka (23.3. SANSKRIT)
pati talabi khoodro poda (18.1. URDU)
patiar sa ragat (45.4. MARANAO)
patience (58.3. FRENCH)
patience dock (58.3. ENGLISH)
patientie (58.3. DUTCH)
patiitasi (62.24. INUKTITUT, EASTERN CANADIAN)
patikan (24.16. INDONESIAN)
patikan cina (24.15. INDONESIAN)
patikim (24.15. JAVANESE)
patilla de golero (38.6. SPANISH)
patison (38.16. CROATIAN)
patisson (38.18. MORISYEN)
patisson (38.18. VLAAMS)
patissoni (38.18. GEORGIAN)
patiujaq (24.11. INUKTITUT, EASTERN CANADIAN)
patkulī (30.27a. MAITHILI)
patla (38.29. KANNADA)
patla kaayi (38.29. TULU)
patladžan (62.19. BULGARIAN)
pătlăgea rosie (62.9. ROMANIAN)
pătlăgea vinătă (62.19. ROMANIAN)
pătlăgele roşii (62.9. ROMANIAN)
pătlăgele vinete (62.19. ROMANIAN)
patlagină (57.1. ROMANIAN)
paṭlakāyi (38.29. TULU)
patlazhùy (62.19. ROMANI, BALKAN)
patlican (62.19. TURKISH)
patlidžan (62.9. SERBO-CROATIAN)
patlidzan (62.19. CROATIAN)
patlidžan (62.19. MACEDONIAN)
patlidžan (62.19. SERBO-CROATIAN)
patlixhan (62.19. ALBANIAN, TOSK)
patma (48.1a. INDONESIAN)
patmage (44.5. KRIO)
patmagi (44.5. MANDING-MANDINKA)
patmajin (44.4. KRIO)
patmɛnji (44.4. KRIO)
patmɛnji (44.5. KRIO)
patnga (38.14. HOPI)
patnga (38.16. HOPI)
patnga (38.17. HOPI)
patnga (38.18. HOPI)
patol (38.29. MORISYEN)
patol (38.31. ENGLISH)
patola (38.21. BICOLANO, CENTRAL)
patola (38.21. CEBUANO)
patola (38.21. TAGALOG)
patola (38.21. WÁRAY-WÁRAY)
patola (38.22. BICOLANO, CENTRAL)
patola (38.22. CEBUANO)
patola (38.22. TAGALOG)
patola (38.22. TAGBANWA)
patola (38.29. PORTUGUESE)
patola (38.30. FRENCH)
patola (38.31. PORTUGUESE)
patola (38.31. SANSKRIT)
patola-ihalas (38.22. WÁRAY-WÁRAY)
patolam (38.31. MALAYALAM)
patolang kastila (38.22. FILIPINO)
patolang Tagalog (38.21. FILIPINO)
patolang-balimbing (38.21. TAGALOG)
patolang-bilog (38.22. TAGALOG)

patolang-ligaw (38.22. TAGALOG)
patolang-uwak (38.22. TAGALOG)
patole (38.31. FRENCH)
patole (38.31. ITALIAN)
patole (38.31. SPANISH)
patoli (38.31. HINDI)
paṭolo (38.31. PALI)
patomelooni (38.7a. HOPI)
patóng (17.2. WÁRAY-WÁRAY)
pâtowi (40.3. PULAAR)
pâtoy (40.3. PULAAR)
patraseli (24.19. JAVANESE)
pătronjelul (24.19. ROMANIAN)
pătruniel cultivat (24.21. ROMANIAN)
pătrunjei cu frunză creată (24.20. ROMANIAN)
pătrunjel (24.19. ROMANIAN)
pătrunjel cu frunză creată (24.5. ROMANIAN)
pătrunjel cutivat (24.4. ROMANIAN)
pătrunjel de grădină (24.19. ROMANIAN)
patrysuintjie (10.1. AFRIKAANS)
patsun (63.1. URDU)
patta (63.1. SANSKRIT)
patta pattak (40.28. INUKTITUT, EASTERN CANADIAN)
pattani (40.28. MALAYALAM)
pattani (40.28. TAMIL)
pattashaka (36.2. SANSKRIT)
patu (38.23. SANSKRIT)
patuasag (36.2. HINDI)
patuasaga (36.2. HINDI)
patula (38.21. HANUNOO)
patula-amu (38.22. TAUSUG)
patula-ba-imbing (38.21. TAUSUG)
patung (17.1. TAGALOG)
patwa (45.4. BENGALI)
patwa (45.4. HINDI)
patwa (45.4. MARATHI)
pau fede (27.24. PORTUGUESE)
pau maia (40.40. NEPALI)
paui (38.26. VLAAMS)
paukena (38.14. MAORI)
paukena (38.16. MAORI)
paukena (38.17. MAORI)
paukena (38.18. MAORI)
paumak (62.9. SUMO-MAYANGNA)
pau-sim-pėh-chhài (30.40. CHINESE, MIN NAN)
pavakka (38.23. MALAYALAM)
pavakka (38.23. TAMIL)
pavakka chedi (38.23. TAMIL)
pavakkaayi (38.23. TAMIL)
pavakkachedi (38.23. TAMIL)
pavakkacheti (38.23. MALAYALAM)
pavakkayi (38.23. TAMIL)
paval (38.23. MALAYALAM)
paval kaipavalli (38.23. MALAYALAM)
pavila (59.2. TELUGU)
pavta (40.9. MARATHI)
paw le$_V$ shi$_-$ (62.9. LAHU)
pawa (17.1. CEBUANO)
pawmuhmp′ (30.52. PAIUTE, SOUTHERN)
paxla (40.26. AZERBAIJANI, NORTHERN)
paxlal ı bitki (40.26. AZERBAIJANI, NORTHERN)
payaru (40.37. TAMIL)
payasvinee (27.45. SANSKRIT)
payaw-payaw (18.1. CEBUANO)
payaz (7.3. PANJABI, EASTERN)
payda (40.9. KANKANAEY)
payong (38.20. BONTOC, CENTRAL)
pazdaba (38.1. ORIYA)
pazhitnik (40.32. RUSSIAN)
pazi (35.3. ALBANIAN, TOSK)
pazi (35.3. TURKISH)
pazienza (58.3. ITALIAN)
pažíka (7.11. SLOVENIAN)
pažitka (7.11. SLOVAK)
pažitka pobřežní (7.11. CZECH)
pažitka zahradni (7.11. CZECH)
pazitnik (40.32. BELARUSAN)
pcctsee (35.9. TIBETAN)
pchpech pačokavor (62.1. ARMENIAN)
pɔ̃kin (38.14. KRIO)
pɔm-*ta*-gbɛŋguɛ (59.2. BULLOM SO)
pɔndɔ-koko (38.27. KONO)
pɔndo-koko (38.28. KONO)
pɔnkin (38.14. KRIO)
pɔnkin (38.16. KRIO)
pɔnkin (38.17. KRIO)
pɔnkin (38.18. KRIO)
pé de cavalo (24.7. PORTUGUESE)
pe' kuĕk (40.21. KHMER, CENTRAL)
pea (40.28. ENGLISH)
pea eggplant (62.23. ENGLISH)
pea nut (40.1. ENGLISH)
pea sair (40.12. GAELIC, SCOTTISH)
peaj (7.3. FARSI, EASTERN)
peaj (7.3. SANTALI)
peāju (7.3. MAITHILI)

pei lan (30.30. CHINESE, XIANG)
peikonkello (33.3. FINNISH)
peila (62.18. MENDE)
peili (62.18. GOLA)
peili (62.18. MENDE)
peiŋga (44.5. KONO)
pein-mohawaya (8.1. BURMESE)
pein-u (8.1. BURMESE)
peirsil (24.19. GAELIC, IRISH)
Péiterséileg (24.19. LUXENBOURGEOIS)
peiulill (27.8. ESTONIAN)
pekaja (48.1a. INDONESIAN)
pekankai (38.21. TAMIL)
pek-hȧp (13.4. CHINESE, MIN NAN)
pek-hȧp-hoe (13.4. CHINESE, MIN NAN)
pekin carami (30.40. UZBEK)
pekin kastusy (30.36. KIRGHIZ)
pekin kastusy (30.40. KIRGHIZ)
Peking cabbage (30.40. ENGLISH)
Pekingkohl (30.40. GERMAN)
pekino bastutis (30.40. LITHUANIAN)
pekinška zelka (30.40. MACEDONIAN)
pela yam (11.3. URDU)
pelampong (27.12. INDONESIAN)
pelampong (27.12. MALAY)
pele (45.2. NIUE)
pele (45.2. SAMOAN)
peleng (35.9. INDONESIAN)
peletrūnas (27.6. LITHUANIAN)
peli (62.18. GOLA)
peli (62.18. LOKO)
pelin (27.6. ROMANIAN)
pe-lin-mway (38.29. BURMESE)
pelo (62.18. KISSI, NORTHERN)
peltokorte (1.1. FINNISH)
peltonätkemä (40.10. FINNISH)
pelyněk (27.6. CZECH)
pēmbere (36.3. FULFULDE, ADAMAWA)
pēmberēji (36.3. FULFULDE, ADAMAWA)
pemento (62.1. GALICIAN)
penantazi (40.32. BURMESE)
penda (45.1. TELUGU)
pendalam (11.1. TELUGU)
pendalam (11.2. TELUGU)
pendalamu (11.1. TELUGU)
pendli (38.4. KONKANI)
pendligummadikaaya (38.1. TELUGU)
peng e shu (20.4. CHINESE, YUE)
pengapi (40.8. BURMESE)
penggaga (24.7. INDONESIAN)
peni sag (30.52. HINDI)
penikeel (17.7. ESTONIAN)
penjelang (24.9. JAVANESE)
penjelangan (24.15. JAVANESE)
penjilang (24.9. MALAY)
penniseto rosso (17.3. ITALIAN)
penniseton the porphyroun (17.3. GREEK)
pennyroyal mint (44.2. ENGLISH)
pensamiento silvestre (68.1. SPANISH)
pensaspaprika (62.5. FINNISH)
pensée sauvage (68.1. FRENCH)
pensieri (68.1. MALTESE)
pentiveduru (17.1. TELUGU)
pentru frunze (24.4. ROMANIAN)
pentru rădăcină (24.21. ROMANIAN)
pentrubulb (7.3. ROMANIAN)
peòk (17.7. NICOBARESE, CAR)
pepa (62.1. NIUE)
pepa (62.2. SAMOAN)
pepanan (63.1. GOURMANCHÉMA)
pepare (38.23. INDONESIAN)
pepasan (38.4. MALAY)
pɛpɛ (62.5. KRIO)
pepe di Cayenna (62.2. ITALIAN)
pepelesta loboda (35.5. MACEDONIAN)
pepene (38.7a. ROMANIAN)
pepene (38.7b. ROMANIAN)
pepene (38.8. ROMANIAN)
pepene (38.9. ROMANIAN)
pepene (38.10. ROMANIAN)
pepene (38.11. ROMANIAN)
pepene galben (38.9. ROMANIAN)
pepene verde (38.3. ROMANIAN)
pepenele verde (38.3. ROMANIAN)
peper (62.1. VLAAMS)
peper van Guinée (62.14a. DUTCH)
pepere e tala (62.1. SOTHO, SOUTHERN)
peperetan (10.2. JAVANESE)
peperkers (30.51. AFRIKAANS)
peperoncino (62.1. ITALIAN)
peperoncino piccante (62.2. ITALIAN)
peperone (62.1. ITALIAN)
peperone cinese (62.4. ITALIAN)
peperone dolce (62.1. ITALIAN)
peperone ornamentale (62.5. ITALIAN)
peperone rabbioso (62.2. ITALIAN)
peperwortel (30.1. AFRIKAANS)
pepiino (62.20. JAPANESE)
pepinela (38.25. PORTUGUESE)
pepiniko (38.13b. LADINO)

perila (44.7. ROMANIAN)
perilë (44.7. ALBANIAN, TOSK)
perilha (44.7. PORTUGUESE)
perilla (44.7. AFRIKAANS)
perilla (44.7. DANISH)
perilla (44.7. ENGLISH)
perilla (44.7. GREEK)
perilla (44.7. ITALIAN)
perilla (44.7. NORWEGIAN)
perilla (44.7. SPANISH)
perilla (44.7. SWEDISH)
perilla (44.7. TURKISH)
perilla mint (44.7. ENGLISH)
perille (44.7. DUTCH)
perille (44.7. FRENCH)
Perille (44.7. GERMAN)
perille de Nankin (44.7. FRENCH)
periok (38.23. MALAY)
peri-ouo (62.18. KISI, SOUTHERN)
perishíl (24.19. LADINO)
peri-uo (62.18. KISSI, NORTHERN)
periya' (38.23. BINUKID)
periya (38.23. MANOBO)
periya mungili (17.1. TAMIL)
periya venkayam (7.3. TAMIL)
periyakarumeikkilhaugu (19.1. TAMIL)
perkoles (40.25. ILOCANO)
perleblomst (13.5. NORWEGIAN)
perlehyacinth (13.5. DANISH)
perleleg (7.1. DANISH)
perleløk (7.1. NORWEGIAN)
Perllauch (7.1. GERMAN)
perllys (60.1. WELSH)
perllys y perthi (24.8. WELSH)
permen (44.3. INDONESIAN)
pero pungente (32.1. ITALIAN)
perperek a amo (27.8. MARANAO)
persi (24.19. MORISYEN)
persi frize (24.20. MORISYEN)
Persian melon (38.7b. ENGLISH)
persil (24.19. FRENCH)
persil (24.19. LADINO)
persil commum (24.19. FRENCH)
persil frisé (24.19. FRENCH)
persil italienne (24.20. FRENCH)
persil racine (24.21. FRENCH)
persil séri (24.17. FRENCH)
persil tubéreux (24.21. FRENCH)
persila (24.19. KOONGO)
persila (24.19. NGUNDI)
persilja (24.19. FINNISH)
persilja (24.19. SWEDISH)
persille (24.19. DANISH)
persille (24.19. NORWEGIAN)
persille (24.20. NORWEGIAN)
peršin (24.19. BOSNIAN)
peršin (24.19. CROATIAN)
persiska dinja (38.7b. MACEDONIAN)
perslee (24.19. GAELIC, MANX)
persli (24.19. WELSH)
persli eidal (24.20. WELSH)
persli gwrraidd naip (24.21. WELSH)
peršun (24.19. SERBO-CROATIAN)
Peru çeltiği (35.6. TURKISH)
peru varaimungil (17.1. TAMIL)
Peruaanse lampionplant (62.12. DUTCH)
Peruaanse spinazie (35.6. VLAAMS)
Peruanische Blasenkirsche (62.12. GERMAN)
Peruanischer Reisspinat (35.6. GERMAN)
peruanskij fizalis (62.12. RUSSIAN)
perugu thotakoora (23.3. TELUGU)
perum jeerakam (24.13. MALAYALAM)
perum ponnaku poondu (63.1. TAMIL)
perumelde (35.6. NORWEGIAN)
perumvalli kizhangu (11.1. TAMIL)
peruna (62.24. FINNISH)
perungeeragam (24.13. TAMIL)
perungunni (62.12. TAMIL)
perunpalsternakka (24.6. FINNISH)
peruskjellfrø (27.22. NORWEGIAN)
perutsiat makpatât (30.27a. ALEUT)
peruu füüsal (62.12. ESTONIAN)
peruvalli (11.1. TAMIL)
peruvallikkilannu (11.1. MALAYALAM)
Peruvian carrot (24.6. ENGLISH)
Peruvian groundcherry (62.12. ENGLISH)
Peruvian parsnip (24.6. ENGLISH)
peruvianski fizalis (62.12. BULGARIAN)
peryanskaja vishnja (62.12. RUSSIAN)
peryatpym (40.8. BURMESE)
pèsali (40.10. BURMESE)
pesalu (40.37. TELUGU)
pesalu pessara (40.37. TELUGU)
pe-saung-ya (40.29. BURMESE)
pèsi (24.19. CREOLE, HAITIAN FRENCH)
pesi (24.19. KWÉYÒL)
peslan (44.5. TURKISH)
pèsol (40.28. CATALAN-VALENCIAN-BALEAR)
pésol (40.28. SPANISH)

pessara (40.38. HINDI)
pestarrot (27.35. ICELANDIC)
pesusienikurkku (38.22. FINNISH)
petaside japonês (27.35. PORTUGUESE)
pétasitès japonais (27.35. FRENCH)
petasitis o iaponikos (27.35. GREEK)
petatas (62.24. BINUKID)
petate (62.24. KURDISH, NORTHERN)
pe-taung-shay (40.26. BURMESE)
pete (35.4. SOTHO, NORTHERN)
pete (35.4. SOTHO, SOUTHERN)
pete (35.4. TSWANA)
pétéoré (27.8. GURO)
peterseli (24.19. INDONESIAN)
peterséli (24.20. INDONESIAN)
peterselie (24.19. DUTCH)
peterselie (24.19. VLAAMS)
peterseliekool (30.35. VLAAMS)
petersell (24.19. ESTONIAN)
pētersīļi (24.19. LATVIAN)
Petersilie (24.19. GERMAN)
pētersīlis (24.19. LATVIAN)
pɛtɛtɛ (36.3. KRIO)
petha (38.1. HINDI)
petha (38.1. PANJABI, EASTERN)
petha kaddoo (38.1. PANJABI, EASTERN)
petit pois (40.28. FRENCH)
petit radis (30.56. FRENCH)
petit riz du Pérou (35.6. FRENCH)
petite bringelle (62.18. FRENCH)
petite fève (40.34b. FRENCH)
petola (38.21. INDONESIAN)
petola balat (38.22. MALAY)
petola buntal (38.22. MALAY)
petola manis (38.22. MALAY)
petola segi (38.21. MALAY)
petola ular (38.29. MALAY)
petonciano (62.19. ITALIAN)
pěťour maloúborný (27.22. CZECH)
petražolė (24.19. LITHUANIAN)
petrezselyem (24.19. HUNGARIAN)
petrishkeh (24.19. YIDDISH)
petroselinon (24.19. GREEK)
petrosilia italkit (24.20. HEBREW)
petroslinon (24.21. HEBREW)
petroslinon hagina (24.19. HEBREW)
petroslinon ha-ginna (24.19. HEBREW)
petršilj (24.19. SLOVENIAN)
pětršilka (24.19. SORBIAN, UPPER)
petrushca (24.19. UZBEK)
petrushka (24.19. KAZAKH)
petrushka (24.19. KIRGHIZ)
petrušhka (24.19. RUSSIAN)
petrushka (24.19. TATAR)
petrushka (24.19. UZBEK)
petrushka agarodnaja (24.20. RUSSIAN)
petrushka karanjaplodnaja (24.21. BELARUSAN)
petrushka karnevaja (24.21. RUSSIAN)
petrushka kucharavaja (24.19. BELARUSAN)
petrushka kudrjavaja (24.5. RUSSIAN)
petrushka kudrjavaja (24.20. RUSSIAN)
petrushka listovaja (24.4. RUSSIAN)
petrushka pasevnaja (24.19. RUSSIAN)
petrushka sadovaja (24.19. RUSSIAN)
petruşka (24.19. TURKMEN)
petruška (24.19. UKRAINIAN)
petruška (24.19. UZBEK)
petržel (24.19. CZECH)
petrzel kadeřava (24.5. CZECH)
petržel kadeřava (24.20. CZECH)
petržel kořenova (24.21. CZECH)
petržel natova (24.4. CZECH)
petržel zahradni (24.19. CZECH)
petržel zahradní kadeřavá (24.19. CZECH)
petržel zahradní kořenová (24.21. CZECH)
petržlen (24.19. SLOVAK)
petržlen záhradný (24.19. SLOVAK)
petsai (30.36. INDONESIAN)
pe-tsai (30.40. AFRIKAANS)
pe-tsai (30.40. CHINESE, MANDARIN)
petsai (30.40. DUTCH)
pe-tsai (30.40. ENGLISH)
pe-tsai (30.40. FILIPINO)
pé-tsai (30.40. FRENCH)
petsai (30.40. INDONESIAN)
petsai (30.40. MALAY)
petsai (30.40. PORTUGUESE)
pe-tsai (30.40. VLAAMS)
petsay (30.40. CEBUANO)
petsay (30.40. ILOCANO)
petsay (30.40. MORISYEN)
petsay (30.40. TAGALOG)
petty morel (62.15. ENGLISH)
peul (40.27. DUTCH)
peutètè (36.3. BUDIK)
pevithi (40.5. GREEK)
pey (40.28. SCOTS)
peyavarai (40.3. TAMIL)
peyazcheh (7.11. FARSI, WESTERN)

pé-yin (40.40. BURMESE)
peyppeechukku (38.21. TAMIL)
peyppeerakam (38.21. TAMIL)
peyupadal (38.31. TAMIL)
pe-zaung-ya (40.29. BURMESE)
pezeda o evosmos (60.1. GREEK)
pezenta (60.1. GREEK)
pézsmatök (38.17. HUNGARIAN)
Pfeffer-Gurke (38.6. GERMAN)
Pfeilwurz (15.2. GERMAN)
Pferdebohne (40.34a. GERMAN)
Pferdebohne Sommerwurz (52.1. GERMAN)
Pferdekorn (40.16. GERMAN)
Pferdesaat (24.17. GERMAN)
pfuh$^{\vee}$ chaw (17.5. LAHU)
pfuh ti$_{\vee}$ mvuh$_{\wedge}$ (59.2. LAHU)
pfunde (36.2. SENA)
pgnenocrven grav (40.24. MACEDONIAN)
pha (7.5. KOREAN)
phai (17.1. THAI)
phai bong yai (17.2. THAI)
phai lueang (17.1. THAI)
phai paa (17.1. THAI)
phai-luang (17.1. THAI)
phai-luong (17.1. THAI)
phai-nam (17.1. THAI)
phai-nual-yai (17.2. THAI)
phai-nuan (17.2. THAI, NORTHERN)
phai-ngachang (17.1. THAI)
phai-pa (17.1. THAI)
phai-pah (17.1. THAI)
phai-pao (17.2. THAI)
phai-sangdoi (17.2. THAI, NORTHERN)
phai-sangnuan (17.2. THAI, NORTHERN)
phai-sangyen (17.2. THAI, NORTHERN)
phai-sisuk (17.1. THAI)
phai-tong (17.2. THAI)
phaix ba:nz (17.1. LAO)
phaix pa:x (17.1. LAO)
phai-zangkum (17.2. THAI, NORTHERN)
phak ho:m pa:nx (24.9. LAO)
phak ho:m po:mz (24.9. LAO)
phak 'i:x tu:x (44.5. LAO)
('phak) kas'ééd (40.19. LAO)
phak khaakhiat (4.1. THAI)
phak kuut (5.1. THAI)
phak kuut kao kwuang (4.1. THAI)
phak kuutnam (4.1. THAI)
phak s 'i (24.13. LAO)
phak sian (34.2. THAI)
phak sian khao (34.2. THAI)
phak sôb (6.1. LAO)
phak som sian (34.2. THAI)
phak waan ban (39.5. THAI)
phakan (7.10. NEWAR)
phakan (24.17. THAI)
phakbang (27.18. THAI)
phakbia-dokluang (59.2. THAI)
phakbia-yai (59.2. THAI)
phakbung (36.2. THAI)
phakbung-pling (27.19. THAI)
phakbung-ruam (27.19. THAI)
phakchi (24.9. THAI)
phakchi-duanha (24.13. THAI, NORTHERN)
phakchi-farang (24.19. THAI)
phakchi-lom (24.17. THAI)
phakchit (40.19. THAI)
phakdaeng (27.18. THAI)
phake (40.12. GREEK)
phakha (38.23. THAI)
phakhaan (27.14. THAI)
phakhin (18.2. THAI)
phakhom (24.9. THAI, NORTHERN)
phakhom-noi (24.9. THAI)
phakkat-daeng (35.4. THAI)
phakkat-farang (35.4. THAI)
phakkat-foi (27.9. THAI)
phakkat-hangkai (27.40. THAI)
phakkat-hom (27.31. THAI)
phakkat-hom (27.32. THAI)
phakkat-hom (27.40. THAI)
phakkat-hua (30.54. THAI)
phakkat-hua (30.56. THAI)
phakkat-kaeo (27.31. THAI)
phakkat-kaeo (27.32. THAI)
phakkat-khaokwangtung (30.36. THAI)
phakkat-khaopli (30.40. THAI)
phakkatkhaoyai (30.40. THAI)
phakkatkheokwangtung (30.37. THAI)
phakkat-khieo (30.17. THAI)
phakkat-khieopli (30.17. THAI)
phakkengkheng (45.4. THAI)
phakkhaep (38.4. THAI)
phakkhana (30.25. THAI)
phakkhangkai (6.1. THAI)
phakkhaopun (24.4. THAI)
phakkhaothong (61.1. THAI)
phakkhikhwai (27.23. THAI)
phakkhithut (30.53. THAI)
phakkhithut (30.54. THAI)

phakkhithut (30.56. THAI)
phakkhom-suan (23.3. THAI)
phakkhrat (27.3. THAI)
phakkhrat (27.4. THAI)
phakkrachet (40.19. THAI)
phakkwangtung (30.37. THAI)
phakpaeng (27.19. THAI)
phakpang (28.1. THAI)
phakpet-nam (23.1. THAI)
phakphet (27.3. THAI)
phakphet (27.4. THAI)
phakphet chaang (27.14. THAI)
phakplang (28.1. THAI)
phakplang-yai (28.1. THAI)
phakpong (18.1. THAI)
phakpum (24.4. THAI)
phaksalat (27.31. THAI)
phaksalat (27.32. THAI)
phaktamlung (38.4. THAI)
phaktang-o (27.23. THAI)
phakthotyot (36.2. THAI)
phaktop (18.1. THAI)
phaktop-thai (18.1. THAI)
phalēj (38.7b. MAITHILI)
phalin (38.20. SANSKRIT)
phanasi (40.26. GUJARATI)
phang-koe (38.7. CHINESE, MIN NAN)
phara (38.13a. SOTHO, NORTHERN)
pha-rai (38.3. BURMESE)
pharasbean (40.26. HINDI)
pharash bin (40.26. BENGALI)
pharashbi (8.3. BENGALI)
phare (38.13a. TSWANA)
pharn (2.1. TAJIKI)
pharsi (38.14. NEPALI)
pharsi (38.16. NEPALI)
pharsi (38.18. NEPALI)
phaséole (40.26. FRENCH)
phasìli (24.19. VENDA)
phatilajon (62.9. TAJIKI)
phat-si (38.14. NEWAR)
phat-si (38.16. NEWAR)
phat-si (38.17. NEWAR)
phat-si (38.18. NEWAR)
phav npoob (36.2. HMONG-DAW)
pha-ye-thi (38.3. BURMESE)
pherefere (62.1. TSWANA)
pherefere yetala (62.1. SOTHO, NORTHERN)
pherick (16.1a. GAELIC, MANX)
phhari pudina (44.3. MARATHI)
phinyu (63.1. LOMWE, MALAWI)
phkaa spéi (30.26. KHMER, CENTRAL)
phlu-kae (61.1. THAI)
phombo (63.1. DIGO)
phool ghobi (30.25. URDU)
phool gobhi (30.26. HINDI)
phool gobhi (30.26. PANJABI, EASTERN)
phool gobhy (30.26. HINDI)
phool kobi (30.26. ASSAMESE)
phopti (62.12. MARATHI)
photlovani salata (27.32. GEORGIAN)
photlovani tsharkhali (35.3. GEORGIAN)
phov ponm (30.27a. KWÉYÒL)
phrik khee nuu (62.5. THAI)
phtii (23.3. KHMER, CENTRAL)
phüak (8.4. LAO)
phuak (8.4. THAI)
phul gobhi (30.26. HINDI)
phul gobhi (30.26. HINDUSTANI)
phul gobhi (30.26. PANJABI, EASTERN)
phul gòbi (30.26. BALTI)
phul gobi (30.26. FIJIAN)
phul gobi (30.26. HINDI)
phul kobi (30.26. ASSAMESE)
phul kobi (30.26. MARATHI)
phulkōbī (30.26. MAITHILI)
phumde (36.2. CHUWABU)
phunjipat (63.1. BENGALI)
phút (38.7b. HINDUSTANI)
phuṭ (38.11. SANTALI)
phut (38.11. SANTALI)
phutika (38.1. HINDI)
phutium (27.8. GUJARATI)
phuttharaksa (9.1. THAI)
phutthason (9.1. THAI)
phūūl qɔ̄pi (30.26. TIBETAN)
phwetekele (62.9. LOMWE, MALAWI)
phwetekele (62.10. LOMWE, MALAWI)
phylla revas (30.41c. GREEK)
Phyllostochis (17.4. GREEK)
Physalie (62.12. GERMAN)
phytolakka (56.1. GREEK)
pi (40.28. SAMOAN)
pi jiu hua (46.1. CHINESE, MANDARIN)
pi jiu hua (46.1. CHINESE, WÚ)
pi lu suan chiang (62.12. CHINESE, MANDARIN)
pi tsi (10.3. FRENCH)
pia (8.1. TOK PISIN)
pia (19.1. HAWAIIAN)

pia (19.1. MANU'A SAMOAN)
pia (19.1. MARQUESAN)
pia (19.1. NIUE)
pia (19.1. RAROTONGAN)
pia (19.1. SAMOAN)
piaha (44.10. DAGBANI)
piaj (7.3. BENGALI)
piaj (7.3. HINDI)
piąj (7.3. SANTALI)
piantaggine (16.1a. ITALIAN)
piasa (44.10. MÒORÉ)
piäz (7.2. PANJABI, EASTERN)
piaz (7.2. URDU)
piaz (7.3. FARSI, WESTERN)
piaz (7.3. HINDI)
piaz (7.3. PANJABI, EASTERN)
piaz (7.3. URDU)
pibbyr caiaan (62.2. GAELIC, MANX)
pibe-løg (7.5. DANISH)
picao branco (27.22. PORTUGUESE)
picão bravo (27.22. PORTUGUESE)
picapica (40.18. SPANISH)
piccinna (38.21. MALAYALAM)
pickerel weed (18.2. ENGLISH)
picki (8.4. SANTALI)
pickling cucumber (38.13b. ENGLISH)
pichay (30.37. IVATAN)
pichenga (38.21. MALAYALAM)
pichha (16.1a. MALAYALAM)
pichhila (28.1. SANSKRIT)
pichhila (36.3. SANSKRIT)
pichhila (45.1. SANSKRIT)
pichhila chhada (28.1. SANSKRIT)
pichinna (38.21. MALAYALAM)
pichinna (38.22. MALAYALAM)
pichkati (62.21. ASSAMESE)
pichukku (38.22. TAMIL)
pidagadali (16.1a. TAMIL)
pie lan (30.30. CHINESE, WÚ)
piēcē (8.3. MÒORÉ)
pièn t'éou (40.12. CHINESE, HAKKA)
pienda (40.1. HINDUSTANI)
pienene (27.45. LATVIAN)
pieplant (58.1. ENGLISH)
pieprz czerwony (62.6. POLISH)
pieprz turecki (62.3. POLISH)
pieprzyca (30.50. POLISH)
pieprzyca siewna (30.51. POLISH)
pièrè (8.4. DAGAARE, SOUTHERN)
piesang (16.1a. AFRIKAANS)
piesangblom (16.1b. AFRIKAANS)
pietersielie (24.19. AFRIKAANS)
pietruszka korzeniowa (24.21. POLISH)
pietruszka lisciowa (24.4. POLISH)
pietruszka liściowa (24.19. POLISH)
pietruszka lisciowa fryzowana (24.5. POLISH)
pietruszka lisciowa fryzowana (24.20. POLISH)
pietruszka włoska (24.20. POLISH)
pievinė kartenė (30.44. LITHUANIAN)
pievinis pūtelis (27.47. LITHUANIAN)
piez (7.3. UZBEK)
piëz (7.10. UZBEK)
pigeon bean (40.34b. ENGLISH)
pigeon pea (40.2. ENGLISH)
pigeon pea (40.2. FILIPINO)
pigoglys (22.2. WELSH)
pigweed (23.3. ENGLISH)
piha (44.10. DAGBANI)
pihala (59.2. HOPI)
pihúraa (40.26. COMANCHE)
pii (62.18. KONO)
piikkiaralia (26.3. FINNISH)
piiman (62.1. JAPANESE)
piimjuur (27.46. ESTONIAN)
pijaz tup (7.3. KIRGHIZ)
pijjipesara (40.35. TELUGU)
pijkruid (6.2. DUTCH)
pijlkruid (6.2. VLAAMS)
pijlwortel (15.2. DUTCH)
pikakku (23.4. IFUGAO)
pikampob (38.18. GOURMANCHÉMA)
pikampolo (38.29. GOURMANCHÉMA)
pikampolo (38.30. GOURMANCHÉMA)
pikaw (15.4. BONTOC, CENTRAL)
pikingské zelí (30.40. CZECH)
pikku (38.22. TAMIL)
pikkujamssi (11.5. FINNISH)
pikpo kremmydi (7.5. GREEK)
pikri (11.3. GREEK)
pikunkaai (38.21. TAMIL)
pikunkai (38.21. TAMIL)
pila kohda (38.14. HINDI)
pila kohda (38.16. HINDI)
pila kohda (38.18. HINDI)
pilanda (17.1. KISSI, NORTHERN)
pilblad (6.2. NORWEGIAN)
pile murgha (23.4. URDU)
pile murghka (23.4. HINDI)
pilea (66.1. ASSAMESE)

pilea (66.1. DANISH)
pilea (66.1. DUTCH)
pilea (66.1. ENGLISH)
pilea (66.1. FRENCH)
Pilea (66.1. GERMAN)
pilea (66.1. ITALIAN)
pílea (66.1. PORTUGUESE)
pilea (66.1. SPANISH)
pilea (66.2. ASSAMESE)
pilea (66.2. DANISH)
pilea (66.2. DUTCH)
pilea (66.2. ENGLISH)
pilea (66.2. FRENCH)
Pilea (66.2. GERMAN)
pilea (66.2. GREEK)
pílea (66.2. PORTUGUESE)
pilea (66.2. SPANISH)
piliadagu kaila (40.18. TELUGU)
pilipili (62.2. SWAHILI)
pilipili (62.5. SWAHILI)
pilipili hoho (62.2. SWAHILI)
pilipili kali (62.2. SWAHILI)
pilipili kichaa (62.1. SWAHILI)
pilipili kichaa (62.5. SWAHILI)
pilipli mbuzi (62.4. SWAHILI)
pilipli tamu (62.1. SWAHILI)
pillert (40.28. NORWEGIAN)
pilli (62.5. FRENCH)
pilliadugu (40.18. TELUGU)
pillisipuli (7.5. FINNISH)
pilliyadagu (40.18. TELUGU)
pilliyuyu (62.15. QUECHUA, CUSCO)
pilpel (62.1. HEBREW)
pilpel cayenne (62.2. HEBREW)
pilpel ḥarif (62.2. HEBREW)
pilpel katan (62.3. HEBREW)
pilpel tobasco (62.5. HEBREW)
pilpelet (62.2. HEBREW)
pilpelet (62.5. HEBREW)
pilpelet ha-ginna (62.1. HEBREW)
pilrotmaranta (15.2. NORWEGIAN)
piludi (62.21. GUJARATI)
pilum kohalum (38.14. GUJARATI)
pilum kohalum (38.16. GUJARATI)
pilum kohalum (38.18. GUJARATI)
piluparni (38.4. SANSKRIT)
pima kari (62.4. MORISYEN)
pima salad (62.1. MORISYEN)
piman bouk (62.2. CREOLE, HAITIAN FRENCH)
piman dou (62.1. CREOLE, HAITIAN FRENCH)
piman dous (62.1. CREOLE, HAITIAN FRENCH)
piman kayenn (62.2. KWÉYÒL)
piman pike (62.2. CREOLE, HAITIAN FRENCH)
piman zwézo (62.2. CREOLE, HAITIAN FRENCH)
piman zwézo (62.3. CREOLE, HAITIAN FRENCH)
piman zwézo (62.5. CREOLE, HAITIAN FRENCH)
pimando (62.1. KWÉYÒL)
piman-sézòné (62.2. KWÉYÒL)
piment (62.1. ESTONIAN)
piment (62.5. FRENCH)
piment antillais (62.4. FRENCH)
piment de Cayenne (62.2. FRENCH)
piment d'Espagne (62.1. FRENCH)
piment d'oiseau (62.2. FRENCH)
piment d'oiseau (62.3. FRENCH)
piment d'oiseaux (62.5. FRENCH)
piment doux (62.1. CAJUN FRENCH)
piment doux (62.1. FRENCH)
piment rouge (62.3. FRENCH)
piment zouézeau (62.3. FRENCH)
pimenta das paredes (37.1. PORTUGUESE)
pimenta e pharmakestiki (62.1. GREEK)
pimenta malagueta (62.2. PORTUGUESE)
pimenta vermelha (62.2. PORTUGUESE)
pimentão (62.1. PORTUGUESE)
pimentão de Caiena (62.2. PORTUGUESE)
pimenteira (62.5. PORTUGUESE)
pimentinha (62.5. PORTUGUESE)
pimento (62.1. PORTUGUESE)
pimento cereja (62.3. PORTUGUESE)
pimento chinês (62.4. PORTUGUESE)
pimento comarim (62.3. PORTUGUESE)
pimento das Caraíbas (62.6. PORTUGUESE)
pimento de Caiena (62.2. PORTUGUESE)
pimento pepper (62.1. ENGLISH)
pimento piripiri (62.3. CRIOULO, CAPE VERDE)
pimenton (62.1. CRIOULO, CAPE VERDE)
pimiento (62.1. GALICIAN)
pimiento (62.1. SPANISH)
pimiento bravo (62.6. SPANISH)
pimiento caribe (62.6. SPANISH)
pimiento chinese (62.4. SPANISH)

pimiento dulce (62.1. SPANISH)
pimiento loco (62.3. SPANISH)
pimiento morron (62.1. SPANISH)
pimiento picante (62.2. SPANISH)
pimiento ruso espinoso (26.4. SPANISH)
pimiento salvaje (26.4. SPANISH)
pimiento silvestre (62.3. SPANISH)
pimienton (62.1. LADINO)
pimyènta (62.1. LADINO)
pimyenton (62.1. LADINO)
pin (62.24. POTAWATOMI)
pin daalu (11.1. MARATHI)
pina (35.9. MÒORÉ)
pina (59.3. BEMBÉ)
pinaatti (35.9. FINNISH)
pinaattihierakka (58.3. FINNISH)
pinaattikiinankaali (30.36. FINNISH)
pinaattikunankaali (30.40. FINNISH)
pinard brou (28.1. JULA)
pinati (40.1. NIUE)
pinati (40.1. SAMOAN)
pincel de amor (27.18. SPANISH)
pinchão (30.48. PORTUGUESE)
pinda (40.1. DUTCH)
pinda (40.1. SUMO-MAYANGNA)
pinda (40.1. VLAAMS)
pindaaloo (62.24. SANSKRIT)
pindaalu (11.1. SANSKRIT)
pindaalu (36.3. SANSKRIT)
pindaphala (38.20. SANSKRIT)
pinḍe (8.3. SANTALI)
pindithaka (36.3. SANSKRIT)
pine (40.26. NIUE)
pine fua loloa (40.42. NIUE)
pinheira (1.1. PORTUGUESE)
pinheirinha de água (42.1. PORTUGUESE)
pini'ae puaka (40.9. TONGAN)
pinino (38.17. ISNAG)
pinipita (16.1a. WÁRAY-WÁRAY)
pinksterbloem (30.44. DUTCH)
pinksterbloem (30.44. VLAAMS)
pinnis (35.9. BENGALI)
pinnis (35.9. HINDI)
pinochez (35.9. BRETON)
piŋpesi (45.1. DANGME-KROBO)
pin-sein (44.5. BURMESE)
pinutian (36.3. WÁRAY-WÁRAY)
pinyak (62.24. POTAWATOMI)
pinzèla (40.28. LADINO)
piobar dearg (62.2. GAELIC, IRISH)
piobar dearg (62.5. GAELIC, IRISH)
piobar glas (62.1. GAELIC, IRISH)
pion (58.1. GREEK)
pionandra (62.7. ITALIAN)
piopo (24.17. INDONESIAN)
piopo (24.17. MALAY)
pioz (7.2. TAJIKI)
pioz (7.3. TAJIKI)
pioz (7.9a. TAJIKI)
pipangay (38.21. MORISYEN)
pipari (62.1. LATVIAN)
piparjuuri (30.1. FINNISH)
piparrohi (44.9. ESTONIAN)
piparrot (30.1. ICELANDIC)
pipeløk (7.5. NORWEGIAN)
pipere (62.1. ALBANIAN, TOSK)
piperia (62.1. GREEK)
piperia e makrdkarpos (62.2. GREEK)
piperia e thamnodis (62.5. GREEK)
piperka (62.1. MACEDONIAN)
pipinela (38.25. SPANISH)
pipinko (38.13b. LADINO)
pipino (38.13a. CEBUANO)
pipino (38.13a. LADINO)
pipino (38.13a. SAMBAL, TINÀ)
pipino (38.13a. TAGALOG)
pipino (38.13b. TAGALOG)
pipino sikwa (38.13. MARANAO)
pipínu (38.13a. HILIGAYNON)
pipingkaai (38.13a. TAMIL)
piplök (7.5. SWEDISH)
pipñña (38.13a. SINHALA)
pippali (62.2. NEWAR)
pipulka (27.2. KONKANI)
pipulka (27.2. MARATHI)
piragadi (62.19. TAMIL)
pirasa (7.8. TURKISH)
pirasa yaprakli teke sakali (27.46. TURKISH)
pirca (27.8. QUECHUA, CUSCO)
pirifol (24.3. GALICIAN)
pirinç fasulyesi (40.40. TURKISH)
piripiri (62.5. RONGA)
piriya halim (30.52. PANJABI, EASTERN)
pirkankai (38.21. TAMIL)
pirkinkody (38.21. TAMIL)
pirshushkha (30.1. GEORGIAN)
pis (40.28. GAELIC, IRISH)
pisang (16.1a. DANISH)
pisang (16.1a. INDONESIAN)
pisang (16.1a. MALAY)

pjepën (38.9. ALBANIAN, TOSK)
pjepën (38.10. ALBANIAN, TOSK)
pjepën (38.11. ALBANIAN, TOSK)
pjepër (38.7a. ALBANIAN, TOSK)
pjepër (38.7b. ALBANIAN, TOSK)
pjepër (38.8. ALBANIAN, TOSK)
pjepër (38.9. ALBANIAN, TOSK)
pjepër (38.10. ALBANIAN, TOSK)
pjepër (38.11. ALBANIAN, TOSK)
pjepjer (62.1. SORBIAN, LOWER)
planhigyn persawrus (24.9. WELSH)
plantagos o koronopous (57.1. GREEK)
plantain (16.1a. ENGLISH)
plantain (16.1a. FRENCH)
plantin (16.1a. KRIO)
plàtan (16.1a. CATALAN-VALENCIAN-BALEAR)
platano (16.1a. SPANISH)
plátano de cocinar (16.1a. SPANISH)
plãtano macho (16.1a. SPANISH)
platdissel (27.45. AFRIKAANS)
plate brush (62.23. ENGLISH)
platlapu cemere (24.23. LATVIAN)
platte-peters (40.34a. AFRIKAANS)
plattvial (40.10. SWEDISH)
platycodon à grandes fleurs (34.3. FRENCH)
pļavas plastbārdis (27.47. LATVIAN)
plawhigyn-ia (22.1. WELSH)
plectranthe (44.8. FRENCH)
plejju (44.2. MALTESE)
pletik apo (24.19. JAVANESE)
plocksallat (27.32. SWEDISH)
plūduriuojantysis agaras (64.3. LITHUANIAN)
pluimhyacint (13.5. DUTCH)
pluksalat (27.32. DANISH)
pluksalat (27.32. DUTCH)
plym hyacint (13.5. SWEDISH)
pó de mico (40.18. PORTUGUESE)
po leng (35.9. CHINESE, MANDARIN)
p'o lo min shin (27.46. CHINESE, MANDARIN)
po tsai suanmo (58.3. CHINESE, MANDARIN)
poagap (40.43. TAGALOG)
poanrey (40.26. GAELIC, MANX)
poashani (38.14. TAMIL)
poashani (38.16. TELUGU)
poashani (38.18. TAMIL)
pob kws (17.7. HMONG-DAW)
pob kws qab zib (17.7. HMONG-DAW)
pob tsawb (16.1b. HMONG-DAW)
pochrzyn biały (11.7. POLISH)
pochrzyn chińsik (11.6. POLISH)
pochrzyn jadalny (11.5. POLISH)
pochrzyn skrzydlaty (11.1. POLISH)
pochrzyn trójdzielny (11.8. POLISH)
pochrzyn ziemniaczany (11.2. POLISH)
pochrzyn żółty (11.3. POLISH)
pock-tsoï (30.36. GREEK)
pod fern (4.1. ENGLISH)
podezřeň (3.1. CZECH)
podina (44.3. URDU)
podolem (38.29. KONKANI)
podu (40.31. IGBO)
podum choka (48.1b. ASSAMESE)
podum sipa (48.1a. ASSAMESE)
podzemna koleraba (30.22. SLOVENIAN)
podzemnica olejná (40.1. SLOVENIAN)
podzemnice olejná (40.1. CZECH)
podzo (40.38. DIGO)
poego (44.2. GALICIAN)
poejo (44.2. PORTUGUESE)
poélé (45.1. GOURMANCHÉMA)
poe-lêng-á-chhài (35.9. CHINESE, MIN NAN)
poenxo (44.2. GALICIAN)
pofadderboontjie (38.29. AFRIKAANS)
poh choy (35.9. MALAY)
pohaua (38.20. MAORI)
poh-pohan (66.1. INDONESIAN)
poh-pohan (66.1. JAVANESE)
poh-pohan (66.2. INDONESIAN)
poh-pohan (66.2. JAVANESE)
pohung (39.4. JAVANESE)
poi (28.1. BENGALI)
poi (28.1. HINDI)
poi (28.1. HINDUSTANI)
poi (28.1. MARATHI)
poi kare (40.29. MORISYEN)
poikivel (28.1. HINDI)
poiko saag (28.1. NEPALI)
poilo (7.8. KOONGO)
poilo (7.8. TEKE)
pointed gourd (38.31. ENGLISH)
poire (35.3. MORISYEN)
poire de terre (27.39. FRENCH)
poire de terre Cochet (27.39. FRENCH)
poire melon (62.20. FRENCH)
poireau (7.8. FRENCH)
poireau (7.8. KOONGO)
poireau (7.8. TEKE)

praavrishayani (40.18. TELUGU)
prakar khada (23.3. ORIYA)
pralbønne (40.24. DANISH)
pralbönne (40.24. NORWEGIAN)
prān (7.3. KASHMIRI)
prāna-bagāra (7.3. KASHMIRI)
prâpiëy (40.29. KHMER, CENTRAL)
pras (7.8. ALBANIAN, TOSK)
prasa (7.8. BOSNIAN)
prása (7.8. GEORGIAN)
prása (7.8. LADINO)
prasini piperia (62.1. GREEK)
praskavec ledencový (40.29. CZECH)
praso (7.8. GREEK)
prassa (7.8. GEORGIAN)
práta (62.24. GAELIC, IRISH)
prátaí céadfhómhair (62.24. GAELIC, IRISH)
prátaí lutaha (62.24. GAELIC, IRISH)
prâtiël prèah 'ângkaôl (20.4. KHMER, CENTRAL)
pravrishayana (40.18. SANSKRIT)
praz (7.8. BULGARIAN)
praz (7.8. ROMANIAN)
praziluk (7.8. BOSNIAN)
praziluk (7.8. SERBO-CROATIAN)
pre (7.8. JAVANESE)
preeria-auringonkukka (27.26. FINNISH)
prei (7.6. INDONESIAN)
prei (7.8. AFRIKAANS)
prei (7.8. DUTCH)
prei (7.8. VLAAMS)
prêle des champs (1.1. FRENCH)
preserving melon (38.2. ENGLISH)
presh (7.8. ALBANIAN, TOSK)
presh kulmak (7.11. ALBANIAN, TOSK)
preslica (1.1. SERBO-CROATIAN)
přeslička rolní (1.1. CZECH)
prexil (24.19. LADINO)
prezzemolo (24.19. ITALIAN)
prezzèmolo napolitanum (24.20. ITALIAN)
prezzèmolo tuberoso (24.21. ITALIAN)
prickly amaranth (23.3. ENGLISH)
prickly comfrey (29.3. ENGLISH)
prickly-pear (32.1. ENGLISH)
priest crown (27.45. ENGLISH)
prijel (24.19. GALICIAN)
primoula (51.1. GREEK)
primrose malanga (8.10. ENGLISH)
prince's feather (23.3. ENGLISH)
princesseboon (40.26. VLAAMS)
pring gesing (17.1. JAVANESE)
pring ori (17.1. JAVANESE)
pring petung (17.2. JAVANESE)
prinsesseboon (40.26. DUTCH)
Prinzeßbohne (40.26. GERMAN)
prishíl (24.19. LADINO)
prithubeejaka (40.12. SANSKRIT)
prixel (24.19. GALICIAN)
prochugaali chine (39.4. MARATHI)
prokelj (30.29. SERBO-CROATIAN)
prokolice (30.31. CZECH)
prokula (30.31. MACEDONIAN)
prokule (30.31. SERBO-CROATIAN)
proletna surepka (30.2. MACEDONIAN)
pronkbone (40.26. STELLINGWERFS)
pronkboon (40.24. DUTCH)
pronkboon (40.24. VLAAMS)
pronker (40.26. STELLINGWERFS)
pronkmeloen (38.18. STELLINGWERFS)
proso perłowe afrykańskie (17.3. POLISH)
prosvirjak (45.5. RUSSIAN)
prosvirnik (45.5. RUSSIAN)
protažitka (27.25. CZECH)
proutnjak (35.2. RUSSIAN)
proutnjak venichny (35.2. BELARUSAN)
prugugurka (38.13b. ICELANDIC)
Prunkbohne (40.24. GERMAN)
prydbønne (40.24. NORWEGIAN)
prydmelon (38.8. NORWEGIAN)
psianka (62.21. POLISH)
psianka amerykańska (62.15. POLISH)
psianka czarna (62.21. POLISH)
psianka etiopska (62.14a. POLISH)
psianka gilo (62.14b. POLISH)
psianka lulo (62.22. POLISH)
pšiŋ (7.3. DAKOTA)
pšiŋ (7.8. DAKOTA)
pšiŋ ikceka (7.10. DAKOTA)
pśiŋ (7.3. LAKOTA)
pśiŋ (7.8. LAKOTA)
psofokarpus (40.29. CZECH)
ptê (8.9. BÉTÉ)
ptèride â l'aigle (2.1. FRENCH)
ptéris aigle (2.1. FRENCH)
pterosilia (24.19. HEBREW)
pû (38.20. CHINESE, MIN NAN)
pu cong ying (27.45. CHINESE, MANDARIN)
pu gong ying (27.45. CHINESE, WÚ)
pu gua (38.20. CHINESE, YUE)
pu kung ying (27.45. CHINESE, MANDARIN)

pu qin (24.8. CHINESE, MANDARIN)
pu qin (24.8. CHINESE, WÚ)
pu tong bai cai (30.36. CHINESE, YUE)
pu tong gang dou (40.43. CHINESE, WÚ)
pu tong jiang dou (40.43. CHINESE, MANDARIN)
pu tong shan yao (11.6. CHINESE, MANDARIN)
pu tong shan yao (11.6. CHINESE, WÚ)
pu tong si gua (38.22. CHINESE, YUE)
pu tou bai cai (30.36. CHINESE, GÀN)
puaileng (35.9. THAI)
pualele (27.43. ENGLISH)
puauhi (11.1. MARQUESAN)
pu-bɔndɔ (38.21. MENDE)
pucai (7.13. JAVANESE)
pucuk labu (38.18. INDONESIAN)
pucuna uchu (62.6. QUECHUA, CUSCO)
pucúo (38.20. MANDJAK)
pudal (38.29. TAMIL)
pudalai (38.31. TAMIL)
pudalangai (38.29. TAMIL)
puddila an maluhadan (40.26. IFUGAO)
puddila an mummool (40.26. IFUGAO)
pudding grass (44.2. ENGLISH)
pudelkõrvits (38.20. ESTONIAN)
pudina (44.3. BENGALI)
pudina (44.3. HINDI)
pudina (44.3. MARATHI)
pudina (44.3. MIZO)
pudina (44.3. NEPALI)
pudina (44.3. SANSKRIT)
pudina (44.3. TAMIL)
pudina (44.3. TELUGU)
pudina hulluj (44.3. TAJIKI)
pudinai dashti (44.2. TAJIKI)
puding mas (39.3. MALAY)
pudol (38.29. TAMIL)
pueixo (44.2. GALICIAN)
pueraria (40.30. CZECH)
puéraria (40.30. FRENCH)
pueraria (40.30. ITALIAN)
pueraria (40.30. PORTUGUESE)
pueraria (40.30. SLOVAK)
pueraria (40.30. TURKISH)
pueraria se'ira (40.30. HEBREW)
pueraria strisciante (40.30. ITALIAN)
puerarie (40.30. ALBANIAN, TOSK)
puerarija (40.30. SERBO-CROATIAN)
puerarya se'jra (40.30. HEBREW)
puero (40.30. PORTUGUESE)
puerro (7.8. LADINO)
puerro agreste (7.1. SPANISH)
puerro salvage (7.1. SPANISH)
puerro silvestre (7.1. SPANISH)
Puffbohne (40.34a. GERMAN)
pu-gbawo (44.5. MENDE)
puggie nut (40.1. SCOTS)
pu-goja (38.3. MENDE)
pugyapa (23.3. SUBANON)
puhi bihnáaʔ (38.3. COMANCHE)
puhi tuhkapu (27.32. COMANCHE)
puhog (40.29. IVATAN)
pui (45.1. DANGME-KROBO)
pui hiɔ (45.1. DANGME-KROBO)
puichenggah (38.21. MALAYALAM)
puileng (35.9. INDONESIAN)
pujɛ (62.1. MENDE)
pujɛ (62.5. MENDE)
pujɛ-mumu (62.5. MENDE)
pujɛ-wawa (62.1. MENDE)
pujɛ-wawa (62.2. MENDE)
puju (27.7. ESTONIAN)
puka (30.27a. MAORI)
puka beunteur (5.1. INDONESIAN)
pukinparta (27.47. FINNISH)
pul (40.26. AZERBAIJANI, NORTHERN)
pul çap edilən yer (44.3. AZERBAIJANI, NORTHERN)
pula'a (8.6. SAMOAN)
pulaakkilanu (39.4. MALAYALAM)
pulachakiri (45.4. KANNADA)
pulaka (8.9. NIUE)
pulakizhanna (20.4. MALAYALAM)
pu-langei (39.4. BULLOM SO)
pulankilhangu (20.4. TAMIL)
pulau (50.2. TAGALOG)
pulaw (50.2. TAGALOG)
pulet (62.12. BONTOC, CENTRAL)
pulezzo (44.2. ITALIAN)
puli (11.2. MENDE)
pulichai-keerai (45.4. TAMIL)
pulichakeere (45.4. KANNADA)
pulichchai (45.4. MALAYALAM)
pulichchai (45.4. TAMIL)
pulichchai kerai (45.4. TAMIL)
pulicheera (45.4. MALAYALAM)
pulikkeerai (59.2. TAMIL)
puliya (38.23. SUBANON)
puliya nataw (38.23. SUBANON)

pullikirai (59.2. TAMIL)
pullokurpitsa (38.20. FINNISH)
pullóóx (39.4. WOLOF)
pullum (38.3. TAMIL)
puloh (39.4. WOLOF)
pulóók (11.1. PULAAR)
pulóók (11.2. PULAAR)
pulóók (11.5. PULAAR)
pulóók (39.4. PULAAR)
puloth (16.1a. BULLOM SO)
pulpmelon (38.24. VLAAMS)
pulugan (11.2. BICOLANO, CENTRAL)
pum hub (44.3. HMONG-DAW)
pum hum (44.3. HMONG-NJUA)
pumalapaggar danggu (7.2. IFUGAO)
pumbela (40.18. TEM)
pumpa (38.16. SWEDISH)
pumpa (38.18. SWEDISH)
pumpkin (38.14. ENGLISH)
pumpkin (38.14. GAELIC, MANX)
pumpkin (38.16. ENGLISH)
pumpkin (38.16. GAELIC, MANX)
pumpkin (38.16. NORWEGIAN)
pumpkin (38.18. ENGLISH)
pumpkin (38.18. GAELIC, MANX)
pumpkin (38.18. NORWEGIAN)
punaikalei (40.18. TAMIL)
punaikkali (40.18. TAMIL)
punaippidukkan (40.18. TAMIL)
punajuuri (35.4. FINNISH)
punakaali (30.27a. FINNISH)
punane kapsas (30.27b. ESTONIAN)
punapipar (62.2. ESTONIAN)
punauba (40.9. ESTONIAN)
pundareeka (48.1a. SANSKRIT)
pundi beeja (45.4. KANNADA)
punḍi sāngā (36.3. MUNDARI)
pundi soppu (45.4. KANNADA)
pundibija (45.4. KANNADA)
pundithappu (45.4. TULU)
puṇḍittu (45.4. TULU)
pundreekamu (48.1a. TELUGU)
pungapong (8.3. TAGALOG)
pungapung (8.2. TAGALOG)
pungapung (8.3. FILIPINO)
pungapung (8.3. TAGALOG)
pungikeerai (23.3. TAMIL)
punguza (58.3. SWAHILI)
punje (17.7. SWAHILI)
punocvjetni grah (40.24. SERBO-CROATIAN)
punti (16.1a. BUHID)
punti (16.1a. PANGASINAN)
punti (16.1a. TAGBANWA)
puntik (16.1a. TAGBANWA)
punto (16.1a. TAGBANWA)
puolanminttu (44.2. FINNISH)
puošnioji džioveklė (52.1. LITHUANIAN)
pupa (23.4. YORUBA)
pupa (40.26. LATVIAN)
pupa (40.34a. LITHUANIAN)
pupak (30.29. SORBIAN, UPPER)
pupalka dvouletá (51.1. CZECH)
pupas (40.34a. LATVIAN)
pupelės (40.26. LITHUANIAN)
pupenda (50.2. MENDE)
puper glâs (62.2. WELSH)
pupinas (40.26. LATVIAN)
pupoljka (51.1. SERBO-CROATIAN)
pupu lele (27.18. NIUE)
pupu lele (27.42. NIUE)
pupur poeth (62.1. WELSH)
pur (7.11. ROMANIAN)
pura cortar (27.32. PORTUGUESE)
puraka (8.6. RAROTONGAN)
purasruoho (29.1. FINNISH)
pura:vare (40.26. BADAGA)
puravi (7.1. LATVIAN)
puravi (7.8. LATVIAN)
purayin (48.1a. HINDI)
purcha (38.3. TELUGU)
purchaq (40.26. UYGHUR)
purchaq (40.28. UYGHUR)
pureh-tsibehleh (7.8. YIDDISH)
púri (11.2. PULAAR)
puri-balé (11.2. PULAAR)
puriket (27.8. BONTOC CENTRAL)
puriket (27.8. ILOCANO)
puring (39.3. INDONESIAN)
purjo (7.8. FINNISH)
purjo (7.8. SWEDISH)
purjo lök (7.8. SWEDISH)
pur-løg (7.11. DANISH)
puroi sak (28.1. ASSAMESE)
purple amaranth (23.3. ENGLISH)
purple brinjal (62.19. ENGLISH)
purpurikit (27.14. ILOCANO)
purre (7.8. NORWEGIAN)
purslaan (59.2. AFRIKAANS)
purslane (59.2. ENGLISH)
purula (38.22. HINDI)

purun (7.3. MIZO)
purun sen (7.3. MIZO)
purun var (7.10. MIZO)
purunisag (59.2. ORIYA)
purunisag (59.4. ORIYA)
purusha (20.1. SANSKRIT)
purvi sak (35.1. ASSAMESE)
purwul (38.29. HINDI)
puryaw (38.23. HANUNOO)
pus sha (40.38. HMONG-DAW)
pusanikkai (38.1. TAMIL)
pusdienziedes (22.1. LATVIAN)
pu-see-nan (44.3. BURMESE)
pushani kai (38.1. TAMIL)
pushini (38.1. TAMIL)
pushkara (48.1a. SANSKRIT)
pushpaphala (38.1. SANSKRIT)
pushpavha (24.13. SANSKRIT)
puskarāmul (48.1a. NEWAR)
puso (16.1b. FILIPINO)
puso ug saging (16.1b. TAGALOG)
puspu-poroto (40.2. QUECHUA, CUSCO)
pusu' (16.1b. BINUKID)
putaakkizhangu (39.4. TAMIL)
putaka (48.1a. SANSKRIT)
putali phool (27.12. NEPALI)
putaque pariu (62.2. KONKANI)
putcha (38.3. TELUGU)
putcha kaaya (38.3. TELUGU)
pute (36.3. PULAAR)
puterseli (24.19. JAVANESE)
puthika (28.1. SANSKRIT)
putri india (59.2. INDONESIAN)
putta sullad (27.32. PANJABI, EASTERN)
puttabaale (16.1a. KANNADA)
puttalpiram (38.21. MALAYALAM)
puttalpiram (38.22. MALAYALAM)
putthari chunta (62.23. MALAYALAM)
putthur (23.2. SANSKRIT)
puttla (38.29. TAMIL)
putulika (38.31. SANSKRIT)
putu-mɛsɛ (62.5. KONO)
putu-wa (62.2. KONO)
putzakova (38.7b. TELUGU)
puuhaa (27.42. MAORI)
puungitangita (66.3. MAORI)
puunitangita (66.3. MAORI)
puutomaatti (62.7. FINNISH)
puuwhaa (27.42. MAORI)
puvalai (9.1. TAMIL)
puvalai (16.1a. TAMIL)
puvanvali (16.1a. TAMIL)
pu-wā (17.7. NEWAR)
puwida (40.25. BONTOC, CENTRAL)
pwa (40.12. KWÉYÒL)
pwa (40.26. CREOLE, HAITIAN FRENCH)
pwa (40.26. KWÉYÒL)
pwa (40.28. KWÉYÒL)
pwa Angòl (40.2. KWÉYÒL)
pwa chous (40.25. CREOLE, HAITIAN FRENCH)
pwa enkoni (40.43. CREOLE, HAITIAN FRENCH)
pwa frans (40.28. CREOLE, HAITIAN FRENCH)
pwa inkoni (40.41. CREOLE, HAITIAN FRENCH)
pwa kongo (40.2. CREOLE, HAITIAN FRENCH)
pwa pijon (40.12. CREOLE, HAITIAN FRENCH)
pwa tann (40.26. CREOLE, HAITIAN FRENCH)
pwa valèt (40.25. CREOLE, HAITIAN FRENCH)
pwad chous (40.25. CREOLE, HAITIAN FRENCH)
pwad souch (40.25. CREOLE, HAITIAN FRENCH)
pwaro (7.8. MORISYEN)
pwa-salad (40.26. KWÉYÒL)
pwev piman (62.1. KWÉYÒL)
pyààl (62.19. BEROM)
pyāj (7.3. HINDI)
pyanj (7.3. BENGALI)
pyaung-bu (17.7. BURMESE)
pyaz (7.2. NEPALI)
pyāz (7.3. KURDISH, NORTHERN)
pyaz (7.3. NEPALI)
pyaz (7.5. NEPALI)
pyâz (7.3. PASHTO, NORTHERN)
pyntekål (30.35. NORWEGIAN)
pyò (10.1. BEROM)
pypsteelboon (38.20. AFRIKAANS)
pys (40.28. WELSH)
pystykeiholehti (6.2. FINNISH)
pythagorean bean (48.1b. ENGLISH)
pytidomeno sinapi (30.12. GREEK)
q ırmızı turp (30.56. AZERBAIJANI, NORTHERN)

qadas (40.12. AFAR)
qalamfur (62.1. TAJIKI)
qamh ir-rum (17.7. MALTESE)
qānra (38.13a. TIBETAN)
qaqocc (27.16. MALTESE)
qaqocc tax-xewk (27.15. MALTESE)
qáquaussaq (40.12. INUKTITUT, WEST GREENLANDIC)
qara hamra (38.18. MALTESE)
qara sudani (38.14. ARABIC)
qara sudani (38.16. ARABIC)
qara sudani (38.17. ARABIC)
qara sudani (38.18. ARABIC)
qara tork (38.4. MALTESE)
qarbız (38.3. KAZAKH)
qarğıdalı (17.7. AZERBAIJANI, NORTHERN)
qargha hamra (38.16. MALTESE)
qarpız (38.3. AZERBAIJANI, NORTHERN)
qärqäḥa (17.1. AMHARIC)
qärqäḥa (17.1. HARARI)
qärqäḥa (17.2. HARARI)
qärqäḥa (17.4. HARARI)
q^{w}asa (38.13a. AMHARIC)
qäsbädäğğ (27.14. AMHARIC)
qàtar (40.28. AFAR)
qatron (30.45. TAJIKI)
qawın (38.7a. KAZAKH)
qawın (38.7b. KAZAKH)
qäyysər (30.56. AMHARIC)
qcqpa (7.10. TIBETAN)
qej (7.10. HMONG-DAW)
qej ntyw (7.12. HMONG-DAW)
qəl (38.20. AMHARIC)
qelbo (17.7. AFAR)
q̱əmuù (20.6. TIBETAN)
qepë (7.3. ALBANIAN, TOSK)
qepë dimrore (7.5. ALBANIAN, TOSK)
qepë shalotë (7.2. ALBANIAN, TOSK)
qera (16.1a. TIBETAN)
qes tsw ha (8.1. HMONG-DAW)
qhau liab (9.1. HMONG-DAW)
qhiav (20.6. HMONG-DAW)
qhiav qus (20.2. HMONG-DAW)
qhiav qus (20.5. HMONG-DAW)
qhūmaà me̱tɔ̀ɔ̀ (27.45. TIBETAN)
qi cai (30.43. CHINESE, MANDARIN)
qi mi (10.2. CHINESE, XIANG)
qi nu ya li (35.6. CHINESE, MANDARIN)
qi nu ya li (35.6. CHINESE, WÚ)
qi zi yu (19.1. CHINESE, GÀN)
qian guan cao (57.1. CHINESE, YUE)
qian ri ju (27.2. CHINESE, MANDARIN)
qian shi (50.1. CHINESE, MANDARIN)
qian shi (50.1. CHINESE, WÚ)
qian tou gan lan (30.32. CHINESE, WÚ)
qiang wei gen jing tian (37.1. CHINESE, MANDARIN)
qiang wei gen jing tian (37.1. CHINESE, WÚ)
qiao (7.4. CHINESE, YUE)
qiao tou (7.4. CHINESE, GÀN)
qiao tou (7.4. CHINESE, MIN BEI)
qiao tou (7.4. CHINESE, XIANG)
qiao tou (7.4. CHINESE, YUE)
qie gua (62.19. CHINESE, YUE)
qie zi (62.19. CHINESE, GÀN)
qie zi (62.19. CHINESE, MANDARIN)
qie zi (62.19. CHINESE, WÚ)
qie zi (62.19. CHINESE, XIANG)
qin cai (24.4. CHINESE, MANDARIN)
qin cai (24.4. CHINESE, WÚ)
qin cai (24.4. CHINESE, YUE)
qin jing hui xiang (24.12. CHINESE, YUE)
qing bo he (44.2. CHINESE, YUE)
qing cai (30.36. CHINESE, GÀN)
qing cai (30.36. CHINESE, WÚ)
qing cai (30.36. CHINESE, YUE)
qing dou (40.8. CHINESE, XIANG)
qing gua (38.13a. CHINESE, YUE)
qing hua cai (30.31. CHINESE, MANDARIN)
qing hua cai (30.31. CHINESE, WÚ)
qing hua cai (30.31. CHINESE, YUE)
qing xing (23.4. CHINESE, WÚ)
qing zi ge (69.1. CHINESE, WÚ)
qiqër (40.5. ALBANIAN, TOSK)
qiu jing gan lan (30.30. CHINESE, MANDARIN)
qiu jing gan lan (30.30. CHINESE, WÚ)
qiu ye wo ju (27.32. CHINESE, YUE)
qiu ying (27.12. CHINESE, WÚ)
qiyar (38.13a. KAZAKH)
qoe (40.26. CHECHEN)
qoe (40.28. CHECHEN)
qoghun (38.7a. UYGHUR)
qoghun (38.7b. UYGHUR)
qolqas (8.4. ARABIC)
qolqas el-mustaukat (8.6. ARABIC)
qömaphatiko (40.25. HOPI)
qonaq (17.7. UYGHUR)
qônco (38.14. SIDAMO)
qônco (38.16. SIDAMO)

qônco (38.17. SIDAMO)
qônco (38.18. SIDAMO)
qondari (44.9. GEORGIAN)
qondo (62.2. HARARI)
qōŋi̲puù (24.11. TIBETAN)
qoqu (27.45. TAJIKI)
qos (11.6. HMONG-DAW)
qos daj (11.3. HMONG-DAW)
qos dej (48.1a. HMONG-DAW)
qos hmab soo (40.21. HMONG-DAW)
qos liab (11.3. HMONG-DAW)
qos liab (36.3. HMONG-DAW)
qos ntoo (27.26. HMONG-DAW)
qos ntoo (39.4. HMONG-DAW)
qos tsov (8.4. HMONG-DAW)
qos tsw ha (8.4. HMONG-DAW)
qos yaj ywm (62.24. HMONG-DAW)
qötsahatiko (40.25. HOPI)
qovun (38.7a. AZERBAIJANI, NORTHERN)
qovun (38.7a. UZBEK)
qovun (38.7b. AZERBAIJANI, NORTHERN)
qovun (38.7b. UZBEK)
qu bai jie (30.57. CHINESE, WÚ)
qu mai cai (27.10a. CHINESE, YUE)
qua bi (38.18. VIETNAMESE)
qua dua hău (38.7a. VIETNAMESE)
quả dua hấu (38.7b. VIETNAMESE)
qua dua hău (38.11. VIETNAMESE)
quail grass (23.4. ENGLISH)
Quaker comfrey (29.3. ENGLISH)
qualuza (8.9. PORTUGUESE)
quantan (40.8. VIETNAMESE)
que dou (40.9. CHINESE, MANDARIN)
que dou (40.9. CHINESE, WÚ)
Queensland arrowroot (9.1. ENGLISH)
quelembe (8.8. SPANISH)
quélimbé (8.8. FRENCH)
quenopódio bom-henrique (35.5. PORTUGUESE)
qui zi jiang (27.26. CHINESE, YUE)
quiabo (45.1. CRIOULO, UPPER GUINEA)
quiabo (45.1. PORTUGUESE)
quimbombó (45.1. SPANISH)
quimgombó (45.1. SPANISH)
quinoa (35.6. CRIOULO, CAPE VERDE)
quinoa (35.6. DANISH)
quinoa (35.6. ENGLISH)
quinoa (35.6. FRENCH)
quinoa (35.6. ITALIAN)
quinoa (35.6. PORTUGUESE)
quinoa (35.6. QUECHUA, CUSCO)
quinoa (35.6. SWEDISH)
quinoa branca (35.6. PORTUGUESE)
quinua (35.6. SPANISH)
quing hsiang (23.4. CHINESE, MANDARIN)
quingombo (45.1. CRIOULO, UPPER GUINEA)
quingombo (45.1. PORTUGUESE)
quitadji (62.14b. FULFULDE)
qulë dimrore (7.5. SLOVENIAN)
qulu' (38.20. HARARI)
qūmüü (13.4. TIBETAN)
qungordlit (30.27a. INUKTITUT, WEST GREENLANDIC)
qunguleq (30.27a. INUKTITUT, WEST GREENLANDIC)
quy[ees]t g[aj]c nai (4.1. VIETNAMESE)
qu[yf] d[oj]i (27.26. VIETNAMESE)
raadi (30.56. ZARMA)
raaja koshaathakee (38.22. SANSKRIT)
raajaadri (23.3. SANSKRIT)
raajabaale (16.1a. KANNADA)
raajagira (23.3. MARATHI)
raajagiri (23.3. SANSKRIT)
raajakooshmaanda (62.19. SANSKRIT)
raajakoshaathakee (38.21. SANSKRIT)
raajaluka (30.56. SANSKRIT)
raajamaasha (40.26. SANSKRIT)
raajamaasha (40.43. SANSKRIT)
raajamudga (40.35. SANSKRIT)
raajanea (23.3. KANNADA)
raajapalaandu (7.3. SANSKRIT)
raajapriya (7.3. SANSKRIT)
raajashakini (23.3. SANSKRIT)
raajashimbi (40.9. HINDI)
raajashimbi (40.9. SANSKRIT)
raajathinisha (38.3. SANSKRIT)
raajeeva (48.1a. SANSKRIT)
raajeshta (16.1a. SANSKRIT)
raajeshtha (7.3. SANSKRIT)
raajgeer (23.3. KANNADA)
raajimathphala (38.21. SANSKRIT)
raajiphal (38.31. SANSKRIT)
raajkiri (23.3. HINDI)
raajputhri (38.20. SANSKRIT)
rààke (17.5. HAUSA)
raama seethalika (23.3. SANSKRIT)
raama thulasi (44.4. KANNADA)
raama thulasi (44.5. MALAYALAM)
raamadana sirukeere (23.3. KANNADA)

ràdika (30.41a. ROMANI, BALKAN)
radiki (27.10a. GREEK)
radikya (27.10a. TURKISH)
radis (30.54. FRENCH)
radis (30.56. DANISH)
radis (30.56. FRENCH)
radis (30.56. INDONESIAN)
radis (30.56. KANKANAEY)
radis (30.56. SUNDA)
radis serpent (30.53. FRENCH)
radisa (30.56. ICELANDIC)
rädisa (30.56. SWEDISH)
radish (30.54. ENGLISH)
radish (30.56. ENGLISH)
radish (30.56. KRIO)
radish (30.56. WOLOF)
radishi (30.56. SWAHILI)
radisi (30.56. TSWANA)
radiska (30.56. SORBIAN, UPPER)
radraka (16.1a. SANSKRIT)
radsâ (58.1. KATI)
radys (30.56. AFRIKAANS)
raeddike (30.54. DANISH)
raeddike (30.56. DANISH)
raeshime gadde (20.1. KANNADA)
rafanis (30.56. GREEK)
rafano (30.54. ITALIAN)
rafano (30.56. ITALIAN)
rafano rusticano (30.1. ITALIAN)
ragadali (40.12. SANSKRIT)
raghyl (30.1. GAELIC, MANX)
ragiyang (8.1. CEBUANO)
ragom a amo (27.8. MARANAO)
raguinikavaja (63.1. BELARUSAN)
rahadri (23.3. SANSKRIT)
rahar (40.2. NEPALI)
rāhaṛi (40.2. MAITHILI)
rahavalli (38.23. SANSKRIT)
rahbahrber (58.1. YIDDISH)
raheṛ (40.2. SANTALI)
rahgyl (30.56. GAELIC, MANX)
rahilaa (40.5. HINDI)
rahuchhana (20.6. SANSKRIT)
rahun (7.10. KASHMIRI)
rahurahu (2.1. MAORI)
rai (30.17. BENGALI)
rai (30.17. HINDI)
rāī (30.17. HINDI)
rāi (30.17. MUNDARI)
rai (30.24. HINDI)
rai (30.24. ORIYA)
rai (30.24. URDU)
ráib (30.21. GAELIC, IRISH)
raidis fhiáin (30.1. GAELIC, IRISH)
raifort japonais (30.58. FRENCH)
raifort sauvage (30.1. FRENCH)
raihon (44.5. TAJIKI)
raikhon (44.5. UZBEK)
raimame (40.25. JAPANESE)
raiponce (33.2. FRENCH)
raipóntico (33.2. SPANISH)
raish (40.43. HINDI)
raisin d'Amérique (56.2. FRENCH)
raithneach (2.1. GAELIC, IRISH)
raíz açucarada (24.23. PORTUGUESE)
raiz forte (30.1. PORTUGUESE)
raj manh (40.34a. PANJABI, EASTERN)
räj mung (40.40. PANJABI, EASTERN)
raja kosa taki (38.22. SANSKRIT)
rajagaro (23.3. GUJARATI)
rajagiri (23.3. SANSKRIT)
rajakere (23.3. KANNADA)
rajakoshataki (38.22. SANSKRIT)
rajapa (30.41a. BULGARIAN)
rajasarsapa (30.17. SANSKRIT)
rajashakini (23.3. SANSKRIT)
rajče (62.9. CZECH)
rajče jedlé (62.9. CZECH)
rajčenka (62.7. CZECH)
rajčiak (62.9. SLOVAK)
rajčica (62.9. BOSNIAN)
rajcica (62.9. CROATIAN)
rajčica (62.9. SERBO-CROATIAN)
rajgaroo (23.3. GUJARATI)
rajgira (23.3. HINDI)
rajgira (23.3. MARATHI)
rajgira (23.3. MARWARI)
raji (30.24. SANSKRIT)
rajika (30.17. SANSKRIT)
rajika (30.24. SANSKRIT)
räjmã (40.26. PANJABI, EASTERN)
rajmah (40.26. HINDI)
raj-rawam (40.34a. URDU)
raj-rawan (40.34a. PANJABI, EASTERN)
rakcacha (24.6. QUECHUA, CUSCO)
rà'ke (17.5. BEROM)
rake (17.5. HAUSA)
raket de frans (32.1. MORISYEN)
rakkyo (7.4. DANISH)
rakkyo (7.4. ENGLISH)

rakkyo (7.4. FRENCH)
rakkyou (7.4. JAPANESE)
raknashak (23.3. BENGALI)
raktakamal (50.2. MARATHI)
raktakanda (7.3. SANSKRIT)
raktamaci (62.9. SANSKRIT)
raktavardhak (62.9. SANSKRIT)
raktha gmjna (35.4. SANSKRIT)
raktha maricha (62.2. SANSKRIT)
rakthaalu (36.3. SANSKRIT)
rakthabeeja (30.51. SANSKRIT)
rakthabeeja (38.3. SANSKRIT)
rakthabeeja (40.1. SANSKRIT)
rakthakanda (7.3. SANSKRIT)
rakthamaricha (62.5. SANSKRIT)
rakthambashtaa (45.4. SANSKRIT)
rakthaphala (38.4. SANSKRIT)
rakthapushpa (20.1. SANSKRIT)
raktharaaji (30.51. SANSKRIT)
raktharenu (20.1. SANSKRIT)
rakthasarshapa (30.24. SANSKRIT)
rakthavrunthaak (62.9. SANSKRIT)
rakthika (30.24. SANSKRIT)
rakthphala (62.19. SANSKRIT)
rakto-kambal (50.2. BENGALI)
ram begun (62.16. BENGALI)
ram dana (23.3. HINDI)
ram jhingạ (45.1. SANTALI)
ram jokhola (40.33. ASSAMESE)
räm tōri (38.21. PANJABI, EASTERN)
räm tōri (38.22. PANJABI, EASTERN)
ram turai (38.21. NEPALI)
Rama (45.4. GERMAN)
ramacha (40.9. MALAYALAM)
ramamulaka (62.9. TELUGU)
ramas (40.43. HINDI)
ramathana champa (44.5. KANNADA)
rambai (62.10. INDONESIAN)
rambha (16.1a. KANNADA)
rambha (16.1a. SANSKRIT)
rambha (16.1a. TELUGU)
rambṛa (40.37. SANTALI)
ramdana (23.3. HINDI)
ramenas (30.54. DUTCH)
ramenas (30.56. DUTCH)
ramgatora (38.20. MAORI)
ramhikolai (40.43. BENGALI)
ramkasturi (44.5. KANNADA)
ramkurthi (40.8. HINDI)
ramma (45.4. HAUSA)
rammenas (30.54. STELLINGWERFS)
rampai (62.15. INDONESIAN)
ramphala (62.9. SANSKRIT)
ramphali (40.6. HINDI)
rampion (33.2. ENGLISH)
rampion (33.2. VLAAMS)
rampon (33.2. FRENCH)
rampon (67.2. FRENCH)
ramsik (40.38. MARANAO)
ramtal urahi (40.3. ASSAMESE)
ramtal urahi (40.4. ASSAMESE)
ramthurai (45.1. HINDI)
ramtorai (45.1. BENGALI)
ramtorai (45.1. NEPALI)
ramturai (45.1. HINDI)
ramturai (45.1. PANJABI, EASTERN)
ramudyu (38.29. HANUNOO)
ramuza (7.2. HAUSA)
ran aalu (8.4. MARATHI)
ranā<u>sh</u> (58.1. PASHTO, NORTHERN)
ranbhendo (45.2. MARATHI)
ran-bhopla (38.20. MARATHI)
randa miang (27.12. SUNDA)
randa midang (27.12. INDONESIAN)
randa midang (27.12. JAVANESE)
randhwni (24.4. BENGALI)
rang he (20.5. CHINESE, MANDARIN)
rang he (20.5. CHINESE, WÚ)
ranga alu (36.3. BENGALI)
ranga sak (23.3. ASSAMESE)
rangam (62.9. INDONESIAN)
rangkain (50.2. ORIYA)
rangoonboon (40.25. VLAAMS)
rani unyu (3.1. NEPALI)
rani unyu (3.2. NEPALI)
ranmaath (23.3. MARATHI)
ranmat (23.3. MARATHI)
ranoun koulos (24.17. GREEK)
ranskalainen rakuuna (27.6. FINNISH)
ransvaalbossie (23.3. AFRIKAANS)
ranthuri (45.1. HINDI)
ranti (62.10. INDONESIAN)
ranti (62.15. INDONESIAN)
ranti (62.15. JAVANESE)
ranti (62.15. MALAY)
ranti bali (62.9. JAVANESE)
ranti gendel (62.9. JAVANESE)
ranti kenong (62.9. JAVANESE)
rao den do (23.3. VIETNAMESE)
raopen (30.41a. STELLINGWERFS)

rapa (30.21. INDONESIAN)
řapa (30.22. CZECH)
rapa (30.41a. CZECH)
rapa (30.41a. ITALIAN)
rapa (30.41a. SOTHO, NORTHERN)
rapa (30.41a. SOTHO, SOUTHERN)
rapa coltivata (30.41a. ITALIAN)
rapa da foglia (30.41b. ITALIAN)
rapa di primavera (30.41a. ITALIAN)
rapa lachanon (30.41b. GREEK)
rapa senza testa (30.41b. ITALIAN)
rapak (36.3. MARANAO)
rapanaki (30.56. GREEK)
rapanidhe e kultivuar (30.54. ALBANIAN, TOSK)
rapanidhe e kultivuar (30.56. ALBANIAN, TOSK)
raparperi (58.1. FINNISH)
rapavo mitrovdenče (27.17. MACEDONIAN)
raperònzolo (33.2. ITALIAN)
raperonzolo (67.2. ITALIAN)
rapiska (30.21. TURKISH)
rapiţa (30.21. ROMANIAN)
rapiţa colza (30.21. ROMANIAN)
rapkol (30.41a. INDONESIAN)
rapodjo (30.27a. CRIOULO, CAPE VERDE)
raponc harangvirág (33.2. HUNGARIAN)
rapónchigo (33.2. SPANISH)
rapôncio (33.2. PORTUGUESE)
rapontico (58.1. ITALIAN)
rapóntico (58.1. SPANISH)
rapontika (51.1. NORWEGIAN)
raponzo (33.2. ITALIAN)
rapònzolo (33.2. ITALIAN)
raps (30.21. ALBANIAN, TOSK)
raps (30.21. DANISH)
raps (30.21. ESTONIAN)
Raps (30.21. GERMAN)
raps (30.21. NORWEGIAN)
raps (30.21. RUSSIAN)
raps (30.21. SWEDISH)
raps (30.21. TURKMEN)
rapsas (30.21. LITHUANIAN)
rapsi (30.21. FINNISH)
rapúncio (33.2. PORTUGUESE)
rapúnculo (33.2. SPANISH)
rapunkelklokke (33.2. NORWEGIAN)
rapunkeltlocta (33.2. SWEDISH)
rapunsel-klokke (33.2. DANISH)
rapuntsel (33.2. RUSSIAN)
rapuntselkellukas (33.2. ESTONIAN)
Rapunzel (33.2. GERMAN)
Rapunzel-Glockenblume (33.2. GERMAN)
rapunzelklokje (33.2. DUTCH)
rapunzelklokje (33.2. VLAAMS)
Rapunzelsalat (59.2. GERMAN)
raquette (32.1. FRENCH)
rara ayam (23.4. SUNDA)
rarauhe (2.1. MAORI)
rarawea (40.16. SUNDA)
rasaayani (62.15. SANSKRIT)
rasabaale (16.1a. KANNADA)
rasālo (17.5. PALI)
rasbhari (62.12. HINDI)
rasbhary (62.12. PANJABI, EASTERN)
rashun (7.10. BENGALI)
r[as]ng g[aj]c nai (4.1. VIETNAMESE)
rãss bēn (40.26. PANJABI, EASTERN)
raštan (30.35. SERBO-CROATIAN)
raštika (30.35. CROATIAN)
rastiteinaja goubka (38.22. RUSSIAN)
rasun (7.10. BENGALI)
rasun (7.10. ORIYA)
ṛasun (7.10. SANTALI)
rasuni (7.10. MUNDARI)
r[as]y (8.1. VIETNAMESE)
r[as]y [aw]n (8.1. VIETNAMESE)
rat tail radish (30.53. ENGLISH)
rataaloo (11.1. HINDI)
rataaloo (11.2. HINDI)
ratago (23.3. QUECHUA, CUSCO)
ratakaju (40.1. SINHALA)
ratalu (11.1. GUJARATI)
ratalu (11.1. HINDI)
ratálú (11.1. HINDUSTANI)
ratalu (11.1. PANJABI, EASTERN)
ratalu (11.2. GUJARATI)
ratalu (11.2. HINDI)
ratálú (11.2. HINDUSTANI)
ratalu (11.2. PANJABI, EASTERN)
ratalu (11.2. URDU)
ratálú (11.3. HINDUSTANI)
ratálú (11.4. HINDUSTANI)
ratálú (11.5. HINDUSTANI)
ratálú (11.6. HINDUSTANI)
ratálú (11.7. HINDUSTANI)
ratálú (11.8. HINDUSTANI)
ratalu (36.3. MARATHI)
ratāru (11.1. MAITHILI)
rateise (30.56. SOTHO, NORTHERN)

rebung (17.1. INDONESIAN)
rebung (17.1. MALAY)
rebung (17.2. MALAY)
rebung (17.4. INDONESIAN)
rebung (17.4. MALAY)
red amaranth (23.3. ENGLISH)
red beet (35.4. ENGLISH)
red cabbage (30.27b. ENGLISH)
red gourd (38.16. ENGLISH)
red gram (40.2. ENGLISH)
red green (23.3. KRIO)
red jute (63.1. ENGLISH)
red root amaranth (23.3. ENGLISH)
red sorrel (45.4. ENGLISH)
reddik (30.54. NORWEGIAN)
reddik (30.56. NORWEGIAN)
redies (30.56. STELLINGWERFS)
redis (30.56. RUSSIAN)
rediscka (30.56. UZBEK)
redīsi (30.56. LATVIAN)
rediski (30.56. RUSSIAN)
redka (30.54. KAZAKH)
red'ka (30.54. RUSSIAN)
red'ka (30.56. RUSSIAN)
red'ka (30.56. UKRAINIAN)
red'kev (30.54. SLOVAK)
ředkev (30.56. CZECH)
redkev (30.56. SLOVAK)
redkev (30.56. SLOVENIAN)
ředkev seté (30.53. CZECH)
red'kovka (30.56. SLOVAK)
redleaf chicory (27.10b. ENGLISH)
red-leaved hibiscus (45.3. ENGLISH)
ředrvička (30.56. CZECH)
redzika pasjannaja (30.56. BELARUSAN)
reefort (30.56. SCOTS)
reevand (58.1. FARSI, WESTERN)
reevase (58.1. FARSI, WESTERN)
refasm'ari (40.17. ICELANDIC)
regamo (44.6. ITALIAN)
regelat haginah (59.2. HEBREW)
regelat ha-ginna (59.2. HEBREW)
regrat (27.45. SLOVAK)
reḥan (44.7. HEBREW)
rehan canum (44.4. HEBREW)
rehan matok (44.5. HEBREW)
rehani (44.5. GEORGIAN)
rehtekh (30.54. YIDDISH)
rehtekh (30.56. YIDDISH)
rehuydinkaali (30.32. FINNISH)
Réidech (30.54. LUXENBOURGEOIS)
Réidech (30.56. LUXENBOURGEOIS)
Reis-Bohne (40.40. GERMAN)
Reismelde (35.6. GERMAN)
rèké (17.5. KANURI)
remayong (28.1. MALAY)
rembayog (27.14. MARANAO)
remdéng (20.1. KHMER, CENTRAL)
remédio do vaquerio (44.9. PORTUGUESE)
remolacha (24.11. TAGALOG)
remolacha de mesa (35.4. SPANISH)
remolacha hortelana (35.4. SPANISH)
remolacha roja (35.4. SPANISH)
remolatsa (35.4. CEBUANO)
remolatsa (35.4. PAMPANGAN)
remolatsa pula (35.4. TAGALOG)
remolatxa (35.4. CATALAN-VALENCIAN-BALEAR)
remuroa (62.15. MAORI)
remuroa (62.21. MAORI)
ren (30.1. MACEDONIAN)
ren (30.1. SERBO-CROATIAN)
renand (58.1. PASHTO, NORTHERN)
renchi (62.12. OROMO, WEST CENTRAL)
rendeng (24.7. JAVANESE)
rendeng (24.15. JAVANESE)
rengamutu (22.2. MAORI)
rengarenga (22.2. MAORI)
renkon (48.1a. JAPANESE)
rentira (40.12. KIRIBATI)
renzekalebas (38.15. VLAAMS)
renzumame (40.12. JAPANESE)
repa (30.36. BELARUSAN)
repa (30.41a. BELARUSAN)
repa (30.41a. SERBO-CROATIAN)
rěpa (30.41a. SORBIAN, LOWER)
rěpa (30.41a. SORBIAN, UPPER)
repa (35.4. BOSNIAN)
řepa burák pravá (35.4. CZECH)
repa karmavaja (30.22. RUSSIAN)
řepa obecná pravá (35.4. CZECH)
repani (30.53. GREEK)
repasta (24.8. SERBO-CROATIAN)
repja (30.21. ICELANDIC)
řepka (30.21. CZECH)
repka olejná (30.22. SLOVENIAN)
repkasto zvonče (33.2. MACEDONIAN)
repkë (30.56. ALBANIAN, TOSK)
repolho (30.27a. PORTUGUESE)
repolho chinês (30.40. PORTUGUESE)

ria (20.6. BUHID)
ria ol (40.1. SERER-SINE)
riantshi (40.43. HINDI)
riaungi (62.21. PANJABI, EASTERN)
ribásá (7.3. JUNKUN TAKUM)
ribbas ha-ginna (58.1. HEBREW)
ribbed gourd (38.21. ENGLISH)
ribbed luffa (38.21. ENGLISH)
ribbekål (30.28. DANISH)
rice bean (40.40. ASSAMESE)
rice bean (40.40. ENGLISH)
richpah (60.1. HEBREW)
ricin batard (39.1. FRENCH)
rid (20.6. BUHID)
ridged gourd (38.21. ENGLISH)
ridiche de grădină (30.54. ROMANIAN)
ridiche de grădină (30.56. ROMANIAN)
ridiche de lună (30.56. ROMANIAN)
ridikas (30.54. LITHUANIAN)
ridikėlis (30.56. LITHUANIAN)
ri[eef]ng n[ees]p (20.1. VIETNAMESE)
Riesen-Kürbis (38.14. GERMAN)
Riesen-Kürbis (38.16. GERMAN)
Riesen-Kürbis (38.18. GERMAN)
riga (30.48. SERBO-CROATIAN)
rigano (44.6. LADINO)
rìgízà (10.1. HAUSA)
rihan (44.5. HINDI)
rihan (44.5. URDU)
ríídín bàrééwáá (55.1. HAUSA)
ríídùwà (40.26. NGIZIM)
riiki (7.8. MAORI)
riímií (23.4. HAUSA)
riisisaniainen (4.1. FINNISH)
riiwai (62.24. MAORI)
rijstboon (40.40. VLAAMS)
rike (7.8. KIRIBATI)
rikhpa reḫanit (60.1. HEBREW)
rimang (62.23. INDONESIAN)
rimo-rimo (62.5. BICOLANO, CENTRAL)
rimskij latuk (27.32. RUSSIAN)
rindeline sibul (7.9a. ESTONIAN)
ringni (62.19. GUJARATI)
rinn (6.2. GAELIC, IRISH)
rinsag (56.1. PANJABI, EASTERN)
rion (58.1. GREEK)
ripolyo (30.27a. MARANAO)
Rippenkohl (30.28. GERMAN)
Rippen-melone (38.7a. GERMAN)
ripstomat (62.10. NORWEGIAN)
risagrasker (38.16. ICELANDIC)
risbønne (40.40. DANISH)
risbønne (40.40. NORWEGIAN)
rishabhajata (40.18. SANSKRIT)
rishabhi (40.18. SANSKRIT)
rishi buz (27.46. TAJIKI)
rishi buz (27.47. TAJIKI)
rishyaprokta (40.18. SANSKRIT)
riso d'acqua (17.8. ITALIAN)
riso d'America (17.8. ITALIAN)
riso indiano (17.8. ITALIAN)
ristmadar (27.7. ESTONIAN)
ritta kanda (19.1. TELUGU)
rivož (58.1. UZBEK)
riya (20.6. GIANGAN)
riz sauvage (17.8. FRENCH)
riz sauvage de Mandchourie (17.8. FRENCH)
rjapa (30.56. BULGARIAN)
rjazanec (7.11. BULGARIAN)
rjetkej (30.56. SORBIAN, UPPER)
rjewjeń (58.1. SORBIAN, UPPER)
roay (40.9. INDONESIAN)
roɓaji (40.3. HAUSA)
robambole (7.12. ROMANIAN)
robiglio (40.10. ITALIAN)
robirohi (23.4. ESTONIAN)
robol (1.1. STELLINGWERFS)
rochaka (7.3. SANSKRIT)
rochani (44.3. SANSKRIT)
rockdu (40.38. KOREAN)
rocket salad (30.48. ENGLISH)
rocoto (62.6. DANISH)
rocoto (62.6. DUTCH)
rocoto (62.6. ENGLISH)
rocoto (62.6. FRENCH)
Rocoto (62.6. GERMAN)
rocoto (62.6. ITALIAN)
rocoto (62.6. PORTUGUESE)
rocoto (62.6. SPANISH)
rocoto (62.6. VLAAMS)
rød cikorie (27.10b. DANISH)
rodal (38.1. BICOLANO, CENTRAL)
rødbede (35.4. DANISH)
rödbeta (35.4. SWEDISH)
rødbete (35.4. NORWEGIAN)
rode bestomaat (62.10. VLAAMS)
rode biet (35.4. DUTCH)
rode biet (35.4. VLAAMS)
rode kool (30.27b. DUTCH)
rode kool (30.27b. VLAAMS)

ruiponce (33.2. SPANISH)
rukevník východní (30.42. CZECH)
rukh tamater (62.7. NEPALI)
rukiew siewna (30.52. POLISH)
rukiew wodna (30.52. POLISH)
rukuɓu (23.3. HAUSA)
rukuɓuho (23.3. FULFULDE, ADAMAWA)
ruku-ruku (44.4. MALAY)
ruku-ruku (44.5. TAUSUG)
rukwiel szpalerowa (30.47. POLISH)
rum'-chäng (50.2. KHMER, CENTRAL)
rumach (58.1. KASHMIRI)
rumbhā (16.1a. PALI)
rumex à écusson (58.4. FRENCH)
rumex oseille (58.2. FRENCH)
rumex patience (58.3. FRENCH)
rumi bânjân (62.9. PASHTO, NORTHERN)
rumi nane (44.3. TURKISH)
rumput beremi (59.2. MALAY)
rumput gajah (17.3. INDONESIAN)
rumput gajah (17.3. MALAY)
rumu (39.4. QUECHUA, CUSCO)
run dou (40.26. CHINESE, XIANG)
rungia (21.2. DANISH)
rungia (21.2. DUTCH)
rungia (21.2. ENGLISH)
rungia (21.2. FRENCH)
Rungia (21.2. GERMAN)
rungia (21.2. ITALIAN)
rungia (21.2. PORTUGUESE)
rungia (21.2. SPANISH)
rũngũ (38.20. GIKUYU)
runkelis (35.4. LITHUANIAN)
ruohosipuli (7.11. FINNISH)
ruoka parsa (13.2. FINNISH)
ruokasipuli (7.3. FINNISH)
ruoti (27.15. FINNISH)
ruotsinraunioyrtti (29.3. FINNISH)
rupjseklu lecas (40.12. LATVIAN)
rũreru (13.2. GIKUYU)
ruridama-no-ki (39.5. JAPANESE)
rūsganā dienziede (13.3. LATVIAN)
rushnut (10.1. ENGLISH)
rusivi (44.4. KWANGALI)
ruski gavez (29.3. MACEDONIAN)
ruskija babi (40.34a. BELARUSAN)
rüssèi kaèw (17.1. KHMER, CENTRAL)
rüssèi khléi (17.1. KHMER, CENTRAL)
rüssèi préi (17.1. KHMER, CENTRAL)
rüssèi roliëk (17.1. KHMER, CENTRAL)
russekål (30.42. NORWEGIAN)
russervin (69.1. DANISH)
Russian comfrey (29.3. ENGLISH)
Rüstiele (30.41b. GERMAN)
ruszpunka jadalna (67.2. POLISH)
rutabaga (30.22. ALBANIAN, TOSK)
rutabaga (30.22. ENGLISH)
rutabaga (30.22. FILIPINO)
rutabaga (30.22. FRENCH)
rutabaga (30.22. ITALIAN)
rutabaga (30.22. JAPANESE)
rutabaga (30.22. PORTUGUESE)
rutabaga (30.22. SPANISH)
rutabaga (30.22. VLAAMS)
rutchyakanda (8.3. SANSKRIT)
Ruten-Kohl (30.17. GERMAN)
rutki (30.53. LATVIAN)
rutki (30.56. LATVIAN)
ruusukaali (30.29. FINNISH)
ruusupapu (40.24. FINNISH)
ružičková kapusta (30.29. CZECH)
rvfo-tvhvyv (38.14. MUSKOGEE)
rvfo-tvhvyv (38.16. MUSKOGEE)
rvfo-tvhvyv (38.17. MUSKOGEE)
rwdins (30.22. WELSH)
ryoku-chiku (17.1. JAPANESE)
ryokutou (40.38. JAPANESE)
ryokuyô-kanran (30.35. JAPANESE)
rysboon (40.40. AFRIKAANS)
rysk rot (26.4. SWEDISH)
ryssgubbe (30.42. SWEDISH)
rytinė engra (30.42. LITHUANIAN)
ryvässipuli (7.2. FINNISH)
rzepa (30.41a. POLISH)
rzepa liściowa (30.41b. POLISH)
rzepa turnips (30.22. POLISH)
rzepa turnips (30.41a. POLISH)
rzepa włoska (30.41c. POLISH)
rzepak (30.21. POLISH)
rzewień (58.1. POLISH)
rzeżucha łąkowa (30.44. POLISH)
rzezucha ogrodowa (30.51. POLISH)
rzodkiew (30.56. POLISH)
rzodkiew chińska (30.54. POLISH)
rzodkiew długa (30.53. POLISH)
rzodkiewka (30.56. POLISH)
sa (38.3. VAI)
sa (38.5. KONO)
sa (38.5. VAI)
sa soo (40.18. MANDING-MANDINKA)

sakonu (28.1. BAATONUM)
sakorab (7.10. MARANAO)
sakpa (45.3. GA)
sakpa (45.4. GA)
sakpa (45.4. KRIO)
sakribonte (38.16. GA)
sakribonte (38.18. GA)
sakrithphalaa (16.1a. SANSKRIT)
saksankirveli (24.16. FINNISH)
saku (15.2. KHMER, CENTRAL)
saku (59.2. GOLA)
sàkuru (11.1. BAATONUM)
sakuyot (63.1. BICOLANO, CENTRAL)
sakuyot (63.1. ILOCANO)
sakuyot (63.1. PAMPANGAN)
salaadi soppu (27.32. KANNADA)
salaat (27.32. KANNADA)
salaati (27.32. ZARMA)
salaattifenkoli (24.13. FINNISH)
salaattikiinankaali (30.40. FINNISH)
salaattikleitonia (59.1. FINNISH)
salaattisikuri (27.10a. FINNISH)
salaattivirmakka (27.21. FINNISH)
salad (27.29. ASSAMESE)
salad (27.29. HINDI)
salad (27.29. ORIYA)
salad (27.29. PANJABI, EASTERN)
salad (27.31. ASSAMESE)
salad (27.31. BENGALI)
salad (27.31. BULU)
salad (27.31. HINDI)
salad (27.31. KRIO)
salad (27.31. MALAY)
salad (27.31. MÒORÉ)
salad (27.31. ORIYA)
salad (27.31. PANJABI, EASTERN)
salad (27.32. ASSAMESE)
salad (27.32. BENGALI)
salad (27.32. BULU)
salad (27.32. HINDI)
salad (27.32. HINDUSTANI)
salad (27.32. KRIO)
salad (27.32. MALAY)
salad (27.32. MÒORÉ)
salad (27.32. ORIYA)
salad (27.32. PANJABI, EASTERN)
salad pat (27.29. ASSAMESE)
salad pat (27.30. ASSAMESE)
salad pat (27.31. ASSAMESE)
salad pat (27.32. ASSAMESE)
salada (27.31. SUNDA)
salada (27.32. SUNDA)
salada (30.52. SUNDA)
salada air (30.52. INDONESIAN)
salada bogor (27.31. SUNDA)
salada bogor (27.32. SUNDA)
salada bokor (27.32. SUNDA)
salada kola (27.32. SINHALA)
salada liangan (24.17. SUNDA)
salade (27.32. PULAR)
saladi (27.31. BEMBÉ)
saladi (27.31. DAGAARE, SOUTHERN)
saladi (27.31. GOURMANCHÉMA)
saladi (27.31. KOONGO)
saladi (27.32. BEMBÉ)
saladi (27.32. DAGAARE, SOUTHERN)
saladi (27.32. GOURMANCHÉMA)
saladi (27.32. GOURMANTCHÉ)
saladi (27.32. KOONGO)
saladi (27.32. NYANJA)
saladi (30.52. SWAHILI)
saladri (24.4. SUNDA)
salagam (30.41a. NEPALI)
salag-salag (38.22. TAGALOG)
salaikat (64.3. JAVANESE)
salakhoe (55.1. SUSU)
salakunda (63.1. MANDING-MANDINKA)
salampauyan (20.2. GIANGAN)
salanggi (36.3. BUHID)
salanoi (44.5. FILIPINO)
salat (27.31. KAZAKH)
salat (27.31. KIRGHIZ)
salat (27.31. TATAR)
salat (27.31. TURKMEN)
salat (27.31. UZBEK)
salat (27.32. KAZAKH)
salat (27.32. KIRGHIZ)
salat (27.32. NORWEGIAN)
salat (27.32. TATAR)
salat (27.32. TURKMEN)
salat (27.32. UZBEK)
salat (27.32. WOLOF)
salát chřestový (27.30. CZECH)
salát hlávkovy (27.31. CZECH)
šalát hlávkový (27.31. SLOVAK)
salat kočannyj (27.31. RUSSIAN)
salat kočhannyi (27.31. RUSSIAN)
salát křezu (27.32. CZECH)
salat latuk (27.32. BELARUSAN)
salat latuk (27.32. RUSSIAN)

salát listový (27.32. CZECH)
šalát listovy (27.32. SLOVAK)
salat litovol (27.32. RUSSIAN)
salat pol'ovij (67.2. UKRAINIAN)
salat rimski (35.3. RUSSIAN)
salát římský (27.32. CZECH)
šalát římsky (27.32. SLOVAK)
salat romain bladsalat (27.32. NORWEGIAN)
šalát záhradný (27.31. SLOVAK)
šalát záhradný (27.32. SLOVAK)
sälata (27.31. AMHARIC)
salata (27.31. ARABIC)
salata (27.31. BELARUSAN)
salata (27.31. CROATIAN)
sälata (27.32. AMHARIC)
salata (27.32. BOSNIAN)
salata (27.32. CROATIAN)
salâ ṯa (27.32. SIDAMO)
saláta bakszakál (27.46. HUNGARIAN)
salată căpătină (27.31. ROMANIAN)
salată de cimp (33.2. ROMANIAN)
salată de foi (27.32. ROMANIAN)
sałata długolistna (27.32. POLISH)
sałata do skubania (27.32. POLISH)
salata endivija (27.9. BULGARIAN)
salata glavičasta (27.31. SERBO-CROATIAN)
salata glavičasta (27.32. SERBO-CROATIAN)
sałata głowiasta (27.31. POLISH)
salata golovčastij (27.31. UKRAINIAN)
sałata indiańska (27.29. POLISH)
salata latuk (27.31. BELARUSAN)
sałata liskowa (27.32. POLISH)
salata romana (27.32. SERBO-CROATIAN)
salata romeni (27.32. GEORGIAN)
sałata rzymska (27.32. POLISH)
sałata szparagowa (27.30. POLISH)
salata vrtna (27.31. SERBO-CROATIAN)
saláta-katáng (27.9. HUNGARIAN)
salatalik (38.13a. TURKISH)
Salatbohne (40.26. GERMAN)
salatfifill (27.9. ICELANDIC)
Salat-Gurke (38.13a. GERMAN)
salatna cichorija (27.10a. BULGARIAN)
salatno cveklo (35.4. BULGARIAN)
salato (27.32. MANINKAKAN, WESTERN)
salat-romen (27.32. UKRAINIAN)
salatsennep (30.48. DANISH)
salatsennep (30.48. NORWEGIAN)
salatsigur (27.10a. ESTONIAN)
salatsikori (27.10a. NORWEGIAN)
salattu (27.31. TAMIL)
salattu (27.32. TAMIL)
salātu baldriņš (67.2. LATVIAN)
salàtu selerija (24.4. LATVIAN)
salatum tikin (24.15. INDONESIAN)
salatun (24.15. INDONESIAN)
sa-lat-ywet (27.31. BURMESE)
sa-lat-ywet (27.32. BURMESE)
salayo (63.1. IBANAG)
salaysay (7.2. BINUKID)
salbissa (7.2. GOURMANTCHÉ)
salbissa-djãg (7.10. GOURMANTCHÉ)
saldais fenhelis (24.12. LATVIAN)
saldais kartupelis (36.3. LATVIAN)
salderi djepang (24.10. MALAY)
salderi Jepang (24.10. INDONESIAN)
saledri (24.4. SUNDA)
saleggia (67.2. ITALIAN)
salgam (30.22. FIJIAN)
salgam (30.22. URDU)
salgam (30.41a. HINDI)
sālgam (30.41a. MUNDARI)
salgam (30.41a. SANTALI)
şalgam (30.41a. TURKISH)
şalgam lahanasi (30.22. TURKISH)
salgham (30.41a. GUJARATI)
salgham (30.41a. HINDI)
salgom (30.41a. ASSAMESE)
salgum (30.22. URDU)
salgum (30.41a. BODO)
salgum (30.41a. MARATHI)
salgum (30.41a. ORIYA)
salgum salgum (30.41a. URDU)
salib l-art (57.1. MALTESE)
saligot (64.3. FRENCH)
saligun (38.23. TAUSUG)
saliiya (38.3. CAROLINIAN)
salit (27.29. GUJARATI)
salit (27.29. MARATHI)
salit (27.31. GUJARATI)
salit (27.31. MARATHI)
salit (27.32. GUJARATI)
salit (27.32. MARATHI)
salk (35.3. ARABIC)
salklë (35.3. ALBANIAN, TOSK)
sallaat (27.31. KHMER, CENTRAL)
sallaat (27.32. KHMER, CENTRAL)
sallad (27.31. PANJABI, EASTERN)
sallad (27.32. PANJABI, EASTERN)
sallada lök (7.5. SWEDISH)

salladu keerai (27.32. TAMIL)
salooalu (62.24. HINDI)
salot (7.2. AFRIKAANS)
šalot (7.2. BULGARIAN)
šalot (7.2. UKRAINIAN)
šalota (7.2. SORBIAN, UPPER)
salota (27.31. LITHUANIAN)
salota (27.32. LITHUANIAN)
šalotes sīpoli (7.2. LATVIAN)
salotinė sultenė (67.2. LITHUANIAN)
šalotka (7.2. CZECH)
šalotka (7.2. SLOVENIAN)
salotos (27.31. LITHUANIAN)
šalott (7.2. ESTONIAN)
salotten sipuli (7.2. FINNISH)
salottisipuli (7.2. FINNISH)
salsa (24.19. CRIOULO, CAPE VERDE)
salsa (24.19. PORTUGUESE)
salsa (24.20. PORTUGUESE)
salsa comum (24.19. PORTUGUESE)
salsa crespa (24.20. PORTUGUESE)
salsa de pastor (24.3. SPANISH)
salsa frisada (24.19. PORTUGUESE)
salsa italiana (24.20. PORTUGUESE)
salsa tuberosa (24.21. PORTUGUESE)
salsão (24.4. PORTUGUESE)
salsefia (27.46. POLISH)
salsefia hiszpańska wezymord czarny korzen (27.37. POLISH)
salsefia łąkowa (27.47. POLISH)
salsèfrica (27.47. ITALIAN)
salsey (27.46. GAELIC, MANX)
salsifi (27.46. FRENCH)
sal'sifi (27.46. RUSSIAN)
salsifi blanc (27.46. FRENCH)
salsifi blanco (27.46. SPANISH)
salsifi branco (27.46. PORTUGUESE)
salsifi negro (27.38. PORTUGUESE)
salsifi negro (27.38. SPANISH)
salsifis (27.46. CZECH)
salsifis (27.46. FRENCH)
salsifis des prés (27.47. FRENCH)
salsifis noir (27.38. FRENCH)
salsify (27.46. ASSAMESE)
salsify (27.46. ENGLISH)
salsinha (24.19. PORTUGUESE)
salsola (35.8. ENGLISH)
salsola (35.8. PORTUGUESE)
salsola (35.8. VLAAMS)
salsola de Komarov (35.7. FRENCH)
salsola de Komarov (35.7. PORTUGUESE)
salsola de Komarov (35.7. SPANISH)
salsola di Komarov (35.7. ITALIAN)
salsola soda (35.8. FRENCH)
šaltmėtė (44.3. LITHUANIAN)
salu (55.1. DAGBANI)
salusifai (27.46. JAPANESE)
saluyot (59.3. TAGBANWA)
saluyot (63.1. FILIPINO)
saluyot (63.1. ILOCANO)
saluyot (63.1. TAGALOG)
saluyot a bunga (45.1. ILOCANO)
saluyot nga bulang (63.1. ILOCANO)
saluyot nga bunga (45.1. ILOCANO)
saluyu (63.1. IBANAG)
saluyuk (45.1. ITAWIT)
saluyut (45.1. ITAWIT)
saluyut (63.1. ILOCANO)
salvinavelya (11.2. GUJARATI)
salyote (38.26. CEBUANO)
salzifi (27.46. BULGARIAN)
sam (40.13. FIJIAN)
sama (20.2. SAMOAN)
samadh (58.2. GAELIC, IRISH)
samag albostan (24.16. ARABIC)
samaiji (44.8. FULFULDE, ADAMAWA)
samaiwol (44.8. FULFULDE, ADAMAWA)
samambaia do campo (2.1. PORTUGUESE)
saman nsusuwa (62.23. TWI)
samana (7.2. DANGME-KROBO)
samana (7.3. DANGME-KROBO)
samana tʃupã (7.10. DANGME-KROBO)
samanatʃupa (7.2. DANGME-KROBO)
samangka (38.3. SUNDA)
samanka (38.3. HINDI)
samanntɔrewa (62.23. TWI)
šämäqqo (17.1. AMHARIC)
šämäqqo (17.2. AMHARIC)
šämäqqo (17.4. AMHARIC)
samara-kokadi (27.8. GUJARATI)
samaray (62.21. BUHID)
samba (8.3. ORIYA)
sambaar moggu (34.1. KANNADA)
sambal poosani (38.1. TAMIL)
sambalpushani kaai (38.1. TAMIL)
sambe (40.3. KANNADA)
sambiru (62.17. BAATONUM)
samboangan (36.3. MARANAO)
sambram gøjah (7.3. BODO)
sambram gufut (7.10. BODO)

sambu (8.4. TAMIL)
samho (30.36. SUNDA)
sami (16.1a. TAMIL)
sámle (40.1. DAGBANI)
samma (67.2. ARABIC)
sammakonputki (24.7. FINNISH)
samouya (39.4. THEMNE)
sampeu (39.4. SUNDA)
samputantastha (17.7. SANSKRIT)
samsam (23.3. TAGBANWA)
samsaq (7.10. UYGHUR)
san (39.4. VIETNAMESE)
San Fernando gabi (8.8. TAGALOG)
San Francisco (39.3. TAGALOG)
san mafé (59.2. BAMANANKAN)
san mafé (59.2. SONGHAY)
san se jin (68.1. CHINESE, YUE)
san ye sheng cai (27.32. CHINESE, YUE)
san ye sheng cai (27.33. CHINESE, YUE)
san ye wo fu (27.32. CHINESE, WÚ)
san yeh kuei jen tsao (27.8. CHINESE, MANDARIN)
sana (17.7. SAMOAN)
sana (40.5. BODO)
sanaadir (17.7. AFAR)
sanab (62.21. BRETON)
sänābi (30.17. HARARI)
sanagalu (40.5. TELUGU)
sanagulu (40.5. TELUGU)
sanal (63.1. TAMIL)
sananjalka (2.1. FINNISH)
sanaraje (11.7. DAGBANI)
sanaray (62.21. BUHID)
sanchi (23.2. BENGALI)
sand leek (7.12. ENGLISH)
sândaèk an gen sar (40.8. KHMER, CENTRAL)
sândaêk ângkât miëhs (40.40. KHMER, CENTRAL)
sândaêk ângkuy (40.43. KHMER, CENTRAL)
sândaèk barang (40.26. KHMER, CENTRAL)
sândaèk ba:y (40.38. KHMER, CENTRAL)
sândaèk dai (40.2. KHMER, CENTRAL)
sândaêk kâng (40.43. KHMER, CENTRAL)
sândaêk khmau (40.41. KHMER, CENTRAL)
sândaèk klöng (40.2. KHMER, CENTRAL)
sândaêk krâhâm (40.41. KHMER, CENTRAL)
sândaèk kroëb sâ (40.2. KHMER, CENTRAL)
sândaèk muul (40.28. KHMER, CENTRAL)
sândaêk riech mieh (40.40. KHMER, CENTRAL)
sândaêk sâ (40.41. KHMER, CENTRAL)
sândaèk sieng (40.8. KHMER, CENTRAL)
sândaêk troeung (40.42. KHMER, CENTRAL)
sandalye sazi (10.2. TURKISH)
sandan birii (19.1. HAUSA)
sandan dutse (19.1. HAUSA)
sandan yan bori (19.1. HAUSA)
sandhi (20.6. HINDI)
sandia (38.3. ILOCANO)
sandia (38.3. SPANISH)
sàndij (30.17. KASHMIRI)
sandika (40.10. SANSKRIT)
sandiya (38.3. BICOLANO, CENTRAL)
sandíya (38.3. HILIGAYNON)
sandiya (38.3. IBANAG)
sandiya (38.3. ILOCANO)
sandiya (38.3. ITAWIT)
sandiya sa raba (38.3. HANUNOO)
sandõ (45.4. KISSI, NORTHERN)
sando (45.4. KONO)
sandusi (40.34a. BURMESE)
sanduyung (17.5. BUHID)
sanely (17.7. BINUKID)
sa-nɛsɛ (38.5. KONO)
s’a:ng (17.2. LAO)
s’a:ng kh’am’ (17.1. LAO)
sāngā (36.3. MUNDARI)
sangai (20.6. TAMIL)
sãŋga-lɛ (38.3. BULLOM SO)
sangchu (27.31. KOREAN)
sangday (39.4. IBALOI)
sanggariya (62.5. BICOLANO, CENTRAL)
sanghata pathrika (24.13. SANSKRIT)
sanghatica (64.3. SANSKRIT)
sanghusip (39.3. HANUNOO)
sangíg (44.5. WÁRAY-WÁRAY)
sanglay kayo (39.4. TAGBANWA)
sangou (17.7. BEMBÉ)
sangou (17.7. KOONGO)
sangseri (23.4. SUNDA)
sanguinaria (23.2. SPANISH)
sanhusip (39.3. HANUNOO)
sankalu (40.21. HINDI)
sankalu (40.21. URDU)
sankalwa (64.3. HAUSA)
šänk$_w$är (17.5. AMHARIC)
saŋk-ɛ (20.6. BULLOM SO)
sanley (17.7. MANOBO)

sânndaèk dèi (40.1. KHMER, CENTRAL)
sano kerau (40.27. NEPALI)
sanpat (63.1. NEPALI)
sansandok (23.6. ILOCANO)
sansoni (40.18. HAUSA)
sansonv (30.24. KONKANI)
santa (40.43. HINDI)
santa-maria (62.21. CRIOULO, CAPE VERDE)
sãntõ (45.4. SUSU)
santoma cimarrona (23.2. SPANISH)
santoreggia (44.9. ITALIAN)
santou (45.1. SUSU)
santra pawra (17.5. HINDUSTANI)
santu (45.1. SUSU)
santu (45.4. SUSU)
santui (45.4. SUSU)
santuméri (16.1a. SUSU)
santumir (16.1a. BULLOM SO)
santuŋ (45.4. MALIMBA)
santu-na (45.4. YALUNKA)
sanyi (27.8. MANDING-MANDINKA)
são caetano (38.23. CRIOULO, CAPE VERDE)
sáóppiso (38.14. BLACKFOOT)
sáóppiso (38.16. BLACKFOOT)
sáóppiso (38.17. BLACKFOOT)
sáóppiso (38.18. BLACKFOOT)
sap kereviz (24.4. TURKISH)
sapan kizhangu (8.4. TAMIL)
sapcha-kovock (38.18. UZBEK)
saperge-erwt (40.13. VLAAMS)
saperis (13.2. INDONESIAN)
sapersis (13.2. INDONESIAN)
saphong (2.1. BLAAN, SARANGANI)
saphuril kumra (38.14. BENGALI)
saphuril kumra (38.16. BENGALI)
saphuril kumra (38.18. BENGALI)
sạpin dare (32.1. SANTALI)
saping na saba (16.1a. TAGALOG)
sapnuan (8.4. BUHID)
sapo (38.22. KRIO)
sapo-ɔkrɔ (38.22. KRIO)
sapodila (38.23. KRIO)
sapor-kpmane (38.22. DAGAARE, SOUTHERN)
sappaatthukkalli (32.1. TAMIL)
sar (44.9. DANISH)
sar (44.9. ICELANDIC)
sar (44.9. NORWEGIAN)
sara (38.3. VAI)
sara (38.5. MANDING-MANDINKA)
sara (38.5. VAI)
sara chema (8.1. TELUGU)
sara kumba (38.3. MANINKAKAN, EASTERN)
sara mriçẖ (62.2. PASHTO, NORTHERN)
saraana (13.4. MONGOLIAN)
sarach hakinamon (3.1. HEBREW)
sarach hamaim (4.1. HEBREW)
sarach yapani (3.2. HEBREW)
saradèni (38.3. JULA)
sarahattika (27.2. SANSKRIT)
sarai-yipun (42.1. THAI)
sarakumba (38.3. MANDING-MANDINKA)
saramago amarelo (30.47. PORTUGUESE)
saramago oriental (30.42. PORTUGUESE)
sara-mɛsɛ (38.5. MANINKAKAN, EASTERN)
sarangan (36.3. BUHID)
sarangun (56.1. HINDI)
saraswataku (24.7. TELUGU)
saraswathi aaku (24.7. TELUGU)
saraswathi soppu (24.7. KANNADA)
saraypati (27.17. TURKISH)
sarbada (9.1. NEPALI)
sarda (38.7b. PASHTO, NORTHERN)
sarɛ (38.5. MANINKAKAN, EASTERN)
sarepta kapsasrohi (30.17. ESTONIAN)
sareptä senap (30.17. SWEDISH)
sarepta sinep (30.17. ESTONIAN)
Sareptamosterd (30.17. DUTCH)
sareptamosterd (30.17. VLAAMS)
sareptansinappi (30.4. FINNISH)
Sarepta-Senf (30.17. GERMAN)
sareptasennep (30.17. NORWEGIAN)
sárgadinnye (38.7a. HUNGARIAN)
sárgadinnye (38.7b. HUNGARIAN)
sárgadinnye (38.8. HUNGARIAN)
sárgadinnye (38.9. HUNGARIAN)
sárgadinnye (38.10. HUNGARIAN)
sárgadinnye (38.11. HUNGARIAN)
sárgaliliom (13.3. HUNGARIAN)
šargarepa (24.11. BOSNIAN)
sárgarépa (24.11. HUNGARIAN)
sarha nsoe (23.3. AKAN-AKUAPEM)
sarhoş otu (62.12. TURKISH)
sari batakseven (14.1. TURKISH)
sari topalak (10.1. TURKISH)
sariboya (20.2. TURKISH)
sarimsak (7.14. TURKISH)

sɔbuɛ (23.3. DANGME-KROBO)
sɔbui (34.2. ÉWÉ)
sɔ̃chal (27.42. PANJABI, EASTERN)
sɔɛ (40.25. KONO)
šćehel (58.2. SORBIAN, UPPER)
sɔgbee (34.2. ÉWÉ)
Schabzierger-Klee (40.32. GERMAN)
schakeri schora (38.16. MALAYALAM)
Schälgurke (38.13a. GERMAN)
Schangen-Melone (38.10. GERMAN)
Schantungkohl (30.40. GERMAN)
scharfer Mauerpfeffer (37.1. GERMAN)
scharlaken aubergine (62.14c. VLAAMS)
scharlottenlök (7.2. SWEDISH)
schav (35.3. YIDDISH)
Scheimkraut (31.1. GERMAN)
scherpe komkommer (38.31. VLAAMS)
schijfcactus (32.1. DUTCH)
Schild-Ampfer (58.4. GERMAN)
schilfartige Pfeilwurz (15.2. GERMAN)
schinoprasson (7.11. GREEK)
Schlangengurke (38.13a. GERMAN)
Schlangengurke (38.29. GERMAN)
Schlangen-Haargurke (38.29. GERMAN)
Schlaugenlauch (7.12. GERMAN)
Schlitz-Wegerich (57.1. GERMAN)
Schnittkohl (30.23. GERMAN)
Schnittlauch (7.11. GERMAN)
schoezole (40.26. STELLINGWERFS)
schopfige Traubenhyazinthe (13.5. GERMAN)
schorseneer (27.38. DUTCH)
schorsoneer (27.38. VLAAMS)
Schorzonere (27.38. GERMAN)
schtor (7.10. ARMENIAN)
Schufflör (30.26. LUXENBOURGEOIS)
Schwammgurke (38.22. GERMAN)
Schwammkürbis (38.22. GERMAN)
schwarze Nachtschatten (62.21. GERMAN)
schwarze Teufelskirsche (62.21. GERMAN)
schwarzer Senf (30.24. GERMAN)
Schwarzwurzel (27.38. GERMAN)
Schwedischer Beinwell (29.3. GERMAN)
Schwertbohne (40.3. GERMAN)
Schwertbohne (40.4. GERMAN)
schwimmende Wassernuß (64.3. GERMAN)
šćir (23.3. SERBO-CROATIAN)
sɔkɛ (62.15. SUSU)
sɔ-kpɔ-mɛsɛ (40.43. VAI)
sɔlwi (34.2. ÉWÉ)
sɔma (23.4. AJA)
sɔma (23.4. FON)
sɔ-mɛsɛ (40.2. KONO)
sɔ-mɛsɛ (40.43. KONO)
sɔndi (23.3. KONO)
scolyme d'Espagne (27.36. FRENCH)
scolyme tache (27.37. FRENCH)
scolymus (27.36. ENGLISH)
scolymus (27.36. VLAAMS)
scoparia (35.2. ITALIAN)
scortionera (27.38. ROMANIAN)
scorzobianca (27.46. ITALIAN)
scorzoner (27.38. DANISH)
scorzonera (27.38. ENGLISH)
scorzonera (27.38. ITALIAN)
scorzonera di spagna (27.38. ITALIAN)
scorzoneră hispanică (27.38. ROMANIAN)
scorzoneră spaniolă (27.38. ROMANIAN)
scorzonère (27.38. FRENCH)
scorzonère d'Espagne (27.38. FRENCH)
scorzonère noire (27.38. FRENCH)
Scotch kale (30.35. ENGLISH)
sɔrel (45.4. KRIO)
sɔrɛl (45.4. KRIO)
sɔrɛŋ (63.1. PULAAR)
sɔso (40.25. MANDING-MANDINKA)
ščvel' morščinistyj (58.2. RUSSIAN)
se'np'odal-e' (50.2. KIOWA)
sea kale (30.45. ENGLISH)
sealbhag (58.2. GAELIC, SCOTTISH)
seallóid (7.2. GAELIC, IRISH)
sēǣgārrha (64.2. PANJABI, EASTERN)
sèb (40.43. PULAAR)
seb (40.43. WOLOF)
sèb u tubab (40.3. WOLOF)
seba (16.1a. PANGASINAN)
sebe (9.1. SUNDA)
sɛbɛ (62.19. GA)
sɛbɛ baa (62.18. GA)
sebijari (40.25. DUTCH)
sebijari (40.25. SRANAN)
sebo (62.14b. BAMILEKE)
seboliat (7.3. ITAWIT)
sebu (40.43. WOLOF)
sebulnusa (62.23. SUBANON)
sebze bezelyesi (40.28. TURKISH)
sebze eğreltisi (5.1. TURKISH)
sedanina coltivata (24.23. ITALIAN)
sedano (24.4. ITALIAN)
sedano da coste (24.4. ITALIAN)
sedano-rapa (24.5. ITALIAN)

sedep malem (13.3. INDONESIAN)
sederi (24.4. JAVANESE)
sedon (37.1. GREEK)
sedum (37.1. ENGLISH)
sedum (37.1. VLAAMS)
sédum âcre (37.1. FRENCH)
see genasu (11.5. KANNADA)
seegumbala (38.16. KANNADA)
seekool (30.45. AFRIKAANS)
seema mullanki (30.41a. MALAYALAM)
seema vankaya (62.9. TELUGU)
seemagonguru (45.4. TELUGU)
seemagummadi (38.16. TELUGU)
seemai avarai (40.26. TAMIL)
seemai averai kody (40.26. TAMIL)
seemaikkasuru (45.4. TAMIL)
seemajeelakarra (24.13. TELUGU)
seema-kathirikkaayi (38.25. TAMIL)
seemamirapakaaya (62.5. TELUGU)
seemay tekkali (62.9. TAMIL)
seeme badane (38.25. KANNADA)
seeme badane kaayi (38.25. KANNADA)
seeme kattirikkai (38.25. TAMIL)
seeme pundi (45.4. KANNADA)
seeme pundi soppu (45.4. KANNADA)
seemegogu (45.4. TELUGU)
seeniavaraikai (40.6. TAMIL)
seenikilangu (36.3. TAMIL)
seepe kumbuda (38.16. TULU)
seer (7.10. FARSI, WESTERN)
seeramsiaq (7.10. CHECHEN)
Seerose (50.2. GERMAN)
sɛɛsu (38.22. DENDI)
seethambel (50.2. MALAYALAM)
sefoorjench (27.38. FARSI, WESTERN)
sega' (66.3. BINUKID)
segadilla (40.29. CEBUANO)
ségal (38.3. SERER-SINE)
segan jantan (59.2. MALAY)
segappuvalli kizhangu kody (36.3. TAMIL)
segapu thambattai (40.4. TAMIL)
segaputham battai (40.4. TAMIL)
sɛgbula-lowa (38.18. MENDE)
segha (40.25. TIV)
segidilya (40.29. AKLANON)
segidilya (40.29. BICOLANO, CENTRAL)
segidilya (40.29. WÁRAY-WÁRAY)
segidilyas (40.29. TAGALOG)
segili (16.1a. TAMIL)
segu betawi (15.2. INDONESIAN)
segurelha (44.9. PORTUGUESE)
şehvet otu (10.1. TURKISH)
sehwete se sesweu (24.19. SOTHO, SOUTHERN)
sę̨i-goup (32.1. KIOWA)
sę̨i-'Hdā'-gɑ (32.1. KIOWA)
seiyô-ha-geito (23.3. JAPANESE)
seiyou (38.16. JAPANESE)
seiyou-kabocha (38.14. JAPANESE)
seiyou-kabocha (38.16. JAPANESE)
seiyou-kabocha (38.18. JAPANESE)
seiyo-wasabi (30.1. JAPANESE)
sėjamasis avinžirnis (40.5. LITHUANIAN)
sėjamasis pelėžirnis (40.10. LITHUANIAN)
sėjamasis žirnis (40.28. LITHUANIAN)
sėjamoji petražolė (24.19. LITHUANIAN)
sėjamoji petražolė (24.21. LITHUANIAN)
sėjamoji petražole (30.51. LITHUANIAN)
sėjamoji pipirnė (30.51. LITHUANIAN)
sējas burkāns (24.11. LATVIAN)
sejas dedestina (40.10. LATVIAN)
sējas lucerna (40.17. LATVIAN)
sējas zirņi (40.28. LATVIAN)
seker kamişi (17.5. TURKISH)
sekhu (45.1. FON)
səkkʷar dənnəč (36.3. AMHARIC)
sekkirai (27.10a. TAMIL)
sekstainā celozija (23.4. LATVIAN)
sekule-gbɛŋgbɛ (62.2. SUSU)
sekunā (36.3. MAITHILI)
sekwa (38.21. BICOLANO, CENTRAL)
sekwa (38.29. CEBUANO)
sel kambang (38.20. TOK PISIN)
selada (27.31. INDONESIAN)
selada (27.31. JAVANESE)
selada (27.31. MALAY)
selada (27.32. INDONESIAN)
selada (27.32. JAVANESE)
selada (27.32. MALAY)
selada air (30.52. INDONESIAN)
selada air (30.52. MALAY)
selada bulat (27.31. MALAY)
selada kerup (27.31. INDONESIAN)
selada kerup (27.32. INDONESIAN)
seladah (30.52. INDONESIAN)
seladeri (24.4. INDONESIAN)
seladeri (24.4. MALAY)
seladi air (30.52. INDONESIAN)
seladren (24.17. JAVANESE)
selae (27.32. SOTHO, NORTHERN)

seminga (16.1a. KONO)
semiz otu (59.3. TURKISH)
semizotu (59.2. TURKISH)
sempi (40.15. DAGBANI)
sen (17.1. GOLA)
sen (48.1a. VIETNAMESE)
senaga (40.5. TELUGU)
senai kizhangu (8.3. TELUGU)
sênâluk (58.1. ALEUT)
senape (30.3. ITALIAN)
senape (30.4. ITALIAN)
senape (30.5. ITALIAN)
senape (30.6. ITALIAN)
senape (30.7. ITALIAN)
senape (30.8. ITALIAN)
senape (30.9. ITALIAN)
senape (30.10. ITALIAN)
senape (30.11. ITALIAN)
senape (30.12. ITALIAN)
senape (30.13. ITALIAN)
senape (30.14. ITALIAN)
senape (30.15. ITALIAN)
senape (30.16. ITALIAN)
senape (30.18. ITALIAN)
senape (30.19. ITALIAN)
senape (30.20. ITALIAN)
senape bianca (30.57. ITALIAN)
senape indiana (30.17. ITALIAN)
senape spinacio (30.39. ITALIAN)
senartok (58.1. ALEUT)
šənbəra (40.5. AMHARIC)
sendapoki (27.38. BASQUE)
sendjap (38.16. MALIMBA)
sendjap (38.17. MALIMBA)
sénegré (40.32. FRENCH)
seneni thogo-na (40.43. YALUNKA)
Senf (30.4. GERMAN)
Senf (30.6. GERMAN)
Senf (30.7. GERMAN)
Senf (30.8. GERMAN)
Senf (30.9. GERMAN)
Senf (30.10. GERMAN)
Senf (30.11. GERMAN)
Senf (30.15. GERMAN)
Senf (30.16. GERMAN)
senf (30.17. BOSNIAN)
Senf (30.20. GERMAN)
Senfkapper (34.2. GERMAN)
Senfkohl (30.24. GERMAN)
sengeri (53.1. ASSAMESE)
senggang bener (23.3. SUNDA)
senggang beureum (23.3. SUNDA)
senggang bodas (23.3. SUNDA)
senggang cucuk (23.3. SUNDA)
senggang itik (23.3. SUNDA)
sengkuang (40.21. INDONESIAN)
sengkuang (40.21. MALAY)
sengkuang (40.22. MALAY)
sengkuwang (40.21. MALAY)
seŋgo (38.5. KISSI, NORTHERN)
seŋgo-pɔmbɔ (38.5. KISSI, NORTHERN)
seŋgu-chielachalo (38.5. KISSI, NORTHERN)
seṉhur (32.1. HINDUSTANI)
seni (17.1. GOLA)
seni (17.1. VAI)
senitra (9.1. JAVANESE)
senku (45.1. FON)
šənkurta (7.3. HARARI)
sennaugaksak (17.7. ALEUT)
sennep (30.24. DANISH)
sennep (30.57. DANISH)
sennep (30.57. NORWEGIAN)
sennepskål (30.4. NORWEGIAN)
sennepskål (30.5. NORWEGIAN)
sennepskål (30.6. NORWEGIAN)
sennepskål (30.7. NORWEGIAN)
sennepskål (30.8. NORWEGIAN)
sennepskål (30.9. NORWEGIAN)
sennepskål (30.10. NORWEGIAN)
sennepskål (30.11. NORWEGIAN)
sennepskål (30.12. NORWEGIAN)
sennepskål (30.13. NORWEGIAN)
sennepskål (30.14. NORWEGIAN)
sennepskål (30.15. NORWEGIAN)
sennepskål (30.16. NORWEGIAN)
sennepskål (30.17. NORWEGIAN)
sennepskål (30.18. NORWEGIAN)
sennepskål (30.19. NORWEGIAN)
sennepskål (30.20. NORWEGIAN)
sennichi-modoki (27.2. JAPANESE)
senninkoku (23.3. JAPANESE)
sensitiva acuatica (40.19. SPANISH)
sente (8.1. INDONESIAN)
sente (8.1. JAVANESE)
sente (8.1. SUNDA)
sɛnutsõe (23.3. ÉWÉ)
senye (17.1. VAI)
s(e)òbhrach (51.1. GAELIC, SCOTTISH)
seoha (30.48. HINDI)
sep (40.43. WOLOF)

sepersi (13.2. INDONESIAN)
seppan kizhangu (8.4. TAMIL)
sepvion oinanihi (24.17. GREEK)
sequidillas (40.13. TAGALOG)
sêr (7.10. FARSI, EASTERN)
serat (17.7. MANOBO)
serawung (44.4. INDONESIAN)
şerbetçi otu (46.1. TURKISH)
şerbetçiotu (46.1. TURKISH)
sĕrébé (44.5. GURO)
sereni (45.4. SPANISH)
sererella (40.29. CEBUANO)
sererelya (40.29. AKLANON)
ser-flodau (27.17. WELSH)
seri (24.17. JAPANESE)
seri (24.17. VLAAMS)
şərik (24.19. AZERBAIJANI, NORTHERN)
serikeerai (23.3. TAMIL)
serikibaba (40.31. HAUSA)
seroja (48.1a. INDONESIAN)
seroja (48.1a. MALAY)
seroja merah (48.1a. INDONESIAN)
serori (24.4. JAPANESE)
serpent japonais (38.29. FRENCH)
serpent melon (38.10. ENGLISH)
serpent végétal (38.29. FRENCH)
serpentona (38.29. ITALIAN)
serpiente vegetal (38.29. SPANISH)
serralha (27.18. CRIOULO, CAPE VERDE)
serralha (27.42. ITALIAN)
serralha (27.42. PORTUGUESE)
serralha macia (27.42. PORTUGUESE)
sertebønne (40.2. DANISH)
seruni (27.2. INDONESIAN)
seruni (27.23. INDONESIAN)
seruriatku (24.5. JAPANESE)
serurii (24.4. JAPANESE)
servilleta del pobre (38.22. SPANISH)
sésamo bravo (55.1. PORTUGUESE)
sésamo selvagem (55.1. PORTUGUESE)
sésamo slavaje (55.1. SPANISH)
sesawi (30.17. INDONESIAN)
se-sawi (30.17. INDONESIAN)
sesawi putih (30.36. INDONESIAN)
sèsé (40.31. YORUBA)
ṣẹṣẹ (40.31. YORUBA)
ṣẹṣẹ erèé (40.31. YORUBA)
sesebebe (38.20. LUYIA)
seskoulo (35.3. GREEK)
sessile alternanthera (23.2. ENGLISH)
sessonge (40.31. BASSARI)
seta (6.2. PORTUGUESE)
se̱ta (20.6. TIV)
seta (27.8. CRIOULO, CAPE VERDE)
seta (38.17. SWAHILI)
seta (38.18. SWAHILI)
'se-tapho (30.27a. MUSKOGEE)
'se-tapho-rakko (30.35. MUSKOGEE)
setha (38.3. SANSKRIT)
setim (27.18. CEBUANO)
seto bojho (20.1. NEPALI)
seto mula (30.54. NEPALI)
setumbak merah (27.18. MALAY)
seturia (17.6. GREEK)
sevali (16.1a. TAMIL)
sevattu mullangi (27.10a. TAMIL)
seviette des pauvres (38.22. FRENCH)
sevketibostan (27.36. TURKISH)
sevlák cukrový (24.23. CZECH)
sevlák zeleninový (24.23. CZECH)
sevojekool (30.34. STELLINGWERFS)
sevoy´u (7.3. PAIUTE, SOUTHERN)
sevóya (7.3. LADINO)
sewejaarboontjie (40.4. AFRIKAANS)
sewejaarsboontjie (40.25. AFRIKAANS)
sewipak (58.1. CREE, PLAINS)
seyagit (24.3. HEBREW)
seyuti (38.25. BINUKID)
sezam (55.1. POLISH)
sezam paprsčitý (55.1. CZECH)
sfeclă de frunze (35.3. ROMANIAN)
sfeclă obisnuită (35.4. ROMANIAN)
sfeclă roşie (35.4. ROMANIAN)
sfecla mangold (35.3. ROMANIAN)
sfedădefruuze (33.3. ROMANIAN)
sferoïdis sinapis (30.4. GREEK)
sgatong (62.19. BLAAN, SARANGANI)
sgourolachano (30.34. GREEK)
sha ge (11.1. CHINESE, YUE)
sha ge (40.21. CHINESE, YUE)
sha ma (17.7. LAHU)
shaamali (38.7a. CHECHEN)
shaame gedde (8.4. KANNADA)
shaave gedde (8.4. KANNADA)
shada natey (23.3. BENGALI)
shadanatey (23.3. BENGALI)
shaenga (40.1. KANNADA)
shaenga daana (40.1. MARATHI)
shafa (7.2. HAUSA)
shah (30.27a. ROMANI, BALKAN)

shahalaim tarovteem (30.51. HEBREW)
shaḥalayim tarbutyim (30.51. HEBREW)
shahe abic (30.52. FARSI, WESTERN)
shahed (38.11. ARABIC)
shahlahtn (27.30. YIDDISH)
shahlahtn (27.31. YIDDISH)
shahlahtn (27.32. YIDDISH)
shakabilva (62.19. SANSKRIT)
shakabilvaka (62.19. SANSKRIT)
shakadevi (27.42. TAMIL)
shakar kand (36.3. HINDI)
shakardand (36.3. HINDI)
shakaria (36.3. GUJARATI)
shakarkandi (36.3. HINDI)
shakarkandi (36.3. PANJABI, EASTERN)
shakashreshtha (62.19. SANSKRIT)
shakayogya (24.9. SANSKRIT)
shakerkand (36.3. HINDI)
shak-stmeshdha (62.9. SANSKRIT)
shakushi-na (30.36. JAPANESE)
shakwesh (7.3. POTAWATOMI)
shalamta (7.2. HAUSA)
shalanta (7.2. HAUSA)
shaleya (24.13. SANSKRIT)
shalga (30.41a. HINDI)
shalgam (30.41a. BENGALI)
shalgam (30.41a. HINDUSTANI)
shalgam (30.41a. PANJABI, EASTERN)
shalgham (30.22. FARSI, WESTERN)
shalgham (30.41a. FARSI, EASTERN)
shalgham (30.41a. FARSI, WESTERN)
shalghama (30.41a. PASHTO, NORTHERN)
shalina (24.13. SANSKRIT)
shaljam (30.41a. HINDUSTANI)
shaljum (30.41a. HINDI)
shallaathu (27.29. TAMIL)
shallaathu (27.31. TAMIL)
shallaathu (27.32. TAMIL)
shallathu (27.31. TAMIL)
shallathu (27.32. TAMIL)
shallot (7.2. ENGLISH)
shalnikavaja strelkslist (6.2. BELARUSAN)
shalqi (38.3. ALBANIAN, TOSK)
shaluk (50.2. BENGALI)
s̱halum (30.41a. KATI)
shamagadde (8.4. KANNADA)
shamakkizhangu (8.4. TAMIL)
shamakoora (8.4. TELUGU)
shamantippu (27.23. TAMIL)
shamari (24.13. SWAHILI)
shamathumpa (8.4. TELUGU)
shambarakanda (11.2. SANSKRIT)
shambe avare (40.3. KANNADA)
shambe kayi (40.29. KANNADA)
shamman (38.7b. HAUSA)
shamukei (27.45. PANJABI, EASTERN)
shan da huang (19.1. CHINESE, WÚ)
shan hu cai (24.14. CHINESE, MANDARIN)
shan hu cai (24.14. CHINESE, WÚ)
shan luo bo (56.3. CHINESE, WÚ)
shan mai leng shui hua (66.2. CHINESE, WÚ)
shan po ts'ai (35.1. CHINESE, MANDARIN)
shan se jin (68.1. CHINESE, WÚ)
shan shu (11.6. CHINESE, WÚ)
shan wo ju (27.29. CHINESE, MANDARIN)
shan wo ju (27.29. CHINESE, WÚ)
shan yao (11.1. CHINESE, WÚ)
shan ye qui zhen cao (27.8. CHINESE, WÚ)
shan ye shu yu (11.8. CHINESE, WÚ)
shâna (30.27a. SIDAMO)
shana dumpa (8.4. TAMIL)
shanapushpika (40.2. SANSKRIT)
shandalj (24.23. MONGOLIAN)
shandong bai (30.40. CHINESE, MIN BEI)
shang hai da tou cai (30.22. CHINESE, MIN BEI)
shang lu (56.1. CHINESE, YUE)
shang lu (56.3. CHINESE, YUE)
shang yu cai (30.58. CHINESE, WÚ)
shanghu (7.3. IBALOI)
shanglu (56.1. FRENCH)
shankhamoola (30.56. SANSKRIT)
shao cai (30.40. CHINESE, YUE)
shapaka (64.3. SANSKRIT)
sharada (48.1a. SANSKRIT)
sharalzagana (27.42. MONGOLIAN)
sharanga (20.6. SANSKRIT)
sharapadma (48.1a. SANSKRIT)
shargundei (30.51. PANJABI, EASTERN)
shargundei (30.58. PANJABI, EASTERN)
sharotto (7.2. JAPANESE)
sharrag (27.9. GAELIC, MANX)
shasha (38.13a. BENGALI)
shatawar (13.1. HINDI)
shathaparna (36.2. SANSKRIT)
shathapathra (48.1a. SANSKRIT)
shathapathrika (24.13. SANSKRIT)
shathaprasuna (24.13. SANSKRIT)
shathapushpa (24.13. SANSKRIT)
shathaputhri (38.21. SANSKRIT)

shathi (20.4. SANSKRIT)
shatlatuvirai (27.31. TAMIL)
shatlatuvirai (27.32. TAMIL)
shawewe (23.2. EDE NAGO)
shawewe (23.2. YORUBA)
shcavel (58.2. KAZAKH)
shɔ̰ɔ̰qɔ̰ɔ̀ (62.24. TIBETAN)
shchane shpinatna (58.3. BELARUSAN)
shchane zvichainie (58.2. BELARUSAN)
shchavel (58.2. RUSSIAN)
shchavel morščinistyj (58.2. RUSSIAN)
shchavelnik (58.2. RUSSIAN)
shɔkɔ (23.4. GA)
shɔkɔt (23.4. EDE NAGO)
shɔkɔt (23.4. YORUBA)
shɔkɔtɔ-yɔkɔtɔ (23.4. KRIO)
shɛ ɛjɛ shɔnrɔ (28.1. EDE NAGO)
shɛ ɛjɛ shɔnrɔ (28.1. YORUBA)
she gua (38.29. CHINESE, WÚ)
she tian gua (38.10. CHINESE, MANDARIN)
she tian gua (38.10. CHINESE, WÚ)
she tou cai' (27.35. CHINESE, WÚ)
she xiang gua (38.26. CHINESE, MANDARIN)
she xiang gua (38.26. CHINESE, WÚ)
sheemaimanathakkaali (62.9. TAMIL)
sheethashiva (24.13. SANSKRIT)
sheinan refuee (27.45. HEBREW)
shĕkar (36.3. KASHMIRI)
shelum (24.17. MALAY)
shemakkalenga (8.4. TAMIL)
shembi avare (40.4. KANNADA)
shembiavari tumbeki (40.4. KANNADA)
shen jiang (20.6. CHINESE, WÚ)
shen qin (24.23. CHINESE, MANDARIN)
shen qin (24.23. CHINESE, WÚ)
shen shu (11.7. CHINESE, YUE)
shên tzu (39.4. CHINESE, MANDARIN)
shenagalu (40.5. TELUGU)
sheng cai (27.31. CHINESE, GÀN)
sheng cai (27.31. CHINESE, XIANG)
sheng cai (27.32. CHINESE, YUE)
sheng cai (27.33. CHINESE, YUE)
sheng jiang (20.6. CHINESE, GÀN)
sheng jiang (20.6. CHINESE, MIN BEI)
sheng jiang (20.6. CHINESE, YUE)
sheonit (54.1. HEBREW)
shephard's purse (30.43. ENGLISH)
shervola (24.23. ITALIAN)
sheuit (40.23. HEBREW)
sheʿuʿit (40.23. HEBREW)
sheuit aduma (40.24. HEBREW)
sheʿuʿit aduma (40.24. HEBREW)
sheuit hagima (40.26. HEBREW)
sheuit haginna (40.26. HEBREW)
sheʿuʿit ha-ginna (40.26. HEBREW)
sheʿuʿit lima (40.25. HEBREW)
sheuit lima (40.25. HEBREW)
shevanti (27.23. KANNADA)
shi (62.5. KORO)
shi diao bai (13.2. CHINESE, WÚ)
shi diao bai (13.2. CHINESE, YUE)
shi yong bin li (35.1. CHINESE, MANDARIN)
shi yong ji (27.15. CHINESE, MANDARIN)
shi yong ji (27.15. CHINESE, WÚ)
shi yong lan hua jiao (14.1. CHINESE, MANDARIN)
shi yong lan hua jiao (15.1. CHINESE, WÚ)
shi yong mei ren jiao (9.1. CHINESE, MANDARIN)
shi yong mei ren jiao (9.1. CHINESE, WÚ)
shi yong zi li (40.41. CHINESE, YUE)
shi young bin li (35.1. CHINESE, WÚ)
shibbolet ha-mayim (25.1. HEBREW)
shibolet hamaim (25.1. HEBREW)
shihomandora (62.7. JAPANESE)
shìitá (62.1. NGIZIM)
shikaku mame (40.29. JAPANESE)
shikhalu (17.7. SANSKRIT)
shikhámúl (30.41a. HINDUSTANI)
shikha-mula (24.11. SANSKRIT)
shikhivardhaka (38.1. SANSKRIT)
shikoli (27.10a. ZULU)
shilarambha (9.1. SANSKRIT)
shilɔt (7.2. KRIO)
shim (40.9. BENGALI)
shima houzuki (62.12. JAPANESE)
shimagonguru (45.4. TELUGU)
shima-tsuru-azuki (40.40. JAPANESE)
shimbe balli (40.3. KANNADA)
shimbi (40.3. KANNADA)
shimbi (40.3. SANSKRIT)
shimbi (40.3. TELUGU)
shimbi (40.9. SANSKRIT)
shimbi (40.18. SANSKRIT)
shimbiphala (30.56. SANSKRIT)
shimul ali (39.4. BENGALI)
shina-kanzô (13.3. JAPANESE)
shinfa (30.17. OROMO, WEST CENTRAL)
shingada (64.3. MARATHI)

shingori (64.2. ASSAMESE)
shinnan refuyi (27.45. HEBREW)
shinu (44.5. JUNKUN TAKUM)
shiny bugleweed (44.1. ENGLISH)
shionlä mirch (62.1. PANJABI, EASTERN)
shirai buginak (35.5. TAJIKI)
shirayama-giku (27.17. JAPANESE)
shirayama-giru (27.17. VLAAMS)
shiri kharak (27.42. TAJIKI)
shirin calampir (62.1. UZBEK)
shirnavrintha (38.3. SANSKRIT)
shiro-garashi (30.57. JAPANESE)
shiro-guwai (6.1. JAPANESE)
shirola (38.21. MARATHI)
shirone (44.1. JAPANESE)
shirone (44.1. VLAAMS)
shiro-uri (38.8. JAPANESE)
shiro-uri (38.9. JAPANESE)
shishi-tôgarashi (62.1. JAPANESE)
shiso (44.7. JAPANESE)
shiso (44.7. VLAAMS)
shi̱ta (20.6. TIV)
shitapú (62.2. JUNKUN TAKUM)
shitunguo (7.3. LUYIA)
shivamaale chedy (9.1. TAMIL)
shivappuvaslakkeerai (28.1. TAMIL)
shiwáákáá (27.24. HAUSA)
shìwaákaá (27.24. HAUSA)
shnghara (64.2. GUJARATI)
shnittluk (7.11. KAZAKH)
shohbulut (10.2. TAJIKI)
shohbulut dashti (10.3. TAJIKI)
shọkọtọ-yọkọtọ (23.4. IGBO)
shokshi' (38.20. CHICKASAW)
shokuyōbīto (35.4. JAPANESE)
shokuyou-giku (27.23. JAPANESE)
shokuyou-houzuki (62.11. JAPANESE)
shokuyou-kanna (9.1. JAPANESE)
shokuyou-tanpopo (27.45. JAPANESE)
sholgom (30.41a. UZBEK)
shombu (24.13. TAMIL)
shona nguo (27.8. SWAHILI)
shonkôra (17.5. SIDAMO)
shọ̀ọ̀ shọ̀ọ̀ (40.2. IGBO)
shōō̃grä (30.53. PANJABI, EASTERN)
shookashimba (40.18. SANSKRIT)
shookashimbika (40.18. SANSKRIT)
shorakkaai (38.20. TAMIL)
shorakkai (38.20. TAMIL)
shorivalli (40.18. MALAYALAM)
shoriyanam (40.18. MALAYALAM)
shornoi (35.1. MONGOLIAN)
short-podded yam bean (40.21. ENGLISH)
shoshan (13.4. HEBREW)
shoshan yom (13.3. HEBREW)
shou gong mu (39.5. CHINESE, MANDARIN)
shou gong mu (39.5. CHINESE, WÚ)
shouga (20.6. JAPANESE)
shoushan (7.11. GAELIC, MANX)
shovul (58.3. UZBEK)
shpargull (13.2. ALBANIAN, TOSK)
shpinaht (35.9. YIDDISH)
shpinat (35.9. KIRGHIZ)
shpinat (35.9. TATAR)
shpinat (35.9. UZBEK)
shpinat ogorodnyj (35.9. KAZAKH)
shpiynat (35.9. KAZAKH)
shpurdhë (38.16. ALBANIAN, TOSK)
shradhashaka (36.2. SANSKRIT)
shravan gevda (40.26. MARATHI)
shravanghevda (40.26. MARATHI)
shri (48.1a. SANSKRIT)
shrigavera (20.6. SANSKRIT)
shringakanda (64.3. SANSKRIT)
shringamoola (64.3. SANSKRIT)
shringaruha (64.3. SANSKRIT)
shringata (64.2. SANSKRIT)
shringata (64.3. SANSKRIT)
shringataka (64.3. SANSKRIT)
shriparna (48.1a. SANSKRIT)
shu_ (7.3. LAHU)
shu (30.27a. BULU)
shu (30.27b. BULU)
shu (36.3. CHINESE, MANDARIN)
shu dou (40.2. CHINESE, MANDARIN)
shu dou (40.2. CHINESE, WÚ)
shu fan qie (62.7. CHINESE, MANDARIN)
shu fan qie (62.7. CHINESE, WÚ)
shu fan qie (62.7. CHINESE, YUE)
shu hua du jing shan (47.1. CHINESE, WÚ)
shu mi (17.7. CHINESE, YUE)
shu shu (39.4. CHINESE, MANDARIN)
shu shu (39.4. CHINESE, WÚ)
shu wei luo bo (30.53. CHINESE, MANDARIN)
shu wei lou bo (30.53. CHINESE, WÚ)
shu wie pu tao (30.53. CHINESE, YUE)
shu yu (11.1. CHINESE, YUE)
shu yu (11.6. CHINESE, WÚ)
shu zai (62.24. CHINESE, YUE)

shu zai yu (6.2. CHINESE, MIN BEI)
shuang gai jue (5.1. CHINESE, WÚ)
shugaken (17.5. KRIO)
shuga-kɛn (17.5. KRIO)
shughlaig (58.2. GAELIC, MANX)
shu_hpu (7.10. LAHU)
shui cong (7.5. CHINESE, YUE)
shui cong (7.14. CHINESE, MIN BEI)
shui gua (38.22. CHINESE, MIN BEI)
shui gua (38.22. CHINESE, YUE)
shui han ts'ai (30.52. CHINESE, MANDARIN)
shui jue (4.1. CHINESE, MANDARIN)
shui jue (4.1. CHINESE, WÚ)
shui kwa (38.22. CHINESE, MANDARIN)
shui qie (62.23. CHINESE, MANDARIN)
shui qie (62.23. CHINESE, WÚ)
shui qin (24.17. CHINESE, MANDARIN)
shui qin (24.17. CHINESE, WÚ)
shui qin (24.17. CHINESE, YUE)
shui tian jie (30.52. CHINESE, YUE)
shukapindi (40.18. SANSKRIT)
shukari (11.2. SANSKRIT)
shukavathi (40.18. SANSKRIT)
shukladugda (64.3. SANSKRIT)
shuklapadma (48.1a. SANSKRIT)
shulgam gɔ̃glɔ̃ɔ̃ (30.41a. PANJABI, EASTERN)
shulkha (58.2. TAJIKI)
shum ashkelon (7.2. HEBREW)
shum dak (7.11. HEBREW)
shum gavoha (7.1. HEBREW)
shum hagina (7.10. HEBREW)
shum ha-ginna (7.10. HEBREW)
shum hakarash (7.8. HEBREW)
shum ha-peka'ot (7.13. HEBREW)
shum kurrat (7.6. HEBREW)
shum sini (7.4. HEBREW)
shumar (24.12. HEBREW)
shumar (24.13. HEBREW)
shumar pashut (24.12. HEBREW)
shumar pashut (24.13. HEBREW)
shumburà (40.5. SIDAMO)
shumgioh (52.1. TAJIKI)
shun jên kua (38.25. CHINESE, MANDARIN)
shunam (7.10. GUJARATI)
shungiku (27.23. ENGLISH)
shungiku (27.23. JAPANESE)
shunkurta (7.8. SIDAMO)
shunkurti (7.3. OROMO, WEST CENTRAL)
shunthi (20.6. TELUGU)
shunti (20.6. KANNADA)
shunti (20.6. SANSKRIT)
shura (35.1. TAJIKI)
shuragioh (35.7. TAJIKI)
shuragioh (35.10. TAJIKI)
shure (63.1. SUSU)
shuti (38.3. KURDISH, NORTHERN)
shvatvat (1.1. HEBREW)
shvatvat ha-sadè (1.1. HEBREW)
shvethapathra (48.1a. SANSKRIT)
shvetharaaji (38.29. SANSKRIT)
shvethashimbika (40.9. SANSKRIT)
shwee (28.1. GA)
shwee (59.3. GA)
shwe-pe (40.28. BURMESE)
shwe-pha-yone (38.14. BURMESE)
shwetsursha (30.48. BENGALI)
shwie (59.3. GA)
shylim (64.3. KAZAKH)
si brèam (38.18. CAJUN FRENCH)
si gua (38.22. CHINESE, GÀN)
si gua (38.22. CHINESE, MANDARIN)
si gua (38.22. CHINESE, MIN BEI)
si gua (38.22. CHINESE, WÚ)
si gua (38.22. CHINESE, YUE)
si ji dou (40.26. CHINESE, GÀN)
si ji dou (40.26. CHINESE, MIN BEI)
si ji dou (40.26. CHINESE, XIANG)
si ji dou (40.26. CHINESE, YUE)
si ji luo bu (40.31. CHINESE, XIANG)
si jiao ling (64.2. CHINESE, WÚ)
si jiao ling (64.3. CHINESE, MANDARIN)
sī kouā (38.3. CHINESE, HAKKA)
Siamese galangal (20.1. ENGLISH)
sian-jîn-chióng (32.1. CHINESE, MIN NAN)
siao suan (7.10. CHINESE, MANDARIN)
sib zamini (62.24. FARSI, WESTERN)
siba (7.3. MENDE)
siba (7.3. VAI)
sibanana (16.1. NDONGA)
sibatse simaron (40.25. TAGALOG)
sibatsi (40.9. TAGALOG)
sibɔla (7.2. BULLOM SO)
sibɔla (7.3. BULLOM SO)
sibé zamine (62.24. FARSI, WESTERN)
Siberian ginseng (26.4. ENGLISH)
Siberian kale (30.23. ENGLISH)
Siberische kool (30.23. VLAAMS)
sibezamihic hendie (11.1. FARSI, WESTERN)
sibir songino (7.11. MONGOLIAN)

sibirgrasløk (7.11. NORWEGIAN)
Sibirischer Ginseng (26.4. GERMAN)
sibito jambo (62.15. MANDING-MANDINKA)
sibolyas (7.3. IBANAG)
sibolyino (7.2. BICOLANO, CENTRAL)
sibo-siboyasan (10.2. TAGALOG)
siboyas (7.3. ILOCANO)
siboyas (7.3. MARANAO)
sibsibit (23.3. BONTOC, CENTRAL)
sibsifit (23.3. BONTOC, CENTRAL)
sibul (7.3. ESTONIAN)
sibulet (7.11. TURKISH)
sibulyas (7.2. BICOLANO, CENTRAL)
sibulyas (7.3. ILOCANO)
sibuyas (7.2. TAGALOG)
sibuyas (7.3. BICOLANO, CENTRAL)
sibuyas (7.3. BONTOC CENTRAL)
sibuyas (7.3. CEBUANO)
sibuyas (7.3. FILIPINO)
sibuyas (7.3. GIANGAN)
sibúyas (7.3. HILIGAYNON)
sibuyas (7.3. IBANAG)
sibuyas (7.3. IFUGAO)
sibuyas (7.3. PAMPANGAN)
sibúyas (7.3. PANGASINAN)
sibuyas (7.3. TAGALOG)
sibuyas (7.3. TAUSUG)
sibuyas (7.3. WÁRAY-WÁRAY)
sibuyas bombay (7.3. TAGALOG)
sibuyas buyas (7.3. TAGALOG)
sibuyas talalog (7.2. TAGALOG)
sibuyas-bisaya (7.8. CEBUANO)
sibuyas-Bisaya (7.8. CEBUANO)
sibuyas-bombay (7.2. WÁRAY-WÁRAY)
sibuyas-bombay (7.3. TAGALOG)
sibuyas-litmo (7.8. CEBUANO)
sibuyas-tagalog (7.2. FILIPINO)
sibuyas-Tagalog (7.2. TAGALOG)
sibwn (7.2. WELSH)
sicimè (10.1. DAGAARE, SOUTHERN)
sicori (27.10a. WELSH)
sidamate (62.9. NDONGA)
sidá-sidâ (59.3. WÁRAY-WÁRAY)
siddhaartha (30.17. SANSKRIT)
siddhattho (30.57. PALI)
sidèldéla (38.12. JOLA-FONYI)
siderolopatho (58.3. GREEK)
sidesalaatti (27.32. FINNISH)
sidhaprayojana (30.51. SANSKRIT)
sidirgh (40.10. TAJIKI)
Sieberiese blaarkool (30.23. AFRIKAANS)
sieboontjjie (40.24. AFRIKAANS)
siēn-gnin-tchòng (32.1. CHINESE, HAKKA)
siènz (34.2. LAO)
siepel (7.3. STELLINGWERFS)
sieq il-hamiema (27.21. MALTESE)
Sierra Leone bologi (27.13. ENGLISH)
sieva bean (40.25. ENGLISH)
sigadillas (40.29. TAGALOG)
sigadilyas (40.29. TAGALOG)
sigal telat goni (68.1. HEBREW)
sigando (7.13. INDONESIAN)
sigaram (20.6. TAMIL)
sigarillas (40.29. PAMPANGAN)
sigarillas (40.29. TAGALOG)
sigarilyas (40.29. PANGASINAN)
sigarilyas (40.29. SAMBAL, TINÀ)
sigarilyas (40.29. TAGALOG)
sigarilyis (40.29. TAGALOG)
sigbut-ta-bato (24.15. GIANGAN)
siggalfut (8.4. BUHID)
siğilli lahana (30.42. TURKISH)
siğirdili (29.1. TURKISH)
sigli (63.1. GBE-VHE)
sigod (24.12. ICELANDIC)
sigorei (27.10a. AFRIKAANS)
sigra tat-tadam (62.7. MALTESE)
siguane (20.6. SENOUFO, CEBAARA)
siguidilyas (40.29. FILIPINO)
siguine (8.7. FRENCH)
sigumchi (35.9. KOREAN)
sigur (27.10a. ESTONIAN)
sigurlaukur (7.9b. ICELANDIC)
sigutung (62.19. MANOBO)
sihi genasu (36.3. KANNADA)
sihi thonde (38.4. KANNADA)
sihutung (62.19. MANOBO)
sii (17.1. LOKO)
s'ii s'aangz (24.4. LAO)
s'ii s'aangz (24.17. LAO)
siikuri (27.9. ARABIC)
siitan (17.1. ILOCANO)
siitan (23.3. ILOCANO)
siiwi (7.3. HOPI)
sikai (38.20. CEBUANO)
sikalignirini (62.5. SENOUFO, CEBAARA)
sikalikpolougui (62.1. SENOUFO, CEBAARA)
sikaliwalih (62.5. SENOUFO, CEBAARA)
sikamas (40.21. PAMPANGAN)

sikandakanda (40.9. LUYIA)
sikaritigue (17.5. SENOUFO, CEBAARA)
sikay (38.20. CEBUANO)
sikhubi (40.43. LUYIA)
sikildjé (38.3. FULFULDE)
sikirre (38.7b. FULFULDE, ADAMAWA)
siklan tera (38.19. TURKISH)
si-koe (38.3. CHINESE, MIN NAN)
sikoi (38.3. MALAY)
sikon (27.10a. TURKISH)
šikorej (27.10a. SORBIAN, UPPER)
sikori (27.10a. NORWEGIAN)
sikori (27.10b. NORWEGIAN)
sikuli gbéngbé (62.2. SUSU)
sikuri (27.10a. FINNISH)
sikuwa' (38.21. BINUKID)
sikuwa' (38.22. BINUKID)
sikwa (38.22. MARANAO)
sikyahatiko (40.25. HOPI)
sikyatotona (24.11. HOPI)
sil (23.4. PANJABI, EASTERN)
sila (43.1. PULAAR)
silá (43.1. PULAAR)
silakadom (11.5. TELUGU)
silam (44.7. NEPALI)
silanro (24.9. HOPI)
silara (23.4. HINDI)
silɔ-na (45.1. YALUNKA)
sileban (17.7. MANOBO)
sili (62.1. BICOLANO, CENTRAL)
sili (62.1. BONTOC CENTRAL)
sili (62.1. CEBUANO)
sili (62.1. GIANGAN)
sili (62.1. IBANAG)
sili (62.1. IFUGAO)
sili (62.1. ILOCANO)
sili (62.1. PAMPANGAN)
síli (62.1. PANGASINAN)
sili (62.1. TAGALOG)
sili (62.1. TAUSUG)
síli (62.2. PANGASINAN)
sili amiling (62.5. ITAWIT)
sili macopa (62.2. ITAWIT)
sili-diyablo (62.5. TAGALOG)
silig (35.4. ARABIC)
siling labuyo (62.5. FILIPINO)
siling labuyo (62.5. TAGALOG)
silingbalang (62.5. PANGASINAN)
siling-bilog (62.1. FILIPINO)
siling-bilog (62.1. TAGALOG)
siling-bundok (62.5. TAGALOG)
siling-diyablo (62.5. TAGALOG)
siling-haba (62.2. TAGALOG)
siling-labuyo (62.2. BICOLANO, CENTRAL)
siling-labuyo (62.2. BONTOC CENTRAL)
siling-labuyo (62.2. CEBUANO)
siling-labuyo (62.2. GIANGAN)
siling-labuyo (62.2. IBANAG)
siling-labuyo (62.2. IFUGAO)
siling-labuyo (62.2. ILOCANO)
siling-labuyo (62.2. PAMPANGAN)
siling-labuyo (62.2. TAGALOG)
siling-labuyo (62.2. TAUSUG)
siling-labuyo (62.5. BICOLANO, CENTRAL)
siling-labuyo (62.5. BONTOC CENTRAL)
siling-labuyo (62.5. CEBUANO)
siling-labuyo (62.5. GIANGAN)
siling-labuyo (62.5. IBANAG)
siling-labuyo (62.5. IFUGAO)
siling-labuyo (62.5. ILOCANO)
siling-labuyo (62.5. PAMPANGAN)
siling-labuyo (62.5. TAGALOG)
siling-labuyo (62.5. TAUSUG)
siling-maanghang (62.5. TAGALOG)
silingpalay (62.5. TAGALOG)
siling-pasiti (62.5. TAGALOG)
sili-sili (62.21. SUBANON)
silit-diyablo (62.5. ILOCANO)
silla (43.1. PULAAR)
silma (40.1. KONKOMBA)
silo-na (45.1. YALUNKA)
silonyi (45.1. SUSU)
silopersilja (24.20. FINNISH)
silosinappi (30.7. FINNISH)
silq (35.4. KURDISH, NORTHERN)
silver spinach (23.4. ENGLISH)
silver-spiked cockscomb (23.4. ENGLISH)
sīm (40.9. MAITHILI)
sim saag (30.52. NEPALI)
sima (40.1. DAGBANI)
simaikkasuru (45.4. TAMIL)
simal tarul (39.4. NEPALI)
simalalu (39.4. ASSAMESE)
simalla (17.1. OROMO, WEST CENTRAL)
simalla (17.2. OROMO, WEST CENTRAL)
simalla (17.4. OROMO, WEST CENTRAL)
siman bēzē (62.1. DAGAARE, SOUTHERN)
simane (62.2. DAGAARE, SOUTHERN)
simani (39.5. MINANGKABAU)
simba (8.3. ORIYA)

simba (40.34a. ORIYA)
simbande (40.39. LUYIA)
simbi (40.9. SANSKRIT)
simbung (8.4. BUHID)
simhi (62.25. SANSKRIT)
simi (40.25. NEPALI)
simindi (17.7. GEORGIAN)
siminɛ (17.1. KONO)
simiyé (40.1. DAGAARE, SOUTHERN)
simla mirch (62.1. HINDI)
simolu aloo (39.4. ASSAMESE)
simolu alu (39.4. ASSAMESE)
simon (38.3. IVATAN)
simsim (55.1. TAJIKI)
sina kitka (52.1. MACEDONIAN)
sinak (40.26. SUMO-MAYANGNA)
sinakkantal (39.4. MALAYALAM)
sinambong (38.16. MARANAO)
sinambong (38.17. MARANAO)
sinap i bardhë (30.57. ALBANIAN, TOSK)
sinap i zi (30.24. ALBANIAN, TOSK)
sinapi (30.57. GREEK)
sinapi (35.9. NIUE)
sinapi me diogkoumenon stelechos (30.19. GREEK)
sinapi me kovdylodis rizes (30.14. GREEK)
sinapi me lefki anthisi (30.11. GREEK)
sinapi me makri mischo (30.13. GREEK)
sinapi me mikra phylla (30.7. GREEK)
sinapi me phyllon jananthos (30.16. GREEK)
sinapi me plati mischo (30.10. GREEK)
sinapis vlastou indo kalamou (30.5. GREEK)
sinapl lachanikon (30.17. GREEK)
sinappi (30.17. FINNISH)
sinappikaali (30.48. FINNISH)
sinarafung (11.1. BUHID)
sinay (40.40. ILOCANO)
sincama (40.22. SPANISH)
síndria (38.3. CATALAN-VALENCIAN-BALEAR)
sindz (27.38. ARMENIAN)
si-në-meh (11.1. NICOBARESE, CAR)
si-në-meh (11.2. NICOBARESE, CAR)
si-në-meh (11.5. NICOBARESE, CAR)
sinep (30.3. ESTONIAN)
sinfito mayor (29.2. SPANISH)
sinfiton refuyi (29.2. HEBREW)
sing ghika (63.1. SANSKRIT)
singaade (64.2. KANNADA)
singaara beeja (64.2. KANNADA)
singapur (8.10. HANUNOO)
singara (64.1. NEPALI)
singara (64.2. NEPALI)
singavaeram (20.6. TAMIL)
singbũlé (40.39. DAGAARE, SOUTHERN)
singcamat (40.21. ITAWIT)
singgamas (40.21. MARANAO)
singgika (63.1. SANSKRIT)
singguliyas malagti (40.25. HANUNOO)
singhaada (64.2. HINDI)
singhaada (64.3. MARATHI)
singhaada balli (64.2. KANNADA)
singhaara (64.2. TAMIL)
singhara (64.2. HINDI)
singhara (64.2. KANNADA)
singhara (64.2. MARATHI)
singhara (64.2. TAMIL)
singhara (64.2. URDU)
singhara nut (64.2. ENGLISH)
singhara water chestnut (64.2. ENGLISH)
singivaeram (20.6. TAMIL)
siṅgiveraṁ (20.6. PALI)
singkamás (30.41a. HILIGAYNON)
singkamas (40.21. BINUKID)
singkamas (40.21. FILIPINO)
singkamas (40.21. TAGALOG)
singkamas (40.21. TAUSUG)
singkong (39.4. INDONESIAN)
singkong (39.4. SUNDA)
singli (63.1. GBE-VHE)
siŋgli (63.1. GBE-VHE)
singra (30.56. URDU)
singri (30.53. HINDI)
sinh kh [uw][ow]ng (20.6. VIETNAMESE)
siṅhāṛ (64.2. MAITHILI)
sinhi (62.19. SANSKRIT)
sinimailanen (40.17. FINNISH)
sininaba (8.1. ILOCANO)
sinir out (57.1. TURKISH)
sinisamfa (16.1a. BODO)
sinji (27.42. SERBO-CROATIAN)
sinjibil (20.6. SOMALI)
sinkade (62.16. BURMESE)
sinkamas (40.21. HANUNOO)
sinkamas (40.21. PAMPANGAN)
sinkamas (40.21. TAGALOG)
sinkamás (40.21. TAGALOG)
sinkamas (40.21. TAUSUG)
sinmeti (62.5. BALANTA-KENTOHE)
sinna keerai (23.3. TAMIL)

slao (27.32. STELLINGWERFS)
slatka pčenka (17.7. MACEDONIAN)
slaz (45.5. POLISH)
slaz kędzierzawy (45.5. POLISH)
slez (45.4. MACEDONIAN)
sléz (45.5. CZECH)
slez (45.5. SERBO-CROATIAN)
slez (45.5. SLOVAK)
slingerboon (40.9. DUTCH)
slnečnica topinambur (27.26. SLOVAK)
sloffenbonen (40.34a. STELLINGWERFS)
slök baahs (38.4. KHMER, CENTRAL)
slök chuu (45.4. KHMER, CENTRAL)
slök lumpaèng (6.1. KHMER, CENTRAL)
slök mchuu (58.2. KHMER, CENTRAL)
sloní tráva (17.3. CZECH)
sluitkool (30.27a. DUTCH)
sluitkool (30.27a. VLAAMS)
šluotelinė kochija (35.2. LITHUANIAN)
s'ma (40.1. KONKOMBA)
smailialapės pupelės (40.23. LITHUANIAN)
smalblaarplantago (57.1. AFRIKAANS)
smalbladige mosterd (30.7. VLAAMS)
small flower galinsoga (27.22. ENGLISH)
small gourd (38.4. ENGLISH)
small leaf mustard (30.7. ENGLISH)
small mallow (45.5. ENGLISH)
small pepper (62.3. ENGLISH)
small radish (30.56. ENGLISH)
smaržīgā kārvele (24.8. LATVIAN)
smaržīgā rezēda (60.1. LATVIAN)
smau reang tük (27.18. KHMER, CENTRAL)
smetanka (27.45. CZECH)
smetanka lékařska (27.45. CZECH)
smooth loofah (38.22. ENGLISH)
smooth pigweed (23.3. ENGLISH)
smörgåskrasse (30.51. SWEDISH)
smorodinovidniy tomat (62.10. RUSSIAN)
smulkiažiedė galinsoga (27.22. LITHUANIAN)
smullish (24.4. GAELIC, MANX)
šmykace (40.26. SORBIAN, LOWER)
snake bean (40.42. ENGLISH)
snake gourd (38.29. ENGLISH)
snake gourd (38.29. FILIPINO)
snake melon (38.10. ENGLISH)
snake palm (8.2. ENGLISH)
snap bean (40.26. ENGLISH)
snèap (30.41a. GAELIC, SCOTTISH)
sneha beejaka (40.1. SANSKRIT)
snèip (30.22. GAELIC, SCOTTISH)
snek tamatis (38.29. KRIO)
snɛk-tamatis (38.29. KRIO)
sniebiet (35.3. DUTCH)
sniedebone (40.26. STELLINGWERFS)
snigdhapathra (35.9. SANSKRIT)
snijbiet (35.3. VLAAMS)
snijboon (40.26. VLAAMS)
snijpeterselie (24.19. VLAAMS)
snijsalat (27.32. DUTCH)
snitbladbede (35.3. DANISH)
snitsalat (27.32. DANISH)
šnitt pijazi (7.11. KIRGHIZ)
snittbladbete (35.3. NORWEGIAN)
šnitt-louk (7.11. RUSSIAN)
šnittluk (7.11. RUSSIAN)
snow pea (40.27. ENGLISH)
snybeet (35.3. AFRIKAANS)
snyboontjie (40.26. AFRIKAANS)
snylterot (52.1. NORWEGIAN)
sò (40.43. JUNKUN TAKUM)
so (41.1. JAVANESE)
soá (24.13. HINDUSTANI)
soaga-nmei (23.3. GA)
soaka (27.44. FULFULDE)
soan (20.5. URDU)
soàn-á (7.8. CHINESE, MIN NAN)
soàn-á (7.10. CHINESE, MIN NAN)
soant (20.6. URDU)
soàn-thâu (7.10. CHINESE, MIN NAN)
soap (24.13. FIJIAN)
söb (40.43. WOLOF)
sobaima (40.26. BODO)
sóballa (35.10. HUNGARIAN)
soblé (7.2. WOLOF)
soblé (7.3. SERER-SINE)
soblé (7.3. WOLOF)
soblé futô (7.2. WOLOF)
soblere (7.3. PULAAR)
soboe (34.2. ÉWÉ)
sobon (63.1. KURANKO)
soch (7.3. ARMENIAN)
soch (7.11. ARMENIAN)
sočivo (40.12. SERBO-CROATIAN)
sockermajs (17.7. SWEDISH)
socoti (62.5. RONGA)
soczewica (40.12. POLISH)
soczewica jadalna (40.12. POLISH)
sød kartoffel (36.3. DANISH)
soda kruid (35.8. DUTCH)
sodaly şora (35.8. TURKMEN)

sodaotu (35.7. TURKISH)
Soda-Salzkraut (35.8. GERMAN)
sodaurt (35.7. NORWEGIAN)
sodaurt (35.8. NORWEGIAN)
sødfennikel (24.12. DANISH)
sødskærm (24.16. DANISH)
soɛ (40.25. KONO)
soekwa (38.21. VLAAMS)
soepui (7.5. AFRIKAANS)
soe-soe (38.25. AFRIKAANS)
soetrissie (62.1. AFRIKAANS)
soffione tarassaco (27.45. ITALIAN)
sogan (7.3. AZERBAIJANI, NORTHERN)
sogan (7.3. BOSNIAN)
soğan (7.3. TURKISH)
sogan (7.3. TURKMEN)
sogan (7.11. AZERBAIJANI, NORTHERN)
soğancik (7.11. TURKISH)
sogho (44.5. LOKO)
sogia (40.8. GREEK)
soging (16.1a. IBALOI)
sogoi (44.5. KURANKO)
sogrī (30.53. MARWARI)
sohās (38.13a. MAITHILI)
sohikeerai (24.13. TAMIL)
soi (11.2. MANU'A SAMOAN)
soi (11.2. SAMOAN)
soia (40.8. BASQUE)
soia (40.8. FRENCH)
soia (40.8. ITALIAN)
soia (40.8. ROMANIAN)
soia de cultură (40.8. ROMANIAN)
soijapapa (40.8. FINNISH)
soilire (24.4. GAELIC, IRISH)
soilire (24.4. GAELIC, SCOTTISH)
soja (40.8. BULGARIAN)
soja (40.8. BULU)
soja (40.8. CREOLE, HAITIAN FRENCH)
soja (40.8. CRIOULO, CAPE VERDE)
sója (40.8. CZECH)
soja (40.8. FRENCH)
soja (40.8. KAZAKH)
soja (40.8. LATVIAN)
soja (40.8. MACEDONIAN)
soja (40.8. NORWEGIAN)
soja (40.8. POLISH)
soja (40.8. PORTUGUESE)
soja (40.8. RUSSIAN)
soja (40.8. SERBO-CROATIAN)
sója (40.8. SLOVAK)
soja (40.8. SLOVENIAN)
soja (40.8. SPANISH)
soja böna (40.8. SWEDISH)
sója luštinatá (40.8. CZECH)
soja vjalikaja (40.8. BELARUSAN)
soja zwyczajna (40.8. POLISH)
sojabaun (40.8. ICELANDIC)
Sojabohne (40.8. GERMAN)
sojabønne (40.8. DANISH)
sojaboon (40.8. AFRIKAANS)
sojaboon (40.8. DUTCH)
sojaboon (40.8. VLAAMS)
sojaboontjie (40.8. AFRIKAANS)
sojas (40.8. LATVIAN)
sojë (40.8. ALBANIAN, TOSK)
sok choy (35.9. MORISYEN)
sok mai (17.7. CHINESE, MANDARIN)
sokerijuuri (24.23. FINNISH)
sokerimaissi (17.7. FINNISH)
sokeriruoko (17.5. FINNISH)
sokkarqaàla (17.5. AFAR)
soko (23.4. ENGLISH)
ṣọkọ (23.4. YORUBA)
soko rahkaʔmiitsaʔ (38.13a. COMANCHE)
soko sikusa (27.45. COMANCHE)
sọkọ yọ̀kọ̀tọ̀ (23.4. YORUBA)
sọkọ yọ̀kọ̀tọ̀ sọkọtọ (23.4. YORUBA)
ṣokudalẹ (62.23. YORUBA)
solanka komarowa (35.7. POLISH)
solanka sodowa (35.8. POLISH)
solano bietolone (62.7. ITALIAN)
solano negro (62.21. SPANISH)
solano nero (62.21. ITALIAN)
solanon to melan (62.15. GREEK)
solanum (62.21. ROMANIAN)
solanum amerikani (62.15. HEBREW)
solanüm ha-pekaʿot (62.24. HEBREW)
solanum he-ḫaẕil (62.19. HEBREW)
solanum shaḫor (62.21. HEBREW)
solasi (44.5. TAGALOG)
solasih (44.5. SUNDA)
solata berivka (27.32. SLOVENIAN)
solata vezivka (27.32. SLOVENIAN)
soldinella sibtorpioide (24.15. ITALIAN)
solena solnenka (35.8. MACEDONIAN)
solésolé (44.5. AKAN-ASANTE)
solid bamboo (17.2. ENGLISH)
sollə (40.26. AZERBAIJANI, NORTHERN)
solmazçiçek (22.2. TURKISH)
solotej (27.32. SORBIAN, UPPER)

solui (34.2. ÉWÉ)
sølvbede (35.3. DANISH)
solvirf (44.6. ROMANIAN)
somaka (38.1. SANSKRIT)
sombo (62.5. YORUBA)
som-chin (59.3. THAI)
som-kaoli (59.3. THAI)
som-khon (59.3. THAI)
sommar cypress (35.2. SWEDISH)
Sommer-Bohnenkraut (44.9. GERMAN)
sommercypres (35.2. DANISH)
Sommer-Endivie (27.32. GERMAN)
sommergresskar (38.18. NORWEGIAN)
sompu (24.13. KANNADA)
sômz ph'oox dii (45.4. LAO)
sômz sien (30.57. LAO)
sona (24.13. PASHTO, NORTHERN)
sona (40.43. TEM)
sonagee beeja (40.18. KANNADA)
sonakaela (16.1a. MARATHI)
sonbol (24.1. FARSI, WESTERN)
sonchala (45.5. HINDI)
sondaegida (62.23. KANNADA)
sonde (62.23. KANNADA)
sonde gida (62.23. KANNADA)
sonf (24.12. PANJABI, EASTERN)
sonf (24.13. HINDI)
sonf (24.13. PANJABI, EASTERN)
song tian gua (38.11. CHINESE, MANDARIN)
songa (38.17. SWAHILI)
songa (38.18. SWAHILI)
songino (7.1. MONGOLIAN)
sóngol (43.1. BALANTA-KENTOHE)
sôngough (17.1. TIV)
sôngough (17.2. TIV)
sôngough (17.4. TIV)
soni bihnáaʔ (17.5. COMANCHE)
sonka (11.7. ÉWÉ)
sonp (24.13. HINDI)
sonson di otot (38.29. KALAGAN)
sont (24.13. HINDI)
són-t'êou (7.10. CHINESE, HAKKA)
sonta (20.6. HINDI)
son̲th (20.6. HINDUSTANI)
sonthi (20.6. TELUGU)
sonti (20.6. KONKANI)
sonti (20.6. TELUGU)
són-ťzè (7.10. CHINESE, HAKKA)
sookshmavalli (38.23. SANSKRIT)
soomukas (52.1. ESTONIAN)
soopshreshta (40.37. SANSKRIT)
soo-qua (38.21. HMONG-DAW)
sooran (8.3. HINDI)
soorana (8.3. SANSKRIT)
soorana gedde (8.3. KANNADA)
soorna (8.3. KONKANI)
soornaka (8.3. SANSKRIT)
sooryaavartha (45.5. SANSKRIT)
sooryakaanthi gedde (27.26. KANNADA)
soosan (13.3. FARSI, WESTERN)
sopéré (11.7. BAOULÉ)
soppu (35.9. BADAGA)
sopropo (38.23. VLAAMS)
sopu (24.13. TELUGU)
sörä bēn (40.8. PANJABI, EASTERN)
sora mame (40.34a. JAPANESE)
sorakaaya (38.20. TELUGU)
sorakaya (38.20. TELUGU)
sorakkaai (38.20. TAMIL)
sora-sora (24.7. BICOLANO, CENTRAL)
sore (63.1. SUSU)
soreballi (38.20. KANNADA)
sorekaayi (38.20. KANNADA)
sore-kayi (38.20. KANNADA)
sorelu (58.2. JAPANESE)
soriah (30.3. ASSAMESE)
soriah (30.4. ASSAMESE)
sorofikhe (63.1. SUSU)
sórrè (63.1. SUSU)
sorrel (45.3. SPANISH)
sorrel (58.2. ENGLISH)
sort mungbønne (40.37. DANISH)
sort natskygge (62.21. DANISH)
sort sennep (30.24. DANISH)
soru manimuni (24.15. ASSAMESE)
sorzycha ziemna (40.39. POLISH)
sos- (38.21. HAUSA)
sosɛ (40.25. KURANKO)
sóska (58.2. HUNGARIAN)
soso (40.43. BAMANANKAN)
soso (40.43. MANDING-MANDINKA)
soso (40.43. MANINKA)
sosolon a gadong a kamo (35.9. MARANAO)
sosolon a gadong a kamo (35.9. MARINDUQUE)
šošovica (40.12. SLOVAK)
sosso (40.26. JULA)
sosso (40.26. MANINKAKAN, WESTERN)
soteen (16.1a. IVATAN)
soteen gayod (16.1a. IVATAN)

søtkjeks (24.23. NORWEGIAN)
sötpotatis (36.3. SWEDISH)
søtpotet (36.3. NORWEGIAN)
soũ (30.27a. GOURMANCHÉMA)
souberi fiyou (59.2. JAPANESE)
souchet comestible (10.1. FRENCH)
soude (35.10. FRENCH)
soude asperge (35.10. FRENCH)
soude commune (35.8. FRENCH)
soué (40.39. TEM)
soufleur (30.26. PULAR)
sougandhikamu (48.1a. TELUGU)
souïdiki kram i rapis (30.22. GREEK)
souka (39.4. GBAYA)
soukonanbourou (44.5. JULA)
soumochnik (30.43. RUSSIAN)
soumou (17.1. KABIYÉ)
souna (40.43. GOURMANCHÉMA)
sou-nave (30.30. MORISYEN)
sounf (24.12. ASSAMESE)
sounf (24.13. ASSAMESE)
sounounga (38.20. TEM)
souou (40.1. BALANTA-KENTOHE)
soupommé (30.27a. MANINKAKAN, WESTERN)
soupommé (30.27a. PULAR)
soupommé (30.27b. MANINKAKAN, WESTERN)
soupommé (30.27b. PULAR)
soũ-poné (30.26. GOURMANCHÉMA)
sourann (8.3. MORISYEN)
sousami (55.1. GREEK)
southern pea (40.43. ENGLISH)
soutslaai (22.1. AFRIKAANS)
sovirf (44.6. ROMANIAN)
šovul (58.3. UZBEK)
sow thistle (27.18. ENGLISH)
sow thistle (27.43. ENGLISH)
sowatou (23.2. KABIYÉ)
sowori (62.5. TSONGA)
sowthistle (27.27. ENGLISH)
sowthistle (27.42. ENGLISH)
soy bean (40.8. ENGLISH)
soya (40.8. ENGLISH)
soya (40.8. GIKUYU)
soya (40.8. HEBREW)
soya (40.8. ITALIAN)
soya (40.8. NORWEGIAN)
soya (40.8. SPANISH)
soya (40.8. TURKISH)
soya bean (40.8. ASSAMESE)
soya bean (40.8. ENGLISH)
soya bean (40.8. KANNADA)
soya beeja (40.8. KANNADA)
soya bonchi (40.8. SINHALA)
soya chikkudu (40.8. TELUGU)
soya fasülyesi (40.8. TURKISH)
soya payaru (40.8. TAMIL)
soyabean (40.8. PANJABI, EASTERN)
soyabin (40.8. GUJARATI)
soyabin (40.8. ORIYA)
soyaboontjie (40.8. AFRIKAANS)
soyauba (40.8. ESTONIAN)
soybean (40.8. ENGLISH)
soybean (40.8. MALAYALAM)
soybean (40.8. URDU)
soybin (40.8. MARATHI)
soyeh (40.8. YIDDISH)
soza (40.8. MORISYEN)
soza kolgo (40.8. MÒORÉ)
spaanse erwt (40.5. DUTCH)
spaanse erwt (40.5. VLAAMS)
spaanse peper (62.2. DUTCH)
spaanse peper (62.2. VLAAMS)
spaanse schorsoneer (27.36. DUTCH)
spaanse zuring (58.4. DUTCH)
spaghettipompoen (38.18. VLAAMS)
spamridgeh (13.2. YIDDISH)
spanać (35.9. BOSNIAN)
spanac (35.9. ROMANIAN)
spanač (35.9. SERBO-CROATIAN)
spanac sălbatic (35.6. ROMANIAN)
spanach (35.9. ARMENIAN)
spanacul (35.9. ROMANIAN)
spanacul ciobanilor (35.5. ROMANIAN)
spanacul de Nova Zeelandă (22.2. ROMANIAN)
spanak (35.9. BULGARIAN)
spanak (35.9. MACEDONIAN)
spanak de Nova Zeelandă (22.2. ROMANIAN)
spanaka (35.9. GREEK)
spanaki keïlanis (28.1. GREEK)
spanaki Neas Zilandias (22.2. GREEK)
spanakoos (35.9. ROMANI, BALKAN)
spanakovidna repka (30.39. MACEDONIAN)
spanarkerfill (24.16. ICELANDIC)
spaniolă (27.37. ROMANIAN)
Spanische Golddistel (27.36. GERMAN)
Spanische Schwarzwurzel (27.38. GERMAN)
Spanischer Pfeffer (62.1. GERMAN)

Spanischer Pfeffer (62.2. GERMAN)
Spanish golden thistle (27.36. ENGLISH)
Spanish needle (27.8. ENGLISH)
Spanisher Golddistel (27.36. GERMAN)
španjolska (27.38. SERBO-CROATIAN)
spansk gold tistle (27.36. SWEDISH)
spansk hvitlök (7.12. SWEDISH)
spansk kjørvel (24.16. NORWEGIAN)
spansk körvel (24.16. SWEDISH)
spansk peber (62.1. DANISH)
spansk peber (62.2. DANISH)
spansk peppar (62.1. SWEDISH)
španska artičoka (27.15. MACEDONIAN)
španska kozička (27.46. MACEDONIAN)
španska kozja brada (27.46. MACEDONIAN)
španska zmijovka (27.38. MACEDONIAN)
španski zlaten oso (27.36. MACEDONIAN)
spanskpepper (62.1. NORWEGIAN)
ṡpaŋ'ṡniyutapi (38.3. DAKOTA)
ṡpaŋ'ṡniyutapi (38.7a. DAKOTA)
ṡpaŋ'ṡniyutapi (38.11. DAKOTA)
spanspek (38.7a. AFRIKAANS)
spanspek (38.7b. AFRIKAANS)
spanspekboompie (62.20. AFRIKAANS)
sparanga (13.2. LADINO)
sparanghel (13.2. ROMANIAN)
sparangi (13.2. GREEK)
sparengj (13.2. ALBANIAN, TOSK)
sparevaja kapusta (30.31. RUSSIAN)
spárga (13.2. HUNGARIAN)
spárgakel (30.31. HUNGARIAN)
spárgatök (38.18. HUNGARIAN)
spargel (13.2. ESTONIAN)
Spargel (13.2. GERMAN)
spärgel (13.2. SWEDISH)
spargel (40.13. ESTONIAN)
špargel'a (13.2. SLOVAK)
Spargelbohne (40.42. GERMAN)
Spargelerbse (40.13. GERMAN)
sparġelis (13.2. LATVIAN)
špargelj (13.2. SLOVENIAN)
Spargelsalat (27.30. GERMAN)
špargl (13.2. CZECH)
spargl (13.2. SORBIAN, UPPER)
špargla (13.2. SERBO-CROATIAN)
sparja (13.2. UZBEK)
sparnuotoji dioskorėja (11.1. LITHUANIAN)
šparoga (13.2. CROATIAN)
sparris (13.1. SWEDISH)
sparris (13.2. SWEDISH)
sparrisärt (40.13. SWEDISH)
sparrisböna (40.42. SWEDISH)
sparrissallat (27.30. SWEDISH)
sparža (13.2. UKRAINIAN)
sparžha aptečnhaja (13.2. RUSSIAN)
sparzha leuavaja (13.2. BELARUSAN)
sparzha obyknovennaja (13.2. RUSSIAN)
sparzha ovoshtnaja (13.2. RUSSIAN)
sparzha sedovnaja (13.2. RUSSIAN)
spatawar jangli asparagus (13.1. URDU)
spearmint (44.3. ENGLISH)
Spearmint (44.3. GERMAN)
spec (62.1. ALBANIAN, TOSK)
spéi barang (30.35. KHMER, CENTRAL)
spéi k'daôb (30.27a. KHMER, CENTRAL)
spéi k'daôb (30.27b. KHMER, CENTRAL)
spéi k'daôb (30.34. KHMER, CENTRAL)
spéi khiew (30.36. KHMER, CENTRAL)
spéi khiew (30.40. KHMER, CENTRAL)
spéi sâå (30.25. KHMER, CENTRAL)
spéi tôôch tôôch (30.29. KHMER, CENTRAL)
Speiserübe (30.41a. GERMAN)
Speisezwiebel (7.3. GERMAN)
špenát (35.9. CZECH)
špenát (35.9. SLOVAK)
spenat (35.9. SWEDISH)
špenát Novozélandksý (22.2. SLOVAK)
špenát setý (35.9. CZECH)
spenót (35.9. HUNGARIAN)
spergilk'al (30.31. ICELANDIC)
spergill (13.2. ICELANDIC)
spergilsalat (27.30. ICELANDIC)
spèrsis (13.2. INDONESIAN)
spersis (13.2. SUNDA)
sperzieboon (40.26. DUTCH)
sperzieboon (40.26. VLAAMS)
spevek (13.2. ARMENIAN)
sphurjathu (23.3. SANSKRIT)
spider flower (34.2. ENGLISH)
spikenard (26.2. ENGLISH)
spilanthe des lieux humides (27.2. FRENCH)
spilanthes o lachanodes (27.2. GREEK)
spina santa cinese (62.8. ITALIAN)
špinača (35.9. SLOVENIAN)
spinaccio d'America (28.1. ITALIAN)
spinaccio del Malabar (28.1. ITALIAN)
spinaccio della Cina (28.1. ITALIAN)
spinach (35.9. ENGLISH)
spinach (35.9. FILIPINO)
spinach (35.9. KANNADA)

Stachys (44.11. GERMAN)
stachys tubereux (44.11. FRENCH)
stalk celery (24.4. ENGLISH)
stambialapis šakya (2.1. LITHUANIAN)
stambiažiedis plačvarpis (34.3. LITHUANIAN)
stamkool (30.35. VLAAMS)
stamsniede bonen (40.26. STELLINGWERFS)
stangbønne (40.26. DANISH)
Stangensellerie (24.4. GERMAN)
stangselleri (24.4. NORWEGIAN)
stapfilo (24.11. GEORGIAN)
staphilo (24.11. GEORGIAN)
stapilo (24.11. GEORGIAN)
star gooseberry (39.5. ENGLISH)
stashkeh (40.1. YIDDISH)
št'av (58.2. SLOVAK)
št'av špenátny (58.3. SLOVAK)
štavel (58.3. MACEDONIAN)
šťavel hliznatý (53.1. CZECH)
steblena salata (27.30. MACEDONIAN)
Steckrübe (30.22. GERMAN)
stedmoderblomst (68.1. DANISH)
steeltjes (30.41b. DUTCH)
steinselja (24.19. ICELANDIC)
stekkool (30.33. VLAAMS)
stem lettuce (27.30. ENGLISH)
stemorsblomst (68.1. NORWEGIAN)
stengelsalat (27.30. NORWEGIAN)
stengelui (7.9b. VLAAMS)
štěrbák (27.9. CZECH)
sterie (58.2. ROMANIAN)
sterkers (30.51. DUTCH)
stevia (58.3. ROMANIAN)
sthirphala (38.1. SANSKRIT)
sthoola chanchu (63.1. SANSKRIT)
sthoolakandaka (8.3. SANSKRIT)
sthoolashimbi (40.3. SANSKRIT)
štiav kyslý (58.2. SLOVAK)
stilkkål (30.36. NORWEGIAN)
stilkselleri (24.4. ICELANDIC)
stilkselleri (24.4. NORWEGIAN)
stinging nettle (66.3. ENGLISH)
štir (23.3. MACEDONIAN)
štítenka Schreberova (31.1. CZECH)
štitest štavej (58.4. MACEDONIAN)
stjälkselleri (24.4. SWEDISH)
stobrainais sīpols (7.5. LATVIAN)
stocchis (44.11. GREEK)
stočna repa (35.4. MACEDONIAN)
stokboon (40.26. DUTCH)
stoksniedebonen (40.26. STELLINGWERFS)
stolova morkva (24.11. UKRAINIAN)
stolovij burjak (35.4. UKRAINIAN)
stolthenrik (35.5. NORWEGIAN)
stolthenriks-ga'sefod (35.5. DANISH)
stonecrop (37.1. ENGLISH)
stopelknol (30.41a. DUTCH)
stor galanga (20.1. DANISH)
stor nælde (66.3. DANISH)
storašaknis katilėlis (33.2. LITHUANIAN)
störböna (40.26. SWEDISH)
storborre (27.5. NORWEGIAN)
stornesle (66.3. NORWEGIAN)
stornetla (66.3. ICELANDIC)
šťovík (58.2. CZECH)
šťovík kyselý (58.2. CZECH)
šťovík štitnatý (58.4. CZECH)
šťovík zahradní (58.3. CZECH)
strandblađka (35.3. ICELANDIC)
strandkål (30.45. DANISH)
strandkål (30.45. NORWEGIAN)
strandkoal (30.45. DANISH)
strandköl (30.45. SWEDISH)
strandkvann (24.1. NORWEGIAN)
Strauchbohne (40.2. GERMAN)
Straucherbse (40.2. GERMAN)
strawberry tomato (62.11. ENGLISH)
stregona giapponese (44.11. ITALIAN)
strėlialapė papliauška (6.2. LITHUANIAN)
strelki zvychainaja (30.43. BELARUSAN)
strelolist strelolistyj (6.2. KAZAKH)
struikkool (30.33. VLAAMS)
struikokra (63.1. VLAAMS)
struiktomaat (62.7. DUTCH)
strumous mustard (30.18. ENGLISH)
struupmoes (30.35. STELLINGWERFS)
strzałka wodna (6.2. POLISH)
studentenkruid (35.2. DUTCH)
studentenkruid (35.2. VLAAMS)
stumbill (27.38. ALBANIAN, TOSK)
styfnos (62.21. GREEK)
styselkanna (9.1. AFRIKAANS)
styvmorsviol (68.1. SWEDISH)
su (30.27a. ZARMA)
su (30.35. WOLOF)
su başaği (25.1. TURKISH)
su cevizi (64.3. TURKISH)
su da zhu mao cai (35.8. CHINESE, MANDARIN)
su da zhu mao cai (35.8. CHINESE, WÚ)

su eğreltisi (4.1. TURKISH)
su h[af]o (30.30. VIETNAMESE)
su kabaği (38.16. TURKISH)
su kabaği (38.20. TURKISH)
su kalkani (31.1. TURKISH)
su kazayaği (24.23. TURKISH)
su keneviri (44.1. TURKISH)
su kerevizi (24.17. TURKISH)
su kestanesi (10.2. TURKISH)
su kestanesi (64.3. TURKISH)
su kildizi (27.8. TURKISH)
su su (38.25. VIETNAMESE)
su teresi (30.44. TURKISH)
su teresi (30.52. TURKISH)
suan (7.1. CHINESE, YUE)
suan (7.10. CHINESE, MIN BEI)
suan (7.10. CHINESE, WÚ)
suan (7.10. CHINESE, YUE)
suan jiang (62.11. CHINESE, MANDARIN)
suan jiang (62.11. CHINESE, WÚ)
suan mo (58.2. CHINESE, MANDARIN)
suan mu (58.2. CHINESE, WÚ)
suan pan zi (40.31. CHINESE, MANDARIN)
suan pan zi (40.31. CHINESE, WÚ)
suaralu (11.2. BENGALI)
suaralu (11.2. HINDI)
suariya (11.2. GUJARATI)
suart mungböna (40.37. SWEDISH)
suartrot (27.38. SWEDISH)
subak (38.3. KOREAN)
subang puteri batek (27.8. MALAY)
su-brocoli (30.31. WOLOF)
subwuyas (7.3. CAROLINIAN)
succory (27.10a. ENGLISH)
suchanchuka (63.1. SANSKRIT)
sucikcikana seca (62.1. DAKOTA)
sucré-dow (17.1. GOURMANCHÉMA)
sucré-dow (17.2. GOURMANCHÉMA)
sucré-dow (17.4. GOURMANCHÉMA)
sucul gece sefasi (36.2. TURKISH)
sucumadeira (27.24. PORTUGUESE)
sudaikai (62.23. TAMIL)
sudhavaasa (38.13a. SANSKRIT)
sudheergha (38.29. SANSKRIT)
sudhimudi (27.18. BENGALI)
sudi (8.4. IVATAN)
sudimirapakaaya (62.5. TELUGU)
sudmirapakaaya (62.2. TELUGU)
sudo (11.5. JAVANESE)
sudu loonu (7.10. SINHALA)
suduwelala (11.7. SINHALA)
suɛ (40.25. KONO)
sufaid murga (23.4. HINDI)
sufed kadsumbal (40.3. HINDI)
sufed rai (30.57. HINDI)
sufedbachla (28.1. HINDI)
su-folor (30.26. WOLOF)
suga ken (17.5. IGBO)
sugaken (17.5. LOKO)
sugama (40.39. YAO)
sugan (7.3. TATAR)
suganda (44.6. TAGALOG)
sugandam (20.1. TAMIL)
sugandha (20.1. SANSKRIT)
sugandha (24.13. SANSKRIT)
sugandha (40.5. SANSKRIT)
sugandham (16.1a. TAMIL)
sugandhavaachi (20.1. KANNADA)
sugandhavacha (20.1. SANSKRIT)
sugandhayoga (20.1. SANSKRIT)
sugandhhalu (16.1a. TELUGU)
sugandhi (10.1. SANSKRIT)
sugandhi (24.9. SANSKRIT)
sugar pea (40.27. ENGLISH)
sugarcane inflorescence (17.5. ENGLISH)
sugasimbi (40.18. TAMIL)
sugɛ-ŋɛŋɛɛgɛ (62.5. LOKO)
sugina (1.1. JAPANESE)
sugreng (64.3. BODO)
sugupta (40.18. SANSKRIT)
suh^ meu_ (39.4. LAHU)
suhkruroog (17.5. ESTONIAN)
suhu tsitsinaʔ (62.24. COMANCHE)
sui dou (40.6. CHINESE, MANDARIN)
sui dou (40.6. CHINESE, WÚ)
sui kwa (38.3. JAPANESE)
suiba (58.2. JAPANESE)
suidissel (27.42. AFRIKAANS)
suika (38.3. JAPANESE)
suikermais (17.7. DUTCH)
suikermais (17.7. VLAAMS)
suikermielie (17.7. AFRIKAANS)
suikerriet (17.5. AFRIKAANS)
suikerriet (17.5. DUTCH)
suikerwortel (24.23. AFRIKAANS)
suippokaali (30.27a. FINNISH)
suizenjina (27.25. JAPANESE)
suizenjina (27.25. VLAAMS)
sujala (48.1a. SANSKRIT)
suka thi (38.25. BURMESE)

sukaattisitruuna (38.2. FINNISH)
sukampo (58.2. JAPANESE)
sukan (48.1a. BICOLANO, CENTRAL)
sukanaasaka (63.1. SANSKRIT)
sukanda (10.1. SANSKRIT)
sukanda (38.23. SANSKRIT)
sukandaka (11.2. SANSKRIT)
sukandaka (38.23. SANSKRIT)
sukandako (7.3. PALI)
sukandi (8.3. SANSKRIT)
sukara (17.5. BUDIK)
sukareshta (10.1. SANSKRIT)
sukaw (48.1a. ILOCANO)
sukerei (27.10a. STELLINGWERFS)
sukeri'j (27.10a. STELLINGWERFS)
sukhasha (38.3. SANSKRIT)
sukhavasa (38.3. SANSKRIT)
sukkerært (40.27. DANISH)
sukkerert (40.27. NORWEGIAN)
sukkermais (17.7. NORWEGIAN)
sukkermays (17.7. DANISH)
sukkerrod (24.23. DANISH)
sukkerrør (17.5. DANISH)
sukkerrør (17.5. NORWEGIAN)
sukku (20.6. BADAGA)
sukku (20.6. TAMIL)
sukola (44.5. MANINKA)
sukola (44.5. MANINKAKAN, EASTERN)
sukora (44.5. PULAAR)
sukosha (38.21. SANSKRIT)
sukrin (38.7a. BRETON)
sukshmapathra (24.9. SANSKRIT)
suku (38.20. MUNDARI)
sukula (44.5. BAMANANKAN)
sukuma (30.35. LUYIA)
sukuma wiki (30.6. SWAHILI)
sukuma wiki (30.35. SWAHILI)
sukumaara (16.1a. SANSKRIT)
sukuru (17.5. WAMEY)
sulɛnyi (45.1. SUSU)
suli (8.4. BINUKID)
sũlou (38.27. DAGAARE, SOUTHERN)
suludjatô (62.15. MANDING-MANDINKA)
sulyom (64.1. HUNGARIAN)
sulyom (64.3. HUNGARIAN)
suma (8.3. KONKANI)
suma (40.1. DAGBANI)
suma (40.39. MÒORÉ)
suma (44.8. JUNKUN TAKUM)
sumalle (38.20. PULAAR)
sumaráy (39.4. WÁRAY-WÁRAY)
šumbəra (40.5. AMHARIC)
sumbul (13.5. TAJIKI)
sumbullbar (46.1. ALBANIAN, TOSK)
sumbura (40.5. AFAR)
šumbura (40.5. HARARI)
sum-kam (40.1. MÒORÉ)
summer cypress (35.2. ENGLISH)
summer savory (44.9. ENGLISH)
summer squash (38.18. ENGLISH)
summersquash (38.18. NORWEGIAN)
sumpa (23.3. WÁRAY-WÁRAY)
sumpae légu (40.31. KONKOMBA)
sumpelegu (40.31. KONKOMBA)
Sumpfparakresse (27.2. GERMAN)
Sumpf-Trichterwinde (36.2. GERMAN)
sumpi-sumping (23.6. PANGASINAN)
sumptaro (8.6. DANISH)
sumranz (45.3. MONGOLIAN)
sumranz (45.4. MONGOLIAN)
sún (17.1. CHINESE, MIN NAN)
sún (17.2. CHINESE, MIN NAN)
sún (17.4. CHINESE, MIN NAN)
sun gua (38.14. CHINESE, GÀN)
sun zi gai (30.5. CHINESE, MANDARIN)
sun zi jie (30.5. CHINESE, WÚ)
sunariysh (27.36. ARABIC)
sunḏ (20.6. PASHTO, NORTHERN)
sundai (62.23. TAMIL)
sundaikaai (62.23. TAMIL)
sundaikeerai (40.19. TAMIL)
sundakkai (62.23. KANNADA)
sundakkayi (62.23. KANNADA)
sundari (62.15. SANSKRIT)
sundi (20.6. TAMIL)
sundrogai (17.1. ORIYA)
sungay-kalabaw (64.3. TAGALOG)
sungot (36.3. BUHID)
sungra (30.53. HINDI)
sunka (11.7. ÉWÉ)
sunkaaru (40.1. FULFULDE, ADAMAWA)
sunkuk buladlad (2.1. HANUNOO)
sunma (44.8. JUNKUN TAKUM)
sunna (40.26. GOURMANCHÉMA)
sunset hibiscus (45.2. ENGLISH)
sunte (20.6. MARATHI)
sunthi (20.6. TULU)
šuṇṭhi (20.6. TULU)
sunti (20.6. TULU)
suntunlɛ (45.4. BULLOM SO)

sununga (38.20. TEM)
suo cao (10.1. CHINESE, WÚ)
suolaheinä (58.4. FINNISH)
suotu (6.2. TURKISH)
supathra (35.9. SANSKRIT)
suphala (16.1a. SANSKRIT)
suphala (38.1. SANSKRIT)
su-pommé (30.27a. SERER-SINE)
su-pommé (30.27a. WOLOF)
su-põnpe (30.27a. MÒORÉ)
suppome (30.27a. PULAAR)
supriya (24.7. SANSKRIT)
süpürge kahya (35.2. TURKISH)
süpürge out (35.2. TURKISH)
supushpa (38.21. SANSKRIT)
supushpika (24.13. SANSKRIT)
supuwiri (24.4. HINDUSTANI)
supya (40.2. SANSKRIT)
suqömhatiko (40.25. HOPI)
suqömvàapu (40.26. HOPI)
sura (40.12. SANSKRIT)
sura (58.3. ICELANDIC)
surabhi (44.5. SANSKRIT)
surage (24.9. INDONESIAN)
surai (38.20. TAMIL)
suraikaai (38.18. TAMIL)
suraikayi (38.18. TAMIL)
Suraim spinach (59.3. ENGLISH)
suraka (35.1. PANJABI, EASTERN)
sura-kai (38.20. TAMIL)
surake (35.1. HINDI)
suran (8.3. GUJARATI)
suran (8.3. HINDI)
suran (8.3. MARATHI)
suran (8.3. SANSKRIT)
suran y cwn (58.2. WELSH)
sūraṇa (8.3. TULU)
suranah (8.3. SANSKRIT)
surangga (23.4. WÁRAY-WÁRAY)
surang-surang (23.4. WÁRAY-WÁRAY)
surasa (44.4. SANSKRIT)
surasa (44.5. SANSKRIT)
surasema (40.3. HINDI)
surashthaja (40.2. SANSKRIT)
su-rav (30.30. WOLOF)
surawung (44.5. INDONESIAN)
surawung (44.5. SUNDA)
surelle (58.2. FRENCH)
surelle (58.4. FRENCH)
surengan (58.2. INDONESIAN)
surengan (58.4. INDONESIAN)
Surinaamse postelein (59.3. DUTCH)
Surinaamse postelein (59.3. VLAAMS)
surinaamse zuring (45.4. DUTCH)
Surinam purslane (59.3. ENGLISH)
Surinam semizotu (59.3. TURKISH)
suriyakamal (48.1a. GUJARATI)
surkhkaram (30.27b. TAJIKI)
surnawi (8.4. BUHID)
suro-na (63.1. YALUNKA)
surosandiya (38.29. CEBUANO)
suruku katama (19.1. MANDING-BAMBARA)
suruma (11.4. GUN)
sürünücü dam koruğu (37.1. TURKISH)
survari (23.4. HINDI)
süs biberi (62.5. TURKISH)
susak (38.20. ALBANIAN, TOSK)
šušanbandžar (24.8. ARMENIAN)
susemiori (38.22. KOREAN)
sushaka (23.3. SANSKRIT)
sushaka (40.6. SANSKRIT)
sushaka (45.1. SANSKRIT)
sushavi (38.23. SANSKRIT)
susheethala (38.13a. SANSKRIT)
sushkaka (20.6. SANSKRIT)
s[us]ng lam (50.2. VIETNAMESE)
susnialu (11.5. BENGALI)
s[us]p l[ow] (30.26. VIETNAMESE)
susruba (62.23. GBE-VHE)
Süsskartoffel (36.3. GERMAN)
susuo (62.14b. BAFIA)
susuruba (62.23. TWI)
su-suu (38.25. KHMER, CENTRAL)
susuvi (38.23. SANSKRIT)
Süßdolde (24.16. GERMAN)
Sußwurz (24.23. GERMAN)
suthi (20.6. NEWAR)
suthni (11.5. BENGALI)
sütótök (38.16. HUNGARIAN)
sutri (40.40. HINDI)
sutri (40.40. URDU)
suuk (40.1. SUNDA)
suureviljalist (38.16. ESTONIAN)
suvariyaa kand (11.2. HINDI)
suvarna gedda (8.3. KANNADA)
suvarna gedde (8.3. KANNADA)
suvarna-guddab (8.3. KANNADA)
suvartula (38.3. SANSKRIT)
suvasura (30.51. SANSKRIT)
suvithra (8.3. SANSKRIT)

szczypiorek chiński (7.4. POLISH)
szeges borsó (40.10. HUNGARIAN)
szeges lednek (40.10. HUNGARIAN)
szegletes lednek (40.10. HUNGARIAN)
szelíd csorbóka (27.42. HUNGARIAN)
szezám (55.1. HUNGARIAN)
szíki ballagófű (35.8. HUNGARIAN)
szivacstök (38.22. HUNGARIAN)
szkarłatka amerykańska (56.2. POLISH)
szójabab (40.8. HUNGARIAN)
szparag (13.2. NORWEGIAN)
szparag dziki (13.1. POLISH)
szparag lekarski (13.2. NORWEGIAN)
szpinak (35.9. POLISH)
szpinak nowozelandzki (22.2. POLISH)
szpinak warzywny (35.9. POLISH)
szpinak wodny (36.2. POLISH)
szümcső (30.42. HUNGARIAN)
szurokfű (44.6. HUNGARIAN)
tà (27.24. GA)
ta (40.1. MANDINKA)
ta chii chi (27.15. CHINESE, MANDARIN)
t'ą̄'-gɑ (38.20. KIOWA)
ta huang (58.1. CHINESE, MANDARIN)
ta kên hsuch wei ts'ai (24.8. CHINESE, MANDARIN)
taa kôngz (59.2. LAO)
t[aaf]m v[oo]ng (17.2. VIETNAMESE)
tààjiwâr (20.6. KANURI)
taakaka (2.1. MAORI)
ta'amu (8.1. SAMOAN)
ta'amu papalagi (8.9. SAMOAN)
taao-hla-chao (40.42. HMONG-DAW)
taara-yaara (19.1. HAUSA)
taaro (8.4. FINNISH)
taasai (30.38. JAPANESE)
taatsunyaa (23.3. HAUSA)
tabadai (34.2. HAUSA)
tabadamashe (34.2. HAUSA)
tabaday (34.2. HAUSA)
tabag (62.17. URDU)
tabagwang (38.1. HANUNOO)
tabagwang (38.16. HANUNOO)
tabagway (38.17. HANUNOO)
tabaibo (32.1. CRIOULO, CAPE VERDE)
tabanburo (62.16. TAGALOG)
tabare (17.1. BADAGA)
tabaru (40.18. BAATONUM)
tabasco (62.5. ITALIAN)
tabasco (62.5. PORTUGUESE)
tabasco (62.5. SPANISH)
tabasco pepper (62.5. ENGLISH)
tabasco pepper (62.5. FRENCH)
Tabasco Pfeffer (62.5. GERMAN)
tabascopaprika (62.5. NORWEGIAN)
tabau-tabau (38.21. ILOCANO)
tabaw-tabaw (38.22. ILOCANO)
tabayag (38.20. HANUNOO)
tabayug (38.20. HANUNOO)
tabinere (11.2. KANNADA)
tabios (40.2. BICOLANO, CENTRAL)
tabios (40.2. CEBUANO)
tabios (40.2. TAGALOG)
tabiyayong (38.20. PANGASINAN)
tabiyos (40.2. BICOLANO, CENTRAL)
tabiyos (40.2. CEBUANO)
table beet (35.4. ENGLISH)
tablug-batug (40.3. HANUNOO)
tabo (17.5. MANOBO)
tabobog (38.21. BICOLANO, CENTRAL)
tabobog (38.21. CEBUANO)
tabobog (38.21. TAGALOG)
tabobog (38.22. TAGALOG)
tabobok (38.22. TAGALOG)
tabogok (38.13. TAGALOG)
tabtabilokong (27.14. BONTOC, CENTRAL)
tabubok (38.29. TAGALOG)
tabubuan (36.3. BUHID)
tabugok (38.29. SUBANON)
tabugok (38.29. TAGALOG)
tabuguk (38.1. AGTA)
tabuka (55.1. VAI)
tabukɔ (55.1. MENDE)
tabulɔgɔ (55.1. LOKO)
tabungau (38.20. ILOCANO)
tabungaw (38.20. BONTOC, CENTRAL)
tabungaw (38.20. ILOCANO)
tabungaw (38.20. KANKANAEY)
tabuyuk (38.22. HANUNOO)
tabyos (40.2. BICOLANO, CENTRAL)
tabyos (40.2. CEBUANO)
tacaltique (17.1. SENOUFO, CEBAARA)
taccola (40.27. ITALIAN)
tachi suberi hiyu (59.2. JAPANESE)
tachi-nata-mame (40.3. JAPANESE)
tadagani (40.43. KANNADA)
tadagunny (40.43. KANNADA)
tadam (62.9. MALTESE)
tadkernit (27.42. ARABIC)
îɑdl-t'ɑn (40.26. KIOWA)

t̂ɑdl-t'ɑn-sɑhyei-gɑ (40.28. KIOWA)
tadtad (23.3. BONTOC CENTRAL)
tadung (62.19. KALAGAN)
tadyaw (16.1a. MARANAO)
táem (8.4. BALANTA-KENTOHE)
taeng-chin (38.3. THAI)
taeng-kariang (38.25. THAI)
taeng-kwa (38.13a. THAI)
taeng-mo (38.3. THAI)
taeng-om (38.13a. THAI)
taeng-ran (38.13a. THAI)
taewa (62.24. MAORI)
taewha (62.24. MAORI)
taf a nuk (59.2. SERER-SINE)
tafal (8.4. GADDANG)
tafannuwa (7.10. HAUSA)
tafanua (7.10. DENDI)
tafárnúúwáá (7.10. HAUSA)
tafarnuwa (7.10. ZARMA)
tafod yr ych (29.1. WELSH)
tafol (58.3. WELSH)
tafvmpe (7.3. MUSKOGEE)
tafvmpe-vhake (7.10. MUSKOGEE)
tâg i mpitöx (59.2. WOLOF)
taga (38.20. BULLOM SO)
tagaban (45.1. HANUNOO)
tagabang (63.1. CEBUANO)
tagabang (63.1. SUBANON)
tagabang diririt marara (63.1. HANUNOO)
tagabang talun (63.1. HANUNOO)
tagabas (5.1. TAGALOG)
tagaditak (24.7. IVATAN)
tàgalê (40.3. PULAAR)
taga-opay (24.17. KALAGAN)
tagatum (62.16. AKLANON)
tagatum (62.16. CEBUANO)
ta-gbeŋgbe-ta-bana (62.1. THEMNE)
ta-gbɛŋgbɛ-*ta*-bana (62.2. THEMNE)
ta-gbɛŋgbɛ-*ta*-bɔli (62.2. THEMNE)
ta-gbɛŋgbɛ-*ta*-but (62.2. THEMNE)
ta-gbɛŋgbɛ-*ta*-lɔl-*ta*-lɔl (62.5. THEMNE)
ta-gbeŋgbe-ta-potho (62.1. THEMNE)
ta-gbɛŋgbɛ-*ta*-potho (62.2. THEMNE)
ta-gbɛŋgbɛ-*ta*-temne (62.5. THEMNE)
ta-gbeŋgbe-ta-wir (62.1. THEMNE)
taglabag dikot (59.2. SAMBAL, BOTOLAN)
Taglilie (13.3. GERMAN)
tago (24.7. MANU'A SAMOAN)
tagogtog (23.4. TAGBANWA)
tagogtog (62.19. TAGBANWA)
tagotong (62.19. MARANAO)
tagotung (62.19. MARANAO)
tagtagu (23.4. BONTOC, CENTRAL)
tagughug (23.4. TAGBANWA)
tagughug (23.6. AKLANON)
tagulinao (27.18. FILIPINO)
tagulinas (27.18. TAGALOG)
tagulinau (27.18. TAGALOG)
tagulinaw (27.18. TAGALOG)
taguntung (62.19. TAGBANWA)
tagutong (62.16. CEBUANO)
tagutung (62.16. BUHID)
tagwamba (8.4. HAUSA)
tagyas (40.2. SUBANON)
ta-hā-ngen (11.1. NICOBARESE, CAR)
ta-hā-ngen (11.2. NICOBARESE, CAR)
ta-hā-ngen (11.5. NICOBARESE, CAR)
tahaunon (40.30. MANOBO)
ta-hɔndi (23.3. MENDE)
tahdun (8.4. BODO)
tahe̱r (38.13a. SANTALI)
tahganda (39.4. BODO)
tahguna (36.3. BODO)
tahiti ararotu (19.1. TURKISH)
Tahiti arrowroot (19.1. ENGLISH)
tahou (38.3. SENOUFO, CEBAARA)
tahung (62.19. AGTA, CASIGURAN DUMAGAT)
tai gai (30.20. CHINESE, MANDARIN)
tai jie (30.20. CHINESE, WÚ)
t'ái vông (58.1. CHINESE, HAKKA)
taiawa (62.24. MAORI)
taibo bihnaaʔ (38.7a. COMANCHE)
taibwo (44.5. CAROLINIAN)
Taigawurzel (26.4. GERMAN)
taikuda (23.3. TAUSUG)
taimiá de cipó (38.19. PORTUGUESE)
tain jiao (62.1. CHINESE, WÚ)
tai-nyuk-ngö (16.1a. NICOBARESE, CAR)
taioba (8.4. PORTUGUESE)
taioba (8.9. PORTUGUESE)
taisi (30.36. JAPANESE)
tai-si (30.36. JAPANESE)
taiuiá de cipó (38.19. PORTUGUESE)
Taiwan giant bamboo (17.2. ENGLISH)
Taiwan-tsunaso (63.1. JAPANESE)
taiwha (62.24. MAORI)
tajer (8.9. DUTCH)
tajerblad (8.9. VLAAMS)
täjihoraz (23.4. TURKMEN)

taka (19.1. CZECH)
taka (63.1. MAGUINDANAO)
taka laut (19.1. INDONESIAN)
taka-ma-gindanaw (63.1. MAGUINDANAO)
takamariya (44.5. GUJARATI)
tàkàndá (17.5. GONJA)
takanda (17.5. GONJA)
takanɗa (17.5. HAUSA)
takapasin (62.23. TAGALOG)
takarmány lucerna (40.17. HUNGARIAN)
takarmány tarlórépa (30.41a. HUNGARIAN)
takarmánykáposzta (30.35. HUNGARIAN)
takenoko (17.1. JAPANESE)
takenoko (17.2. JAPANESE)
takenoko (17.4. JAPANESE)
takere foondu (30.51. FULFULDE, ADAMAWA)
taket (27.18. SUBANON)
takio-kohol (24.7. FILIPINO)
takip-kohol (24.7. TAGALOG)
takip-kuhol (24.7. TAGALOG)
takip-kuto (24.7. TAGALOG)
takip-suso (24.7. TAGALOG)
takittukaujaq (24.11. INUKTITUT, EASTERN CANADIAN)
takjas (27.5. ESTONIAN)
takka (19.1. TURKISH)
takkali (62.9. TAMIL)
takkeklap (30.42. DANISH)
tako la hassan (59.2. SWAHILI)
ta-kobo (62.19. BAGA)
takokak (62.23. INDONESIAN)
takokak (62.23. SUNDA)
takolô (40.39. MANDINKA)
takompv (38.13a. MUSKOGEE)
takoudiakatou (45.1. PULAR)
taksāl (44.3. PASHTO, NORTHERN)
taktôjk (40.26. ALEUT)
taku (45.1. PULAAR)
takup-suso (24.7. WÁRAY-WÁRAY)
takuri (40.40. SUBANON)
takwara (43.1. DAGBANI)
tala (38.20. KISSI, NORTHERN)
talad (17.5. AGTA)
talailo (50.2. CEBUANO)
tálan (40.1. BAGA)
talaparni (24.13. SANSKRIT)
talapias (9.1. BUHID)
talas (8.4. INDONESIAN)
talas belitung (8.9. INDONESIAN)
talas belitung (8.10. INDONESIAN)
talas bogor (8.4. INDONESIAN)
tala-suruntu-na (62.2. YALUNKA)
talcódja (40.18. PULAAR)
tâlebi (38.7a. FARSI, WESTERN)
talebie (38.7b. FARSI, WESTERN)
talembak (58.1. MADURA)
Talerkürbis (38.28. GERMAN)
tales (8.4. JAVANESE)
tales (8.4. TAGBANWA)
taletale (24.13. NIUE)
taleus balang (8.4. SUNDA)
taleus balitung (8.9. SUNDA)
taleus hideung (8.9. SUNDA)
taleus lahun indung (8.9. SUNDA)
taleus sente (8.1. SUNDA)
talewort (29.1. ENGLISH)
talgami (30.41b. GEORGIAN)
tàlhánà (36.2. SONGHAY)
talhpakha' (32.1. CHICKASAW)
tali jiwo (27.25. INDONESIAN)
talia (59.3. ITALIAN)
talilong (23.4. TAGALOG)
talilong (59.3. TAGALOG)
talingisag (45.4. SUBANON)
Talinum (59.3. GERMAN)
talinum (59.3. TAGALOG)
talinum (59.4. DANISH)
talinum (59.4. TAGALOG)
talipan (8.1. BICOLANO, CENTRAL)
talise (36.3. BUHID)
talisibul (7.5. ESTONIAN)
talitis (23.3. WÁRAY-WÁRAY)
talkha (24.23. TAJIKI)
tallow gourd (38.1. ENGLISH)
talo (8.4. BEMBÉ)
talo (8.4. NIUE)
talo (8.4. NORWEGIAN)
talo (8.4. SAMOAN)
talo (8.4. TONGAN)
talom (62.19. MANOBO)
talom (62.19. SAMBAL, BOTOLAN)
talom (62.19. SAMBAL, TINÀ)
talon (62.19. AGTA, CASIGURAN DUMAGAT)
talon (62.19. PANGASINAN)
talón (62.19. PANGASINAN)
talon (62.19. SAMBAL, BOTOLAN)
talong (62.19. BICOLANO, CENTRAL)
talong (62.19. BONTOC CENTRAL)

talong (62.19. FILIPINO)
talong (62.19. TAGALOG)
talong (62.19. TAGBANWA)
talonga (62.19. TAGALOG)
talong-ayam (62.16. BICOLANO, CENTRAL)
talong-buru (62.16. TAGALOG)
talong-gubat (62.16. TAGALOG)
talong-na-puti (62.25. TAGALOG)
talong na puti (62.25. TAGALOG)
talongon (62.23. TAGALOG)
talong-pipit (62.25. TAGALOG)
talong-talong (62.16. TAGALOG)
talong-talongan (62.23. TAGALOG)
talote (38.25. SPANISH)
talugbati (28.1. HANUNOO)
talum (62.19. MAGUINDANAO)
talum (62.19. SAMA, CENTRAL)
talum (62.19. TAUSUG)
talun (62.19. HANUNOO)
talun balubun (62.19. HANUNOO)
talúng (62.19. HILIGAYNON)
talun-kiya (62.23. TAUSUG)
talus-talus (8.4. PALAWANO, CENTRAL)
taluyot (45.1. IFUGAO)
talvari sem (40.4. URDU)
talviportulakka (59.1. FINNISH)
talvispuli (7.5. FINNISH)
talvisquash (38.14. FINNISH)
talyang (8.1. CEBUANO)
talyon meshulash (59.3. HEBREW)
tam bul bawk (30.41a. MIZO)
tam ling (38.4. LAO)
tam nhè (40.18. LAO)
tam nin (38.4. LAO)
tama (40.33. AFRIKAANS)
tama chisha (27.31. JAPANESE)
tamaatar (62.9. HINDI)
tamaate (62.9. WOLOF)
tamaàtim (62.9. AFAR)
tamados (62.9. KWÉYÒL)
tamahiba (20.4. TAGALOG)
tamahilan (20.4. BICOLANO, CENTRAL)
tamahli (62.9. CHEROKEE)
tamaka (44.10. HAUSA)
tamane (7.3. JAPANESE)
tamanegi (7.3. JAPANESE)
tamangyan (20.2. CEBUANO)
ta-mank (17.7. THEMNE)
tamantasi (62.9. KONO)
ta:mara (48.1a. BADAGA)
tamarillo (62.7. ENGLISH)
tamarillo (62.7. FRENCH)
tamarillo (62.7. NORWEGIAN)
tamasis (62.9. LIMBA)
tamasis (62.9. SUSU)
tamasisio (62.9. KISSI, NORTHERN)
tamatar (62.9. BENGALI)
tamatar (62.9. FIJIAN)
tamatar (62.9. HINDI)
ṭamāṭar (62.9. MAITHILI)
tamatar (62.9. PANJABI, EASTERN)
tamatar (62.9. URDU)
tamätarē (62.10. PANJABI, EASTERN)
tamaté (62.9. MANINKAKAN, WESTERN)
tamaté (62.9. WOLOF)
tamatere (62.9. PULAAR)
tamati (62.9. MANDING-BAMBARA)
tamati (62.9. SOTHO, NORTHERN)
tamati (62.9. SOTHO, SOUTHERN)
tamati (62.9. TSWANA)
tamatie (62.9. AFRIKAANS)
tamatis (62.9. CEBUANO)
tamatis (62.9. HERERO)
tamatis (62.9. KRIO)
tamatìsi (62.9. VENDA)
tamato (62.9. TOK PISIN)
tamattar (62.9. PANJABI, EASTERN)
tamatum (62.9. ARABIC)
tamatum el-sholak (62.11. ARABIC)
tamaylan (20.4. BICOLANO, CENTRAL)
tambadamaath (23.3. MARATHI)
tambadepavade (40.9. MARATHI)
tambalyun (38.1. HANUNOO)
tambasisi (62.9. LOKO)
tambasu (40.40. HANUNOO)
tambatas (62.9. GOLA)
tambatasi (62.9. VAI)
tambati (62.9. MARATHI)
tambda math (23.3. MARATHI)
tambdamath (23.3. MARATHI)
tambɛn (17.3. PULAAR)
tambo (17.1. TAGALOG)
tambobon (19.1. SAMBAL, TINÀ)
támbu' (17.1. HILIGAYNON)
tambulok (38.1. TAGBANWA)
tambuta (62.9. KONKANI)
tambuta (62.9. MARATHI)
tamelang (17.1. MALAY)
tameta (62.9. GUJARATI)
tamguma (39.4. BAATONUM)

taminɔghɔ (55.1. LOKO)
tam-is (11.5. CEBUANO)
tamis malaopi (17.5. BUHID)
tamis malayung (17.5. BUHID)
tamis mariruk (17.5. BUHID)
tamis sigbo (17.5. BUHID)
tamlueng (38.4. THAI)
tammani (40.33. AFRIKAANS)
tammatikaya (40.3. TELUGU)
tammatikaya (40.4. TELUGU)
tamme gnetumboom (41.1. DUTCH)
tamo (20.4. PAMPANGAN)
tamo (20.4. TAGALOG)
tamokansi (20.4. TAGALOG)
tamole (59.2. SAMOAN)
tamole (59.2. TONGAN)
tampala (23.3. ENGLISH)
tampopo (27.45. JAPANESE)
tampoyang (20.4. SUBANON)
tampuranga (9.1. CEBUANO)
tamuyu (38.20. HANUNOO)
tamyakhrua (40.30. THAI)
tan ge (11.1. CHINESE, YUE)
tànà nītúó (36.2. IZON)
tànàïn túó (36.2. IZON)
tanaman hitam (27.38. INDONESIAN)
tanap (30.41a. CHICKASAW)
tanap homma' (35.4. CHICKASAW)
tanaposi (30.22. LOMWE, MALAWI)
tanaposi (30.36. NYANJA)
tanaposi (30.41a. LOMWE, MALAWI)
tanchi' (17.7. CHICKASAW)
tan-chu (17.4. CHINESE, MANDARIN)
tàndà (38.20. SONGHAY)
tandalajo (23.3. GUJARATI)
tandalja (23.3. GUJARATI)
tandaljo (23.3. GUJARATI)
tandang-aso (62.23. TAGALOG)
tandauje (8.4. FULFULDE, ADAMAWA)
tandaure (8.4. FULFULDE, ADAMAWA)
tandula (23.3. SANSKRIT)
taṇḍuleyyo (23.3. PALI)
tanduliya (23.3. SANSKRIT)
taṇḍulīyo (23.3. PALI)
tandulja (23.3. GUJARATI)
tandulja (23.3. MARATHI)
tandur (38.3. PANJABI, EASTERN)
tang da huang (58.1. CHINESE, MANDARIN)
tang da huang (58.1. CHINESE, WÚ)
tang i mpiteurh (59.2. WOLOF)
tang jie (30.49. CHINESE, MANDARIN)
tang 'ôô (27.23. LAO)
t'ang ts'ai (30.29. CHINESE, MANDARIN)
tanga (39.4. KONO)
tangal (62.5. SAMBAL, BOTOLAN)
tangale (40.18. YAO)
tangawissi (20.6. BEMBÉ)
tangawissi (20.6. KOONGO)
tangawizi (20.2. SWAHILI)
tangawizi (20.6. DIGO)
tangawizi (20.6. SWAHILI)
tange (38.16. NDALI)
tangei (39.4. MENDE)
tangho (27.23. ENGLISH)
tanghoy (38.1. IBANAG)
tangila a elepanti (8.1. MARANAO)
tangila a lopa (24.7. MARANAO)
tangkay (38.1. IBANAG)
tangki (40.19. MALAY)
tangkil (41.1. SUNDA)
tangkil sake (41.1. SUNDA)
tangkong (36.2. BUTUANON)
tangkong (36.2. CEBUANO)
tangkong (36.2. ILOCANO)
tangkong (36.2. WÁRAY-WÁRAY)
tangkoy (38.1. ILOCANO)
tangkoy (38.1. ITAWIT)
tangkue (38.1. INDONESIAN)
tangkung (36.2. BINUKID)
tangkung (36.2. HILIGAYNON)
tangkung (36.2. TAUSUG)
tangnan (17.1. WÁRAY-WÁRAY)
tang-o (24.19. CHINESE, MIN NAN)
tango (27.23. FILIPINO)
tango (27.23. JAVANESE)
tango (27.23. TAGALOG)
tango (38.3. SWAHILI)
tango (38.13a. SWAHILI)
tangoh (24.9. SUNDA)
tangoũmbo (39.4. GOURMANTCHÉ)
tangoy (8.4. BICOLANO, CENTRAL)
tang-sún (17.1. CHINESE, MIN NAN)
tang-sún (17.2. CHINESE, MIN NAN)
tang-sún (17.4. CHINESE, MIN NAN)
tanguipeter (59.2. WOLOF)
tanhangui (45.4. SENOUFO, CEBAARA)
tani (21.2. TOK PISIN)
tània (8.9. HUNGARIAN)
tanier (8.9. FRENCH)
taniku takana (30.17. JAPANESE)

tanilang (17.1. MARANAO)
tanipu (30.41a. SWAHILI)
tanje (38.16. TONGA)
tãnk i mpétah (59.2. WOLOF)
tankari (62.12. SANSKRIT)
tankasi (62.12. SANSKRIT)
tankoru (43.1. DAGBANI)
tankun (36.2. HANUNOO)
tankwa (62.2. HAUSA)
tanmadòz (62.9. KWÉYÒL)
tanmanan (20.2. WÁRAY-WÁRAY)
tanndawre (8.4. FULFULDE)
tannia (8.9. DANISH)
tannia (8.9. ENGLISH)
tannia (8.9. FIJIAN)
Tannia (8.9. GERMAN)
tannie (8.9. FRENCH)
tannier spinach (8.8. ENGLISH)
tantan kpwen (10.1. BAATONUM)
tantandok (34.2. ILOCANO)
tantandok-a-dadakel (34.2. ILOCANO)
tantgesiesebloommetjie (27.8. AFRIKAANS)
ta'o (8.4. MARQUESAN)
tao tao (38.25. HMONG-DAW)
taogotan (62.23. AKLANON)
taogotan (62.23. CEBUANO)
taokiat (6.1. THAI)
tao-tue (38.1. HMONG-DAW)
tapasapriya (24.13. SANSKRIT)
tāpătăt' (62.24. MICMAC)
tāpesāoomakāwāāl (30.52. MICMAC)
taphla gogra (38.17. GEORGIAN)
tapilan (40.40. TAGALOG)
tapingan-daga (24.7. TAGALOG)
tapioca (39.4. FIJIAN)
tapioca (39.4. HINDI)
tapioca (39.4. MANU'A SAMOAN)
tapioca (39.4. MARATHI)
tapioca (39.4. PANJABI, EASTERN)
tapiocă (39.4. ROMANIAN)
tapioca (39.4. SAMOAN)
tapioca (39.4. SPANISH)
tapioca (39.4. VLAAMS)
tapioca plant (39.4. ENGLISH)
tapioka (39.4. BULGARIAN)
tapioka (39.4. HEBREW)
tápióka (39.4. HUNGARIAN)
Tapiokapflanze (39.4. GERMAN)
tapole (62.24. SOTHO, SOUTHERN)
tappuah adama (62.24. HEBREW)
tappud (62.24. HEBREW)
tapsoy (30.52. IBALOI)
taptapingar (23.6. ILOCANO)
taptu (40.9. BUHID)
taptu (40.25. BUHID)
tapuranga (9.1. CEBUANO)
tapyoka (39.4. TURKISH)
tar (38.7a. PANJABI, EASTERN)
tar (38.10. PANJABI, EASTERN)
tara (30.48. PANJABI, EASTERN)
tara (38.13a. PASHTO, NORTHERN)
tarabi (62.16. BURMESE)
tarada (38.22. ORIYA)
tára-dar (17.1. DAGAARE, SOUTHERN)
tarai (38.13a. PASHTO, NORTHERN)
tarakaankadee (38.29. MARATHI)
taralông (62.8. ESTONIAN)
tarambulo (62.16. TAGALOG)
taramira (30.24. HINDI)
taramira (30.48. HINDI)
taramira (30.48. URDU)
tarampay (62.23. WÁRAY-WÁRAY)
tarani kaayi (40.43. KANNADA)
taranoki (26.3. JAPANESE)
taratai (48.1a. INDONESIAN)
tarate (48.1a. JAVANESE)
tarate (48.1a. SUNDA)
taraware simi (40.4. NEPALI)
taráxaco (27.45. GALICIAN)
taraxaco (27.45. PORTUGUESE)
taraxacónn (27.45. SPANISH)
taraxakon to pharmakeftikon (27.45. GREEK)
tārbhuj (38.3. MAITHILI)
tarbooj (38.3. HINDI)
tarbooz (38.3. PANJABI, EASTERN)
tarboz (38.3. FARSI, EASTERN)
tarbuc (38.3. HINDUSTANI)
tarbuge (38.3. FIJIAN)
tarbuj (38.3. PANJABI, EASTERN)
tarbuja (38.7b. NEPALI)
tarbuz (38.3. BENGALI)
tarbuz (38.3. GUJARATI)
tarbuz (38.3. HINDI)
tarbuz (38.3. MARATHI)
tarbuz (38.3. PANJABI, EASTERN)
tarbuz (38.3. URDU)
tarbuz (38.3. UZBEK)
tarbúz (38.7b. HINDUSTANI)
tarbuza (38.3. HINDUSTANI)
tarbūza (38.3. PASHTO, NORTHERN)

tarch koohie (30.44. FARSI, WESTERN)
tarché farangy (7.8. FARSI, WESTERN)
tarçin eğreltisi (3.1. TURKISH)
tarçini kral eğreltisi (3.1. TURKISH)
tare bodi (40.42. NEPALI)
tare tezak (30.52. FARSI, WESTERN)
tarfo (36.2. HAUSA)
tarhaherne (40.28. FINNISH)
tarhamaltsa (35.1. FINNISH)
tarhapapu (40.26. FINNISH)
tarhasaurikki (27.22. FINNISH)
tarhasuolaheinä (58.2. FINNISH)
tarhun (27.6. TURKISH)
tarifuan (36.3. BUHID)
tarimal luuvan (30.54. MONGOLIAN)
tarimal luuvan (30.56. MONGOLIAN)
tarimal songino (7.10. MONGOLIAN)
tarimal tömördöö (40.10. MONGOLIAN)
tarimal tsargas (40.17. MONGOLIAN)
tarimal vandui (40.27. MONGOLIAN)
tarimal vandui (40.28. MONGOLIAN)
tarḵẖa kāoṛa (38.23. PASHTO, NORTHERN)
tärkhämäk (38.13a. UYGHUR)
tarkhun (27.6. TAJIKI)
tarkhuna (27.6. GEORGIAN)
tárkony (27.6. HUNGARIAN)
tarla atkuyruğu (1.1. TURKISH)
tarlórépa (30.41a. HUNGARIAN)
tarmuj (38.3. BENGALI)
tarmuj (38.3. HINDI)
tarmuj (38.3. MARATHI)
tarmus (40.14. URDU)
tarna:psu (30.30. BADAGA)
táró (8.1. HUNGARIAN)
taro (8.4. CZECH)
taro (8.4. DANISH)
taro (8.4. DUTCH)
taro (8.4. ENGLISH)
taro (8.4. ESTONIAN)
taro (8.4. FRENCH)
Taro (8.4. GERMAN)
táró (8.4. HUNGARIAN)
taro (8.4. LATVIAN)
taro (8.4. MAORI)
taro (8.4. NORWEGIAN)
taro (8.4. PORTUGUESE)
taro (8.4. RUSSIAN)
taro (8.4. SERBO-CROATIAN)
taro (8.4. SLOVAK)
taro (8.4. SWEDISH)
taro (8.4. TAGALOG)
taro (8.4. VLAAMS)
taro (8.6. AFRIKAANS)
taro géant (8.1. FRENCH)
taroi (38.31. FIJIAN)
taron (62.19. WÁRAY-WÁRAY)
tarong (62.19. ILOCANO)
taróng (62.19. WÁRAY-WÁRAY)
tarongtarong (62.16. WÁRAY-WÁRAY)
tarong-tarong (62.23. WÁRAY-WÁRAY)
tarra (38.10. TAJIKI)
tarrelzaad naaldekruid (27.8. DUTCH)
tartaar (30.46. VLAAMS)
tartar bread-plant (30.46. ENGLISH)
tartarian sea kale (30.46. ENGLISH)
tar-tszak (30.51. TAJIKI)
tartufo di prato (40.11. ITALIAN)
taru kala (40.18. BENGALI)
tarubhuj (38.3. ORIYA)
tarubhuja (38.3. ORIYA)
tarukala (40.18. BENGALI)
tarul (11.1. NEPALI)
tarul (11.5. NEPALI)
tarum (40.6. INDONESIAN)
tarung (62.19. ATTA, PAMPLONA)
tarung (62.19. TAGALOG)
tarung (62.19. TAGBANWA)
taruyin-pabo (23.6. IVATAN)
tarvuz (38.3. UZBEK)
tarwi (40.14. QUECHUA, CUSCO)
tas sarmisaği (7.12. TURKISH)
tasbih (9.1. INDONESIAN)
tasbulja repchataja (7.5. BELARUSAN)
ta-sɛŋa (20.6. THEMNE)
ta-seŋa (20.6. THEMNE)
ta-seŋa-*ta*-thənk (20.6. THEMNE)
tasije tokotje (62.5. FULFULDE, ADAMAWA)
taškuotoji mėta (44.2. LITHUANIAN)
tassel flower (23.3. ENGLISH)
tassel grapehyacinth (13.5. ENGLISH)
tasshi (62.1. HAUSA)
tasznik pospolity (30.43. POLISH)
tata (36.3. WAMEY)
tàtaci (62.2. KORO)
tatai (27.8. SUBANON)
tatami (17.1. SUSU)
tatami-na (17.1. YALUNKA)
tatar (30.46. TURKISH)
tatarisk strandkål (30.46. DANISH)
tatase (62.1. YORUBA)

tɔmati (62.9. BAOULÉ)
tɔmbega (9.1. LOKO)
tɔndɔ wuli (59.2. MENDE)
tɔnŋgɔ (62.1. GA)
tɔslwɛn (23.2. GBE, WACI)
tɔw (40.25. MENDE)
tɔwɔ (40.25. GOLA)
tɔwɔ (40.25. MENDE)
tɔwe (40.25. BULLOM SO)
tɔwɛlikan (36.2. GUN)
tɛ (11.7. ÉWÉ)
te (17.7. ÉWÉ)
te anian (7.2. KIRIBATI)
te anian (7.3. KIRIBATI)
te anian (7.5. KIRIBATI)
te babai (8.6. JAPANESE)
te babai (8.6. KIRIBATI)
te bambin (38.14. KIRIBATI)
te bambin (38.16. KIRIBATI)
te bambin (38.17. KIRIBATI)
te bambin (38.18. KIRIBATI)
te banana (16.1a. KIRIBATI)
te bangke (38.14. KIRIBATI)
te bangke (38.16. KIRIBATI)
te bangke (38.17. KIRIBATI)
te bangke (38.18. KIRIBATI)
te baukin (38.14. KIRIBATI)
te baukin (38.16. KIRIBATI)
te baukin (38.17. KIRIBATI)
te baukin (38.18. KIRIBATI)
te beneka (62.2. KIRIBATI)
te beneka (62.5. KIRIBATI)
te boteto (62.24. KIRIBATI)
te iam (11.1. KIRIBATI)
te iam (11.2. KIRIBATI)
te iam (11.3. KIRIBATI)
te iam (11.4. KIRIBATI)
te iam (11.5. KIRIBATI)
te iam (11.6. KIRIBATI)
te iam (11.7. KIRIBATI)
te iam (11.8. KIRIBATI)
te kaibaba (17.1. KIRIBATI)
te kaibaba (17.2. KIRIBATI)
te kaibaba (17.4. KIRIBATI)
te kaipapa (17.1. KIRIBATI)
te kaipapa (17.2. KIRIBATI)
te kaipapa (17.4. KIRIBATI)
te kumara (36.3. KIRIBATI)
te papai (8.6. KIRIBATI)
te poteto (62.24. KIRIBATI)
te rarai (16.1a. KIRIBATI)
te tabicka (39.4. KIRIBATI)
te taororo (8.4. KIRIBATI)
te tomato (62.9. KIRIBATI)
te touru (16.1a. KIRIBATI)
tea (38.18. LOKO)
tebo (17.5. MAGUINDANAO)
tebo (17.5. MARANAO)
tebtebbek (10.2. BONTOC, CENTRAL)
tebu (17.5. BICOLANO, CENTRAL)
tebu (17.5. INDONESIAN)
tebu (17.5. MALAY)
tebu (17.5. TAGALOG)
tebu bertelur (17.5. MALAY)
tebu endog (17.5. JAVANESE)
tebu ikan (17.5. INDONESIAN)
tebu telur (17.5. INDONESIAN)
tebu trubuk (17.5. JAVANESE)
teddu tekkunde (59.2. FULFULDE, ADAMAWA)
tɛ-dze (11.7. ÉWÉ)
teeagomgura (45.4. TELUGU)
teegabatchali (35.3. TELUGU)
teegabatchali (35.9. TELUGU)
teehibiskus (45.4. FINNISH)
tèèng (38.13a. LAO)
tèèng (38.13b. LAO)
tèèng môô (38.3. LAO)
teftlo (35.4. GREEK)
te-gba (11.7. ÉWÉ)
tege (63.1. EWONDO)
tegudtung (62.19. BINUKID)
teh hka¯ shi_ (38.3. LAHU)
tehénborsó (40.43. HUNGARIAN)
tehũ (40.1. KISSI, NORTHERN)
tēi (40.1. KISSI, NORTHERN)
teibwo (44.5. CAROLINIAN)
tek-á (17.1. CHINESE, MIN NAN)
tek-á (17.2. CHINESE, MIN NAN)
tek-á (17.4. CHINESE, MIN NAN)
tekesakali (27.46. TURKISH)
teketera (27.22. SHONA)
teki (10.2. INDONESIAN)
tekkarè (38.13a. TULU)
téklo (11.1. ÉWÉ)
tekokak (62.23. INDONESIAN)
tek-sún (17.1. CHINESE, MIN NAN)
tek-sún (17.2. CHINESE, MIN NAN)
tek-sún (17.4. CHINESE, MIN NAN)
tekvica (38.18. SLOVAK)

t'éou (40.26. CHINESE, HAKKA)
tĕpanā (24.18. MICMAC)
ţepari (40.23. ROMANIAN)
tepari (62.12. BENGALI)
tepariya (62.12. HINDI)
tepary (40.23. HUNGARIAN)
tepary bean (40.23. ENGLISH)
Tepary-Bohne (40.23. GERMAN)
tepary-bønne (40.23. DANISH)
teparybønne (40.23. NORWEGIAN)
teparyboon (40.23. AFRIKAANS)
təpə (40.26. AZERBAIJANI, NORTHERN)
t̲epër (30.41a. PASHTO, NORTHERN)
tépösaláta (27.32. HUNGARIAN)
teppô-yurine (13.4. JAPANESE)
teratai (48.1a. INDONESIAN)
teratai (48.1a. MALAY)
teratai (48.1a. SUNDA)
teratai (50.2. INDONESIAN)
teratai (50.2. JAVANESE)
terayag (38.20. BINUKID)
tered (35.9. HEBREW)
tered hagina (35.9. HEBREW)
tered ha-ginna (35.9. HEBREW)
tered new-zelandy (22.2. HEBREW)
terere (23.3. GIKUYU)
terfai (36.2. HAUSA)
teroi (38.21. HINDUSTANI)
terong (62.19. INDONESIAN)
terong (62.19. JAVANESE)
terong (62.19. MALAY)
terong (62.19. SUNDA)
terong asam (62.16. INDONESIAN)
terong asam (62.16. MALAY)
terong belanda (62.7. INDONESIAN)
terong belut (62.23. JAVANESE)
terong betanda (62.7. INDONESIAN)
terong cina (62.18. INDONESIAN)
terong dayak (62.16. MALAY)
terong engkol (62.18. INDONESIAN)
terong engkol (62.18. SUNDA)
terong gayung (62.18. SUNDA)
terong glatik (62.18. INDONESIAN)
terong hitam (62.15. INDONESIAN)
terong iban (62.16. MALAY)
terong kalapa (62.18. SUNDA)
terong kelapa (62.18. INDONESIAN)
terong kopek (62.18. SUNDA)
terong menen (62.7. INDONESIAN)
terong menen (62.7. SUNDA)
terong meranti (62.15. MALAY)
terong perat (62.15. MALAY)
terong perat (62.16. INDONESIAN)
terong perus (62.7. INDONESIAN)
terong pipit (62.23. INDONESIAN)
terong pipit (62.23. JAVANESE)
terong pipit (62.23. MALAY)
terong pipit (62.23. MINANGKABAU)
terong pipit (62.25. MALAY)
terong pipit putch (62.23. MALAY)
terong rapoh (62.18. MALAY)
terong rembang (62.23. MALAY)
terong sabrang (62.9. JAVANESE)
terong santan (62.18. MALAY)
terong susu (62.23. INDONESIAN)
terong telunjuk (62.21. MALAY)
terongan berries (62.23. ENGLISH)
teroung hitam (62.15. INDONESIAN)
terpary boon (40.23. VLAAMS)
tertufa (27.26. ARABIC)
terubuk (17.5. INDONESIAN)
tes-ɛ (40.25. BULLOM SO)
tesi-lɛ (40.25. BULLOM SO)
tespong (24.17. INDONESIAN)
tespong (24.17. SUNDA)
tespong (24.23. INDONESIAN)
tessurom (40.39. MANDINKA)
testi kabaği (38.20. TURKISH)
tesu-tu (17.5. NEWAR)
tetambak merah (27.18. INDONESIAN)
tɛtɛ (23.3. BAATONUM)
tẹ̀tẹ̀ (23.3. YORUBA)
tɛtɛ (34.2. AKAN-AKUAPEM)
tɛtɛ (34.2. AKAN-ASANTE)
tɛtɛ (34.2. DANGME)
tɛtɛ (34.2. DANGME-KROBO)
tɛtɛ (34.2. GA)
tɛtɛ (34.2. TWI)
tẹ̀tẹ̀ atẹle danji (23.3. YORUBA)
tẹ̀tẹ̀ àtètèdáyé (23.3. YORUBA)
tête de lettue (27.31. CAJUN FRENCH)
tẹ̀tẹ̀ ẹ̀gún (23.3. YORUBA)
tẹ̀tẹ̀ eléègún (23.3. YORUBA)
tẹ̀tẹ̀ kékeré (23.3. YORUBA)
tɛtɛ kereke (23.3. EDE NAGO)
tɛtɛ kereke (23.3. YORUBA)
tɛtɛ moto (23.3. FON)
tɛtɛ moto (23.3. GUN)
tɛtɛ n'la (23.3. EDE NAGO)
tɛtɛ n'la (23.3. YORUBA)

tɛtɛ wuli wuli (23.3. GEN)
tɛtɛma (23.3. GBE, WACI)
tɛtɛtɛrɛ (34.2. TWI)
tɛtɛtɛtɛ (34.2. AKAN-ASANTE)
tetimun (38.13a. INDONESIAN)
teto (21.1. URDU)
tetragónia (22.2. PORTUGUESE)
tetragonia della Nuova Zelanda (22.2. ITALIAN)
tétragonne (22.2. FRENCH)
tétragonne cornue (22.2. FRENCH)
tetragonolobus purpurnyj (40.13. RUSSIAN)
tetrocgonolovos oorphyros (40.13. GREEK)
tettu (21.1. ENGLISH)
tettu (21.1. YORUBA)
Teufelsbusch (26.4. GERMAN)
teve (8.3. MANU'A SAMOAN)
teve (8.3. MARQUESAN)
teve (8.3. NIUE)
teve (8.3. RAROTONGAN)
teve (8.3. SAMOAN)
teve (8.3. TONGAN)
tɛvɛ (21.1. MENDE)
tevi (11.1. FON)
tevi gami (8.9. FON)
tevu (17.5. IBANAG)
tewm tal-icheli (7.10. MALTESE)
tewm ta'ras hoxna (7.12. MALTESE)
tɛyɛ (21.1. MENDE)
tezak (30.51. PANJABI, EASTERN)
tezak (30.58. PANJABI, EASTERN)
tʃakatʃaka (40.18. DANGME)
tfief (27.42. MALTESE)
thaamara (48.1a. KANNADA)
thaamara (48.1a. MALAYALAM)
thaamara (48.1a. TELUGU)
thaamara (48.1a. TULU)
thaamara puvvu (48.1a. TELUGU)
thaamarai (48.1a. TAMIL)
thaambeta (62.9. MARATHI)
thaavare (48.1a. KANNADA)
thaavare beeja (48.1a. KANNADA)
thaavare gedde (48.1a. KANNADA)
thabai (66.3. GIKUYU)
thabinera (11.6. KANNADA)
t^{h}ada (8.4. PUMÉ)
thageti (34.2. GIKUYU)
thagiti (34.2. GIKUYU)
thai kumbalom (38.7a. MALAYALAM)
thai kumbalom (38.7b. MALAYALAM)
thaihbeng (38.13a. BODO)
thakhwa (38.13a. BURMESE)
thakhwa (38.13b. BURMESE)
thakkaali (62.9. KANNADA)
thakkaali (62.9. TAMIL)
thakkaali pandu (62.9. TELUGU)
thakkali (62.9. MALAYALAM)
thakkali (62.9. SINHALA)
tha-kwa (38.13a. BURMESE)
tha-kwa-hmwe (38.7a. BURMESE)
thali chi khat (40.12. MIZO)
thalooin (40.1. GAELIC, MANX)
thamarai (48.1a. TAMIL)
thamarasa (48.1a. SANSKRIT)
thamate balli (40.3. KANNADA)
thamattan kodi (40.3. TAMIL)
thamba (40.3. TELUGU)
thambada maath (23.3. MARATHI)
thambattai (40.3. TAMIL)
thamboolaraaga (40.12. SANSKRIT)
thamma (40.3. TELUGU)
thamma kaaya (40.4. TELUGU)
thampala (23.3. SINHALA)
thandaar (48.1a. MALAYALAM)
thandu keerai (23.3. TAMIL)
thandukkeerai (23.3. TAMIL)
thandula (23.3. SANSKRIT)
thandulajaa (23.3. MARATHI)
thandulanama (23.3. SANSKRIT)
thanduleraka (23.3. SANSKRIT)
thandulibeeja (23.3. SANSKRIT)
thanduliya (23.3. MARATHI)
thanduliya (23.3. SANSKRIT)
thankeerai (23.3. TAMIL)
thanmori (62.12. MARATHI)
thannir mathan (38.3. MALAYALAM)
thantawan-hua (27.26. THAI)
thanthuvigraha (16.1a. SANSKRIT)
thantikada (36.2. TELUGU)
thanuwen (20.4. BURMESE)
thaoyaimom (19.1. THAI)
tharabooja (38.3. SANSKRIT)
tharbooj (38.3. HINDI)
tharbooj (38.3. MARATHI)
tharbooz (38.3. HINDI)
tharbooza (38.3. HINDI)
tharbuza (38.3. TAMIL)
tharoi (38.21. HINDI)
thatachhadamu (16.1a. TELUGU)
thatapayaru (40.43. TAMIL)

thatapayuru (40.43. TAMIL)
thathpathri (16.1a. SANSKRIT)
thattaam payiru (40.9. TAMIL)
thattapayir (40.43. TAMIL)
thause (38.13a. KONKANI)
thavaikachchu (11.2. TAMIL)
thavaksheera (15.2. SANSKRIT)
thavasai murungai (39.5. TAMIL)
thavase (38.13a. MARATHI)
thaviani salata (27.31. GEORGIAN)
theekshagandha (30.24. SANSKRIT)
theekshna (40.18. SANSKRIT)
theekshna (62.2. SANSKRIT)
theekshna (62.5. SANSKRIT)
theekshna moola (20.1. SANSKRIT)
theekshnagandha (30.17. SANSKRIT)
theeta kogial (40.18. TELUGU)
theevrakantha (8.3. SANSKRIT)
theevrashakti (62.5. SANSKRIT)
thei kawr sak (64.3. MIZO)
thekkare (38.13a. TULU)
thelele (36.2. LOMWE, MALAWI)
thelele (45.1. LOMWE, MALAWI)
thelele la amwenye (45.1. NYANJA)
thelele lobzala (45.1. NYANJA)
thella dumpa (36.3. TELUGU)
thella thaamara (48.1a. TELUGU)
thellachikkudu (40.9. TELUGU)
thellachikurkai (40.9. TELUGU)
thelladooradagondi (40.18. TELUGU)
thellakaakara (38.23. TELUGU)
thellane padmanu (48.1a. TELUGU)
thellatotakura (23.3. TELUGU)
thendle (38.4. KONKANI)
thendli (38.4. MARATHI)
theneyarathi (16.1a. TELUGU)
thenipe (30.41a. TSWANA)
thes (40.25. BULLOM SO)
thevu (8.4. TULU)
thian-klaep (24.13. THAI, NORTHERN)
thibinici (35.3. GIKUYU)
thibinici (35.9. GIKUYU)
thick head (27.14. TOK PISIN)
thierëzë (40.12. ALBANIAN, TOSK)
thikthabimbi (38.4. SANSKRIT)
thikthaka (38.20. SANSKRIT)
thikthakhya (38.4. SANSKRIT)
thikthathumbi (38.20. SANSKRIT)
thikthathundi (38.4. SANSKRIT)
thikthika (62.15. SANSKRIT)
thimare (24.7. TULU)
thindisa (45.1. SANSKRIT)
thingalavare (40.26. KANNADA)
thippe tomato hannu (62.10. KANNADA)
thipper (30.41a. PANJABI, EASTERN)
thirkola (38.4. HINDI)
thirnupatchi (44.5. TAMIL)
thirumaalundi (48.1a. TAMIL)
thirunithru (44.5. MALAYALAM)
thirunithru (44.5. TAMIL)
thirunuthpatchi (44.5. TAMIL)
thitalau (38.20. HINDI)
thiyya gummadi (38.16. TELUGU)
thlai hnah nei chi (35.6. MIZO)
thlaizung ei mi (15.2. MIZO)
thmate balli (40.3. KANNADA)
thmate balli (40.4. KANNADA)
thô͘-tāu (40.1. CHINESE, MIN NAN)
th[of] l[of] (30.30. VIETNAMESE)
thogari (40.2. KANNADA)
thogari (40.2. TULU)
thogari bele (40.2. KANNADA)
thogari uddu (40.2. KANNADA)
thogo hongkure-na (40.43. YALUNKA)
thol-kuri (24.7. BENGALI)
tholoo-pany (38.23. TAMIL)
tholthakkali (62.12. TAMIL)
thom (7.10. PANJABI, EASTERN)
thô-má-toh (62.9. CHINESE, MIN NAN)
thondai (38.4. TAMIL)
thondali (38.4. MARATHI)
thonde balli (38.4. KANNADA)
thonde kaayi (38.4. KANNADA)
thonde kayi (38.4. KANNADA)
thondi (38.4. MALAYALAM)
thoor (40.2. HINDI)
th[oor] nh[aa]n s[aa]m (59.3. VIETNAMESE)
thoora (40.2. MARATHI)
thoori (40.2. MARATHI)
thootikoora (36.2. TELUGU)
thoovar (40.2. MARATHI)
thora paeru (40.2. MALAYALAM)
thora parippu (40.2. SINHALA)
thorai (38.21. HINDI)
thori (40.2. KONKANI)
thornless prickly pear (32.1. ENGLISH)
thorny amaranth (23.3. ENGLISH)
thoroko (40.41. GIKUYU)
thoroko (40.43. GIKUYU)
thosani (27.23. RONGA)

thotadoolagondi (40.18. TELUGU)
thotakura (23.3. TELUGU)
thousand-headed kale (30.33. ENGLISH)
thovarai (40.2. TAMIL)
thovaray (40.2. TAMIL)
thovari (40.2. KANNADA)
thrapukarkati (38.13a. SANSKRIT)
thrapusa (38.7b. SANSKRIT)
thrapusha (38.13a. SANSKRIT)
threebulka (7.11. BELARUSAN)
thribeeja (40.1. SANSKRIT)
thridax e imeros me kotsani (27.30. GREEK)
thrika (64.3. SANSKRIT)
thrikaalamalli (45.5. TELUGU)
thrikonaphala (64.3. SANSKRIT)
thrikota (64.3. SANSKRIT)
thrinethra (11.2. SANSKRIT)
thriputa (40.10. SANSKRIT)
throny bamboo (17.1. ENGLISH)
thua daeng (40.12. THAI)
thua daeng (40.40. THAI)
thua din (40.1. THAI)
thua hua chang (40.5. THAI)
thua khaao (40.41. THAI)
thua khaek (40.26. THAI)
thua khieo (40.38. THAI)
thua lisong (40.1. THAI)
thua lueang (40.8. THAI)
thua maetaai (40.2. THAI)
thua pae (40.40. THAI)
thua paep (40.9. THAI)
thua phra lueang (40.8. THAI)
thua phum (40.26. THAI)
thua plu (40.29. THAI)
thua rachamat (40.25. THAI)
thua rae (40.2. THAI)
thua rae (40.8. THAI)
thua rai (40.41. THAI)
thua thong (40.38. THAI)
thua yang (40.34a. THAI)
thua-duang (44.11. THAI)
thua-phra (40.4. THAI)
thua-phu (40.29. THAI)
th[uf] l[uf] d[uwj]c (62.15. VIETNAMESE)
thúhar (32.1. HINDUSTANI)
thukhmerihan (44.5. HINDI)
thulaathipathi (62.12. HINDI)
thulasi (62.21. MALAYALAM)
thulasi aarjaki (44.4. HINDI)
thulasidvesha (44.5. SANSKRIT)
thulo bihin (62.21. NEPALI)
thumba (38.20. SANSKRIT)
thumbaka (38.20. SANSKRIT)
thumbaru (24.9. SANSKRIT)
ṱhumbe (40.38. VENDA)
thumbe kanji (40.3. KANNADA)
thumbee (38.20. SANSKRIT)
thumri (38.20. HINDI)
th'u:n (8.10. LAO)
thunagenasu (11.1. KANNADA)
thundi (38.4. SANSKRIT)
thundika (38.4. SANSKRIT)
thundikeri (38.4. SANSKRIT)
thundilaphala (38.13a. SANSKRIT)
thundiparyyayaga (38.4. SANSKRIT)
thunga masthe (10.1. KANNADA)
thungaksheeri (15.2. SANSKRIT)
thungi (44.5. SANSKRIT)
thupari balli (38.22. KANNADA)
thupi jolokia (62.5. ASSAMESE)
thupi mah (40.6. ASSAMESE)
thupi urahi (40.6. ASSAMESE)
thupiurahi (40.6. ASSAMESE)
thuppa heera kavi (38.22. KANNADA)
thuppa heere kayi (38.22. KANNADA)
thuppeere kaayi (38.22. KANNADA)
thupri kaayi (38.22. KANNADA)
thuraivalai (16.1a. TAMIL)
thuraka hesaru (40.35. KANNADA)
thuri (38.21. HINDI)
ṭhu̱sam (58.3. TIBETAN)
thuvar (40.2. HINDI)
thuvara (40.2. KANNADA)
thuvara (40.2. MALAYALAM)
thuvarai (40.2. TAMIL)
thuvaran (40.2. MALAYALAM)
thuvari (40.2. SANSKRIT)
thuvarika (40.2. SANSKRIT)
th[uw][ow]ng l[uj]c (56.1. VIETNAMESE)
th[uw][ow]ng l[uj]c (56.2. VIETNAMESE)
thvashti (24.7. SANSKRIT)
thwàx din (40.1. LAO)
thwàx do (40.43. LAO)
thwàx fàlangx (40.26. LAO)
thwàx h'ê (40.2. LAO)
thwàx ho (40.1. LAO)
thwàx khiêw (40.38. LAO)
thwàx khôn (40.8. LAO)
thwâx la:ng tê:k (40.40. LAO)
thwàx ngo:k (40.38. LAO)

thwàx phi (40.40. LAO)
thwàx ph'uu (40.29. LAO)
thwàx sadê:k (40.38. LAO)
thwàx sadê:t pa:x (40.40. LAO)
thwàx siênx (40.41. LAO)
thwàx tê (40.8. LAO)
ti¯ law^ pfuh¯ (40.21. LAHU)
t'í- t'éou (40.1. CHINESE, HAKKA)
tia fana (40.1. SERER-SINE)
tia kolèñ (40.15. MANDINKA)
tia kolõ (40.15. MANDINKA)
tiaf a nuk (59.2. SERER-SINE)
tiagué dabbii (40.39. PULAAR)
t'iâm kouā (38.7b. CHINESE, HAKKA)
t'iâm ts'ói-t'êou (35.4. CHINESE, HAKKA)
tia-mimak (27.32. MORISYEN)
tian cai (35.3. CHINESE, MANDARIN)
tian cai (35.3. CHINESE, WÚ)
tian cai wan dou (40.27. CHINESE, MANDARIN)
tian cai wan dou (40.27. CHINESE, WÚ)
tian cai wan dou (40.27. CHINESE, YUE)
tian hu sui (24.10. CHINESE, WÚ)
tian hui xiang (24.12. CHINESE, MANDARIN)
tian jiao (62.1. CHINESE, MANDARIN)
tian jiao (62.1. CHINESE, YUE)
tian jui (27.25. CHINESE, YUE)
tian shu (11.1. CHINESE, YUE)
tian shu (11.5. CHINESE, MANDARIN)
tian shu (11.5. CHINESE, WÚ)
tian shu (11.7. CHINESE, MIN BEI)
tian shu (11.7. CHINESE, YUE)
tiare aru ra (15.2. RAROTONGAN)
tibbatu (62.25. SINHALA)
tibbu (17.5. SAMA, CENTRAL)
tibiayon (38.1. CEBUANO)
tibiyayon (38.1. CEBUANO)
tibo (17.5. MARANAO)
tibu (17.5. MANOBO)
tibu (17.5. SUBANON)
tichunia (11.4. KOONGO)
tick pea (40.34b. ENGLISH)
tiɔ̃ŋ (40.1. KISSI, NORTHERN)
ticsau (65.1. QUECHUA, CUSCO)
ticuna (11.1. QUECHUA, CUSCO)
tidek (62.5. BINUKID)
tidi (17.5. MANOBO)
tidik (17.5. MANOBO)
tido (17.5. MANOBO)
tidog (17.5. ATTA, PAMPLONA)
ti[eer]uh[oof]i h[uw][ow]ng (24.13. VIETNAMESE)
tiɛlɛbdo (34.2. MÒORÉ)
tien kua (38.7b. CHINESE, MANDARIN)
t'ien ts'ai (35.3. CHINESE, MANDARIN)
tifalia (36.3. BUHID)
tife (44.10. PULAAR)
tiga (40.1. BAMANANKAN)
tiga (40.1. JULA)
tiga (40.1. MANDING-MANDINKA)
tiga (40.1. MANDINKA)
tiga (40.1. MANDJAK)
tiga (40.1. MANINKA)
tiga (40.1. PULAAR)
tiga ba (40.39. MANDING-MANDINKA)
tiga bélé (40.39. MANINKA)
tiga bélé (40.39. MANINKAKAN, WESTERN)
tiga kolô (40.39. MANDINKA)
tiga kolõ (40.39. MANINKA)
tiga ni kuélé (40.39. MANDING-MANDINKA)
tiga ni kuru (40.39. MANDING-MANDINKA)
tiga ni nguélé (40.39. BAMANANKAN)
tiga ni nkourou (40.39. BAMANANKAN)
tiga ouélé (40.39. BAMANANKAN)
tigafun (5.1. BUHID)
tigage (40.1. FULFULDE, ADAMAWA)
tigagué (40.1. PULAAR)
tigan binio (40.39. KISI, SOUTHERN)
tigare (40.1. PULAAR)
tige (40.1. KURANKO)
tiger bamboo (17.1. ENGLISH)
tigerblomst (12.1. DANISH)
tigerblomst (12.1. NORWEGIAN)
tīgerlilijas (13.4. LATVIAN)
Tigerlilie (12.1. GERMAN)
tiger-nuts (10.1. ENGLISH)
tigi (17.7. BONTOC, CENTRAL)
tigi-nga magmanto (8.3. ILOCANO)
tigrídia (12.1. PORTUGUESE)
tigrisblom (12.1. ICELANDIC)
tigrisvirág (12.1. HUNGARIAN)
tigsi (36.3. CEBUANO)
tigudtung (62.19. MANOBO)
tigué (40.1. BAMANANKAN)
tigurskruo (39.3. ICELANDIC)
tigvă (38.20. ROMANIAN)
tihon̠ (40.3. SANTALI)
tihu (38.13a. ASSAMESE)

tinkshottham (20.6. MALAYALAM)
tinta-de-macaco (28.1. CRIOULO, CAPE VERDE)
tinumbaga (16.1a. WÁRAY-WÁRAY)
tinyēre (7.3. FULFULDE, ADAMAWA)
tinyẽre (7.3. FULFULDE, ADAMAWA)
tio (40.1. MANDING-MANDINKA)
tiô (40.1. MANDING-MANDINKA)
tío (40.1. MANDINKA)
tio ba (40.1. MANDING-MANDINKA)
tio kolon (40.39. MANDING-MANDINKA)
tioeuhs (40.4. KHMER, CENTRAL)
tion (40.1. KISI, SOUTHERN)
tion (40.1. KISSI, NORTHERN)
tipari (62.12. HINDI)
tipari (62.12. ORIYA)
tipariya (62.12. BENGALI)
tipima (62.2. MORISYEN)
tipoi (40.28. MORISYEN)
tipoi manz-tou (40.27. MORISYEN)
tippa tiga (11.5. TELUGU)
ti'psina (30.41a. DAKOTA)
ti'psinażiżi (30.22. DAKOTA)
tipula (7.3. BASQUE)
tipulatxi (7.5. BASQUE)
tipŋsiŋla śaśa (35.4. LAKOTA)
tira (30.24. HINDI)
tírā (40.1. BAMANANKAN)
tíra (40.1. MANDINKA)
tira (40.43. YOM)
tiras (17.7. HEBREW)
tiridinkitinyi (59.2. SUSU)
tírìì (20.2. HAUSA)
tirnirupachai (44.5. TAMIL)
tirnutpatchi (44.5. TAMIL)
tirṛạ da (40.30. SANTALI)
tirrica amarela (10.1. RONGA)
tirtil (53.1. TURKISH)
tīruma kosa (1.1. LATVIAN)
tirunitru (44.5. MALAYALAM)
tirunitru (44.5. TAMIL)
t[is]a t[oo] (44.7. VIETNAMESE)
tistel (27.11. NORWEGIAN)
tita alu (11.4. ASSAMESE)
tita bhekuri (62.17. ASSAMESE)
tita kerela (38.23. ASSAMESE)
titbaigun (62.23. BENGALI)
tite karela (38.23. NEPALI)
titikukɔḍɛ (62.2. GBE-VHE)
titina (62.23. AKLANON)
titina (62.23. HILIGAYNON)
tito (40.10. SPANISH)
titthiram (20.1. TAMIL)
ti-tu (17.5. NEWAR)
tiukula (44.5. BAMANANKAN)
tiung (62.19. MINANGKABAU)
tiuri (40.10. HINDI)
tivu (17.5. MANOBO)
tivvitiga (11.5. TELUGU)
tiwu (17.5. SUNDA)
tiwu endog (17.5. SUNDA)
tiwu tirubus (17.5. SUNDA)
tiya (40.1. MANDING-MANDINKA)
tiya (40.1. MANDINKA)
tiya (40.1. SERER-SINE)
tiyoh (38.13a. ASSAMESE)
tiŋpaiŋla śaśa (30.54. LAKOTA)
tiŋpaiŋla śaśa (30.56. LAKOTA)
tiŋpsila (30.41a. LAKOTA)
tiŋpsiŋla skaska (30.41a. LAKOTA)
tiŋpsiŋlazizi (24.11. LAKOTA)
tjiekoriebos (27.10a. AFRIKAANS)
tjikare (38.20. TEM)
tjongô (10.1. BÉTÉ)
tjongô (10.1. JULA)
tł'ohchin (7.3. NAVAJO)
tō (17.5. NIUE)
to (17.5. TONGAN)
to (38.20. ZUNI)
to (40.1. MANDINKA)
to hua tou (40.24. CHINESE, MANDARIN)
to indiko arrapoúte (20.4. GREEK)
to ofioides (38.29. GREEK)
toa (40.43. DAGBANI)
toa yamman (38.20. BAOULÉ)
tōa-chhang (7.5. CHINESE, MIN NAN)
toal (8.3. KHMER, CENTRAL)
toan th[ar]o (58.2. VIETNAMESE)
toårig natlys (51.1. DANISH)
Toassa-Jute (63.1. GERMAN)
tōa-tāu (40.8. CHINESE, MIN NAN)
tobalõ soso (40.3. MANDINKA)
tobalô-sosso (40.3. MANDING-MANDINKA)
tobi (40.30. SUNDA)
tobo (17.5. SAMBAL, BOTOLAN)
tobok (17.5. SAMBAL, BOTOLAN)
tobu (17.5. SUBANON)
todagunny (40.43. KANNADA)
todza (27.8. DIGO)
t'oexka (40.28. CHECHEN)

tomate (62.9. GALICIAN)
Tomate (62.9. GERMAN)
tomate (62.9. GREEK)
tomate (62.9. PORTUGUESE)
tomaté (62.9. PULAR)
tomate (62.9. SPANISH)
tomate (62.9. STELLINGWERFS)
tomate (62.9. WOLOF)
tomaté (62.9. WOLOF)
tomate amère (62.14a. FRENCH)
tomate arbol (62.7. SPANISH)
tomate arbóreo (62.7. SPANISH)
tomate arbustive (62.7. FRENCH)
tomate cereja (62.10. PORTUGUESE)
tomate cereza (62.10. SPANISH)
tomate cerise (62.10. FRENCH)
tomate cherry (62.10. FRENCH)
tomate cherry (62.10. SPANISH)
tomate chimango (62.7. PORTUGUESE)
tomate cimarrón (62.10. SPANISH)
tomate de árvore (62.7. PORTUGUESE)
tomate de cáscara (62.13. SPANISH)
tomate en arbre (62.7. FRENCH)
tomate groseille (62.10. FRENCH)
tomate inglés (62.11. SPANISH)
tomate salvage (62.10. SPANISH)
tomate selvagem (62.10. PORTUGUESE)
tomateiro (62.9. CRIOULO, CAPE VERDE)
tomateiro (62.9. GALICIAN)
tomateiro (62.9. PORTUGUESE)
tomateiro arbisto (62.7. PORTUGUESE)
tomateiro arbóreo (62.7. PORTUGUESE)
tomateiro bravo (62.17. PORTUGUESE)
tomateiro en arbre (62.7. FRENCH)
tomateiro francês (62.20. PORTUGUESE)
tomatele (62.9. ROMANIAN)
tomatera (62.9. CATALAN-VALENCIAN-BALEAR)
tomates (62.9. AKAN-FANTE)
tomatese (62.9. AKAN-ASANTE)
tomatese (62.9. TWI)
tomatez (62.9. BRETON)
tomati (62.9. BAOULÉ)
tomati (62.9. CRIOULO, CAPE VERDE)
tomati (62.9. DAGAARE, SOUTHERN)
tomati (62.9. HINDUSTANI)
tomati (62.9. JULA)
tomāti (62.9. LATVIAN)
tomati (62.9. MANDING-BAMBARA)
tomati (62.9. YORUBA)
tomati (62.10. DAGAARE, SOUTHERN)
tomatillo (62.10. SPANISH)
tomatillo (62.13. DANISH)
tomatillo (62.13. ENGLISH)
tomatillo (62.13. FINNISH)
tomatillo (62.13. FRENCH)
tomatilló (62.13. HUNGARIAN)
tomatillo (62.13. SPANISH)
tomatillo (62.13. VLAAMS)
tomatillos del diablo (62.21. SPANISH)
tomatim (62.9. AMHARIC)
tomatin (62.13. ARABIC)
tomatinho (62.9. CRIOULO, CAPE VERDE)
tomatinho de capuz (62.11. PORTUGUESE)
tomatinho preto (17.5. RONGA)
tomatitos (62.21. SPANISH)
tòmátò (62.9. AKPA)
tòmátò (62.9. BAFUT)
tomáto (62.9. CHINESE, HAKKA)
tomato (62.9. ENGLISH)
tomato (62.9. GAELIC, MANX)
tomàto (62.9. GAELIC, SCOTTISH)
tomato (62.9. GBE-VHE)
tòmatò (62.9. IGBO)
tomato (62.9. JAPANESE)
tomato (62.9. KANNADA)
tomato (62.9. KOREAN)
tomato (62.9. MALAY)
tomato (62.9. MIZO)
tomato (62.9. MÒORÉ)
tomato (62.9. NIUE)
tomato (62.9. TELUGU)
tomato (62.9. TWI)
tomato (62.9. WELSH)
tomato (62.10. MÒORÉ)
tomato eggplant (62.14c. ENGLISH)
tòmátò eléjò (38.29. YORUBA)
tomato elejo (38.30. YORUBA)
tomato hannu (62.9. KANNADA)
Tomatobaum (62.7. GERMAN)
tomatom (38.5. CRIOULO, UPPER GUINEA)
tomatos (62.9. ÉWÉ)
tomatoshi (62.9. BEMBA)
tomāts (62.9. LATVIAN)
tomátum (62.9. KANURI)
tomatur (62.9. ICELANDIC)
tomatv (62.9. MUSKOGEE)
tomaty (62.9. RUSSIAN)
tombo ya vɔlɔ (9.1. MENDE)
tombo-maga (27.8. MENDE)

torshak (58.2. FARSI, WESTERN)
tọrụbẹ́ín (36.2. IZON)
torwv-lvstēcv (40.43. MUSKOGEE)
toshkaram (30.30. TAJIKI)
tosonzoro (27.24. SHONA)
tosre (23.2. GBE-VHE)
tossa jute (63.1. ENGLISH)
tota kura (23.3. TELUGU)
totakura (23.3. TELUGU)
tôtigue (45.1. SENOUFO, CEBAARA)
totoe (23.3. SAMOAN)
toubab sarro (38.9. MANINKAKAN, WESTERN)
toubabou soso (40.25. MANINKAKAN, WESTERN)
toubanio (17.7. MANINKAKAN, WESTERN)
tougan (38.1. JAPANESE)
tougara-shi (62.5. JAPANESE)
toũm bani (40.26. GOURMANTCHÉ)
toũng kouā (38.1. CHINESE, HAKKA)
touretskii garoh (40.5. RUSSIAN)
toutou (38.21. GOURMANCHÉMA)
tovi aku (9.1. ÉWÉ)
towa (38.16. MENDE)
towa (38.18. MENDE)
towa-wawa (38.18. MENDE)
towɛi (38.16. MENDE)
toyavalli (38.23. SANSKRIT)
töyhtöhelmililja (13.5. FINNISH)
toyita (15.1. QUECHUA, CUSCO)
toyu uphol-svān (50.2. NEWAR)
trâb put lonhoong (62.19. KHMER, CENTRAL)
trâb vèèng (62.19. KHMER, CENTRAL)
trạbhuj (38.11. SANTALI)
trädgårdsböna (40.26. SWEDISH)
trädgårdsmålla (35.1. SWEDISH)
trädgärdsportlak (59.2. SWEDISH)
trädgårdsrättika (30.54. SWEDISH)
trädgårdssyra (58.2. SWEDISH)
trädtomat (62.7. SWEDISH)
tragopono o leimonios (27.47. GREEK)
tragoponon o procsophylos (27.46. GREEK)
trai su (38.25. VIETNAMESE)
trailing amaranth (23.3. ENGLISH)
traista-ciobanului (30.43. ROMANIAN)
trâkuön (36.2. KHMER, CENTRAL)
trâllaach (38.1. KHMER, CENTRAL)
trangkil (41.1. JAVANESE)
trangull (38.13a. ALBANIAN, TOSK)
trâsâk (38.13a. KHMER, CENTRAL)
trâsâk (38.13b. KHMER, CENTRAL)
tras̱ẖtu (7.3. KATI)
tráta (62.9. GAELIC, IRISH)
trate (48.1a. JAVANESE)
tra:w (8.4. KHMER, CENTRAL)
trætomat (62.7. DANISH)
trbulja (24.17. SERBO-CROATIAN)
tre (38.20. GEN)
tre gai (17.1. VIETNAMESE)
tre gai r [uw]ng (17.1. VIETNAMESE)
tre l[af] ng[af] (17.1. VIETNAMESE)
tre ta[uf] (17.2. VIETNAMESE)
trebul'ka pravá siata (24.3. SLOVAK)
tree kale (30.35. ENGLISH)
tree onion (7.9a. ENGLISH)
tree tomato (62.7. ENGLISH)
trei fratipatati (68.1. ROMANIAN)
trejkrāsu vijolīte (68.1. LATVIAN)
tremoceiro (40.14. PORTUGUESE)
tremoço (40.14. PORTUGUESE)
tremoco (40.14. TSONGA)
tres-marias (55.1. CRIOULO, CAPE VERDE)
triaglog (27.21. WELSH)
tribolo acquatico (64.3. ITALIAN)
tribulo acuático (64.3. SPANISH)
trifoliate yam (11.4. ENGLISH)
triganella (40.32. RUSSIAN)
trigonella e frekocho-rios (40.32. GREEK)
trigonelle (40.32. FRENCH)
trigonelle fenugrec (40.32. FRENCH)
trikertė žvaginė (30.43. LITHUANIAN)
trikonaphalam (64.3. SANSKRIT)
trikosantes (38.29. DANISH)
trilliw (68.1. WELSH)
triñchin (58.2. BRETON)
triñchin-gad (58.4. BRETON)
triphyli (40.17. GREEK)
trishahara (24.13. SANSKRIT)
trispalvė našlaitė (68.1. LITHUANIAN)
trivoles o cherseos (64.3. GREEK)
trobojna (68.1. SERBO-CROATIAN)
trobojna temjanuška (68.1. MACEDONIAN)
troell (36.1. BRETON)
trolisten strelolist (6.1. MACEDONIAN)
Trompetenkürbis (38.20. GERMAN)
trompetillas (27.21. SPANISH)
tronchuda (30.28. ENGLISH)
Tronchudakohl (30.28. GERMAN)
tronchudakool (30.28. VLAAMS)

tronhude (30.28. RUSSIAN)
tropæolum (65.1. DANISH)
tropeolë (65.1. ALBANIAN, TOSK)
tropeolum gadul (65.1. HEBREW)
tropical kudzu (40.30. ENGLISH)
tropiki amerikaniki dioscorea (11.8. GREEK)
trosboon (40.6. VLAAMS)
třtina cukrová (17.5. CZECH)
trubus (17.5. INDONESIAN)
true caper berry (34.1. ENGLISH)
true mustard (30.24. ENGLISH)
truffette acide (53.1. FRENCH)
trukwa egipska (38.22. POLISH)
trukwa ostrokątna (38.21. POLISH)
trybula ogrodowa (24.3. POLISH)
trzcina cukrowa (17.5. POLISH)
tsabar matsuee (32.1. HEBREW)
tsabola (62.1. LOMWE, MALAWI)
tsabola (62.2. LOMWE, MALAWI)
tsabola (62.3. LOMWE, MALAWI)
tsabola (62.5. LOMWE, MALAWI)
tsabulevaja (7.1. BELARUSAN)
tsabulja batun (7.5. BELARUSAN)
tsabulja shalot (7.2. BELARUSAN)
tsabulja skarada (7.11. BELARUSAN)
tsabulja tabchataja (7.5. BELARUSAN)
tsabulji repchatui (7.3. BELARUSAN)
tsabulji zepchatyi (7.3. BELARUSAN)
ts'ai tou (40.26. CHINESE, MANDARIN)
tsaka (39.4. BEMBÉ)
tsaki̱tsetu (38.14. COMANCHE)
tsaki̱tsetu (38.16. COMANCHE)
tsaki̱tsetu (38.17. COMANCHE)
tsaki̱tsetu (38.18. COMANCHE)
tsalaph kotsani (34.1. HEBREW)
tsan tou (40.34a. CHINESE, MANDARIN)
tsanguu (30.51. MONGOLIAN)
ts'ao wan tou (40.10. CHINESE, MANDARIN)
tsara (38.3. BAMANANKAN)
tsatsai (30.18. CHINESE, MANDARIN)
tsatsai (30.18. ENGLISH)
tsatsai (30.18. FRENCH)
tsatsai (30.18. VLAAMS)
tsatsanija shirakolistnaja (17.8. BELARUSAN)
tsatsaymori (40.23. HOPI)
tsattsafi (7.2. HAUSA)
tsavtawaktsi (17.7. HOPI)
tsayote (38.25. BONTOC CENTRAL)
tsayote (38.25. CEBUANO)
tsayote (38.25. GIANGAN)
tsayote (38.25. IBANAG)
tsayote (38.25. ILOCANO)
tsayote (38.25. PAMPANGAN)
tsayote (38.25. TAGALOG)
tsayote (38.25. TAUSUG)
tschangei (62.5. TEM)
tschangei-kubong (62.2. TEM)
tschanguilou (40.31. KABIYÉ)
tschibilli (20.6. DAGBANI)
tschocho (38.20. DAGBANI)
tse tin tsiu (62.1. CHINESE, MANDARIN)
tseidlei (62.2. KIOWA)
tseidlei (62.5. KIOWA)
tsene (38.20. GA)
tseque (23.3. RONGA)
tsereriya (40.43. HAUSA)
tsértsvi (40.26. GEORGIAN)
tsertsvi (40.34a. GEORGIAN)
tsharkhali (35.4. GEORGIAN)
tshilài (27.32. VENDA)
tshipinìtshi (35.9. VENDA)
tshlacvi (7.5. GEORGIAN)
tsi (61.1. ENGLISH)
tsi lu (62.19. HMONG-DAW)
tsibehleh (7.3. YIDDISH)
tsibekhes (7.5. YIDDISH)
tsichungu (40.1. LUYIA)
tsiili (62.1. HOPI)
tsiili (62.2. HOPI)
tsiiraʔ (62.2. COMANCHE)
tsikoriah (27.10a. YIDDISH)
tsimbande (40.39. LUYIA)
tsimboka (23.3. LUYIA)
tsindashe (38.4. LUYIA)
tsing'oli (40.43. LUYIA)
tsintsimir (20.6. AKAN-ASANTE)
tsintsinmin (20.6. AKAN-ASANTE)
tsirebika (38.18. MALAGASY)
tsisaka (34.2. LUYIA)
tsisoya (40.8. LUYIA)
tsisus (69.1. MONGOLIAN)
tsitsaca (62.2. GEORGIAN)
tsitsmati (30.51. GEORGIAN)
tsnon (30.53. HEBREW)
tsnonit (30.56. HEBREW)
tsobola (62.1. NYANJA)
tsobola (62.2. NYANJA)
tsobola (62.5. NYANJA)
ts'ói-t'êou (30.41a. CHINESE, HAKKA)
ts'ói t'êou (30.54. CHINESE, HAKKA)

tsokombéle (40.18. AKWA)
tsōn (7.3. TIBETAN)
tson:dzoro (27.24. SHONA)
tsong (7.3. BALTI)
tsong (17.1. HMONG-DAW)
tsoorgono (24.1. MONGOLIAN)
tsoorgono (24.2. MONGOLIAN)
ts'oūng-t'êou (7.3. CHINESE, HAKKA)
ts'oūng-tzè (7.3. CHINESE, HAKKA)
tsubo-kusa (24.7. JAPANESE)
tšuki (40.26. SORBIAN, LOWER)
tsukushi (1.1. JAPANESE)
tsumamono (24.19. JAPANESE)
ts'ung (7.5. CHINESE, MANDARIN)
tsuro-no-geitô (23.2. JAPANESE)
tsuru reishi (38.23. JAPANESE)
tsuru-mannen-gusa (37.1. JAPANESE)
tsuru-murasaki (28.1. JAPANESE)
tsuruna (22.2. JAPANESE)
tsuru-ninjin (33.3. JAPANESE)
tsuwa (10.1. SHONA)
tsuwabuki (27.20. JAPANESE)
tsuwabuki (27.20. POLISH)
tsuwabuki (27.20. VLAAMS)
tsymjanets clubnjanosny (24.8. BELARUSAN)
ttangdureup (26.1. KOREAN)
tł̄ohchinií (7.3. APACHE, JICARILLA)
ttōshchi̱nii (7.3. APACHE, JICARILLA)
tu (17.5. NEWAR)
tu dang qui (26.2. CHINESE, MANDARIN)
tu dang qui (26.2. CHINESE, WÚ)
tu dou (40.35. CHINESE, WÚ)
tu dou (62.24. CHINESE, WÚ)
tu dou (62.24. CHINESE, XIANG)
tu dou (62.24. CHINESE, YUE)
tu hsing ts'ai (30.51. CHINESE, MANDARIN)
tu jen san (59.4. CHINESE, MANDARIN)
tu ren shen (59.3. CHINESE, YUE)
tu ren shen (59.4. CHINESE, WÚ)
tua dam (40.43. THAI)
tua fak yaow (40.42. THAI)
tua Phnom (40.42. THAI)
tubaabuɲambi (11.2. MANDING-MANDINKA)
tubabu soso (40.25. MANDING-MANDINKA)
tubabu soso (40.25. MANDINKA)
tubayan (17.5. BUHID)
tubbo (17.5. ILOCANO)
tuberina (44.11. ITALIAN)
tuberous chervil (24.8. ENGLISH)
tuberous nasturtium (65.2. ENGLISH)
tuberous yam bean (40.22. ENGLISH)
tuberous-rooted mustard (30.14. ENGLISH)
tubo (17.5. CEBUANO)
tubo' (17.5. FILIPINO)
tubo (17.5. TAGALOG)
tubó (17.5. WÁRAY-WÁRAY)
tubos (17.5. SAMBAL, TINÀ)
tubú (17.5. HILIGAYNON)
tubu (17.5. TAUSUG)
tubuh (17.5. TAUSUG)
tubuk (17.5. TAGBANWA)
tučnica (59.2. MACEDONIAN)
tue-chien (62.21. FRENCH)
tufănică (27.23. ROMANIAN)
tuffed hyacinth (13.5. ENGLISH)
tuffieh tal-fanali (62.12. MALTESE)
tugabang (63.1. CEBUANO)
tuggong (62.16. TAGBANWA)
tugi (11.1. BONTOC, CENTRAL)
tugi (11.5. FILIPINO)
tugi (11.5. HILIGAYNON)
tugi (11.5. ILOCANO)
tugi (11.5. TAGALOG)
tugi (36.3. BONTOC CENTRAL)
tugi (36.3. IBANAG)
tugmachagul (45.5. TAJIKI)
tugo-na (40.25. YALUNKA)
tugutong (62.16. CEBUANO)
tuh´dus (62.24. PAIUTE, SOUTHERN)
tuimeruca (30.48. DUTCH)
tuinbonekruid (44.9. VLAAMS)
tuinboon (40.26. AFRIKAANS)
tuinboon (40.34a. DUTCH)
tuinboon (40.34a. VLAAMS)
tuinert (40.28. AFRIKAANS)
tuinkers (30.51. AFRIKAANS)
tuinkers (30.51. DUTCH)
tuinkers (30.51. VLAAMS)
tuinmelde (35.1. DUTCH)
tuinmelde (35.1. VLAAMS)
tuinzuring (58.2. DUTCH)
tuk (39.4. MANKANYA)
tukas-tukas (9.1. TAGALOG)
tukhamariya (44.5. MARATHI)
tuki phool (27.45. NEPALI)
tukm-i-kahu (27.31. FARSI, WESTERN)
tukm-i-kahu (27.32. FARSI, WESTERN)
tukog-banog (63.1. CEBUANO)
tukol (16.1a. ILOCANO)

turfed stone leek (7.9b. ENGLISH)
tùrgúnùùwáá (63.1. HAUSA)
turi (38.21. HINDI)
turi (38.22. URDU)
turi (40.2. MARATHI)
turi bhol (38.22. MARATHI)
turia (38.21. GUJARATI)
turia (38.22. GUJARATI)
túrìì (20.2. HAUSA)
tuřin (30.22. CZECH)
turksvy (32.1. AFRIKAANS)
turku ķirbis (38.20. LATVIAN)
turmas (40.14. HINDI)
turmeric (20.2. ENGLISH)
turmuj (38.3. ASSAMESE)
turmus (40.14. HEBREW)
turmuz (40.14. BENGALI)
turmuz (40.14. PANJABI, EASTERN)
turn sole (27.42. ENGLISH)
turneps (30.22. ROMANIAN)
turneps (30.22. RUSSIAN)
turneps (30.41a. KAZAKH)
turneps (30.41a. ROMANIAN)
turneps (30.41a. RUSSIAN)
turnepsas (30.22. LITHUANIAN)
turnip (30.41a. ENGLISH)
turnip (30.41a. FILIPINO)
turnip (30.41a. INDONESIAN)
turnip (30.41a. KANNADA)
turnip (30.41a. MALAY)
turnip (30.41a. TELUGU)
turnip broccoli (30.41c. ENGLISH)
turnip greens (30.41b. ENGLISH)
turnip kale (30.30. ENGLISH)
turnip root (30.41a. ENGLISH)
turnip-rooted celery (24.5. ENGLISH)
turnip-rooted chervil (24.8. ENGLISH)
turnip-rooted parsley (24.21. ENGLISH)
turnips (30.22. POLISH)
turnips (30.41a. POLISH)
turnips (30.41a. SERBO-CROATIAN)
turnips (30.41a. SWEDISH)
turnipsi (30.41a. FINNISH)
turp (30.54. UZBEK)
turp (30.56. KIRGHIZ)
turp (30.56. PASHTO, NORTHERN)
turp (30.56. TURKISH)
turp (30.56. UZBEK)
turshak (53.1. TAJIKI)
turshak (58.2. TAJIKI)
tursin (24.19. MALTESE)
tursin taljan (24.20. MALTESE)
tursiphal (27.23. MARATHI)
turški fižol (40.24. SLOVENIAN)
turubus (17.5. SUNDA)
turuchi avare (40.18. KANNADA)
turuku thogari (40.2. KANNADA)
turumja (38.3. BODO)
tururuwa (63.1. HAUSA)
tuscarora (17.8. ITALIAN)
tuščialaiškis česnakas (7.5. LITHUANIAN)
tusima (27.24. TEM)
tüskés kápricserje (34.1. HUNGARIAN)
tuszkarora (17.8. HUNGARIAN)
tutavia hmel zvychainie (46.1. BELARUSAN)
tutikura (36.2. TELUGU)
tutubo (35.1. GEORGIAN)
tuutae-ika-moana (22.2. MAORI)
tuva (40.1. HOPI)
tuvarai (40.2. TAMIL)
tuvari (40.2. HINDI)
tuvari (40.2. KANNADA)
tuvari (40.2. SANSKRIT)
tuvarika (40.2. HINDI)
tuvarika (40.2. SANSKRIT)
tuved (8.4. BINUKID)
tuver (40.2. BENGALI)
tuver (40.2. GUJARATI)
tuver (40.2. HINDI)
tuver (40.2. MARATHI)
tuwari (40.2. SANSKRIT)
t[uwf] c[oo] (6.1. VIETNAMESE)
t[uwf] gai (11.5. VIETNAMESE)
t[uwr] t[oo] (44.7. VIETNAMESE)
tuya (40.26. CHEROKEE)
tuya (40.43. DAGBANI)
tuyapak (2.1. HANUNOO)
tüzbab (40.24. HUNGARIAN)
tűzliliom (13.3. HUNGARIAN)
tvhvyv (38.14. MUSKOGEE)
tvhvyv (38.16. MUSKOGEE)
tvhvyv (38.17. MUSKOGEE)
tvhvyv (38.18. MUSKOGEE)
tvis boloki (30.56. GEORGIAN)
tvlak' encakhv (40.26. MUSKOGEE)
tvlak-cvpko (40.26. MUSKOGEE)
tvlak-cvpko (40.43. MUSKOGEE)
tvlak-espokhv (40.26. MUSKOGEE)
tvlak-lane (40.26. MUSKOGEE)
tvlak-lvste (40.26. MUSKOGEE)

upêtche (40.1. MANDJAK)
upiate (40.1. MANDJAK)
upland cress (30.2. ENGLISH)
Uplandse smeerwortel (29.3. DUTCH)
upo (38.20. FILIPINO)
upo (38.20. TAGALOG)
upó (38.20. WÁRAY-WÁRAY)
upodaki (28.1. SANSKRIT)
upodika (28.1. SANSKRIT)
upong-ahas (38.29. TAGALOG)
upoti (28.1. SANSKRIT)
upperi-parnki (62.2. MALAYALAM)
uppländsk vallört (29.3. SWEDISH)
uppukkadalai (40.5. TAMIL)
ŭpsĕskw' (44.3. MICMAC)
upu (38.20. HANUNOO)
upupu (40.18. SWAHILI)
upupu (66.3. SWAHILI)
uqalo (17.1. XHOSA)
uqalo (17.1. ZULU)
uqalo (17.2. XHOSA)
uqalo (17.2. ZULU)
uqalo (17.4. XHOSA)
uqalo (17.4. ZULU)
ura da colorare (56.2. ITALIAN)
urabã (43.1. JOLA-FONYI)
urabi (11.1. BUHID)
urad (40.35. GUJARATI)
urad (40.37. HINDI)
urad (40.38. HINDI)
uradishi (30.56. ZULU)
uraga (24.7. KANNADA)
urahi (40.34a. ASSAMESE)
urakakara (38.23. TELUGU)
urala kizangu (62.24. MALAYALAM)
uralagadda (62.24. TELUGU)
uralakilangu (62.24. TAMIL)
urambing (40.25. BUHID)
urard (15.2. TAGALOG)
uraru (15.4. BICOLANO, CENTRAL)
uraru (15.4. HANUNOO)
uray (23.3. BICOLANO, CENTRAL)
uray (23.3. TAGALOG)
urd (40.37. ENGLISH)
urd (40.37. FIJIAN)
urd (40.37. HINDI)
urd (40.37. PANJABI, EASTERN)
urd-bab (40.37. HUNGARIAN)
Urdbohne (40.37. GERMAN)
urdbønne (40.37. DANISH)
urdbønne (40.37. NORWEGIAN)
urdhvasita (38.23. SANSKRIT)
uṛdi (40.37. MUNDARI)
urdu (40.37. TULU)
uré (36.3. SUSU)
uré firke (36.3. SUSU)
uré gbuéli (36.3. SUSU)
urgǣd (30.1. PANJABI, EASTERN)
urhi-mah (40.43. ASSAMESE)
uri (11.1. BINUKID)
uri (11.2. BINUKID)
uri (11.5. BINUKID)
uri (23.3. TAGALOG)
uri (38.9. PORTUGUESE)
uri (38.10. PORTUGUESE)
uṛic alaṅ (59.2. SANTALI)
urid (40.38. HINDI)
uridschŏ (36.3. BASSARI)
urigaddalu (7.3. TELUGU)
uriri (7.3. KWANGALI)
uri:ri (40.18. SENA)
úritök (38.18. HUNGARIAN)
urlagadda (62.24. TELUGU)
urngol (44.5. FULFULDE, ADAMAWA)
urobi-mahor-pat (40.43. ASSAMESE)
urohi (40.34a. ASSAMESE)
urohi (40.43. ASSAMESE)
uroslam archichoke (27.26. URDU)
ursică mare (66.3. ROMANIAN)
ürt-allikkerss (30.52. ESTONIAN)
urtiga japonesa (44.11. PORTUGUESE)
urtiga maior (66.3. PORTUGUESE)
urtiga mansa (66.3. PORTUGUESE)
urtigão (66.3. PORTUGUESE)
ùrud (20.2. AFAR)
urud (40.37. URDU)
urulaikizhangu (62.24. TAMIL)
urulakizhangu (62.24. MALAYALAM)
urustambhhaa (16.1a. SANSKRIT)
uruzhai kilangu (62.24. TAMIL)
u-ryakán (43.1. WAMEY)
urzicuţe negre (44.7. ROMANIAN)
usan (30.48. PANJABI, EASTERN)
useleri (24.4. ZULU)
ushnon (35.8. TAJIKI)
usi (11.8. SUMO-MAYANGNA)
usĩ (17.3. IZON)
usimilagaay (62.2. TAMIL)
usipag (45.2. ASSAMESE)
usjevna (30.48. SERBO-CROATIAN)

usnes (27.11. LATVIAN)
ussu-nifing (44.10. BAMANANKAN)
ussu-ni-fing (44.10. BAMANANKAN)
ussu-ni-gue (44.10. BAMANANKAN)
ussurian thorny pepperbush (26.4. ENGLISH)
ustekaya (62.23. TELUGU)
usti (62.23. TELUGU)
usturoi cultivat (7.10. ROMANIAN)
usu beninigana (27.18. JAPANESE)
usu ni fin (44.10. BAMANANKAN)
usu ni gé (44.10. BAMANANKAN)
usub (8.4. JOLA-FONYI)
u-suri (40.43. WAMEY)
usuujaq (24.11. INUKTITUT, EASTERN CANADIAN)
uszyca (6.2. POLISH)
uszyca trójlistna (6.1. POLISH)
utamatisi (62.9. ZULU)
utaw (40.8. FILIPINO)
utaw (40.8. TAGALOG)
utheniphu (30.41a. ZULU)
uthpala (50.2. SANSKRIT)
uthpalaka (50.2. SANSKRIT)
uthulu (40.38. TELUGU)
utí (38.18. IZON)
ụ̀-tīb (45.1. BETE-BENDI)
utiu (38.7b. SAMOAN)
utong (40.26. SAMBAL, BOTOLAN)
utong (40.42. ILOCANO)
utong-utongan (11.2. TAGALOG)
utumatisi(la) (62.9. ZULU)
ụtụta (7.3. IZON)
ututo ogiri (43.1. IGBO)
utuutu (10.2. MANU'A SAMOAN)
uurh (16.1a. CAROLINIAN)
uus-meremaa spinat (22.2. ESTONIAN)
uva-de-santa maria (62.21. CRIOULO, CAPE VERDE)
uva-macaco (56.2. CRIOULO, CAPE VERDE)
uvanitsoq (7.3. INUKTITUT, WEST GREENLANDIC)
uvi (11.1. BUGHOTU)
uvi (11.1. FIJIAN)
uvi (11.1. IBANAG)
uvi (11.2. BUGHOTU)
uvi (11.5. BUGHOTU)
uwhi (11.1. MAORI)
uwhi (11.2. MAORI)
uwhi (11.3. MAORI)
uwhi (11.4. MAORI)
uwhi (11.5. MAORI)
uwhi (11.6. MAORI)
uwhi (11.7. MAORI)
uwhi (11.8. MAORI)
uwhikaaho (11.1. MAORI)
uwhikaaho (11.2. MAORI)
uwhikaaho (11.3. MAORI)
uwhikaaho (11.4. MAORI)
uwhikaaho (11.5. MAORI)
uwhikaaho (11.6. MAORI)
uwhikaaho (11.7. MAORI)
uwhikaaho (11.8. MAORI)
uwi (11.1. INDONESIAN)
uwi gandul (11.1. JAVANESE)
uwi kata (11.1. JAVANESE)
uwi kentang (62.24. JAVANESE)
uwi singapur (11.1. INDONESIAN)
uwimbi (38.13a. TUMBUKA)
u-wong (62.5. WAMEY)
uwua nor (17.3. TIV)
uzhunnu (40.37. MALAYALAM)
üzümsü bitkiler (62.23. TURKISH)
uŋjiŋjiŋtka (62.9. LAKOTA)
va bvuh-eh (17.2. LAHU)
va k'aw (17.2. LAHU)
va meu (17.2. LAHU)
va meu hpu (17.2. LAHU)
va sha (17.2. LAHU)
va$^{\vee}$ shi (17.1. LAHU)
va'a (40.30. MANU'A SAMOAN)
vaajibhakshya (40.5. SANSKRIT)
vaajimantha (40.5. SANSKRIT)
vaalamara (40.3. MALAYALAM)
vaanari (40.18. SANSKRIT)
vaaraahi (11.2. SANSKRIT)
vaarana (16.1a. SANSKRIT)
vaarthaaki (62.19. SANSKRIT)
vaartthaagam (62.19. TAMIL)
vaasapushpa (30.51. SANSKRIT)
vaasu preshpa (30.51. SANSKRIT)
vàát (44.8. BEROM)
vaayasi (62.15. SANSKRIT)
vaayingana (62.19. KONKANI)
vad árticsóka (27.15. HUNGARIAN)
vad spárga (13.1. HUNGARIAN)
va-diegon (40.18. BASSARI)
vadoekplant (38.22. AFRIKAANS)
vadzjany arech plavanchi (64.3. BELARUSAN)
vaenti (8.4. KONKANI)

vaepudupachha (44.5. TELUGU)
vaerkkadalai (40.1. TAMIL)
vaerushanagalu (40.1. TELUGU)
vafo (62.5. GUN)
vafo daho (62.1. GUN)
vaggarane saasive (30.24. KANNADA)
vago (38.1. FIJIAN)
vahakurpitsa (38.1. FINNISH)
vahosa (62.19. IVATAN)
vainagu krizantēma (27.23. LATVIAN)
vaine 'initia (38.23. TONGAN)
vaingim (62.19. KONKANI)
väinönputki (24.1. FINNISH)
vaistinė aqurklė (29.1. LITHUANIAN)
vaistinė ožragė (40.32. LITHUANIAN)
vaistinė taukė (29.2. LITHUANIAN)
vaistinis smidras (13.2. LITHUANIAN)
vajrangi (27.26. SANSKRIT)
vakrashimbi (40.6. SANSKRIT)
val (40.9. GUJARATI)
val (40.9. MARATHI)
vala (16.1a. MALAYALAM)
valai (16.1a. TAMIL)
valavarai (40.3. TAMIL)
valchibhagi (28.1. KONKANI)
valeriaan (27.21. AFRIKAANS)
valeriaan (27.21. VLAAMS)
valeriana (27.21. GREEK)
valeriana (27.21. PORTUGUESE)
valeriana (27.21. ROMANIAN)
valeriana (27.21. RUSSIAN)
valeriana (67.2. SPANISH)
valeriana de Itália (67.1. PORTUGUESE)
valériane (67.2. BRETON)
valériane d'Alger (27.21. FRENCH)
valériane petite (67.2. FRENCH)
valerianela (67.2. SPANISH)
valerianela de Italia (67.1. SPANISH)
valerianella (67.2. GREEK)
valerianella (67.2. ITALIAN)
valérianella á fruits velus (67.1. FRENCH)
valerianelle potagère (67.2. FRENCH)
valeriani e keratoïdis (27.21. GREEK)
valeriánka pol'ná (67.2. SLOVAK)
valerian-palderjan (67.2. ESTONIAN)
valerijana (27.21. MACEDONIAN)
valerijanavaja (67.1. BELARUSAN)
valerjana (27.21. RUSSIAN)
valerjanela malaja (67.2. BELARUSAN)
valeryanit (67.1. HEBREW)
valeryanit (67.2. HEBREW)
valge peakapsas (30.27a. ESTONIAN)
valgomasis batatas (36.3. LITHUANIAN)
valgomasis česnakas (7.10. LITHUANIAN)
valgomasis krienas (30.1. LITHUANIAN)
valgomasis lęšis (40.12. LITHUANIAN)
valgomasis manijokas (39.4. LITHUANIAN)
valgomasis pūtelis (27.46. LITHUANIAN)
valgomasis salieras (24.4. LITHUANIAN)
valgomasis salieras (24.5. LITHUANIAN)
valgomasis svoqūnas (7.3. LITHUANIAN)
valgomoji gelteklė (27.38. LITHUANIAN)
valgomoji rūgštynė (58.2. LITHUANIAN)
valgomoji viksvuolė (10.1. LITHUANIAN)
valipodaki (28.1. SANSKRIT)
valkoinen guineanjamssi (11.7. FINNISH)
valkokaali (30.27a. FINNISH)
valksipuli (7.10. FINNISH)
vallaarai (24.7. TAMIL)
vallai keerai (36.2. TAMIL)
vallai kodi (36.2. TAMIL)
vallai-pundu (7.10. TAMIL)
vallaka (40.9. SANSKRIT)
vallareekeeri (24.7. TAMIL)
vallarei (24.7. TAMIL)
vallari (24.7. TAMIL)
vallari (40.32. SANSKRIT)
vallaringa (40.3. MALAYALAM)
vallaringa (40.4. MALAYALAM)
valli phala (38.1. SANSKRIT)
vallia capo molago (62.5. MALAYALAM)
vallikilangu (36.3. TAMIL)
vallikilanna (36.3. MALAYALAM)
vallimudga (40.35. SANSKRIT)
valm (40.9. HINDI)
valse yam (43.1. VLAAMS)
valsk bønne (40.34a. DANISH)
valudalai (62.19. TAMIL)
valudunai (62.19. TAMIL)
valutina (62.19. MALAYALAM)
valverde (35.2. PORTUGUESE)
vambèyon (39.4. TIV)
van kavunu (38.7a. TURKISH)
van nalu (45.2. NEPALI)
vana (27.2. KANNADA)
vanabhhushaa (16.1a. SANSKRIT)
vanakadali (9.1. SANSKRIT)
vanalakshmi (16.1a. SANSKRIT)
vanamaalini (11.2. SANSKRIT)
vanamudga (40.35. SANSKRIT)

vanamugali (27.2. KANNADA)
vaṇaṅṅana (23.2. TULU)
vanapushpa (24.13. SANSKRIT)
vanashobhana (48.1a. SANSKRIT)
vanashukari (40.18. SANSKRIT)
vanashunti (20.6. KANNADA)
vanavaasi (11.2. SANSKRIT)
vandelaga (24.7. KANNADA)
vandeninė zizanija (17.8. LITHUANIAN)
vandisi (17.7. VENDA)
vandmelon (38.3. DANISH)
vandmimose (40.19. DANISH)
vandnavle (24.16. DANISH)
vanga chettu (62.19. TELUGU)
vangachettu (62.19. TELUGU)
va-ngal (40.1. WAMEY)
vangam (62.19. TAMIL)
vangamu (62.19. TELUGU)
vangana (62.19. SANSKRIT)
vange (62.19. MARATHI)
va-ngond (8.4. WAMEY)
vankaaya (62.19. TELUGU)
vankaya (62.19. TAMIL)
vankaya (62.19. TELUGU)
vanlig agurkurt (29.1. NORWEGIAN)
vanlig blomkarse (65.1. NORWEGIAN)
vanlig flatklokke (34.3. NORWEGIAN)
vanlig kepaløk (7.3. NORWEGIAN)
vannan (16.1a. MALAYALAM)
vannmelon (38.3. NORWEGIAN)
vannsuy (24.9. KHMER, CENTRAL)
vannsuy baraing (24.19. KHMER, CENTRAL)
vansh (17.1. SANSKRIT)
vanta (45.1. MALAYALAM)
vantak (62.19. GUJARATI)
vanzylsesuikerboontjie (40.26. AFRIKAANS)
vao sosolo (23.2. MANU'A SAMOAN)
vao sosolo (23.2. SAMOAN)
va-péna (40.3. BASSARI)
va-pèna (40.3. WAMEY)
va̱r inyaregh (44.3. TIV)
vara (20.6. SANSKRIT)
varaahi (11.2. MARATHI)
varaahi (11.6. SANSKRIT)
varaahikanda (11.2. SANSKRIT)
varahi (11.2. SANSKRIT)
varahika (40.18. SANSKRIT)
varahikanda (11.2. GUJARATI)
varahikanda (11.2. SANSKRIT)
varaka (40.35. SANSKRIT)
varam (27.45. HINDI)
varanabusa (16.1a. SANSKRIT)
varanavallabha (16.1a. SANSKRIT)
varariksha (16.1a. SANSKRIT)
vardakačača (27.10a. GEORGIAN)
varemerohi (29.2. ESTONIAN)
varesputk (24.3. ESTONIAN)
variali (24.13. GUJARATI)
variari (24.13. GUJARATI)
varikantaka (64.3. SANSKRIT)
varikubshaka (64.3. SANSKRIT)
varisoha (48.1a. SANSKRIT)
varivalli (38.23. SANSKRIT)
variyali (24.13. GUJARATI)
varjúháj (37.1. HUNGARIAN)
varkkos (59.2. AFRIKAANS)
vårklynne (67.2. SWEDISH)
varkoortjies (24.7. AFRIKAANS)
varot (24.18. ARMENIAN)
vårsalat (67.2. DANISH)
vårsalat (67.2. NORWEGIAN)
varsseller (24.4. ESTONIAN)
vartaikaai (38.3. TAMIL)
vartaku peetaphala (62.19. SANSKRIT)
vartaku vatigama (62.19. SANSKRIT)
vartha (62.19. SANSKRIT)
varthaka (62.19. SANSKRIT)
vartoj (27.26. TAJIKI)
varvara (44.5. SANSKRIT)
varvarea e prolïmi (30.2. GREEK)
varvaree (44.5. SANSKRIT)
varyaali (24.13. HINDI)
varză albă (30.27a. ROMANIAN)
varză creaţă (30.34. ROMANIAN)
varză chinezească (30.36. ROMANIAN)
varză chinezească (30.40. ROMANIAN)
varză cu căpăţină (30.27a. ROMANIAN)
varză de Bruxelles (30.29. ROMANIAN)
varză de frunze (30.35. ROMANIAN)
varză de iarnă (30.27a. ROMANIAN)
varza de mare (30.46. ROMANIAN)
varză furajeră (30.35. ROMANIAN)
vasaiyilaikkeerai (35.9. TAMIL)
vasaras eglīte (35.2. LATVIAN)
vasayleykiray (35.9. TAMIL)
vasilikos (44.5. GREEK)
vaslakkeerai (28.1. TAMIL)
vastukakara (35.9. SANSKRIT)
vasunta (40.35. SANSKRIT)
vatana (40.28. GUJARATI)

vataun (62.19. PANJABI, EASTERN)
vathaari (8.3. SANSKRIT)
va-tienega (45.1. BASSARI)
va-tieneka (45.1. BASSARI)
vatigana (62.19. SANSKRIT)
va-tiitien (40.18. WAMEY)
vātiṅgano (62.19. PALI)
va-tiur (40.43. WAMEY)
va-tiuri (40.43. WAMEY)
vatnakarsi (30.52. ICELANDIC)
vatnalilja (50.2. ICELANDIC)
vatnshneta (64.3. ICELANDIC)
vatnsmelona (38.3. ICELANDIC)
vatrachion (24.17. GREEK)
vāttākī (62.19. PALI)
vattakka (38.3. MALAYALAM)
vattenkrasse (30.52. SWEDISH)
vattenmelon (38.3. SWEDISH)
vavad (23.3. HINDI)
vavanala (17.7. SANSKRIT)
vavibomanga (36.3. MALAGASY)
vay (15.4. IVATAN)
vayasavha (62.15. SANSKRIT)
vazhai (16.1a. TAMIL)
vazhukkaikeerai (59.2. TAMIL)
vazhuthana (62.19. MALAYALAM)
vazirak (30.2. TAJIKI)
vce (17.7. MUSKOGEE)
vebaotu (27.35. TURKISH)
veccia (40.43. ITALIAN)
veccia pisello (40.13. ITALIAN)
vedhaka (24.9. SANSKRIT)
vedhani (40.32. SANSKRIT)
vedhya (20.4. SANSKRIT)
vedru (17.1. TELUGU)
veduru (17.1. TELUGU)
vegetable fern (5.1. ENGLISH)
vegetable mustard (30.17. ENGLISH)
vegetable oyster (27.46. ENGLISH)
vegetable pear (38.25. ENGLISH)
vegetable rape (30.21. ENGLISH)
vegetable sponge (38.22. ENGLISH)
Veilchen (68.1. GERMAN)
vejbred (16.1a. DANISH)
vejiga de perro (62.11. SPANISH)
vel vangi (62.9. MARATHI)
vela pandu (38.7b. TELUGU)
vela tikva (38.16. SERBO-CROATIAN)
velapalam (38.7b. TAMIL)
veldsla (67.2. DUTCH)
veldsla (67.2. VLAAMS)
veldzuring (58.2. DUTCH)
vele (17.7. VENDA)
vella poodu (7.10. TAMIL)
vella tamma (40.3. TELUGU)
vella thambattankai (40.3. TAMIL)
vella thambattankai (40.4. TAMIL)
vellaaraku (24.7. TELUGU)
vellai tambattai (40.3. TAMIL)
vellaikeerai (36.2. TAMIL)
vellaitambattai (40.3. TAMIL)
vellaitthambattai (40.3. TAMIL)
vellambal (50.2. TAMIL)
vellapayaru (40.43. MALAYALAM)
vellari (38.13a. TAMIL)
vellarikai (38.13a. TAMIL)
vellarikka (38.13a. MALAYALAM)
vellarikkai (38.13a. TAMIL)
vellatamma (40.3. TELUGU)
vella-vengayam (7.3. TAMIL)
vellu (17.1. KONKANI)
velluli (7.10. TELUGU)
vellulli (7.10. MALAYALAM)
vellulli (7.10. TELUGU)
veloudinos faslolos (40.18. GREEK)
veluthulli (7.10. MALAYALAM)
velvet bean (40.18. ENGLISH)
velvet simba (40.18. ORIYA)
venäjänjuuri (26.4. FINNISH)
venanachine (10.1. GOURMANCHÉMA)
venda (45.1. MALAYALAM)
venda (45.1. TELUGU)
vendai (45.1. TAMIL)
vendai chedy (45.1. TAMIL)
vendaikkaai (45.1. TAMIL)
vendaikkay (45.1. TAMIL)
vendayam (40.32. TAMIL)
vendi (45.1. ORIYA)
vendi (45.1. TAMIL)
vendiyam (40.32. TAMIL)
vengaayam eerulli (7.3. TAMIL)
vengan (62.19. GUJARATI)
vengayam (7.3. TAMIL)
vengni (62.19. GUJARATI)
venilai valli (11.2. TAMIL)
venkel (24.13. DUTCH)
ventagam (40.32. TAMIL)
venthaiam (40.32. MALAYALAM)
venthayam (40.32. MALAYALAM)
venthayom (40.32. MALAYALAM)

venthiam (40.32. MALAYALAM)
Venüs otu (10.1. TURKISH)
vepudupachha (44.5. TELUGU)
verconia (27.24. TURKISH)
verdolaga (59.2. CATALAN-VALENCIAN-BALEAR)
verdolaga (59.2. SPANISH)
verdolaga de Cuba (59.1. SPANISH)
verdolaga de invierno (59.1. SPANISH)
verdolaga de playa (59.3. SPANISH)
verkkala (40.1. MALAYALAM)
verkkolehti (25.1. FINNISH)
verkkomeloni (38.7b. FINNISH)
vernonia (27.24. FRENCH)
Vernonia (27.24. GERMAN)
vernonia (27.24. ITALIAN)
vernonia (27.24. PORTUGUESE)
vernonia (27.24. SPANISH)
vernonie (27.24. DANISH)
vernonie (27.24. FRENCH)
verrkkombu (20.6. TAMIL)
verrolaivalli (11.6. TAMIL)
verusenagalu (40.1. TELUGU)
veshana (24.9. SANSKRIT)
veshmakula (38.29. SANSKRIT)
vesikastanja (10.2. FINNISH)
vesikastanja (10.3. FINNISH)
vesikupp (50.2. ESTONIAN)
vesimeloni (38.3. FINNISH)
vesipähkel (10.2. ESTONIAN)
vesipähkel (64.3. ESTONIAN)
vesipähkinä (64.1. FINNISH)
vesipähkinä (64.3. FINNISH)
vesiputk (24.17. ESTONIAN)
vesiroos (50.2. ESTONIAN)
vestindisk pilrod (15.2. DUTCH)
Vest-Indisk-gurke (38.6. SWEDISH)
veteménybab (40.26. HUNGARIAN)
vetési lednek (40.10. HUNGARIAN)
větevna tá kapusta (30.33. CZECH)
vethayam (40.32. TAMIL)
vetrargrasler (38.14. ICELANDIC)
vettiiaivalli kizhangu (11.6. TAMIL)
vettilaivalli (11.2. TAMIL)
vhv (62.24. MUSKOGEE)
vhv-cvmpv (11.1. MUSKOGEE)
vhv-cvmpv (11.2. MUSKOGEE)
vhv-cvmpv (11.5. MUSKOGEE)
vhv-cvmpv (36.3. MUSKOGEE)
via (8.1. FIJIAN)
via (8.1. TOK PISIN)
via (8.6. TONGAN)
via kana (8.6. FIJIAN)
viadranu (8.1. FIJIAN)
viazi (35.9. SWAHILI)
viazi (62.24. DIGO)
viazi muiringo (62.24. SWAHILI)
viazi vikuu (11.1. SWAHILI)
viazi vitamu (36.3. SWAHILI)
viboodipathri (44.5. TELUGU)
vibudipatri (44.5. TELUGU)
vibuyu (38.20. SWAHILI)
vibuyu (38.27. SWAHILI)
vibuyu (38.29. SWAHILI)
viči bob (40.14. CZECH)
victory onion (7.14. ENGLISH)
videshivrunthaak (62.9. SANSKRIT)
vie cu cā-röt (24.11. VIETNAMESE)
vienadienė (13.3. LITHUANIAN)
vienmetė paprika (62.1. LITHUANIAN)
vierhoruk (22.2. VLAAMS)
vīges kaktuss (32.1. LATVIAN)
vigna (40.35. CZECH)
vigna adzuki (40.36. CZECH)
vigna cinese (40.41. ITALIAN)
vigna e imitripodios (40.42. GREEK)
vigna e kylindiki (40.41. GREEK)
vigna e siniki (40.43. GREEK)
vigna mungo (40.37. CZECH)
vigna rýžová (40.40. CZECH)
vigna zlatá (40.38. CZECH)
vignabønne (40.43. DANISH)
vignaboon (40.43. DUTCH)
vigne japonaise (40.30. FRENCH)
vihanneskrassi (30.51. FINNISH)
vihanneskrysanteemi (27.23. FINNISH)
vihannespaprika (62.1. FINNISH)
vihannesportulakka (59.2. FINNISH)
vihannessinappi (30.39. FINNISH)
viherminttu (44.3. FINNISH)
viigikaktus (32.1. ESTONIAN)
viikunaopuntia (32.1. FINNISH)
vikke (40.10. NORWEGIAN)
vilaayathi pundi (45.4. KANNADA)
vilaayathi pundrike (45.4. KANNADA)
vilaaythi bengar (62.9. HINDI)
vilaaythimug (40.1. MARATHI)
vilaaythimung (40.1. HINDI)
vilaiti ghavat (40.17. MARATHI)
vilaiti wagan (62.9. GUJARATI)

vilayati kaddu (38.14. HINDI)
vilayati kaddu (38.16. HINDI)
vilayati kaddu (38.17. HINDI)
vilayati kaddu (38.18. HINDI)
vilayati mung (40.1. HINDI)
vilayati palak (35.9. HINDI)
vilayati wagan (62.9. GUJARATI)
vilay-ati-gavat (40.17. MARATHI)
vilayati-hullu (40.17. KANNADA)
vilayithi (62.9. BENGALI)
vilayithi baingan (62.9. HINDI)
vilayithi vengan (62.9. GUJARATI)
vilaythibaigun (62.9. HINDI)
vilayti palak (35.9. HINDI)
vilayti-ghas (40.17. GUJARATI)
vilbag (27.8. MORISYEN)
vildris (17.8. DANISH)
villaty kaddu (38.14. URDU)
villaty kaddu (38.16. URDU)
villgulrot (24.11. NORWEGIAN)
villucchio d'acqua (36.2. ITALIAN)
vilucchio giapponese (36.1. ITALIAN)
vinătă (62.19. ROMANIAN)
vindel (36.1. NORWEGIAN)
vindelklokke (33.3. NORWEGIAN)
vine Initia (38.23. NIUE)
vineveh (16.1a. IVATAN)
vingeert (40.13. NORWEGIAN)
vingerblaartee (34.2. AFRIKAANS)
vingëz e koltivuar (40.10. ALBANIAN, TOSK)
vinha virgem (69.1. PORTUGUESE)
vinivih (16.1a. IVATAN)
vinivik (16.1a. IVATAN)
vinkel (24.13. AFRIKAANS)
vinteereddik (30.56. NORWEGIAN)
vinterasters (27.23. DANISH)
vintergresskar (38.16. NORWEGIAN)
vintergresskar (38.17. NORWEGIAN)
vinterkarse (30.2. DANISH)
vintermelon (38.7b. NORWEGIAN)
vintermelon (38.11. NORWEGIAN)
vinterportulak (59.1. DANISH)
vinterportulakk (59.1. NORWEGIAN)
vintersallat (67.2. SWEDISH)
vioelen (68.1. STELLINGWERFS)
vioeltien (68.1. STELLINGWERFS)
viola (68.1. ASSAMESE)
viola (68.1. ITALIAN)
viola (68.1. PORTUGUESE)
viola (68.1. SPANISH)
viola del pensiero (68.1. ITALIAN)
viola tricolore (68.1. ITALIAN)
violet (68.1. ENGLISH)
violeta (68.1. PORTUGUESE)
violeta (68.1. SPANISH)
violetta (68.1. ITALIAN)
violette (68.1. FRENCH)
violette tricolore (68.1. FRENCH)
violier (68.1. FRENCH)
violka trojbarevná (68.1. CZECH)
viooltje (68.1. DUTCH)
vịọ̀vịo (40.2. IGBO)
vira (23.3. SANSKRIT)
virágkel (30.26. HUNGARIAN)
virágnád (9.1. HUNGARIAN)
virdhida (11.2. SANSKRIT)
virzinkāposti (30.34. LATVIAN)
visalacheera (35.3. MALAYALAM)
visalacheera (35.9. MALAYALAM)
visha thogari (40.10. KANNADA)
vishaari (63.1. SANSKRIT)
vishaghana (23.3. SANSKRIT)
vishaghna (23.3. SANSKRIT)
vishakantaki (38.23. SANSKRIT)
vishala (28.1. SANSKRIT)
vishani (64.3. SANSKRIT)
vishvabheshaj (20.6. SANSKRIT)
vishvaksenapriya (11.2. SANSKRIT)
vishvarochanna (36.2. SANSKRIT)
vishvatulasi (28.1. SANSKRIT)
visi (38.3. SHONA)
visoki (40.24. CROATIAN)
vista (10.1. CRIOULO, CAPE VERDE)
vite indiana (38.23. ITALIAN)
vitenu lobija (40.9. LATVIAN)
vithunnaka (24.9. SANSKRIT)
vitkål (30.27a. SWEDISH)
vitlök (7.10. SWEDISH)
vìtluf (27.10a. UKRAINIAN)
vitlupin (40.14. SWEDISH)
vitsenap (30.57. SWEDISH)
vitunguu liki (7.8. SWAHILI)
vitunguu saumu (7.11. SWAHILI)
vitunguu shaloti (7.2. SWAHILI)
vitunna (23.4. SANSKRIT)
vìvsjanij korìn (27.46. UKRAINIAN)
vizijátszint (18.1. HUNGARIAN)
vizililiom (50.2. HUNGARIAN)
vizitorma (30.52. HUNGARIAN)
vizizsázsa (30.52. HUNGARIAN)

vkeçi biberi (62.5. TURKISH)
vlašac (7.11. SERBO-CROATIAN)
vlasstal (27.16. GAELIC, MANX)
vlcí bôb biely (40.14. SLOVAK)
voandzeja podzemní (40.39. CZECH)
voandzou (40.39. FRENCH)
voden kreson (30.52. BULGARIAN)
voden spanak (36.2. MACEDONIAN)
vodena (24.17. SERBO-CROATIAN)
Vodianoi orekh dvurogii (64.1. RUSSIAN)
vodna mimoza (40.19. MACEDONIAN)
vodna oreška (64.3. MACEDONIAN)
vodna paprat (4.1. MACEDONIAN)
vodní kaštan (10.2. CZECH)
vodnica (30.41a. SLOVAK)
vodnice (30.22. CZECH)
vodnice (30.41a. CZECH)
vodopija (27.10a. SERBO-CROATIAN)
vodový melón (38.3. SLOVAK)
vodunga (27.8. FON)
voem (40.42. MORISYEN)
voem (40.43. MORISYEN)
Vogel Pfeffer (62.2. GERMAN)
voikukat (27.45. FINNISH)
voikukka (27.45. FINNISH)
võilill (27.45. ESTONIAN)
voitonlaukka (7.14. FINNISH)
vojtěška (40.17. CZECH)
voksagurk (38.1. DANISH)
volantincillo (34.2. SPANISH)
volatín (34.2. SPANISH)
volchii bob (40.14. RUSSIAN)
volstruisslaai (22.1. AFRIKAANS)
vondio (45.4. SERER-SINE)
vông-p'oû (38.14. CHINESE, HAKKA)
vông-p'oû (38.16. CHINESE, HAKKA)
voodoo lily (8.2. ENGLISH)
vork en lepel (30.43. STELLINGWERFS)
vörös cikória (27.10b. HUNGARIAN)
vöröshagyma (7.3. HUNGARIAN)
vörösjagyma (7.3. HUNGARIAN)
vöröskáposzta (30.27b. HUNGARIAN)
vorraginon to pharmakeftico (29.1. GREEK)
vorsalat (67.2. ICELANDIC)
vortel (24.11. INDONESIAN)
vošni sok (38.14. BOSNIAN)
vošni sok (38.16. BOSNIAN)
vošni sok (38.17. BOSNIAN)
vošni sok (38.18. BOSNIAN)
vosočna tikvička (38.1. MACEDONIAN)
voú-tzè (11.1. CHINESE, HAKKA)
voú-tzè (11.2. CHINESE, HAKKA)
voú-tzè (11.3. CHINESE, HAKKA)
voú-tzè (11.4. CHINESE, HAKKA)
voú-tzè (11.5. CHINESE, HAKKA)
voú-tzè (11.6. CHINESE, HAKKA)
voú-tzè (11.7. CHINESE, HAKKA)
voú-tzè (11.8. CHINESE, HAKKA)
vowa (23.3. VENDA)
vpe-cvmpv (17.5. MUSKOGEE)
vraski e lachanodis (30.26. GREEK)
vrassiki (kramvi) e goggylodis (30.30. GREEK)
vrihikanchana (40.12. SANSKRIT)
vrinthaak (62.19. SANSKRIT)
vrischikali (66.3. SANSKRIT)
vrishchikapriya (28.1. SANSKRIT)
vrishya (40.18. SANSKRIT)
vrithabeeja (40.2. SANSKRIT)
vritthabeeja (45.1. SANSKRIT)
vritthaphala (38.3. SANSKRIT)
vritthaphala (62.19. SANSKRIT)
vroeg barbarakruid (30.2. DUTCH)
vroeg barbarakruid (30.2. VLAAMS)
vrtna (35.1. SERBO-CROATIAN)
vrtna kislaca (58.3. SLOVENIAN)
vrtna loboda (35.1. SLOVENIAN)
vrtni bob (40.34a. SLOVENIAN)
vudi (16.1a. BUGHOTU)
vudi (16.1a. FIJIAN)
vulli (7.3. TELUGU)
vulligaddalu (7.3. TELUGU)
vullipayalu (7.3. TELUGU)
vuononkaali (67.2. FINNISH)
vürtsbasiilik (44.5. ESTONIAN)
vusavyeley (35.3. TAMIL)
vusavyeley (35.9. TAMIL)
vusavyeley keerai (35.9. TAMIL)
vusovuso (38.22. FIJIAN)
vwende (38.3. NYANJA)
vyaaghra (40.18. SANSKRIT)
vyadhihanta (11.2. SANSKRIT)
vyakur (11.8. NEPALI)
vyanga (40.18. SANSKRIT)
vyo (17.1. JUNKUN TAKUM)
vzduš (11.2. CZECH)
wa (8.3. BURMESE)
wā (17.7. NEWAR)
wa la-phru (24.11. BALTI)
wa macaba (8.9. GBAYA)
wa ngagu (62.18. GBAYA)

waa' (35.9. NAVAJO)
waa baw (17.1. KAREN)
waaákén na Makka (40.2. HAUSA)
waa-chuu (17.1. KAREN)
waa-gler (17.2. KAREN)
waa-kahlue-pho (17.2. KAREN)
wáákéé (40.25. HAUSA)
wáákee (40.43. HAUSA)
wáákée (40.43. HAUSA)
wáákéé ankwai (40.25. HAUSA)
wáákéé bauchi (40.25. HAUSA)
wáákéé bisa (40.25. HAUSA)
wáákén ankwai (40.3. HAUSA)
wáákén awaki (40.43. HAUSA)
wáákén baibààyíí (40.9. HAUSA)
wáákén bareya (40.43. HAUSA)
wáákén bísáá (40.3. HAUSA)
wáákén daamánáá (40.43. HAUSA)
wáákén damfami (40.43. HAUSA)
wáákén dànfámíí (40.2. HAUSA)
wáákén dànfàmíí (40.9. HAUSA)
wáákén dàrníí (40.9. HAUSA)
wááken gizo (40.43. HAUSA)
wáákén gwààrí (40.9. HAUSA)
wáákén Másàr (40.2. HAUSA)
wáákén Màsàr (40.9. HAUSA)
wáákén Sàntànbûl (40.2. HAUSA)
wáákén Sàntànbûl (40.9. HAUSA)
wáákén tàntábàráá (40.2. HAUSA)
wáákén tàntábàráá (40.9. HAUSA)
wáákén tuuraawáá (40.2. HAUSA)
wáákén tuuraawaa (40.9. HAUSA)
wáákén yán máátáá (40.2. HAUSA)
wáákén yán-máátáá (40.9. HAUSA)
waa-khlue (17.2. KAREN)
waa-khru (17.2. KAREN)
waa-klu (17.2. KAREN)
waa-kwaa (17.2. KAREN)
waambar (20.6. CHECHEN)
waa-mee (17.2. KAREN)
wàa-phru (24.11. BALTI)
waata-kirihi (30.52. MAORI)
waatlemoen (38.3. AFRIKAANS)
waáwán-goónaá (38.16. HAUSA)
wabasi'binik (6.2. POTAWATOMI)
wabem (62.19. AKAN-ASANTE)
wabo (17.2. BURMESE)
waccuure (10.1. FULFULDE, ADAMAWA)
Wachskürbis (38.1. GERMAN)
wad gourd (38.1. ENGLISH)
wade rawan (40.25. PANJABI, EASTERN)
wadi (30.56. KWÉYÒL)
waerut (15.2. JAVANESE)
wagmeza (17.7. DAKOTA)
wagmu (38.14. DAKOTA)
wagmu (38.16. DAKOTA)
wagmu (38.17. DAKOTA)
wagmu (38.18. DAKOTA)
wãgun (62.19. KASHMIRI)
wahlea (28.1. MARATHI)
wa'hnakapizi' (30.17. DAKOTA)
waȟpeiŋkpajiji (27.31. LAKOTA)
waȟpeiŋkpajiji (27.32. LAKOTA)
waȟpetaŋka (30.27a. DAKOTA)
waȟpeyutapi (27.32. DAKOTA)
wahpeyutapi (30.27a. LAKOTA)
wahuwapa (17.7. DAKOTA)
wahyakhahon (62.9. MOHAWK)
wai (8.4. KISSI, NORTHERN)
waigōre (38.14. FULFULDE, ADAMAWA)
waigōre (38.16. FULFULDE, ADAMAWA)
waigōre (38.17. FULFULDE, ADAMAWA)
waigōre (38.18. FULFULDE, ADAMAWA)
waijoré (38.14. FULFULDE)
waijoré (38.16. FULFULDE)
waijoré (38.17. FULFULDE)
waijoré (38.18. FULFULDE)
wait-yam (11.7. KRIO)
waka (16.1a. SUMO-MAYANGNA)
wâkâs (16.1a. CREE, PLAINS)
wakay (36.3. IVATAN)
wāke (40.26. HAUSA)
waké (40.43. HAUSA)
wakegi (7.9b. JAPANESE)
waken Turawa (40.28. HAUSA)
waki (16.1a. SUMO-MAYANGNA)
wakmaheza (17.7. DAKOTA)
wakmu (38.14. DAKOTA)
wakmu (38.16. DAKOTA)
wakmu (38.17. DAKOTA)
wakmu (38.18. DAKOTA)
wakourouhou (11.3. SENOUFO, CEBAARA)
wąkrotka zwyczajna (24.7. POLISH)
wal (40.9. GUJARATI)
wal (40.9. MARATHI)
walalto (32.1. AFAR)
walati pälak (35.9. PANJABI, EASTERN)
waleriana (27.21. POLISH)
walia bocio (23.3. SONGHAY)
wall (40.9. HINDI)

wallrocket (30.47. ENGLISH)
waloh (38.17. JAVANESE)
walub kenti (38.20. JAVANESE)
waluh (38.15. INDONESIAN)
waluh (38.17. INDONESIAN)
waluh (38.17. JAVANESE)
waluh (38.17. SUNDA)
waluh bodas (38.20. SUNDA)
waluh deleg (38.20. JAVANESE)
waluh gambas (38.25. SUNDA)
waluh jepang (38.25. INDONESIAN)
waluh jepang (38.25. JAVANESE)
waluh panjang (38.20. SUNDA)
waluh siyem (38.25. INDONESIAN)
waluh siyem (38.25. JAVANESE)
waluh siyem (38.25. SUNDA)
walur (8.3. INDONESIAN)
walur (8.3. JAVANESE)
wambotu (62.19. SINHALA)
wamde (38.20. MÒORÉ)
wamgaso (43.1. MANDING-MANDINKA)
wa-mi (17.2. KAREN)
wamla (17.7. KABIYÉ)
wamnaheza (17.7. DAKOTA)
wamnu (38.14. DAKOTA)
wamnu (38.16. DAKOTA)
wamnu (38.17. DAKOTA)
wamnu (38.18. DAKOTA)
wamnuha (38.20. DAKOTA)
wamu (17.2. BURMESE)
wa-mu (17.2. KAREN)
wan (38.14. BALTI)
wan (38.16. BALTI)
wan (38.17. BALTI)
wan (38.18. BALTI)
wan dou (40.28. CHINESE, GÀN)
wan dou (40.28. CHINESE, MANDARIN)
wan dou (40.28. CHINESE, WÚ)
wan dou (40.28. CHINESE, YUE)
wan dou shu (40.11. CHINESE, MANDARIN)
wan dou shu (40.11. CHINESE, WÚ)
wan tou (40.28. CHINESE, MANDARIN)
wandoe (40.2. DUTCH)
wandu (40.28. KOREAN)
wanet (17.1. BURMESE)
wang gua (38.13a. CHINESE, WÚ)
wang gua (38.29. CHINESE, MANDARIN)
wang gua (38.30. CHINESE, WÚ)
wang wen tian gua (38.7b. CHINESE, MANDARIN)
wang wen tian gua (38.7b. CHINESE, WÚ)
wangaso (43.1. MANDING-MANDINKA)
wani (17.2. BURMESE)
wan-prachim (11.2. THAI)
wans (17.1. GUJARATI)
wansui (24.9. MALAY)
wansuy (24.9. TAGALOG)
wapikoskatask (30.56. CREE, PLAINS)
wâpisk oskatask (24.18. CREE, PLAINS)
wâpiskaskosiy (7.3. CREE, PLAINS)
wâpiski oskatask (24.18. CREE, PLAINS)
wapyu (17.2. BURMESE)
wara karis (30.52. TOK PISIN)
warabi (2.1. JAPANESE)
warrigal cabbage (22.2. ENGLISH)
waru (62.24. GIKUYU)
warzywo (7.8. POLISH)
wasabi (30.58. DANISH)
wasabi (30.58. DUTCH)
wasabi (30.58. ENGLISH)
wasabi (30.58. FRENCH)
wasabi (30.58. JAPANESE)
wasabi (30.58. PORTUGUESE)
wasabi (30.58. SPANISH)
wasabi (30.58. VLAAMS)
wasabi daikon (30.1. JAPANESE)
wasabia japońska (30.58. POLISH)
wasakûtcä'pûk (27.45. POTAWATOMI)
wasalde (7.3. PULAAR)
wasgagIn (62.2. POTAWATOMI)
wasgagIn (62.5. POTAWATOMI)
waskalebas (38.1. VLAAMS)
waskōre (45.1. FULFULDE, ADAMAWA)
waskuyecaśaśa (62.9. DAKOTA)
Wasserlilie (50.2. GERMAN)
Wassermelone (38.3. GERMAN)
Wassernuß (64.1. GERMAN)
Wassernuß (64.3. GERMAN)
Wasser-Rebendolde (24.17. GERMAN)
Wasseryam (11.1. GERMAN)
wasure-gusa (13.3. JAPANESE)
wasvadoek (38.22. AFRIKAANS)
wasvadoekkalbas (38.22. AFRIKAANS)
wata-bitas (27.45. KRIO)
watāɔ̃ (62.17. PANJABI, EASTERN)
watakrɛs (30.52. KRIO)
watalo-binch (40.43. KRIO)
watamɛlɔn (38.3. KRIO)
watani (40.28. GUJARATI)
watani (40.28. MARATHI)

wēnaangnié (8.3. DAGAARE, SOUTHERN)
wenewene (38.20. MAORI)
weng cai (36.2. CHINESE, MANDARIN)
weng cai (36.2. CHINESE, WÚ)
weng cai (36.2. CHINESE, YUE)
wĕnjooesŭgŭbŭn (30.41a. MICMAC)
wèpè (40.18. YORUBA)
werekpe (40.18. EDE NAGO)
werekpe (40.18. YORUBA)
werekpe (40.21. KRIO)
wɛrɛkpɛ (40.21. KRIO)
wèrepe (40.18. YORUBA)
werk-bemeda (38.4. AMHARIC)
werke (16.1a. OROMO, WEST CENTRAL)
wertel (24.11. JAVANESE)
wertol (24.11. JAVANESE)
Wes-Indiese agurkie (38.6. AFRIKAANS)
West Indian arrowroot (15.2. ENGLISH)
West Indian gherkin (38.6. ENGLISH)
western radish (30.56. ENGLISH)
westerse karmozijnbes (56.2. DUTCH)
West-Indische augurk (38.6. VLAAMS)
West-indische pijlwortel (15.2. VLAAMS)
Westindische-Gurke (38.6. GERMAN)
wetakolu (38.21. SINHALA
wètè (34.2. BAOULÉ)
wété wété (11.1. BAOULÉ)
wet-egusi (38.5. KRIO)
wet-ɛgusi (38.5. KRIO)
wet-yam (11.7. KRIO)
wewehan (18.1. INDONESIAN)
wewehan (18.2. INDONESIAN)
wężowata japońska (38.29. POLISH)
wężyk (7.12. POLISH)
whaangai (38.20. MAORI)
white cabbage (30.27a. ENGLISH)
white carrot (24.6. ENGLISH)
white flowered mustard (30.11. ENGLISH)
white gourd (38.1. ENGLISH)
white gourd melon (38.1. ENGLISH)
white Guinea yam (11.7. ENGLISH)
white Guinea yam (11.7. FILIPINO)
white mustard (30.57. ENGLISH)
white pumpkin (38.1. ENGLISH)
white pumpkin (38.20. ENGLISH)
white quinoa (35.6. ENGLISH)
white vetch (40.10. ENGLISH)
white-flowered gourd (38.20. ENGLISH)
white-seeded melon (38.5. ENGLISH)
wibab (58.1. KWÉYÒL)
wide petiole mustard (30.10. ENGLISH)
wiekschellen (40.26. STELLINGWERFS)
Wiesen-Bocksbart (27.47. GERMAN)
Wiesen-Schaumkraut (30.44. GERMAN)
wiesiołek dwuletni (51.1. POLISH)
wihcekaskosiy (7.3. CREE, PLAINS)
wiisch (16.1a. CAROLINIAN)
wiiwa (23.3. HOPI)
wijen (55.1. INDONESIAN)
wilayti-gawuth (40.17. HINDI)
wild asparagus (13.1. ENGLISH)
wild blite (23.3. ENGLISH)
wild cucumber (38.19. ENGLISH)
wild eggplant (62.16. ENGLISH)
wild leek (7.1. ENGLISH)
wild lettuce (27.33. ENGLISH)
wild look (7.1. DUTCH)
wild marjoram (44.6. ENGLISH)
wild pansey (68.1. ENGLISH)
wild pea eggplant (62.16. ENGLISH.)
wild spiny cucumber (38.6. ENGLISH)
wild violet (24.7. ENGLISH)
wild violet (68.1. ENGLISH)
wild water chestnut (10.3. ENGLISH)
Wild Yamswurzel (11.4. GERMAN)
wilde asperge (13.1. VLAAMS)
wilde aspersie (13.1. AFRIKAANS)
wilde basielkruid (44.4. AFRIKAANS)
wilde gesiggie (68.1. AFRIKAANS)
wilde komkommer (38.19. VLAAMS)
wilde marjolein (44.6. DUTCH)
wilde marjolein (44.6. VLAAMS)
wilde prei (7.1. DUTCH)
wilde-ertappel (23.3. AFRIKAANS)
wildegrenadella (23.3. AFRIKAANS)
wildenastagal (62.21. AFRIKAANS)
wilder Majoran (44.6. GERMAN)
wildes Stiefmuetterchen (68.1. GERMAN)
wilis (8.9. SUMO-MAYANGNA)
wimbiona (40.2. LOKO)
winbati (38.14. POTAWATOMI)
winbati (38.16. POTAWATOMI)
winbati (38.17. POTAWATOMI)
winbati (38.18. POTAWATOMI)
winged bean (40.29. ENGLISH)
winged pea (40.13. ENGLISH)
winged yam (11.1. ENGLISH)
winobluszcz (69.1. POLISH)
wintaa melon (38.11. JAPANESE)
winter cherry (62.12. ENGLISH)

xiang wan dou (40.10. CHINESE, WÚ)
xiao bai cai (30.36. CHINESE, MANDARIN)
xiao bai cai (30.36. CHINESE, YUE)
xiao bai cai ya zhong (30.36. CHINESE, WÚ)
xiao guo suan jiang (62.12. CHINESE, MANDARIN)
xiao hsiao yu ts'ai (30.36. CHINESE, MANDARIN)
xiao hua sui mi ji (30.44. CHINESE, WÚ)
xiao hua sui mi qi (30.44. CHINESE, MANDARIN)
xiao hui xiang (24.13. CHINESE, WÚ)
xiao luo bu (40.31. CHINESE, XIANG)
xiao mi jiao (62.5. CHINESE, MANDARIN)
xiao mi jiao (62.5. CHINESE, WÚ)
xiao pak choi (30.36. CHINESE, MANDARIN)
xiao quo suan jiang (62.12. CHINESE, WÚ)
xiao shu (11.6. CHINESE, MIN BEI)
xiao suan (7.12. CHINESE, MANDARIN)
xiao suan (7.12. CHINESE, WÚ)
xiao xi gua (38.2. CHINESE, MANDARIN)
xiao xi gua (38.2. CHINESE, WÚ)
xiao ye gai (30.7. CHINESE, MANDARIN)
xiao ye jie (30.7. CHINESE, WÚ)
xicoira amarga (27.10a. CATALAN-VALENCIAN-BALEAR)
xie (7.4. CHINESE, MANDARIN)
xie (7.4. CHINESE, WÚ)
xie bai (7.7. CHINESE, WÚ)
xie wie cai (24.3. CHINESE, YUE)
xikore (27.10a. BASQUE)
xin cai (23.3. CHINESE, GÀN)
xin dou jiao (40.41. CHINESE, GÀN)
xing cai (23.3. CHINESE, MIN BEI)
xinxebra (20.6. GALICIAN)
xirivia (24.18. CATALAN-VALENCIAN-BALEAR)
xiyar (38.13a. AZERBAIJANI, NORTHERN)
xorbaz (38.3. CHECHEN)
xòs-sor (38.20. BALTI)
xox (7.3. CHECHEN)
xrika (17.6. MALTESE)
xuan cao (13.3. CHINESE, WÚ)
xue cai (30.6. CHINESE, YUE)
xue li hong (30.6. CHINESE, WÚ)
xue li hong (30.6. CHINESE, YUE)
xu-xu (38.25. VIETNAMESE)
xwb kuab (38.21. HMONG-DAW)
xwb kuab (38.22. HMONG-DAW)
xyong (17.2. HMONG-DAW)
ya (40.1. KISSI, NORTHERN)
ya (40.1. KONO)
ya baw (17.2. AKHA)
ya cai (40.19. CHINESE, MANDARIN)
ya cai (40.19. CHINESE, WÚ)
ya cong (27.38. CHINESE, WÚ)
ya er qin (24.10. CHINESE, MANDARIN)
ya er qin (24.10. CHINESE, WÚ)
yä ƒäräñğ dubba (38.13a. AMHARIC)
ya lai jia li ya (24.6. CHINESE, MANDARIN)
ya lai jia li ya (24.6. CHINESE, WÚ)
ya she cao (18.2. CHINESE, WÚ)
ya shu (17.1. AKHA)
ya ts'ung (27.38. CHINESE, MANDARIN)
ya˄ yi¯ shi_ (62.24. LAHU)
ya yu (8.9. CHINESE, MANDARIN)
ya yu (8.9. CHINESE, WÚ)
yaa kho on (27.14. THAI)
yaaknonkaawaa' (40.1. CHICKASAW)
yaakon (27.39. JAPANESE)
yáála (40.1. KONO)
yáàpá (40.1. BEROM)
yàbaàsi (7.3. IGBO)
yäbahər mašəlla (17.7. AMHARIC)
yäbahəŕ qùlqwal (32.1. AMHARIC)
yäbahər qulqwal (38.7a. AMHARIC)
yàbákó (16.1a. NUPE-NUPE-TAKO)
yaban turpu (33.2. TURKISH)
yabani çeltik (17.8. TURKISH)
yabani enginar (27.15. TURKISH)
yabani havuç (24.18. TURKISH)
yabani hindiba (27.10a. TURKISH)
yabani ispanak (35.5. TURKISH)
yabani kuşkonmaz (13.1. TURKISH)
yabani şalgam (30.22. TURKISH)
yabani sarmisak (7.2. TURKISH)
yabani şeker havucu (24.23. TURKISH)
yabani su kestanesi (10.3. TURKISH)
yábas (7.3. BULLOM SO)
yabas (7.3. KRIO)
yabas (7.3. MENDE)
yabasi (7.3. IGBO)
yabásị (7.3. IZON)
yabasi (7.3. KRIO)
yabasi (7.3. MENDE)
yabasi-laa (7.2. MENDE)
yabatian (45.4. WAMEY)
yabɛrɛ (8.4. MALIMBA)
yabéré (8.4. MANDINKA)
yabɛre (8.4. MANINKAKAN, EASTERN)

yɔɔ (40.43. GA)
yɔɔ (40.43. TWI)
yɔgɔtɔ (23.4. MENDE)
yɔkhɔmi (20.6. SUSU)
yɔ̀ŋ (11.2. BEROM)
yɔ-*n*gengeyo (63.1. KISSI, NORTHERN)
ydrocharis (36.2. GREEK)
ydrokotyli asiatiki (24.7. GREEK)
ydrokotyli tou chortotapitos (24.14. GREEK)
ydropepon (38.3. GREEK)
yɛ (11.7. DANGME)
ye (38.18. MANDING-MANDINKA)
ye cai (30.27a. CHINESE, YUE)
ye cai hua (30.26. CHINESE, YUE)
ye ci gu (6.1. CHINESE, WÚ)
ye ci qu (6.1. CHINESE, MANDARIN)
ye hai jiao (62.21. CHINESE, WÚ)
ye hu lu ba (40.34a. CHINESE, MANDARIN)
ye jiang (20.5. CHINESE, YUE)
ye ju (67.2. CHINESE, MANDARIN)
ye ju (67.2. CHINESE, WÚ)
ye kai hua (38.20. CHINESE, GÀN)
ye kai hua (38.20. CHINESE, WÚ)
ye ku jii (27.10a. CHINESE, WÚ)
ye ku qu (27.10a. CHINESE, MANDARIN)
ye lai xiang (51.1. CHINESE, YUE)
ye liu jiu (30.18. CHINESE, WÚ)
ye lu sun (13.1. CHINESE, WÚ)
ye she gua (38.31. CHINESE, MANDARIN)
ye she gua (38.31. CHINESE, WÚ)
ye tian cai (35.3. CHINESE, WÚ)
ye wo ju (27.32. CHINESE, MANDARIN)
ye wo ju (27.33. CHINESE, WÚ)
ye yong jie cai (30.7. CHINESE, YUE)
ye yong jie cai (30.16. CHINESE, YUE)
yebese (62.1. GEN)
yɛ̀ɛ́ (38.18. BEROM)
yɛɛku (62.5. WAAMA)
yeg (39.4. BULLOM SO)
ye-gbɛ (38.18. MANINKAKAN, EASTERN)
yeh chü (67.2. CHINESE, MANDARIN)
yeḥḥ (38.7b. HARSŪSI)
yɛhwɛshi (62.14a. AJA)
yehwlen (44.10. FON)
yeiŋgei (63.1. MENDE)
yek (39.4. BULLOM SO)
yeke (39.4. BULLOM SO)
yɛkɛ-a-potho (39.4. BULLOM SO)
yɛkɛ-lɛ (39.4. BULLOM SO)
yekola-kulkual (32.1. AMHARIC)
yɛku boo (62.1. BAATONUM)
yɛku yaki (62.5. BAATONUM)
ye-kumba (38.18. MANINKAKAN, EASTERN)
yɛlɛ (11.7. GA)
yele kosu (30.27a. KANNADA)
yella gadda (11.6. TELUGU)
yellagadda (11.2. TELUGU)
yellow dhal (40.2. ENGLISH)
yellow ginger (20.2. ENGLISH)
yellow Guinea yam (11.3. ENGLISH)
yellow mustard (30.57. ENGLISH)
yellow ocumo (8.9. ENGLISH)
yellow velvetleaf (14.1. ENGLISH)
yellow weed (27.22. ENGLISH)
yellow yam (11.3. ENGLISH)
yellow yautia (8.9. ENGLISH)
yen zhi cai (45.4. CHINESE, YUE)
yenebilir diplazyum (5.1. TURKISH)
yenen yahudi kirazi (62.12. TURKISH)
yeŋgei (63.1. MENDE)
yeŋgelesi (38.18. LOKO)
yengi (62.9. NUPE-NUPE-TAKO)
Yeni Zelanda ispanaği (22.2. TURKISH)
yenilen yam (11.5. TURKISH)
yenkisa (36.3. MALIMBA)
yenkisa ka (36.3. MANDING-MANDINKA)
yɛntɛ (38.3. BULLOM SO)
yɛntɛ (38.5. BULLOM SO)
yɛnthɛ-pokan-dɛ (55.1. BULLOM SO)
ye-nyɛ̃ (38.18. MANINKAKAN, EASTERN)
yer baddemi (10.1. TURKISH)
yer bezelyesi (40.39. TURKISH)
yer elmasi (27.26. TURKISH)
yer fistiği (40.1. TURKISH)
yer fistiği (40.15. TURKISH)
yer lahanasi (30.22. TURKISH)
ýeralma (62.24. TURKMEN)
yerba mora (62.15. SPANISH)
yeregh (17.5. TIV)
yerelmasi (27.26. TURKISH)
yèrepe (40.18. YORUBA)
yerf ındığı (40.1. AZERBAIJANI, NORTHERN)
yerfistiği (40.1. TURKISH)
yerkökü (24.11. AZERBAIJANI, NORTHERN)
yerpənək (38.13a. AZERBAIJANI, NORTHERN)
yerra gogu (45.4. TELUGU)
yerra tamma (40.4. TELUGU)

ysmanak (35.9. TURKMEN)
ysplant (22.1. AFRIKAANS)
yu (8.4. CHINESE, GÀN)
yu (8.4. CHINESE, MANDARIN)
yu (8.4. CHINESE, WÚ)
yu dou (40.11. CHINESE, WÚ)
yu dou (40.25. CHINESE, GÀN)
yu dou (40.25. CHINESE, MIN BEI)
yu dou (40.26. CHINESE, YUE)
yu gu zi (17.7. CHINESE, XIANG)
yu jin (20.2. CHINESE, YUE)
yü kwa (38.16. CHINESE, MANDARIN)
yu mi (17.7. CHINESE, MANDARIN)
yu mi (17.7. CHINESE, WÚ)
yu mi (17.7. CHINESE, YUE)
yu nai (8.4. CHINESE, WÚ)
yu qian bo cai (35.1. CHINESE, WÚ)
yu tou (7.13. CHINESE, YUE)
yü tou (8.4. CHINESE, MANDARIN)
yu tou (8.4. CHINESE, MIN BEI)
yu tou (8.4. CHINESE, WÚ)
yu tou (8.4. CHINESE, YUE)
yu yi gan lan (30.23. CHINESE, WÚ)
yu yi gan lan (30.35. CHINESE, MANDARIN)
yu yi gan lan (30.35. CHINESE, WÚ)
yu yi gan lan (30.35. CHINESE, YUE)
yu zi (8.4. CHINESE, XIANG)
yuan cong (7.3. CHINESE, XIANG)
yuan dem qui (24.1. CHINESE, YUE)
yuan gen (30.22. CHINESE, WÚ)
yuan gen (30.41a. CHINESE, WÚ)
yuan jiao (62.1. CHINESE, YUE)
yuan jiao ling (64.3. CHINESE, WÚ)
yuan xing shu yu (11.7. CHINESE, MANDARIN)
yuan xing shu yu (11.7. CHINESE, WÚ)
yuan ye jin kui (45.5. CHINESE, MANDARIN)
yuan ye jin kui (45.5. CHINESE, WÚ)
yuan ye juan mo (58.4. CHINESE, MANDARIN)
yuan ye suan mu (58.4. CHINESE, WÚ)
yuan zhui hjua tu ren shen (59.4. CHINESE, YUE)
yuca (39.4. ENGLISH)
yuca (39.4. SPANISH)
yuco de bejuco (40.21. SPANISH)
yue gua (38.9. CHINESE, MANDARIN)
yue gua (38.9. CHINESE, WÚ)
yue gua (38.9. CHINESE, YUE)
yue jian cao (51.1. CHINESE, WÚ)
yue jian cao (51.1. CHINESE, YUE)
yue si gua (38.21. CHINESE, YUE)
yugao (38.20. JAPANESE)
yuka (39.4. TURKISH)
yukɛnɛ (36.3. KONO)
yü-kwa (38.14. CHINESE, MANDARIN)
yulonguga (44.5. MÒORÉ)
yulyungu (44.5. MÒORÉ)
yumɛn (9.1. BULLOM SO)
yumrulu ekşi üçgül (53.1. TURKISH)
yumrulu frenk maydanozu (24.8. TURKISH)
yumrulu latin çiçeği (65.2. TURKISH)
yumrulu mürdümük (40.11. TURKISH)
yumrulu yam (11.2. TURKISH)
yun dou (40.26. CHINESE, GÀN)
yun dou (40.26. CHINESE, MIN BEI)
yun dou (40.26. CHINESE, WÚ)
yung (20.2. BALTI)
yuquilla (15.2. QUECHUA, CUSCO)
yurine (13.4. JAPANESE)
yuro blé (40.39. MANDING-MANDINKA)
yute (63.1. SPANISH)
yute de fruto alargado (63.1. SPANISH)
yute de hojas comestibles (63.1. SPANISH)
yuvarlak-küçük yaprakli (45.5. TURKISH)
yvlonkv-cate (35.4. MUSKOGEE)
yvlonkv-catuce (30.56. MUSKOGEE)
za (42.1. CHINESE, WÚ)
za baw (17.2. AKHA)
za cui (17.2. AKHA)
za ju (17.1. AKHA)
za khao (17.1. LAHU)
za pi-euh (17.2. AKHA)
za shaw (17.1. AKHA)
zaailathyrus (40.10. DUTCH)
zaákiíbánzaá zarangade (23.3. HAUSA)
zaandknollen (30.41a. STELLINGWERFS)
zabache (40.25. TAGALOG)
ẕabbar (32.1. HEBREW)
zabibi (20.2. HAUSA)
zacharna carevica (17.7. BULGARIAN)
zaenil (27.23. PANJABI, EASTERN)
zafferano delle Indie (20.2. ITALIAN)
zagaĩ zagagbé (27.8. BAOULÉ)
zaher (30.26. ARABIC)
zaher el-mooz (16.1b. ARABIC)
zail (7.10. CAJUN FRENCH)
zail el-hosam (1.1. ARABIC)
za-jeh (17.2. AKHA)

Chapter 3. Latin binomials and synonyms[1]

Aalius androgynus Kuntze– see *Sauropus androgynus*
Aalius sumatranus Kuntze – see *Sauropus androgynus*
Abelmoschus bammia Webb. – see *A. esculentus*
Abelmoschus cruentus Walp. – see *Hibiscus sabdariffa*
***Abelmoschus esculentus* (L.) Moench (45.1)**
Abelmoschus esculentus Mey. – see *A. esculentus* (L.) Moench
Abelmoschus luzonensis Merrill – see *A. manihot*
Abelmoschus magnificus Wall. – see *A. manihot*
***Abelmoschus manihot* (L.) Medikus (45.2)**
Abelmoschus manihot (L.) Medikus subsp. *manihot* – see *A. manihot*
Abelmoschus mindanaensis Warb. – see *A. manihot*
Abelmoschus multiformis Wall – see *A. esculentus*
Abelmoschus warreensis Dalz. – see *A. manihot*
Acacia lacustris Desf. – see *Neptunia prostrata*
Acanthopanax eleutherococcus Makino – see *Eleutherococcus seoulensis*
Acanthopanax senticosus (Rupr. & Maxim.) Harms – see *Eleutherococcus seoulensis*
Acetosa agrestis Rafin. – see *Rumex acetosa*
Acetosa amplexicaulis Rafin. – see *Rumex acetosa*
Acetosa angustata Rafin. – see *Rumex acetosa*
Acetosa bidentula Rafin. – see *Rumex acetosa*
Acetosa hastifolia Schur. – see *Rumex acetosa*
Acetosa hastulata Rafin. – see *Rumex acetosa*
Acetosa hortensis Garsault – see *Rumex scutatus*
Acetosa magna Gilib. – see *Rumex acetosa*
Acetosa officinalis Gueldenst. ex Ledeb. – see *Rumex acetosa*
Acetosa olitoria Rafin. – see *Rumex acetosa*
Acetosa patientia (L.) Gomez de la Maza – see *Rumex patientia*
Acetosa pratensis Garsault – see *Rumex acetosa*
Acetosa pratensis Mill. – see *Rumex acetosa*
Acetosa scutata (L.) Mill. – see *Rumex scutatus*
Acetosa scutata subsp. *hastifolia* (Bieb.) A. Löve & Kapoor – see *Rumex scutatus*
Acetosa subalpina Schur – see *Rumex acetosa*
Acetosella crassicaulis (Zucc.) Kuntze – see *Oxalis tuberosa*
Acetosella crenata (Jacq.) Kuntze – see *Oxalis tuberosa*
Acetosella tuberosa (Molina) Kuntze – see *Oxalis tuberosa*
Achyranthes philoxeroides (Mart.) Standl. – see *Alternanthera philoxeroides*
Achyranthes sessilis (L.) Besser – see *Alternanthera sessilis*
Achyranthes triandra Roxb. – see *Alternanthera sessilis*

---

[1] There is often more than one scientific name (Latin binomial) used for a vegetable crop; these alternative names are considered synonyms. In some instances, the name that is currently used is not universally accepted. In addition, binomials are subject to change, especially with the increasing use of molecular genetics for assessing relatedness. As a consequence, synonyms can greatly complicate the identification of a species of interest. Since the common names are listed taxonomically in Chapter 1, having the correct scientific name is essential for finding a crop. This chapter lists synonyms for the scientific names of the vegetable crops in the text. The correct name is in bold, followed by a number that indicates the family and the specific crop within the family in Chapter 1 (e.g. 24.1. indicates the 24th family and the 1st crop). Each synonym is keyed back to the correct scientific name, which should be used to find the crop of interest.

Allium foliosum Clarion, ex DC. – see *A. schoenoprasum*
Allium glaucum hort. Par. ex Ledeb. – see *A. schoenoprasum*
Allium grayi Regel – see *A. macrostemon*
Allium grayi Regel var. *chanetii* (H. Lév.) H. Lév. – see *A. macrostemon*
Allium gredense Rivas – see *A. schoenoprasum*
Allium iatasen H. Lév. – see *A. macrostemon*
Allium idzuense Hara – see *A. schoenoprasum*
***Allium kurrat* Schweinf. ex K. Krause (7.6)**
Allium latissimum Prokhanov – see *A. victorialis*
Allium ledebourianum Schult. – see *A. schoenoprasum*
Allium ledebourianum var. *intermedium* Krylov – see *A. schoenoprasum*
Allium lusitanicum Link, ex Regel – see *A. schoenoprasum*
***Allium macrostemon* Bunge (7.7)**
Allium margaritaceum Moench – see *A. scorodoprasum*
Allium martini H. Lév. & Vaniot – see *A. chinense*
Allium microdictyon Prokh. – see *A. victorialis*
Allium montanum Schrank – see *A. schoenoprasum*
Allium montanum Schrank, non F.W. Schmidt – see *A. schoenoprasum*
Allium multitabulatum S. Cic. – see *A. × proliferum*
Allium neglectum Wender. – see *A. scorodoprasum*
Allium nereidum Hance – see *A. macrostemon*
Allium nipponicum Franch. & Sav. – see *A. macrostemon*
Allium odorum auct. non L. – see *A. tuberosum*
Allium odorum L. – see *A. tuberosum*
Allium oliganthum Kar. & Kir. – see *A. schoenoprasum*
Allium ouensanense Nakai – see *A. macrostemon*
Allium palustre Chaix – see *A. schoenoprasum*
Allium palustre Pourr. ex Lag. – see *A. schoenoprasum*
Allium pekinense Prokh. – see *A. sativum*
Allium plantagineum Lam. – see *A. victorialis*
***Allium porrum* L. (7.8)**
Allium potaninii Regel – see *A. tuberosum*
***Allium × proliferum* (Moench) Schrad. ex Willd (7.9)**
Allium punctulatum Schlecht. – see *A. schoenoprasum*
Allium purpurascens Losa. – see *A. schoenoprasum*
Allium raddeanum Regel – see *A. schoenoprasum*
Allium reflexum F.G. Dietr. – see *A. schoenoprasum*
Allium riparium Opiz – see *A. schoenoprasum*
Allium roxburghii Kunth – see *A. tuberosum*
Allium salota Dostál – see *A. cepa* var. *aggregatum*
***Allium sativum* L. (7.10)**
Allium sativum L. form *pekinense* Makino – see *A. sativum*
Allium sativum L. subsp. *asiae-mediae* Kazakova – see *A. sativum*
Allium sativum L. var. *controversum* (Schrad.) Regel – see *A. sativum*
Allium sativum L. var. *ophioscorodon* (Link) Doll – see *A. sativum*
Allium sativum L. var. *pekinense* F. Maek. – see *A. sativum*
Allium schoenoprasioides Fries, ex Nym. – see *A. schoenoprasum*
***Allium schoenoprasum* L. (7.11)**
Allium schoenoprasum L. subsp. *alpinum* (DC) Čelak – see *A. schoenoprasum*
Allium schoenoprasum L. var. *alpinum* DC. – see *A. schoenoprasum*

Allium schoenoprasum L. var. *folissum* Regel – see *A. schoenoprasum*
***Allium scorodoprasum* L. (7.12)**
Allium scorodoprasum L. var. *viviparum* Regel – see *A. scorodoprasum*
Allium senescens Miq. – see *A. tuberosum*
Allium senescens Miq. non L. – see *A. tuberosum*
Allium sibiricum auct. non L. – see *A. schoenoprasum*
Allium sibiricum L. – see *A. schoenoprasum*
Allium splendens Miq. – see *A. chinense*
Allium sulvia Buch.-Ham. ex G. Don – see *A. tuberosum*
Allium tartaricum L. f. – see *A. tuberosum*
Allium tenuifolium Salisb. – see *A. schoenoprasum*
Allium triquetrum Lour. – see *A. chinense*
***Allium tuberosum* Rottler, ex Sprengel (7.13)**
Allium tuberosum Roxb. – see *A. tuberosum* Rottler, ex Sprengel
Allium udinicum Antsup – see *A. schoenoprasum*
Allium uliginosum G. Don – see *A. tuberosum*
Allium uliginosum G. Don, non Ledeb. – see *A. tuberosum*
Allium uratense Franchet – see *A. macrostemon*
Allium utilissimum Baranov & Skvortz – see *A. macrostemon*
***Allium victorialis* L. (7.14)**
Allium victorialis L. subsp. *platyphyllum* Hultén – see *A. victorialis*
Allium wakegi Araki – see *A.* × *proliferum*
Allium yedoense Nakai – see *A. tuberosum*
Alocasia alba Schott – see *A. macrorrhizos*
Alocasia indica (Lour.) Spach – see *A. macrorrhizos*
Alocasia macrorrhiza (L.) Schott – see *A. macrorrhizos*
***Alocasia macrorrhizos* (L.) G. Don. (8.1)**
Alocasia odora (Roxb.) Koch – see *A. macrorrhizos*
***Alpinia galanga* (L.) Sw. (20.1)**
Alpinia galanga (L.) Willd. – see *A. galanga*
Alternanthera achyranthoides Forsk. – see *A. sessilis*
Alternanthera glabra Moq. – see *A. sessilis* (L.) R. Br.
Alternanthera nodiflora R. Br. – see *A. sessilis*
***Alternanthera philoxeroides* (C. Martius) Griseb. (23.1)**
Alternanthera repens Gmel. – see *A. sessilis*
***Alternanthera sessilis* (L.) R. Br. (23.2)**
Alternanthera triandra Lam. – see *A. sessilis*
Althaea verticillata Alef. – see *Malva verticillata*
Amaranthus abyssinicus hort. – see *A. caudatus*
Amaranthus abyssinicus hort. ex L.H. Bailey – see *A. caudatus*
Amaranthus alopecurus Hochst. ex A. Br. & Bouche – see *A. caudatus*
Amaranthus ascendens Loisel. – *A. blitum* subsp. *oleraceus*
Amaranthus ascendens var. *oleraceus* (L.) Priszter – see *A. blitum* subsp. *oleraceus*
Amaranthus blitum L. – see *A. blitum* subsp. *oleraceus*
Amaranthus blitum L. cv. group Oleraceus – see *A. blitum* subsp. *oleraceus*
***Amaranthus blitum* L. subsp. *oleraceus* (L.) Costea (23.3)**
Amaranthus blitum L. var. *oleraceus* (L.) Hook. f. – see *A. blitum* subsp. *oleraceus*
Amaranthus blitum var. *oleraceus* (L.) Hook – see *A. blitum* subsp. *oleraceus*
Amaranthus caudatus auct. nonnull. Afr. – see *A. cruentus*
***Amaranthus caudatus* L. (23.3)**

Amaranthus caudatus L. subsp. *mantegazzianus* (Pass.) – see *A. caudatus*
Amaranthus caudatus L. subsp. *saueri* Jehlik – see *A. caudatus*
Amaranthus caudatus L. var. *alopecurus* Moq. – see *A. caudatus*
Amaranthus chlorostachys Willd. – see *A. hybridus*
***Amaranthus cruentus* L. (23.3)**
Amaranthus cruentus Willd., non hort. – see *A. caudatus*
Amaranthus cruentus Willd., non L. – see *A. caudatus*
Amaranthus delilei Richter & Loret – see *A. retroflexus*
***Amaranthus dubius* C. Martius, ex Thell. (23.3)**
Amaranthus edulis Speg. – see *A. caudatus*
Amaranthus frumentaceus Ham, non Buch. – see *A. caudatus*
Amaranthus gangeticus L. – see *A. tricolor*
Amaranthus gracilis Desf. – see *A. viridis*
Amaranthus hybridus 'chlorostachys' – see *A. hybridus*
***Amaranthus hybridus* L. (23.3)**
Amaranthus hybridus L. subsp. *cruentus* (L.) Thell. – see *A. cruentus*
Amaranthus hybridus L. subsp. *cruentus* Thell. var. *paniculatus* Thell. – see *A. hybridus*
Amaranthus hybridus L. var. *paniculatus* (L.) Thell. – see *A. cruentus*
Amaranthus hybridus 'paniculatus' – see *A. cruentus*
Amaranthus hybridus subsp. *cruentus* (L.) Thell. – see *A. caudatus*
Amaranthus inamoenus Willd. – see *A. mangostanus*
Amaranthus incurvatus Tim. ex Gren. & Godr. – see *A. hybridus*
Amaranthus leucocarpus S. Wats. – see *A. spinosus*.
Amaranthus lividus L. – see *A. blitum* subsp. *oleraceus*
***Amaranthus mangostanus* L. (23.3)**
Amaranthus mantegazzianus Pass. – see *A. caudatus*
Amaranthus maximus Mill. – see *A. caudatus*
Amaranthus melancholicus L. – see *A. tricolor*
Amaranthus oleraceus L. – see *A. blitum* subsp. *oleraceus*
Amaranthus paniculatus L. – see *A. cruentus*
Amaranthus paniculatus Wall., non L. – see *A. caudatus*
Amaranthus patulus auct. nonnull. – see *A. retroflexus*
Amaranthus patulus Bertol. – see *A. hybridus*
Amaranthus polystachyus Blume – see *A. viridis*
Amaranthus polystachyus Willd. – see *A. viridis*
***Amaranthus retroflexus* L. (23.3)**
Amaranthus salicifolius Veitch – see *A. tricolor*
Amaranthus sanguineus L. – see *A. cruentus*
Amaranthus speciosus Sims – see *A. cruentus*
***Amaranthus spinosus* L. (23.3)**
***Amaranthus* spp. (23.3)**
***Amaranthus sylvestris* Desf. nom. nud. (23.3)**
***Amaranthus tricolor* L. (23.3)**
Amaranthus tricolor L. subsp. *tricolor* – see *A. tricolor*
Amaranthus tricolor subsp. *mangostanus* (L.) Aell. – see *A. mangostanus*
Amaranthus tristis L. – see *A. tricolor*
Amaranthus tristis Moq. – see *A. dubius*
Amaranthus tristis Roxb. – see *A. mangostanus*
***Amaranthus viridis* L. (23.3)**
Amatula flava Medik. – see *Lycopersicon esculentum* var. *esculentum*

Amatula rubra Medik. – see *Lycopersicon esculentum* var. *esculentum*
Amomum galanga (L) Lour. – see *Alpinia galanga*
Amomum latifolium Lam. – see *Curcuma zedoaria*
Amomum medium Lour. – see *Alpinia galanga*
Amomum mioga Thunb. – see *Zingiber mioga*
Amomum zedoaria Bergius – see *Curcuma zedoaria*
Amomum zedoaria Christmann – see *Curcuma zedoaria*
Amomum zerumbeth Koenig – see *Curcuma zedoaria*
Amomum zingiber L. – see *Zingiber officinale*
Amorphophallus campanulatus Decne. – see *A. paeoniifolius* var. *campanulatus*
Amorphophallus campanulatus (Roxb.) Blume, ex Decne. – see *A. paeoniifolius* var. *campanulatus*
Amorphophallus chatty André – see *A. paeoniifolius* var. *campanulatus*
Amorphophallus gigantiflorus Hayata – see *A. paeoniifolius* var. *campanulatus*
***Amorphophallus konjac* Koch. (8.2)**
Amorphophallus paeonifolius (Dennst.) Nicolson – see *A. paeoniifolius* var. *campanulatus*
***Amorphophallus paeoniifolius* (Dennst.) Nicolson var. *campanulatus* (Decne.) Sivad. (8.3)**
Amorphophallus rex Prain – see *A. paeoniifolius* var. *campanulatus*
Amorphophallus rivieri Dorieu, ex Carriere – see *A. konjac*
Amorphophallus sativus Blume – see *A. paeoniifolius* var. *campanulatus*
Amorphophallus virosus N.E. Br. – see *A. paeoniifolius* var. *campanulatus*
Ampelosicyos scandens Thou. – see *Telfairia pedata*
Andropogon dulce Burm.f. – see *Eleocharis dulcis*
Anethum dulce DC. – see *Foeniculum vulgare* subsp. *vulgare* var. *dulce*
Anethum foeniculum L. var. *azoricum* Schkuhr – see *Foeniculum vulgare* subsp. *vulgare* var. *azoricum*
Anethum foeniculum L. var. *dulce* Schkuhr – see *Foeniculum vulgare* subsp. *vulgare* var. *dulce*
Anethum foeniculum romanum Hoffm. – see *Foeniculum vulgare* subsp. *vulgare* var. *dulce*
***Angelica archangelica* L. (24.1)**
***Angelica keiskei* (Miq.) Koidz. (24.2)**
Angelica litoralis Fr. – see *A. archangelica*
Angelica major Gilib. – see *A. archangelica*
Angelica officinalis Moench – see *A. archangelica*
Angelica utilis Makino, ex Yabe – see *A. keiskei*
Anguina cucumerina Kuntze – see *Trichosanthes cucumerina*
Anguina sinensis Mill. – see *Trichosanthes cucumerina*
Anguria citrullus (L.) Mill. – see *Citrullus lanatus* var. *lunatus*
Anguria pedatisecta Nees & Mart. – see *Cyclanthera pedata*
Anisogonium esculentum (Retz.) Presl – see *Diplazium esculentum*
Anneslea spinosa Andr. – see *Euryale ferox*
Anserina bonus-henricus (L.) Dumort. – see *Chenopodium bonus-henricus*
Anthemis artemisiifolia Willd. – see *Chrysanthemum × morifolium*
Anthemis coronaria hort. – see *Chrysanthemum coronarium*
***Anthriscus cerefolium* (L.) Hoffm. (24.3)**
Anthriscus chaerophyllus St.-Lag. – see *A. cerefolium*
Apium celeri Gaertn. – see *A. graveolens* var. *dulce*
Apium crispum Mill. – see *Petroselinum crispum* var. *crispum*
Apium dulce Mill. – see *A. graveolens* var. *dulce*
Apium graveolens L. – see *A. graveolens* var. *dulce*
Apium graveolens L. *sativum* Gaudin – see *A. graveolens* var. *dulce*
Apium graveolens L. var. *dulce* (Mill.) DC. – see *A. graveolens* L. var. *dulce* (Mill.) Pers.
***Apium graveolens* var. *dulce* (Mill.) Pers. (24.4)**

Apium graveolens L. var.-gr. *rapaceum* Alef. – see *A. graveolens* var. *rapaceum*
Apium graveolens L. var. *rapaceum* DC. – see *A. graveolens* L. var. *rapaceum* (Mill.) Gaud-Beaup.
***Apium graveolens* var. *rapaceum* (Mill.) Gaudin (24.5)**
Apium latifolium Mill. – see *Petroselinum crispum* var. *tuberosum*
Apium lusitanicum Miller – see *A. graveolens* var. *dulce*
Apium petroselinum L. – see *Petroselinum crispum* var. *crispum*
Apium petroselinum L. var. *latifolium* Hayne – see *Petroselinum crispum* var. *tuberosum*
Apium petroselinum L. var. *tuberosum* Noisette – see *Petroselinum crispum* var. *tuberosum*
Apium petroselinum L. var. *vulgare* Noisette – see *Petroselinum crispum* var. *neapolitanum*
Apium rapaceum Mill. – see *A. graveolens* var. *rapaceum*
Apium sisarum Calest. – see *Sium sisarum*
Apium tuberosum Bernh. ex Rchb. – see *Petroselinum crispum* var. *tuberosum*
***Aponogeton distachyos* L. f. (25.1)**
Aponogeton distachyos Thunb. – see *A. distachyos* L. f.
Aponogeton distachys Edgew. – see *A. distachyos* L. f.
Aponogeton distachyum Thunb. – see *A. distachyos*
Aponogeton lagrangei Ed. Andre – see *A. distachyos*
Arabis chinensis Rottl. ex Wight – see *Lepidium sativum*
Arabis nasturtium [Clairv.] – see *Nasturtium officinale*
Arabis sylvestris Scop. – see *Diplotaxis muralis*
Arachis africana Burm. f. – see *Vigna subterranea*
***Arachis hypogaea* L. (40.1)**
Arachnida hypogaea (L.) Moench – see *A. hypogaea* L.
Arachnida quadrifolia Trew – see *A. hypogaea*
Aralia canescens Siebold & Zucc. – see *A. elata*
Aralia chinensis var. *canescens* (Siebold & Zucc.) Koehne – see *A. elata*
Aralia chinensis var. *mandshurica* (Rupr. & Maxim.) Rehd. – see *A. elata*
***Aralia continentalis* Kitag. (26.1)**
***Aralia cordata* Thunb. (26.2)**
Aralia edulis Siebold & Zucc. – see *A. cordata*
Aralia elata (Miq.) Seeman – see *A. elata* Seeman
***Aralia elata* Seeman (26.3)**
Aralia mandshurica Rupr. & Maxim. – see *A. elata*
Aralia nudicaulis Blume– see *A. cordata*
Aralia schmidtii Pojark. – see *A. cordata*
Aralia spinosa L. var. *canescens* Franch. & Sav. – see *A. elata*
Aralia spinosa L. var. *glabrescens* – see *A. elata*
Archangelica archangelica H. Karst. – see *Angelica archangelica*
Archangelica keiskei Miq. – see *Angelica keiskei*
Archangelica officinalis (Moench) Hoffm – see *Angelica archangelica*
Arctium adhaereus Gilib. – see *A. lappa*
Arctium intermedium Bab. – see *A. lappa*
***Arctium lappa* L. (27.5)**
Arctium majus Bernh. – see *Arctium lappa*
Arctium ruderale Salisb. – see *A. lappa*
Arisacontis chamissonis Schott – see *Cyrtosperma merkusii*
Armoracia lapathifolia Gilib. – see *A. rusticana*
Armoracia lapathifolia Usteri. – see *A. rusticana*
***Armoracia rusticana* P. Gaertn., B. Mey., Scherb. (30.1)**
Armoracia sativa Bernh. – see *A. rusticana*

Arnica tussilaginea Burm. f. – see *Farfugium japonicum*
Arnoglossum subulatum S.F. Gray – see *Plantago coronopus* var. *sativa*
Arracacia esculenta DC. – see *A. xanthorrhiza*
***Arracacia xanthorrhiza* Bancroft (24.6)**
***Artemisia dracunculus* L. (27.6)**
Artemisia dracunculus L. var. *indica* Maxim. – see *A. indica* var. *maximowiczii*
Artemisia indica Willd. – see *A. indica* var. *maximowiczii*
***Artemisia indica* Willd. var. *maximowiczii* (Nakai) H. Hara (27.7)**
Artemisia inodora Willd. – see *A. dracunculus*
Artemisia princes Pamp. – see *A. indica* var. *maximowiczii*
Artemisia redowskii Ledeb. – see *A. dracunculus*
Arum campanulatum Roxb. – see *Amorphophallus paeoniifolius* var. *campanulatus*
Arum chinense L. – see *Colocasia esculenta*
Arum colocasia L. – see *Colocasia esculenta*
Arum esculentum L. – see *Colocasia esculenta*
Arum indicum Lour. – see *Alocasia macrorrhizos*
Arum lividum Salisb. – see *Colocasia esculenta*
Arum macrorrhiza L. – see *Alocasia macrorrhizos*
Arum macrorrhizon L. – see *Alocasia macrorrhizos*
Arum mucronatum Lam. – see *Alocasia macrorrhizos*
Arum nigrum Vell. – see *Xanthosoma violaceum*
Arum odorum Roxb. – see *Alocasia macrorrhizos*
Arum paeoniaefolium Dennst. – see *Amorphophallus paeoniifolius* var. *campanulatus*
Arum peltatum Lam. – see *Colocasia esculenta*
Arum peregrinum L. – see *Alocasia macrorrhizos*
Arum rumphii Gaud. – see *Amorphophallus paeoniifolius* var. *campanulatus*
Arum sagittaefolium L. – see *Xanthosoma sagittifolium*
Arum sagittifolium Link – see *Xanthosoma brasiliense*
Arum xanthorrhizon Jacq. – see *Xanthosoma sagittifolium*
Arundo bambos L. – see *Bambusa bambos*
Ascalonicum schoenoprasum Renault – see *Allium schoenoprasum*
Asparagopsis javanica Kunth – see *A. officinalis*
***Asparagus acutifolius* L. (13.1)**
Asparagus caspius Hohen. – see *A. officinalis*
Asparagus collinus Schur – see *A. officinalis*
Asparagus dubius Decne. – see *A. officinalis*
Asparagus esculentus Salisb. – see *A. officinalis*
Asparagus hedecarpus Andr. ex Baker – see *A. officinalis*
Asparagus hortensis Mill. ex Baker – see *A. officinalis*
Asparagus maritimus Mill. – see *A. officinalis*
***Asparagus officinalis* L. (13.2)**
Asparagus officinalis L. var. *altilis* L. – see *A. officinalis*
Asparagus oxycarpus Stev. – see *A. officinalis*
Asparagus paragus Gueldenst. ex Ledeb. – see *A. officinalis*
Asparagus polyphyllus Stev. – see *A. officinalis*
Asparagus prostratus Dumort. – see *A. officinalis*
Asparagus sativus Mill. – see *A. officinalis*
Asparagus schoberioides Kunth – see *A. officinalis*
Asparagus tenuifolius Gilib. – see *A. officinalis*
Asparagus vulgaris Gueldenst. ex Ledeb. – see *A. officinalis*

Asplenium esculentum (Retz.) C. Presl – see *Diphazium esculentum*
Aster scaber Thunb. – see *Doellingeria scabra*
Aster yomena Kitam. – see *Kalimeris yomena*
Asterogeum laciniatum S.F. Gray – see *Plantago coronopus* var. *sativa*
Athyrium esculentum (Retz.) Copel. – see *Diphazium esculentum*
Atrema testiculatum Miq. – see *Coriandrum sativum*
Atriplex ameghinoi Speg. – see *A. hortensis*
Atriplex benghalensis Lam. – see *A. hortensis*
Atriplex bonus-henricus (L.) Crantz – see *Chenopodium bonus-henricus*
Atriplex heterantha Wight – see *A. hortensis*
***Atriplex hortensis* L. (35.1)**
Atriplex microtheca Moq. – see *A. hortensis*
Atriplex scoparia (L.) Crantz – see *Bassia scoparia*
Atriplex virgata Roth – see *A. hortensis*
Azukia angularis (Willd.) Ohwi – see *Vigna angularis*
Azukia mungo (L.) Masam. – see *Vigna mungo* var. *mungo*
Azukia radiata (L.) Ohwi – see *Vigna radiata* var. *radiata*
Azukia subtrilobata Takahashi – see *Vigna angularis*
Azukia typica Miura – see *Vigna radiata* var. *radiata*
Azukia umbellata (Thunb.) Ohwi – see *Vigna umbellata*
Baeobotrys latifolia Blume – see *Maesa latifolia*
Baeumerta nasturtium P. Gaertn. B. Mey. & Scherb. – see *Nasturtium officinale*
Bambos stricta Roxb. – see *Dendrocalamus strictus*
Bambus arundo Blanco – see *Bambusa blumeana*
Bambusa arundinacea (Retz.) Willd. – see *B. bambos*
Bambusa aspera Schultes f. – see *Dendrocalamus asper*
Bambusa aspera Schult. & Schult. f. – see *Dendrocalamus asper*
Bambusa atrovirens Wen – see *B. oldhamii*
***Bambusa bambos* (L.) Voss (17.1)**
Bambusa bambusa Huth. – see *B. bambos*
***Bambusa beecheyana* (Munro) McClure (17.1)**
***Bambusa blumeana* J.A. & J.H. Schultes (17.1)**
Bambusa brandisii Munro – see *Dendrocalamus brandisii*
Bambusa edulis Carrière – see *Phyllostachys edulis*
Bambusa edulis (non Poiret) Carriêre – see *P. pubescens*
Bambusa edulis (Odash.) Keng. – see *Phyllostachys edulis*
Bambusa gigantea Wall. – see *Dendrocalamus giganteus*
Bambusa heterocycla Carrière – see *Phyllostachys edulis*
Bambusa latiflora (Munro) Kurz – see *Dendrocalamus latiflorus*
Bambusa mitis hort. ex Carr. – see *Phyllostachys viridis*
Bambusa mitis Poiret – see *Phyllostachys pubescens*
Bambusa monogyna Griff. – see *Dendrocalamus hamiltonii*
***Bambusa oldhamii* Munro (17.1)**
Bambusa pungens Blanco – see *B. blumeana*
Bambusa spinosa Blume, ex Nees – see *B. blumeana*
Bambusa spinosa Roxb. – see *B. bambos*
***Bambusa* spp. (17.1)**
Bambusa surinamensis Ruprecht – see *B. vulgaris*
Bambusa thouarsii Kunth – see *B. vulgaris*
Bambusa verticillata Benth. – see *Dendrocalamus latiflorus*

***Bambusa vulgaris* Schrad. ex J.C. Wendl. (17.1)**
Bamia magnifica Wall. – see *Abelmoschus manihot*
Bamia manihot (L.) Wall. – see *Abelmoschus manihot*
Bancroftia xanthorrhiza Billb. – see *Arracacia xanthorrhiza*
Barbarea praecox (Sm.) R. Br. – see *B. verna*
***Barbarea verna* (Mill.) Asch. (30.2)**
***Basella alba* L. (28.1)**
Basella alba L. cv Rubra – see *B. alba*
Basella cordifolia Lam. – see *B. alba*
Basella crassifolia Salisb. – see *B. alba*
Basella japonica Burm.f. – see *B. alba*
Basella lucida L. – see *B. alba*
Basella nigra Lour. – see *B. alba*
Basella oleracea Alef. – see *B. alba*
Basella rubra L. – see *B. alba*
Basella tuberosa (Lozano) Kunth – see *Ullucus tuberosus*
Basella volubilis Salisb. – see *B. alba*
Basilicum polystachion Moench – see *Ocimum basilicum*
***Bassia scoparia* (L.) A.J. Scott (35.2)**
Batatas edulis (Thunb.) Choisy – see *Ipomoea batatas* var. *batatas*
Batatas xanthorhiza Boj. – see *Ipomoea batatas* var. *batatas*
Battata tuberosa Hill – see *Solanum tuberosum*
Bauhinia burkeana Benth. ex Harv. & Sond. – see *Tylosema esculentum*
Bauhinia esculenta Burchell – see *Tylosema esculentum*
Benincasa cerifera Savi – see *B. hispida*
***Benincasa hispida* (Thunb.) Cogn. (38.1)**
Benincasa vacua (F. Muell.) F. Muell. – see *B. hispida*
Beta cicla L. – see *B. vulgaris* Cicla group
Beta esculenta Salisb. – see *B. vulgaris* Crassa group
Beta hortensis Mill. – see *B. vulgaris* Cicla group
Beta maritima campestris Lange – see *B. vulgaris* Crassa group
Beta maritima hortensis Lange – see *B. vulgaris* Crassa group
Beta maritima saccharifera Lange – see *B. vulgaris* Crassa group
Beta rapa Dum. – see *B. vulgaris* Crassa group
Beta rapacea Hegetschw. – see *B. vulgaris* Crassa group
Beta vulgaris altissima Döll – see *B. vulgaris* Crassa group
Beta vulgaris crassa (Alef.) Voss – see *B. vulgaris* Crassa group
***Beta vulgaris* L. Cicla group (35.3)**
Beta vulgaris L. convar. *crassa* (Alef.) J. Helm – see *Beta vulgaris* Crassa group
Beta vulgaris L. convar. *crassa* (Alef.) J. Helm provar. *conditiva* – see *B. vulgaris* Crassa group
Beta vulgaris L. convar. *vulgaris* – see *B. vulgaris* Cicla group
Beta vulgaris L. convar. *vulgaris* provar. *flavescens* DC. sensu Helm – see *B. vulgaris* Cicla group
***Beta vulgaris* L. Crassa group (35.4)**
Beta vulgaris L. cv. group Swiss chard – see *B. vulgaris* Cicla group
Beta vulgaris L. subsp. *alba* (DC.) Döll – see *B. vulgaris* Cicla group
Beta vulgaris L. subsp. *cicla* (L.) – see *B. vulgaris* Cicla group
Beta vulgaris L. subsp. *cicla* (L.) Döll – see *B. vulgaris* Cicla group
Beta vulgaris L. subsp *cicla* (L.) Koch sensu Ford-Lloyd – see *B. vulgaris* Cicla group
Beta vulgaris L. subsp. *crassa* Alef. – see *B. vulgaris* Crassa group
Beta vulgaris L. subsp. *esculenta* Becker-Dillingen – see *B. vulgaris* Crassa group

Beta vulgaris L. subsp. *rapacea* (Koch) Döll – see *B. vulgaris* Crassa group
Beta vulgaris L. subsp. *vulgaris* convar. *vulgaris* var. *vulgaris* sensu Mansfeld – see *B. vulgaris* Crassa group
Beta vulgaris L. subsp. *vulgaris* sensu Ford-Lloyd – see *B. vulgaris* Crassa group
Beta vulgaris L. var. *altissima* DC. – see *B. vulgaris* Crassa group
Beta vulgaris L. var. *ameliora* Becker-Dillingen – see *B. vulgaris* Cicla group
Beta vulgaris L. var. *cicla* Koch – see *B. vulgaris* Cicla group
Beta vulgaris L. var. *cicla* L. – see *B. vulgaris* Cicla group
Beta vulgaris L. var. *cicla* Mag. – see *B. vulgaris* Cicla group
Beta vulgaris L. var. *communis* Becker-Dillingen – see *B. vulgaris* Cicla group
Beta vulgaris L. var. *crassa* (Alef.) J. Helm – see *B. vulgaris* Crassa group
Beta vulgaris L. var. *esculenta* (Salisb.) Gürke – see *B. vulgaris* Crassa group
Beta vulgaris L. var. *flavescens* DC. – see *B. vulgaris* Cicla group
Beta vulgaris L. var. *rapa* (Dum.) Asch. – see *B. vulgaris* Crassa group
Beta vulgaris L. var. *rapa* Dumont – see *B. vulgaris* Crassa group
Beta vulgaris L. var. *rapa* Dumont form *rubra* DC. – see *B. vulgaris* Crassa group
Beta vulgaris L. var. *rapacea* Koch – see *B. vulgaris* Crassa group
Beta vulgaris L. var. *rubra* Moq. – see *B. vulgaris* Crassa group
Beta vulgaris L. var. *vulgaris* covar. *cicla* (L.) Alef. var. *flavesecens* DC. sensu Mansfeld – see *B. vulgaris* Cicla group
Beta vulgaris (Poirée) DC. – see *B. vulgaris* Cicla group
Beta vulgaris (Poirée) Moq. – see *B. vulgaris* Cicla group
Beta vulgaris L. var.-gr. *crassa* Alef. – see
B. vulgaris Crassa group
Beta vulgaris rapacea W.D.J. Koch – see *B. vulgaris* Crassa group
Beta vulgaris vulgaris form *cicla* (L.) Beck – see *B. vulgaris* Cicla group
Beta vulgaris vulgaris form *esculenta* (Salisb.) Beck – see *B. vulgaris* Crassa group
Bidens abortiva Schum & Thonn. – see *B. pilosa*
Bidens acmella Lam. – see *Acmella oleracea*
Bidens acmelloides Berg. – see *Acmella oleracea*
Bidens chinensis auct., non Willd. – see *B. pilosa*
Bidens fervida Lam. – see *Acmella oleracea*
Bidens fusca Lam. – see *Acmella oleracea*
Bidens leucantha Griseb. – see *B. pilosa*
Bidens oleracea Cav. ex Steud. – see *Acmella oleracea*
***Bidens pilosa* L. (27.8)**
Bifora loureirii Kostel. – see *Coriandrum sativum*
Blitum bonus-henricus (L.) C.A. Mey. – see *Chenopodium bonus-henricus*
Boberella alkekengi E.H.L. Krause – see *Physalis alkekengi*
Boberella peruviana E.H.L. Krause – see *Physalis peruviana*
Bonnania officinalis Presl – see *Sinapis alba*
***Borago officinalis* L. (29.1)**
Botor tetragonoloba Kuntze – see *Psophocarpus tetragonolobus*
Botor tetragonoloba (L.) Kuntze – see *Psophocarpus tetragonolobus*
Botor tetragonoloba (Stickm.) DC. – see *Psophocarpus tetragonolobus*
Bracheilema paniculatum R. Br. – see *Gymnanthemum amygdalinum*
Brasenia nymphoides Baill. – see *B. schreberi*
Brasenia peltata Pursh – see *B. schreberi*
Brasenia purpurea Casp. – see *B. schreberi*
***Brasenia schreberi* J.P. Gmel. (31.1)**

Brassica acephala auct. – see *B. oleracea* var. *viridis*
Brassica alba (L.) Rabenh. – see *Sinapis alba*
Brassica alboglabra L.H. Bailey – see *B. oleracea* var. *alboglabra*.
Brassica atrovirens Mao – see *B. rapa* subsp. *narinosa*
Brassica besseriana Andrz. – see *B. juncea* var. *rugosa*
Brassica botrytis (L.) Mill. – see *B. oleracea* var. *botrytis*
Brassica botrytis subsp. *italica* Lizg. – see *B. oleracea* var. *italica*
Brassica botrytis var. *cauliflora* Presl – see *B. oleracea* var. *botrytis*
Brassica brachycarpa Candargy – see *B. nigra* Koch.
Brassica bullata (DC.) Pasquale – see *B. oleracea* var. *sabauda*
Brassica bullata Pasquale var. *sabauda* – see *B. oleracea* var. *sabauda*
Brassica campestris DC. – see *B. napus* L. var. *napobrassica*
Brassica campestris L. – see *B. napus* var *napus*
Brassica campestris L. var. *pablaria* DC. – see *Brassica napus* var. *pabularia*
Brassica campestris L. Chinensis group – see *B. rapa* subsp. *chinensis*
Brassica campestris L. Japonica group – see *B. napus* var *napus*
Brassica campestris L. Nariosa group – see *B. rapa* subsp. *narinosa*
Brassica campestris L. Pekinensis group – see *B. rapa* subsp. *pekinensis*
Brassica campestris L. Perviridis group – see *Brassica rapa* subsp. *nipposinica* var. *perviridis*
Brassica campestris L. Rapifera group – see *B. rapa* subsp. *rapa*
Brassica campestris L. subsp. *chinensis* (L.) Makino – see *B. rapa* subsp. *chinensis*
Brassica campestris L. subsp. *chinensis* Makino var. *amplexicaulis* Makino – see *B. rapa* subsp. *pekinensis*
Brassica campestris L. subsp. *chinensis* var. *parachinensis* L.H. Bailey – see *B. rapa* subsp. *chinensis* var. *parachinensis*
Brassica campestris L. subsp. *chinensis* var. *rosularis* Tsen & Lee – see *B. rapa* subsp. *narinosa*
Brassica campestris L. subsp. *chinensis* var. *utilis* Tsen & Lee – see *B. rapa* subsp. *chinensis* var. *parachinensis*
Brassica campestris L. subsp. *napobrassica* (L.) Schübl. & Mart. – see *B. napus* var. *napobrassica*
Brassica campestris L. subsp. *napus* Hook. & T. Anderson var. *parachinensis* Makino – see *B. rapa* subsp. *chinensis* var. *parachinensis*
Brassica campestris L. subsp. *napus* Hook. f. & T. Anderson – see *B. napus* var *napus*
Brassica campestris L. subsp. *parachinensis* – see *B. rapa* subsp. *pekinensis*
Brassica campestris L. subsp. *pekinensis* (Lour.) Olsson – see *B. rapa* subsp. *pekinensis*
Brassica campestris L. subsp. *rapa* (L.) Hook.f. & T. Anderson – see *B. rapa* subsp. *rapa*
Brassica campestris L. subsp. *rapa* Hook. f. & T. Anderson – see *B. rapa* subsp. *rapa*
Brassica campestris L. subsp. *rapifera* (Metzg.) Sinskaya – see *B. rapa* subsp. *rapa*
Brassica campestris L. var. *chinensis* Ito – see *B. rapa* subsp. *chinensis*
Brassica campestris L. var. *napobrassica* DC. – see *B. napus* var. *napobrassica*
Brassica campestris L. var. *nippo-oleifera* Makino – see *B. napus* var *napus*
Brassica campestris L. var. *oleifera* DC. – see *B. napus* var *napus*
Brassica campestris L. var. *pabularia* DC. – see *B. napus* var. *pabularia*
Brassica campestris L. var. *rapa* (L.) Hartm. – see *B. rapa* subsp. *rapa*
Brassica campestris L. var. *rapifera* Metz. – see *B. rapa* subsp. *rapa*
Brassica campestris Ruvo group – see *B. rapa* subsp. *rapa*
Brassica campestris var. *chinensis* (L.) Makino – see *B. rapa* subsp. *chinensis*
Brassica campestris var. *chinensis* (L.) T. Ito – see *B. rapa* subsp. *chinensis*
Brassica campestris var. *napus* (L.) Babingt. – see *B. napus* var. *napus*
Brassica campestris var. *rapa* (L.) Hartm. – see *B. rapa* subsp. *rapa*
Brassica capitata (L.) Lizg. – see *B. oleracea* var. *capitata*

Brassica capitata auct. – see *B. oleracea* var. *capitata*
Brassica capitata Lévl. – see *B. oleracea* var. *capitata*
Brassica capitata subsp. *costata* (DC.) Lizg. – see *B. oleracea* var. *costata*
***Brassica carinata* A. Braun (30.3)**
Brassica cauliflora Gars. – see *B. oleracea* var. *botrytis*
Brassica cauliflora subsp. *abortiva* Lizg. – see *B. oleracea* var. *botrytis*
Brassica cauliflora subsp. *simplex* Lizg. – see *B. oleracea* var. *italica*
Brassica caulorapa DC. – see *B. oleracea* var. *gongylodes*
Brassica caulorapa (DC.) Pasquale – see *B. oleracea* var. *gongylodes*
Brassica caulorapa Pasq. – see *B. oleracea* var. *gongylodes*
Brassica cernua Forb. & Hemsl. var. *chirimenna* auct. non Makino – see *B. juncea* var. *multiceps*
Brassica cernua Forbes & Hemsley – see *B. juncea* var. *rugosa*
Brassica cernua Hemsl. – see *B. juncea* var. *rugosa*
Brassica cernua Thunb. – see *B. juncea* var. *rugosa*
Brassica cernua var. *hirimenna* Makino – see *B. juncea* var. *crispifolia*
Brassica chinensis Jusl. – see *B. rapa* subsp. *chinensis*
Brassica chinensis L. – see *B. rapa* subsp. *chinensis*
Brassica chinensis L. subsp. *laminata* Titov – see *B. rapa* subsp. *pekinensis*
Brassica chinensis L. subsp. *petiolata* Titou – see *B. rapa* subsp. *chinensis*
Brassica chinensis L. var. *communis* Tsen & Lee – see *B. rapa* subsp. *chinensis*
Brassica chinensis L. var. *oleifera* Makino – see *B. napus* var *napus*
Brassica chinensis L. var. *oleifera* Tsen & Lee – see *B. napus* var *napus*
Brassica chinensis L. var. *parachinensis* (L.H. Bailey) Tsen & Lee – see *B. rapa* subsp. *chinensis* var. *parachinensis*
Brassica chinensis L. var. *pekinensis* (Rupr.) Sun. – see *B. rapa* subsp. *pekinensis*
Brassica chinensis L. var. *rosularis* Tsen & Lee – see *B. rapa* subsp. *narinosa*
Brassica chinensis Regel – see *B. rapa* subsp. *pekinensis*
Brassica chinensis var. *pekinensis* (Lour.) Sun – see *B. rapa* subsp. *pekinensis*
Brassica cibaria Dierb. – see *B. rapa* subsp. *rapa*
Brassica cretica var. *cauliflora* (DC.) Schwarz – see *B. oleracea* var. *botrytis*
Brassica decumbens Bub. – see *Diplotaxis muralis*
Brassica eruca L. – see *Eruca sativa*
Brassica erucoides Roxb. – see *Eruca sativa*
Brassica fimbriata DC. – see *B. napus* var. *pabularia*
Brassica gemmifera auct. – see *B. oleracea* var. *gemmifera*
Brassica gemmifera Lév. – see *B. oleracea* var. *gemmifera*
Brassica gemmifera Lizg. – see *B. oleracea* var. *gemmifera*
Brassica gongylodes (L.) Mill. – see *B. oleracea* var. *gongylodes*
Brassica gongylodes Lizg. subsp. *gongylodes* – see *B. oleracea* var. *gongylodes*
Brassica gongylodes subsp. *asiatica* Lizg. – see *B. oleracea* var. *gongylodes*
Brassica hirta Moench – see *Sinapis alba*
Brassica hispida Tenore – see *Eruca sativa*
Brassica integrifolia (West) Rupr. – see *B. juncea* var. *rugosa*
Brassica integrifolia (West) Thell. var. *carinata* (A. Braun) O.E. Schulz – see *B. carinata*
Brassica japonica hort. non Thunb. – see *B. juncea* (L.) Czern. var. *crispifolia* L.H. Bailey
Brassica japonica Makino – see *B. juncea* var. *rugosa*
***Brassica juncea* (L.) Czern. var. *capitata* Hort. (30.4)**
***Brassica juncea* (L.) Czern. var. *crassicaulis* Chen and Yang (30.5)**
***Brassica juncea* (L.) Czern. var. *crispifolia* L.H. Bailey (30.6)**
***Brassica juncea* (L.) Czern. var. *foliosa* L.H. Bailey (30.7)**

***Brassica juncea* (L.) Czern. var. *gemmifera* Lee & Lin (30.8)**
***Brassica juncea* (L.) Czern. var. *involuta* Yang & Chen (30.9)**
***Brassica juncea* (L.) Czern. var. *latipa* Li (30.10)**
***Brassica juncea* (L.) Czern. var. *leucanthus* Chen & Yang (30.11)**
***Brassica juncea* (L.) Czern. var. *linearifolia* Sun (30.12)**
***Brassica juncea* (L.) Czern. var. *longepetiolata* Yang & Chen (30.13)**
***Brassica juncea* (L.) Czern. var. *megarrhiza* N. Tsen & S.N. Lee (30.14)**
***Brassica juncea* (L.) Czern. var. *multiceps* N. Tsen & S.N. Lee (30.15)**
***Brassica juncea* (L.) Czern. var. *mutisecta* L.H. Bailey (30.16)**
***Brassica juncea* (L.) Czern. var. *rugosa* (Roxb.) N. Tsen & S.N. Lee (30.17)**
***Brassica juncea* (L.) Czern. var. *strumata* N. Tsen & S.N. Lee (30.18)**
***Brassica juncea* (L.) Czern. var. *tumida* N. Tsen & Lee (30.19)**
***Brassica juncea* (L.) Czern. var. *utilis* Li (30.20)**
Brassica juncea Coss. – see *B. juncea* var. *rugosa*
Brassica juncea HK. F. & Thoms. – see *B. juncea* var. *rugosa*
Brassica juncea subsp. *integrifolia* (West) Thell. – see *B. juncea* var. *rugosa*
Brassica juncea subsp. *rugosa* Prain – see *B. juncea* var. *rugosa*
Brassica juncea var. *agrestis* Prain – see *B. juncea* var. *rugosa*
Brassica juncea var. *atropurpurea* Makino – see *B. juncea* var. *rugosa*
Brassica juncea var. *capitata* hort. ex Li – see *B. juncea* (L.) Czern. var. *capitata* hort.
Brassica juncea var. *capitata* Li – see *B. juncea* (L.) Czern. var. *capitata* hort.
Brassica juncea var. *chinensis* Busch – see *B. juncea* var. *rugosa*
Brassica juncea var. *chirimenna* Makino – see *B. juncea* var. *crispifolia*
Brassica juncea var. *crassicaulis* Chen & Yang – see *B. juncea* var. *crassicaulis*
Brassica juncea var. *crispifolia* auct., non L.H. Bailey – see *B. juncea* var. *multiceps*
Brassica juncea var. *crispifolia* L.H. Bailey – see *B. juncea* var. *crispifolia*
Brassica juncea var. *foliosa* Bailey – see *B. juncea* var. *foliosa*
Brassica juncea var. *gemmifera* Lee & Lin – see *B. juncea* var. *gemmifera*
Brassica juncea var. *involuta* Yang & Chen – see *B. juncea*. var. *involuta*
Brassica juncea var. *latipa* Li – see *B. juncea* var. *latipa*
Brassica juncea var. *leucanthus* Chen & Yang – see *B. juncea*. var. *leucanthus*
Brassica juncea var. *linearifolia* Sun – see *B. juncea* var. *linearifolia*
Brassica juncea var. *longepetiolata* Yang & Chen – see *B. juncea* var. *longepetiolata*
Brassica juncea var. *megarrhiza* Tsen & Chen – see *B. juncea* var. *megarrhiza*
Brassica juncea var. *napiformis* (Paill. & Bois) Kitam. – see *B. juncea* var. *megarrhiza*
Brassica juncea var. *napiformis* Pai & Bois – see *B. juncea* var. *megarrhiza*
Brassica juncea var. *rugosa* L.H. Bailey – see *B. juncea* var. *rugosa*
Brassica juncea var. *rugosa* Makino – see *B. juncea* var. *rugosa*
Brassica juncea var. *rugosa* (Roxb.) Tsen & Lee – see *B. juncea* var. *rugosa*
Brassica juncea var. *sabellica* Kitam. – see *B. juncea* var. *crispifolia*
Brassica lanceolata Lange – see *B. juncea* var. *rugosa*
Brassica lanceolata Schweinf. – see *B. nigra*
Brassica napiformis L.H. Bailey – see *B. juncea* var. *megarrhiza*
Brassica napiformis (Paill. & Bois) L.H. Bailey – see *B. juncea* var. *megarrhiza*
Brassica napiformis L.H. Bailey var. *multisecta* Barano – see *B. juncea* var. *mutisecta*
Brassica napiformis Pailleux & Bois – see *B. juncea* var. *megarrhiza*
Brassica napiformis var. *multisecta* Baranov – see *B. juncea* var. *megarrhiza*
Brassica napobrassica DC. – see *B. napus* var. *napobrassica*
Brassica napobrassica (L.) Mill. – see *B. napus* var. *napobrassica*
Brassica napobrassica (L.) Mill. var. *solidiflora* L.H. Bailey – see *B. napus* var. *napobrassica*

Brassica napobrassica Mill. – see *B. napus* var. *napobrassica*
Brassica napobrassica subsp. *flava* Sinskaya – see *B. napus* var. *napobrassica*
Brassica napobrassica subsp. *primitiva* Sinskaya – see *B. napus* var. *napobrassica*
Brassica napus form *annua* (Schübl. & G Martens). Thell. – see *B. napus* var. *napus*
Brassica napus form *biennis* (Schübl. & G Martens) Thell. – see *B. napus* var. *napus.*
Brassica napus L. – see *B. napus* var. *napus*
Brassica napus L. Napobrassica group – see *B. napus* var. *napobrassica*
Brassica napus L. Pabularia group. – see *B. napus* var. *pabularia*
Brassica napus L. subsp. *napobrassica* (L.) Jafri – see *B. napus* var. *napobrassica*
Brassica napus L. subsp. *oleifera* (Delile) Sinskaya – see *B. napus* var. *napus*
Brassica napus L. subsp. *pabularia* (DC.) Janchen – see *B. napus* var. *pabularia*
Brassica napus L. var. *chinensis* (L.) O.E. Schulz – see *B. rapa* subsp. *chinensis*
Brassica napus L. var. *edulis* Delile – see *B. napus* var. *napobrassica*
Brassica napus L. var. *napobrassica* (L.) Rchb. – see *B. napus* var. *napobrassica*
***Brassica napus* L. var. *napobrassica* (L.) Reichb. (30.22)**
***Brassica napus* L. var *napus* (30.21)**
Brassica napus L. var. *oleifera* (Moench) Delile – see *B. napus* var. *napus*
***Brassica napus* L. var. *pabularia* (DC.) Reichb. (30.23)**
Brassica napus L. var. *rapifera* L. – see *B. napus* var. *napobrassica*
Brassica napus L. var. *rapifera* Metzg. – see *B. napus* var. *napobrassica*
Brassica narinosa L.H. Bailey – see *B. rapa* subsp. *narinosa*
***Brassica nigra* (L.) W.D.J. Koch. (30.24)**
Brassica oleifera Moench – see *B. napus* var. *napus*
Brassica oleracea botrytis DC. – see *B. oleracea* var. *botrytis*
Brassica oleracea botrytis var. *italica* Plenck. – see *B. oleracea* var. *italica*
Brassica oleracea capitata DC. – see *B. oleracea* var. *capitata*
Brassica oleracea convar. *caulorapa* (DC.) Alef. – see *B. oleracea* var. *gongylodes*
Brassica oleracea costata DC. – see *B. oleracea* var. *costata*
Brassica oleracea form *ramosa* (DC.) Thell. – see *B. oleracea* var. *ramosa*
Brassica oleracea gemmifera DC. – see *B. oleracea* var. *gemmifera*
Brassica oleracea L. Acephala group – see *B. oleracea* var. *viridis*
Brassica oleracea L. Alboglabra group – see *B. oleracea* var. *alboglabra.*
Brassica oleracea L. *asparogoides* – see *B. oleracea* var. *italica*
Brassica oleracea L. Botrytis group – see *B. oleracea* var. *botrytis*
Brassica oleracea L. Capitata group – see *B. oleracea* var. *capitata*
Brassica oleracea L. *cauliflora* – see *B. oleracea* var. *botrytis*
Brassica oleracea L. cv. group Broccoli – see *B. oleracea* var. *italica*
Brassica oleracea L. cv. group Brussels sprouts – see *B. oleracea* var. *gemmifera*
Brassica oleracea L. cv. group Cauliflower – see *B. oleracea* var. *botrytis*
Brassica oleracea L. cv. group Chinese kale – see *B. oleracea* var. *alboglabra*
Brassica oleracea L. cv. group Kohlrabi – see *B. oleracea* var. *gongylodes*
Brassica oleracea L. Gemmifera group – see *B. oleracea* var. *gemmifera*
Brassica oleracea L. Gongylodes group – see *B. oleracea* var. *gongylodes*
Brassica oleracea L. Italica group – see *B. oleracea* var. *italica*
Brassica oleracea (L.) Lizg. subsp. *oleracea* – see *B. oleracea* var. *viridis*
Brassica oleracea L. subsp. *capitata* (L.) Schübl. & Mart. – see *B. oleracea* var. *capitata*
Brassica oleracea L. subsp. *caulo-rapa* Metzg. – see *B. oleracea* var. *gongylodes*
Brassica oleracea L. subsp. *gemmifera* (DC.) Lizg. – see *B. oleracea* var. *gemmifera*
Brassica oleracea L. subsp. *gongylodes* (L.) Schübl. & Mart. – see *B. oleracea* var. *gongylodes*
Brassica oleracea L. subsp. *sabauda* L. – see *B. oleracea* var. *sabauda*

Brassica oleracea L. subvar. *albo-glabra* Burkill – see *B. oleracea* var. *alboglabra*
Brassica oleracea L. subvar. *cymosa* Thell. – see *B. oleracea* var. *italica*
Brassica oleracea L. subvar. *gemmifera* (DC.) Schulz – see *B. oleracea* var. *gemmifera*
Brassica oleracea L. subvar. *medullosa* (Thell.) Lamprecht – see *B. oleracea* var. *medullosa*
Brassica oleracea L. var. *acephala* DC. – see *B. oleracea* var. *viridis*
Brassica oleracea L. var. *acephala* L. f. var. *acephala* DC. – see *B. oleracea* var. *viridis*
Brassica oleracea L. var. alboglabra L.H. Bailey (Musil) (30.25)
Brassica oleracea L. var. *asparagoides* Gmel. – see *B. oleracea* var. *italica*
Brassica oleracea L. var. botrytis L. (30.26)
Brassica oleracea L. var. *bullata* DC. – see *B. oleracea* var. *sabauda*
Brassica oleracea L. var. *bullata* DC. subvar. *gemmifera* DC. – see *B. oleracea* var. *gemmifera*
Brassica oleracea L. var. capitata L. (30.27)
Brassica oleracea L. var. *caulorapa communis* DC. – see *B. oleracea* var. *gongylodes*
Brassica oleracea L. var. *caulorapa* DC. – see *B. oleracea* var. *gongylodes*
Brassica oleracea L. var. *caulorapa* L. – see *B. oleracea* var. *gongylodes*
Brassica oleracea L. var. *caulorapa* Pasq. – see *B. oleracea* var. *gongylodes*
Brassica oleracea L. var. *chinensis* (L.) Prain – see *B. rapa* subsp. *chinensis*
Brassica oleracea L. var. *costata* Bailey – see *B. oleracea* var. *costata*
Brassica oleracea L. var. costata DC (30.28)
Brassica oleracea L. var. *fimbriata* Mill. – see *B. napus* var. *pabularia*
Brassica oleracea L. var. *frutescens* Vis. – see *B. oleracea* var. *ramosa* Alef.
Brassica oleracea L. var. *gemmifera* DC. – see *B. oleracea* var. *gemmifera*
Brassica oleracea L. var. gemmifera Zenk. (30.29)
Brassica oleracea L. var. gongylodes L. (30.30)
Brassica oleracea L. var. italica Plenck (30.31)
Brassica oleracea L. var. *luteola* Alef. – see *B. oleracea* var. *costata*
Brassica oleracea L. var. medullosa Thell. (30.32)
Brassica oleracea L. var. *millecapitata* (Lev.) Helm – see *B. oleracea* var. *gemmifera*
Brassica oleracea L. var. *napobrassica* L. – see *B. napus* var. *napobrassica*
Brassica oleracea L. var. ramosa DC. (30.33)
Brassica oleracea L. var. *rubra* L. – see *B. oleracea* var. *capitata*
Brassica oleracea L. var. sabauda L. (30.34)
Brassica oleracea L. var. *tronchuda* L.H. Bailey – see *B. oleracea* var. *costata*
Brassica oleracea L. var. viridis L. (30.35)
Brassica parachinensis L.H. Bailey – see *B. rapa* subsp. *chinensis* var. *parachinensis*
Brassica pekinensis (Lour.) Rupr. – see *B. rapa* subsp. *pekinensis*
Brassica personii Rouy & Foucaud – see *B. nigra*
Brassica pe-tsai L.H. Bailey – see *B. rapa* subsp. *pekinensis*
Brassica radice-rapi Tillandz, ex Ahokas – see *B. napus* var. *napobrassica*
Brassica rapa convar. *rapifera* (Metzg.) Alef. – see *B. rapa* subsp. *rapa*
Brassica rapa L. Chinensis group – see *B. rapa* subsp. *chinensis*
Brassica rapa L. cv. group Caisin – see *B. rapa* subsp. *chinensis* var. *parachinensis*
Brassica rapa L. Pekinensis group – see *B. rapa* subsp. *pekinensis*
Brassica rapa L. Perviridis group – see *B. rapa* subsp. *nipposinica* var. *perviridis*
Brassica rapa L. Ruvo group – see *B. rapa* subsp. *rapa*
Brassica rapa L. subsp. chinensis (L.) Hanelt (30.36)
Brassica rapa L. subsp. chinensis (L.) Hanelt var. parachinensis (L.H. Bailey) Hanelt (30.37)
Brassica rapa L. subsp. *napus* (L.) Briq. – see *B. napus* var. *napus*
Brassica rapa L. subsp. *napus* (L.) Schübl. & Mart. – see *B. napus* var. *napus*
Brassica rapa L. subsp. narinosa (L.H. Bailey) Hanelt (30.38)

***Brassica rapa* L. subsp. *nipposinica* (L.H. Bailey) Hanelt var. *perviridis* L.H. Bailey (30.39)**
Brassica rapa L. subsp. *pekinensis* (Lour.) Hanelt – see *B. rapa* subsp. *pekinensis*
***Brassica rapa* L. subsp. *pekinensis* (Lour.) Hanelt (30.40)**
***Brassica rapa* L. subsp. *rapa* (30.41)**
Brassica rapa L. subsp. *rapifera* Metzg. – see *B. rapa* subsp. *rapa*
Brassica rapa L. var. *amplexicaulis* Tanak & Ono – see *B. rapa* subsp. *pekinensis*
Brassica rapa L. var. *campestris* (L.) Clapham – see *B. napus* var *napus*
Brassica rapa L. var. *chinensis* Kitam. – see *B. rapa* subsp. *chinensis*
Brassica rapa L. var. *glabra* Regel – see *B. rapa* subsp. *pekinensis*
Brassica rapa L. var. *komatsuna* Hara – see *Brassica rapa* subsp. *nipposinica* var. *perviridis*
Brassica rapa L. var. *narinosa* Kitamura – see *B. rapa* subsp. *narinosa*
Brassica rapa L. var. *perviridis* L.H. Bailey – see *Brassica rapa* subsp. *nipposinica* var. *perviridis*
Brassica rapa L. var. *piaorh-tsai* Kitamura – see *B. rapa* subsp. *narinosa*
Brassica rapa L. var. *septiceps* L.H. Bailey – see *B. rapa* subsp. *rapa*
Brassica rapeuropaea Sinskaya – see *B. rapa* subsp. *rapa*
Brassica rapifera (Metzg.) Dalla Torre & Sarnth. – see *B. rapa* subsp. *rapa*
Brassica rapoasiatica Sinskaya – see *B. rapa* subsp. *rapa*
Brassica rugosa hort. – see *B. juncea* var. *rugosa*
Brassica rugosa L.H. Bailey – see *B. juncea* var. *rugosa*
Brassica rupestris subsp. *gongylodes* (L.) Janchen – see *B. oleracea* var. *gongylodes*
Brassica ruvo L.H. Bailey – see *B. rapa* subsp. *rapa*
Brassica sinapoides Roth – see *B. nigra*
Brassica sinensis L. – see *B. rapa* subsp. *chinensis*
Brassica sinensis L. var. *pe tsai* Pault – see *B. rapa* subsp. *pekinensis*
Brassica taquetii Lev. – see *B. juncea* var. *rugosa*
Brassica tenuifolia (L.) Fries – see *Diplotaxis muralis*
Brassica tuberosa Salisb. – see *B. rapa* subsp. *rapa*
Brassica turgida Pers. – see *Eruca sativa*
Brassica urbaniana O.E. Schulz – see *B. juncea* var. *rugosa*
Brassica willdenowii Boiss. – see *B. juncea* var. *rugosa*
Brunyera biennis Bub. – see *Oenthera biennis*
Bryonia alceifolia Willd. – see *Coccinia grandis*
Bryonia cheirophylla Wall. – see *Luffa aegyptiaca*
Bryonia cucumeroides Ser. – see *Trichosanthes cucumeroides*
Bryonia grandis L. – see *Coccinia grandis*
Bryonia umbellata Wall. – see *Trichosanthes cucumerina*
Buceras foenum-graecum (L.) All. – see *Trigonella foenum-graecum*
Buceras odoratissima Medik. – see *Trigonella foenum-graecum*
Bucholzia philoxeroides Mart. – see *Alternanthera philoxeroides*
***Bunias orientalis* L. (30.42)**
Bunias verrucosa Moench – see *B. orientalis*
Buphthalmum oleraceum Lour. – see *Glebionis coronarium*
Bursa pastoris Weber – see *Capsella bursa-pastoris*
Bushiola scoparia (L.) Nieuw. – see *Bassia scoparia*
Cacalia amygdalina Kuntze – see *Gymnanthemum amygdalinum*
Cacalia bicolor Roxb. ex Willd. – see *Gynura bicolor*
Cacalia cusimbua D. Don– see *Gynura bicolor*
Cacalia glabra Heyne – see *Emilia sonchifolia*
Cacalia prenanthoides Sieber, ex DC. – see *Emilia sonchifolia*
Cacalia sonchifolia L. – see *Emilia sonchifolia*

Cacara erosa (L.) Kuntze – see *Pachyrhizus erosus*
Cacara erosa Safford – see *Pachyrhizus erosus*
Cacara tuberosa Britt. – see *Pachyrhizus tuberosus*
Cactus elongatus Willd. – see *Opuntia ficus-indica*
Cactus ficus-indica L. – see *Opuntia ficus-indica*
Cactus lanceolatus Haw. – see *Opuntia ficus-indica*
Cactus opuntia Gussone – see *Opuntia ficus-indica*
Cajan cajan (L.) Huth – see *Cajanus cajan*
Cajan inodorum Medik. – see *Cajanus cajan*
Cajanum thora Raf. – see *C. cajan*
Cajanus bicolor DC. – see *C. cajan*
Cajanus cajan (L.) Huth. – see *C. cajan* (L.) Millsp.
***Cajanus cajan* (L.) Millsp. (40.2)**
Cajanus flavus DC. – see *C. cajan*
Cajanus indicus Spreng. – see *C. cajan*
Cajanus luteus Bello – see *C. cajan*
Cajanus obcordifolia Singh – see *C. cajan*
Cajanus pseudocajan (Jacq.) Schinz & Guillaumin. – see *C. cajan*
Cajanus striatus Boj. – see *C. cajan*
Caladium brasiliense Desf. – see *Xanthosoma brasiliense*
Caladium colocasia (L.) W.F. Wight, ex Safford – see *Colocasia esculenta*
Caladium edule G.F.W. Mey. – see *Xanthosoma sagittifolium*
Caladium esculentum hort. – see *Colocasia esculenta*
Caladium esculentum Vent. – see *Colocasia esculenta*
Caladium glycirrhizum Fraser – see *Alocasia macrorrhizos*
Caladium macrorrhizon R. Br. – see *Alocasia macrorrhizos*
Caladium odorum Lindl. – see *Alocasia macrorrhizos*
Caladium sagittifolium (L.) Vent. – see *Xanthosoma sagittifolium*
Caladium utile hort. ex Engl. – see *Xanthosoma sagittifolium*
Caladium xanthorrhizon (Jacq.) Willd. – see *Xanthosoma sagittifolium*
Calamintha hortensis hort. – see *Satureja hortensis*
Calandrinia andrewsii (Sweet) Sweet – see *Talinum fruticosum*
Calathea allouga Steud. – see *C. allouia*
***Calathea allouia* (Aubl.) Lindley (15.1)**
Calathea cylindrica K. Schum. – see *C. allouia*
Calathea grandifolia Lindl. – see *C. allouia*
Calathea macrosepala K. Schum. – see *C. allouia*
Calathea semperflexus Horan. – see *C. allouia*
Calathea violacea Lindl.– see *C. allouia*
Caloscordum exsertum Lindl. – see *Allium chinense*
Calystegia japonica Choisy – see *Convolvulus japonicus*
Calystegia pubescens Lindl. – see *Convolvulus japonicus*
Calyxhymenia expansa Ruiz & Pav. – see *Mirabilis expansa*
Campanula decurrens Thore. – see *C. rapunculus*
Campanula esculenta Salisb. – see *C. rapunculus*
Campanula gentianoides Lam. – see *Platycodon grandiflorus*
Campanula glauca Thunb. – see *Platycodon grandiflorus*.
Campanula grandiflora Jacq. – see *Platycodon grandiflorus*
Campanula lambertiana A. DC. – see *C. rapunculus*
***Campanula rapunculus* L. (33.2)**

Campanula tetraphylla Thunb. – see *Adenophora triphylla*
Campanula triphylla Thunb. – see *Adenophora triphylla*
Campanula verruculosa Hoffgg. & Link – see *C. rapunculus*
Campanula verticillata Pallas – see *Adenophora triphylla*
Campanula virgata DC. – see *C. rapunculus*
Campanumoea lanceolata Siebold & Zucc. – see *Codonopsis lanceolata*
Campe praecox (R. Br.) Dulac – see *Barbarea verna*
Canavalia africana Dunn. – see *C. ensiformis*
***Canavalia ensiformis* (L.) DC. (40.3)**
Canavalia ensiformis (L.) DC. var. *gladiata* (Jacq.) Kuntze – see *C. gladiata*
Canavalia ensiformis auct. non (L.) DC. – see *C. gladiata*
Canavalia ensiformis var. *alba* Makino – see *C. gladiata*
Canavalia ensiformis var. *gladiata* (Jacq.) Kuntze – see *C. gladiata* (Jacq.) DC.
Canavalia gladiata DC. var. *ensiformis* DC. – see *C. ensiformis*
***Canavalia gladiata* (Jacq.) DC. (40.4)**
Canavalia gladiata var. *leucosperma* Voigt – see *C. ensiformis*
Canavalia incurva (Thunb.) DC. – see *C. gladiata*
Canavalia loureirii G. Don – see *C. gladiata*
Canavalia lunareti Carr. – see *C. gladiata*
Canavalia machaeroides (DC.) Steud. – see *C. gladiata*
Canavalia maxima Thou. – see *C. gladiata*
Canavalia regalis Dunn. – see *C. ensiformis*
Canna achiras Gill – see *C. indica*
Canna aureavittata Lodd. – see *C. indica*
Canna bidentata Bertol. – see *C. indica*
Canna coccinea P. Miller – see *C. indica*
Canna edulis Ker-Gawl. – see *C. indica*
Canna ehrenbergii Bouché – see *C. indica*
Canna ellipticifolia Stokes – see *C. indica*
Canna esculenta Lodd. & Cat. ex Loud. – see *C. indica*
***Canna indica* L. (9.1)**
Canna limbata Roscoe – see *C. indica*
Canna meridensis Kränzlin, ex Engl. – see *C. indica*
Canna nepalensis Wall. – see *C. indica*
Canna orientalis Roscoe – see *C. indica*
Canna patens Roscoe – see *C. indica*
Canna rubricaulis Link – see *C. indica*
Canna speciosa Roscoe – see *C. indica*
Canna surinamensis Bouché – see *C. indica*
Cannabis lupulus Scop. – see *Humulus lupulus*
Cannacorus indicus Medik. – see *Canna indica*
Cannacorus ovatus Moench – see *Canna indica*
Capparis aegyptia Lam. – see *C. spinosa*
Capparis cordifolia Lam. – see *C. spinosa*
Capparis herbacea Willd. – see *C. spinosa*
Capparis leucophylla DC. – see *C. spinosa*
Capparis mariana Jacq. – see *C. spinosa*
Capparis murrayana J. Grah. – see *C. spinosa*
Capparis obovata Royle – see *C. spinosa*
Capparis ovato Desuaux – see *C. spinosa*

Capparis ovata M. Bieb. – see *C. spinosa*
Capparis rupestris Sibth. & Sm. – see *C. spinosa*
Capparis sandwichiana DC. – see *C. spinosa*
Capparis sicula Duham. – see *C. spinosa*
***Capparis spinosa* L. (34.1)**
Capparis spinosa L. var. *inermis* – see *C. spinosa*
Capparis spinosa var. *mariana* (Jacq.) K. Schum. – see *C. spinosa*
Capparis spinosa var. *rupestris* Boiss. – see *C. spinosa*
***Capsella bursa-pastoris* (L.) Medikus (30.43)**
Capsella bursa-pastoris Moench – see *C. bursa-pastoris* (L.) Medikus
Capsella bursa-pastoris var. *auriculata* Makino – see *C. bursa-pastoris*
Capsella polymorpha Cav. – see *C. bursa-pastoris*
Capsella praecox Jord. – see *C. bursa-pastoris*
Capsella ruderalis Jord. – see *C. bursa-pastoris*
Capsella sabulosa Jord. – see *C. bursa-pastoris*
Capsella virgata Jord. – see *C. bursa-pastoris*
Capsicum annuum L. var. *angulosum* Miller – see *C. annuum* var. *annuum*
***Capsicum annuum* L. var. *annuum* (62.1)**
Capsicum annuum L. var. *annuum* Grossum group – see *C. annuum* var. *annuum*
Capsicum annuum L. var. *grossum* (L.) Sendt. – see *C. annuum* var. *annuum*
***Capsicum annuum* L. var. *longum* Bailey (62.2)**
Capsicum annuum L. var. *longum* (DC.) Sendt. – see *C. annuum* L. var. *longum* Bailey
Capsicum baccatum auct., non L. – see *C. annuum* var. *annuum*
***Capsicum baccatum* L. var. *baccatum* (62.3)**
***Capsicum chinense* Jacq. (62.4)**
Capsicum fastigiatum Blume – see *C. frutescens*
Capsicum frutescens auct., non L. – see *C. frutescens* L.
***Capsicum frutescens* L. (62.5)**
Capsicum frutescens L. var. *grossum* (L.) L.H. Bailey – see *C. annuum* var. *annuum*
Capsicum frutescens L. var. *longum* L.H. Bailey – see *C. annuum* var. *longum*
Capsicum frutescens L. var. *longum* Sendt. – see *C. annuum* var. *longum*
Capsicum frutescens Willd. – see *C. frutescens* L.
Capsicum grossum L. – see *C. annuum* var. *annuum*
Capsicum guatemalense Bitter – see *C. pubescens*
Capsicum luteum Lam. – see *C. chinense.*
Capsicum microcarpum Cav. – see *C. baccatum* var. *baccatum*
Capsicum minimum Roxb. – see *C. annuum* var. *annuum*
***Capsicum pubescens* Ruiz & Pavon (62.6)**
Caraca tuberosa (Lam.) Britt. – see *Pachyrhizus tuberosus*
Cardamindum majus Moench – see *Tropaeolum majus*
Cardamine aquatica Nieuwl. – see *Nasturtium officinale*
Cardamine fontana Lam. – see *Nasturtium officinale*
Cardamine nasturtium Kuntze – see *Nasturtium officinale*
Cardamine nasturtium-aquaticum (L.) Borbás – see *Nasturtium officinale*
***Cardamine pratensis* L. (30.44)**
Cardaminum nasturtium Moench – see *Nasturtium officinale*
Cardamon sativum (L.) Fourr. – see *Lepidium sativum*
Cardamon sativum G. Beck – see *Lepidium sativum*
Cardiocrinum giganteum (Wall.) Makino – see *Lilium cordifolium*
Carex tuberosa Blanco – see *Eleocharis dulcis*

Carigola hastata Rafin. – see *Monochoria hastata*
Carota sativa (Hoffm.) Rupr. – see *Daucus carota* subsp. *sativus*
Carota sylvestris Rupr. – see *Daucus carota* subsp. *sativus*
Carpopogon pruriens Roxb. – see *Mucuna pruriens*
Carum petroselinum (L.) Benth. & Hook. f. – see *Petroselinum crispum* var. *crispum*
Carum sisarum Baill. – see *Sium sisarum*
Castalia pubescens (Willd.) Blume – see *Nymphaea nouchali*
Celeri graveolens Britt. – see *Apium graveolens* var. *dulce*
***Celosia argentea* L. (23.4)**
Celosia argentea L. var. *cristata* (L.) O. Kuntze – see *C. argentea*
Celosia cristata L. – see *C. argentea*
Celosia cristata L. var. *splendens* Moq. – see *C. argentea*
Celosia debilis S. Moore – see *C. argentea*
Celosia splendens Schum. & Thonn. – see *C. argentea*
***Centella asiatica* (L.) Urban (24.7)**
Chenopodium scoparia L. – see *Bassia scoparia*
Cepa fistulosa (L.) S.F. Gray – see *Allium fistulosum*
Cepa prolifera Moench – see *Allium* × *proliferum*
Cepa schoenoprasa (L.) Moench – see *Allium schoenoprasum*
Cepa tenuifolia S.F. Gray – see *Allium schoenoprasum*
Cepa ventricosa Moench – see *Allium fistulosum*
Cepa victorialis (L.) Moench – see *Allium victorialis*
Cephalandra indica Naud. – see *Coccinia grandis*
Ceratocephalus acmella Kuntze – see *Acmella oleracea*
Ceratocephalus ciliatus Kuntze – see *Acmella ciliata*
Ceratocephalus debilis Kuntze – see *Acmella oleracea*
Ceratocephalus grandiflora Kuntze – see *Acmella oleracea*
Ceratocephalus tenellus Kuntze – see *Acmella oleracea*
Ceratopteris siliquosa (L.) Copel. – see *C. thalictroides*
***Ceratopteris thalictroides* (L.) Brongn. (4.1)**
Cerefolium cerefolium Britt. – see *Anthriscus cerefolium*
Chaerefolium cerefolium (L.) Schinz & Thell. – see *Anthriscus cerefolium*
Chaerophyllum bobrovii Schischk. – see *C. bulbosum*
***Chaerophyllum bulbosum* L. (24.8)**
Chaerophyllum bulbosum L. var. *caucasicum* Hoffm. – see *C. bulbosum*
Chaerophyllum caucasicum (Hoffm.) Schischk. – see *C. bulbosum*
Chaerophyllum cerefolium (L.) Crantz – see *Anthriscus cerefolium*
Chaerophyllum laevigatum Vis. – see *C. bulbosum*
Chaerophyllum odorata Lam. – see *Myrrhis odorata*
Chaerophyllum rapaceum Alef. – see *C. bulbosum*
Chaerophyllum rapaceum var. *hortense* Alef. – see *C. bulbosum*
Chaerophyllum rapaceum var. *silvestre* Alef. – see *C. bulbosum*
Chaetochloa palmifolia (Willd.) Hitchc. & Chase – see *Setaria palmifolia*
Chamaeraphis palmifolia (Willd.) Kuntze – see *Setaria palmifolia*
Chataea tacca Soland. ex Seem. – see *Tacca leontopetaloides*
Chayota edulis (Jacq.) Jacq. – see *Sechium edule*
Cheliusia abyssinica Sch. Bip. ex A. Rich. – see *Gymnanthemum amygdalinum*
Chenopodina glauca Moq. – see *Suaeda glauca*
Chenopodium album subsp. *quinoa* (Willd.) Kuntze – see *C. quinoa* subsp. *quinoa*
Chenopodium album var. *purpurascens* (Jacq.) Alef. – see *C. quinoa* subsp. *quinoa*

Chenopodium album var. *quinoa* Alef. – see *C. quinoa* subsp. *quinoa*
***Chenopodium bonus-henricus* L. (35.5)**
Chenopodium esculentum Salisb. – see *C. bonus-henricus*
Chenopodium guinoa Krocker – see *C. quinoa* subsp. *quinoa*
Chenopodium hastatum St.-Lag. – see *C. bonus-henricus*
Chenopodium hircinum var. *quinoa* Aellen – see *C. quinoa* subsp. *quinoa*
Chenopodium leucospermum Schrad. – see *C. quinoa* subsp. *quinoa*
Chenopodium millmi J.T. del Granado – see *Amaranthus caudatus*
Chenopodium purpurascens var. *punctulatum* Moq. – see *C. quinoa* subsp. *quinoa*
***Chenopodium quinoa* Willd. subsp. *quinoa* (35.6)**
Chenopodium ruderale St. Lag. – see *C. bonus-henricus*
Chenopodium sagittatum Lam. – see *C. bonus-henricus*
Chenopodium scoparia L. – see *Bassia scoparia*
Chenopodium spinacifolium Stokes – see *C. bonus-henricus*
Chenopodium triangulare Dulac – see *C. bonus-henricus*
Chenopodium triangularifolia Gilib. – see *C. bonus-henricus*
Chlorocyperus aureus Palla – see *Cyperus esculentus*
Chondrocarpus sibthorpioides (Lam.) Sweet. – see *Hydrocotyle sibthorpioides*
Chrysanthemum coronarium L. – see *Glebionis coronarium*
Chrysanthemum roxburghii Desf. – see *Glebionis coronarium*
Chrysanthemum segetum L. – see *Glebionis coronarium*
Chrysanthemum spatiosum L.H. Bailey – see *Glebionis coronarium*
Chymocarpus tuberosus (Ruiz & Pavón) Heynh. – see *Tropaeolum tuberosum*
***Cicer arietinum* L. (40.5)**
Cicer grossum Salisb. – see *C. arietinum*
Cicer kabulicum Dixit – see *C. arietinum*
Cicer lens (L.) Willd. – see *Lens culinaris*
Cicer physodes Reichb. – see *C. arietinum*
Cicer rotundum Jord. ex Alef. – see *C. arietinum*
Cicer sativum Schkuhr – see *C. arietinum*
Cicercula alata Moench – see *Lathyrus sativus*
Cicercula alba Medik. – see *Lathyrus sativus*
Cicercula caerulea Medik. – see *Lathyrus sativus*
Cicercula sativa Medik. – see *Lathyrus sativus*
Cicerula sativa (L.) Alef. – see *Lathyrus sativus*
***Cichorium endivia* L. subsp. *endiva* (27.9)**
Cichorium endivia var. *crispum* Lam. – see *C. endivia* subsp. *endiva*
Cichorium endivia var. *latifolium* Lam. – see *C. endivia* subsp. *endiva*
***Cichorium intybus* L. (27.10)**
Cichorium intybus var. *foliosum* Hegi – see *C. intybus*
Cichorium intybus var. *sativum* (Bisch.) Janch. – see *C. intybus*
Cieca glabrata M. Roem. – see *Passiflora biflora*
***Cirsium dipsacolepis* (Maxim.) Matsum. (27.11)**
***Cissus discolor* Blume (69.1)**
Cissus javana DC. – see *C. discolor*
Citrullus battich Forssk. – see *Citrullus lanatus* var. *lunatus*
Citrullus caffer Schrad. – see *Citrullus lanatus* var. *lunatus*
Citrullus caffrorum Schrad. – see *Citrullus lanatus* var. *lunatus*
Citrullus colocynthoides Pangalo – see *C. lanatus* var. *citroides*
Citrullus colocynthoides var. *citroides* (L.H. Bailey) Millan – see *C. lanatus* var. *citroides*

Citrullus edulis Spach – see *C. lanatus* var. *lunatus*
Citrullus fistulosus Stocks – see *Praecitrullus fistulosus*
Citrullus lanatus (Thunb.) Mansf. – see *C. lanatus* var. *lanatus*
***Citrullus lanatus* (Thunb.) Matsum. & Nakai var. *citroides* (L. H. Bailey) Mansf. (38.2)**
***Citrullus lanatus* (Thunb.) Matsum. & Nakai var. *lanatus* (38.3)**
Citrullus lanatus (Thunb.) Natsum & Nakai – see *C. lanatus* var. *lanatus*
Citrullus lanatus var. *caffer* (Schrad.) Mansfeld – see *C. lanatus* var. *lunatus*
Citrullus lanatus var. *caffrorum* (Schrad.) Fosberg – see *C. lanatus* var. *lunatus*
Citrullus lanatus var. *capensis* (Alef.) Fursa – see *C. lanatus* var. *lunatus*
Citrullus lanatus var. *citroides* (L.H. Bailey) Mansf. – see *C. lanatus* var. *citroides*
Citrullus pasteca Sag. – see *C. lanatus* var. *lunatus*
Citrullus vulgaris L. – see *C. lanatus* var. *lunatus*
Citrullus vulgaris Schrad. – see *C. lanatus* var. *lunatus*
Citrullus vulgaris Schrad. ex Eckl. & Zeyh. – see *C. lanatus* var. *lunatus*
Citrullus vulgaris Schrad. var. *citroides* Bailey – see *C. lanatus* var. *citroides*
Citrullus vulgaris var. *citroides* L.H. Bailey – see *C. lanatus* var. *citroides*
Citrullus vulgaris var. *fistulosus* (Stocks) Duthie & Fuller – see *Praecitrullus fistulosus*
Cjoleus salangensis Gürke – see *Solenostemon rotundifolius*
Cladosicyos edulis Hook. f. – see *Cucumeropsis mannii*
Claytonia cubensis Bonpl. – see *Montia perfoliata*
Claytonia perfoliata Dougl. ex Hook. – see *Montia perfoliata*
Claytonia perfoliata J. Donn. – see *Montia perfoliata*
Claytonia triangularis (Jacq.) Kuntze – see *Talinum fruticosum*
Cleome acuta Schum. & Thonn. – see *C. gynandra*
***Cleome gynandra* L. (34.2)**
Cleome pentaphylla L. – see *C. gynandra*
Clinopodium hortense Kuntze – see *Satureja hortensis*
Clutia androgyna L. – see *Sauropus androgynus*
***Cnidoscolus aconitifolius* (Miller) I.M. Johnston (39.1)**
***Cnidoscolus chayamansa* McVaugh (39.2)**
Cnidoscolus napaeifolius (A. Juss.) Pohl – see *C. aconitifolius*
Coccinia cordifolia (L.) Cogn. – see *C. grandis*
Coccinia corifolia auct. non (L.) Cogn. – see *C. grandis*
***Coccinia grandis* (L.) Voigt (38.4)**
Coccinia indica Wight & Arn. – see *C. grandis*
Cochlearia armoracia L. – see *Armoracia rusticana*
Cochlearia japonica Franch & Sav. – see *Wasabia japonica*
Cochlearia maritima (L.) Crantz – see *Crambe maritime*
Cochlearia rusticana Lam. – see *Armoracia rusticana*
Cochlearia wasabi Siebold – see *Wasabia japonica*
***Codiaeum variegatum* (L.) Blume (39.3)**
***Codonopsis lanceolata* (Siebold & Zucc.) Trautv. (33.3)**
Coleus dazo A. Chev. – see *Plectranthus esculentus*
Coleus dysentericus Bak. – see *Solenostemon rotundifolius*
Coleus esculentus (N.E. Brown) G. Tayl. – see *Plectranthus esculentus*
Coleus floribundus (N.E. Br.) Robyns & Lebrun – see *Plectranthus esculentus*
Coleus langouassiensis A. Chev. – see *Plectranthus esculentus*
Coleus pallidiflorus A. Chev. – see *Solenostemon rotundifolius*
Coleus rotundifolius (Poir.) A. Chev. & Perrot – see *Solenostemon rotundifolius*
Coleus rotundifolius (Blume) Benth. – see *Plectranthus esculentus*

Coleus salagensis Gürke – see *Solenostemon rotundifolius*
Coleus tuberosus (Blume) Bentham – see *Plectranthus esculentus*
Colocasia antiquorum Schott, ex Schott & Endl. – see *C. esculenta*
Colocasia antiquorum Schott var. *esculenta* (L.) Schott – see *C. esculenta*
Colocasia antiquorum var. *esculenta* Engl. – see *C. esculenta*
Colocasia antiquorum var. *globulifera* Engl. & Krause – see *C. esculenta*
***Colocasia esculenta* (L.) Schott (8.4)**
Colocasia esculenta (L.) Schott var. *antiquorum* (Schott) Hubb. & Rehder – see *C. esculenta*
***Colocasia gigantea* (Blume) Hook. f. (8.5)**
Colocasia indica (Lour.) Hassk. – see *C. gigantea*
Colocasia indica (Lour.) Kunth – see *Alocasia macrorrhizos*
Colocasia macrorrhiza (L.) Schott, ex Schott & Endl. – see *Alocasia macrorrhizos*
Colocasia macrorrhiza Schott – see *Alocasia macrorrhizos*
Colocasia mucronata Kunth – see *Alocasia macrorrhizos*
Colocasia odora Brongn. – see *Alocasia macrorrhizos*
Colocasia peregrina Raf. – see *C. esculenta*
Colocasia peregrina Schott – see *Alocasia macrorrhizos*
Colocasia vulgaris Raf. – see *C. esculenta*
Colocynthis citrullus (L.) O. Kuntze – see *Citrullus lunatus* var. *lunatus*
Colocynthis citrullus var. *fistulosus* (Stocks) Chakrav. – see *Praecitrullus fistulosus*
Conium arracacha Hook. – see *Arracacia xanthorrhiza*
Consolida major Garsault – see *Symphytum officinale*
Consolida major Gilib. – see *Symphytum officinale*
Convolvulus adansonii Desr. – see *Ipomoea aquatica*
Convolvulus batatas L. – see *Ipomoea batatas* var. *batatas*
Convolvulus batatas Lam. – see *Ipomoea batatas* var. *batatas*
Convolvulus candicans Soland. ex Sims – see *Ipomoea batatas* var. *batatas*
Convolvulus chrysorhizus Forst. ex Wien. – see *Ipomoea batatas* var. *batatas*
Convolvulus chrysorrhizus Soland. ex G. Forster – see *Ipomoea batatas* var. *batatas*
Convolvulus cordatifolius Vell. – see *Ipomoea batatas* var. *batatas*
Convolvulus edulis Thunb. – see *Ipomoea batatas* var. *batatas*
Convolvulus esculentus Salisb. – see *Ipomoea batatas* var. *batatas*
***Convolvulus japonicus* Thunb. (36.1)**
Convolvulus repens Vahl. – see *Ipomoea aquatica*
Convolvulus reptans L. – see *Ipomoea aquatica*
Convolvulus rostratus Zipp. ex Span. – see *Ipomoea aquatica*
Convolvulus septangularis Steud. – see *Ipomoea batatas* var. *batatas*
Convolvulus tuberifer Steud. – see *Ipomoea batatas* var. *batatas*
Convolvulus tuberosus Vell. – see *Ipomoea batatas* var. *batatas*
Convolvulus varius Vell. – see *Ipomoea batatas* var. *batatas*
***Corchorus olitorius* L. (63.1)**
Corchous catharticus Blanco – see *Corchorus olitorius*
Corchous decemangularis Roxb. – see *Corchorus olitorius*
Corchous lobatus Wildem. – see *Corchorus olitorius*
Corchous quinquelocularis Moench – see *Corchorus olitorius*
Cordaea fabaeformis (L'Hérit.) Spreng. – see *Cyamopsis tetragonoloba*
Coreopsis acmella E.H.L. Krause – see *Acmella oleracea*
Coriandrum diversifolium Gilib. – see *C. sativum*
Coriandrum globosum Salisb. – see *C. sativum*
Coriandrum majus Gouan – see *C. sativum*

Coriandrum melphitense Ten. & Guss. – see *C. sativum*
***Coriandrum sativum* L. (24.9)**
Coriandrum testiculatum auct. – see *C. sativum*
Coronopus vulgaris Fourr. – see *Plantago coronopus* var. *sativa*
Corynosicyos edulis (Hook. f.) F. Muell. – see *Cucumeropsis mannii*
Cosmos bipinnatus Ridley – see *C. caudatus*
***Cosmos caudatus* Kunth (27.12)**
Cotula conica Wall. – see *Acmella oleracea*
Crambe aspera M. Bieb. – see *C. tataria*
***Crambe maritima* L. (30.45)**
Crambe pontica Steven, ex Rupr. – see *C. maritima*
***Crambe tataria* Sebeók (30.46)**
Crambe tatarica Jacq. – see *C. tataria*
Crambe tatarica Pall. ex Güldenstädt – see *C. tataria*
Crambe buschii (O.E. Schulz) Stankev. ex Wul'f – see *C. tataria*
Crambe gibberosa Rupr. – see *C. tataria*
Crambe litwinowii Gross – see *C. tataria*
Crambe pinnatifida R. Br. – see *C. tataria*
***Crassocephalum biafrae* (Oliv. & Hiern) S. Moore (27.13)**
***Crassocephalum crepidioides* (Benth.) S. Moore (27.14)**
Crassocephalum sonchifolium (L.) Less. – see *Emilia sonchifolia.*
Crepidiastrum sonchifolium (Maxim.) Pak & Kawano – see *Youngia sonchifolia*
Crepis taraxacum Stokes – see *Taraxacum officinale*
Croton variegatum L. – see *Codiaeum variegatum*
Crucifera eruca E.H.L. Krause – see *Eruca sativa*
Crucifera fontana E.H.L.Krause – see *Nasturtium officinale*
Crucifera nasturtium E.H.L. Krause – see *Lepidium sativum*
Crucifera sinapis E.H.L. Krause – see *Brassica nigra*
Cryophytum crystallinum (L.) N.E. Br. – see *Mesembryanthemum crystallinum*
Cryophytum crystallinum N.E. Br. ex E.P Phillips – see *Mesembryanthemum crystallinum*
Cryphiospermum repens P. Beauv. – see *Enydra fluctuans*
Cryptolobus africanus (Burm. f.) Spreng. – see *Vigna subterranea*
Cryptolobus subterraneus (L.) Spreng. – see *Vigna subterranea*
Cryptotaenia canadensis auct., non DC. – see *C. japonica*
Cryptotaenia canadensis (L.) DC. – see *C. japonica*
Cryptotaenia canadensis (L.) DC. subsp. *japonica* (Hassk.) Hand.-Mazz. – see *C. japonica*
Cryptotaenia canadensis (L.) DC. var. *japonica* (Hassk.) Makino – see *C. japonica*
Cryptotaenia canadensis Pers. – see *C. japonica*
Cryptotaenia canadensis var. *japonica* (Hassk.) Makino – see *C. japonica*
***Cryptotaenia japonica* Hassk. (24.10)**
Cucumeropsis edulis (Hook. f.) Cogn. – see *C. mannii*
Cucumeropsis mannii A. Chev. – see *C. mannii* Naudin
***Cucumeropsis mannii* Naudin (38.5)**
Cucumis acutangula L. – see *Luffa acutangula*
Cucumis acutangulus Descorut – see *Sechium edule*
Cucumis acutangulus L. – see *Luffa acutangula*
Cucumis acutangulus Wall. – see *Luffa aegyptiaca*
Cucumis africanus Lindl. – see *Momordica charantia*
Cucumis anguinus L. – see *Trichosanthes cucumerina*
Cucumis anguria A. Meeuse var. *anguria* – see *C. anguria* var. *anguria*

Cucumis anguria L. – see *C. anguria* var. *anguria*
***Cucumis anguria* L. var. *anguria* (38.6)**
Cucumis argyi Lév. – see *Momordica charantia*
Cucumis aromaticus Royle – see *C. melo* subsp. *melo* var. *chito*
Cucumis cantalupensis Haberle, ex M.J. Roem. – see *C. melo* subsp. *melo* var. *cantalupensis*
Cucumis cantalupo (Ser.) Haberle, ex Rchb. – see *C. melo* subsp. *melo* var. *cantalupensis*
Cucumis chinensis Pangalo – see *C. melo* subsp. *melo* var. *conomon*
Cucumis chito Morren – see *C. melo* subsp. *melo* var. *chito*
Cucumis citrullus (L.) Ser. – see *Citrullus lanatus* var. *lunatus*
Cucumis conomon Thunb. – see *C. melo* subsp. *melo* var. *conomon*
Cucumis dissectus Decne. – see *Citrullus lanatus* var. *lunatus*
Cucumis dudaim L. – see *C. melo* subsp. *melo* var. *chito*
Cucumis erinaceus hort. – see *C. anguria* var. *anguria*
Cucumis esculentus Salisb. – see *C. sativus*
Cucumis eumelo var. *cantalupa* Pangalo – see *C. melo* subsp. *melo* var. *cantalupensis*
Cucumis flexuosus L. – see *C. melo* subsp. *melo* var. *flexuosus*
Cucumis flexuosus Ser. var. *tarra* Pangalo – see *C. melo* subsp. *melo* var. *flexuosus*
Cucumis grossulariiformis hort. – see *C. anguria* var. *anguria*
Cucumis intermedius M.J. Roem. – see *Momordica charantia*
Cucumis lineatus Bosc – see *Luffa aegyptiaca*
Cucumis mairei Lévl. – see *Lagenaria siceraria*
Cucumis megacarpus G. Don – see *Luffa aegyptiaca*
Cucumis melo convar. *dudaim* (L.) Grebenšč. – see *C. melo* subsp. *melo* var. *chito*
Cucumis melo convar. *flexuosus* (L.) Grebenšč. – see *C. melo* subsp. *melo* var. *flexuosus*
Cucumis melo E. Morr. – see *C. melo* subsp. *melo* var. *chito*
Cucumis melo J. Schultze-Motel convar. *melo* – see *C. melo* subsp. *melo* var. *cantalupensis*
Cucumis melo L. Cantaloupensis group – see *C. melo* subsp. *melo* var. *cantalupensis*
Cucumis melo L. *cantalupensis* Munger & Robinson – see *C. melo* subsp. *melo* var. *cantalupensis*
Cucumis melo L. Chito group – see *C. melo* subsp. *melo* var. *chito*
Cucumis melo L. *chito* Munger & Robinson – see *C. melo* subsp. *melo* var. *chito*
Cucumis melo L. Conomon group – see *C. melo* subsp. *melo* var. *conomon*
Cucumis melo L. *conomon* Munger & Robinson – see *C. melo* subsp. *melo* var. *conomon*
Cucumis melo L. convar. *cantalupa* (Pangalo) Grebenšč. – see *C. melo* subsp. *melo* var. *cantalupensis*
Cucumis melo L. convar. *conomon* (Thunb.) Grebenšč. – see *C. melo* subsp. *melo* var. *conomon*
Cucumis melo L. *dudaim* Munger & Robinson – see *C. melo* subsp. *melo* var. *chito*
Cucumis melo L. Flexuosus group – see *C. melo* subsp. *melo* var. *flexuosus*
Cucumis melo L. *flexuosus* Munger & Robinson – see *C. melo* subsp. *melo* var. *flexuosus*
Cucumis melo L. Inodorus group – see *C. melo* subsp. *melo* var. *inodorus*
Cucumis melo L. subsp. *chinensis* (Pangalo) Filov – see *C. melo* subsp. *melo* var. *conomon*
Cucumis melo L. subsp. *melo* Chito group – see *C. melo* subsp. *melo* var. *chito*
Cucumis melo L. subsp. *melo* Conomon group – see *C. melo* subsp. *melo* var. *conomon*
Cucumis melo L. subsp. *melo* Flexuosus group – see *C. melo* subsp. *melo* var. *flexuosus*
Cucumis melo L. subsp. *melo* Inodorus group – see *C. melo* subsp. *melo* var. *inodorus*
Cucumis melo L. subsp. *melo* Reticulatus group – see *C. melo* subsp. *melo* var. *cantalupensis*
***Cucumis melo* L. subsp. *melo* var. *cantalupensis* Naudin (38.7)**
***Cucumis melo* L. subsp. *melo* var. *chito* (C. Morren) Naudin (38.8)**
***Cucumis melo* L. subsp. *melo* var. *conomon* (Thunb.) Makino (38.9)**
***Cucumis melo* L. subsp. *melo* var. *flexuosus* (L.) Naudin (38.10)**
***Cucumis melo* L. subsp. *melo* var. *inodorus* H. Jacq. (38.11)**
Cucumis melo L. subsp. *spontaneus* Filov – see *C. melo* subsp. *melo* var. *chito*

Cucumis melo L. var. *acidulus* Naud. – see *C. melo* subsp. *melo* var. *conomon*
Cucumis melo L. var. *bitariana* (Sickenberger) Hassib – see *C. melo* subsp. *melo* var. *chito*
Cucumis melo L. var. *cantalupa* Sag. ex Filov– see *C. melo* subsp. *melo* var. *cantalupensis*
Cucumis melo L. var. *cantalupensis* Naud. – see *C. melo* subsp. *melo* var. *cantalupensis*
Cucumis melo L. var. *cantalupo* Ser. – see *C. melo* subsp. *melo* var. *cantalupensis*
Cucumis melo L. var. *chito* (E. Morr.) Naud. – see *C. melo* subsp. *melo* var. *chito*
Cucumis melo L. var. *conomon* form *albus* Makino – see *C. melo* subsp. *melo* var. *conomon*
Cucumis melo L. var. *conomon* (Thunb.) Makino – see *C. melo* subsp. *melo* var. *conomon*
Cucumis melo L. var. *cultus* – see *C. melo* subsp. *melo* var. *chito*
Cucumis melo L. var. *erythraeus* Naud – see *C. melo* subsp. *melo* var. *chito*
Cucumis melo L. var. *flexuosus* Naud. – see *C. melo* subsp. *melo* var. *flexuosus*
Cucumis melo L. var. *indicus* Malin. – see *C. melo* subsp. *melo* var. *conomon*
Cucumis melo L. var. *inodorus* Naud. – see *C. melo* subsp. *melo* var. *inodorus*
Cucumis melo L. var. *reticulatus* Naud. – see *C. melo* subsp. *melo* var. *cantalupensis*
Cucumis melo L. var. *reticulatus* Ser. – see *C. melo* subsp. *melo* var. *cantalupensis*
Cucumis melo L. var. *utilissimus* Duthie & Fuller – see *C. melo* subsp. *melo* var. *conomon*
Cucumis melo L. var.-gr. *cantalupensis* Alef. – see *C. melo* subsp. *melo* var. *cantalupensis*
Cucumis melo L. var.-gr. *microcarpus* Alef – see *C. melo* subsp. *melo* var. *chito*
Cucumis melo subsp. *dudaim* (L.) Grebenšč. – see *C. melo* subsp. *melo* var. *chito*
Cucumis melo subsp. *flexuosus* (L.) Grebenšč. – see *C. melo* subsp. *melo* var. *flexuosus*
Cucumis melo subsp. *flexuosus* (L.) Pangalo– see *C. melo* subsp. *melo* var. *flexuosus*
Cucumis melo var. *dudaim* (L.) Naud. – see *C. melo* subsp. *melo* var. *chito*
Cucumis melo var. *flexuosus* (L.) Naud. – see *C. melo* subsp. *melo* var. *flexuosus*
Cucumis melo var. *utilissimus* (Roxb.) Duthie & Fuller – see *C. melo* subsp. *melo* var. *conomon*
***Cucumis metuliferus* E. Meyer, ex Naudin (38.12)**
Cucumis microcarpus (Alef.) Pangalo – see *C. melo* subsp. *melo* var. *chito*
Cucumis missionis Wall. – see *Trichosanthes cucumerina*
Cucumis muricatus Willd. – see *C. sativus*
Cucumis odoratissimus Moench – see *C. melo* subsp. *melo* var. *chito*
Cucumis operculatus Roxb. ex wight & Arn. – see *Luffa acutangula*
Cucumis pentandrus Roxb. ex Wight & Arn. – see *Luffa aegyptiaca*
Cucumis pictus Jacq. – see *C. melo* subsp. *melo* var. *chito*
***Cucumis sativus* L. (38.13)**
Cucumis tinneanus Kotschy & Peyr. – see *C. metuliferus*
Cucumis utilissimus Roxb. – see *C. melo* subsp. *melo* var. *conomon*
Cucurbita acutangula (L.) Blume – see *Luffa acutangula*
Cucurbita anguria Duch. ex Lam. – see *Citrullus lanatus* var. *lunatus*
***Cucurbita argyrosperma* C. Huber. (38.14)**
Cucurbita cerifera Fisch. – see *Benincasa hispida*
Cucurbita citrullus L. – see *Citrullus lanatus* var. *lunatus*
Cucurbita esculenta S.F. Gray – see *C. pepo*
Cucurbita evodicarpa Hassk. – see *Sicana odorifera*
Cucurbita farinosa Blume – see *Benincasa hispida*
Cucurbita fastuosa Salisb. – see *C. pepo*
***Cucurbita ficifolia* Bouché (38.15)**
Cucurbita hippopera Ser. – see *C. moschata*
Cucurbita hispida Thunb. ex Murray – see *Benincasa hispida*
Cucurbita idolatrica Willd. – see *Lagenaria siceraria*
Cucurbita lagenaria L. – see *Lagenaria siceraria*
Cucurbita leucantha Duch. ex Lam. – see *Lagenaria siceraria*

Cucurbita leucantha Lam. – see *Lagenaria siceraria*
Cucurbita littoralis Hassk – see *Benincasa hispida.*
Cucurbita macrocarpa Gaspar. – see *C. moschata*
Cucurbita mammeata Molina – see *C. pepo*
Cucurbita maxima Duch. ex. Lam – see *Cucurbita maxima* Duchesne
***Cucurbita maxima* Duchesne (38.16)**
Cucurbita maxima oblonga courgero (Ser.) Ser. – see *C. pepo*
Cucurbita maxima var. *courgero* Ser. – see *C. pepo*
Cucurbita melanosperma A. Braun – see *C. ficifolia*
Cucurbita melanosperma Gaspar. – see *C. ficifolia*
Cucurbita melopepo L. – see *C. pepo*
Cucurbita melopepo Lour. – see *C. moschata*
Cucurbita mixta Pang. – see *C. argyrosperma*
Cucurbita moschata (Duch.) Poir. – see *C. moschata*
Cucurbita moschata Duch. var. *melonaeformis* Makino – see *C. moschata*
Cucurbita moschata Duch. var. *toonas* Makino – see *C. moschata*
***Cucurbita moschata* Duchesne (38.17)**
Cucurbita moschata Poir. – see *C. moschata* Duchesne
Cucurbita moschata subsp. *argyrosperma* (Huber) Filov – see *C. argyrosperma*
Cucurbita oblonga (Duch. ex Lam.) Link– see *C. pepo*
Cucurbita odorifera Vell. – see *Sicana odorifera*
Cucurbita ovifera L. – see *C. pepo*
Cucurbita pepo DC. – see *C. pepo* L.
***Cucurbita pepo* L. (38.18)**
Cucurbita pepo L. *moschata* Duch. ex Lam. – see *C. moschata*
Cucurbita pepo L. *polymorpha* Duch. ex Lam. – see *C. pepo*
Cucurbita pepo L. var. *meloniformis* (Carriere) Makino – see *C. moschata*
Cucurbita pepo L. var. *toonas* Makino – see *C. moschata*
Cucurbita pepo-aspera Blanco – see *Benincasa hispida*
Cucurbita polymorpha Duchesne – see *C. pepo*
Cucurbita potiro Pers. – see *C. maxima*
Cucurbita siceraria Molina – see *Lagenaria siceraria*
Cucurbita spathularis Schrad. – see *C. moschata*
Cucurbita subverrucosa Willd. – see *C. pepo*
Cucurbita vacua F. Muell. – see *Benincasa hispida*
Cucurbita venosa Descourt. – see *C. pepo*
Cucurbita verrucosa L. – see *C. pepo*
Cucurbita villosa Blume – see *Benincasa hispida*
Curcuma americana Lam. – see *Calathea allouia*
Curcuma domestica Valeton – see *C. longa*
Curcuma longa Koenig. – see *C. longa* L.
***Curcuma longa* L. (20.2)**
Curcuma longifolia Wall. – see *Zingiber officinale*
***Curcuma mangga* Valeton & Van Zijp (20.3)**
Curcuma pallida Lour. – see *C. zedoaria*
Curcuma speciosa Hort. ex Link – see *C. zedoaria*
Curcuma zedoaria (Bergius) Roxb. – see *C. zedoaria* (Christm.) Roscoe
***Curcuma zedoaria* (Christm.) Roscoe (20.4)**
Curcuma zerumbet (Berg.) Roxb. – see *C. zedoaria*
Curcuma zerumbet (Koen.) Roxb. – see *C. zedoaria*

Cyamopsis psoraloides (Lam.) DC. – see *C. tetragonoloba*
***Cyamopsis tetragonoloba* (L.) Taubert (40.6)**
Cyamus nelumbo (L.) Sm. – see *Nelumbo nucifera*
Cyclanthera digitata Arn. – see *C. pedata*
Cyclanthera edulis Naud. – see *C. pedata*
***Cyclanthera pedata* (L.) Schrader (38.19)**
Cyclanthera pedata var. *edulis* (Naudin) Cogn. – see *C. pedata*
Cymopterus litoralis A. Gray – see *Glehnia littoralis*
Cynara cardunculus form *scolymus* Halac. – see *C. scolymus*
Cynara cardunculus Hayek subsp. *cardunculus* – see *C. cardunculus* subsp. *cardunculus*
***Cynara cardunculus* L. subsp. *cardunculus* (27.15)**
Cynara cardunculus L. subsp. *cardunculus* Hegi – see *C. cardunculus* subsp. *cardunculus*
Cynara cardunculus L. subsp. *flavescens* Wiklund – see *C. cardunculus* subsp. *cardunculus*
Cynara cardunculus L. subsp. *scolymus* (L.) Hegi. – see *C. scolymus*
Cynara cardunculus L. var. *sativa* Moretti – see *C. scolymus*
Cynara cardunculus L. var. *scolymus* L. – see *C. scolymus*
Cynara horrida Aiton – see *C. scolymus*
Cynara humilis Viv. – see *C. scolymus*
***Cynara scolymus* L. (27.16)**
Cynara spinosissima Presl – see *C. scolymus*
Cynara sylvestris Lam. – see *C. scolymus*
Cyperus aureus Ten. – see *C. esculentus*
Cyperus buchananii Boeck. – see *C. esculentus*
Cyperus esculentus Gouan – see *C. esculentus* L.
***Cyperus esculentus* L. (10.1)**
Cyperus hydra H.B.K. – see *C. esculentus*
Cyperus melanorrhizus Delile – see *C. esculentus*
Cyperus nervosus Roem. & Schult. – see *C. esculentus*
Cyperus pallidus Savi – see *C. esculentus*
Cyperus phymatodes Mubl. – see *C. esculentus*
Cyperus repens Ell. – see *C. esculentus*
Cyperus retusus Nees – see *C. esculentus*
Cyperus rotundus Benth. – see *C. esculentus*
Cyperus rotundus L. – see *C. esculentus*
Cyperus sieberianus Link – see *C. esculentus*
Cyperus tenorianus Schult. f. – see *C. esculentus*
Cyperus tenorii Presl – see *C. esculentus*
Cyperus tuberosus Pursh – see *C. esculentus*
***Cyphomandra betacea* (Cav.) Sendtner (62.7)**
Cyphomandra crassifolia (Ortega) Kuntze – see *C. betacea*
Cyphomandra procera Wawra. – see *C. betacea*
***Cyrtosperma chamissonis* (Schott) Merr. (8.6)**
Cyrtosperma cuspidilobum Schott – see *C. merkusii*
Cyrtosperma dubium Schott – see *C. merkusii*
Cyrtosperma edule Schott, ex Seem. – see *C. chamissonis*
Cyrtosperma intermedium Schott – see *C. merkusii*
Cyrtosperma lasioides Griff. – see *C. chamissonis*
***Cyrtosperma merkusii* (Hassk.) Schott. (8.7)**
Cyrtosperma nadeaudianum Moore – see *C. merkusii*
Cyssopetalum javanicum (Blume) Turcz. – see *Oenanthe javanica* subsp. *javanica*

Cytisus cajan L. – see *Cajanus cajan*
Cytisus pseudocajan Jacq. – see *Cajanus cajan*
Damasonium flavum Mill. – see *Limnocharis flava*
Dasyloma benghalensis DC. – see *Oenanthe javanica* subsp. *javanica*
Dasyloma javanicum (Blume) Miq. – see *Oenanthe javanica* subsp. *javanica*
Dasyloma subbipinnatum Miq. – see *Oenanthe javanica* subsp. *javanica*
Daucus abyssinicus C.A. Mey. – see *D. carota* subsp. *sativus*
Daucus carota convar. *afganicus* Setchkarev – see *D. carota* subsp. *sativus*
Daucus carota convar. *sativus* Setchkarev – see *D. carota* subsp. *sativus*
***Daucus carota* L. subsp. *sativus* (Hoffm.) Arcang. (24.11)**
Daucus carota L. subsp. *sativus* (Hoffm.) Thell. – see *D. carota* L. subsp. *sativus* (Hoffm.) Arcang.
Daucus carota L. var. *atrorubens* Alef. – see *D. carota* subsp. *sativus*
Daucus carota L. var. *sativus* DC. – see *D. carota* subsp. *sativus*
Daucus carota L. var. *sativus* Hoffm. – see *D. carota* subsp. *sativus*
Daucus carota var. *sativus* Hoffm. – see *D. carota* subsp. *sativus*
Daucus gingidium L. – see *D. carota* subsp. *sativus*
Daucus sativa (Hoffm.) Pass. – see *D. carota* subsp. *sativus*
Decaloba biflora M. Roem. – see *Passiflora biflora*
Decaneurum amygdalinum (Del.) DC. – see *Gymnanthemum amygdalinum*
Demidovia tetragoniodes Pall. – see *Tetragonia tetragoniodes*
Dendranthema × *grandiflorum* (Ramat.) Kitamura – see *Chrysanthemum* × *morifolium*
Dendranthema morifolium (Ramat.) Tzvel. - see *Chrysanthemum* × *morifolium*
Dendrocalamopsis atrovirens (Wen) Keng f. ex W.T. Lin – see *Bambusa oldhami*
Dendrocalamopsis oldhamii (Munro) Keng f. – see *Bambusa oldhamii*
***Dendrocalamus asper* (Schultes f.) Backer, ex Heyne (17.2)**
***Dendrocalamus brandisii* (Munro) Kurz (17.2)**
Dendrocalamus flagellaris Ridley – see *D. asper*
Dendrocalamus flagellifer Munro – see *D. asper*
***Dendrocalamus giganteus* Wall. ex Munro (17.2)**
***Dendrocalamus hamiltonii* Nees & Arn. ex Munro (17.2)**
***Dendrocalamus latiflorus* Munro (17.2)**
***Dendrocalamus membranaceus* Munro (17.2)**
Dendrocalamus merrillianus (Elmer) Elmer – see *D. asper*
***Dendrocalamus strictus* (Roxb.) Nees (17.2)**
Dentidia nankinensis Lour. – see *Perilla frutescens* var. *crispa*
Dentidia purpurascens Pers. – see *Perilla frutescens* var. *crispa*
Dentidia purpurascens Poir. – see *Perilla frutescens* var. *crispa*
Deringa japonica (Hassk.) Koso-Pol. – see *Cryptotaenia japonica*
Desmanthus lacustris Willd. – see *Neptunia prostrata*
Desmanthus natans Willd. – see *Neptunia prostrata*
Desmanthus stolonifer DC. – see *Neptunia prostrata*
Dimorphanthus elatus Miq. – see *Aralia elata*
Dimorphanthus mandshuricus (Rupr. & Maxim.) Maxim. – see *Aralia elata*
Dioscorea aculeata Balb. ex Kunth – see *D. cayenensis*
Dioscorea aculeata F. Ham. ex Wall. – see *D. esculenta*
Dioscorea aculeata L. – see *D. esculenta*
Dioscorea aculeata Wight – see *D. alata*
Dioscorea acutangula F. Ham. ex Wall. – see *D. alata*
Dioscorea affinis Kunth – see *D. trifida*
***Dioscorea alata* L. (11.1)**

Dioscorea alata L. var. *purpurea* hort. ex Bull. – see *D. alata*
Dioscorea alata L. var. *purpurea* (Roxb.) M. Pouch. – see *D. alata*
Dioscorea angustiflora Rendle – see *D. cayenensis*
Dioscorea anthropophagorum A. Chev. – see *D. bulbifera*
Dioscorea atropurpurea Roxb. – see *D. alata*
Dioscorea batatas Decne. – see *D. oppositifolia*
Dioscorea berteroana Kunth – see *D. cayenensis*
Dioscorea brasiliana Poir. – see *D. trifida*
Dioscorea brasiliensis Willd. – see *D. trifida*
Dioscorea buchholziana Engl. – see *D. dumetorum*
***Dioscorea bulbifera* L. (11.2)**
Dioscorea bulbifera L. form *domestica* Makino & Nemoto – see *D. bulbifera*
Dioscorea camerunensis Knuth – see *D. cayenensis*
***Dioscorea cayenensis* Lam. (11.3)**
Dioscorea cayenensis subsp. *rotundata* (Poir.) J. Miège – see *D. rotundata*
Dioscorea cliffortiana Lam. – see *D. bulbifera*
Dioscorea colocasiaefolia Pax – see *D. alata*
Dioscorea colocasiifolia Dalz. non Pax – see *D. cayenensis*
Dioscorea combilium F. Ham. ex Wall. – see *D. esculenta*
Dioscorea combilium Prain & Burk. – see *D. esculenta*
Dioscorea coriacea Wight, ex Wall. – see *D. oppositifolia*
Dioscorea crispata Roxb. – see *D. bulbifera*
Dioscorea cylindrica Burm. f. – see *Ipomoea batatas* var. *batatas*
Dioscorea daemona Hook. – see *D. dumetorum*
Dioscorea decaisneana Carrière – see *D. oppositifolia*
Dioscorea demeusei De Wild. & Th. Dur. – see *D. cayenensis*
Dioscorea diacantha Zipp. ex Span. – see *D. esculenta*
Dioscorea doryphora Hance – see *D. oppositifolia*
***Dioscorea dumetorum* (Kunth) Pax (11.4)**
Dioscorea eburina Lour. – see *D. alata*
Dioscorea eburnea Willd. – see *D. alata*
Dioscorea echinata F. Ham. ex Wall. – see *D. esculenta*
Dioscorea elliptica Thunb. ex Prain & Burkill – see *D. oppositifolia*
***Dioscorea esculenta* (Lour.) Burkill (11.5)**
Dioscorea fasciculata Roxb. – see *D. esculenta*
Dioscorea glabra Wall. – see *D. oppositifolia*
Dioscorea globosa Roxb. – see *D. alata*
Dioscorea goyazensis Griseb. – see *D. trifida*
Dioscorea heterophylla Roxb. – see *D. bulbifera*
Dioscorea hoffa Cordem. – see *D. bulbifera*
Dioscorea javanica Queva – see *D. alata*
Dioscorea lanata Bail. f. – see *D. esculenta*
Dioscorea lanceolata Heyne, ex Hook. f. – see *D. oppositifolia*
Dioscorea latifolia Benth. – see *D. bulbifera*
Dioscorea lutea Griseb. – see *D. bulbifera*
Dioscorea martinicensis Spreng. – see *D. alata*
Dioscorea moma De Wild. – see *D. cayenensis*
Dioscorea nummularia Lam. – see *D. oppositifolia*
Dioscorea occidentalis Knuth – see *D. cayenensis*
Dioscorea opposita Thunb. – see *D. oppositifolia*

Dioscorea oppositifolia Campbell – see *D. bulbifera*
***Dioscorea oppositifolia* L. (11.6)**
Dioscorea palmata Juss. ex Pers. – see *D. trifida*
Dioscorea papillaris Blanco – see *D. bulbifera*
Dioscorea papuana Warb. – see *D. esculenta*
Dioscorea praehensilis Baker. – see *D. cayenensis*
Dioscorea preangeriana Uline, ex Prain & Burkill – see *D. oppositifolia*
Dioscorea prehensilis Benth. – see *D. cayenensis*
Dioscorea pruinosa Chev. – see *D. cayenensis*
Dioscorea pulchella Roxb. – see *D. bulbifera*
Dioscorea purpurea Roxb. – see *D. alata*
Dioscorea pyrifolia Kunth – see *D. oppositifolia*
Dioscorea quinquelobata Vell. – see *D. trifida*
***Dioscorea rotundata* Poiret (11.7)**
Dioscorea rubella Roxb. – see *D. alata*
Dioscorea ruiziana Klotzsch, ex Kunth – see *D. trifida*
Dioscorea sapini De Wild. – see *D. alata*
Dioscorea sativa auct., non L. – see *D. bulbifera*
Dioscorea sativa Beatson. – see *D. cayenensis*
Dioscorea sativa Blanco – see *D. esculenta*
Dioscorea sativa Del. – see *D. alata*
Dioscorea sativa form *domestica* Makino – see *D. esculenta*
Dioscorea sativa L. – see *D. esculenta*
Dioscorea sativa Munro – see *D. alata*
Dioscorea sativa Thunb. – see *D. bulbifera*
Dioscorea sativa var. *domestica* Makino – see *D. esculenta*
Dioscorea spiculata Blume – see *D. alata*
Dioscorea spinosa Roxb. ex Wall. – see *D. esculenta*
Dioscorea swinhoei Rolfe – see *D. oppositifolia*
Dioscorea tamnifolia Salisbury – see *D. bulbifera*
Dioscorea tiliifolia Kunth – see *D. esculenta*
Dioscorea tredecimnervis Pierre, ex Kunth – see *D. esculenta*
Dioscorea triandra Hort. ex Griseb. – see *D. bulbifera*
***Dioscorea trifida* L. f. (11.8)**
Dioscorea triloba Lam. – see *D. trifida*
Dioscorea triphylla Hochst. – see *D. dumetorum*
Dioscorea triphylla Schimp. ex Kunth – see *D. dumetorum*
Dioscorea triphylla var. *dumetorum* (Kunth) R. Knuth – see *D. dumetorum*
Dioscorea tuberosa Vell. – see *D. trifida*
Dioscorea tugui Blanco – see *D. esculenta*
Dioscorea tunga F. Ham. ex Hook. f. – see *D. bulbifera*
Dioscorea tunga Hamilton – see *D. bulbifera*
Dioscorea tunga Wall. – see *D. bulbifera*
Dioscorea versicolor F. Ham. ex Wall. – see *D. esculenta*
Dioscorea violacea Baudon – see *D. bulbifera*
Dioscorea vulgaris Miq. – see *D. alata*
***Diplazium esculentum* (Retz.) Swartz. (5.1)**
Diplazium malabaricum Spreng. – see *D. esculentum*
***Diplotaxis muralis* (L.) DC. (30.47)**
Diplotaxis tenuifolia (L.) DC. – see *D. muralis*

***Doellingeria scabra* (Thunb.) Nees (27.17)**
Dolichos acinaciformis Jacq. – see *Canavalia ensiformis*
Dolichos ahipa Wedd – see *Pachyrhizus ahipa*
Dolichos albus Lour. – see *Lablab purpureus*
Dolichos altissimus Lour. – see *Lablab purpureus*
Dolichos amoenus Salisb. – see *Lablab purpureus*
Dolichos angularis Willd. – see *Vigna angularis*
Dolichos articulatus Lam. – see *Pachyrhizus erosus*
Dolichos bengalensis Jacq. – see *Lablab purpureus*
Dolichos biflorus auct. mult,. non L. – see *Macrotyloma uniflorum*
Dolichos biflorus L. – see *Vigna unguiculata* subsp. *cylindrica*
Dolichos bulbosus L. – see *Pachyrhizus erosus*
Dolichos catjang Burm. f. – see *Vigna unguiculata* subsp. *cylindrica.*
Dolichos culturatus Thunb. – see *Lablab purpureus*
Dolichos dissectus Lam. – see *Vigna aconitifolia*
Dolichos ensiformis auct. non (L.) DC. – see *Canavalia gladiata*
Dolichos ensiformis L. – see *Canavalia ensiformis*
Dolichos ensiformis Thunb. – see *Lablab purpureus*
Dolichos erosus L. – see *Pachyrhizus erosus*
Dolichos fabaeformis L'Hér. – see *Cyamopsis tetragonoloba*
Dolichos fabaeformis Willd. – see *Cyamopsis tetragonoloba*
Dolichos gladiatus Jacq. – see *Canavalia gladiata*
Dolichos hirsutus Thunb. – see *Pueraria montana* var. *lobata*
Dolichos incurvus Thunb. – see *Canavalia gladiata*
Dolichos japonicus hort. – see *Pueraria montana* var. *lobata*
Dolichos lablab L. – see *Lablab purpureus*
Dolichos lobatus Willd. – see *Pueraria montana* var. *lobata*
Dolichos monachalis Brot. – see *Vigna unguiculata* subsp. *cylindrica*
Dolichos ovatus Graham – see *Psophocarpus tetragonolobus*
Dolichos psoraloides Lam. – see *Cyamopsis tetragonoloba*
Dolichos pubescens L. – see *Macrotyloma uniflorum*
Dolichos pugioniformis Gmel. – see *Canavalia ensiformis*
Dolichos purpureus L. – see *Lablab purpureus*
Dolichos sesquipedalis L. – see *Vigna unguiculata* subsp. *sesquipedalis*
Dolichos sinensis L. – see *Vigna unguiculata* subsp. *unguiculata*
Dolichos soja L. – see *Glycine max*
Dolichos soudanensis hort. – see *Lablab purpureus*
Dolichos stenocarpa Hochst. ex A. Rich – see *Sphenostylis stenocarpa*
Dolichos tetragonolobus L. – see *Psophocarpus tetragonolobus*
Dolichos tetragonolobus Stickm. – see *Psophocarpus tetragonolobus*
Dolichos tranquebaricus Jacq. – see *Vigna unguiculata* subsp. *cylindrica*
Dolichos trilobus L. – see *Pueraria montana* var. *lobata*
Dolichos tuberosus Lam – see *Pachyrhizus tuberosus*
Dolichos umbellatus Thunb. – see *Vigna umbellata*
Dolichos unguiculatus L. – see *Vigna unguiculata* subsp. *unguiculata*
Dolichos uniflorus Lam. – see *Macrotyloma uniflorum*
Dolichos vestitus Graham – see *Flemingia vestita*
Donax arundinacea auct., non Lour. – see *Maranta arundinacea*
Dracontium polyphyllum Forst. – see *Amorphophallus paeoniifolius* var. *campanulatus*
Dubrueilia melastomoides (Poir.) Gaud. – see *Pilea melastomoides*

Ecbolium insulare Kuntze – see *Justicia insularis*
Eclipta filicaulis Schum. & Thonn. – see *Acmella oleracea*
***Eleocharis dulcis* (Burm. f.) Trin. ex Henschel (10.2)**
Eleocharis dulcis var. *tuberosa* (Roem. & Schult.) T. Koyama – see *E. dulcis*
Eleocharis equisetina Presl. – see *E. dulcis*
Eleocharis indica (Lour.) Druce – see *E. dulcis*
***Eleocharis kuroguwai* Ohwi (10.3)**
Eleocharis plantaginea R. Br. – see *E. dulcis*
Eleocharis plantaginea R. Br. var. *tuberosa* Makino – see *E. dulcis*
Eleocharis plantaginea Roemer & Schultes – see *E. dulcis*
Eleocharis plantaginoides (Retz.) Domin – see *E. dulcis*
Eleocharis tuberosa (Roxb.) Roem. & Schult. – see *E. dulcis*
Eleocharis tumida Roem. & Schult. – see *E. dulcis*
Eleutherococcus koreanus Nakai – see *E. seoulensis*
***Eleutherococcus senticosus* (Nakai) S.Y. Hu (26.4)**
Eleutherococcus senticosus (Rupr. & Maxim.) Maxim – see *E. senticosus* (Nakai) S.Y. Hu
Emilia flammea Cass. – see *E. sonchifolia*
Emilia mucronata Wall. – see *E. sonchifolia*
Emilia purpurea Cass. – see *E. sonchifolia*
Emilia rigidula DC. – see *E. sonchifolia*
Emilia sagittata DC. – see *E. sonchifolia*
Emilia scabra DC. – see *E. sonchifolia*
Emilia sinica Miq. – see *E. sonchifolia*
Emilia sonchifolia hort. – see *E. sonchifolia* (L.) DC.
***Emilia sonchifolia* (L.) DC. (27.18)**
Emilia sonchifolia (L.) DC. var. *javanica* (Burm.) Mattfeld – see *E. sonchifolia*
Englerastrum floribundum (N.E. Br.) Th. Fries. – see *Plectranthus esculentus*
***Enydra fluctuans* Lour. (27.19)**
Enydra longifolia (Blume) DC. – see *E. fluctuans*
Enydra paludosa (Reinw.) DC. – see *E. fluctuans*
Enydria aquatica Vell. – see *Myriophyllum aquaticum*
***Equisetum arvense* L. (1.1)**
Erebennus canaranus Alef. – see *Abelmoschus manihot*
Eruca alba (L.) Noulet – see *Sinapis alba*
Eruca cappadocica Reut. – see *E. sativa*
Eruca eruca (L.) Asch. & Graebn. – see *E. sativa*
Eruca foetida Moench – see *E. sativa*
Eruca glabrescens Jord. – see *E. sativa*
Eruca grandiflora Cav. – see *E. sativa*
Eruca hispida DC. – see *E. sativa*
Eruca lativalvis Boiss. – see *E. sativa*
Eruca longirostris Uechtr. – see *E. sativa*
Eruca muralis Bess. – see *Diplotaxis muralis*
Eruca oleracea J. St.-Hil. – see *E. sativa*
Eruca perennis Mill. – see *Diplotaxis muralis*
Eruca permixta Jord. – see *E. sativa*
Eruca ruchetta Spach – see *E. sativa*
Eruca sativa Garsault – see *E. sativa* Miller
Eruca sativa Hill – see *E. sativa* Miller
***Eruca sativa* Miller (30.48)**

Eruca sinapis-alba DC. – see *Sinapis alba*
Eruca stenocarpa Boiss. & Reut. – see *E. sativa*
Eruca sylvestris Bub. – see *E. sativa*
Eruca tenuifolia (L.) Moench – see *Diplotaxis muralis*
Eruca vesicaria (L.) Cav. – see *E. sativa*
Eruca vesicaria (L.) Cav. subsp. *sativa* (Mill.) Thell. – see *E. sativa*
Eruca vesicaria subsp. *sativa* (Mill.) Thell. – see *E. sativa*
Eruca vesicaria var. *genuina* Rouy & Fouc. – see *E. sativa*
Eruca vesicaria var. *sativa* (Mill.) Thell. – see *E. sativa*
Ervum lens L. – see *Lens culinaris*
***Erysimum amurense* Kitag. (30.49)**
Erysimum glabrum Presl – see *Brassica nigra*
Erysimum praecox L. – see *Barbarea verna*
Erysimum praecox R. Br. – see *Barbarea verna*
Erysimum praecox Sm. – see *Barbarea verna*
Erysimum vernum Mill. – see *Barbarea verna*
Ethulia sparganophora L. – see *Struchium sparganophorum*
Ethulia struchium Sw. – see *Struchium sparganophorum*
***Euryale ferox* Salisb. (50.1)**
Euryale indica Planch. – see *E. ferox*
Eutrema japonica (Miq.) Koidz. – see *Wasabia japonica*
Eutrema wasabi L. – see *Wasabia japonica*
Eutrema wasabi Maxim. – see *Wasabia japonica*
Eutrema wasabi Mix. – see *Wasabia japonica*
Eutrema wasabi (Siebold) Maxim. – see *Wasabia japonica*
Euzomum hispidum Link – see *Eruca sativa*
Euzomum sativum Link – see *Eruca sativa*
Exolus ascendens Hara – see *Amaranthus lividus*
Exolus caudatus sensu auct. Taiwan – see *Amaranthus viridis*
Exolus viridis (L.) Miq. – see *Amaranthus viridis*
Faba bona Medik. – see *Vicia faba*
Faba equina Medik. – see *Vicia faba*
Faba faba (L.) H.D. House – see *Vicia faba*
Faba major Desf. – see *Vicia faba*
Faba minor Roxb. – see *Vicia faba*
Faba sativa (L.) Bernh. – see *Vicia faba*
Faba vulgaris Mill. – see *Vicia faba*
Faba vulgaris Moench – see *Vicia faba*
Falcaria javanica (Blume) DC. – see *Oenanthe javanica* subsp. *javanica*
Falcaria laciniata (Blume) DC. – see *Oenanthe javanica* subsp. *javanica*
Farfugium grande Lindl. – see *F. japonicum*
***Farfugium japonicum* (L.) Kitam. (27.20)**
Farfugium kaempferi (Siebold & Zucc.) Benth. – see *F. japonicum*
Fedia campanulata Steud. – see *Valerianella eriocarpa*
***Fedia cornucopiae* Gaertn. (27.21)**
Fedia eriocarpa Roem. & Schult. – see *Valerianella eriocarpa*
Fedia graciliflora Fisch. & Mey. – see *F. cornucopiae*
Fedia graciliflora Willk. & Lange – see *F. cornucopiae*
Fedia graeca Dufr. – see *F. cornucopiae*
Fedia incrassata Moench – see *F. cornucopiae*

Fedia langei Pomel – see *F. cornucopiae*
Fedia locusta Rchb. – see *Valerianella locusta*
Fedia olitoria Vahl – see *Valerianella locusta*
Fedia paniculata Colla – see *Valerianella locusta*
Fedia rugulosa Spreng. – see *Valerianella eriocarpa*
Fedia scorpioides Dufr. – see *F. cornucopiae*
Ferraria pavonia L. f. – see *Tigridia pavonia*
Ferraria tigridia Sims – see *Tigridia pavonia*
Feuillca pedata Hook. – see *Telfairia pedata*
Fevillea pedata J. Sm. – see *Telfairia pedata*
Fevillea pedata Sm. ex Sims – see *Telfairia pedata*
Flemingia procumbens Roxburgh – see *F. vestita*
Flemingia procumbens Wight – see *F. vestita*
***Flemingia vestita* Benth. ex Bak. (40.7)**
Foeniculum azoricum Mill. – see *F. vulgare* subsp. *vulgare* var. *azoricum*
Foeniculum capillaceum Gilib. – see *F. vulgare* subsp. *vulgare* var. *azoricum*
Foeniculum dulce DC. – see *F. vulgare* subsp. *vulgare* var. *dulce*
Foeniculum dulce DC., non Mill. – see *F. vulgare* subsp. *vulgare* var. *azoricum*
Foeniculum dulce Mill. – see *F. vulgare* subsp. *vulgare* var. *dulce*
Foeniculum officinale All. var. *dulce* Alef. – see *F. vulgare* subsp. *vulgare* var. *dulce*
Foeniculum panmorium DC. – see *F. vulgare* subsp. *vulgare* var. *azoricum*
Foeniculum rigidum Brot. ex Steud. – see *F. vulgare* subsp. *vulgare* var. *azoricum*
Foeniculum sativum Bertol. – see *F. vulgare* subsp. *vulgare* var. *dulce*
Foeniculum vulgare Gaertn. var. *azoricum* (Mill.) Thell. – see *F. vulgare* subsp. *vulgare* var. *azoricum*
Foeniculum vulgare Mill. convar. *azoricum* (Mill.) J. Holub. – see *F. vulgare* subsp. *vulgare* var. azoricum
***Foeniculum vulgare* Mill. subsp. *vulgare* var. *azoricum* (Mill.) Thell. (24.12)**
***Foeniculum vulgare* Mill. subsp. *vulgare* var. *dulce* (Mill.) Batt. (24.13)**
Foeniculum vulgare Mill. var. *azoricum* (Miller) Thell. – see *F. vulgare* subsp. *vulgare* var. *azoricum*
Foeniculum vulgare Mill. var. *dulce* Alef. – see *F. vulgare* subsp. *vulgare* var. *dulce*
Foeniculum vulgare Mill. var. *dulce* Alef. – see *F. vulgare* subsp. *vulgare* var. *dulce*
Foeniculum vulgare Mill. var. *dulce* Batt. & Trab. – see *F. vulgare* subsp. *vulgare* var. *dulce*
Foeniculum vulgare Mill. var. *dulce* Fiori – see *F. vulgare* subsp. *vulgare* var. *dulce*
Foeniculum vulgare Mill. var. *dulce* (Mill.) Battand. & Trabut – see *F. vulgare* subsp. *vulgare* var. dulce
Foeniculum vulgare Mill. var. *sativum* Presl – see *F. vulgare* subsp. *vulgare* var. *dulce*
Foenum-graecum officinale Moench – see *Trigonella foenum-graecum*
Foenum-graecum officinale var. *cultum* Alef. – see *Trigonella foenum-graecum*
Foenum-graecum sativum Medik. – see *Trigonella foenum-graecum*
Folliculigera graveolens Pasq. – see *Trigonella foenum-graecum*
Furcaria sabdariffa Ulbr. – see *Hibiscus sabdariffa*
Galanga major Garsault – see *Alpinia galanga*
Galanga officinalis Salisb. – see *Alpinia galanga*
Galinsoga acmella Steud. – see *G. parviflora*
Galinsoga hirsuta Baker – see *G. parviflora*
***Galinsoga parviflora* Cav. (27.22)**
Galinsoga quinqueradiata Ruiz & Pav. – see *G. parviflora*
Gandola nigra (Lour.) Raf. – see *Basella alba*
Gandola peruviana Moq. – see *Ullucus tuberosus*
Gandola tuberosa Moq. – see *Ullucus tuberosus*

Geolobus flavus Raf. – see *Vigna subterranea*
Germanea rotundifolia Poir. – see *Solenostemon rotundifolius*
Geropogon hirsutum Urv. – see *Tragopogon porrifolius*
Gigantochloa aspera (Schult.) Kurz. – see *Dendrocalamus asper*
***Glebionis coronarium* (L.) C. Jeffrey (27.23)**
Glehnia leiocarpa Mathias – see *G. littoralis*
***Glehnia littoralis* F. Schmidt (24.14)**
Glehnia littoralis F. Schmidt, ex Miq. – see *G. littoralis* F. Schmidt
Gloriosa luxurians Lour. ex Gomes – see *Hemerocallis fulva*
Glycine gracilis Skvorc. – see *G. max*
Glycine hispida (Moench) Maxim. – see *G. max*
***Glycine max* (L.) Merr. (40.8)**
Glycine orex (L.) Merrill – see *G. max*
Glycine soja Bentham – see *G. max*
Glycine soja hort. – see *G. max*
Glycine subterranea L. – see *Vigna subterranea*
Glycine uniflora (Lam.) Dalz. – see *Macrotyloma uniflorum*
Glycine ussuriensis Kegel – see *G. max*
Gnemon brunoniana Kuntze – see *Gnetum gnemon*
Gnemon griffitii Kuntze – see *Gnetum gnemon*
Gnemon ovalifolia Kuntze – see *Gnetum gnemon*
Gnetum acutum Miq. – see *G. gnemon*
Gnetum brunonianum Griff. – see *G. gnemon*
***Gnetum gnemon* L. (41.1)**
Gnetum griffithii Parl. – see *G. gnemon*
Gnetum ovalifolium Poir. – see *G. gnemon*
Gnetum polystachyum Reinw. – see *G. gnemon*
Gnetum sylvestre Brongn. – see *G. gnemon*
Gnetum vinosum Elm. – see *G. gnemon* .
Gomphima vaginalis (Burman f.) Rafin – see *Monochoria vaginalis*
Gomphrena sessilis L. – see *Alternanthera sessilis*
Gymnanthemum abyssinicum Sch. Bip. ex Walp. – see *G. amygdalinum*
***Gymnanthemum amygdalinum* (Delile) Sch. Bip. ex Walp. (27.24)**
Gymnopetalum calyculatum Miq. – see *Cucurbita moschata*
Gymnopetalum septemlobum Miq. – see *Benincasa hispida*
Gynandropsis denticulata DC. – see *Cleome gynandra*
Gynandropsis gynandra (L.) Briq. – see *Cleome gynandra*
Gynandropsis gynandra (L.) Merr. – see *Cleome gynandra*
Gynandropsis pentaphylla (L.) DC. – see *Cleome gynandra*
Gynandropsis viscida Bunge – see *Cleome gynandra*
Gynura angulosa DC. – see *G. bicolor*
***Gynura bicolor* DC. (27.25)**
Gynura bicolor Willd. – see *G. bicolor* DC.
Gynura crepidioides Benth. – see *Crassocephalum crepidioides*
Gynura ecalyculata DC. – see *Emilia sonchifolia*
Gynura longifolia Kerr – see *G. bicolor*
Gynura microcephala Vatke – see *Crassocephalum crepidioides*
Hedera senticosa Rupr. & Maxim. – see *Eleutherococcus seoulensis*
Hedypnois taraxacum Scop. – see *Taraxacum officinale*
Heleocharis dulcis Trin. ex Henschel – see *Eleocharis dulcis*

Heleocharis dulcis var. *tuberosa* Schult. – see *Eleocharis dulcis*
Heleocharis plantaginoidea W.F. Wight. – see *Eleocharis dulcis*
Helianthus doronicoides Torr. & Gray – see *H. tuberosus*
Helianthus mollissimus E.E. Wats – see *H. tuberosus*
***Helianthus tuberosus* L. (27.26)**
Helmia bulbifera (L.) Kunth – see *Dioscorea bulbifera*
Helmia dumetorum Kunth. – see *Dioscorea dumetorum*
Hemerocallis crocea Lam. – see *H. fulva*
Hemerocallis disticha Donn. – see *H. fulva*
***Hemerocallis flava* L. (13.3)**
Hemerocallis flava L. var. *minor* (Mill.) M. Hotta – see *H. minor*
Hemerocallis flava Suter– see *H. fulva*
***Hemerocallis fulva* L. (13.3)**
Hemerocallis graminea Andr. – see *H. minor*
Hemerocallis graminifolia Schltdl. – see *H. minor*
Hemerocallis kwanso hort. ex Regel – see *H. fulva*
Hemerocallis lilio-asphodelus L. – see *H. flava*
Hemerocallis lilio-asphodelus L. var. *fulvus* L. – see *H. fulva*
Hemerocallis longituba Miq. – see *H. fulva*
***Hemerocallis minor* Mill. (13.3)**
Hemerocallis pumila Salisb. – see *H. minor*
Hemionitis esculenta Retz. – see *Diphazium esculentum*
Herschelia edulis (Sims) Bowdich – see *Physalis peruviana*
***Hibiscus acetosella* Wel. ex Hiern (45.3)**
Hibiscus acetosus Noronha – see *H. sabdariffa*
Hibiscus bartlingianus Steud – see *Abelmoschus manihot*
Hibiscus canaranus Miq. ex Mast – see *Abelmoschus manihot*
Hibiscus digitatus Cav. – see *H. sabdariffa*
Hibiscus eetveldianus De Willd. & Th. Durand – see *H. acetosella*
Hibiscus esculentus L. – see *Abelmoschus esculentus*
Hibiscus ficifolius Mill. – see *Abelmoschus esculentus*
Hibiscus fraternus L. – see *H. sabdariffa*
Hibiscus gossypifolius Mill. – see *H. sabdariffa*
Hibiscus hispidissimus A. Chev. – see *Abelmoschus esculentus*
Hibiscus hostilis Wall. ex Mast. – see *Abelmoschus manihot*
Hibiscus japonicus Miq. – see *Abelmoschus manihot*
Hibiscus longifolius Roxb. – see *Abelmoschus esculentus*
Hibiscus maculatus Bartl. – see *Abelmoschus manihot*
Hibiscus manihot L. – see *Abelmoschus manihot*
Hibiscus notho-manihot F. Muell. – see *Abelmoschus manihot*
Hibiscus palmatilobus Baill. – see *H. sabdariffa*
Hibiscus palmatus Cav. – see *Abelmoschus manihot*
Hibiscus papyriferus Salisb. – see *Abelmoschus manihot*
Hibiscus pentaphyllus Roxb. – see *Abelmoschus manihot*
Hibiscus praecox Forssk. – see *Abelmoschus esculentus*
Hibiscus pseudo-manihot DC. – see *Abelmoschus manihot*
Hibiscus pyramidatus Garcke – see *Abelmoschus manihot*
Hibiscus ricinoides Zipp. ex Span. – see *Abelmoschus manihot*
Hibiscus rosella hort. – see *H. sabdariffa*
***Hibiscus sabdariffa* L. (45.4)**

Ipomoea batatas Lam. – see *I. batatas* var. *batatas*
Ipomoea batatas Lam. var. *edulis* Makino – see *I. batatas* var. *batatas*
Ipomoea batatas Poir. – see *I. batatas* var. *batatas*
Ipomoea batatas var. *edulis* (Thunb.) Kuntze – see *I. batatas* var. *batatas*
Ipomoea catesbaei G.F.W. Mey. – see *I. batatas* var. *batatas*
Ipomoea chrysorrhiza (Soland. ex G. Forster) Peter – see *I. batatas* var. *batatas*
Ipomoea edulis Makino – see *I. batatas* var. *batatas*
Ipomoea fastigiata Sweet – see *I. batatas* var. *batatas*
Ipomoea repens Poir. – see *I. aquatica*
Ipomoea repens Roth – see *I. aquatica*
Ipomoea reptans (L.) Poir. – see *I. aquatica*
Ischurochloa spinosa Büse – see *Bambusa blumeana*
***Ixeridium dentatum* (Thunb.) Tzvelev (27.27)**
Ixeris dentata (Thunb.) Nakai – see *Ixeridium dentatum*
Ixeris sonchifolia (Maxim.) Hance – see *Youngia sonchifolia*
Jacea tricolor (L.) Opiz – see *Viola tricolor*
Jaegeria uliginosa Spreng. – see *Acmella oleracea*
Janipha manihot H.B.K. – see *Manihot esculenta* subsp. *esculenta*
Jasminoides rhombifolium Moench – see *Lycium chinense*
Jatropha aconitifolia Mill. – see *Cnidoscolus aconitifolius*
Jatropha dulcis J.F. Gmel. – see *Manihot esculenta* subsp. *esculenta*
Jatropha janipha Lour. – see *Manihot esculenta* subsp. *esculenta*
Jatropha loureiri Steud. – see *Manihot esculenta* subsp. *esculenta*
Jatropha manihot L. – see *Manihot esculenta* subsp. *esculenta*
Jatropha napaeifolia Desr. ex A. Juss. – see *Cnidoscolus aconitifolius*
Jatropha stipulata Vell. – see *Manihot esculenta* subsp. *esculenta*
Joliffia africana Delile – see *Telfairia pedata*
***Justicia insularis* T. Anderson (21.1)**
Justicia karschiana Buettner – see *J. insularis*
Justicia rostellarioides Lindau – see *J. insularis*
Kali inermis Moench – see *Salsola soda*
Kali soda Scop. – see *Salsola soda*
***Kalimeris yomena* (Kitam.) Kitam. (27.28)**
Keringa amygdalina Rafin. – see *Gymnanthemum amygdalinum*
Kerstingiella biflora Lackey – see *Macrotyloma uniflorum*
Kerstingiella geocarpa Harms – see *Macrotyloma geocarpum*
Kerstingiella tisserantii Pellegrin – see *Macrotyloma geocarpum*
Kerstingiella uniflora (Lam.) Lackey – see *Macrotyloma uniflorum*
Kleinia cusimbua (D. Don) Less. – see *Gynura bicolor*
Kochia childsii hort. ex anon. – see *Bassia scoparia*
Kochia parodii Aellen – see *Bassia scoparia*
Kochia scoparia form *trichophylla* (hort. ex Voss) Schinz & Thell. – see *Bassia scoparia*
Kochia scoparia (L.) Schrad. – see *Bassia scoparia*
Kochia trichophylla hort. ex Voss. – see *Bassia scoparia*
Kochia virgata Kostel. – see *Bassia scoparia*
Lablab benghalensis Medik. – see *L. purpureus*
Lablab ferrugineus Medik. – see *L. purpureus*
Lablab leucocarus Savi – see *L. purpureus*
Lablab nankinicus Savi – see *L. purpureus*
Lablab niger Medik. – see *L. purpureus*

Lablab perennans DC. – see *L. purpureus*
***Lablab purpureus* (L.) Sweet (40.9)**
Lablab rufus Medik. – see *L. purpureus*
Lablab vulgaris Savi – see *L. purpureus*
Labradia pruriens Swediaur – see *Mucuna pruriens*
Lactuca amurensis Regel, ex Maxim. – see *L. indica*
Lactuca angustana hort. ex Vilmorin – see *L. sativa* var. *angustana*
Lactuca angustana Vilm., non All. – see *L. sativa* var. *angustana*
Lactuca brevirostris Champ. ex Benth. – see *L. indica*
Lactuca bungeana Nakai – see *Youngia sonchifolia*
Lactuca capitata (L.) Garsault – see *L. sativa* var. *capitata*
Lactuca capitata DC. – see *L. sativa* var. *capitata*
Lactuca crispa (L.) DC. – see *L. sativa* var. *longifolia*
***Lactuca indica* L. (27.29)**
Lactuca kotschyi Sch. Bip. – see *Launaea taraxacifolia*
Lactuca laciniata (Houtt.) Makino – see *L. indica*
Lactuca laciniata Roth – see *L. sativa* var. *longifolia*
Lactuca mauritiana Poir. – see *L. indica*
Lactuca pentaphylla Sch. Bip. – see *Launaea taraxacifolia*
Lactuca romana Gars. – see *L. sativa* var. *longifolia*
Lactuca saligna – see *L. indica*
Lactuca sativa asparagus L. – see *L. sativa* var. *angustana*
Lactuca sativa capitata L. – see *L. sativa* var. *capitata*
Lactuca sativa foliosa Brehmer – see *L sativa* var. *longifolia*
Lactuca sativa L. provar. *angustana* (L.H. Bailey) Helm – see *L. sativa* var. *angustana*
Lactuca sativa L. provar. *crispa* (L.) Helm – see *L. sativa* var. *longifolia*
Lactuca sativa L. var. *angustana* hort. – see *L. sativa* var. *angustana*
Lactuca sativa L. var. *angustana* hort. ex L.H. Bailey – see *L. sativa* var. *angustana*
Lactuca sativa L. var. *angustana* Irish – see *L. sativa* var. *angustana*
Lactuca sativa L. var. *angustana* Irish, ex Bremer – see *L. sativa* var. *angustana*
***Lactuca sativa* L. var. *angustana* L.H. Bailey (27.30)**
Lactuca sativa L. var. *augustana* hort. – see *L. sativa* var. *asparagina*
Lactuca sativa L. var. *aurescens* Hegi – see *L sativa* var. *longifolia*
Lactuca sativa L. var. *capitata* DC. – see *L. sativa* var. *capitata*
Lactuca sativa L. var. *capitata* hort. – see *L. sativa* var. *capitata*
Lactuca sativa L. var. *capitata* (hort. ex L.H. Bailey) St. Cycl. – see *L. sativa* var. *capitata*
Lactuca sativa L. var. *capitata* L. – see *L. sativa* var. *capitata*
***Lactuca sativa* L. var. *capitata* L. (27.31)**
Lactuca sativa L. var. *crispa* L. – see *L sativa* var. *longifolia*
Lactuca sativa L. var. *integrifolia* Bisch. ex Stankov & Taliev – see *L. sativa* var. *angustana*
Lactuca sativa L. var. *intybacea* hort. ex L.H. Bailey – see *L. sativa* var. *longifolia*
Lactuca sativa L. var. *longifolia* L. – see *L. sativa* var. *longifolia*
***Lactuca sativa* L. var. *longifolia* Lam. (27.32)**
Lactuca sativa L. var. *longifolia* (Lam.) Hegi – see *L sativa* var. *longifolia*
Lactuca sativa L. var. *romana* Gars. – see *L. sativa* var. *longifolia*
Lactuca sativa L. var. *romana* hort. – see *L. sativa* var. *longifolia*
Lactuca sativa L. var.-gr. *secalina* Alef. – see *L. sativa* var. *longifolia*
Lactuca sativa romana L. – see *L. sativa* var. *longifolia*
Lactuca sativa subsp. *capitata* (L.) Alef. – see *L. sativa* var. *capitata*
Lactuca sativa tenerrima Helm – see *L. sativa L.* var. *capitata*

Lactuca squarrosa Miq. – see *L. indica*
Lactuca taraxacifolia (Willd.) Schum. ex Hornem. – see *Launaea taraxacifolia*
Laelia orientalis (L.) Desv. – see *Bunias orientalis*
Lagenaria idolatrica (Willd.) Ser. – see *L. siceraria*
Lagenaria lagenaria (L.) Cockerell – see *L. siceraria*
Lagenaria leucantha (Duch.) Rusby – see *L. siceraria*
Lagenaria microcarpa Naud. – see *L. siceraria*
Lagenaria siceraria convar. *clavatina* Grebenšč. – see *L. siceraria*
Lagenaria siceraria convar. *cugurda* Grebenšč. – see *L. siceraria*
Lagenaria siceraria Grebenšč. convar. *siceraria* – see *L. siceraria*
***Lagenaria siceraria* (Molina) Standl. (38.20)**
Lagenaria siceraria Standley var. *hispida* Hara – see *L. siceraria*
Lagenaria siceraria var. *depressa* (Ser.) Hara – see *L. siceraria*
Lagenaria siceraria var. *hispida* Hara – see *L. siceraria*
Lagenaria siceraria var. *microcarpa* (Naud.) Hara – see *L. siceraria*
Lagenaria vulgaris Ser. – see *L. siceraria*
Lagenaria vulgaris Ser. subsp. *asiatica* Kobyakova – see *L. siceraria*
Lagenaria vulgaris Ser. var. *hispida* Nakai – see *L. siceraria*
Languas galanga (L.) Merr. – see *Alpinia galanga*
Languas galanga (L.) Stuntz – see *Alpinia galanga*
Languas galanga Stuntz – see *Alpinia galanga*
Languas vulgare (L) Stuntz – see *Alpinia galanga*
Lapathum acetosa Scop. – see *Rumex acetosa*
Lapathum acutatum (L.) Lam. – see *Rumex scutatus*
Lapathum alpestre Scop. – see *Rumex scutatus*
Lapathum hortense Garsault – see *Rumex patientia*
Lapathum hortense Lam. – see *Rumex patientia*
Lapathum hortense Moench – see *Rumex patientia*
Lapathum pratense Lam. – see *Rumex acetosa*
Lapathum scutatum Lam. – see *Rumex scutatus*
Lappa amplissima Kotschy – see *Arctium lappa*
Lappa arctium Hill – see *Arctium lappa*
Lappa bardana Moench – see *Arctium lappa*
Lappa communis Coss. & Germ. – see *Arctium lappa*
Lappa communis var. *major* Neilr – see *Arctium lappa*
Lappa edulis hort. – see *Arctium lappa*
Lappa edulis Sieb. – see *Arctium lappa*
Lappa glabra Lam. – see *Arctium lappa*
Lappa kotschyi Boiss. – see *Arctium lappa*
Lappa major Gaertn. – see *Arctium lappa*
Lappa officinalis All. – see *Arctium lappa*
Lappa vulgaris Hill – see *Arctium lappa*
Lasia merkusii Hassk. – see *Cyrtosperma chamissonis*
Lathyrus abyssinicus A. Br. ex Chiov. – see *L. sativus*
Lathyrus asiaticus (Zalk.) Kudrj. – see *L. sativus*
Lathyrus attenuatus Viv. – see *L. tuberosus*
Lathyrus lens (L.) Bernh. – see *Lens culinaris*
***Lathyrus sativus* L. (40.10)**
***Lathyrus tuberosus* L. (40.11)**
***Launaea taraxacifolia* (Willd.) Amin, ex C. Jeffrey (27.33)**

Limnocharis caduciflora Turcz. ex Trinius – see *Zizania latifolia*
Limnocharis emarginata Humb. & Bonpl. – see *L. flava*
Limnocharis emarginata Kunth – see *L. flava*
Limnocharis flava **(L.) Buchenau (14.1)**
Limnocharis laforesti Duchass. – see *L. flava*
Limnocharis plumieri Rich. – see *L. flava*
Limnochloa caduciflora Turcz. ex Trin. – see *Zizania latifolia*
Lindera odorata Aschers – see *Myrrhis odorata*
Lipusa coccinea (L.) Alef. – see *Phaseolus coccineus*
Lipusa multiflora (Lam.) Alef. – see *Phaseolus coccineus*
Locusta communis Delarbre – see *Valerianella locusta*
Lotus tetragonolobus **L. (40.13)**
Luffa acutangula **(L.) Roxb. (38.21)**
Luffa acutangula (L.) Roxb. var. *subangulata* (Miq.) Cogn. – see *L. acutangula* (L.) Roxb.
Luffa acutangula Roxb. – see *L. acutangula* (L.) Roxb.
Luffa aegyptiaca **P. Miller (38.22)**
Luffa aegyptiaca Roxb. – see *L. aegyptiaca*
Luffa amara Roxb. – see *L. acutangula*
Luffa arabum Steud. – see *L. aegyptiaca*
Luffa cattu-picinna Ser. – see *L. aegyptiaca*
Luffa clavata Roxb. – see *L. aegyptiaca*
Luffa cordata Meissn. – see *L. aegyptiaca*
Luffa cylindrica auct., non M.J. Roem. – see *L. aegyptiaca*
Luffa cylindrica (L.) M.J. Roem. – see *L. aegyptiaca*
Luffa cylindrica (L.) T. Durand & H. Durand – see *L. aegyptiaca*
Luffa cylindrica M.J. Roem. – see *L. aegyptiaca*
Luffa drastica Mart. – see *L. acutangula*
Luffa fluminensis M.J. Roem. – see *L. acutangula*
Luffa foetida Cav. – see *L. acutangula*
Luffa foetida (Jacq. f.) hort. Vindob. ex Schrad. – see *L. aegyptiaca*
Luffa gigantea hort. – see *L. aegyptiaca*
Luffa gosa Buch.-Ham. – see *L. acutangula*
Luffa hederacea Wall. – see *L. aegyptiaca*
Luffa hermaphrodita Singh & Bandhari – see *L. acutangula*
Luffa insularum A. Gray – see *L. aegyptiaca*
Luffa jacquinii Schrad. – see *L. aegyptiaca*
Luffa leiocarpa F. Muell. – see *L. aegyptiaca*
Luffa leucosperma M.J. Roem. – see *L. aegyptiaca*
Luffa macrocarpa hort. – see *L. aegyptiaca*
Luffa marylandica hort. – see *L. aegyptiaca*
Luffa parvala Buch.-Ham. – see *L. aegyptiaca*
Luffa pentandra Roxb. – see *L. aegyptiaca*
Luffa petola Ser. – see *L. aegyptiaca*
Luffa plukenetiana Ser. – see *L. acutangula*
Luffa racemosa Roxb. – see *L. aegyptiaca*
Luffa satpatia Buch.-Ham. – see *L. aegyptiaca*
Luffa scabra Schum. & Thonn. – see *L. aegyptiaca*
Luffa striata Schrad. – see *L. aegyptiaca*
Luffa subangulata Miq. – see *L. aegyptiaca*
Luffa sylvestris Miq. – see *L. acutangula*

Luffa veitchii Naud. – see *L. aegyptiaca*
Luffa vittata Spanoghe – see *L. aegyptiaca*
Luffa vittata Zipp. ex Span. – see *L. aegyptiaca*
Lunaria japonica Miq. – see *Wasabia japonica*
***Lupinus albus* L. (40.14)**
Lupinus albus L. subsp. *termis* (Forssk.) Caruel – see *L. albus*
Lupinus albus L. var. *albus* – see *L. albus*
Lupinus albus L. var. *termis* (Forssk.) Alef. – see *L. albus*
***Lupinus angustifolius* L. (40.14)**
Lupinus bivonii Presl – see *L. termis*
Lupinus cryptanthus Shuttlew. ex Campb. – see *L. angustifolius*
Lupinus hirsutus Eichw. – see *L. albus*
Lupinus leucospermus Boiss. – see *L. angustifolius*
Lupinus linifolius Roth – see *L. angustifolius*
***Lupinus mutabilis* Sweet. (40.14)**
Lupinus opsianthus Atabek. & Maissurjan – see *L. angustifolius*
Lupinus philistaeus Boiss. – see *L. angustifolius*
Lupinus prolifer Desr. – see *L. termis*
Lupinus reticulatus Desv. – see *L. angustifolius*
Lupinus sativus Gaertn. – see *L. albus*
***Lupinus* spp. (40.14)**
Lupinus sylvestris Lam. – see *L. angustifolius*
***Lupinus termis* Forskal. (40.14)**
Lupinus varius Gaertn. – see *L. albus*
Lupinus varius L. – see *L. angustifolius*
Lupulus amarus Gilib. – see *Humulus lupulus*
Lupulus communis Gaertn. – see *Humulus lupulus*
Lupulus humulus Mill. – see *Humulus lupulus*
Lupulus scandens Lam. – see *Humulus lupulus*
Lupulus scandens St. Lager. – see *Humulus lupulus*
Lycium barbarum Lam. – see *L. chinense*
Lycium carnosum hort. Tonelle, ex Dun. – see *L. chinense*
Lycium carnosum hort., non Poir. – see *L. chinense*
***Lycium chinense* Miller (62.8)**
Lycium chinense var. *ovatum* C.K. Schneid. – see *L. chinense*
Lycium elongatum Salisb. – see *L. chinense*
Lycium europaeum Merat. ex Dun. – see *L. chinense*
Lycium lanceolatum Loisel. – see *L. chinense*
Lycium megistocarpim Dunal var. *ovatum* Dunal. – see *L. chinense*
Lycium megistocarpum Dunal – see *L. chinense*
Lycium ovatum Loisel. – see *L. chinense*
Lycium rhombifolium (Moench) Dippel – see *L. chinense*
Lycium sinenses Grenier – see *L. chinense*
Lycium trewianum Roem. & Schult. – see *L. chinense*
Lycium turbinatum Loisel. – see *L. chinense*
Lycopersicon esculentum subsp. *cultum* Brezh. – see *L. esculentum* var. *esculentum*
Lycopersicon esculentum subsp. *pimpinellifolium* Brezh. – see *L. pimpinellifolium*
Lycopersicon esculentum subsp. *subspontaneum* Brezh. – see *L. esculentum* var. *esculentum*
***Lycopersicon esculentum* var. *esculentum* P. Miller (62.9)**
Lycopersicon galeni Mill. – see *L. esculentum* var. *esculentum*

Moghania vestita Benth. – see *Flemingia vestita*
Moghania vestita (Benth. ex Bak.) Kuntze – see *Flemingia vestita*
Moghania vestita Kuntze – see *Flemingia vestita*
Momordica anthelmintica Schum. & Thonn. – see *M. charantia*
Momordica balsamina Descourt. – see *M. charantia*
Momordica bicolor Blume – see *Coccinia grandis*
Momordica carinata Vell. – see *Luffa aegyptiaca*
***Momordica charantia* L. (38.23)**
Momordica charantia L. subsp. *abbreviata* (Ser.) Grebenšč. – see *M. charantia*
Momordica charantia L. var. *abbreviata* Ser. – see *M. charantia*
Momordica charantia L. var. *minor* Naud. – see *M. charantia*
Momordica chinensis Spreng. – see *M. charantia*
Momordica cylindrica Blanco – see *M. charantia*
Momordica cylindrica L. – see *Luffa aegyptiaca*
Momordica elegans Salisb. – see *M. charantia*
Momordica humilis Wall. – see *M. charantia*
Momordica indica L. – see *M. charantia*
Momordica jagorana C. Koch – see *M. charantia*
Momordica lanata Thunb. – see *Citrullus lanatus* var. *lunatus*
Momordica luffa L. – see *Luffa aegyptiaca*
Momordica luffa Vell. – see *Luffa acutangula*
Momordica macropetala Mart. – see *M. charantia*
Momordica monadelpha Roxb. – see *Coccinia grandis*
Momordica muricata Willd. – see *M. charantia*
Momordica operculata Blanco – see *Luffa aegyptiaca*
Momordica operculata Vell. – see *M. charantia*
Momordica papillosa Peckolt, ex Rosenth. – see *M. charantia*
Momordica pedata L. – see *Cyclanthera pedata*
Momordica procera A. Chev. – see *Cucumeropsis mannii*
Momordica reticulata Salisb. – see *Luffa aegyptiaca*
Momordica roxburghiana G. Don – see *M. charantia*
Momordica senegalensis Lam. – see *M. charantia*
Momordica sinensis Spreng. – see *M. charantia*
Momordica tubiflora Wall. – see *Luffa acutangula*
Momordica zeylanica Mill. – see *M. charantia*
Monochoria dilatata Kunth. – see *M. hastata*
Monochoria dubia Miq. – see *M. vaginalis*
Monochoria hastaefolia C. Presl – see *M. hastata*
***Monochoria hastata* (L.) Solms (18.1)**
Monochoria hastata Solms – see *M. hastata*
Monochoria junghuhniana Hassk. – see *M. vaginalis*
Monochoria korsakowii Regel & Maack – see *M. vaginalis*
Monochoria linearis Miq. – see *M. vaginalis*
Monochoria loureirii Kunth – see *M. vaginalis*
Monochoria pauciflora Kunth – see *M. vaginalis*
Monochoria plantaginea (Roxb.) Kunth. – see *M. hastifolia*
Monochoria sagittata Kunth. – see *M. hastata*
***Monochoria vaginalis* (Brum. f.) C. Presl. (18.2)**
***Montia perfoliata* (Donn, ex Willd.) Howell (59.1)**
Moraea pavonia Thunb. – see *Tigridia pavonia*

Moraea tigridia Baker – see *Tigridia pavonia*
Mucuna minima Haines – see *M. pruriens* var. *utilis*
Mucuna pauciflora Blume, ex Burck – see *M. pruriens* var. *utilis*
***Mucuna pruriens* (L.) DC. var. *utilis* (Wall, ex Wight) Baker, ex Burck. (40.18)**
Mucuna pruriens subsp. *deeringiana* (Bort) Hanelt – see *M. pruriens* var. *utilis*
Mucuna pruriens var. *utilis* auct. mult. – see *M. pruriens* var. *utilis*
Mucuna prurita Hook. – see *M. pruriens* var. *utilis*
Mucuna utilis Wall. ex Wight – see *M. pruriens* var. *utilis*
***Musa* × *paradisiaca* L. var. *paradisiaca* (16.1)**
Musa × *sapientum* L. – see *M.* × *paradisiaca* var. *paradisiaca*
Musa alphurica Miq. – see *M.* × *paradisiaca* var. *paradisiaca*
Musa berteroi Colla – see *M.* × *paradisiaca* var. *paradisiaca*
Musa berteroniana Steud. – see *M.* × *paradisiaca* var. *paradisiaca*
Musa chaparra Perr. – see *M.* × *paradisiaca* var. *paradisiaca*
Musa paradisiaca var. *normalis* Kuntze – see *M.* × *paradisiaca* var. *paradisiaca*
Musa paradisiaca var. *sapientum* Kuntze – see *M.* × *paradisiaca* var. *paradisiaca*
Musa sapidisiaca Jacob – see *M.* × *paradisiaca* var. *paradisiaca*
Musa sapientum var. *paradisiaca* Baker – see *M.* × *paradisiaca* var. *paradisiaca*
***Muscari comosum* (L.) Mill. (13.5)**
Mutarda nigra Bernh. – see *Brassica nigra* Koch.
Myagrum taraxacifolium Lam. – see *Bunias orientalis*
***Myriophyllum aquaticum* (Vell.) Verdc. (42.1)**
Myriophyllum brasiliense Camb. – see *M. aquaticum*
Myriophyllum proserpinacoides Gillies, ex Hook. & Arn. – see *M. aquaticum*
***Myrrhis odorata* (L.) Scop. (24.16)**
Myscolus hispanicus D. Dietr. – see *Scolymus hispanicus*
Myscolus hispanicus (L.) Cass. ex Dietr. – see *Scolymus hispanicus*
Myscolus hispanicus (L.) Endl. – see *Scorzonera hispanica*
Myscolus microcephalus Cass. – see *Scolymus hispanicus*
Nardosmia japonica Siebold & Zucc. – see *Petasites japonicus*
Nardosmia japonicus (Siebold & Zucc.) Maxim – see *Petasites japonicus*
Nasturtium aquaticum Garsault – see *N. officinale*
Nasturtium aquaticum Wahlenb. – see *N. officinale*
Nasturtium armoracia (L.) Fries – see *Armoracia rusticana*
Nasturtium bursa-pastoris (L.) Roth – see *Capsella bursa-pastoris*
Nasturtium crispum Medik. – see *Lepidium sativum*
Nasturtium fontanum Asch. – see *N. officinale*
Nasturtium fontanum (Lam.) Asch. – see *N. officinale*
Nasturtium hortense Garsault – see *Lepidium sativum*
Nasturtium insipidum Reverch. ex Fouc. & Rouy – see *N. officinale*
Nasturtium microphyllum Reichb. – see *N. officinale*
Nasturtium nasturtium Cockerell – see *N. officinale*
Nasturtium nasturtium-aquaticum Karst. – see *N. officinale*
Nasturtium nasturtium-aquaticum (L.) H. Karst. – see *N. officinale*
***Nasturtium officinale* R. Br. (30.52)**
Nasturtium officinale R. Br. var. *siifolium* (Rchb.) Alef. – see *N. officinale*
Nasturtium officinale R. Br. var. *vulgare* Alef. – see *N. officinale*
Nasturtium sativum Kuntze – see *Lepidium sativum*
Nasturtium sativum (L.) Medik. – see *Lepidium sativum*
Nasturtium sativum Moench – see *Lepidium sativum*

Nasturtium siifolium Phil. – see *N. officinale*
Nasturtium siifolium Rchb. – see *N. officinale*
Nasturtium siifolium Reichb. – see *N. officinale*
Nasturtium spinosum Desv. ex Thell. – see *Lepidium sativum*
Negretia pruriens Blanco – see *Mucuna pruriens*
Negretia pruriens Naves, ex Villar – see *Mucuna pruriens*
Nelumbium asiaticum Rich. – see *Nelumbo nucifera*
Nelumbium caspicum Fisch. ex DC. – see *Nelumbo nucifera*
Nelumbium caspium Eichw. – see *Nelumbo nucifera*
Nelumbium nelumbo (L.) Druce – see *Nelumbo nucifera*
Nelumbium nicifera Gaertn – see *Nelumbo nucifera*
Nelumbium nuciferum auct. – see *Nelumbo nucifera*
Nelumbium speciosum Willd. – see *Nelumbo nucifera*
Nelumbo alba Berchtold & J. Presl – see *N. nucifera*
Nelumbo asiatica Rich. – see *N. nucifera*
Nelumbo caspica Eichwald – see *N. nucifera*
Nelumbo indica Pers. – see *N. nucifera*
Nelumbo indica Poir. – see *N. nucifera*
Nelumbo javanica Poir. – see *N. nucifera*
Nelumbo nelumbo (L.) H. Karst. – see *N. nucifera*
***Nelumbo nucifera* Gaertner (48.1)**
Nelumbo nucifera Gaertn. var. *speciosa* – see *N. nucifera*
Nelumbo rheedei C. Presl – see *N. nucifera*
Nelumbo speciosa Willd. – see *N. nucifera*
Nelumbo speciosa Willd. var. *tamara* DC. – see *N. nucifera*
Nelumbo tamara (DC.) Sweet – see *N. nucifera*
Nelumbo transversa C. Presl – see *N. nucifera*
Nelumbo turbinata Blanco – see *N. nucifera*
Neptunia natans (L. f.) Druce – see *N. prostrata*
Neptunia oleracea Lour. – see *N. prostrata*
Neptunia plena Lindl. – see *N. prostrata*
***Neptunia prostrata* (Lam.) Baill. (40.19)**
Neptunia stolonifera Guill. & Perr. – see *N. prostrata*
Neustanthus chinensis Benth. – see *Pueraria montana* var. *lobata*
Nochotta oleracea S.G. Gmel. – see *Cicer arietinum*
Nymphaea nelumbo L. – see *N. nucifera*
***Nymphaea nouchali* Burm. f. (50.2)**
Nymphaea stellata Willd – see *N. nouchali*
Ocimum acutum Thunb. – see *Perilla frutescens* var. *crispa*
Ocimum africanum Lour. – see *O. americanum*
Ocimum album L. – see *O. basilicum*
Ocimum americanum Jacq. – see *O. basilicum* L.
***Ocimum americanum* L. (44.4)**
Ocimum americanum var. *pilosum* (Willd.) A.J. Paton – see *O. americanum*
Ocimum anisatum hort. ex Benth. – see *O. basilicum*
Ocimum barrelieri Roth – see *O. basilicum*
***Ocimum basilicum* L. (44.5)**
Ocimum basilicum L. subsp. *basilicum* – see *O. basilicum*
Ocimum basilicum L. var. *bullatum* Alef. – see *O. basilicum*
Ocimum basilicum L. var. *difforme* Benth. – see *O. basilicum*

Ocimum basilicum L. var. *glabratum* Benth. – see *O. basilicum*
Ocimum basilicum L. var. *majus* Benth. – see *O. basilicum*
Ocimum basilicum L. var. *minimum* L. – see *O. basilicum*
Ocimum basilicum L. var. *pelvifolium* Alef. – see *O. basilicum*
Ocimum basilicum L. var. *pilosum* (Willd.) Benth. – see *O. basilicum*
Ocimum basilicum L. var. *purpurascens* Benth. – see *O. basilicum*
Ocimum basilicum L. var. *thyrsiflorum* (L.) Benth. – see *O. basilicum*
Ocimum basilicum L. var. *violaceum* Alef. – see *O. basilicum*
Ocimum basilicum L. var. *vulgare* Alef. – see *O. basilicum*
Ocimum brachiatum Blume – see *O. americanum*
Ocimum bullatum Lam. – see *O. basilicum*
Ocimum canum Sims – see *O. americanum*
Ocimum caryophyllatum Roxb – see *O. basilicum*
Ocimum ciliare Heyne, ex Hook. f. – see *O. basilicum*
Ocimum ciliatum Hornem. – see *O. basilicum*
Ocimum citrodorum Blanco – see *O. basilicum*
Ocimum cochleatum Desf. – see *O. basilicum*
Ocimum crispum Thunb. – see *Perilla frutescens* var. *crispa*
Ocimum fimbriatum hort. ex Benth. – see *O. basilicum*
Ocimum frutescens L. – see *Perilla frutescens* var. *crispa*
Ocimum frutescens L. var. *crispum* (Benth.) Decne. – see *Perilla frutescens* var. *crispa*
Ocimum gratissimum Seem. – see *O. basilicum*
Ocimum hispidum Lam. – see *O. basilicum*
Ocimum integerrimum Willd. – see *O. basilicum*
Ocimum lacerum [Heyne], ex Wall. – see *O. basilicum*
Ocimum lanceolatum Schum. & Thonn. – see *O. basilicum*
Ocimum laxum Vahl, ex Benth. – see *O. basilicum*
Ocimum majus Garsault – see *O. basilicum*
Ocimum majus hort. ex Benth. – see *O. basilicum*
Ocimum medium Mill. – see *O. basilicum*
Ocimum menthaefolium Benth. – see *O. basilicum*
Ocimum minimum L. – see *O. basilicum*
Ocimum minus Garsault – see *O. basilicum*
Ocimum nigrum Thou. ex Benth. – see *O. basilicum*
Ocimum odorum Salisb. – see *O. basilicum*
Ocimum pilosum L. – see *O. basilicum*
Ocimum pilosum Willd. – see *O. basilicum*
Ocimum salinum Molina – see *O. basilicum*
Ocimum scabrum Wight, ex Hook. f. – see *O. basilicum*
Ocimum stamineum Sims – see *O. americanum*
Ocimum thyrsiflorum L. – see *O. basilicum*
Ocimum urticaefolium hort. ex Benth. – see *O. basilicum*
Oenanthe benghalensis Benth. & J.H. Hook. – see *O. javanica* subsp. *javanica*
Oenanthe decumbens Koso-Pol. – see *O. javanica* subsp. *javanica*
Oenanthe japonica (Miq.) Drude – see *O. javanica* subsp. *javanica*
Oenanthe javanica (Blume) DC. subsp. *javanica* – see *O. javanica* subsp. *javanica*
***Oenanthe javanica* (Blume) DC. subsp. *javanica* (Batalin) H. Hara (24.17)**
Oenanthe javanica (Blume) DC. subsp. *stolonifera* (Wall. ex DC.) Murata – see *O. javanica* subsp. *javanica*
Oenanthe javanica DC. – see *O. javanica* (Blume) DC. subsp. *javanica*

Oenanthe laciniata (Blume) Zoll. – see *O. javanica* subsp. *javanica*
Oenanthe linearis DC. – see *O. javanica* subsp. *javanica*
Oenanthe rivularis Dunn – see *O. javanica* subsp. *javanica*
Oenanthe schlechteri Wolff – see *O. javanica* subsp. *javanica*
Oenanthe stolonifera (Roxb.) DC. – see *O. javanica* subsp. *javanica*
Oenanthe stolonifera Wall. – see *O. javanica* subsp. *javanica*
Oenanthe subbipinnata (Miq.) Drude – see *O. javanica* subsp. *javanica*
Oenothera angustifolia Mill. – see *O. biennis*
***Oenothera biennis* L. (51.1)**
Oenothera communis Lév. – see *O. biennis*
Oenothera cruciata Nutt. ex G. Don – see *O. biennis*
Oenothera gauroides Hornem. – see *O. biennis*
Oenothera glabra Mill. – see *O. biennis*
Oenothera grandiflora [Soland.] – see *O. biennis*
Oenothera graveolens Gilib. – see *O. biennis*
Oenothera hirsuta D. Dietr. – see *O. biennis*
Oenothera hirsutissima De Vries – see *O. biennis*
Oenothera hookeri Torr. & Gray – see *O. biennis*
Oenothera lamarkiana Ser. – see *O. biennis*
Oenothera muricata L. – see *O. biennis*
Oenothera odorata Hook. & Arn. – see *O. biennis*
Oenothera parviflora L. – see *O. biennis*
Oenothera pubescens Nees – see *O. biennis*
Oenothera suaveolens Desf. – see *O. biennis*
Oligosporus condimentarius Cass. – see *Artemisia dracunculus*
Oligosporus dracunculus (L.) Polj. – see *Artemisia dracunculus*
Onagra biennis (L.) Scop. – see *Oenothera biennis*
Onagra chrysantha Spach – see *Oenothera biennis*
Onagra cruciata Small – see *Oenothera biennis*
Onagra europaea Spach – see *Oenothera biennis*
Onagra hookeri Small – see *Oenothera biennis*
Onagra muricata Moench – see *Oenothera biennis*
Onagra vulgaris Spach – see *Oenothera biennis*
Oncus esculentus Lour. – see *Dioscorea esculenta*
Opuntia compressa Macbride – see *O. ficus-indica*
Opuntia crassa Haw. – see *O. ficus-indica*
Opuntia elongata (Willd.) Haw. – see *O. ficus-indica*
Opuntia engelmannii Salm-Dyck – see *O. ficus-indica*
Opuntia ficus-barbarica Berger – see *O. ficus-indica*
***Opuntia ficus-indica* (L.) Miller (32.1)**
Opuntia gymnocarpa A. Web. – see *O. ficus-indica*
Opuntia lanceolata (Haw.) Haw. – see *O. ficus-indica*
Opuntia megacantha Salm-Dyck – see *O. ficus-indica*
Opuntia occidentalis Engelm & Bigel. – see *O. ficus-indica*
Opuntia parvula Salm-Dyck – see *O. ficus-indica*
Opuntia undosa Griff. – see *O. ficus-indica*
Opuntia undulata Griff. – see *O. ficus-indica*
Origanum anglicum Hill – see *O. vulgare*
Origanum barcense Simonkai – see *O. vulgare*
Origanum capitatum Willd. ex Benth – see *O. vulgare*

Origanum creticum L. – see *O. vulgare*
Origanum creticum Lour. – see *O. vulgare*
Origanum decipiens Wallr. ex Benth. – see *O. vulgare*
Origanum elegans Sennen – see *O. vulgare*
Origanum floridum Salisb. – see *O. vulgare*
Origanum garcile C. Koch – see *O. vulgare*
Origanum heracleoticum auct., non L. – see *O. vulgare*
Origanum heracleoticum Reichb. – see *O. vulgare*
Origanum hirtum Noe, ex Nym. – see *O. vulgare*
Origanum humile Mill. – see *O. vulgare*
Origanum latifolium Mill. – see *O. vulgare*
Origanum laxiflorum Royle, ex Benth. – see *O. vulgare*
Origanum macrostachyum Hoffmgg. & Link – see *O. vulgare*
Origanum majus Garsault – see *O. vulgare*
Origanum megastachyum Link – see *O. vulgare*
Origanum minus Garsault – see *O. vulgare*
Origanum normale D. Don – see *O. vulgare*
Origanum nutans Willd. ex Benth. – see *O. vulgare*
Origanum oblongatum Link – see *O. vulgare*
Origanum orientale Mill. – see *O. vulgare*
Origanum pruinosum C. Koch – see *O. vulgare*
Origanum purpurescens Gilib. – see *O. vulgare*
Origanum semiglaucum Boiss. & Reut. ex Briq. – see *O. vulgare*
Origanum stoloniferum Bess. ex Reichb. – see *O. vulgare*
Origanum thymiflorum Reichb. – see *O. vulgare*
Origanum venosum Willd. ex Benth. – see *O. vulgare*
Origanum viride Willd. ex Benth. – see *O. vulgare*
***Origanum vulgare* L. (44.6)**
Origanum vulgare L. subsp. *genuinum* Gaudin – see *O. vulgare*
Origanum vulgare L. subsp. *prismaticum* Gaudin – see *O. vulgare*
Origanum vulgare L. subsp. *vulgare* – see *O. vulgare*
Origanum vulgare L. var. *formosanum* Hayata – see *O. vulgare*
Origanum vulgare L. var. *tauricum* Boriss. – see *O. vulgare*
Origanum watsoni Schmidt & Schlagintweit – see *O. vulgare*
Orignaum officinale Gueldenst. – see *O. vulgare*
Orignaum venosum Wild. ex Benth. – see *O. vulgare*
Orignaum virens C.A. Mey. – see *O. vulgare*
Orignaum viridulum Martr. – see *O. vulgare*
Orobanche alba Bieb. – see *O. crenata*
Orobanche altissima Noe, ex Nym. – see *O. crenata*
Orobanche canescens F. Schultz – see *O. crenata*
Orobanche congesta Reichb. f. – see *O. crenata*
***Orobanche crenata* Forsskal (52.1)**
Orobanche foetida Viv. – see *O. crenata*
Orobanche grandiflora Bory & Chaub. – see *O. crenata*
Orobanche gussonii Nym. – see *O. crenata*
Orobanche pruinsoa Lapeyr. – see *O. crenata*
Orobanche segetum Spurn. – see *O. crenata*
Orobanche speciosa DC. – see *O. crenata*
Orobanche viciaefabae F. Schultz – see *O. crenata*

Orobus bimarginatus Stokes – see *Lathryus sativus*
Orobus faba Brot. – see *Vicia faba*
Oroga heracleotica Rafin. – see *Origanum vulgare*
Orthospermum bonus-henricus (L.) Schur – see *Chenopodium bonus-henricus*
Orthosporum bonus-henricus (L.) T. Nees – see *Chenopodium bonus-henricus*
Orthosporum unctuosum Montandon – see *Chenopodium bonus-henricus*
Osmunda asiatica (Fernald) Ohw – see *O. cinnamomea*
Osmunda biformis (Benth.) Makino – see *O. japonica*
***Osmunda cinnamomea* L. (3.1)**
Osmunda cinnamomea (L.) Pr. var. *fokiense* (Copel.) Tagawa – see *O. cinnamomea*
Osmunda cinnamomea L. var. *asiatica* Fernald – see *O. cinnamomea*
Osmunda cinnamomea L. var. *fokiensis* Copel. – see *O. cinnamomea*
***Osmunda japonica* Th. (3.2)**
Osmunda nipponica Makino – see *O. japonica*
Osmunda regalis L. var. *biformis* Benth. – see *O. japonica*
Osmunda regalis L. var. *japonica* Milde – see *O. japonica*
Osmunda siliquosa (L.) E. Copel. – see *Ceratopteris thalictroides*
Osmundastrum cinnamomea var. *fokiense* (Copel.) Tagawa – see *Osmunda cinnamomea*
Osmundastrum japonicum (Thunb.) K. Presl – see *Osmunda japonica*
Oxalis arracacha G. Don – see *O. tuberosa*
Oxalis crassicaulis Zucc. – see *O. tuberosa*
Oxalis crenata Jacq. – see *O. tuberosa*
***Oxalis tuberosa* Molina (53.1)**
Oxybaphus expansus (Ruiz & Pav.) Vahl – see *Mirabilis expansa*
***Pachyrhizus ahipa* (Wedd.) L. Parodi (40.20)**
Pachyrhizus angulatus Rich. ex DC. – see *P. erosus*
Pachyrhizus articulatus (Lam.) Duch. & Walp. – see *P. erosus*
Pachyrhizus bulbosus (L.) Kurz. – see *P. erosus*
***Pachyrhizus erosus* (L.) Urban (40.21)**
Pachyrhizus jicamas Blanco – see *P. erosus*
Pachyrhizus palmatilobus Benth. & Hook.f. – see *P. erosus*
Pachyrhizus strigosus Clausen – see *P. erosus*
Pachyrhizus thunbergianus Siebold & Zucc. – see *Pueraria montana* var. *lobata*
Pachyrhizus trilobus DC. – see *Pueraria montana* var. *lobata*
***Pachyrhizus tuberosus* (Lam.) Sprengel (40.22)**
Panicum palmaefolium Koenig – see *Setaria palmifolia*
Panicum palmaefolium Willd. ex Poir. – see *Setaria palmifolia*
Panicum plicatum Willd. – see *Setaria palmifolia*
Paraixeris sonchifolia (Maxim.) Tzvelev – see *Youngia sonchifolia*
Parmentiera edulis Rafin. – see *Solanum tuberosum*
***Passiflora biflora* Lam. (54.1)**
Passiflora brighami Wats. – see *P. biflora*
Passiflora glabrata H.B.K. – see *P. biflora*
Passiflora lunata Willd. – see *P. biflora*
Passiflora spathulata Mast. – see *P. biflora*
Passiflora transversa Mast. – see *P. biflora*
Pastinaca divaricata Desf. – see *P. sativa*
Pastinaca fleischmannii Hladnik, ex D. Dietr. – see *P. sativa*
Pastinaca opaca Bernh. ex Hornem. – see *P. sativa*
Pastinaca pratensis (Pers.) H. Mart. – see *P. sativa*

***Pastinaca sativa* L. (24.18)**
Pastinaca sativa L. subsp. *divaricata* (Desf.) Rouy & Camus – see *P. sativa*
Pastinaca sativa L. subsp. *pratensis* (Pers.) Čelak. – see *P. sativa*
Pastinaca sativa L. subsp. *sylvestris* (Mill.) Rouy & Camus – see *P. sativa*
Pastinaca sativa L. subsp. *umbrosa* (Steven, ex DC.) Bondar. ex O.N. Korovina – see *P. sativa*
Pastinaca sativa L. subsp. *urens* (Req. ex Godr.) Čelak. – see *P. sativa*
Pastinaca sativa L. var. *brevis* Alef. – see *P. sativa*
Pastinaca sativa L. var. *edulis* DC. – see *P. sativa*
Pastinaca sativa L. var. *hortensis* Ehrh. ex Hoffm. – see *P. sativa*
Pastinaca sativa L. var. *longa* Alef. – see *P. sativa*
Pastinaca sativa L. var. *pratensis* Pers. – see *P. sativa*
Pastinaca sativa L. var. *siamensis* Roem. & Schult. ex Alef. – see *P. sativa*
Pastinaca sylvestris Mill. – see *P. sativa*
Pastinaca teretiuscula Boiss. – see *P. sativa*
Pastinaca umbrosa Steven, ex DC. – see *P. sativa*
Pastinaca urens Req. ex Godr. – see *P. sativa*
Pedicellaria gynandra (L.) Chiov. – see *Cleome gynandra*
Pedicellaria pentaphylla (L.) Schrank – see *Cleome gynandra*
Pennisetum benthamii Steud. – see *P. purpureum*
Pennisetum flavicomum Leeke – see *P. purpureum*
Pennisetum flexispica K. Schum. – see *P. purpureum*
Pennisetum macrostachyum Benth. ex Hook. – see *P. purpureum*
Pennisetum nitens Hackel – see *P. purpureum*
Pennisetum pruinosum Leeke – see *P. purpureum*
***Pennisetum purpureum* Schumacher (17.3)**
Pepo citrullus Sag. – see *Cucurbita pepo*
Pepo eximius Sag. – see *Cucurbita moschata*
Pepo ficifolia Britton – see *Cururbita ficifolia*
Pepo ficifolius (Bouche) Britton – see *Cucurbita ficifolia*
Pepo indicus Burm. – see *Cucurbita moschata*
Pepo lagenarius Moench – see *Lagenaria siceraria*
Pepo macrocarpus Rich. ex Spach – see *Cucurbita maxima*
Pepo maximus (Duch. ex Lam.) Peterm. – see *Cucurbita maxima*
Pepo melopepo (L.) Moench – see *Cucurbita pepo*
Pepo moschatus (Duch. ex Lam.) Sag. – see *Cucurbita moschata*
Pepo potiro (Pers.) Sag. – see *Cucurbita maxima*
Pepo verrucosus (L.) Moench – see *Cucurbita pepo*
Pepo vulgaris Moench – see *Cucurbita pepo*
Perilla acuta (Thunb.) Nakai – see *P. frutescens* var. *crispa*
Perilla albiflora Odashima – see *P. frutescens* var. *crispa*
Perilla arguta Benth. – see *P. frutescens* var. *crispa*
Perilla avium Dunn – see *P. frutescens* var. *crispa*
Perilla cavaleriei H. Lév. – see *P. frutescens* var. *crispa*
Perilla citriodora (Makino) Nakai – see *P. frutescens* var. *crispa*
Perilla crispa (Thunb.) Tanaka – see *P. frutescens* var. *crispa*
Perilla frutescens (L.) Britton – see *P. frutescens* var. *crispa*
***Perilla frutescens* (L.) Britton var. *crispa* (Thunb.) W. Deane (44.7)**
Perilla frutescens cv. Crispa – see *P. frutescens* var. *crispa*
Perilla frutescens var. *crispa* form *crispa* Makino – see *P. frutescens* var. *crispa*
Perilla frutescens var. *nankinensis* Decne. ex Bailey – see *P. frutescens* var. *crispa*

Perilla frutescens var. *nankinensis* (Lour.) Brill – see *P. frutescens* var. *crispa*
Perilla hirtella Nakai – see *P. frutescens* var. *crispa*
Perilla laciniata hort. ex W. Mill & L.H. Bailey – see *P. frutescens* var. *crispa*
Perilla macrostachya Benth – see *P. frutescens* var. *crispa*
Perilla nankinensis Decne. – see *P. frutescens* var. *crispa*
Perilla nankinensis (Lour.) Decne. – see *P. frutescens* var. *crispa*
Perilla ocimoides L. – see *P. frutescens* var. *crispa*
Perilla ocimoides L. var. *crispa* Benth. – see *P. frutescens* var. *crispa*
Perilla shimadai Kudo – see *P. frutescens* var. *crispa*
Perilla urticaefolia Salisb. – see *P. frutescens* var. *crispa*
Petasites japonica Miq. – see *P. japonicus* (Siebold & Zucc.) Maxim
Petasites japonicus Fr. Schmidt – see *P. japonicus* (Siebold & Zucc.) Maxim
Petasites japonicus Miq. – see *P. japonicus* (Siebold & Zucc.) Maxim
***Petasites japonicus* (Siebold & Zucc.) Maxim (27.35)**
Petasites japonicus (Siebold & Zucc.) R. Schmidt – see *P. japonicus* (Siebold & Zucc.) Maxim.
Peterium aquilinum Kuhn – see *P. aquilinum*
Petroselinum crispum convar. *radicosum* (Alef.) Danert – see *P. crispum* var. *tuberosum*
Petroselinum crispum (Mill.) Hill – see *P. crispum* var. *crispum*
Petroselinum crispum (Mill.) Nyman – see *P. crispum* var. *crispum*
Petroselinum crispum (Mill.) Nyman var. *tuberosum* Crovetto – see *P. crispum* (Mill.) Nyman, ex A.W. Hill var. *tuberosum* (Bernh.) Mart. Crov.
***Petroselinum crispum* (P. Mill.) Nyman, ex A.W. Hill var. *crispum* (24.19)**
***Petroselinum crispum* (P. Mill.) Nyman, ex A.W. Hill var. *neapolitanum* Danert (24.20)**
***Petroselinum crispum* (P. Mill.) Nyman, ex A.W. Hill var. *tuberosum* (Bernh.) Mart. Crov. (24.21)**
Petroselinum crispum subsp. *tuberosum* (Bernh. ex Rchb.) Soó – see *P. crispum* var. *tuberosum*
Petroselinum crispum var. *erfurtense* Danert – see *P. crispum* var. *tuberosum*
Petroselinum crispum var. *latifolium* auct., non (Mill.) Airy-Shaw – see *P. crispum*. var. *neapolitanum*
Petroselinum crispum var. *neapolitanum* Danert – see *P. crispum* var. *neapolitanum*
Petroselinum crispum var. *radicosum* (Alef.) L.H. Bailey – see *P. crispum* var. *tuberosum*
Petroselinum crispum var. *radicosum* Bailey – see *P. crispum* var. *tuberosum*
Petroselinum crispum var. *silvestre* (Alef.) Danert – see *P. crispum* var. *neapolitanum*
Petroselinum crispum var. *tuberosum* Crovetto – see *P. crispum* var. *tuberosum*
Petroselinum crispum var. *vulgare* (Noisette) Danert – see *P. crispum* var. *neapolitanum*
Petroselinum hortense Hoffm. – see *P. crispum* var. *crispum*
Petroselinum hortense subsp. *macrocarpum* Mackevič – see *P. crispum* var. *crispum*
Petroselinum hortense subsp. *microcarpum* Mackevič – see *P. crispum* var. *crispum*
Petroselinum hortense var. *crispum* Bailey – see *P. crispum* var. *crispum*
Petroselinum hortense var. *foliosum* (Alef.) Thell. – see *P. crispum* var. *neapolitanum*
Petroselinum hortense var. *radicosum* (Alef.) L.H. Bailey – see *P. crispum* var. *tuberosum*
Petroselinum hortense var. *tuberosum* (Bernh. ex Rchb.) Thell. – see *P. crispum* var. *tuberosum*
Petroselinum petroselinum H. Karst. – see *P. crispum* var. *crispum*
Petroselinum sativum Hoffm. – see *P. crispum* var. *crispum*
Petroselinum sativum var. *latifolium* Airy-Shaw – see *P. crispum* var. *tuberosum*
Petroselinum sativum var. *silvestre* Alef. – see *P. crispum* var. *neapolitanum*
Petroselinum sativum var. *tuberosum* (Bernh. ex Rchb.) Schübl. & G. Martens – see *P. crispum* var. tuberosum
Petroselinum sativum var.-gr. *foliosum* Alef. – see *P. crispum* var. *neapolitanum*
Petroselinum sativum var.-gr. *radicosum* Alef. – see *P. crispum* var. *tuberosum*
Petroselinum vulgare Hill – see *P. crispum* var. *crispum*
Petroselinum vulgare J. Hill – see *P. crispum* var. crispum

Petroselinum vulgare S.F. Gray – see *P. crispum* var. crispum
Phaseolus abyssinicus Savi – see *Vigna radiata* var. *radiata*
Phaseolus aconitifolius Jacq. – see *Vigna aconitifolia*
***Phaseolus acutifolius* A. Gray var. *acutifolius* (40.23)**
Phaseolus acutifolius var. *latifolius* G.F. Freeman – see *P. acutifolius* var. *acutifolius*
Phaseolus acutifolius var. *tenuifolius* A. Gray – see *P. acutifolius* var. *acutifolius*
Phaseolus amazonicus Benth. – see *P. lunatus*
Phaseolus amoenus Fingerh. – see *P. vulgaris* var. *vulgaris*
Phaseolus angularis (Willd.) W.F. Wight – see *Vigna angularis*
Phaseolus angulosus Schübl. & Mart. – see *P. vulgaris* var. *vulgaris*
Phaseolus asparagoides Schur – see *P. vulgaris* var. *vulgaris*
Phaseolus aureus Roxb. – see *Vigna radiata* var. *radiata*
Phaseolus bipunctatus Jacq. – see *P. lunatus*
Phaseolus calcaratus Roxb. – see *Vigna umbellata*
Phaseolus chrysanthos Savi. – see *Vigna angularis*
***Phaseolus coccineus* L. (40.24)**
Phaseolus compressus DC. – see *P. vulgaris* var. *vulgaris*
Phaseolus cruentus hort. ex Schur – see *P. vulgaris* var. *vulgaris*
Phaseolus cylindricus L. – see *Vigna unguiculata* subsp. *cylindrica*
Phaseolus derasus Schrank – see *P. lunatus*
Phaseolus dimidiatus Haberle, ex Schübl. & Martens – see *P. vulgaris* var. *vulgaris*
Phaseolus esculentus Salisb. – see *P. vulgaris* var. *vulgaris*
Phaseolus foecundus Macf. – see *P. lunatus*
Phaseolus glaber Roxb. – see *Vigna umbellata*
Phaseolus gonospermus Savi – see *P. vulgaris* var. *vulgaris*
Phaseolus haematocarpus Savi – see *P. vulgaris* var. *vulgaris*
Phaseolus hernandezii Savi – see *Vigna mungo* var. *mungo*
Phaseolus hirtus Retz. – see *Vigna radiata* var. *radiata*
Phaseolus ilocanus Blanco – see *P. lunatus*
Phaseolus inamoenus L. – see *P. lunatus*
Phaseolus ionocarpus Fingerh. – see *P. vulgaris* var. *vulgaris*
Phaseolus latifolius Freeman – see *P. acutifolius* var. *acutifolius*
Phaseolus latisiliquus Macf. – see *P. lunatus*
Phaseolus lilac Zucc. – see *P. vulgaris* var. *vulgaris*
Phaseolus limensis Macf. – see *P. lunatus*
***Phaseolus lunatus* L. (40.25)**
Phaseolus lunatus L. var. *macrocarpus* Bonth – see *P. lunatus*
Phaseolus lupinoides Fingerh. – see *P. vulgaris* var. *vulgaris*
Phaseolus macrocarpus Moench – see *P. lunatus*
Phaseolus max L. – see *Glycine max*
Phaseolus maximus Roxb. – see *P. lunatus*
Phaseolus melanospermus Fingerh. – see *P. vulgaris* var. *vulgaris*
Phaseolus mexicanus Mart. – see *P. vulgaris* var. *vulgaris*
Phaseolus montanus Brandegee – see *P. acutifolius* var. *acutifolius*
Phaseolus multiflorus Lam. – see *P. coccineus*
Phaseolus multiflorus Willd. – see *P. coccineus*
Phaseolus mungo L. – see *Vigna mungo* var. *mungo*
Phaseolus mungo L. var. *radiatus* Bak. – see *Vigna radiata* var. *radiata*
Phaseolus mungo Roxb. – see *Vigna radiata* var. *radiata*
Phaseolus nanus Juslen. – see *P. vulgaris* var. *vulgaris*

Phytolacca decandra L. – see *P. americana*
Phytolacca decandra var. *acinosa* (Roxb.) Moq. – see *P. acinosa*
***Phytolacca esculenta* Van Houtte (56.3)**
Phytolacca kaempferi A. Gray – see *P. esculenta*
Phytolacca longespica Moq. – see *P. octandra*
Phytolacca mexicana Crantz – see *P. octandra*
***Phytolacca octandra* L. (56.4)**
Phytolacca pekinensis Hance – see *P. esculenta*
Phytolacca rugosa A. Br. & Bouche – see *P. octandra*
Phytolacca vulgaris Crantz – see *P. americana*
Phytosalpinx lucida Lunell. – see *Lycopus lucidus*
Pilea elongata (Blume) Miq. – see *P. melastomoides*
***Pilea glaberrima* (Blume) Blume (66.1)**
***Pilea melastomoides* (Poir.) Wedd. (66.2)**
Pilea oreophila (Miq.) Blume – see *P. melastomoides*
Pilea pellucida Blume – see *P. melastomoides*
Pilea smilacifolia (Wallich, ex Voigt) Weddell – see *P. glaberrima*
Pilea trinervia Wight – see *P. melastomoides*
***Pimpinella brachycarpa* (Komarov) Nakai (24.22)**
Pimpinella calycina var. *brachycarpa* Kom. – see *P. brachycarpa*
Pimpinella sisarum Jess. – see *Sium sisarum*
Pimpinella sisarum Koso-Poljansky – see *Sium sisarum*
Pinardia coronaria (L.) Less. – see *Glebionis coronarium*
Pionandra betacea (Cav.) Miers – see *Cyphomandra betacea*
Pircunia esculenta (van Houtte) Moq. – see *Phytolacca esculenta*
Pircunia latbenia Moq. – see *Phytolacca acinosa*
Pisum arvense L. – see *P. sativum* subsp. *sativum* var. *sativum*
Pisum biflorum Stokes – see *P. sativum* subsp. *sativum* var. *sativum*
Pisum commune Clav. – see *P. sativum* subsp. *sativum* var. *sativum*
Pisum gullosum Risso, ex Alef. – see *P. sativum* subsp. *sativum* var. *sativum*
Pisum humile Mill. – see *P. sativum* subsp. *sativum* var. *sativum*
Pisum lathyrus E.H.L. Krause – see *Lathyrus sativus*
Pisum leptolobum Rchb. – see *P. sativum* subsp. *sativum* var. *sativum*
Pisum macrocarpum Ser. ex Schur – see *P. sativum* subsp. *sativum* var. *sativum*
Pisum pachylobum Dierb. ex Alef. – see *P. sativum* subsp. *sativum* var. *sativum*
Pisum quadratum Rchb. – see *P. sativum* subsp. *sativum* var. *sativum*
Pisum ramulare Rchb. – see *P. sativum* subsp. *sativum* var. *sativum*
Pisum saccharatum Ser. – see *P. sativum* subsp. *sativum* var. *sativum*
Pisum sativum Alef. – see *P. sativum* subsp. *sativum* var. *sativum*
Pisum sativum L. – see *P. sativum* L. subsp. *sativum* var. *sativum*
Pisum sativum L. subsp. *arvense* Asch. & Graebn. – see *P. sativum* subsp. *sativum* var. *sativum*
Pisum sativum L. subsp. *commune* (Clav.) Govorov – see *P. sativum* subsp. *sativum* var. *sativum*
Pisum sativum L. subsp. *hortense* (Nielr.) Asch. & Graebn. – see *P. sativum* subsp. *sativum* var. *sativum*
***Pisum sativum* L. subsp. *sativum* var. *macrocarpon* (40.27)**
***Pisum sativum* L. subsp. *sativum* var. *sativum* (40.28)**
Pisum sativum L. var. *macrocarpon* – see *P. sativum* subsp. *sativum* var. *macrocarpon*
Pisum sativum L. var. *saccharatum* – see *P. sativum* subsp. *sativum* var. *macrocarpon*
Pisum speciosum Dierb. ex Alef. – see *P. sativum* subsp. *sativum* var. *sativum*
Pisum tetragonum Hort. ex Pasqu. – see *P. sativum* subsp. *sativum* var. *sativum*

Pisum tuberosum E.H.L. Krause – see *Lathyrus tuberosus*
Pisum tuberosum (L.) E.H.L. Krause – see *Lathyrus tuberosus*
Pisum umbellatum Mill. – see *P. sativum* subsp. *sativum* var. *sativum*
Pisum umbellatum Rchb. – see *P. sativum* subsp. *sativum* var. *sativum*
Pisum uniflorum Moench – see *P. sativum* subsp. *sativum* var. *sativum*
Pisum vulgare Fingerh. – see *P. sativum* subsp. *sativum* var. *sativum*
Pisum vulgare Jundz. – see *P. sativum* subsp. *sativum* var. *sativum*
Plantago columnae Gouan – see *P. coronopus* var. *sativa*
Plantago commutata Guss. – see *P. coronopus* var. *sativa*
Plantago cornuti Jacq. – see *P. coronopus* var. *sativa*
Plantago coronopifolia Murr. – see *P. coronopus* var. *sativa*
Plantago coronopoda St.-Lag. – see *P. coronopus* var. *sativa*
Plantago coronopus L. – see *P. coronopus* var. *sativa*
***Plantago coronopus* L. var. *sativa* Fiori (57.1)**
Plantago cupani Guss. – see *P. coronopus* var. *sativa*
Plantago erubescens Shuttlew. ex Costa – see *P. coronopus* var. *sativa*
Plantago filiformis C. Koch – see *P. coronopus* var. *sativa*
Plantago jacquini Roem. & Schult. – see *P. coronopus* var. *sativa*
Plantago lacustris De Cresp. – see *P. coronopus* var. *sativa*
Plantago neglecta Guss. – see *P. coronopus* var. *sativa*
Plantago parishii Macbr. – see *P. coronopus* var. *sativa*
Plantago stellaris Salisb. – see *P. coronopus* var. *sativa*
Platycodon autumnale Decne. – see *P. grandiflorus*
Platycodon chinense Lindl. & Paxt. – see *P. grandiflorus*
Platycodon glaucum (Thunb.) Nakai – see *P. grandiflorus*
Platycodon grandiflorum A. DC. – see *P. grandiflorus*
***Platycodon grandiflorus* (Jacq.) A. DC. (34.3)**
Platycodon sinensis Lem. – see *P. grandiflorus*
Platygonia kaempferi Naud. – see *Trichosanthes cucumeroides*
Plectranthus barrelieri Spreng. – see *Ocimum basilicum*
Plectranthus coppinii Cornu – see *Solenostemon rotundifolius*
***Plectranthus esculentus* N.E. Br. (44.8)**
Plectranthus floribundus N.E. Br. – see *P. esculentus*
Plectranthus nankinensis (Lour.) Spreng. – see *Perilla frutescens* var. *crispa*
Plectranthus rotundifolius (Poiret) Sprengel – see *P. esculentus*
Plectranthus tuberosus Blume – see *Solenostemon sonchifolius*
Polymnia edulis Wedd. – see *Smallanthus sonchifolius*
Polymnia sonchifolia Poeppig & Endl. – see *Smallanthus sonchifolius*
Polynome alata Salisb. – see *Dioscorea alata*
Polynome bulbifera Salisb. – see *Dioscorea bulbifera*
Polypara cochinchinensis Lour. – see *Houttuynia cordata*
Polypara cordata (Thunb.) Kuntze – see *Houttuynia cordata*
Pontederia alba Buch.-Ham. ex Wall. – see *Monochoria vaginalis*
Pontederia cordata Lour. – see *Monochoria vaginalis*
Pontederia dilatata [Buch.-Ham.] – see *Monochoria hastata*
Pontederia dubia Blume – see *Monochoria vaginalis*
Pontederia hastata L. – see *Monochoria hastata*
Pontederia lanceolata Wall. ex Kunth – see *Monochoria vaginalis*
Pontederia linearis Hassk. – see *Monochoria vaginalis*
Pontederia loureiriana Schult. f. – see *Monochoria vaginalis*

Pontederia pauciflora Blume – see *Monochoria vaginalis*
Pontederia plantaginea Roxb. – see *Monochoria vaginalis*
Pontederia racemosa Buch.-Ham. ex Wall. – see *Monochoria vaginalis*
Pontederia sagittata Roxb. – see *Monochoria hastata*
Pontederia sagittifolia Heyne, ex Wall. – see *Monochoria hastata*
Pontederia vaginalis Blanco – see *Monochoria hastata*
Pontederia vaginalis Burm. f. – see *Monochoria vaginalis*
Pontederia vaginalis N.L. Burman – see *Monochoria vaginalis*
Pontederia vaginata Royle – see *Monochoria vaginalis*
Poppya fabiana C. Koch – see *Luffa aegyptiaca*
Porophyllum cusimbua (D. Don) DC. – see *Gynura bicolor*
Porrum commune Rchb. – see *Allium porrum*
Porrum fistulosum Schur. – see *Allium fistulosum*
Porrum sativum Mill. – see *Allium porrum*
Porrum scorodoprasum (L.) Rchb. – see *Allium scorodoprasum*
Portulaca aurea hort. ex DC. – see *P. oleracea*
Portulaca crassicaulis Jacq. – see *Talinum fruticosum*
Portulaca crassifolia Jacq. – see *Talinum fruticosum*
Portulaca dichotoma Gueldenst. – see *P. oleracea*
Portulaca fruticosa Linn. – see *Talinum fruticosum*
Portulaca hortensis Rupr. – see *P. oleracea*
Portulaca laevis Buch.-Ham. – see *P. oleracea*
Portulaca latifolia Hornem. – see *P. oleracea*
Portulaca maritima S.G. Gmel. – see *P. oleracea*
Portulaca officinarum Crantz – see *P. oleracea*
***Portulaca oleracea* L. (59.2)**
Portulaca oleracea L. subsp. *oleracea* – see *P. Oleracea*
Portulaca oleracea L. subsp. *sativa* (Haw.) Schübl. & Mart. – see *P. oleracea*
Portulaca olitoria Pall. – see *P. oleracea*
Portulaca paniculata Jacq. – see *Talinum paniculatum*
Portulaca paniculata L. – see *Talinum fruticosum*
Portulaca parvifolia Haw. – see *P. oleracea*
Portulaca patens L. – see *Talinum paniculatum*
Portulaca racemosa L. – see *Talinum fruticosum*
Portulaca repens Noronha – see *P. oleracea*
Portulaca sativa Haw. – see *P. oleracea*
Portulaca suffruticosa Thwaites – see *P. oleracea*
Portulaca sylvestris Bub. – see *P. oleracea*
Portulaca sylvestris Garsault – see *P. oleracea*
Portulaca sylvestris Mont. – see *P. oleracea*
Portulaca triangularis Jacq. – see *Talinum fruticosum*
Portulaca viridis hort. ex DC. – see *P. oleracea*
Potamogeton bifolius Lapeyr. – see *Vicia faba*
***Praecitrullus fistulosus* (Stocks) Pang. (38.24)**
Praecitrullus indicus Pangalo, ex Vul'f & Maleeva – see *P. fistulosus*
Prenanthes dentata Thunb. – see *Ixeridium dentatum*
Prenanthes sonchifolia Bunge – see *Youngia sonchifolia*
Prenathes sarmentosa Wall. – see *Emilia sonchifolia*
Proteinophallus rivieri Hook. f. – see *Amorphophallus konjac*
Pseudocapsicum torulosum Moench – see *Solanum aethiopicum*

Psophocarpus tetragonolobus **(L.) DC. (40.29)**
Psophocarpus tetragonolobus (Stickm.) DC. – see *P. tetragonolobus* (L.) DC.
Psoralea tetragonoloba L. – see *Cyamopsis tetragonoloba* Taubert
Pteridium aquilina L. – see *P. aquilinum*
Pteridium aquilinum **(L.) Kuhn (2.1)**
Pteridium esculentum (Forst.) Nakai – see *P. aquilinum*
Pteris esculenta Forst. – see *Pteridium aquilinum*
Pteris thalictroides (L.) Sw. – see *Ceratopteris thalictroides*
Pterocypsela indica (L.) C. Shih – see *Lactuca indica*
Pueraria argyi Lév. & Vaniot – see *P. montana* var. *lobata*
Pueraria bodinieri Lév. & Vaniot – see *P. montana* var. *lobata*
Pueraria caerulea Lév. & Vaniot – see *P. montana* var. *lobata*
Pueraria harmsii Rech. – see *P. montana* var. *lobata*
Pueraria hirsuta (Thunb.) C.K. Schneid. – see *P. montana* var. *lobata*
Pueraria hirsuta (Thunb.) Matsumura – see *P. montana* var. *lobata*
Pueraria hirsuta Matsum. – see *P. montana* var. *lobata*
Pueraria hirsuta Schneid – see *P. montana* var. *lobata*
Pueraria koten Lév. – see *P. montana* var. *lobata*
Pueraria lobata (Willd.) Ohwi – see *P. montana* var. *lobata*
Pueraria montana **(Lour.) Merr. var. *lobata* (Willd.) Maesen & S. M. Almeida, ex Sanjappa & Predeep (40.30)**
Pueraria neo-caledonica Harms – see *P. montana* var. *lobata*
Pueraria novo-guineensis Warburg – see *P. montana* var. *lobata*
Pueraria pseudo-hirsuta Tang & Wang – see *P. montana* var. *lobata*
Pueraria thunbergiana (Siebold & Zucc.) Benth. – see *P. montana* var. *lobata*
Pueraria triloba (Houtt.) Makino – see *P. montana* var. *lobata*
Pueraria triloba (L.) Makino – see *P. montana* var. *lobata*
Pueraria triloba (Lour.) Makino – see *P. montana* var. *lobata*
Pueraria volkensii Hosokawa – see *P. montana* var. *lobata*
Pulegium aromaticum S.F. Gray – see *Mentha pulegium*
Pulegium erectum Mill. – see *Mentha pulegium*
Pulegium heterophyllum Opiz, ex Boenn. – see *Mentha pulegium*
Pulegium micranthum Claus – see *Mentha pulegium*
Pulegium pubescens Opiz, ex Boenn. – see *Mentha pulegium*
Pulegium pulegium Karst. – see *Mentha pulegium*
Pulegium tomentella Presl – see *Mentha pulegium*
Pulegium vulgare Mill. – see *Mentha pulegium*
Pycreus esculentus (L.) Hayek – see *Cyperus esculentus*
Pyretheum indicum Roxb. – see *Chrysanthemum coronarium*
Pyrethrum acmella Medik. – see *Acmella oleracea*
Radicula armoracia (L.) B.L. Rob. – see *Armoracia rusticana*
Radicula nasturtium-aquaticum Britt. & Rendle – see *Nasturtium officinale*
Radicula nasturtium-aquaticum (L.) Rendle & Britten – see *Nasturtium officinale*
Radicula officinalis Groves – see *Nasturtium officinale*
Radicula officinalis (R. Br.) H. Groves – see *Nasturtium officinale*
Rajania flexuosa Bello – see *Dioscorea alata*
Raphanis magna Moench – see *Armoracia rusticana*
Raphanus acanthiformis Morel – see *R. sativus* Dakion group
Raphanus albus (L.) Crantz – see *Sinapis alba*
Raphanus caudatus L. f. – see *R. sativus* Caudatus group

Raphanus chinensis Mill. – see *R. sativus* Oleiformis group
Raphanus rapa (L.) Crantz – see *Brassica rapa* subsp. *rapa*
Raphanus raphanistroides Nakai – see *R. sativus* Dakion group
Raphanus raphanistrum form *chinensis* (Mill.) Thell. – see *R. sativus* Oleiformis group
Raphanus raphanistrum form *longirostris* Thell. – see *R. sativus* Oleiformis group
Raphanus raphanistrum subsp. *caudatus* (L. f.) Thell. – see *R. sativus* Caudatus group
Raphanus sativus convar. *oleiferus* (Stokes) Sazon. & Stankev. – see *R. sativus* Oleiformis group
Raphanus sativus form *raphanistroides* Makino – see *R. sativus* Dakion group
***Raphanus sativus* L. Caudatus group (30.53)**
Raphanus sativus L. cv. Caudatus – see *R. sativus* Caudatus group
Raphanus sativus L. cv. group Chinese radish – see *R. sativus* Dakion group
Raphanus sativus L. cv. group Leaf radish – see *R. sativus* Oleiformis group
Raphanus sativus L. cv. group Rat-tail radish – see *R. sativus* Caudatus group
Raphanus sativus L. cv. group Small radish – see *R. sativus* radicula group
Raphanus sativus L. cv. Longipinnatus – see *R. sativus* Dakion group
***Raphanus sativus* L. Dakion group (30.54)**
Raphanus sativus L. Hortensis group – see *Raphanus sativus* Radicula group
Raphanus sativus L. Japonicus group – see *R. sativus* Dakion group
Raphanus sativus L. Longipinnatus group – see *R. sativus* Dakion group
***Raphanus sativus* L. Oleiformis group (30.55)**
***Raphanus sativus* L. Radicula group (30.56)**
Raphanus sativus L. subsp. *acanthiformis* (Blanch.) A.K. Stankevich – see *R. sativus* Dakion group
Raphanus sativus L. subsp. *acanthiformis* (Morel) A.K. Stankevich – see *R. sativus* Dakion group
Raphanus sativus L. subsp. *oleiferus* (Stokes) Metzg. – see *R. sativus* Oleiformis group
Raphanus sativus L. var. *acanthiformis* Makino – see *R. sativus* Dakion group
Raphanus sativus L. var. *caudatus* Alef. – see *R. sativus* Caudatus group
Raphanus sativus L. var. *caudatus* L. – see *R. sativus* Caudatus group
Raphanus sativus L. var. *caudatus* (L. f.) Vilmorin – see *R. sativus* Caudatus group
Raphanus sativus L. var. *caudatus* (L.) Vilmorin – see *R. sativus* Caudatus group
Raphanus sativus L. var. *chinensis* (Mill.) Alef. – see *R. sativus* Oleiformis group
Raphanus sativus L. var. *hortensis* Baker – see *R. sativus* Radicula group
Raphanus sativus L. var. *hortensis* Baker form *acanthiformis* Makino – see *R. sativus* Dakion group
Raphanus sativus L. var. *longipinnatus* L.H. Bailey – see *R. sativus* Dakion group
Raphanus sativus L. var. *mougri* Helm – see *R. sativus* Caudatus group
Raphanus sativus L. var. *niger* (Miller) Persoon – see *R. sativus* Dakion group
Raphanus sativus L. var. *oleiferus* Stokes – see *R. sativus* Oleiformis group
Raphanus sativus L. var. *oleiformis* Pers. – see *R. sativus* Oleiformis group
Raphanus sativus L. var. *radicula* DC. – see *R.* Radicula group
Raphanus sativus L. var. *radicula* DC. – see *R. sativus* Radicula group
Raphanus sativus L. var. *radicula* Pers. – see *R. sativus* Radicula group
Raphanus sativus L. var. *sativus* – see *R. sativus* Radicula group
Raphanus sativus var. *raphanistroides* (Makino) Sinsk. – see *R. sativus* Dakion group
Raphanus sinapis-officinalis Crantz – see *Brassica nigra* Koch.
Raphanus taquetii Lévl. – see *R. sativus* Dakion group
Rapistrum glandulosum Berger. – see *Bunias orientalis*
Rapunculus verus Fourr. – see *Campanula rapunculus*
Reseda arborea hort. ex Vilm. – see *R. odorata*
Reseda neilgherrensis Müll. Arg – see *R. odorata*
***Reseda odorata* L. (60.1)**
***Rheum rhabarbarum* L. (58.1)**

Rheum rhaponticum L. – see *R. rhabarbarum*
Rheum rhubarban L. – see *R. rhabarbarum*
Rheum sibiricum Pall. – see *R. rhabarbarum*
Rheum undulatum L. – see *R. rhabarbarum*
Robynsia macrophylla Mart. & Gal. – see *Pachyrhizus erosus*
Roripa armoracia (L.) Hitchc. – see *Armoracia rusticana*
Roripa nasturtium Beck – see *Nasturtium officinale*
Roripa rusticana (P. Gaertn., B. Mey. & Scherb.) Gren. & Godr. – see *Armoracia rusticana*
Rorippa armoracia (L.) A.S. Histche – see *Armoracia rusticana*
Rorippa nasturtium Beck – see *Nasturtium officinale*
Rorippa nasturtium (L.) Hayek. – see *Nasturtium officinale*
Rorippa nasturtium-aquaticum (L.) Hayek – see *Nasturtium officinale*
Rorippa officinalis (R. Br.) P. Royen – see *Nasturtium officinale*
Rudua aurea (Roxb.) F. Maek. – see *Vigna radiata* var. *radiata*
Rudua aurea (Roxb.) Maekawa – see *Vigna radiata* var. *radiata*
Rudua mungo (L.) F. Maek. – see *Vigna mungo* var. *mungo*
Ruelingia triangularis (Jacq.) Ehrh. – see *Talinum fruticosum*
***Rumex acetosa* L. (58.2)**
Rumex acetosus St Lag. – see *R. acetosa*
Rumex acidus Salisb. – see *R. acetosa*
Rumex acuminatus Campd. – see *R. acetosa*
Rumex acutatus subsp. *glaucus* (Jacq.) Gaud. ex Wulf – see *R. scutatus*
Rumex aetnensis Presl – see *R. scutatus*
Rumex agrestis Rafn. – see *R. acetosa*
Rumex alpestris Jacq. – see *R. scutatus*
Rumex ambiguus Gren. – see *R. acetosa*
Rumex amplexicaulis Rafin. – see *R. acetosa*
Rumex angustatus Rafin. – see *R. acetosa*
Rumex bidentula Rafin. – see *R. acetosa*
Rumex commersonnii Campd. – see *R. acetosa*
Rumex domesticus Schur – see *R. patientia*
Rumex glaucus Jacq. – see *R. scutatus*
Rumex haplorhizus Czern. ex Turcz. – see *R. acetosa*
Rumex hastaefolius Campd. – see *R. scutatus*
Rumex hastatus Link, ex Meissn. – see *R. scutatus*
Rumex hastifolius M. Bieb. – see *R. scutatus*
Rumex hastulatus Rafin. – see *R. acetosa*
Rumex hispanicus Koch – see *R. acetosa*
Rumex hortensis Vis. ex Nym. – see *R. acetosa*
Rumex induratus Boiss. & Reut. – see *R. scutatus*
Rumex intermedius Sturm ex Meissn. – see *R. acetosa*
Rumex luxurians hort. ex Meissn. – see *R. scutatus*
Rumex micranthus Campd. ex Meissn. – see *R. acetosa*
Rumex mutabilis Campd. ex Meissn. – see *R. acetosa*
Rumex oblongus Campd. – see *R. acetosa*
Rumex olitoria Rafin. – see *R. acetosa*
Rumex olympicus Boiss. – see *R. patientia*
Rumex oxyotus Campd. – see *R. acetosa*
Rumex papillaris Boiss. & Reut. – see *R. acetosa*
***Rumex patientia* L. (58.3)**

Rumex patientia var. *kurdicus* Boiss. – see *R. patientia*
Rumex pratensis Dulac – see *R. acetosa*
Rumex pseudo-acetosa Bertol. – see *R. acetosa*
Rumex pubescens C. Koch – see *R. scutatus*
Rumex rugosus Campd. – see *R. acetosa*
Rumex sagittifolius Campd. – see *R. acetosa*
***Rumex scutatus* L. (58.4)**
Rumex scutatus subsp. *hastifolius* (Bieb.) Borod. – see *R. scutatus*
Rumex scutatus var. *hortensis* DC. – see *R. scutatus*
Rumex stenophyllus Timb. & Jeanb. – see *R. acetosa*
Rumex thyrsiflorus Fingerh. – see *R. acetosa*
Rumex triangularis DC. – see *R. acetosa*
Rumex tuberosus Poir. – see *R. acetosa*
Rumex vinealis Timb. & Jeanb. – see *R. acetosa*
***Rungia klossii* S. Moore (21.2)**
Sabdariffa digitata Kostel. – see *Hibiscus sabdariffa*
Sabdariffa rubra Kostel. – see *Hibiscus sabdariffa*
***Saccharum edule* Hassk. (17.5)**
Saccharum officinarum L. var. *edule* (Hassk.) Backer – see *S. edule* Hassk.
Sagitta aquatica "Pline" St.-Leg. – see *Sagittaria sagittifolia* subsp. *sagittifolia*
Sagitta maior Scop. – see *Sagittaria sagittifolia* subsp. *sagittifolia*
Sagitta palustris Bub. – see *Sagittaria sagittifolia* subsp. *sagittifolia*
Sagittaria acuminata Sm. – see *S. sagittifolia* subsp. *sagittifolia*
Sagittaria alpina Willd. – see *S. sagittifolia* subsp. *sagittifolia*
Sagittaria aquatica Lam. – see *S. sagittifolia* subsp. *sagittifolia*
Sagittaria arifolia Nutt. ex J.G. Sm. – see *S. sagittifolia* subsp. *sagittifolia*
Sagittaria doniana Sweet – see *S. sagittifolia* subsp. *sagittifolia*
Sagittaria edulis Schlecht. – see *S. sagittifolia* subsp. *sagittifolia*
Sagittaria gigantea hort. ex Vilm. – see *S. sagittifolia* subsp. *sagittifolia*
Sagittaria gracilis Pursh – see *S. sagittifolia* subsp. *sagittifolia*
Sagittaria hastata Pursh – see *S. sagittifolia* subsp. *sagittifolia*
Sagittaria hermaphrodita Buch.-Ham. – see *S. sagittifolia* subsp. *sagittifolia*
Sagittaria hermaphrodita Wall. – see *S. sagittifolia* subsp. *sagittifolia*
Sagittaria heterophylla Schreb. – see *S. sagittifolia* subsp. *sagittifolia*
Sagittaria hirundinacea Blume – see *S. sagittifolia* subsp. *sagittifolia*
Sagittaria japonica hort. ex Vilm. – see *S. sagittifolia* subsp. *sagittifolia*
Sagittaria lancifolia hort. ex Vilm. – see *S. sagittifolia* subsp. *sagittifolia*
Sagittaria latifolia Willd. – see *S. sagittifolia* subsp. *sagittifolia*
Sagittaria longerostris J.G. Sm. – see *S. sagittifolia* subsp. *sagittifolia*
Sagittaria longiloba Engelm. – see *S. sagittifolia* subsp. *sagittifolia*
Sagittaria macrophylla Bunge – see *S. sagittifolia* subsp. *sagittifolia*
Sagittaria major Scop. – see *S. sagittifolia* subsp. *sagittifolia*
Sagittaria minor Mill. – see *S. sagittifolia* subsp. *sagittifolia*
Sagittaria monoeca Gilib. – see *S. sagittifolia* subsp. *sagittifolia*
Sagittaria natans Pall. – see *S. sagittifolia* subsp. *sagittifolia*
Sagittaria natans Pall. non Michx. – see *S. sagittifolia* subsp. *sagittifolia*
Sagittaria obtusa Thunb. – see *S. sagittifolia* subsp. *sagittifolia*
Sagittaria pygmaea Miq. – see *S. sagittifolia* subsp. *sagittifolia*
Sagittaria sagittaefolia Auct. – see *S. sagittifolia* subsp. *sagittifolia*
Sagittaria sagittata Thunb. – see *S. sagittifolia* subsp. *sagittifolia*

Sagittaria sagittifolia L. – see *S. sagittifolia* subsp. *sagittifolia*
***Sagittaria sagittifolia* L. subsp. *leucopetala* (Miq.) Hartog (6.1)**
***Sagittaria sagittifolia* L. subsp. *sagittifolia* (6.2)**
Sagittaria sagittifolia L. var. *edulis* Siebold, ex Miq. – see *S. sagittifolia* subsp. *leucopetala*
Sagittaria sagittifolia L. var. *leucopetala* Miq. – see *S. sagittifolia* subsp. *leucopetala*
Sagittaria sagittifolia L. var. *sinensis* Sims – see *S. sagittifolia* subsp. *sagittifolia*
Sagittaria sagittifolia Lour., non L. – see *S. sagittifolia* subsp. *leucopetala*
Sagittaria simplex Pursh – see *S. sagittifolia* subsp. *sagittifolia*
Sagittaria sinensis Sims – see *S. sagittifolia* subsp. *leucopetala*
Sagittaria trifolia L. – see *S. sagittifolia* subsp. *leucopetala*
Sagittaria trifolia L. var. *edulis* (Siebold) Ohwi – see *S. sagittifolia* subsp. *leucopetala*
Sagittaria trifolia L. var. *typica* Mak. – see *S. sagittifolia* subsp. *leucopetala*
Sagittaria trifolia var. *sinensis* (Sims) Makino – see *S. sagittifolia* subsp. *leucopetala*
Sagittaria variabilis Engelm. – see *S. sagittifolia* subsp. *sagittifolia*
Sagittaria vulgaris Gueldenst. – see *S. sagittifolia* subsp. *sagittifolia*
Salsola asparagoides Miq. – see *S. glauca*
***Salsola komarovi* Iljin (35.7)**
Salsola longifolia Lam. – see *S. soda*
Salsola scoparia (L.) M. Bieb. – see *Bassia scoparia*
Salsola sieversiana Pall. ex Steud. – see *Bassia scoparia*
Salsola soda F. Schmidt – see *S. komarovii* Iljin
***Salsola soda* L. (35.8)**
Salsola soda sensu auct. Japon., non L. – see *S. komarovii* Iljin
Salsola songarica Siev. ex Pall. – see *Bassia scoparia*
Sarcoca acinosa (Roxb.) Skalický – see *Phytolacca acinosa*
Sarcoca esculenta (van Houtte) Skalický – see *Phytolacca esculenta*
Satureia hortensis L. – see *Satureja hortensis*
Satureja brachiata Stokes – see *S. hortensis*
***Satureja hortensis* L. (44.9)**
Satureja laxiflora K. Koch – see *S. hortensis*
Satureja officinarum Crantz – see *S. hortensis*
Satureja pachyphylla K. Koch – see *S. hortensis*
Satureja viminea Burm. f. – see *S. hortensis*
Sauropus albicans Blume – see *S. androgynus*
***Sauropus androgynus* (L.) Merrill (39.5)**
Sauropus gardnerianus Wight – see *S. androgynus*
Sauropus sumatranus Miq. – see *S. androgynus*
Sauropus zeylanicus Wight – see *S. androgynus*
Scandalida rubra Medik. – see *Lotus tetragonolobus*
Scandalida tetragonoloba Medik. – see *Lotus tetragonolobus*
Scandix cerefolium L. – see *Anthriscus cerefolium*
Scandix odorata L. – see *Myrrhis odorata*
Scandix tenuifolia Salisb. – see *Anthriscus cerefolium*
Schaberia maritime C.A. Gray var. *asparagoides* Franch & Sav. – see *Suaeda glauca*
Schoberia glauca Bunge – see *Suaeda glauca*
Schoenoprasum vulgare Fourr. – see *Allium schoenoprasum*
Scirpus plantagineus Retz. – see *Eleocharis dulcis*
Scirpus plantaginoides Rottb. – see *Eleocharis dulcis*
Scirpus tuberosus Roxb. – see *Eleocharis dulcis*
Scirpus tumidus Roxb. – see *Eleocharis dulcis*

Scolymus angiospermus Gaertn. – see *S. hispanicus*
Scolymus annuus Gerard, ex Steud. – see *S. maculatus*
Scolymus clusii Bub. – see *S. maculatus*
Scolymus congestus Lam. – see *S. hispanicus*
Scolymus gymnospermus Gaertn. – see *S. maculatus*
***Scolymus hispanicus* L. (27.36)**
***Scolymus maculatus* L. (27.37)**
Scolymus pectinatus Cass. – see *S. maculatus*
Scolymus perennis Gerard, ex Steud. – see *S. hispanicus*
Scolymus theophrasti Bub. – see *S. hispanicus*
Scolymus vivax Sauvages, ex DC. – see *S. hispanicus*
Scorzonera acaulis Willd. – see *Taraxacum officinale*
Scorzonera angustifolia Wohlleb. ex Wallr. – see *S. hispanica*
Scorzonera asphodeloides Wallr. – see *S. hispanica*
Scorzonera crispatula Boiss. – see *S. hispanica*
Scorzonera dentata Kit. – see *S. hispanica*
Scorzonera denticulata Lam. – see *S. hispanica*
Scorzonera edulis Moench – see *S. hispanica*
Scorzonera glastifolia Willd. – see *S. hispanica*
Scorzonera graminifolia Hoffm. – see *S. hispanica*
***Scorzonera hispanica* L. (27.38)**
Scorzonera montana Mutel – see *S. hispanica*
Scorzonera sativa Gater. – see *S. hispanica*
Scorzonera taurica M. Bieb. – see *S. hispanica*
Sececio kaempferi DC. – see *Farfugium japonicum*
Sechium americanum Lam. – see *S. edule*
Sechium americanum Poir. – see *S. edule*
Sechium chayota Hemsl. – see *S. edule*
***Sechium edule* (Jacq.) Swartz (38.25)**
Sechium edule Swartz. – see *S. edule* (Jacq.) Swartz.
***Sedum sarmentosum* Burge (37.1)**
Sedum sheareri S. Moore – see *S. sarmentosum*
Selinum cerefolium E.H.L. Krause – see *Anthriscus cerefolium*
Selinum coriandrum E.H.L. Krause – see *Coriandrum sativum*
Selinum sisarum E.H.L. Krause – see *Sium sisarum*
Senecio biafrae Oliv. & Hiern – see *Crassocephalum biafrae*
Senecio ecalyculatus Sch. Bip. – see *Emilia sonchifolia*
Senecio kaempferi DC. – see *Farfugium japonicum*
Senecio kaempferi (Siebold & Zucc.) Benth. – see *Farfugium japonicum*
Senecio sonchifolius (L.) Moench – see *Emilia sonchifolia*
Senecio splendens Lev. & Van – see *Ligularia fischeri*
Sesamopteris radiata DC. – see *Sesamum radiatum*
Sesamopteris radiata DC. ex Meissn. – see *Sesamum radiatum*
Sesamum foetidum Afz. ex Engl. – see *S. radiatum*
Sesamum occidentale Regel & Heer – see *S. radiatum*
***Sesamum radiatum* Schumacker (55.1)**
Seseli sisarum (L.) Crantz – see *Sium sisarum*
***Setaria palmifolia* (Koenig) Stapf (17.6)**
Sicana odorifera Naudin – see *S. odorifera* (Vell.) Naudin
***Sicana odorifera* (Vell.) Naudin (38.26)**

Sicyos edulis Jacq. – see *Sechium edule*
Sicyos fauriei H. Lev. – see *Momordica charantia*
Sicyos laciniata Descourt. – see *Sechium edule*
Sicyos laciniatus Descourt. – see *Sechium edule*
***Sinapis alba* L. (30.57)**
Sinapis alba L. subsp. *alba* – see *S. alba*
Sinapis arensis var. *alba* – see *S. alba*
Sinapis arensis var. *nigra* – see *Brassica nigra*
Sinapis brassicata Griseb. – see *Brassica juncea* var. *rugosa*
Sinapis cernua Thunb. – see *Brassica juncea* var. *rugosa*
Sinapis chinensis var. *integrifolia* Stokes – see *Brassica juncea* var. *rugosa*
Sinapis cuneifolia Roxb. – see *Brassica juncea* var. *rugosa*
Sinapis eruca [Clairv.] – see *Eruca sativa*
Sinapis exotica hort. ex DC. – see *Eruca sativa*
Sinapis hispida Ten. – see *S. alba*
Sinapis integrifolia West – see *Brassica juncea* var. *rugosa*
Sinapis juncea L. – see *Brassica juncea* var. *rugosa*
Sinapsis juncea var. *napiformis* Pailet & Bois – see *Brassica juncea* var. *megarrhiza*
Sinapis lanceolata DC. – see *Brassica juncea* var. *rugosa*
Sinapis muralis R. Br. – see *Diplotaxis muralis*
Sinapis nigra Descourt. – see *Brassica juncea* var. *rugosa*
Sinapis nigra L. – see *Brassica nigra*
Sinapis patens Roxb. – see *Brassica juncea* var. *rugosa*
Sinapis pekinensis Lour – see *Brassica rapa* var. *pekinensis*
Sinapis rugosa Roxb. – see *Brassica juncea* var. *rugosa*
Sinapis tuberosa Poir. – see *Brassica rapa* subsp. *rapa*
Sinapistrum album (L.) Chev. – see *Sinapis alba*
Sinocalamus beecheyanus (Munro) McClure – see *Bambusa beecheyana*
Sinocalamus beecheyanus var. *pubescens* Li. – see *Bambusa beecheyana*
Sinocalamus brandisii (Munro) Keng f. – see *Dendrocalamus brandisii*
Sinocalamus giganteus (Munro) Keng f. – see *Dendrocalamus giganteus*
Sinocalamus latiflorus (Munro) McClure – see *Dendrocalamus latiflorus*
Sinocalamus oldhamii (Munro) McClure – see *Bambusa oldhamii*
Siphonoglossa macleodiae S. Moore – see *Justicia insularis*
Sisymbrium acre Lam. – see *Diplotaxis muralis*
Sisymbrium amarum Salisb. – see *Nasturtium officinale*
Sisymbrium barrelieri Thuill. – see *Diplotaxis muralis*
Sisymbrium cardaminefolium Gilib. – see *Nasturtium officinale*
Sisymbrium erucastrum Gouan – see *Diplotaxis muralis*
Sisymbrium fluviatile Vell. – see *Nasturtium officinale*
Sisymbrium minus Lam. – see *Diplotaxis muralis*
Sisymbrium monense Thuill. – see *Diplotaxis muralis*
Sisymbrium murale L. – see *Diplotaxis muralis*
Sisymbrium nasturtium L. – see *Nasturtium officinale*
Sisymbrium nasturtium Thunb. – see *Nasturtium officinale*
Sisymbrium nasturtium-aquaticum L. – see *Nasturtium officinale*
Sisymbrium nigrum (L.) Prantl – see *Brassica nigra*
Sisymbrium tenuifolium L. – see *Diplotaxis muralis*
Sium brevifolium hort. ex DC. – see *S. sisarum*
Sium javanicum Blume – see *Oenanthe javanica* subsp. *javanica*

Sium laciniatum Blume – see *Oenanthe javanica* subsp. *javanica*
Sium podolicum hort. ex G. Don – see *S. sisarum*
***Sium sisarum* L. (24.23)**
***Smallanthus sonchifolius* (Poepp. & Endl.) H. Rob. (27.39)**
Soda inermis Fourr. – see *Salsola soda*
Soja angustifolia Miquel – see *Glycine max*
Soja hispida Moench – see *Glycine max*
Soja japonica Savi – see *Glycine max*
Soja max (L.) Piper – see *Glycine max*
Soja soja Karst. – see *Glycine max*
Solanum acanthifolium hort. Par. ex Dun. – see *S. torvum*
Solanum acutifolium Kit. – see *S. nigrum*
Solanum adventitium Polgár – see *S. americanum*
Solanum aegyptiacum Forsk. – see *S. nigrum*
Solanum aethiopicum Aculeatum Group – see *S. aethiopicum*
Solanum aethiopicum Gilo Group – see *S. aethiopicum*
Solanum aethiopicum Jacq. – see *S. aethiopicum*
***Solanum aethiopicum* L. (62.14)**
Solanum aethiopicum L. var. *aculeatum* Dunal – see *S. aethiopicum*
Solanum agreste Roth – see *S. indicum*
Solanum alatum Moench – see *S. nigrum*
Solanum album Noronha – see *S. melongena*
***Solanum americanum* Miller (62.15)**
Solanum amoenum Jungh. – see *S. torvu*
Solanum anguivi Hook. – see *S. indicum*
Solanum anguivi var. *multiflorum* (Roem. & Schult.) Vajr. – see *Solanum violaceum*
Solanum angulatum Ruiz & Pav. – see *S. quitoense*
Solanum aracatscha Bess. – see *S. tuberosum*
Solanum arenarium Schur – see *S. nigrum*
Solanum asperum Hornem. ex Walp. – see *S. nigrum*
Solanum astroites Forst. f – see *S. nigrum*
Solanum atriplicifolium Desp. ex Dun. – see *S. nigrum*
Solanum atropo Schum. – see *S. macrocarpon*
Solanum besseri Weinm. ex Roem. & Schult. – see *S. nigrum*
Solanum betaceum Cav – see *Cyphomandra betacea*
Solanum brieyi De Wild. – see *S. aethiopicum*
Solanum calvum Bitter – see *S. americanum*
Solanum campechiense hort. Par. ex Dun. – see *S. torvum*
Solanum canescens Blume – see *S. indicum*
Solanum canescens Kit. – see *S. nigrum*
Solanum caribaeum Dunal – see *S. americanum*
Solanum cestrifolium Jacq. ex Weinw. – see *S. nigrum*
Solanum chenopodioides Lam. – see *S. nigrum*
Solanum chlorocarpum Schur – see *S. nigrum*
Solanum coagulans Forsskal – see *S. melongena*
Solanum columbianum Dunal – see *S. tuberosum*
Solanum crassifolium Ortega – see *Cyphomandra betacea*
Solanum crassifolium Salisb. – see *S. macrocarpon*
Solanum crenato-dentatum Dunal – see *S. nigrum*
Solanum crotonoides Michx. ex Dun. – see *S. torvum*

Solanum cultum (A. DC.) Berth. – see *S. tuberosum*
Solanum cumingii Dunal – see *S. melongena*
Solanum cuneatum Moench – see *S. indicum*
Solanum cuneifolium hort. Avens. ex Dunal – see *S. nigrum*
Solanum curtipes Bitter – see *S. americanum*
Solanum dasyphyllum Thonn. ex Schum. – see *S. macrocarpon*
Solanum decurrens Wall. ex Dunal – see *S. nigrum*
Solanum demissum Lindl. – see *S. tuberosum*
Solanum deppei hort. ex Dunal – see *S. nigrum*
Solanum diffusum Vell. – see *S. nigrum*
Solanum dilleni Schult. – see *S. nigrum*
Solanum dimorphum Matsum. – see *S. macrocarpon*
Solanum douglasii Dunal – see *S. nigrum*
Solanum edule Schum. & Thonn. – see *S. melongena*
Solanum elskensi De Wild. – see *S. aethiopicum*
Solanum erosum Heurck & Muell. Arg. – see *S. indicum*
Solanum erythraeum Dunal – see *S. nigrum*
Solanum erythraeum G.F.W. Mey. – see *S. nigrum*
Solanum esculentum Dunal – see *S. melongena*
Solanum esculentum Neck. – see *S. tuberosum*
Solanum fendleri A. Gray – see *S. tuberosum*
Solanum ferox Jungh. ex Miq. – see *S. indicum*
***Solanum ferox* L. (62.16)**
Solanum ferrugineum Jacq. – see *S. torvum*
Solanum ficifolium Ortega – see *S. torvum*
Solanum fistulosum Dunal, ex Poir. – see *S. nigrum*
Solanum flavum Kit. – see *S. nigrum*
Solanum foliosum Link – see *Lycopersicon esculentum* var. *esculentum*
Solanum geminifolium Thonn. – see *S. aethiopicum*
Solanum gilo Requien, ex Dunal – see *S. integrifolium*
Solanum gollmeri Bitter – see *S. americanum*
Solanum guatemalense hort. ex Bitter – see *S. muricatum*
Solanum guatemalense Rous – see *S. muricatum*
Solanum guineense (L.) Lam. – see *S. nigrum*
Solanum heteracanthum Dunal – see *S. melongena*
Solanum heterogonum hort. ex Dunal – see *S. nigrum*
Solanum heterophyllum Balb. ex Dunal – see *S. torvum*
Solanum heynii Roem. & Schult. – see *S. indicum*
Solanum hierochuntinum Dunal – see *S. incanum*
Solanum himalense Dunal – see *S. indicum*
Solanum hirsutum Dunal – see *S. nigrum*
Solanum hirsutum Roxb. – see *S. aethiopicum*
Solanum humboldtii Willd. – see *Lycopersicon esculentum* var. *esculentum*
Solanum humile Salisb. – see *S. nigrum*
Solanum immane Hance. ex Walp. – see *S. aethiopicum*
Solanum immite Dunal – see *S. tuberosum*
Solanum incanum auct. – see *S. melongena*
Solanum incanum Kit. ex Schult. – see *S. nigrum*
***Solanum incanum* L. (62.17)**
Solanum incertum Dunal – see *S. nigrum*

Solanum inconspicuum Bitter – see *S. americanum*
Solanum indicum auct., non L. – see *S. violaceum*
Solanum indicum form *album* C.Y. Wu & S.C. Huang – see *S. violaceum*
Solanum indicum Roxb. – see *S. violaceum*
Solanum indicum var. *eroso-pinnatifidum* Duna – see *S. violaceum*
Solanum indicum var. *multiflorum* (Roem. & Schult.) C.B. Clarke – see *S. violaceum*
Solanum indicum var. *pubescens* Dunal – see *S. violaceum*
Solanum indicum var. *recurvatum* C.Y. Wu & S.C. Huang – see *S. violaceum*
Solanum indicum var. *sinuato-lobatum* Dunal – see *S. violaceum*
Solanum inops Dunal – see *S. americanum*
Solanum insanum L. – see *S. melongena*
Solanum insigne Lowe – see *Cyphomandra betacea*
Solanum integrifolium Poir. – see *S. aethiopicum*
Solanum judaicum Bess. – see *S. nigrum*
Solanum kitaibeli Schult. – see *S. nigrum*
Solanum kraussianum Buching. ex Krauss – see *S. nigrum*
Solanum kurzii Brace, ex Prain – see *S. violaceum*
Solanum largiflorum C. White – see *S. torvum*
Solanum lasiocarpum Dunal – see *S. aethiopicum*
Solanum lasiocarpum Dunal var. *velutinum* Callery & Gaudich. – see *S. aethiopicum*
Solanum lobelii Tenore – see *S. aethiopicum*
Solanum longum Roxb. – see *S. melongena*
Solanum luridum Salisb. – see *Lycopersicon esculentum* var. *esculentum*
Solanum luteo-virescens C.C. Gmel – see *S. nigrum*
Solanum luteum Mill. – see *S. nigrum*
Solanum lycopersicon Blanco – see *Lycopersicon esculentum* var. *esculentum*
Solanum lycopersicum L. – see *Lycopersicon esculentum* var. *esculentum*
Solanum macaonense Dunal – see *S. torvum*
Solanum maccai Bert. ex Dunal – see *S. torvum*
***Solanum macrocarpon* L. (62.18)**
Solanum macrocarpon Pav. ex Dunal – see *S. quitoense*
Solanum mammosum Herb. ex Dunal – see *S. torvum*
Solanum mannii Wright – see *S. torvum*
Solanum mayanum Lundell – see *S. torvum*
Solanum melaniferum Moric. ex Dunal – see *S. muricatum*
Solanum melanocarpum Dunal – see *S. melongena*
Solanum melanocerasum Willd. – see *S. nigrum*
***Solanum melongena* L. (62.19)**
Solanum melongena var. *depressum* L.H. Bailey – see *S. melongena*
Solanum melongena var. *esculentum* Nees – see *S. melongena*
Solanum melongena var. *incanum* (L.) Kuntze – see *S. incanum*
Solanum melongena var. *inerme* C.H. Wright – see *S. aethiopicum*
Solanum melongena var. *inerme* (DC.) Hiern – see *S. melongena*
Solanum melongena var. *ovigerum* Nees – see *S. melongena*
Solanum melongena var. *serpentinum* (Desf.) L.H. Bailey – see *S. melongena*
Solanum melongenum St. Lag. – see *S. melongena*
Solanum memphiticum J.F. Gmel. – see *S. nigrum*
Solanum microtatanthum Bitter – see *S. americanum*
Solanum miniatum Bemh. ex Willd. – see *S. nigrum*
Solanum minutibaccatum Bitter – see *S. americanum*

Solanum minutibaccatum subsp. *curtipedunculatum* Bitter – see *S. americanum*
Solanum minutiflorum Dunal – see *S. nigrum*
Solanum monteiroi C.H. Wright – see *S. aethiopicum*
Solanum morselephantum Dammann – see *S. macrocarpon*
Solanum moschatum J. & C. Presl – see *S. nigrum*
***Solanum muricatum* Aiton (62.20)**
Solanum naumannii Engl. – see *S. aethiopicum*
Solanum nigrum auct. non L. – see *S. americanum*
***Solanum nigrum* L. (62.21)**
Solanum nigrum var. *dillenii* (Schult.) A. Gray – see *S. americanum*
Solanum nigrum var. *nodiflorum* (Jacq.) A. Gray – see *S. americanum*
Solanum nivalomontanum C.Y. Wu & S.C Huang – see *Solanum violaceum*
Solanum nodiflorum Jacq. – see *S. americanum*
Solanum nodiflorum var. *acuminatum* Dunal – see *S. americanum*
Solanum nodiflorum var. *macrophyllum* Dunal – see *S. americanum*
Solanum nodiflorum var. *petiolastrum* Dunal – see *S. americanum*
Solanum nodiflorum var. *puberulum* Dunal – see *S. americanum*
Solanum obliquum Bert. ex Dunal – see *Cyphomandra betacea*
Solanum obtusifolium Hartw. ex Sendt. – see *S. torvum*
Solanum ochroleucum Bast. – see *S. nigrum*
Solanum oleraceum Dunal, ex Poir. – see *S. nigrum*
Solanum olivare Pailleux & Bois – see *S. aethiopicum*
Solanum ovatifolium De Wild. – see *S. aethiopicum*
Solanum oviferum Salisb. – see *S. melongena*
Solanum ovigerum Dunal – see *S. melongena*
Solanum ovoideum Zipp. ex Miq. – see *S. aethiopicum*
Solanum pachystylum Polgár – see *S. americanum*
Solanum paludosum Dunal – see *S. nigrum*
Solanum parmentieri Molina, ex Walp. – see *S. tuberosum*
Solanum parviflorum Moris, ex Badaro – see *S. nigrum*
Solanum patulum Roth – see *S. nigrum*
Solanum peruvianum Lour. – see *Lycopersicon esculentum* var. *esculentum*
Solanum photeinocarpum Nakamura & Odashima – see *S. americanum*
Solanum pierreanum Pailleux & Bois – see *S. aethiopicum*
Solanum pimpinellifolium Jusl. – see *Lycopersicon pimpinellifolium*
Solanum pimpinellifolium L. – see *Lycopersicum pimpinellifolium*
Solanum pinnatifidum Roth – see *S. indicum*
Solanum plumieri Dunal – see *S. melongena*
Solanum poggei Dammer – see *S. aethiopicum*
Solanum pomiferum Cav. – see *Lycopersicon esculentum* var. *esculentum*
Solanum pressum Dunal – see *S. melongena*
Solanum pseudo-lycopersicum Jacq. – see *Lycopersicon esculentum* var. *esculentum*
Solanum pseudomelongena Tenore – see *S. aethiopicum*
Solanum pseudo-saponaceum Blume – see *S. torvum*
Solanum pseudo-undatum Blume – see *S. melongena*
Solanum pseudo-villosum Schur – see *S. nigrum*
Solanum pterocaulum Dunal – see *S. nigrum*
Solanum ptychanthum Dunal – see *S. nigrum*
Solanum puniceum C.C. Gmel. – see *S. nigrum*
Solanum quadriloculare Spreng. – see *S. aethiopicum*

Solanum quitense H.B.K. – see *S. quitoense*
***Solanum quitoense* Lam. (62.22)**
Solanum racemiflorum Dunal – see *Lycopersicon pimpinellifolium*
Solanum racemiflorum Vilmor. – see *Lycopersicum pimpinellifolium*
Solanum racemigerum Zodda – see *Lycopersicum pimpinellifolium*
Solanum repens Noronha – see *S. nigrum*
Solanum retroflexum Dunal – see *S. nigrum*
Solanum rhinozerothis Blume – see *S. nigrum*
Solanum roxburghii Dunal – see *S. nigrum*
Solanum rubrum Mill. – see *S. nigrum*
Solanum rumphii Dunal – see *S. nigrum*
Solanum saccianum Naud. – see *S. muricatum*
Solanum sanctum Jan, ex Dun. – see *S. torvum*
Solanum sanctum L. – see *S. incanum*
Solanum sanitwongsei Craib – see *S. violaceum*
Solanum sapinii De Wild. – see *S. macrocarpon*
Solanum sativum Dunal – see *S. melongena*
Solanum scabrum Lam. – see *S. muricatum*
Solanum scabrum Mill. – see *S. nigrum*
Solanum sciaphilum Bitter – see *S. americanum*
Solanum serpentinum Desf. – see *S. melongena*
Solanum silarium Buch.-Ham. ex Wall. – see *S. torvum*
Solanum sinense Blanco – see *S. tuberosum*
Solanum sinuatorepandum Dammer, ex K. Braun – see *S. Aethiopicum*
Solanum sodomaeum L. – see *S. violaceum*
Solanum sodomeum L. – see *S. violaceum*
Solanum sodomeum Russ. ex Nees – see *S. indicum*
Solanum sparsespinosum De Wild. – see *S. aethiopicum*
Solanum speciosum hort. ex Dunal – see *S. nigrum*
Solanum spurium J.F. Gmel. – see *Lycopersicon esculentum* var. *esculentum*
Solanum stenopetalum A. Br. – see *S. nigrum*
Solanum stramonifolium auct. non Jacq. – see *S. aethiopicum*
Solanum stramonifolium Dunal – see *S. aethiopicum*
Solanum stramonifolium Lam. – see *S. torvum*
Solanum strictum Zuccagni – see *S. nigrum*
Solanum suffruticosum Schousb. ex Willd. – see *S. nigrum*
Solanum sylvestre Audib. ex Dunal – see *S. tuberosum*
Solanum tenellum Bitter – see *S. americanum*
Solanum tenuiflorum Steud. – see *S. nigrum*
Solanum texanum Dunal – see *S. aethiopicum*
Solanum texanum hort. ex Tenore – see *S. integrifolium*
Solanum thonningianum Jacq.f. – see *S. macrocarpon*
Solanum tomentosum Hesb. Hasselt, ex Miq. – see *S. melongena*
***Solanum torvum* Swartz (62.23)**
Solanum triangulare Lam. – see *S. nigrum*
Solanum trongum Poir. – see *S. melongena*
***Solanum tuberosum* L. (62.24)**
Solanum tuberosum L. subsp. *tuberosum* [(L.) Hawkes] – see *S. tuberosum*
Solanum uliginosum Blume – see *S. nigrum*
Solanum umbelliferum Torr. – see *S. nigrum*

Solanum undatum Lam. – see *S. melongena*
Solanum undatum Poiret sensu Ochse – see *S. melongena*
Solanum undulatum Poir. – see *S. melongena*
Solanum variegatum Ruiz & Pav. – see *S. muricatum*
Solanum verapazense Standl. & Steyerm. – see *S. torvum*
Solanum verrucosum Schlecht. – see *S. tuberosum*
Solanum villosum Mill. – see *S. nigrum*
Solanum violaceum DC. ex Dun. – see *S. melongena*
***Solanum violaceum* Ortega (62.25)**
Solanum violaceum subsp. *multiflorum* (C.B. Clarke) K.M. Matthew – see *S. violaceum*
Solanum virginianum Pers. – see *S. nigrum*
Solanum virginianum Russ. ex Wall. – see *S. indicum*
Solanum viride Schur – see *S. nigrum*
Solanum viridescens hort. ex Reichb. – see *S. nigrum*
Solanum vulgare Hegetschw. – see *S. nigrum*
Solanum vulgatum Willd. ex Steud. – see *S. nigrum*
Solanum wightii Miq. ex C.B. Clarke – see *S. torvum*
Solanum zanoni Gouan – see *S. macrocarpon*
Solanum zeylanicum Blanco – see *S. aethiopicum*
Solanum zeylanicum Scop. – see *S. melongena*
***Solenostemon rotundifolius* (Poir.) J. K. Morton (44.10)**
Sonchus arvensis auct. mult. – see *S. wightianus*
Sonchus arvensis L. – see *S. wightianus*
***Sonchus asper* (L.) Hill (27.40)**
Sonchus asper P. Gaertn., B. Mey. & Scherb. – see *S. oleraceus*
Sonchus ciliatus Lam. – see *S. oleraceus*
Sonchus cuspidatus Blume – see *S. asper*
Sonchus glaber Gilib. – see *S. oleraceus*
Sonchus javanicus Jungh. – see *S. malaianus*
Sonchus lacerus Willd. – see *S. oleraceus*
Sonchus laevis Garsault – see *S. oleraceus*
Sonchus laevis Vill. – see *S. oleraceus*
Sonchus macrotus Fenzl – see *S. oleraceus*
***Sonchus malaianus* Miquel (27.41)**
Sonchus maritimus L. var. *malaianus* (Miquel) Hochr. – see *S. malaianus*
***Sonchus oleraceus* L. (27.42)**
Sonchus oleraceus var. *asper* L. – see *S. asper*
Sonchus oreophilus Miquel – see *S. malaianus*
Sonchus parviflorus Lej. ex Reichb. – see *S. oleraceus*
Sonchus pentaphyllus (Sch. Bip.) Sch. Bip. ex Schweinf. – see *Launaea taraxacifolia*
Sonchus reversus E. Mey. ex DC. – see *S. oleraceus*
Sonchus roseus Besser – see *S. oleraceus*
Sonchus royleanus DC. – see *S. oleraceus*
Sonchus schimperi A. Br. & Bouche – see *S. oleraceus*
Sonchus schmidianus C. Koch – see *S. oleraceus*
Sonchus sundaicus Blume – see *S. oleraceus*
Sonchus taraxacifolius Willd. – see *Launaea taraxacifolia*
Sonchus tenerrimus Schur – see *S. oleraceus*
Sonchus umbellifer Thunb. – see *S. oleraceus*
***Sonchus wightianus* DC. (27.43)**

Sonchus zacinthoides DC. – see *S. oleraceus*
Sparganophorus sparganophora (L.) C. Jeffrey – see *Struchium sparganophorum*
Sparganophorus struchium Poir. – see *Struchium sparganophorum*
Sparganophorus vaillantii Crantz – see *Struchium sparganophorum*
Sphenostylis ornata A. Chev. – see *S. stenocarpa*
Sphenostylis ornata (Welw. ex Baker) A. Chev. – see *S. stenocarpa*
***Sphenostylis stenocarpa* (Hochst. ex. A. Rich.) Harms. (40.31)**
Spilanthes acmella auct. – see *Acmella oleracea*
Spilanthes ciliata HBK – see *Acmella ciliata*
Spilanthes oleracea L. – see *Acmella oleracea*
Spilanthes paniculata Wall. ex DC. – see *Acmella paniculata*
Spilanthes pseudo-acmella auct., non (L.) Murr. – see *Acmella paniculata*
Spilanthes uliginosa Sw. – see *Acmella uliginosa*
Spinacia domestica Borkh. – see *S. oleracea*
***Spinacia oleracea* L. (35.9)**
Spinacia oleracea Miller – see *S. oleracea* L.
Spuriopimpinella calycina (Maxim.) Kitag. – see *Pimpinella brachycarpa*
***Stachys affinis* Bunge (44.11)**
Stachys palustris auct., non L. – see *S. affinis*
Stachys sieboldii Miq. – see *S. affinis*
Stachys tuberifera Naudin. – see *S. affinis*
Stizolobium bulbosum (L.) Spreng. – see *Pachyrhizus erosus*
Stizolobium deeringianum (Small) Bort. – see *Mucuna pruriens*
Stizolobium domingense Spreng. – see *Pachyrhizus erosus*
Stizolobium pruriens Medik. – see *Mucuna pruriens*
Stizolobium tuberosum (Lam.) Spreng. – see *Pachyrhizus tuberosus*
Stizolobium utile subsp. *deeringianum* (Bort) Ditm. – see *Mucuna pruriens*
Stizolobium utile Wall. ex Wight – see *Mucuna pruriens*
Struchium africanum P. Beauv. – see *S. sparganophorum*
Struchium herbaceum J. St.-Hil. – see *S. sparganophorum*
***Struchium sparganophorum* (L.) O. Kuntze (27.44)**
Suaeda asparagoides Miq. – see *S. glauca*
Suaeda cinerea Schur – see *Bassia scoparia*
***Suaeda glauca* (Bunge) Bunge (35.10)**
Suaeda sieversiana Pall. – see *Bassia scoparia*
Suaeda stauntonii Moq. – see *S. glauca*
***Symphytum × uplandicum* Nyman (29.3)**
Symphytum album hort. ex Steud. – see *S. officinale*
Symphytum ambiguum Pau – see *S. officinale*
Symphytum bohemicum F.W. Schmidt – see *S. officinale*
Symphytum coeruleum Petitmengin, ex Thell. – see *S. × uplandicum*
Symphytum consolida Gueldenst. – see *S. officinale*
Symphytum elatum Tausch – see *S. officinale*
Symphytum majus Bubani – see *S. officinale*
Symphytum microcalyx Opiz – see *S. officinale*
***Symphytum officinale* L. (29.2)**
Symphytum orientale auct. – see *S. × uplandicum*
Symphytum patens Sibth. – see *S. officinale*
Symphytum peregrinum Hook. f. – see *S. × uplandicum*
Symphytum stenophyllum Beck – see *S. officinale*

Symphytum tanaicense Stev. – see *S. officinale*
Symphytum uliginosum Kern. – see *S. officinale*
Symphytum variegatum hort. – see *S. officinale*
Tacca abyssinica Hochst. ex Baker – see *T. leontopetaloides*
Tacca artocarpifolia Seem. – see *T. leontopetaloides*
Tacca brownii Seem. – see *T. leontopetaloides*
Tacca dubia J.A. Schult. & J.H. Schult. – see *T. leontopetaloides*
Tacca gaogao Blanco – see *T. leontopetaloides*
Tacca guineensis G. Don, ex Loud. – see *T. leontopetaloides*
Tacca hawaiiensis H.W. Limpr. – see *T. leontopetaloides*
Tacca involucrata (Limpr.) Schum. & Thonn. – see *T. leontopetaloides*
***Tacca leontopetaloides* (L.) O. Kuntze (19.1)**
Tacca maculata Seem. – see *T. leontopetaloides*
Tacca madagascariensis Bojer – see *T. leontopetaloides*
Tacca oceanica Seem. – see *T. leontopetaloides*
Tacca phallifera J.H. Schult. – see *T. leontopetaloides*
Tacca pinnatifida J.R. Forst. & G. Forst. – see *T. leontopetaloides*
Tacca pinnatifolia Gaertn. – see *T. leontopetaloides*
Tacca quanzensis Welw. – see *T. leontopetaloides*
Tacca samoensis Reinecke – see *T. leontopetaloides*
Tacca umbrarum Jum. & Perr. – see *T. leontopetaloides*
Tacca viridis Hemsl. – see *T. leontopetaloides*
Taeniocarpum articulatum (Lam.) Desv. – see *Pachyrhizus erosus*
Talinum andrewsii Sweet – see *T. fruticosum*
Talinum attenuatum Rose & Standl. – see *T. fruticosum*
Talinum confusum Rose & Standl. – see *T. fruticosum*
Talinum crassifolium (Jacq.) Willd. – see *T. fruticosum*
Talinum crassifolium var. *albiflorum* DC. – see *T. fruticosum*
Talinum diffusum Rose & Standl. – see *T. fruticosum*
***Talinum fruticosum* (L.) Juss. (59.3)**
Talinum fruticosum Willd. – see *T. fruticosum* (L.) Juss.
***Talinum paniculatum* (Jacq.) Gaertner (59.4)**
Talinum paniculatum Moench – see *T. fruticosum*
Talinum patens (L.) Juss. – see *T. paniculatum*
Talinum racemosum (L.) Rohrb. – see *T. fruticosum*
Talinum triangulare (Jacq.) Willd. – see *T. fruticosum*
Tapeinophallus rivieri Baill. – see *Amorphophallus konjac*
Taraxacum affine Jord. – see *T. officinale*
Taraxacum alpestre DC. – see *T. officinale*
Taraxacum alpinum C. Koch – see *T. officinale*
Taraxacum arcuatum Dum. – see *T. officinale*
Taraxacum aureum Fisch. ex Herd. – see *T. officinale*
Taraxacum bellidifolium Fisch. ex Herd. – see *T. officinale*
Taraxacum caucasicum DC. – see *T. officinale*
Taraxacum ceratophorum DC. – see *T. officinale*
Taraxacum commune Dum.-Cours. – see *T. officinale*
Taraxacum commutatum Jord. – see *T. officinale*
Taraxacum corniculatum DC. – see *T. officinale*
Taraxacum dens-leonis Desf. – see *T. officinale*
Taraxacum erectum Schrank – see *T. officinale*

Taraxacum eriopodum DC. – see *T. officinale*
Taraxacum erythrospermum Andrz. ex Bess. – see *T. officinale*
Taraxacum glaciale Huet, ex Nym. – see *T. officinale*
Taraxacum glaucanthum DC. – see *T. officinale*
Taraxacum glaucescens Reichb. – see *T. officinale*
Taraxacum halophilum Trautv. ex Beck. – see *T. officinale*
Taraxacum hoppeanum Griseb. & Schenk – see *T. officinale*
Taraxacum laciniatum Martr. – see *T. officinale*
Taraxacum laevigatum DC. – see *T. officinale*
Taraxacum lanceolatum Poir. – see *T. officinale*
Taraxacum leontodon Gueldenst. – see *T. officinale*
Taraxacum leucospermum Jord. – see *T. officinale*
Taraxacum levigatum Poir. – see *T. officinale*
Taraxacum maculatum Jord. – see *T. officinale*
Taraxacum mexicanum DC. – see *T. officinale*
Taraxacum microcephalum Schur – see *T. officinale*
Taraxacum minimum Heldr. ex Nym. – see *T. officinale*
Taraxacum montanum Nutt. – see *T. officinale*
Taraxacum nigricans Reichb. – see *T. officinale*
Taraxacum obovatum DC. – see *T. officinale*
Taraxacum officinale auct. – see *T. officinale*
***Taraxacum officinale* G.H. Weber, ex Wiggers (27.45)**
Taraxacum officinale Wiggers var. *glaucescens* Koch. – see *T. officinale*
Taraxacum officinarum Rupr. – see *T. officinale*
Taraxacum parvulum DC. – see *T. officinale*
Taraxacum platycarpum H. Dahlst. – see *T. officinale*
Taraxacum pratense Hegetschw. – see *T. officinale*
Taraxacum prupurascens Schur – see *T. officinale*
Taraxacum pseudo-officinale Schur – see *T. officinale*
Taraxacum pyrenaicum Timb. ex Nym. – see *T. officinale*
Taraxacum rubinerve Jord. – see *T. officinale*
Taraxacum runcinatum Fisch. ex Herd. – see *T. officinale*
Taraxacum salinum Schur – see *T. officinale*
Taraxacum sinense Poir. – see *T. officinale*
Taraxacum stevenii DC. – see *T. officinale*
Taraxacum taraxacoides Willk. – see *T. officinale*
Taraxacum taraxacum Karst. – see *T. officinale*
Taraxacum triviale Hegetschw. – see *T. officinale*
Taraxacum vulgare auct. – see *T. officinale*
Taraxacum vulgare Schrank – see *T. officinale*
Taraxacum wallichii DC. – see *T. officinale*
Telanthera philoxeroides (Mart.) Moq. – see *Alternanthera philoxeroides*
***Telfairia occidentalis* Hook. f. (38.27)**
Telfairia pedata Hook. f. – see *T. pedata* (Smith, ex Sims) Hook.
***Telfairia pedata* (Sm. ex Sims) Hook. (38.28)**
Telis foenum-graeca (L.) Kuntze – see *Trigonella foenum-graecum*
Teremis turbinata Rafin. – see *Lycium chinense*
Ternanthera philoxeroides (Mart.) Moq. – see *Alternanthera philoxeroides*
Tetragonia cornuta Gaertn. – see *T. tetragoniodes*
Tetragonia expansa Murr. – see *T. tetragoniodes*

Tetragonia halimifolia Forst. f. – see *T. tetragoniodes*
Tetragonia japonica Thunb. – see *T. tetragoniodes*
Tetragonia quadricornis Stokes – see *T. tetragoniodes*
***Tetragonia tetragoniodes* (Pallas) O. Kuntze (22.2)**
Tetragonolobus biflorus var. *purpureus* (Moench) Alef. – see *Lotus tetragonolobus*
Tetragonolobus edulis Link – see *Lotus tetragonolobus*
Tetragonolobus purpureus Moench – see *Lotus tetragonolobus*
Tetrodea subterranea (L.) Raf. – see *Vigna subterranea*
Thalysia mays Kuntze – see *Zea mays*
Thlaspi bursa (Weber) Thunb. – see *Capsella bursa-pastoris*
Thlaspi bursa-pastoris L. – see *Capsella bursa-pastoris*
Thlaspi nasturtium Bergeret, ex Steud. – see *Lepidium sativum*
Thlaspi sativum Desf. – see *Lepidium sativum*
Thlaspi sativum (L.) Crantz – see *Lepidium sativum*
Thlaspidium sativum Spach – see *Lepidium sativum*
Thymus bidentatus Stokes – see *Mentha pulegium*
Thymus origanum Carnel – see *Orignaum vulgare*
Thymus origanum (L.) Kuntze – see *Origanum vulgare*
Tigridia conchiiflora Sweet – see *T. pavonia*
Tigridia grandiflora Salisb. – see *T. pavonia*
Tigridia oxypetala R. Morris – see *T. pavonia*
***Tigridia pavonia* (L. f.) Ker.-Gawl. (12.1)**
Tigridia speciosa Poit. – see *T. pavonia*
Tragopogon australis Jord. – see *T. porrifolius*
Tragopogon brachyphyllus Nym. – see *T. porrifolius*
Tragopogon carinatus Gilib. – see *T. pratensis*
Tragopogon grandiflorus Saut. – see *T. pratensis*
Tragopogon leiocarpos Saut. – see *T. pratensis*
Tragopogon minor Fries – see *T. pratensis*
Tragopogon pomifolius J.F. Gmel. – see *T. porrifolius*
***Tragopogon porrifolius* L. (27.46)**
***Tragopogon pratensis* L. (27.47)**
Tragopogon sativus Gater. – see *T. porrifolius*
Tragopogon shuttleworthii Godet – see *T. pratensis*
Tragopogon sinuatus Ave-Lall. – see *T. pratensis*
Tragopogon sylvestris Garsault – see *T. pratensis*
Tragopogon tortilis Pritz. – see *T. pratensis*
Tragopogon undulatus Thuill. – see *T. pratensis*
Trapa bicornis L. f. – see *T. natans* var. *natans*
***Trapa bicornis* Osbeck (64.1)**
Trapa bicornis var. *bispinosa* (Roxb.) Nakano – see *T. natans* var. *bispinosa*
Trapa bicornuta L. – see *T. bicornis*
Trapa bispinosa Roxb. – see *T. natans* var. *bispinosa*
Trapa bispinosa Roxb. var. *iinumai* Makino – see *T. natans* var. *bispinosa*
Trapa chinensis Lour. – see *T. bicornis*
Trapa cochinchinensis Lour. – see *T. bicornis*
Trapa japonica Flerov. – see *T. natans* var. *natans*
Trapa japonica var. *iinumai* Makino – see *T. natans* var. *natans*
Trapa natans L. var. *bicornis* Makino, ex Iinuma – see *T. bicornis*
***Trapa natans* L. var. *bispinosa* (Roxb.) Makino (64.2)**

Trapa natans L. var. *japonica* – see *T. natans* var. *natans*
***Trapa natans* L. var. *natans* (64.3)**
Trapa natans L. var. *quadrispinosa* – see *T. natans* var. *natans*
Trapa quadrispinosa Roxb. – see *T. natans* var. *natans*
Trapa quadrispinosa Wall. – see *T. bicornis*
Trichosanthes amara Blanco – see *Luffa acutangula*
Trichosanthes anguina L. – see *T. cucumerina*
Trichosanthes cavalerei Lévl. – see *T. cucumeroides*
Trichosanthes chinensis Ser. – see *T. cucumeroides*
Trichosanthes cochinchinesis M. Roem. – see *T. cucumerina*
Trichosanthes colubrina Jacq. f. – see *T. cucumerina*
***Trichosanthes cucumerina* L. (38.29)**
Trichosanthes cucumerina L. var. *anguina* (L.) Haines – see *T. cucumerina*
Trichosanthes cucumerina Miq. – see *T. cucumeroides*
Trichosanthes cucumerina subsp. *anguina* (L.) Grebenšč. – see *T. cucumerina*
Trichosanthes cucumerina var. *anguina* (L.) Haines – see *T. cucumerina*
***Trichosanthes cucumeroides* (Ser.) Maxim. (38.30)**
Trichosanthes cucumeroides (Ser.) Maxim var. *formosana* (Hayata) Kitamura – see *T. cucumeroides*
***Trichosanthes dioica* Roxb. (38.31)**
Trichosanthes laciniosa Klein, ex Willd. – see *T. cucumerina*
Trichosanthes lobata Roxb. – see *T. cucumerina*
Trichosanthes officinalis Buch.-Ham. ex Wall. – see *T. dioica*
Trichosanthes officinalis M.J. Roem – see *T. dioica*
Trichosanthes olida F. Muell. – see *T. cucumerina*
Trichosanthes ovigera subsp. *cucumeroides* (Ser.) C. Jeffrey – see *T. cucumeroides*
Trichosanthes pedatifolia Miq. – see *T. cucumerina*
Trichosanthes pilosa Buch.-Han. ex Wall. – see *T. cucumerina*
Trichosanthes quadricirrha Miq. – see *T. cucumeroides*
Trichosanthes reniformis Miq. – see *T. cucumerina*
Trichosanthes turolata Steud. – see *T. cucumerina*
Trichosanthes wightiana M. Roem. – see *T. cucumerina*
***Trigonella foenum-graecum* L. (40.32)**
Trigonella foenum-graecum provar. *foenum-graecum* – see *T. foenum-graecum*
Trigonella foenum-graecum subsp. *culta* (Alef.) Gams – see *T. foenum-graecum*
Trigonella graeca St. Lag. – see *T. foenum-graecum*
Trigonella jemenensis (Serp.) Sinskaya – see *T. foenum-graecum*
Trisanthus cochinchinensis Lour. – see *Centella asiatica*
Tropaeolum cubio Bukasov – see *T. tuberosum*
Tropaeolum elatum Salisb. – see *T. majus*
Tropaeolum hybridum L. – see *T. majus*
***Tropaeolum majus* L. (65.1)**
Tropaeolum mucronatum Meyen – see *T. tuberosum*
Tropaeolum quinquelobium Bergius – see *T. majus*
Tropaeolum repandifolium Stokes – see *T. majus*
Tropaeolum scheuerianum E. Morr. – see *T. majus*
Tropaeolum schillingii Vilmorin – see *T. majus*
***Tropaeolum tuberosum* Ruiz & Pavon (65.2)**
Trophaeum majus Kuntze – see *Tropaeolum majus*
Trophaeum tuberosum (Ruiz & Pavon) Kuntze – see *Tropaeolum tuberosum* Ruiz & Pavon
Turia cordata J.F. Gmel. – see *Luffa aegyptiaca*

Turia cylindrica J.F. Gmel. – see *Luffa aegyptiaca*
Turia sativa Forsk. – see *Luffa aegyptiaca*
Tussilago japonica L. f. – see *Farfugium japonicum*
Tussilago petasites L. – see *Petasites japonicus*
Tylosema esculentum **(Burch.) A. Schreib. (40.33)**
Ubium altissimum Desf. ex Steud. – see *Dioscorea alata*
Ullucus kunthii Moq. – see *U. tuberosus*
Ullucus tuberosus **Caldas (28.2)**
Ullucus tuberosus Lozano – see *U. tuberosus* Caldas
Urtica angustifolia Fisch. ex Hornem. – see *U. dioica*
Urtica aquatica Liebm. – see *U. dioica*
Urtica bollae Kanitz – see *U. dioica*
Urtica cataractae Steud. – see *Pilea melastomoides*
Urtica dioecia Hill – see *U. dioica*
Urtica dioica **L. (66.3)**
Urtica diplotricha Phil. – see *U. dioica*
Urtica eckloniana Blume – see *U. dioica*
Urtica elongata Blume – see *Pilea melastomoides*
Urtica galeopsifolid Wierzb. ex Opiz – see *U. dioica*
Urtica gibbosa Zipp. ex Blume – see *Pilea melastomoides*
Urtica glabrata Clem. ex Vis. – see *U. dioica*
Urtica gracilis Aiton – see *U. dioica*
Urtica hispida DC. – see *U. dioica*
Urtica hispidula Cariot – see *U. dioica*
Urtica kioviensis Bogowitch – see *U. dioica*
Urtica major Garsault – see *U. dioica*
Urtica major Kanitz – see *U. dioica*
Urtica melastomoides Poir. – see *Pilea melastomoides*
Urtica mexicana Blume – see *U. dioica*
Urtica oreophila Miq. – see *Pilea melastomoides*
Urtica peduncularis Sm. – see *Pilea melastomoides*
Urtica platyphylla Wedd. – see *U. dioica*
Urtica procera Muhl. ex Willd. – see *U. dioica*
Urtica procera Willd. – see *U. dioica*
Urtica pseudo-dioica Steud. – see *U. dioica*
Urtica pubescens Ledeb. – see *U. dioica*
Urtica radicans Bolla – see *U. dioica*
Urtica sicula Gasp. ex Guss. – see *U. dioica*
Urtica submitis Boiss. – see *U. dioica*
Urtica trinervia Roxb. – see *Pilea melastomoides*
Urtica viridis Rydb. – see *U. dioica*
Utica melastomoides Poir. – see *Pilea melastomoides*
Valeriana cornucopiae L. – see *Fedia cornucopiae*
Valeriana incrassata Bory & Chaub. – see *Valerianella eriocarpa*
Valeriana locusta subsp. *olitoria* L. – see *Valerianella locusta*
Valeriana locusta Ucria – see *Fedia cornucopiae*
Valeriana ringens Salisb. – see *Fedia cornucopiae*
Valerianella campanulata Bivona – see *V. eriocarpa*
Valerianella cornucopiae Loisel. – see *Fedia cornucopiae*
Valerianella cucullata Dufour, ex Krok – see *V. eriocarpa*

***Valerianella eriocarpa* Desv. (67.1)**
Valerianella incrassata Nym. – see *V. eriocarpa*
Valerianella locusta Betcke var. *olitoria* L. – see *V. locusta*
Valerianella locusta (L.) Betcke – see *V. locusta* (L.) Laterrade
***Valerianella locusta* (L.) Laterrade (67.2)**
Valerianella macrocyathus Pomel – see *V. eriocarpa*
Valerianella muricata Heldr. ex Nym. – see *V. eriocarpa*
Valerianella olitoria (L.) Pollich – see *V. locusta*
Valerianella olitoria Moench – see *V. locusta*
Valerianella plagiocyanthus Pomel – see *V. eriocarpa*
Valerianella puberula Reichb. ex Nym. – see *V. eriocarpa*
Valerianella pusilla Miégev. – see *V. locusta*
Valerianella rugulosa Fisch. – see *V. eriocarpa*
Valerianella striata Stev. ex Fisch. – see *V. locusta*
Valerianella truncata Pett. ex Nym. – see *V. eriocarpa*
Vallisneria bulbosa Poir. – see *Sagittaria sagittifolia* subsp. *sagittifolia*
Verbesina pseudo-acmella Linn. – see *Acmella oleracea*
Vernonia adenosticta Fenzl, ex Walp. – see *Gymnanthemum amygdalinum*
Vernonia amygdalina Delile. – see *Gymnanthemum amygdalinum*
Vicia equina Steud. – see *V. faba*
Vicia esculenta Salisb. – see *V. faba*
***Vicia faba* L. (40.34)**
Vicia faba L. form *anacarpa* Makino – see *V. faba*
Vicia lens (L.) Coss. & Germ. – see *Lens culinaris*
Vicia vulgaris S.F. Gray – see *V. faba*
***Vigna aconitifolia* (Jacq.) Maréchal (40.35)**
***Vigna angularis* (Willd.) Ohwi & Ohashi (40.36)**
Vigna angularis (Willd.) Wight – see *V. angularis* (Willd.) Ohwi & Ohashi
Vigna aureus (Roxb.) Hepper – see *V. radiata* var. *radiata*
Vigna baoulensis A. Chev. – see *V. unguiculata* subsp. *cylindrica*
Vigna calcarata (Roxb.) Kurz – see *V. umbellata*
Vigna calcaratus Roxb. – see *V. umbellata*
Vigna catjang (Burm. f.) Walp. – see *V. unguiculata* subsp. *cylindrica*
Vigna catjang Endl. – see *V. unguiculata* subsp. *unguiculata*
Vigna catjang Endl. var. *sinensis* King – see *V. unguiculata* subsp. *unguiculata*
Vigna catjang var. *sinensis* form *sesquipedalis* Makino – see *V. unguiculata* subsp. *sesquipedalis*
Vigna cylindrica (L.) Skeels – see *V. unguiculata* subsp. *cylindrica*
Vigna lobata (Willd.) Endl. ex Miq. – see *Pueraria montana* var. *lobata*
***Vigna mungo* (L.) Hepper var. *mungo* (40.37)**
Vigna mungo (L.) Hepper var. *radiatus* Bak. – see *V. radiata* var. *radiata*
Vigna ornata Welw. ex Baker – see *Sphenostylis stenocarpa*
Vigna radiata (L.) R. Wilczek – see *V. radiata* var. *radiata*
***Vigna radiata* (L.) R. Wilczek var. *radiata* (40.38)**
Vigna sesquipedalis Koern – see *V. unguiculata*. subsp. *sesquipedalis*
Vigna sesquipedalis L. – see *V. unguiculata* subsp. *sesquipedalis*
Vigna sesquipedalis (L.) Fruhwirth – see *V. unguiculata* subsp. *sesquipedalis*
Vigna sesquipedalis var. *melanophtalmis* Nakai – see *V. unguiculata* subsp. *unguiculata*
Vigna sesquipedalis W.F. Wight – see *V. unguiculata* subsp. *sesquipedalis*
Vigna sinensis Endl. – see *V. unguiculata*. subsp. *unguiculata*
Vigna sinensis Endl. var. *sesquipedalis* Kern – see *V. unguiculata* subsp. *sesquipedalis*

Vigna sinensis Hassk. – see *V. unguiculata* subsp. *unguiculata*
Vigna sinensis L. – see *V. unguiculata* subsp. *unguiculata*
Vigna sinensis (L.) Hassk. – see *V. unguiculata* subsp. *unguiculata*
Vigna sinensis (L.) Hassk. subsp. *sesquipedalis* (L.) Van Eseltine – see *V. unguiculata* subsp. *sesquipedalis*
Vigna sinensis (L.) Savi, ex Hassk. – see *V. unguiculata* subsp. *unguiculata*
Vigna sinensis (L.) Savi, ex Hassk subsp. *sesquipedalis* (L.) Van Eselt – see *V. unguiculata* subsp. *sesquipedalis*
Vigna sinensis (L.) Savi, ex Hassk. subsp. *unguiculata* – see *V. unguiculata* subsp. *unguiculata*
Vigna sinensis (L.) Sevi, ex Hassk var. *cylindricus* – see *V. unguiculata* subsp. *cylindrica*
Vigna sinensis (L.) Walp. subsp. *sesquipedalis* (L.) Fruw. – see *V. unguiculata* subsp. *sesquipedalis*
Vigna sinensis (L.) Walp. subsp. *sinensis* – see *V. unguiculata* subsp. *unguiculata*
Vigna sinensis Savi, ex Hassk. subsp. *cylindrica* – see *V. unguiculata* subsp. *unguiculata*
Vigna sinensis Savi form *dschiuroku* Ito – see *V. unguiculata* subsp. *sesquipedalis*
Vigna sinensis Savi var. *catjang* (Burm.) T. Ito – see *V. unguiculata* subsp. *cylindrica*
Vigna sinensis Savi var. *elongata* Miquel – see *V. unguiculata* subsp. *sesquipedalis*
Vigna sinensis Savi var. *sesquipedalis* Koern. – see *V. unguiculata* subsp. *sesquipedalis*
Vigna sinensis (Stickm) Savi, ex Hassk – see *V. unguiculata* subsp. *unguiculata*
Vigna sinensis (Stickm) Savi, ex Hassk var. *sesquipedalis* Koern – see *V. unguiculata* subsp. *sesquipedalis*
Vigna sinensis subsp. *cylindrica* Hedrick – see *V. unguiculata* subsp. *cylindrica*
Vigna sinensis subsp. *cylindrica* (L.) van Eseltine – see *V. unguiculata* subsp. *cylindrica*
Vigna sinensis subsp. *sesquipedalis* (L.) van Eseltine – see *V. unguiculata* subsp. *sesquipedalis*
Vigna sinensis unguiculata (L.) Walp. – see *V. unguiculata* subsp. *cylindrica*
Vigna sinensis var. *catjang* (Burm. f.) Chiov. – see *V. unguiculata* subsp. *cylindrica*
Vigna sinensis var. *sesquipedalis* (L.) Asch. & Schweinf. – see *V. unguiculata* subsp. *sesquipedalis*
Vigna sinensis var. *sesquipedalis* (L.) Koern – see *V. unguiculata*. subsp. *sesquipedalis*
***Vigna subterranea* (L.) Verdc. (40.39)**
***Vigna umbellata* (Thunb.) Ohwi & Ohashi (40.40)**
Vigna unguiculata (L.) Walp. – see *V. unguiculata* subsp. *unguiculata*
Vigna unguiculata (L.) Walp. cv. group Biflora – see *V. unguiculata* subsp. *cylindrica*
Vigna unguiculata (L.) Walp. cv. group Sesquipedalis – see *V. unguiculata* subsp. *sesquipedalis*
Vigna unguiculata (L.) Walp. cv. group Unguiculata – see *V. unguiculata* subsp. *unguiculata*
***Vigna unguiculata* (L.) Walp. subsp. *cylindrica* (L.) Van Eselt. ex Verdr. (40.41)**
***Vigna unguiculata* (L.) Walp. subsp. *sesquipedalis* (L.) Verdr. (40.42)**
***Vigna unguiculata* (L.) Walp. subsp. *unguiculata* (40.43)**
Vigna unguiculata (L.) Walp. subsp. *unguiculata* (Unguiculata Group) Westphal – see *V. unguiculata* subsp. *unguiculata*
Vigna unguiculata subsp. *catjang* (Burm. f.) Chiov. – see *V. unguiculata* subsp. *cylindrica*
Vigna unguiculata subsp. *unguiculata* (cv group Unguiculata) Marechal – see *V. unguiculata* subsp. *unguiculata*
Vigna unguiculata var. *catjang* (Burm. f.) Ohashi – see *V. unguiculata* subsp. *cylindrica*
Vigna unguiculata var. *cylindrica* (L.) Ohashi – see *V. unguiculata* subsp. *cylindrica*
Vigna unguiculata var. *sesquipedalis* (L.) Ohashi – see *V. unguiculata* subsp. *sesquipedalis*
Vigolina acmella Poir. – see *Galinsoga parviflora*
Villa lloydii Jord. ex Bor. – see *Viola tricolor*
Viola agrestis Jord. – see *V. tricolor*
Viola alpestris Jord. – see *V. tricolor*
Viola arvatica Jord. – see *V. tricolor*
Viola arvensis Murr. – see *V. tricolor*

Viola atlantica Pomel – see *V. tricolor*
Viola banatica Kit. ex Roem. & Schult. – see *V. tricolor*
Viola bella Gren. ex Gren. & Godr. – see *V. tricolor*
Viola bicolor Gilib. – see *V. tricolor*
Viola bubanii Timb. – see *V. tricolor*
Viola calaminaria Lej. – see *V. tricolor*
Viola calcarata Vill. – see *V. tricolor*
Viola cenisia Schangin – see *V. tricolor*
Viola confinis Jord. ex Nym. – see *V. tricolor*
Viola contempta Jord. – see *V. tricolor*
Viola curtisii E. Forst. – see *V. tricolor*
Viola demetria Prolongo, ex Boiss. – see *V. tricolor*
Viola deminuta Schur – see *V. tricolor*
Viola derelicta Jord. – see *V. tricolor*
Viola deseglisei Jord. ex Bor. – see *V. tricolor*
Viola flavida Jord. – see *V. tricolor*
Viola foucaudi Sav. – see *V. tricolor*
Viola garganica Strobl – see *V. tricolor*
Viola gracilescens Jord. – see *V. tricolor*
Viola gracilis Martr. ex Nym. – see *V. tricolor*
Viola grandiflora L. – see *V. tricolor*
Viola hymettia Boiss. & Heldr. – see *V. tricolor*
Viola kitaibeliana Roem. & Schult. – see *V. tricolor*
Viola lejeunii Jord. – see *V. tricolor*
Viola lepida Jord. – see *V. tricolor*
Viola leptoceras Schur – see *V. tricolor*
Viola lilascens Heldr. ex Boiss. – see *V. tricolor*
Viola lutea Huds. – see *V. tricolor*
Viola luteola Jord. – see *V. tricolor*
Viola maritima Spreng. ex Mert. & Koch – see *V. tricolor*
Viola media Chaillet, ex Roem. & Schult. – see *V. tricolor*
Viola meduanensis Bor. – see *V. tricolor*
Viola mentita Jord. – see *V. tricolor*
Viola micrantha Bertol. – see *V. tricolor*
Viola monticola Jord. – see *V. tricolor*
Viola mucuzii Orph. ex Nym. – see *V. tricolor*
Viola mutabilis Rochel – see *V. tricolor*
Viola nana Godr. – see *V. tricolor*
Viola nemausensis Jord. – see *V. tricolor*
Viola occulta hort. Argent. ex Otto – see *V. tricolor*
Viola olonnensis Genev. – see *V. tricolor*
Viola olyssiponensis Rouy. – see *V. tricolor*
Viola paillouxi Jord. – see *V. tricolor*
Viola pallescens Jord. – see *V. tricolor*
Viola parviflora Kit. ex Roem. & Schult. – see. *V. tricolor*
Viola peregrina Jord. ex Bor. – see *V. tricolor*
Viola perennis Vilm. – see *V. tricolor*
Viola perrobusta Borb. – see *V. tricolor*
Viola provostii Bor. – see *V. tricolor*
Viola pseudo-lutea Schur – see *V. tricolor*

Viola ruralis Jord. ex Bor. – see *V. tricolor*
Viola sabulosa Bor. – see *V. tricolor*
Viola sagoti Jord. – see *V. tricolor*
Viola saxatilis F.W. Schmidt – see *V. tricolor*
Viola segetalis Jord. – see *V. tricolor*
Viola soongorica Fisch. ex Ledeb. – see *V. tricolor*
Viola subtilis Jord. – see *V. tricolor*
Viola sudetica Willd. – see *V. tricolor*
Viola symei Baker – see *V. tricolor*
Viola tenella Muhl. – see *V. tricolor*
Viola timbali Jord. – see *V. tricolor*
***Viola tricolor* L. (68.1)**
Viola variata Jord. – see *V. tricolor*
Viola versicolor Salisb. – see *V. tricolor*
Viola villarsiana Roem. & Schult. – see *V. tricolor*
Viola virgata Gren. – see *V. tricolor*
Viola vivariensis Jord. – see *V. tricolor*
Vitis discolor (Blume) Dalz. Blume – see *Cissus discolor*
Voandzeia geocarpa (Harms) A. Chev. – see *Macrotyloma geocarpum*
Voandzeia poissoni A. Chev. – see *Macrotyloma geocarpum*
Voandzeia subterranea (L.) Thouars, ex DC. – see *Vigna subterranea*
***Wasabia japonica* (Miq.) Matsum. (30.58)**
Wasabia pungens Matsum. – see *W. japonica*
Wasabia wasabi (Siebold) Makino – see *W. japonica*
Wilborgia acmelea Roth – see *Galinsoga parviflora*
Wilborgia parviflora H.B. & K. – see *Galinsoga parviflora*
Witheringia chenopodioides Remy – see *Solanum nigrum*
Witheringia rubra Remy – see *Solanum nigrum*
***Xanthosoma brasiliense* (Desf.) Engler (8.8)**
Xanthosoma edule C.F. Myer – see *X. sagittifolium*
Xanthosoma edule Schott, ex Schott & Endl. – see *X. sagittifolium*
Xanthosoma ianthinum K. Koch – see *X. violaceum*
Xanthosoma jacquinii Schott, ex Schott & Endl. – see *X. sagittifolium*
Xanthosoma mucronatum Schott. – see *Alocasia macrorrhiza*
Xanthosoma nigrum (Vell.) Mansf. – see *X. violaceum*
***Xanthosoma sagittifolium* (L.) Schott (8.9)**
Xanthosoma utile C. Koch & Bouché – see *X. sagittifolium*
***Xanthosoma violaceum* Schott (8.10)**
Xanthosoma xanthorrhizon (Jacq.) K. Koch – see *X. sagittifolium*
Xanthoxalis crassicaulis (Zucc.) Small – see *Oxalis tuberosa*
***Youngia sonchifolia* Maxim. (27.48)**
Zea amylea-saccharata Sturtev. – see *Z. mays* subsp. *mays*
Zea amylea-saccharata Sturtev. ex L.H. Bailey – see *Z. mays* subsp. *mays*
Zea mays convar. *saccharata* Koern. – see *Zea mays* subsp. *mays*
***Zea mays* L. subsp. *mays* (17.7)**
Zea mays rugosa Bonaf. – see *Zea mays* subsp. *mays*
Zea mays subsp. *saccharata* (Sturt.) Zhuk. – see *Zea mays* subsp. *mays*
Zea mays var. *amylea-saccharata* Bailey – see *Z. mays* subsp. *mays*
Zea mays var. *saccharata* (Koern.) Asch. & Graebn. – see *Zea mays* subsp. *mays*
Zea saccharata Sturt. – see *Zea mays* subsp. *mays*

Zea saccharata Sturt. ex L.H. Bailey – see *Z. mays* subsp. *mays*
Zingiber aromaticum Noronha – see *Z. officinale*
Zingiber delte – see *Z. officinale*
Zingiber galanga Stokles – see *Alpinia galanga*
Zingiber mijooka Siebold – see *Z. mioga*
***Zingiber mioga* (Thunb.) Roscoe (20.5)**
Zingiber missionis Wall. – see *Z. officinale*
***Zingiber officinale* Roscoe (20.6)**
Zingiber oligophyllum K. Schum. – see *Z. mioga*
Zingiber zingiber Karst. – see *Z. officinale*
Zizania aquatica auct. Japon., non L. – see *Z. latifolia*
Zizania aquatica Hack. – see *Z. latifolia*
Zizania aquatica L. – see *Z. latifolia*
Zizania aquatica var. *angustifolia* A.S. Hitchc. – see *Z. latifolia*
Zizania aquatica var. *latifolia* Kom. – see *Z. latifolia*
Zizania caduciflora L. – see *Z. latifolia*
Zizania caduciflora (Turcz.) Hand.-Mazz. – see *Z. latifolia*
Zizania dahurica Turcz. ex Steud. – see *Z. latifolia*
***Zizania latifolia* (Griseb.) Turcz. ex Stapf. (17.8)**
Zizania latifolia Turcz. – see *Z. latifolia*
Zizania palustris Sieb. – see *Z. latifolia*

Chapter 4a. Edible plant part utilized, listed by English common name

Young plants (aerial parts)

For additional information on plant parts consumed, see Chapter 1.

Alligator weed (23.1)
Chombujiggangi (30.49)[1]
Common seepweed (35.10)
Deodeog (33.3)[2]
Kalimeris (27.28)
Marjoram (44.6)
Palm-grass (17.6)
Sessile alternanthera (23.2)

Shoots

Alfalfa (40.17)
Amaranthus (23.3)
American black nightshade (62.15)
Angelica (24.1)
Asparagus (13.2)
Bamboo shoots (17.1, 17.2, 17.4)
Barilla plant (35.7)
Bitter leaf (27.24)
Bottle gourd (38.20)[3]
Buffalo spinach (27.19)
Buko (41.1)
Broomrape (52.1)
Bur marigold (27.8)
Cape pondweed (25.1)
Cat's whiskers (34.2)
Ceylon spinach (28.1)
Chayote (38.25)[3]
Cockscomb (23.4)
Common comfrey (29.2)
Cosmos (27.12)
Edible chrysanthemum (27.23)
Elephant grass (17.3)
Euryale (50.1)[3]
Evening primrose (51.1)
False roselle (45.3)
Fenugreek (40.32)
Flameflower (59.4)
Fluted gourd (38.27)[3]
Galinsoga (27.22)
Gilo (62.14b)
Ginger (20.6)[3]
Gogoro (55.1)
Greater galangal (20.1)[3]
Hawksbeard velvetplant (27.14)
Hops (46.1)
Hydrocotyle (24.15)
Indian poke (56.1)
Inkweed (56.4)
Ivy gourd (38.4)[3]
Japanese angelica (24.2)
Japanese wild ginger (20.5)[3]
Jew's marrow (63.1)
Kangaroo vine (69.1)
Kipeit (47.1)[4]
Kudzu (40.30)[3]
Mallow (45.5)
Mango ginger (20.3)[3]
Meadow salsify (27.47)
New Zealand spinach (22.2)
Parrot's feather (42.1)
Passion flower (54.1)
Pea (40.28)[3]
Pea eggplant (62.23)[3]
Pointed gourd (38.31)[3]
Poke salat (56.2)
Pokeweed (56.3)
Prickly comfrey (29.3)
Salsola (35.8)
Sierra Leone bologi (27.13)
Snake gourd (38.29)[3]
Sow thistle (27.18)
Spikenard (26.2)
Summer cypress (35.2)
Summer savory (44.9)
Sunset hibiscus (45.2)
Tettu (21.1)

---

[1] Korean name
[2] Japanese name
[3] Secondary use
[4] Javanese name
[5] Indonesian name

Wasabi (30.58)[3]
Water bamboo (17.8)
Water bitter leaf (27.44)
Water dropwort (24.17)
Water mimosa (40.19)
Water spinach (36.2)
Wild asparagus (13.1)
Wonderberry (62.21)[3]

Stems

Bamboo shoot mustard (30.5)
Celtuce (27.30)
Coastal glehnia (24.14)
Common seepweed (35.10)
Gemmiferous mustard (30.8)
Kohlrabi (30.30)
Siberian ginseng (26.4)
Strumous mustard (30.18)
Swollen stem mustard (30.19)
Water bamboo (17.8)

Pseudostem

Kurrat (7.6)
Leek (7.8)
Shallot (7.2)
Welsh onion (7.5)

Leaves

Abyssinian mustard (30.3)
African eggplant (62.18)
African valerian (27.21)
Alfalfa (40.17)
Amaranthus (23.3)
American black nightshade (62.15)
Angelica (24.1)
Asiatic pennywort (24.7)
Aster (27.17)
Barilla plant (35.7)
Beet (35.4)
Bitter gourd (38.23)[3]
Black mustard (30.24)
Bottle gourd (38.20)[3]
Broad beaked mustard (30.38)
Buckhorn plantain (57.1)
Buffalo spinach (27.19)
Buko (41.1)
Bur marigold (27.8)
Capitata mustard (30.4)
Carrot (24.11)[3]
Cassava (39.4)[3]
Cat's whiskers (34.2)
Celeriac (24.5)[3]
Celery (24.4)[3]
Ceylon spinach (28.1)
Chamnamul (24.22)[1]
Chard (35.3)
Chaya (30.2, 39.1)
Chayote (38.25)[3]
Chervil (24.3)
Chicory (27.10a)
Chimapaya (27.4)[1]
Chinese bellflower (34.3)
Chinese cabbage (30.40)
Chinese chive (7.13)
Chinese kale (30.25)
Chinese wolfberry (62.8)
Chive (7.11)
Coastal glehnia (24.14)
Cockscomb (23.4)
Common cabbage (30.27a)
Common comfrey (29.2)
Common sauropus (39.5)
Common seepweed (35.10)
Coriander (24.9)
Cornsalad (67.2)
Cosmos (27.12)
Croton (39.3)
Cuckoo-flower (30.44)
Cut leaf mustard (30.6)
Dandelion (27.45)
Dock (58.3)
Edible chrysanthemum (27.23)
Endive (27.9)
European pennyroyal (44.2)
Evening primrose (51.1)
False roselle (45.3)
Fennel (24.13)
Fenugreek (40.32)
Florence fennel (24.12) leaf base
Flower-like leaf mustard (30.16)
Fluted gourd (38.27)
French sorrel (58.4)
Garden cress (30.51)
Garden myrrh (24.16)
Garlic (7.10)[3]
Giant garlic (7.12)
Giant taro (8.1)[3]

Godulbbaegi (27.48)[1]
Golden apple (62.14a)
Gomchwi (27.34)[1]
Good King Henry (35.5)
Great-headed garlic (7.1)[3]
Guasca (27.1)
Gynura (27.25)
Hastate-leaved pondweed (18.1)
Hawksbeard velvetplant (27.14)
Head lettuce (27.31)
Hill mustard (30.42)
Hoary basil (44.4)
Horseradish (30.1)[3]
Hydrocotyle (24.15)
Ice plant (22.1)
Indian lettuce (27.29)
Indian poke (56.1)
Inkweed (56.4)
Involute mustard (30.9)
Italian corn-salad (67.1)
Italian parsley (24.20)
Ivy gourd (38.4)[3]
Japanese angelica (24.2)
Japanese aralia (26.3)
Japanese garlic (7.7)
Japanese hornwort (24.10)
Japanese wild ginger (20.5)[3]
Jew's marrow (63.1)
Johnny-jump-up (68.1)
Kale (30.35)
Kalimeris (27.28)
Kangaroo vine (69.1)
Kipeit (47.1)[4]
Kudzu (40.30)[3]
Kumindelan (27.41)[5]
Kurrat (7.6)
Ladybell (33.1)
Large leaf mustard (30.17)
Leaf lettuce (27.32)
Leaf radish (30.55)
Leek (7.8)
Line mustard (30.12)
Long petiole mustard (30.13)
Longroot onion (7.14)[3]
Lotus root (48.1a)[3]
Mallow (45.5)
Marrow stem kale (30.32)
Mignonette (60.1)
Miner's lettuce (59.1)
Mock pak choy (30.37)
Mugwort (27.7)
Nasturtium (65.1)
New Zealand spinach (22.2)
Orach (35.1)
Oval-leaved pondweed (18.2)
Pak-choi (30.36)
Para cress (27.2)
Parsley (24.19)
Parsnip (24.18)[3]
Passion flower (54.1)
Perilla (44.7)
Pigeon pea (40.2)[3]
Pilea (66.1, 66.2)
Poke salat (56.2)
Pokeweed (56.3)
Prickly comfrey (29.3)
Prickly-pear (32.1)
Pualele (27.43)
Purslane (59.2)
Quinoa (35.6)
Rampion (33.2)[3]
Red cabbage (30.27b)
Redleaf chicory (27.10b)
Roquette salad (30.48)
Roselle (45.4)[3]
Rungia (21.2)
Rutabaga (30.22)[3]
Salsify (27.46)
Salsola (35.8)
Saururis (61.1)
Savoy cabbage (30.34)
Scolymus (27.36)[3]
Scorzonera (27.38)[3]
Sea kale (30.45)[3]
Sedum (37.1)
Shallot (7.2)
Shephard's purse (30.43)
Siberian ginseng (26.4)
Siberian kale (30.23)
Sierra Leone bologi (27.13)
Small leaf mustard (30.7)
Smooth loofah (38.22)[3]
Snake gourd (38.29)[3]
Sorrel (58.2)
Sow thistle (27.18, 27.27, 27.42)
Spearmint (44.3)
Spinach (35.9)
Spinach mustard (30.39)
Spiny sow thistle (27.40)
Spot flower (27.3)

Spotted garden thistle (27.37)
Stinging nettle (66.3)
Summer savory (44.9)
Sunset hibiscus (45.2)
Sweet basil (44.5)
Sweetpotato (36.3)[3]
Swollen stem mustard (30.19)
Tannia (8.9)[3]
Tannier spinach (8.8)
Tartar bread-plant (30.46)
Taro (8.4)[3]
Tarragon (27.6)
Tettu (21.1)
Thousand-headed kale (30.33)
Tillered mustard (30.15)
Tronchuda (30.28)
Turfed stone leek (7.9b)
Turnip greens (30.41b)
Turnip-rooted parsley (24.21)[3]
Upland cress (30.2)
Vegetable fern (5.1)
Vegetable rape (30.21)
Wallrocket (30.47)
Water dropwort (24.17)
Water fern (4.1)
Water mimosa (40.19)
Water spinach (36.2)
Watercress (30.52)
Waterleaf (59.3)
Watershield (31.1)
Welsh onion (7.5)
White flowered mustard (30.11)
White mustard (30.57)
Wide petiole mustard (30.10)
Wild lettuce (27.33)
Winged bean (40.29)[3]
Wonderberry (62.21)
Yellow velvetleaf (14.1)

Petioles

Borage (29.1)
Butterbur (27.35)
Cardoon (27.15)
Celery (24.4)
Edible burdock (27.5)[3]
Giant elephant's ear (8.5)
Giant taro (8.1)[3]
Japanese farfugium (27.20)
Rhubarb (58.1)
Sea kale (30.45)
Tartar bread-plant (30.46)
Yellow velvetleaf (14.1)

Flower/inflorescence/flower stalk/spike

Banana inflorescence (16.1b)
Broccoli (30.31)
Broccoli raab (30.41c)
Cape pondweed (25.1)
Capper (34.1)
Cauliflower (30.26)
Chimapaya (27.4)[1]
Chinese chive (7.13)
Chinese kale (30.25)
Day lily (13.3)
Globe artichoke (27.16)
Greater galangal (20.1)
Japanese wild ginger (20.5)[3]
Johnny-jump-up (68.1)
Nasturtium (65.1)[3]
Marjoram (44.6)
Mignonette (60.1)
Passion flower (54.1)
Peduncled mustard (30.20)
Spearmint (44.3)
Spot flower (27.3)
Sugarcane inflorescence (17.5)
Tronchuda (30.28)
Vegetable rape (30.21)
Water lily (50.2)
White mustard (30.57)
Yellow velvetleaf (14.1)

Immature seed pods

Asparagus pea (40.13)
Black gram (40.37)
Black-eyed pea (40.43)
Catjang (40.41)
Cluster bean (40.6)
Edible podded pea (40.27)
Fenugreek (40.32)
Grass pea (40.10)
Green bean (40.26)
Jicama (40.21)[3]
Lentil (40.12)
Marama bean (40.33)
Moth bean (40.35)
Mung bean (40.38)

Pigeon pea (40.2)
Potato bean (40.22)
Rat tail radish (30.53)
Scarlet runner bean (40.24)
Tepary bean (40.23)
Winged bean (40.29)
Yard-long bean (40.42)

Fruit

African eggplant (62.18)
African horned cucumber (38.12)
American black nightshade (62.15)
Angled loofah (38.21)
Armonian cucumber (38.10)
Bell pepper (62.1)
Bitter gourd (38.23)
Bonnet pepper (62.4)
Bottle gourd (38.20)
Buko (41.1)
Cantaloupe (38.7a)
Cape gooseberry (62.12)
Casabanana (38.26)
Cat's whiskers (34.2)
Cayenne pepper (62.2)
Chayote (38.25)
Citron (38.2)
Cucumber (38.13a)
Currant tomato (62.10)
Cyclanthera (38.19)
Eggplant (62.19)
Garden egg (62.17)
Golden apple (62.14a)
Hairy-fruited eggplant (62.16)
Indian nightshade (62.25)
Ivy gourd (38.4)
Japanese snake gourd (38.30)
Malabar gourd (38.15)
Mango melon (38.8)
Muskmelon (38.7b)
Musky pumpkin (38.17)
Naranjillo (62.22)
Okra (45.1)
Oriental pickling melon (38.9)
Pea eggplant (62.23)
Pepino (62.20)
Pickling cucumber (38.13b)
Plantain (16.1a)
Pointed gourd (38.31)
Prickly-pear (32.1)
Pumpkin (38.14)
Rocoto (62.6)
Scarlet eggplant (62.14c)
Small pepper (62.3)
Smooth loofah (38.22)
Snake gourd (38.29)
Squash melon (38.24)
Strawberry tomato (62.11)
Summer squash (38.18)
Tabasco pepper (62.5)
Tomatillo (62.13)
Tomato (62.9)
Tree tomato (62.7)
Watermelon (38.3)
Wax gourd (38.1)
West Indian gherkin (38.6)
White-seeded melon (38.5)
Winter melon (38.11)
Winter squash (38.16)
Wonderberry (62.21)

Seed

For details on seed maturity, see Chapter 1.

Adzuki bean (40.36)
African yam bean (40.31)[3]
Alfalfa (40.17) sprouts
Amaranthus (23.3) sprouts
Asian water chestnut (64.1)
Bambara groundnut (40.39)
Black gram (40.37)
Black-eyed pea (40.43)
Broad bean (40.34a)
Catjang (40.41)
Cluster bean (40.6)
Coriander (24.9)
Euryale (50.1)
Fluted gourd (38.27)
Garbanzo bean (40.5)
Garden myrrh (24.16)[3]
Grass pea (40.10)
Green bean (40.26)
Horse gram (40.16)
Horseradish (30.1) sprouts[3]
Hyacinth bean (40.9)
Jack bean (40.3)
Jicama (40.21)[3]
Kersting's groundnut (40.15)
Lentil (40.12) and sprouts[3]

Chapter 4b. Edible part utilized, listed by abbreviated scientific name

Young plants (aerial parts)

For additional details on the plant part consumed, see Chapter 1.

Alternanthera philoxeroides (23.1)
Alternanthera sessilis (23.2)
Codonopsis lanceolata (33.3)
Erysimum amurense (30.49)[1]
Kalimeris yomena (27.28)
Origanum vulgare (44.6)
Setaria palmifolia (17.6)
Suaeda glauca (35.10)

Shoots

Abelmoschus manihot (45.2)
Alpinia galanga (20.1)[1]
Amaranthus spp. (23.3)
Angelica archangelica (24.1)
Angelica keiskei (24.2)
Aponogeton distachyos (25.1)
Aralia cordata (26.2)
Asparagus acutifolius (13.1)
Asparagus officinalis (13.2)
Bambusa spp. (17.1)
Basella alba (28.1)
Bassia scoparia (35.2)
Bidens pilosa (27.8)
Celosia spp. (23.4)
Cissus discolor (69.1)
Cleome gynandra (34.2)
Coccinia grandis (38.4)[1]
Corchorus olitorius (63.1)
Cosmos caudatus (27.12)
Crassocephalum biafrae (27.13)
Crassocephalum crepidioides (27.14)
Curcuma mangga (20.3)[1]
Dendrocalamus spp. (17.2)
Emilia sonchifolia (27.18)
Enydra fluctuans (27.19)
Euryale ferox (50.1)[1]
Galinsoga parviflora (27.22)
Glebionis coronarium (27.23)
Gnetum gnemon (41.1)
Gymnanthemum amygdalinum (27.24)
Hibiscus acetosella (45.3)
Humulus lupulus (46.1)
Hydrocotyle sibthorpioides (24.15)
Ipomoea aquatica (36.2)
Justicia insularis (21.1)
Lagenaria siceraria (38.20)[1]
Maesa latifolia (47.1)
Malva rotundifolia (45.5)
Medicago sativa subsp. *sativa* (40.17)
Myriophyllum aquaticum (42.1)
Neptunia prostrata (40.19)
Oenanthe javanica subsp. *javanica* (24.17)
Oenothera biennis (51.1)
Orobanche crenata (52.1)
Passiflora biflora (54.1)
Pennisetum purpureum (17.3)
Phyllostachys spp. (17.4)
Phytolacca acinosa (56.1)
Phytolacca americana (56.2)
Phytolacca esculenta (56.3)
Phytolacca octandra (56.4)
Pisum sativum subsp. *sativum* var. *sativum* (40.28)[1]
Pueraria montana var. *lobata* (40.30)[1]
Salsola komarovii (35.7)
Salsola soda (35.8)
Satureja hortensis (44.9)
Sechium edule (38.25)[1]
Sesamum radiatum (55.1)
Solanum aethiopicum (62.14)
Solanum americanum (62.15)
Solanum nigrum (62.21)[1]
Solanum torvum (62.23)[1]
Struchium sparganophorum (27.44)
Symphytum officinale (29.2)
Symphytum × *uplandicum* (29.3)
Talinum paniculatum (59.4)
Telfairia occidentalis (38.27)[1]
Tetragonia tetragoniodes (22.2)
Tragopogon pratensis (27.47)

[1] Secondary use

Trichosanthes cucumerina (38.29)[1]
Trichosanthes dioica (38.31)[1]
Trigonella foenum-graecum (40.32)
Wasabia japonica (30.58)[1]
Zingiber mioga (20.5)[1]
Zingiber officinale (20.6)[1]
Zizania latifolia (17.8)

Stems

Brassica juncea var. *crassicaulis* (30.5)
Brassica juncea var. *gemmifera* (30.8)
Brassica juncea var. *strumata* (30.18)
Brassica juncea var. *tumida* (30.19)
Brassica oleracea var. *gongylodes* (30.30)
Eleutherococcus senticosus (26.4)
Glehnia littoralis (24.14)
Lactuca sativa var. *angustana* (27.30)
Suaeda glauca (35.10)
Zizania latifolia (17.8)

Pseudostem

Allium cepa var. *aggregatum* (7.2)
Allium fistulosum (7.5)
Allium kurrat (7.6)
Allium porrum (7.8)

Leaves

Abelmoschus manihot (45.2)
Acmella ciliata (27.1)
Acmella paniculata (27.3)
Acmella uliginosa (27.4)
Adenophora triphylla (33.1)
Allium ampeloprasum var. *ampeloprasum* (7.1)[1]
Allium cepa var. *aggregatum* (7.2)
Allium fistulosum (7.5)
Allium kurrat (7.6)
Allium macrostemon (7.7)
Allium porrum (7.8)
Allium × proliferum (7.9)
Allium sativum (7.10)[1]
Allium schoenoprasum (7.11)
Allium scorodoprasum (7.12)
Allium tuberosum (7.13)
Allium victorialis (7.14)[1]
Alocasia macrorrhizos (8.1)[1]
Amaranthus spp. (23.3)
Angelica archangelica (24.1)
Angelica keiskei (24.2)
Anthriscus cerefolium (24.3)
Apium graveolens var. *dulce* (24.4)[1]
Apium graveolens var. *rapaceum* (24.5)[1]
Aralia elata (26.3)
Armoracia rusticana (30.1)[1]
Artemisia dracunculus (27.6)
Artemisia indica var. *maximowiczii* (27.7)
Atriplex hortensis (35.1)
Barbarea verna (30.2)
Basella alba (28.1)
Beta vulgaris Cicla group (35.3)
Beta vulgaris Crassa group (35.4)
Bidens pilosa (27.8)
Brasenia schreberi (31.1)
Brassica carinata (30.3)
Brassica juncea var. *capitata* (30.4)
Brassica juncea var. *crispifolia* (30.6)
Brassica juncea var. *foliosa* (30.7)
Brassica juncea var. *involuta* (30.9)
Brassica juncea var. *latipa* (30.10)
Brassica juncea var. *leucanthus* (30.11)
Brassica juncea var. *linearifolia* (30.12)
Brassica juncea var. *longepetiolata* (30.13)
Brassica juncea var. *multiceps* (30.15)
Brassica juncea var. *mutisecta* (30.16)
Brassica juncea var. *rugosa* (30.17)
Brassica juncea var. *tumida* (30.19)
Brassica napus var. *napobrassica* (30.22)[1]
Brassica napus var. *napus* (30.21)
Brassica napus var. *pabularia* (30.23)
Brassica nigra (30.24)
Brassica oleracea var. *alboglabra* (30.25)
Brassica oleracea var. *capitata* (30.27)
Brassica oleracea var. *costata* (30.28)
Brassica oleracea var. *medullosa* (30.32)
Brassica oleracea var. *ramosa* (30.33)
Brassica oleracea var. *sabauda* (30.34)
Brassica oleracea var. *viridis* (30.35)
Brassica rapa subsp. *chinensis* (30.36)
Brassica rapa subsp. *chinensis* var. *parachinensis* (30.37)
Brassica rapa subsp. *narinosa* (30.38)
Brassica rapa subsp. *nipposinica* var. *perviridis* (30.39)
Brassica rapa subsp. *pekinensis* (30.40)
Brassica rapa subsp. *rapa* (30.41)
Bunias orientalis (30.42)
Cajanus cajan (40.2)[1]
Campanula rapunculus (33.2)[1]

Capsella bursa-pastoris (30.43)
Cardamine pratensis (30.44)
Celosia spp. (23.4)
Centella asiatica (24.7)
Ceratopteris thalictroides (4.1)
Chenopodium bonus-henricus (35.5)
Chenopodium quinoa subsp. *quinoa* (35.6)
Cichorium endivia subsp. *endiva* (27.9)
Cichorium intybus (27.10)
Cissus discolor (69.1)
Cleome gynandra (34.2)
Cnidoscolus aconitifolius (39.1)
Cnidoscolus chayamansa (39.2)
Coccinia grandis (38.4)[1]
Codiaeum variegatum (39.3)
Colocasia esculenta (8.4)[1]
Corchorus olitorius (63.1)
Coriandrum sativum (24.9)
Cosmos caudatus (27.12)
Crambe maritima (30.45)[1]
Crassocephalum biafrae (27.13)
Crassocephalum crepidioides (27.14)
Cryptotaenia japonica (24.10)
Daucus carota subsp. *sativus* (24.11)[1]
Diplazium esculentum (5.1)
Diplotaxis muralis (30.47)
Doellingeria scabra (27.17)
Eleutherococcus senticosus (26.4)
Emilia sonchifolia (27.18)
Enydra fluctuans (27.19)
Eruca sativa (30.48)
Fedia cornucopiae (27.21)
Foeniculum vulgare subsp. *vulgare* var. *azoricum* (24.12)
Foeniculum vulgare subsp. *vulgare* var. *dulce* (24.13)
Glebionis coronarium (27.23)
Glehnia littoralis (24.14)
Gnetum gnemon (41.1)
Gynura bicolor (27.25)
Hibiscus acetosella (45.3)
Hibiscus sabdariffa (45.4)[1]
Houttuynia cordata (61.1)
Hydrocotyle sibthorpioides (24.15)
Ipomoea aquatica (36.2)
Ipomoea batatas var. *batatas* (36.3)[1]
Ixeridium dentatum (27.27)
Justicia insularis (21.1)
Kalimeris yomena (27.28)
Lactuca indica (27.29)
Lactuca sativa var. *capitata* (27.31)
Lactuca sativa var. *longifolia* (27.32)
Lagenaria siceraria (38.20)[1]
Launaea taraxacifolia (27.33)
Lepidium sativum (30.51)
Ligularia fischeri (27.34)
Limnocharis flava (14.1)
Luffa aegyptiaca (38.22)
Lycium chinense (62.8)
Maesa latifolia (47.1)
Malva rotundifolia (45.5)
Manihot esculenta subsp. *esculenta* (39.4)[1]
Medicago sativa subsp. *sativa* (40.17)
Mentha pulegium (44.2)
Mentha spicata (44.3)
Mesembryanthemum crystallinum (22.1)
Momordica charantia (38.23)[1]
Monochoria hastata Solms (18.1)
Monochoria vaginalis (18.2)
Montia perfoliata (59.1)
Myrrhis odorata (24.16)
Nasturtium officinale (30.52)
Nelumbo nucifera (48.1)[1]
Neptunia prostrata (40.19)
Ocimum americanum (44.4)
Ocimum basilicum (44.5)
Oenanthe javanica subsp. *javanica* (24.17)
Oenothera biennis (51.1)
Opuntia ficus-indica (32.1)
Passiflora biflora (54.1)
Pastinaca sativa (24.18)[1]
Perilla frutescens var. *crispa* (44.7)
Petroselinum crispum var. *crispum* (24.19)
Petroselinum crispum var. *neapolitanum* (24.20)
Petroselinum crispum var. *tuberosum* (24.21)[1]
Phytolacca acinosa (56.1)
Phytolacca americana (56.2)
Phytolacca esculenta (56.3)
Phytolacca octandra (56.4)
Pilea glaberrima (66.1)
Pilea melastomoides (66.2)
Pimpinella brachycarpa (24.22)
Plantago coronopus var. *sativa* (57.1)
Platycodon grandiflorus (34.3)
Portulaca oleracea (59.2)
Psophocarpus tetragonolobus (40.29)[1]
Pueraria montana var. *lobata* (40.30)[1]
Raphanus sativus Oleiformis group (30.55)
Reseda odorata (60.1)
Rumex acetosa (58.2)

Rumex patientia (58.3)
Rumex scutatus (58.4)
Rungia klossii (21.2)
Salsola komarovii (35.7)
Salsola soda (35.8)
Satureja hortensis (44.9)
Sauropus androgynus (39.5)
Scolymus hispanicus (27.36)[1]
Scolymus maculatus (27.37)
Scorzonera hispanica (27.38)[1]
Sechium edule (38.25)[1]
Sedum sarmentosum (37.1)
Sinapis alba (30.57)
Solanum aethiopicum (62.14)
Solanum americanum (62.15)
Solanum macrocarpon (62.18)
Solanum nigrum (62.21)
Sonchus asper (27.40)
Sonchus malaianus (27.41)
Sonchus oleraceus (27.42)
Sonchus wightianus (27.43)
Spinacia oleracea (35.9)
Suaeda glauca (35.10)
Symphytum officinale (29.2)
Symphytum × uplandicum (29.3)
Talinum fruticosum (59.3)
Taraxacum officinale (27.45)
Telfairia occidentalis (38.27)
Tetragonia tetragoniodes (22.2)
Tragopogon porrifolius (27.46)
Trichosanthes cucumerina (38.29)[1]
Trigonella foenum-graecum (40.32)
Tropaeolum majus (65.1)
Urtica dioica (66.3)
Valerianella eriocarpa (67.1)
Valerianella locusta (67.2)
Viola tricolor (68.1)
Xanthosoma brasiliense (8.8)
Xanthosoma sagittifolium (8.9)[1]
Youngia sonchifolia (27.48)
Zingiber mioga (20.5)[1]

Petioles

Alocasia macrorrhizos (8.1)[1]
Apium graveolens var. *dulce* (24.4)
Arctium lappa (27.5)[1]
Borago officinalis (29.1)
Colocasia gigantea (8.5)
Crambe maritima (30.45)
Cynara cardunculus subsp. *cardunculus* (27.15)
Farfugium japonicum (27.20)
Limnocharis flava (14.1)
Petasites japonicus (27.35)
Rheum rhabarbarum (58.1)

Flower/inflorescence/flower stalk/spike

Acmella paniculata (27.3)
Acmella uliginosa (27.4)
Allium tuberosum (7.13)
Alpinia galanga (20.1)
Aponogeton distachyos (25.1)
Brassica juncea var. *utilis* (30.20)
Brassica napus var. *napus* (30.21)
Brassica oleracea var. *alboglabra* (30.25)
Brassica oleracea var. *botrytis* (30.26)
Brassica oleracea var. *costata* (30.28)
Brassica oleracea var. *italica* (30.31)
Brassica rapa subsp. *rapa* (30.41)
Capparis spinosa (34.1)
Cynara cardunculus (27.16)
Hemerocallis spp. (13.3)
Limnocharis flava (14.1)
Mentha spicata (44.3)
Musa × paradisiaca var. *paradisiaca* (16.1)
Nymphaea nouchali (50.2)
Origanum vulgare (44.6)
Passiflora biflora (54.1)
Reseda odorata (60.1)
Saccharum edule (17.5)
Sinapis alba (30.57)
Tropaeolum majus (65.1)[1]
Viola tricolor (68.1)
Zingiber mioga (20.5)[1]

Immature seed pods

Cajanus cajan (40.2)
Cyamopsis tetragonoloba (40.6)
Lathyrus sativus (40.10)
Lens culinaris (40.12)
Lotus tetragonolobus (40.13)
Pachyrhizus erosus (40.21)[1]
Pachyrhizus tuberosus (40.22)
Phaseolus acutifolius var. *acutifolius* (40.23)
Phaseolus coccineus (40.24)
Phaseolus vulgaris var. *vulgaris* (40.26)
Pisum sativum subsp. *sativum* var. *macrocarpon* (40.27)

Psophocarpus tetragonolobus (40.29)
Raphanus sativus Caudatus group (30.53)
Trigonella foenum-graecum (40.32)
Tylosema esculentum (40.33)
Vigna aconitifolia (40.35)
Vigna mungo var. *mungo* (40.37)
Vigna radiata var. *radiata* (40.38)
Vigna unguiculata subsp. *cylindrica* (40.41)
Vigna unguiculata subsp. *sesquipedalis* (40.42)
Vigna unguiculata subsp. *unguiculata* (40.43)

Fruit

Abelmoschus esculentus (45.1)
Benincasa hispida (38.1)
Capsicum annuum var. *annuum* (62.1)
Capsicum annuum var. *longum* Bailey (62.2)
Capsicum baccatum var. *baccatum* (62.3)
Capsicum chinense (62.4)
Capsicum frutescens (62.5)
Capsicum pubescens (62.6)
Citrullus lanatus var. *citroides* (38.2)
Citrullus lanatus var. *lanatus* (38.3)
Cleome gynandra (34.2)
Coccinia grandis Voigt (38.4)
Cucumeropsis mannii Naudin (38.5)
Cucumis anguria var. *anguria* (38.6)
Cucumis melo subsp. *melo* var. *cantalupensis* (38.7)
Cucumis melo subsp. *melo* var. *chito* (38.8)
Cucumis melo subsp. *melo* var. *conomon* (38.9)
Cucumis melo subsp. *melo* var. *flexuosus* (38.10)
Cucumis melo subsp. *melo* var. *inodorus* (38.11)
Cucumis metuliferus (38.12)
Cucumis sativus (38.13)
Cucurbita argyrosperma (38.14)
Cucurbita ficifolia (38.15)
Cucurbita maxima (38.16)
Cucurbita moschata (38.17)
Cucurbita pepo (38.18)
Cyclanthera pedata (38.19)
Cyphomandra betacea (62.7)
Gnetum gnemon (41.1)
Lagenaria siceraria (38.20)
Luffa acutangula (38.21)
Luffa aegyptiaca (38.22)
Lycopersicon esculentum var. *esculentum* (62.9)
Lycopersicon pimpinellifolium (62.10)
Momordica charantia (38.23)
Musa × paradisiaca var. *paradisiaca* (16.1)
Opuntia ficus-indica (32.1)
Physalis alkekengi (62.11)
Physalis peruviana (62.12)
Physalis philadelphica (62.13)
Praecitrullus fistulosus (38.24)
Sechium edule (38.25)
Sicana odorifera (38.26)
Solanum aethiopicum (62.14)
Solanum americanum (62.15)
Solanum ferox (62.16)
Solanum incanum (62.17)
Solanum macrocarpon (62.18)
Solanum melongena (62.19)
Solanum muricatum (62.20)
Solanum nigrum (62.21)
Solanum quitoense (62.22)
Solanum torvum (62.23)
Solanum violaceum (62.25)
Trichosanthes cucumerina (38.29)
Trichosanthes cucumeroides (38.30)
Trichosanthes dioica (38.31

Seed

Amaranthus spp. (23.3)
Arachis hypogaea (40.1)
Aralia continentalis (26.1)
Armoracia rusticana (30.1)[1]
Canavalia ensiformis (40.3)
Canavalia gladiata (40.4)
Cicer arietinum (40.5)
Citrullus lanatus var. *lanatus* (38.3)[1]
Coriandrum sativum (24.9)
Cucumeropsis mannii (38.5)
Cucurbita argyrosperma (38.14)[1]
Cucurbita maxima (38.16)[1]
Cyamopsis tetragonoloba (40.6)
Euryale ferox (50.1)
Glycine max (40.8.)
Lablab purpureus (40.9)
Lathyrus sativus (40.10)
Lens culinaris (40.12)
Lupinus spp.(40.14)
Macrotyloma geocarpum (40.15)
Macrotyloma uniflorum (40.16)
Medicago sativa subsp. *sativa* (40.17)
Mucuna pruriens var. *utilis* (40.18)
Myrrhis odorata (24.16)[1]
Nelumbo nucifera (48.1)
Nymphaea nouchali (50.2)[1]

Pachyrhizus erosus (40.21)[1]
Perilla frutescens var. *crispa* (44.7)[1]
Phaseolus acutifolius var. *acutifolius* (40.23)
Phaseolus coccineus (40.24)
Phaseolus lunatus (40.25)
Phaseolus vulgaris var. *vulgaris* (40.26)
Pisum sativum subsp. *sativum* var. *sativum* (40.28)
Psophocarpus tetragonolobus (40.29)
Sphenostylis stenocarpa (40.31)[1]
Telfairia occidentalis (38.27)
Telfairia pedata (38.28)
Trapa bicornis (64.1)
Trapa natans var. *bispinosa* (64.2)
Trapa natans var. *natans* (64.3)
Vicia faba (40.34)
Vigna aconitifolia (40.35)
Vigna angularis (40.36)
Vigna mungo var. *mungo* (40.37)
Vigna radiata var. *radiata* (40.38)
Vigna subterranea (40.39)
Vigna umbellata (40.40)
Vigna unguiculata subsp. *cylindrica* (40.41)
Vigna unguiculata subsp. *sesquipedalis* (40.42)
Vigna unguiculata subsp. *unguiculata* (40.43)
Zea mays subsp. *mays* (17.7)

Roots

Adenophora triphylla (33.1)
Apium graveolens var. *rapaceum* (24.5)
Arctium lappa (27.5)
Armoracia rusticana (30.1)
Arracacia xanthorrhiza (24.6.)
Beta vulgaris Crassa group (35.4)
Brassica juncea var. *megarrhiza* (30.14)
Brassica napus var. *napobrassica* (30.22)
Brassica rapa subsp. *rapa* (30.41)
Campanula rapunculus (33.2)
Chaerophyllum bulbosum (24.8)
Cirsium dipsacolepis (27.11)
Convolvulus japonicus (36.1)
Crambe tatarica (30.46)[1]
Daucus carota subsp. *sativus* (24.11)
Euryale ferox (50.1)[1]
Glehnia littoralis (24.14)
Ipomoea batatas var. *batatas* (36.3)
Justicia insularis (21.1)[1]
Lepidium meyenii (30.50)
Manihot esculenta subsp. *esculenta* (39.4)
Myrrhis odorata (24.16)[1]
Pachyrhizus ahipa (40.20)
Pachyrhizus erosus (40.21)
Pachyrhizus tuberosus (40.22)
Pastinaca sativa (24.18)
Petroselinum crispum var. *tuberosum* (24.21)
Psophocarpus tetragonolobus (40.29)[1]
Pueraria montana var. *lobata* (40.30)
Raphanus sativus Dakion group (30.54)
Raphanus sativus Radicula group (30.56)
Scolymus hispanicus (27.36)
Scorzonera hispanica (27.38)
Sium sisarum (24.23)
Smallanthus sonchifolius (27.39)
Taraxacum officinale (27.45)[1]
Tragopogon porrifolius (27.46)
Tragopogon pratensis (27.47)
Tylosema esculentum (40.33)[1]
Youngia sonchifolia (27.48)

Tubers

Calathea allouia (15.1)
Cyperus esculentus (10.1)
Dioscorea alata (11.1)
Dioscorea bulbifera (11.2)
Dioscorea cayenensis (11.3)
Dioscorea dumetorum (11.4)
Dioscorea esculenta (11.5)
Dioscorea oppositifolia (11.6)
Dioscorea rotundata (11.7)
Dioscorea trifida (11.8)
Flemingia vestita (40.7)
Helianthus tuberosus (27.26)
Icacina oliviformis (43.1)
Lathyrus tuberosus (40.11)
Mirabilis expansa (49.1)
Oxalis tuberosa (53.1)
Plectranthus esculentus (44.8)
Solenostemon rotundifolius (44.10)
Solanum tuberosum (62.24)
Sphenostylis stenocarpa (40.31)
Stachys affinis (44.11)
Tropaeolum tuberosum (65.2)
Ullucus tuberosus (28.2)

Corms

Alocasia macrorrhizos (8.1)
Amorphophallus konjac (8.2)
Amorphophallus paeoniifolius var. *campanulatus* (8.3)
Colocasia esculenta (8.4)
Cyrtosperma chamissonis (8.6)
Cyrtosperma merkusii (8.7)
Eleocharis dulcis (10.2)
Eleocharis kuroguwai (10.3)
Sagittaria sagittifolia subsp. *leucopetala* (6.1)
Sagittaria sagittifolia subsp. *sagittifolia* (6.2)
Xanthosoma sagittifolium (8.9)
Xanthosoma violaceum (8.10)

Bulbs

Allium ampeloprasum var. *ampeloprasum* (7.1)
Allium cepa var. *cepa* (7.3)
Allium chinense (7.4)
Allium × proliferum (7.9)
Allium sativum (7.10)
Allium scorodoprasum (7.12)
Allium victorialis (7.14)
Lilium spp. (13.4)
Muscari comosum (13.5)
Tigridia pavonia (12.1)

Rhizomes

Alpinia galanga (20.1)[1]
Canna indica (9.1)
Codonopsis lanceolata (33.3)
Curcuma longa (20.2)
Curcuma mangga (20.3)
Curcuma zedoaria (20.4)
Lycopus lucidus (44.1)
Maranta arundinacea (15.2)
Nelumbo nucifera (48.1)
Nymphaea nouchali (50.2)
Tacca leontopetaloides (19.1)
Wasabia japonica (30.58)
Zingiber mioga (20.5)
Zingiber officinale (20.6)

Miscellaneous structures

Brassica juncea var. *gemmifera* (30.8)
Brassica oleracea var. *gemmifera* (30.29)
Centella asiatica (24.7)[1]
Equisetum arvense (1.1)
Hibiscus sabdariffa (45.4)
Osmunda cinnamomea (3.1)
Osmunda japonica (3.2)
Pteridium aquilinum (2.1)
Zea mays subsp. *mays* (17.7)

Chapter 5a. Method of preparation, listed by English common name

Vegetables consumed cooked

For additional information on the plant part consumed, see Chapter 1.

Abyssinian mustard (30.3)
Adzuki bean (40.36)
African eggplant (62.18)
African valerian (27.21)
African yam bean (40.31)
Alfalfa (40.17)
Alligator weed (23.1)
Amaranthus (23.3)
American black nightshade (62.15)
Angelica (24.1)
Angled loofah (38.21)
Arracacha (24.6.)
Arrowhead (6.2)
Asian water chestnut (64.1)
Asiatic pennywort (24.7)
Asparagus (13.2)
Asparagus pea (40.13)
Aster (27.17)
Bambara groundnut (40.39)
Bamboo shoots (17.1, 17.2, 17.4)
Bamboo shoot mustard (30.5)
Banana inflorescence (16.1b)
Barilla plant (35.7)
Beet (35.4)
Bell pepper (62.1)
Bitter gourd (38.23)
Bitter leaf (27.24)
Bitter yam (11.4)
Black gram (40.37)
Black mustard (30.24)
Black-eyed pea (40.43)
Bonnet pepper (62.4)
Borage (29.1)
Bottle gourd (38.20)
Bracken (2.1)
Broad beaked mustard (30.38)
Broad bean (40.34a)
Broccoli (30.31)
Broccoli raab (30.41c)
Broomrape (52.1)
Brussels sprouts (30.29)
Buckhorn plantain (57.1)
Buffalo spinach (27.19)
Buko (41.1)
Bur marigold (27.8)
Butterbur (27.35)
Cape gooseberry (62.12)
Cape pondweed (25.1)
Capitata mustard (30.4)
Capper (34.1)
Cardoon (27.15)
Carrot (24.11)
Casabanana (38.26)
Cassava (39.4)
Catjang (40.41)
Cat's whiskers (34.2)
Cauliflower (30.26)
Cayenne pepper (62.2)
Celeriac (24.5)
Celery (24.4)
Celtuce (27.30)
Ceylon spinach (28.1)
Chard (35.3)
Chaya (39.1, 39.2)
Chayote (38.25)
Chervil (24.3)
Chicory (27.10)
Chimapaya (27.4)[1]
Chinese arrowhead (6.1)
Chinese bellflower (34.3)
Chinese cabbage (30.40)
Chinese chive (7.13)
Chinese kale (30.25)
Chinese radish (30.54)
Chinese water chestnut (10.2)
Chinese wolfberry (62.8)
Chinese yam (11.6)
Chombujiggangi (30.49)[1]
Chufa (10.1)
Cinnamon fern (3.1)
Cluster bean (40.6)
Coastal glehnia (24.14)
Cockscomb (23.4)

[1] Korean name
[2] Japanese name
[3] Javanese name
[4] Indonesian name

Kangaroo vine (69.1)
Kersting's groundnut (40.15)
Kipeit (47.1)[3]
Kohlrabi (30.30)
Konjac (8.2)
Kudzu (40.30)
Kumindelan (27.41)[4]
Kurrat (7.6)
Ladybell (33.1)
Large leaf mustard (30.17)
Leaf radish (30.55)
Leek (7.8)
Lentil (40.12)
Lesser yam (11.5)
Lily (13.4)
Lima bean (40.25)
Line mustard (30.12)
Long petiole mustard (30.13)
Long zedoary (20.4)
Longroot onion (7.14)
Lotus root (48.1a)
Lotus seed (48.1b)
Lupin (40.14)
Maca (30.50)
Malabar gourd (38.15)
Mallow (45.5)
Manchurian spikenard (26.1)
Mango ginger (20.3)
Marama bean (40.33)
Maravilla bulb (12.1)
Marjoram (44.6)
Marrow stem kale (30.32)
Mauka (49.1)
Meadow salsify (27.47)
Mignonette (60.1)
Miner's lettuce (59.1)
Mock pak choy (30.37)
Moth bean (40.35)
Mugwort (27.7)
Mung bean (40.38)
Musky pumpkin (38.17)
New Zealand spinach (22.2)
Oca (53.1)
Okra (45.1)
Onion (7.3)
Orach (35.1)
Oriental pickling melon (38.9)
Oval-leaved pondweed (18.2)
Oyster nut (38.28)
Pak-choi (30.36)
Palm-grass (17.6)
Para cress (27.2)
Parrot's feather (42.1)
Parsley (24.19)
Parsnip (24.18)
Passion flower (54.1)
Pea (40.28)
Pea eggplant (62.23)
Peanut (40.1)
Peduncled mustard (30.20)
Perilla (44.7)
Pigeon bean (40.34b)
Pigeon pea (40.2)
Pilea (66.1, 66.2)
Plantain (16.1a)
Pointed gourd (38.31)
Poke salat (56.2)
Pokeweed (56.3)
Potato (62.24)
Potato bean (40.22)
Potato yam (11.2)
Prickly comfrey (29.3)
Prickly-pear (32.1)
Pualele (27.43) 1
Pumpkin (38.14)
Purslane (59.2)
Quinoa (35.6)
Radish (30.56)
Rakkyo (7.4)
Rampion (33.2)
Rat tail radish (30.53)
Red cabbage (30.27b)
Rhubarb (58.1)
Rice bean (40.40)
Rocoto (62.6)
Roquette salad (30.48)
Rose glorybind (36.1)
Roselle (45.4)
Rungia (21.2)
Rutabaga (30.22)
Salsify (27.46)
Salsola (35.8)
Saururis (61.1)
Savoy cabbage (30.34)
Scarlet eggplant (62.14c)
Scarlet runner bean (40.24)
Scolymus (27.36)
Scorzonera (27.38)
Sea kale (30.45)
Sedum (37.1)

Vegetables consumed raw

African horned cucumber (38.12)
Alfalfa (40.17)
Alligator weed (23.1)
Amaranthus (23.3)
American black nightshade (62.15)
Angelica (24.1)
Armonian cucumber (38.10)
Asiatic pennywort (24.7)
Beet (35.4)
Bell pepper (62.1)
Black mustard (30.24)
Broccoli (30.31)
Buckhorn plantain (57.1)
Buffalo spinach (27.19)
Cantaloupe (38.7a)
Cape gooseberry (62.12)
Cape pondweed (25.1)
Capper (34.1)
Carrot (24.11)
Cauliflower (30.26)
Celeriac (24.5)
Celery (24.4)
Celtuce (27.30)
Chervil (24.3)
Chicory (27.10a)
Chimapaya (27.4)
Chinese chive (7.13)
Chinese radish (30.54)
Chinese water chestnut (10.2)
Chive (7.11)
Chufa (10.1)
Common cabbage (30.27a)
Common sauropus (39.5)
Coriander (24.9)
Cornsalad (67.2)
Cosmos (27.12)
Country potato (44.8)
Cuckoo-flower (30.44)
Cucumber (38.13a)
Currant tomato (62.10)
Dandelion (27.45)
Deodeog (33.3)[1]
Edible podded pea (40.27)
Endive (27.9)
Evening primrose (51.1)
Fennel (24.13)
Florence fennel (24.12)
Garden cress (30.51)
Garden egg (62.17)
Garden myrrh (24.16)
Garlic (7.10)
Giant garlic (7.12)
Ginger (20.6)
Great-headed garlic (7.1)
Greater galangal (20.1)
Guasca (27.1)
Hairy-fruited eggplant (62.16)
Hawksbeard velvetplant (27.14)
Head lettuce (27.31)
Hoary basil (44.4)
Horseradish (30.1)
Hydrocotyle (24.15)
Ice plant (22.1)
Indian nightshade (62.25)
Italian corn-salad (67.1)
Italian parsley (24.20)
Japanese artichoke (44.11)
Japanese garlic (7.7)
Jerusalem artichoke (27.26)
Jew's marrow (63.1)
Johnny-jump-up (68.1)
Kohlrabi (30.30)
Kumindelan (27.41)
Kurrat (7.6)
Leaf lettuce (27.32)
Leek (7.8)
Lentil (40.12)
Mango ginger (20.3)
Mango melon (38.8)
Marjoram (44.6)
Meadow salsify (27.47)
Miner's lettuce (59.1)
Muskmelon (38.7b)
Naranjillo (62.22)
Nasturtium (65.1)
New Zealand spinach (22.2)
Onion (7.3)
Oriental pickling melon (38.9)
Palm-grass (17.6)
Para cress (27.2)
Parsley (24.19)
Parsnip (24.18)
Passion flower (54.1)
Peanut (40.1)
Pepino (62.20)
Perilla (44.7)
Pickling cucumber (38.13b)
Pilea (66.1, 66.2)

Plantain (16.1a)
Prickly-pear (32.1)
Pualele (27.43)
Pumpkin (38.14)
Purslane (59.2)
Radish (30.56)
Rampion (33.2)
Red cabbage (30.27b)
Redleaf chicory (27.10b)
Roquette salad (30.48)
Salsify (27.46)
Saururis (61.1)
Scorzonera (27.38)
Sessile alternanthera (23.2)
Shallot (7.2)
Sierra Leone bologi (27.13)
Sorrel (58.2)
Sow thistle (27.18)
Sowthistle (27.42)
Spearmint (44.3)
Spinach (35.9)
Spiny sow thistle (27.40)
Spot flower (27.3)
Strawberry tomato (62.11)
Summer squash (38.18)
Sunset hibiscus (45.2)
Sweet basil (44.5)
Sweet corn (17.7)
Swollen stem mustard (30.19)
Tarragon (27.6)
Tartar bread-plant (30.46)
Tettu (21.1)
Tomato (62.9)
Tomatillo (62.13)
Tree onion (7.9a)
Turnip-rooted parsley (24.21)
Upland cress (30.2)
Wallrocket (30.47)
Water dropwort (24.17)
Water lily (50.2)
Watercress (30.52)
Watermelon (38.3)
Watershield (31.1)
Welsh onion (7.5)
White mustard (30.57)
White-seeded melon (38.5)
Wild water chestnut (10.3)
Winged bean (40.29)
Winter melon (38.11)

Vegetables consumed preserved

Armonian cucumber (38.10)
Cabbage (30.27)
Capitata mustard (30.4)
Capper (34.1)
Chinese cabbage (30.40)
Chinese radish (30.54)
Citron (38.2)
Country potato (44.8)
Cucumber (38.13a)
Ginger (20.6)
Gobouazami (27.11)[1]
Greater galangal (20.1)
Horsetail (1.1)
Involute mustard (30.9)
Japanese artichoke (44.11)
Okra (45.1)
Oriental pickling melon (38.9)
Pickling cucumber (38.13b)
Rat tail radish (30.53)
Shiny bugleweed (44.1)
Strumous mustard (30.18)
Swollen stem mustard (30.19)
Tabasco pepper (62.5)
Tillered mustard (30.15)
Wax gourd (38.1)
West Indian gherkin (38.6)
White flowered mustard (30.11)

Chapter 5b. Method of preparation, listed by abbreviated scientific name

Vegetables consumed cooked

For additional information on plant part consumed method, see Chapter 1

Abelmoschus esculentus (45.1)
Abelmoschus manihot (45.2)
Acmella ciliata (27.1)
Acmella paniculata (27.3)
Acmella uliginosa (27.4)
Adenophora triphylla (33.1)
Allium ampeloprasum var. *ampeloprasum* (7.1)
Allium cepa var. *aggregatum* (7.2)
Allium cepa var. *cepa* (7.3)
Allium chinense (7.4)
Allium fistulosum (7.5)
Allium kurrat (7.6)
Allium macrostemon (7.7)
Allium porrum L. (7.8)
Allium × proliferum (7.9)
Allium sativum (7.10)
Allium scorodoprasum (7.12)
Allium tuberosum (7.13)
Allium victorialis (7.14)
Alocasia macrorrhizos (8.1)
Alpinia galanga (20.1)
Alternanthera philoxeroides (23.1)
Alternanthera sessilis (23.2)
Amaranthus spp. (23.3)
Amorphophallus konjac (8.2)
Amorphophallus paeoniifolius (8.3)
Angelica archangelica (24.1)
Angelica keiskei (24.2)
Anthriscus cerefolium (24.3)
Apium graveolens var. *dulce* (24.4)
Apium graveolens var. *rapaceum* (24.5)
Aponogeton distachyos (25.1)
Arachis hypogaea (40.1)
Aralia continentalis (26.1)
Aralia cordata (26.2)
Aralia elata (26.3)
Arctium lappa (27.5)
Armoracia rusticana (30.1)
Arracacia xanthorrhiza (24.6.)
Artemisia dracunculus (27.6)
Artemisia indica var. *maximowiczii* (27.7)
Asparagus acutifolius (13.1)
Asparagus officinalis (13.2)
Atriplex hortensis (35.1)
Bambusa spp. (17.1)
Barbarea verna (30.2)
Basella alba (28.1)
Bassia scoparia (35.2)
Benincasa hispida (38.1)
Beta vulgaris Cicla group (35.3)
Beta vulgaris Crassa group (35.4)
Bidens pilosa (27.8)
Borago officinalis (29.1)
Brassica carinata (30.3)
Brassica juncea var. *capitata* (30.4)
Brassica juncea var. *crassicaulis* (30.5)
Brassica juncea var. *crispifolia* (30.6)
Brassica juncea var. *foliosa* (30.7)
Brassica juncea var. *gemmifera* (30.8)
Brassica juncea var. *involuta* (30.9)
Brassica juncea var. *latipa* (30.10)
Brassica juncea var. *leucanthus* (30.11)
Brassica juncea var. *linearifolia* (30.12)
Brassica juncea var. *longepetiolata* (30.13)
Brassica juncea var. *megarrhiza* (30.14)
Brassica juncea var. *mutisecta* (30.16)
Brassica juncea var. *rugosa* (30.17)
Brassica juncea var. *strumata* (30.18)
Brassica juncea var. *utilis* (30.20)
Brassica napus var. *napobrassica* (30.22)
Brassica napus var. *napus* (30.21)
Brassica napus var. *pabularia* (30.23)
Brassica nigra (30.24)
Brassica oleracea var. *alboglabra* (30.25)
Brassica oleracea var. *botrytis* (30.26)
Brassica oleracea var. *capitata* (30.27)
Brassica oleracea var. *costata* (30.28)
Brassica oleracea var. *gemmifera* (30.29)
Brassica oleracea var. *gongylodes* (30.30)
Brassica oleracea var. *italica* (30.31)
Brassica oleracea var. *medullosa* (30.32)
Brassica oleracea var. *ramosa* (30.33)
Brassica oleracea var. *sabauda* (30.34)
Brassica oleracea var. *viridis* (30.35)
Brassica rapa subsp. *chinensis* (30.36)
Brassica rapa subsp. *chinensis* var. *parachinensis* (30.37)

Raphanus sativus Oleiformis group (30.55)
Raphanus sativus Radicula group (30.56)
Reseda odorata (60.1)
Rheum rhabarbarum (58.1)
Rumex acetosa (58.2)
Rumex patientia (58.3)
Rumex scutatus (58.4)
Rungia klossii (21.2)
Saccharum edule (17.5)
Sagittaria sagittifolia subsp. *leucopetala* (6.1)
Sagittaria sagittifolia subsp. *sagittifolia* (6.2)
Salsola komarovii (35.7)
Salsola soda (35.8)
Satureja hortensis (44.9)
Sauropus androgynus (39.5)
Scolymus hispanicus (27.36)
Scolymus maculatus (27.37)
Scorzonera hispanica (27.38)
Sechium edule (38.25)
Sedum sarmentosum (37.1)
Sesamum radiatum (55.1)
Setaria palmifolia (17.6)
Sicana odorifera (38.26)
Sinapis alba (30.57)
Sium sisarum (24.23)
Smallanthus sonchifolius (27.39)
Solanum americanum (62.15)
Solanum ferox (62.16)
Solanum incanum (62.17)
Solanum macrocarpon (62.18)
Solanum melongena (62.19)
Solanum muricatum (62.20)
Solanum nigrum (62.21)
Solanum torvum (62.23)
Solanum tuberosum (62.24)
Solanum violaceum (62.25)
Solenostemon rotundifolius (44.10)
Sonchus asper (27.40)
Sonchus malaianus (27.41)
Sonchus oleraceus (27.42)
Sonchus wightianus (27.43)
Sphenostylis stenocarpa (40.31)
Spinacia oleracea (35.9)
Stachys affinis (44.11)
Struchium sparganophorum (27.44)
Suaeda glauca (35.10)
Symphytum officinale (29.2)
Symphytum × uplandicum (29.3)
Tacca leontopetaloides (19.1)
Talinum fruticosum (59.3)
Talinum paniculatum (59.4)
Taraxacum officinale (27.45)
Telfairia occidentalis (38.27)
Telfairia pedata (38.28)
Tetragonia tetragoniodes (22.2)
Tigridia pavonia (12.1)
Tragopogon porrifolius (27.46)
Tragopogon pratensis (27.47)
Trapa bicornis Osbeck (64.1)
Trapa natans var. *bispinosa* (64.2)
Trapa natans var. *natans* (64.3)
Trichosanthes cucumerina (38.29)
Trichosanthes cucumeroides (38.30)
Trichosanthes dioica (38.31)
Trigonella foenum-graecum (40.32)
Tropaeolum tuberosum (65.2)
Tylosema esculentum (40.33)
Ullucus tuberosus (28.2)
Urtica dioica (66.3)
Valerianella eriocarpa (67.1)
Valerianella locusta (67.2)
Vicia faba (40.34)
Vigna aconitifolia (40.35)
Vigna angularis (40.36)
Vigna mungo var. *mungo* (40.37)
Vigna radiata var. *radiata* (40.38)
Vigna subterranea (40.39)
Vigna umbellata (40.40)
Vigna unguiculata subsp. *cylindrica* (40.41)
Vigna unguiculata subsp. *sesquipedalis* (40.42)
Vigna unguiculata subsp. *unguiculata* (40.43)
Wasabia japonica (30.58)
Xanthosoma brasiliense (8.8)
Xanthosoma sagittifolium (8.9)
Xanthosoma violaceum (8.10)
Youngia sonchifolia (27.48)
Zea mays subsp. *mays* (17.7)
Zingiber mioga (20.5)
Zingiber officinale (20.6)
Zizania latifolia (17.8)

Vegetables consumed raw

Abelmoschus manihot (45.2)
Acmella ciliata (27.1)
Acmella paniculata (27.3)
Allium ampeloprasum var. *ampeloprasum* (7.1)
Allium cepa var. *aggregatum* (7.2)
Allium cepa var. *cepa* (7.3)
Allium fistulosum (7.5)

Petroselinum crispum var. *crispum* (24.19)
Petroselinum crispum var. *neapolitanum* (24.20)
Petroselinum crispum var. *tuberosum* (24.21)
Physalis alkekengi (62.11)
Physalis peruviana (62.12)
Physalis philadelphica (62.13)
Pilea glaberrima (66.1)
Pilea melastomoides (66.2)
Pisum sativum subsp. *sativum* var. *macrocarpon* (40.27)
Plantago coronopus var. *sativa* (57.1)
Plectranthus esculentus (44.8)
Portulaca oleracea (59.2)
Psophocarpus tetragonolobus (40.29)
Pueraria montana var. *lobata* (40.30)
Raphanus sativus Dakion group (30.54)
Raphanus sativus Radicula group (30.56)
Rumex acetosa (58.2)
Sauropus androgynus (39.5)
Scorzonera hispanica (27.38)
Setaria palmifolia (17.6)
Sinapis alba (30.57)
Solanum aethiopicum (62.14)
Solanum americanum (62.15)
Solanum ferox (62.16)
Solanum incanum (62.17)
Solanum quitoense (62.22)
Solanum violaceum (62.25)
Sonchus asper (27.40)
Sonchus malaianus (27.41)
Sonchus oleraceus (27.42)
Sonchus wightianus (27.43)
Spinacia oleracea (35.9)
Stachys affinis (44.11)
Taraxacum officinale (27.45)
Tetragonia tetragoniodes (22.2)
Tragopogon porrifolius (27.46)
Tragopogon pratensis (27.47)
Tropaeolum majus (65.1)
Valerianella eriocarpa (67.1)
Valerianella locusta (67.2)
Viola tricolor (68.1)
Zea mays subsp. *mays* (17.7)
Zingiber officinale (20.6)

Vegetables consumed preserved

Abelmoschus esculentus (45.1)
Alpinia galanga (20.1)
Brassica juncea var. *capitata* (30.4)
Brassica juncea var. *involuta* (30.9)
Brassica juncea var. *leucanthus* (30.11)
Brassica juncea var. *multiceps* (30.15)
Brassica juncea var. *strumata* (30.18)
Brassica juncea var. *tumida* (30.19)
Brassica oleracea var. *capitata* (30.27)
Brassica rapa subsp. *pekinensis* (30.40)
Capparis spinosa (34.1)
Capsicum frutescens (62.5)
Cirsium dipsacolepis (27.11)
Citrullus lanatus var. *citroides* (38.2)
*Cucumis anguria*var. *anguria* (38.6)
Cucumis melo subsp. *melo* var. *conomon* (38.9)
Cucumis melo subsp. *melo* var. *flexuosus* (38.10)
Cucumis sativus (38.13)
Equisetum arvense (1.1)
Lycopus lucidus (44.1)
Plectranthus esculentus (44.8)
Raphanus sativus Caudatus group (30.53)
Raphanus sativus Dakion group (30.54)
Stachys affinis (44.11)
Zingiber officinale (20.6)

Chapter 6. Languages[1]

ABUA – abn, (syn. Abuan). Abua is spoken by 25,000 people in Nigeria. Classification: Niger Congo, Atlantic-Congo, Volta-Congo, Benue-Congo, Cross River, Delta Cross, Central Delta, Abua-Odual. Dialects: Central Abuan, Emughan, Otapha, and Okpeden.

AFAR – aar, (syn. Afaraf, Danakil, Denkel, Afar Af, Adal). Afar is spoken by 1,400,000-1,500,000 people in Ethiopia, Somalia, Djibouti, and Eritrea. Classification: Afro-Asiatic, Cushitic, East, Saho-Afar. Dialects: Northern Afar, Central Afar, Aussa, and Baadu (Ba'adu).

AFRIKAANS – afr. Afrikaans is spoken by 6,000,000 people in South Africa and 10,000,000 second language speakers. It is also spoken by immigrants in Australia, Botswana, Canada, Lesotho, Malawi, Namibia, New Zealand, Zambia, and Zimbabwe. Classification: Indo-European, Germanic, West, Low Saxon-Low Franconian, Low Franconian. Dialects: Cape Afrikaans (West Cape Afrikaans), Orange River Afrikaans, and East Cape Afrikaans.

AGTA – *afa*. Agta is a cluster of 8 languages spoken on Luzon, the Philippines, one of which is extinct. There are a very limited number of speakers i.e. ~ 2,000. Classification: Austronesian, Malayo-Polynesian.

AGTA, CASIGURAN DUMAGAT – dgc, (syn. Casiguran Dumagat). Casiguran Dumagat Agta is spoken by 600 people in the Aurora province of Luzon, the Philippines. Classification: Austronesian, Malayo-Polynesian, Northern Philippine, Northern Luzon, Northern Cordilleran, Dumagat, Northern. Casiguran Dumagat Agta is similar to Paranan.

AJA – aja, (syn. Adja, Ajja). Aja is spoken by a very limited number of people in southern Sudan, western Bahr el Ghazal Province and in the western Central African Republic. Classification: Nilo-Saharan, Central Sudanic, West, Kresh. Aja is thought to be midway between Banda and Kresh.

AKAN – aka. Akan is a language spoken by 8,300,000 in Ghana and eastern Côte d'Ivoire; there are 1,000,000 second language speakers. Classification: Niger-Congo, Atlantic-Congo, Volta-Congo, Kwa, Nyo, Potou-Tano, Tano, Central, Akan. Dialects: Fante, Akuapem, Asante, Agona, Dankyira, Asen, Akyem Bosome, Kwawu, and Ahafo.

AKAN-AKUAPEM – aka-*aub*. Akuapem is a dialect of Akan. There are 8,300,000 speakers of Akan in Ghana with 555,000 Akuapem in the southeast, north of Accra.

AKAN-ASANTE – aka-*aag*, (Asante syn. Ashante Twi, Asanti, Achanti). Asante is a dialect of Akan spoken in Ghana by the Asante people. The Asante population numbers 2,800,000 people that are located in south central Ghana.

AKAN-FANTE – aka-*fat*, (Fante syn. Fanti, Mfantse). Fante is a dialect of Akan spoken in Ghana by the Fante. The Fante population numbers 1,900,000 people that are located in south central Ghana.

AKHA – ahk, (syn. Kaw, Ekaw, Ko, Aka, Ikaw, Ak'a, Ahka, Khako, Kha Ko, Khao Kha Ko, Ikor, Aini, Yani). Akha is spoken by 1,000,000 people in the Chiangmai, Chiangrai, and Maehongson provinces in Thailand, the Yunnan province of China, and in Burma. Classification: Sino-Tibetan, Tibeto-Burman, Lolo-Burmese, Loloish, Southern Loloish, Akha, Hani, Ha-Ya. Dialects: Ako, Asong.

AKLANON – akl, (syn. Aklan, Aklano, Panay, Aklanon-Bisayan). There are 394,500 speakers of Aklanon, primarily in the Aklan Province of Northern Panay, the Philippines. Classification: Austronesian, Malayo-Polynesian, Meso Philippine, Central Philippine, Bisayan, West, Aklan. Aklanon is similar to Hiligaynon.

AKPA – akf, (syn. Akweya). Akpa is spoken by 27,000 people in Nigeria. Classification: Niger-Congo, Atlantic-Congo, Volta-Congo, Benue-Congo, Idomoid, Yatye-Akpa. Akpa is similar to Ekpari.

[1] Primary scource – Gordon, 2005. Each language has a three letter ISO/DIS 639-3 identity code, established by the International Organization for Standardization, listed after its name. In instances where the identity of the language is not definitive, a temporary code has been assigned and is in italics.

AKWA – akw. Akwa is spoken by 24,000 people in the Cuvette Region, Makoua District of Congo. Classification: Niger-Congo, Atlantic-Congo, Volta-Congo, Benue-Congo, Bantoid, Southern, Narrow Bantu, Northwest, C, Mbosi.

ALBANIAN, TOSK – als, (syn. Albanian, Tosk, Arnaut, Shkip, Shqip, Skchip, Shqiperë, Zhgabe). Tosk Albanian is spoken by 2,980,000 people, primarily in south Albania to the Shkumbi River. It is also spoken by a small number of immigrants in Belgium, Egypt, Germany, Greece, Turkey, Ukraine, and the United States. Classification: Indo-European, Albanian, Tosk. Dialects: Arbanasi, Srem, Camerija, and Korca.

ALEUT – ale, (syn. Unangam Tunuu). Only about 300 Aleut speakers remain; Western Aleut is spoken on Atka Island and Eastern Aleut on the eastern Aleutian Islands, Pribilofs, Siberian Commander Islands and Alaskan Peninsula. Classification: Eskimo-Aleut, Aleut. Dialects: Western Aleut and Eastern Aleut.

AMHARIC – amh, (syn. Abyssinian, Ethiopian, Amarinya, Amarigna). Amharic is a Semitic language spoken by 17,373,000 of the Amhara people in North Central Ethiopia, the Amhara Region and in Addis Ababa. A small number of speakers are found in Egypt and Sweden. In Israel it is used by around 40,000 Jews of Ethiopian origin. Worldwide there are approximately 4,000,000 second language users. Classification: Afro-Asiatic, Semitic, South Ethiopian, South, Transversal, Amharic-Argobba.

ANDO – *ang*. Ando is reported by Burkill as a language in the Côte d'Ivoire. There are 78 languages spoken there. The correct language of the words listed is not known.

ANUAK – anu, (syn. Anywak, Anywa, Yambo, Jambo, Nuro, Anyuak, Dho Anywaa). Anuak is a Nilotic language of the Nilo-Saharan language family. It is spoken by 45,000 of the Anuak people primarily in the western part of Ethiopia. It is also spoken by 52,000 people in the Sudan with a total of 97,000 speakers in all countries. Classification: Nilo-Saharan, Eastern Sudanic, Nilotic, Western, Luo, Northern, Anuak. Dialects: Adoyo, Coro, Lul, and Opëno.

ANYIN – any, (syn. Anyi, Agni). There are 610,000 Anyin speakers in Côte d'Ivoire and 200,000 in Ghana, plus anywhere from 10,000 to 100,000 second language speakers. Classification: Niger-Congo, Atlantic-Congo, Volta-Congo, Kwa, Nyo, Potou-Tano, Tano, Central, Bia, Northern. Dialects: Sanvi, Indenie, Bini, Bona, Moronou, Djuablin, Ano, Abe, Barabo and Alangua. Baoulé and Nzema are the closest languages to Anyin.

APACHE, JICARILLA – apj. There are 2,000 Jicarilla Apaches in Northern New Mexico primarily in Rio Arriba and Sandoval counties, though the number of speakers is significantly lower. Use of the language is declining. Classification: Na-Dene, Nuclear Na-Dene, Athanpaskan-Eyak, Athapaskan, Apachean, Navajo-Apache, Eastern Apache.

ARABIC – arb. Arabic is a Central Semitic language that is spoken by more than 280,000,000 people as a first language and 250,000,000 as a second language, mostly in the Middle East and North Africa. Classification: Afro-Asiatic, Semitic, Central, South, Arabic. A diverse cross-section of dialects are spoken, not all of which are mutually intelligible. Standard Arabic is the official language that is widely taught in schools and used for all written materials and formal speeches. Dialects: Egyptian, Gulf, Iraqi, Levantine, Maghrebi, Andalusi, Bahrani, Central Asia, Hassaniya, Hejazi, Judeo-Arabic dialects, Maltese, Najdi, Shuwa, Siculo, Sudanese, and Yemeni.

ARMENIAN – hye, (syn. Haieren, Somkhuri, Ermenice, Ermeni Dili, Armjansik Yazyk). There are a total of 6,723,000 Armenian speakers in all countries, of which 3,400,000 live in Armenia. Remaining speakers are found in a wide range of countries from India to the United States. Classification: Indo-European, Armenian. Dialects: Eastern Armenian, Erevan, Tbilisi, Karabagh, Shamakhi, Astrakhan, Dzhulfa, Agulis, Khvoy-Salmst, Urmia-Maragheh, Artvin, Karin, Mus, Van, Tigranakert, Kharberd, Shabin-Karahissar, Trabzon, Hamshen, Malatya, Kilikien, Syria, Arabkir, Akn, Sebaste, Ewdokia, Smyrna, North Komedia, Constantinople, Rodosto, Crimea, and Ashkharik. All of the dialects are intelligible. Eastern Armenian is the most widely used and is spoken in Armenia and its Turkish and Iranian borderlands.

ASSAMESE – asm, (syn. Asambe, Asami, Asamiya). There are 15,334,000 speakers of Assamese, most of whom are in the Assam state in northern India. Classification: Indo-European, Indo-Iranian, Indo-Aryan, Eastern zone, Bengali-Assamese. Dialects: Jharwa, Mayang, Standard Assamese, and Western Assamese.

ATTA, PAMPLONA – att, (syn. Northern Cagayan Negrito). Pamplona Atta is spoken by 1000 people in the Northwestern Cagayan Province of Luzon, the Philippines. Classification: Austronesian, Malayo-Polynesian, Northern Philippine, Northern Luzon, Northern Cordilleran, Ibanagic, Ibanag. Pamplona Atta is similar to Ibanag North.

AZERBAIJANI, NORTHERN – azj, (syn. Azeri Turk, Azerbaijani, Azerbaydzhani). Azerbaijani is spoken by 7,000,000 people, the majority of which (6,069,000) are in Azerbaijan and southern Dagestan. It is also spoken in a wide range of other countries, e.g. Georgia, Germany, Russia, Estonia, the United Kingdom, the United States, Uzbekistan, Syria, Iraq, Turkey, Turkmenistan, Ukraine, Canada. Classification: Altaic, Turkic, Southern, Azerbaijani. Dialects: Quba, Derbend, Baku, Shamakhi, Salyan, Lenkaran, Qazakh, Airym, Borcala, Terekeme, Qyzylbash, Nukha, Zaqatala, Qabala, Yerevan, Nakhchivan, Ordubad, Ganja, Shusha, and Karapapak.

BAATONUM – bba, (syn. Baatonu, Baatombu, Baruba, Bargu, Burgu, Berba, Barba, Bogung, Bargawa, Barganchi, Bariba). Baatonum is spoken by 560,000 people, 460,000 of whom are located in Benin, primarily in the central, north, and Borgou Provience. It is also spoken (~100,000) in some areas of Nigeria (Kwara and Niger states and in the Borgu local government area). Classification: Niger-Congo, Atlantic-Congo, Volta-Congo, North, Gur, Bariba.

BADAGA – bfq, (syn. Badag, Badagu, Badugu, Baduga, Vadagu). Badaga is spoken by 50,000 people in the Nilgiri Hills of southern India. Classification: Dravidian, Southern, Tamil-Kannada, Kannada.

BAFIA – ksf, (syn. Rikpa, Ripey, Rikpa', Bekpak). Bafia is spoken by 60,000 people in Cameroon. Classification: Niger-Congo, Atlantic-Congo, Volta-Congo, Benue-Congo, Bantoid, Southern, Narrow Bantu, Northwest, A, Bafia (A.50). Dialects: Kpa and Bape.

BAFUT – bfd, (syn. Bufe, Fu, Fut, Befe). Bafut is spoken in Cameroon by 50,000 people. Classification: Niger-Congo, Atlantic-Congo, Volta-Congo, Benue-Congo, Bantoid, Southern, Wide Grassfields, Narrow Grassfields, Mbam-Nkam, Ngemba. Bafut is related to Bamum. Dialects: Bafut and Buge.

BAGA – *bad*. Baga is a group of closely related languages that are spoken by 30,000 people in the coastal region of Guinea. Approximately one-half speak Landoma, a closely related language; many also speak the Mande language Susu. Classification: Niger-Congo, Atlantic-Congo, Atlantic, Southern, Mel, Temne. The Baga languages: Baga Binari (bcg), Baga Koga (bgo), Baga Manduri (bmd), Baga Mboteni (bgm), Baga Sitemu (bsp), Baga Sobane (bsv), are often considered dialects of Baga. The specific dialect of the crops listed was not indicated.

BALANTA-KENTOHE – ble, (syn. Balanta, Balant, Balante, Balanda, Ballante, Belante, Bulanda, Brassa, Alante, Frase). Blanta-Kentohe is spoken by 393,000 people, mostly in Guinea-Bissau, along the north central and central coast. It is also spoken in Gambia. Classification: Niger-Congo, Atlantic-Congo, Atlantic, Northern, Bak, Balant-Ganja. Dialects: Fora, Kantohe (Kentohe, Queuthoe), Naga, and Mane.

BALTI – bft, (syn. Sbalti, Baltistani, Bhotia of Baltistan). Balti is spoken by 337,000 people in Baltistan, northeastern Pakistan (270,000) and in parts of India (Ladakh). Classification: Sino-Tibetan, Tibeto-Burman, Himalayish, Tibeto-Kanauri, Tibetic, Tibetan, Western. Dialects: Chorbat, Purik, and others.

BAMANANKAN – bam, (syn. Bambara, Bamanakan). Bamanankan is spoken by 2,700,000 people in Mali; including second language speakers, there are a total of 6,000,000 speakers in all countries. Bamanankan is similar to Dioula. Classification: Niger-Congo, Mande, Western, Central-Southwestern, Central, Manding-Jogo, Manding-Vai, Manding-Mokole, Manding, Manding-East, Northeastern Manding, Bamana. Dialects: Standard Bambara, Somono, Segou, San, Beledugu, Ganadugu, Wasulu and Sikasso.

BAMILEKE – *bai*. The Bamileke are a collection of Semi-Bantu ethnic groups found in the highlands of Cameroon's West Province. They comprise of more than 2,120,000 members within over 100 individual groups that are linguistically related, i.e. from the Bantoid branch of the Niger-Congo language family. The specific language of the crops listed was not indicated.

BAOULÉ – bci, (syn. Baule, Bawule). There are 2,130,000 Baoulé speakers found primarily in southern Cote d'Ivorie. Classification: Niger-Congo, Atlantic-Congo, Volta-Congo, Kwa, Nyo, Potou-Tano, Tano, Central, Bia, Northern. Dialects: similar to Anyin; many subgroups but largely mutually intelligible.

BASQUE – eus, (syn. Vascuense, Euskera). Basque is spoken by 1,063,000 people (approximately 665,700 first language speakers) in the Spanish Basque country and in southwest France. Classification: Basque. Dialects: Bizkaian, Gipuzkoan, and Upper Navarrese (in Spain), and Lower Navarrese, Lapurdian, and Zuberoan (in France).

BASSARI – bsc, (syn. Basari, Tenda Basari, Biyan, Onëyan, Onian, Ayan, Wo). There are approximately 8,800 Bassari speakers in Senegal. It is also spoken in Guinea and Guinea-Bissau with a total of 18,000 speakers in all countries. Classification: Niger-Congo, Atlantic-Congo, Atlantic, Northern, Eastern Senegal-Guinea, Tenda. Dialects: closely related to Budik.

BELARUSAN – bel, (syn. Belarusian, Belorussian, Bielorussian, White Russian, White Ruthenian, Byelrussian). Belarusan is spoken by 6,715,000 people in Belarus with an estimated total of 9,081,000 speakers in all countries, the majority being in Russia, Ukraine, Poland, Azerbaijan, Canada, Estonia, Kazakhstan, Latvia, Lithuania, Moldova, Tajikistan, and Turkmenistan. Classification: Indo-European, Slavic, East. Dialects: Northeast Belarusan, Southwest Belarusan, and Central Belarusan. Linguistically, Belarusan is between Russian and Ukrainian.

BEMBA – bem, (syn. Chibemba, Cibemba, Ichibemba, Icibemba, Wemba and Chiwemba). Bemba is spoken by 3,300,000 people in Zambia with a total of 3,600,000 speakers in all countries, the balance being primarily in Botswana, Democratic Republic of Congo and Malawi. Classification: Niger-Congo, Atlantic-Congo, Volta-Congo, Benue-Congo, Bantoid, Southern, Narrow Bantu, Central, M, Bemba. Dialects: Lembue, Lomotua, Ngoma, Nwesi, Town Bemba, Luunda, Chishinga, Kabende, Mukulu, Ng'umbo, Twa of Bangweulu, and Unga.

BEMBÉ – bmb, (syn. Beembe, Ebembe, Ibembe, Kibembe, Mbeembe). Bembé is spoken by 252,000 people in Democratic Republic of Congo; it is also used in some parts of Burundi and Tanzania. Classification: Niger-Congo, Atlantic-Congo, Volta-Congo, Benue-Congo, Bantoid, Southern, Narrow Bantu, Central, D, Bembe. Bembé is similar to several other languages: Lega-Mwenga dialects, Buyu, Lega-Shabunda, Zimba, Enya, Nyanga, Tembo, and Lengola.

BENGALI – ben, (syn. Banga-Bhasa, Bangala, Bangla). Bengali evolved from the Magadhi Prakrit and Sanskrit languages and is spoken in the eastern Indian subcontinent by nearly 230,000,000 people worldwide, ranking it fifth or sixth among the most spoken languages of the world. There are an estimated 100,000,000 speakers in Bangladesh; it is also spoken in India, Malawi, Nepal, Saudi Arabia, Singapore, United Arab Emirates, United Kingdom, and the United States. Classification: Indo-European, Indo-Iranian, Indo-Aryan, Eastern zone, Bengali-Assamese. Dialects or languages in the Bengali group: Central Bengali, Western Bengali, Southwestern Bengali, Northern Bengali, Rajbanshi, Bahe, Eastern Bengali, Haijong, Southeastern Bengali, Ganda, Vanga, and Chittagonian.

BEROM – bom, (syn. Birom, Berum, Gbang, Kibo, Kibbo, Kibbun, Kibyen, Shosho, Aboro, Boro-Aboro, Afango, Chenberom, Cen Berom). Berom is spoken by 300,000 people primarily within Nigeria's Plateau State. Classification: Niger-Congo, Atlantic-Congo, Volta-Congo, Benue-Congo, Plateau, Beromic. Dialects: Gyell-Kuru-Vwang (Ngell-Kuru-Vwang), Fan-Foron-Heikpang, Bachit-Gashish, Du-Ropp-Rim, Hoss, and Cen.

BÉTÉ – *byf*. There are three Bété languages spoken in Côte d'Ivoire – Daloa, Gagnoa, and Guiberoua with 130,000, 150,000 and 130,000 speakers, respectively. Each is classified as a Niger-Congo, Atlantic-Congo, Volta-Congo, Kru, Eastern, Bete language. The specific language for the vegetable names listed in the text was not identified.

BETE-BENDI – btt, (syn. Bette-Bendi, Dama). Bete-Bendi is spoken by 36,800 people in the Cross River State and Obudu local government area of Nigeria. Classification: Niger-Congo, Atlantic-Congo, Volta-Congo, Benue-Congo, Cross River, Bendi. Dialects: Bete (Bette, Mbete) and Bendi. Bete-Bendi is not the same as the language Bete spoken in the Nigerian state of Gongola, Bete spoken in Cameroon, or Bété of Côte d'Ivoire.

BICOLANO, CENTRAL – bcl, (syn. Bikol, Biko, Bicalano). Central Bicolano is spoken by 2,500,000 people in the Philippines, primarily in Southern Catanduanes, Northern Sorsogon, Albay, Camarines, Norte and Sur, and Luzon. Classification: Austronesian, Malayo-Polynesian, Meso Philippine, Central Philippine, Bikol, Coastal, Naga. Dialects: Naga and Legaspi.

BINUKID – bkd, (syn. Manobo, Binukid Manobo, Binokid, Bukidnon, Western Bukidnon Manobo). Spoken by 100,000 people in north central Mindanao, southern Bukidnon, northeastern Cotabato and Agusan del Sur Proviences in the Philippines. Classification: Austronesian, Malayo-Polynesian, Southern Philippine, Manobo, North. Dialects: Kalabugao, Bukidnon, Minalowang, Esperanza, and Agusan del Sur. Binukid is similar to Higaonon.

BLAAN, SARANGANI – bps, (syn. Bilaan, Balud, Tumanao). Sarangani Blaan is spoken by 90,700 people on Mindanao, the Philippines. Classification: Austronesian, Malayo-Polynesian, South Mindanao, Bilic, Blaan.

BLACKFOOT – bla, (syn. Pikanii, Siksika, Blackfeet). Blackfoot is spoken by 5,000 people in Alberta and northern Montana. Classification: Algic, Algonquian, Plains. Dialects: Piegan and Blood.

BOBO MADARÉ, SOUTHERN – bwq, (syn. Bobo Fing, Bobo Fi, Black Bobo, Bobo). The language is listed in the text as Bobo. Southern Bobo Madaré is spoken by 311,000 people primarily in the Houet Province of Burkino Faso. Classification: Niger-Congo, Mande, Western, Northwestern, Soninke-Bobo, Bobo. Dialects: Benge, Sogokiré, Voré, Syabéré (Sya), and Zara (Bobo Dioula, Bobo Jula).

BODO – brx, (syn. Boro, Bodi, Bara, Boroni, Mechi, Meche, Mech, Meci, Kachari). Bodo is spoken by 600,000 of the Bodo people in the state of Assam in India and to some extent in Nepal and Bangladesh. Classification: Sino-Tibetan, Tibeto-Burman, Jingpho-Konyak-Bodo, Konyak-Bodo-Garo, Bodo-Garo, Bodo. Dialects: Chote and Mech. Bodo is an official language in Assam and is one of 22 languages with constitutional status in India.

BOKYI – bky, (syn. Boki, Nki, Okii, Uki, Nfua, Osikom, Osukam, Vaaneroki). Bokyi is spoken by 140,000 people in Nigeria and 4,000 in Cameroon. Classification: Niger-Congo, Atlantic-Congo, Volta-Congo, Benue-Congo, Cross River, Bendi. Dialects: Basua (Bashua), Irruan (Erwan, Eerwee), Boje (Bojie), Kwakwagom, Nsadop, Osokom, Wula (Baswo, Okundi, Kecwan), Oku, Boorim, Oyokom, Abo (Abu), and Eastern Bokyi (Eastern Boki).

BONTOC, CENTRAL – bnc, (syn. Igorot, Bontok, Bontoc). Central Bontoc is spoken by 40,000 people in the Central Mountain Province of Luzon in the Philippines. Classification: Austronesian, Malayo-Polynesian, Northern Philippine, Northern Luzon, South-Central Cordilleran, Central Cordilleran, Nuclear Cordilleran, Bontok-Kankanay, Bontok. Dialects: Sadanga, Guinaang Bontoc, and Bayyu.

BOSNIAN – bos. Bosnian is a South Slavic language spoken by 2,700,000 people primarily in Bosnia and Herzegovina. It is also spoken in Serbia, Montenegro, Croatia, Albania, Kosova, and by immigrants in a number of countries. Classification: Indo-European, Balto-Slavic, Slavic, South, Western South Slavic.

BRETON – bre, (syn. Brezhoneg). Breton is spoken by 500,000 people in France, primarily in Western Brittany with additonal speakers in Eastern Brittany and among immigrants in other countries. Classification: Indo-European, Celtic, Insular, Brythonic. Dialects: Leonais, Tregorrois, Vannetais, and Cornouaillais. Cornish and Welsh are closely related to Breton and are generally mutually comprehensible.

BUDIK – tnr, (syn. Bedik, Tandanke, Tendanke, Tenda, Bande, Basari du Bandemba). Budik is spoken by 3,000 people in Senegal. Classification: Niger-Congo, Atlantic-Congo, Atlantic, Northern, Eastern Senegal-Guinea, Tenda. Budik closely resembles Bassari.

BUGHOTU – bgt, (syn. Bugotu, Bugoto, Bugota, Mbughotu, Mahaga). Bughotu is spoken by 4,000 people in the Solomon Islands, primarily on Santa Isabel and Furona Islands. Classification: Austronesian, Malayo-Polynesian, Central-Eastern, Eastern Malayo-Polynesian, Oceanic, Central-Eastern Oceanic, Southeast Solomonic, Gela-Guadalcanal, Bughotu. Dialects: Hageulu and Vulava.

BUHID – bku, (syn. Bukil, Bangon, Batangan). Buhid is spoken by 8,000 people in southern Mindoro, the Philippines. Classification: Austronesian, Malayo-Polynesian, Meso Philippine, South Mangyan, Buhid-Taubuid.

BULGARIAN – bul, (syn. Balgarski). There are 8,000,000 Bulgarian speakers in Bulgaria and another 1,000,000 among immigrants in Canada, Greece, Hungary, Israel, Moldova, Montenegro, Romania, Serbia, Turkey, and the United States. Classification: Indo-European, Slavic, South, Eastern. Dialects: Palityan (Palitiani, Bogomil) and Pomak.

BULLOM SO – buy, (syn. Northern Bullom, Bolom, Bulem, Bullun, Bullin, Bulom, Mmani, Mandingi, Mandenyi). Bullom So is spoken by approximately 500 people in Sierra Leone and along its border with Guinea. Classification: Niger-Congo, Atlantic-Congo, Atlantic, Southern, Mel, Bullom-Kissi, Bullom, Northern. Dialects: Mmani and Kafu.

BULU – bum, (syn Boulou). Bulu is a Bantu language with 174,000 speakers in Cameroon, predominately in the Southern, Central, and Eastern Provinces and 800,000 second language speakers. Classification: Niger-Congo, Atlantic-Congo, Volta-Congo, Benue-Congo, Bantoid, Southern, Narrow Bantu, Northwest, A, Yaunde-Fang. Dialects: Yelinda, Yembana, Yengono, Zaman, and Bene.

BURMESE – mya, (syn. Bama, Bamachaka, Myen, Myanmar). Burmese is spoken by 32,000,000 people predominately in Myanmar. It is also spoken by a relatively small number (ca. 300,000) in Bangladesh, Malaysia, and Thailand. Classification: Sino-Tibetan, Tibeto-Burman, Lolo-Burmese, Burmish, Southern. Dialects: Merguese (Mergui, Beik), Yaw, Danu (Trauw), Burmese, and Palaw.

BUTUANON – btw. Butuanon is spoken by 34,500 people on Mindanao in the Philippines. Classification: Austronesian, Malayo-Polynesian, Meso Philippine, Central Philippine, Bisayan, South, Butuan-Tausug. Butuanon is similar to Kamayo and Surigaonon.

CAJUN FRENCH – frc, (syn. Français Acadien, Acadian, Cajun, Cajan, Cadien). Cajun French is a dialect of French spoken by approximately 1,000,000 people in Southern Louisiana west of the Mississippi. It differs from Metropolitan French in pronunciation, vocabulary and intonation. The use of Cajun French continues to decline. Classification: Indo-European, Italic, Romance, Italo-Western, Western, Gallo-Iberian, Gallo-Romance, Gallo-Rhaetian, Oïl, French. Dialects: Marsh French, Prairie French, and Big Woods French.

CAROLINIAN – cal, (syn. Saipan Carolinian, Southern Carolinian). Carolinian is an Austronesian language spoken by 3,000 people on Saipan, Anatahan and Agrihan islands of the Northern Mariana Islands. Along with English and Chamorro, it is one of the three official languages of the islands. Classification: Austronesian, Malayo-Polynesian, Central-Eastern, Eastern Malayo-Polynesian, Oceanic, Central-Eastern Oceanic, Remote Oceanic, Micronesian, Micronesian Proper, Ponapeic-Trukic, Trukic. Carolinian is similar to Satawalese, Woleaian, Puluwatese, Mortlockese, Chuukese, and Ulithian.

CATALAN-VALENCIAN-BALEAR – cat, (syn. Català, Catalán, Bacavès, Catalonian). Catalan-Valencian-Balear is spoken by 6,473,000 people in Northeastern Spain and Catalonia and Valencia Provinces, Balearic Islands and certain other areas in Spain and by immigrants in a number of countries around the world. Classification: Indo-European, Italic, Romance, Italo-Western, Western, Gallo-Iberian, Ibero-Romance, East Iberian. Dialects: Catalan-Rousillonese (Northern Catalán), Valencian (Valenciano, Valencià), Balearic (Balear, Insular Catalan, Mallorqui, Menorqui,

Eivissenc), Central Catalan, Algherese, and Northwestern Catalan (Pallarese, Ribagorçan, Lleidatà, Aiguavivan).

CEBUANO – ceb, (syn. Sugbuhanon, Sugbuanon, Visayan, Bisayan, Binisaya, Sebuano). Cebuano is an Austronesian language spoken by 20,000,000 people in the Philippines, predominately in Negros, Cebu, Bohol, Visayas and parts of Mindanao. Classification: Austronesian, Malayo-Polynesian, Meso Philippine, Central Philippine, Bisayan, Cebuan. Dialects: Cebu, Boholano, Leyete, and Mindanao Visayan.

CHECHEN – che, (syn. Nokhchiin, Nokchiin Muott, Galancho). Chechan is spoken by 1,500,00 people, primarily in Chechnya. Classification: Northeast Caucasian, Nakh, Vainakh, Chechen. Dialects: Akkhiy, Chiantiy, Chiebarloy, Mialkhiy, Nokhchmakhkakhoy, Orstkhoy, Sharoy, Shuotoy and Terloy.

CHEROKEE – chr, (syn. Tsalagi, Tslagi). Cherokee is an Iroquoian language spoken by 15,000 to 22,500 people in Oklahoma and North Carolina. Classification: Iroquoian, Southern Iroquoian. Dialects: Elati (Lower Cherokee, Eastern Cherokee), Kituhwa (Middle Cherokee), Otali (Upper Cherokee, Western Cherokee, Overhill Cherokee) and Overhill Middle Cherokee.

CHICKASAW – cic, (syn. Chikashshanompa). Chickasaw is a Muskogean language. Of an ethnic population of 35,000 to 37,000, there are only around 1000 Chickasaw speakers. They are found primarily in south central Oklahoma. Classification: Muskogean, Western. Chickasaw is very similar to Choctaw, such that the two languages are considered by some as dialects of the same language.

CHINESE, GÀN – gan, (syn. Gan, Kan, Jiangxihua). Gàn is spoken by 20,580,000 people in Jiangxi, the southeastern corner of Hubei and parts of Anhui, Hunan and Fujian provinces of China. Classification: Sino-Tibetan, Chinese. Dialects: Chang-Jing, Yi-Liu, Ji-Cha, Fu-Guang, and Ying-Yi.

CHINESE, HAKKA – hak, (syn. Hakka, Hokka, Kejia, Kechia, Ke, Xinminhua, Majiahau, Tu Guangdonghau). Hakka is a primary subdivision of Chinese spoken by the Hakka ethnic group in southern China and immigrants and their descendents in many areas of the world. It is spoken by 25,725,000 in China with a total of 30,000,000 speakers worldwide. Classification: Sino-Tibetan, Chinese. Dialects: Yue-Tai (Meixian, Raoping, Taiwan, Kejia), Yuezhong (Central Guangdong), Huizhou, Yuebei (Northern Guangdong), Tingzhou (Min-Ke), Ning-Long (Longnan), Yugui, and Tonggu. Yue-Tai is the standard dialect.

CHINESE, MANDARIN – cmn, (syn. Mandarin, Guanhua, Beifang Fangyan, Northern Chinese, Guoyu, Standard Chinese). Mandarin Chinese is the most widely spoken language in the world with 867,200,000 speakers on mainland China and an estimated total of 885,000,000 worldwide. Classification: Sino-Tibetan, Chinese. Dialects: Huabei Guanhua (Northern Mandrin), Xibei Guanhua (Northwestern Mandarin), Xinan Guanhua (Southwestern Mandarin), and Jinghuai Guanhua (Jiangxia Guanhua, Lower Yangze Mandarin).

CHINESE, MIN BEI – mnp, (syn. Northern Min, Min Pei). Min Bei is spoken by 10,290,000 people in China, primarily in northern Fujian Province. Classification: Sino-Tibetan, Chinese. Dialects: Min Nan, Min Bei, Min Dong, Min Zhong, and Pu-Xian.

CHINESE, MIN NAN – nan, (syn. Southern Min, Minnan). Min Nan is spoken by 25,725,000 people on the mainland of China. It is also spoken in Brunei, Indonesia, Malaysia, Philippines, Singapore, Taiwan, Thailand, and the United States for a total of 46,228,000 speakers in all countries. Classification: Sino-Tibetan, Chinese. Dialects: Xiamen (Amoy), Leizhou (Lei Hua, Li Hua), Chao-Shan (Choushan, Chaozhou), Hainan (Hainanese, Qiongwen Hua, Wenchang), Longdu, and Zhenan Min.

CHINESE, WÚ – wuu. Wú is spoken by 77,000,000 people, second among Chinese languages to Mandarin and tenth among all languages worldwide. It is primarily spoken in Shanghai, much of Zhejiang province, and southern Jiangsu province. Classification: Sino-Tibetan, Chinese. Dialects: Taihu, Jinhua (Kinhwa), Taizhou, Oujiang, Wuzhou, Chuqu, and Xuanzhou.

CHINESE, XIANG – hsn, (syn. Hunan, Hunanese, Xiang, Hsiang). Xiang is spoken by 36,015,000 people in central and southwestern Hunan, Sichuan, and northern Guangxi provinces. Classification: Sino-Tibetan, Chinese. Dialects: Changyi, Luoshao, and Jishu.

CHINESE, YUE – yue, (syn. Yuet Yue, Gwong Dung Waa, Cantonese, Yue, Yueh, Yueyu, Baihua). Yue, commonly referred to as Cantonese in the United Sates, is spoken by an estimated 52,000,000 people in China, primarily in Guangdong and parts of Guangxi in the southern mainland of China, Hong Kong and Macau. It is also spoken by immigrants in Southeast Asia, Canada and the United States. Classification: Sino-Tibetan, Chinese. Yue is distinct and mutually unintelligible from the other major Chinese languages.

CHUWABU – chw, (syn. Chuwabo, Chwabo, Cuwabo, Cuabo, Chuabo, Chichwabo, Cicuabo, Txuwabo, Echuwabo, Echuabo). Chuwabu is spoken by 786,000 people in Mozambique along the central coast. Classification: Niger-Congo, Atlantic-Congo, Volta-Congo, Benue-Congo, Bantoid, Southern, Narrow Bantu, Central, P, Makua (P.30). Dialects: Central Chuwabo, Nyaringa, Marale, Karungu, and Maindo.

COMANCHE – com. Commache is a Uto-Aztecan language now spoken by only about 200 of the Comanche people in western Oklahoma. The language split off from the Shoshone in the early 1700s. The name comes from a Ute word for enemy or stranger. Classification: Uto-Aztecan, Northern Uto-Aztecan, Numic, Central. Comanche is close to Shoshoni. During World War II, Comanche troops in the U.S. Army conveyed sensitive operational messages in Comanche that the enemy was unable to decipher.

CREE, PLAINS – crk, (syn. Western Cree). Plains Cree is an Algonquian language spoken by 34,000 people in the plains region of Canada, primarily in north central Manitoba, westward across Saskatchewan and Central Alberta to the Rocky Mountains. Classification: Algic, Algonquian, Central, Cree-Montagnais-Naskapi. Dialects: Plains Cree, Western York Cree, and Northern Alberta Cree.

CREOLE, HAITIAN FRENCH – hat, (syn. Kreyol, Haitian, Haitian Creole, Aiysyen). Haitian Creole is spoken by 6,965,000 people in Haiti and another 400,000 in the Caribbean region, Canada and the United States. It the most widely spoken Creole language in the world. Classification: Creole, French based. Dialects: Fablas and Plateau Haitian Creole.

CRIOULO, CAPE VERDE – kea, (syn. Kabuverdianu, Caboverdiano). Cape Verde Crioulo is a Portuguese Creole spoken by 926,000 people in the Republic of Cape Verde off the western coast of Africa and by a limited number of people in Angola, Brazil, France, Germany, Guinea-Bissau, Italy, Luxembourg, Netherlands, Portugal, Senegal, Spain, and the United States. It is considered the oldest Creole language. Classification: Portuguese based Creole. Dialects: Sotavento and Baravento.

CRIOULO, UPPER GUINEA – pov, (syn. Kiryol, Portuguese Creole, Kriulo, Guinea-Bissau Creole). Upper Guinea Crioulo is a Portuguese Creole spoken by 190,000 people in Guinea-Bissau and a total of 392,000 worldwide. Classification: Portuguese base Creole. Dialects: Bissau-Bolama Creole, Bafatá Creole, and Cacheu-Ziguinchor Creole.

CROATIAN – hrv, (syn. Hrvatski). Croatian is a South Slavic language spoken by 4,800,000 people in Croatia and 6,214,000 worldwide, many of which are in Bosnia and Herzegovina, Serbia (Vojvodina), and Montenegro but also in Austria, Germany, Hungary, Italy, Slovakia and Slovenia. Classification: Indo-European, Slavic, South, Western. Dialects: Kaykayski, Chakavski, and Shtokavski (Ijekavski), the latter being the official dialect.

CZECH – ces, (syn. Cestina). Czech is a West Slavic language spoken by 10,004,000 people in the Czech Republic and in a number of other countries (e.g. Austria, Bulgaria, Canada, Israel, Poland, Romania, United States, Germany, Croatia, Serbia, Slovakia, and Ukraine) for a total of 11,525,000 speakers worldwide. Classification: Indo-European, Slavic, West, Czech-Slovak. Dialects: Central Bohemian, Czecho-Morvian, Hanak, Lach (Yalach), Northeast Bohemian, and Southwest Bohemian.

DAGAARE, SOUTHERN – dga, (Southern Dagari, Dagari, Dagara, Degati, Dagati, Dogaari, Dagaare). Southern Dagaare is spoken by 700,000 of the Dagaaba people in Ghana; including Northern Dagara in Burkina Faso, there is a total of 1,000,000 speakers. Classification: Niger-Congo, Atlantic-Congo, Volta-Congo, North, Gur, Central, Northern, Oti-Volta, Western, Northwest, Dagaari-Birifor, Dagaari. Dialects: Dagaare and Birifor.

DAGBANI – dag, (syn. Dagbane, Dagomba, Dagbamba). Dagbani is Gur language spoken by 800,000 people in Northeast Ghana. Classification: Niger-Congo, Atlantic-Congo, Volta-Congo, North, Gur, Central, Northern, Oti-Volta, Western, Southeast. Dialect: Nanuni (Nanumba). Dagbani is similar to Mampruli, Talni, and Kusaal.

DAKOTA – dak, (syn. Sioux). Dakota is spoken by 5,000 Dakota Sioux in the western United States and southern Canada. Dakota is one of forty-five languages that make-up the Siouan family; Dakota and Lakota are thought by some to be dialects of the same language. Classification: Siouan, Siouan Proper, Central, Mississippi Valley, Dakota. Dialects: Dakota (Santee) and Nakota (Yankton).

DANGME – ada, (syn. Adangme). Dangme is a Kwa language spoken by 800,000 people in southeast Ghana, coastal areas east of Accra and inland. Classification: Niger-Congo, Atlantic-Congo, Volta-Congo, Kwa, Nyo, Ga-Dangme. Dialects: Ada, Ningo, Osu, Shai, Gbugbla, and Krobo.

DANGME-KROBO – ada-*kro*. Dangme-Krobo is a dialect of Dangme (Adangme) that is spoken in Ghana. It is listed by Burkill as Adangme-Krobo.

DANISH – dan, (syn. Dansk, Central Danish, Sjaelland). Danish is a Scandinavian language spoken by 5,000,000 people in Denmark and an additional 300,000 in Canada, Faroe Islands, Germany, Greenland, Iceland, Norway, Sweden, United Arab Emirates and the United States. Classification: Indo-European, Germanic, North, East Scandinavian, Danish-Swedish, Danish-Riksmal, Danish. Danish is divided into three distinct dialect groups – Eastern Danish (including Bornholm, Scanian and Halland dialects), Island Danish (including Zealand, Funen, Lolland, Falster, and Møn), and Jutlandic (including North, East, West and South Jutlandic dialects).

DENDI – ddn, (syn. Dandawa). Dendi is a Songhai language spoken by 30,000 people in Benin and a few thousand more in other countries. Classification: Nilo-Saharan, Songhai, Southern. Dendi is closely related to Zarma and Songai.

DIGO – dig, (Kidigo, Chidigo). Digo is a Bantu language spoken by 217,000 of the Digo people in Kenya. It is also spoken in Tanzania with a total of 305,000 speakers in all countries. Classification: Niger-Congo, Atlantic-Congo, Volta-Congo, Benue-Congo, Bantoid, Southern, Narrow Bantu, Central, E, Nyika, Mijikenda. Dialects: Chinondo (Northern Digo), Ungu (Lungu, Southern Digo), Ts'imba, and Tsw'aka (Chw'aka).

DITAMMARI – tbz, (syn. Ditamari, Tamari, Somba). Ditammari is spoken by 20,000 people in Benin. It is also spoken in Togo with at total of 47,500 speakers in all countries. Classification: Niger-Congo, Atlantic-Congo, Volta-Congo, North, Gur, Central, Northern, Oti-Volta, Eastern. Dialects: Eastern Ditammari and Western Ditammari.

DOONDO – dde, (syn. Dondo, Kidoondo). Doondo is spoken by 3,000 people in Congo. Classification: Niger-Congo, Atlantic-Congo, Volta-Congo, Benue-Congo, Bantoid, Southern, Narrow Bantu, Central, H, Kongo (H.10). It is a member of the Kongo cluster of languages.

DUTCH – nld, (syn. Nederlands, Hollands). Dutch is a West Germanic language spoken by 12,360,000 people in the Netherlands and 22,000,000 native speakers worldwide. It is also spoken in Belgium, Suriname, Aruba, and to a lesser extent in parts of France (Flanders), Germany (lower Rhine) and several former Dutch colonies and by immigrant populations throughout the world. Classification: Indo-European, Germanic, West, Low Saxon-Low Franconian. Dialects: Brabanitc, East Flemish, West Flemish, Hollandic, Lumburgish, South Geldric, and Zealandic.

EDE NAGO – nqg, (syn. Nago, Nagots, Nagot). Ede Nago is spoken by 200,000 people in southeastern Benin. Classification: Niger-Congo, Altantic-Congo, Volta-Congo, Benue-Congo, Defoid, Yoruboid, Edekiri. Dialects: Ede Nago is a member of the Ede language cluster.

EDO – bin, (syn. Bini, Benin, Addo, Oviedo, Ovioba). Edo is Benue-Congo language spoken by 1,000,000 people in Nigeria, primarily in the Bendel State and the Ovia, Oredo and Orthionmwon local government areas. Classification: Niger-Congo, Atlantic-Congo, Volta-Congo, Benue-Congo, Edoid, North-Central, Edo-Esan-Ora.

EFIK – efi, (syn. Calabar). Efik is a member of the Ibibio dialect cluster spoken by 400,000 people in the Cross River and Akwa Ibom states of Nigeria. Classification: Niger-Congo, Atlantic-Congo, Volta-Congo, Benue-Congo, Cross-River, Delta Cross, Lower Cross, Oholo, Efik.

ENGLISH – eng. English is a West Germanic language that evolved in England. It is spoken by 55,000,000 people in the United Kingdom, 309,000,000 to 400,000,000 as a first language worldwide, and an estimated 199,000,000 to 1,400,000,000 people as a second language. Classification: Indo-European, Germanic, West, English. There are a diverse range of dialects in the United Kingdom from Cockney to Lower Scottish. English is now generally considered the universal language of science and it is estimated that one half of the world population will be fluent in English by 2015.

ESTONIAN – est, (syn. Eesti, Viro). Estonian is a Finno-Ugric language spoken by 953,000 people in Estonia and a total of 1,075,000 in all countries. Classification: Uralic, Finnic. Language groups and (dialects): Northeastern Coastal Estonian (Tallinn and Narva), North Estonian (island, western, central and eastern), and South Estonian (Mulgi, Tartu, and Vôru). The language is most closely related to Finnish.

ÉWÉ – ewe, (syn. Eibe, Ebwe, Eve, Efe, Eue, Vhe, Gbe, Krepi, Krepe, Popo). Éwé is a Niger-Congo language spoken in Ghana by 2,250,000 people and in Togo and Benin with a total of over 3,000,000 in all countries. Classification: Niger-Congo, Atlantic-Congo, Volta-Congo, Kwa, Left Bank, Gbe. Dialects: Anglo (Anlo), Awuna, Hudu, and Kotafoa.

EWONDO – ewo, (syn. Ewundu, Jaunde, Yaounde, Yaunde). Ewondo is spoken by 577,700 people in Cameroon. Classification: Niger-Congo, Atlantic-Congo, Volta-Congo, Benue-Congo, Bantoid, Southern, Narrow Bantu, Northeast, A, Yaunde-Fang (A.70). Dialects: Badjia (Bakjo), Bafeuk, Bamvele (Mvele, Yezum, Yesoum), Bane, Beti, Fong, Mbida-Bani, Mvete, Mvog-Niengue, Omvang, Yabekolo (Yebekolo), Yabeka, Yabekanga, Enoah, Evouzok.

FARSI, EASTERN – prs, (syn. Persian, Dari, Parsi). There are 5,600,000 Eastern Farsi speakers in Afghanistan. It is also spoken in Iran and Pakistan with a total of 7,600,000 speakers in all countries. Classification: Indo-European, Indo-Iranian, Iranian, Western, Southwestern, Persian. Dialects: Dari (Afghan, Farsi, Herati, Tajiki, Kaboli, Kabuli, Khorasani), and Parsiwan.

FARSI, WESTERN – pes, (syn. Persian, New Persian, Parsi, Irani). Western Farsi is spoken by 22,000,000 people in Iran and a total of 24,000,000 speakers in all countries. Classification: Indo-Iranian, Iranian, Western, Southwestern, Persian. Dialects: Sedehi, Shirazi, Ketabi, Mashadi, Shahrudi, Kazeruni, Esfahani, Kermani, Araki, Basseri, Old Shirazi, Shirazjahromi, Yazdi, Qazvini, Tehrani, Bandari, Kashani, and Mahalhamadani.

FIJIAN – fij, (syn. Fiji, Standard Fijian, Eastern Fijian, Nadroga, Nadronga). Fijian is an Austronesian language spoken by 330,000 people in Fiji and a total of 334,000 speakers in all countries. Classification: Austronesian, Malayo-Polynesian, Central-Eastern Eastern Malayo-Polynesain, Oceanic, Remote Oceanic, Central Pacific, East Fijian-Polynesian, East Fijian. Dialects: Kadavu, Southeast Viti Levu, Bau, Northeast Viti Levu, Central Vanua Levu, Northeast Vanua Levu, Southeast Vanua Levu, and West Vanua Levu.

FILIPINO – fil, (syn. Pilipino). Filipino is based on Tagalog and is spoken throughout the Philippines. Classification: Austronesian, Malayo-Polynesian, Meso Philippine, Central Philippine, Tagalog.

FINNISH – fin, (syn. Suomi, Suomea). Finnish is a Finno-Ugric language spoken by 4,700,000 people in Finland. It is also spoken in Canada, Estonia, Norway, Russia, Republic of Karelia, Sweden and the United States, with a total of 5,233,000 speakers in all countries. Classification: Uralic, Finnic. Dialects: Southwestern Finnish, Häme (Tavast), south Pohjanmaa, Central and North Pohjanmaa, Peräpohja, Savo (Savolax), and Southeastern Finish (Finnish Karjala, Finnish Karelian).

FON – fon, (syn. Fo, Fongbe, Fonnu, Fogbe, Dahomeen, Djedji). Fon is part of the Gbe language cluster spoken by 1,700,000 people in Benin and 1,735,000 in all countries (e.g. Togo). Classification: Niger-Congo, Atlantic-Congo, Volta-Congo, Kwa, Left Bank, Gbe, Fon. Dialects: Agbome, Arohun, Gbekon, and Kpase.

FRENCH – fra, (syn. Français). French is spoken by 51,000,000 people in France. It is also spoken in a diverse range of other countries for a total of 64,858,000 speakers in all countries. Classification: Indo-European, Italic, Romance, Italo-Western, Western, Gallo-Iberian, Gallo-Romance, Gallo-Rhaetian, Oïl. Dialects: Standard French, Norman (Normand), Angevin, Berrichon, Bourbonnais, Bourguignon, Franc-Comtois, Gallo, Poitevin, Santongeais, and Lorraine.

FULFULDE – *ful*. The eastern, east central, and west central dialects of the Fula languages are called Fulfulde. They include Adamawa (fub), Bagirmi (fui), Maasina (ffm), and Borgu (fub) Fulfulde. The language for the crops listed was not specified though based on location (Nigeria and Cameroon) the names may represent both Borgu and Adamawa Fulfulde.

FULFULDE, ADAMAWA – fub, (synonyms for Adamawa Fulfulde vary with location. In Cameroon they are: Adamawa, Fulani, Peul, Peulh, Ful, Fula, Fulbe, Boulbe, Eastern Fulani, Fulfulde, Foulfoulde, Pullo, Gapelta, Pelta Hay, Domona, Pladina, Palata, Paldena, Dzemay, Zemay, Zaakosa, Pule, Taareyo, Sanyo, and Biira while in Nigeria common synonyms are: Eastern Fulfulde, Fulatanchi, Fulani, Fula, Fillanci, Fula-Fulfulde). Adamawa Fulfulde is a Fula language spoken by the Fula people over a relatively extensive area of West Africa. The western dialects are called Pulaar (Pular) and the eastern, east central, and west central dialects are called Fulfulde. Adamawa Fulfulde is an east central form of Fula that is spoken in parts of Cameroon (700,000 speakers), Chad (128,000), Sudan (90,000) and Nigeria where the total number of Fulfulde (Nigerian Fulfulde) speakers is 7,611,000. Classification: Niger-Congo, Atlantic-Congo, Atlantic, Northern, Senegambian, Fulani-Wolof, Fula, Eastern.

GA – gaa, (syn. Amina, Gain, Accra, Acra). Ga is a Kwa language spoken by 600,000 people in southeastern Ghana around Accra and is one of 16 languages the Ghanaian government uses in publishing. Classification: Niger-Congo, Atlantic-Congo, Volta-Congo, Kwa, Nyo, Ga-Dangme.

GADDANG – gad, (syn. Cagayan). Gaddang is spoken by 30,000 people on Luzon, the Philippines. Classification: Austronesian, Malayo-Polynesian, Northern Philippine, Northern Luzon, Northern Cordilleran, Ibanagic, Gaddang. Many Gaddang speakers also use Ilocano.

GAELIC, IRISH – gle, (syn. Irish, Erse, Gaeilge). Irish is a Goidelic language spoken by 260,000 people in Ireland and a total of 355,000 speakers in all countries. Classification: Indo-European, Celtic, Insular, Goidelic. Dialects: Munster-Leinster (Southern Irish), Connacht (Western Irish), and Donegal (Ulster, Northern Irish). It is now spoken by only a small minority of the population.

GAELIC, MANX – glv, (syn. Manx, Gaelg, Gailck). Manx is a Goidelic language previously spoken on the Isle of Man. The language essentially died out in the early 1970s but has been the subject of revival efforts such that there are currently around 100 speakers. Classification: Indo-European, Celtic, Insular, Goidelic. Dialects: Manx is close to Scottish Gaelic.

GAELIC, SCOTTISH – gla, (syn. Gàidhlig, Gaelic, Scots Gaelic, Albannach, Gaidhlig, Erse). Scottish is a Goidelic language spoken by 58,000 people in the United Kingdom, in particular around Glasgow, the north and central counties of the Ross region of Scotland, and on Skye and the Hebrides. Classification: Indo-European, Celtic, Insular, Goidelic. Dialects: East Sutherlandshire.

GALICIAN – glg, (syn. Galego, Gallego). Galician is spoken by 3,173,000 people in the Galicia region of Spain and neighboring areas and in northern Portugal. Its close similarity to Portuguese has simulated an ongoing debate as to whether it is simply a dialect of Portuguese or a distinct language. Classification: Indo-European, Italic, Romance, Italo-Western, Western, Gallo-Iberian, Ibero-Romance, West Iberian, Portuguese-Galician. Dialects: Eastern, Central, and Western.

GANDA – lug, (syn. Luganda, Ugandan). Ganda, also commonly called Luganda, is spoken by 3,016,000 people in Uganda, especially in southern Uganda in the Buganda region. There are an additional 1,000,000 second language users. Classification: Niger-Congo, Atlantic-Congo, Volta-

Congo, Benue-Congo, Bantoid, Southern, Narrow Bantu, Central, J, Nyoro-Ganda. Dialects: Kooki (Olukooki), Sese (Olusese), Luvuma, and Ludiopa.

GBAYA – gya, (syn. Baya, Gbaya Nord-Ouest, Northwest Gbaya). Gbaya is branch of 12 Ubangian languages spoken in the Central African Republic and to a lesser extent in Democratic Republic of Congo, the Republic of the Congo and Cameroon. The crops listed (from Cameroon) appear to be in the Northwest Gbaya language. The following classification is not considered definitive: Niger-Congo, Ubangian, Gbaya. Dialects of Northwest Gbaya are: Banginda, Mbai, Gaymona, Lai (Lay), Lombu, Mbere, Mombe, and Yaáyuwee (Yaiwe, Kalla).

GBE-VHE – *gbe-vhe*. Gbe is a cluster of approximately 20 languages spoken in eastern Ghana and western Nigeria. The crops listed as Gbe-Vhe remain under that listing, though both Gbe and Vhe are synonyms of Éwé. There are 3,112,000 speakers of Éwé in Ghana and Togo. Éwé is classified as: Niger-Congo, Atlantic-Congo, Volta-Congo, Kwa, Left Bank, Gbe. Dialects of Éwé include Anglo (Anlo), Awuna, Hudu, and Kotafoa.

GBE, WACI – wci, (syn. Ouatchi, Waci, Waci-Gbe, Wachi, Watyi). Waci Gbe is spoken by 365,500 people in Togo. It is also spoken in Benin with a total of 475,500 speakers in all countries. Classification: Niger-Congo, Atlantic-Congo, Volta-Congo, Kwa, Left Bank, Gbe. Dialects: Waci is one of 10 languages that make up the Gbe language cluster.

GEN – gej, (syn. Ge, Gen-Gbe, Mina-Gen, Mina, Guin, Gegbe, Popo). Gen is a Gbe language spoken by 358,900 people in Benin and Togo. Classification: Niger-Congo, Atlantic-Congo, Volta-Niger, Gbe. Dialects: Anexo, Agoi, Gen, and Gliji.

GEORGIAN – kat, (syn. Kartuli, Gruzinski, Common Kartvelian). Georgian is the official language of Georgia, spoken by 3,981,000 inhabitants and by a limited number of people in Iran, Azerbaijan, Turkey, Russia, Europe and the United States for a total of 4,178,000 speakers in all countries. Classification: Kartvelian, Georgian. Dialects: Imeretian, Racha-Lexchxum (Lechkhum), Gurian, Adzhar (Acharian), Imerxev Kartlian, Kaxetian (Kakhetian), Ingilo, Tush, Xevsur (Kheysur), Moxev (Mokhev), Pshav, Mtiul, Ferejdan, and Meskhur-Javakhuri.

GERMAN – deu, (syn. Deutsch, Tedesco, Standard German). German is a West Germanic language that is spoken by 75,300,000 people in Germany. It is the most widely spoken language in the European Union and used by a number of people in Austria, Switzerland, Italy, Liechtenstein, Luxembourg, Denmark, and Belgium. It is also used by immigrants around the world. There are 105,000,000 native and 80,000,000 non-native speakers worldwide. Classification: Indo-European, Germanic, West, High German, German. Dialects: Bavarian, Schwäbisch, Allemannisch, Mainfränkisch, Hessisch, Palatinian, Rheinfränkisch, Westfälisch, Saxonian, Thuringian, Brandenburgisch, and Low Saxon.

GIANGAN – bgi, (syn. Bagobo, Jangan, Guanga, Gulanga, Clata, Eto, Atto). Giangan is an Austronesian language spoken by 55,000 people on Mindanao in the Philippines. Classification: Austronesian, Malayo-Polynesian, South Mindanao, Bagobo. Dialects: Giangan is similar to Tagabawa and Obo Manobo.

GIKUYU – kik, (syn. Kikuyu, Gekoyo, Gigikuyu). Gikuyu is member of the Niger-Congo language family spoken by 5,347,000 of the Kikuyu people in Kenya. Classification: Niger-Congo, Atlantic-Congo, Benue-Congo, Bantoid, Southern, Bantu, Central Bantu, Northeast Bantu, Kikuyu-Kamba. Dialects: Southern Gikuyu (Kiambu, Southern Murang'a), Ndia (Southern Kirinyaga), Gichugu (Northern Kirinyaga), Mathira (Karatina), and Northern Gikuyu (Northern Murang'a, Nyeri).

GOLA – gol, (syn. Gula). Gola is spoken by 8,000 people in Sierra Leone and 99,000 in Liberia. Classification: Niger-Congo, Atlantic-Congo, Atlantic, Southern, Mel, Gola. Dialects in Sierra Leone: De (Deng), Managobla (Gobla), Kongbaa, Kpo, Senje (Sene), Tee (Tege), and Toldil (Toodii); in Liberia: Deng (Todii), Kongba, and Senje. Distinctly different forms of Gola are spoken in Nigeria, Guinea-Bissau and Guinea.

GONJA – gjn, (syn. Ngbanyito). Gonja is spoken by 230,000 people in northern Ghana. Classification: Niger-Congo, Atlantic-Congo, Volta-Congo, Kwa, Nyo, Potou-Tano, Tano, Guang, North Guang. Dialects: Gonja, Choruba (Choroba).

GOURMANCHÉMA – gux, (syn. Gourma, Gourmantche, Gurma, Migulimancema, Goulmacema, Gulmancema, Gulimancema, Gourmantché). Gourmanchéma is spoken by 600,000 of the Gourounsi people of Burkina Faso and there are 812,000 speakers in all countries. Classification: Niger-Congo, Atlantic-Congo, Volta-Congo, North, Gur, Central, Northern, Oti-Volta, Gurma. Dialects: Northern Gourmanchéma, Central Gourmanchéma, and Southern Gourmanchéma.

GOURMANTCHÉ – fuh, (syn. Northeastern Burkina Faso Fulfulde). Gourmantché is a dialect of Fulfulde spoken by 750,000 people the northeastern corner of Burkina Faso. Classification: Niger-Congo, Atlantic-Congo, Atlantic, Northern, Senegambian, Fulani-Wolof, Fula, East Central. Dialects: Barani (Barain, Baraniire), Gourmantche, Bogandé, Jelgoore, Liptaakoore, Barkoundouba, Seeba-Yaga (Yaaga) Ouhiguyua, and Fada Ngurma.

GREEK – ell, (syn. Ellinika, Grec, Greco, Graecae, Romaic, Neo-Hellenic). Greek is an Indo-European language with written documents dating back 34 centuries. It is spoken by 9,860,000 people in Greece and a total of 12,258,000 for all countries. Classification: Indo-European, Greek, Attic. Dialects: Katharevousa, Dimotiki, and Saracatsan.

GUJARATI – guj, (syn. Gujrathi, Gujerati, Gujerathi). Gujarati is an Indo-Aryan language spoken by 45,479,000 people in India, primarily in the states of Gujarat, Maharashtra, Rajasthan, Karnataka and Madhya Pradesh. Worldwide there are 46,106,000 speakers. Classification: Indo-European, Indo-Iranian, Indo-Aryan, Central Zone, Gujarati. Dialects: Standard Gujarati, Gamadia, Parsi, Kathiyawadi, Kharwa, Kakari, and Tarimuki.

GUN – guw, (syn. Alada, Alada-Gbe, Gun-Alada, Gun-Gbe, Goun, Egun, Gu, Gugbe, Seto-Gbe, Toli-Gbe). Gun is spoken by 243,000 people in Benin; it is also spoken in Nigeria with a total for all countries of 501,000 speakers. Classification: Niger-Congo, Atlantic-Congo, Volta-Congo, Kwa, Left Bank, Gbe, Aja. Dialects: Ajra, Alada, Seto, and Toli.

GURO – goa, (syn. Gouro, Kweni, Lo, Kwéndré). Guro is spoken by 332,000 people in the West Central and Central departments of Côte d'Ivoire. Classification: Niger-Congo, Mande, Eastern, Southeastern, Guro-Tura, Guro-Yaoure.

HANUNOO – hnn, (syn. Hanonoo). Hanunoo is spoken by 13,000 people in southern Oriental Mindoro Province, the Philippines. Classification: Austronesian, Malayo-Polynesian, Meso Philippine, South Mangyan, Hanunoo. Dialects: Gubatnon (Gubat, Sorsogonon), Binli, Kagankan, Waigan, Wawan, and Bulalakawnon.

HARARI – har, (syn. Hararri, Adare, Adere, Aderinya, Adarinnya, Gey Sinan). Harari is spoken by 21,000 of the Harari people in the Harari Region of Ethiopia. Classification: Afro-Asiatic, Semitic, South, Ethiopian, South, Transversal, Harari-East Gurage.

HARSŪSI – hss, (syn. Hersyet, Harsi 'Aforit). Harsūsi is a Semitic language spoken by 1000 to 2,000 of the Harasis people in the Dhofar Province of Oman. Classification: Afro-Asiatic, Semitic, South, South Arabian. Dialects: Similar to Mehri.

HAUSA – hau, (syn. Haussa, Hausawa, Haoussa, Abakwariga, Mgbakpa, Habe, Kado). Hausa is a Chadic language spoken by 18,525,000 people in Nigeria. It also spoken in parts of Benin, Burkino Faso, Cameroon, Ghana, Niger, Sudan and Togo for a total of 24,000,000 first language speakers and 15,000,000 second language speakers in all countries. Classification: Afro-Asiatic, Chadic, West Chadic, West Chadic A, Hausa-Gwandara (A.1). Dialects: Kano, Katagum, Hadejiya, Sokoto, Gobirawa, Adarawa, Kebbawa, Zamfarawa, Katsina, and Arewa.

HAWAIIAN – haw, (syn. 'Olelo Hawai'i, 'Olelo Hawai'i Makuahine). Hawaiian is a Polynesian language spoken by 1000 people primarily on Hawai'i and Ni'ihau but also throughout the Hawaiian Islands. Use of the language has declined appreciably from 37,000 first language speakers in 1900. Classification: Austronesian, Malayo-Polynesian, Oceanic, Polynesian, Nuclear Polynesian,

Eastern Polynesian, Tahitic, Nuclear Tahitic. Dialects: similar to Rarotongan, Tuamotuan, Tahitian, Maori, Marquesan, and Rapa Nui.

HEBREW – heb, (syn. Ivrit). Hebrew is a Semitic language spoken by 4,847,000 people in Israel and a total of 5,055,000 in all countries. Classification: Afro-Asiatic, Semitic, Central, South, Canaanite. Dialects: Standard and Oriental Hebrew.

HERERO – her, (syn. Otjiherero, Ochiherero). Herero is a language of the Bantu family spoken by 113,000 people in Namibia and to a lesser extent in Botswana for a total of 133,000 speakers in all countries. Classificaiton: Niger-Congo, Atlantic-Congo, Benue-Congo, Bantoid, Southern Bantoid, Bantu, Central Bantu, Zone R.30. Dialects: Mbandieru and Kuvale.

HILIGAYNON – hil, (syn. Ilonggo, Illogo, Hiligainon). Hiligaynon is an Austronesian language spoken by 7,000,000 people in the Philippines, primarily in the Iloilo, Negros Occidental, and Capiz provinces of Visayas. There are 11,000,000 total speakers in all countries. Classification: Austronesian, Malayo-Polynesian, Meso Philippine, Central Philippine, Bisayan, Central, Peripheral. Dialects: Hiligaynon, Kawayan, Bantayan, and Kari.

HINDI – hin, (syn. Khari Boli, Khadi Boli). Hindi is a Indo-Aryan language spoken by people in India, primarily in the central and northern parts of the country. There are 500,000,000 to 680,000,000 first language speakers and 120,000,000 to 225,000,000 second language speakers in all countries. Classification: Indo-European, Indo-Iranian, Indo-Aryan, Central zone, Western Hindi, Hindustani. Subdivisions of Hindi include: Western Hindi, Eastern Hindi, Bihari, Pahari, and Rajasthani. Dialects: Kumaoni (Kumauni), Garhwali, Braj, Haryanvi (Bangaru), Bundeli (Bundelkhandi), Kannauji, Kauravi (Vernacular Hindustani), Khari boli, Awadhi, Bagheli, and Chattisgarhi.

HINDUSTANI – hns. Hindustani is spoken in Pakistan, India, Fiji, Guyana, Malaysia, Suriname, Trinidad, Tobago, and Myanmar with 541,000,000 native and 904,000,000 total speakers. The vegetable names listed are in Caribbean Hindustani from Suriname where there are 150,000 speakers. Classification: Indo-European, Indo-Iranian, Indo-Aryan, Eastern zone, Bihari. Dialects of Caribbean Hindustani: Trinidad Bhojpuri and Sarnami Hindustani (Sarnami Hindi, Aili Gaili).

HMONG-DAW – mww, (syn. Hmong-Der, Hmong-White, White Meo, White Miao, Meo Kao, White Lum, Peh Miao, Pe Miao, Chuan Miao, Bai Miao). Hmong-Daw is a form of Hmong spoken by 232,700 of the Hmong people in China. It is also spoken in France, Laos, Thailand and the United States for a total of 514,000 speakers in all countries. Classification: Hmong-Mien, Hmongic, Chuanqiandian. Dialects: Hmong Gu Mba (Hmong Qua Mba, Striped Hmong), Mong Leng, and Petchabun Miao.

HMONG-NJUA – blu, (syn. Chuanqiandian Miao, Chuanchientien Miao, Sichuan-Guizhou-Yunnan Hmong, Tak Miao, Meo, Miao, Western Miao, Western Hmong). Hmong-Njua is a Hmong language spoken by 1,000,000 of the Hmong people in China, primarily where the Guizhou, Sichuan, and Yunnan provinces meet. It is also spoken in French Guiana, Laos, Myanmar, Thailand, Vietnam, and the United States with 1,290,000 speakers in all countries. Classification: Hmong-Mien, Hmongic, Chuanqiandian. Dialects: Xiao Hua Miao (Atse, Small Flowery Miao), Tak Miao (Ching Miao, Green Miao, Blue Miao).

HOPI – hop, (syn. Hopilàvayi, Hopìikwa). Hopi is spoken by 5,000 of the Hopi people in northeastern Arizona. Classification: Uto-Aztecan, Northern Uto-Aztecan, Hopi. Dialects: First Mesa, Mishongnovi, Shipaulovi, and Third Mesa.

HUNGARIAN – hun, (syn. Magyar). Hungarian is a Uralic language spoken by 10,299,000 people in Hungary and by Hungarian minorities in Romania, Serbia, Slovakia, Israel, Ukraine, Croatia, Austria, and Slovenia for a total of 13,611,000 speakers in all countries. Classification: Uralic, Finno-Ugric, Ugric, Hungarian. Dialects: Alföld, West Danube, Danube-Tisza, King's Pass Hungarian, Northeast Hungarian, Northwest Hungarian, Székely, and West Hungarian.

IBALOI – ibl, (syn. Ibaloy, Ibadoy, Inibaloi, Nabaloi, Benguet-Igorot, Igodor). Ibaloi is spoken by 111,000 people primarily in the Benguet and Nueva Vizcaya Provinces of Luzon, the Philippines. Classification: Austronesian, Malayo-Polynesian, Northern Philippine, Northern Luzon, South-

Central Cordilleran, Southern Cordilleran, Pangasinic, Benguet, Ibaloi-Karao. Dialects: Daklan, Kabayan, and Bokod.

IBANAG – ibg, (syn. Ybanag). Ibanag is spoken by 500,000 of the Ibanag people in northern Luzon, the Philippines. It is also spoken by immigrants in the middle East, United Kingdom and the United States. Classification: Austronesian, Malayo-Polynesian, Northern Philippine, Northern Luzon, Northern Cordilleran, Ibanagic, Ibanag. Dialects: Northern Ibanag and Southern Ibanag.

ICELANDIC – isl, (syn. Íslenska). Icelandic is spoken by 230,000 people in Iceland. It is also spoken in Canada, Denmark, Norway and the United States for a total of 320,000 speakers in all countries. Classification: Indo-European, Germanic North, West Scandinavian. Dialects: none.

ICEVE-MACI – bec, (syn. Icheve, Becheve, Bechere, Bacheve, Ochebe, Ocebe, Ocheve, Utse, Utser, Utseu). Iceve-Maci is spoken by 7,000 people in Cameroon and a total of 12,000 in all countries. Classification: Niger-Congo, Atlantic-Congo, Volta-Congo, Benue-Congo, Bantoid, Southern, Tivoid. Dialects: Icheve (Bacheve) and Oliti (Maci, Matchi, Oliti-Akwaya, Olithi, Olit, Kwaya, Akwaya, Motom, Motomo, Ihekwot).

IFUGAO – *ifc*, Ifugao is a cluster of 4 languages spoken by about 130,000 people on Luzon, the Philippines. Classification: Austronesian, Malayo-Polynesian, Northern Philippine, Northern Luzon, South-Central Cordilleran, Nuclear Cordilleran, Ifugao. The four languages are Amganad, Batad, Mayoyao, and Tuwali Ifugao. The precise language was not specified.

IGBO – ibo, (syn. Ibo). Igbo is spoken predominately by the Igbo people in Nigeria where there are 18,000,000 speakers in the southeastern and southern areas of the country. It is also spoken in parts of Cameroon, Equatorial Guinea, and Gabon with a total of 20 to 25 million speakers in all countries. Classification: Niger-Congo, Atlantic-Congo, Volta-Congo, Benue-Congo, Igboid, Igbo. Dialects: Owerri (Isuama), Onitsha, Umuahia, Orlu, Ngwa, Afikpo, Nsa, Oguta, Aniocha, Eche, Egbema, Oka (Awka), Bonny-Opobo, Mbaise, Nsuka, Ohuhu, and Unwana.

ILOCANO – ilo, (syn. Iloko, Ilokano). Ilocano is an Austronesian language spoken by 8,000,000 people predominately in northwestern Luzon. It is also spoken in the United States with a total of 10,000,000 speakers in all countries. Classification: Austronesian, Malayo-Polynesian, Northern Philippine, Northern Luzon, Ilocano.

INDONESIAN – ind, (syn. Bahasa Indonesia). Indonesian is the national language and is spoken by 240,000,000 Indonesians. It is one of the most widely spoken languages in the world with 250,000,000 speakers in all countries. Indonesian is also spoken in East Timor and by immigrants in the Philippines, Malaysia, Singapore, Thailand, Brunei, Australia, Netherlands and Saudi Arabia. Classification: Austronesian, Malayo-Polynesian, Malayic, Malayan, Local Malay.

INUKTITUT – ike, (syn. Eastern Canadian Eskimo, Eastern Arctic Eskimo, Inuit). Inuktitut is a macrolanguage comprised of two languages, Eastern Canadian Inuktitut (ike) and Western Canadian Inuktitut (ikt), within which there are a number of dialects. Eastern Canadian Inuktitut is spoken by 14,000 people in Quebec and Labrador. Collectively there are 35,000 Inukitut speakers (both languages). Classification (Eastern): Eskimo-Aleut, Eskimo, Inuit. Dialects: Baffinland Eskimo, Labrador Eskimo and Quebec Eskimo.

INUKTITUT, WEST GREENLANDIC – kal, (syn. West Greenlandic Eskimo, Greenlandic, Kalaallisut). There are 54,000 speakers of the three dialects of Greenlandic Inuktitut, predominately in Greenland but also in Denmark; approximately 44,000 speak the West dialect. Classification: Eskimo-Aleut, Eskimo, Inuit, Greenlandic. Dialects: West Greenlandic, East Greenlandic, Polar Eskimo (North Greenlandic, Thule Inuit).

ISNAG – isd, (syn. Dibagat-Kabugao-Isneg, Isneg, Maragat). There are 30,000 speakers of Isnag in the Apayao Province of Luzon, the Philippines. Classification: Austronesian, Malayo-Polynesian, Northern Philippine, Northern Luzon, Northern Cordilleran, Ibanagic, Isnag. Dialects: Bayag, Dibagat-Kabugao, Calanasan, and Karagawan (Daragawan), Talifugu-Ripang (Tawini).

ITALIAN – ita, (syn. Italiano). Italian is a Romance language spoken by 60,000,000 people in Italy and a total of 70,000,000 worldwide. Other locations where Italian is spoken are Switzerland,

San Marino, Vatican City, Slovenia and Croatia and to a lesser extent in a wide range of countries around the world. Classification: Indo-European, Italic, Romance, Italo-Western, Italo-Dalmatian. Dialects: Tuscan, Abruzzese, Pugliese, Umbrian, Laziale, Central Marchigiano, Cicolano-Reatino-Aquilano, and Molisano.

ITAWIT – itv, (syn. Itawis, Tawit, Itawes). Itawit is spoken by 134,000 people on Luzon, the Philippines. Classification: Austronesian, Malayo-Polynesian, Northern Philippine, Northern Luzon, Northern Cordilleran, Ibanagic, Ibanag. Dialects: Malaweg (Malaueg), and Itawis.

IVATAN – ivv. Ivatan is spoken by 35,000 people primarily on the Batanes Islands in the Philippines but also in other areas. Classification: Austronesian, Malayo-Polynesian, Northern Philippine, Bashiic-Central Luzon-Northern Mindoro, Bashiic, Ivatan. Dialects: Itbayaten, Basco Ivatan, and Southern Ivatan.

IZON – ijc, (syn. Izo, Uzo, Ijo, Ijaw, Central-Western Ijo, Ijo-Izon). Izon is spoken by 1,000,000 people in Nigeria. Classification: Niger-Congo, Atlantic-Congo, Ijoid, Ijo, West Ijo. Dialects: Arogbo, Furupagha, Egbema, West Olodiama, Oporoza (Gbaranmatu), Ogulagha, Iduwini, Ikibiri, Ogboin, West Tarakiri, Kabo (Kabowei), Kumbo (Kumbowei), Mein, Operemo, Tuomo, Ogbe Ijo, Gbarain, Kolokuma-Opokuma, Ekpetiama, Apoi, Koluama, Basan (Bassan), East Olodiama, East Tarakiri, Oyiakiri, Oporomo (Oporoma), and Bumo (Boma).

JAPANESE – jpn, Japanese is spoken by 121,050,000 people in Japan and a total of 130,000,000 in all countries. Classification: Japanese. There are dozens of dialects that have been separated into three primary categories: Tokyo-type, Kyoto-Osaka-type, and Kyūshū-type dialects.

JAVANESE – jav, (syn. Jawa, Djawa). Javanese is spoken by 75,200,000 people on the island of Java in Indonesia. It is also spoken in Peninsular Malaysia, Suriname, and New Caledonia for a total of 75,508,000 speakers in all countries. Classification: Austronesian, Malayo-Polynesian, Javanese. Dialects: Jawa Halus, Cirebon (Tjirebon, Cheribon), Tegal, Indramayu, Solo, Tembung, Pasisir, Surabaya, Malang-Pasuruan, Banten, and Manuk.

JOLA-FONYI – dyo, (syn. Diola, Kújoolaak Kati Fooñi, Kujamataak, Jola-Fogny, Diola-Fogny, Dyola, Jóola, Jola, Yola). Jola-Fonyi is spoken by 292,000 people in Senegal with a total of 358,000 speakers in all countries. Classification: Niger-Congo, Atlantic-Congo, Atlantic, Northern, Bak, Jola, Jola Proper, Jola Central, Jola-Fonyi. Dialects: Buluf, Fonyi, Kombo, Kalounaye, Narang, Gusilay, Kwatay, Karon, Mlomp, Kerak, Ejamat, and Bayot.

JULA – dyu, (syn. Dyula, Dyoula, Diula, Dioula, Djula). Jula is spoken by 1,000,000 people in Burkina Faso; it is also spoken in Côte d'Ivoire with a total of 1,229,000 speakers in all countries. Classification: Niger-Congo, Mande, Western, Central-Southwestern, Central, Manding-Jogo, Manding-Vai, Manding-Mokole, Manding, Manding-East, Northeastern Manding, Bamana.

JUNKUN TAKUM – jbu, (syn. Diyi, Njikum, Jukun). Junkun Takum is spoken as a second language by 40,000 people in Nigeria. Classification: Niger-Congo, Atlantic-Congo, Volta-Congo, Benue-Congo, Platoid, Benue, Jukunoid, Central, Jukun-Mbembe-Wurbo, Jukun. Dialects: Takum and Donga (Akpanzhi).

KABIYÉ – kbp, (syn. Kabre, Cabrai, Kabure, Kabye, Cabrais). Kabiyé is a Grusi language spoken by 700,000 people in northern Togo and around 30,000 in Benin and Ghana. Classification: Niger-Congo, Atlantic-Congo, Volta-Congo, North, Gur, Central, Southern, Grusi, Eastern. Dialects: Kewe, Kijang, Lama-Tissi, and Boufale.

KALAGAN – kqe. Kalagan is spoken by 21,000 people in Davao del Sur and Davao Oriental Provinces, the Philippines. Classification: Austronesian, Malayo-Polynesian, Meso Philippine, Central Philippine, Mansakan, Western. Dialects: Isamal, Piso, Tumuaong, Lactan.

KALINGA, BUTBUT – kyb, (syn. Butbut). Butbut Kalinga is spoken by 8,000 people on Luzon, the Philippines. Classification: Austronesian, Malayo-Polynesian, Northern Philippine, Northern Luzon, South-Central Cordilleran, Central Cordilleran, Kalinga-Itneg, Kalinga.

KANKANAEY – kne, (syn. Central Kankanaey, Kankanai, Kankanay). Kankanaey is spoken by 150,000 people on Luzon, the Philippines. Classification: Austronesian, Malayo-Polynesian,

Northern Philippine, Northern Luzon, South-Central Cordilleran, Central Cordilleran, Nuclear Cordilleran, Bontok-Kankanay, Kankanay. Dialects: Mankayan-Buguias, Kapangan, Bakun-Kibungan, and Guinzadan.

KANNADA – kan, (syn. Kanarese, Canarese, Banglori, Madrassi). Kannada is a Dravidian language spoken by 35,346,000 people primarily in the Karnataka state of India. It is also spoken by immigrants in Australia, Singapore, the United Arab Emirates, and the United States for a total of 44,000,000 speakers in all countries. Classification: Dravidian, Southern, Tamil-Kannada, Kannada. There are around 20 dialects spoken, e.g. Kundagannada, Nadavar-Kannada, Havigannada, Are Bhashe, Soliga, Badaga, Gulbarga Kannada, Dharawad Kannada, Chitradurga Kannada.

KANURI – kby, (syn. Manga, Kanouri, Kanoury). Kanuri is a group of languages spoken by approximately 4,000,000 people in Nigeria, Niger, Chad, Cameroon and to a very limited extent elsewhere. The specific language was not designated, though it is probably Manga Kanuri (kby). Classification: Nilo-Saharan, Saharan, Western, Kanuri. Dialects: Manga and Dagara.

KAREN – *kar*. Karen represents a group of languages spoken by 3,000,000 people in Myanmar and along the border in Thailand. There are 3 main branches of Karen (Sgaw, Pwo, and Pa'o) and 17 Karen languages are spoken in Thailand. The exact language of the vegetables listed was not indicated. Classification: Sino-Tibetan, Tibeto-Burman, Karen.

KASHMIRI – kas, (Keshur, Kaschemiri, Cashmiri, Cashmeeree, Kacmiri). Kashmiri is spoken by 4,391,000 people, primarily the Kashmiri in India but also a small number in Pakistan for a total of 4,611,000 speakers in all countries. Classification: Indo-European, Indo-Iranian, Indo-Aryan, Northwestern zone, Dardic, Kashmiri. Dialects: Bakawali, Bunjwali, Standard Kashmiri, Kishtwari (Kashtawari, Kistwali, Kashtwari, Kathiawari), Miraski, Poguli, Rambani, Riasi, Shah-Mansuri, Siraji of Doda, Siraji-Kashmiri, Zayoli, and Zirak-Boli.

KATI – bsh, (syn. Bashgali, Kativiri, Nuristani). Kati is spoken by 15,000 people in Afghanistan and a small number elsewhere for a total of 18,700 in all countries. Classification: Indo-European, Indo-Iranian, Indo-Aryan, Nuristani. Dialects: Eastern Kativiri (Shekhani), Western Kativiri, and Mumviri.

KAZAKH – kaz, (syn. Kazak, Kaisak, Kosach, Qazaq). Kazakh is spoken by 5,293,000 people in Kazakhstan where it is the official state language and by a total of 10,000,000 people in all countries, especially China, Mongolia, Afghanistan, Tajikistan, Turkey, Turkmenistan, Ukraine, Uzbekistan, Russia, and Iran. Classification: Altaic, Turkic, Western, Aralo-Caspian. Dialects: Northeastern Kazakh, Southern Kazakh and Western Kazakh.

KHMER, CENTRAL – khm, (syn. Khmer, Cambodian). Central Khmer is spoken by 12,110,000 Cambodian people and a total of 13,276,000 in all countries. Classification: Austro-Asiatic, Mon-Khmer, Eastern Mon-Khmer, Khmer. Dialects: Central Khmer is distinct from Northern Khmer spoken in Thailand.

KIOWA – kio, Kiowa is a Native American language of the Southern Plains spoken by 1000 of the Kiowa Tribe in Oklahoma. It is considered an endangered language with the number of speakers declining. Classification: Kiowa Tanoan, Kiowa-Towa, Kiowa.

KIRGHIZ – kir, (syn. Kara-Kirgiz, Kirgiz). Kirghiz is a Turkic language spoken by 2,448,000 people in Kyrgyzstan and by a total of 4,000,000 in all countries, in particular Afghanistan, Kazakhstan, Tajikistan, Turkey, Uzbekistan, Pakistan, Russia and Xinjiang, China. Classification: Altaic, Turkic, Western, Aralo-Caspian. Dialects: Northern and Southern Kirghiz.

KIRIBATI – gil, (syn. I-Kiribati, Gilbertese, Ikiribati). Kiribati is spoken by 58,320 people in Kiribati. It is also spoken in Fiji, Marshall Islands, Nauru, Solomon Islands, Tuvalu and Vanuatu for a total 102,000 speakers in all countries. Classification: Austronesian, Malayo-Polynesian, Central-Eastern, Eastern Malayo-Polynesian, Oceanic, Central-Eastern Oceanic, Remote Oceanic, Micronesian, Micronesian Proper, Ikiribati. Dialects: Banaban.

KISI, SOUTHERN – kss, (Kisi, Gissi, Kissien). Southern Kisi is spoken by 85,000 people in Sierra Leone and 115,000 in Liberia. Classification: Niger-Congo, Atlantic-Congo, Atlantic, Southern,

Mel, Bullom-Kissi, Kissi. Dialects: Luangkori, Tengia, and Warn. It differs from Northern Kissi which is spoken in Guinea.

KISSI, NORTHERN – kqs, (syn. Gizi, Kisi, Kissien, Kisie). Northern Kissi is spoken by 40,000 people in Sierra Leone and 286,500 in Guinea. Classification: Niger-Congo, Atlantic-Congo, Atlantic, Southern, Bullom-Kissi, Kissi. Dialects: Liaro, Kama, Teng, and Tung. Similar to Southern Kissi spoken in Liberia and Sierra Leone.

KONKANI – knn, (syn. Konkan Standard, Bankoti, Kunabi, North Konkan, Central Konkan, Concorinum, Cugani, Konkanese). Konkain is Indo-Aryan language spoken by 4,000,000 people in India. It is the official language of the state of Goa. Konkani is considered to be a macrolanguage comprising Konkani (knn) and Goan Konkani (gom). Classification: Indo-European, Indo-Iranian, Indo-Aryan, Southern zone, Konkani. Dialects: Agari of Kolaba, Parabhi (Kayasthi, Damani), Koli, Kiristav, Dhanagari, Bhandari, Thakuri (Thakari, Thakri, Thakua, Thakura), Karhadi, Sangamesvari (Bakoti, Bankoti), Ghati (Maoli), and Mahari (Dhed, Holia, Parvari).

KONKOMBA – xon, (syn. Likpakpaln, Kpankpam, Kom Komba). Konkomba is spoken by 500,000 of the Konkomba people in Ghana and in Togo for a total of 550,000 speakers in all countries. Classification: Niger-Congo, Atlantic-Congo, Volta-Congo, North, Gur, Central, Northern, Oti-Volta, Gurma. Dialects: Lichabool-Nalong, Limonkpel, Linafiel, Likoonli, and Ligbeln.

KONO – kno, (syn. Konnoh). Kono is spoken by 190,000 of the Kono people in the Eastern Province of Sierra Leone. Classification: Niger-Congo, Mande, Western, Central-Southwestern, Central, Manding-Jogo, Manding-Vai, Vai-Kono. Dialects: Northern Kono (Sando), and Central Kono (Fiama, Gbane, Gbane Kando, Gbense, Gorama Kono, Kamara, Lei, Mafindo, Nimi Koro, Nimi Yama, Penguia, Soa, Tankoro, Toli).

KOONGO – kng, (syn. Kongo, Kikongo, Congo, Kikoongo). Koongo is spoken by 1,000,000 people in the Democratic Republic of the Congo, predominately in the Bas-Congo Province. Classification: Niger-Congo, Atlantic-Congo, Volta-Congo, Benue-Congo, Bantoid, Southern, Narrow Bantu, Central, H, Kongo (H.10). Dialects: South Congo, Central Kongo, West Kongo (Fiote, Fioti), Bwende (Buende), East Kongo, Southeast Kongo, and Nzamba (Dzamba).

KOREAN – kor, (syn. Hanguohua, Hanguk Mal). Korean is spoken by 42,000,000 people in South Korea, 20,000,000 in North Korea and a total of 67,019,000 speakers in all countries. Classification: Language isolate (no discernable genealogical relationship with other languages). Dialects: Seoul (Kangwondo, Kyonggido), Ch'ungch'ongdo (North Ch'ungch'ong, South Ch'ungch'ong), Kyongsangdo (North Kyongsangdo, South Kyongsangdo), Chollado (North Chollado, South Chollado), and Cheju Island.

KORO – kfo, (syn. Koro Jula). Koro is spoken by 40,000 people in Cote d'Ivorie. Classification: Niger-Congo, Mande, Western, Central-Southwestern, Central, Manding-Jogo, Manding-Vai, Manding-Mokole, Manding, Manding-East, Southeastern Manding, Maninka-Mori.

KRIO – kri, (syn. Creole, Patois). Krio is spoken by 472,000 of the Krio people in Sierra Leone and a total of 481,000 speakers in all countries. Krio is a Creole language of the descendants of freed slaves from the West Indies, Britain and the United States. Classification: Creole, English based, Atlantic, Krio. Dialects: Aku.

KUNG-EKOKA – knw, (syn. Ekoka-!Xû, Kung, !Kung, !Ku, !Xu, !Hu, Qxü, !Xun, !Khung, !Xung). Kung-Ekoka is spoken by 1,757 of the Kung Bushmen in Namibia and a total of 6,899 in all countries. Classification: Khoisan, Southern Africa, Northern. Dialect: Akhoe.

KURANKO – knk, (syn. Koranko). Kuranko is spoken by 250,000 of the Kuranko people in the Northern Province of Sierra Leone. It is also spoken in Guinea for a total of 305,000 speakers in all countries. Classification: Niger-Congo, Mande, Western, Central-Southwestern, Central, Manding-Jogo, Manding-Vai, Manding-Mokole, Mokole. Dialects: Barrawa, Nieni, Mankaliya, Sambaya, Ney, Sengbe, and Mongo.

KURDISH, NORTHERN – kmr, (syn. Kurmanji, Kurmancî, Kirmancî, Kermancî, Kurdi, Kurdî). Northern Kurdish is spoken by 3,950,000 people in Turkey. It is also spoken in Iraq, Iran and to a lesser extent elsewhere for a total of 9,113,000 speakers in all countries. Classification: Indo-European, Indo-Iranian, Iranian, Western, Northwestern, Kurdish. Dialects: Boti (Botani), Marashi, Ashiti, Bayezidi, Hekari, and Shemdinani.

KWANGALI – kwn, (syn. Sikwangali, Rukwangali, Kwangari, Kwangare). Kwangali is a Bantu language spoken by 73,000 of people in the Okavango Region of Namibia. It is also spoken in Angola for a total of 84,000 in all countries. Classification: Niger-Congo, Atlantic-Congo, Volta-Congo, Benue-Congo, Bantoid, Southern, Narrow Bantu, Central, K, Kwangwa (K.40) Dialect: Sambyu (Shisambyu, Sambiu, Sambio).

KWANYAMA – kua, (syn. Ochikwanyama, Kuanyama, Ovambo, Humba, Kwanjama, Kwancama, Otjiwambo, Owambo). Kwanyama is spoken by 421,000 people in Angola and 250,000 in Namibia. Classification: Niger-Congo, Atlantic-Congo, Benue-Congo, Bantoid, Southern, Bantu, Central, Zone R, Kwanyama. Dialects: similar to Ndonga and Kwambi.

KWÉYÒL – acf, (Patwa, Patois, Saint Lucian Creole French, Lesser Antillean Creole French). Kwéyòl is a French Creole spoken by 158,000 people in Saint Lucia. It is also spoken in Dominica, France, Grenada, Guyana, Trinidad, and Tobago for a total of 357,000 speakers in all countries. Classification: Creole, French based.

LADINO – lad, (syn. Judeo Spanish, Judæo-Spanish, Judæo-Espagnol, Judeoespañol, Sefardi, Dzhudezmo, Judezmo, Spanyol, Haquetiya). Ladino or Judæo-Spanish is a Romance language spoken by 100,000 Sephardic Jews in Israel and in Greece, Turkey and a number of countries around the world for a total of 110,000 speakers in all countries. Classification: Indo-European, Italic, Romance, Italo-Western, Western, Gallo-Iberian, Ibero-Romance, West Iberian, Castilian. Dialects: Judezmo (Judyo, Jidyo), Ladino, and Haquetiya (Haketia, Haketiya, Hakitia).

LAHU – lhu, (Lohei, Lahuna, Laku, Kaixien, Namen, Mussuh, Muhso, Musso, Mussar, Moso). Lahu is a Tibeto-Burman language spoken by 411,000 of primarily the Laho people in the Yunnan Province of China. It is also spoken in Myanmar, Thailand, and Laos for a total of 577,000 speakers in all countries. Classification: Sino-Tibetan, Tibeto-Burman, Lolo-Burmese, Loloish, Southern, Akha, Lahu. Dialects: Na (Black Lahu, Musser Dam, Northern Lahu, Loheirn), Nyi (Red Lahu, Southern Lahu, Musseh Daeng, Luhishi, Luhushi), and Shehleh.

LAKOTA – lkt, (Lakhota, Teton, Teton Sioux). Lakota is a Siouan language spoken by 6,000 of the Lakota people found in North and South Dakota, northern Nebraska and Montana and southern Minnesota. Classification: Siouan, Siouan Proper, Central, Mississippi Valley, Dakota. Dialect: Brulé.

LAO – lao, (syn. Laotian Tai, Laotian, Phou Lao, Eastern Thai, Lum Lao, Lao Wiang, Lao Kao, Rong Kong, Tai Lao, Lao-Tai, Lào, Lao-Lum, Lao, Noi). Lao is a member of the Kradai language family spoken primarily by 3,000,000 the Lao people in Laos. It is also spoken in the Isan Region in northeastern Thailand, a number of other countries (e.g. France, Canada, China, Australia, United States) and as a second language by a number of ethnic groups in Thailand for a total of 3,188,000 speakers. Classification: Tai-Kadai, Kam-Tai, Be-Tai, Tai-Sek, Tai, Southwestern, Lao-Phutai. Dialects: Luang Prabang, Vientiane (Wiang Jan), Savannakhet (Suwannakhet), Pakse, Lao-Kao, Lao-Khrang.

LATVIAN – lav, (syn. Lettish, Lettisch). Latvian is a Baltic Language spoken by 1,394,000 people in Latvia and in a number of other countries for a total of 1,544,000 speakers. Classification: Indo-European, Baltic, Eastern. Dialects: West Latvian (Central Latvian) and East Latvian (High Latvian, Latgalian).

LIMBA– lia, (syn. West Central Limba, Yimba, Yumba). West Central Limba is spoken by 335,000 people in Sierra Leone. Other West African Limba languages include East Limba (lma) spoken in Sierra Leone and Limba or Malimba (mzd) spoken in Cameroon. Classification (West Central Limba): Niger-Congo, Atlantic-Congo, Atlantic, Southern, Limba. Dialects: Western Limba (Tonko, Sela) and Central (Tamiso, Gbongogbo).

LISU – lis, (syn. Lissu, Lisaw, Li-Shaw, Li-Hsaw, Lu-Tzu, Lesuo, Li, Lishu, Liso, Leisu, Leshuoopa, Loisu, Southern Lisu, Yao Yen, Yaw-Yen, Yaw Yin, Yeh-Jen, Chung, Cheli, Chedi, Lip'a, Lusu, Khae). Lisu is spoken by 580,000 people in China and 723,000 in all countries. Classification: Sino-Tibetan, Tibeto-Burman, Lolo-Burmese, Loloish, Northern, Lisu. Dialects: Hau Lisu (Flowery Lisu), Pai Lisu (White Lisu), and Lu Shi Lisu.

LITHUANIAN – lit, (syn. Lietuviskai, Lietuvi, Litovskiy, Litewski, Litauische). Lithuanian is a Baltic language related to Latvian and Old Prussian that is spoken by 2,955,000 people in Lithuania and a total of 3,125,000 in all countries. Classification: Indo-European, Baltic, Eastern. Dialects: Aukshtaitish (Aukshtaichiai, Aukstaitiskai, Highland Lithuanian), Dzukish (Dzukiskai), Shamaitish (Samogitian, Zhemaitish, Zemaitis, Zemaitiskai, Zemachiai, Lowland Lithuanian), and Suvalkietishai.

LOKO – lok, (syn. Landogo). Loko is a Mande language spoken by 115,000 of the Loko people in Sierra Leone. Classification: Niger-Congo, Mande, Western, Central-Southwestern, Southwestern, Mende-Loma, Mende-Bandi, Mende-Loko. Dialects: Magbiambo, Gbendembu, Ngoahu, Nagbanmba, Sanda, Laia, Libisegahun, Koya, Ribbi, and Buya.

LOMWE, MALAWI – lon, (syn. Nguru, Anguru). Lomwe is spoken by 250,000 people in Malawi. Classification: Mixed language, Makhua-Nyanja. Dialects: Different (not intelligible) from the Lomwe spoken in Mozambique.

LUO – *luo*. Luo is a group of Nilo-Saharan langauges spoken over a wide area of Africa. The vegetable names listed are for a currently undetermined Southern Luo language spoken in Uganda. Possible languages include Adhola (adh – syn. Dhopadhola, Japadhola, Ludama), Kumam (kdi – syn. Kuman, Kumum, Ikumama, Akum, Kwania) and several Luo-Acholi languages [i.e. Acholi – ach (syn. Acoli, Atscholi, Shuli, Gang, Lwo, Lwoo, Akoli, Acooli, Log, Acoli, Dok Acoli), Alur – alz (syn. Lur, Luri, Aloro, Alua, Alulu, Dho Alur, Jo Alur), and Lango – laj (syn. Lwo, Lwoo, Leb-Lano).

LUXENBOURGEOIS – ltz, (syn. Lëtzebuergesch, Luxemburgish, Luxemburgian, Luxembourgish, Letzhurgisch, Moselle Franconian, Frankish). Luxenbourgeois is a Germanic language spoken by 300,000 people in Luxembourg and a total of 390,000 in all countries (e.g. Belgium, France, Germany, United States). Classification: Indo-European, Germanic, West, High German, German, Middle German, West Middle German, Moselle Franconian.

LUYIA – luy, (syn. Luluyia, Luhya). Luyia is spoken by 3,418,000 people in Kenya. It is used by the Kabras, Wanga, Marama and a number of people in the Lake Victoria area of the Western province with a total of 3,643,000 speakers in all countries. Classification: Niger-Congo, Atlantic-Congo, Volta-Congo, Benue-Congo, Bantoid, Southern, Narrow Bantu, Central, J, Masaba-Luyia (J.30). Dialects: Kisa (Shisa, Lushisa), Marama, Wanga (Hanga, Luhanga, Oluhanga, Kawanga, Oluwanga), Tsotso, Saamia (Samia, Olusamia, Lusamia, Lusaamia, Samya), West Nyala (Nyala-B), Khayo, Songa, Marachi, and Kabras..

MACEDONIAN – mkd, (syn. Makedonski, Slavic, Macedonian Slavic). Macedonian is a South Slavic language spoken by 1,386,000 people in Macedonia and a total of 1,598,000 in all countries (e.g. Greece, Australia, Serbia, Albania, Germany, France, Italy, United States, Canada, Turkey). Classification: Indo-European, Slavic, South, Eastern. Dialects: Northern Macedonian, Southeastern Macedonian, and Western Macedonian.

MADURA – mad, (syn. Madurese, Madhura, Basa Mathura). Madura is spoken by 13,695,000 Madurese on the Island of Madura, Sapudi Islands and the northern coastal area of east Java in Indonesia. Classification: Austronesian, Malayo-Polynesian, Madurese. Dialects: Bawean (Boyanese), Bangkalan (Bangkalon), Pamekesan (Pamekasan), Sampang, Sapudi, and Sumenep.

MAGUINDANAO – mdh, (syn. Magindanaon, Magindanaw). There are 1,000,000 speakers of Maguindanao, predominately the Magindanaon people in southwest Mindanao, the Philippines. Classification: Austronesian, Malayo-Polynesian, Southern Philippine, Danao, Magindanao. Dialects: Laya, Ilud, Biwangan, Sibugay, Iranun (Ilanon, Illanon, Ilanum, Iranon), and Tagakawanan.

MAITHILI – mai, (Maitli, Maitili, Methli, Tirahutia, Bihari, Tirhuti, Tirhutia, Apabhramsa). Maithili is an Eastern Indic language spoken by 22,000,000 people in the Indian state of Bihar and the eastern Terai region of Nepal. There are a total of 24,797,000 speakers in all countries. Classification: Indo-European, Indo-Iranian, Indo-Aryan, Eastern zone, Bihari. Dialects: Standard Maithili, Southern Standard Maithili, Eastern Maithili (Khotta, Kortha, Kortha Bihari), Western Maithili, Jolaha, Central Colloquial Maithili (Sotipura), Kisan, and Dehati.

MAKHUWA – vmw, (syn. Central Makhuwa, Makhuwa-Makhuwana, Macua, Emakua, Makua, Makoane, Maquoua, Makhuwwa of Nampula, Emakhuwa, Emakchu). Makhuwa is a Bantu language, the most important indigenous language spoken in Mozambique. It is spoken by 2,500,000 of the Makua people primarily in the Nampula province. Classification: Niger-Congo, Atlantic-Congo, Volta-Congo, Benue-Congo, Bantoid, Southern, Narrow Bantu, Central, P, Makua. Dialects: Emwaja, Enaharra (Maharra, Nahara, Emathipane), Enyara, Central Makua (Makhuwana, Makuana, Emakhuwana), Empamela (Nampamela), and Enlai (Mulai).

MALAGASY – plt, (syn. Plateau Malagasy). Malagasy is spoken by 5,940,000 people in Madagascar and a total of 5,948,000 speakers in all countries (e.g. Comoros, Réunion, Mayotte). Classification: Austronesian, Malayo-Polynesian, Barito, East, Malagasy. Dialects: Merina, Betsileo, Sihanaka, Bezanozano, and Tanala.

MALAY – mly, (syn. Bahasa Malaysia, Bahasa Malayu, Malayu, Melaju, Melayu, Standard Malay). Malay is from the Malayo-Polynesian branch of the Austronesia language family and is spoken by 7,181,000 people in Malaysia and to a considerable extent in other countries in the region (e.g. Brunei, Indonesia, Malaysia, Singapore) for a total of 17,604,000 speakers in all countries. Classification: Austronesian, Malayo-Polynesian, Malayic, Malayan, Local Malay. Dialects: Trengganu, Kelantan, Kedah, Perak (Southern Malay), Sarawak Malay, and Bazaar Malay (Low Malay, Pasar Malay, Pasir Malay, Trade Malay).

MALAYALAM – mal, (syn. Alealum, Malayalani, Malayali, Malean, Maliyad, Mallealle, Mopla). Malayalam is a Dravidian language spoken by 35,351,000 people in south India, predominately in the Kerala state. It is also spoken in Bahrain, Fiji, Israel, Malaysia, Qatar, Singapore, United Arab Emirates and the United Kingdom for a total of 35,757,000 speakers in all countries. Classification: Dravidian, Southern, Tamil-Kannada, Tamil-Kodagu, Tamil-Malayalam, Malayalam. Dialects: Malabar, Nagari-Malayalam, Malayalam, Southern Kerala, Central Kerala, Northern Kerala, Kayavar, Namboodiri, Moplah, Pulaya, Nasrani, and Nayar.

MALIMBA – mzd, (syn. Mulimba, Mudima, Limba, Lemba). Malimba is spoken by 2,230 people in Cameroon. Classification: Niger-Congo, Atlantic-Congo, Volta-Congo, Benue-Congo, Bantoid, Southern, Narrow Bantu, Northwest, A, Duala (A.20). Dialects: similar to Duala.

MALTESE – mlt, (syn. Malti). Maltese is spoken by 300,000 people on Malta and along with English are the official languages of the Republic of Malta. There are a total of 371,000 speakers in all countries. Classification: Afro-Asiatic, Semitic, Central, South, Arabic. Dialects: Standard Maltese, Port Maltese, Rural West Maltese, Rural East Maltese, Rural Central Maltese, Zurrieq, and Gozo.

MANDING-BAMBARA – *m-bam*. Burkill lists a series of vegetables used in Mali under the language title of Manding-Bambara. Since Bambara or Bamananka is a Manding language spoken by 2,700,000 people in Mali, the assumption is that the names are Bamananka, though they have been left designated as Manding-Bambara in the text. The classification for Bamananka (Bambara) is: Niger-Congo, Mande, Western, Central-Southwestern, Central, Manding-Jogo, Manding-Vai, Manding-Mokole, Manding, Manding-East, Northeastern Manding, Bamana.

MANDING-MANDINKA – *m-mnk*, (syn. Mandinque, Mandingo, Manding, Mandé, Socé). Mandinka is a Manding language that is spoken by 453,000 of the Mandina people in the western half of Gambia. It is also spoken in Senegal, Mali, Guinea, Burkina Faso, Sierra Leone, Liberia, Guinea-Bissau and Chad for a total of 5,300,000 speakers in all countries. The language is listed as Manding-Mandinka by Burkill and is so designated in the text. The classification for Mandinka remains in question: Niger-Congo, Mande, Western Mande, Manding, West Manding, Mandinka.

MANDING-MANINKA – *m-emk*, (syn. Maninka, Mande, Southern Maninka, Kankan Maninka, Eastern Malinke). Maninka or Eastern Maninkakan is a member of the Manding language family that is spoken in Guinea by 1,890,000 of the Malinké people and a total of 2,012,000 people in all countries. Eastern Maninkakan is classified as: Niger-Congo, Mande, Western, Central-Southwestern, Central, Manding-Jogo, Manding-Vai, Manding-Mokole, Manding, Manding-East, Southwestern-Manding.

MANDINKA – mnk, (syn. Manding, Mandingo, Mandingue, Mandinque, Mande, Socé). Mandinka is a Mandé language spoken by 96,000 of the Mandinaka people in southwest Senegal. It is also spoken in Mali, The Gambia, Guinea, Ivory Coast, Burkina Faso, Sierra Leone, Liberia, and Guinea-Bissau and Chad for a total of 5,300,000 speakers in all countries. Classification: Niger-Congo, Mande, Western, Central, Southwestern, Central, Manding-Jogo, Manding-Vai, Manding-Mokole, Manding, Manding-West.

MANDJAK – mfv, (syn. Mandjaque, Manjaca, Manjaco, Manjak, Manjaku, Manjack, Mandyak, Majak, Ndjak, Kanyop). Mandjak is spoken by 95,670 people in southwest Senegal. It is also spoken in Guinea-Bissau (170,000), Gambia and France for a total of 285,000 in all countries. Classification: Niger-Congo, Altantic-Congo, Atlantic, Northern, Bak, Manjaku-Papel. Dialects: Bok (Babok, Sarar, Teixiera, Pinto, Tsaam), Likes-Utsia (Baraa, Kalkus), Cur (Churo), Lund, and Yu (Pecixe, Siis, Pulhilh).

MANINKA – *man*, (syn. Maninkakan, Malinke, Malinka). Maninka is a group of closely related languages spoken by the Malinké people in Senegal, Guinea, Mali, Liberia, Sierra Leone and Côte d'Ivoire. There are four languages: Konyanka, Sankaran and Forest Maninka, and Eastern Maninkakan, the latter being the largest. The crops names listed are from Senegal and were designated simply as Maninka by Burkill.

MANINKAKAN, EASTERN – emk, (syn. Maninka, Mande, Southern Maninka, Kankan Maninka, Eastern Malinke). Eastern Maninkakan is spoken by 1,890,000 people in Guinea. It is also spoken in Liberia and Sierra Leone for a total of 2,014,000 speakers in all countries. Classification: Niger-Congo, Mande, Western, Central-Southwestern, Central, Manding-Jogo, Manding-Vai, Manding-Mokole, Manding, Manding-East, Southwestern Manding. Dialects: Bö, Amana (Kourousa), Koulounkalan, Maninka-Mori (Mori), and Wasulu (Wassulu, Wassulunka, Wassulunke).

MANINKAKAN, WESTERN – mlq, (syn. Maninka-Western, Maninga, Malinka, Malinke, Western Malinke). Western Maninkakan is spoken in Senegal by 382,000 people. It is also spoken in Gambia and Mali for a total 525,000 speakers in all countries. Classification: Niger-Congo, Mande, Western, Central-Southwestern, Central, Manding-Jogo, Manding-Vai, Manding-Mokole, Manding, Manding-West. Dialects: Jahanka (Jahanque, Jahonque, Diakkanke, Diakhanke, Kyakanke).

MANKANYA – knf, (syn. Mankanha, Mancanha, Mancagne, Mancang, Bola). Mankanya is spoken by 41,000 people in Guinea-Bissau. It is also spoken in Senegal and Gambia with total of 70,000 speakers in all countries. Classification: Niger-Congo, Atlantic-Congo, Atlantic, Northern, Bak, Manjaku-Papel. Dialects: Burama (Bulama, Buram, Brame) and Shadal (Sadar).

MANOBO – *mno*. Manobo is a cluster of 11 languages spoken by 386,000 people on Mindanao, the Philippines. The specific language of the crops listed was not given. Classification: Austronesian, Malayo-Polynesian, Southern Philippine, Manobo.

MANU'A SAMOAN – *m*-smo. Manu'a Samoan is a variation of Samoan spoken in the Manua Islands. There are 370,000 speakers of Samoan, an Austronesia language which is also spoken in New Zealand, Australia and the United States for a total of 870,000 speakers in all countries. Samoan is classified as: Austronesian, Malayo-Polynesian, Central-Eastern, Eastern Malayo-Polynesian, Oceanic, Central-Eastern Oceanic, Remote Oceanic, Central Pacific, East Fijian-Polynesian, Polynesian, Nuclear, Samoic-Outlier, Samoan.

MAORI – mri, (syn. New Zealand Maori). Maori is an Eastern Polynesian language spoken by 50,000 to 70,000 Maori in New Zealand. Classification: Austronesian, Malayo-Polynesian, Central-Eastern, Eastern Malayo-Polynesian, Oceanic, Central-Eastern Oceanic, Remote Oceanic,

Central Pacific, East Fijian-Polynesian, Polynesian, Nuclear, East, Central, Tahitic. Dialects: North Auckland, South Island, Taranaki, Wanganui, Bay of Plenty, Rotorua-Taupo, and Moriori

MARANAO – mrw, (syn. Ranao, Maranaw). Maranao is spoken by 776,000 people in the Lanao del Norte and Lanao del Sur provinces of Mindanao, the Philippines. Classification: Austronesian, Malayo-Polynesian, Southern Philippine, Danao, Maranao-Iranon.

MARATHI – mar, (syn. Maharashtra, Maharathi, Malhatee, Marthi, Muruthu). Marathi is an Indo-Aryan language spoken by 68,030,000 of the Marathi people in southwestern India. It is also spoken in Mauritius and Israel for a total of 90,000,000 speakers worldwide of which 20,000,000 are second language speakers. Classification: Indo-European, Indo-Iranian, Indo-Aryan, Southern zone. There are 42 dialects of Marathi, the major divisions of which are Ahirani, Khandeshi, Varhadi, Wadvali, Samavedi and Are Marathi.

MARINDUQUE – tgl, (syn. Maeinduque). Marinduque is one of the four main dialects of Tagalog spoken on Marinduque, an island in the Mimaropa region near the main island of Luzon. Tagalog is spoken by 14,486,000 people in the Philippines and a total of 15,900,000 people worldwide. There are 229,000 people on Marinduque; the number of those speaking the Marinduque dialect was not available. Tagalog is classified as: Austronesian, Malayo-Polynesian, Meso Philippine, Central Philippine, Tagalog.

MARQUESAN – mrq/mqm. There are two Marquesan languages spoken on the Marquesan Islands of French Polynesia, North Marquesan (mrq) and South Marquesan (mqm), which are sufficiently distinct to be classed as separate languages. There are 5,500 speakers roughly split between the two languages. Classification: Austronesian, Malayo-Polynesian, Central-Eastern, Eastern Malayo-Polynesian, Oceanic, Central-Eastern Oceanic, Remote Oceanic, Central Pacific, East Fijian-Polynesian, Polynesian, Nuclear, East, Central, Marquesic. Dialects for North Marquesan: Hatutu, Nuku Hiva, Ua Huka, and Ua Pou; South Marquesan: Hiva Oa, Tahuta, and Fatu Hiva.

MARWARI – mve, (syn. Merwari, Rajasthani, Marwari Meghwar, Jaiselmer, Marawar, Marwari Bhil). Marwari is spoken by 13,200,000 people in the Indian states of Rajasthan and Guijarat and a small number (200,000) in Eastern Pakistan. Classification: Indo-European, Indo-Iranian, Indo-Aryan, Central zone, Rajasthani, Marwari. Dialects: Northern Marwari, Southern Marwari, Marwari Bhil, Marwari Meghwar, and Marwari Bhat.

MEHRI – gdq, (syn. Mahri). Mehri is a Modern South Arabian language spoken by 70,643 people in the eastern part of Yemen and western Oman. Classification: Afro-Asiatic, Semitic, South, South Arabian. Dialects: Western Mehri (Mehriyet) and Eastern Mehri (Mehriyot).

MENDE – men, (syn. Boumpe, Hulo, Kossa, Kosso). Mende is spoken by 1,460,000 of the Mende people and other ethnic groups in Sierra Leone and a small number of people in Liberia and elsewhere for a total of 1,480,000 speakers in all countries. Classification: Niger-Congo, Mande, Western, Central-Southwestern, Southwestern, Mende-Loma, Mende-Bandi, Mende-Loko. Dialects: Kpa, Ko, Waanjama, and Sewawa.

MICMAC – mic, (syn. Mi'gmaq, Miigmao, Mi'kmaq, Restigouche). Micmac is an Eastern Algonquian language spoken by 7,310 Míkmaq in eastern Canada. It is also spoken in the United States for a total of 8,510 in both countries. Classification: Algic, Algonquian, Eastern. Dialects: Northern Micmac and Southern Micmac.

MINANGKABAU – min, (syn. Minang, Padang). There are 6,500,000 speakers of Minangkabau predominantly in Sumatra but also found elsewhere in the Indonesian Archipelago. Classification: Austronesian, Malayo-Polynesian, Malayic, Malayan, Para-Malay. Dialects: Agam, Pajokumbuh, Tanah, Si Junjung, Batu Sangkar-Pariangan, Singkarak, Orang Mamak, Ulu, Kerinci-Minangkabau, Aneuk Jamee (Jamee), and Penghulu.

MIZO – lus, (syn. Dulien, Duhlian Twang, Lusai, Lushai, Lusei, Lushei, Lukhai, Lusago, Sailau, Hualngo, Whelngo). Mizo is spoken by 529,000 people in the Mizoram state in eastern India. It is also spoken in Bangladesh and Myanmar with a total of 542,000 speakers in all countries.

Classification: Sino-Tibetan, Tibeto-Burman, Kuki-Chin-Naga, Kuki-Chin, Central. Dialects: Fannai, Mizo, Ngente, Tlau, and Le.

MOHAWK – moh. Mohawk is a member of the Iroquoian family of languages spoken by 3,350 members of the Mohawk nation in the United States and Canada. Classification: Iroquoian, Northern Iroquoian, Five Nations, Mohawk-Oneida. Dialects: Western (Six Nations and Tyendinaga), Central (Ahkwesáhsne), and Eastern (Kahnawà:ke and Kanehsatà:ke).

MONGOLIAN – khk, (syn. Halh, Halh Mongolian, Khalkha Mongolian, Mongol, Central Mongolian). Mongolian is member of the Mongolic language family spoken by 2,329,000 people in Mongolia. It is also spoken in the Inner Mongolia region of China for a total of 2,337,000 speakers in all countries. Classification: Altaic, Mongolian, Eastern, Oirat-Khalkha, Khalkha-Buriat, Mongolian Proper. Dialects: Halh (Khalkha), Dariganga, Khotogoit, Sartul, and Tsongol.

MÒORÉ – mos, (syn. Mole, Moose, More, Moshi, Mossi). Mòoré is spoken by 5,000,000 of the Mossi people in Burkina Faso. It is also spoken in Benin, Côte d'Ivoire, Ghana, Mali, and Togo with a total of 5,036,000 speakers in all countries. Classification: Niger-Congo, Atlantic-Congo, Volta-Congo, North, Gur, Central, Northern, Oti-Volta, Western, Northwest. Dialects: Saremdé, Taolendé, Yaadré, Ouapadoupou, Yaande, Zaore (Joore), and Yana (Yanga, Jaan).

MORISYEN – mfe, (syn. Mauritius Creole French, Kreole, Kreol, Mauritian, Maurysen). Morisyen is a French based Creole spoken by 600,000 people in Mauritius with a total of 604,000 speakers in all countries. Classification: Creole, French based. Dialect: Rodrigues Creole.

MUNDARI – muw, (syn. Mandari, Munari, Munda, Mondari, Horo, Colh, Killi). Mundari is a Munda language spoken by 2,069,000 of the Munda people in east India, Bangladesh, and Nepal. Classification: Austro-Asiatic, Munda, North Munda, Kherwari, Mundari. Dialects: Hasada', Latar, Naguri, Kera', and Bhumij (Sadar Bhumij, Bhumij Munda, Bhumij Thar).

MUSKOGEE – mus, (syn. Muscogee, Creek). Muskogee is a Muskogean language of the American southeast spoken by 4,300 of the Muskogee Nation in Oklahoma, Alabama, Georgia and Florida. Classification: Muskogean, Eastern. Dialects: Creek and Seminole.

NAMA-DAMARA – naq, (syn. Naman, Namakwa, Namaqua, Maqua, Tama, Tamma, Tamakwa, Berdama, Bergdamara, Kakuya Bushman Nasie, Rooi Nasie, Hottentot, Klipkaffer, Klipkaffern, Khoekhoegowap, Khoekhoegowab). Nama-Damara is a dialect of Nama. Nama is spoken by 176,000 people in Namibia. It is also spoken in Botswana and South Africa for a total of 233,000 speakers in all countries. Classification: Khoisan, Southern Africa, Central, Nama. Dialects of Nama: Damara, Sesfontein Damara, Namidama, Central Damara, and Nama.

NAVAJO – nav, (syn. Diné Bizaad, Navaho). Navajo is an Athabaskan language spoken by 149,000 of the Navajo people in the southwestern United States. It is the most widely spoken Native American language in the United States and was used by the Navajo code talkers in the military during World War II to send secure messages. Classification: Na-Dene, Nuclear Na-Dene, Athapaskan-Eyak, Athapaskan, Apachean, Navajo-Apache, Western Apache-Navajo.

NDALI – ndh, (syn. Chindali, Kindali, Ici-Ndali). Ndali is spoken by 150,000 people in Tanzania. It is also spoken in Malawi for a total of 220,000 speakers in all countries. Classification: Niger-Congo, Atlantic-Congo, Volta-Congo, Benue-Congo, Bantoid, Southern, Narrow Bantu, Central, M, Nyika-Safwa (M.20) Dialect: Sukwa.

NDONGA – ndo, (syn. Ochindonga, Oshindonga, Osindonga, Otjiwambo, Owambo, Ambo). Ndonga is a Bantu language spoken by 429,000 people in Namibia. Classification: Niger-Congo, Atlantic-Congo, Volta-Congo, Benue-Congo, Bantoid, Southern, Narrow Bantu, Central, R, Ndonga (R.20). Dialects: Ngandyera, Eunda Kolonkadhi, and Kwaludhi.

NEPALI – nep, (syn. Nepalese, Gorkhali, Gurkhali, Khaskura, Parbatiya, Eastern Pahari). Napali, an Indo-Aryan language closely related to Hindi, is spoken by 11,053,000 people in Nepal. It is also spoken in Bhutan and parts of India and Myanmar. Classification: Indo-European, Indo-Iranian, Indo-Aryan, Northern zone, Eastern Pahari. Dialects: Baitadi, Bajhangi, Bajurali (Bajura), Doteli (Dotali, Gaunle), Soradi, Acchami, and Darjula.

NEWAR – new, (syn. Nepal Bhasa, Newari). Newar is spoken by 825,000 of the Newar or Newah people, a multiple ethnicity group in the Kathmandu valley and in towns and villages throughout Nepal. Classification: Sino-Tibetan, Tibeto-Burman, Himalayish, Mahakiranti, Newari. Dialects: Dolkhali (Dolakha), Sindhupalchok Pahri (Pahri, Pahari), Totali, Citlang, Kathmandu-Pathan-Kirtipur, Baktapur, and Baglung.

NGIZIM - ngi, (syn. Ngizmawa, Ngezzim). Ngizim is spoken by 80,000 of the Ngizim people in the Yobe State of Nigeria. Classification: Afro-Asiatic, Chadic, West, B, Bade-Ngizim (B.1), Bade Proper.

NGUNDI – ndn, (syn. Ingundi, Ngondi). Ngundi is spoken by 3,000 people in the Sangha Region of the Republic of Congo. The original listing was Nsundi, which was believed to be a misspelling of Ngundi. Classication: Niger-Congo, Atlantic-Congo, Volta-Congo, Benue-Congo, Bantoid, Southern, Narrow Bantu, Northern, C, Ngundi.

NICOBARESE, CAR – caq, (syn. Pu, Car). Car Nicobarese is one of the Nicobarese languages spoken by 30,000 people in the Nicobar Islands of India. Classification: Austro-Asiatic, Mon-Khmer, Nicobar, Nancowry. Other Nicobarese langauages are: Chaura, Teressa, Bompoka, Southern Nicobarese (dialects: Condul, Great Nicobar, Little Nicobar, Milo, Sambelong, Tafwap), and Central Nicobarese (dialects: Camorta, Katchal, Nancowry, Trinkut).

NIUE – niu, (syn. Niuean, Niuefekai). Niue is spoken by 2,000 people in Niue. It is also spoken in Cook Islands, New Zealand, and Tonga for a total of 8,000 speakers in all countries. Classification: Austronesian, Malayo-Polynesian, Central-Eastern, Eastern Malayo-Polynesian, Oceanic, Central-Eastern Oceanic, Remote Oceanic, Central Pacific, East Fijian-Polynesian, Polynesian, Tongic. Dialects: similar to Tongan.

NORWEGIAN – nob, p. 549; (syn. Norwegian Bokmål, Bokmål, Bokmaal). Norwegian is a Northern Germanic language spoken by 6,300,000 people in Norway and a few in Norwegian settlements in the United States. There are two forms of written Norwegian: Norwegian Bokmål (nob) and Norwegian Nynorsk (nno). Classification: Indo-European, Germanic, North, East Scandinavian, Danish-Swedish, Danish- Bokmål. Dialects: there is such a diverse range of dialects of Norwegian that is difficult to estimate the total number, some of which are virtually unintelligible.

NUPE-NUPE-TAKO – nup, (syn. Nupe, Nufawa, Nupeci, Nupenchi, Nupecidji, Nupencizi). Nupe-Nupe-Tako is spoken by 800,000 people in Nigeria. It is also a trade language with 200,000 second language speakers. Classification: Niger-Congo, Atlantic-Congo, Volta-Congo, Benue-Congo, Nupoid, Nupe-Gbagyi, Nupe. Dialects: Nupe Central (Nife, Anupe, Nupecizi, Nupencizi, Ampeyi, Anupecwayi, Anuperi, Tapa, Tappah, Takpa), and Nupe Tako (Bassa Nge).

NYAKYUSA-NGONDE – nyy, (syn. Ikinyakyusa, Mombe, Ngonde, Ikingonde, Konde, Nkonde, Nyakusa, Nyikyusa, Sochile, Sokile, Sokili, Kukwe, Nyekyosa, Kinyakyusa). Nyakyusa-Ngonde is spoken 750,000 people in Tanzania and a total of 1,050,000 in all countries. Classification: Niger-Congo, Atlantic-Congo, Volta-Congo, Benue-Congo, Bantoid, Southern, Narrow Bantu, Central, M, Nyakyusa (M.30). Dialects: Nyakyusa, Kukwe, Mwamba (Lungulu), Ngonde, and Selya (Salya, Seria).

NYANJA – nya, (syn. Chinyanja, Chewa). Nyanja, also called Chinyanja or Chichewa after the Chewas and other tribes who use the language, is spoken by 7,000,000 people in Malawi. Nyanja is also spoken in Mozambique, Zambia, and Zimbabwe for a total of 9,300,000 speakers in all countries. Classification: Niger-Congo, Atlantic-Congo, Volta-Congo, Benue-Congo, Bantoid, Southern, Narrow Bantu, Central, N, Nyanja (N.30). Dialects: Chewa (Chichewa, Cheva, Sheva), Ngoni, Manganja (Waganga, Cimanganja), Nyasa, and Peta (Cipeta, Maravi, Marave, Malawi).

OGBIA – ogb, (syn. Ogbinya). Ogbia is spoken by 200,000 people in the Bayelsa state and the Brass local government area of Nigeria. Classification: Niger-Congo, Atlantic-Congo, Volta-Congo, Benue-Congo, Cross River, Delta Cross, Central Delta. Dialects: Agholo (Kolo), Oloibiri, and Anyama.

OGBIA-AGHOLO – ogb-*hlo*, (syn. Ogiba-Kolo). Ogbia-Agholo is a dialect of Ogbia spoken in Nigeria. See Ogbia.

ORIYA – ori, (syn. Uriya, Utkali, Odri, Odrum, Oliya, Orissa, Vadiya, Yudhia). Oriya or Odia is language in the Indo-Aryan branch of the Indo-European family spoken by 31,666,000 people in the Indian state of Orissa. Classification: Indo-European, Indo-Iranian, Info-Aryan, Eastern zone, Oriya. Dialects: Mughalbandi (Oriya Proper, Standard Oriya), Southern Oriya, Northwestern Oriya, Western Oriya (Sambalpuri), North Balasore Oriya, Midnapore Oriya, and Halbi.

OROMO, WEST CENTRAL – gaz, (syn. Galla, Afan Oromo, Oromiffa, Oromoo). West Central Oromo is spoken by 8,920,000 people in Ethiopia. Classification: Afro-Asistic, Cushitic, East, Oromo. Dialects: Western Oromo and Central Oromo.

PAIUTE, SOUTHERN – ute, (syn. Piute) Southern Paiute, a Ute language of the Numic branch of the Uto-Aztecan language family, is spoken by 2,000 people in Arizona, southeastern California and Nevada. Classification: Uto-Aztecan, Northern Uto-Aztecan, Numic, Southern. Dialects: Southern Paiute, Ute, and Chemehuevi.

PALAWANO, CENTRAL – plc, (syn. Quezon Palawano, Palawanen, Palaweño). Central Palawano is spoken by 12,000 people in Palawan Province, the Philippines. Classification: Austronesian, Malayo-Polynesian, Meso Philippine, Palawano.

PALI – pli. Pali is an extinct literary and liturgical language used in Cambodia, Bangladesh, India, Laos, Burma, Nepal, Sri Lanka, Thailand, and Vietnam. Classification: Indo-European, Indo-Iranian, Indo-Aryan.

PAMPANGAN – pam, (syn. Pampango, Pampangueño, Kapampangan, Amanung Sisuan, Capampañgan). Pampangan is spoken by 1,897,000 people, predominantly in the Pampanga province in Central Luzon of the Philippines. There are a total of 2,900,000 speakers in all countries. Classification: Austronesian, Malayo-Polynesian, Northern Philippine, Bashiic-Central Luzon-Northern Mindoro, Central Luzon, Pampangan.

PANGASINAN – pag, Pangasinan is spoken by 1,164,000 people, primarily in the Ilocos Region and Central Luzon in the Philippines. It is the primary language of a half million of the Pangasinan people in the Pangasinan province. Classification: Austronesian, Malayo-Polynesian, Northern Philippine, Northern Luzon, South-Central Cordilleran, Southern Cordilleran, Pangasinic.

PANJABI, EASTERN – pan, (syn. Punjabi, Gurmukhi, Gurumukhi). Eastern Panjabi is spoken by 27,109,000 people primarily in the Punjab of India. It is also spoken by immigrants in Kenya, United Kingdom, Mauritius, United States, Fiji, Malaysia, United Arab Emirates, Bangladesh, Canada, and Singapore for a total of 28,006,000 speakers in all countries. Classification: Indo-European, Indo-Iranian, Indo-Aryan, Central zone, Panjabi. Dialects: Panjabi Proper, Majhi, Doab, Bhatyiana (Bhatneri, Bhatti), Powadhi, Malwa, and Bathi.

PASHTO, NORTHERN – pbu, (syn, Pakhto, Pashtu, Pushto, Yusufzai, Pashto). Northern Pashto is spoken by 9,585,000 people in Pakistan. It is also spoken along the Afghanistan border with a total of 9,700,000 speakers in all countries. Classification: Indo-European, Indo-Iranian, Iranian, Eastern, Southeastern, Pashto. Dialects: Ningraharian Pashto and Northeastern Pashto.

POLISH – pol, (syn. Polski, Polnisch). Polish is a West Slavic language spoken by 36,554,000 people in Poland and a total of 42,708,000 speakers in all countries. It is spoken by immigrants in Belarus, Ukraine, Lithuania, Latvia, United Kingdom, Romania, Czech Republic, Russia, Brazil, United States, Canada, Germany, France, Australia, Ireland, Israel and elsewhere. Classification: Indo-European, Slavic, West, Lechitic. Dialects: Greater Polish, Lesser Polish, Mazovian (Mazur) and Silesian in Poland and a number of minor dialects spoken in various areas of the world.

PORTUGUESE – por, (syn. Português). Portuguese is a Romance language spoken by 10,000,000 people in Portugal and 177,457,000 people in all countries, ranking it 6th among the languages of the world. It is spoken in parts of Africa, Asia and South America and has official language status in Angola, Brazil, Cape Verde, East Timor, Equatorial Guinea, Macau, Mozambique, São Tomé and Príncipe. Classification: Indo-European, Italic, Romance, Italo-Western, Western,

Gallo-Iberian, Ibero-Romance, West-Iberian, Portuguese-Galician. There are a diverse range of dialects of Portuguese spoken, even within individual countries. For example the dialects in Brazil are: Caipira, Cearense, Baiano, Fluminense, Gaúcho, Mineiro, Nordestino, Nortista, Paulistano, Sertanejo, Sulista, and Carioca. Similar diversity is found in other countries as well as a number of Portuguese-based Creoles.

POTAWATOMI – pot, (syn. Pottawotomie, Bodéwadmimwen, Bodéwadmi Zheshmowen, Neshnabémwen). Potawatomi is a Central Algonquian language spoken by only 50 members of the Potawatomi tribe in North America which are found primarily in Michigan, Wisconsin, southern Ontario and Kansas. Classification: Algic, Algonquian, Central.

PULAAR – fuc, (syn. Pulaar Fulfulde, Peul, Peulh). Pulaar is a language spoken by the Fula people in West Africa from Senegal to Sudan. Pulaar represents the western dialects of the Fula language and Fulfulde the central and eastern dialects. There are 2,387,000 Pulaar speakers in Senegal and a total of 3,244,000 in all countries. Classification: Niger-Congo, Atlantic-Congo, Atlantic, Northern, Senegambian, Fulani-Wolof, Fula, Western. Dialects: Toucouleur (Tukolor, Tukulor, Tokilor, Pulaar, Haalpulaar, Fulbe Jeeri), and Fulacunda (Fulakunda, Fulkunda).

PULAR – fuf, (syn. Fuuta Jalon, Futa Jallon, Fouta Dyalon, Fulbe, Fullo Fuuta, Futa Fula, Foula Fouta, Fulfulde Jalon, Jalon). Pular is a Fula language spoken by 2,550,000 of the Fula people in Guinea and is distinct from Pulaar. It is also spoken in Guinea Bissau, Sierra Leone and by a few people in Mali for a total 2,915,000 speakers in all countries. Classification: Niger-Congo, Atlantic-Congo, Atlantic, Northern, Senegambian, Fulani-Wolof, Fula, West Central. Dialects: Kebu Fula and Fula Peta.

PUMÉ – yae, (syn. Llaruro, Yaruro, Yaruru, Yuapín). Pumé is spoken by 3,396 of the Pumé Indian tribe in lowland southwestern Venezuela. It is currently unclassificed.

QUECHUA, CUSCO – quz, (syn. Cuzco Quechua, Quechua, Qosqo-Qollaw, Runasimi Qusqu Qullaw, Quechua de Cusco-Collao, Qheswa, Quechua Cusco, Quechua de Cuzco). Cusco Quechua is one of at least seven Quechua languages spoken by 300,000 to 500,000 people in Peru. There are a total of 3,500,000 to 4,400,000 Quechua speakers in South America, principally in Argentina, Bolivia, Chile, Colombia, Ecuador and Peru. Classification: Quechuan, Quechua II, C. Dialects: Caylloma Quechua, Eastern Apurímac Quechua, and Puno Quechua.

RAROTONGAN – rar, (syn. Rarotongan-Mangaian, Māori, Kūki ʻĀirani, Cook Island, Cook Islands Maori, Te reo Ipukarea). Rarotongan is spoken by 16,800 people in the Cook Islands and a total of 42,669 speakers in all countries. Classification: Austronesian, Malayo-Polynesian, Central-Eastern, Eastern Malayo-Polynesian, Oceanic, Central-Eastern Oceanic, Remote Oceanic, Central Pacific, East Fijian-Polynesian, Polynesian, Nuclear, East, Central, Tahitic. Dialects: Rakahanga-Manihiki, Tongareva (Penrhyn), Atiu, Mitiaro, Mauke, Aitutaki, Rarotonga, and Mangaia.

ROMANI, BALKAN – rmn, (syn. Gypsy). Balkan Romani is spoken by 187,900 people in Bulgaria, 120,000 in Serbia and Montenegro and in Macedonia with a total of 523,900 speakers in all countries. Classification: Indo-European, Indo-Iranian, Indo-Aryan, Central zone, Romani, Balkan. Dialects: Arlija, Tinners Romani, Greek Romani, Dzambazi, East Bulgarian Romani, Paspatian, and Ironworker Romani.

ROMANIAN – ron, (syn. Rumanian, Moldavian, Daco-Rumanian). Romanian is a Romance language spoken by 19,741,000 people in Romania and a total of 23,498,000 people in all countries. It is spoken primarily in Romania and Moldovia but also in Italy, Spain, Russia, Ukraine, Israel, Portugal, United Kingdom, the United States, Canada, France, and Germany. Classification: Indo-European, Italic, Romance, Eastern. Dialects: Moldavian, Muntenian (Walachian), Transylvanian, Banat, and Bayash.

RONGA – rng, (syn. Shironga, Xironga, Gironga). Ronga is a Southeastern Bantu language spoken by 641,000 people in Mozambique and a total of 727,565 speakers in all countries. Classification: Niger-Congo, Atlantic-Congo, Volta-Congo, Benue-Congo, Bantoid, Southern, Narrow Bantu, Central, S, Tswa-Ronga (S.50). Dialects: Konde, Putru, and Kalanga.

RUSSIAN – rus, (syn. Russki). Russian is a Slavic language that is the most widely spoken native language in Europe. It is spoken by 117,863,000 people in Russia and is also spoken by immigrant communities around the world for a total of 278,000,000 speakers worldwide. It is an official language in Russia, Belarus, Kazakhstan and Kyrgyzstan. Classification: Indo-European, Slavic, East. Dialects: North Russian and South Russian.

SAMA, CENTRAL – sml, (syn. Siasi Sama, Central Sinama, Samal, Sinama) Central Sama is spoken by 90,000 people in the Sulu Province of the Philippines and in Malaysia. Classification: Austronesian, Malayo-Polynesian, Sama-Bajaw, Sulu-Borneo, Inner Sulu Sama. Dialect: Dilaut-Badjao.

SAMBAL, BOTOLAN – sbl, (syn. Aeta Negrito, Botolan Zambal, Ayta Hambali). There are 33,000 Botolan Sambal speakers in Central Luzon, the Philippines. Classification: Austronesian, Malayo-Polynesian, Northern Philippine, Bashiic-Central Luzon-Northern Mindoro, Central Luzon, Sambalic. Dialects: Ayta Hambali (Hambali Botolan), Sambali Botolan.

SAMBAL, TINÀ – xsb, (syn. Tina, Sambali). Tinà Sambal is spoken by 70,000 people in the northern Zambales Province of Luzon, the Philippines. Classification: Austronesian, Malayo-Polynesian, Northern Philippine, Bashiic-Central Luzon-Northern Mindoro, Central Luzon, Sambalic. Dialects: Santa Cruz, Masinloc, and Iba.

SAMOAN – smo. Samoan is an Austronesian language spoken by 199,000 people in Samoa; it is also spoken in American Samoa, Fiji, New Zealand, Tonga and United States for a total of 370,000 speakers in all countries. Classification: Austronesian, Malayo-Polynesian, Central-Eastern, Eastern Malayo-Polynesian, Oceanic, Central-Eastern Oceanic, Remote Oceanic, Central Pacific, East Fijian-Polynesian, Polynesian, Nuclear, Samoic-Outlier, Samoan.

SANGA – xsn, (syn. Isanga, Asanga). Sanga is spoken by 20,000 people in the Bauchi State and Toro local government area in Nigeria. Classification: Niger-Congo, Atlantic-Congo, Volta-Congo, Benue-Congo, Kainji, Eastern, Northern Jos, Jera. Dialect: Bujiyel.

SANSKRIT – san. Sanskrit is the ancient language of India, the oral use of which has greatly declined though it continues to be widely used as a ceremonial language in Hindu religious rituals. There are currently only 14,000 speakers in India and 194,000 second language speakers. Classification: Indo-European, Indo-Iranian, Indo-Aryan.

SANTALI – sat, (syn. Hor, Har, Satar, Santhali, Sandal, Sangtal, Santal, Sentali, Samtali, Santhiali, Sonthal). Santali is spoken by 5,959,000 people in the Indian states of Jharkhand, Assam, Bihar, Orissa, Tripura and West Bengal. Is also spoken in Bangladesh, Nepal, and Bhutan for a total of 6,050,000 speakers in all countries. Classification: Austro-Asiatic, Munda, North Munda, Kherwari, Santali. Dialects: Karmali (Khole), Kamari-Santali, Lohari-Santali, Manjhi, Paharia, and Mahali (Mahili, Mahli).

SCOTS – sco, (syn. Lowland Scots, Lallans). In its structure, a dialect of English, Scots has a distinctive vocabulary unintelligible to many speakers of English. Its 200,000 speakers are mostly in Scotland and Ulster. Classification: Indo-European, Germanic, West, English. Dialects: Insular, Northern Scots, Southern Scots, and Ulster.

SENA – seh, (syn. Cisena, Chisena). Sena is spoken by 876,000 people in Mozambique, principally in the Tete, Sofala, Zambezia and Manica provinces. Including second language speakers there are a total of 1,500,000 speakers. Classification: Niger-Congo, Altantic-Congo, Volta-Congo, Benue-Congo, Bantoid, Southern, Narrow Bantu, Central, N, Senga-Sena (N.40), Sena. Dialects: Caia (Care, Sare, North Sena), Bangwe (South Sena), Rue (Chirue), Gombe, Sangwe, and Podzo (Phodzo, Chipodzo, Cipodzo, Puthsu, Shiputhsu).

SENOUFO, CEBAARA – sef, (syn. Sénoufo, Senadi, Senari, Syenere, Tiebaara, Tyebala). The Senoufo or Senufic languages comprise 15 languages spoken by the Senufo in the north of Côte d'Ivoire, the south of Mali and the southwest of Burkina Faso. Cebaara, Djimini, Nyarafolo, Palaka, Shempire, and Tagwana are the most common in Côte d'Ivoire with Cebaara Senoufo being spoken by 862,000 people. Classification: Niger-Congo, Atlantic-Congo, Volta-Congo, North, Gur, Senufo,

Senari. Dialects: Kafire, Kasara, Kufuru, Tagbari (Mbengui-Niellé), Patara, Pogara, Tyebara, Tagara, Tenere, Takpasyeeri (Messeni), Southwest Senari, Kandere (Tengrela), Papara, Fodara, Kulere, Nafara.

SERBO-CROATIAN – hbs. Standard Serbian, modern Croatian, and Bosnian are mutually intelligible and have been unified as Serbo-Croatian or Croato-Serbian. It encompasses all the dialects spoken in Serbia, Croatia, Montenegro, and Bosnia and Herzegovina, and was the official standard language of Yugoslavia from 1954 to 1991. It is spoken by 23,000,000 people predominately in these countries. Classification: Indo-European, Balto-Slavic, Slavic, South, Western, Serbo-Croatian. Dialects: Štokavian, Čakavian, and Kajkavian.

SERER-SINE – srr, (syn. Sérère-Sine, Serer, Serrer, Sereer, Seereer, Serer-Sin, Sine-Saloum, Seex, Sine-Sine). Serer-Sine is spoken by 1,155,000 people in Senegal. It is also spoken in Gambia with a total of 1,183,000 speakers in all countries. Classification: Niger-Congo, Atlantic-Congo, Atlantic, Northern, Senegambian, Serer. Dialects: Segum, Fadyut-Palmerin, Sine, Dyegueme (Gyegem), and Niominka.

SHAN – shn, (Sha, Tai Shan, Sam, Thai Yai, Tai Yai, Great Thai, Tai Luang, Mau, Ngio, Ngiow, Ngiaw, Ngiao, Ngeo). Shan is a Kadai language spoken by 3,200,000 of the Shan people in northeast Myanmar. It is also spoken in some areas of northern Thailand and the Yunnan province of southwestern China for a total of 3,300,000 speakers in all countries. Classification: Tai-Kadai, Kam-Tai, Be-Tai, Tai-Sek, Tai, Southwestern, Northwest. Dialects: Kokant Shan and Tai Mao (Mao, Maw, Mau, Tai Long, Northern Shan).

SHONA – sna, (Swina, Chishona, chiShona). Shona is a Bantu language spoken by 10,663,000 of the Shona people in Zimbabwe. It is also spoken in Zambia, Mozambique, and Botswana for a total of 10,704,000 speakers in all countries. Classification: Niger-Congo, Atlantic-Congo, Volta-Congo, Benue-Congo, Bantoid, Southern, Narrow Bantu, Central, S, Shona (S.10). Dialects: Toko, Hwesa, Karanga (Chikaranga), Zezuru (Chizezuru, Bazezuru, Bazuzura, Mazizuru, Vazezuru, Wazezuru), and Korekore (Northern Shona, Goba, Gova, Shangwe).

SIDAMO – sid, (syn. Sidámo 'Afó, Sidaminya). Sidamo is an Afro-Asiatic language spoken by 1,876,000 people in southern Ethiopia. Classification: Afro-Asiatic, Cushitic, East, Highland. Dialects: similar to Allaaba, Kambaata, and Hadiyya.

SINHALA – sin, (syn. Sinhalese, Singhalese, Singhala, Cingalese). Sinhala is an Indo-European language spoken by 13,190,000 Sinhalese in Sri Lanka and to a limited extent elsewhere (e.g. Canada, Maldives, Singapore, Thailand, United Arab Emirates) with a total of 13,220,000 speakers in all countries. Classification: Indo-European, Indo-Iranian, Indo-Aryan, Sinhalese-Maldivian. Dialect: Rodiya.

SLOVAK – slk, (syn. Slovakian). Slovak is a West Slavic language spoken by 4,865,000 people in Slovakia and a total of 5,011,000 people in all countries (e.g. United States, Czech Republic, Serbia, Ireland, Romania, Hungary, Poland, Canada). Classification: Indo-European, Slavic, West, Czech-Slovak. Dialect groups: Western Slovak, Central Slovak, Eastern Slovak and Lowlands Slovak.

SLOVENIAN – slv, (syn. Slovene, Slovenscina). Slovenian is a South Slavic language spoken by 1,727,000 people in Slovenia and a total of 1,984,000 speakers in all countries (e.g. Italy, Austria, Hungary, Croatia). Classification: Indo-European, Slavic, South, Western. Dialects: there are 32 main dialects in Slovenia.

SOMALI – som, (syn. Af-Soomaali, Af-Maxaad Tiri, Common Somali, Standard Somali). Somali, a member of the Afro-Asiatic language family, is spoken by 7,784,000 people in Somalia. It is also spoken in Djibouti, Yemen, and Kenya and in immigrant communities in a number of countries for a total of 12,653,000 speakers in all countries. Classification: Afro-Asiatic, Cushitic, East, Somali. Dialects: there are three main groups: Northern, Benaadir, and Maay.

SONGHAY – hmb, (syn. Songay, Senni, Songoy, Songai, Songhai, Songay, Songoi, Sonrai, Sonrhai, Central Songai). Songhay is a group of closely related dialects along the Niger River in Mali, Niger, Benin, Burkino Faso, and Nigeria spoken by 125,000 people in Burkino Faso and a total of 140,000 speakers in all countries. Classification: Nilo-Saharan, Songhai, Southern. Dialect: Marensé (Maransé, Koroboré).

SORBIAN, LOWER – dsb, (syn. Lower Wendish, Niedersorbisch, Bas Sorabe, Wendish, Lusatian, Lower Lusatian, Dolnoserbski, Delnoserbski). Lower Sorbian is a Slavic language spoken by 14,000 people in the Brandenburg region of Germany. Classification: Indo-European, Slavic, West, Sorbian.

SORBIAN, UPPER – hsb, (syn. Obersorbisch, Haut Sorabe, Upper Lusatian, Wendish, Hornjoserbski, Hornoserbski). Upper Sorbian is a Slavic language spoken by 15,000 people in the Saxony and Brandenburg regions of Germany. Classification: Indo-European, Slavic, West, Sorbian. Dialects: Bautzen and Kamenz.

SOTHO, NORTHERN – nso, (syn. Pedi, Sepedi, Transvaal Sotho). Northern Sotho is spoken by 3,696,000 people in Gauteng, Limpopo, and Mpumalanga provinces of South Africa. It is also spoken in Botswana with a total of 3,710,000 speakers in all countries. Classification: Niger-Congo, Atlantic-Congo, Volta-Congo, Benue-Congo, Bantoid, Southern, Narrow Bantu, Central, S, Sotho-Tswana (S.30), Sotho, Northern. Dialects: Masemola (Masemula, Tau), Kgaga (Kxaxa, Khaga), Koni (Kone), Tswene (Tsweni), Gananwa (Xananwa, Hananwa), Pulana, Phalaborwa (Phalaburwa, Thephalaborwa), Khutswe (Khutswi, Kutswe), Lobedu (Lubedu, Lovedu, Khelobedu), Tlokwa (Tlokoa, Tokwa, Dogwa), Pai, Dzwabo (Thabine-Roka-Nareng), Kopa, and Matlala-Moletshi.

SOTHO, SOUTHERN – sot, (syn. Suto, Suthu, Souto, Sesotho, Sisutho). Southern Sotho is a Bantu language spoken by 3,104,000 people in South Africa. It is also spoken by 1,493,000 people in Lesotho for a total of 5,000,000 speakers in all countries. Classification: Niger-Congo, Atlantic-Congo, Benue-Congo, Bantoid, Southern Bantoid, Bantu, Central Bantu, Southern Bantu, Sotho-Tswana, Sotho. Dialects: Taung and Phuthi.

SPANISH – spa, (syn. Español, Castellano, Castilian). Spanish is spoken by 28,173,000 people in Spain and in a wide range of other countries (e.g. Argentina, Bolivia, Chile, Colombia, Costa Rica, Cuba, Dominican Republic, Ecuador, El Salvador, Equatorial Guinea, Guatemala, Honduras, Mexico, Nicaragua, Panama, Paraguay, Peru, Puerto Rico, Uruguay, Venezuela and by a significant number of people in Belize, Gibraltar, United States, Philippines and Andorra) for a total of 350,000,000 first language speakers in all countries. Classification: Indo-European, Italic, Romance, Italo-Western, Western, Gallo-Iberian, Ibero-Romance, West Iberian, Castilian. Dialects: Andalusian (Andalú, Andaluz, Andalusí), Murcian, Aragonese, Navarrese, Castilian, Canary Islands Spanish, and American Spanish.

SRANAN – srn, (syn. Sranan Tongo, Taki-Taki, Surinaams, Surinamese, Suriname Creole English). Sranan is a Creole language spoken by 120,000 people in Suriname and a total of 126,400 in all countries. Classification: Creole, English based, Atlantic, Suriname. Dialect: similar to Ndyuka.

STELLINGWERFS – stl, (syn. Stellingwarfs). Stellingwerfs is a Dutch-Saxon dialect spoken in the southeastern part of the Friesland province in the Netherlands. Classification: Indo-European, Germanic, West, Low Saxon-Low Franconian, Low Saxon.

SUBANON – suc, (syn. Siocon). Western Subanon is spoken by 75,000 people in the Zamboanga Peninsula region of Mindanao, the Philippines. Classification: Austronesian, Malayo-Polynesian, Southern Philippine, Subanun, Kalibugan. Dialect: Western Kolibugan (Western Kalibugan).

SUMO-MAYANGNA – sum, (syn. Sumo, Sumu, Soumo, Sumoo, Woolwa, Sumo Tawahka, Taguaca). Sumo-Mayangna is spoken by 6,700 people in southeastern Nicaragua. It is also spoken in the northern part of Honduras with a total of 7,400 speakers in all countries. Classification: Misumalpan. Dialects: Panamahka, Nicaraguan Tawahka, Ulwa, Bawihka, and Kukra.

SUNDA – sun, (syn. Sundanese, Priangan). Sunda is spoken by 27,000,000 people in the western third of Java. Classificatiori Austronesian, Malayo-Polynesian, Sundanese. Dialects: Banten, Bogor (Krawang), Pringan, and Cirebon.

SUSU – sus, (syn. Soso, Susoo, Soussou, Sose). Susu is a Mande language spoken by 905,500 of the Susu people in Guinea, West Africa. It is also spoken in Guinea-Bissau and Sierra Leone with a total of 1,029,000 speakers in all countries. Classification: Niger-Congo, Mande, Western, Central-Southwestern, Central, Susu-Yalunka. Dialects: Susu is distinct from Yalunka.

SWAHILI – swh, (syn. Kiswahili, Kisuaheli). Swahili is spoken in a number of countries in East Africa and the Democratic Republic of Congo. In Tanzania it is spoken by 541,000 people and is estimated to be spoken by a total 5 to 10 million people in all countries. Swahili is also spoken by varying numbers of people in Burundi, Kenya, Mayotte, Mozambique, Oman, Rwanda, Somalia, South Africa, Uganda, United Arab Emirates and the United States. Classification: Niger-Congo, Atlantic-Congo, Volta-Congo, Benue-Congo, Bantoid, Southern, Narrow Bantu, Central, G, Swahili (G.40). Dialects: Mrima, Unguja (Kiunguja, Zanzibar), Pemba, and Mgao.

SWEDISH – swe, (syn. Svenska, Ruotsi). Swedish is a North Germanic language spoken by 7,825,000 people in Sweden. It is also spoken in parts of Finland, Northern Europe and the United States with a total of 8,790,000 speakers in all countries. Classification: Indo-European, Germanic, North, East Scandinavian, Danish-Swedish, Swedish. Dialects: major groups include Northern Swedish (Norrland), Eastern Swedish (Finland Swedish, Estonian Swedish), Svea, and Gutniska (Gutamal, Gotlandic, Gutnic). There are thought to be several hundred rural dialects that vary widely with location in Sweden and are often so localized they are limited to individual parishes.

TAGALOG – tgl. Tagalog is an Austronesian language spoken by 22,000,000 people in the Philippines, principally in Central and South Luzon. It is also spoken in Australia, Bahrain, Brunei, Canada, Guam, Hong Kong, Japan, Kuwait, Malaysia, Northern Mariana Islands, Palau, Qatar, Saudi Arabia, Switzerland, United Arab Emirates, United Kingdom and United States with a total of 90,000,000 speakers in all countries. Classification: Austronesian, Malayo-Polynesian, Meso Philippine, Central Philippine, Tagalog. There are believed to be four main dialects: Northern, Central, Southern, and Marinduque.

TAGBANWA – tbw, (syn. Tagbanua, Apurawnon, Aborlan Tagbanwa). Tagbanwa is an Austronesian language spoken by 10,000 people in the central and northern regions of the Palawan province in the Philippines. The use of Tagbanwa is declining in favor of Cuyonon and Tagalog. Classification: Austronesian, Malayo-Polynesian, Meso Philippine, Palawano. Dialects: similar to Quezon Palawano and Cuyonon.

TAJIKI – tgk, (syn. Tadzhik, Tajiki Persian, Galcha). Tajiki or Tajik Persian is a variety of Persian spoken by 3,344,000 people in Tajikistan. It is also spoken in Iran, Kazakhstan, Kyrgyzstan, Russia, Turkmenistan, Ukraine and Uzbekistan with a total of 4,380,000 speakers in all countries. Classification: Indo-European, Indo-Iranian, Iranian, Western, Southwestern, Persian. The dialects are separated into North, Central, Southern, and Southeastern groups.

TAMIL – tam, (Tamalsan, Tambul, Tamili, Tamal, Damulian). Tamil is a Dravidian language spoken by 61,527,000 people in India and is the official administrative language in the state of Tamil Nadu. It is also spoken in Sri Lanka, Singapore, Canada, Malaysia, Mauritius, Réunion and by emigrant communities around the world with a total of 66,020,000 speakers in all countries. Classification: Dravidian, Southern, Tamil-Kannada, Tamil-Kodagu, Tamil-Malayalam, Tamil. Dialects: Adi Dravida, Aiyar, Aiyangar, Arava, Burgandi, Kongar, Madrasi, Pattapu Bhasha, Tamil, Sri Lanka Tamil, Malaya Tamil, Burma Tamil, South Africa Tamil, Tigalu, Harijan, Sanketi, Hebbar, Mandyam Brahmin, and Secunderabad Brahmin.

TATAR – tat, (syn. Tartar). Tatar is a Turkic language spoken by 465,000 of the Tatar people in Russia. It is the official language in Tatarstan and is spoken by a total of 1,610,000 people in all countries. Classification: Altaic, Turkic, Western, Uralian. Dialects: Middle Tatar (Kazan), Western Tatar (Misher), and Eastern Tatar (Siberian Tartar).

TAUSUG – tsg, (syn. Taw Sug, Sulu, Suluk, Tausog, Moro Joloano, Jolohano, Sinug Tausug). Tausug is a Visayan language spoken by 900,000 of the Tausūg people in the Philippines. It is also spoken in Malaysia and Indonesia with at total of 1,022,000 speakers in all countries. Classification: Austronesian, Malayo-Polynesian, Meso Philippine, Central Philippine, Bisayan, South, Butuan-Tausug.

TEKE – *tec*. There are nine Teke languages spoken in the Republic of Congo (Teke-Eboo, Teke-Fuumu, Teke-Ibali, Teke-Kukuya, Teke-Laali, Teke-Nzikou, Teke-Tege, Teke-Tsaayi, Teke-Tyee); however, the specific language for the vegetable names listed was not given. There are 490,000 speakers for all of the Teke languages spoken in Congo. It is also spoken in Gabon and Congo-Kinshasa. The Teke languages are classified as: Niger-Congo, Atlantic-Congo, Volta-Congo, Benue-Congo, Bantoid, Southern, Narrow Bantu, Northwest, B, Teke (B.70).

TELUGU – tel, (syn. Telegu, Andhra, Gentoo, Tailangi, Telangire, Telgi, Tengu, Terangi, Tolangan). Telugu is a Dravidian language spoken by 69,634,000 people in India, primarily in the state of Andhra Pradesh where it is the official language, but also in states of Tamil Nadu, Karnataka, Maharashtra and Jharkhand. Classification: Dravidian, South-Central, Telugu. Dialects: Berad, Dasari, Dommara, Golari, Kamathi, Komtao, Konda-Reddi, Salewari, Telangana, Telugu, Vadaga, Srikakula, Vishakhapatnam, East Godavari, Rayalseema, Nellore, Guntur, Vadari, and Yanadi (Yenadi).

TEM – kdh, (syn. Kotokoli, Cotocoli, Tim, Timu, Temba). Tem is spoken by 204,000 people in Togo. It is also spoken in Benin and Ghana with a total of 307,000 speakers in all countries. Classification: Niger-Congo, Atlantic-Congo, Volta-Congo, North, Gur, Central, Southern, Grusi, Eastern.

THAI – tha, (syn. Central Tai, Standard Thai, Thaiklang, Siamese). Thai is a Kradai language spoken by 20,182,000 people in Thailand where it is the official language. It is also spoken in Northern Malaysia, Cambodia, Southern Myanmar, and Laos with a total of 20,230,000 speakers in all countries. Classification: Tai-Kadai, Kam-Tai, Be-Tai, Tai-Sek, Tai, Southwestern, East Central, Chiang Saeng. Dialect: Khorat Thai (Korat, Thaikorat).

THAI, NORTHERN – nod, (syn. Lanna, Lan Na, Lanatai, Yuan, Phyap, Phayap, Payap, Kammüang, Kammyang, Myang, Kam Mu'ang, Mu'ang, Khon Mung, Khon Myang, Tai Nya, La Nya, Northern Thai, Western Laotian). Northern Thai or Lanna is a Tai language closely related to Thai and Lao that is spoken by 6,000,000 people in northern Thailand and a few thousand speakers in northwestern Laos. Classification: Tai-Kadai, Kam-Tai, Be-Tai, Tai-Sek, Tai, Southwestern, East Central, Ching Saeng. Dialects: Nan, Bandu, and Tai Wang.

THEMNE – tem, (Temne, Timne, Timene, Timmannee, Temen). Themne is spoken by 1,200,000 people in Sierra Leone, primarily in the Northern Province. Classification: Niger-Congo, Atlantic-Congo, Atlantic, Southern, Mel, Temne, Temne-Banta. Dialects: Banta, Konike, Yoni, Bombali, Western Temne (Pil), Sanda, Ribia, Kholifa, Koya, Masingbi, and Malal.

TIBETAN – bod, (syn. Central Tibetan, Wei, Weizang, Bhotia, Zang, Phoke, Dbus, Dbusgtsang, U). Tibetan is the official language and is spoken by 1,066,000 people in the Tibet Autonomous Region of China. It is also spoken in Kashmir region and the states of Sikkim and Arunachal Pradesh in India. Classification: Sino-Tibetan, Tibeto-Burman, Himalayish, Tibeto-Kanauri, Tibetic, Tibetan, Central. Dialects: Gtsang (Tsang, Lhasa), Dbus, and Mngahris (Ngari).

TIGRIGNA – tir, (syn. Tigrinya, Tigray). Tigrigna is spoken by 3,225,000 people in Ethiopia. It is also spoken by 1,200,000 people in Eritrea, and in German and Israel with a total of 4,450,000 speakers in all countries. Classification: Afro-Asiatic, Semitic, South, Ethiopian, North.

TIRURAY – tiy, (syn. Tirurai, Teduray). Tiruray is spoken by 50,000 people on Mindanao, the Philippines. Classification: Austronesian, Malayo-Polynesian, South Mindanao, Tiruray.

TIV – tiv, (syn. Munshi). Tiv is spoken by 2,212,000 people in Nigeria, primarily in the state of Benue but also in the states of Plateau, Taraba, and Nasarawa and to a limited extent in Cameroon. Classification: Niger-Congo, Atlantic-Congo, Volta-Congo, Benue-Congo, Bantoid, Southern, Tivoid.

TOK PISIN – tpi, (syn. Pisin, Pidgin, Neomelanesian, New Guinea Pidgin English, Melanesian English). Tok Pisin is an English based Creole spoken as a primary language by perhaps 1,000,000 people in Papua New Guinea, principally in the northern half of the country. Classification: Creole, English Creole, Pacific, Tok Pisin.

TONGA – tog, (syn. Chitonga, Siska, Sisya, Western Nyasa, Kitonga). Tonga is spoken by 170,000 people in Mali. It differs from the Tonga language (toi) spoken in Zambia and Zimbabwe. Classification: Niger-Congo, Atlantic-Congo, Volta-Congo, Benue-Congo, Bantoid, Southern, Narrow Bantu, Central, N, Manda (N.10).

TONGAN – ton. Tongan is a Austronesian language spoken by 96,000 people in Tonga. It is also spoken in American Samoa, Australia, Canada, Fiji, New Zealand, Niue, the United States, and Vanuatu with a total of 105,000 speakers in all countries. Classification: Austronesian, Malayo-Polynesian, Central-Eastern, Eastern Malayo-Polynesian, Oceanic, Central-Eastern Oceanic, Remote Oceanic, Central Pacific, East Fijian-Polynesian, Polynesian, Tongic. Dialects: similar to Niue.

TSONGA – tso, (Shitsonga, Xitsonga, Thonga, Tonga, Gwamba, Xichangana). Tsonga is spoken by 223,000 of the Tsonga people in Namibia. It is also spoken in Mozambique, South Africa, Swaziland, and Zimbabwe with a total of 3,275,000 speakers in all countries. Classification: Niger-Congo, Atlantic-Congo, Benue-Congo, Bantoid, Southern Bantoid, Bantu, Central Bantu, Southern Bantu, Tswa-Ronga, Tsonga. Dialects: Bila (Vila), Changana (Xichangana, Changa, Shangaan, Hlanganu, Hanganu, Langanu, Shilanganu, Shangana), Jonga (Djonga, Dzonga), and Ngwalungu (Shingwalungu).

TSWANA – tsn, (syn. Tsiwaha, Beetjuans, Chuana, Coana, Cuana, Setswana). Tswana is a Bantu language spoken by 3,301,000 people in South Africa. It is also spoken in Botswana, Zimbabwe and Namibia with 4,407,000 speakers in all countries. Classification: Niger-Congo, Atlantic-Congo, Volta-Congo, Benue-Congo, Bantoid, Southern, Narrow Bantu, Central, S, Sotho-Tswana (S.30). Dialects in South Africa: Tlahaping (Tlapi), Rolong, Kwena, Kgatla, Ngwatu (Ngwato), Tawana, Lete, Ngwaketse, and Tlokwa.

TULU – tcy, (syn. Tal, Tallu, Thalu, Tilu, Tuluva Bhasa, Tullu, Thulu). Tulu is a Dravidian language spoken by 1,949,000 of the Tuluvas people primarily in the states of Karnataka, Kerala, and Maharashtra in India. Classification: Dravidian, Southern, Tulu. Dialects: Brahmin, Jain, Common, and Tribal.

TUMBUKA – tum, (syn. Tumboka, Chitumbuka, Tamboka, Tambuka, Timbuka, Tombucas). Tumbuka is a Bantu language spoken by 940,000 people in Malawi. It is also spoken in parts of Zambia and Tanzania with a total of 1,332,000 speakers in all countries. Classification: Niger-Congo, Atlantic-Congo, Benue-Congo, Bantoid, Southern, Narrow Bantu, Central, N, Tumbuka (N.20). Dialects: Chitumbuka (Kamanga, Henga), Nenya, Poka (Chipoka, Phoka), Yombe, Senga, Nthali, Fungwe, Wenya, and Hewe (Hewa).

TUPURI – tui, (syn. Toupouri, Tuburi, Toubouri, Ndore, Ndoore, Wina, Tongoyna, Honya, Dema, Mata). Tupuri is spoken by 125,000 of the Tupuri people in parts of northern Cameroon and 90,785 in southern Chad for a total of 215,785 speakers in all countries. Classification: Niger-Congo, Atlantic-Congo, Volta-Congo, North, Adamawa-Ubangi, Adamawa, Mbum-Day, Mbum, Northern, Tupuri-Mambai.

TURKISH – tur, (syn. Türkçe, Türkisch, Anatolian). Turkish is spoken by 46,278,000 people in Turkey. It is also spoken in Albania, Azerbaijan, Bosnia and Herzegovina, Bulgaria, Croatia, Cyprus, Greece, Hungary, Iraq, Jordon, Kosovo, Lebanon, Republic of Macedonia, Moldova, Montenegro, Palestine, Romania, Russia, Serbia, Syria, Turkmenistan, Uzbekistan and by immigrant communities in a number of countries for a total of 50,625,000 speakers in all countries. Classificaton: Altaic, Turkic, Southern, Turkish. Dialects: Danubian, Eskisehir, Razgrad, Dinler, Rumelian, Karamanli, Edirne, Gaziantep, and Urfa.

TURKMEN – tuk, (syn. Turkomans, Turkmenler, Turkmanian, Trukhmen, Trukhmeny, Turkmani). Turkmen is spoken by 3,430,000 people in Turkmenistan. It is also spoken in Iran, Iraq, Afghanistan and Turkey with a total of 6,403,000 speakers in all countries. Classification: Altaic, Turkic, Southern, Turkmenian. Dialects: Nokhurli, Anauli, Khasarli, Nerezim, Yomud, Teke (Tekke), Goklen, Salyr, Saryq, Esari, and Cawdur.

TWI – twi. Twi is a member of the Kwa language family and one of the three dialects of the Akan language (Twi, Akuapem, and Fante). Twi is spoken by 15,800,000 people in Ghana. Classification: Niger-Congo, Atlantic-Congo, Kwa, Akan languages, Akan, Twi. Dialects: different tribes of the Akan people (e.g. Ashantis, Fantis, Akuapems, Kwawus, Akyems, Brongs and Dankyiras) use the same grammar but have a different tonation that is mutually intelligable.

UKRAINIAN – ukr. Ukrainian is a East Slavic language spoken by 31,058,000 people in the Ukraine. It is the official language in both Ukraine and Moldova and is spoken by Ukrainan immigrants in a number of other countries for a total of 39,441,000 speakers worldwide. Classification: Indo-European, Slavic, East. Ukrainian dialects are separated into three main groups – Northern dialects: Eastern Polissian, Central Polissian and West Polissian; Southeastern dialects: Middle Dnieprian, Slobodan, Steppe, and Kuban; and Southwestern dialects: Boyko, Hutsul, Lemko, Podillian, Volynian, Pokuttia (Bukovynian), Upper Dniestrian, and Upper Sannian.

URDU – urd. Urdu is an Indo-European language spoken by 10,719,000 people in Pakistan. It is also spoken in five Indian states and by immigrants in a wide range of countries around the world with a total of 60,503,000 speakers in all countries. Classification: Indo-European, Indo-Iranian, Indo-Aryan, Central zone, Western Hindi, Hindustani. Indian dialects include Dakhini (Dakani, Deccan, Desia, Mirgan), Pinjari, and Rekhta (Rekhti).

UYGHUR – uig, (syn. Uighur, Uygur, Uigur, Uighuir, Uiguir, Weiwuer, Wiga). Uyghur is a Turkic language spoken by 7,214,000 of the Uyghur people in the Xinjiang Uyghur Autonomous Region administered by China and in Kazakhstan. There are a total of 7,601,000 speakers in all countries. Classification: Altaic, Turkic, Eastern. Dialects: Central Uyghur, Hotan (Hetian), and Lop (Luobu).

UZBEK – uzn, (syn. Özbek, Northern Uzbek) Northern Uzbek is a Turkic language spoken by 16,539,000 people in Uzbekistan. It is also spoken in Kyrgyzstan, Afghanistan, Pakistan, Kazakhstan, Turkmenistan, Tajikistan, Russia, and China for a total of 18,795,000 speakers in all countries. Classification: Altaic, Turkic, Eastern. Dialects: Karluk (Qarlug), Kipchak (Kypchak), and Oghuz.

VAI – vai, (syn. Vei, Vy, Gallinas, Gallines). Vai is a Mande language spoken by 89,500 people in Liberia and 15,500 in Sierra Leone. Classification: Niger-Congo, Mande, Western, Central-Southwestern, Central, Manding-Jogo, Manding-Vai, Vai-Kono.

VENDA – ven, (syn. Chivenda, Tshivenḓa, Luvenḓa). Venda is a Bantu language spoken by 876,000 people in South Africa. It is also spoken in Zimbabwe with a total of 960,000 speakers in all countries. Classification: Niger-Congo, Atlantic-Congo, Volta-Congo, Benue-Congo, Bantoid, Southern, Narrow Bantu, Central, S, Venda (S.20). Dialects: Phani, Tavha-Tsindi, Ilafuri, Manda, Guvhu, Mbedzi, and Lembetu.

VIETNAMESE – vie, (syn. Kinh, Gin, Jing, Ching, Viet, Annamese). Vietnamese, a member of the Austroasiatic language family, is spoken by 65,795,000 people in Vietnam and by immigrants in the United States, Canada, France and a number of other countries. It is also spoken as a second language by ethnic minorities in Vietnam and a total of 67,439,000 speakers in all countries. Classification: Austro-Asiatic, Mon-Khmer, Viet-Muong, Vietnamese. There are a number of dialects that are separated into three regions: Northern Vietnamese (Tonkinese, Hanoi), Central Vietnamese (Hue), and Southern Vietnamese.

VLAAMS – vls, (syn. Flamand, Flemish, Vlaemsch). Vlaams is spoken by 1,070,000 people in Belgium, in particular the West Flanders province. It is also spoken in France and the Netherlands with a total of 1,202,000 speakers in all countries. Classification: Indo-European, Germanic, West, Low Saxon-Low Franconian, Low Franconian. Dialects: West Vlaams (Vlaemsch), Oostvlaams, Antwerps, Limburgs, and Brabants.

WAAMA – wwa, (syn. Yoabu, Yoabou). Waama is spoken by 50,000 people in Benin, principally in the Atakora province but also in some parts of western Nigeria. Classification: Niger-Congo, Atlantic-Congo, Volta-Congo, North, Gur, Central, Northern, Oti-Volta, Eastern. Dialects: Waama and Tangamma.

WAMEY – cou, (syn. Wamay, Wamei, Konyagi, Coniagui, Conhague, Koniagui). Wamey is spoken by 16,700 people in Senegal. It is also spoken in parts of Guinea with a total of 21,970 speakers in all countries. Classification Niger-Congo, Atlantic-Congo, Atlantic, Northern, Eastern Senegal-Guinea, Tenda.

WÁRAY-WÁRAY – war, (syn. Samareño, Samaran, Samar-Leyte, Waráy, Winaray, Binisaya). Wáray-Wáray is spoken by 2,437,000 people in the provinces of Samar, North Samar, Eastern Samar, Leyte and Biliran in the Philippines. Classification: Austronesian, Malayo-Polynesian, Meso Philippine, Central Philippine, Bisayan, Central, Warayan, Samar-Waray. Dialects: Waray, Samar-Leyte, and Northern Samar.

WELSH – cym, (syn. Cymraeg). Welsh is a Celtic language spoken by 611,000 people in Wales. It is also spoken in England along the Welsh border, in the Chubut Valley of Patagonia and among immigrants in Canada, United States, New Zealand, and Australia with a total of 750,000 speakers in all countries. Classification: Indo-European, Celtic, Insular, Brythonic. Dialects: North Walian and South Walian.

WOLOF – wol, (syn. Ouolof, Yallof, Walaf, Volof, Waro-Waro). Wolof is a member of the Niger-Congo language family spoken in Senegal by 3,568,000 the Wolof people and other Senegalese. It is also spoken in Mauritania and Gambia with a total of 3,612,000 speakers in all countries. Classification: Niger-Congo, Atlantic-Congo, Atlantic, Northern, Senegambian, Fula-Wolof, Wolof. Dialects: Baol, Cayor, Dyolof (Djolof, Jolof), Lebou (Lebu), and Jander.

XHOSA – xho, (syn. Isixhosa, Xosa, Koosa, Kaffer, Kaffir, Caffre, Cafre, Cauzuh). Xhosa is a Bantu language spoken by 7,196,000 people in South Africa where it is an official language. Classification: Niger-Congo, Atlantic-Congo, Volta-Congo, Benue-Congo, Bantoid, Southern, Narrow Bantu, Central, S, Nguni (S.40). Dialects: Gcaleka, Ndlambe, Gaika (Ncqika), Thembu, Bomvana, Mpondomse (Mpondomisi), Mpondo, Xesibe, Rhathabe, Bhaca, Cele, Hlubi, and Mfengu.

YALA – yba, (syn. Iyala). Yala is a member of the Niger-Congo language family that is spoken by 50,000 people in Nigeria, primarily in the Cross River State and the Ogoja, Obubra, and Ikom local government areas. Classification: Niger-Congo, Atlantic-Congo, Volta-Congo, Benue-Congo, Idomoid, Akweya, Etulo-Idoma, Idoma. Dialects: Nkum (Yala Ikom), Nkum Akpambe (Yala Obubra), and Yala Ogoja.

YALUNKA – yal, (syn. Yalunke, Djallonke, Kjalonke, Dialonke, Jalonke, Jallonka). Yalunka, a Mande language, is spoken by 30,000 of the Yalunka people in Sierra Leone. It is also spoken in Guinea, Mali, and Senegal with a total of 55,000 speakers in all countries. Classification: Niger-Congo, Mande, Western, Central-Southwestern, Central, Susu-Yalunka. Dialects: Sulima, Firia, and Musaia.

YAO – yao, (syn. Chiyao, Achawa, Adsawa, Adsoa, Ajawa, Ayawa, Ayo, Ayao, Djao, Haiao, Hiao, Hyao, Jao, Veiao, Wajao). Yao is spoken by 1,000,000 people in Malawi. It is also spoken in Mozambique, Tanzania and Zambia with a total of 1,942,000 speakers in all countries. Classification: Niger-Congo, Atlantic-Congo, Volta-Congo, Benue-Congo, Bantoid, Southern, Narrow Bantu, Central, P, Yao (P.20). Dialect: Mangoche.

YIDDISH – ydd, (syn. Judeo-German, Eastern Yiddish). Yiddish is a Germanic language spoken by 215,000 people in Israel. It is also spoken in Argentina, Australia, Belarus, Belgium, Brazil, Canada, Costa Rica, Estonia, Hungary, Latvia, Lithuania, Moldova, Panama, Poland, Puerto Rico, Romania, Russia, South Africa, Ukraine, Uruguay, United Kingdom, United States and elsewhere by a total of 3,142,000 speakers in all countries. Written Yiddish uses the Hebrew alphabet. Classification: Indo-European, Germanic, West, High German, Yiddish. Dialects: Southeastern Yiddish, Mideastern Yiddish, and Northeastern Yiddish.

YOM – pil, (syn. Pila, Pilapila, Kpilakpila). Yom is spoken by 74,000 people in Benin. Classification: Niger-Congo, Atlantic-Congo, Volta-Congo, North, Gur, Central, Northern, Oti-Volta, and Yom-Nawdm.

YORUBA – yor, (syn. Yooba, Yariba). Yoruba is spoken by 18,850,000 of primarily the Yoruba people in Nigeria. It is also spoken in Benin and Togo in West Africa and by immigrants in the United Kingdom and United States with a total of 19,327,000 speakers in all countries. Classification: Niger-Congo, Atlantic-Congo, Volta-Congo, Benue-Congo, Defoid, Yoruboid, Edekiri. Dialects: Oyo, Ijesha, Ila, Ijebu, Ondo, Wo, Owe, Jumu, Iworro, Igbonna, Yagba, Gbedde, Egba, Akono, Aworo, Bunu (Bini), Ekiti, Ilaje, Ikale, and Awori.

ZARMA – dje, (syn. Dyerma, Dyarma, Dyabarma, Adzerma, Djerma, Zabarma, Zarbarma, Zarmaci). Zarma, a member of the Songhay language grouping, is spoken by 2,100,000 people in Niger, primarily in the southwestern part of the country and a limited number elsewhere. Classification: Nilo-Saharan, Songhai, Southern. Dialect: Kaado.

ZULU – zul, (syn. Isizulu, Zunda). Zulu is a Bantu language spoken by 9,200,000 of the Zulu people in South Africa, especially in Zululand and northern Natal. Is is also spoken in Botswana, Lesotho, Malawi, Mozambique, and Swaziland. Classification: Niger-Congo, Atlantic-Congo, Volta-Congo, Benue-Congo, Bantoid, Southern, Narrow Bantu, Central, S, Nguni (S.40). Dialects: Lala and Qwabe; similar to Swazi and Xhosa.

ZUNI – zun, (syn. Zunni, Shiwi'ma). Zuni is spoken by 9,651 of the Zuni Indian tribe in western New Mexico and eastern Arizona. Classification: language isolate.

Chapter 7. Language synonyms[1]

Abakwariga – see Hausa
Aborlan Tagbanwa – see Tagbanwa
Aboro – see Berom
ABUA – abn
Abuan – see Abua
Abyssinian – see Amharic
Acadian – see Cajun French
Accra – see Ga
Achawa – see Yao
Acra – see Ga
Adal – see Afar
Adamawa – see Fulfulde, Adamawa
Adangme – see Dangme
Adare – see Harari
Adarinnya – see Harari
Addo – see Edo
Adere – see Harari
Aderinya – see Harari
Adja – see Aja
Adsawa – see Yao
Adsoa – see Yao
Adzerma – see Zarma
Aeta Negrito – see Sambal, Botolan
Afan Oromo – see Oromo, West Central
Afango – see Berom
AFAR – aar
Afar Af – see Afar
Afaraf – see Afar
Af-Maxaad Tiri – see Somali
AFRIKAANS – afr
Af-Soomaali – see Somali
AGAT – *afa*
Agni – see Anyin
AGTA, CASIGURAN DUMAGAT – dgc
Ahka – see Akha
Aini – see Akha
Aiysyen – see Creole, Haitian French
AJA – aja
Ajawa – see Yao
Ajja – see Aja
Ak'a – see Akha
Aka – see Akha
AKAN – aka
AKAN-AKUAPEM – aka-*aub*
AKAN-ASANTE – aka-*aag*
AKAN-FANTE – aka-*fat*
AKHA – ahk
Aklan – see Aklanon
Aklano – see Aklanon
AKLANON – akl
Aklanon-Bisayan – see Aklanon
AKPA – akf
AKWA – akw
Akweya – see Akpa
Alada – see Gun
Alada-Gbe – see Gun
Alante – see Balanta-Kentohe
Albanian – see Albanian, Tosk
ALBANIAN, TOSK – als
Albannach – see Gaelic, Scottish
Alealum – see Malayalam
ALEUT – ale
Amanung Sisuan – see Pampangan
Amarigna – see Amharic
Amarinya – see Amharic
Ambo – see Ndonga
AMHARIC – amh
Amina – see Ga
Anatolian – see Turkish
Andhra – see Telugu
ANDO – *ang*
Anguru – see Lomwe, Malawi
Annamese – see Vietnamese
ANUAK – anu
Anyi – see Anyin
ANYIN – any
Anyuak – see Anuak
Anywa – see Anuak
Anywak – see Anuak
Apabhramsa – see Maithili
APACHE, JICARILLA – apj
Apurawnon – see Tagbanwa
ARABIC – arb
ARMENIAN – hye
Armjansik Yazyk – see Armenian
Arnaut – see Albanian, Tosk
Asambe – see Assamese
Asami – see Assamese

[1] The names of the languages listed in bold capitals are indicated as correct in *Ethnologue – Languages of the World*, 15th edition. Since many languages have more than one name, their synonyms are also listed and keyed back to the appropriate name. Each language has a three letter ISO/DIS 639-3 identity code; in instances where the identity of the language is not definitive, a temporary code has been assigned and is in italics.

Asamiya – see Assamese
Asanga – see Sanga
ASSAMESE – asm
ATTA, PAMPLONA – att
Atto – see Giangan
Ayan – see Bassari
Ayao – see Yao
Ayawa – see Yao
Ayo – see Yao
Ayta Hambali – see Sambal, Botolan
Azerbaijani – see Azerbaijani, Northern
AZERBAIJANI, NORTHERN – azj
Azerbaydzhani – see Azerbaijani, Northern
Azeri Turk – see Azerbaijani, Northern
Baatombu – see Baatonum
Baatonu – see Baatonum
BAATONUM – bba
Bacavès – see Catalan-Valencian-Balear
Bacheve – see Iceve-Maci
Badag – see Badaga
BADAGA – bfq
Badagu – see Badaga
Baduga – see Badaga
Badugu – see Badaga
BAFIA – ksf
BAFUT – bfd
BAGA – *bad*
Bagobo – see Giangan
Bahasa Indonesia – see Indonesian
Bahasa Malaysia – see Malay
Bahasa Malayu – see Malay
Bai Miao – see Hmong-Daw
Baihua – see Chinese, Yue
Balanda – see Balanta-Kentohe
Balant – see Balanta-Kentohe
Balanta – see Balanta-Kentohe
BALANTA-KENTOHE – ble
Balante – see Balanta-Kentohe
Balgarski – see Bulgarian
Ballante – see Balanta-Kentohe
BALTI – bft
Baltistani – see Balti
Balud – see Blaan, Sarangani
Bama – see Burmese
Bamachaka – see Burmese
Bamana – see Bamanankan
BAMANANKAN – bam
Bambara – see Bamanankan
BAMILEKE – *bai*
Bande – see Budik
Banga-Bhasa – see Bengali
Bangala – see Bengali
Bangla – see Bengali
Banglori – see Kannada
Bangon – see Buhid
Bankoti – see Konkani
BAOULÉ – bci
Bara – see Bodo
Barba – see Baatonum
Barganchi – see Baatonum
Bargawa – see Baatonum
Bargu – see Baatonum
Bariba – see Baatonum
Baruba – see Baatonum
Bas Sorabe – see Sorbian, Lower
Basa Mathura – see Madura
Basari – see Bassari
Basari du Bandemba – see Budik
Bashgali – see Kati
BASQUE – eus
BASSARI – bsc
Batangan – see Buhid
Baule – see Baoulé
Bawule – see Baoulé
Baya – see Gbaya
Bechere – see Iceve-Maci
Becheve – see Iceve-Maci
Bedik – see Budik
BEEMBE – beq
Beembe – see Bembé
Beetjuans – see Tswana
Befe – see Bafut
Beifang Fangyan – see Chinese, Mandarin
Bekpak – see Bafia
Belante – see Balanta-Kentohe
BELARUSAN – bel
Belarussian – see Belarusan
Belorussian – see Belarusan
BEMBA – bem
BEMBÉ – bmb
Bembe – see Beembe
BENGALI – ben
Benguet-Igorot – see Ibaloi
Benin – see Edo
Berba – see Baatonum
Berdama – see Nama-Damara
Bergdamara – see Nama-Damara
BEROM – bom
Berum – see Berom
BÉTÉ – *byf*

BETE-BENDI – btt
Bette-Bendi – see Bete-Bendi
Bhotia – see Tibetan
Bhotia of Baltistan – see Balti
Bicalano – see Bicolano, Cental
BICOLANO, CENTRAL – bcl
Bielorussian – see Belarusan
Bihari – see Maithili
Biira – see Fulfulde, Adamawa
Biko – see Bicolano, Cental
Bikol – see Bicolano, Cental
Bilaan – see Blaan, Sarangani
Bini – see Edo
Binisaya – see Cebuano
Binisaya – see Wáray-Wáray
Binokid – see Binukid
BINUKID – bkd
Binukid Manobo – see Binukid
Birom – see Berom
Bisayan – see Cebuano
Biyan – see Bassari
BLAAN, SARANGANI – bps
Black Bobo – see Bodo Madaré, Southern
Blackfeet – see Blackfoot
BLACKFOOT – bla
Bobo – see Bodo Madaré, Southern
Bobo Fi – see Bodo Madaré, Southern
Bobo Fing – see Bodo Madaré, Southern
BOBO MADARÉ, SOUTHERN – bwq
Bodéwadmi Zheshmowen – see Potawatomi
Bodéwadmimwen – see Potawatomi
Bodi – see Bodo
BODO – brx
Bogung – see Baatonum
Boki – see Bokyi
Bokmaal – see Norwegian
Bokmål – see Norwegian
BOKYI – bky
Bola – see Mankanya
Bolom – see Bullom So
Bontoc – see Bontoc, Central
BONTOC, CENTRAL – bnc
Bontok – see Bontoc, Central
Boro – see Bodo
Boro-Aboro – see Berom
Boroni – see Bodo
BOSNIAN – bos
Botolan Zambal – see Sambal, Botolan
Boulbe – see Fulfulde, Adamawa
Boulou – see Bulu
Boumpe – see Mende
Brassa – see Balanta-Kentohe
BRETON – bre
Brezhoneg – see Breton
BUDIK – tnr
Bufe – see Bafut
BUGHOTU – bgt
Bugota – see Bughotu
Bugoto – see Bughotu
Bugotu – see Bughotu
BUHID – bku
Bukidnon – see Binukid
Bukil – see Buhid
Bulanda – see Balanta-Kentohe
Bulem – see Bullom So
BULGARIAN – bul
Bullin – see Bullom So
BULLOM SO – buy
Bullun – see Bullom So
Bulom – see Bullom So
BULU – bum
Burgu – see Baatonum
BURMESE – mya
Butbut – see Kalinga, Butbut
BUTUANON – btw
Byelrussian – see Belarusan
Caboverdiano – see Crioulo, Cape Verde
Cabrai – see Kabiyé
Cabrais – see Kabiyé
Cadien – see Cajun French
Caffre – see Xhosa
Cafre – see Xhosa
Cagayan – see Gaddang
Cajan – see Cajun French
Cajun – see Cajun French
CAJUN FRENCH – frc
Calabar – see Efik
Cambodian – see Khmer, Central
Canarese – see Kannada
Cantonese – see Chinese, Yue
Capampañgan – see Pampangan
Car – see Nicobarese, Car
CAROLINIAN – cal
Cashmeeree – see Kashmiri
Cashmiri – see Kashmiri
Casiguran Dumagat – see Agta, Casiguran Dumagat
Castellano – see Spanish
Castilian – see Spanish
Català – see Catalan-Valencian-Balear

Danakil – see Afar
Dandawa – see Dendi
DANGME – ada
DANGME-KROBO – ada-*kro*
DANISH – dan
Dansk – see Danish
Dari – see Farsi, Eastern
Dbus – see Tibetan
Dbusgtsang – see Tibetan
Degati – see Dagaare, Southern
Delnoserbski – see Sorbian, Lower
Dema – see Tupuri
DENDI – ddn
Denkel – see Afar
Deutsch – see German
Dho Anywaa – see Anuak
Dialonke – see Yalunka
Dibagat-Kabugao-Isneg – see Isnag
DIGO – dig
Diné Bizaad – see Navajo
Diola – see Jola-Fonyi
Diola-Fogny – see Jola-Fonyi
Dioula – see Jula
Ditamari – see Ditammari
DITAMMARI – tbz
Diula – see Jula
Diyi – see Junkun Takum
Djallonke – see Yalunka
Djao – see Yao
Djawa – see Javanese
Djedji – see Fon
Djerma – see Zarma
Djula – see Jula
Dogaari – see Dagaare, Southern
Dolnoserbski – see Sorbian, Lower
Domona – see Fulfulde, Adamawa
Dondo – see Doondo
DOONDO – dde
Duhlian Twang – see Mizo
Dulien – see Mizo
DUTCH – nld
Dyabarma – see Zarma
Dyarma – see Zarma
Dyerma – see Zarma
Dyola – see Jola-Fonyi
Dyoula – see Jula
Dyula – see Jula
Dzemay – see Fulfulde, Adamawa
Dzhudezmo – see Ladino
Eastern Arctic Eskimo – see Inuktitut
Eastern Canadian Eskimo – see Inuktitut
Eastern Fijian – see Fijian
Eastern Fulani – see Fulfulde, Adamawa
Eastern Fulfulde – see Fulfulde, Adamawa
Eastern Malinke – see Manding-Maninka
Eastern Malinke – see Maninkakan, Eastern
Eastern Pahari – see Nepali
Eastern Thai – see Lao
Eastern Yiddish – see Yiddish
Ebembe – see Bembé
Ebwe – see Éwé
Echuabo – see Chuwabu
Echuwabo – see Chuwabu
EDE NAGO – nqg
EDO – bin
Eesti – see Estonian
Efe – see Éwé
EFIK – efi
Egun – see Gun
Eibe – see Éwé
Ekaw – see Akha
Ekoka-!Xû – see Kung-Ekoka
Ellinika – see Greek
Emakchu – see Makhuwa
Emakhuwa – see Makhuwa
Emakua – see Makhuwa
ENGLISH – eng
Ermeni Dili – see Armenian
Ermenice – see Armenian
Erse – see Gaelic, Irish
Erse – see Gaelic, Scottish
Eskimo – see Inupiatun
Español – see Spanish
ESTONIAN – est
Ethiopian – see Amharic
Eto – see Giangan
Eue – see Éwé
Euskera – see Basque
Eve – see Éwé
ÉWÉ – ewe
EWONDO – ewo
Ewundu – see Ewondo
FARSI, EASTERN – prs
FARSI, WESTERN – pes
Fiji – see Fijian
FIJIAN – fij
FILIPINO – fil
Fillanci – see Fulfulde, Adamawa
FINNISH – fin
Flamand – see Vlaams

Flemish – see Vlaams
Fo – see Fon
Fogbe – see Fon
FON – fon
Fongbe – see Fon
Fonnu – see Fon
Foula Fouta – see Pular
Foulfoulde – see Fulfulde, Adamawa
Fouta Dyalon – see Pular
Français – see French
Français Acadien – see Cajun French
Frankish – see Luxenbourgeois
Frase – see Balanta-Kentohe
FRENCH – fra
Fu – see Bafut
Ful – see Fulfulde, Adamawa
Fula – see Fulfulde, Adamawa
Fula-Fulfulde – see Fulfulde, Adamawa
Fulani – see Fulfulde, Adamawa
Fulatanchi – see Fulfulde, Adamawa
Fulbe – see Fulfulde, Adamawa
Fulbe – see Pular
FULFULDE – *ful*
Fulfulde – see Fulfulde, Adamawa
Fulfulde Jalon – see Pular
FULFULDE, ADAMAWA – fub
Fullo Fuuta – see Pular
Fut – see Bafut
Futa Fula – see Pular
Futa Jallon – see Pular
Fuuta Jalon – see Pular
GA – gaa
GADDANG – gad
Gaeilge – see Gaelic, Irish
Gaelg – see Gaelic, Manx
Gaelic – see Gaelic, Scottish
GAELIC, IRISH – gle
GAELIC, MANX – glv
GAELIC, SCOTTISH – gla
Gaidhlig – see Gaelic, Scottish
Gàidhlig – see Gaelic, Scottish
Gailck – see Gaelic, Manx
Gain – see Ga
Galancho – see Chechen
Galcha – see Tajiki
Galego – see Galician
GALICIAN – glg
Galla – see Oromo, West Central
Gallego – see Galician
Gallinas – see Vai
Gallines – see Vai
Gan – see Chinese, Gàn
GANDA – lug
Gapelta – see Fulfulde, Adamawa
Gbang – see Berom
GBAYA – gya
Gbaya Nord-Ouest – see Gbaya
Gbe – see Éwé
GBE, WACI – wci
GBE-VHE – *gbe-vhe*
Ge – see Gen
Gegbe – see Gen
Gekoyo – see Gikuyu
GEN – gej
Gen-Gbe – see Gen
Gentoo – see Telugu
GEORGIAN – kat
GERMAN – deu
Gey Sinan – see Harari
GIANGAN – bgi
Gigikuyu – see Gikuyu
GIKUYU – kik
Gilbertese – see Kiribati
Gin – see Vietnamese
Gironga – see Ronga
Gissi – see Kisi, Southern
Gizi – see Kisi, Southern
Gizi – see Kissin Northern
GOLA – gol
GONJA – gjn
Gorkhali – see Nepali
Goulmacema – see Gourmanchéma
Goun – see Gun
Gourma – see Gourmanchéma
GOURMANCHÉMA – gux
GOURMANTCHÉ – fuh
Gourmantche – see Gourmanchéma
Gourmantché – see Gourmanchéma
Gouro – see Guro
Graecae – see Greek
Great Thai – see Shan
Grec – see Greek
Greco – see Greek
GREEK – ell
Greenlandic, Kalaallisut – see Inuktitut, West Greenlandic
Gruzinski – see Georgian
Gu – see Gun
Guanga – see Giangan
Guanhua – see Chinese, Mandarin

Gugbe – see Gun
Guin – see Gen
Guinea-Bissau Creole – see Crioulo, Upper Guinea
GUJARATI – guj
Gujerathi – see Gujarati
Gujerati – see Gujarati
Gujrathi – see Gujarati
Gula – see Gola
Gulanga – see Giangan
Gulimancema – see Gourmanchéma
Gulmancema – see Gourmanchéma
GUN – guw
Gun-Alada – see Gun
Gun-Gbe – see Gun
Guoyu – see Chinese, Mandarin
Gurkhali – see Nepali
Gurma – see Gourmanchéma
Gurmukhi – see Panjabi, Eastern
GURO – goa
Gurumukhi – see Panjabi, Eastern
Gwamba – see Tsonga
Gwong Dung Waa – see Chinese, Yue
Gypsy – see Romani, Balkan
Habe – see Hausa
Haiao – see Yao
Haieren – see Armenian
Haitian – see Creole, Haitian French
Haitian Creole – see Creole, Haitian French
Hakka – see Chinese, Hakka
Halh – see Mongolian
Halh Mongolian – see Mongolian
Hanguk Mal – see Korean
Hanguohua – see Korean
Hanonoo – see Hanunoo
HANUNOO – hnn
Haoussa – see Hausa
Haquetiya – see Ladino
Har – see Santali
HARARI – har
Hararri – see Harari
Harsi ʻAforit – see Harsūsi
HARSŪSI – hss
HAUSA – hau
Hausawa – see Hausa
Haussa – see Hausa
Haut Sorabe – see Sorbian, Upper
HAWAIIAN – haw
HEBREW – heb
HERERO – her
Hersyet – see Harsūsi
Hiao – see Yao
Hiligainon – see Hiligaynon
HILIGAYNON – hil
HINDI – hin
HINDUSTANI – hns
HMONG-DAW – mww
Hmong-Der – see Hmong-Daw
HMONG-NJUA – blu
Hmong-White – see Hmong-Daw
Hokka – see Chinese, Hakka
Hollands – see Dutch
Honya – see Tupuri
HOPI – hop
Hopìikwa – see Hopi
Hopilàvayi – see Hopi
Hor – see Santali
Hornjoserbski – see Sorbian, Upper
Hornoserbski – see Sorbian, Upper
Horo – see Mundari
Hottentot – see Nama-Damara
Hrvatski – see Croatian
Hsiang – see Chinese, Xiang
!Hu – see Kung-Ekoka
Hualngo – see Mizo
Hulo – see Mende
Humba – see Kwanyama
Hunan – see Chinese, Xiang
Hunanese – see Chinese, Xiang
HUNGARIAN – hun
Hyao – see Yao
Ibadoy – see Ibaloi
IBALOI – ibl
Ibaloy – see Ibaloi
IBANAG – ibg
Ibembe – see Bembé
Ibo – see Igbo
ICELANDIC – isl
ICEVE-MACI – bec
Icheve – see Iceve-Maci
Ichibemba – see Bemba
Icibemba – see Bemba
Ici-Ndali – see Ndali
IFUGAO – *ifc*
IGBO – ibo
Igodor – see Ibaloi
Igorot – see Bontoc, Central
Ijaw – see Izon
Ijo – see Izon
Ijo-Izon – see Izon

Ikaw – see Akha
Ikingonde – see Nyakyusa-Ngonde
Ikinyakyusa – see Nyakyusa-Ngonde
Ikiribati – see Kiribati
I-Kiribati – see Kiribati
Ikor – see Akha
Illogo – see Hiligaynon
ILOCANO – ilo
Ilokano – see Ilocano
Iloko – see Ilocano
Ilonggo – see Hiligaynon
INDONESIAN – ind
Ingundi – see Ngundi
Inibaloi – see Ibaloi
Inuit – see Inuktitut
INUKTITUT – ike
INUKTITUT, WEST GREENLANDIC – kal
Iñupiak – see Inupiatun
Inupiaq – see Inupiatun
Inupiat – see Inupiatun
INUPIATUN – esi/esk
Inupiatun – see Inupiatun
Irani – see Farsi, Western
Irish – see Gaelic, Irish
Isanga – see Sanga
Isixhosa – see Xhosa
Isizulu – see Zulu
Íslenska – see Icelandic
ISNAG – isd
Isneg – see Isnag
ITALIAN – ita
Italiano – see Italian
Itawes – see Itawit
Itawis – see Itawit
ITAWIT – itv
IVATAN – ivv
Ivrit – see Hebrew
Izo – see Izon
IZON – ijc
Jaiselmer – see Marwari
Jallonka – see Yalunka
Jalon – see Pular
Jalonke – see Yalunka
Jambo – see Anuak
Jangan – see Giangan
Jao – see Yao
JAPANESE – jpn
Jaunde – see Ewondo
JAVANESE – jav
Jawa – see Javanese
Jiangxihua – see Chinese, Gàn
Jing – see Vietnamese
Jola – see Jola-Fonyi
Jola-Fogny – see Jola-Fonyi
JOLA-FONYI – dyo
Jolohano – see Tausug
Jóola – see Jola-Fonyi
Judæo-Espagnol – see Ladino
Judæo-Spanish – see Ladino
Judeo Spanish – see Ladino
Judeoespañol – see Ladino
Judeo-German Pila – see Yiddish
Judezmo – see Ladino
Jukun Diyi – see Junkun Takum
JULA – dyu
JUNKUN TAKUM – jbu
KABIYÉ – kbp
Kabre Diyi – see Kabiyé
Kabure – see Kabiyé
Kabuverdianu – see Crioulo, Cape Verde
Kabye – see Kabiyé
Kachari – see Bodo
Kacmiri – see Kashmiri
Kado – see Hausa
Kaffer – see Xhosa
Kaffir – see Xhosa
Kaisak – see Kazakh
Kaixien – see Lahu
Kakuya Bushman Nasie – see Nama-Damara
KALAGAN – kqe
KALINGA, BUTBUT – kyb
Kam Mu'ang – see Thai, Northern
Kammüang – see Thai, Northern
Kammyang – see Thai, Northern
Kan – see Chinese, Gàn
Kanarese – see Kannada
Kankan Maninka – see Manding-Maninka
Kankan Maninka – see Maninkakan, Eastern
KANKANAEY – kne
Kankanai – see Kankanaey
Kankanay – see Kankanaey
KANNADA – kan
Kanouri – see Kanuri
Kanoury – see Kanuri
KANURI – kby
Kanyop – see Mandjak
Kapampangan – see Pampangan
Kara-Kirgiz – see Kirghiz
KAREN – *kar*
Kartuli – see Georgian

Kaschemiri – see Kashmiri
KASHMIRI – kas
KATI – bsh
Kativiri – see Kati
Kaw – see Akha
Kazak – see Kazakh
KAZAKH – kaz
Ke – see Chinese, Hakka
Kechia – see Chinese, Hakka
Kejia – see Chinese, Hakka
Kermancî – see Kurdish, Northern
Keshur – see Kashmiri
Kha Ko – see Akha
Khadi Boli – see Hindi
Khae – see Lisu
Khako – see Akha
Khalkha Mongolian – see Mongolian
Khao Kha Ko – see Akha
Khari Boli – see Hindi
Khaskura – see Nepali
Khmer – see Khmer, Central
KHMER, CENTRAL – khm
Khoekhoegowab – see Nama-Damara
Khoekhoegowap – see Nama-Damara
Khon Mung – see Thai, Northern
Khon Myang – see Thai, Northern
!Khung – see Kung-Ekoka
Kibbo – see Berom
Kibbun – see Berom
Kibeembe – see Beembe
Kibembe – see Bembé
Kibo – see Berom
Kibyen – see Berom
Kidigo – see Digo
Kidoondo – see Doondo
Kikongo – see Koongo
Kikoongo – see Koongo
Kikuyu – see Gikuyu
Killi – see Mundari
Kindali – see Ndali
Kinh – see Vietnamese
Kinyakyusa – see Nyakyusa-Ngonde
KIOWA – kio
KIRGHIZ – kir
Kirgiz – see Kirghiz
KIRIBATI – gil
Kirmancî – see Kurdish, Northern
Kiryol – see Crioulo, Upper Guinea
Kisi – see Kisi, Southern
Kisi – see Kissi, Northern
KISI, SOUTHERN – kss
Kisie – see Kissi, Northern
KISSI, NORTHERN – kqs
Kissien – see Kisi, Southern
Kissien – see Kissi, Northern
Kisuaheli – see Swahili
Kiswahili – see Swahili
Kitonga – see Tonga
Kjalonke – see Yalunka
Klipkaffer – see Nama-Damara
Klipkaffern – see Nama-Damara
Ko – see Akha
Kom Komba – see Konkomba
Konde – see Nyakyusa-Ngonde
Kongo – see Koongo
Koniagui – see Wamey
Konkan Standard – see Konkani
Konkanese – see Konkani
KONKANI – knn
KONKOMBA – xon
Konnoh – see Kono
KONO – kno
Konyagi – see Wamey
KOONGO – kng
Koosa – see Xhosa
Koranko – see Kuranko
KOREAN – kor
KORO – kfo
Koro Jula – see Koro
Kosach – see Kazakh
Kossa – see Mende
Kosso – see Mende
Kotokoli – see Tem
Kpankpam – see Konkomba
Kpilakpila Pila – see Yom
Kreol – see Morisyen
Kreole – see Morisyen
Krepe – see Éwé
Krepi – see Éwé
Kreyol – see Creole, Haitian French
KRIO – kri
Kriulo – see Crioulo, Upper Guinea
!Ku – see Kung-Ekoka
Kuanyama – see Kwanyama
Kujamataak – see Jola-Fonyi
Kújoolaak Kati Fooñi – see Jola-Fonyi
Kūki ʻĀirani – see Rarotongan
Kukwe – see Nyakyusa-Ngonde
Kunabi – see Konkani
!Kung – see Kung-Ekoka

KUNG-EKOKA – knw
KURANKO – knk
Kurdi – see Kurdish, Northern
Kurdî – see Kurdish, Northern
KURDISH, NORTHERN – kmr
Kurmancî – see Kurdish, Northern
Kurmanji – see Kurdish, Northern
Kwancama – see Kwanyama
KWANGALI – kwn
Kwangare – see Kwangali
Kwangari – see Kwangali
Kwanjama – see Kwanyama
KWANYAMA – kua
Kwéndré – see Guro
Kweni – see Guro
KWÉYÒL – acf
La Nya – see Thai, Northern
LADINO – lad
LAHU – lhu
Lahuna – see Lahu
Lakhota – see Lakota
LAKOTA – lkt
Laku – see Lahu
Lallans – see Scots
Lan Na – see Thai, Northern
Lanatai – see Thai, Northern
Landogo – see Loko
Lanna – see Thai, Northern
LAO – lao
Lao – see Lao
Lào – see Lao
Lao Kao – see Lao
Lao Wiang – see Lao
Lao-Lum – see Lao
Lao-Tai – see Lao
Laotian – see Lao
Laotian Tai – see Lao
LATVIAN – lav
Leisu – see Lisu
Lemba – see Malimba
Leshuoopa – see Lisu
Lesser Antillean Creole French – see Kwéyòl
Lesuo – see Lisu
Lettisch – see Latvian
Lettish – see Latvian
Lëtzebuergesch – see Luxenbourgeois
Letzhurgisch – see Luxenbourgeois
Li – see Lisu
Lietuvi – see Lithuanian
Lietuviskai – see Lithuanian
Li-Hsaw – see Lisu
Likpakpaln – see Konkomba
Limba – see Malimba
LIMBA – lia
Lip'a – see Lisu
Lisaw – see Lisu
Li-Shaw – see Lisu
Lishu – see Lisu
Liso – see Lisu
Lissu – see Lisu
LISU – lis
Litauische – see Lithuanian
Litewski – see Lithuanian
LITHUANIAN – lit
Litovskiy – see Lithuanian
Llaruro – see Pumé
Lo – see Guro
Lohei – see Lahu
Loisu – see Lisu
LOKO – lok
LOMWE, MALAWI – lon
Lower Lusatian – see Sorbian, Lower
Lower Wendish – see Sorbian, Lower
Luganda – see Ganda
Luhya – see Luyia
Lukhai – see Mizo
Luluyia – see Luyia
Lum Lao – see Lao
LUO – *luo*
Lusago – see Mizo
Lusai – see Mizo
Lusatian – see Sorbian, Lower
Lusei – see Mizo
Lushai – see Mizo
Lushei – see Mizo
Lusu – see Lisu
Lu-Tzu – see Lisu
Luvenḓa – see Venda
Luxembourgish – see Luxenbourgeois
Luxemburgian – see Luxenbourgeois
Luxemburgish – see Luxenbourgeois
LUXENBOURGEOIS – ltz
LUYIA – luy
MACEDONIAN – mkd
MACEDONIAN – mkd
Macedonian Slavic – see Macedonian
Macua – see Makhuwa
Madhura – see Madura
Madrassi – see Kannada
MADURA – mad

Madurese – see Madura
Maeinduque – see Marinduque
Magindanaon – see Maguindanao
Magindanaw – see Maguindanao
MAGUINDANAO – mdh
Magyar – see Hungarian
Mahaga – see Bughotu
Maharashtra – see Marathi
Maharathi – see Marathi
Mahri – see Mehri
MAITHILI – mai
Maitili – see Maithili
Maitli – see Maithili
Majak – see Mandjak
Majiahau – see Chinese, Hakka
Makedonski – see Macedonian
MAKHUWA – vmw
Makhuwa-Makhuwana – see Makhuwa
Makhuwwa of Nampula – see Makhuwa
Makoane – see Makhuwa
Makua – see Makhuwa
MALAGASY – plt
MALAY – mly
MALAYALAM – mal
Malayalani – see Malayalam
Malayali – see Malayalam
Malayu – see Malay
Malean – see Malayalam
Malhatee – see Marathi
MALIMBA – mzd
Malinka – see Maninka
Malinka – see Maninkakan, Western
Malinke – see Maninka
Malinke – see Maninkakan, Western
Maliyad – see Malayalam
Mallealle – see Malayalam
MALTESE – mlt
Malti – see Maltese
Mancagne – see Mankanya
Mancang – see Mankanya
Mancanha – see Mankanya
Mandari – see Mundari
Mandarin – see Chinese, Mandarin
Mande – see Manding-Maninka
Mandé – see Mandinka
Mande – see Maninkakan, Eastern
Mandenyi – see Bullom So
Manding – see Mandinka
MANDING-BAMBARA – *m-bam*
Mandingi – see Bullom So
MANDING-MANDINKA – *m*-mnk
MANDING-MANINKA – *m-emk*
Mandingo – see Mandinka
Mandingue – see Mandinka
MANDINKA – mnk
Mandinque – see Mandinka
MANDJAK – mfv
Mandjaque – see Mandjak
Mandyak – see Mandjak
Manga – see Kanuri
Maninga – see Maninkakan, Western
MANINKA – *man*
Maninka – see Manding-Maninka
Maninka – see Maninkakan, Eastern
Maninkakan – see Maninka
MANINKAKAN, EASTERN – emk
MANINKAKAN, WESTERN – mlq
Maninka-Western – see Maninkakan, Western
Manjaca – see Mandjak
Manjack – see Mandjak
Manjaco – see Mandjak
Manjak – see Mandjak
Manjaku – see Mandjak
Mankanha – see Mankanya
MANKANYA – knf
MANOBO – *mno*
Manobo – see Binukid
MANU'A SAMOAN – *m*-smo
Manx – see Gaelic, Manx
MAORI – mri
Māori – see Rarotongan
Maqua – see Nama-Damara
Maquoua – see Makhuwa
Maragat – see Isnag
MARANAO – mrw
Maranaw – see Maranao
MARATHI – mar
Marawar – see Marwari
MARINDUQUE – tgl
MARQUESAN – mrq/mqm
Marthi – see Marathi
MARWARI – mve
Marwari Bhil – see Marwari
Marwari Meghwar – see Marwari
Mata – see Tupuri
Mau – see Shan
Mauritian – see Morisyen
Mauritius Creole French – see Morisyen
Maurysen – see Morisyen
Mbeembe – see Bembé

Mbughotu – see Bughotu
Mech – see Bodo
Meche – see Bodo
Mechi – see Bodo
Meci – see Bodo
MEHRI – gdq
Melaju – see Malay
Melanesian English – see Tok Pisin
Melayu – see Malay
MENDE – men
Meo – see Hmong-Njua
Meo Kao – see Hmong-Daw
Merwari – see Marwari
Methli – see Maithili
Mgbakpa – see Hausa
Mi'gmaq – see Micmac
Mi'kmaq – see Micmac
Miao – see Hmong-Njua
MICMAC – mic
Migulimancema – see Gourmanchéma
Miigmao – see Micmac
Min Pei – see Chinese, Min Bei
Mina – see Gen
Mina-Gen – see Gen
Minang – see Minangkabau
MINANGKABAU – min
Minnan – see Chinese, Min Nan
MIZO – lus
Mmani – see Bullom So
MOHAWK – moh
Moldavian – see Romanian
Mole – see Mòoré
Mombe – see Nyakyusa-Ngonde
Mondari – see Mundari
Mongol – see Mongolian
MONGOLIAN – khk
Mongolian – see Mongolian
MÒORÉ – mos
Moose – see Mòoré
Mopla – see Malayalam
More – see Mòoré
MORISYEN – mfe
Moro Joloano – see Tausug
Moselle Franconian – see Luxenbourgeois
Moshi – see Mòoré
Moso – see Lahu
Mossi – see Mòoré
Mu'ang – see Thai, Northern
Mudima – see Malimba
Muhso – see Lahu
Mulimba – see Malimba
Munari – see Mundari
Munda – see Mundari
MUNDARI – muw
Munshi – see Tiv
Muruthu – see Marathi
Muscogee – see Muskogee
MUSKOGEE – mus
Mussar – see Lahu
Musso – see Lahu
Mussuh – see Lahu
Myang – see Thai, Northern
Myanmar – see Burmese
Myen – see Burmese
Nabaloi – see Ibaloi
Nadroga – see Fijian
Nadronga – see Fijian
Nago – see Ede Nago
Nagot – see Ede Nago
Nagots – see Ede Nago
NAMA-DAMARA – naq
Namakwa – see Nama-Damara
Naman – see Nama-Damara
Namaqua – see Nama-Damara
Namen – see Lahu
Navaho – see Navajo
NAVAJO – nav
NDALI – ndh
Ndjak – see Mandjak
NDONGA – ndo
Ndoore – see Tupuri
Ndore – see Tupuri
Nederlands – see Dutch
Neo-Hellenic – see Greek
Neomelanesian – see Tok Pisin
Nepal Bhasa – see Newar
Nepalese – see Nepali
NEPALI – nep
Neshnabémwen – see Potawatomi
New Guinea Pidgin English – see Tok Pisin
New Persian – see Farsi, Western
New Zealand Maori – see Maori
NEWAR – new
Newari – see Newar
Nfua – see Bokyi
Ngbanyito – see Gonja
Ngeo – see Shan
Ngezzim – see Ngizim
Ngiao – see Shan
Ngiaw – see Shan

Ngio – see Shan
Ngiow – see Shan
NGIZIM – ngi
Ngizmawa – see Ngizim
Ngonde – see Nyakyusa-Ngonde
Ngondi – see Ngundi
NGUNDI – ndn
Nguru – see Lomwe, Malawi
NICOBARESE, CAR – caq
Niedersorbisch – see Sorbian, Lower
NIUE – niu
Niuean – see Niue
Niuefekai – see Niue
Njikum Diyi – see Junkun Takum
Nki – see Bokyi
Nkonde – see Nyakyusa-Ngonde
Noi – see Lao
Nokchiin Muott – see Chechen
Nokhchiin – see Chechen
North Alaskan Inupiat – see Inupiatun
North Alaskan Inupiatun – see Inupiatun
North Konkan – see Konkani
Northeastern Burkina Faso Fulfulde – see Gourmantché
Northern Bullom – see Bullom So
Northern Cagayan Negrito – see Atta, Pamplona
Northern Chinese – see Chinese, Mandarin
Northern Min – see Chinese, Min Bei
Northern Thai – see Thai, Northern
Northern Uzbek – see Uzbek
Northwest Alaskan Inupiatun – see Inupiatun
Northwest Alsaka Inupiat – see Inupiatun
Northwest Gbaya – see Gbaya
NORWEGIAN – nob
Norwegian Bokmål – see Norwegian
Nufawa – see Nupe-Nupe-Tako
Nupe – see Nupe-Nupe-Tako
Nupeci – see Nupe-Nupe-Tako
Nupecidji – see Nupe-Nupe-Tako
Nupenchi – see Nupe-Nupe-Tako
Nupencizi – see Nupe-Nupe-Tako
NUPE-NUPE-TAKO – nup
Nuristani – see Kati
Nuro – see Anuak
Nyakusa – see Nyakyusa-Ngonde
NYAKYUSA-NGONDE – nyy
NYANJA – nya
Nyekyosa – see Nyakyusa-Ngonde
Nyikyusa – see Nyakyusa-Ngonde
Obersorbisch – see Sorbian, Upper
Ocebe – see Iceve-Maci
Ochebe – see Iceve-Maci
Ocheve – see Iceve-Maci
Ochiherero – see Herero
Ochikwanyama – see Kwanyama
Ochindonga – see Ndonga
Odri – see Oriya
Odrum – see Oriya
OGBIA – ogb
OGBIA-AGHOLO – ogb-*hlo*
Ogbinya – see Ogbia
Ogiba-Kolo – see Ogbia-Agholo
Okii – see Bokyi
'Olelo Hawai'i – see Hawaiian
'Olelo Hawai'i Makuahine – see Hawaiian
Oliya – see Oriya
Onëyan – see Bassari
Onian – see Bassari
Orissa – see Oriya
ORIYA – ori
Oromiffa – see Oromo, West Central
OROMO, WEST CENTRAL – gaz
Oromoo – see Oromo, West Central
Oshindonga – see Ndonga
Osikom – see Bokyi
Osindonga – see Ndonga
Osukam – see Bokyi
Otjiherero – see Herero
Otjiwambo – see Kwanyama
Otjiwambo – see Ndonga
Ouatchi – see Gbe, Waci
Ouolof – see Wolof
Ovambo – see Kwanyama
Oviedo – see Edo
Ovioba – see Edo
Owambo – see Kwanyama
Owambo – see Ndonga
Özbek – see Uzbek
Padang – see Minangkabau
PAIUTE, SOUTHERN – ute
Pakhto – see Pashto, Northern
Palata – see Fulfulde, Adamawa
Palawanen – see Palawano, Central
PALAWANO, CENTRAL – plc
Palaweño – see Palawano, Central
Paldena – see Fulfulde, Adamawa
PAMPANGAN – pam
Pampango – see Pampangan
Pampangueño – see Pampangan

Panay – see Aklanon
PANGASINAN – pag
Panjabi – see Panjabi, Eastern
PANJABI, EASTERN – pan
Parbatiya – see Nepali
Parsi – see Farsi, Eastern
Parsi – see Farsi, Western
Pashto – see Pashto, Northern
PASHTO, NORTHERN – pbu
Pashtu – see Pashto, Northern
Patois – see Krio
Patois – see Kwéyòl
Patwa – see Kwéyòl
Payap – see Thai, Northern
Pe Miao – see Hmong-Daw
Pedi – see Sotho, Northern
Peh Miao – see Hmong-Daw
Pelta Hay – see Fulfulde, Adamawa
Persian – see Farsi, Eastern
Persian – see Farsi, Western
Peul – see Fulfulde, Adamawa
Peul – see Pulaar
Peulh – see Fulfulde, Adamawa
Peulh – see Pulaar
Phayap – see Thai, Northern
Phoke – see Tibetan
Phou Lao – see Lao
Phyap – see Thai, Northern
Pidgin – see Tok Pisin
Pikanii – see Blackfoot
Pila – see Yom
Pilapila Pila – see Yom
Pilipino – see Filipino
Pisin – see Tok Pisin
Piute – see Paiute, Southern
Pladina – see Fulfulde, Adamawa
Plateau Malagasy – see Malagasy
POLISH – pol
Polnisch – see Polish
Polski – see Polish
Popo – see Éwé
Popo – see Gen
Português – see Portuguese
PORTUGUESE – por
Portuguese Creole – see Crioulo, Upper Guinea
POTAWATOMI – pot
Pottawotomie – see Potawatomi
Priangan – see Sunda
Pu – see Nicobarese, Car
PULAAR – fuc
Pulaar Fulfulde – see Pulaar
PULAR – fuf
Pule – see Fulfulde, Adamawa
Pullo – see Fulfulde, Adamawa
PUMÉ – yae
Pushto – see Pashto, Northern
Qazaq – see Kazakh
Quezon Palawano – see Palawano, Central
Qxü – see Kung-Ekoka
Rajasthani – see Marwari
Ranao – see Maranao
RAROTONGAN – rar
Rarotongan-Mangaian – see Rarotongan
Restigouche – see Micmac
Rikpa – see Bafia
Rikpa' – see Bafia
Ripey – see Bafia
Romaic – see Greek
ROMANI, BALKAN – rmn
ROMANIAN – ron
Rong Kong – see Lao
RONGA – rng
Rooi Nasie – see Nama-Damara
Rukwangali – see Kwangali
Rumanian – see Romanian
Ruotsi – see Swedish
RUSSIAN – rus
Russki – see Russian
Sailau – see Mizo
Saint Lucian Creole French – see Kwéyòl
Saipan Carolinian – see Carolinian
Sam – see Shan
SAMA, CENTRAL – sml
Samal – see Sama, Central
Samaran – see Wáray-Wáray
Samareño – see Wáray-Wáray
Samar-Leyte – see Wáray-Wáray
SAMBAL, BOTOLAN – sbl
SAMBAL, TINÀ – xsb
Sambali – see Sambal, Tinà
SAMOAN – smo
Samtali – see Santali
Sandal – see Santali
SANGA – xsn
Sangtal – see Santali
SANSKRIT – san
Santal – see Santali
SANTALI – sat
Santhali – see Santali
Santhiali – see Santali

Sanyo – see Fulfulde, Adamawa
Satar – see Santali
Sbalti – see Balti
SCOTS – sco
Scots Gaelic – see Gaelic, Scottish
Sebuano – see Cebuano
Seereer – see Serer-Sine
Seex – see Serer-Sine
Sefardi – see Ladino
SENA – seh
Senadi – see Senoufo, Cebaara
Senari – see Senoufo, Cebaara
Senni – see Songhay
Sénoufo – see Senoufo, Cebaara
SENOUFO, CEBAARA – sef
Sentali – see Santali
Sepedi – see Sotho, Northern
SERBO-CROATIAN – hbs
Sereer – see Serer-Sine
Serer – see Serer-Sine
Sérère-Sine – see Serer-Sine
Serer-Sin – see Serer-Sine
SERER-SINE – srr
Serrer – see Serer-Sine
Sesotho – see Sotho, Southern
Seto-Gbe – see Gun
Setswana – see Tswana
Sha – see Shan
SHAN – shn
Shironga – see Ronga
Shitsonga – see Tsonga
Shiwi'ma – see Zuni
Shkip – see Albanian, Tosk
SHONA – sna
Shosho – see Berom
Shqip – see Albanian, Tosk
Shqiperë – see Albanian, Tosk
Siamese – see Thai
Siasi Sama – see Sama, Central
Sichuan-Guizhou-Yunnan Hmong – see Hmong-Njua
Sidaminya – see Sidamo
SIDAMO – sid
Sidámo 'Afó – see Sidamo
Siksika – see Blackfoot
Sikwangali – see Kwangali
Sinama – see Sama, Central
Sine-Saloum – see Serer-Sine
Sine-Sine – see Serer-Sine
Singhala – see Sinhala
Singhalese – see Sinhala
SINHALA – sin
Sinhalese – see Sinhala
Sinug Tausug – see Tausug
Siocon – see Subanon
Siska – see Tonga
Sisutho – see Sotho, Southern
Sisya – see Tonga
Sjaelland – see Danish
Skchip – see Albanian, Tosk
Slavic – see Macedonian
SLOVAK – slk
Slovakian – see Slovak
Slovene – see Slovenian
SLOVENIAN – slv
Slovenscina – see Slovenian
Socé – see Mandinka
Sochile – see Nyakyusa-Ngonde
Sokile – see Nyakyusa-Ngonde
Sokili – see Nyakyusa-Ngonde
SOMALI – som
Somba – see Ditammari
Somkhuri – see Armenian
Songai – see Songhay
Songay – see Songhay
Songhai – see Songhay
SONGHAY – hmb
Songoi – see Songhay
Songoy – see Songhay
Sonrai – see Songhay
Sonrhai – see Songhay
Sonthal – see Santali
SORBIAN, LOWER – dsb
SORBIAN, UPPER – hsb
Sose – see Susu
Soso – see Susu
SOTHO, NORTHERN – nso
SOTHO, SOUTHERN – sot
Soumo – see Sumo-Mayangna
Soussou – see Susu
Southern Carolinian – see Carolinian
Southern Dagari – see Dagaare, Southern
Southern Lisu – see Lisu
Southern Maninka – see Manding-Maninka
Southern Maninka – see Maninkakan, Eastern
Southern Min – see Chinese, Min Nan
Southern Paiute – see Paiute, Southern
Souto – see Sotho, Southern
SPANISH – spa
Spanyol – see Ladino

SRANAN – srn
Sranan Tongo – see Sranan
Standard Chinese – see Chinese, Mandarin
Standard Fijian – see Fijian
Standard German – see German
Standard Malay – see Malay
Standard Somali – see Somali
Standard Thai – see Thai
STELLINGWERFS – stl
Stellingwarfs – see Stellingwerfs
SUBANON – suc
Sug – see Tausug
Sugbuanon – see Cebuano
Sugbuhanon – see Cebuano
Sulu – see Tausug
Suluk – see Tausug
Sumo – see Sumo-Mayangna
Sumo Tawahka – see Sumo-Mayangna
SUMO-MAYANGNA – sum
Sumoo – see Sumo-Mayangna
Sumu – see Sumo-Mayangna
SUNDA – sun
Sundanese – see Sunda
Suomea – see Finnish
Suomi – see Finnish
Surinaams – see Sranan
Suriname Creole English – see Sranan
Surinamese – see Sranan
Susoo – see Susu
SUSU – sus
Suthu – see Sotho, Southern
Suto – see Sotho, Southern
Svenska – see Swedish
SWAHILI – swh
SWEDISH – swe
Swina – see Shona
Syenere – see Senoufo, Cebaara
Taareyo – see Fulfulde, Adamawa
Tadzhik – see Tajiki
TAGALOG – tgl
Tagbanua – see Tagbanwa
TAGBANWA – tbw
Taguaca – see Sumo-Mayangna
Tai Lao – see Lao
Tai Luang – see Shan
Tai Nya – see Thai, Northern
Tai Shan – see Shan
Tai Yai – see Shan
Tailangi – see Telugu
TAJIKI – tgk
Tajiki Persian – see Tajiki
Tak Miao – see Hmong-Njua
Taki-Taki – see Sranan
Tal – see Tulu
Tallu – see Tulu
Tama – see Nama-Damara
Tamakwa – see Nama-Damara
Tamal – see Tamil
Tamalsan – see Tamil
Tamari – see Ditammari
Tamboka – see Tumbuka
Tambuka – see Tumbuka
Tambul – see Tamil
TAMIL – tam
Tamili – see Tamil
Tamma – see Nama-Damara
Tandanke – see Budik
Tartar – see Tatar
TATAR – tat
Tausog – see Tausug
TAUSUG – tsg
Taw – see Tausug
Tawit – see Itawit
Te reo Ipukarea – see Rarotongan
Tedesco – see German
Teduray – see Tiruray
TEKE – *tec*
Telangire – see Telugu
Telegu – see Telugu
Telgi – see Telugu
TELUGU – tel
TEM – kdh
Temba – see Tem
Temen – see Themne
Temne – see Themne
Tenda – see Budik
Tenda Basari – see Bassari
Tendanke – see Budik
Tengu – see Telugu
Terangi – see Telugu
Teton – see Lakota
Teton Sioux – see Lakota
THAI – tha
Thai Yai – see Shan
THAI, NORTHERN – nod
Thaiklang – see Thai
Thalu – see Tulu
THEMNE – tem
Thonga – see Tsonga
Thulu – see Tulu

TIBETAN – bod
Tibetan – see Tibetan
Tiebaara – see Senoufo, Cebaara
Tigray – see Tigrigna
TIGRIGNA – tir
Tigrinya – see Tigrigna
Tilu – see Tulu
Tim – see Tem
Timbuka – see Tumbuka
Timene – see Themne
Timmannee – see Themne
Timne – see Themne
Timu – see Tem
Tina – see Sambal, Tinà
Tirahutia – see Maithili
Tirhuti – see Maithili
Tirhutia – see Maithili
Tirurai – see Tiruray
TIRURAY – tiy
TIV – tiv
TOK PISIN – tpi
Tolangan – see Telugu
Toli-Gbe – see Gun
Tombucas – see Tumbuka
Tonga – see Tsonga
TONGA – tog
TONGAN – ton
Tongoyna – see Tupuri
Tosk – see Albanian, Tosk
Toubouri – see Tupuri
Toupouri – see Tupuri
Transvaal Sotho – see Sotho, Northern
Trukhmen – see Turkmen
Trukhmeny – see Turkmen
Tsalagi – see Cherokee
Tshivenḓa – see Venda
Tsiwaha – see Tswana
Tslagi – see Cherokee
TSONGA – tso
TSWANA – tsn
Tu Guangdonghau – see Chinese, Hakka
Tuburi – see Tupuri
Tullu – see Tulu
TULU – tcy
Tuluva Bhasa – see Tulu
Tumanao – see Blaan, Sarangani
Tumboka – see Tumbuka
TUMBUKA – tum
TUPURI – tui
Türkçe – see Turkish
Türkisch – see Turkish
TURKISH – tur
Turkmani – see Turkmen
Turkmanian – see Turkmen
TURKMEN – tuk
Turkmenler – see Turkmen
Turkomans – see Turkmen
TWI – twi
Txuwabo – see Chuwabu
Tyebala – see Senoufo, Cebaara
U – see Tibetan
Ugandan – see Ganda
Uighuir – see Uyghur
Uighur – see Uyghur
Uiguir – see Uyghur
Uigur – see Uyghur
Uki – see Bokyi
UKRAINIAN – ukr
Unangam Tunuu – see Aleut
Upper Lusatian – see Sorbian, Upper
URDU – urd
Uriya – see Oriya
Utkali – see Oriya
Utse – see Iceve-Maci
Utser – see Iceve-Maci
Utseu – see Iceve-Maci
UYGHUR – uig
Uygur – see Uyghur
UZBEK – uzn
Uzo – see Izon
Vaaneroki – see Bokyi
Vadagu – see Badaga
Vadiya – see Oriya
VAI – vai
Vascuense – see Basque
Vei – see Vai
Veiao – see Yao
VENDA – ven
Vhe – see Éwé
Viet – see Vietnamese
VIETNAMESE – vie
Viro – see Estonian
Visayan – see Cebuano
VLAAMS – vls
Vlaemsch – see Vlaams
Volof – see Wolof
Vy – see Vai
WAAMA – wwa
Wachi – see Gbe, Waci
Waci – see Gbe, Waci

Waci-Gbe – see Gbe, Waci
Wajao – see Yao
Walaf – see Wolof
Wamay – see Wamey
Wamei – see Wamey
WAMEY – cou
Waráy – see Wáray-Wáray
WÁRAY-WÁRAY – war
Waro-Waro
Watyi – see Gbe, Waci
Wei – see Tibetan
Weiwuer – see Uyghur
Weizang – see Tibetan
WELSH – cym
Wemba – see Bemba
Wendish – see Sorbian, Lower
Wendish – see Sorbian, Upper
West Central Limba – see Limba
West Greenlandic Eskimo – see Inuktitut, West Greenlandic
Western Bukidnon Manobo – see Binukid
Western Cree – see Cree, Plains
Western Hmong – see Hmong-Njua
Western Laotian – see Thai, Northern
Western Malinke – see Maninkakan, Western
Western Miao – see Hmong-Njua
Western Nyasa – see Tonga
Whelngo – see Mizo
White Lum – see Hmong-Daw
White Meo – see Hmong-Daw
White Miao – see Hmong-Daw
White Russian – see Belarusan
White Ruthenian – see Belarusan
Wiga – see Uyghur
Wina – see Tupuri
Winaray – see Wáray-Wáray
Wo – see Bassari
WOLOF – wol
Woolwa – see Sumo-Mayangna
XHOSA – xho
Xiang – see Chinese, Xiang
Xichangana – see Tsonga
Xinminhua – see Chinese, Hakka
Xironga – see Ronga
Xitsonga – see Tsonga
Xosa – see Xhosa
!Xu – see Kung-Ekoka
!Xun – see Kung-Ekoka
!Xung – see Kung-Ekoka
Yallof – see Wolof
YALUNKA – yal
Yalunke – see Yalunka
Yambo – see Anuak
Yani – see Akha
YAO – yao
Yao Yen – see Lisu
Yaounde – see Ewondo
Yariba – see Yoruba
Yaruro – see Pumé
Yaruru – see Pumé
Yaunde – see Ewondo
Yaw Yin – see Lisu
Yaw-Yen – see Lisu
Ybanag – see Ibanag
Yeh-Jen – see Lisu
YIDDISH – yds
Yimba – see Limba
Yoabou – see Waama
Yoabu – see Waama
Yola – see Jola-Fonyi
YOM – pil
Yooba – see Yoruba
YORUBA – yor
Yuan – see Thai, Northern
Yuapín – see Pumé
Yudhia – see Oriya
Yue – see Chinese, Yue
Yueh – see Chinese, Yue
Yuet Yue – see Chinese, Yue
Yueyu – see Chinese, Yue
Yumba – see Limba
Yusufzai – see Pashto, Northern
Zaakosa – see Fulfulde, Adamawa
Zabarma – see Zarma
Zang – see Tibetan
Zarbarma – see Zarma
ZARMA – dje
Zarmaci – see Zarma
Zemay – see Fulfulde, Adamawa
Zhgabe – see Albanian, Tosk
ZULU – zul
Zunda – see Zulu
Zuni – see Zunni
ZUNI – zun

Chapter 8. Dialects[1]

Abo – see Bokyi
Abruzzese – see Italian
Abu – see Bokyi
Acchami – see Nepali
Ada – see Dangme
Adarawa – see Hausa
Adi Dravida – see Tamil
Adoyo – see Anuak
Adzhar – see Georgian
Afghan – see Farsi, Eastern
Afikpo – see Igbo
Agam – see Minangkabau
Agari of Kolaba – see Konkani
Agbome – see Fon
Agholo – see Ogbia
Agoi – see Gen
Agona – see Akan
Agulis – see Armenian
Agusan del Sur – see Binukid
Ahafo – see Akan
Ahirani – see Marathi
Ahkwesáhsne – see Mohawk
Aiguavivan – see Catalan-Valencian-Balear
Aili Gaili – see Hindustani
Airym – see Azerbaijani, Northern
Aitutaki – see Rarotongan
Aiyangar – see Tamil
Aiyar – see Tamil
Ajra – see Gun
Akhoe – see Kung-Ekoka
Akkhiy – see Chechen
Akn – see Armenian
Ako – see Akha
Akono – see Yoruba
Akpanzhi – see Junkun Takum
Aku – see Krio
Akuapem – see Akan
Akwaya – see Icheve-Maci
Akyem Bosome – see Akan
Alada – see Gun
Alföld – see Hungarian
Algherese – see Catalan-Valencian-Balear
Allemannisch – see German
Amana – see Maninkakan, Eastern
American Spanish – see Spanish
Amganad – see Ifugao
Amoy – see Chinese, Min Nan
Ampeyi – see Nupe-Nupe-Tako
Anauli – see Turkmen
Andalú – see Spanish
Andalusí – see Spanish
Andalusi Arabic – see Arabic
Andalusian – see Spanish
Andaluz – see Spanish
Aneuk Jamee – see Minangkabau
Anexo – see Gen
Angevin – see French
Anglo – see Éwé
Anglo – see Gbe-Vhe
Aniocha – see Igbo
Anlo – see Éwé
Anlo – see Gbe-Vhe
Antwerps – see Vlaams
Anupe – see Nupe-Nupe-Tako
Anupecwayi – see Nupe-Nupe-Tako
Anuperi – see Nupe-Nupe-Tako
Anyama – see Ogbia
Apoi – see Izon
Arabkir – see Armenian
Aragonese – see Spanish
Araki – see Farsi, Western
Arava – see Tamil
Arbanasi – see Albanian, Tosk
Are Bhashe – see Kannada
Are Marathi – see Marathi
Arewa – see Hausa
Arlija – see Romani, Balkan
Arogbo – see Izon
Arohun – see Fon
Artvin – see Armenian
Asante – see Akan
Asen – see Akan
Ashiti – see Kurdish, Northern
Ashkharik – see Armenian
Asong – see Akha
Astrakhan – see Armenian
Atiu – see Rarotongan
Atse – see Hmong-Njua
Aukshtaichiai – see Lithuanian
Aukshtaitish – see Lithuanian

[1] A number of languages are known by a dialect name. This chapter lists the correct language; for additional details, see Chapter 6.

Aukstaitiskai – see Lithuanian
Aussa – see Afar
Awadhi – see Hindi
Awka – see Igbo
Awori – see Yoruba
Aworo – see Yoruba
Awuna – see Éwé
Awuna – see Gbe-Vhe
Ayta Hambali – see Sambal, Botolan
Ba'adu – see Afar
Baadu – see Afar
Babok – see Mandjak
Bacheve – see Icheve-Maci
Bachit-Gashish – see Berom
Badaga – see Kannada
Badjia – see Ewondo
Bafatá Creole – see Crioulo, Upper Guinea
Bafeuk – see Ewondo
Baffinland Eskimo – see Inuktitut
Bafut – see Bafut
Bagheli – see Hindi
Baglung – see Newar
Bahe – see Bengali
Bahrani Arabic – see Arabic
Baiano – see Portuguese
Baitadi – see Nepali
Bajhangi – see Nepali
Bajura – see Nepali
Bajurali – see Nepali
Bakawali – see Kashmiri
Bakjo – see Ewondo
Bakoti – see Konkani
Baktapur – see Newar
Baku – see Azerbaijani, Northern
Bakun-Kibungan – see Kankanaey
Balear – see Catalan-Valencian-Balear
Balearic – see Catalan-Valencian-Balear
Bamvele – see Ewondo
Banaban – see Kiribati
Banat – see Romanian
Bandari – see Farsi, Western
Bandu – see Thai, Northern
Bane – see Ewondo
Bangaru – see Hindi
Banginda – see Gbaya
Bangkalan – see Madura
Bangkalon – see Madura
Bangwe – see Sena
Bankoti – see Konkani
Banta – see Themne
Bantayan – see Hiligaynon
Banten – see Javanese
Banten – see Sunda
Baol – see Wolof
Bape – see Bafia
Baraa – see Mandjak
Barain – see Gourmantché
Barani – see Gourmantché
Baraniire – see Gourmantché
Baravento – see Crioulo, Cape Verde
Barkoundouba – see Gourmantché
Barrawa – see Kuranko
Basan – see Izon
Basco Ivatan – see Ivatan
Bashua – see Bokyi
Bassa Nge – see Nupe-Nupe-Tako
Bassan – see Izon
Basseri – see Farsi, Western
Basua – see Bokyi
Baswo – see Bokyi
Batad – see Ifugao
Bathi – see Panjabi, Eastern
Batu Sangkar-Pariangan – see Minangkabau
Bau – see Fijian
Bautzen – see Sorbian, Upper
Bavarian – see German
Bawean – see Madura
Bawihka – see Sumo-Mayangna
Bay of Plenty – see Maori
Bayag – see Isnag
Bayash – see Romanian
Bayezidi – see Kurdish, Northern
Bayot – see Jola-Fonyi
Bayyu – see Bontoc, Central
Bazaar Malay – see Malay
Bazezuru – see Shona
Bazuzura – see Shona
Beik – see Burmese
Beledugu – see Bamanankan
Benaadir – see Somali
Bendi – see Bete-Bendi
Bene – see Bulu
Benge – see Bobo Madaré, Southern
Berad – see Telugu
Berrichon – see French
Bete – see Bete-Bendi
Beti – see Ewondo
Betsileo – see Malagasy
Bette – see Bete-Bendi
Bezanozano – see Malagasy

Bhaca – see Xhosa
Bhandari – see Konkani
Bhatneri – see Panjabi, Eastern
Bhatti – see Panjabi, Eastern
Bhatyiana – see Panjabi, Eastern
Bhumij – see Mundari
Bhumij Munda – see Mundari
Bhumij Thar – see Mundari
Big Woods French – see Cajun French
Bila – see Tsonga
Bini – see Yoruba
Binli – see Hanunoo
Birifor – see Dagaare, Southern
Bissau-Bolama Creole – see Crioulo, Upper Guinea
Biwangan – see Maguindanao
Bizkaian – see Basque
Black Lahu – see Lahu
Blood – see Blackfoot
Blue Miao – see Hmong-Njua
Bö – see Maninkakan, Eastern
Bobo Dioula – see Bobo Madaré, Southern
Bobo Jula – see Bobo Madaré, Southern
Bogandé – see Gourmantché
Bogomil – see Bulgarian
Bogor – see Sunda
Boholano – see Cebuano
Boje – see Bokyi
Bojie – see Bokyi
Bok – see Mandjak
Bokod – see Ibaloi
Boma – see Izon
Bombali – see Themne
Bompoka – see Nicobarese, Car
Bomvana – see Xhosa
Bonny-Opobo – see Igbo
Boorim – see Bokyi
Borcala – see Azerbaijani, Northern
Bornholm – see Danish
Botani – see Kurdish, Northern
Boti – see Kurdish, Northern
Boufale – see Kabiyé
Bourbonnais – see French
Bourguignon – see French
Boyanese – see Madura
Boyko – see Ukrainian
Brabanitc – see Dutch
Brabants – see Vlaams
Brahmin – see Tulu
Braj – see Hindi
Brame – see Mankanya
Brandenburgisch – see German
Brulé – see Lakota
Buende – see Koongo
Buge – see Bafut
Bujiyel – see Sanga
Bukidnon – see Binukid
Bukovynian – see Ukrainian
Bulalakawnon – see Hanunoo
Bulama – see Mankanya
Buluf – see Jola-Fonyi
Bumo – see Izon
Bundeli – see Hindi
Bundelkhandi – see Hindi
Bunjwali – see Kashmiri
Bunu – see Yoruba
Buram – see Mankanya
Burama – see Mankanya
Burgandi – see Tamil
Burma Tamil – see Tamil
Burmese – see Burmese
Buya – see Loko
Bwende – see Koongo
Cacheu-Ziguinchor Creole – see Crioulo, Upper Guinea
Caia – see Sena
Caipira – see Portuguese
Čakavian – see Serbo-Croatian
Calanasan – see Isnag
Camerija – see Albanian, Tosk
Camorta – see Nicobarese, Car
Canary Islands Spanish – see Spanish
Cape Afrikaans – see Afrikaans
Care – see Sena
Carioca – see Portuguese
Castilian – see Spanish
Catalan-Rousillonese – see Catalan-Valencian-Balear
Cawdur – see Turkmen
Caylloma Quechua – see Quechua, Cusco
Cayor – see Wolof
Cearense – see Portuguese
Cebu – see Cebuano
Cele – see Xhosa
Cen – see Berom
Central Abuan – see Abua
Central Afar – see Afar
Central Asia Arabic – see Arabic
Central Belarusan – see Belarusan
Central Bengali – see Bengali

Central Bohemian – see Czech
Central Catalan – see Catalan-Valencian-Balear
Central Chuwabo – see Chuwabu
Central Colloquial Maithili – see Maithili
Central Damara – see Nama- Damara
Central Estonian – see Estonian
Central Galician – see Galician
Central Gourmanchéma – see Gourmanchéma
Central Guangdong – see Chinese, Hakka
Central Kerala – see Malayalam
Central Kongo – see Koongo
Central Kono – see Kono
Central Latvian – see Latvian
Central Limba – see Limba
Central Makua – see Makhuwa
Central Marchigiano – see Italian
Central Mohawk – see Mohawk
Central Nicobarese – see Nicobarese, Car
Central Oromo – see Oromo, West Central
Central Pohjanmaa – see Finnish
Central Polissian – see Ukrainian
Central Slovak – see Slovak
Central Tagalog – see Tagalog
Central Tajiki – see Tajiki
Central Uyghur – see Uyghur
Central Vanua Levu – see Fijian
Central Vietnamese – see Vietnamese
Chakavsk – see Croatian
Changa – see Tsonga
Changana – see Tsonga
Chang-Jing – see Chinese Gàn
Changyi – see Chinese, Xiang
Chao-Shan – see Chinese, Min Nan
Chaozhou – see Chinese, Min Nan
Chattisgarhi – see Hindi
Chaura – see Nicobarese, Car
Cheju Island – see Korean
Chemehuevi – see Paiute, Southern
Cheribon – see Javanese
Cheva – see Nyanja
Chewa – see Nyanja
Chiantiy – see Chechen
Chichewa – see Nyanja
Chiebarloy – see Chechen
Chikaranga – see Shona
Ching Miao – see Hmong-Njua
Chinondo – see Digo
Chipodzo – see Sena
Chipoka – see Tumbuka
Chirue – see Sena
Chishinga – see Bemba
Chitradurga Kannada – see Kannada
Chittagonian – see Bengali
Chitumbuka – see Tumbuka
Chizezuru – see Shona
Chollado – see Korean
Chorbat – see Balti
Choroba – see Gonja
Choruba – see Gonja
Chote – see Bodo
Choushan – see Chinese, Min Nan
Ch'ungch'ongdo – see Korean
Chuqu – see Chinese, Wú
Churo – see Mandjak
Chw'aka – see Digo
Cicolano-Reatino-Aquilano – see Italian
Cimanganja – see Nyanja
Cipeta – see Nyanja
Cipodzo – see Sena
Cirebon – see Javanese
Cirebon – see Sunda
Citlang – see Newar
Common – see Tulu
Condul – see Nicobarese, Car
Connacht – see Gaelic, Irish
Constantinople – see Armenian
Cornouaillais – see Breton
Coro – see Anuak
Creek – see Muskogee
Crimea – see Armenian
Cur – see Mandjak
Czecho-Morvian – see Czech
Dagaare – see Dagaare, Southern
Dagara – see Kanuri
Dakani – see Urdu
Dakhini – see Urdu
Daklan – see Ibaloi
Dakota – see Dakota
Damani – see Konkani
Damara – see Nama- Damara
Dankyira – see Akan
Danu – see Burmese
Danube-Tisza – see Hungarian
Danubian – see Turkish
Daragawan – see Isnag
Dari – see Farsi, Eastern
Dariganga – see Mongolian
Darjula – see Nepali
Dasari – see Telugu
Dbus – see Tibetan

Haijong – see Bengali
Hainan – see Chinese, Min Nan
Hainanese – see Chinese, Min Nan
Haketia – see Ladino
Haketiya – see Ladino
Hakitia – see Ladino
Halbi – see Oriya
Halh – see Mongolian
Halland – see Danish
Hambali Botolan – see Sambal, Botolan
Häme – see Finnish
Hamshen – see Armenian
Hanak – see Czech
Hananwa – see Sotho, Northern
Hanga – see Luyia
Hanganu – see Tsonga
Hanoi – see Vietnamese
Haquetiya – see Ladino
Harijan – see Tamil
Haryanvi – see Hindi
Hasada' – see Mundari
Hassaniya Arabic – see Arabic
Hatutu – see Marquesan
Hau Lisu – see Lisu
Havigannada – see Kannada
Hebbar – see Tamil
Hejazi Arabic – see Arabic
Hekari – see Kurdish, Northern
Henga – see Tumbuka
Herati – see Farsi, Eastern
Hessisch – see German
Hetian – see Uyghur
Hewa – see Tumbuka
Hewe – see Tumbuka
High Latvian – see Latvian
Highland Lithuanian – see Lithuanian
Hiligaynon – see Hiligaynon
Hiva Oa – see Marquesan
Hlanganu – see Tsonga
Hlubi – see Xhosa
Hmong Gu Mba – see Hmong-Daw
Hmong Qua Mba – see Hmong-Daw
Holia – see Konkani
Hollandic – see Dutch
Hoss – see Berom
Hotan – see Uyghur
Huabei Guanhua – see Chinese, Mandarin
Hudu – see Éwé
Hudu – see Gbe-Vhe
Hue – see Vietnamese
Huizhou – see Chinese, Hakka
Hutsul – see Ukrainian
Hwesa – see Shona
Iba – see Sambal, Tinà
Icheve – see Icheve-Maci
Iduwini – see Izon
Igbonna – see Yoruba
Ihekwot – see Icheve-Maci
Ijebu – see Yoruba
Ijekavski – see Croatian
Ijesha – see Yoruba
Ikale – see Yoruba
Ikibiri – see Izon
Ila – see Yoruba
Ilafuri – see Venda
Ilaje – see Yoruba
Ilanon – see Maguindanao
Ilanum – see Maguindanao
Illanon – see Maguindanao
Ilud – see Maguindanao
Imeretian – see Georgian
Imerxev Kartlian – see Georgian
Indramayu – see Javanese
Ingilo – see Georgian
Insular – see Scots
Insular Catalan – see Catalan-Valencian-Balear
Iranon – see Maguindanao
Iranun – see Maguindanao
Iraqi Arabic – see Arabic
Ironworker Romani – see Romani, Balkan
Irruan – see Bokyi
Isamal – see Kalagan
Island Estonian – see Estonian
Isuama – see Igbo
Itawis – see Itawit
Itbayaten – see Ivatan
Iworro – see Yoruba
Jaan – see Mòoré
Jahanka – see Maninkakan, Western
Jahanque – see Maninkakan, Western
Jahonque – see Maninkakan, Western
Jain – see Tulu
Jamee – see Minangkabau
Jander – see Wolof
Jawa Halus – see Javanese
Jelgoore – see Gourmantché
Jharwa – see Assamese
Jiangxia Guanhua – see Chinese, Mandarin
Ji-Cha – see Chinese Gàn
Jidyo – see Ladino

Jinghuai Guanhua – see Chinese, Mandarin
Jinhua – see Chinese, Wú
Jishu – see Chinese, Xiang
Jolaha – see Maithili
Jolof – see Wolof
Jonga – see Tsonga
Joore – see Mòoré
Judezmo – see Ladino
Judyo – see Ladino
Jumu – see Yoruba
Kaado – see Zarma
Kabayan – see Ibaloi
Kabende – see Bemba
Kabo – see Izon
Kaboli – see Farsi, Eastern
Kabowei – see Izon
Kabras – see Luyia
Kabuli – see Farsi, Eastern
Kadavu – see Fijian
Kafire – see Senoufo, Cebaara
Kafu – see Bullom So
Kagankan – see Hanunoo
Kahnawà:ke – see Mohawk
Kajkavian – see Serbo-Croatian
Kakari – see Gujarati
Kakhetian – see Georgian
Kalabugao – see Binukid
Kalanga – see Ronga
Kalkus – see Mandjak
Kalla – see Gbaya
Kalounaye – see Jola-Fonyi
Kama – see Kissi, Northern
Kamanga – see Tumbuka
Kamara – see Kono
Kamari-Santali – see Santali
Kamathi – see Telugu
Kamenz – see Sorbian, Upper
Kandere – see Senoufo, Cebaara
Kanehsatà:ke – see Mohawk
Kangwondo – see Korean
Kannauji – see Hindi
Kano – see Hausa
Kantohe – see Balanta-Kentohe
Kapangan – see Kankanaey
Karabagh – see Armenian
Karagawan – see Isnag
Karamanli – see Turkish
Karanga – see Shona
Karapapak – see Azerbaijani, Northern
Karatina – see Gikuyu
Karhadi – see Konkani
Kari – see Hiligaynon
Karin – see Armenian
Karluk – see Uzbek
Karmali – see Santali
Karon – see Jola-Fonyi
Karungu – see Chuwabu
Kasara – see Senoufo, Cebaara
Kashani – see Farsi, Western
Kashtawari – see Kashmiri
Kashtwari – see Kashmiri
Katagum – see Hausa
Katchal – see Nicobarese, Car
Katharevousa – see Greek
Kathiawari – see Kashmiri
Kathiyawadi – see Gujarati
Kathmandu-Pathan-Kirtipur – see Newar
Katsina – see Hausa
Kauravi – see Hindi
Kawanga – see Luyia
Kawayan – see Hiligaynon
Kaxetian – see Georgian
Kayasthi – see Konkani
Kayavar – see Malayalam
Kaykayski – see Croatian
Kazan – see Tatar
Kazeruni – see Farsi, Western
Kebbawa – see Hausa
Kebu Fula – see Pular
Kecwan – see Bokyi
Kedah – see Malay
Kejia – see Chinese, Hakka
Kelantan – see Malay
Kentohe, Queuthoe – see Balanta-Kentohe
Kera' – see Mundari
Kerak – see Jola-Fonyi
Kerinci-Minangkabau – see Minangkabau
Kermani – see Farsi, Western
Ketabi – see Farsi, Western
Kewe – see Kabiyé
Kgaga – see Sotho, Northern
Kgatla – see Tswana
Khaga – see Sotho, Northern
Khalkha – see Mongolian
Khandeshi – see Marathi
Kharberd – see Armenian
Khari boli – see Hindi
Kharwa – see Gujarati
Khasarli – see Turkmen
Khayo – see Luyia

Khelobedu – see Sotho, Northern
Kheysur – see Georgian
Khole – see Santali
Kholifa – see Themne
Khorasani – see Farsi, Eastern
Khorat Thai – see Thai
Khotogoit – see Mongolian
Khotta – see Maithili
Khutswe – see Sotho, Northern
Khutswi – see Sotho, Northern
Khvoy-Salmst – see Armenian
Kiambu – see Gikuyu
Kijang – see Kabiyé
Kilikien – see Armenian
King's Pass Hungarian – see Hungarian
Kinhwa – see Chinese, Wú
Kipchak – see Uzbek
Kiristav – see Konkani
Kisa – see Luyia
Kisan – see Maithili
Kishtwari – see Kashmiri
Kistwali – see Kashmiri
Kituhwa – see Cherokee
Kiunguja – see Swahili
Ko – see Mende
Kokant Shan – see Shan
Koli – see Konkani
Kolo – see Ogbia
Kolokuma-Opokuma – see Izon
Koluama – see Izon
Kombo – see Jola-Fonyi
Komtao – see Telugu
Konda-Reddi – see Telugu
Konde – see Ronga
Kone – see Sotho, Northern
Kongar – see Tamil
Kongba – see Gola
Kongbaa – see Gola
Koni – see Sotho, Northern
Konike – see Themne
Konyanka – see Maninka
Kooki – see Ganda
Kopa – see Sotho, Northern
Korat – see Thai
Korca – see Albanian, Tosk
Korekore – see Shona
Koroboré – see Songhay
Kortha – see Maithili
Kortha Bihari – see Maithili
Kotafoa – see Éwé
Kotafoa – see Gbe-Vhe
Koulounkalan – see Maninkakan, Eastern
Kourousa – see Maninkakan, Eastern
Koya – see Loko
Koya – see Themne
Kpa – see Bafia
Kpa – see Mende
Kpase – see Fon
Kpo – see Gola
Krawang – see Sunda
Krobo – see Dangme
Kuban – see Ukrainian
Kufuru – see Senoufo, Cebaara
Kukra – see Sumo-Mayangna
Kukwe – see Nyakyusa-Ngonde
Kulere – see Senoufo, Cebaara
Kumaoni – see Hindi
Kumauni – see Hindi
Kumbo – see Izon
Kumbowei – see Izon
Kundagannada – see Kannada
Kutswe – see Sotho, Northern
Kuvale – see Herero
Kwakwagom – see Bokyi
Kwaludhi – see Ndonga
Kwatay – see Jola-Fonyi
Kwawu – see Akan
Kwaya – see Icheve-Maci
Kwena – see Tswana
Kxaxa – see Sotho, Northern
Kyakanke – see Maninkakan, Western
Kyonggido – see Korean
Kyongsangdo – see Korean
Kyoto-Osaka-type – see Japanese
Kypchak – see Uzbek
Kyūshū-type – see Japanese
Labrador Eskimo – see Inuktitut
Lach – see Czech
Lactan – see Kalagan
Ladino – see Ladino
Lai – see Gbaya
Laia – see Loko
Lala – see Zulu
Lama-Tissi – see Kabiyé
Langanu – see Tsonga
Lao-Kao – see Lao
Lao-Khrang – see Lao
Lapurdian – see Basque
Latar – see Mundari
Latgalian – see Latvian

Lay – see Gbaya
Laya – see Maguindanao
Laziale – see Italian
Le – see Mizo
Lebou – see Wolof
Lebu – see Wolof
Lechkhum – see Georgian
Legaspi – see Bicolano, Central
Lei – see Kono
Lei Hua – see Chinese, Min Nan
Leizhou – see Chinese, Min Nan
Lembetu – see Venda
Lembue – see Bemba
Lemko – see Ukrainian
Lenkaran – see Azerbaijani, Northern
Leonais – see Breton
Lesser Polish – see Polish
Lete – see Tswana
Levantine Arabic – see Arabic
Leyete – see Cebuano
Lhasa – see Tibetan
Li Hua – see Chinese, Min Nan
Liaro – see Kissi, Northern
Libisegahun – see Loko
Lichabool-Nalong – see Konkomba
Ligbeln – see Konkomba
Likes-Utsia – see Mandjak
Likoonli – see Konkomba
Limburgs – see Vlaams
Limonkpel – see Konkomba
Linafiel – see Konkomba
Liptaakoore – see Gourmantché
Little Nicobar – see Nicobarese, Car
Lleidatà – see Catalan-Valencian-Balear
Lobedu – see Sotho, Northern
Lohari-Santali – see Santali
Loheirn – see Lahu
Lolland – see Danish
Lombu – see Gbaya
Lomotua – see Bemba
Longdu – see Chinese, Min Nan
Longnan – see Chinese, Hakka
Lop – see Uyghur
Lorraine – see French
Lovedu – see Sotho, Northern
Low Malay – see Malay
Low Saxon – see German
Lower Cherokee – see Cherokee
Lower Navarrese – see Basque
Lower Yangze Mandarin – see Chinese, Mandarin
Lowland Lithuanian – see Lithuanian
Lowlands Slovak – see Slovak
Lu Shi Lisu – see Lisu
Luang Prabang – see Lao
Luangkori – see Kisi, Southern
Lubedu – see Sotho, Northern
Ludiopa – see Ganda
Luhanga – see Luyia
Luhishi – see Lahu
Luhushi – see Lahu
Lul – see Anuak
Lumburgish – see Dutch
Lund – see Mandjak
Lungu – see Digo
Lungulu – see Nyakyusa-Ngonde
Luobu – see Uyghur
Luoshao – see Chinese, Xiang
Lusaamia – see Luyia
Lusamia – see Luyia
Lushisa – see Luyia
Luunda – see Bemba
Luvuma – see Ganda
Maay – see Somali
Maci – see Icheve-Maci
Madrasi – see Tamil
Mafindo – see Kono
Magbiambo – see Loko
Maghrebi Arabic – see Arabic
Mahalhamadani – see Farsi, Western
Mahali – see Santali
Mahari – see Konkani
Maharra – see Makhuwa
Mahili – see Santali
Mahli – see Santali
Maindo – see Chuwabu
Mainfränkisch – see German
Majhi – see Panjabi, Eastern
Makhuwana – see Makhuwa
Makuana – see Makhuwa
Malabar – see Malayalam
Malal – see Themne
Malang-Pasuruan – see Javanese
Malatya – see Armenian
Malaueg – see Itawit
Malaweg – see Itawit
Malawi – see Nyanja
Malaya Tamil – see Tamil
Malayalam – see Malayalam
Mallorqui – see Catalan-Valencian-Balear
Maltese – see Arabic

Malwa – see Panjabi, Eastern
Managobla – see Gola
Manda – see Venda
Mandyam Brahmin – see Tamil
Mane – see Balanta-Kentohe
Manga – see Kanuri
Mangaia – see Rarotongan
Manganja – see Nyanja
Mangoche – see Yao
Maninka-Mori – see Maninkakan, Eastern
Manjhi – see Santali
Mankaliya – see Kuranko
Mankayan-Buguias – see Kankanaey
Manuk – see Javanese
Mao – see Shan
Maoli – see Konkani
Marachi – see Luyia
Marale – see Chuwabu
Marama – see Luyia
Maransé – see Songhay
Marashi – see Kurdish, Northern
Marave – see Nyanja
Maravi – see Nyanja
Marensé – see Songhay
Marinduque – see Tagalog
Marsh French – see Cajun French
Marwari Bhat – see Marwari
Marwari Bhil – see Marwari
Marwari Meghwar – see Marwari
Masemola – see Sotho, Northern
Masemula – see Sotho, Northern
Mashadi – see Farsi, Western
Masingbi – see Themne
Masinloc – see Sambal, Tinà
Matchi – see Icheve-Maci
Mathira – see Gikuyu
Matlala-Moletshi – see Sotho, Northern
Mau – see Shan
Mauke – see Rarotongan
Maw – see Shan
Mayang – see Assamese
Mayoyao – see Ifugao
Mazizuru – see Shona
Mazovian – see Polish
Mazur – see Polish
Mbai – see Gbaya
Mbaise – see Igbo
Mbandieru – see Herero
Mbedzi – see Venda
Mbengui-Niellé – see Senoufo, Cebaara
Mbere – see Gbaya
Mbete – see Bete-Bendi
Mbida-Bani – see Ewondo
Mech – see Bodo
Mehriyet – see Mehri
Mehriyot – see Mehri
Mein – see Izon
Meixian – see Chinese, Hakka
Menorqui – see Catalan-Valencian-Balear
Merguese – see Burmese
Mergui – see Burmese
Merina – see Malagasy
Meskhur-Javakhuri – see Georgian
Messeni – see Senoufo, Cebaara
Mfengu – see Xhosa
Mgao – see Swahili
Mialkhiy – see Chechen
Middle Cherokee – see Cherokee
Middle Dnieprian – see Ukrainian
Middle Tatar – see Tatar
Mideastern Yiddish – see Yiddish
Midnapore Oriya – see Oriya
Milo – see Nicobarese, Car
Min Bei – see Chinese, Min Bei
Min Dong – see Chinese, Min Bei
Min Nan – see Chinese, Min Bei
Min Zhong – see Chinese, Min Bei
Minalowang – see Binukid
Mindanao Visayan – see Cebuano
Mineiro – see Portuguese
Min-Ke – see Chinese, Hakka
Miraski – see Kashmiri
Mirgan – see Urdu
Misher – see Tatar
Mishongnovi – see Hopi
Mitiaro – see Rarotongan
Mizo – see Mizo
Mlomp – see Jola-Fonyi
Mmani – see Bullom So
Mngahris – see Tibetan
Mokhev – see Georgian
Moldavian – see Romanian
Molisano – see Italian
Mombe – see Gbaya
Møn – see Danish
Mong Leng – see Hmong-Daw
Mongo – see Kuranko
Moplah – see Malayalam
Mori – see Maninkakan, Eastern
Moriori – see Maori

Motom – see Icheve-Maci
Motomo – see Icheve-Maci
Moxev – see Georgian
Mpondo – see Xhosa
Mpondomisi – see Xhosa
Mpondomse – see Xhosa
Mrima – see Swahili
Mtiul – see Georgian
Mughalbandi – see Oriya
Mukulu – see Bemba
Mulai – see Makhuwa
Mulgi – see Estonian
Mumviri – see Kati
Munster-Leinster – see Gaelic, Irish
Muntenian – see Romanian
Murcian – see Spanish
Mus – see Armenian
Musaia – see Yalunka
Musseh Daeng – see Lahu
Musser Dam – see Lahu
Mvele – see Ewondo
Mvete – see Ewondo
Mvog-Niengue – see Ewondo
Mwamba – see Nyakyusa-Ngonde
Na – see Lahu
Nadavar-Kannada – see Kannada
Nafara – see Senoufo, Cebaara
Naga – see Balanta-Kentohe
Naga – see Bicolano, Central
Nagari-Malayalam – see Malayalam
Nagbanmba – see Loko
Naguri – see Mundari
Nahara – see Makhuwa
Najdi – see Arabic
Nakhchivan – see Azerbaijani, Northern
Nakota – see Dakota
Nama – see Nama- Damara
Namboodiri – see Malayalam
Namidama – see Nama- Damara
Nampamela – see Makhuwa
Nan – see Thai, Northern
Nancowry – see Nicobarese, Car
Nanumba – see Dagbani
Nanuni – see Dagbani
Narang – see Jola-Fonyi
Narva – see Estonian
Nasrani – see Malayalam
Navarrese – see Spanish
Nayar – see Malayalam
Ncqika – see Xhosa
Ndia – see Gikuyu
Ndlambe – see Xhosa
Nellore – see Telugu
Nenya – see Tumbuka
Nerezim – see Turkmen
Ney – see Kuranko
Ng'umbo – see Bemba
Ngandyera – see Ndonga
Ngari – see Tibetan
Ngell-Kuru-Vwang – see Berom
Ngente – see Mizo
Ngoahu – see Loko
Ngoma – see Bemba
Ngonde – see Nyakyusa-Ngonde
Ngoni – see Nyanja
Ngwa – see Igbo
Ngwaketse – see Tswana
Ngwalungu – see Tsonga
Ngwato – see Tswana
Ngwatu – see Tswana
Nicaraguan Tawahka – see Sumo-Mayangna
Nieni – see Kuranko
Nife – see Nupe-Nupe-Tako
Nimi Koro – see Kono
Nimi Yama – see Kono
Ning-Long – see Chinese, Hakka
Ningo – see Dangme
Ningraharian Pashto – see Pashto, Northern
Niominka – see Serer-Sine
Nkum – see Yala
Nkum Akpambe – see Yala
Nokhchmakhkakhoy – see Chechen
Nokhurli – see Turkmen
Nordestino – see Portuguese
Norman – see French
Normand – see French
Norrland – see Swedish
North Auckland – see Maori
North Balasore Oriya – see Oriya
North Ch'ungch'ong – see Korean
North Chollado – see Korean
North Estonian – see Estonian
North Greenlandic – see Inuktitut, West Greenlandic
North Jutlandic Danish – see Danish
North Komedia – see Armenian
North Kyongsangdo – see Korean
North Pohjanmaa – see Finnish
North Russian – see Russian
North Sena – see Sena

North Tajiki – see Tajiki
North Walian – see Welsh
Northeast Belarusan – see Belarusan
Northeast Bohemian – see Czech
Northeast Hungarian – see Hungarian
Northeast Vanua Levu – see Fijian
Northeast Viti Levu – see Fijian
Northeastern Coastal Estonian – see Estonian
Northeastern Kazakh – see Kazakh
Northeastern Pashto – see Pashto, Northern
Northeastern Yiddish – see Yiddish
Northern – see Somali
Northern Afar – see Afar
Northern Alberta Cree – see Cree, Plains
Northern Bengali – see Bengali
Northern Catalán – see Catalan-Valencian-Balear
Northern Digo – see Digo
Northern Gikuyu – see Gikuyu
Northern Gourmanchéma – see Gourmanchéma
Northern Guangdong – see Chinese, Hakka
Northern Ibanag – see Ibanag
Northern Irish – see Gaelic, Irish
Northern Kerala – see Malayalam
Northern Kirghiz – see Kirghiz
Northern Kirinyaga – see Gikuyu
Northern Kono – see Kono
Northern Lahu – see Lahu
Northern Macedonian – see Macedonian
Northern Mandrin – see Chinese, Mandarin
Northern Marwari – see Marwari
Northern Micmac – see Micmac
Northern Murang'a – see Gikuyu
Northern Samar – see Wáray-Wáray
Northern Scots – see Scots
Northern Shan – see Shan
Northern Shona – see Shona
Northern Swedish – see Swedish
Northern Tagalog – see Tagalog
Northern Vietnamese – see Vietnamese
Northwest Hungarian – see Hungarian
Northwestern Catalan – see Catalan-Valencian-Balear
Northwestern Mandarin – see Chinese, Mandarin
Northwestern Oriya – see Oriya
Nortista – see Portuguese
Nsa – see Igbo
Nsadop – see Bokyi
Nsuka – see Igbo
Nthali – see Tumbuka
Nukha – see Azerbaijani, Northern
Nuku Hiva – see Marquesan
Nupe Central – see Nupe-Nupe-Tako
Nupe Tako – see Nupe-Nupe-Tako
Nupecizi – see Nupe-Nupe-Tako
Nupencizi – see Nupe-Nupe-Tako
Nwesi – see Bemba
Nyakyusa – see Nyakyusa-Ngonde
Nyala-B – see Luyia
Nyaringa – see Chuwabu
Nyasa – see Nyanja
Nyeri – see Gikuyu
Nyi – see Lahu
Nzamba – see Koongo
Ogbe Ijo – see Izon
Ogboin – see Izon
Oghuz – see Uzbek
Ogulagha – see Izon
Oguta – see Igbo
Ohuhu – see Igbo
Oka – see Igbo
Okpeden – see Abua
Oku – see Bokyi
Okundi – see Bokyi
Old Shirazi – see Farsi, Western
Olit – see Icheve-Maci
Olithi – see Icheve-Maci
Oliti – see Icheve-Maci
Oliti-Akwaya – see Icheve-Maci
Oloibiri – see Ogbia
Oluhanga – see Luyia
Olukooki – see Ganda
Olusamia – see Luyia
Olusese – see Ganda
Oluwanga – see Luyia
Omvang – see Ewondo
Ondo – see Yoruba
Onitsha – see Igbo
Oostvlaams – see Vlaams
Opëno – see Anuak
Operemo – see Izon
Oporoma – see Izon
Oporomo – see Izon
Oporoza – see Izon
Orang Mamak – see Minangkabau
Orange River Afrikaans – see Afrikaans
Ordubad – see Azerbaijani, Northern
Oriental Hebrew – see Hebrew
Oriya Proper – see Oriya
Orlu – see Igbo
Orstkhoy – see Chechen

Osokom – see Bokyi
Osu – see Dangme
Otali – see Cherokee
Otapha – see Abua
Ouapadoupou – see Mòoré
Ouhiguyua – see Gourmantché
Oujiang – see Chinese, Wú
Overhill Cherokee – see Cherokee
Overhill Middle Cherokee – see Cherokee
Owe – see Yoruba
Owerri – see Igbo
Oyiakiri – see Izon
Oyo – see Yoruba
Oyokom – see Bokyi
Pahari – see Newar
Paharia – see Santali
Pahri – see Newar
Pai – see Sotho, Northern
Pai Lisu – see Lisu
Pajokumbuh – see Minangkabau
Pakse – see Lao
Palatinian – see German
Palaw – see Burmese
Palitiani – see Bulgarian
Palityan – see Bulgarian
Pallarese – see Catalan-Valencian-Balear
Pamekasan – see Madura
Pamekesan – see Madura
Panamahka – see Sumo-Mayangna
Panjabi Proper – see Panjabi, Eastern
Papara – see Senoufo, Cebaara
Parabhi – see Konkani
Parsi – see Gujarati
Parsiwan – see Farsi, Eastern
Parvari – see Konkani
Pasar Malay – see Malay
Pasir Malay – see Malay
Pasisir – see Javanese
Paspatian – see Romani, Balkan
Patara – see Senoufo, Cebaara
Pattapu Bhasha – see Tamil
Paulistano – see Portuguese
Pecixe – see Mandjak
Pemba – see Swahili
Penghulu – see Minangkabau
Penguia – see Kono
Penrhyn – see Rarotongan
Perak – see Malay
Peräpohja – see Finnish
Peta – see Nyanja
Petchabun Miao – see Hmong-Daw
Phalaborwa – see Sotho, Northern
Phalaburwa – see Sotho, Northern
Phani – see Venda
Phodzo – see Sena
Phoka – see Tumbuka
Phuthi – see Sotho, Southern
Piegan – see Blackfoot
Pil – see Themne
Pinjari – see Urdu
Pinto – see Mandjak
Piso – see Kalagan
Plains Cree – see Cree, Plains
Plateau Haitian Creole – see Creole, Haitian French
Podillian – see Ukrainian
Podzo – see Sena
Pogara – see Senoufo, Cebaara
Poguli – see Kashmiri
Poitevin – see French
Poka – see Tumbuka
Pokuttia – see Ukrainian
Polar Eskimo – see Inuktitut, West Greenlandic
Pomak – see Bulgarian
Port Maltese – see Maltese
Powadhi – see Panjabi, Eastern
Prairie French – see Cajun French
Pringan – see Sunda
Pshav – see Georgian
Pugliese – see Italian
Pulaar – see Fulfulde, Adamawa
Pulaar – see Pulaar
Pulana – see Sotho, Northern
Pular – see Fulfulde, Adamawa
Pulaya – see Malayalam
Pulhilh – see Mandjak
Puno Quechua – see Quechua, Cusco
Purik – see Balti
Puthsu – see Sena
Putru – see Ronga
Pu-Xian – see Chinese, Min Bei
Qabala – see Azerbaijani, Northern
Qarlug – see Uzbek
Qazakh – see Azerbaijani, Northern
Qazvini – see Farsi, Western
Qiongwen Hua – see Chinese, Min Nan
Quba – see Azerbaijani, Northern
Quebec Eskimo – see Inuktitut
Qwabe – see Zulu
Qyzylbash – see Azerbaijani, Northern

Racha-Lexchxum – see Georgian
Rajbanshi – see Bengali
Rakahanga-Manihiki – see Rarotongan
Rambani – see Kashmiri
Raoping – see Chinese, Hakka
Rarotonga – see Rarotongan
Rayalseema – see Telugu
Razgrad – see Turkish
Red Lahu – see Lahu
Rekhta – see Urdu
Rekhti – see Urdu
Rhathabe – see Xhosa
Rheinfränkisch – see German
Riasi – see Kashmiri
Ribagorçan – see Catalan-Valencian-Balear
Ribbi – see Loko
Ribia – see Themne
Rodiya – see Sinhala
Rodosto – see Armenian
Rodrigues Creole – see Morisyen
Rolong – see Tswana
Rotorua-Taupo – see Maori
Rue – see Sena
Rumelian – see Turkish
Rural Central Maltese – see Maltese
Rural East Maltese – see Maltese
Rural West Maltese – see Maltese
Saamia – see Luyia
Sadanga – see Bontoc, Central
Sadar – see Mankanya
Sadar Bhumij – see Mundari
Salewari – see Telugu
Salya – see Nyakyusa-Ngonde
Salyan – see Azerbaijani, Northern
Salyr – see Turkmen
Samar-Leyte – see Wáray-Wáray
Samavedi – see Marathi
Sambali Botolan – see Sambal, Botolan
Sambalpuri – see Oriya
Sambaya – see Kuranko
Sambelong – see Nicobarese, Car
Sambio – see Kwangali
Sambiu – see Kwangali
Sambyu – see Kwangali
Samia – see Luyia
Samogitian – see Lithuanian
Sampang – see Madura
Samya – see Luyia
San – see Bamanankan
Sanda – see Loko
Sanda – see Themne
Sando – see Kono
Sangamesvari – see Konkani
Sangwe – see Sena
Sankaran – see Maninka
Sanketi – see Tamil
Santa Cruz – see Sambal, Tinà
Santee – see Dakota
Santongeais – see French
Sapudi – see Madura
Saracatsan – see Greek
Sarar – see Mandjak
Sarawak Malay – see Malay
Sare – see Sena
Saremdé – see Mòoré
Sarnami Hindi – see Hindustani
Sarnami Hindustani – see Hindustani
Sartul – see Mongolian
Saryq – see Turkmen
Savannakhet – see Lao
Savo – see Finnish
Savolax – see Finnish
Saxonian – see German
Scanian – see Danish
Schwäbisch – see German
Sebaste – see Armenian
Secunderabad Brahmin – see Tamil
Sedehi – see Farsi, Western
Seeba-Yaga – see Gourmantché
Segou – see Bamanankan
Segum – see Serer-Sine
Sela – see Limba
Selya – see Nyakyusa-Ngonde
Seminole – see Muskogee
Sene – see Gola
Senga – see Tumbuka
Sengbe – see Kuranko
Senje – see Gola
Senje – see Gola
Seoul – see Korean
Seria – see Nyakyusa-Ngonde
Sertanejo – see Portuguese
Sese – see Ganda
Sesfontein Damara – see Nama- Damara
Seto – see Gun
Sewawa – see Mende
Shabin-Karahissar – see Armenian
Shadal – see Mankanya
Shah-Mansuri – see Kashmiri
Shahrudi – see Farsi, Western

Southwestern Finnish – see Finnish
Southwestern Mandarin – see Chinese, Mandarin
Srem – see Albanian, Tosk
Sri Lanka Tamil – see Tamil
Srikakula – see Telugu
Standard Assamese – see Assamese
Standard Bambara – see Bamanankan
Standard French – see French
Standard Gujarati – see Gujarati
Standard Hebrew – see Hebrew
Standard Kashmiri – see Kashmiri
Standard Maithili – see Maithili
Standard Maltese – see Maltese
Standard Oriya – see Oriya
Steppe – see Ukrainian
Štokavian – see Serbo-Croatian
Striped Hmong – see Hmong-Daw
Sudanese – see Arabic
Sukwa – see Ndali
Sulima – see Yalunka
Sulista – see Portuguese
Sumenep – see Madura
Surabaya – see Javanese
Suvalkietishai – see Lithuanian
Suwannakhet – see Lao
Svea – see Swedish
Sya – see Bobo Madaré, Southern
Syabéré – see Bobo Madaré, Southern
Syria – see Armenian
Székely – see Hungarian
Tafwap – see Nicobarese, Car
Tagakawanan – see Maguindanao
Tagara – see Senoufo, Cebaara
Tagbari – see Senoufo, Cebaara
Tahuta – see Marquesan
Tai Long – see Shan
Tai Mao – see Shan
Tai Wang – see Thai, Northern
Taihu – see Chinese, Wú
Taiwan – see Chinese, Hakka
Taizhou – see Chinese, Wú
Tajiki – see Farsi, Eastern
Tak Miao – see Hmong-Njua
Takpa – see Nupe-Nupe-Tako
Takpasyeeri – see Senoufo, Cebaara
Takum – see Junkun Takum
Talifugu-Ripang – see Isnag
Tallinn – see Estonian
Tamil – see Tamil
Tamiso – see Limba
Tanah – see Minangkabau
Tanala – see Malagasy
Tangamma – see Waama
Tankoro – see Kono
Taolendé – see Mòoré
Tapa – see Nupe-Nupe-Tako
Tappah – see Nupe-Nupe-Tako
Taranaki – see Maori
Tarimuki – see Gujarati
Tartu – see Estonian
Tau – see Sotho, Northern
Taung – see Sotho, Southern
Tavast – see Finnish
Tavha-Tsindi – see Venda
Tawana – see Tswana
Tawini – see Isnag
Tbilisi – see Armenian
Tee – see Gola
Tegal – see Javanese
Tege – see Gola
Tehrani – see Farsi, Western
Teixiera – see Mandjak
Teke – see Turkmen
Tekke – see Turkmen
Telangana – see Telugu
Telugu – see Telugu
Tembung – see Javanese
Tenere – see Senoufo, Cebaara
Teng – see Kissi, Northern
Tengia – see Kisi, Southern
Tengrela – see Senoufo, Cebaara
Terekeme – see Azerbaijani, Northern
Teressa – see Nicobarese, Car
Terloy – see Chechen
Thabine-Roka-Nareng – see Sotho, Northern
Thaikorat – see Thai
Thakari – see Konkani
Thakri – see Konkani
Thakua – see Konkani
Thakura – see Konkani
Thakuri – see Konkani
Thembu – see Xhosa
Thephalaborwa – see Sotho, Northern
Third Mesa – see Hopi
Thule Inuit – see Inuktitut, West Greenlandic
Thuringian – see German
Tigalu – see Tamil
Tigranakert – see Armenian
Tingzhou – see Chinese, Hakka
Tinners Romani – see Romani, Balkan

Tjirebon – see Javanese
Tlahaping – sec Tswana
Tlapi – see Tswana
Tlau – see Mizo
Tlokoa – see Sotho, Northern
Tlokwa – see Sotho, Northern
Tlokwa – see Tswana
Todii – see Gola
Tokilor – see Pulaar
Toko – see Shona
Tokwa – see Sotho, Northern
Tokyo-type – see Japanese
Toldil – see Gola
Toli – see Gun
Toli – see Kono
Tongareva – see Rarotongan
Tonggu – see Chinese, Hakka
Tonkinese – see Vietnamese
Tonko – see Limba
Toodii – see Gola
Totali – see Newar
Toucouleur – see Pulaar
Town Bemba – see Bemba
Trabzon – see Armenian
Trade Malay – see Malay
Transylvanian – see Romanian
Trauw – see Burmese
Tregorrois – see Breton
Trengganu – see Malay
Tribal – see Tulu
Trinidad Bhojpuri – see Hindustani
Trinkut – see Nicobarese, Car
Ts'imba – see Digo
Tsaam – see Mandjak
Tsang – see Tibetan
Tsongol – see Mongolian
Tsotso – see Luyia
Tsw'aka – see Digo
Tswene – see Sotho, Northern
Tsweni – see Sotho, Northern
Tukolor – see Pulaar
Tukulor – see Pulaar
Tumuaong – see Kalagan
Tung – see Kissi, Northern
Tuomo – see Izon
Tuscan – see Italian
Tush – see Georgian
Tuwali Ifugao – see Ifugao
Twa of Bangweulu – see Bemba
Tyebara – see Senoufo, Cebaara
Tyendinaga – see Mohawk
Ua Huka – see Marquesan
Ua Pou – see Marquesan
Ulster – see Gaelic, Irish
Ulster – see Scots
Ulu – see Minangkabau
Ulwa – see Sumo-Mayangna
Umbrian – see Italian
Umuahia – see Igbo
Unga – see Bemba
Ungu – see Digo
Unguja – see Swahili
Unwana – see Igbo
Upper Cherokee – see Cherokee
Upper Dniestrian – see Ukrainian
Upper Navarrese – see Basque
Upper Sannian – see Ukrainian
Urfa – see Turkish
Urmia-Maragheh – see Armenian
Ute – see Paiute, Southern
Vadaga – see Telugu
Vadari – see Telugu
Valencià – see Catalan-Valencian-Balear
Valencian – see Catalan-Valencian-Balear
Valenciano – see Catalan-Valencian-Balear
Van – see Armenian
Vanga – see Bengali
Vannetais – see Breton
Varhadi – see Marathi
Vazezuru – see Shona
Vernacular Hindustani – see Hindi
Vientiane – see Lao
Vila – see Tsonga
Vishakhapatnam – see Telugu
Vlaemsch – see Vlaams
Volynian – see Ukrainian
Voré – see Bobo Madaré, Southern
Vôru – see Estonian
Vulava – see Bughotu
Waama – see Waama
Waanjama – see Mende
Wadvali – see Marathi
Waganga – see Nyanja
Waigan – see Hanunoo
Walachian – see Romanian
Wanga – see Luyia
Wanganui – see Maori
Waray – see Wáray-Wáray
Warn – see Kisi, Southern
Wassulu – see Maninkakan, Eastern

References

Abaase, T.A., A. Gouzaye, L. Woltering and D. Pasternak. 2007. The role of indigenous leafy vegetables on daily diet and rural and urban economy of Niger. In: *Indigenous Vegetables and Legumes Prospectus for Fighting Poverty, Hunger and Malnutrition*. T.A. Lumpkin and J. Hughes (eds.), Acta Hort. 752:35-39.

Agbegha, M.L. 1996. *Ịzọn-English Dictionary.* Riverside Communications, Port Harcourt, Nigeria.

Albert, R. and D.L. Shaul. 1985. *A Concise Hopi and English Lexicon.* John Benjamins Pub. Co., Philadelphia, Pennsylvania, USA.

Aldave, A. and J. Mostacero. 1988. *Botánica Farmaceútica*. Editorial Libertad, Trujillo, Peru.

Allen, B.M. 1967. *Malayan Fruits: An Introduction to the Cultivated Species*. Donald Moore LTD, Singapore.

Alpern, S.B. 2008. Exotic plants of Western Africa: Where they came from and when. *History of Africa* 35:63-102.

Anderson, E.F. 1993. *Plants and People of the Golden Triangle.* Dioscorides Press, Portland, Oregon.

Anon. 1994. *Catalan Dictionary*. Routledge, New York, New York, USA.

Anon. 1991. *South African Multi-language Dictionary and Phrase Book. English, Afrikaans, Northern Sotho, Sesotho, Tswana, Xhosa, Zulu*. Readers Digest Assoc. South Africa, Cape Town, South Africa.

Anon. 1997. *Index Kewensis* (disc), Oxford University Press, Oxford, UK.

Asfaw, Z. and M. Tadesse. 2001. Prospects for sustainable use and development of wild food plants in Ethiopia. *Econ. Bot*. 55:47-62.

Asher, R.E. (ed.). 1994. *The Encyclopedia of Language and Linguistics*. Pergamon Press, Oxford, UK.

Austin, M. and R. Lynch (eds.). 1986. *Saad Ahaah Sinil*. Rough Rock Demonstration School, Rough Rock, Arizona, USA.

Awde, N. 1996. *Hausa-English, English-Hausa Dictionary*. Hippocrene Books, New York, New York, USA.

Awde, N., A. Sarwan, S. Davatolhagh and S. Aziz. 2002. *Dari. Dari-English, English- Dari Dictionary and Phrasebook*. Hippocrene Books, New York, New York, USA.

Awde, N. and C. Shahribaf. *Farsi. Farsi-English, English-Farsi Dictionary & Phrasebook*. Hippocrene Books, New York, New York, USA.

Backes, A. 1999. *Nomes Populares e Científicos de Plantas do Rio Grande do Sul.* Editora Unisinos, São Leopoldo RS, Brazil.

Balashev, L.L. (ed.). 1970. *Slovar 'poleznykh rastenii na dvadtsati evropeiskikh iazykakh*. [Dictionary of Useful Plants in Twenty European Languages]. Nauka Pub. House, Moscow, Russia.

Banfield, A.W. 1916. *Dictionary of the Nupe Language*. Niger Press, Shonga, Nigeria.

Baret, John and A. Fleming. 1580. *An aluearie or quadruple dictionarie containing foure* sundrie tongues: namelie, English, Latine, Greeke, and French. Newlie enriched with varietie of wordes, phrases, prouerbs, and diuers lightsome obseruations of grammar. By the tables you may contrairwise finde out the most necessarie wordes placed after the alphabet, whatsoeuer are to be found in anie other dictionarie: which tables also serue for lexicons, to lead the learner vnto the English of such hard wordes as are often read in authors, being faithfullie examined, are truelie numbered. Verie profitable for such as be desirous of anie *of those languages*. Excudebat Henricus Denhamus typographus, Gulielmi Seresij vnicus assignatus, Londini, England.

Barigozzi, C. (ed.). 1986. *The Origin and Domestication of Cultivated Plants*. Elsevier, Amsterdam, the Netherlands.

Beever, J. 1991. *Dictionary of Maori Plant Names*. Auckland Botanical Society, Auckland, New Zealand.

Benton, R.A. 1971. *Pangasinan Dictionary*. University of Hawaii Press, Honolulu, Hawaii, USA.

Bianco, V.V., F. Pimpini and L. Basoccu. 1990. *Orticoltura.* Pàtron, Bologna, Italy.

Biswas, S., S.C. Sen Gupta, S. Sengupta, B. Dasgupta and S. Banerjee. 2006. *Samsad English-Bengali Dictionary*. Sahitya Samsad, Kolkata, India.

Biggs, B. 1981. *The Complete English-Maori Dictionary*. Auckland University Press, Auckland, New Zealand.

Bingham, H. 1908. *Gilbertese-English Dictionary*. American Board of Commissioners for the Foreign Missions, Boston, Massachusetts, USA.

Blench, R.M., K. Williamson and B. Connell. 1994. The diffusion of maize in Nigeria: A historical and linguistic investigation. *Sprache und Geschichte in Afrika* 15:38-39.

Bose, T.K. and M.G. Som. 1986. *Vegetable Crops in India*. Nava Prokash, Calcutta, India.

Bright, W. (ed.). 1992. *International Encyclopedia of Linguistics*. Oxford University Press, Oxford, UK.

Brookes, R. 1763. *The Natural History of Vegetables*, London, UK.

Brücher, H. 1989. *Useful Plants of Neotropical Origin and Their Wild Relatives.* Springer-Verlag, New York, New York, USA.

Brummitt, R.K. and C.E. Powell (eds.). 1992. *Authors of Plant Names*. Royal Botanical Garden, Kew, UK.

Buck, C.D. 1949. *A Dictionary of Selected Synonyms in the Principal Indo-European Languages.* University of Chicago Press, Chicago, Illinois, USA.

Buechel, E. 1970. *A Dictionary of the Teton Dakota Sioux Language*. P. Manhart (ed.). Red Cloud Indian School, Holy Rosary Mission, Pine Ridge, South Dakota, USA.

Burkill, H.M. 1985-1997. *The Useful Plants of West Tropical Africa*. Royal Botanic Gardens, Kew, UK.

Busson, F., P. Jaeger, P. Lunven and M. Pinta. 1965 *Plantes Alimentaires de l'Ouest Africain. Etude Botanique, Biologique et Chimiqu*e. M. Leconte, Marseille, France.

Camargo, L. de S. 1992. *As Hortaliças e Sci Cultivo*. 3rd ed. Fundaçã Cargill, Campinas, Brazil.

Cerrate, E. 1979. *Vegetación del valle de Chiquián*. Univ. Nac. Mayor de San Marcos, Lima, Peru.

Chauhan, D.V.S. 1972. *Vegetable Production in India*. Ram Praasad and Sons, Agra U.P., India.

Childers, R.C. 1979. *A Dictionary of the Pali Language.* Cosmo Publications, New Delhi, India.

Christophersen, E. 1935. *Flowering Plants of Samoa*. Bull. 128, Bernice P. Bishop Museum, Honolulu, Hawaii, USA.

Christophersen, E. 1938. *Flowering Plants of Samoa*. Bull. 154, Bernice P. Bishop Museum, Honolulu, Hawaii, USA.

Christophory, Jules. 1995. *English-Luxembourgish Dictionary*. Esch/Alzette, Luxembourg.

Chittenden, F.J. (ed.). 1951. *Dictionary of Gardening*. Clarendon Press, Oxford, UK.

Chung, K.-F., Y. Kono, C.-M. Wang and C.-I. Peng. 2008. Notes on *Acemella* (Asteraceae: Heliantheae) in Taiwan. *Botanical Studies* 49:73-82.

Churchward, C.M. 1959. *Tongan Dictionary*. Government Printing Press, Nuku'alofa, Tonga.

Coe, R.G. and G.J. Anderson. 1999. Ethnobotany of the Sumu (Ulwa) of southeastern Nicaragua and comparisons with Miskitu plant lore. *Econ. Bot*. 53:363-386.

Colon de Reguero, L. and SM. Rodriguez de Santiago. 1978. *Tabla de Composicion de Alimentos de Uso Corriente en Puerto Rico*. University of Puerto Rico, San Juan, Puerto Rico.

Corlett, J.L., E.A. Dean and L.E. Grivetti. 2003. Hmong gardens: Botanical diversity in an urban setting. *Econ. Bot*. 57:365-379.

Corrêa, M. Pio and Leonam de Azeredo Pena.1926-1978. *Dicionário das Plantas Uteis do Brasil e das Exoticas Cultivadas.* Imprensa Nacional, Rio de Janeiro, Brazil.

Council of Scientific and Industrial Research. 1952-1988. *The Wealth of India: A Dictionary of Indian Raw Materials and Industrial Products.* Council of Scientific and Industrial Research, New Delhi, India.

Cox, P.A. and S.A. Banack. 1991. *Islands, Plants, and Polynesians, An Introduction to Polynesian Ethnobotany*. Diocorides Press, Portland, Oregon, USA.

Craven, T. 1932. *The New Royal Dictionary. English into Hindustani and Hindustani into English.* Methodist Pub. House, Lucknow, India.

Cronquist, A. 1988. *The Evolution and Classification of Flowering Plants.* New York Botanical Garden, Bronx, New York, New York, USA.

Crosby, A.W., Jr. 1972. *The Columbian Exchange*. Greenwood Press, Westport, Connecticut.

Crystal, D. 2000. *Language Death*. Cambridge University Press, Cambridge, UK.

Dalziel, J.M. 1937. *The Useful Plants of Tropical Africa*. Crown Agents for Overseas Governments, London, UK.

Deighton, F.C. 1957. *Vernacular Botanical Vocabulary for Sierra Leone*. Crown Agents for Overseas Governments on behalf of the Government of Sierra Leone, London, UK.

de Koning, J. 1993. *Checklist of Vernacular Plants in Mozambique*. Wageningen Agricultural University, Wageningen, the Netherlands.

de Larajasse, E. 1897. *Somali-English and English-Somali Dictionary.* Kegan Paul, Trench, Trübner & Co., London, UK.

de Malherbe, F. 1862. *Oeuvres Completes de Malherbe / Recueillies et Aannotees par M.L. Lalanne.* Hachette, Paris, France.

de Winter, W.P. and V.B. Amoroso (eds.). 1989. *Plant Resources of South-East Asia*. No. 15(2), *Cryptograms: Ferns and Fern Allies*. Pudoc Scientific, Leiden, the Netherlands.

Diamond, J.M. 2005. *Guns, Germs, and Steel: The Fates of Human Societies*. W.W. Norton, New York, New York, USA.

Dobring, M. 2002. *Lowerwendish-English and English-Lowerwendish Dictionary*. Published by the author, Port Huron, Michigan, USA.

Doke, C.M., D.M. Malcolm, J.M.A. Sikakana and B.W. Vilakazi. 1990. *English-Zulu, Zulu-English Dictionary*. Witwatersrand University Press, Johannesburg, South Africa.

Dransfield, S. and E.W. Widjaja (eds.). 1995. *Plant Resources of South-East Asia*. No. 7, *Bamboos*. Pudoc Scientific, Leiden, the Netherlands.

Drury, H. 1978. *The Useful Plants of India; with Notices of their Chief Value in Commerce, Medicine, and the Arts*. International Book Distributors, Dehradun, India.

Duke, J.A. 1986. *Isthmian Ethnobotanical Dictionary*. Scientific Publishers, Jodhpur, India.

Dupriez, H. and P. De Leener. 1989. *African Gardens and Orchards. Growing Vegetables and Fruits.* Macmillan, London, UK.

Elkins, R.E. 1968. *Manobo-English Dictionary*. Univ. Hawaii Press, Honolulu, Hawaii, USA.

Ertuğ, F. 2000. An ethnobotanical study in Central Anatolia (Turkey). *Econ. Bot.* 54:155-182.

Fargher, D.C., B. Stowell and I. Faulds. 1979. *Fargher's English-Manx dictionary*. Shearwater Press, Douglas, Isle of Man, Great Britain.

Fisher, A., E. Weiss, E. Mdala and S. Tshabe. *English-Xhosa Dictionary*. Oxford University Press, Cape Town, South Africa.

Flach, M., and R. Rumawas (eds.). 1989. *Plant Resources of South-East Asia*. No. 9, *Plants Yielding Non-seed Carbohydrates.* Pudoc Scientific, Leiden, the Netherlands.

Foot, E.C. 1913. *Galla-English, English-Galla Dictionary*. Cambridge University Press, Cambridge, UK.

Franquemont, C., T. Plowman, E. Franquemont, S. King, C. Niezgoda, W. Davis and C. Sperling. 1990. The ethnobotany of Chinchero, an Andean community in southern Peru. *Fieldiana* 24:1-126.

Frantz, D.G. and N.J. Russell. 1995. *Blackfoot Dictionary of Stems, Roots, and Affixes.* University of Toronto Press, Toronto, Canada.

Frawley, W. 2003. *International Encyclopedia of Linguistics.* Oxford University Press, Oxford, UK.

Fritz, D., W. Stolz, F. Venter, J. Weichmann and C. Wonneberger. 1989. *Gemüsebau*. Eugen Ulmer, Stuttgart, Germany.

Fyle, C.N. and E.D. Jones. 1980. *A Krio-English Dictionary*. Oxford University Press, Oxford, UK.

Gasparini, A. 1983. *Sidamo-English Dictionary*. E.M.I., Bologna, Italy.

Gestwicki, Ronald, Rev. and Mrs. 1966. *An English-Herero, Herero-English Dictionary*. Anglican Church, Windhoek, South West Africa.

Goddard, M. 2007. *English-Kurdish, Kurdish-English Sorani Dictionary*. Wallenberg Press, London, UK.

Goodwill, J., A. Kotzé, L. Namaseb, A. Tjouṭuku, E. Kaura, P. Hasheela, L. Mbenzi, D. Nakare, H. Nakare, M. !Goraseb and V. ≠Eiases. 1991. *First School Dictionary for Namibia*. Oxford University Press, Cape Town, South Africa.

Gordon, R.G., Jr. (ed.). 2005. *Ethnologue Languages of the World*. 15th edition. SIL International, Dallas, Texas, USA.

Gove, P.B. (ed.). 1971. *Webster's Third New International Dictionary*. G. and C. Merriam, Springfield, Massachusetts, USA.

Gragson, T.L. 1997. The use of underground plant organs and its relation to habitat selection among the Pumé Indians of Venezuela. *Econ. Bot.* 51:377-384.

Gunn, G.C. 2003. *First Globalization. The Eurasian Exchange, 1500-1800*. Rowman & Littlefield, Oxford, UK.

Gurudeva, M.R. 2001. *Botanical and Vernacular Names of South Indian Plants*. Divyachandra Prakashana, Bangalore, India.

Guzmann, C.C. de, and J.S. Siemonsma (eds.). 1999. *Plant Resources of South-East Asia*. No. 13. *Spices*. Backhuys Publishers, Leiden, the Netherlands.

Hannan, M. 1973. *Standard Shona Dictionary*. Macmillan, Johannesburg, South Africa.

Harlan, J.R. 1975. Our vanishing genetic resources. *Science* 188:617-621.

Harrington, J.P. 1928. Vocabulary of the Kiowa language. *Smithsonian Inst. Bull.* 34, 255 pp.

Heemkerk, W. 1987. *Especies e Variedades de Feijoes em Mocambique*. National Institute of Agricultural Research, Maputo, Mozambique.

Hill, K.C. (ed.). 1998. *Hopi Dictionary*. University Arizon Press, Tucson, Arizona, USA.

Hoch, E. 1960. *Bemba Pocket Dictionary*. Kipalapala Tanganyika, Tanzania.

Hockings, P. and C. Pilot-Raichoor. 1992. *A Badaga-English Dictionary*. Mouton de Gruyter, Berlin, Germany.

Hung, L., H. Huang and H.-F. Yen (eds.). 1992. *The Nomenclarture of Vegetable Crops*. National Taiwan University, Taipei, Republic of China.

Huxley, A. (ed.). 1992. *Dictionary of Gardening*. Macmillian, London, UK.

Indian Council of Agricultural Research. 1961. *Handbook of Agriculture*. ICAR, New Delhi.

Ivens, W.G. 1998. *Bugotu-English, English-Bugotu Concise Dictionary*. Hippocrene Books, New York, New York, USA.

Jackson, F.H. and J.C. Marck. 1991. *Carolinian-English Dictionary*. University Hawaii Press, Honolulu, Hawaii, USA.

Jaeger, P., P. Lunven and M. Pinta. 1965. *Plantes Alimentaires de L'Ouest Africain. Étude Botanique, Biologique et Chimaque*. M Leconte, Marseille, France.

Jansen, R.K. 1985. The systematics of *Acemlla* (Asteraceae-Heliantheae). *Syst. Bot. Monogr*. 8:1-115.

Jayaweera, D.M.A. 1981. *Medical Plants (Indigenous and Exotic) Used in Ceylon*. National Science Council of Sri Lanka, Colombo, Sri Lanka.

Jeeni, C.E. 2001. *Hindi-English-Mizo Dictionary*. Vani Prakasana, New Delhi, India.

Jerome, S., W. Caxton, W. de Worde and R. Curzon. 1495. *Vitas patrum*. Wynkyn de Worde, Westmynstre, UK.

Jha, Govinda. 1999. *Kalyani Kosh (A Maithili-English Dictionary)*. Maharajadhiraja Kamesvara Simha Kalyani Phaundesana, Darabhanga, India.

Johnstone, T.M. 1977. *Harsūsi Lexicon and English-Harsūsi Word-List*. Oxford University Press, London, UK.

Johnstone, T.M. 1987. *Mehri Lexicon and English-Mehri Word-List.* School of Oriental and African Languages, University of London, London, UK.

Jones, S.B., Jr. and A.E. Luchsinger. 1986. *Plant Systematics*. McGraw-Hill, New York, New York, USA.

Jørgensen, H. 1936. *A Dictionary of the Classical Newārī.* Kevin & Munksgaard, Copenhagen.

Kabithe, D.W. 1980. *Afrikan Language Dictionary. Swahili-English, English-Swahili*. Understanding International, Toledo, Ohio, USA.

Kachroo, P. (ed.). 1970. *Pulse Crops of India*. Indian Council of Agri. Res., New Delhi, India.

Kane, T.L. 1990. *Amharic-English Dictionary*. Harrassowitz, Wiesbaden, Germany.

Kays, S.J. and S.F. Nottingham. 2007. *Biology and Chemistry of Jerusalem Artichoke*: (*Helianthus tuberosus* L.). Taylor and Francis, Boca Raton, Florida, USA.

Kays, S.J. and J.C. Silva Dias. 1996. *Cultivated Vegetables of the World: Latin Binomial, Common Name in 15 Languages, Edible Part, and Method of Preparation*. Exon Press, Athens, Georgia, USA.

Koenon, E. von. 2001. *Medicinal, Poisonous, and Edible Plants in Namibia.* Klaus Hess Publishers, Windhoek, Namibia.

Kohen, E. and D. Kohen-Gordon. 2000. *Ladino-English, English-Ladino, Concise Encyclopedia Dictionary.* Hippocrene Books, New York, New York, USA.

Kroll, S. and D. Zahirović. 1998. *Bosnian-English / English-Bosnian Dictionary and Phrasebook.* Hippocrene Books, New York, New York, USA.

Lamberts, M. 1992. Production trends for specialty Asian vegetables in Dade County, Florida. *Acta Hort.* 318:79-85.

Lemmens, R.H.M.J. and N. Bunyapraphatsara (eds.). 2003. *Plant Resources of South-East Asia.* No. 12(3). *Medicinal and Poisonous Plants*. Backhuys Publishers, Leiden, the Netherlands.

Leslau, W. 1963. *Etymological Dictionary for Harari.* University of California Press, Berkeley, California, USA.

Leslau, W. 1976. *Concise Amharic Dictionary*. University of California Press, Berkeley, California, USA.

Levang, P. and H. de Foresta. 1991. *Economic Plants of Indonesia*. Orstom, Bogor, Indonesia.

Linné, Carl von. 1753. *Caroli Linnæi Species plantarum, exhibentes plantas rite cognitas, ad genera relatas: cum differentiis specificis, nominibus trivialibus, synonymis selectis, locis natalibus, secundum systema sexuale digestas*. Impensis Laurentii Salvii, Holmiæ (Stockholm, Sweden).

Lorenzi, H. 1991. *Plantas Daninhas do Brasil: Terrestres, Aquáticas, Parasitas, Tóxicase Medicinais.* 2nd ed. Plantarum Press, Nova Odessa, Brazil.

Luomala, K. 1953. *Ethnobotany of the Gilbert Islands.* Bishop Museum Bul. 213, 129p.

MacPhail, R.M. 1984. *Campbell's English-Santali Dictionary*. Eastern Books, Calcutta, India.

Malherbe, F. de. 1862. *Oeuvres completes de Malherbe / recueillies et aannotees par M.L. Lalanne*. Hachette, Paris, France.

Mamedov, S. 1995. *Azerbaijani. English-Azerbaijani / Azerbaijani-English*. Hippocrene Books, New York, New York, USA.

Man, Ssu-ch'ien. 1958. *English-Hakka Dictionary.* Qi Guang Chubanshe, Hong Kong.

Manner, A. 1983. *Tulu-English Dictionary*. Asian Educational Services, New Delhi, India.

Maracle, D.K. 1992. *One Thousand Useful Mohawk Words*. Audio-Forum, Guilford, Connecticut, USA.

Marijana, Kiš. 1996. *Dijalektnata leksika od oblasta na rastitelniot svet.* Institut za Makedonski Jazik "Krste Misirkov", Skopje, Macedonia.

Martin, J.B. and M.M. Mauldin. 2000. *A Dictionary of Creek/Muskogee.* University of Nebraska Press, Lincoln, Nebraska, USA.

Martineau, L. 1992. *The Southern Paiutes. Legends, Lore, Language and Lineage.* KC Publications, Las Vegas, Nevada, USA.
McNeill, J., F.R. Barrie, H.M. Burdet, V. Demoulin, D.L. Hawksworth, K. Marhold, D.H. Nicolson, J. Prado, P.C. Silva, J.E. Skog and J.H. Wiersema (eds.). 2005. *International Code of Botanical Nomenclature* (*Vienna Code*): adopted by the seventeenth International Botanical Congress, Vienna, Austria. A.R.G. Ganter, Königstein, Germany.
Mertz, O., A.M. Lykke and A. Reenberg. 2001. Importance and seasonality of vegetable consumption and marketing in Burkina Faso. *Econ. Bot.* 55:276-289.
Milner, G.B. 1993. *Samoan Dictionary.* Polynesian Press, Aotearoa, New Zealand.
Mintz, M.W. 1971. *Bikol Dictionary*. University of Hawaii Press, Honolulu, Hawaii, USA.
Mochari, M. (ed.). 1985. *Bodo-English Dictionary*. The Bodo Catholic Youth Association, Bengtol, Kakrajhar, Assam, India.
Molina, F.S., H. Valenzuela and D.L. Shaul. 1999. *Yoeme-English, English-Yoeme Standard Dictionary.* Hippocrene Books, New York, New York, USA.
Motus, C. 1971. *Hiligaynon Dictionary*. Hawaii Press, Honolulu, Hawaii, USA.
Munro, P. and C. Willmond. 1994. *Chickasaw. An Analytical Dictionary*. University Oklahoma Press, Norman, Oklahoma.
Newman, S. 1958. *Zuni Dictionary*. Intern. J. Amer. Linguistics. 24:117p.
Niang, M. 1997. *Pulaar-English / English-Pulaar.* Hippocrene Books, New York, New York, USA.
Nichols, J. and A. Vagapov. 2004. *Chechen-English and English-Chechen Dictionary*. RoutledgeCurzon, New York, New York, USA.
Nguyen, M.L.T. 2005. Bạc hà (*Colocasia gigantean* [Blume] Hook. f.) in the culinary history of Vietnamese-Americans. *Econ. Bot.* 59:185-196.
Oideachais, A.R. 1978. *Ainmneacha Plandaí Agus Ainmhithe*. Baile Atha Cliath, Oifig an tSolathair, Ireland.
Oideachais, A.R. 1978. *Focloir Talmhaíochta*. An Gum, Baile Atha Cliath, Ireland.
Öztopçu, K., Z. Abuov, N. Kambarov and Y. Azemoun. 1996. *Dictionary of the Turkic Languages*. Routledge, New York, New York, USA.
Parham, B.E.V. 1957. Some Samoan plant names, *Laufasi Ola, Dept. Agri. Circ.*, 2(5 and 7), Apia, Samoa.
Parham, B.E.V. 1958. Some Samoan plant names, *Laufasi Ola, Dept. Agri. Circ.*, 3(5, 8 and 11), Apia, Samoa.
Parham, B.E.V. 1972. *Plants of Samoa*. New Zealand Dept. Scientific and Industrial Res., Wellington, New Zealand.
Parker, E.M. and R.J. Hayward. 1985. *An Afar-English-French Dictionary*. School of Oriental and African Studies, University of London, London, UK.
Parker, S.P. 1989. *McGraw-Hill Dictionary of Scientific and Technical Terms*. McGraw-Hill, New York, New York, USA.
Parrotta, J.A. 2001. *Healing Plants of Peninsular India*. CABI Publishing, New York, New York, USA.
Passy, A.M. 1999. *Sephardic Folk Dictionary, English to Ladino / Ladino to English*. Published by the author, Los Angeles, California, USA.
Pereira, C. and F.V. Agarez. 1980. *Botânica: Taxonomia e Organografia das Angiospermas*. Interamericana, Rio de Janeiro, Brazil.
Phone, W. 2007. *Dictionary of Jicarilla Apache*. University of New Mexico Press, Albuquerque, New Mexico, USA.
Pikia, M. 2006. *African Traditional Plant Knowledge Today.* Lit Verlag, Berlin, Germany.
Plants of the Philippines Work Group. 1996. *Plants of the Philippines*. Island Pub. House, Manila.
Pulte, W. (ed.). 1975. *Cherokee-English Dictionary*. Cherokee Nation of Oklahoma, Tahlequah, Oklahoma, USA.

Ramser, J., K. Weising, C. Lopez-Peralta, W. Terhalle, R. Terauchi and G. Kahl. 1997. Molecular marker based taxonomy and phylogeny of Guinea yam (*Dioscorea rotundata – D. cayenensis*). *Genome* 40:903-915.

Rand, S.T., 1888. *Dictionary of the Language of the Micmac Indians.* Nova Scotia Printing, Halifax, Nova Scotia, Canada.

Raven, P.H., R.F. Evert and S.E. Eichhorn. 1992. *Biology of Plants*. Worth, New York, New York, USA.

Reh, M. 1999. *Anywa-English and English-Anywa Dictionary*. Rüdiger, Köppe Verlag, Köln, Germany.

Reid, L.A. (ed.). 1971. *Philippine Minor Languages: Word Lists and Phonologies*. University of Hawaii Press, Honolulu, Hawaii, USA.

Reitz, R. 1972. *Flora Ilustrada Catarinense*. Herbário Barbosa Rodrigues, Itajaí, Santa Catarina, Brazil.

Robinson, L.W. and J. Armagost. 1990. *Comanche Dictionary and Grammar.* The Summer Institute of Lingustics, Dallas, Texas, USA.

Ross, D. and G.D. Smith. 1998. *Scots-English, English-Scots Dictionary*. Hippocrene Books, New York, New York, USA.

Royal Society, Great Britain. 1790. Ancient Cookery. *Society of Antiquaries of London.*Collection of ordinances and regulations for the government of the Royal Household ... London, England.

Rubino, C.R.G. 2000. *Ilocano-English Dictionary.* University of Hawaii Press, Honolulu, Hawaii, USA.

Rytz, O. (ed.). 1960. *Gonja-English Dictionary and Spelling Book*. 1960. Institute of African Studies, University of Ghana, Legon, Ghana.

Sahrmüller, R., H.-S. Banholzer, I. Galler, F. Bummeltenberg, D. Kulewatz, I. Lepiksarr and K. Semrau. 1990. *Mehrsprachiges Verzeichnis der Gemüsenamen. Teil 1: Allphabetisches Verzeichnis* [Index of Vegetable Names in Several Languages. Part 1. Alphabetical List]. Institut für Gemüseproduktion, Grossbeeren, Germany.

Sahrmüller, R., H.-S. Banholzer, I. Galler, F. Bummeltenberg, D. Kulewatz, I. Lepiksarr, K. Semrau, H. Voigt and G. Voll. 1990. *Mehrsprachiges Verzeichnis der Gemüsenamen. Teil 2: 100 Gemüsearten in 50 Sprachen* [Index of Vegetable Names in Several Languages. Part 2. 100 Vegetable Species 50 Languages]. Institut für Gemüseproduktion, Grossbeeren, Germany.

Salaman, R.N. 1940. Why "Jerusalem" artichoke? *Royal Hort. Soc.* LXVI:338-348, 376-383.

Sánchez-Monge y Parellada, E. 1981. *Diccionario de Plantas Agrícolas*. Ministry of Agriculture, Sercicio de Publ., Agrarias, Madrid, Spain.

Sapir, E. 1930. Southern Paiute, a Shoshonean language. *Proceedings of the American Academy of Arts and Sciences* 65:1-296.

Schuh, R.G. 1981. *Dictionary of Ngizim*. University of California Press, Berkeley, California, USA.

Setchell, W.A. 1924. *American Samoa*. Carnegie Inst. Pub. 341, Washington, DC, USA.

Seymour, E.L.O. 1990. *The Wise Garden Encyclopedia*. Harper Collins, New York, New York, USA.

Siemonsma, J.S. and K. Piluek (eds.). 1993. *Plant Resources of South-East Asia*. No. 8, *Vegetables.* Pudoc Scientific Publishers, Leiden, the Netherlands.

Simpson, J.A. and Weiner, E.S.C. 1989. *The Oxford English Dictionary*, 20 vols, Claredon Press, Oxford, UK.

Skinner, N. 1996. *Hausa Comparative Dictionary.* Rüdiger, Köppe Verlan, Köln, Germany.

Slavov, A. 1999. *Gypsy-English, English-Gypsy, Concise Dictionary*. Hippocrene Books, New York, New York, USA.

Smith, H.H. 1933. *Ethnobotany of the Forest Potawatomi Indians*. Bull. Pub. Museum of the City of Milwaukee, Milwaukee, Wisconsin, USA.

Soukup, J. 1980. *Vocabulario de los nombres vulgares de la Flora Peruana y Catálogo de los Géneros.* Editorial Salesiana. Segunda Edición, Lima, Peru.

Soule, J. 1985. *Glossary for Horticultural Crops*. John Wiley & Sons, New York, New York, USA.

Sperlich, W.B. (ed.). 1997. *Tohi Vagahau Niue / Niue Language Dictionary*. Government Press, Niue.

Spring, R.K. 2002. *Balti-English, English-Balti Dictionary*. RoutledgeCurzon, New York, New York, USA.

Sprinkle, R. 1976. *Zhonggou Min nan yu Ying yu zi dian* (*Amoy-English Dictionary*). Maryknoll Fathers, Taichung, Taiwan.

Stevels, J.M.C. 1990. *Légumes Traditionnels du Cameroun une Étude Agro-botanique*. Wageningen Agricultural University Papers, Wageningen, the Netherlands.

Stone, G. 2002. *Upper Sorbian-English Dictionary.* Domowina Verlag GmbH, Bautzen, Germany.

Strømme, E. (ed.) 1993. *Navn på grønnsaker og krydderplanter*. Norsk plantenomenklaturråd for hagebruk. Faginfo/Statens fagtjeneste for landbruket, Ås, Norway.

Subhadrabandhu, S. 1999. Ethnomedical uses of common plants in Thailand. *Chronica Horticulturae* 39(1):7-9.

Svenson, H.K. 1939. Monographic studies in the genus Eleocharis – V. *Rhodora* 41:1-19.

Tanaka, Y. 1976. *Tanaka's Cyclopedia of Edible Plants of the World.* Sasuke Nakao (ed.), Keigaku Pub. Co., Tokyo, Japan.

Taylor, F.W. 1995. *Fulani-English*. Hippocrene Books, New York, New York, USA.

Terpstra, G. 1968. *English-Tiv Dictionary*. Institute for African Studies, University of Ibadan, Ibadan, Nigeria.

Terrell, E., S.R. Hill, J.H. Wiersema and W.E. Rice. 1986. *A Check List of Names for 3,000 Vascular Plants of Economic Importance*. United States Department of Agriculture, Agricultural Handbook 505, Washington, DC, USA.

Thibert, A. 1997. *Eskimo-English, English-Eskimo Dictionary.* AES + Laurier Pub., Ottawa, Canada.

Thomson, D.S. 1994. *The New English-Gaelic Dictionary*. GAIEM, Glasgow, Scotland, UK.

Tominaga, T. 1992. Fruits and vegetables in markets in Bintulu, Sarawak, Malaysia. *J. Fac. Agric. Shinshu University* 29(2):137-147.

Torikashvili, J.J. 1992. *Georgian-English, English-Georgian Dictionary.* Hippocrene Books, New York, New York, USA.

Tovar, O. 1990. *Tipos de Vegetación, Diversidad Florística y Estado de Conservación de la Cuenca del Mantaro*. Centro de Datos para la Conservación, Univ. Nac. Agraria La Molina, Lima, Peru.

Uzicani, N.S. 1995. *Bosnian-English, English-Bosnian Dictionary.* Hippocrene Books, New York, New York, USA.

Van der Maesen, L.J.G. and S. Somaatmadja (eds.). 1989. *Plant Resources of South-East Asia.* No. 1, *Pulses*. Pudoc Scientific, Leiden, the Netherlands.

Van Warmelo, N.J. 1989. *Venda Dictionary, Tsguvenḓa-English.* J.L. van Schaik, Pretoria, South Africa.

Van Wyk, B.-E. 2005. *Food Plants of the World.* Timber Press, Portland, Oregon, USA.

Vavilov, N.I. 1992. *Origin and Geography of Cultivated Plant*s. Cambridge University Press, Cambridge, UK.

Verheij, E.W.M. and R.E. Coronel (eds.). 1991. *Plant Resources of South-East Asia*. No. 2, *Edible Fruits and Nuts*. Pudoc Scientific, Leiden, the Netherlands.

Vestal, P.A. and R.E. Schultes. 1939. *The Economic Botany of the Kiowa Indians.* Botanical Museum, Cambridge, Massachusetts, USA.

Wells, J.C. 2010. Orthographic diacritics and multilingual computing. <http://www.phon. ucl.ac.uk/ home/wells/dia/diacritics-revised.htm>.

Whistler, W.A. 1991. Polynesian plant introductions. pp. 41-66, In: *Islands, Plants, and Polynesians, An Introduction to Polynesian Ethnobotany*. P.A. Cox and S.A. Banack (eds.), Dioscorides Press, Portland, Oregon, USA.

Whistler, W.A. 1991. *The Ethnobotany of Tonga: The Plants, Their Tongan Names, and Their Uses*. Bishop Museum Botany Bull. 2, Bishop Museum, Honolulu, Hawaii, USA.

Wilder, G.P. 1931. *Flora of Rarotonga*. Bishop Museum Bull. 86, Honolulu, Hawaii, USA.
Williamson, K. 1993. Linguistic evidence for the use of some tree and tuber food plants in Southern Nigeria. pp. 150-51. In: Shaw, T., P. Sinclair, B. Andah, and A. Okpoka, *The Archaeology of Africa: Food, Metals, and Towns*. London, UK.
Williamson, J. 1960. *Useful Plants of Malawi*. Government Printers, Zomba, Malawi.
Williamson, J.P. 1992. *An English-Dakota Dictionary*. Minnesota Historical Soc. Press, St Paul, Minnesota, USA.
Wilson, P.M. nd. *English-Swahili*. East African Literature Bureau, Nairobi, Kenya.
Yuncker, T.G. 1943. *The Flora of Niue Island*. Bishop Museum Bull. 178, Honolulu, Hawaii, USA.
Yang, Yigeng, Lianci Liu, Xuequn Chen, Cailin Chen, Guangfan Zhou and Keli Niao. 1988. Classificational research for vegetables in mustard (*Brassica juncea* Coss.). pp. 342-347 In: *International Symposium on Horticultural Germplasm, Cultivated and Wild. Part II vegetables*. International Academic Publishers, Beijing, China.
Yawalkar, K.S. 1965. *Vegetable Crops of India*. Agri-Horticultural Pub. House, Nagpur, India.
Young, R.W. and W. Morgan. 1980. *The Navajo Language*. University of New Mexico Press, Albuquerque, New Mexico, USA.
Yuncker, T.G. 1945. *Plants of the Manua Islands*. Bishop Museum Bull. 184, Honolulu, Hawaii.
Yuncker, T.G. 1959. *Plants of Tonga*. Bishop Museum Bull. 220, Honolulu, Hawaii, USA.
Zohary, D. and M. Hopf. 2000. *Domestication of Plants in the Old World.* Oxford University Press, Oxford, UK.

Appendix I. Diacritics

English is one of the few languages that utilizes the standard 26 letters of the Latin alphabet. Many other Latin alphabet-based languages supplement the sounds that can be represented by the letters using diacritics (e.g. á, ã, â, ä, å, ā, ă, ą, ȁ, ȧ, ạ, ả, a̱, ǻ, ấ, ầ, ẩ, ẫ, ậ, ắ, ằ, ẳ, ẵ, ặ) (Wells, 2010). Diacritics used in the text are listed below; however, their effect on pronunciation is not consistent across languages. As a consequence, it is best to consult an explanation of the use of diacritics for each language.

Acute accents (e.g. á, ǽ, ć, é, í, ó, ĺ, ń, ŕ, ś, ú, ź) have a wide range of applications depending upon the language. For example, in certain languages (e.g. Lakota, Portuguese, Spanish) they represent stressed vowels. In certain other languages, they indicate a long vowel (e.g. Czech, Hungarian, Irish) and in Polish palatalization. In some African languages, an acute accent over a vowel typically indicates a rising tone, while over *d*, *k*, or *t* in Senegalese languages, it hardens and shortens the letter. A double acute (e.g. ő, ű) is primarily used in Hungarian and acts as a combination of an acute and a diaeresis, giving a longer sound to the respective letters.

A grave accent (e.g. à, è, ì, ò, ù, ỳ), like an acute accent, has various effects on pronunciation depending upon the language. In French a grave accent indicates a half-open vowel; in Italian and Catalan it is a stressed vowel; in Scottish Gaelic a long vowel; and in Vietnamese it denotes a low tone. In a number of the African languages listed, a grave accent over a vowel denotes a falling tone.

In certain African languages, a double grave accent (e.g. ȁ, ȅ, ȉ, ȍ, ȑ, ȕ) over a vowel indicates a deeper falling tone when following a syllable already under a falling tone. Likewise, in Serbian, Croatian and Slovenian, a double grave accent indicates a falling tone.

A circumflex accent (e.g. â, ê, î, ô, û, ŵ, ŷ) in Welsh denotes a long vowel including ŵ and ŷ which are also vowels. In Chichewa, ŵ is a bilabial fricative, while in Portuguese â, ê, î, and ô are nasalized vowels. In Turkish, a circumflex appears over an *a* or *u* when a preceding consonant (e.g. k, g, l), is pronounced as a palatal plosive, while in certain African languages a circumflex accent over a vowel indicates a rising tone. In Potawatomi, ê is pronounced as in the English word b*e*t, î as in b*i*t and û as in l*u*ck (Smith, 1933), while in Kiowa the circumflex (î, t̂) indicates a combination of an acute and a grave accent.

A inverted circumflex, also known as a hachek (e.g. č, ě, š, ž), is represented by a relatively sharp wedge over the letter. In Czech, Slovak, Croation, Slovene, and certain other languages, š and ž indicate palato-alveolar fricatives. In Amharic, the grapheme č denotes the voiceless postalveolar affricate consonant, similar to *ch* in the English word *ch*ocolate, ǧ signals [dʒ] and š signals [ʃ]. In Czech, č is a voiceless affricate while a hachek over an *e* indicates palatalization of the preceding consonant. In several of the African languages listed, a hachek over a consonant indicates a hardened aspiration, a fricative or sibilant; however, when over a vowel it denotes a falling tone in the Nigerian languages.

The breve (e.g. ă, ĕ, ĭ, ŏ, ŭ) differs from the hachek in that it is more rounded. In several languages, it denotes the shortness of a vowel. For example, in Romanian ă indicates a mid central vowel. In several of the African languages the presence of a breve over a vowel shortens it.

A macron over a vowel (e.g. ā, ē, ī, ō, ū) indicates a long vowel in Maori, Hawaiian, and Latvian, while in Yoruba it indicates a mid tone. In certain African languages, a macron lengthens a vowel, though more commonly the vowel is doubled to show length. In Pinyin, ā, ē, ī, ō, ū, ǖ are used to indicate the first tone of Mandarin Chinese. In contrast, a macron under a vowel (e.g. a̲, e̲, i̲, o̲, u̲) in several of the African languages indicates a low level pitch.

An overdot (e.g. ȧ, ḃ, ċ, ḋ, ė, ḟ, ġ, ḣ, ṁ, ṅ, ȯ, ṗ, ṙ, ṡ, ṫ, ẇ, ẋ, ẏ, ż) when used as a diacritic (in contrast to the letters i and j where the dot over the letter is a nondistinctive mark) has several functions. In Lithuanian, ė stands for a half closed vowel. In Maltese, ċ is a voiceless postalveolar affricate, while in Lakota it is sometimes used to indicate an explosive stop. In some of the African languages, ṅ indicates *ng* as in the English si*ng*.

A dot under a letter (e.g. ạ, ḅ, ḍ, ẹ, ḥ, ḳ, ḷ, ṃ, ṇ, ọ, ṛ, ṣ, ṭ, ụ, ṿ, ẉ, ẓ) in Vietnamese indicates a low, glottal tone when under a vowel. In Igbo, it indicates a reduction of the height of the vowel while in Yoruba, when used under a consonant, an underdot indicates a sibilant sound as in ṣ = *sh*. In several of the African languages a dot under a vowel makes it more open.

A diaeresis or trema is created using two dots over the letter (e.g. ä, ë, ï, ö, ü). When placed over one of two adjacent vowels, a diaeresis indicates the vowels are sounded separately (i.e. split into two syllables rather than pronounced as a digraph). A variation of this is the German umlaut where ä, ö, and ü represent front vowels. This changes the sound as in ä = *ae*.

A tilde is written over a letter (e.g. ã, ĩ, ñ, õ, ũ) to indicate a change in pronunciation. In Spanish, Basque, Filipino, Galacian, Guarani and certain other languages it represents a palatal nasal consonant. In other letters, a tilde indicates nasalization of the base vowel in Portuguese and Guarani. In Estonian, õ stands for a close-mid back unrounded vowel; in Guarani, g̃ stands for the nasalized upper central rounded vowel; and in Vietnamese a tilde over a vowel represents a dipping tone.

A ring over a letter (e.g. å, ů, ẘ, ẙ), especially the letter *a*, is a distinctive feature in Danish, Norwegian and Swedish. The letter *ů* is used in Czech where it has the same pronunciation as *ú* but changes to a short *o* in declension. A ring under a letter (e.g. ḁ) may also be used in certain Indo-European languages to indicate syllabicity.

A hook (e.g. ả, ẻ, ỉ, ỏ, ủ, ỷ) appears as a small question mark placed on the top of vowels in the Vietnamese language, causing them to be pronounced with a falling, then rising tone.

A stroke (e.g. ø), also called a slash, is a diacritic that is placed through the letters *o* in Danish and Norwegian creating a half-close front rounded vowel. A velarized *l* (e.g. ł) in Polish is pronounced as *w*.

A bar (e.g. đ, ħ, ŧ) is a diacritic the effect of which varies with the language in question. The Croatian letter đ represents a palatalized *d* which can also be written as the digraph *dj* (Wells, 2010). In the Vietnamese alphabet, however, đ stands for [d], while in Sámi đ and ŧ are fricatives. In Maltese ħ represents a voiceless pharyngeal fricative.

The horn (e.g, ơ, ư) is a Vietnamese diacritic that creates unrounded variants of each vowel and are considered as separate letters from *o* and *u*.

A cedilla (e.g. ç, ş) originated from Spanish and is represented by a miniature curved structure at the base of the letters *c* and *s*. With ç it represents a soft sound /s/ rather than a hard sound /k/ before *a*, *o*, *u*, or at the end of a word in Basque, Catalan, French and Portuguese. However, in Albanian, Azerbaijani, Kurdish, Tatar, Turkish and Turkmen, ç represents a voiceless affricate. In Azerbaijani, Kurdish, Tatar, Turkish and Turkmen, ş represents a voiceless postalveolar fricative /ʃ/.

A comma at the base of a letter (ķ, ļ, ņ, ŗ, ţ) is used as a diacritic in several languages (e.g. Latvian, Romanian) and represents various soft consonants. With the letter *g*, the comma is reversed and placed out of necessity over the lower case letter (ģ).

A tail or ogonek (e.g. ą, ę, į, ų) is used in several eastern European languages. In Polish, ą and ę denote nasalized vowels. In Lithuanian the tail simply means a long vowel. In transcriptions of Navajo, Kiowa and certain other Amerindian languages, the diacritic is used to indicate nasalization (Harrington, 1928).

A dot (e.g. ạ, ẹ, ị, ọ, ụ, ỵ) in Vietnamese indicates a low tone and glottalized or tense voice while in Yoruba, ạ and ẹ denote a half-open vowel. A dotted *s* represents a palatoalveolar fricative /ʃ/. A dot is also sometimes used in Romanization of languages using non-Latin alphabets. For example, several languages from Southeast Asia use a dot under *d, n, s* and *t* to represent retroflex consonants while under *d, s* or *t* in the Romanization of Arabic, it represents an emphatic consonant.

An apostrophe (') has an exceptionally wide range of uses as a diacritic. In Hausa it indicates a glottalisation of the following consonant or when between two vowels, a glottal stop. In the Romanization of Mandarin Chinese, it marks aspiration of the preceding consonant sound as in ts'ê where ts' is aspirated. In English it denotes possessive as in the common name giant elephant's ear (*Colocasia gigantea* (Blume) Hook. f.). In Swahili, when placed after *ng* (ng'), the *g* is silent as in the English word si*ng*er, while in Italian it is occasionally used for a grave accent after a final vowel. In some languages (e.g. Guarani, Hawaiian, Samoan, Tahitian, Tongan) and transliterations, the apostrophe is used to indicate a glottal stop. Likewise, it may indicate the presence or lack of palatalization in Belarusian, Russian and Ukrainian depending upon where it is used. In Southern Paiute, -k'w indicates the *w* is unvoiced at the end of the word. With -ng'w the *w* at the end of the word is unvoiced (Martineau, 1992). In some languages an apostrophe is used to separate the root word from its affixes, especially if the root word is foreign and unassimilated.

The effect of two diacritics on a single letter (e.g. ǻ, ấ, ầ, ẩ, ẫ, ậ, ắ, ằ, ẳ, ẵ, ặ, ế, ề, ể, ễ, ệ, ǖ, ǘ, ǚ, ǜ) varies with the language in question. In Vietnamese one may relate to the vowel quality and the other to the tone.

Appendix II. Unique letters/symbols and alphabetization

Unique letters and symbols

In some instances, unique letters are required for representing distinctive sounds that cannot be reproduced using Latin letters and diacritics. For example the palatoalveolar fricative (ʃ) may be represented by *sh* in English or š, ŝ or ş, depending upon the language. In English certain sounds can be represented by combining two letters [sh or ng], called a digraph. In other languages, however, these adjacent letters may sometimes be pronounced separately. In other instances, it is not possible to accurately represent the required sound with existing letters or diacritics, in which case, additional letters are required. For example in the Khoisan languages clicking sounds are used which cannot be reproduced using the Latin alphabet. In addition to new letters, ligatures (the fusion of two letters such as æ) are routinely used in some languages.

Several West African languages use additional letters (e.g. ŋ, ɲ, ə, ɛ, ɔ). Where possible the pronunciation indicated by the letter is listed below, typically by language or language groups. In a few instances, symbols used from several sources lack descriptions of their pronunciation and an explanation could not be found. For more detailed explanation of unique characters, one should utilize a dictionary for the language in question.

Afro-roman characters used in a number of African languages

ɓ – a glottalised *b*
ḅ – an implosive *b*
ḏ – a post-alveolar *d*
ɗ – a glottalised *d*
ḍ – an implosive *d*
ɛ – an open *e*
ə – a central vowel (schwa), as at the end of English soda
ƒ – a bilabial *f*
ɣ – a voiced velar fricative
ƙ – a glottalized *k*
ɲ – a palatal nasal *ny*
ŋ – a velar nasal as the *ng* in English si*ng*
ɔ – a short back vowel as in English brought
ọ – a short back vowel in Nigerian languages
ʃ – *sh*
ƴ – a glottalised y

Amharic (see Kane, 1990)

g^w – the superscript w indicates a voiced labiovelar stop.
ʔ – represents a voiceless glottal stop.

Azerbaijani, Kazakh, Kirghiz, Tatar

ı – indicates a retracted i sound.

Herero is a Bantu language and like the Khoisan languages of southern Africa, uses clicks as speech sounds. The primary sounds are symbolized using |, ≠ (or ǂ), and !. Many of the Bantu languages (e.g. Zulu) use the letters c, x, and q for dental, lateral, and alveolar clicks.

! – represents an alveolar click which is made using the tip of the tongue which is pulled down abruptly from the roof of the mouth.
≠ – or more correctly ǂ, is a palatal click that is made with a flat tongue that is pulled down abruptly from the roof of the mouth. It makes a softer sound clicking sound than the !.

Comanche

ʔ – represents a glottal stop.

German

ß – the German eszett or scharfes symbol has been replaced in modern German by *ss*.

Khmer, Central

: – when following a vowel, it indicates a lengthening of the vowel sound.

Kung-Ekoka is synonymous with !KUNG, a member of the Khoisan language family that uses click sounds. The sound is created by sucking on the molars in the back of the mouth to create a sound (*tchick*) somewhat like that used in English to get a horse moving.

// or ǂ – a pulmonic click.
! – click consonant.

Lahu

Diacritical marks in some versions of Lahu are placed one space after each vowel.

^ – represents a rising tone.
˅ – represents a falling tone.
ˉ – represents a very low tone.

Lakota

η – The lower case Greek letter η (eta) is used in popular Lakota orthography to indicate the nasality of a preceding vowel.

Lao

: – when following a vowel, it indicates a lengthening of the vowel sound.

Nama-Damara

! – represents a post-alveolar click release (sharp click).
// – represents affricated lateral clicks.
/ – represents affricated dental clicks.
- – represents a stem that occurs with a prefix.

Paiute, Southern (Martineau, 1992)

'– an apostrophe indicates a glottal stop.
´– an accute accent mark is used to distinguish a long vowel from a short one.

Potawatomi (Smith, 1933)

' – an apostrophe indicates a glottal stop.

Pume

t^h – a superscript *h* indicates aspiration of the sound.

Samoan and **Raratongan**

' – indicates a glottal unaspirated voiceless stop.

Somali

ď – indicates an implosive dental stop.

Tibetan

∂ – though not specified in the source, it probably represents a low mid-back rounded vowel / alternatively, a high central unrounded vowel.

Vietnamese

[ ] – Letters within brackets represent transcriptions of Vietnamese characters (e.g. [aws] = ắ, [oos] = ố, [oox] = ỗ, [oo] = ô, etc.).

Warma

δ – indicates the voiced *th* sound in the English word thus.

Yom

μ – is thought to represent a syllabic nasal sound as in the *om* of the English word bottom.
τ – though not indicated by the source, it probably indicates a partially voiced *t* sound when adjacent to an *m* or *r*.

Zuni

/ – a slash represents a glottal stop in Zuni that is unaspirated and voiceless (Newman, 1958).

Italicization

Certain prefixes are italicized in several African languages, the use of which varies depending upon the language and source. In a number of cases, the meaning of italicization in vernacular name lists is not indicated. In some instances, the use of italics indicates singular versus plural definite for the plant or for the useful part of the plant. For example, in Themne the prefixes *a-*, *am-*, *an-* indicate singular while *ɛ-* indicates plural; *ka-* singular for the useful part of the plant and *ta-* the plural; *ma-* the fruit (Deighton, 1957). In various Niger-Congo languages the vernacular name may begin with an italicized *n*, *m*, *g* or *k* followed by a specific letter (e.g., *m*b, *n*d, *n*g, *n*j, *g*b or *k*p) which indicates an altered pronunciation of the second letter. When the name is modified with a prefix, the initial consonant is commonly altered for euphony (e.g., t becomes l; k becomes g). In some sources prefixes are ignored in alphabetization; however, herein the prefix is treated as part of the word.

Alphabetization

In the Latin alphabet used by the English language, individual letters are alphabetized in the following order: a, b, c, d, e, f, g, h, i, j, k, l, m, n, o, p, q, r, s, t, u, v, w, x, y, z. In some languages certain digraphs (a two letter combinations such as sh, ng, nh) are treated as distinct letters and their alphabetization varies depending upon the language. In addition, letters with diacritics over them in some languages may be alphabetized using distinctly different methods, as illustrated by Estonian vowels with diacritics which are found in the following order at the end of the alphabet: p, r, s, š, z, ž, t, u, v, õ, ä, ö, ü. The Kiowa alphabet exhibits the following order: a, au, e, i, o, u, ai, aui, oi, ui, b, f, p, v, g, c, k, q, d, j, t, th, ch, x, s, z, l, w, y, h, m, n, while the Zuni alphabet, comprised of only 18 letters, is ordered as follows: a, b, d, ch, e, h, i, k, l, ł, m, n, o, p, s, t, u, w, y.

In the text, diacritics are ignored with regard to alphabetization. The following alphabetical sequence is used: a/ɑ, æ/ae, b/ɓ, c/ɔ, d/ɗ/ɖ, e/ɛ/ʃ/ə, f, ƒ, g, h/ɣ, i/ı, j, k/ƙ, l, m, n/ɲ/ŋ, o, p, q, r, s, ß, t, u, v, w, x, y/ƴ, z, δ, η, where slashes (/) indicate the letters are handled alphabetically as equals.